Nonlinear Functional Analysis and its Applications

IV : Applications to Mathematical Physics

Springer Science+Business Media, LLC

Carl Friedrich Gauss (1777–1855)

Eberhard Zeidler

Nonlinear Functional Analysis and its Applications

IV: Applications to Mathematical Physics

Translated by Juergen Quandt

With 201 Illustrations

Springer

Eberhard Zeidler
Max-Planck-Institut für Mathematik
 in den Naturwissenschaften
Inselstraße 22-26
D-04103 Leipzig
Germany

Juergen Quandt (*Translator*)
Department of Mathematics
University of Central Florida
Orlando, FL 32816
U.S.A.

Mathematics Subject Classification (1980): 46-XX

Library of Congress Cataloging-in-Publication Data
(Revised for vol. 4)
Zeidler, Eberhard.
  Nonlinear functional analysis and its applications.
  Vol. 3: Translated by Leo L. Boron; vol. 4:
Translated by Juergen Quandt.
  Includes bibliographies and indexes.
  Contents: v. 1. Fixed point theorems —
v. 3. Variational methods and optimization — v. 4.
Applications to mathematical physics.
  1. Nonlinear functional analysis. I. Title.
QA321.5.Z4513   1985   515.7   83-20455
ISBN 978-1-4612-8926-5   ISBN 978-1-4612-4566-7 (eBook)
DOI 10.1007/978-1-4612-4566-7

Previous edition, *Vorlesungen über nichtlineare Funktionalanalysis*, Vols. I–III, published by BSB B.G. Teubner Verlagsgesellschaft, 7010 Leipzig, Sternwartenstrasse 8, Deutsche Demokratische Republik.

© 1988 by Springer Science+Business Media New York
Originally published by Springer-Verlag New York Inc. in 1988
Softcover reprint of the hardcover 1st edition 1988

All rights reserved. This work may not be translated or copied in whole or in part without the written permission of the publisher (Springer Science+Business Media, LLC),
             except for brief excerpts in connection with reviews or scholarly analysis. Use in connection with any form of information storage and retrieval, electronic adaptation, computer software, or by similar or dissimilar methodology now known or hereafter developed is forbidden.

Typeset by Asco Trade Typesetting Ltd., Hong Kong.
Printed and bound by R. R. Donnelley & Sons, Harrisonburg, Virginia.

9 8 7 6 5 4 3 2 (Corrected second printing, 1997)

Dedicated in love to my sister's family

Edith, Günter, Jürgen, and Ulrich

So teach us to number our days, that we may apply our hearts unto wisdom.
Psalm 90, 12

And though I have the gift of prophecy, and understand all mysteries, and all knowledge; and though I have all faith, so that I could remove mountains, and have not charity, I am nothing.
I. Corinthians 13, 2

Preface

> The main concern in all scientific work must be the human being himself. This, one should never forget among all those diagrams and equations.
> Albert Einstein

This volume is part of a comprehensive presentation of nonlinear functional analysis, the basic content of which has been outlined in the Preface of Part I. A Table of Contents for all five volumes may also be found in Part I. The present Part IV and the following Part V contain applications to mathematical physics. Our goals are the following:

(i) A detailed motivation of the basic equations in important disciplines of theoretical physics.
(ii) A discussion of particular problems which have played a significant role in the development of physics, and through which important mathematical and physical insight may be gained.
(iii) A combination of classical and modern ideas.
(iv) An attempt to build a bridge between the language and thoughts of physicists and mathematicians.

We shall always try to advance as soon as possible to the heart of the problem under consideration and to concentrate on the basic ideas.

The treatment of mathematical physics in the mathematical literature is often as follows. A mathematical problem is discussed and then followed only by a short remark that physical interpretations exist. But for mathematicians it is also very important to have a profound knowledge of the *physical background*, because it is possible that a mathematical result lies in a parameter space which is physically meaningless. Furthermore, it might happen that mathematicians will, with a great deal of effort, determine solutions which are actually unstable, i.e., which are never realized in nature. Finally, we should mention that every mathematical model of processes in nature is based on simplifications and approximations. Thereby it is important to learn the particular model assumptions. For example, the addition of a small viscosity term may greatly simplify the mathematical study, and this viscosity term might actually yield a much better model of physical reality than the previous idealization without viscosity.

It is therefore important that mathematicians are concerned about the physical motivation of the problems they study. Our presentation might help in this direction. In order to avoid confusion we clearly distinguish between the mathematical formulation of the basic equations, their physical motivation, and the proof of purely mathematical results. The word "proof" is always understood in the sense of a rigorous mathematical proof.

Although we emphasize a detailed discussion of the basic equations, this is never a goal in itself. We therefore do *not* use an oversophisticated axiomatic approach. Our main concern is the study of typical problems. We have tried to select interesting and important problems in order to mediate the fascination which results from the interrelation between mathematics and physics. At the same time, we have tried to cover a broad spectrum which might lead to a diversified and colorful picture. We think that a broad scope in the education of students is very important, so that later on, the researcher will be able to apply different ways of thinking to the solutions of his very special problems. Our main goal is also to emphasize the relation between classical theories and modern developments in physics, including presently exciting open problems. For a student, it is important to know such relations; otherwise, modern theories may appear to him as unrelated. In the previous volumes we have emphasized the *unity of pure and applied mathematics*. Here we want to complement this picture by trying to show the *unity of mathematics and physics*.

The recognition process in theoretical physics can very roughly be described by the following scheme:

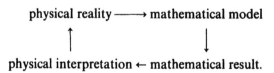

The treatment of mathematical problems may therefore be very difficult. As a mathematician, however, one should not forget that the important intellectual contribution in the finding of new insights into nature's secrets lies in the process of passing from the physical reality to the mathematical model. This, for example, can be seen from Einstein's struggle with the general theory of relativity. He began with physical observations, and for their realization he then tried to find a suitable mathematical setting.

A mathematician who wants to occupy himself with physics is confronted with the following *difficulty*.

(a) In the physical literature the mathematical apparatus is not always correctly handled. Even the best textbooks in theoretical physics contain mathematical statements which are wrong or proofs which are not correct. This has the consequence that many mathematicians restrict themselves to purely mathematical considerations, or if they do treat physical subjects, then physicists have difficulty in recognizing the familiar physical material.

(b) On the other hand, creative physicists have a number of objections to mathematics. "The rigor of mathematics is a luxury. We need concrete answers to our problems. The general results of mathematicians are often useless for our purposes, or the formulation is so abstract that we cannot recognize whether a mathematical result is useful to us or not. Thus we have to develop our own methods and cannot wait until they are mathematically justified. By experience, this takes much too long."

For a student who wants to learn both mathematics and physics, a schizophrenic situation may occur, and he might be caught on the horns of a dilemma.

Actually, both the above views are justified: the critique of the mathematician and the critique of the physicist. Until today, for example, it has not been possible to develop a mathematically rigorous quantum field theory. For about 40 years, however, physicists have worked with dubious mathematical methods and have obtained results which are in fantastic coincidence with experiment. On the other hand, in cases where rigorous and transparent mathematical theories exist, they are not always used by physicists. This applies, for example, to the calculus of variations. The difficulties which occurred during the creation of differential and integral calculus and variational calculus in the eighteenth century, with concepts like "infinitesimally small," "variation," and "virtual displacement," are still preserved in many present-day physics textbooks and cause a great deal of trouble for the student. Here, the words of the Austrian musician Gustav Mahler (1860–1911) are appropriate: tradition = slovenliness. Some particular cases will be critically discussed in this volume. This is not meant as a cheap polemic but rather we hope to help the student. That many such difficulties exist is evident to the author in his daily contact with mathematics and physics students.

By choosing particular examples, we also try to explain that specific forms of mathematical and physical *thinking* exist, and that both their differences and synthesis are important for effective research. Many important physical theories were obtained, not by way of mathematical deduction but instead, by ingenious physicists, in close contact with observed physical phenomena, who developed their specific physical ideas and, from a mathematical point of view, employed some very questionable mathematical tools. At the same time, however, they developed a number of very fruitful mathematical ideas. The different forms of mathematical and physical thinking make it very difficult to find a common language for both disciplines. A student, for example, might become desperate in trying to translate the standard ten-volume work in theoretical physics of Landau and Lifšic (1962) into a precise mathematical language. Experience shows that it is more useful for a student to take a pragmatic point of view and get acquainted with *both* languages, i.e., the language of mathematics and the language of physics, and to look for as many relations between them as possible.

In this volume we hope that the *mathematician* feels comfortable, because his precise mathematical language is used whenever a purely mathematical subject is treated, but also that, at the same time, he learns something about the physical interpretation. On the other hand, we hope that the *physicist* recognizes his familiar physical ideas and sees how these are related to important modern mathematical concepts and also how these concepts can fruitfully be applied.

Only the reader can decide whether we have succeeded in this or not. For the future, however, it will be important that mathematicians and natural scientists make greater efforts towards a better understanding, so that ideas may flow in both directions. No abstract statements are needed here, but rather the *personal* contact between mathematicians and natural scientists, in order to help demolish barriers.

This volume has the same formal structure as the previous Parts I–III. Details may be found in the Preface of Part I. We mention that at the end of this volume a listing of all theorems may be found. The Appendix contains a survey of the international system of units and the values of important universal constants.

Mathematicians should bear in mind that, at the end of even the most abstract theory, physicists need numbers which they can compare with experiment. It is also very important for physicists to have a precise knowledge of the orders of magnitude in which the various effects lie, so that they know which of them to neglect. In order to give mathematicians a feeling for this, we have added several tables throughout this volume.

In the Introduction to Part I we have already mentioned that in the study of many problems the following steps are used:

(i) Translation of the problem into the language of functional analysis.
(ii) Applications of abstract functional analytic theorems.
(iii) To verify the assumptions in (ii) it is often necessary to apply deep and very specific analytical tools.

According to our title "Nonlinear Functional Analysis," we concentrate on (i) and (ii). The tools in (iii), which usually require very long proofs (e.g., *a priori* estimates for linear elliptic systems in spaces of Hölder continuously differentiable functions), are often used without proof, but we always say where the corresponding proof may be found.

The sciences today are characterized by an increased splitting. In this presentation we try to emphasize unifying *principles* and relations. If this is done by a single author, then surely only with imperfection. Critical remarks and suggestions are therefore very welcome.

For assistance in typing parts of this manuscript I am very thankful to Ursula Abraham, Helga Hedwig, Kristina Friedrich, Hiltraud Lehmann, Rita Löffler, Karin Quasthoff, Karla Rietz, Stefan Ackermann, Frank Benkert, Werner Berndt, Stefan Friedrich, Bernd Fritzsche, Matthias Günther, Uwe Heide, Jürgen Herrler, Jürgen Janassary, Bernd Kirstein, Wolfgang Kliesch, Klaus Schenk, Rainer Schumann, and Friedemann Schuricht.

Also extremely helpful was the understanding and knowledgable assistance of the librarian of our Institute, Mrs. Ina Letzel, for which I wish to express my sincerest thanks.

I would heartily like to thank Professor Friedrich Hirzebruch for his repeated generous hospitality at the Max-Planck Institute for Mathematics in Bonn.

I would also like to thank Juergen Quandt for his English translation. Finally, my special thanks are due to Springer-Verlag for the harmonious collaboration and the beautiful layout of the book.

Leipzig
Summer 1987

Eberhard Zeidler

Preface to the Corrected Second Printing

I am very pleased that Springer-Verlag is publishing a corrected reprint of my book. In this edition, I have made minor revisions and added a fairly comprehensive list of recent references.

Nowadays, nonlinear mathematical techniques play a fundamental role in all areas of natural sciences. The additional references should help the reader to discover new ways of applying mathematics to the fascinating phenomena of our real world. I hope that this volume will contribute to a better understanding between mathematicians and natural scientists.

Leipzig
Spring 1997

Eberhard Zeidler

Translator's Preface

I would like to take this opportunity to thank Professor Eberhard Zeidler for much advice during the course of the translation. I am very grateful to him for a pleasant time of collaboration. I also want to thank Steve Kane for his help with stylistic questions.

What follows are a few observations about the process of discovery. The most difficult form of thinking seems to be the thinking about thinking. This may be related to the problem of self-reference in logic. It appears to be in principle impossible to completely analyze one's own process of thought, since such an analysis interferes with the process to be analyzed. Nevertheless, the conscious mind can look at the subconscious mind and with the aid of memory also at the conscious mind.

To visualize the mental state of thought, one could use the analogy

$$\text{awakenness} = \text{consciousness}$$
$$\text{dream} = \text{subconsciousness}.$$

The conscious mind is subject to our will, as elements of thought serve certain more or less vague images and signs, which are continuously combined by the subconscious mind (i.e., involuntarily). According to logical rules and controlled by the conscious mind, they are formed into certain chains. Those which appeal to the conscious mind are kept as further elements of thought. In this selection process the conscious mind is guided by a certain feeling of harmony and simplicity and the desire to arrive at logically connected entities. Discovery itself occurs suddenly (like a flash of lightning), usually after a period of mental exertion (and possibly preceded by some incubation time).

What are the necessary prerequisites for discovery?

(1) Character. The most important factor seems to be one's character. The better it is, the more likely the conscious mind will be guided by a feeling of harmony.
(2) Intelligence. The higher the intelligence, the faster the subconscious mind is able to combine the elements of thought.
(3) Knowledge. The broader the knowledge, the more elements of thought the subconscious mind can combine.
(4) Inspiration. Without being inspired by other peoples' ideas, one's thoughts often proceed in circles.
(5) Persistence. To arrive at deep results (see below), a certain persistence over long periods of time seems to be required.

How can the value of a discovery be judged?

A result is deep if it is basic (the most basic results being the universal laws of nature). Such results assume a simple form ($F = ma$, $E = mc^2$, etc.). It seems to be the hardest task to arrive at those discoveries.

Winter Park, FL
Spring 1987

Juergen Quandt

Contents

Preface	vii
Translator's Preface	xiii

INTRODUCTION
Mathematics and Physics — 1

APPLICATIONS IN MECHANICS — 7

CHAPTER 58
Basic Equations of Point Mechanics — 9

§58.1.	Notations	10
§58.2.	Lever Principle and Stability of the Scales	14
§58.3.	Perspectives	17
§58.4.	Kepler's Laws and a Look at the History of Astronomy	22
§58.5.	Newton's Basic Equations	25
§58.6.	Changes of the System of Reference and the Role of Inertial Systems	28
§58.7.	General Point System and Its Conserved Quantities	32
§58.8.	Newton's Law of Gravitation and Coulomb's Law of Electrostatics	35
§58.9.	Application to the Motion of Planets	38
§58.10.	Gauss' Principle of Least Constraint and the General Basic Equations of Point Mechanics with Side Conditions	45
§58.11.	Principle of Virtual Power	48
§58.12.	Equilibrium States and a General Stability Principle	50
§58.13.	Basic Equations of the Rigid Body and the Main Theorem about the Motion of the Rigid Body and Its Equilibrium	52
§58.14.	Foundation of the Basic Equations of the Rigid Body	55

§58.15.	Physical Models, the Expansion of the Universe, and Its Evolution after the Big Bang	57
§58.16.	Legendre Transformation and Conjugate Functionals	65
§58.17.	Lagrange Multipliers	67
§58.18.	Principle of Stationary Action	69
§58.19.	Trick of Position Coordinates and Lagrangian Mechanics	70
§58.20.	Hamiltonian Mechanics	72
§58.21.	Poissonian Mechanics and Heisenberg's Matrix Mechanics in Quantum Theory	77
§58.22.	Propagation of Action	81
§58.23.	Hamilton–Jacobi Equation	82
§58.24.	Canonical Transformations and the Solution of the Canonical Equations via the Hamilton–Jacobi Equation	83
§58.25.	Lagrange Brackets and the Solution of the Hamilton–Jacobi Equation via the Canonical Equations	84
§58.26.	Initial-Value Problem for the Hamilton–Jacobi Equation	87
§58.27.	Dimension Analysis	89

CHAPTER 59
Dualism Between Wave and Particle, Preview of Quantum Theory, and Elementary Particles — 98

§59.1.	Plane Waves	99
§59.2.	Polarization	101
§59.3.	Dispersion Relations	102
§59.4.	Spherical Waves	103
§59.5.	Damped Oscillations and the Frequency–Time Uncertainty Relation	104
§59.6.	Decay of Particles	105
§59.7.	Cross Sections for Elementary Particle Processes and the Main Objectives in Quantum Field Theory	106
§59.8.	Dualism Between Wave and Particle for Light	107
§59.9.	Wave Packets and Group Velocity	110
§59.10.	Formulation of a Particle Theory for a Classical Wave Theory	111
§59.11.	Motivation of the Schrödinger Equation and Physical Intuition	112
§59.12.	Fundamental Probability Interpretation of Quantum Mechanics	113
§59.13.	Meaning of Eigenfunctions in Quantum Mechanics	114
§59.14.	Meaning of Nonnormalized States	116
§59.15.	Special Functions in Quantum Mechanics	117
§59.16.	Spectrum of the Hydrogen Atom	118
§59.17.	Functional Analytic Treatment of the Hydrogen Atom	121
§59.18.	Harmonic Oscillator in Quantum Mechanics	122
§59.19.	Heisenberg's Uncertainty Relation	123
§59.20.	Pauli Principle, Spin, and Statistics	125
§59.21.	Quantization of the Phase Space and Statistics	126
§59.22.	Pauli Principle and the Periodic System of the Elements	127
§59.23.	Classical Limiting Case of Quantum Mechanics and the WKB Method to Compute Quasi-Classical Approximations	129
§59.24.	Energy–Time Uncertainty Relation and Elementary Particles	130
§59.25.	The Four Fundamental Interactions	134
§59.26.	Strength of the Interactions	136

APPLICATIONS IN ELASTICITY THEORY 143

CHAPTER 60
Elastoplastic Wire 145
§60.1. Experimental Result 147
§60.2. Viscoplastic Constitutive Laws 149
§60.3. Elasto-Viscoplastic Wire with Linear Hardening Law 151
§60.4. Quasi-Statical Plasticity 154
§60.5. Some Historical Remarks on Plasticity 155

CHAPTER 61
Basic Equations of Nonlinear Elasticity Theory 158
§61.1. Notations 166
§61.2. Strain Tensor and the Geometry of Deformations 168
§61.3. Basic Equations 176
§61.4. Physical Motivation of the Basic Equations 180
§61.5. Reduced Stress Tensor and the Principle of Virtual Power 184
§61.6. A General Variational Principle (Hyperelasticity) 190
§61.7. Elastic Energy of the Cuboid and Constitutive Laws 198
§61.8. Theory of Invariants and the General Structure of Constitutive Laws and Stored Energy Functions 202
§61.9. Existence and Uniqueness in Linear Elastostatics (Generalized Solutions) 209
§61.10. Existence and Uniqueness in Linear Elastodynamics (Generalized Solutions) 212
§61.11. Strongly Elliptic Systems 213
§61.12. Local Existence and Uniqueness Theorem in Nonlinear Elasticity via the Implicit Function Theorem 215
§61.13. Existence and Uniqueness Theorem in Linear Elastostatics (Classical Solutions) 221
§61.14. Stability and Bifurcation in Nonlinear Elasticity 221
§61.15. The Continuation Method in Nonlinear Elasticity and an Approximation Method 224
§61.16. Convergence of the Approximation Method 227

CHAPTER 62
Monotone Potential Operators and a Class of Models with Nonlinear Hooke's Law, Duality and Plasticity, and Polyconvexity 233
§62.1. Basic Ideas 234
§62.2. Notations 242
§62.3. Principle of Minimal Potential Energy, Existence, and Uniqueness 244
§62.4. Principle of Maximal Dual Energy and Duality 245
§62.5. Proofs of the Main Theorems 247
§62.6. Approximation Methods 252
§62.7. Applications to Linear Elasticity Theory 255
§62.8. Application to Nonlinear Hencky Material 256
§62.9. The Constitutive Law for Quasi-Statical Plastic Material 257

§62.10.	Principle of Maximal Dual Energy and the Existence Theorem for Linear Quasi-Statical Plasticity	259
§62.11.	Duality and the Existence Theorem for Linear Statical Plasticity	262
§62.12.	Compensated Compactness	264
§62.13.	Existence Theorem for Polyconvex Material	273
§62.14.	Application to Rubberlike Material	277
§62.15.	Proof of Korn's Inequality	278
§62.16.	Legendre Transformation and the Strategy of the General Friedrichs Duality in the Calculus of Variations	284
§62.17.	Application to the Dirichlet Problem (Trefftz Duality)	288
§62.18.	Application to Elasticity	289

CHAPTER 63
Variational Inequalities and the Signorini Problem for Nonlinear Material 296

§63.1.	Existence and Uniqueness Theorem	296
§63.2.	Physical Motivation	298

CHAPTER 64
Bifurcation for Variational Inequalities 303

§64.1.	Basic Ideas	303
§64.2.	Quadratic Variational Inequalities	305
§64.3.	Lagrange Multiplier Rule for Variational Inequalities	306
§64.4.	Main Theorem	308
§64.5.	Proof of the Main Theorem	309
§64.6.	Applications to the Bending of Rods and Beams	311
§64.7.	Physical Motivation for the Nonlinear Rod Equation	315
§64.8.	Explicit Solution of the Rod Equation	317

CHAPTER 65
Pseudomonotone Operators, Bifurcation, and the von Kármán Plate Equations 322

§65.1.	Basic Ideas	322
§65.2.	Notations	325
§65.3.	The von Kármán Plate Equations	326
§65.4.	The Operator Equation	327
§65.5.	Existence Theorem	332
§65.6.	Bifurcation	332
§65.7.	Physical Motivation of the Plate Equations	334
§65.8.	Principle of Stationary Potential Energy and Plates with Obstacles	339

CHAPTER 66
Convex Analysis, Maximal Monotone Operators, and Elasto-Viscoplastic Material with Linear Hardening and Hysteresis 348

§66.1.	Abstract Model for Slow Deformation Processes	349
§66.2.	Physical Interpretation of the Abstract Model	352
§66.3.	Existence and Uniqueness Theorem	355
§66.4.	Applications	358

APPLICATIONS IN THERMODYNAMICS 363

CHAPTER 67
Phenomenological Thermodynamics of Quasi-Equilibrium and Equilibrium States 369
§67.1. Thermodynamical States, Processes, and State Variables 371
§67.2. Gibbs' Fundamental Equation 374
§67.3. Applications to Gases and Liquids 375
§67.4. The Three Laws of Thermodynamics 378
§67.5. Change of Variables, Legendre Transformation, and Thermodynamical Potentials 385
§67.6. Extremal Principles for the Computation of Thermodynamical Equilibrium States 387
§67.7. Gibbs' Phase Rule 391
§67.8. Applications to the Law of Mass Action 392

CHAPTER 68
Statistical Physics 396
§68.1. Basic Equations of Statistical Physics 397
§68.2. Bose and Fermi Statistics 402
§68.3. Applications to Ideal Gases 403
§68.4. Planck's Radiation Law 408
§68.5. Stefan–Boltzmann Radiation Law for Black Bodies 409
§68.6. The Cosmos at a Temperature of 10^{11} K 411
§68.7. Basic Equation for Star Models 412
§68.8. Maximal Chandrasekhar Mass of White Dwarf Stars 412

CHAPTER 69
Continuation with Respect to a Parameter and a Radiation Problem of Carleman 422
§69.1. Conservation Laws 422
§69.2. Basic Equations of Heat Conduction 423
§69.3. Existence and Uniqueness for a Heat Conduction Problem 425
§69.4. Proof of Theorem 69.A 426

APPLICATIONS IN HYDRODYNAMICS 431

CHAPTER 70
Basic Equations of Hydrodynamics 433
§70.1. Basic Equations 434
§70.2. Linear Constitutive Law for the Friction Tensor 436
§70.3. Applications to Viscous and Inviscid Fluids 438
§70.4. Tube Flows, Similarity, and Turbulence 439
§70.5. Physical Motivation of the Basic Equations 441
§70.6. Applications to Gas Dynamics 444

CHAPTER 71
Bifurcation and Permanent Gravitational Waves — 448
- §71.1. Physical Problem and Complex Velocity — 451
- §71.2. Complex Flow Potential and Free Boundary-Value Problem — 454
- §71.3. Transformed Boundary-Value Problem for the Circular Ring — 456
- §71.4. Existence and Uniqueness of the Bifurcation Branch — 459
- §71.5. Proof of Theorem 71.B — 462
- §71.6. Explicit Construction of the Solution — 464

CHAPTER 72
Viscous Fluids and the Navier–Stokes Equations — 479
- §72.1. Basic Ideas — 480
- §72.2. Notations — 485
- §72.3. Generalized Stationary Problem — 486
- §72.4. Existence and Uniqueness Theorem for Stationary Flows — 490
- §72.5. Generalized Nonstationary Problem — 491
- §72.6. Existence and Uniqueness Theorem for Nonstationary Flows — 494
- §72.7. Taylor Problem and Bifurcation — 495
- §72.8. Proof of Theorem 72.C — 500
- §72.9. Bénard Problem and Bifurcation — 505
- §72.10. Physical Motivation of the Boussinesq Approximation — 512
- §72.11. The Kolmogorov 5/3-Law for Energy Dissipation in Turbulent Flows — 513
- §72.12. Velocity in Turbulent Flows — 515

MANIFOLDS AND THEIR APPLICATIONS — 527

CHAPTER 73
Banach Manifolds — 529
- §73.1. Local Normal Forms for Nonlinear Double Splitting Maps — 531
- §73.2. Banach Manifolds — 533
- §73.3. Strategy of the Theory of Manifolds — 535
- §73.4. Diffeomorphisms — 537
- §73.5. Tangent Space — 538
- §73.6. Tangent Map — 540
- §73.7. Higher-Order Derivatives and the Tangent Bundle — 541
- §73.8. Cotangent Bundle — 545
- §73.9. Global Solutions of Differential Equations on Manifolds and Flows — 546
- §73.10. Linearization Principle for Maps — 550
- §73.11. Two Principles for Constructing Manifolds — 554
- §73.12. Construction of Diffeomorphisms and the Generalized Morse Lemma — 560
- §73.13. Transversality — 563
- §73.14. Taylor Expansions and Jets — 566
- §73.15. Equivalence of Maps — 571
- §73.16. Multilinearization of Maps, Normal Forms, and Castastrophe Theory — 572

Contents   xxi

§73.17.	Applications to Natural Sciences	579
§73.18.	Orientation	582
§73.19.	Manifolds with Boundary	584
§73.20.	Sard's Theorem	587
§73.21.	Whitney's Embedding Theorem	588
§73.22.	Vector Bundles	589
§73.23.	Differentials and Derivations on Finite-Dimensional Manifolds	595

CHAPTER 74
Classical Surface Theory, the Theorema Egregium of Gauss, and Differential Geometry on Manifolds   609

§74.1.	Basic Ideas of Tensor Calculus	615
§74.2.	Covariant and Contravariant Tensors	617
§74.3.	Algebraic Tensor Operations	621
§74.4.	Covariant Differentiation	623
§74.5.	Index Principle of Mathematical Physics	625
§74.6.	Parallel Transport and Motivation for Covariant Differentiation	626
§74.7.	Pseudotensors and a Duality Principle	627
§74.8.	Tensor Densities	630
§74.9.	The Two Fundamental Forms of Gauss of Classical Surface Theory	631
§74.10.	Metric Properties of Surfaces	634
§74.11.	Curvature Properties of Surfaces	636
§74.12.	Fundamental Equations and the Main Theorem of Classical Surface Theory	639
§74.13.	Curvature Tensor and the Theorema Egregium	642
§74.14.	Surface Maps	644
§74.15.	Parallel Transport on Surfaces According to Levi-Civita	645
§74.16.	Geodesics on Surfaces and a Variational Principle	646
§74.17.	Tensor Calculus on Manifolds	648
§74.18.	Affine Connected Manifolds	649
§74.19.	Riemannian Manifolds	651
§74.20.	Main Theorem About Riemannian Manifolds and the Geometric Meaning of the Curvature Tensor	653
§74.21.	Applications to Non-Euclidean Geometry	655
§74.22.	Strategy for a Further Development of the Differential and Integral Calculus on Manifolds	663
§74.23.	Alternating Differentiation of Alternating Tensors	664
§74.24.	Applications to the Calculus of Alternating Differential Forms	664
§74.25.	Lie Derivative	673
§74.26.	Applications to Lie Algebras of Vector Fields and Lie Groups	676

CHAPTER 75
Special Theory of Relativity   694

§75.1.	Notations	699
§75.2.	Inertial Systems and the Postulates of the Special Theory of Relativity	699
§75.3.	Space and Time Measurements in Inertial Systems	700
§75.4.	Connection with Newtonian Mechanics	702
§75.5.	Special Lorentz Transformation	706

§75.6.	Length Contraction, Time Dilatation, and Addition Theorem for Velocities	708
§75.7.	Lorentz Group and Poincaré Group	710
§75.8.	Space–Time Manifold of Minkowski	713
§75.9.	Causality and Maximal Signal Velocity	714
§75.10.	Proper Time	717
§75.11.	The Free Particle and the Mass–Energy Equivalence	719
§75.12.	Energy Momentum Tensor and Relativistic Conservation Laws for Fields	723
§75.13.	Applications to Relativistic Ideal Fluids	726

CHAPTER 76
General Theory of Relativity — 730

§76.1.	Basic Equations of the General Theory of Relativity	730
§76.2.	Motivation of the Basic Equations and the Variational Principle for the Motion of Light and Matter	732
§76.3.	Friedman Solution for the Closed Cosmological Model	736
§76.4.	Friedman Solution for the Open Cosmological Model	741
§76.5.	Big Bang, Red Shift, and Expansion of the Universe	742
§76.6.	The Future of our Cosmos	745
§76.7.	The Very Early Cosmos	747
§76.8.	Schwarzschild Solution	756
§76.9.	Applications to the Motion of the Perihelion of Mercury	758
§76.10.	Deflection of Light in the Gravitational Field of the Sun	765
§76.11.	Red Shift in the Gravitational Field	766
§76.12.	Virtual Singularities, Continuation of Space–Time Manifolds, and the Kruskal Solution	767
§76.13.	Black Holes and the Sinking of a Space Ship	771
§76.14.	White Holes	775
§76.15.	Black–White Dipole Holes and Dual Creatures Without Radio Contact to Us	775
§76.16.	Death of a Star	776
§76.17.	Vaporization of Black Holes	780

CHAPTER 77
Simplicial Methods, Fixed Point Theory, and Mathematical Economics — 794

§77.1.	Lemma of Sperner	797
§77.2.	Lemma of Knaster, Kuratowski, and Mazurkiewicz	798
§77.3.	Elementary Proof of Brouwer's Fixed-Point Theorem	799
§77.4.	Generalized Lemma of Knaster, Kuratowski, and Mazurkiewicz	800
§77.5.	Inequality of Fan	801
§77.6.	Main Theorem for n-Person Games of Nash and the Minimax Theorem	802
§77.7.	Applications to the Theorem of Hartman–Stampacchia for Variational Inequalities	803
§77.8.	Fixed-Point Theorem of Kakutani	804
§77.9.	Fixed-Point Theorem of Fan–Glicksberg	805

Contents   xxiii

§77.10.	Applications to the Main Theorem of Mathematical Economics About Walras Equilibria and Quasi-Variational Inequalities	806
§77.11.	Negative Retract Principle	808
§77.12.	Intermediate-Value Theorem of Bolzano–Poincaré–Miranda	808
§77.13.	Equivalent Statements to Brouwer's Fixed-Point Theorem	810

CHAPTER 78
Homotopy Methods and One-Dimensional Manifolds   817

§78.1.	Basic Idea	818
§78.2.	Regular Solution Curves	818
§78.3.	Turning Point Principle and Bifurcation Principle	821
§78.4.	Curve Following Algorithm	822
§78.5.	Constructive Leray–Schauder Principle	823
§78.6.	Constructive Approach for the Fixed-Point Index and the Mapping Degree	824
§78.7.	Parametrized Version of Sard's Theorem	828
§78.8.	Theorem of Sard–Smale	829
§78.9.	Proof of Theorem 78.A	830
§78.10.	Parametrized Version of the Theorem of Sard–Smale	832
§78.11.	Main Theorem About Generic Finiteness of the Solution Set	834
§78.12.	Proof of Theorem 78.B	834

CHAPTER 79
Dynamical Stability and Bifurcation in B-Spaces   840

§79.1.	Asymptotic Stability and Instability of Equilibrium Points	841
§79.2.	Proof of Theorem 79.A	843
§79.3.	Multipliers and the Fixed-Point Trick for Dynamical Systems	846
§79.4.	Floquet Transformation Trick	848
§79.5.	Asymptotic Stability and Instability of Periodic Solutions	851
§79.6.	Orbital Stability	852
§79.7.	Perturbation of Simple Eigenvalues	853
§79.8.	Loss of Stability and the Main Theorem About Simple Curve Bifurcation	856
§79.9.	Loss of Stability and the Main Theorem About Hopf Bifurcation	860
§79.10.	Proof of Theorem 79.F	863
§79.11.	Applications to Ljapunov Bifurcation	867

Appendix	883
References	885
List of Symbols	933
List of Theorems	943
List of the Most Important Definitions	946
List of Basic Equations in Mathematical Physics	953
Index	959

INTRODUCTION
Mathematics and Physics

The more I have learned about physics, the more convinced I am that physics provides, in a sense, the deepest applications of mathematics. The mathematic problems that have been solved, or techniques that have arisen out of physics in the past, have been the lifeblood of mathematics.... The really deep questions are still in the physical sciences. For the health of mathematics at its research level, I think it is very important to maintain that link as much as possible.
Sir Michael Atiyah (1984)

The theory of general relativity is a good example for the basic character in a modern development of a theory. The original hypotheses become more and more abstract and remote from experience, but thereby the goal of deriving a maximum of results of our experience from a minimum of hypotheses has become closer.
Albert Einstein (1955)

The nuclear physicist Niels Bohr (1885–1962) did not—as did Arnold Sommerfeld (1868–1951)—start from precisely defined mathematical assumptions, but, starting from phenomena and guided by intuition, felt his way towards a new physics. This form of research was also characteristic for the founder of modern quantum mechanics, Werner Heisenberg (1901–1976).
Bartel Leendert van der Waerden (1981)

If one does not sometimes think the illogical, one will never discover new ideas in science.
Max Planck (1945)

"I think this is so," says Cicha, "in the fight for new insights, the breaking brigades are marching in the front row. The vanguard that does not look to left nor to right, but simply forges ahead—those are the physicists. And behind them there are following the various canteen men, all kinds of stretcher bearers, who clear the dead bodies away or, simply put, get things in order. Well, those are the mathematicians."
From the criminal roman "Dead loves poetry" of the
Czech physicist Jan Klimá (born in 1938)

The most vitally characteristic fact about mathematics, in my opinion, is its quite peculiar relationship to the natural sciences, or, more generally, to *any* science which interprets experience on a higher more than on a purely descriptive level....

I think that this is a relatively good approximation to truth—which is much too complicated to allow anything but approximations—that mathematical ideas originate in empirics, although the genealogy is sometimes long and obscure. But, once they are so conceived, the subject begins to live a peculiar life of its own and is better, compared to a creative one, governed by almost entirely aesthetic motivations, than to anything else and, in particular, to an empirical science....

But there is a grave danger that the subject will develop along the line of *least* resistance, that the stream, so far from its source, will separate into a multitude of insignificant tributaries, and that the discipline will become a disorganized mass of details and complexities. In other words, at a great distance from its empirical sources, or after much "abstract" inbreeding a mathematical object is in danger of degeneration. At the inception the style is usually classical; when it shows signs of becoming *baroque*, then the danger signal is up....

Whenever this stage is reached, the only remedy seems to be a rejuvenating return to the source: the reinjection of more or less directly empirical ideas. I am convinced that this was a necessary condition to conserve the freshness and the vitality of the subject and that this will remain equally true in the future.

John von Neumann (1947)

For a mentor of Ph.D. candidates it would be most easy to educate a poor applied mathematician. The next simplest thing would be to educate a poor pure mathematician. Then an entire quantum gap lies between the education of a good pure mathematician, and finally, an enormous quantum gap, the education of a good applied mathematician. For the latter task (especially after the death of John von Neumann) I would consider no one sufficiently qualified. The knowledge and abilities which are nowadays required of a really successful applied mathematician, presume an extraordinary high intellectual standard, and, even for the career of our present-day students, it is almost impossible to predict which parts of mathematics will prove most suited for applications.

Peter Hilton (1973)

Mathematics is not a deductive science—that's a cliché. When you try to prove a theorem, you don't just list the hypotheses, and then start to reason. What you do is trial-and-error, experimentation, and guesswork.

Paul Halmos (1985)

Between the ages of 12–16, I familiarized myself with the elements of mathematics. In doing so I had the good fortune of discovering books which were not too particular in their logical rigor.

At the age of 17, I entered the Polytechnic Institute of Zurich. There I had excellent teachers (for example, Hurwitz and Minkowski), so that I really could have obtained a sound mathematical education. However, most of the time I worked in the physical laboratory, fascinated by the direct contact with experience. The rest of the time I used, in the main, to study at home the works of Kirchhoff, Helmoltz, Hertz, etc. The fact that I neglected mathematics to a certain extent had its cause not merely in my stronger interest in the natural sciences than in mathematics, but also in the following strange experience. I saw that mathematics was split up into numerous specialities, each of which could easily absorb the short life granted to us. Consequently, I saw myself in the position of Buridan's ass which was unable to decide upon any specific bundle of hay. This was obviously due to the fact that my intuition was not strong enough in the field of mathematics in order to differentiate clearly that which was fundamentally important, and that which is really basic, from the rest of the more or less dispensable erudition, and it was not clear to me as a student that the approach to a more profound knowledge of the basic principles of physics is tied up with the most intricate mathematical methods. This only dawned upon me gradually after years of independent scientific work. True enough, physics was also divided into separate fields. In this field, however, I soon learned to scent out that which was able to lead to fundamentals.

Albert Einstein (1955)

Dick Feynman was a profoundly original scientist. He refused to take anybody's word for anything. This meant that he was forced to rediscover or reinvent for himself almost the whole of physics. It took him five years of concentrated work to reinvent quantum mechanics. At the end, he had a version of quantum mechanics that he could understand. The calculations I did for Hans Bethe, using the orthodox theory, took me several months of work and several hundred sheets of paper. Dick could get the same answer, calculating on a blackboard, in half an hour.

In orthodox physics it can be said: Suppose an electron is in this state at a certain time, then you calculate what it will do next by solving a certain differential equation (the Schrödinger equation). Instead of this, Dick said simply: "The electron does whatever it likes." A history of the electron is any possible path in space and time. The behavior of the electron is just the result of adding together all the histories according to some simple rules that Dick worked out. I had the enormous luck to be there at Cornell in 1948 when the idea was newborn, and to be for a short time Dick's sounding board.

Dick distrusted my mathematics and I distrusted his intuition. Dick fought against my scepticism, arguing that Einstein had failed because he stopped thinking in concrete physical images and became a manipulator of equations. I had to admit that was true. The great discoveries of Einstein's earlier years were all based on direct physical intuition. Einstein's later unified theories failed because they were only sets of equations without physical meaning.

Nobody but Dick could use his theory. Without success I tried to understand him .... For two weeks I had not thought about physics, and then it came bursting into my consciousness like an explosion. Feynman's pictures and Schwinger's equations began sorting themselves out in my head with a clarity they have never had before. I had no pencil or paper, but everything was so clear I did not need to write it down. Feynman and Schwinger were just looking at the same set of ideas from two different sides. Putting their methods together, you would have a theory of quantum electrodynamics that combined the mathematical precision of Schwinger with the practical flexibility of Feynman.

Freeman J. Dyson (1979)

There are mathematicians who reject a binding of mathematics to physics, and who justify mathematical work solely by aesthetical satisfaction which, besides all the difficulty of the material, mathematics is able to offer. Such mathematicians are more likely to regard mathematics as a form of art than a science, and this point of view of mathematical unselfishness can be characterized by the slogan "l'art pour l'art".

On the other hand, there are physicists who regret that their science is so much related to mathematics. They fear a loss of intuition in the natural sciences. They consider the intimate relation with nature, the finding of ideas in nature itself, which was given to Goethe (1749–1832) in such a high degree, as being destroyed by mathematics, and their anger or sorrow is the more serious the more they are forced to realize the inevitability of mathematics.

Both points of view deserve serious consideration; because not only people with narrow minds have expressed such opinions. Yes, one can say that such a radical inclination to one side or the other, if not caused by a lack of talent, is sometimes evidence of a deeper perception of science, as if someone is interested in both sciences, but at the same time is satisfied with obvious connections between mathematics and physics....

Mathematics is an organ of knowledge and an infinite refinement of language. It grows from the usual language and world of intuition as does a plant from the soil, and its roots are the numbers and simple geometrical intuitions. We do not know which kind of content mathematics (as the only adequate language) requires; we cannot imagine into what depths and distances this spiritual eye (mathematics) will lead us.

Erich Kähler (1941)

Relations between mathematics and physics vary with time. Right now, and for the past few years, harmony reigns and a honeymoon blossoms. However, I have seen other times, times of divorce and bitter battles, when the sister sciences declared each other as useless—or worse. The following exchange between a famous theoretical physicist and an equally famous mathematician might have been typical, some fifteen or twenty years ago:

Says the physicist: "I have no use for mathematics. All the mathematics I ever need, I invent in one week."
Answers the mathematician: "You must mean the seven days it took the Lord to create the world."

A slightly more reliable document is found in the preface of the first edition of Hermann Weyl's book on group theory and quantum mechanics from 1928. He writes: "I cannot abstain from playing the role of an (often unwelcome) intermediary in this drama between mathematics and physics, which fertilize each other in the dark, and deny and misconstrue one another when face to face."

This dramatic situation, described here by one of the great masters in both sciences, is a result of recent times. At the time of Newton (1642–1727) disharmony between mathematics and physics seemed unthinkable and unnatural, since both were his brainchildren; and close symbiosis persisted through the whole of the eighteenth century. The rift arose around 1800 and was caused by the development of pure mathematics (represented by number theory) on the one hand, and of a new kind of physics, independent of mathematics, which developed out of chemistry, electricity and magnetism on the other. This rift was widened in Germany under the influence of Goethe (1749–1832) and his followers, Schelling (1775–1854) and Hegel (1770–1831) and their "Naturphilosophie."...

Our protagonists are Carl Friedrich Gauss (1777–1855), as the creator of modern number theory, and Michael Faraday (1791–1867) as the inventor of physics without mathematics (in the strict sense of the word).

It would be foolish, of course, to claim the nonexistence of number theory before Gauss. An amusing document may illustrate the historical development. Erich Hecke's famous "Lectures on the Theory of Algebraic Numbers" has on its last page a "timetable," which chronologically lists the names and dates of the great number theoreticians, starting with Euclid (300 B.C.) and ending with Hermann Minkowski (1864–1909). As a physicist, I am impressed to find so many familiar names in this Hall of Fame: Fermat (1601–1665), Euler (1707–1783), Lagrange (1736–1813), Legendre (1752–1833), Fourier (1768–1830), and Gauss (1777–1855). In fact, we cannot find a single great number theoretician before Gauss, whom we would not count among the great physicists, provided we disregard antiquity. Specialization starts after 1800 with names like Kummer, Galois, and Eisenstein; who were all under the great influence of Gauss' "Disquisitiones Arithmeticae." In this specific sense Gauss' book marks the dividing line between mathematics as a universal science and mathematics as a union of special disciplines, and between the "géomètre" as a universal "savant" in the sense of the eighteenth century and the specialized "mathématicien" of modern times. As is typical for a man of transition, Gauss does not belong to either category, he was universal and specialized. The struggle raged within him—and made him suffer.

Res Jost (1984)
(Mathematics and Physics Since 1800: Discord and Sympathy)

Mathematics is an ancient art, and from the outset it has been both the most highly esoteric and the most intensely practical of human endeavors. As long ago as 1800 B.C., the Babylonians investigated the abstract properties of numbers; and in Athenian Greece, geometry attained the highest intellectual status. Alongside this theoretical understanding, mathematics blossomed as a day-to-day tool for surveying lands, for navigation, and for the engineering of public works. The practical problems and theoretical pursuits stimulated one another; it would be impossible to disentangle these two strands.

Much the same is true today. In the twentieth century, mathematics has burgeoned in scope and in diversity and has been deepened in its complexity and abstraction. So profound has this explosion of research been that entire areas of mathematics may seem unintelligible to laymen—and frequently to mathematicians working in other subfields. Despite this trend towards—indeed because of it—mathematics has become more concrete and vital than ever before.

In the past quarter of a century, mathematics and mathematical techniques have become an integral, pervasive, and essential component of science, technology, and business. In our technically oriented society, "innumeracy" has replaced illiteracy as our principal educational gap. One could compare the contributions of mathematics to our society with the necessity of air and food for life. In fact, we could say that we live in the age of mathematics—that our culture has been "mathematicized." No reflection of mathematics around us is more striking than the omnipresent computer....

There is an exciting development taking place right now, *reunification* of mathematics with theoretical physics....

In the last ten or fifteen years mathematicians and physicists realized that modern geometry is in fact the natural mathematical framework for gauge theory. The gauge potential of gauge theory is the connection of mathematics. The gauge field is the mathematical curvature defined by the connection; certain "charges" in physics are the topological invariants studied by mathematicians. While the mathematicians and physicists worked separately on similar ideas, they did *not* just duplicate each other's efforts. The mathematicians produced general, far-reaching theories and investigated their ramifications. Physicists worked out details of certain examples which turned out to describe nature beautifully and elegantly. When the two met again, the results were more powerful than either anticipated....

In mathematics we now have a new motivation to use specific insights from the examples worked out by physicists. This signals the return to an ancient tradition....

Mathematical research should be as *broad* and as *original* as possible, with very *long-range goals*. We expect history to repeat itself: we expect that the most profound and useful future applications of mathematics cannot be predicted today, since they will arise from mathematics yet to be discovered.

Arthur M. Jaffe (1984)
(Ordering the Universe: the Role of Mathematics)

The mathematician is a very lonely man. Certainly, he has family and friends, whose company he enjoys and whose understanding he needs; but his work, his mathematical problems—i.e., that which makes up the essence of his conscious life—he cannot share with anyone.

Krysztof Maurin (1981)

APPLICATIONS IN MECHANICS

> The book of nature is written in the language of mathematics.
> Galileo Galilei (1564–1642)

> Indolence destroys knowledge, let us live and work.
> Johannes Kepler (1571–1630)

> In 1605, Kepler recognized the orbit of the planet Mars as an ellipse, and, in 1638, the triumphal march of the natural sciences began with Galileo's law of falling bodies. Today we start to ask questions, as we see how little help the sciences are able to offer in questions of humanity, and also, because their child, technology, which gave rise to so many hopes, can so terribly be misused.
> Wilhelm Blaschke (1946)

In the following two chapters we discuss the basic ideas of mechanics and the dualism between wave and particle. This is fundamental for a deeper understanding of all problems, considered in Parts IV and V.

CHAPTER 58

Basic Equations of Point Mechanics

> Lex prima: A stationary body will remain motionless, and a moving body will continue to move in the same direction with unchanging speed unless it is acted on by some force.
> Lex secunda: The time-rate-of-change of the momentum of a body is proportional to the force.
> Lex tertia: If any object exerts a force on another object, then the second object also exerts an equal and opposite force on the first.
> Lex quarta: Forces are added like vectors.
> Isaac Newton, *Philosophiae Naturalis Principia Mathematica* (London, 1687)

> The same applies to the concept of force as does to any other physical concept: Verbal definitions are meaningless; real definitions are given through a measuring process.
> Arnold Sommerfeld (1954)

Mechanics is the oldest physical discipline. Its ideas, however, have influenced many other branches of physics. The *goal* in this chapter is to present some general principles of point mechanics which are necessary to understand elasticity theory, hydrodynamics, and many other branches of physics (statistical physics, theory of relativity, electrodynamics, quantum mechanics and quantum field theory, etc.). We will try to explain the close relation between the results about variational problems of Part III and the basic principles of mechanics. In particular, we explain the connection between Lagrange's multiplier rule and the principle of least constraint and the least (stationary) action. To introduce the reader to the basic ideas, we consider in Section 58.2 a simple, but typical example: equilibrium state and motion of a balance, and its stability. Many modern expositions begin with the principle of stationary action. This principle, however, does not explicitly contain the most important physical concept—the force. Also, the principle of the stationary action, other than the principle of the least constraint, does not admit the most general side conditions, with nonlinear relations for the velocities. We therefore choose to present the basic principles in the following order:

(i) Principle of the least constraint of Gauss (most general principle in mechanics).
(ii) Principle of virtual power.
(iii) General stability principle for equilibrium states and, as a special case, the principle of minimal potential energy.
(iv) Lagrange function and the principle of stationary action.
(v) Hamiltonian formalism (canonical equations and the partial differential equation of Hamilton–Jacobi).
(vi) Poisson brackets and Poisson's mechanics.

Since the principle of virtual work often leads to misunderstandings, we replace it with the principle of virtual power. For a discussion of this, see Section 58.3. We hope our approach will highlight the relation between elegant mathematical theories (Lagrangian and Hamiltonian formalism) on the one hand, and physical reality on the other. Especially emphasized are stability questions. Points (iv)–(vi) play an important role in the formulation of physical theories other than classical mechanics.

We consider the following interesting applications:

(a) Motion of planets.
(b) Motion of a rigid body.
(c) Expanding universe in the context of classical mechanics.

In Section 58.21 we show that Poisson's formulation of classical mechanics allows a simple deduction of quantum mechanics. As an application we consider the quantum mechanical oscillator in the context of Heisenberg's matrix mechanics. This implies Planck's famous formula about the quantization of energy for the harmonic oscillator.

We choose here a presentation of classical mechanics, which later on might help to understand modern generalizations such as the special and general theory of relativity and quantum field theory. The most elegant mathematical fundament for mechanics is formed by the geometry on symplectic manifolds. This will be shown in Part V.

58.1. Notations

In classical mechanics one often works in the real linear three-dimensional space V_3 of vectors $x, y, \ldots$. Intuitively, x is an arrow in the usual three-dimensional Euclidean space, and arrows obtained by translations are identified (Fig. 58.1). In the usual way, let

$$xy \quad \text{and} \quad x \times y$$

58.1. Notations

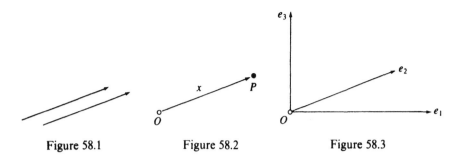

Figure 58.1   Figure 58.2   Figure 58.3

denote the scalar product and the vector product, respectively. If we specify a point O as the coordinate origin and attach x at O, then x is called the *position vector* and we identify the endpoint P with x (Fig. 58.2). A coordinate system consists of a point O and three linearly independent vectors b_1, b_2, b_3 (basis vectors). We can write

$$x = \xi^1 b_1 + \xi^2 b_2 + \xi^3 b_3,$$

where the numbers ξ^1, ξ^2, ξ^3 are called the coordinates of the point x. In classical physics we shall use *Einstein's sum convention*, i.e., equal upper and lower indices are summed from 1 to 3. In particular, we write

$$x = \xi^i b_i.$$

In Cartesian coordinates the basis vectors are denoted by e_1, e_2, e_3, i.e., all e_i are unit vectors, pairwise orthogonal, and oriented as in Figure 58.3.

The trajectory of a point is given by a map

$$t \mapsto x(t)$$

with real time parameter t and position vector $x(t)$. A dot denotes derivatives with respect to time.

The vectors

$$\dot{x}(t) \quad \text{and} \quad \ddot{x}(t)$$

are called *velocity vector* and *acceleration vector* at time t, respectively. If $x(t) = \xi^i(t) b_i$, then

$$\dot{x}(t) = \dot{\xi}^i(t) b_i \quad \text{and} \quad \ddot{x}(t) = \ddot{\xi}^i(t) b_i.$$

The absolute values $|\dot{x}(t)|$ and $|\ddot{x}(t)|$ are called velocity and acceleration at time t, respectively.

EXAMPLE 58.1 (Rotation About an Axis). Let b be a unit vector. Consider a rotation $y = Tx$ about the point O and the axis of rotation b with an angle φ, which is measured clockwise (Fig. 58.4(a)).

We choose Cartesian coordinates with $e_3 = b$, and obtain

$$y = Tx = (xe_1)f_1 + (xe_2)f_2 + (xb)b,$$

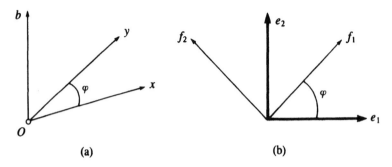

Figure 58.4

with vectors

$$f_1 = (\cos \varphi)e_1 + (\sin \varphi)e_2, \qquad f_2 = -(\sin \varphi)e_1 + (\cos \varphi)e_2$$

(Fig. 58.4(b)). If we fix x and choose $\varphi = \varphi(t)$ with $\varphi(0) = 0$, then we obtain a rotational motion about the axis of rotation b with $y(0) = x$. Differentiation with respect to t gives

$$\dot{f}_1 = \dot{\varphi}f_2, \qquad \dot{f}_2 = -\dot{\varphi}f_1, \qquad \dot{b} = 0.$$

For the velocity vector of this rotational motion we obtain

$$\dot{y}(t) = \dot{\varphi}(t)b \times y(t).$$

The number $\omega = \dot{\varphi}(t)$ is called the *angular velocity* at time t. Often, one writes

$$\omega(t) = \dot{\varphi}(t)b$$

and calls ω the angular velocity vector.

Expansion of Tx with respect to φ gives

$$Tx = \varphi b \times x + O(\varphi^2), \qquad \varphi \to 0.$$

The linearized map $y = \varphi b \times x$ is called an *infinitesimal rotation* about the angle φ. Furthermore,

$$y = (\boldsymbol{\varphi} \times x) + c$$

with $\boldsymbol{\varphi} = \varphi b$ and the constant vector c is called an *infinitesimal motion*. It is composed of an infinitesimal rotation and a translation.

The space V_3 is an H-space with scalar product

$$(x|y) \stackrel{\text{def}}{=} xy.$$

For $x = \xi^i b_i$ and $y = \eta^i b_i$ we have

$$xy = \xi^i \eta^j b_i b_j.$$

Let $A: V_3 \to V_3$ be a linear operator. In a fixed coordinate system equation

$$y = Ax$$

means that relation

$$\eta^j = A^j_i \xi^i$$

holds for the coordinates. Such linear operators appear in the form of inertia tensors, strain tensors, or stress tensors. The adjoint operator of A is denoted by $A^*\colon V_3 \to V_3$. By definition, we have $(x|Ay) = (A^*x|y)$, i.e.,

$$x(Ay) = y(A^*x) \quad \text{for all} \quad x, y \in V_3.$$

Thus in Cartesian coordinates this gives

$$(A^*)^j_i = A^i_j \quad \text{for all } i, j.$$

For the F-derivative of a function $U\colon V_3 \to \mathbb{R}$ at a point x we find

$$U'(x)h = h^i D_i U(x) \quad \text{for all} \quad h \in V_3,$$

where $h = h^i b_i$ and $D_i = \partial/\partial \xi^i$. If we define linear functionals $b^i \in V_3^*$ through $\langle b^i, h \rangle = h^i$, then

$$U'(x) = D_i U(x) b^i.$$

One also writes

$$U'(x) = \operatorname{grad}(U(x))$$

for this. Obviously, $\{b^i\}$ is a basis for the dual space V_3^*. We call $\{b^i\}$ the dual basis of $\{b_i\}$. We agree to identify V_3^* with V_3. More precisely, for every $b^* \in V_3^*$ there exists a unique $b \in V_3$ with

$$\langle b^*, x \rangle = (b|x) = bx$$

and we identify b^* with b. From

$$\langle b^i, b_j \rangle = \delta^i_j \quad \text{for all } i, j$$

we obtain $b^i \in V_3$ and

$$b^i b_j = \delta^i_j \quad \text{for all } i, j.$$

These equations uniquely determine b^i. Explicitly, we have

$$b^1 = (b_2 \times b_3)/b_1(b_2 \times b_3).$$

The formulas for b^2 and b^3 are obtained through cyclic permutations. In this sense $U'(x) \in V_3$ holds. In Cartesian coordinates it follows that $b^i = b_i = e_i$. Then

$$U'(x) = \operatorname{grad} U(x) = D_i U(x) e_i,$$

where i is summed from 1 to 3.

Using the previous considerations we now may apply functional analytic methods in classical mechanics. This enables us to work coordinate free. In particular, we can use the coordinate free differential calculus for the F-derivative of Chapter 4. This greatly simplifies formulas, and calculations.

58.2. Lever Principle and Stability of the Scales

In the history of physics many general results have been obtained in studying concrete phenomena. Archimedes' lever principle (statics) and Kepler's laws about the planetary motion together with Galilei's study of the free fall (dynamics) were of great significance for the development of mechanics. Using the example of the scales we try to show in this section how one is naturally led to the fundamental concepts of torque, work, power, constraining force, potential energy, kinetic energy, and total energy. Moreover, we consider the central principle of virtual power and discuss stability criteria. In the following section we explain connections with other general principles in mechanics, which later on will be studied in greater detail.

We begin with the lever principle. Figure 58.5 shows the typical configuration of a pair of scales, i.e., we consider the simplest model of a weighing machine. It consists of three firmly connected points with masses m_i and corresponding vectors y_i, which are attached at the point Q. At the endpoint y_i, the force K_i is applied, $i = 1, 2, 3$, where K_i is a vector. We first assume that, except for the firm connection, all three points are able to move freely (freely moving scales). In Section 58.13 we derive the following equilibrium condition for this example of a rigid body: total *force* and total *torque* with respect to Q must be zero, i.e.,

$$K_1 + K_2 + K_3 = 0, \tag{1}$$

$$(y_1 \times K_1) + (y_2 \times K_2) + (y_3 \times K_3) = 0, \tag{2}$$

and the body must be at rest at some fixed time. Condition (1) is satisfied with $K_3 = -K_1 - K_2$. From (2) follows

$$|K_1||y_1| = |K_2||y_2|. \tag{3}$$

This relation is called the *lever principle* and more suggestively it reads:

Force times force arm equals weight times weight arm.

For the gravitational force $K_i = -m_i g e_3$, $i = 1, 2$, with acceleration due to gravity $g = 9.81 \text{ ms}^{-2}$, formula (3) becomes

$$m_1|y_1| = m_2|y_2|. \tag{4}$$

Now we consider a balance, which is not freely moving and for the remainder of this chapter make the natural assumption that the third point y_3 is the fixed point of rotation O. We choose O as in Figure 58.6 as coordinate origin. This way we are able to neglect the motion of the third mass point m_3. The other two mass points with masses m_1 and m_2 and with position vectors x_1 and x_2

58.2. Lever Principle and Stability of the Scales

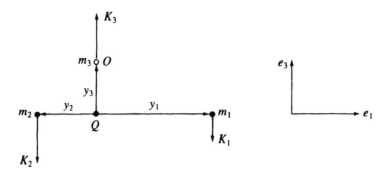

Figure 58.5

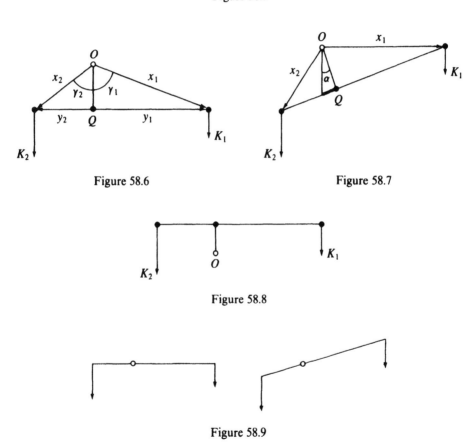

Figure 58.6 Figure 58.7

Figure 58.8

Figure 58.9

are affected by the gravitational forces K_1 and K_2, respectively. The system thus obtained only has one degree of freedom, characterized by the angle of rotation α of Figure 58.7. We are now looking for a *general principle*, which yields equilibrium conditions and the stability of the equilibrium positions. To engineers, only the stable equilibrium positions are of interest in constructing buildings, bridges, etc. In order to find such a principle, we use the concept of work. For an arbitrary motion of the system

$$x_i = x_i(t)$$

which for $t = 0$ is shown by Figure 58.6, we define the *work* during the time interval $[0, t]$ as

$$W(t) = \int_0^t K_1 \dot{x}_1 + K_2 \dot{x}_2 \, dt.$$

Stability Principle 58.2. *Let $i = 1, 2$. If*

$$W(t) < 0$$

for all $|t| > 0$ which are sufficiently small and all motions $x_i = x_i(t)$ of the system which satisfy $x_i(t) \neq x_i(0)$ for $|t| > 0$, then the system is in a stable equilibrium for the position $(x_1(0), x_2(0))$.

This principle will be formulated in a general form in Section 58.12. Intuitively, it means the following. If we set the system of Figure 58.6 into the motion $x_i = x_i(t)$, then for $W(t) > 0$ the gravitational force has to do work, i.e., in this case we gain work, whereas for $W(t) < 0$ we have to do work ourselves. Thus we expect the stability of the equilibrium position to increase, if $W(t) < 0$ decreases. We now assume the lever condition (4). In the following we will see that the stability of the configuration in Figure 58.6 increases, the higher the point of rotation O lies above the rocking beam, i.e., the larger $d = |y_3|$ is. The configuration in Figure 58.8, on the other hand, satisfies $W(t) > 0$ for $|t| > 0$, i.e., as expected, this equilibrium position is unstable. If the point of rotation O lies on the rocking beam, then this will imply that $W(t) \equiv 0$. Here, every configuration of Figure 58.9 is in an equilibrium position. But all these equilibrium positions are unstable, which everybody knows who once sat on a swing.

Principle of Virtual Power 58.3. *Since $W(0) = 0$, the Stability Principle 58.2 implies that $\dot{W}(0) = 0$ for a stable equilibrium i.e.,*

$$\dot{W}(0) = K_1 \dot{x}_1(0) + K_2 \dot{x}_2(0) = 0. \tag{5}$$

By definition, $\dot{W}(0)$ is the *power* of the forces K_1 and K_2 at time $t = 0$. Therefore (5) is called the principle of virtual power. The attribute "virtual" comes from the fact that in (5) we consider all mathematically possible motions $x_i = x_i(t)$ of the system, which are compatible with the side conditions, not only those which are physically possible, under the influence of the forces. The equations for the physically possible motions will be studied in (12) below.

Proposition 58.4. *If the point of rotation O of the scales lies properly above the rocking beam, and if equilibrium condition (4) is satisfied, then Figure 58.6 corresponds to a stable equilibrium position.*

PROOF. Let $x_i = \xi_i e_1 + \zeta_i e_3$. We define the *potential energy* as
$$U(x_1, x_2) = m_1 g \zeta_1 + m_2 g \zeta_2 + U_0,$$
where U_0 is an arbitrary constant. We choose U_0 such that $U(x_1(0), x_2(0)) = 0$. Because of $K_i = -m_i g e_3 = -U_{x_i}$ we have
$$W(t) = \int_0^t K_1 \dot{x}_1 + K_2 \dot{x}_2 \, dt = -\int_0^t \frac{dU}{dt} dt = -U(x_1(t), x_2(t)).$$

For the configuration in Figure 58.7 we obtain
$$\begin{aligned} x_1 &= |x_1|(\sin(\gamma_1 + \alpha)e_1 - \cos(\gamma_1 + \alpha)e_3), \\ x_2 &= -|x_2|(\sin(\gamma_2 - \alpha)e_1 + \cos(\gamma_2 - \alpha)e_3). \end{aligned} \quad (6)$$

Figure 58.6 illustrates the meaning of the angle γ_i. All mathematically possible motions are then obtained from all possible functions $\alpha = \alpha(t)$ with $\alpha(0) = 0$. One finds
$$W(t) = -V(\alpha(t)) \quad (7)$$
with $U(x_1, x_2) = V(\alpha)$, where
$$\begin{aligned} V(\alpha) &= m_1 g |x_1|(\cos \gamma_1 - \cos(\gamma_1 + \alpha)) + m_2 g |x_2|(\cos \gamma_2 - \cos(\gamma_2 - \alpha)) \\ &= \alpha V'(0) + \alpha^2 V''(0)/2 + O(\alpha^3), \qquad \alpha \to 0, \end{aligned}$$
and
$$V'(0) = m_1 g |y_1| - m_2 g |y_2|,$$
$$V''(0) = d(m_1 g + m_2 g) \stackrel{\text{def}}{=} \beta, \qquad d \stackrel{\text{def}}{=} |y_3|.$$

Equation $\dot{W}(0) = 0$ implies
$$\dot{W}(0) = -V'(0)\dot{\alpha}(0) = 0.$$

Since $|\dot{\alpha}(0)|$ can be chosen arbitrarily, it follows that $V'(0) = 0$. This is equilibrium condition (4). Moreover, for $d > 0$ we get $W(t) < 0$ if $|\alpha(t)| > 0$ is sufficiently small. □

58.3. Perspectives

The simple example of the previous section illustrates already a number of important ideas in mechanics, which in a general form will be discussed in the following sections. Here we begin with several observations and hope that this detailed discussion will help to a better understanding of relations in physics.

(i) *Principle of virtual power*. The corresponding condition
$$\dot{W}(0) = 0$$
for the power has been discussed already in (5) above.

(ii) *Principle of virtual work.* The work $W(t)$ depends only on the angle $\alpha(t)$. It is

$$W = -V'(0)\alpha + O(\alpha^2), \qquad \alpha \to 0.$$

The linearization

$$\delta W \stackrel{\text{def}}{=} -V'(0)\alpha$$

is called virtual work. Often one writes $\delta\alpha$ instead of α, i.e.,

$$\delta W = -V'(0)\,\delta\alpha.$$

Instead of $\dot{W}(0) = 0$ in (i) one may also require that

$$\delta W = 0.$$

This also leads to the equilibrium condition $V'(0) = 0$.

In this volume we prefer (i) over (ii), since (i) allows an immediate physical interpretation, while (ii) is only an auxiliary mathematical construction. Interestingly, physicists prefer (ii). The reason for this is simple. In order to really understand the principle, (i) is most appropriate. But in many simple situations, one can use (ii), together with intuitive geometrical arguments about the nature of the virtual displacements, to obtain faster results than through exact computations via (i).

In many textbooks, however, one finds formulations, that are not very elucidating. In the classical textbook by Sommerfeld (1954) one reads: "A virtual displacement is an arbitrary and infinitely small change of the position of the system, compatible with the side conditions.... The virtual work of the reactions needed is equal to zero." In the language of Sommerfeld, which accidentally was the language of the mathematicians at the beginning of differential and integral calculus during the eighteenth century, one would say: If the real function $y = f(x)$ has a minimum at x_0, then the "virtual displacement"

$$\delta y = f'(x_0)\,\delta x,$$

is equal to zero for "infinitely small" displacements δx. This means that the Taylor expansion is only considered up to the linear term

$$f(x_0 + h) - f(x_0) = f'(x_0)h$$

and that this expression is set equal to zero with $h = \delta x$. Today we simply write

$$f'(x_0) = 0.$$

In Budó (1954) one reads the obscure remark: "While the real displacement always takes place during a certain time dt, the virtual displacement can be regarded as timeless $\delta t = 0$. The velocity of the virtual displacement is infinitely large."

58.3. Perspectives

(iii) *Principle of minimal potential energy.* Using the relation

$$W(t) = -U(x_1(t), x_2(t)),$$

with $U(x_1(0), x_2(0)) = 0$ and $W(t) < 0$ for $|t| > 0$, we obtain from the Stability Principle 58.2: If the potential energy U has a strict minimum at $(x_1(0), x_2(0))$ with respect to all possible positions $(x_1(t), x_2(t))$, then $(x_1(0), x_2(0))$ corresponds to a stable equilibrium position.

(iv) *Dynamics and the principle of virtual power.* We now want to study the actual physical motion of our weighing machine. The reader who is familiar with Newton's law "Force equals mass times acceleration" should be warned that here the equations of motions are *not* of the form

$$m_i \ddot{x}_i = K_i, \qquad i = 1, 2. \tag{8}$$

The reason for this is that, besides the forces K_i, additional so-called constraining forces act, which are caused by the side conditions (linkings) and which guarantee that during the motion the rigid fixings are preserved. For time-independent side conditions, as in our case, the actual physical motion

$$x_i = x_i(t)$$

of the system is not determined by (8), but by the dynamical principle of virtual power, which will be discussed in Section 58.11. This principle states that

$$\sum_{i=1}^{2} (m_i \ddot{x}_i(t_0) - K_i) v_i = 0 \tag{9}$$

for all t_0. Here $v_i = \dot{y}_i(t_0)$ are arbitrary velocity vectors, which correspond to all mathematically possible motions $x_i = y_i(t)$ of the system, which are compatible with the side conditions and satisfy

$$y_i(t_0) = x_i(t_0)$$

at some fixed time t_0. One refers to $x_i = y_i(t)$ as the virtual motion. Among all these virtual motions, the actual motion $x_i = x_i(t)$ of the system is determined by (9).

If, in particular, we choose $t_0 = t$ and $v_i = \dot{x}_i(t)$, then we obtain from (9) that

$$\dot{T}(t) = K_1 \dot{x}_1(t) + K_2 \dot{x}_2(t) \equiv \dot{W}(t), \tag{10}$$

where, by definition,

$$T = (m_1 \dot{x}_1^2 + m_2 \dot{x}_2^2)/2,$$

is the *kinetic energy*. Because of $W(t) = -V(\alpha(t))$ we obtain from (5) the fundamental energy conservation law

$$T(t) + V(\alpha(t)) = E_0, \tag{11}$$

with E_0 a constant which is called the total energy or simply the *energy* of the system. From (6) follows

$$T = \mu \dot{\alpha}^2/2$$

with $\mu = m_1 x_1^2 + m_2 x_2^2$. The equation of motion for $\alpha = \alpha(t)$ therefore becomes

$$2^{-1}\mu\dot\alpha^2 + V(\alpha) = E_0, \qquad (12)$$

with $V(\alpha) = \beta\alpha^2/2 + O(\alpha^3)$, $\alpha \to 0$, $\beta > 0$. The solution of (12) with $\alpha(0) = 0$ is

$$\sqrt{\frac{\mu}{2}} \int_0^\alpha \frac{d\alpha}{\sqrt{E_0 - V(\alpha)}} = t.$$

If, approximately, we set $V(\alpha) = \beta\alpha^2/2$, then we obtain

$$\alpha(t) = (\dot\alpha(0)/\omega)\sin \omega t, \qquad \omega = \sqrt{\beta/\mu}. \qquad (13)$$

These are the oscillations of the scales with angular frequency ω.

In Figures 58.8 and 58.9 we have $\beta < 0$ and $V(\alpha) \equiv 0$, respectively. This changes the qualitative behavior of the motion. For example, Figure 58.9 yields rotations about O with constant angular velocity $\dot\alpha$.

If we use approximations of $V(\alpha)$ up to order three or four, then we obtain elliptic functions in (13).

(v) *Dynamic stability.* In Stability Principle 58.2 only the statics is taken into consideration. A general stability analysis, however, needs to take the dynamical nature of the problem into account, i.e., changes of the system under small time-dependent perturbations of position and velocity. This corresponds to the *Ljapunov stability* for equations of motion, which has been discussed in Section 3.6. In many cases, stability results are difficult to obtain. In our case, however, the conservation of energy admits a simple analysis. If $\alpha(0)$, $\dot\alpha(0)$ are the initial data, then E_0 follows from (12). Furthermore, (12) implies that the motion is possible only for

$$E_0 - V(\alpha) \geq 0.$$

In Figure 58.10 this corresponds to the dark printed α-interval. This implies the following:

If $|\alpha(0)|$ and $|\dot\alpha(0)|$ are sufficiently small, then the total energy $|E_0|$ is small and hence $|\alpha(t)|$ remains small for all times.

This is stability of the system in the sense of Ljapunov. If the slope of V increases, i.e., β and $d = |y_3|$ in Figure 58.5 increase, then the interval in Figure

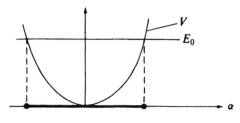

Figure 58.10

58.3. Perspectives

58.10, in which α varies, decreases. This implies:

The higher the point of rotation O lies above the rocking beam, the more stable the equilibrium state of the scales becomes.

Finally, we will use our simple example to explain the basic ideas of Lagrangian and Hamiltonian formalism.

(vi) *Lagrangian formalism.* We consider the Lagrange function

$$L(\alpha, \dot{\alpha}) = T - U = 2^{-1}\mu\dot{\alpha}^2 - V(\alpha).$$

The equations of motions are then the so-called Lagrangian equations

$$\frac{d}{dt}L_{\dot{\alpha}} = L_{\alpha}, \tag{14}$$

i.e.,

$$\mu\ddot{\alpha} + V'(\alpha) = 0.$$

Hence multiplication with $\dot{\alpha}$ and integration imply the energy conservation law (12).

(vii) *Hamiltonian formalism.* We define the generalized momentum $p = L_{\dot{\alpha}} = \mu\dot{\alpha}$ and the Hamilton function

$$H(\alpha, p) = p\dot{\alpha} - L = (p^2/2\mu) + V(\alpha).$$

One immediately observes that H corresponds to the energy. The equation of motion (14) then assumes the symmetric form

$$\dot{p} = -H_{\alpha}, \qquad \dot{\alpha} = H_p. \tag{15}$$

This is a so-called Hamiltonian system.

(viii) *Geometrization of mechanics.* In Figure 58.6 we describe the position of the scales through (x_1, x_2). This is a point in a six-dimensional space

$$V_6 = V_3 \times V_3.$$

This form of description is very convenient if, for example, one wants to define the work $W(t)$ and the kinetic energy T. In fact, formula (6) implies that all possible positions only depend on a single parameter α. One may think of it as the angle, which describes the position of points on the unit circle. Hence, the so-called configuration space C, i.e., the set of all possible positions (x_1, x_2) forms a curve in V_6. More precisely, C is a one-dimensional C^{∞}-manifold, which is diffeomorphic to the unit circle. A motion

$$x_1 = x_1(t), \qquad x_2 = x_2(t), \qquad t_1 \le t \le t_2$$

corresponds to a curve in C, where $(\dot{x}_1(t), \dot{x}_2(t))$ is the tangent vector to this curve. The vectors (v_1, v_2), which appear in the principle of the virtual power (9) are precisely the tangent vectors to C at the point $(x_1(t_0), x_2(t_0))$. The space

V_6 only plays the role of an auxiliary space. Important for the motion of the system is the configuration space C. From a satisfactory mathematical theory for mechanics one would expect that it is a geometrical theory for the manifold C, i.e., the theory is independent of the choice of the parametrization α. The most elegant geometrical formulation of mechanics is obtained by formulating and studying the Hamiltonian system (15) in an invariant, i.e., parameter independent way, in the context of symplectic geometry on the cotangent bundle of C. This will be discussed in Part V.

In all parts of physics one observes a trend towards a geometrization. This is the realization of Einstein's program of finding a unification of physics. An immediate and more formal reason for this is that physicists perform their measurements in coordinate systems. But, of course, the physical essence of the phenomena should be independent of the choice of the coordinate system. A second and more profound reason is the following. The general theory of relativity, as well as the modern gauge field theories, have led to the consequence that all physical interactions can be described through the curvature of suitable manifolds, and a unified theory of matter should be sought in this direction.

The methods, demonstrated in this section by using the scales as an example, can equally well be applied to all static constructions in technology. Naturally, the computations become more complex if the number of degrees of freedom of the system increases.

58.4. Kepler's Laws and a Look at the History of Astronomy

> It is a sign for the power of mathematics that it is able to reproduce such a fine and superior process as the motion of celestial bodies, that it has words and symbols enough to determine the orbit of Jupiter for the next one hundred years from amongst all these infinitely many orbits which one can imagine, and this with an impressive accuracy.
>
> Erich Kähler (1941)

It seems that a look at the stellar sky has fascinated people since the early days and that it was one of the main reasons which led them to think about the meaning of the universe and its scientific and mathematical description. The ingenuity of astronomers and astrophysicists, who have gathered our present knowledge about the universe, is admirable, and should shame all master detectives in the world's literature. A first culminating point was the discovery of the laws of planetary motions by the Prague astronomer and mathematician Johannes Kepler (1571–1639) during the years from 1609 to 1619. At this time Kepler studied a tremendous amount of numerical data, which had been collected by Tycho Brahe (1546–1601). The laws are:

58.4. Kepler's Laws and a Look at the History of Astronomy

(i) The planets move on elliptic orbits with the sun at one focus.
(ii) The line segment joining a planet and the sun sweeps out equal areas in equal times.
(iii) The squares of the periods of revolution of two planets about the sun are proportional to the cubes of the semimajor axes of the ellipses.

These laws are true with respect to a Cartesian coordinate system Σ with the sun at its origin and whose axes are firmly connected with the stellar sky. A vague impression of Kepler's tremendeous scientific achievement is obtained by noting that all the numerical data originated from the moving planet Earth, i.e., in order to obtain (i) to (iii), Kepler first had to transform them to Σ. This step was never taken by the great astronomers of antiquity. For example, it was important to observe that Kopernikus' (1473–1543) hypothesis about a circular orbit of the planet Mars led to an average error of 8 minutes of arc. Laws (i) to (iii) are of a kinematic nature, i.e., they describe the motion, but not its cause. Isaac Newton's (1643–1727) path-breaking idea was then to recognize that (i) to (iii) follow from a universal law (law of gravity) and a general equation of motion (see Section 58.9). Newton based his work on Kepler's results and Galilei's (1564–1642) observation that all bodies fall at the same rate, i.e., receive a constant acceleration. This last observation led Newton to assume that on the surface of the earth a gravitational force exists which causes the free fall. Because of the interactions between the planets, (i) to (iii) are only approximately true. Until 1781 the only known planets were Mercury, Venus, Earth, Mars, Jupiter, and Saturn. In 1781, Herschel discovered Uranus with a telescope and in 1811, in connection with a prize problem of the Paris Academy, Delambre collected numerical data about the motion of Uranus. It was noticed that for several times during the previous one hundred years, Uranus had been registered as a fixed star. But the observed and the calculated data did not completely match each other. It was expected that the deviations were caused by a still unknown planet. In 1845 and 1846, after long and complicated computations two young astronomers, the Englishman Adams (1819–1892) and the Frenchman Leverrier (1811–1877), independently found the orbit of a new planet, called Neptune. Later, in 1846, Galle (1812–1910) at the Berlin astronomical observatory discovered it by following the numerical data contained in a letter by Leverrier. Jacobi then wrote: "One can only admire, how it is possible to obtain such precise results from so few and uncertain results. Those who call this discovery accidental, should also be encouraged to make such accidental discoveries themselves."

In 1930, the planet Pluto was discovered at the Flagstaff astronomical observatory in Arizona (U.S.A.) as a result of a perturbation calculation for Neptune. The following numbers should illustrate the orders of magnitude. To get some idea of our solar system, we consider in Table 58.1 a model to the scale of $1:10^9$. The sun then has a diameter of 1.4 m. The distance from the moon to the earth equals about 30 diameters of the earth, i.e., 0.4 m in this

Table 58.1. Model of the Solar System to the Scale of $1 \text{ m} \triangleq 10^6$ km.

Planet	Diameter	Comparison	Distance from the sun
Mercury	5 mm	pea	58 m
Venus	12 mm	cherry	108 m
Earth	13 mm	cherry	149 m
Mars	7 mm	pea	229 m
Jupiter	143 mm	coconut	778 m
Saturn	121 mm	coconut	1400 m
Uranus	50 mm	apple	2900 m
Neptune	53 mm	apple	4500 m
Pluto	10 mm	cherry	5900 m

model. The mass of the earth is

$$m_{\text{earth}} = 6 \cdot 10^{24} \text{ kg}.$$

Moreover, one roughly has

$$m_{\text{sun}} : m_{\text{earth}} : m_{\text{moon}} = 3 \cdot 10^5 : 1 : 10^{-2}.$$

Table 58.2 contains a survey about the huge distances in the universe. Our Milky Way has the form of a lens with a diameter of 10^5 light years and a thickness of maximally $1.5 \cdot 10^4$ light years. The sun is located almost at the boundary of the Milky Way and travels around its center with 285 km/s. This is about ten times as fast as the earth travels around the sun. For one revolution about the center of the galaxy the sun takes about $200 \cdot 10^6$ years. Since its creation $5 \cdot 10^9$ years ago, the sun has traveled only 25 times around the center of the galaxy. Our galaxy consists of approximately 10^{11} suns with a total mass of $2.5 \cdot 10^{11}$ sun masses, whereby 80 percent is located in the center, and recently probably a black hole has been discovered there. The

Table 58.2

Object	Travel time of light from the sun to the object
Earth	$8\frac{1}{3}$ minutes
Pluto	$5\frac{1}{2}$ hours
Next fixed star (α Centauri)	4 years
Sirius	9 years
Center of the Milky Way	$3 \cdot 10^4$ years
Next major galaxy (Andromeda nebulus)	$7 \cdot 10^5$ years
Quasars	10^{10} years (approximate age of the universe)

number of galaxies is estimated as 10^{11} if the closed cosmological model with finite volume is assumed. In the open model the volume is infinite (see Chapter 76).

Already in 1802 Newton's theory of gravity was a great triumph. One year earlier Piazzi, in Palermo, discovered the planetoid Ceres as a star of magnitude eight and was able to follow its orbit for 9 degrees before losing it. The young Gauss (1777–1855) then computed the entire orbit by employing new methods of the calculus of observations; and using this result, Olbers rediscovered Ceres in 1802.

As a consequence of the great success perturbation calculus has had in the discovery of Neptune in 1846, a perturbation of the orbit of Mercury in the form of a motion of the Perihelion of $\Delta\varphi = 43''$ during 100 years greatly intrigued astronomers, and at first led to the hypothesis of a planet Volcano. But it could never be observed. Today we know that the motion of the perihelion cannot be explained with Newton's theory of gravity, but is a consequence of the general theory of relativity, which was developed by Einstein in 1915. From this theory the above value follows very accurately (see Section 76.9).

The development of mechanics has greatly influenced the development of mathematics. Newton (1643–1727) himself, in searching for a mathematical formulation of his theory, created the differential and integral calculus. He did this independently of Leibniz (1646–1716). Analytical mechanics was further developed by Euler (1707–1783) and Lagrange (1736–1813), who both also developed variational calculus as a mathematical discipline, and by Hamilton (1805–1865) and Jacobi (1804–1851).

The famous n-body problem is the subject of Section 58.9c, and cosmological questions, which presently are at the center of interest, will be discussed in Section 58.15 and, more thoroughly, in Chapter 76.

58.5. Newton's Basic Equations

Galilei (1564–1642), in formulating his law of falling bodies, could do this by using the old algebra. But for Newton, who created the general principles of mechanics from Galilei's dynamics, including his now famous law of gravity, it was clear from the beginning that only a mathematics, containing the concept of infinitesimal small changes, would be able to draw conclusions from this new mechanics.

Such a form of mathematics was known to Newton (1643–1727) since his early days, but his notes on this remained in the hands of his closest friends, and even when, during the height of his life, he published his new concepts about the universe "*Principles of Natural Philosophy*", he carefully avoided this new calculus. He probably feared that his ideas would lose some of their strength by formulating them in a completely new mathematical language. In this regard he underestimated his fellow scientists; because when Leibniz (1646–1716), independently of Newton, formulated the infinitesimal calculus, using a more elegant formalism than Newton, this mathematics received an enthusiastic reception; however, "Newton's principles", despite the brash presentation, were well received on the European continent and translated back into the language of differential and integral calculus.

Erich Kähler (1941)

According to Newton, the motion of n mass points with position vectors x_i, $i = 1, \ldots, n$ and masses m_i is described by the *fundamental differential equations*

$$m_i \ddot{x}_i = K_i, \qquad i = 1, \ldots, n, \tag{16}$$

with forces $K_i = K_i(x_1, \ldots, x_n, \dot{x}_1, \ldots, \dot{x}_n, t)$. These are vector functions. In concrete situations it is then up to the physicist to determine the force field K_i. Let position and velocity of the system at time t_0 be given, i.e.,

$$x_i(t_0) = a_i, \qquad \dot{x}_i(t_0) = b_i, \qquad i = 1, \ldots, n, \tag{17}$$

and if every K_i is of class C^1 in a neighborhood of $(a_1, \ldots, a_n, b_1, \ldots, b_n, t_0)$, then Theorem 3.A implies that in a neighborhood of t_0, equation (16) has a unique solution with (17).

EXAMPLE 58.5 (Free Fall). Let O be a point on the surface of the earth and e_3 a unit vector perpendicular to the surface of the earth at O. We choose a Cartesian coordinate system e_1, e_2, e_3 with its origin at O. From experience we know that in a neighborhood of O, a *gravitational force*

$$K = -mge_3$$

acts on a point of mass m. The equation of motion

$$m\ddot{x} = K, \qquad x(0) = he_3, \qquad \dot{x}(0) = -ve_3$$

has the unique solution

$$x(t) = -2^{-1}gt^2 e_3 + x(0) + t\dot{x}(0). \tag{18}$$

For $\dot{x}(0) = 0$ this is the law of the free fall, discovered by Galilei. Conversely, by using $m\ddot{x} = K$, one obtains the form of the gravitational force K from (18). For arbitrary $x(0)$ and $\dot{x}(0)$, formula (18) yields the motion of a stone, thrown from an initial position $x(0)$ with the initial velocity $\dot{x}(0)$. From experiments one finds for the gravitational acceleration $g = 9.81$ m/s^2.

Let $x = \xi^i e_i$ and $\zeta = \xi^3$. The function

$$U(x) = mg\zeta$$

is called the *potential energy* of the gravitational force. It is equal to weight mg times height ζ above the surface of the earth. For every motion $x = x(t)$ in the gravitational field, the energy conservation law holds

$$\tfrac{1}{2}m\dot{x}^2(t) + U(x(t)) = \text{const} = E_0.$$

This follows, because differentiation with respect to t of the left-hand side

58.5. Newton's Basic Equations

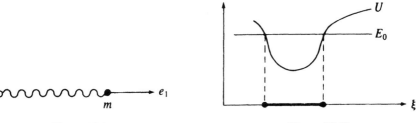

Figure 58.11   Figure 58.12

yields

$$m\ddot{x}\dot{x} + U_x \dot{x} = (m\ddot{x} - K)\dot{x} = 0.$$

In Section 58.8 we will show that in first-order approximation the force $K = -mge_3$ follows from Newton's general force of gravity.

EXAMPLE 58.6 (Harmonic Oscillator). In first-order approximation consider the most general motion

$$x(t) = \xi(t)e_1$$

of a mass point of mass m on a straight line through the vector e_1 with force $K(\xi) = k(\xi)e_1$ (Fig. 58.11). The equation of motion $m\ddot{x} = K$ here becomes

$$m\ddot{\xi} = k(\xi). \tag{19}$$

We choose a function U with

$$k(\xi) = -U'(\xi)$$

and call U the potential energy. We find that $K = -U_x$, i.e., $K = -\operatorname{grad} U$. For every solution of (19) we obtain

$$\tfrac{1}{2}m\dot{\xi}^2(t) + U(\xi(t)) = \text{const} = E_0. \tag{20}$$

This is easily checked by differentiating (20). We call $\tfrac{1}{2}m\dot{\xi}^2$ the kinetic energy and E_0 the energy. Equation (20) shows that the motion is possible only for

$$U(\xi) \leq E_0$$

(Fig. 58.12). Thereby E_0 is determined by prescribing $\xi(t_0)$ and $\dot{\xi}(t_0)$. Series expansion yields

$$k(\xi) = k_0 - k_1 \xi + O(\xi^2), \quad \xi \to 0.$$

By only considering linear terms, we obtain

$$m\ddot{\xi}(t) - k_0 + k_1 \xi(t) = 0.$$

We now assume that K is a repelling force as in the case of a spring, i.e., $k_1 > 0$. Moreover, we assume that the force is equal to zero if the deflection ξ is equal to zero, i.e., $k_0 = 0$. Then the force K and the potential energy U are equal to

$$K = -k_1 \xi e_1, \quad U = \tfrac{1}{2}k_1 \xi^2, \quad k_1 > 0.$$

This important physical system is called an *harmonic oscillator*. The equation of motion becomes

$$\ddot{\xi} + \omega^2 \xi = 0, \qquad \omega = \sqrt{k_1/m},$$

and the solution is

$$\xi(t) = \xi(0)\cos\omega t + \frac{\dot{\xi}(0)}{\omega}\sin\omega t. \tag{21}$$

These are periodic oscillations with so-called angular frequency ω. Since the trigonometric functions have period 2π, the time needed for one oscillation is equal to

$$T = 2\pi/\omega.$$

The number $\nu = 1/T$ is called the *frequency*. It is equal to the number of oscillations during one unit of time.

The harmonic oscillator is the most simple oscillating system. More complicated oscillating systems are often modeled by superposition of harmonic oscillators (cf. Problem 58.11).

The mass m describes the inertia of the body and is measured in kilograms. By definition, 10^3 cm^3 of water have the mass of 1 kg at a temperature of 4° Celsius. On the earth, masses are measured by comparison on scales using the lever principle (4). Masses of celestial bodies are determined from their motions using the gravitational law (see Section 58.8).

In Newtonian mechanics one assumes that the mass is constant and does not depend on the motion and on the system of reference. An exception is formed by rockets, which continuously eject mass, and whose equations of motion are

$$\dot{p} = K \tag{22}$$

with momentum $p = m\dot{x}$. One should also note that the basic equation (16) only describes motions of points which are not subject to constraints. The case of constraints will be discussed in Section 58.10 (principle of the least constraint of Gauss).

Forces, that depend on velocities are called forces of friction. The unit of the force is the Newton N. It is

$$1 \text{ N} = 1 \text{ kg m/s}^2.$$

58.6. Changes of the System of Reference and the Role of Inertial Systems

Physical processes are described by space and time coordinates, i.e., relative to systems of reference. For a physical law one requires that it is precisely

58.6. Changes of the System of Reference and the Role of Inertial Systems

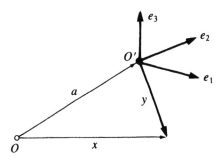

Figure 58.13

stated for which systems of reference it is valid and how it is changed under changes of the system of reference. Moreover, physicists are very much interested in *equivalent systems of reference*. Thereby two systems Σ and Σ' are called equivalent if and only if the form of the physical processes is the same for both systems. More precisely, this means: If two physicists build the same measuring device in Σ and Σ' and if for both systems the same initial and boundary conditions are satisfied, then the physicists obtain the same results of measurement. Not all systems of reference are equivalent. Consider, for example, a physicist P in a closed box B, who drops a stone with initial velocity zero. Then P will obtain very different results depending on whether B stands on the surface of the earth, rides in an accelerated lift or travels in a spacelab (weightlessness).

In the following we discuss the difficulties connected with the system of reference for the basic equation (16). To this end we introduce an auxiliary mathematical construction, the *absolute space*. This means that, as in Section 58.1, we choose a fixed three-dimensional vector space V_3 as the absolute space together with a fixed coordinate origin O. Every motion $x = x(t)$ with $x(t) \in V_3$ for all t, where $x(t)$ is attached at O, is then called an absolute motion. Following Newton, we assume that the basic equation (16) holds true for the absolute space. We then consider the situation of Figure 58.13, where Σ' is a moving Cartesian coordinate system with origin O' and axial vectors $e_i(t)$, $i = 1, 2, 3$, whose absolute motion is given by the equation

$$\dot{e}_i(t) = \omega(t) \times e_i(t).$$

Suppose the absolute motion of O' is $x = a(t)$. We want to find out how equation (16) looks in Σ'.

A physicist P' in Σ' cannot observe the absolute motion $x = x(t)$ of a given point, but only its relative motion $y = y(t)$ with

$$x(t) = a(t) + y(t). \tag{23}$$

We set $y(t) = \eta^i e_i$. For P' the axes e_i are fixed. Thus P' measures the velocity

$$\dot{y}(t) = \dot{\eta}^i(t) e_i,$$

and the acceleration
$$\ddot{y}(t) = \ddot{\eta}^i(t)e_i.$$

Differentiation of (23) yields $\dot{x} = \dot{a} + \dot{\eta}^i e_i + \eta^i \dot{e}_i$, hence

$$\dot{x} = \dot{a} + \dot{y} + \omega \times y, \tag{24}$$

and

$$\ddot{x} = \ddot{a} + \ddot{y} + L(y, \dot{y}, \omega, \dot{\omega}), \tag{25}$$

with

$$L = \omega \times (2\dot{y} + \omega \times y) + \dot{\omega} \times y.$$

From $m_i \ddot{x}_i = K_i$ in the absolute space, we find as the equation of motion in the new system Σ'

$$m_i \ddot{y}_i = K'_i, \quad i = 1, \ldots, n, \tag{26}$$

with

$$K'_i = K_i - m_i \ddot{a} - m_i L(y_i, \dot{y}_i, \omega, \dot{\omega}).$$

For constant ω the motion (24) corresponds to a translation about a and a rotation about the axis ω with angular velocity $|\omega|$ (see Example 58.1). If $\omega(t)$ is time-dependent, then the axis of rotation and the angular velocity both change in time. According to Section 58.14, equation (24) describes the most general motion of a Cartesian coordinate system in the absolute space. We obtain the important statement:

The transformed basic equations (26) have the same structure in an arbitrarily moving Cartesian coordinate system as do Newton's basic equations (16) in the absolute space. Only the forces need to be changed. In addition to K_i we have that K'_i contains so-called inertial forces $-m_i \ddot{a} - m_i L$. Thus the basic equations (16) can be used in every system of reference if the forces are suitably chosen.

If, for example, Σ' performs a constant rotational motion, i.e., $a(t) \equiv 0$ and $\omega(t) = \text{const}$, then we obtain in Σ' the equation of motion

$$m_i \ddot{y}_i = K_i - m_i \omega \times (\omega \times y_i) - 2m_i(\omega \times \dot{y}_i)$$

for $i = 1, \ldots, n$. The second (or third) term on the right-hand side is the centrifugal force (or Coriolis force). Those are inertial forces. In contrast to the centrifugal force the Coriolis force vanishes for a body which is at rest in Σ', i.e., for $\dot{y}_i = 0$.

We now consider an important special case. We call Σ' an *inertial system* if and only if Σ' performs only a purely translationary motion in the absolute space with constant velocity, i.e., $\dot{a}(t) = \text{const}$ and $\omega(t) \equiv 0$. Then (26) becomes

$$m_i \ddot{y}_i = K_i, \quad i = 1, \ldots, n, \tag{27}$$

where, at the same time, K_i is the force which acts on the particle in the absolute space. We obtain: In all inertial systems the same force is applied to the particle, and the equation of motion has the same form (principle of relativity in the restriced sense). In classical mechanics one then assumes that the following stronger postulate holds.

Galileian Principle of Relativity 58.7. *For mechanical processes all inertial systems are equivalent.*

We will show that this principle greatly restricts the number of possible force fields. Consider the motion of two mass points in the absolute space under the influence of a force $K_i = K_i(x_1, x_2)$, i.e.,

$$m_i \ddot{x}_i = K_i(x_1, x_2), \qquad i = 1, 2. \tag{28}$$

After a transformation we find in the inertial system Σ' the equation

$$m_i \ddot{y}_i = K_i(y_1 + a, y_2 + a). \tag{29}$$

The Principle of Relativity 58.7 requires then that the two differential equations (28) and (29) have the same solutions for identical initial conditions

$$x_i(0) = y_i(0), \qquad \dot{x}_i(0) = \dot{y}_i(0), \qquad i = 1, 2.$$

This is the case if

$$K_i = A_i(x_1 - x_2),$$

i.e., if the forces only depend on the difference $x_1 - x_2$.

For the previous coordinate transformation we have either explicitly or implicitly used the following three facts:

(a) There exists an absolute space.
(b) There exists an absolute time, i.e., there exists a rule of measurement to set all clocks in such a way that under changes between two moving coordinate systems the time remains unchanged.
(c) The mass m_i remains unchanged under changes between two moving systems of reference.

The correctness of these hypotheses can only be tested in physical experiments. In the context of the special theory of relativity, (a) to (c) are not correct. In fact, physicists were not able to find an absolute motion, because physical motion can only be observed relative to a system of reference. Therefore, the construction of the absolute space is only a useful mathematical tool. Actually, only inertial systems exist in nature. Experience in classical celestial mechanics shows that, as a good approximation of an inertial system, Σ_{sun} can be chosen where the origin is the center of mass of our solar system and the axes are firmly connected with the stellar sky (see Section 58.9). Then Σ' is an inertial system precisely if it performs a translatory motion with constant velocity with respect to Σ_{sun}. The special theory of relativity which will be discussed in Chapter 75, begins with Einstein's principle of relativity: All inertial systems are equivalent, not only with respect to mechanical processes but with respect to *all* possible physical processes. Furthermore, the velocity of light in the vacuum is the same for all inertial systems. This implies that (b) is no longer valid. It also follows that in relativistical mechanics, which is based on Einstein's principle of relativity, (c) is no longer true. But the effects of relativity theory only occur for velocities which are close to the velocity of light.

In atomic regions, classical mechanics must be replaced with quantum mechanics. There, the concept of the classical orbit and the classical velocity is replaced with probability theoretical expectation values (see Section 59.12). But in this chapter we will work classically. In particular, we assume (a) to (c).

58.7. General Point System and its Conserved Quantities

We consider the motion of n mass points in an arbitrary coordinate system, i.e., we consider the Newtonian basic equation

$$m_i \ddot{x} = K_i(x_1, \ldots, x_n, \dot{x}_1, \ldots, \dot{x}_n, t), \qquad i = 1, \ldots n, \tag{30}$$

for given forces K_i and masses m_i. To study this differential equation we introduce a number of fundamental quantities. For a fixed motion $x_i = x_i(t)$ we define the kinetic energy

$$T = \sum_i m_i \dot{x}_i^2 / 2,$$

the total momentum

$$P = \sum_i m_i \dot{x}_i,$$

the angular momentum

$$N = \sum_i m_i (x_i \times \dot{x}_i),$$

the torque

$$M = \sum_i x_i \times K_i,$$

the work during the time interval $[t_0, t]$

$$W(t) = \int_{t_0}^{t} \sum_i K_i \dot{x}_i \, dt,$$

and the power

$$\dot{W} = \sum_i K_i \dot{x}_i.$$

58.7. General Point System and its Conserved Quantities

Moreover, we call

$$m = \sum_i m_i$$

the total mass and

$$y = \sum_i m_i x_i / m$$

the center of mass of the system. It is always summed from $i = 1$ to n.

We now assume that the forces admit a decomposition of the form

$$K_i = K_i^{(p)} + K_i^{(q)}$$

whereby a function U exists with

$$K_i^{(p)}(x_1, \ldots, x_n) = -U_{x_i}(x_1, \ldots, x_n)$$

for all i. Thereby U is called the potential or potential energy, and U_x denotes the F-derivative of U with respect to x. In Cartesian coordinates $x = \xi^i e_i$ we have

$$U_x = \sum_{j=1}^{3} \frac{\partial U}{\partial \xi^j} e_j.$$

Note that in a simply connected region the C^1-function U is determined only up to an additive constant by the C^2-forces $K_i^{(p)}$. This constant may be specified through a gauge. *Gauge invariance* in classical mechanics means that the gauge does not cause any observable effects. This is because in the equations of motion there occur only forces which are independent of the gauge. In modern elementary particle theory however, different gauges of the fields are of central interest (gauge field theories). For example, the photon can be interpreted as the elementary particle, which carries the gauge information of the electron–positron field. This naturally implies that the photon has no rest energy. This will be discussed in Part V.

We call $K_i^{(p)}$ a *conservative force*. The knowledge of the potential energy U greatly simplifies the computation of the work W, because then

$$W(t) = U(X(t_0)) - U(X(t)) + \int_{t_0}^{t} \sum_i K_i^{(q)} \dot{x}_i \, dt,$$

with $X = (x_1, \ldots, x_n)$. This yields the following important interpretation of the potential energy:

If all forces are conservative, i.e., $K_i^{(q)} \equiv 0$ for all i, then the work W done by these forces is path independent and equal to the difference between the values of the potential energy at the initial point and at the endpoint of the motion.

This motivates the expression potential energy in the sense of accumulated energy. In particular, according to Example 58.5, the potential energy in the gravitational field of the earth is equal to weight times height. During the falling of a stone we gain work, while during its lifting we do work. This work is always equal to weight mg times height difference.

Theorem 58.A (Balance and Conservation Laws). *For a solution of the basic equation (30) we have the following:*

(i) *Energy balance:* $(d/dt)(T + U) = \sum_i K_i^{(q)} \dot{x}_i$.
(ii) *Angular momentum balance:* $\dot{N} = M$.
(iii) *Momentum balance:* $\dot{P} = \sum_i K_i$.
(iv) *Motion of the center of mass:* $m\ddot{y} = \sum_i K_i$.

PROOF. All claims are easily checked by differentiating. For example, we have

$$\frac{d}{dt}(T(t) + U(X(t))) = \sum_i m_i \ddot{x}_i(t) \dot{x}_i(t) + U_{x_i}(X(t))\dot{x}_i(t)$$
$$= \sum_i (K_i - K_i^{(p)})\dot{x}_i = \sum_i K_i^{(q)}\dot{x}_i,$$

and also

$$\dot{N} = \sum m_i \dot{x}_i \times \dot{x}_i + \sum m_i x_i \times \ddot{x}_i = \sum x_i \times K_i. \qquad \square$$

Theorem 58.A has a number of important consequences which are not only of importance for mechanics. The following results can be extended to much more general physical situations, as we shall see in later chapters. In modern physics, the generalization to fields plays an important role and will be discussed in Section 75.12 in connection with the relativistic energy–momentum tensor.

Energy conservation law. The basic quantity

$$E(t) = T(t) + U(X(t))$$

is called total energy or simply energy of the motion at time t. The energy balance (i) means that the change of energy in time is equal to the power of the nonconservative forces. Integration yields

$$E(t) - E(t_0) = \int_{t_0}^{t} \sum_i K_i^{(q)} \dot{x}_i \, dt,$$

which shows that for positive work of nonconservative forces, energy is added to the system. In the case that no nonconservative forces are present, i.e., if all $K_i^{(q)}$ are identically zero, then the most important physical theorem—the energy conservation law

$$E(t) = \text{const},$$

is valid for the motion.

Momentum conservation law. If the total force is identically zero, then the motion satisfies

$$P(t) = \text{const.}$$

Angular momentum conservation law. If the total torque is identically zero, i.e., $M(t) \equiv 0$, then the motion satisfies

$$N(t) = \text{const.}$$

Motion of the center of mass. Obviously, (iii) and (iv) in Theorem 58.A are identical. According to (iv), the center of mass moves as if the entire mass were in concentrated there and as if all forces were acting on it. In the case of momentum conservation the center of mass moves on a straight line with constant velocity.

Necessary equilibrium condition. From (ii) and (iii) in Theorem 58.A we obtain: If the system is in an equilibrium position, i.e.,

$$x_i(t) = \text{const} \quad \text{for all } i,$$

is a solution of the basic equation (30), then the total force and total torque vanish for this configuration.

As some important applications of the conservation laws above, we consider:

(a) Planetary motion (Section 58.9).
(b) The rigid body (Section 58.13).

58.8. Newton's Law of Gravitation and Coulomb's Law of Electrostatics

58.8a. The Gravitational Law

We consider two mass points P_1 and P_2 in the absolute space with masses m_1 and m_2 and corresponding position vectors x_1 and x_2. According to Newton's fundamental observation, a *gravitational force* K acts from P_1 onto P_2 with

$$K = -\frac{Gm_1 m_2 (x_1 - x_2)}{|x_1 - x_2|^3}.$$

The direction of this force shows that it is an attracting force (Fig. 58.14). Because of

$$|K| = Gm_1 m_2 / r^2$$

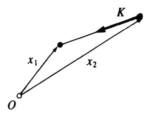

Figure 58.14

with $r = |x_1 - x_2|$ we find that this force is proportional to the product of the masses and indirectly proportional to the square of the distance. The quantity G is a universal constant and is called *gravitational constant*. We have

$$G = 6.7 \cdot 10^{-11} \text{ Nm}^2/\text{kg}^2.$$

For fixed x_1 the gravitational force K has the potential

$$U(x_2) = -\frac{Gm_1 m_2}{|x_2 - x_1|},$$

i.e., $U = -Gm_1 m_2/r$. In fact, we have

$$K = -U_{x_2}.$$

According to Newton's principle of

$$\text{actio} = \text{ractio}$$

we find that, conversely, the gravitational force $-K$ acts from P_2 onto P_1.

Proposition 58.8 (First-Order Approximation of the Gravitational Force). *Let x_1 and x_2^0 be fixed with $x_1 \neq x_2^0$ and let $x_2 = x_2^0 + x$. Moreover, we choose the unit vector $e_3 = (x_2^0 - x_1)/|x_2^0 - x_1|$.*

Then for $x \to 0$ we have

$$K = K_0 + O(|x|), \qquad U = U_0(x) + o(|x|)$$

with

$$K_0 = -gm_2 e_3, \qquad g = Gm_1/|x_2^0 - x_1|^2,$$
$$U_0 = \text{const} + gm_2(xe_3).$$

Moreover, U_0 is the potential of K_0, i.e., $K_0 = -(U_0)_x$.

PROOF. This immediately follows from the Taylor expansion. □

Since K only depends on the difference $x_1 - x_2$, the gravitational law above is valid for any system of reference, according to Section 58.6. If, however, Σ' is no inertial system, then additional inertial forces occur.

58.8. Newton's Law of Gravitation and Coulomb's Law of Electrostatics   37

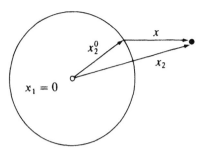

Figure 58.15

EXAMPLE 58.9 (Special Case Earth and the Determination of Masses of Celestial Bodies). We apply Proposition 58.8 to the earth (Fig. 58.15). According to Problem 58.4, the gravitational force of the earth onto a point x_2 with mass m_2 outside the earth is obtained by computing the gravitational force, which is exerted from the center of the earth x_1, assuming the total mass of the earth m_1 is concentrated there. We set $x_1 = 0$. Moreover, we choose x_2^0 as in Figure 58.15 and use the first-order approximations K_0 and U_0 from Proposition 58.8 in a neighborhood of the surface of the earth. This corresponds to Example 58.5.

According to Example 58.5 the acceleration of the earth g can be determined from gravitational experiments. One obtains $g = 9.81$ m/s^2. The gravitational constant G can be determined experimentally by using the Cavendish torsion balance in order to measure the gravitational effect of balls. From

$$g = Gm_1/|x_2^0|^2,$$

and the radius of the earth $|x_2^0| = 6378$ km one obtains the mass of the earth $m_1 = 6 \cdot 10^{24}$ kg.

The mass of the sun can be determined from the orbit of the earth and Kepler's third law (40) below. If one knows the mass of the sun, then analogously the mass of a planet can be determined from the planetary orbits and (40). The mass of the earth can also be derived from (40) and the motion of artificial satellites.

The rotating earth with angular velocity $\omega = 2\pi/\text{day} = 1.2 \cdot 10^{-5}$ s^{-2} is not an inertial system (strictly speaking ω above is somewhat larger because of the motion of the earth around the sun). Therefore, according to Section 58.6, additional inertial forces occur at the surface of the earth, namely the centrifugal force and the Coriolis force. These, however, are very small compared with the gravitational force, and in first-order approximation they can be neglected. Actually, the centrifugal force at the surface of the earth satisfies the estimate

$$|m_2 \omega \times (\omega \times x_2^0)| \leq m_2 \omega^2 |x_2^0|.$$

Thus, at the surface of the earth we obtain for the ratio between the maximal centrifugal force and the gravitational force $m_2 \omega^2 |x_2^0|/m_2 g = 10^{-4}$.

58.8b. Coulomb's Law

If similarly as above P_1 and P_2 are two points with charges Q_1 and Q_2 and corresponding position vectors x_1 and x_2, then an electrostatic force K acts from P_1 onto P_2 with

$$K = \frac{Q_1 Q_2 (x_1 - x_2)}{4\pi\varepsilon_0 |x_1 - x_2|^3},$$

and the dielectricity constant ε_0 with

$$\frac{1}{4\pi\varepsilon_0} = 8.988 \cdot 10^9 \text{ Nm}^2/\text{A}^2\text{s}^2.$$

This law was formulated by Coulomb in 1775, i.e., about one hundred years after Newton's gravitational law.

The main difference between Coulomb's law and the gravitational law is that in Coulomb's law for equal charges $Q_1 Q_2 > 0$ a repelling force occurs, and for different charges $Q_1 Q_2 < 0$ an attracting force occurs. The similarity of both laws on the other hand is the reason for the close relation between the motion of planets around the sun and electrons around the atomic nucleus, which in 1913 led to Bohr's atomic model (see Section 59.16). In Section 59.26 we show that for the motion of electrons in the atom the gravitational force is very small compared to the electrostatic force and therefore plays no role.

The potential U of K is obtained similarly as in the previous section through

$$U(x_2) = \frac{Q_1 Q_2}{4\pi\varepsilon_0 |x_1 - x_2|}$$

for fixed x_1.

58.9. Application to the Motion of Planets

We want to show which differential equation governs the planetary motion in the Solar System. Also, we want to derive Kepler's laws by explicitly solving the two-body problem (sun and one planet).

58.9a. The n-Body Problem

We begin with the motion of n mass points which are subject only to the gravitational force (sun x_1 and $n - 1$ planets $x_2, \ldots, x_n$). According to Newton this motion is described in the absolute space by the equation

$$m_i \ddot{x}_i = K_i(x_1, \ldots, x_n), \qquad i = 1, \ldots, n, \tag{31}$$

58.9. Application to the Motion of Planets

with gravitational force

$$K_i = \sum_{j=1, j \neq i}^{n} K_{ij}$$

and

$$K_{ij} = -Gm_i m_j (x_i - x_j)/|x_i - x_j|^3.$$

This means that the force K_i acting on the ith point is equal to the sum of the gravitational forces acting from all points on the ith point. All forces are conservative. The potential is

$$U = \sum_{j=1, j \neq i}^{n} U_{ij}$$

with

$$U_{ij} = -Gm_i m_j /|x_i - x_j|.$$

In fact, we have $K_i = -U_{x_i}$.

If, according to Section 58.6, we pass from the absolute space to any abitrary inertial system with $x_i = y_i + a$, then we obtain (31) with y_i instead of x_i, since K_i only depends on $x_j - x_i$. Hence (31) holds in every inertial system, i.e., Newton's law of gravitation satisfies Galilei's Principle of Relativity 58.7. From Theorem 58.A we find for the center of mass

$$m\ddot{y} = \sum_i K_i = 0,$$

thus y moves with constant velocity. Therefore there exists an inertial system Σ_0 whose origin is the center of mass. We set

$$x_i = y_i + y$$

and find that in this system

$$\sum_i m_i y_i = \sum_i m_i (x_i - y) = 0.$$

Also we obtain (31) with y_i instead of x_i.

58.9b. Solution of the Two-Body Problem and Kepler's Laws

We now consider the important special case of the two-body problem, i.e., we set $n = 2$. We have

$$m_1 \ddot{y}_1 = K, \quad m_2 \ddot{y}_2 = -K, \quad K = Gm_1 m_2 (y_2 - y_1)/|y_1 - y_2|^3, \quad (32)$$

$$m_1 y_1 + m_2 y_2 = 0, \quad (33)$$

where y_1 and y_2 describe the motion of the sun and one planet, respectively. We want to show that by using the conservation laws of Section 58.7, equations (32) and (33) can be reduced to the simple system (37) below, which can explicitly be solved. The much more difficult case of the general n-body problem will be discussed a little further in Section 58.9c.

For the relative motion of the planet with respect to the sun

$$x = y_2 - y_1$$

we have

$$m_2 \ddot{x} = -Gm_2 mx/|x|^3, \quad m = m_1 + m_2. \tag{34}$$

This follows from (32) and (33). If one knows the motion of x, then one obtains from (33) that

$$y_1 = -m_2 x/m, \quad y_2 = m_1 x/m. \tag{35}$$

By using Theorem 58.A one can easily integrate (34). The angular momentum conservation law and the energy conservation law immediately imply

$$m_2 x(t) \times \dot{x}(t) = \text{const} = N,$$
$$U(x(t)) + m_2 \dot{x}(t)^2/2 = \text{const} = E, \tag{36}$$

with $U(x) = -Gmm_2/|x|$. Note that $U'(x) = Gmm_2 x/|x|^3$ for $x \neq 0$.

Given $y_i(0)$, $\dot{y}_i(0)$ for $i = 1, 2$, we find $x(0)$, $\dot{x}(0)$ and hence from (36) the angular momentum N and the energy E for the relative motion. Let $N \neq 0$, and hence $x(0) \neq 0$.

Then the initial-value problem which corresponds to (34) has a unique solution which, according to Theorem 58.A, satisfies equations (36). From (36) follows

$$Nx(t) \equiv 0,$$

i.e., the motion occurs on a plane perpendicular to N. For a suitable choice of basis vectors e_1, e_2, and $e_3 = N/|N|$ we therefore have

$$x = re_r, \quad e_r \stackrel{\text{def}}{=} (\cos \varphi)e_1 + (\sin \varphi)e_2$$

with polar coordinates $r = r(t)$, $\varphi = \varphi(t)$. Let $\varphi(0) = 0$. If we set

$$e_\varphi \stackrel{\text{def}}{=} -(\sin \varphi)e_1 + (\cos \varphi)e_2,$$

then we find $e_r e_\varphi = 0$, $e_r \times e_\varphi = e_3$ and

$$\dot{x} = \dot{r}e_r + r\dot{\varphi}e_\varphi.$$

From (36) we obtain the desired form of the equation of motion

$$r^2 \dot{\varphi} = |N|/m_2,$$
$$\dot{r}^2 + r^2 \dot{\varphi}^2 - 2Gm/r = 2E/m_2. \tag{37}$$

Theorem 58.B (Two-Body Problem). *The solution of* (37) *is*

$$r = p/(1 + \varepsilon \cos \varphi) \tag{38}$$

58.9. Application to the Motion of Planets

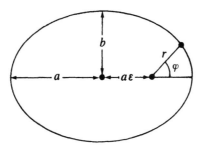

Figure 58.16

with

$$p = N^2/\alpha^2 m_2, \quad \varepsilon = \sqrt{1 + 2EN^2/m_2\alpha^2}, \quad \alpha = Gmm_2,$$

and

$$t = \frac{m_2}{|N|} \int_0^\varphi r^2(\varphi)\, d\varphi. \tag{39}$$

PROOF. From (37) follows $d\varphi/dr = \dot\varphi/\dot r = F(r)$. Integration yields (38). □

For $\varepsilon < 1, = 1, > 1$, the orbit (38) is an ellipse, parabola, or hyperbola, respectively, with corresponding energies $E < 0, = 0, > 0$. The sun, i.e., $r = 0$ always corresponds to a focus.

We now consider the case of the ellipse $\varepsilon < 1$ (Fig. 58.16). We want to show that for the relative motion $x = x(t)$ all three Kepler laws of Section 58.4 are satisfied.

(i) *First law.* Using Cartesian coordinates $\xi = r\cos\varphi$, $\eta = r\sin\varphi$, $r = \sqrt{\xi^2 + \eta^2}$ we obtain from (38) the equation of an ellipse

$$\frac{(\xi + \varepsilon a)^2}{a^2} + \frac{\eta^2}{b^2} = 1$$

with $a = p/(1 - \varepsilon^2)$ and $b = p/\sqrt{1 - \varepsilon^2}$. The sun corresponds to the focus $\xi = 0, \eta = 0$.

(ii) *Second law.* From (37) follows that during the time interval $[t_1, t_2]$, the position vector x sweeps out the surface

$$2^{-1} \int_{t_1}^{t_2} r^2 \dot\varphi\, dt = 2^{-1}(t_2 - t_1)|N|/m_2.$$

(iii) *Third law.* If a and b are the great and small semiaxes of the ellipse, respectively, then the area of the ellipse is equal to πab and for the period of revolution T we obtain $\pi ab = T|N|/2m_2$, i.e.,

$$T^2/a^3 = 4\pi^2/G(m_1 + m_2). \tag{40}$$

Since the mass of the planet m_2 is small compared to the mass of the sun m_1 we obtain approximately $T^2/a^3 = $ const for all planets.

Experience shows that, as a good approximation for an inertial system Σ_0, one can choose a system, whose origin is the center of mass of the solar system and whose axes are firmly connected with the stellar sky.

58.9c. Discussion of the n-Body Problem

Is the solar system stable? Properly speaking, the answer is still unknown, and yet this question has led to very deep results which probably are more important than the answer to the original question.
  Jürgen Moser, *Neue Zürcher Zeitung* (May 14, 1975)

Conserved quantities. According to the previous discussion, the two-body problem can be solved in two steps:

(i) Newton's second-order system of differential equations (31) is reduced to a first-order system by using conservation laws.
(ii) This first-order system is solved by quadrature (computation of integrals).

The n-body problem, one of the most famous mathematical problems of the eighteenth and nineteenth centuries, consisted in the problem of extending the solution procedure for $n = 2$ to arbitrary n. This, however, already *failed* for $n = 3$ (three-body problem). For $n = 3$ the general solution of (31) contains exactly 18 constants. The goal of the mathematicians then was to find 18 conserved quantities

$$C_i(t, x_1, x_2, x_3, \dot{x}_1, \dot{x}_2, \dot{x}_3) = \text{const}, \qquad i = 1, \ldots, 18$$

from which the orbital motion $x_i = x_i(t)$ could have been determined by solving for $\dot{x}_i$ and using quadrature. But the energy law, the angular momentum law, and the center of mass theorem of Section 58.7, only yield 10 conserved quantities E, N, $y(0)$, and $\dot{y}(0)$. Observe that except for the energy E all these quantities are vectors. In 1887, Bruns showed that no other conserved quantities C_i exist, which depend algebraically on x_i and $\dot{x}_i$. This result then was generalized by Poincaré in 1889, who also excluded a number of cases with analytic dependence.

Stability of the planetary system. The stability problem in general form is as follows:

(a) Is it possible that the planets collide, fall into the sun, drastically change their orbits, or as an extremal situation leave the solar system?
(b) Are small perturbations, e.g., caused by cosmic dust, able to cause one of the cases in (a)?

This is a question of existential consequences, because already small perturbations of the orbit of the earth would have catastrophic consequences for the life on our planet. The great interest in explicit analytic solutions for the n-body problem was motivated by the hope that through this the stability problem could be solved. The complete answer is still open.

Lagrange (1736–1813) and Laplace (1749–1827), by using their perturbation theory, as well as later astronomers were able to show that the great semiaxes of the planetary orbits only slightly change during the course of many hundreds of years (theorem of secular stability). Thereby Lagrange's method of the variation of constants played an important role. Today, simple variants of this method are taught in every course about ordinary differential equations. In perturbation theory it is an important task to find suitable coordinates for which a simple description of the problem is possible. For example, one would like to separate the orbital motion during one revolution and the secular change of the orbit. Usually, Cartesian coordinates are of no use. A very important instrument in this direction is the theory created by Hamilton (1805–1865) and Jacobi (1804–1851) which will be discussed in Section 58.23. In 1858, Dirichlet (1805–1859) told his colleague Kronecker (1823–1891) that he had found a proof for the stability of the planetary system. After his death, however, no such notes could be found. Initiated by Weierstrass and Mittag-Leffler, the Norwegian–Swedish King Oscar II offered, in 1885, a prize for the solution of this problem, and in 1889 it went to the young, ingenious Henri Poincaré (1854–1912). His prize memoir can be found in Acta Mathematica, Vol. 13 (1890), pp. 5–270. He did not find a solution, but developed a great number of new ideas, which led to the creation of a qualitative theory for dynamical systems, algebraic topology, and differential topology. Thereby Poincaré opened new dimensions in mathematics and greatly influenced the mathematics of our century. The important work of Ljapunov (1857–1918) on stability theory at the end of the last century was also influenced by the n-body problem.

Series expansion. The prize problem on which Poincaré worked, is described by Mittag-Leffler in Vol. 7 (1885) of the Acta Mathematica as follows: Given an arbitrary system of material points, all of which attract each other according to Newton's law, and assuming that no collisions of any two points occur the problem is to expand the coordinates of each single point into infinite series, which are composed of known functions of the time and which are uniformly convergent for all times.

An important, special result in this direction was obtained in 1912 by the Finlander K. S. Sundman. For the three-body problem he introduced a new uniforming variable u and was able to expand the coordinates with respect to this new variable. His series converge in the unit circle. For $-1 \leq u \leq 1$ these series yield real solutions for the three-body problem which exist for all times $-\infty < t < \infty$. The case of possible collisions is also included. A proof can be found in Siegel and Moser (1971, M). Unfortunately, Sundman's method cannot be generalized to the case $n > 3$.

Principial mathematical difficulties. Two typical difficulties occur in the study of the n-body problem:

(a) The right-hand side of (31) becomes singular for $x_i = x_j$ (collisions).
(b) The formally computed perturbation series contain dangerously small divisors.

If the existence theorem for ordinary differential equations (Theorem 3.A) is applied to (31) with initial condition $x_i(0) \neq x_j(0)$ for $i \neq j$, then one obtains the existence of a unique solution for a sufficiently *small* time interval. In order to guarantee existence for all times, a priori estimates such as in Section 3.3, are needed to guarantee the boundedness of the solutions for all times and to exclude collisions. The singularities in (a) above would then become meaningless.

The problem of small divisors has already been discussed in Section 3.8. They greatly complicate the convergence proofs for the perturbation series in celestial mechanics. In his prize memoir to the problem of King Oscar II, Poincaré proved that the formally constructed series in celestial mechanics may diverge. This, however, did not disturb him, because he found a way out by inventing the concept of asymptotic series. Those are series which possibly diverge but for any given degree of accuracy yield approximate results. By definition it is

$$f(x) \sim a_0 + \frac{a_1}{x} + \frac{a_2}{x^2} + \cdots, \qquad x \to +\infty,$$

i.e., this series is an asymptotic expansion of f for $x \to +\infty$ if and only if, for each fixed n, we have

$$\left| f(x) - \left(a_0 + \frac{a_1}{x} + \cdots + \frac{a_n}{x^n} \right) \right| \leq \frac{b_n(x)}{x^n}$$

for all $x > R$ with $b_n(x) \to 0$ as $x \to +\infty$. For example, one has

$$\int_x^\infty t^{-1} e^{x-t} dt \sim \frac{1}{x} - \frac{1!}{x^2} + \frac{2!}{x^3} \cdots$$

as $x \to +\infty$.

Today, other than to the times of Poincaré, we are able to give convergence proofs for series with small divisors by using the Kolmogorov–Arnold–Moser theory. Thereby the technique of the hard implicit function theorem is used (see Chapter 5). Very roughly, the following result holds:

(a) The formally constructed series for the perturbation of quasi-periodic motions, which, e.g., occur in celestial mechanics, can diverge or converge.
(b) The convergence depends very sensibly on the particular parameter value. In every neighborhood of a good-natured parameter, no matter how small, there exist bad-natured parameters with divergence.
(c) In a certain sense good-natured situations are far more common than bad-natured ones.

All those are typical difficulties for perturbations of quasi-periodic motions, which are related to resonance phenomena. And mathematicians have to live with this.

Explicitly solvable cases of the three-body problem. For a number of special cases the solution of the three-body system has been known for some time.

(i) *First special case of Lagrange.* The three mass points form an equilateral triangle at time $t = 0$. Then these three points move on similar ellipses, whereby the triangle remains equilateral. This case is very closely-related to the two-body problem. It approximately occurs for the configuration sun, Jupiter, Trojan group of asteroids.

(ii) *Second special case of Lagrange.* The three mass points with masses m_1, m_2, and m_3 lie on a rotating straight line.

(iii) *Restringent three-body problem.* Here, in addition to (ii), it is assumed that m_3 is very small. Strictly speaking, the limiting case $m_3 = 0$ is considered. The other two points move on circular orbits around a common center of mass. One may think of the sun, a planet, and an asteroid. This case has already been discussed by Euler (1707–1783) and was rediscovered by Jacobi, 1836. In great detail, Poincaré studied the special case $m_2 \ll m_1$ (see Fig. 3.12 in Chapter 3).

(iv) *Hill's moon theory.* At the end of the last century, Hill gave a number of stability proofs for a class of planar three-body problems which model the configuration of the sun, earth, and moon. Thereby he constructed the so-called Hill's limiting curve, and showed that it cannot be crossed by any body.

For the questions of this section we recommend the literature at the end of this chapter under the heading "celestial mechanics".

58.10. Gauss' Principle of Least Constraint and the General Basic Equations of Point Mechanics with Side Conditions

58.10a. Special Case

In order to explain the simple basic idea we begin with a special case. In an arbitrary system of reference we consider the motion of a point with mass m and position vector x under the influence of a force K. If the point is able to move freely, then equation

$$m\ddot{x} = K \tag{41}$$

Figure 58.17

holds. Now let us consider the following general side conditions

$$f(x, t) = 0, \qquad g(x, \dot{x}, t) = 0. \tag{42}$$

Then equation (41) needs to be modified since, in general, the solutions of (41) do not satisfy (42). Our goal is the modification

$$m\ddot{x} = K + Z,$$

where Z is a so-called *constraining force*, whose task it is to guarantee (42) for the motion. In technology such constraining forces occur for motions of machines. There, they have a very real meaning. If Z is too large, then the machine can be torn apart. Let us consider, for example, a pendulum under the influence of the gravitational force K. Then Z acts as in Figure 58.17, i.e., it keeps the pendulum mass m on the circular orbit. According to Gauss, the *basic equations* are

$$m\ddot{x} = K + \lambda f_x + \mu g_{\dot{x}}, \tag{43}$$

i.e., the constraining force is equal to $Z = \lambda f_x + \mu g_{\dot{x}}$. Thereby we have

$$K = K(x, \dot{x}, t), \qquad \lambda = \lambda(x, \dot{x}, t), \qquad \mu = \mu(x, \dot{x}, t).$$

As we shall see below, (43) can be obtained from the general principle of least constraint. We begin, however, by studying the mathematical consistence of (43). Sometimes, in more formally written textbooks on theoretical physics this is not included. We need to answer two questions:

(i) How can the functions λ and μ be determined?
(ii) When do solutions of (43) satisfy the side conditions (42)?

We assume:

(H) f and g are C^2 and f_x, $g_{\dot{x}}$ are linearly independent for all considered arguments.

By considered arguments, we mean all arguments in a fixed open, nonempty set in the $(x, \dot{x}, t)$-space, i.e., in $V_3 \times V_3 \times \mathbb{R}$. We also allow the case that g does not appear. Then we assume that $f_x \neq 0$ for all considered arguments.

Let $x = x(t)$ be a motion which satisfies (42). Differentiation of (42) with respect to t yields

$$f_x \dot{x} + f_t = 0. \tag{44}$$

58.10. Gauss' Principle of Least Constraint

Further differentiation with respect to t implies

$$f_x \ddot{x} + \dot{x} f_{xx} \dot{x} + f_{tt} = 0,$$
$$g_{\dot{x}} \ddot{x} + g_x \dot{x} + g_t = 0.$$
(45)

Inserting the basic equation (43) into (45), we obtain for fixed arguments $(x, \dot{x}, t)$ a linear system of equations for λ and μ, which can uniquely be solved, since the corresponding homogeneous system has only the trivial solution $\lambda = \mu = 0$. In fact, the solution of the homogeneous system consists in the problem of finding the point of intersection between the vector $\lambda f_x + \mu g_{\dot{x}}$ and the orthogonal subspace

$$\{h : f_x h = 0, g_{\dot{x}} h = 0\}.$$

Proposition 58.10. *Assume* (H). *If $x = x(t)$ is a solution of (43) which satisfies the side conditions (42), (44) for some fixed time, then this solution satisfies the side conditions for all times.*

PROOF. From the construction of λ, μ it follows that $x = x(t)$ satisfies equations (45). Integration of (45) yields (44) and (42). □

Corollary 58.11. *The general existence and uniqueness theorem (Theorem 3.A) about initial-value problems can be applied to (43), since*

$$\lambda = \lambda(x, \dot{x}, t) \quad \text{and} \quad \mu = \mu(x, \dot{x}, t)$$

are C^1-functions as unique solutions of a linear system of equations with C^1-coefficients.

The minimization problem

$$m(\ddot{x} - K/m)^2 = \min!$$
(46)

with side conditions (45) is called the *principle of least constraint*. Thereby *only $\ddot{x}$ varies*, while $(x, \dot{x}, t)$ is considered a fixed parameter. Lagrange's multiplier rule for real functions of Section 43.10 immediately implies (43).

58.10b. General Case

In an analogous fashion we now consider the general case of a motion of n mass points with position vectors x_i, masses m_i, and forces K_i acting on x_i. Let the side conditions be given through the real functions

$$f^{(r)}(X, t) = 0, \quad r = 1, \ldots, R,$$
$$g^{(s)}(X, \dot{X}, t) = 0, \quad s = 1, \ldots, S,$$
(47)

with $X = (x_1, \ldots, x_n)$ and $R + S < 3n$. We assume:

(R) The side conditions (47) are regular, i.e., all $f^{(r)}$ and $g^{(s)}$ are C^2 for all considered arguments, and the linear system of equations for $\ddot{X}$, which similarly as (45), is obtained from (47) by differentiation with respect to time t, has maximal rank $R + S$ for all considered arguments.

By considered arguments we mean all arguments in a fixed, open, nonempty set in the $(X, \dot{X}, t)$-space, i.e., in $\prod_{i=1}^{2n} V_3 \times \mathbb{R}$. We also allow the case that g in (47) does not appear. Then the maximal rank in (R) must be equal to R. If g does not appear in (47), then for every t, equations (47) describe an R-dimensional C^2-manifold C in the X-space $\prod_{i=1}^{n} V_3$ (see the preimage theorem of Section 4.18). Thereby C is called the configuration space if, in addition, every $f^{(r)}$ is independent of t.

Similarly as in (43), Gauss' basic equations are

$$m_i \ddot{x}_i = K_i + \sum_r \lambda_r f_{x_i}^{(r)} + \sum_s \mu_s g_{\dot{x}_i}^{(s)} \qquad (48)$$

for $i = 1, \ldots, n$ with

$$K_i = K_i(X, \dot{X}, t), \qquad \lambda_r = \lambda_r(X, \dot{X}, t), \qquad \mu_s = \mu_s(X, \dot{X}, t).$$

Again, the functions λ_r and μ_s are uniquely determined by the linear system for $\ddot{X}$ which, as in the case of (45), is obtained from (47) by differentiation with respect to t. Then a result analogous to Proposition 58.10 and Corollary 58.11 is valid. Finally, we must replace the principle of least constraint (46) with

$$\sum_i m_i (\ddot{x}_i - K_i/m_i)^2 = \min!.$$

Experience shows that, in fact, Gauss' basic equations yield a correct representation of processes in nature with the corresponding constraining forces. Equations (47) represent the most general mechanical side conditions. The principle of least constraint is very natural, because, in the absence of side conditions, one immediately obtains Newton's equations $m\ddot{x}_i = K_i$, $i = 1, \ldots, n$.

58.11. Principle of Virtual Power

We now discuss the important special case of *holonomic constraints*, i.e., the side conditions (47) do not depend on time and velocity. Our goal is, by using a simple trick, to eliminate the constraining forces, which in many instances are very disturbing.

Definition 58.12. All possible positions $X = (x_1, \ldots, x_n)$ of the system, which are compatible with the side conditions, form the *configuration space* C. By a

58.11. Principle of Virtual Power

virtual motion

$$x_i = y_i(t), \quad i = 1, \ldots, n$$

we mean any C^1-motion which satisfies the side conditions. Then

$$v_i = \dot{y}_i(t)$$

is called the corresponding *virtual velocity*.

By an arbitrary virtual velocity $(v_1, \ldots, v_n)$ at point X we mean the virtual velocity $v_i = \dot{y}_i(t)$ of a virtual motion $Y = Y(t)$, which for the fixed time t is at the point X.

Geometrically, the virtual motions correspond precisely to the C^1-curves in C. The virtual velocities at the point $X \in C$ are precisely the tangent vectors to C at X. The virtual C^1-motions, which physically are possible under the influence of given forces will be simply called *motions* in the following.

The difference between virtual motions and motions will play an important role in our further discussion.

Principle of Virtual Power 58.13. *We consider n points with position vectors x_i and masses m_i. The force $K_i = K_i(X, \dot{X}, t)$ with $X = (x_1, \ldots, x_n)$ acts on x_i. For holonomic constraints we then obtain for the motion $X = X(t)$ of the points the equation*

$$\sum_{i=1}^{n} (m_i \ddot{x}_i(t) - K_i) v_i = 0 \tag{49}$$

for all t and arbitrary virtual velocities $(v_1, \ldots, v_n)$ at each point $X(t)$.

This general principle, which goes back to d'Alembert (1717–1783), is very useful, because of its invariant formulation, i.e., the concrete analytical form of the side conditions does not appear and all concepts involved have a physical interpretation. From (49), for example, it is possible to derive in a simple form the basic equations for the rigid body (see Section 58.14).

In the regular case (R) of Section 58.10, relation (49) follows from Gauss' equations. In fact, by differentiating with respect to t one immediately obtains from (47) that

$$\sum_i f_{x_i}^{(r)} v_i = 0.$$

Multiplication of (48) with v_i and summation over i yields (49).

If one chooses especially $v_i = \dot{x}_i(t)$ in (49), then one obtains for the motion the energy balance

$$\dot{T} = \sum_i K_i \dot{x}_i \tag{50}$$

with kinetic energy $T = \sum_i m_i \dot{x}_i^2 / 2$. This equation is the key for the following stability discussion.

58.12. Equilibrium States and a General Stability Principle

In this section we consider the same situation as in the Principle of Virtual Power 58.13. We say the system is in an *equilibrium state* (equilibrium position) at X_0 if and only if every C^2-motion of the system $X = X(t)$ with $X(t_0) = X_0$ and $\dot{X}(t_0) = 0$ satisfies

$$X(t) \equiv X_0.$$

We call

$$W(t) = \int_{t_0}^{t} \sum_i K_i(Y(t), \dot{Y}(t), t) \dot{Y}(t) \, dt$$

the *virtual work*. Thereby $X = Y(t)$ is an arbitrary virtual C^1-motion with $Y(t_0) = X_0$. We call $W(\cdot)$ *nonpositive* if and only if

$$W(t) \leq 0 \quad \text{for all } t.$$

Moreover, $W(\cdot)$ is called *strictly negative* if and only if

$$W(t) < 0 \quad \text{for all } t \neq t_0$$

and for all virtual C^1-motions $X = Y(t)$ with

$$Y(t) \neq X_0 \quad \text{for all } t \neq t_0.$$

Proposition 58.14 (Equilibrium Condition). *If X_0 is an equilibrium state, then*

$$\sum_i K_i(X_0, 0, t) v_i = 0 \quad \text{for all } t \tag{51}$$

and all virtual velocities $(v_1, \ldots, v_n)$ at X_0.

If conversely, the virtual work $W(\cdot)$ is nonpositive in the sense of the above definition, then X_0 is an equilibrium state.

PROOF. From (49) with $\ddot{x}_i = 0$ follows (51). Conversely, suppose that $W(t) \leq 0$ with $Y(t) \equiv X(t)$, $X(t_0) = X_0$ and $\dot{X}(t_0) = 0$. By integration it follows from the energy balance (50) that

$$T(t) - T(t_0) \leq 0 \quad \text{and} \quad T(t_0) = 0.$$

This means that $\dot{x}_i(t) \equiv 0$. □

Definition 58.15. If $W(\cdot)$ is strictly negative, then X_0 is called a *statically stable equilibrium state*.

We want to motivate this definition and show: If the system is at rest at time t_0, then it remains at rest for all times $t \geq t_0$, i.e., from $X(t_0) = X_0$ and $\dot{X}(t_0) = 0$ follows that the system remains at X_0. More precisely, we show that

58.12. Equilibrium States and a General Stability Principle

there exists *no* C^1-motion $X = X(t)$ with

$$X(t) \neq X_0 \quad \text{for all} \quad t > t_0.$$

Suppose such a motion exists. From $\dot{X}(t_0) = 0$ follows $T(t_0) = 0$ and the energy balance (50) yields

$$T(t) = W(t) < 0 \quad \text{for all} \quad t > t_0$$

after integration. But this contradicts $T(t) \geq 0$.

As in Section 58.7 we now assume that the forces can be decomposed in the form

$$K_i = K_i^{(p)} + K_i^{(q)}$$

with $K_i^{(p)}(X) = -U_{x_i}(X)$ for all i and all X, i.e., $K_i^{(p)}$ has the C^1-potential U. This implies

$$W(t) = U(X_0) - U(Y(t)) + W^{(q)}(t).$$

The condition $W(t) < 0$ for statical stability can now be written in the form

$$-\Delta U + \Delta W^{(q)} < 0. \tag{52}$$

Definition 58.16. Suppose the function $U: C \to \mathbb{R}$ on the metric space C has a local minimum at the point X_0. This minimum is called *regular* if and only if it is strict and there exists a neighborhood $V(X_0)$ such that the diameter of the set

$$\{X \in V(X_0): U(X) \leq E\}$$

for fixed, real E tends to zero as $E \to U(X_0)$.

For example, the real function $U = x^{2k}$, $k = 1, 2, \ldots$ has a regular minimum at $x = 0$, while for $U(x) \equiv 0$ this is not the case.

Theorem 58.C (Stability Principle of the Minimal Potential Energy). *We assume holonomic constraints which only depend on the position X and suppose that all forces have a potential, i.e., there exists a C^1-function U with*

$$K_i(X) = -U_{x_i}(X) \quad \text{for all } i \text{ and all } X.$$

Then the following is true.

(a) *If the system is in an equilibrium state at X_0, then U on the configuration space C has a critical point at X_0.*
(b) *If U on the configuration space C has a local, regular minimum at X_0, then X_0 is an equilibrium state, which is locally statically stable and stable in the sense of Ljapunov.*

By "locally" statically stable we mean a local variant of Definition 58.15, for which only virtual motions $X = Y(t)$ in a spatial neighborhood of X_0 and a time neighborhood of t_0 are considered. By *stability in the sense of Ljapunov* we mean that small deflections from the equilibrium state with small initial velocities only cause small motions of the system. More precisely, we require that for every $\varepsilon > 0$ there exists a $\delta > 0$ such that

$$|X(t_0) - X_0| + |\dot{X}(t_0)| < \delta$$

implies

$$|X(t) - X_0| + |\dot{X}(t)| < \varepsilon$$

for all C^1-motions $X = X(t)$ and all t. Thereby we have $|X| = \sum_i |x_i|$.

PROOF. Ad(a). Let $X = Y(t)$ be a virtual C^1-motion with $Y(t_0) = X_0$. From (51) follows that

$$\frac{d}{dt} U(Y(t)) = -\sum_i K_i(Y(t)) \dot{Y}_i(t) = 0 \quad \text{for} \quad t = t_0.$$

Ad(b). The local statical stability follows from

$$W(t) = U(X_0) - U(Y(t)) < 0 \quad \text{for} \quad Y(t) \neq X_0.$$

In order to prove the stability in the sense of Ljapunov we choose a motion $X = X(t)$ with $|X(t_0) - X_0| + |\dot{X}(t_0)| < \delta$. From the energy balance (50) follows the energy conservation theorem

$$T(t) + U(X(t)) = \text{const} = E \quad \text{for all } t.$$

From $T(t) \geq 0$ it follows that

$$U(X(t)) \leq E \quad \text{for all } t.$$

Since $E \to U(X_0)$ as $\delta \to 0$, Definition 58.16 implies that the diameter of the region, in which the motion occurs tends to zero as $\delta \to 0$. Moreover, it follows from

$$T(t) + U(X_0) \leq E \quad \text{for all } t$$

that $\sup_t |\dot{X}(t)| \to 0$ as $\delta \to 0$. □

58.13. Basic Equations of the Rigid Body and the Main Theorem about the Motion of the Rigid Body and its Equilibrium

We want to apply the previous observations, and, in particular, the principle of virtual power, to one of the most important mechanical models, namely the rigid body. Many objects in technology, can in first-order approximation be considered as rigid bodies, e.g., buildings, bridges, ships, planes, space ships, etc.

We consider $n \geq 3$ points, which do not all lie on a straight line, with masses

58.13. Basic Equations of the Rigid Body and the Main Theorem

m_i and position vectors x_i. Supppose the force

$$K_i = K_i(X, \dot{X}, t)$$

with $X = (x_1, \ldots, x_n)$ acts on x_i. This system is called a *rigid body* if and only if the side conditions

$$(x_i - x_j)^2 = c_{ij}, \qquad i, j = 1, \ldots, n \tag{53}$$

are satisfied with fixed numbers $c_{ij} \neq 0$ for all $i \neq j$. This means that the distances between the points remain the same for all motions. In the following section we will see that the configuration space C, i.e., the set of all points with (53) forms a six-dimensional C^∞-manifold in the X-space $\prod_{i=1}^n V_3$. In order to separate the motion of the center of mass

$$y = \sum_i m_i x_i / m$$

with the total mass $m = \sum_i m_i$ we set

$$x_i = y + z_i.$$

The angular momentum N with respect to the center of mass is, by definition, equal to

$$N = \sum_i m_i z_i \times \dot{z}_i.$$

Moreover, we define the *inertia tensor* θ through

$$\theta \omega = \sum_i m_i (z_i^2 \omega - (\omega z_i) z_i).$$

In an arbitrary coordinate system with the center of mass as the origin we have

$$\theta \omega = (\theta_s^r \omega^s) b_r$$

with the components

$$\theta_s^r = \sum_{k=1}^n m_k (z_k^2 \delta_s^r - b_s b_j z_k^j z_k^r), \qquad r, s = 1, 2, 3.$$

Here we set $\omega = \omega^s b_s$ and $z_k = z_k^j b_j$. In Cartesian coordinates we obtain for the basis vectors $b_s b_j = \delta_{sj}$. Hence there we have $\theta_s^r = \theta_r^s$ for all r, s and therefore

$$\theta = \theta^*,$$

i.e., $\theta: V_3 \to V_3$ is a self-adjoint linear operator. Thus there exist three orthonormal eigenvectors e_1, e_2, e_3 of θ with

$$\theta e_i = \theta_i e_i, \qquad i = 1, 2, 3. \tag{54}$$

If we choose a Cartesian coordinate system with the center of mass as the origin and axes e_i, then these axes are called *principal axes of inertia* and the eigenvalues θ_i of θ are called *principal moments of inertia*. In this coordinate system θ has a diagonal form, i.e.,

$$\theta_s^r = \theta_s \delta_s^r \quad \text{for all } r, s.$$

We shall see in the following section, by using the principle of virtual power, that the *basic equations for the motion of the rigid body* are equal to

$$m\ddot{y} = \sum_i K_i, \tag{55}$$

$$\dot{N} = \sum_i z_i \times K_i, \tag{56}$$

$$\dot{z}_i = \omega \times z_i, \quad i = 1, \ldots, n, \tag{57}$$

$$N = \theta \omega. \tag{58}$$

These equations have a very intuitive meaning. If

$$\omega(t) = \text{const},$$

then, according to Example 58.1, relation (57) corresponds to a rotational motion about the center of mass with axis of rotation ω and angular velocity $\omega = |\omega|$. In the general case, ω varies with time and we call $\omega(t)$ and $\omega(t)$ the momentary axis of rotation or the momentary angular velocity, respectively. According to (58), the angular momentum N is obtained from the momentary axis of rotation ω by an application of the linear operator θ, i.e., the inertia tensor. Relation (55) shows that the center of mass moves as if all forces would act on it. Altogether, the basic equations (55) to (58) represent the following statement.

The motion of the rigid body is the combination of the motion of the center of mass and a "rotational motion" about the center of mass, whereby, however, the axis of rotation and the rotational velocity depend on time. Such a motion is called a screw motion.

If $T = \sum_i m_i \dot{z}_i^2/2$ is the kinetic energy with respect to the center of mass, then

$$T = \sum_i m_i (\omega \times z_i)^2/2 = \omega \theta \omega.$$

Therefore θ^{-1} exists. This follows, because $\theta \omega = 0$ implies $\omega \times z_i = 0$ for all i, hence $\omega = 0$.

The following theorem and its corollary answer the central question: Which quantities uniquely determine the motion of the rigid body, and through which conditions are equilibrium states described? Recall that $X = (x_1, \ldots, x_n)$.

Theorem 58.D (Existence and Uniqueness Theorem for the Motion of the Rigid Body). *Assume that we are given C^1-forces*

$$K_i = K_i(X, \dot{X}, t), \quad i = 1, \ldots, n, \quad n \geq 3,$$

the initial position

$$x_1(t_0), \ldots, x_n(t_0),$$

the initial velocity $\dot{y}(t_0)$ of the center of mass, and the initial position $\omega(t_0)$ of the momentary axis of rotation together with the initial angular velocity $|\omega(t_0)|$.

Then for all times t in a neighborhood of t_0 there exists a unique C^2-motion $X = X(t)$ for the rigid body.

PROOF. From $x_i(t_0)$ we obtain $y(t_0)$ and thereby $z_i(t_0)$ and $\theta(t_0)$. Moreover, we have

$$N(t_0) = \theta(t_0)\omega(t_0).$$

If we replace ω in (57) with $\theta^{-1}N$, then, according to Theorem 3.A, equations (55) to (57) have a unique solution. □

Corollary 58.17 (Equilibrium Condition). *Assume the C^1-forces $K_i = K_i(X, \dot{X}, t)$ are given. For the positions $x_{10}, \ldots, x_{n0}$, the rigid body is in an equilibrium state if and only if the following three conditions are satisfied:*

(i) *The total force is zero, i.e.,*

$$\sum_i K_i(X_0, 0, t) = 0 \quad \text{for all } t.$$

(ii) *The total torque is zero, i.e.,*

$$\sum_i x_{i0} \times K_i(X_0, 0, t) = 0 \quad \text{for all } t.$$

(iii) *The body is at rest at time t_0, i.e., $\dot{x}_i(t_0) = 0$ for all i.*

PROOF. If one inserts

$$x_i(t) \equiv x_{i0}$$

into (55) and (56), then one obtains

$$\ddot{y} = \dot{N} = 0$$

and (i) to (iii) follow. Note that

$$\sum_i x_i \times K_i = \sum_i (y + z_i) \times K_i.$$

Conversely, assume (i) to (iii). Then $x_i(t) \equiv x_{i0}$ is a solution of (55) to (58) and, according to Theorem 58.D, unique. □

58.14. Foundation of the Basic Equations of the Rigid Body

Lemma 58.18. *Assume that $n \geq 3$ different points $x_1, \ldots, x_n$ are given, whereby not all of them lie on a straight line. Then one obtains precisely all solutions of* the system of linear equations

$$(x_i - x_j)(h_i - h_j) = 0, \qquad i,j = 1, \ldots, n \qquad (59)$$

through

$$h_i = a + \omega \times x_i$$

with arbitrary $a, \omega \in V_3$.

The proof will be given in Problem 58.1. From the subimmersion theorem (Theorem 4.I in Section 4.16) it follows that the configuration space C of the rigid body, described by (53), is a six-dimensional C^∞-manifold. One can show that $C = SO(3) \times \mathbb{R}^3$, where $SO(3)$ is the Lie group of all orthogonal (3×3)-matrices with determinant one (see Chapter 84).

Proposition 58.19. *The most general possible motion of the rigid body is described by the differential equation*

$$\dot{x}_i(t) = a(t) + (\omega(t) \times x_i(t)), \qquad i = 1, \ldots, n \qquad (60)$$

for arbitrary but fixed vector functions $a(\cdot)$ and $\omega(\cdot)$.

PROOF. Equation (59) with $h_i = \dot{x}_i$ immediately follows from $(x_i - x_j)^2 = c_{ij}$ by differentiating with respect to t. Hence (60) follows from Lemma 58.18.

Conversely, (60) satisfies equation (59) with $h_i = \dot{x}_i$. Integration of (59) yields $(x_i - x_j)^2 = \text{const}$. □

In order to obtain the basic equations for the rigid body we now use the principle of virtual power from Section 58.11. This principle tells us that

$$\sum_i (m_i \ddot{x}_i - K_i) v_i = 0.$$

From Proposition 58.19, an arbitrary virtual velocity has the form $v_i = a + (\omega \times x_i)$. This implies

$$a \sum_i m_i \ddot{x}_i - K_i + \omega \sum_i x_i \times (m_i \ddot{x}_i - K_i) = 0.$$

In what follows we write $\sum$ instead of $\sum_i$.

Since a and ω are arbitrary, it follows that

$$\sum m_i \ddot{x}_i = \sum K_i, \qquad (61)$$

$$\sum m_i x_i \times \ddot{x}_i = \sum x_i \times K_i. \qquad (62)$$

From this one easily derives the basic equations in the form of (55) to (58). Indeed, because of (61), we get

$$m\ddot{y} = \sum K_i.$$

From (62) follows

$$\sum (y + z_i) \times (m\ddot{y} + m_i \ddot{z}_i) = \sum (y + z_i) \times K_i.$$

Because of $\sum m_i z_i = 0$ we obtain $\sum m_i \ddot{z}_i = 0$, hence

$$\sum z_i \times m_i \ddot{z}_i = \sum z_i \times K_i, \quad \text{i.e.,} \quad \dot{N} = \sum z_i \times K_i,$$

and $\dot{x}_i = a + (\omega \times x_i)$ implies

$$\dot{z}_i = a - \dot{y} + (\omega \times y) + (\omega \times z_i).$$

Because of $\sum m_i z_i = 0$ and $\sum m_i \dot{z}_i = 0$ we obtain

$$a - \dot{y} + (\omega \times y) = 0,$$

hence $\dot{z}_i = \omega \times z_i$. This implies

$$N = \sum_i m_i z_i \times (\omega \times z_i) = \sum_i m_i \omega z_i^2 - m_i z_i(\omega z_i) = \theta \omega. \qquad \square$$

58.15. Physical Models, the Expansion of the Universe, and its Evolution after the Big Bang

Objects in the real world are very complicated and a complete mathematical description, i.e., taking all possible effects into account, is principially impossible. Physicists, however, have made the positive experience that many phenomena can be described with relatively simple models. In this section we describe this important sort of physical thought by looking at the interesting example of the expansion of the universe. A complete understanding of this phenomenon is possible only from the general theory of relativity, and will be discussed in Chapter 76. Many effects, however, can be understood in the context of a simple model, which only uses the electromagnetic black-body radiation and further results in statistical physics. This will now be discussed, and will perhaps create some interest in the reader for statistical physics and the general theory of relativity of Chapters 68 and 76. The key question is:

How did the universe evolve?

To answer this question in the context of the so-called standard model we need two experimental results:

(a) Hubble's law about the red shift in the spectrum of galaxies (Hubble, 1929).
(b) Discovery of the 3 K-radiation in 1965 for which Penzias and Wilson received the Nobel prize.

Looking at the spectrum of the galaxies, one observes a red shift, which, if interpreted as a Doppler effect (Problem 58.6), leads to an escape velocity V for the galaxies of

$$V = HR. \tag{63}$$

Thereby R is the distance between the galaxy and our Milky Way, and H denotes the Hubble constant. According to presently available data, we have $H = H_0$ with

$$H_0 = (15 \text{ km/s}) \text{ per } 10^6 \text{ light years} = 1/(2 \cdot 10^{10} \text{ years}).$$

This means that for all 10^6 light years the escape velocity of the galaxies increases by 15 km/s. This, however, is a very rough value.

The expansion of the universe described by (63) can be explained with the hypothesis that at time $t = 0$ an enormous explosion, called the Big Bang, occurred and created our universe. We shall see that the weak 3 K-radiation from the universe can be regarded as an important experimental proof for the hypothesis of the Big Bang. This radiation has its origin in the photon energy which at an early stadium of the hot cosmos existed. Roughly speaking, this photon energy has become very thin as a consequence of the expansion of the universe and today we can only observe a very weak radiation which requires sensitive antenna systems. As one would expect, this radiation is isotropic, i.e., no direction in the universe is preferred.

Another sign for the correctness of the Big Bang hypothesis is the currently observed ratio of

$$\text{hydrogen:helium} = 3:1.$$

This also follows from the standard model.

One important question, however, can presently not be experimentally determined, namely: Is the volume of the universe finite or infinite? For an answer one would have to know a very precise value for the mass density ρ_0 of the present universe, but today the error margins for ρ_0 are too large. We shall show, however, that the age of the universe can be estimated by using the Hubble constant H_0.

The 3 K-radiation is an electromagnetic radiation density in the universe, which satisfies Planck's law about the black-body radiation with energy density

$$E/V = \int_0^\infty 8\pi hc \, d\lambda/\lambda^5 (e^{hc/kT\lambda} - 1) \tag{64}$$
$$= 7.6 \cdot 10^{-16} [T]^4 \quad (\text{J/m}^3).$$

Thereby $T = 3$ K is the so-called temperature of the radiation. As usual, $[T]$ denotes the value without dimension, i.e., $[T] = 3$. Experimentally, E/V is measured and T is calculated from (64). Equation (64) also determines the distribution of the energy onto the wavelengths λ. In Section 68.4, we derive (64) from the Bose statistics for photons, and show that the particle density of the photons is given by

$$N/V = 60.4 \quad (kT/hc)^3. \tag{65}$$

This corresponds to a density of

$$2 \cdot 10^7 \, [T]^3 \quad \text{photons/m}^3.$$

Thus the mean energy of a photon is equal to

$$\varepsilon = E/N = 3.7 \cdot 10^{-23} \, [T] \, \text{J}$$

58.15. Physical Models, Expansion of the Universe

and hence
$$\varepsilon = 2.7kT. \tag{66}$$

The universal constant k is called the Boltzmann constant and has the value
$$k = 1.380 \cdot 10^{-23} \text{ J/K}.$$

In the following the electromagnetic radiation in the universe will simply be called photon gas. Section 68.4 implies that the entropy density of the photon gas is equal to
$$S/V = 4\,aT^3/3, \tag{67}$$

where a is a constant given by (68.25).

Now let us show that relations (63)–(67) above yield the main results about the evolution of our universe. In this connection the key equations below will be the equation of motion of a galaxy (70) and equation (74), which relates the time after the Big Bang to the corresponding energy density.

58.15a. Equation of Motion of the Galaxies

We consider a galaxy Γ of mass m located at a distance $R(t)$ from our Milky Way. Very important is that, according to Problem 58.5, the only force acting on Γ is the gravitational force caused by the mass M which is contained inside a ball of radius $R(t)$, i.e.,
$$M = 4\pi R(t)^3 \rho(t)/3.$$

Thereby $\rho(t)$ is the mass density at time t. According to Problem 58.4, the actual motion is the same as if the entire mass M of the ball would be located at the center $R = 0$. The expansion of the universe can only be completely understood in the context of the general theory of relativity (see Chapter 76). But it is quite interesting that important results can already be obtained by using only simple facts from classical mechanics. To this end we choose $R(t)$ not too large, so that the space curvature can be neglected. Further, we take relativistic effects into account by choosing a mass density $\rho(t)$ which is related to the energy density $e(t)$ by the Einstein relation
$$e(t) = \rho(t)c^2.$$

Here c denotes the velocity of light (see Section 75.11). This energy density $e(t)$ also contains all kinds of radiation. We assume that the escape motion
$$R = R(t)$$

of the galaxy Γ satisfies the classical energy conservation law, i.e.,
$$\frac{m\dot{R}^2}{2} - \frac{GmM}{R} = \text{const} = \frac{mE_0}{2}. \tag{68}$$

The first term is the kinetic energy of the escape motion and, the second term, according to Section 58.8, is the potential energy of the gravitation. Let t_0 denote the *present time* and $t = 0$ the time of the Big Bang. We set

$$H_0 = H(t_0), \qquad R_0 = R(t_0)$$

and

$$\rho_0 = \rho(t_0).$$

From (63) follows

$$E_0 = R_0^2 H_0^2 - 2GM/R_0 = R_0^2(H_0^2 - 8\pi G\rho_0/3). \tag{69}$$

For $R = R(t)$, equation (68) yields the *equation of motion* for the galaxy Γ:

$$\int_{R(t)}^{R_0} \frac{dR}{\sqrt{E_0 + 2GM/R}} = t_0 - t. \tag{70}$$

From this *first key equation* we draw several important conclusions.

At time $t = 0$ of the Big Bang the galaxy Γ is at the origin, i.e., $R(0) = 0$. Consequently, the *age* of the universe t_0 is given by

$$t_0 = \int_0^{R_0} \frac{dR}{\sqrt{E_0 + 2GM/R}} \leq \frac{R_0}{\sqrt{E_0 + 2GM/R_0}} \tag{71}$$

$$= 1/H_0 = 2 \cdot 10^{10} \text{ years}.$$

Thus it follows that the universe is less than 20 milliard years old. Note that the total mass M of the ball of radius $R(t)$ remains constant for all times $t > 0$. At present we approximately have

$$E_0 = 0,$$

as is shown below. Thus with (71) we obtain for the age of the universe

$$t_0 = \int_0^{R_0} \sqrt{\frac{R}{2GM}}\, dR = \frac{2}{3}\sqrt{\frac{R_0^3}{2GM}} = \frac{2}{3H_0}.$$

This gives nearly 13 milliard years. Estimates obtained from the general theory of relativity will be given in Chapter 76.

We define the critical mass density

$$\rho_{\text{crit}} = 3H_0^2/8\pi G = 5 \cdot 10^{-27} \text{ kg/m}^3.$$

It follows that

$$E_0 = 8\pi G R_0^2 (\rho_{\text{crit}} - \rho_0)/3,$$

where $\rho_0 c^2$ is the value of the energy density of the present universe, and ρ_0 is the present mass density. From (70) we derive the following important *alternative* about the structure of our universe:

If $E_0 \geq 0$, i.e., $\rho_0 \leq \rho_{\text{crit}}$, then the integrand in (70) has the form $O(1)$ or $O(\sqrt{R})$ for $R \to \infty$. Consequently, we have

$$R(t) \to +\infty \quad \text{as} \quad t \to +\infty,$$

58.15. Physical Models, Expansion of the Universe

i.e., the galaxy may escape into infinity. Note that $t'(R) > 0$ for all $R > 0$ according to (70). Consequently the function $t \mapsto R(t)$ is strictly monotone increasing for $t > 0$. (72a)

If $E_0 < 0$, i.e., $\rho_0 > \rho_{\text{crit}}$, then the integrand in (70) becomes imaginary for large R. Therefore, $R(t)$ must remain bounded for $t \to +\infty$, i.e., the galaxy Γ cannot escape into infinity. (72b)

In the context of the general theory of relativity one finds that for $\rho_0 < \rho_{\text{crit}}$ the universe is infinite, whereas for $\rho_0 > \rho_{\text{crit}}$ its volume is finite (see Chapter 76). We will show that $\rho_0 = 1.7 \cdot 10^{-27}$ kg/m³, i.e.,

$$\rho_0 \sim \rho_{\text{crit}}$$

and $E_0 \sim 0$. The uncertainty for the value of ρ_0, however, is currently so large that it is impossible to decide whether (72a) or (72b) is true, i.e., whether the universe is infinite or finite. If additional masses would be discovered (interstellar matter, small hypothetical mass of the large number of neutrinos in the universe, etc.), then this would favor the finite universe.

58.15b. Energy Density of the Present Universe

In one cubic meter, there are presently about N_n nucleons (protons or neutrons) with $0.03 \leq N_n \leq 6$. For say $N_n = 1$, this yields a matter density for the present universe of

$$\rho_n = 1.7 \cdot 10^{-27} \text{ kg/m}^3.$$

From (65) with $T = 3$ K we find that $5.4 \cdot 10^8$ photons are contained in one cubic meter. The ratio of the numbers of photons and the number of nucleons in one cubic meter is therefore presently about $10^9 : 1$. This is a very important consequence of the 3 K-radiation, as we shall see in Section 58.15f below. It is assumed that the photon–nucleon ratio has remained constant at all times. Formula (66) implies that the energy density of the photons is presently about $10^9 \cdot 2.7 \, kT$. Dividing by c^2 we obtain the corresponding "photon mass density" of

$$\rho_p = 10^{-30} \text{ kg/m}^3.$$

For the mass density of the present cosmos we find

$$\rho_0 = \rho_n + \rho_p = 1.7 \cdot 10^{-27} \text{ kg/m}^3.$$

We assume that the entropy S of the photon gas has remained constant since the Big Bang. Hence, formula (67) implies that $R^3(t)T^3 = \text{const}$, i.e., the temperature T of the photon gas is given by

$$T(t) = \text{const}/R(t). \quad (73)$$

At an early stage in its development the universe was therefore very hot. Currently the universe is not in a thermodynamical equilibrium, because there exist enormous temperature differences between the 3 K of the photon gas and the very hot matter of galaxies. But one assumes that at an early stage, there existed a common temperature for the universe, i.e., the universe was in a thermodynamical equilibrium. Actually, only a very gradually changing quasi-equilibrium was possible, because of the expansion of the universe (see Section 67.1).

The above considerations show that the ratio of the energy densities of nucleons and photons is currently of magnitude 10^3, i.e., the nucleon energy dominates (matter cosmos). Since, according to (66), the mean energy of a photon depends linearly on the photon temperature, and the ratio of nucleons and photons has remained constant, it follows that for a photon temperature of approximately 3000 K, the ratio of the nucleon energy and the photon energy has been equal to one. For higher temperatures, i.e., at an early stage of the cosmos, the photon energy dominated (radiation cosmos).

58.15c. Critical Temperature

Table 58.3 shows the so-called critical temperature. It is defined as

$$T_{\text{crit}} = m_0 c^2 / k,$$

where m_0 is the rest mass of the particle. One assumes that, in a thermodynamical equilibrium, there exists a great number of particles with rest mass m_0 and hence rest energy $m_0 c^2$ above T_{crit} (see Section 75.11). This is motivated as follows. Let $m_0 \neq 0$. For high temperatures experience shows us that we can use quasi-classical statistical physics. The particles in the early cosmos are regarded as an oscillating system. This is a fairly general assumption as will be shown in Problem 58.11. The law of equipartition then implies that the mean energy of each vibrational degree of freedom is equal to

$$\bar{E} = kT$$

(see Problem 68.4). Moreover, the mean energy must be greater than the rest energy, i.e., $\bar{E} \geq m_0 c^2$. This implies that

$$T \geq T_{\text{crit}}.$$

For $m_0 = 0$ these observations show that T has no lower bound. Thus we set

Table 58.3

Particle	Spin	Critical temperature T_{crit}
proton p, neutron n	$\pm\frac{1}{2}$	10^{13} K
π-mesons π^0, $\pi^\pm$	0	$1.5 \cdot 10^{12}$ K
muons $\mu^\pm$	$\pm\frac{1}{2}$	10^{12} K
electron e^-, positron e^+	$\pm\frac{1}{2}$	$6 \cdot 10^9$ K
neutrinos v_e, $\bar{v}_e$, v_μ, $\bar{v}_\mu$	$\frac{1}{2}$	0 K
photon γ	± 1	0 K

58.15. Physical Models, Expansion of the Universe

$T_{crit} = 0$. In Table 58.3 photons and neutrinos have no rest mass, however, it is possible that neutrinos have a rest mass of 20 eV/c². In comparison the electron mass is $5 \cdot 10^5$ eV/c².

58.15d. Development in Time of the Early Cosmos

At an early stage of the universe, time t and mass density $\rho(t)$ were related like this:

$$t = \sqrt{\frac{3}{32\pi G \rho(t)}}. \tag{74}$$

We prove this *second key formula*. From (68) follows

$$\dot{R}^2 = E_0 + 2GM/R.$$

Because of $\dot{R} = HR$ we have

$$H^2 = \frac{E_0}{R^2} + \frac{8\pi G \rho}{3}.$$

At the early stage of the cosmos ($R \to 0$, $t \to 0$, $\rho \to +\infty$), the energy density was mainly determined by the radiation energy of the photons and the other elementary particles. From (64) and (73) we find for the corresponding mass density $\rho = e/c^2$ the relation

$$\rho = \text{const} \cdot T^4 = \text{const} \cdot R^{-4}.$$

This implies

$$H^2 = \frac{8\pi G \rho}{3} + O(\rho^{1/2}), \qquad \rho \to +\infty.$$

In order to simplify the computations, we neglect the term $O(\rho^{1/2})$. We obtain

$$H = CR^{-2}, \qquad C = \text{const}$$

and $\dot{R} = HR = C/R$. Integration yields

$$C(t - t_1) = 2^{-1}(R(t)^2 - R(t_1)^2)$$

and hence

$$t - t_1 = 2^{-1}(H(t)^{-1} - H(t_1)^{-1}).$$

For $t_1 \to 0$ we have $\rho(t_1) \to +\infty$ (Big Bang). This implies (74).

58.15e. The Universe at a Temperature of 10^{11} K

This temperature is chosen, because, as a look at Table 58.3 shows, at this temperature essentially only photons, neutrinos, electrons, and positrons exist.

In Section 68.6 we shall use methods of statistical physics and results about the spin of Table 58.3, to show that the energy density of the universe at 10^{11} K is 9/2 times that of the photon gas. It follows therefore from (64) that the energy density is equal to

$$e(t) = 3.5 \cdot 10^{29} \text{ J/m}^3.$$

From (74) we obtain the corresponding time

$$t = (3c^2/32\pi Ge(t))^{1/2} = 10^{-2} \text{ s},$$

i.e., one hundredth of a second after the Big Bang the universe had the temperature 10^{11} K.

58.15f. Helium Synthesis

Computations in nuclear physics show that at a temperature of 900 million K the helium synthesis began. This mainly follows from the photon–nucleon ratio of about $10^9:1$. At this temperature the ratio of neutrons and protons is equal to 13:87 as a consequence of nuclear reactions. A hydrogen nucleus

Table 58.4

Object	Approximate age
Cosmos	maximal 20 milliard years
Our galaxy (Milky Way)	10 milliard years
Sun and Earth	5 milliard years
Life on Earth	4 milliard years
Earth continents	3 milliard years
Earth atmosphere	1 milliard years
Explosion of life on Earth (fishes, plants, insects)	600 million years ago
Dinosauriers appear	200 million years ago
Dinosauriers disappear; strong development of mammals	60 million years ago
Man	1–3 million years
Human history	5,000 years
Formation of modern science through Galilei and Newton	400 years ago
Einstein's general theory of relativity	1915
Friedman's model of the expanding universe in the context of the general theory of relativity	1922
Hubble's law about the red shift	1929
Atomic bomb	1945
Discovery of the 3 K-radiation	1965
Humans on the moon	1969

(or a helium nucleus) consists of one proton (or two protons and two neutrons). Furthermore, the mass of the proton is approximately equal to the mass of the neutron. Let us assume that all neutrons have been used in the formation of helium and the remaining protons are being used to form hydrogen. This yields

a mass ratio of helium and hydrogen of 26:74,

i.e., approximately 1:3. This is about the ratio which is currently observed in the universe, and is thus another reason to assume that the standard model, discussed above, is a correct model for the evolution of the universe.

A popular and elementary exposition about the development of the cosmos may be found in Weinberg (1977, M), and a thorough discussion of the mathematical–physical theory in the context of the general theory of relativity is contained in Weinberg (1972, M). We merely mention here that the helium synthesis took place 225 s after the Big Bang. Further interesting numbers can be found in Table 58.4.

58.16. Legendre Transformation and Conjugate Functionals

In mechanics and thermodynamics the change of dependent and independent variables plays an important role. This is called the Legendre transformation. We want to explain the basic idea. Important applications will be discussed in Section 58.20 (Hamiltonian formalism) and in Section 67.5 (thermodynamical potentials). Let the real function

$$E = E(S, V)$$

be C^2 on $\mathbb{R}^2$. For the total differential we obtain

$$dE = T\,dS + P\,dV \tag{75}$$

with

$$T = E_S(S, V), \qquad P = E_V(S, V). \tag{76}$$

Our goal is to choose T and V as independent variables. More precisely, we assume that the first equation in (76) can be solved for S, i.e.,

$$S = S(T, V).$$

Locally, as a consequence of the implicit function theorem (Theorem 4.B), this is possible if $E_{SS}(S, V) \neq 0$. Globally, this is possible if $S \mapsto E_S(S, V)$ is strictly monotone increasing with $E_S(S, V) \to \pm\infty$ for $S \to \pm\infty$. In the thermodynamics of Chapter 67 we have $V =$ volume, $T =$ absolute temperature, $E =$ inner energy, $S =$ entropy, and $P =$ pressure.

The trick of the Legendre transformation is to find a new function

$$F = F(V, T)$$

which is related to E in such a way that the total differential dF assumes a particularly simple form. Formally, F can easily be obtained from the product rule

$$d(ab) = a\,db + b\,da.$$

Equation (75) implies that

$$dE = d(TS) - S\,dT + P\,dV.$$

If we choose $F = E - TS$, then

$$dF = -S\,dT + P\,dV, \tag{77}$$

and hence

$$S = -F_T(V, T), \qquad P = F_V(V, T). \tag{78}$$

The function F is called free energy. In order to justify this formal computation we consider the C^1-curves

$$S = S(t), \qquad V = V(t),$$

and compute $T(t)$, $P(t)$ from (76). For the time derivatives we find

$$\dot{F} = \dot{E} - \dot{T}S - T\dot{S} = T\dot{S} + P\dot{V} - \dot{T}S - T\dot{S} = -S\dot{T} + P\dot{V}.$$

If, in particular, we choose the curves in such a way that

$$T(t) = t, \qquad V = \text{const},$$

then we obtain

$$\dot{F} = F_T = -S.$$

If $T(t) = \text{const}$, $V(t) = t$ we obtain $\dot{F} = F_V = P$. This is (78).

Finally, we explain the connection between conjugate functionals, which have been discussed in Chapter 51 of Part III. We begin with

$$E = E(S, V),$$

consider V as a parameter, and set

$$E^*(S^*, V) = \sup_{S \in \mathbb{R}} S^*S - E(S, V). \tag{79}$$

From Section 51.1 it then follows that E^* is the conjugate function to E, and S^* is the conjugate variable to S. We want to show that for a sufficiently regular function E in the above sense, the following holds:

$$E^* = -F, \qquad S^* = T.$$

In fact, the existence of a unique maximum in (79) and differentiation with respect to S immediately imply $S^* = E_S(S, V)$, hence $T = S^*$ and

$$E^*(S^*, V) = TS - E(S, V) = -F(T, V).$$

In Example 51.4 we showed that the following two conditions are sufficient for the existence of a unique maximum in (79):

strict convexity

$$E_{SS}(S, V) > 0 \quad \text{for all } S, V$$

and coercivity

$$E(S, V)/|S| \to +\infty \quad \text{as} \quad |S| \to \infty \quad \text{for all } S, V.$$

Transformation (79), however, can also be performed if E is not sufficiently regular.

58.17. Lagrange Multipliers

Our goal in the following sections is this. In special situations, we want to use the general principle of Gauss' least constraint to obtain more convenient formulations of the basic equations of mechanics which can also be applied to other physical theories, namely:

(i) Principle of stationary action and Lagrangian mechanics.
(ii) Canonical equations and Hamiltonian mechanics.
(iii) Poisson brackets and Poissonian mechanics.
(iv) Propagation of action and the Hamilton–Jacobi theory about the relations between the canonical equations (ordinary differential equations) and the Hamilton–Jacobi equation (first-order partial differential equation).

To recognize the connection between mechanics and a variational principle, in the following section we consider the variational problem

$$\int_{t_1}^{t_2} L(y, \dot{y}, t) \, dt = \text{stationary!} \tag{80}$$

$$y(t_1) = a, \quad y(t_2) = b,$$

with the side condition

$$M_j(y) = 0, \quad j = 1, \ldots, J, \tag{81}$$

where $y = (y^1, \ldots, y^n) \in \mathbb{R}^n$. The real numbers t_1 and t_2 and $a, b \in \mathbb{R}^n$ are given. Moreover, let $y(t) \in G$ for all $t \in [t_1, t_2]$. We set

$$\bar{L}(y, \dot{y}, t) = L(y, \dot{y}, t) + \sum_{j=1}^{J} \lambda_j(t) M_j(y)$$

and $P(t) = (y(t), \dot{y}(t), t)$. The necessary conditions for a solution $y = y(t)$ of (80),

(81) are then

$$\frac{d}{dt}\bar{L}_{\dot{y}^i}(P(t)) - \bar{L}_{y^i}(P(t)) = 0, \quad i = 1, \ldots, n \tag{82}$$

for all $t \in \,]t_1, t_2[$. The functions λ_j are called *Lagrange multipliers*. Our assumptions are:

(H1) G is a nonempty, open set in $\mathbb{R}^n$ and $1 \leq J < n$.
(H2) The functions $M_j: G \to \mathbb{R}$, $j = 1, \ldots, J$ are C^2, and the matrix $(\partial M_j/\partial y^i)$ has maximal rank J at each point of G.

According to the preimage theorem of Section 4.18 this assumption guarantees that the set of all $y \in G$, which satisfy the side conditions (81), form a $(n - J)$-dimensional C^2-manifold M which is called the configuration space. This fact is the key to our proof.

(H3) The function $L: G \times \mathbb{R}^{n+1} \to \mathbb{R}$ is C^2.

Theorem 58.E. *Assume* (H1)–(H3). *If* $y: [t_1, t_2] \to G$ *is a C^2-solution of* (80), (81), *then there exist functions* $\lambda_j: [t_1, t_2] \to \mathbb{R}$ *such that* (82) *holds.*
Without side conditions (81) *one obtains* (82) *with* $\bar{L} = L$.

The importance of this theorem is that the case of side conditions can formally be reduced to the case *without* side conditions by replacing L with $\bar{L}$.

PROOF.

(I) No side conditions. For this case, (82) with $L = \bar{L}$ has already been proved in Section 37.4b. There we had $n = 1$. But for $n > 1$ the proof is the same if one only varies y^i and leaves all other y^k fixed.

(II) Side conditions. The proof idea is to eliminate the side conditions by using $y = \varphi(q)$ and then to apply (I).

(II-1) Localization. Let $y = y(t)$ be a solution of (80), (81). We choose a fixed solution point $y_0 = y(t_0)$ and vary $y(\cdot)$ only in a neighborhood of y_0. Therefore it suffices to consider the case that $[t_1, t_2]$ is a sufficiently small interval which contains t_0 in its interior and where G is a small neighborhood of y_0 in which the solution-manifold M of (81) has the parameter representation

$$y = \varphi(q) \tag{83}$$

with $q = (q^1, \ldots, q^m)$ and $\varphi(q_0) = y_0$. Because of $\dim M = m$, the matrix $(\partial \varphi^i(q)/\partial q^k)$ has rank $m = n - J$.

(II-2) Elimination. Using transformation (83) we obtain from (80), (81) the new variational problem

$$\int_{t_1}^{t_2} N(q, \dot{q}, t)\, dt = \text{stationary!}, \tag{84}$$

$$q(t_1) = \bar{a}, \quad q(t_2) = \bar{b},$$

where
$$N(q, \dot{q}, t) = L(\varphi(q), \varphi'(q)\dot{q}, t).$$

According to (I) the necessary conditions for (84) are equal to

$$\frac{d}{dt} N_{\dot{q}^k} - N_{q^k} = 0, \quad k = 1, \ldots, m. \tag{85}$$

The chain rule implies for the partial F-derivatives

$$N_q h = L_y \varphi'(q) h + L_{\dot{y}} \varphi''(q) h \dot{q},$$
$$N_{\dot{q}} h = L_{\dot{y}} \varphi'(q) h,$$
$$\frac{d}{dt} N_{\dot{q}} h = \left(\frac{d}{dt} L_{\dot{y}}\right) \varphi'(q) h + L_{\dot{y}} \varphi''(q) \dot{q} h.$$

From (85), i.e., from $(d/dt) N_{\dot{q}} - N_q = 0$ we therefore obtain

$$\left(\frac{d}{dt} L_{\dot{y}} - L_y\right) \varphi'(q) = 0. \tag{86a}$$

Differentiation of (81) with respect to q yields

$$M'_j(\varphi(q)) \varphi'(q) = 0, \quad j = 1, \ldots, J. \tag{86b}$$

(III-3) **Lagrange multipliers.** In component notation equation (86b) is equal to

$$\sum_{r=1}^{n} \frac{\partial M_j}{\partial y^r} \frac{\partial \varphi^r}{\partial q^k} = 0, \quad k = 1, \ldots, m.$$

At the point q_0 this is a linear homogeneous system with coefficient matrix $(\partial \varphi^r(q_0)/\partial q^k)$. This matrix has rank $m = n - J$. Thus the J linearly independent vectors M'_j span the entire solution space. Relation (86b) therefore implies the linear dependence

$$\frac{d}{dt} L_{\dot{y}} - L_y = \sum_j \lambda_j M'_j.$$

This is (82) for $t = t_0$. □

58.18. Principle of Stationary Action

We now want to show that for a typical physical situation the equations of motion which have been obtained from the principle of least constraint, can also be derived from a variational principle. This is an important result.

We consider here the following situation. Let N mass points x_i with masses m_i be given. Suppose the force K_i acts on x_i whereby a potential $U = U(x_1, \ldots, x_N, t)$ exists, i.e.,

$$K_i = -U_{x_i}, \quad i = 1, \ldots, N,$$

and assume that the motion satisfies the side conditions

$$f^r(x_1,\ldots,x_N) = 0, \qquad r = 1,\ldots,R \tag{87}$$

with $R < 3N$. Under suitable regularity conditions one obtains from the principle of least constraint the equations of motion

$$m_i\ddot{x}_i = -U_{x_i} + \sum_{r=1}^{R} \lambda_r f_{x_i}^{(r)}, \qquad i = 1,\ldots,N. \tag{88}$$

The following observation is now important. We set

$$L = T - U \tag{89}$$

with kinetic energy

$$T = \sum_{i=1}^{N} m_i \dot{x}_i^2/2$$

and potential energy U. Moreover, we consider the *variational problem*

$$\int_{t_1}^{t_2} L\,dt = \text{stationary!}, \tag{90a}$$

$$x_i(t_1) = a_i, \qquad x_i(t_2) = b_i, \qquad i = 1,\ldots,N$$

with side conditions

$$f^{(r)}(x_1,\ldots,x_N) = 0, \qquad r = 1,\ldots,R. \tag{90b}$$

Under suitable regularity conditions a solution of (90) then satisfies the equations

$$\frac{d}{dt}\bar{L}_{\dot{x}_i} - \bar{L}_{x_i} = 0, \qquad i = 1,\ldots,N \tag{91}$$

with

$$\bar{L} = L + \sum_{r=1}^{R} \lambda_r f^{(r)},$$

according to Theorem 58.E. A comparison implies the key result that (91) is equal to the equations of motion (88). Therefore (88) follows from the variational principle (90), which is called the *principle of stationary action*.

The physical significance of the Lagrange multipliers λ_r is the fact that they yield constraining forces.

58.19. Trick of Position Coordinates and Lagrangian Mechanics

> By generalizing Euler's method, Lagrange got the idea for his remarkable formulas, where, in a single line, there is contained the solution of all problems of analytic mechanics.
>
> Carl Gustav Jakob Jacobi (1804–1851)

58.19. Trick of Position Coordinates and Lagrangian Mechanics

Lagrangian mechanics starts from the *variational problem*

$$\int_{t_1}^{t_2} L(q, \dot{q}, t)\, dt = \text{stationary!,} \tag{92}$$

$$q(t_1) = a, \qquad q(t_2) = b,$$

where $q = (q^1, \ldots, q^m) \in \mathbb{R}^m$ and $q(t) \in Q$ for all $t \in [t_1, t_2]$. We assume:

(H) The Lagrange function $L: Q \times \mathbb{R}^m \times \mathbb{R} \to \mathbb{R}$ is C^2, where Q is a nonempty, open set in $\mathbb{R}^m$.

For a C^2-solution $q: [t_1, t_2] \to Q$ of (92), Theorem 58.E implies the famous *Lagrangian equation of motion*

$$\frac{d}{dt} L_{\dot{q}}(P(t)) - L_q(P(t)) = 0 \tag{93}$$

for all $t \in [t_1, t_2]$ with $P(t) = (q(t), \dot{q}(t), t)$ or in component notation

$$\frac{d}{dt} L_{\dot{q}^k}(P(t)) - L_{q^k}(P(t)) = 0, \qquad k = 1, \ldots, m.$$

We want to show how (92) follows from the previous section.

(T) *Trick of Lagrangian position coordinates q.* Let a system of mass points be given such as in Section 58.18. We set

$$L = T - U,$$

where T is the kinetic energy and U the potential energy. The important idea is to introduce coordinates q such that with respect to q no side conditions occur. Moreover, we transform T and U into these coordinates and use (92), (93).

If, for example, the equation of motion takes place on a circle, then the angle φ can be introduced as position coordinate q.

Method (T) precisely corresponds to our proof of Theorem 58.E, whereby (93) corresponds to (85). Thus the basic idea is as follows. The side conditions of Section 58.18 yield a configuration space M, which is a manifold, and which is parametrized by the coordinate q. Thereby the side conditions vanish automatically. In the language of manifolds this means that we pass to a chart. Thereby m is called the number of degrees of freedom of the system. The variational problem (90) then becomes (92).

If $q = q(t)$ is a C^2-solution of the Lagrangian equations (93) on $[t_1, t_2]$, and if L does not depend on t, then

$$A = \dot{q} L_{\dot{q}} - L$$

is a conserved quantity, i.e., it is constant along $q = q(t)$. In fact, we have

$$\frac{d}{dt}(\dot{q} L_{\dot{q}} - L) = \ddot{q} L_{\dot{q}} + \dot{q}\frac{d}{dt} L_{\dot{q}} - L_q \dot{q} - L_{\dot{q}} \ddot{q} = 0.$$

Here we use the convention (102) below. In many cases we have $A = T + U$, i.e., A is the energy. The precise conditions for this are given in Remark 58.22 below.

Remark 58.20 (General Manifolds). We now use several important concepts from the theory of manifolds which will be introduced in Chapter 73. By restricting ourselves to open sets Q, our formulation above is local. In fact, we can consider the more general case that the configuration space M is a real m-dimensional C^3-manifold which generally cannot be described through a single chart. The formulation for the general M, however, is completely analogous to (92), (93). In this case q is a point on M and $\dot{q}$ is a point in the tangent space TM_q, since the tangent vector $\dot{q}(t)$ to the curve $t \mapsto q(t)$ at the fixed point $q = q(t)$ lies in TM_q. Therefore $(q, \dot{q})$ lies in the tangent bundle TM which is a C^2-manifold. The Lagrange function

$$L: TM \times \mathbb{R} \to \mathbb{R}$$

must then be C^2. Furthermore, L_q and $L_{\dot{q}}$ in (93) are the corresponding tangent maps, i.e.,

$$L_q(q, \dot{q}, t): TM_q \to \mathbb{R}, \tag{94}$$

$$L_{\dot{q}}(q, \dot{q}, t): TM_q \to \mathbb{R}. \tag{95}$$

More precisely, $q \mapsto L(q, \dot{q}, t)$ is a C^2-map from M into $\mathbb{R}$ with tangent map (94) at the point q, and $\dot{q} \mapsto L(q, \dot{q}, t)$ is a C^2-map from TM_q into $\mathbb{R}$ with tangent map (95). Since, by definition, tangent maps are linear and continuous, it follows that

$$L_q(q, \dot{q}, t), \quad L_{\dot{q}}(q, \dot{q}, t) \in TM_q^*, \tag{96}$$

i.e., these are cotangent vectors (elements of the cotangent space TM_q^*).

The formulation of classical mechanics on manifolds in the context of sympletic geometry will be discussed in Part V.

As an application of the Lagrangian formalism we consider in Problem 58.7 the circlular pendulum. Lagrangian mechanics, i.e., the principle of stationary action (92) is a model, for many field theories in physics. In Chapter 76 we will show that the equations of the general theory of relativity can be derived from a variational principle. Moreover, Maxwell's electrodynamics and all quantum field theories can also be obtained from such variational principles. This will be discussed in Part V.

58.20. Hamiltonian Mechanics

Hamiltonian mechanics starts from the *canonical equations*

$$\begin{aligned} \dot{q}(t) &= H_p(q(t), p(t), t), \\ \dot{p}(t) &= -H_q(q(t), p(t), t), \end{aligned} \tag{97}$$

58.20. Hamiltonian Mechanics

or more simply

$$\dot{q} = H_p, \qquad \dot{p} = -H_q.$$

In component notation this means

$$p = (p_1, \ldots, p_m), \qquad q = (q^1, \ldots, q^m) \qquad \text{with} \quad p, q \in \mathbb{R}^m$$

and

$$\dot{q}^k = H_{p^k}, \qquad \dot{p}_k = -H_{q^k}, \qquad k = 1, \ldots, m. \tag{98}$$

We will see that in many cases H is the energy. Moreover, q describes the position of the system and the components of p are called *generalized momenta*. Our assumptions are:

(H) The Hamilton function $H: G \times \mathbb{R} \to \mathbb{R}$ is C^1, where G is a nonempty, open set in $\mathbb{R}^{2m}$.

G is called the *phase space*, and (97) is also called a Hamiltonian system.

STANDARD EXAMPLE 58.21 (Harmonic Oscillator). Newton's equations of motion for the harmonic oscillator are

$$m\ddot{q} = -kq \tag{99}$$

with $q \in \mathbb{R}$ and $k, m > 0$. The solution is

$$q(t) = C \sin(\omega t + \alpha)$$

with the angular frequency $\omega = \sqrt{k/m}$. The Lagrange function $L = T - U$ is

$$L = (m\dot{q}^2 - kq^2)/2.$$

Lagrange's equation $(d/dt)L_{\dot{q}} - L_q = 0$ implies (99). In order to obtain the form (97) we set

$$p = L_{\dot{q}},$$

i.e., $p = m\dot{q}$ is the momentum. Moreover, let

$$H = p\dot{q} - L,$$

hence $H = (m\dot{q}^2 + kq^2)/2 = T + U$. This is the energy. Transformation onto p yields

$$H(q, p) = \frac{p^2}{2m} + \frac{kq^2}{2}.$$

Thus the Hamiltonian equations $\dot{q} = H_p$, $\dot{p} = -H_q$ mean

$$\dot{q} = p/m, \qquad \dot{p} = -kq.$$

This is equivalent to (99).

Similarly as in this simple example we now want to show how the canonical equations can be obtained from the Lagrangian mechanics of Section 58.19.

The key is the Legendre transformation

$$p = L_{\dot{q}}(q, \dot{q}, t), \tag{100}$$

$$H = \langle p, \dot{q} \rangle - L, \tag{101}$$

i.e., the change from $(q, \dot{q})$ and L to (q, p) and H. In component notation this means that

$$p_k = L_{\dot{q}^k}, \quad k = 1, \ldots, m,$$

$$H = p_j \dot{q}^j - L,$$

where j is summed from 1 to m.

In the following we often use the convenient notation

$$ab \stackrel{\text{def}}{=} \langle a, b \rangle \quad \text{for} \quad a \in X^* \text{ and } b \in X. \tag{102}$$

In the case that $X = X^* = \mathbb{R}^m$ this means that $ab = a_j b^j$.

We assume that L is C^2 and that (100) can be uniquely solved for $\dot{q}$, whereby a C^2-map

$$\dot{q} = \dot{q}(q, p, t)$$

occurs. Then (101) becomes

$$H(q, p, t) = p\dot{q}(q, p, t) - L(q, \dot{q}(q, p, t), t).$$

Let $q = q(t)$ be a C^2-solution of the Lagrangian equation

$$\frac{d}{dt} L_{\dot{q}}(P(t)) - L_q(P(t)) = 0 \tag{103}$$

with $P(t) = (q(t), \dot{q}(t), t)$. Using the Legendre transformation (100) we obtain $p(t)$, whereby

$$\dot{q}(t) = \dot{q}(q(t), p(t), t).$$

In order to obtain (97) we compute the partial F-derivatives

$$H_q h = p(\dot{q}_q h) - L_q h - L_{\dot{q}}(\dot{q}_q h) = -L_q h,$$

$$H_p k = k\dot{q} + p(\dot{q}_p k) - L_{\dot{q}}(\dot{q}_p k) = \dot{q}k,$$

for all $h, k \in \mathbb{R}^m$ and observe (102). It follows that

$$H_q = -L_q, \quad H_p = \dot{q},$$

i.e., more precisely

$$H_q(q, p, t) = -L_q(q, \dot{q}, t), \quad H_p(q, p, t) = \dot{q}$$

with $\dot{q} = \dot{q}(q, p, t)$. From (103), i.e., from $\dot{p} = L_q = -H_q$ and $\dot{q} = H_p$ we immediately obtain (97).

Remark 58.22 (Energy). If $L = T - U$ with kinetic energy $T = T(q, \dot{q}, t)$ and potential energy $U = U(q, t)$ and if T is homogeneous of degree 2 with respect

to $\dot{q}$, then
$$H = T + U,$$
i.e., H is equal to the energy.

For a proof we differentiate the identity
$$T(q, \alpha\dot{q}, t) = \alpha^2 T(q, \dot{q}, t) \quad \text{for all} \quad \alpha \in \mathbb{R},$$
with respect to α at the point $\alpha = 1$. This implies $T_{\dot{q}}\dot{q} = 2T$. Moreover, we have $p = L_{\dot{q}} = T_{\dot{q}}$, hence
$$H = p\dot{q} - L = 2T - T + U.$$

Remark 58.23 (Legendre Transformation and Differentials). The Legendre transformation corresponds precisely to the natural transformation method discussed in Section 58.16 where $\dot{q}$ is replaced with a new variable p. In fact, we have
$$dL = L_q \, dq + L_{\dot{q}} \, d\dot{q} + L_t \, dt = L_q \, dq + p \, d\dot{q} + L_t \, dt,$$
with $L_q \, dq = L_{q^j} \, dq^j$, etc. From
$$d(p\dot{q}) = (dp)\dot{q} + p \, d\dot{q}$$
we obtain
$$d(p\dot{q} - L) = -L_q \, dq + \dot{q} \, dp - L_t \, dt.$$
Letting $H = p\dot{q} - L$, we obtain $H_q = -L_q$, $H_p = \dot{q}$ and $H_t = L_t$.

Remark 58.24 (General Manifolds). We now consider the same situation as in Remark 58.20, i.e., q varies on the configuration space M, where M is a real, m-dimensional C^3-manifold. The canonical equations (97) can then be naturally formulated for this more general situation. From $p = L_{\dot{q}}(q, \dot{q}, t)$ and (96) follows
$$p \in TM_q^*.$$
Thus (q, p) is a point of the cotangent bundle TM^*, which is a C^2-manifold. We now assume that the Hamilton function
$$H: TM^* \times \mathbb{R} \to \mathbb{R}$$
is C^1. The corresponding tangent maps are
$$H_q(q, p, t): TM_q \to \mathbb{R}, \qquad (104)$$
$$H_p(q, p, t): TM_q^* \to \mathbb{R}. \qquad (105)$$
More precisely, $q \mapsto H(q, p, t)$ is a C^1-map from M into $\mathbb{R}$ with tangent map (104), and $p \mapsto H(q, p, t)$ is a C^1-map from TM_q^* into $\mathbb{R}$ with tangent map (105). As an m-dimensional, real vector space the tangent space TM_q is reflexive, i.e.,

$TM_q^{**} = TM_q$. In this sense we have

$$H_q(q, p, t) \in TM_q^*, \qquad H_p(q, p, t) \in TM_q,$$

and the canonical equations (97) are valid.

For fixed time t the Legendre transformation (101) corresponds to the change from

$$(q, \dot{q}) \in TM \qquad \text{to} \quad (q, p) \in TM^*,$$

i.e., to a transformation $\varphi\colon TM \to TM^*$ which maps the tangent bundle onto the cotangent bundle. Behind the Legendre transformation a duality principle is hidden. The precise relation with conjugate functionals in the context of duality theory of convex analysis has already been discussed in Section 51.1 of Part III.

Hamiltonian mechanics has a number of important advantages.

(i) The solution of the canonical equations (97) can often be greatly simplified by using canonical transformations (Section 58.24).
(ii) By using Poisson brackets one can give simple criteria for the conserved quantities (see Section 58.21).
(iii) In contrast to the Lagrangian equations, the canonical equations (97) are first-order differential equations, i.e., they represent a dynamical system for which the entire theory of dynamical systems is available.
(iv) The canonical equations are closely related to symplectic geometry. In particular, this implies the following. If $q = q(t)$, $p = p(t)$ are solutions of (97), then they can be interpreted as trajectories of a flow in the phase space, this flow is volume preserving (Liouville's theorem), and builds the starting point for classical statistical physics (see Part V).
(v) The solution theory for the canonical equations is closely related to the solution theory of a partial differential equation, namely the Hamilton–Jacobi differential equation for the action function (see Section 58.23).
(vi) The canonical formalism can be generalized to infinite-dimensional systems. Thereby field theories can be studied in the language of canonical equations (see Part V).

In summary, one can say the following. Up to now, physical experience shows that the canonical formalism is a very flexible instrument to formulate and study physical theories. The reason for this probably lies in the fact that the connection with two fundamental physical quantities, namely energy H and action S, is very close and explicit.

In Section 75.11 we will use the Hamiltonian formalism to develop the mechanics of a free particle in the context of the special theory of relativity. Thereby we naturally obtain Einstein's fundamental relation

$$E = mc^2$$

between energy E, mass m, and velocity of light c.

In the following sections of this chapter and in Section 74.26 we will study the classical brackets of Poisson, Lagrange, and Lie and explain their meaning. These brackets show that physics, in its essential parts, has a noncommutative structure, which is especially obvious in quantum theory.

The difficulties, however, which occur in the mathematical formulation of quantum field theory show that at present the true character of physical phenomena is mathematically not completely understood. We are looking at a puzzle which in parts has already been solved with an astounding mathematical beauty, but the entire picture is still hidden.

58.21. Poissonian Mechanics and Heisenberg's Matrix Mechanics in Quantum Theory

Poissonian mechanics starts from the equation

$$\frac{d}{dt} F(P(t)) = \{H, F\}(P(t)) + F_t(P(t)), \tag{106}$$

with $P(t) = (q(t), p(t), t)$ and the so-called *Poisson brackets*

$$\{H, F\} = F_q H_p - F_p H_q; \tag{107}$$

recall our convention $ab = \langle a, b \rangle$. More precisely, we have

$$\{H, F\}(P) = F_q(P) H_p(P) - F_p(P) H_q(P)$$

with $P = (q, p, t)$. In component notation this becomes

$$\{H, F\} = F_{q^j} H_{p_j} - F_{p_j} H_{q^j},$$

where j is summed from 1 to m. Thus, in particular, we obtain

$$\{p_i, q^j\} = \delta_i^j, \qquad \{p_i, p_j\} = \{q_i, q_j\} = 0 \tag{108}$$

for all $i, j = 1, \ldots, m$, and moreover, we find

$$\{H, F\} = -\{F, H\}.$$

Equation (106) means that H is given and (106) holds for an arbitrarily smooth F. Precisely, we assume the following:

(H) The functions $F, H: G \times \mathbb{R} \to \mathbb{R}$ are C^1, where G is an open, nonempty set in $\mathbb{R}^{2m}$.

We now want to show that Poissonian mechanics and Hamiltonian mechanics are equivalent. In fact, it immediately follows from the canonical equations (97) that

$$\frac{d}{dt} F = F_q \dot{q} + F_p \dot{p} + F_t = F_q H_p - F_p H_q + F_t.$$

Therefore every solution $q = q(t)$, $p = p(t)$ of (97) is also a solution of (106).

Conversely, (97) follows from (106) for $F = q^k$ and $F = p_k$, since
$$\{H, q^k\} = H_{p_k} \quad \text{and} \quad \{H, p_k\} = -H_{q^k}.$$
The same observation is valid for manifolds, i.e., for the general situation of Remark 58.24.

The advantage of Poissonian mechanics is the following:

(i) Observables. If the functions $F = F(q, p, t)$ are called observables, then (106) is a uniform formula for the evolution in time of all observables.
(ii) Conserved quantities. If
$$\{H, F\} + F_t \equiv 0$$
holds, then F is a conserved quantity, because for every solution $q = q(t)$, $p = p(t)$ of the canonical equations, (106) holds and thereby $F(P(t)) = $ const.
(iii) Generalization. One can use (106) for the formulation of the evolution in time of more general physical systems, by replacing the classical Poisson brackets (107) with other expressions. As we now want to show, one thereby obtains a very elegant approach to the matrix mechanics, developed by Heisenberg (1925), which marked the beginning of quantum mechanics. The problem which was facing the physicists then was the following. Required was a physical theory which would lead to a quantization of classical mechanics, i.e., this theory should contain many features of classical mechanics but at the same time take quantum effects into account, i.e., the quantization of energy of the harmonic oscillator, which was postulated by Planck in 1900 in order to obtain the correct radiation law.

EXAMPLE 58.25 (Matrix Mechanics of Heisenberg (1925)). We begin with the following formal observations. The key idea is to define the Poisson brackets as
$$\{H, F\} = \frac{i}{\hbar}(HF - FH),$$
where $h = 6.625 \cdot 10^{-34}$ J s is Planck's quantum of action and $\hbar = h/2\pi$. From (106) it follows that the equations of motion for the observables $F = F(q, p, t)$ are given by
$$\frac{d}{dt} F = \{H, F\} + F_t.$$
In particular, the equations of motion are
$$\dot{p}_j = \{H, p_j\}, \quad \dot{q}^j = \{H, q^j\} \tag{109}$$
for $j = 1, \ldots, m$. Analogously to (108), we assume
$$\{p_i, q^j\} = \delta_i^j I, \quad \{p_i, p_j\} = \{q^i, q^j\} = 0 \tag{109a}$$

58.21. Poissonian Mechanics and Heisenberg's Matrix Mechanics

for $i, j = 1, \ldots, m$. These are the fundamental *commutation relations* of quantum mechanics. What is the meaning of q^i and p_j? Because of (109a), these quantities do not commute. Following Heisenberg, we require that q^i, p_j and $F(q, p, t)$ are self-adjoint infinite-dimensional matrices. Self-adjointness is assumed in order to assure that the observables F have real eigenvalues. Because of (109a) we cannot use finite-dimensional matrices, since in that case we have

$$\text{trace}\{p_i, q^j\} = 0 \quad \text{and} \quad \text{trace}(\delta_i^j I) \neq 0 \quad \text{for} \quad i = j,$$

where trace denotes the sum of the diagonal elements.

In an exact functional analytic theory, which was formulated by John von Neumann (1932), one requires that q^i, p_j and $F(q, p, t)$ are self-adjoint operators on a complex H-space. The theory is somewhat complicated, because the operators are unbounded. Thus in computations one carefully has to watch the domains of the operators.

We consider an important application.

STANDARD EXAMPLE 58.26 (Quantum Mechanical Harmonic Oscillator). Our goal is the formula

$$E = \hbar\omega(n + \tfrac{1}{2}), \quad n = 0, 1, \ldots \tag{110}$$

for the possible energy values of a quantum mechanical harmonic oscillator with angular frequency ω. In order to clearly present the simple basic idea, we begin with a formal observation. The Hamilton function of the harmonic oscillator is

$$H(q, p) = \frac{p^2}{2m} + \frac{m}{2}\omega^2 q^2,$$

according to Example 58.21. The previous example yields the equations of motion

$$\dot{p} = \frac{i}{\hbar}(Hp - pH), \quad \dot{q} = \frac{i}{\hbar}(Hq - qH). \tag{111}$$

Moreover, we assume that the commutation relation

$$\frac{i}{\hbar}(pq - qp) = I \tag{112}$$

is valid for the solution $q = q(t)$, $p = p(t)$ of (111), and

$$p = p^*, \quad q = q^*, \quad H = H^*.$$

From (111) and (112) follows

$$\dot{p} = -m\omega^2 q, \quad p = m\dot{q}, \tag{113}$$

since

$$\dot{p} = \frac{im\omega^2}{2\hbar}(q^2p - pq^2) = -m\omega^2 q,$$

$$\dot{q} = \frac{i}{2m\hbar}(p^2q - qp^2) = p/m.$$

A solution of (113), i.e., of $\ddot{q} + \omega^2 q = 0$ is

$$q(t) = \gamma(ae^{-i\omega t} + a^*e^{i\omega t})/\sqrt{2},$$
$$p(t) = m\dot{q} = im\omega\gamma(a^*e^{i\omega t} - ae^{-i\omega t})/\sqrt{2}$$

with $\gamma = (\hbar/m\omega)^{1/2}$, where a^* denotes the adjoint matrix to a. Then $q = q^*$, $p = p^*$ and $H = H^*$ is automatically satisfied and (112) yields

$$aa^* - a^*a = 1. \tag{114}$$

Moreover, we have

$$a = \frac{1}{\sqrt{2}}\left(\frac{q(0)}{\gamma} - \frac{i\gamma p(0)}{\hbar}\right),$$

$$a^* = \frac{1}{\sqrt{2}}\left(\frac{q(0)}{\gamma} - \frac{i\gamma p(0)}{\hbar}\right). \tag{115}$$

Letting

$$N = a^*a$$

we obtain from (114) that

$$H(q(t), p(t)) = \hbar\omega(N + \tfrac{1}{2}).$$

We now show

$$N\varphi_n = n\varphi_n, \quad n = 0, 1, \ldots \tag{116}$$

with

$$\varphi_n = (a^*)^n \varphi_0.$$

Therefore H has the eigenvalues (110), which are interpreted as the possible energy states of the oscillator, i.e.,

$$H\varphi_n = \hbar\omega(n + \tfrac{1}{2})\varphi_n, \quad n = 0, 1, \ldots. \tag{117}$$

We prove (116) by induction. For $n = 0$ we postulate the existence of φ_0 with $a\varphi_0 = 0$. This implies

$$N\varphi_0 = a^*a\varphi_0 = 0.$$

From (114) we obtain

$$N\varphi_{k+1} = a^*aa^*\varphi_k = a^*\varphi_k + a^*N\varphi_k = \varphi_{k+1} + a^*N\varphi_k.$$

Hence (116) follows by induction.

In order to rigorously justify the formal computations, we choose the complex H-space $X = L_2^{\mathbb{C}}(\mathbb{R})$ and define the symmetric operators

$$q_0, p_0 \colon C_0^{\infty}(\mathbb{R}) \subseteq X \to X$$

through

$$q_0 \varphi = x\varphi, \qquad p_0 \varphi = \frac{\hbar}{i}\frac{d\varphi}{dx}.$$

Then we choose $q(0)$ and $p(0)$ as the uniquely determined self-adjoint extensions of q_0 and p_0, respectively. From (115) one finds then a and a^* and thereby all other quantities. Here a^* denotes the adjoint operator to a. Moreover, we choose

$$\varphi_0 = c_0 e^{-(\gamma x)^2/2},$$

where the constant c_0 is obtained from the normalization condition $(\varphi_0 | \varphi_0) = 1$. This actually gives $a\varphi_0 = 0$.

This realization of p and q through differential operators is not accidental. It is closely related to the wave mechanical treatment of the harmonic oscillator which we will consider in Section 59.18. In Part V we will show that Heisenberg's matrix mechanics and Schrödinger's wave mechanics are only two different realizations of the same abstract Hilbert space theory.

In quantum field theory, which will also be discussed in Part V, the quantity φ_n is interpreted as a state with n particles, and N is called the particle number operator. The ground state φ_0 with $N\varphi_0 = 0$ is called the vacuum. Moreover, a^* and a are called creation operator and anihilation operator, respectively. This is motivated by the fact that the n-particle state φ_n is obtained from the vacuum φ_0 by an n-fold application of a^*, i.e., $\varphi_n = (a^*)^n \varphi_0$. Furthermore, we have

$$a\varphi_{n+1} = aa^* \varphi_n = (I + N)\varphi_n = (1 + n)\varphi_n,$$

i.e., an application of a to an $(n + 1)$-particle state yields an n-particle state.

This example shows that already purely algebraical operations imply important quantum theoretical results. It is the starting point for applications of the theory of operator algebras in modern quantum theory. In this direction we recommend Bratteli and Robinson (1979, M) and Part V.

58.22. Propagation of Action

If $q = q(t)$ is a solution of the Lagrangian equation

$$\frac{d}{dt} L_{\dot{q}} - L_p = 0,$$

then the action ΔS which during the time interval $[t_1, t_2]$ is transported by this solution is defined as

$$\int_{t_1}^{t_2} L(q(t), \dot{q}(t), t)\, dt. \tag{118}$$

Since $L = T - U$ we obtain that L has the dimension of an energy and the action ΔS therefore has the dimension energy × time, hence J s (joule second). This definition makes the expression "principle of the stationary action" for (92) plausible.

If $q = q(t)$, $p = p(t)$ is a solution of the canonical equations

$$\dot{q} = H_p, \qquad \dot{p} = -H_q,$$

then the action ΔS, which during the time interval $[t_1, t_2]$ is transported by this solution is defined as

$$\Delta S = \int_{t_1}^{t_2} p(t)\dot{q}(t) - H(q(t), p(t), t)\, dt. \tag{119}$$

The Legendre transformation yields $L = p\dot{q} - H$. Therefore (119) and (118) are consistent.

58.23. Hamilton–Jacobi Equation

Besides the canonical equations

$$\dot{q} = H_p, \qquad \dot{p} = -H_q \tag{120}$$

with $H = H(q, p, t)$ and $q, p \in \mathbb{R}^m$, $t \in \mathbb{R}$ we consider the so-called Hamilton–Jacobi equation

$$S_t + H(q, S_q, t) = 0 \tag{121}$$

for $S = S(q, t)$. We shall show that (121) is closely related to the concept of action. The simultaneous study of (120) and (121) has the following advantages:

(i) If an m-parametric solution of (121) is known, then one can obtain a $2m$-parametric solution of (120) by using a canonical transformation, i.e., one obtains the general solution of the canonical equations. Thereby only differentiation and elimination are used.
(ii) If, conversely, an m-parametric solution of (120) is known and, in addition, Lagrange's bracket conditions are satisfied, then one obtains a solution S of (121) by evaluating a line integral. Thereby S is the action which corresponds to the solutions of (120).
(iii) The solution of the initial-value problem for (121) is a special case of (ii).

These points will be discussed in the following three sections. In the context of the general theory of first-order partial differential equations, (120) is the characteristic system of (121). In another form, equation (121) has already been discussed in Section 37.4 of Part III. There, we gave an intuitive interpretation of S in connection with geometrical optics. Here we want to emphasize the role of the Lagrange brackets. In Part V we will show that this important concept is naturally related to symplectic geometry and leads to Lagrangian manifolds in the context of this geometry.

58.24. Canonical Transformations and the Solution of the Canonical Equations via the Hamilton–Jacobi Equation

An important method for solving the canonical equations (120) is the use of a variables transformation

$$a = a(q,p,t), \qquad b = b(q,p,t) \tag{122}$$

in order to obtain the new canonical equations

$$\dot{a} = \bar{H}_b, \qquad \dot{b} = -\bar{H}_a \tag{123}$$

with a modified Hamilton function $\bar{H} = \bar{H}(a,b,t)$. Such transformations (122) are called *canonical transformations*. One tries to find the simplest form of $\bar{H}$. Then a solution of the original canonical equations (120) is obtained by finding a solution of (123) and using a back transformation. The simplest form of (123) exists for $\bar{H} = 0$. Then

$$a = \text{const} \quad \text{and} \quad b = \text{const}$$

is a solution of (123). In fact, following Jacobi, such canonical transformations can be constructed by setting

$$\begin{aligned} S_a(q,t,a) &= b, \\ S_q(q,t,a) &= p. \end{aligned} \tag{124}$$

Thereby $S = S(q,t,a)$ is a solution of the Hamilton–Jacobi equation (121), which also depends on the m-parameters $a = (a^1, \ldots, a^m)$. Our assumptions are:

(H1) In a neighborhood of (q_0, p_0, t) we are given a real C^1-function $H = H(q,p,t)$ with $q, p \in \mathbb{R}^m$ and $t \in \mathbb{R}$.

(H2) The real function $S = S(q,t,a)$ with $a \in \mathbb{R}^m$ is a complete integral of the Hamilton–Jacobi equation (121), i.e., in a neighborhood of (q_0, t_0, a_0) the function S is C^2 and there a solution of (121) which satisfies the additional property

$$\det S_{aq}(q_0, t_0, a_0) \neq 0. \tag{125}$$

We set $b_0 = S_a(q_0, t_0, a_0)$ and $p_0 = S_q(q_0, t_0, a_0)$.

Because of (125) we can solve the first equation in (124) locally for q by using the implicit function theorem (Theorem 4.B).

Theorem 58.F (Jacobi). *Assume* (H1) *and* (H2). *By solving equation* (124) *one obtains, for every fixed* $(a, b) \in \mathbb{R}^{2m}$ *in a neighborhood of* (a_0, b_0), *a solution*

$$q = q(t), \qquad p = p(t)$$

of the canonical equations (120) *in a neighborhood of* t_0.

PROOF. From (124) we obtain $q = q(t, a, b)$ and $p = p(t, a, b)$. We insert these C^1-functions into (124). Furthermore, we insert $S = S(q, t, a)$ into (121). Differentiation of (124) with respect to t yields

$$S_{aq}\dot{q} + S_{at} = 0, \qquad \dot{p} = S_{qq}\dot{q} + S_{qt}.$$

Differentiation of (121) with respect to a and q yields

$$S_{ta} + H_p S_{qa} = 0, \qquad S_{tq} + H_q + H_p S_{qq} = 0$$

and hence $\dot{q} = H_p$ and $\dot{p} = -H_q$. □

The importance of this method is that in a number of significant cases one can easily obtain a complete integral of the Hamilton–Jacobi equation by using the separation ansatz

$$S(q, t) = \sum_{i=1}^{m} S_i(q_i) + S_{m+1}(t).$$

In the case of the harmonic oscillator we already showed this in Example 37.8.

Remark 58.27 (Global Version). Our formulation of Theorem 58.F was local. But the same proof yields global solutions if equation (124) can be solved globally for q, whereby $q = q(t, a, b)$ is C^1. We then have to assume that (125) holds along $(q(t, a, b), t, a)$ and that the functions H and S are C^1 and C^2, respectively.

58.25. Lagrange Brackets and the Solution of the Hamilton–Jacobi Equation via the Canonical Equations

Suppose we are given a family

$$q = q(t, a), \qquad p = p(t, a) \tag{126}$$

of solutions of the canonical equations

$$\dot{q} = H_p, \qquad \dot{p} = -H_q, \tag{127}$$

which depend on the parameter $a \in \mathbb{R}^m$. Thereby we set $\dot{q} = q_t$ and $\dot{p} = p_t$.

58.25. Lagrange Brackets and the Solution of the Hamilton–Jacobi Equation

The question then is, under which conditions this yields a solution of the Hamilton–Jacobi equation

$$S_t + H(q, S_q, t) = 0. \tag{128}$$

The key assumption is:

(L) The Lagrange brackets satisfy

$$[a^i, a^j](t_0, a) = 0$$

for all $i, j = 1, \ldots, n$ and all $a \in U$.

Thereby we define

$$[a^i, a^j] = p_{a^i} q_{a^j} - p_{a^j} q_{a^i}.$$

Recall that $pq = p_k q^k$, where k is summed from 1 to m. Because of (126) these brackets depend on (t, a). The idea is as follows. Consider the line integral

$$\bar{S}(t, a) = \int_{(t_0, a_0)}^{(t, a)} (p\dot{q} - H) dt + (pq_{a^i}) da^i \tag{129}$$

with $q = q(t, a)$ and $p = p(t, a)$. We solve $q = q(t, a)$ for a and set

$$S(q, t) = \bar{S}(t, a(q, t)). \tag{130}$$

Condition (L) guarantees that the line integral in (129) is path independent. In the following assumptions, we mean by "local" more precisely "in a neighborhood of the point $(t_0, a_0) \in \mathbb{R}^{m+1}$". We set

$$q_0 = q(t_0, a_0) \quad \text{and} \quad p_0 = p(t_0, a_0)$$

and make the following assumptions:

(H1) The function $H: U(q_0, p_0, t_0) \subseteq \mathbb{R}^{2m+1} \to \mathbb{R}$ is C^k, $k \geq 1$.
(H2) The functions (126) are local C^1-functions from $\mathbb{R}^{m+1}$ into $\mathbb{R}^m$ and, satisfy, locally, the canonical equations (127).

Furthermore, we have

$$\det q_a(t_0, a_0) \neq 0$$

and (L) holds in a neighborhood U of a_0.

The condition for the determinant is needed in order to solve locally the equation $q = q(t, a)$ for a.

Theorem 58.G. *Assume (H1) and (H2). Then (130) is a local C^k-solution of the Hamilton–Jacobi equation.*

PROOF.

(I) The Lagrange brackets have the important property that for solutions $q = q(t, a)$, $p = p(t, a)$ of the canonical equations the relation

$$\frac{\partial}{\partial t}[a^i, a^j] = 0 \tag{131}$$

holds. In order to simplify the notation we set

$$\alpha = a^i, \quad \beta = a^j.$$

Moreover, we write $p_\alpha = \partial p/\partial \alpha$, etc. Then (131) follows from

$$\frac{\partial}{\partial t}[\alpha, \beta] = \dot{p}_\alpha q_\beta + p_\alpha \dot{q}_\beta - \dot{p}_\beta q_\alpha - p_\beta \dot{q}_\alpha,$$

and

$$\dot{p}_\alpha = -H_{qq}q_\alpha - H_{qp}p_\alpha,$$
$$\dot{q}_\beta = H_{pq}q_\beta + H_{pp}p_\beta.$$

Note that the symmetry of the matrix of the second-order partial derivatives $H_{p_i q^j}$ implies that

$$(H_{qp}p_\alpha)q_\beta = p_\alpha(H_{pq}q_\beta).$$

From (131) and (L) we therefore obtain locally

$$[a^i, a^j](t, a) = 0. \tag{132}$$

(II) The integral (129) is path independent, since the integrability conditions

$$(pq_\alpha)_t = (p\dot{q} - H)_\alpha, \tag{133}$$

$$(pq_\alpha)_\beta = (pq_\beta)_\alpha \tag{134}$$

are satisfied. Actually, (134) follows immediately from $p_\beta q_\alpha = p_\alpha q_\beta$. This is (132). On the other hand, (133) follows from (127), since

$$(pq_\alpha)_t = \dot{p}q_\alpha + p\dot{q}_\alpha,$$
$$(p\dot{q} - H)_\alpha = p_\alpha \dot{q} + p\dot{q}_\alpha - H_q q_\alpha - H_p p_\alpha.$$

(III) We show that S in (130) is a solution of the Hamilton–Jacobi equation. From (129) follows

$$\bar{S}_t = p\dot{q} - H, \quad \bar{S}_a = pq_a.$$

Note that $S_{a^i} = pq_{a^i}$, and hence $S_a h = p(q_a h)$. Formula (130) implies

$$S_t = \bar{S}_t + \bar{S}_a \dot{a}, \quad S_q = \bar{S}_a a_q.$$

Differentiation of the identity $q = q(t, a(q, t))$ with respect to t and q gives

$$0 = \dot{q} + q_a \dot{a}, \quad I = q_a a_q.$$

This implies $S_q = p$ and $S_t = -H$. □

The physical interpretation of S is obtained by inserting $a = a_0$ into (129). We find

$$S(q(t, a_0), t) = \int_{t_0}^{t} (p\dot{q} - H) dt$$

along the solutions $q = q(t, a_0)$ and $p = p(t, a_0)$ of the canonical equations, i.e., S corresponds to the action which propagates along the solution of the canonical equation.

Remark 58.28 (Global Version). Our proof also yields a global version of Theorem 58.G. In order to obtain the path independence of the integral in (129), condition (L) must be valid in a simply connected region U. Moreover, we need the fact that equation $q = q(t, a)$ can be solved globally for a and yields a C^k-map $a = a(t, q)$.

Remark 58.29 (Lagrangian Manifolds). In Part V we will give an invariant definition for Lagrangian manifolds in connection with symplectic geometry. There we will also show that Lagrange's bracket condition is equivalent to the fact that for $t = t_0$ the family (126) yields a Lagrangian manifold in the (q, p)-phase space. The key thereby is the following formal observation. Consider the equation

$$dp_k \wedge dq^k = 0, \tag{135}$$

where k is summed from 1 to m. The symbol "$\wedge$" is used in the same way as a product sign, where $\alpha \wedge \beta = -\beta \wedge \alpha$. From (135) and $q = q(t, a), p = p(t, a)$ it follows formally that for fixed t the relation

$$0 = \frac{\partial p_k}{\partial a^i} \frac{\partial q^k}{\partial a^j} da^i \wedge da^j$$
$$= \tfrac{1}{2}(p_{a^i} p_{a^j} - p_{a^j} p_{a^i}) da^i \wedge da^j$$
$$= \tfrac{1}{2}[a^i, a^j] da^i \wedge da^j$$

holds, and hence

$$[a^i, a^j] = 0 \quad \text{for all } i, j.$$

The differential form on the left-hand side of (135) is the reason for the fact that the canonical formalism can be formulated in the context of symplectic geometry.

58.26. Initial-Value Problem for the Hamilton–Jacobi Equation

As an application of Theorem 58.G we consider the initial-value problem for the Hamilton–Jacobi equation

$$S_t + H(q, S_q, t) = 0,$$
$$S(t_0, q) = S_0(q), \tag{136}$$

where the function S_0 is given. We have to find the function S. By passing to $S - S_0$ we can always assure that $S_0 \equiv 0$ for a changed function H. Thus we can restrict ourselves to the case $S_0 \equiv 0$. For the solution of (136) we use the canonical equations

$$\dot{q} = H_p, \qquad \dot{p} = -H_q,$$
$$q(t_0) = a, \qquad p(t_0) = 0. \tag{137}$$

Proposition 58.30. *Let the function* $H: U(q_0, 0, t_0) \subseteq \mathbb{R}^{m+1} \to \mathbb{R}$ *be* C^3. *Then the initial-value problem* (136) *with* $S_0 \equiv 0$ *has a unique* C^2-*solution in a neighborhood of* (q_0, t_0).

PROOF.
(I) Existence. Using Theorem 4.D we solve (137) and obtain a C^2-family

$$q = q(t, a), \qquad p = p(t, a).$$

Because of

$$p(t_0, a) \equiv 0,$$

the Lagrange brackets are identically zero. Furthermore, from $q(t_0, a) = a$ follows that

$$q_a(t_0, a) = I.$$

Theorem 58.G yields the existence of a solution of (136), and (130) implies that $S(q, t_0) \equiv 0$.

(II) Uniqueness. Let $S = S(q, t)$ be a C^2-solution of (136). We set

$$p(t) = S_q(q(t), t)$$

and determine $q = q(t)$ from

$$\dot{q}(t) = H_p(q(t), p(t), t), \qquad q(t_0) = a.$$

Differentiation with respect to t gives

$$\dot{p} = S_{qq}\dot{q} + S_{qt}.$$

Differentiation with respect to q in (136) with $S = S(q, t)$ yields the equation

$$S_{tq} + H_q + H_p S_{qq} = 0.$$

Consequently,

$$\dot{p}(t) = -H_q(q(t), p(t), t), \qquad p(t_0) = 0,$$

i.e., $q(\cdot)$ and $p(\cdot)$ are the unique local solutions of (137).
We set $\sigma(t) = S(q(t), t)$. This implies

$$\dot{\sigma} = S_q \dot{q} + S_t$$

and hence

$$\dot{\sigma}(t) = p(t)\dot{q}(t) - H(q(t), p(t), t), \quad \sigma(t_0) = 0.$$

Therefore, $\sigma(t)$ is locally unique.

Since $\det q_a(t, a) \neq 0$ in a neighborhood of (t_0, a_0) it follows that there exists a unique solution $q(\cdot)$ of (137) through each point $a \in U(q_0)$ for fixed $t \in U(t_0)$. Therefore, S is locally uniquely determined. □

58.27. Dimension Analysis

Every physical quantity has a dimension. A survey about the international system of units and the numerical values of important universal constants can be found in the Appendix.

The fact that physical quantities have a dimension greatly restricts the number of possible physical laws. We explain this with an example.

EXAMPLE 58.31. Consider a pendulum of length l and mass M. We are looking for a formula for the period of oscillation T of the pendulum. We expect that T depends on l, M and the gravitational acceleration g. Thus we begin with the ansatz

$$T = \delta l^\alpha M^\beta g^\gamma,$$

where δ is a dimensionless constant. Passing to dimensions we obtain

$$s = m^\alpha kg^\beta m^\gamma s^{-2\gamma}.$$

This implies $\beta = 0$, $\gamma = -\frac{1}{2}$, and $\alpha = -\gamma = \frac{1}{2}$, i.e.,

$$T = \delta \sqrt{\frac{l}{g}}. \tag{138}$$

The constant δ has to be determined from experiments. In Problem 58.7 we show that in first-order approximation, for small pendulum motions, (138) actually follows from the Newtonian equations of motions, whereby $\delta = 2\pi$.

In this example we obtained a maximum of information from a minimum of assumptions by using dimension analysis.

We want to mention here an important point. Often in physics one uses approximation calculations, since exact computations are too complicated or cannot be performed at all. Thereby one exactly needs to know which terms in an equation are "small." To this end one uses variables for which the equations become dimensionless. Only those equations allow a precise comparison of the relative magnitude of the particular terms. We will use this method, for example, in connection with the motion of the Perihelion of Mercury in Section 76.9.

Problems

58.1. *Proof of Lemma 58.18.*

Solution: Obviously, the vectors

$$h_i = a + \omega \times x_i, \qquad i = 1, \ldots, n$$

form a solution of (59). Conversely, every solution of (59) has this form. For $n = 3$ this follows from the fact that the solution space is six-dimensional, since the rank of the system of equations is three.

We use induction. Suppose the assertion of Lemma 58.18 is true for fixed $n \geqslant 3$. The trick is then to set

$$y_i = x_i - x_1 \qquad \text{and} \qquad g_i = h_i - h_1 \qquad \text{for} \quad i = 2, \ldots, n+1.$$

Formula (59) implies

$$(y_i - y_j)(g_i - g_j) = 0, \qquad i, j = 2, \ldots, n+1,$$

so that

$$g_i = \bar{a} + \omega \times y_i, \qquad i = 2, \ldots, n$$

follows from the induction assumption. Moreover, (59) implies that $y_i g_i = 0$ for $i = 2, \ldots, n+1$, and hence

$$y_i \bar{a} = 0, \qquad i = 2, \ldots, n+1.$$

Consequently, $\bar{a} = 0$. Note that by hypothesis span $\{y_2, \ldots, y_{n+1}\}$ has the dimension three. If we now choose the vector a so that

$$h_1 = a + \omega \times x_1,$$

we obtain $h_i = a + \omega \times x_i$ for $i = 1, \ldots, n+1$, i.e., the assertion of Lemma 58.18 is true for $n+1$.

58.2. *Gravitational potential of a mass point.* Assume a point of mass M is at y. It exerts a gravitational force

$$K(x) = \frac{GMm(y-x)}{|y-x|^3}$$

to an arbitrary point of mass m at x with $x \neq y$. Compute the potential U of K.

Solution: We must have $U'(x) = -K(x)$ for all $x \neq y$. This is satisfied for

$$U(x) = -\frac{GMm}{|y-x|}.$$

In order to verify this we use Cartesian coordinates and observe that

$$U'(x) = \frac{\partial U(x)}{\partial \xi^i} e_i.$$

58.3. *Gravitational potential of a body.* A body in a bounded region Ω of $\mathbb{R}^3$ with continuous density $\rho \colon \bar{\Omega} \to \mathbb{R}$ exerts a gravitational force

$$K(x) = \int_\Omega \frac{Gm\rho(y)(y-x)}{|y-x|^3} dy$$

on a point x of mass m. This expression follows from Problem 58.2 by summing the forces whose origin is in the volume element Δy with mass $M = \rho \Delta y$, and then passing from the sum to the integral. Prove that the function

$$U(x) = -\int_\Omega \frac{Gm\rho(y)}{|y-x|} dy$$

represents the corresponding potential in $\mathbb{R}^3$, i.e., $K(x) = -U'(x)$.
Hint: See Günter (1957, M), p. 78.

58.4. *Gravitational potential of a spherical shell.* Let $\Omega = \{y \in \mathbb{R}^3 : 0 < r < |y| < R\}$. For the gravitational potential U of the previous problem with $\rho = $ const prove the following results:
 (i) In the interior, i.e., for $|x| \leq r$ we have $U(x) = $ const. Hence, because of $K = 0$ there exists no gravitational force.
 (ii) In the exterior, i.e., for $|x| \geq R$, the potential $U(x)$ behaves as if the entire mass would be located at the center $x = 0$. Hence, it follows that

$$U(x) = -\frac{Gm\int_\Omega \rho(y) dy}{|x|}$$

Hint: See, e.g., Mangoldt and Knopp (1957, M), Vol. 3, p. 429.

Statements (i) and (ii) remain true for rotationally symmetric, continuous densities ρ. To see this, decompose Ω into small spherical shells of constant density and use an approximation argument.

58.5. *Gravitational force in a mine.* If R is radius of the earth and one goes into a mine at a distance r from the center of the earth, then the gravitational effect is the same as if the entire mass of the partial ball of radius r would be located at the center of the earth. This same important phenomenon will be used in Section 58.15 in connection with the expansion of the universe. More precisely, prove the following. If Ω is a ball of radius R with center $x = 0$ and the continuous, radially symmetric density $\rho : \bar{\Omega} \to \mathbb{R}$, then the gravitational potential is equal to

$$U(x) = -\frac{G\int_{|y| \leq r} \rho(y) dy}{|x|}$$

for $|x| \leq r \leq R$.
Solution: Decompose Ω into a ball of radius r and a spherical shell. The assertion then follows from Problem 58.4.

58.6. *Doppler effect.* A source S moves on a straight line away from the observer P with constant velocity V. At times $t = 0$ and $t = T$ the source S emits signals which travel with the velocity c. Then P receives the two signals at a timely distance of $T + TV/c$. Now suppose that S emits light continuously. What is the relation between the wave lengths, observed by S and P?
Solution: P receives the two signals at a timely distance $T + TV/c$. If λ_S and λ_P is the wave length, which S and P observe, respectively, then we obtain

$\lambda_S = Tc$ for a suitable choice of T, and

$$\lambda_P = Tc + TV = \lambda_S(1 + V/c).$$

Therefore P observes a red shift. For the relative change in the wave length we obtain

$$\frac{\lambda_P - \lambda_S}{\lambda_S} = \frac{V}{c} = Ht,$$

where t is the running time of the signal and H the so-called Hubble constant, i.e., $H = V/R$ with $R = ct$. Here R is the distance between S and P at the time of emission of the signal.

58.7. Plane mathematical pendulum. As in Figure 58.18 we consider a pendulum of length l and mass m. Compute the motion of the pendulum and the period of oscillation by using the Lagrangian formalism of Section 58.19. Furthermore, compute the effective constraining forces.

Solution: The orbital motion $x = x(t)$ occurs on a circle, i.e.,

$$x(t) = l(e_1 \sin \varphi(t) - e_3 \cos \varphi(t)). \tag{139}$$

This motion is given by the angle $\varphi = \varphi(t)$. No side conditions appear in this description. The kinetic energy is

$$E_{\text{kin}} = \tfrac{1}{2} m \dot{x}^2 = \tfrac{1}{2} m l^2 \dot{\varphi}^2.$$

The gravitational force $K = -mge_3$ is applied to the pendulum. It has the potential $U = mg\xi_3$, since $K = -U_x$. The Lagrange function is therefore

$$L = E_{\text{kin}} - U = \tfrac{1}{2} m l^2 \dot{\varphi}^2 + mgl \cos \varphi.$$

The Lagrangian equation

$$\frac{d}{dt} L_{\dot{\varphi}} - L_{\varphi} = 0$$

yields

$$ml^2 \ddot{\varphi} + mgl \sin \varphi = 0. \tag{140}$$

As an initial condition we choose

$$\varphi(t_0) = \varphi_0, \qquad \dot{\varphi}(t_0) = 0 \tag{141}$$

with $-\pi < \varphi_0 < \pi$, i.e., φ_0 corresponds to the turning point of the pendulum.

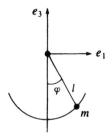

Figure 58.18

Since L does not explicitly depend on time t, it follows from Section 58.19 that $\dot\varphi L_\varphi - L = E_{kin} + U$ is a conserved quantity. This is the energy conservation law

$$\tfrac{1}{2} ml^2 \dot\varphi^2 - mgl \cos \varphi = \text{const} = E.$$

The fact that this expression is constant, also follows directly by differentiation keeping (140) in mind. From (141) we obtain $E = -mgl \cos \varphi_0$, hence

$$\dot\varphi^2 = \frac{4g}{l} \left(\sin^2 \frac{\varphi_0}{2} - \sin^2 \frac{\varphi}{2} \right).$$

We set

$$\sin \frac{\varphi}{2} = k \sin \psi, \qquad k = \sin \frac{\varphi_0}{2}.$$

This way we obtain the elliptic integral

$$\sqrt{\frac{g}{l}} \, t = \int_0^\psi \frac{d\psi}{\sqrt{1 - k^2 \sin^2 \psi}}.$$

This integral yields a periodic motion with $\varphi = 0$ for $t = 0$. The period of oscillation T satisfies $\varphi(T/4) = \varphi_0$, hence

$$\sqrt{\frac{g}{l}} \, \frac{T}{4} = \int_0^{\pi/2} \frac{d\psi}{\sqrt{1 - k^2 \sin^2 \psi}}.$$

For small motions around the equilibrium position $\varphi = 0$, we have that $|\varphi_0|$ is small, i.e., k^2 is small. Expansion of the integrand yields

$$T = 2\pi \sqrt{\frac{l}{g}} \left(1 + \frac{k^2}{4} + O(k^4) \right), \qquad k \to 0.$$

If, for small motions around the equilibrium position, one replaces the function $\sin \varphi = \varphi + O(\varphi^3)$ with φ, then one obtains from (140) the equation of motion

$$\ddot\varphi + \frac{g}{l} \varphi = 0,$$

which has the solution

$$\varphi = \varphi_0 \sin \left(\sqrt{\frac{g}{l}} \, t + \alpha \right).$$

Here the period of oscillation satisfies

$$T = 2\pi \sqrt{\frac{l}{g}}.$$

In order to compute the effective constraining forces Z which are acting on the pendulum, we insert $\varphi = \varphi(t)$ into (139). This gives

$$m\ddot{x} = -mge_3 + Z.$$

See Fig. 58.17 in Section 58.10.

The following Problems 58.8–58.10 serve as a preparation for Problem 58.11 about oscillating systems.

58.8. Complete system of eigenvectors for self-adjoint operators in finite-dimensional H-spaces. Let $A: X \to X$ be a linear self-adjoint operator in an H-space X over $\mathbb{K} = \mathbb{R}, \mathbb{C}$ with $\dim X = N$, $1 \leq N < \infty$. Show that there exists a system $\{x_1, \ldots, x_N\}$ of eigenvectors of A with

$$(x_i | x_j) = \delta_{ij} \quad \text{for all } i, j = 1, \ldots, N.$$

Solution: This is a special case of Theorem 19.B. In order to obtain an entirely elementary approach to oscillating systems we give a full proof.
(I) We begin with the eigenvalue equation

$$Ax = \lambda x, \quad x \in X, \quad \lambda \in \mathbb{K}. \tag{142}$$

Because of

$$\lambda = \frac{(x|Ax)}{(x|x)} \quad \text{and} \quad \overline{(x|Ax)} = (Ax|x) = (x|Ax).$$

all eigenvalues of A are real. The number λ is an eigenvalue of A if and only if the inverse operator $(A - \lambda I)^{-1}$ does not exist, i.e.,

$$\det(A - \lambda I) = 0.$$

According to the fundamental theorem of algebra this polynomial equation in λ has a solution $\lambda = \lambda_1$. Hence it follows that an eigensolution

$$Ax_1 = \lambda_1 x_1, \quad x_1 \in X, \quad \lambda_1 \in \mathbb{R}$$

exists.
(II) Letting

$$Y = \{y \in X : (y|x_1) = 0\},$$

we find that $\dim Y = N - 1$. Moreover, we obtain the key result

$$y \in Y \Rightarrow Ay \in Y. \tag{143}$$

In fact, if $y \in Y$, then

$$(Ay|x_1) = (y|Ax_1) = \lambda_1(y|x_1) = 0.$$

Because of (143) we can now apply the same argument as in (I) to the operator $A: Y \to Y$. This yields the existence of an eigensolution

$$Ax_2 = \lambda_2 x_2, \quad x_2 \in Y, \quad \lambda_2 \in \mathbb{R}.$$

(III) Using N analogous steps we find N eigensolutions

$$Ax_i = \lambda_i x_i, \quad x_i \in X, \quad \lambda_i \in \mathbb{R}, \quad i = 1, \ldots, N.$$

58.9. Normal form for symmetric matrices. Let A^T in the following denote the transposed matrix to A. Suppose we are given a real symmetric $(N \times N)$-matrix $A = (a_{ij})$, i.e., $A = A^T$. Show that there exists a real orthogonal $(N \times N)$-

matrix U with
$$U^T A U = \mathrm{diag}(\lambda_1, \ldots, \lambda_N),$$
where the λ_i's are the eigenvalues of A.

Solution: We choose $X = \mathbb{R}^N$ with $(x|y) = \sum_{i=1}^N \xi_i \eta_i$. Notice that $(x|y) = x^T y$ and hence
$$(Ax|y) = (Ax)^T y = x^T A^T y = x^T A y = (x|Ay),$$
i.e., the linear operator $A: X \to X$ is self-adjoint. According to Problem 58.8 there exists a system of eigenvectors of A with $Ax_i = \lambda_i x_i$ and $x_i^T x_j = \delta_{ij}$ for all $i, j = 1, \ldots, N$. The matrix
$$U = (x_1, \ldots, x_N)$$
satisfies $U^T U = I$ because of
$$U^T U = \begin{pmatrix} x_1 \\ \vdots \\ x_N \end{pmatrix} (x_1, \ldots, x_N) = (x_i^T x_j) = I,$$
i.e., U is orthogonal. Finally, we get
$$U^T A U = \begin{pmatrix} x_1 \\ \vdots \\ x_N \end{pmatrix} (\lambda_1 x_1, \ldots, \lambda_N x_N) = \begin{pmatrix} \lambda_1 & & 0 \\ & \ddots & \\ 0 & & \lambda_N \end{pmatrix}.$$

Because of $\det U^T U = 1$ we find that
$$\det(U^T A U - \lambda I) = \det(U^T (A - \lambda I) U) = \det(A - \lambda I).$$
Hence the solutions of $\det(A - \lambda I) = 0$ are precisely all the λ_i's.

58.10. *Diagonalization of two symmetric matrices.* Suppose we are given two real symmetric $(N \times N)$-matrices A and B with $x^T A x > 0$ for all $x \neq 0$. Prove that there exists a nonsingular real $(N \times N)$-matrix C with
$$C^T A C = I, \qquad C^T B C = \mathrm{diag}(\omega_1, \ldots, \omega_N). \tag{144}$$

Moreover, show that the ω_i's are precisely the solutions of the generalized characteristic equation
$$\det(B - \omega A) = 0. \tag{145}$$

The ω_i's are all positive if and only if $x^T B x > 0$ for all $x \neq 0$.

Solution: From Problem 58.9 there exists an orthogonal matrix U with
$$U^T A U = \mathrm{diag}(\lambda_1, \ldots, \lambda_N).$$
We set $V = \mathrm{diag}(1/\sqrt{\lambda_1}, \ldots, 1/\sqrt{\lambda_N})$, so that
$$V^T U^T A U V = I.$$
The matrix $B_1 = V^T U^T B U V$ is symmetric. Using Problem 58.9 once again we obtain that there exists an orthogonal matrix W with
$$W^T B_1 W = \mathrm{diag}(\omega_1, \ldots, \omega_N).$$
Setting $C = UVW$. formula (144) follows.

The ω_i's are precisely the eigenvalues of B_1, i.e., the solutions of $\det(B_1 - \omega I) = 0$. Observe then that $\det V = \det V^T$ and hence

$$\det(B_1 - \omega I) = \det(V^T U^T [B - \omega A] UV)$$
$$= (\det V)^2 (\det U)^2 \det(B - \omega A).$$

This implies (145).

Notice, furthermore, that $x^T B x = y^T \operatorname{diag}(\omega_1, \ldots, \omega_N) y$ for $y = Cx$.

58.11. *General systems with small oscillations.* We want to find a general model for a mechanical system near an equilibrium state. Suppose that the system has f degrees of freedom, i.e., we can describe the motion of the system by

$$q_i = q_i(t), \qquad i = 1, \ldots, f,$$

where $q_1, \ldots, q_f$ are suitable real coordinates. Suppose that the equilibrium state corresponds to the origin $q_1 = \cdots = q_f = 0$.

58.11a. *Definition.* An oscillating system with f degrees of freedom is given by the equations of motion

$$\sum_{j=1}^{N} a_{ij} \ddot{q}_j + \sum_{j=1}^{N} b_{ij} q_j = 0, \qquad i = 1, \ldots, f. \tag{146}$$

Here the real matrices $A = (a_{ij})$ and $B = (b_{ij})$ are symmetric and the corresponding quadratic forms are positive definite.

58.11b. *Normal coordinates.* Show that there exist new coordinates $r_1, \ldots, r_N$ so that (146) becomes

$$\ddot{r}_i + \omega_i r_i = 0, \qquad i = 1, \ldots, f \tag{147}$$

with $\omega_i > 0$ for all i. Moreover, the frequencies ω_i are precisely the solutions of the generalized characteristic equation

$$\det(B - \omega A) = 0.$$

Solution: We write (147) as the matrix equation

$$A\ddot{q} + Bq = 0.$$

Setting $q = Cr$ we obtain $C^T AC\ddot{r} + C^T BCr = 0$. According to Problem 58.10 this is (147).

58.11c. *Motivation.* We like to show that (146) holds under fairly general hypotheses about the system. We start with a Lagrangian function $L = L(q, \dot{q}, t)$ and assume that the system is homogeneous in time, i.e.,

$$L(q, \dot{q}, t + a) = L(q, \dot{q}, t) \qquad \text{for all} \quad a \in \mathbb{R}$$

and all arguments. This implies $L_t \equiv 0$. From Taylor's theorem,

$$L(q, \dot{q}) = L(0, 0) + \sum_{i=1}^{N} a_i \dot{q}_i + b_i q_i$$
$$+ \frac{1}{2} \sum_{i,j=1}^{N} a_{ij} \dot{q}_i \dot{q}_j - b_{ij} q_i q_j + \cdots.$$

The Lagrangian equations

$$\frac{d}{dt}L_{\dot q_i} - L_{q_i} = 0, \qquad i = 1, \ldots, f \tag{148}$$

coincide with (146) up to a constant term b_i. Since $q = 0$ should be a solution of (148), we need

$$b_i = 0 \qquad \text{for all } i.$$

Moreover, we set

$$a_i = 0 \qquad \text{for all } i$$

because this does not effect (148). The term

$$T = \tfrac{1}{2}\sum a_{ij}\dot q_i \dot q_j = \tfrac{1}{2}\dot q^T A \dot q$$

represents the kinetic energy. Thus, it is reasonable to postulate that $\dot q^T A \dot q > 0$ if $\dot q \neq 0$.

If $\omega_i \leq 0$, then (147) has unbounded solutions. Thus, we also require that $\omega_i > 0$ for all i. This means that $q^T B q > 0$ if $q \neq 0$. By the way, the potential energy of the system is

$$U = \tfrac{1}{2}\sum b_{ij} q_i q_j = \tfrac{1}{2} q^T B q,$$

which has a strict minimum at $q = 0$. This implies the stability of the equilibrium state $q = 0$. Notice that $L = T - U$.

58.11d. *Physical insight.* Our considerations show that, under fairly general assumptions, an oscillating system with f degrees of freedom can be reduced to f *independent harmonic oscillators* in the sense of (147). This result plays an important role in physics. In Section 58.15c, for example, we used this result in the study of the early cosmos.

A detailed mathematical investigation of small oscillations may be found in Gantmacher and Krein (1960, M).

58.12. *Further problems.* Numerous other problems may be found in Sommerfeld (1954, M), Vol. 1 and Landau and Lifšic (1962, M), Vol. 1.

References to the Literature

Classical works: Kepler (1609, M), (1618, M), Galilei (1638, M), Newton (1687, M).
Collected works: Kepler (1939), Galilei (1890/1909), Newton (1779), (1967).
Essay about Galilei and Kepler: Blaschke (1957).
Physical mechanics: Sommerfeld (1954, M), Vol. 1, Landau and Lifšic (1962, M), Vol. 1, Landau, Achieser and Lifšic (1970, M).
Mathematical mechanics: Frank and von Mises (1962, M), Arnold (1978, M), Abraham and Marsden (1978, M).
Celestial mechanics: Wintner (1947, M), Sternberg (1969, L), Vol. 1–2, Siegel and Moser (1971, M), Stumpff (1973, M), Vol. 1–3, Hagihara (1976, M), Vol. 1–5, Abraham and Marsden (1978, M), Arnold (1987, S), Vol. III.
Similarity theory: Sedov (1959, M), Massey (1971, M).
Physics of gravitating systems: Fridman and Polyačenko (1984, M).
Small oscillations: Gantmacher and Krein (1960, M).
History of mechanics: Szábo (1987, M).

CHAPTER 59

Dualism Between Wave and Particle, Preview of Quantum Theory, and Elementary Particles

> Except for atoms and emptyness nothing exists.
> Demokrit (460 B.C.–371 B.C.)

> There exists a limiting case of quantum theory which corresponds to classical particle physics, and there exists another which corresponds to classical wave mechanics. The alternatives which the limiting cases represent are not compatible. Bohr was therefore right when he called the duality between the two "pictures"—wave and particle—an example of complementarity.
> Carl Friedrich von Weizsäcker (1973)

> The last significant turn in quantum theory occurred after de Broglie's discovery of matter waves in 1924, Heisenberg's formulation of quantum mechanics in 1925, and Schrödinger's general wave mechanical equation in 1926.
> Wolfgang Pauli (1958)

> Quantum theory so perfectly illustrates the fact that one might have understood a certain subject with complete clarity, yet at the same time knows that one can speak of it only allegorically and in pictures.
> Werner Heisenberg (1901–1976)

In the previous chapter we considered the concept of particles. This chapter we begin by introducing a number of basic concepts, which are essential for an understanding of wave phenomena in all parts of physics. We then will discuss the relation between waves and particles, which has played an important role in the development of modern physics. In 1925, Werner Heisenberg formulated his matrix mechanics. This is a quantum mechanics, which is derived from classical mechanics by introducing particle quantization. This theory has already been discussed in Section 58.21. Independently, in 1926, Erwin Schrödinger formulated an equivalent wave mechanics which is derived from wave quantization. The main objective of this chapter is to present a survey. This, together with the previous chapter, might help the reader to better understand many of the problems discussed later on. We thereby follow the fascinating line of development, which leads to the central problem of modern physics—the creation of a unified theory for all four interactions in nature. Quantum theory will be discussed in greater detail in Part V. Only a minimal program is presented here. Some interesting problems that we consider are:

(i) Spectrum of the hydrogen atom.
(ii) Quantum mechanical treatment of the harmonic oscillator in the context of Schrödinger's wave mechanics.
(iii) Functional analytical deduction of Heisenberg's uncertainty relation.

Modern quantum theory is not conceivable without the special theory of relativity, because for many elementary particle processes, which occur under extreme conditions, relativistic effects are essential. The special and the general theory of relativity are discussed in Chapters 75 and 76. Here we will use the following results:

(a) Every free particle has a rest mass $m_0 \geq 0$. Its energy is equal to
$$E = \sqrt{m_0^2 c^4 + p^2 c^2}, \tag{1}$$
where p is the momentum vector and c the velocity of light.

(b) For $m_0 > 0$ every free particle with velocity vector v has the mass
$$m = m_0/\sqrt{1 - v^2/c^2} \tag{2}$$
and the energy
$$E = mc^2. \tag{3}$$

(c) Physical effects can travel with, at most, the velocity of light. The velocity of particles with rest mass $m_0 > 0$ is always less than the velocity of light.

At several places in this chapter we will take the opportunity to introduce the reader to the peculiarities and usefulness of physical thought.

59.1. Plane Waves

We begin with a function
$$y = A(t)$$
of the time variable t, which has period T. Then T is called the *period of oscillation*. Moreover, we define the *frequency*
$$v = 1/T$$
and the *angular frequency*
$$\omega = 2\pi v.$$

The frequency v is equal to the number of oscillations during one unit of time. We have

$$T = 2\pi/\omega = 1/v.$$

Definition 59.1. Let $y = W(\alpha)$ be a function of period 2π. By a *plane wave* we mean

$$y = W(kx - \omega t), \tag{4}$$

where x denotes the position vector and t the time. Moreover, k is a non-vanishing vector, which will be called the *wave vector*. The direction of k is called the *direction of propagation*, and $|k|$ is called the *wave number*. The wave length λ and the *propagation velocity* c_p (phase velocity) are, by definition, equal to

$$\lambda = 2\pi/|k|, \quad c_p = \omega/|k|.$$

We call ω the angular frequency and define the period of oscillation T and the frequency v as above through $T = 2\pi/\omega$ and $v = 1/T$.

The function W might be a real function. One may think, for example, of the fluctuations of pressure or density of sound waves in fluids or gases. But W can also be a vector function. In this case, one may think of electric and magnetic fields of electromagnetic waves (light) or the displacement of a body under elastic waves. Such waves are of particular physical importance, since they allow a transport of energy and momentum without a transport of mass.

The intuitive meaning and the motivation of the definition above becomes clear if we lay the e_1-axis of a Cartesian coordinate system in the direction of k and set

$$x = \xi e_1 + \eta e_2 + \zeta e_3.$$

We then obtain $kx = |k|\xi$, and moreover

$$y = W(|k|\xi - \omega t) = W(|k|(\xi - c_p t)).$$

Since W has period 2π this last equation yields a function $y = y(\xi)$ of period $\lambda = 2\pi/|k|$ at every fixed time t. Furthermore, one obtains a function $y = \bar{y}(t)$ of period T for every fixed space point x. The number

$$\alpha = kx - \omega t = |k|(\xi - c_p t)$$

is called the *phase*. All space–time points (x, t) with equal phase yield the same value $W(\alpha)$. If one moves along k with velocity c_p, then W has always the same value. This justifies the expression phase velocity for c_p. From the specific form of a physical model one often obtains a relation of the form

$$\omega = \omega(|k|), \tag{5}$$

which is called a *dispersion relation*. This is equivalent to $c_p = c_p(\lambda)$, i.e., represents a connection between wave length and propagation velocity. The

derivative
$$c_g = \omega'(|k|)$$
is called *group velocity* c_g for $|k|$. We shall motivate this expression below. Actually, from a physical point of view, the group velocity is much more important than the phase velocity. Dispersion relations of the more general form
$$\omega = \omega(k, x, t)$$
are also used.

EXAMPLE 59.2. An important special case of plane waves are the *harmonic* or monochromatic waves
$$y = W_0 \sin(kx - \omega t + \alpha_0).$$
Here α_0 is called the *phase displacement* and $|W_0|$ is called the *amplitude*. The quantity W_0 may be a real number or a vector, e.g., the vector of the electric field strength. Often, one works with complex harmonic waves
$$y = W_0 e^{i(kx - \omega t + \alpha_0)}. \tag{6}$$
Sometimes they are studied directly, as in the case of quantum mechanics, but in other instances only the imaginary part, which is equal to the sinusoidal oscillation above, is used.

59.2. Polarization

If W in (4) is a vector function and W_0 in (6) is a vector, as in the case of electromagnetic waves (light), then one can introduce a number of further concepts. The plane wave is called *transversal* (or *longitudinal*) if and only if W is perpendicular to the direction of propagation k (or parallel to k). Maxwell's equations of Part V show that light waves are always transversal. Elastic waves may be transversal or longitudinal. The direction of W is called the *direction of polarization*.

We now consider transversal waves and classify their polarization. Let P be a fixed plane, perpendicular to k. We move along k with velocity c_p, whereby the projection from W onto P describes a curve C with time period T. The wave is called *circular polarized* (or *elliptic*, *linear*) if and only if C is a circle (or an ellipse, straight line). For a linear polarization, the plane through k and W is called a polarization plane. The polarization is important for an understanding of a number of light phenomena (e.g., anomalous refraction of Iceland spar, intensity of reflected light, birefringence). Circular polarized light has an angular momentum. In the context of quantum field theory this means that the photon has spin $+1$ or -1. This plays an important role in quantum statistics, which in Chapter 68 will be applied to Planck's radiation law and the state of the cosmos following the Big Bang.

59.3. Dispersion Relations

We want to discuss (5). As we shall see in Sections 59.9 and 59.10 below, the group velocity corresponds to the propagation velocity of wave packets and to the propagation velocity of particle rays. Such rays correspond to the first-order approximation of wave theories (e.g., geometrical optics). Therefore the knowledge of the dispersion relation (5) is of particular interest for the computation of the group velocity. This is done by solving the field equations of the corresponding physical theory. In Chapter 71 we will compute the dispersion relation for water waves. Thereby we solve a nonlinear, free boundary-value problem in the context of bifurcation theory. In Part V we show that from Maxwell's equations one obtains the linear dispersion relation

$$c = \lambda v$$

for the propagation of electromagnetic waves in vacuum (light), where c is a universal constant, namely the velocity of light in the vacuum. We therefore have

$$\omega = c|k|.$$

This leads to the important fact that

$$c = c_p = c_g,$$

i.e., group velocity and phase velocity coincide here.

Light Quantum Hypothesis 59.3 (Einstein (1905b)). *Light consists of photons of energy $E = hv$ with momentum $|p| = E/c$ and rest mass $m_0 = 0$. In other terms, we have*

$$E = \hbar\omega, \qquad p = \hbar k, \tag{7}$$

where h is Planck's quantum of action and $\hbar = h/2\pi$.

The meaning of these notations will be motivated in Section 75.11 in the context of the relativistic mechanics of free particles. Einstein used this hypothesis in order to explain the photo effect, for which in 1921 he received the Nobel prize. If a metal plate is exposed to light, then one obtains cathode rays, i.e., electrons are ejected. In the context of classical physics it cannot be understood why the energy of the electrons does not depend on the intensity of light, but only on its frequency v. However, this immediately becomes clear if with Einstein one assumes that the electron is ejected following a collision with a photon, whereby a maximal photon energy of $E = hv$ occurs.

In fact, (7) represents a fundamental relation between the wave and particle picture of microscopical quantum objects, which is valid not only for light.

The universal applicability of (7) was postulated in 1924 by de Broglie in connection with his theory of matter waves. This occurred before Heisenberg (1925) and Schrödinger (1926) gave their formulations of quantum mechanics.

In the nonrelativistic quantum theory of free particles (first quantization) these free particles are described by complex wave functions

$$\psi = \psi_0 e^{i(kx-\omega t)}.$$

For particles without spin (e.g., π-mesons) ψ_0 is a complex number. For the electron one has $\psi_0 \in \mathbb{C}^4$ (see the Klein–Gordon equation and the Dirac equation in Part V). Energy and momentum of the particles follow here from the same relation (7) as in the case of the photon. The number

$$\lambda = 2\pi/|k|$$

is called the de Broglie wave length of the corresponding matter wave. The relativistic equation

$$E^2 = m_0^2 c^4 + p^2 c^2$$

with rest mass m_0, together with (7), implies the dispersion relation for matter waves

$$\omega = c\sqrt{(m_0 c/\hbar)^2 + k^2}.$$

In fact, one observes phenomena of diffraction of electron rays at crystal lattices which are compatible with this wave picture.

59.4. Spherical Waves

Instead of plane waves, one often uses so-called spherical waves

$$y = W(K|x| - \omega t)$$

with $k = Kx/|x|$. This means that the direction of propagation k of the wave is radial and $|k| = K$. For fixed time t, the space points x with constant phase and constant W, lie on spheres around the origin. For variable t, these spheres travel with velocity $c_p = \omega/K$ (Fig. 59.1). The quantities λ, T, and ν are defined as in the case of plane waves.

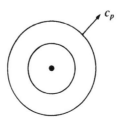

Figure 59.1

59.5. Damped Oscillations and the Frequency–Time Uncertainty Relation

By a damped oscillation which is switched on at time $t = 0$ we mean

$$y = W_0(x)w(t) \tag{8}$$

with

$$w(t) = \begin{cases} e^{-i\omega_0 t + \alpha_0 i - \gamma t} & \text{for } t \geq 0, \\ 0 & \text{for } t < 0 \end{cases}$$

with real numbers ω_0, α_0, and $\gamma > 0$. If W_0 is real, then for $t > 0$ we obtain the imaginary part

$$y = e^{-\gamma t} W_0(x) \sin(\alpha_0 - \omega_0 t),$$

which represents a damped sinusoidal oscillation. For great $\gamma > 0$ these oscillations tend quickly to zero as $t \to +\infty$ (see Fig. 59.2). By definition, the *mean life-time* Δt of the damped oscillation is

$$\Delta t = 1/2\gamma.$$

This is the time, during which $e^{-\gamma t}$ has decreased from 1 at $t = 0$ to the value $e^{-1/2} = 0.6$. The time $t_{1/2}$, during which $e^{-\gamma t}$ decreases from 1 to $\tfrac{1}{2}$ is called *half-life period*. Because of $t_{1/2} = 1.4\Delta t$ one often finds that in the literature there is no distinction made between $t_{1/2}$ and Δt. We call

$$\Delta\omega = \gamma \tag{9}$$

the *half-width of the spectrum* of (8). This will be motivated below. Hence the important so-called frequency–time uncertainty relation

$$\Delta\omega \Delta t = \tfrac{1}{2} \tag{10}$$

is valid. If, according to (7), we assign the energy $E_0 = \hbar\omega_0$ to the frequency ω_0 and set $\Delta E = \hbar\Delta\omega$, then we obtain the so-called *energy–time uncertainty relation*

$$\Delta E \Delta t = \frac{\hbar}{2} \tag{11}$$

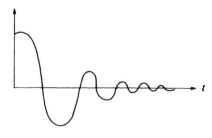

Figure 59.2

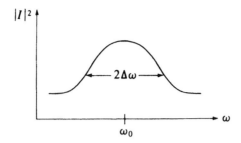

Figure 59.3

whose fundamental importance for elementary particle physics we will discuss in Section 59.24.

Now we shall motivate (9). For this, we let

$$I(\omega) = \frac{1}{2\pi} \int_0^\infty e^{i(\omega-\omega_0)t + \alpha_0 i - \gamma t}\, dt$$

$$= \frac{e^{i\alpha_0}}{2\pi[\gamma - i(\omega - \omega_0)]}.$$

Using a Fourier transformation we obtain the frequency representation

$$y = \int_{-\infty}^\infty I(\omega) W_0(x) e^{-i\omega t}\, d\omega$$

for (8). In many physical theories the squares of the amplitudes are a measure for the intensity of the processes. In our special case

$$|I(\omega)|^2 = \frac{1}{4\pi^2(\gamma^2 + (\omega - \omega_0)^2)}$$

is then a measure for the intensity of the angular frequency ω in the spectrum (Fig. 59.3). According to its definition we now determine $\Delta\omega$ as

$$|I(\omega_0 + \Delta\omega)|^2 / |I(\omega_0)|^2 = \tfrac{1}{2},$$

i.e., for $\omega_0 + \Delta\omega$ the intensity has decreased about one-half compared with the maximum at ω_0. This gives (9). Therefore, roughly, one may say that only the angular frequencies ω with

$$\omega_0 - \Delta\omega \leq \omega \leq \omega_0 + \Delta\omega$$

provide a significant contribution to the damped oscillation.

59.6. Decay of Particles

In connection with decay processes we also have the concept of the mean life-time of particles. What precisely does this mean? We think, for example, of a β-decay (radioactive decay) of the neutron

$$n \to p + e^- + \bar{v}$$

into a proton, an electron, and an antineutrino. This is a typical process of weak interaction. Theoretically, this was studied for the first time by Fermi in 1934. Let $N(t)$ be the number of particles at time t that do not decay. We assume that $w(0) = 0$. Then the *decay probabilty* $w(t)$ for a particle during the time interval $[0, t]$ is given by

$$w(t) = \gamma t + o(t), \qquad t \to 0, \tag{12}$$

where t is sufficiently small in order to guarantee that $w(t) < 1$. It follows that

$$N(t + h) = N(t) - N(t)w(h) = N(t)(1 - \gamma h + o(h))), \qquad h \to 0$$

and this implies the differential equation

$$N' = -\gamma N.$$

Hence, we obtain the well-known *decay law*

$$N(t) = N(0)e^{-\gamma t}. \tag{13}$$

Analogously, as in Section 59.5, we call

$$\Delta t = 1/2\gamma$$

the mean life-time of a particle and $t_{1/2} = 1.4 \cdot \Delta t$ the half-life period of the substance. Again, one finds that in the literature there is often no distinction made between Δt and $t_{1/2}$. We have $N(\Delta t) = 0.6 \cdot N(0)$ and $N(t_{1/2}) = 0.5 \cdot N(0)$. Moreover, γ is called the *decay coefficient*.

59.7. Cross Sections for Elementary Particle Processes and the Main Objectives in Quantum Field Theory

In the case of cosmic rays or particle accelerators one has the following idealized situation. A homogeneous flow of bombarding particles with particle density ρ (number of particles per volume) and constant velocity vector v hits on target particles, whereby a number of N_{target} of such target particles is contained in a fixed given volume. Let N_{reaction} denote the number of certain reactions which during the time interval $[0, t]$ take place there. One may think of traces of emulsions for cosmic rays. The vector $j = \rho v$ is called the vector of the particle current density. The *effective cross section* is then defined as

$$\sigma = \frac{N_{\text{reaction}}}{N_{\text{target}} \cdot t \cdot |j|}.$$

Letting
$$N_{\text{effective}} = \sigma|j|,$$
we obtain that
$$N_{\text{reaction}} = N_{\text{effective}} \cdot N_{\text{target}} \cdot t. \tag{14}$$

In physics, the quantity σ is the most important characterization of scattering processes and has the dimension of a surface. One has the following intuitive interpretation:

The number of reactions with one target particle during a fixed time interval is equal to the number of bombarding particles which penetrate a surface σ during this time interval. Here, the surface is perpendicular to the homogeneous flow of bombarding particles.

The larger σ is, the bigger is the number of reactions. In Section 59.26 we will give numerical values for σ. In Problem 59.3 we compute the effective cross section for the classical Rutherford scattering of α-particles at atomic nucleuses. Using these scattering experiments, Rutherford in 1911 was able to determine the atomic structure.

By
$$w(t) = N_{\text{effective}} \cdot t = \sigma|j|t \tag{15}$$

we define the *reaction probability* $w(t)$ during the time interval $[0, t]$. Strictly speaking, one needs to add the term $o(t)$ as $t \to 0$ to (14) and (15) similarly as in the case of (12).

The first main task of quantum field theory is to compute the decay and reaction probabilities, i.e., the decay coefficients γ and the effective cross sections σ. For this, the formalism of the S-matrix (scattering matrix) is used by physicists without a rigorous mathematical justification. A second main task would be to predict the specific properties of all elementary particles (mass, charge, spin, isospin, strangeness, etc.). This, however, lies in the distant future.

59.8. Dualism Between Wave and Particle for Light

In connection with the previous observations let us make some brief remarks about the physical meaning of the dualism between wave and particle. The mathematical theory will be considered in Part V. For our discussion here we choose light as a typical example. As in the case of elementary particles, physicists and mathematicians are still puzzled.

(i) *Fermat's principle of geometrical optics of 1662.* In connection with reflection and refraction, light behaves like a particle ray, whose motion in geometrical optics is determined by Fermat's variational principle of the least (stationary) light time. From a mathematical point of view, this principle is completely analogous to the principle of the least (stationary) action of classical mechanics. At the same time, the principle of stationary action is the most general variational principle in physics and is true for numerous field theories; in particular, it determines the mathematical structure of the relativistic theories for elementary particles (see Part V).

(ii) *Electromagnetic waves.* In Maxwell's electrodynamics of 1873 light is described by waves of coupled electric and magnetic fields. Thereby diffraction, interference, and polarization effects can be explained. Geometrical optics in this context is obtained by using asymptotic expansions with respect to the small wave length λ and taking only the first-order approximation into consideration. The light rays are here the stream lines of the vector field of the energy density (see Chapter 40).

Maxwell's theory, which will be discussed in Part V, is an ingenious theory. First, up until today no corrections have been necessary either in connection with the special theory of relativity or in connection with quantum electrodynamics. Second, it is the first physical theory, in which two apparently very different phenomena, namely electricity and magnetism, are connected with each other by using the notion of a field. Third, it is a gauge field theory and hence a model for the modern gauge field theories of elementary particles, in connection with which one tries to develop a unified theory for all interactions using the idea of gauge fields. Finally, it enabled Maxwell to predict that light is an electromagnetic wave. The experimental confirmation came in 1888 through Heinrich Hertz. This was nine years after Maxwell's death.

(iii) *Planck's radiation law and quantum theory.* At the end of the nineteenth century it was a famous physical problem to find the correct radiation law to describe the energy distribution in the spectrum of radiating bodies (see Section 68.4). To find this law by means of thermodynamic methods, Planck in 1900 made the revolutionary assumption that the energy of the harmonic oscillator cannot take on all values, but only discrete ones with

$$\Delta E = \hbar\omega.$$

In 1925, Heisenberg showed that the precise values are

$$E = \hbar\omega(n + \tfrac{1}{2}), \qquad n = 0, 1, \ldots$$

(see Example 58.26). For quantum field theory it is of particular interest that there exists a nonzero energy for the ground state $n = 0$. The physical interpretation of this is that the vacuum itself possesses physical properties, which can be observed experimentally, e.g., through the fine structure (Lamb-shift) in the hydrogen spectrum or through the magnetic anomaly of the electron. Moreover, the vaporization of black holes of Section 76.17 is based on these quantum phenomena.

(iv) *Photons and statistical physics.* Other than in geometrical optics, light of all wave lengths behaves during the photoelectric effect like a particle. This light quantum hypothesis, which was formulated by Einstein in 1905, following Planck's quantum hypothesis, can easily be used together with the Bose statistics to derive Planck's radiation law (see Chapter 68).

(v) *Einstein's special theory of relativity* (1905). The fact that the velocity of light is constant for all inertial systems leads to the relativistic structure of time and the development of relativistic mechanics (see Chapter 75).

(vi) *Einstein's general theory of relativity* (1916). Light is deflected by masses, i.e., it behaves like a particle with respect to relativistic gravitation (see Chapter 76).

(vii) *Quantum electrodynamics.* This theory, which was created by Feynman, Schwinger, and Tomonaga during the late 1940s, uses a quantum field to describe the interactions between electrons, positrons, and photons. It is obtained from an equation which results from a combination of Maxwell's equations with Dirac's spinor equation for the electron and the positron. The formalism of quantum theory consists of two steps. The first quantization yields relativistic field equations, which, in connection with an abstract Hilbert space theory, contain particle as well as wave aspects. A probability theoretical interpretation of physical processes is important. The second quantization yields quantum fields. A consistent mathematical theory for this has yet to be found. Because of very strong singularities, the formalism which is presently used leads to mathematically meaningless expressions. Physicists, however, have found a regularization procedure (renormalization of charge and mass), which, formally applied, leads to accurate agreements with the experiment. In this theory, the photon is a *quant*, which can be regarded as a simultaneous generalization of wave and particle. Precisely speaking, the quant is the primary physical phenomenon, and our pictures of particles and waves, which come from our macroscopical experience, are possible approximations for a description of microscopical phenomena.

The unusual structure of the quant is reflected in the historical development of quantum mechanics. Heisenberg in 1925 used particle quantization, i.e., a quantization of classical mechanics, to arrive at quantum mechanics. Schrödinger in 1926, on the other hand, obtained his quantum mechanics completely independent of Heisenberg, by using the idea of matter waves. Actually, both theories are only different realizations of the same abstract Hilbert space theory (Heisenberg picture and Schrödinger picture; see Part V).

(viii) *Gauge field theory.* This modern theory starts from the Dirac equation for electron and positron. According to the principle of greatest simplicity, one automatically obtains the electromagnetic field and the photon as a quant of this field by assuming the gauge invariance of the theory. Roughly speaking, the photon in gauge field theory is obtained for nothing. It is needed to carry the information about the gauge of the electron–positron field. This results in electromagnetic interaction (see Part V).

Quantum electrodynamics and the gauge field theory for the photon form the model for the modern theories which describe all interactions in the microcosmos. All this leads to the same mathematical difficulties.

The previous points (i)–(viii) show the very interesting phenomenon that the light, which is necessary for our biological existence, is also the light of our physical knowledge. Any significant physical theory is in an essential way connected with light.

59.9. Wave Packets and Group Velocity

We want to show that wave packets propagate with group velocity and set $K = |k|$. By definition, a wave packet is obtained by superposition of harmonic waves

$$y = \int_{K_0}^{K_0+\Delta K} A(K)e^{i(K\xi-\omega(K)t)}\,dK,$$

where $\Delta K > 0$ is small. We content ourselves with a rough argument, and approximate the integral with the trapezoid formula

$$y = 2^{-1}[f(K_0) + f(K_0 + \Delta K)]\,\Delta K,$$

where

$$f(K) = A(K)e^{i(K\xi-\omega(K)t)}.$$

Taylor expansion at the point K_0 yields

$$K\xi - \omega(K)t = K_0\xi - \omega(K_0)t + (\xi - \omega'(K_0)t)\,\Delta K$$
$$+ o(\Delta K), \quad \Delta K \to 0.$$

Using the phase velocity $c = \omega(K_0)/K_0$ and the group velocity $c_g = \omega'(K_0)$ we obtain in first-order approximation

$$y = W_0 e^{i(\xi-ct)K_0} \tag{16}$$

with

$$W_0 = A(K_0)[1 + 2^{-1}e^{i(\xi-c_g t)\Delta K}]\,\Delta K. \tag{17}$$

This can be viewed as a harmonic wave, for which, likewise, the amplitude varies according to a harmonic wave law. Denoting the wave lengths of K_0 and ΔK by

$$\lambda_0 = 2\pi/K_0 \quad \text{and} \quad \lambda = 2\pi/\Delta K,$$

respectively, and letting $\Delta K \ll K_0$, we obtain

$$\lambda_0 \ll \lambda.$$

This means that the wave length of the harmonic wave (16) is significantly smaller than the wave length of the amplitude change (17). This situation is pictured in Figure 59.4.

Motivated by this argument, we define the propagation velocity of the wave packet as $c_g = \omega'(K_0)$.

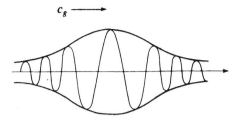

Figure 59.4

59.10. Formulation of a Particle Theory for a Classical Wave Theory

Our starting point is a wave theory with the dispersion relation

$$\omega = \omega(k, x, t) \tag{18}$$

to which we assign the so-called eikonal equation

$$-S_t = \omega(S_x, x, t) \tag{19}$$

and the canonical equations

$$\dot{x} = \omega_k, \qquad \dot{k} = -\omega_x. \tag{20}$$

Here ω has to be replaced with (18). By definition, the particle rays $x = x(t)$ are obtained as a solution of (20).

Let us motivate this general procedure. We will be guided by the geometrical optics of Chapter 40, and begin with the function

$$y = W_0(x, t) e^{iS(x,t)},$$

where S is called an eikonal. In first-order approximation Taylor expansion of S at the point (x_0, t_0) yields

$$S(x, t) = \alpha_0 + k(x - x_0) - \omega(t - t_0) + \cdots$$

with

$$\alpha_0 = S(x_0, t_0), \qquad k = S_x(x_0, t_0), \qquad \omega = -S_t(x_0, t_0). \tag{21}$$

Thus in first-order approximation we obtain the harmonic wave

$$y = W_0 e^{i(k(x - x_0) - \omega(t - t_0) + \alpha_0)} + \cdots.$$

Equations (18) and (21) then yield the eikonal equation (19), which has the form of the Hamilton–Jacobi equations. According to Section 58.23, equations (20) are the corresponding canonical equations.

EXAMPLE 59.4. We consider the important special case

$$\omega = \omega(|k|)$$

which, for example, occurs for light. From (20) we immediately obtain $k = $ const and the particle rays

$$x = c_g \frac{k}{|k|} t + x(0)$$

with $c_g = \omega'(|k|)$. This corresponds to particles which propagate with group velocity c_g.

59.11. Motivation of the Schrödinger Equation and Physical Intuition

> In particular, I would like to mention that I was mainly inspired by the thoughtful dissertation of Mr. Louis de Broglie (Paris, 1924). The main difference here lies in the following. De Broglie thinks of travelling waves, while, in the case of the atom, we are led to standing waves.... I am most thankful to Hermann Weyl with regard to the mathematical treatment of the equation (of the hydrogen atom).
>
> Erwin Schrödinger (1926)

The basic equation of quantum mechanics is the *Schrödinger equation*

$$i\hbar \psi_t = -\frac{\hbar^2}{2m} \Delta \psi + U(x)\psi. \qquad (22)$$

More precisely, Schrödinger (1926) began with the stationary equation

$$E\varphi = -\frac{\hbar^2}{2m} \Delta \varphi + U(x)\varphi \qquad (23)$$

which follows from (22) through

$$\psi(x, t) = \varphi(x) e^{-iEt/\hbar}.$$

From (23) Schrödinger derived the spectrum of the hydrogen atom. This will be discussed in Section 59.16. In this case U is the potential of the electrostatic attracting force of the atomic nucleus, and from the eigenvalues E of (23) one obtains the energy levels of the electron of the hydrogen atom.

We now want to motivate (23) by using an argument in the spirit of Schrödinger's ideas. We begin with the classical energy relation

$$E = \frac{p^2}{2m} + U(x) \qquad (24)$$

for a particle of mass m and with momentum vector p, which is located in a force field with potential U. Formally, we consider a *de Broglie matter wave*

$$\psi = \psi_0 e^{i(kx - \omega t)} \qquad (25)$$

with
$$E = \hbar\omega, \quad p = \hbar k. \tag{26}$$

From (24) we obtain the dispersion relation
$$\hbar\omega = \frac{\hbar^2 k^2}{2m} + U(x). \tag{27}$$

Our goal is the following:

(S) We are looking for a partial differential equation, which has ψ in (25) with (26), (27) as a solution.

Most easily such an equation is obtained by substituting
$$E \Rightarrow i\hbar\frac{\partial}{\partial t}, \quad p \Rightarrow -i\hbar\frac{\partial}{\partial x} \tag{28}$$

in (24). This immediately yields the Schrödinger equation (22).

It is, however, quite surprising that it is possible to obtain such a fundamental equation in such a simple and formal way. The discovery of the Schrödinger equation is an example for the power of physical intuition. It is not the mathematical formalism which plays the important role, but rather the use of physical images and concepts which have been tested in connection with other physical phenomena.

As we shall see in Part V, substitution (28) also directly implies the Schrödinger equation for many-particle systems and the relativistic equation for the electron, i.e., the Dirac equation. In the same way we obtain, in Section 59.24, the Klein–Gordon equation for mesons. Substitution (28) represents a general quantization procedure (first quantization).

59.12. Fundamental Probability Interpretation of Quantum Mechanics

It is remarkable that Schrödinger used his equation (23), without knowing the right interpretation of the complex-valued wave function ψ. This interpretation was discovered by Max Born (1926) in a fundamental paper, in which he used the time-dependent Schrödinger equation (22) to develop a quantum mechanical scattering theory. This statistical interpretation of ψ, which has led to a fundamental change in the way of physical thinking, will now be discussed.

We use the scalar product
$$(\varphi|\psi) = \int_{\mathbb{R}^3} \bar{\varphi}\psi \, dx.$$

In the following only complex-valued functions $\psi = \psi(x, t)$ are considered which belong to the complex H-space $L_2^C(\mathbb{R}^3)$ at all times $t \in \mathbb{R}$, and satisfy the normalization condition

$$(\psi|\psi) = 1.$$

We interpret such functions as the particle state at time t.

(i) *Probability of presence for particles.* The number

$$\int_G |\psi(x, t)|^2 \, dx$$

is the probability of finding a particle at time t in the region G.

(ii) *Expectation value and dispersion of physical quantities in state ψ*. If we choose $A\psi$ either equal to $\xi_j \psi$ or equal to $-i\hbar \, \partial\psi/\partial\xi_j$, then

$$\bar{A} = (\psi|A\psi),$$
$$(\Delta A)^2 = (\psi|(A - \bar{A})^2 \psi) \tag{29}$$

is the expectation value $\bar{A}$ and the dispersion $(\Delta A)^2$ for the position component ξ_j and the momentum component p_j at time t, respectively. Thereby ψ must be chosen at time t. Furthermore, ΔA is called the mean fluctuation or the *mean error* for the physical quantity at state ψ which belongs to A.

This coincides with the correspondence principle

$$\xi_j \Rightarrow \xi_j, \qquad p_j \Rightarrow -i\hbar \, \partial/\partial\xi_j, \tag{30}$$

which has already been used in (28). In (29) one naturally assumes that the right-hand sides exist.

Other than in classical mechanics, a particle state need not correspond to a well-defined position state and momentum state. Instead, only expectation values exist.

59.13. Meaning of Eigenfunctions in Quantum Mechanics

Equations (29) describe a general quantum mechanical principle. Similarly as in (30) one assigns operators A to physical quantities and then computes $\bar{A}$ and ΔA according to (29). The expectation value $\bar{A}$ is called *sharp* if and only if

$$\Delta A = 0.$$

States with sharp expectation values are of particular physical interest.

The mathematical meaning of $\bar{A}$ and ΔA is clear from the *Chebychev*

59.13. Meaning of Eigenfunctions in Quantum Mechanics

inequality

$$p(|A - \bar{A}| \leq \alpha) \geq 1 - \frac{(\Delta A)^2}{\alpha^2} \quad \text{for all} \quad \alpha > 0,$$

where the left-hand side denotes the probability of measuring a value A with $|A - \bar{A}| \leq \alpha$. The simple proof of this may be found in Section 68.1. In particular, if $\Delta A = 0$, then

$$p(A = \bar{A}) = 1,$$

and is the case that $\Delta A > 0$, we obtain, for example,

$$p(|A - \bar{A}| \leq 4\Delta A) \geq \tfrac{15}{16}.$$

In general, the Chebychev inequality tells us that the measurement value A is closer to the mean value $\bar{A}$ the smaller the dispersion $(\Delta A)^2$ is.

The great importance of eigenfunctions and eigenvalues in quantum physics results from the following. If

$$A\psi = \lambda\psi$$

with $(\psi|\psi) = 1$, then, at all times t, the sharp expectation value $\bar{A} = \lambda$ corresponds to the state ψ, since

$$\bar{A} = (\psi|A\psi) = \lambda,$$
$$(\Delta A)^2 = (\psi|(A - \lambda I)^2\psi) = 0.$$

EXAMPLE 59.5 (Hamilton Operator). According to (30) the Hamilton operator

$$H = -\frac{\hbar^2}{2m}\Delta + U$$

is assigned to the classical Hamilton function $\tilde{H} = (p^2/2m) + U$. Since the classical Hamilton function represents the energy of the particle, we interpret $\bar{H}$ as the energy expectation value. If $\varphi = \varphi(x)$, as in the case of the hydrogen atom and the harmonic oscillator below, is an eigenfunction of H with $(\varphi|\varphi) = 1$, i.e., $H\varphi = E\varphi$, then

$$\psi(x, t) = e^{-iEt/\hbar}\varphi(x)$$

is a normalized solution of the Schrödinger equation $i\hbar\psi_t = H\psi$ and we obtain

$$H\psi = E\psi.$$

Thus for the state ψ there exists the sharp expectation value $\bar{H} = E$ at all times t.

EXAMPLE 59.6 (Angular Momentum). We assign the operator $A = N$ to the classical angular momentum $\tilde{N} = x \times p$ with

$$N = x \times p \quad \text{and} \quad p = -i\hbar\, \partial/\partial x.$$

Explicitly, we have $N = \sum_{i=1}^{3} N_i e_i$ and

$$N_3 = \xi_1 p_2 - \xi_2 p_1, \qquad p_j = -i\hbar \, \partial/\partial \xi_j,$$

where N_1 and N_2 are found from cyclic permutations. The operator of the square of the angular momentum is given by $N^2 = N_1^2 + N_2^2 + N_3^2$. In spherical coordinates this means

$$N_3 = -i\hbar \, \partial/\partial \varphi,$$

$$N^2 = -\frac{\hbar^2}{\sin \vartheta} \frac{\partial}{\partial \vartheta} \left(\sin \vartheta \frac{\partial}{\partial \vartheta} \right) + \frac{N_3^2}{\sin^2 \vartheta}.$$

For the Hamilton operator of Example 59.5 we obtain

$$H = \frac{1}{2m} \left[\frac{N^2}{r^2} - \frac{\hbar^2}{r^2} \frac{\partial}{\partial r} \left(r^2 \frac{\partial}{\partial r} \right) \right] + U.$$

These expressions are important for the treatment of the hydrogen atom.

Let Y_l^m denote the surface harmonics which are discussed in the following section. Then, in spherical coordinates, we obtain for every differentiable function

$$\psi(x, t) = R(r, t) Y_l^m(\varphi, \vartheta),$$

the relations

$$N^2 \psi = \hbar^2 l(l + 1) \psi,$$

$$N_3 \psi = \hbar m \psi,$$

(31)

i.e., for these states the square of the angular momentum N^2 has the sharp expectation value $\hbar^2 l(l + 1)$ and N_3 has the sharp expectation value $\hbar m$, where $l = 0, 1, \ldots$ and $m = l, l - 1, \ldots, -l$. In this way one obtains the quantization of the angular momentum, which is important for the understanding of so many physical processes.

59.14. Meaning of Nonnormalized States

For the plane wave

$$\psi = \psi_0 e^{i(kx - \omega t)}$$

with $E = \hbar \omega$ and $p = \hbar k$ the previous interpretation fails, since $(\psi | \psi) = \infty$. The interpretation now is as follows. One uses ψ in scattering experiments and assigns to this function the current density vector

$$j = \rho v$$

of a parallel particle current with density ρ and velocity vector v, where

$$\rho = |\psi_0|^2, \qquad v = p/m.$$

During the time interval $[0, t]$ exactly N particles with

$$N = |j|Ft$$

penetrate a surface F perpendicular to the direction of propagation. The physical meaning of current density vectors will be discussed more thoroughly in Section 69.1 (cf. also Chapter 87).

59.15. Special Functions in Quantum Mechanics

In preparation of the quantum mechanical treatment of the harmonic oscillator and the hydrogen atom, we consider the following special functions of the real variable ξ. Moreover, we set $D = d/d\xi$.

(i) *Hermitian functions (harmonic oscillator functions)*

$$H_n(\xi) = \frac{(-1)^n}{\sqrt{2^n n! \sqrt{\pi}}} e^{\xi^2/2} D^n e^{-\xi^2}.$$

For $n = 0, 1, \ldots$ they form a complete orthonormal system in $L_2(-\infty, \infty)$ and solve the differential equation

$$-y'' + \xi^2 y = (2n + 1)y.$$

(ii) *Legendre polynomials*

$$P_n(\xi) = \sqrt{\frac{2n+1}{2^{2n+1}(n!)^2}} D^n (1 - \xi^2)^n.$$

For $n = 0, 1, \ldots$ they form a complete orthonormal system in $L_2(-1, 1)$ and solve the differential equation

$$-((1 - \xi^2) y')' = n(n + 1) y.$$

(iii) *Generalized Legendre polynomials*

$$P_l^m(\xi) = \sqrt{\frac{(l - m)!}{(l + m)!}} (1 - \xi^2)^{1/2} D^m P_l(\xi).$$

For fixed $m = 0, 1, \ldots$ and $l = m, m + 1, \ldots$ they form a complete orthonormal system in $L_2(-1, 1)$ and solve the differential equation

$$-((1 - \xi^2) y')' + m^2 (1 - \xi^2)^{-1} y = l(l + 1) y.$$

(iv) *Surface harmonics*

$$Y_l^m(\varphi, \vartheta) = \frac{1}{\sqrt{2\pi}} P_l^{|m|}(\cos \vartheta) e^{im\varphi}.$$

For $l = 0, 1, \ldots$ and $m = l, l - 1, \ldots, -l$ they form a complete orthonormal system in $L_2(S^2)$, where S^2 is the surface of the unit ball in $\mathbb{R}^3$ and φ, ϑ denote spherical coordinates. These functions solve the differential equations (31).

(v) *Laguerre functions*

$$L_n^\alpha(\xi) = c_n^\alpha e^{\xi/2} \xi^{-\alpha/2} D^n(e^{-\xi}\xi^{n+\alpha}).$$

The positive constants c_n^α are chosen such that

$$\int_0^\infty (L_n^\alpha)^2 \, d\xi = 1.$$

For fixed $\alpha > -1$ and $n = 0, 1, \ldots$, these functions form a complete orthonormal system in $L_2(0, \infty)$ and solve the differential equation

$$-4(\xi y')' + (\xi + \alpha^2/\xi)y = 2(2n + 1 + \alpha)y.$$

The proofs for this may be found in Triebel (1972, M).

59.16. Spectrum of the Hydrogen Atom

The spectrum of the hydrogen atom consists of discrete lines with wave lengths $\lambda_{mn} = 2\pi c/\omega_{mn}$ and corresponding angular frequencies

$$\omega_{mn} = R\left(\frac{1}{n^2} - \frac{1}{m^2}\right), \tag{32}$$

where $m, n = 1, 2, \ldots$ and $m > n$. This was empirically discovered by Balmer in 1885. Later on Rydberg experimentally determined the so-called Rydberg frequency $R = 2.07 \cdot 10^{16}$ s^{-1}. Thus the problem was to find a theoretical explanation for this surprisingly simple relation.

The first step was taken by Niels Bohr (1913). The hydrogen atom consists of a proton with charge $|e|$ and an electron with charge $e = -1.602 \cdot 10^{-19}$ A s. Bohr postulated that the electron can move only on discrete circular orbits (Fig. 59.5). He determined these orbits similarly as in the Kepler problem for the planetary motion, by replacing Newton's gravitational law with Coulomb's law of electrostatics. Moreover, he added the important nonclassical condition that the angular momentum has the quantized form

$$|N| = n\hbar, \qquad n = 1, 2, \ldots.$$

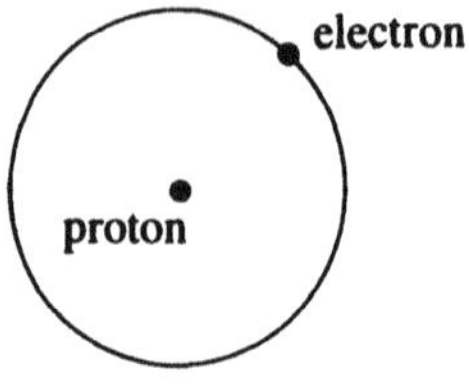

Figure 59.5

59.16. Spectrum of the Hydrogen Atom

This led to the orbital energies

$$E_n = -\frac{\gamma}{n^2}, \quad n = 1, 2, \ldots \tag{33}$$

with

$$\gamma = e^4 m_e / 8\varepsilon_0^2 h^2,$$

where $m_e = 9.1 \cdot 10^{-31}$ kg is the mass of the electron and ε_0 is the dielectric constant. This gives

$$E_1 = -13.6 \text{ eV}.$$

With this energy the electron of the lowest orbit is bound to the nucleus. This energy is needed in order to ionize the hydrogen atom, i.e., to eject the electron. Moreover, this gives roughly the order of magnitude for the energies which occur in chemical reactions per atom. For the radius of the lowest orbit Bohr obtained

$$r_0 = 4\pi\varepsilon_0 \hbar^2 / m_e e^2 = 5 \cdot 10^{-11} \text{ m}.$$

The velocity of the electron in the lowest orbit $n = 1$ is

$$V_0 = c(e^2 / 4\pi\varepsilon_0 \hbar c) = c/137.$$

This velocity is small compared with the velocity of light c. This explains why in this problem relativistic effects can be neglected. Moreover, in Section 59.26, we show that gravitational effects play no role. We will prove all these relations above in Problem 59.1.

A comparison of (33) and (32) yields the simple relation

$$\omega_{mn} = (E_m - E_n)/\hbar \tag{34}$$

with the correct, i.e., experimentally observed, Rydberg frequency R in (32). This fundamental relation of atom spectroscopy can be explained like this. Through outer stimulation the electron is caused to jump into a higher orbit and thereby emits the energy difference ΔE in the form of a photon of energy

$$\Delta E = \hbar \omega_{mn}.$$

This precisely corresponds to Einstein's photon hypothesis.

The fact, however, that the electron should travel in fixed stable orbits was not clear at all, because then, according to Maxwell's theory, the electron should radiate as an accelerated charge and thereby lose energy and consequently tumble into the nucleus. The solution to this famous problem was found, independently, by Pauli and Schrödinger in 1926. Pauli thereby used Heisenberg's matrix mechanics of 1926, which corresponded to a particle quantization, while Schrödinger used his wave equation.

We now discuss the hydrogen atom by using the Schrödinger equation

$$i\hbar \psi_t = -\frac{\hbar^2}{2m_e} \Delta \psi + U\psi. \tag{35}$$

As potential we choose the potential of the Coulomb force between the nucleus and the electron, i.e.,

$$U = -\frac{e^2}{4\pi\varepsilon_0 r}.$$

In spherical coordinates we consider the functions

$$\psi = e^{-iE_n t/\hbar}\varphi_{nlm}(r,\varphi,\vartheta) \tag{36}$$

with

$$\varphi_{nlm} = \frac{1}{r}\sqrt{\frac{2}{nr_0}} L_{n-l-1}^{2l+1}\left(\frac{2r}{nr_0}\right) Y_l^m(\varphi,\vartheta),$$

where E_n and r_0 have the same meaning as in Bohr's atomic model. Moreover, we have $n = 1, 2, \ldots, l = 0, 1, \ldots, n-1$ and $m = l, l-1, \ldots, -l$.

From Section 59.15 one finds through explicit computations that ψ is a normalized solution of (35) with

$$H\psi = E_n\psi,$$
$$N^2\psi = \hbar^2 l(l+1)\psi,$$
$$N_3\psi = \hbar m\psi,$$

where the Hamilton operator H corresponds to the right-hand side of (35). Therefore ψ corresponds to an electron state, which has the sharp enery value E_n for all times t, the sharp value $\hbar^2 l(l+1)$ for the square of the angular momentum and the sharp value $\hbar m$ of the N_3-angular momentum component. The numbers n, l, and m are called quantum numbers. They completely characterize the states (36). Experimentally, these quantum numbers can be observed in the spectrum if U is perturbed by an electric or magnetic field. One then obtains energy values E_{nlm}, which are perturbations of E_n and depend on l and m. For the spectrum this leads to a splitting of the spectral lines, according to (34). In Part V we discuss this by using group-theoretical methods.

In quantum mechanics the classical electron orbits vanish. The number

$$w = \int_G |\psi(x,t)|^2\,dx$$

is the probability of finding the electron at time t in the region G. We choose ψ as in (36). Then this probability is time independent. For $n = 1$ and $G = \{x \in \mathbb{R}^3 : r \leq |x| \leq R\}$ we have

$$w = \int_r^R W(r)\,dr.$$

The function $W(r) = (rr_0/2)^2 e^{-2r/r_0}$ has a maximum for $r = r_0$. Therefore, roughly speaking, for $n = 1$ the probability of a presence for the electron is maximal in the neighborhood of Bohr's classical electron orbit.

The energy values E_n are all negative. For every $E > 0$ one finds solutions of the Schrödinger equation with

$$\psi = e^{-iEt/\hbar}\varphi(x)$$

and $H\varphi = E\varphi$ by using Gauss' hypergeometrical function (see Landau and Lifšic (1962, M), §36). These, however, cannot be normalized. They correspond to free electrons in the field of the hydrogen nucleus. Using the picture of the planetary motion, $E_n < 0$ corresponds to orbital ellipses (planets) and $E_n > 0$ corresponds to orbital hyperbolas (comets).

59.17. Functional Analytic Treatment of the Hydrogen Atom

Quantum mechanics has had a significant influence on the development of linear functional analysis, especially the spectral theory of unbounded, self-adjoint operators, which in 1929 was created by John von Neumann. In 1932 his classical monograph "The Mathematical Foundations of Quantum Mechanics" appeared. Interestingly, in his spectral theory of bounded, self-adjoint operators at the turn of the century, Hilbert intuitively used the notion of the spectrum, without knowing that this mathematical concept was closely related to the theory of atomic spectra.

The functional analytic treatment, which will now be discussed, allows us, among others things, to show in which sense the explicit solutions of the previous section describe *all* solutions. We use the complex H-space $X = L_2^\mathbb{C}(\mathbb{R}^3)$. It consists of all complex-valued, measurable functions $\varphi \colon \mathbb{R}^3 \to \mathbb{C}$ with scalar product

$$(\varphi|\psi) = \int_{\mathbb{R}^3} \bar\varphi \psi \, dx.$$

Every $\varphi \in X$ with $(\varphi|\varphi) = 1$ will be interpreted as a stationary state of the electron. The Hamilton operator

$$H = -\frac{\hbar^2}{2m_e}\Delta + U$$

of the hydrogen atom is a self-adjoint, half-bounded operator, whose domain $D(H)$ is the Sobolev space $W_2^2(\mathbb{R}^3)$ which is dense in $L_2^\mathbb{C}(\mathbb{R}^3)$. The eigenvalues of H are precisely all E_n. The corresponding eigenvectors φ_{nlm} form a complete orthonormal system in a subspace X_0 of X. Every $\varphi \in X_0$ with $(\varphi|\varphi) = 1$ can therefore be written as

$$\varphi = \sum_{nlm} c_{nlm} \varphi_{nlm}.$$

This series converges in X, where

$$1 = \sum_{nlm} |c_{nlm}|^2.$$

We interpret φ as the electron state, in which the electron will be in φ_{nlm} with probability $|c_{nlm}|^2$. The evolution in time of φ is given by

$$\psi = e^{-itH/\hbar}\varphi, \tag{37}$$

i.e., $\psi = \varphi$ for $t = 0$. The exponential function is here to be understood in the sense of functional calculus. Formal differentiation of (37) yields the Schrödinger equation

$$i\hbar\psi_t = H\psi, \tag{38}$$

which, however, will not be used for our development of the theory. The reason is the following. While (37) is meaningful for all initial states $\varphi \in X$, equation (38) only holds for all ψ with $\varphi \in D(H)$. Also the operator H has the continuous spectrum $[0, \infty[$, which corresponds to the free electron of the previous section. All proofs of this can be found in Triebel (1972, M).

59.18. Harmonic Oscillator in Quantum Mechanics

Parallel to the classical harmonic oscillator of Section 58.5 we consider here the one-dimensional Schrödinger equation

$$i\hbar\psi_t = -\frac{\hbar^2}{2m}\psi_{\xi\xi} + U\psi \tag{39}$$

with the potential of the harmonic oscillator $U = m\omega^2\xi^2/2$. This equation follows from (22) if we restrict ourselves to only one space coordinate, i.e., if we are looking for $\psi = \psi(\xi, t)$. With regard to the probability interpretation of Section 59.12 one has to replace then all space integrals with integrals over $\mathbb{R}$.

We consider the function

$$\psi = e^{-iE_n t/\hbar} H_n(\xi/\xi_0)/\sqrt{\xi_0} \tag{40}$$

with $n = 0, 1, \ldots, \xi_0 = \sqrt{\hbar/m\omega}$ and

$$E_n = \hbar\omega(n + \tfrac{1}{2}). \tag{41}$$

With results of Section 59.15 one explicitly verifies that these are solutions of (39). The quantized energy values E_n have been computed for the first time by Heisenberg (1925) using his matrix mechanics (see Example 58.26). These values correspond to Planck's quantum hypothesis $\Delta E = \hbar\omega$ of 1900.

According to Section 59.12 one obtains the expectation values

$$\bar{\xi} = 0 \quad \text{and} \quad \bar{p} = 0$$

for the position and momentum component of (40). The mean error is

$$\Delta\xi = \xi_0\sqrt{(1 + 2n)/2},$$

and
$$\Delta p \, \Delta \xi = \hbar(1 + 2n)/2 \geq \hbar/2.$$

This is a special case of Heisenberg's uncertainty relation of the following section.

The functional analytic treatment is similar to the previous section. We choose the complex H-space $X = L_2^C(\mathbb{R})$. The Hamilton operator

$$H = -\frac{\hbar^2}{2m}\frac{d^2}{d\xi^2} + U$$

with domain $D(H) = C_0^\infty(\mathbb{R})$ is essentially self-adjoint, i.e., it has a self-adjoint closure $\bar{H}$. This operator has precisely all E_n as eigenvalues. The functions ψ in (40) with $t = 0$ are the corresponding eigenvectors in X and form a complete orthonormal system. The proof can be found in Triebel (1972, M).

59.19. Heisenberg's Uncertainty Relation

We begin with the proof of a general functional analytic result and start from the commutation relation

$$AB\psi - BA\psi = iC\psi \qquad \text{for all} \quad \psi \in D. \tag{42}$$

Our goal is

$$2\Delta A \, \Delta B \geq |\bar{C}|. \tag{43}$$

Our assumptions are:

(H1) X is a complex H-space with a dense linear subspace D.
(H2) The operators $A, B, C: D \to D$ are linear and symmetric, i.e., $(A\psi|\psi) = (\psi|A\psi)$ for all $\psi \in D$, etc.

We assign the following numbers to each $\psi \in D$:

$$\bar{A} = (\psi|A\psi), \qquad (\Delta A)^2 = (\psi|(A - \bar{A})^2\psi) = \|(A - \bar{A})\psi\|^2$$

and analogously for B, C.

Theorem 59.A (Abstract Uncertainty Relation). *If* (H1), (H2) *and* (42) *hold, then* (43) *is valid for every* $\psi \in D$ *with* $(\psi|\psi) = 1$.

PROOF. For $a, b \in \mathbb{R}$ and $\psi \in D$ one has

$$i\bar{C} = (\psi|(AB - BA)\psi)$$
$$= (\psi|(A - aI)(B - bI)\psi) - ((A - aI)(B - bI)\psi|\psi)$$
$$= 2i \, \text{Im}(\psi|(A - aI)(B - bI)\psi).$$

The Schwarz inequality yields

$$|\bar{C}| \leq 2\|(A - aI)\psi\| \, \|(B - bI)\psi\|.$$

For $a = \bar{A}$ and $b = \bar{B}$ we obtain (43). □

EXAMPLE 59.7 (Uncertainty of Momentum and Position). We choose

$$X = L_2^{\mathbb{C}}(\mathbb{R}^3) \quad \text{and} \quad D = C_0^\infty(\mathbb{R}^3).$$

By $\psi \in X$ we mean $\psi = \psi(x, t)$ for fixed t. Moreover, let A be the momentum operator

$$p_j = -i\hbar \, \partial/\partial \xi_j,$$

and B the position operator ξ_j. This implies the commutation relation (42) with $C = -\hbar I$, i.e., in short

$$p_j \xi_j - \xi_j p_j = -i\hbar I. \tag{44}$$

From Theorem 59.A we obtain the famous *Heisenberg uncertainty relation* from 1927:

$$\Delta p_j \, \Delta \xi_j \geq \hbar/2. \tag{45}$$

Since D is dense in $W_2^1(\mathbb{R}^3)$, relation (45) is valid also for all normalized $\psi \in W_2^1(\mathbb{R}^3)$ by passing to limits.

The fundamental relation (45) means that for normalized quantum mechanical states, position component and momentum component cannot exactly be measured at the same time. There will always be a mean error which satisfies (45).

According to Section 59.14 we assign the energy $E = \hbar\omega$ and the sharp momentum $p = \hbar k$ to the state

$$\psi = \psi_0 e^{i(kx - \omega t)}.$$

The impossibility of the normalization $(\psi|\psi) = 1$ can now be interpreted as follows. Because of the sharp momentum it is not possible to localize the free particle. Because of $(\psi|\psi) = \infty$ it is not possible to define a probability for the particle to be in the region G.

EXAMPLE 59.8 (Uncertainty of the Angular Momentum Components). We choose X and D as in the previous example. Let N_j denote the jth component of the angular momentum operator $N = x \times p$ with $p = -i\hbar \, \partial/\partial x$. Then the commutation rule (42) holds with $A = N_1$, $B = N_2$, $C = \hbar N_3$, in short

$$N_1 N_2 - N_2 N_1 = i\hbar N_3. \tag{46}$$

This implies

$$\Delta N_1 \, \Delta N_2 \geq \hbar |\bar{N}_3|/2$$

for all normalized $\psi \in W_2^1(\mathbb{R}^3)$. In 1925, during the formulation of his matrix mechanics, Heisenberg noticed that the commutation rules form the essential parts of quantum mechanics. Mathematicians were aware of such commutation rules for a long time in connection with the theory of Lie algebras (see Section 74.26). In fact, behind (46) hides the Lie algebra of the rotation group SO(3) which is responsible for the angular momentum, and also the Lie algebra of the group SU(2) which is responsible for spin and isospin of the elementary particles. As we shall see in Part V, one can therefore obtain important results in quantum theory by using groups and their representation theory. For example, the quark model for elementary particles follows from the representation theory of the group SU(3).

59.20. Pauli Principle, Spin and Statistics

We assign a spin quantum number S to all elementary particles which may assume the values $S = n/2$ with $n = 0, 1, \ldots$. Furthermore, we distinguish between the following spin positions

$$S_z = S, S - 1, \ldots, -S$$

for a fixed elementary particle. Physically, this means that the particle has a spin. Specifying a fixed e_3-axis it may assume states for which the spin vector

$$\vec{S} = \sum_{i=1}^{3} S_i e_i$$

has the sharp value $\hbar S_z$ for S_3 and the sharp value $\hbar^2 S(S + 1)$ for $\vec{S}^2$.

In 1925, the electron spin was hypothetically introduced by Goudsmit and Uhlenbeck, in order to explain the fine structure in the splitting of the spectral lines caused by a magnetic field. Experimentally, this then was confirmed by Gerlach and Stern in 1927. They sent hydrogen atoms through an inhomogeneous magnetic field. If the electron in this experiment is in the ground state of the hydrogen atom, then one has $n = 1$, $l = 0$, and $m = 0$. It follows that $N_3 = 0$. According to classical views of electrodynamics, such an atom without electron angular momentum cannot have a magnetic dipole moment. The ray therefore should not be effected by the magnetic field. But, actually, Gerlach and Stern observed a splitting into two particle rays. This corresponds to an electron spin of $S = \frac{1}{2}$ with both spin positions $S_z = \pm\frac{1}{2}$. In Part V we will show that the electron spin is a typical effect in the theory of relativity and automatically follows from the relativistic equation of the electron, which in 1928 was formulated by Dirac. Moreover, this equation also implies the existence of the positron, the antiparticle of the electron.

For photons, one obtains $S = 1$ with the two spin positions $S_z = \pm 1$. The possible value $S = 0$ does not occur. This follows from the fact that light corresponds to transversal waves. In Section 68.4 we will derive the correct radiation law by making essential use of the fact that photons can only have two spin positions.

Particles with integer spin (or half-numberly) spin are called bosons (or fermions). The electron is a fermion, the photon is a boson. The following principle is a basic natural law.

Pauli Principle 59.9. *In a system of fermions two particles can never be in the same quantum state.*

In Chapter 68 we show how from this principle follows that bosons (resp. fermions) satisfy the Bose statistics (resp. Fermi statistics). As we shall see this yields, e.g., the radiation law, the critical mass for white dwarfs, as well as results about the structure of the cosmos following the Big Bang. In the context of axiomatic quantum field theory, this principle can also be mathematically deduced from suitable axioms. This may be found in Streater and Wightman (1964, M).

59.21. Quantization of the Phase Space and Statistics

In Chapter 68 we consider quantum statistics. There we use, besides the Pauli principle, another very successful general principle. In order to explain this, we consider in a fixed region G of $\mathbb{R}^3$, with volume $V(G)$, elementary particles which are of the same kind. Let p denote the vector of the particle momentum. Moreover, let $\Omega = G \times G_p$ be a region in the six-dimensional phase space. A point $(x, p) \in \Omega$ describes the state of a particle with position $x \in G$ and momentum $p \in G_p$.

Principle of Quantization of the Phase Space 59.10. *The maximal number of particles which can be in Ω is*

$$N = gV(\Omega)/h^3.$$

Thereby g is the number of possible spin positions of a particle described by the spin quantum number S_z.

For example, it is $g = 2$ for electrons and photons. One can also write

$$N = \frac{g}{h^3} \int_\Omega dx\, dp = \frac{gV(G)}{h^3} \int_{G_p} dp.$$

This principle can be extended to arbitrary $2n$-dimensional phase spaces. Intuitively, it states that in a cell of volume h^n there can never be two particles at the same time, having the same quantum state.

We shall use Principle 59.10 in the form of a postulate. We want to motivate this principle. As in Example 58.6, we consider the classical harmonic oscillator with motion

$$q = C\sin(\omega t + \alpha)$$

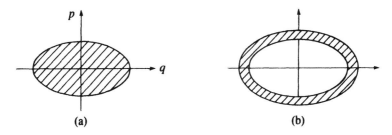

Figure 59.6

and momentum
$$p = m\dot{q} = m\omega C \cos(\omega t + \alpha).$$

The energy of this motion is
$$E = \frac{p^2}{2m} + \frac{m}{2}\omega^2 q^2 = \frac{m}{2}C^2\omega^2.$$

In the (q, p)-phase space, this motion corresponds to an ellipse which is the boundary of a region with measure
$$\int dq\, dp = \pi m\omega C^2 = 2\pi E/\omega$$

(Fig. 59.6(a)). All motions with an energy $E \in [E_0, E_0 + \Delta E]$ cover the surface
$$\int dq\, dp = 2\pi \Delta E/\omega$$

(Fig. 59.6(b)). If according to the principle above we set this surface equal to h, then we obtain
$$\Delta E = \hbar\omega. \tag{47}$$

This is precisely Planck's condition for the quantization of the energy of the harmonic oscillator. We therefore may think of Principle 59.10 as a generalization of (47).

59.22. Pauli Principle and the Periodic System of the Elements

In 1869, great progress was made in chemistry, when, independently, Mendelejev and Meyer were able to systematically order the chemical elements according to phenomenological criteria. Table 59.1 shows the beginning of the periodic system. In horizontal direction the atomic number Z increases, while in vertical direction elements one below the other behave similarly. In 1925, Pauli formulated his Principle 59.10, in order to explain the shell struc-

Table 59.1

1 H 1s							2 He $1s^2$ = K
3 Li K 2s	4 Be K $2s^2$	5 B K $2s^2$, $2p$	6 C K $2s^2$, $2p^2$	7 N K $2s^2$, $2p^3$	8 O K $2s^2$, $2p^4$	9 F K $2s^2$, $2p^5$	10 Ne K $2s^2$, $2p^6$ = L
11 Na K L 3s	12 Mg K L $3s^2$	13 Al K L $3s^2$, $3p$	14 Si K L $3s^2$, $3p^2$	15 P K L $3s^2$, $3p^3$	16 S K L $3s^2$, $3p^4$	17 Cl K L $3s^2$, $3p^5$	18 Ar K L $3s^2$, $3p^6$ = M

ture of the atoms, in the context of Bohr's atomic model and its further development by Sommerfeld. This was even before the discovery of quantum mechanics. Thereby one roughly obtains the following picture.

(i) The atomic number Z is equal to the number of electrons and equal to the number of protons in the nucleus.
(ii) The number of neutrons in the nucleus may vary for fixed Z. Thereby isotopes occur.
(iii) An electron state is characterized by four quantum numbers $n = 1, 2, \ldots$ (orbit), $l = 0, 1, \ldots, n - 1$ (angular momentum), $m = l, l - 1, \ldots, -l$ and $S_z = \pm\frac{1}{2}$ (spin). Two electrons cannot coincide in all four quantum numbers (Pauli in 1925 did not use the spin).
(iv) For energetic reasons the orbits with $n = 1, 2, \ldots$ are filled successively.
(v) The chemical behavior is mainly determined by the outer electrons. The similarity of chemical elements is a consequence of the same number of the outer electrons. Table 59.1 shows the number of electrons in different orbits. In horizontal direction, new electrons are added continuously. Thereby s and p stand for $l = 0$ and $l = 1$, respectively. Furthermore, ns^k means that the number of s-electrons in the nth orbit is equal to k. The maximal number of s-electrons in one orbit is equal to 2 because of

$$l = 0, \quad m = 0, \quad S_z = \pm\frac{1}{2}.$$

The maximal number of p-electrons in one orbit is equal to six because of

$$l = 1, \quad m = -1, 0, 1 \quad \text{and} \quad S_z = \pm\frac{1}{2}.$$

In the vertical lines of Table 59.1 we have the same number of outer s- and p-electrons. This results in a similar chemical behavior of the corresponding elements. The inert gases 2 He, 10 Ne, and 18 Ar have only closed shells denoted by K, L, M, respectively. This is the reason for their chemical inactivity.

Starting with calium with $Z = 19$, the irregularities in the filling of the orbits begin. This follows from the fact that for large Z the electron interaction becomes stronger and the energetic situations more complicated.

For a mathematical treatment in the context of quantum mechanics one has to consider the Schrödinger equation for many-particle systems (see Part V). The Pauli principle then corresponds to the fact that only such wave functions are allowed, which are skew-symmetric with respect to the space variables and the spin variables. The solution of many-electron equations, taking the electron interaction into account, is only approximately possible by using Ritz' method for eigenvalue problems (see Chapter 22). The situation becomes even more complicated for molecular calculations. Today, with the fastest computers, one can only approximately compute relatively small molecules by using the Schrödinger equation. The numerical treatment of large molecules is presently impossible. The main problem in quantum chemistry is to find more effective methods and to build faster computers.

59.23. Classical Limiting Case of Quantum Mechanics and the WKB Method to Compute Quasi-Classical Approximations

As in Section 59.10 we are looking for the solution of the Schrödinger equation (22) in the form

$$\psi = e^{iS/\hbar}.$$

From (22) we obtain the eikonal equation

$$S_t + \frac{S_x^2}{2m} + U - \frac{i\hbar^2}{2m}\Delta S = 0. \tag{48}$$

For $\hbar = 0$ this is the Hamilton–Jacobi equation, which corresponds to the classical energy

$$H(q, p) = \frac{p^2}{2m} + U.$$

The method of Wentzel, Kramers, and Brillouin (WKB method) consists in writing S as an expansion of the form

$$S = S_0 + \hbar S_1 + \hbar^2 S_2 + \cdots$$

and to determine the S_k successively from (48). This is a very effective method of determining quasi-classical approximate solutions of the Schrödinger equation. Note that $\hbar = 1.054 \cdot 10^{-34}$ J s is very small. Actually, one has to use dimensionless quantities in order to determine the smallness of the perturbations in concrete problems. The 0th approximation S_0 is the solution of (48) with $\hbar = 0$ (classical Hamilton–Jacobi equation).

This method illustrates the fact that quantum mechanics becomes classical mechanics as $\hbar \to 0$.

As we have already discussed in Problem 40.11, Maslov succeeded (a few years ago) in developing a global form of this method by finding a procedure to continue the asymptotic expansion beyond singularities (caustics).

59.24. Energy–Time Uncertainty Relation and Elementary Particles

The main difference between classical mechanics and quantum mechanics is that the position component ξ_j and the momentum component p_j cannot be precisely determined at the same time. According to Section 59.19 we have the following estimate for the mean errors $\Delta\xi_j$ and Δp_j:

$$\Delta\xi_j \Delta p_j \geq \hbar/2. \tag{49}$$

If the energy of the quantum system is not sharply determined, then, analogously to (49), we postulate the so-called *energy–time uncertainty relation*

$$\Delta E \, \Delta t \geq \hbar/2. \tag{50}$$

This inequality states that during the time Δt the energy E can only be determined up to a mean error of ΔE, whereby (50) holds. As a motivation, physicists point to the fact that in the special theory of relativity, position x and momentum p are replaced by the contravariant four vectors (x, ct) and $(p, E/c)$ (see Section 75.11). Hence not only x and p, but also ct and E/c correspond to each other. In the sense of this correspondence, (50) follows from (49).

If one considers concrete physical processes, then one often finds, in place of the inequality (49), a relation

$$\Delta\xi_j \Delta p_j \sim \hbar, \tag{51}$$

i.e., the product on the left-hand side has an order of magnitude of $\hbar$. Analogously, one has

$$\Delta E \, \Delta t \sim \hbar. \tag{52}$$

The following two examples should illustrate this more precisely.

EXAMPLE 59.11 (Life-Time of Particles Which are Generated in Accelerators). Reactions in particle accelerators are described by writing the effective cross sections, which are determined experimentally, as a function of the particle energy E. Thereby one often observes resonances such as in Figure 59.7. One refers to them as excited quantum states (one particle or several bound particles) with energy $2\Delta E$, mass $m = 2\Delta E/c^2$ and a life-time

$$\Delta t = \hbar/2\Delta E. \tag{53}$$

This formula is motivated by our observations about quasi-stationary pro-

59.24. Energy–Time Uncertainty Relation and Elementary Particles

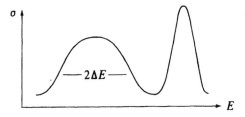

Figure 59.7

cesses (damped oscillations) of Section 59.5. There we saw in (10) that the intensity of a damped oscillation is mainly restricted to frequencies $\omega \in [\omega_0 - \Delta\omega, \omega_0 + \Delta\omega]$ with

$$2\Delta\omega \, \Delta t = 1.$$

The energy formula (53) is obtained from the relation

$$E = \hbar\omega$$

for the de Broglie matter waves.

EXAMPLE 59.12 (Field Quantums and π-Mesons). We want to explain the physical ideas behind Yukawa's meson theory for the nuclear forces. Our starting point is the relativistic equation

$$E^2 = m_0^2 c^4 + p^2 c^2 \tag{54}$$

between energy E, rest mass m_0, and momentum vector p of a free particle. Here c is the velocity of light. In order to obtain a corresponding quantum mechanical wave equation, we use the same quantization procedure as in Section 59.11, i.e., we introduce the substitution

$$E \Rightarrow i\hbar \frac{\partial}{\partial t}, \qquad p \Rightarrow -i\hbar \frac{\partial}{\partial x},$$

and obtain from (54) the so-called Klein–Gordon equation

$$\left(\frac{1}{c^2} \frac{\partial^2}{\partial t^2} - \Delta + \mu^2 \right) \psi = 0 \tag{55}$$

with

$$\mu = m_0 c / \hbar.$$

This equation has the radially symmetric, stationary solution

$$\psi(x) = Q\gamma \frac{e^{-\mu r}}{r} \tag{56}$$

with $r = |x|$. For a suitable choice of the constants Q and γ this, actually, is a solution of the equation

$$(-\Delta + \mu^2)\psi = \delta,$$

in the sense of distributions where δ is Dirac's delta distribution, which is concentrated at the origin (see $A_2(62)$).

We now want to develop the theory of nuclear forces analogously to classical electrodynamics. In Maxwell's theory, which will be discussed in Part V, equation (55) with $\mu = 0$ is the differential equation for the electrical potential ψ. The mechanical potential U of the electrostatic Coulomb force, which is applied from one charge Q to another charge Q_1, is equal to

$$U = Q_1 \psi = Q_1 Q \gamma \frac{e^{-\mu r}}{r} \tag{57}$$

with $\mu = 0$ and $\gamma = 1/4\pi\varepsilon_0$, according to Section 58.8b. An atomic nucleus consists of nucleons, i.e., uncharged neutrons and protons with positive elementary charge. One assumes that the atomic nucleus is kept together by strong nuclear forces which exist between the nucleons. Scattering experiments show that this nuclear force has an extremely short radius of action which is approximately the nuclear radius, i.e.,

$$R = 1.4 \text{ fm} = 1.4 \cdot 10^{-15} \text{ m.} \tag{58}$$

This nuclear radius was determined by Ernest Rutherford in 1911. He used scattering experiments with α-particles (helium nucleuses). In order to describe the nuclear force in analogy to the electrostatic force, we use (57) with free parameters Q_1, Q, and γ. The radius of action of the force, corresponding to the potential U, is defined as $R = 1/\mu$. This is motivated by the fact that $e^{-\mu r}$ has already greatly decreased for $r > 1/\mu$. Together with (58) we obtain the fundamental relation

$$R = 1/\mu = \hbar/m_0 c = 1.4 \cdot 10^{-15} \text{ m.} \tag{59}$$

Formula (57) represents the so-called Yukawa potential. In the case of the electromagnetic force we have $\mu = 0$. Physically, this corresponds to the fact that, as in the case of the gravitational force, the electrostatic Coulomb force has an infinite radius of action R.

So far, we have only considered the forces themselves, without looking for the mechanism through which these forces are transformed. This will be done right now in the context of the idea of exchange forces. In classical electrodynamics one assumes that electromagnetic interactions are transmitted through the electromagnetic field. In quantum field theory, on the other hand, one starts from the assumption that this transmission is done by field quanta which are different for every interaction. The field quantum for the electromagnetic interaction is the photon. According to Einstein's photon hypothesis of Section 59.3, we obtain for the photon $E = c|p|$. A comparison with (54) yields $m_0 = 0$. This coincides with our observation above. There we have $\mu = 0$, and hence $m_0 = 0$. We now pose the question: Which field quantum transmits the nuclear force? To answer this question we interpret the mass m_0 in (55) as the rest mass of the field quantum. From (59) we obtain the rest

59.24. Energy–Time Uncertainty Relation and Elementary Particles

energy

$$E_0 = m_0 c^2 = \hbar c/R = 2 \cdot 10^{-11} \text{ J} = 130 \text{ MeV}.$$

Therefore m_0 corresponds to about 260 rest masses of the electron. The life-time of the field quantum can be computed from (53) as

$$\Delta t_0 = \hbar/E_0 = 0.5 \cdot 10^{-23} \text{ s}. \tag{60}$$

In general, the life-time of an unstable particle gets smaller, the greater the energy gets. In 1935, Yukawa formulated the hypothesis that the nuclear force is transmitted through mesons with the rest mass above. Actually, in 1947, such particles were discovered in cosmic rays and then, in 1948, artificially generated in laboratories at Berkeley. Until 1950, the mesons π^+, π^0, and π^- with corresponding charges $|e|$, 0, and e were discovered in accordance with Yukawa's prediction. In 1949, Yukawa received the Nobel prize for his meson theory.

We shall now look at a picture which illustrates the mechanism for nuclear forces. This also will show why there must be π-mesons with three different charges. It is a consequence of the different charges of the neutron and the proton. We consider the interaction between two protons in the atomic nucleus, as schematically pictured in Figure 59.8(a). The proton p emits a π^+-meson at A. Since π^+ carries a positive elementary charge, the proton p changes into an electrically neutral neutron n. At B, the π^+-meson is captured by another neutron n, which thereby changes into a proton. Analogously, (b) and (c) in Figure 59.8 can be understood. For all these processes, the charge conservation is strictly satisfied, but not so the energy conservation. One therefore calls these π-mesons virtual particles. It is important then that because of the *energy–time uncertainty relation* the classical energy conservation is not required, but instead we give the following interpretation of Figure 59.8(a). The energy of the proton p at A cannot be determined as a sharp value, but, according to (50), may vary in the time interval Δt about the value ΔE with

$$\Delta E \, \Delta t \geq \hbar/2. \tag{61}$$

An analogous result holds for the neutron n at B. This is why during a time interval Δt a π^+-meson may fly from A to B with total energy ΔE and

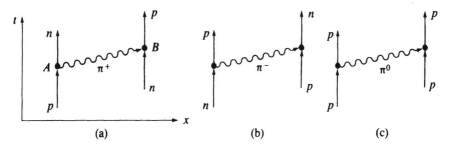

Figure 59.8

momentum vector p_0. According to the relativistic energy formula we have

$$\Delta E = \sqrt{m_0^2 c^4 + c^2 p_0^2}. \tag{62}$$

In this sense the nuclear force is an exchange force which is caused by the exchange of π-mesons.

We want to test the consistency of this picture. According to the theory of relativity of Chapter 75 a physical effect can travel at most with the velocity of light c. The π-meson, which flies through the nucleus, can do this with at most the velocity c, i.e., $R/\Delta t \leq c$. From (59) this means

$$\Delta t \geq R/c = \hbar/m_0 c^2$$

and (62) implies $\Delta E \geq m_0 c^2$, hence

$$\Delta E \, \Delta t \geq \hbar.$$

This is a sharpening of (61).

According to modern views, protons and neutrons consist of quarks, and the nuclear force is a consequence of the quark interaction, which is caused by the exchange of eight gluons.

The diagrams in Figure 59.8 are special cases of the so-called Feynman diagrams. We shall show in Part V that, in the context of quantum field theory, one obtains a mathematical formalism for the computation of quantum processes. Thereby the Feynman diagrams are graphical representations to give an intuitive description of purely analytical expressions, which appear in connection with perturbation calculus. For higher-order perturbations much more complicated diagrams than those in Figure 59.8 appear. Roughly speaking, several particles are involved.

In comparing physical and mathematical thinking, it is interesting that Feynman based his development of quantum electrodynamics, during the late 1940s, on some intuitive physical ideas, which he described by his diagrams. Thereby, he used these diagrams in a virtuous fashion in order to compute physical effects. Dyson wrote in his book "Disturbing the Universe," that he had great difficulty in following Feynman's intuition and in translating it back into a mathematical language which he could understand.

Physical thinking does not seem to follow the lines of a mathematical formalism, but rather proceeds in images, which are closely related to real phenomena. Only thereafter, a mathematical formulation is used. It is known that during the last century, Faraday used the image of electric and magnetic fields, long before Maxwell gave a precise mathematical description.

59.25. The Four Fundamental Interactions

There exist four fundamental interactions in nature: strong, electromagnetic, weak, and gravitational. In Tables 59.2 and 59.3 we give a survey.

59.25. The Four Fundamental Interactions

Table 59.2

Interaction	Relative strength	Radius of action	Example	Particle onto which the force is exerted	Field quanta
Strong	10	10^{-15} m	proton as bound state of three quarks	quarks (hadrons)	8 gluons
Electromagnetic	10^{-2}	∞	atomic force, chemical binding	electrically charged particles	photon
Weak	10^{-5}	$\ll 10^{-16}$ m	β-decay of the neturon (radioactive decay)	hadrons, leptons, e.g., electron, neutrino	vector bosons $W^{\pm}, Z^0$
Gravitational	10^{-39}	∞	planetary motion	particles with mass	graviton

The most significant achievement of modern elementary particle physics is the unified theory for the weak and the electromagnetic interaction, the theory of the electroweak interaction. For this, Glashow, Salam, and Weinberg received the Nobel prize in 1979. This theory can be viewed as a continuation of Maxwell's theory, in which the electric and magnetic interaction is combined to an electromagnetic interaction. The mathematical apparatus is also analogous to Maxwell's theory. This theory can be viewed as a U(1)-gauge field theory, whereas the electroweak theory is a gauge field theory with the group SU(2) × U(1) (see Part V). The field quanta of the electroweak interaction are three heavy vector bosons W^+, W^-, Z^0 with approximately 100 proton masses and the photon. These predicted heavy particles were detected in 1983 at CERN (Geneva) with the 270/270 GeV proton–antiproton collider.

According to quantum chromodynamics, protons, neutrons, and π-mesons consist of quarks, which may carry the three so-called color charges which are called red, green, and blue. Similarly to the photon, which describes the interaction between electrically charged particles, one assumes here the existence of eight gluons, which transmit the interaction between quarks with color charge. One expects that this causes the strong interaction. The quantum chromodynamics is a SU(3)-gauge field theory. Since the Lie algebra of SU(3) is eight-dimensional, this formalism yields the existence of eight gluons.

In order to unify the strong interaction with the electroweak interaction, gauge field theories with larger groups G have been proposed, e.g., with $G = $ SU(5) or $G = $ SO(10). These theories (grand unification theories) imply a proton decay and the existence of magnetic monopoles. For the proton decay one has a half-life period of approximately 10^{30} years, i.e., during this

Table 59.3

Field quanta	Charge	Mass/Proton mass	Spin	Mean life-time
Gluon	0	0	1	∞?
Photon	0	0	1	∞
$W^{\pm}$	$\pm\|e\|$	80	1	$< 10^{-23}$ s
Z^0	0	90	1	$< 10^{-23}$ s

time half of the proton has decayed. At present, serious efforts are being made to find an experimental proof. The field quanta of the unified theory—the so-called Lepto quarks—have approximately 10^{14} proton masses. The magnetic monopoles should have 10^{16} proton masses. This already is in the order of the magnitude of bacteria. Only during the first 10^{-35} s after the Big Bang, the energy needed for creating such heavy particles was large enough. Also one expects that these magnetic monopoles catalyze the proton decay.

The unified theories assume that, for large energies of greater than 10^{15} GeV per particle, only one interaction exists. Such high energies were present only during an extremely short time after the Big Bang. Then, during the cooling-off process of the cosmos, the different interactions crystallized analogously to the crystallization of substances from a fluid. This crystallization goes along with a loss of symmetry. Therefore the present interactions have different symmetry groups (symmetry breaking) (see Section 76.7 and Part V).

59.26. Strength of the Interactions

First we compare the gravitational and electromagnetic interaction. Using classical methods, we compute the gravitational force and the electromagnetic force between two protons. From (63a) and (63b) below we will obtain that the ratio between these two forces is approximately equal to $10^{-39}:10^{-2}$. We use the formula

$$|K| = \alpha \hbar c r^{-2}. \tag{63}$$

The value of the gravitational force between two protons of mass $m = 1.7 \cdot 10^{-27}$ kg at a distance r is equal to (63) with the dimensionless quantity

$$\alpha = Gm^2/\hbar c = 5 \cdot 10^{-39}. \tag{63a}$$

According to Coulomb's law of Section 58.8, the value of the electrostatic repelling force between two protons with positive elementary charge $|e| = 1.7 \cdot 10^{-19}$ A s at a distance r is equal to (63) with

$$\alpha = e^2/4\pi\varepsilon_0 \hbar c = 1/137. \tag{63b}$$

This is the so-called Sommerfeld fine structure constant. The effects of quantum electrodynamics are obtained from perturbation computations with respect to powers of $\alpha = 1/137$. For the strong interaction, such a perturbation formalism fails, because the parameters involved are substantially larger.

Since there exists no precise mathematical quantum field theory, it is not possible to find exact characteristics for the strengths of the strong and weak interaction. One heuristic method is to calculate certain effects by using rough quantum field theoretical approximation procedures, and to form dimensionless quantities analogous to Sommerfeld's fine structure constant [see Bogol-

Table 59.4

Interaction	Typical effective cross sections σ in m²	Typical energy fluctuations of resonances in MeV	Typical mean life-times of resonances in seconds
Strong	10^{-30}	10^2	10^{-23}
Electromagnetic	10^{-33}	10^{-3}	10^{-18}
Weak	10^{-42}	10^{-14}	10^{-7}

jubov and Širkov (1980, M), §10, Landau and Lifšic (1962, M), Vol. 4b, §145 (Fermi theory of the weak interaction) and Becher and Böhm (1981, M) (gauge field theories)]. This procedure, which is still quite uncertain and arbitrary, yields the relative strengths of the interactions given in Table 59.4.

In order to get an idea of the physical effects that are actually observed, we look at the two most important quantities which can be directly determined in experiments with particle accelerators: the *effective cross section* σ with the dimension of a surface, and the *energy fluctuation* $2\Delta E$ of resonances. The concept of effective cross sections has already been discussed in Section 59.7. If, according to (53), we assign mean life-times to the resonances we obtain Table 59.4. It shows that, in fact, the interactions have different strengths.

PROBLEMS

59.1. *Bohr's atomic model*. Prove the formulas of Section 59.16.

Solution: The motion of the electron on a circular orbit of radius r around the proton is described by

$$x = r(\cos \omega t \, e_1 + \sin \omega t \, e_2)$$

with the orthonormal vectors e_1, e_2 and $e_3 = e_1 \times e_2$. This implies

$$\dot{x} = \omega r(-\sin \omega t \, e_1 + \cos \omega t \, e_2),$$

$$\ddot{x} = -\omega^2 x.$$

For the angular momentum of the electron we obtain

$$N = m_e(x \times \dot{x}) = m_e \omega r^2 e_3.$$

The equation of motion for the electron is

$$m_e \ddot{x} = K \tag{64}$$

with Coulomb force

$$K = -\frac{e^2 x}{4\pi\varepsilon_0 |x|^3}$$

according to Section 58.8. From (64) follows

$$m_e \omega^2 r = e^2/4\pi\varepsilon_0 r^2. \tag{65}$$

59. Dualism Between Wave and Particle, Preview of Quantum Theory

The energy is equal to

$$E = \tfrac{1}{2}m_e \dot{x}^2 + U = \frac{m_e}{2}\omega^2 r^2 - \frac{e^2}{4\pi\varepsilon_0 r}$$

$$= -\frac{e^2}{8\pi\varepsilon_0 r}.$$

The angular momentum quantization $|N| = n\hbar$ gives

$$m_e \omega r^2 = n\hbar, \qquad n = 1, 2, \ldots.$$

From (65) we obtain the orbit radii

$$r_n = r_0 n^2$$

with $r_0 = 4\pi\varepsilon_0 \hbar^2/m_e e^2$. The energy is

$$E_n = -e^2/8\pi\varepsilon_0 r_0 n^2,$$

and the orbital velocity of the electron is equal to

$$V_n = |\dot{x}| = \omega r_n = \alpha c/n$$

with $\alpha = e^2/4\pi\varepsilon_0 \hbar c = 1/137$.

59.2. *Scattering of charged particles.* A particle with positive charge Q, mass m, and initial velocity V_∞ is scattered at a fixed particle with positive charge Q_0, as shown in Figure 59.9. Prove that

$$q = (\alpha/mV_\infty^2)^2 \cot\frac{\vartheta}{2}, \qquad (66)$$

where ϑ is the scattering angle and q the so-called collision parameter. Thereby we have $\alpha = QQ_0/4\pi\varepsilon_0$.

Solution: From Section 58.9b we obtain for the Kepler problem

$$m\ddot{x} = \alpha x/|x|^3 \qquad (67)$$

with constant positive energy

$$E = \frac{m}{2}\dot{x}^2 + \alpha/r,$$

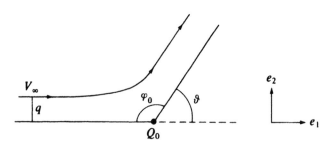

Figure 59.9

the orbital hyperbola

$$\frac{p}{r} = 1 + \varepsilon \cos(\varphi + 2\pi - \varphi_0/2)$$

with

$$\varepsilon = \sqrt{1 + 2EN^2/\alpha^2 m}, \tag{68}$$

and constant angular momentum

$$N = m(x \times \dot{x}).$$

Here r and φ are polar coordinates. Note that on the right-hand side of (67) we have the Coulomb force.

The angle $\varphi_0 \in]0, \pi[$ is chosen so that $r \to \infty$ as $\varphi \to -\pi$, i.e.,

$$1 + \varepsilon \cos(\pi \pm \varphi_0/2) = 0.$$

This implies that $r \to \infty$ as $\varphi \to -\pi + \varphi_0$. The scattering angle is $\vartheta = \pi - \varphi_0$, and hence

$$\sin \vartheta/2 = 1/\varepsilon. \tag{69}$$

For $t \to -\infty$ we obtain $\dot{x}(t) \to V_\infty e_1$. This gives

$$E = mV_\infty^2/2, \quad |N| = mqV_\infty. \tag{70}$$

From (68)–(70) follows (66).

59.3. *Rutherford's scattering formula.* Instead of one particle in Problem 59.2 we now consider a homogeneous particle current with velocity vector $V_\infty e_1$ and particle density ρ of particles with positive charge Q and mass m. This current is scattered at a fixed particle of positive charge Q_0. Compute the number of particles ΔN which, during the time interval $[0, t]$, are scattered about angles which lie in the interval $[\vartheta, \vartheta + \Delta\vartheta]$.

Solution: According to Figure 59.9 and (66), the number ΔN is equal to the number of particles which, during the time interval $[0, t]$, pass through the circular ring surface $\Delta \sigma$, perpendicular to the vector e_1 with

$$\Delta\sigma = \pi[(q + \Delta q)^2 - q^2]$$

$$= \pi(\alpha/mV_\infty^2)^2 \left(\cot^2 \frac{\vartheta + \Delta\vartheta}{2} - \cot^2 \frac{\vartheta}{2} \right).$$

This gives

$$\Delta N = \Delta\sigma(\rho V_\infty t). \tag{71}$$

The surface $\Delta\sigma$ is called the effective cross section. For $\Delta\vartheta \to 0$ we obtain

$$\frac{d\sigma}{d\vartheta} = \pi \left(\frac{\alpha}{mV_\infty^2} \right)^2 \frac{\cos(\vartheta/2)}{\sin^3(\vartheta/2)}.$$

The value $d\sigma$ is called the differential effective cross section.

References to the Literature

Classical works: Planck (1900), Einstein (1905b), Heisenberg (1925), Schrödinger (1926).

Physical quantum theory: Heisenberg (1937, M), Pauli (1958, S), Landau and Lifšic (1962, M), Vols. 3, 4, Bogoljubov and Širkov (1973, M), (1980, M), Itzykson and Zuber (1980, M), Lee (1981, M), Frampton (1987, M).

Mathematical quantum theory: von Neumann (1932, M), Van der Waerden (1980, M) (recommended as an introduction), Streater and Wightman (1964, M), Reed and Simon (1972, M), Vols. 1–4, Triebel (1972, M).

General References to the Literature

General physics: Feynman, Leighton and Sands (1963, L) (Feynman Lectures, Vols. 1–3) (recommended as an introduction), Kittel (1965, L) (Berkeley Physics Course, Vols. 1–5), Orear (1966, M), Hänsel and Neumann (1972, M), Vols. 1–7, Lüscher (1980, M).

Experimental physics: Bergmann and Schaefer (1979, M), Vols. 1–4, Gerthsen (1982, M).

Standard works in theoretical physics: Sommerfeld (1954, M), Vols. 1–6, Landau and Lifšic (1962, M), Vols. 1–10.

Theoretical physics: Pauli (1973, M), Vols. 1–6, Macke (1962, M), Vols. 1–6, Kompanejec (1961, M), Ludwig (1978, M), Vols. 1–4, Weller and Winkler (1974, M), Vols. 1–2, Greiner and Müller (1986, M), Vols. 1–10 (including recent advances in theoretical physics).

Mathematical physics: Courant and Hilbert (1959, M), Vols. 1–2 and Frank and von Mises (1962, M), Vols. 1–2 (classical standard works), Morse and Feshbach (1953, M), Vols. 1–2, Maurin (1967, M), (1976, M), Vols. 1–2, Reed and Simon (1971, M), Vols. 1–4 (emphasis on quantum physics), Triebel (1981, S), Thirring (1983, M), Vols. 1–4.

Handbook of physics: Geiger and Scheel (1926, M), Vols. 1–24 (classic), Flügge (1956, M), Vols. 1–∞.

Encyclopedia: Encyclopedia of Mathematics and its Applications (1976), Vols. 1–∞, Russian Encyclopedia of Mathematics (1977, M), Handbook of Applicable Mathematics (1980, M), Vols. 1–6, Encyclopedia of Astronomy and Space (1976), Van Nostrands Scientific Encyclopedia (1976). Encyclopedic Dictionary of Mathematics (1977) (especially recommended), Encyclopedic Dictionary of Physics (1977), Encyclopedia of Mathematics (1987, M), Vols. 1–∞.

Lexica: Dictionary of Mathematics (1961, M), Vols. 1–2, Brockhaus ABC of Physics (1971, M), Vols. 1, 2, Brockhaus ABC of Chemistry (1965, M), Vols. 1–2.

Four language dictionary of mathematics: Eisenreich and Sube (1982, M), Vols. 1–2 (35,000 termini).

Four language dictionary of physics: Eisenreich and Sube (1973) (75,000 termini).

Qualitative analysis of physical systems: Gitterman and Halpern (1981, M).

Rational mechanics: Truesdell (1977, M), Wang (1979, M).

International system of units: Oberdorfer (1969, M), Massey (1971, M).

Collections of problems in physics: Hajko and Schilling (1975, M), Vols. 1–6 (physics in examples), Vogel (1977, M), Greiner and Müller (1986, M), Vols. 1–10.

Popular expositions about the development of modern physics, cosmology and biology: Riedl (1976, M), Bresch (1978, M), Sullivan (1979, M), Kippenhahn (1980, M), Unsöld (1981, M), Fritzsch (1982, M), (1983, M), Sexl (1982, M), Ivanov (1983, M) (cybernetical methods in neuro-physiology, biology, cultural sciences, and the humanities), Henbest (1984, M), Trefil (1983, M), (1984), Taube (1985, M) (cf. p. 793).

Philosophical problems in mathematics and natural sciences: Kähler (1941), (1979, M), Weyl (1952, M), (1966, M), Born (1957, M), Blaschke (1957, M), Einstein (1965, M), Planck (1945, M), (1967, M), Kline (1972, M), Monod (1970, M) and Bresch (1978, M) (biology), von Weizsäcker (1973), (1976, M), (1976a), (1979, M), (1979a, M), Heisenberg (1977, M), (1977a, M), (1978, M), (1980, M), (1981, M), Dyson (1979, M), Hofstadter (1979, M), Prigogine (1979, M), Prigogine and Stengers (1981, M), Manin (1981, M), Cronin (1981, M), Maurin (1981), (1982), Treder (1983, M), Albers (1985, M), Beckert (1985), Tymoczko (1985, M), Hildebrandt and Tromba (1986, M).

History of physics: von Laue (1950, M), Mehra (1982, M).

History of natural sciences: Wussing (1983, M).

Nobel lectures: The Nobel prizes (1954ff, M).

Unsolved problems in mathematical physics: Simon (1984, S).

Unsolved problems in mathematics: Browder (1976a).

The journals *Nature* and *Scientific American* inform about recent developments in science.

Survey about modern developments in mathematics: Jaffe (1984).

About current results in mathematics and the natural sciences one can consult the Lecture Notes in mathematics, biomathematics, chemistry, computer science, control and information sciences, economics, physics, and statistics. These lecture notes appear by Springer-Verlag.

Current developments in physics may also be found in the two series "Progress in Physics", Birkhäuser, Boston and "Frontiers in Physics", Benjamin, New York. Pursue the summer institute series "Les Houches (1951ff)".

Fundamental formulas in physics: Menzel (1955, M).

Numerical recipies: Press (1986, M) (the art of scientific computing).

APPLICATIONS IN ELASTICITY THEORY

> As any human activity needs goals, mathematical research needs problems.
> David Hilbert (1932)

> Most mathematicians have an idea of the influence of hydrodynamics and electromagnetism on the theory of complex functions and harmonic potentials. The influence of elasticity is less well known. Elasticity led to a vast range of mathematical problems involving linear algebra, differential geometry, ordinary and partial differential equations (mostly nonlinear), elliptic functions, and the calculus of variations.
> Clifford Truesdell (1983)

In Chapters 60 to 66 we want to show that the following mathematical concepts are closely related to problems in nonlinear elasticity and plasticity theory:

convex functionals,
monotone potential operators,
pseudomonotone operators,
maximal monotone operators,
subgradients,
variational inequalities,
duality,
bifurcation.

The corresponding theories, which have been developed in Parts II and III, will here be applied to solve a number of problems in elasticity and plasticity theory. In Chapter 62, for instance, we will see that there exists a duality between strain and stress which allows us to apply the duality theory of Part III. In fact, the duality between strain and stress was one of the reasons for the creation of a general duality theory. Strain and stress are the two fundamental physical quantities of elasticity and plasticity theory.

The two most important concepts used in an elegant and effective description of convex problems in elasticity and hydrodynamics are:

(i) conjugate functionals (duality); and
(ii) subgradients (elastoplastic and viscoplastic material).

In case the considered functionals are not convex, one might be able to apply the modern method of compensated compactness. In this direction we will consider:

(a) polyconvex material in elasticity; and
(b) transonic flow in gas dynamics.

Most difficulties in elasticity and hydrodynamics arise from the fact that in realistic situations in nature no convexity is available.

We emphasize that the application of methods of convex analysis and the theory of monotone operators is only a first step. For a general theory of elastic and plastic phenomena, which presently does not exist, it will be necessary to substantially broaden the mathematical spectrum. Today, one is convinced that the true character of elasticity lies beyond the concepts of convexity and monotonicity.

CHAPTER 60
Elastoplastic Wire

Ut tensio sic vis.[1]

                Robert Hooke, *De Potentia Restitutiva*, (London, 1678)

The first mathematician to consider the nature of resistance of solids to rupture was Galileo (1638).... He endeavoured to determine the resistance of a beam, one end of which is built into a wall, when the tendency to break arises from its own or an applied weight; and he concluded that the beam tends to turn about an axis perpendicular to its length, and in the plane of the wall. This problem and, in particular, the determination of this axis is known as Galileo's problem.

The history of the theory of elasticity started from Galileo's question. Undoubtedly, the two great landmarks are the discovery of Hooke's law in 1660 (published in 1678), and the formulation of the general equations by Navier (1821). Hooke's law provided the necessary experimental foundation for the theory....

In the interval between the discovery of Hooke's law and that of the general differential equations of elasticity by Navier, the attention of those mathematicians who occupied themselves with our science was chiefly directed to the solution and extension of Galileo's problem, and the related theories of the vibrations of bars and plates and the stability of columns.

The first investigation of any importance is that of the elastic line or elastica by Jacob Bernoulli (1705), in which the resistance to bending is a number proportional to the curvature of the rod when bent....

David Bernoulli suggested to Euler (by letter in 1742) that the differential equation of the elastica could be found by making the square of the curvature taken along the rod a minimum; and Euler (1744) was able to obtain the differential equation and to classify various solutions of it (an early study of elliptic integrals and elliptic functions)....

---

[1] The force of a spring is directly proportional to its relative extension (strain). Robert Hooke published this in the anagram

                ceiiinossstuu.

Navier (1821) was the first to investigate the general equations of equilibrium and vibration of elastic solids. He set out from the Newtonian conception of the constitution of bodies, i.e., bodies are made up of small parts called "molecules" which act upon each other by means of central forces....

The studies of Cauchy (1789–1857) in elasticity were first prompted by his being a member of the commission appointed to report upon a memoir by Navier on elastic plates which was presented to the Paris Academy in August, 1820. By the autumn of 1822, Cauchy had discovered most of the elements of the pure theory of elasticity (published in 1827). He had introduced the notion of stress. He had also shown how to introduce both the stress tensor and the strain tensor.... He had determined the equations of motion (or equilibrium) by which the stress components are connected with the forces.... By means of relations between stress components and strain components, he had eliminated the stress components from the equations of motion and equilibrium, and had arrived at equations in terms of the displacement.... Cauchy obtained his stress–strain relation (constitutive law) for isotropic materials by means of two assumptions:

(i) that the relations in question are linear; and
(ii) that the principal planes of stress are normal to the principal axes of strain.

The experimental basis on which these assumptions can be made to rest is the same as that on which Hooke's law rests, but Cauchy did not refer to it. The methods used in these investigations are quite different from those of Navier's memoir (1821). In particular, no use is made of material points and central forces. The resulting equations differ from Navier's in one important respect: Navier's equations contain a single constant to express the elastic behavior of an isotropic body, while Cauchy's contain two such constants (today called the constants of Lamé (1795–1870))....

Green (1793–1841) was dissatisfied with the hypothesis on which the theory of elasticity was based, and he sought a new foundation in his paper (1839). Starting from what is now called the "principle of minimal elastic potential energy" he propounded a new method of obtaining the basic equations. The revolution which Green effected in the elements of the theory is comparable in importance with that produced by Navier's discovery of the basic equations. Green supposed the stored energy function (density of the elastic potential energy) to be capable of being expanded in powers and products of the components of strain.... From this principle Green deduced the equations of elasticity for anisotropic bodies, containing in the general case 21 constants. In the case of isotropy there are two constants, and the equations are the same as those of Cauchy.... (Green followed the pattern of the famous "La Mécanique Analytique" of Lagrange (1788). Green's stored energy function corresponds to the Lagrangian function in mechanics.)

The history of the mathematical theory of elasticity shows clearly that the development of the theory has not been guided exclusively by considerations of its utility for technical mechanics. Most of the men by whose researches it has been founded and shaped have been more interested in natural philosophy than in material progress, in trying to understand the world rather than in trying to make it more comfortable.

A. Love (1906)

The mechanics of continua, which is based on Cauchy's (1827) general notion of stress, has been applied so far only to liquid and solid elastic bodies. In regard to plastic deformations, Saint Venant (1864) (based on experiments of Tresca (1864)) has sketched a theory which, however, does not yield the necessary number of equations in order to completely determine the motion.

But this paper leads to a complete system of equations of motion for plastic bodies.

Richard von Mises (1913)

The mathematical theory of plasticity owes its development to the demand for more realistic methods to determine the safety factors of structures or machine parts, and to the need for better control in technological forming processes such as rolling, drawing, and extruding.

William Prager and Philip Hodge (1951)

In order to understand the basic ideas of elasticity and plasticity theory, we discuss in this chapter the simplest situation by choosing the wire as an example. Special emphasis will be placed on a comparison of important constitutive laws (stress–strain relations). We stress the possibility that plastic behavior may be described by multivalued constitutive equations, i.e., more precisely, by *subgradients*. The calculus of subgradients has been discussed in Part III.

In Chapters 61 to 66 we shall generalize the results of this chapter to three-dimensional bodies and give applications to some important problems. The dynamic plasticity model of Section 60.3 will be substantially generalized in Chapter 66. This model is based on the use of internal state variables.

60.1. Experimental Result

Consider a wire of length l and cross section C, which consists of steel and is expanded under the influence of a tensile force $K > 0$ about Δl (Fig. 60.1). We define

$$u = \Delta l \quad \text{(displacement),}$$

$$\gamma = \Delta l/l \quad \text{(strain),}$$

$$\sigma = K/C \quad \text{(stress).}$$

By definition, the strain γ is equal to the relative change of length. For a compressive force $K < 0$, the stress σ is negative. Experimentally, the stress–strain relation

$$\sigma = \sigma(\gamma)$$

exhibits the qualitative behavior of Figure 60.2. The points I to IV in Figure 60.2 carry the notations: I = proportional limit, II = elastic limit, III = hardening limit, and IV = strength limit. If we slowly increase the tensile force

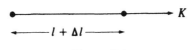

Figure 60.1

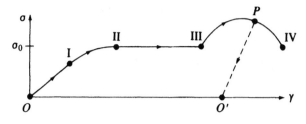

Figure 60.2

K from zero upwards, then the tension σ increases, and the wire is expanded, i.e., in Figure 60.2 we run through the curve from O to IV.

Between O and I there exists a linear relation between stress and strain

$$\sigma = E\gamma, \tag{1}$$

which is called *Hooke's law*. The material constant E is called the elasticity module. In fact, the wire is also subject to a reduction Δd of its thickness d, which is given by

$$\Delta d/d = -\mu \Delta l/l. \tag{2}$$

The material constant μ is called the Poisson number. In this chapter we concentrate on the longitudinal deviation (1). The transverse contraction (2), however, is important for the behavior of three-dimensional bodies, as we shall see in Section 61.7.

Between I and II there exists the nonlinear relation

$$\sigma = \sigma(\gamma), \tag{3}$$

which is called a nonlinear Hooke's law. At point II a plastic behavior can be observed. First, the wire begins to flow, i.e., one observes strains without significant increase of stress. At point III, the so-called hardening limit, the flow process slows down. One says the material has hardened. At point IV the material breaks.

Typical for the plastic behavior is a so-called hysteresis effect. In order to explain this let us consider a point P in Figure 60.2 beyond the elastic limit II. If at this point the force is diminished i.e., the stress is decreased, one does not run backwards through the diagram, but instead along the dotted line. If during this process one reaches the force $K = 0$, and hence the stress $\sigma = 0$, then we are at point O'. This means, although no force is present any more, there still exists a nonzero rest deformation, which is called a plastic rest deformation. While in the elastic region O to II there exists a unique relation between stress σ and strain γ, this is no longer true in the plastic region beyond II. There only a multivalued relation exists between σ and γ, which depends on the chosen path, i.e., on the process. The objective of mathematical plasticity theory is to model such multivalued constitutive laws. This will be discussed during the following sections. Thereby the subgradient of convex analysis of Part III will be the main tool to describe hysteresis effects.

Our observations have a purely phenomenological character, i.e., we do not study the microphysical mechanisms, which are responsible for elasticity. In 1912, Max von Laue's experiments with X-rays proved that many solid materials, especially all metals, have a crystalline structure. During the 1930s one came to the conclusion that plastic behavior is caused by curve-like defects in the lattice structure. Detailed information about this may be found in Sommerfeld (1970, M, H), Chapter IX and in Kleinert (1987, P).

60.2. Viscoplastic Constitutive Laws

If relation

$$\sigma = \sigma(\gamma) \tag{4}$$

holds, then we speak of linear and nonlinear elastic material if $\sigma(\cdot)$ is linear and nonlinear, respectively (Figs. 60.3 and 60.4). We now want to show that plastic, viscous, and viscoplastic behavior can uniformly be described through

$$\dot{\gamma}(t) \in \partial F(\sigma(t)) \tag{5}$$

with time parameter t. More precisely, we consider the processes

$$\gamma = \gamma(t), \qquad \sigma = \sigma(t) \qquad \text{for} \quad 0 \le t \le t_0,$$

which occur slowly (quasi-dynamical processes). Relation (5) holds for almost all $t \in [0, t_0]$.

EXAMPLE 60.1 (Ideal Plastic Behavior). We choose a fixed $\sigma_0 > 0$. Let

$$C = \{\sigma \in \mathbb{R}: |\sigma| \le \sigma_0\}$$

and let χ be the indicator function of C, i.e., $\chi(\sigma) = 0$ or $= +\infty$ for $\sigma \in C$ or $\sigma \notin C$, respectively. Set $F = \chi$. According to Section 47.3 we have

$$\partial \chi(\sigma) = \begin{cases} \{0\} & \text{if } |\sigma| < \sigma_0, \\ \mathbb{R}_\pm & \text{if } \sigma = \pm \sigma_0, \\ \varnothing & \text{if } |\sigma| > \sigma_0. \end{cases}$$

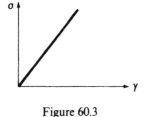

Figure 60.3   Figure 60.4

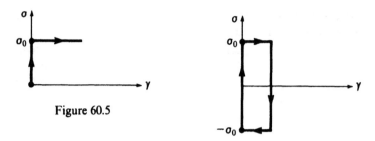

Figure 60.5

Figure 60.6

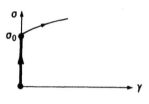

Figure 60.7

Therefore, (5) is equivalent to

$$\dot{\gamma}(t) = 0 \quad \text{if} \quad |\sigma(t)| < \sigma_0,$$
$$\pm\dot{\gamma}(t) \geq 0 \quad \text{if} \quad \sigma(t) = \pm\sigma_0, \tag{6}$$

and $|\sigma(t)| \leq \sigma_0$ for all t. Figures 60.5 and 60.6 show two possibilities. In Figure 60.5 an increase of stress σ, at first, does not yield any strain γ. After reaching the critical stress σ_0, the material flows. Figure 60.6 shows a typical hysteresis effect. It is now important that no unique relation of the form (4) exists, but according to (5) a number of different processes are possible.

EXAMPLE 60.2 (Viscous Behavior). Let $F_1: \mathbb{R} \to \mathbb{R}$ be a convex C^1-function. Then $\partial F_1(\sigma) = \{F_1'(\sigma)\}$. Therefore (5) with $F = F_1$ corresponds to the constitutive law

$$\dot{\gamma}(t) = F_1'(\sigma(t)) \tag{7}$$

which models viscous flows. In contrast to (4), the strain velocity here depends on the stress. If we choose especially

$$F_1(\sigma) = \begin{cases} 0 & \text{if } |\sigma| \leq \sigma_0, \\ (\sigma - \sigma_0)^2/4\mu & \text{if } |\sigma| > \sigma_0, \end{cases}$$

with $\mu > 0$, then (7) corresponds to the process of Figure 60.7 with $\sigma(t) = t$. Formally, F_1 becomes χ as $\mu \to 0$. Hence, formally, ideal plastic behavior can be viewed as a limiting case of viscous behavior (Fig. 60.8).

EXAMPLE 60.3 (Viscoplastic Behavior). For $F = \chi + F_1$, Theorem 47.B implies that

$$\partial F = \partial \chi + \partial F_1.$$

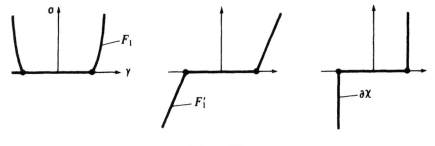

Figure 60.8

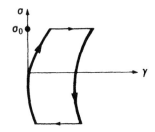

Figure 60.9

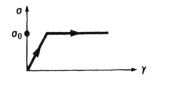 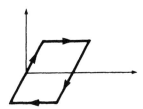

Figure 60.10

In this case (5) describes a superposition of Examples 60.1 and 60.2 of the form

$$\dot{\gamma}(t) \in \partial\chi(\sigma(t)) + F'_1(\sigma(t)).$$

For $F_1(\sigma) = \sigma^2/4\mu$ with $\mu > 0$ and piecewise constant $\dot{\sigma}$ one obtains, e.g., Figure 60.9. Combining (4) with (5) gives

$$\dot{\gamma}(t) \in E^{-1}\dot{\sigma}(t) + \partial F(\sigma(t)). \tag{8}$$

For $F = \chi$, one obtains processes of the form in Figure 60.10.

60.3. Elasto-Viscoplastic Wire with Linear Hardening Law

In Examples 60.1 and 60.3 the plastic limit σ_0 cannot be exceeded, since we have $\partial\chi(\sigma) = \emptyset$ for $|\sigma| > \sigma_0$. However, such an exceeding by hardening effects is actually observed in Figure 60.2. We want to show how one can describe such hardening effects by introducing so-called *internal* state variables e, p and q, r. We set

$$\gamma = e + p, \qquad \sigma = q + r, \tag{9}$$

whereby e is the elastic strain, p is the viscoelastic strain, q is the viscoplastic stress, and r is the hardening stress. The constitutive laws are given by linear Hooke's laws

$$\sigma = Ae, \qquad r = Bp \tag{10}$$

with positive constants A and B and the viscoplastic law

$$\dot{p}(t) \in \partial F(q(t)) \tag{11}$$

for almost all $t \in [0, t_0]$ and fixed $t_0 > 0$. The constants A and B are called elasticity module and hardening module, respectively. Our assumptions are:

(H1) The function $F: \mathbb{R} \to \,]-\infty, \infty]$ is convex and lower semicontinuous with $F \not\equiv +\infty$.
(H2) We are given a function $t \mapsto K(t)$ describing the change in time for the outer force with $K \in W_2^1(0, t_0)$. This implies the stress

$$\sigma(t) = K(t)/C.$$

Thereby $\sigma(0)$ is known. Recall that C is the cross section of the wire.

(H3) The initial data for γ, e, p, r, q at time $t = 0$ are chosen so that (9) and (10) are satisfied at $t = 0$ and $F(q(0)) < \infty$ holds. It is enough to know the initial strain $\gamma(0)$, because all other initial data are then uniquely determined.

Theorem 60.A. *If* (H1) *to* (H3) *hold, then there exists a unique solution* γ, e, p, $r \in W_2^1(0, t_0)$ *which satisfies the given initial data.*

PROOF. This is a special case of Theorem 66.A of Section 66.3 with $U = \Gamma = \mathbb{R}$. Moreover, we have

$$\gamma = Du$$

with $u = \Delta l$ and $D = D^* = 1/l$, and also

$$D^*\sigma = K/Cl.$$

We therefore replace K in Section 66.1 with K/Cl. □

Recall that for any $f \in W_2^1(0, t_0)$ there exists a uniquely determined continuous function $g: [0, t_0] \to \mathbb{R}$, which is equal to f almost everywhere. Moreover, f has a generalized derivative $\dot{f}$ with $\int_0^{t_0} \dot{f}^2 \, dt < \infty$. On the other hand, one can modify f on a set of measure zero, whereby f remains unchanged as an element of $W_2^1(0, t_0)$. Therefore we may always assume that $f: [0, t_0] \to \mathbb{R}$ is continuous. In this sense $f(0)$ is well defined.

60.3. Elasto-Viscoplastic Wire with Linear Hardening Law

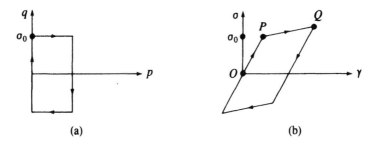

Figure 60.11

EXAMPLE 60.4. (Hardening Effect). Let $F = \chi$ as in Example 60.1. We want to show that there occurs a hardening effect in our model. If p, q travel on the given curve in Figure 60.11(a), then γ, σ run through the curve in Figure 60.11(b). The slope of the line segment $\overrightarrow{OP}$ is equal to A. This is a linear Hooke's law. A so-called linear hardening law corresponds to the line segment $\overrightarrow{PQ}$ with slope $B > 0$, for which the plastic limit σ_0 is exceeded. For $B = 0$, however, $\overrightarrow{PQ}$ is horizotal, i.e., we have a pure flow process. Hence one obtains the two unknown parameters A, B in the constitutive law for the internal state variables from the diagram for the experimentally measurable quantities γ and σ in Figure 60.11(b). This figure is an idealization of the experimental result in Figure 60.2.

As in Figure 60.12, model (9) to (11) may be visualized as a parallel connection of an elastic element B with a viscoelastic element F, which both are serially connected to an elastic element A. Thereby A corresponds to the elastic strain e, while B and F correspond to the viscoelastic strain p. The sum of the quantities e and p gives the total strain

$$\gamma = e + p.$$

At the points A, B, F of Figure 60.12, the strains e, p, p produce the stresses σ, r, q, respectively. We have that

$$\sigma = r + q.$$

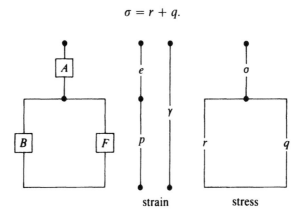

Figure 60.12

60.4. Quasi-Statical Plasticity

We distinguish here between:

(i) quasi-dynamical plasticity,[1]
(ii) quasi-statical plasticity, and
(iii) statical plasticity.

In case (i), time-dependent processes for strain and stress of the form

$$\gamma = \gamma(t), \qquad \sigma = \sigma(t), \qquad 0 \le t \le t_0$$

with constitutive law

$$\dot\gamma(t) \in \partial F(\sigma(t)) \tag{12}$$

will be considered. The quasi-statical case (ii), on the other hand, corresponds to the constitutive law

$$\gamma \in \partial F(\sigma) \tag{13}$$

with the important special case

$$\gamma \in f(\sigma) + \partial\chi(\sigma), \tag{14}$$

where χ denotes the indicator function for the set

$$C = \{\sigma \in \mathbb{R} : |\sigma| \le \sigma_0\}, \qquad \sigma_0 > 0.$$

According to Example 60.1, equation (14) is equivalent to the constitutive law:

$$\begin{aligned}\gamma &= f(\sigma) &&\text{if } |\sigma| < \sigma_0, \\ \gamma &\ge f(\sigma_0) &&\text{if } \sigma \ge \sigma_0, \\ \gamma &\le f(-\sigma_0) &&\text{if } \sigma \le -\sigma_0,\end{aligned} \tag{14*}$$

i.e., the strain γ becomes undetermined if the critical stress $\pm \sigma_0$ has been reached (Fig. 60.13). Note that in the quasi-statical case the history of the

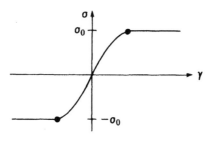

Figure 60.13

---

[1] Note that this terminology is not uniform in the literature. Sometimes quasi-dynamical plasticity is called quasi-statical plasticity.

material does not play any role, which, in fact, represents a strong idealization of the actual behavior of plastic materials. In Chapters 62 and 66, we will discuss generalizations of (i) and (ii) to three-dimensional bodies and prove the following:

quasi-statical plasticity $\Rightarrow$ elliptic variational inequality,

quasi-dynamical plasticity $\Rightarrow$ evolution variational inequality of first order.

From the mathematical point of view, the quasi-statical case is much simpler than the quasi-dynamical case.

The basic ideas of statical plasticity will be discussed in Section 62.11, and the following will be proved:

dual variational problems between strain and stress

+ plasticity condition

$\Rightarrow$ statical plasticity $\Rightarrow$ elliptic variational inequality.

60.5. Some Historical Remarks on Plasticity

The basic experiments on the behavior of plastic material were performed by Tresca (1864), nearly 200 years after the discovery of Hooke's law. Tresca also formulated a plasticity condition (yield condition) which has the following form:

$$\text{Tresca}(\tau) < \tau^* \quad \text{no plasticity,}$$
$$\text{Tresca}(\tau) \geq \tau^* \quad \text{plasticity occurs,}$$
(15)

whereby Tresca(τ) is equal to $\max\{|\tau_1 - \tau_2|, |\tau_2 - \tau_3|, |\tau_1 - \tau_3|\}$ and the τ_1, τ_2, τ_3 denote the eigenvalues of the stress tensor τ. This will be discussed more thoroughly in the following chapter. The numbers $\lambda = \tau_1, \tau_2, \tau_3$ are the solutions of the characteristic equation

$$\det(\tau_{ij} - \lambda \delta_{ij}) = 0,$$

where the τ_{ij} denote the components of τ in a Cartesian coordinate system. Note that the matrix (τ_{ij}) is symmetric. The critical value τ^* thereby depends on the material under consideration. Suppose n_1, n_2, n_3 are eigenvectors of τ which correspond to τ_1, τ_2, τ_3. Following Cauchy (1827), the planes with normal vectors n_1, n_2, n_3 are called the principal planes of stress, and the τ_1, τ_2, τ_3 the principal stresses. Their physical meaning will be explained in Section 61.3. The plasticity condition (15) postulates the following fundamental law of plasticity:

plasticity occurs if the differences between the principal stresses are large enough.

Theoretical work on plasticity goes back to the papers of Saint Venant (1871) and Lévy (1871). In 1909, Haar and von Kármán proposed a so-called statical model in plasticity, which was based on a variational principle for the stresses and a side condition, namely, a plasticity condition. A modern formulation of this, together with existence and uniqueness theorems, will be given in Section 62.11.

The present flow theory in plasticity was founded by von Mises. In his paper (1913) he

(i) replaced the Tresca plasticity condition with a new condition; and
(ii) formulated the equations of motion for an ideal plastic liquid (von Mises liquid).

Von Mises' plasticity condition reads as follows:

$$\begin{aligned} \text{Mises}(\tau) &< \tau_0^2 \quad \text{no plasticity,} \\ \text{Mises}(\tau) &\geq \tau_0^2 \quad \text{plasticity occurs,} \end{aligned} \tag{16}$$

where

$$\text{Mises}(\tau) = \tfrac{1}{3}((\tau_1 - \tau_2)^2 + (\tau_2 - \tau_3)^2 + (\tau_1 - \tau_3)^2). \tag{17}$$

This condition is easier to handle than the old Tresca condition, since, for one thing, the function $\tau \mapsto \text{Mises}(\tau)$ is differentiable, and, moreover, $\text{Mises}(\tau)$ can be expressed in terms of τ. In Section 62.9 we will show that

$$\text{Mises}(\tau) = [\bar{\tau}, \bar{\tau}] = \sum_{i,j=1}^{3} (\tau_{ij} - \tfrac{1}{3}(\tau_{11} + \tau_{22} + \tau_{33}))^2.$$

This is a consequence of the fact that $[\bar{\tau}, \bar{\tau}]$ is an invariant expression, i.e., a special Cartesian coordinate system can be chosen in which $\tau_{ij} = \tau_i \delta_{ij}$. This implies (17).

Throughout this volume we will use the von Mises plasticity condition. The convex structure of (16) enables us to use the methods of convex analysis in plasticity theory.

The basic equations of motion for von Mises liquids and also the more general basic equations for viscoplastic flows will be discussed in Part V (basic equations of rheology for non-Newtonian fluids).

Linear quasi-static plasticity theory was proposed by Hencky (1924), and linear quasi-dynamical plasticity theory dates back to Prandtl (1924) and was further developed by Reuss (1930).

A decisive stride towards a rigorous mathematical theory of plasticity was made by Duvaut and Lions (1972, M), using the modern theory of variational inequalities in order to prove existence theorems. In doing so, they paved the way for applications of nonlinear functional analysis in plasticity and related subjects.

The very fruitful idea of internal state variables was introduced by Nguyen (1973), and has been used already in Section 60.3 in order to describe hardening effects.

We consider the following problems related to plasticity:

(i) Existence theorem in linear quasi-statical plasticity (Section 62.10).
(ii) Existence and uniqueness theorem in statical plasticity (Section 62.11).
(iii) A general quasi-dynamical model in plasticity based on internal state variables (Chapter 66).
(iv) Plastic torsion (Part V).
(v) The basic equations for viscoplastic flows of non-Newtonian liquids, existence and uniqueness for tube flows, and general flows (Part V). We consider, for example, viscous Navier–Stokes liquids and more general viscous Williamson liquids, viscoplastic Bingham liquids, and ideal plastic von Mises liquids.

References to the Literature

Classical works in elasticity: Galilei (1638), Hooke (1678) (basic experiments), Bernoulli (1705) and Euler (1744) (bending of beams), Navier (1821), Cauchy (1827), (1828) (foundation of the general theory), Green (1839) (energy principle).

History of the theory of elasticity: Love (1906), Gurtin (1972, S), Truesdell (1968, M), (1983, S), and Antman (1983, S).

Classical works in plasticity: Tresca (1864) (basic experiments), Saint Venant (1871), Lévy (1871), Haar and von Kármán (1909) (statical plasticity theory), von Mises (1913) (fundamental paper: quadratic plasticity condition and equations of motion for ideal plastic liquids), Hencky (1924) (quasi-statical plasticity theory), Prandtl (1924) and Reuss (1930) (quasi-dynamical plasticity theory), Moreau (1968), Duvaut and Lions (1972, M) (variational inequalities), Nguyen (1973) (internal state variables).

History of plasticity theory: Hill (1950, M), Prager and Hodge (1951, M), Geiringer (1972, S) (handbook article). (See also the detailed References to the Literature for Chapter 66 on plasticity.)

CHAPTER 61

Basic Equations of Nonlinear Elasticity Theory

The geometers who have investigated the equations of equilibrium or motion of thin plates or surfaces have distinguished two kinds of forces, one produced by extension or contraction and the other by the bending of surfaces.... It has seemed to me that these two kinds of forces could be reduced to a single one, which always ought to be called tension or pressure; this force acts upon each element of a section, chosen at will, not only in a flexible surface but also in a solid.

Augustin Louis Cauchy (1827)

The principal creator of three-dimensional elasticity is Cauchy (1789–1857). Mainly for use in three-dimensional hydrodynamics, Euler (1707–1783) had introduced general mappings of regions and had created the associated calculus of partial derivatives, chain rules, Jacobian determinants, etc; he also had formulated the general principles of linear and angular momentum and had shown how to apply them to fluids.

Cauchy mastered all this and turned it to use in elasticity.... It is fair to say that much of the algebra of vectors, matrices, and tensors grew out of Cauchy's work on the strain, local rotation, and stress in elastic bodies.

Clifford Ambrose Truesdell (1983)

For a sufficiently small volume element, the general small change in the position of a deformable body can be represented as the sum of a translation, a rotation, and an extension or contraction in three orthogonal directions.

Hermann Helmholtz (1858)

The following principle, which goes back to Cauchy, is fundamental in stress analysis. If one imagines a volume element which is taken from an elastic body, then the outer forces, the inertial forces and the stress forces which act on the surface of the volume element, are in an equilibrium.

Erich Trefftz (1928)

If methods can be found, which admit a complete understanding of a given system, starting from a single atom and ending with the entire body, only then will a deeper knowledge in elasticity theory be gained.

Adolf Busemann and Otto Föppl (1928)

There are many reasons why nonlinear elasticity is not widely known in the scientific community:

(i) It is basically a new science whose mathematical structure is only now becoming clear.
(ii) Reliable expositions of the theory often take a couple of hundred pages to get to the heart of the matter.
(iii) Many expositions are written in a complicated indicial notation that boggles the eye and turns the stomach.

Stuart Antman (1984)

The goal of elasticity theory is the computation of deformations of elastic bodies and the corresponding stress forces. These deformations need not necessarily be small. Unfortunately, at present, there is no general nonlinear existence theory available. This makes the study of the literature quite difficult. Lacking this comprehensive general theory, a great number of models are used which are based on different approximation assumptions. These assumptions, however, are often not explicitly formulated and their foundation seems doubtful. Difficulties arise mainly from the fact that often there is *no* strict distinction between the different regions which correspond to the undeformed and deformed body.

We will try to present here an approach which might help the reader to understand the different models and approximation assumptions from a *general* and *rigorous* point of view. In what follows, we want to describe our general strategy, which is represented schematically in Figure 61.1.

Basic Equations and Typical Difficulties

In Section 61.3 we formulate the general basic equations of nonlinear elasticity theory. They consist of:

(a) *time-dependent equations of motion for the deformed region of the elastic body*; and
(b) *constitutive laws which describe the connection between deformation and the crucial stress tensor τ*.

The constitutive laws reflect the specific properties of the different elastic materials. The two main difficulties are the following:

(i) The stress tensor τ refers to the *unknown* deformed region of the body.
(ii) The nonlinear constitutive laws are by no means uniquely determined by general physical principles.

In order to avoid *difficulty* (i) we use, in Section 61.5, the following crucial procedure. We replace the stress tensor τ in the deformed region with the *reduced* stress tensor σ in the undeformed region which is sometimes also called the first Piola–Kirchhoff stress tensor. Other than in Section 61.3 we thereby obtain basic equations in the *undeformed* region of the body. The key observation is that τ in the defomed region can be computed from σ in the

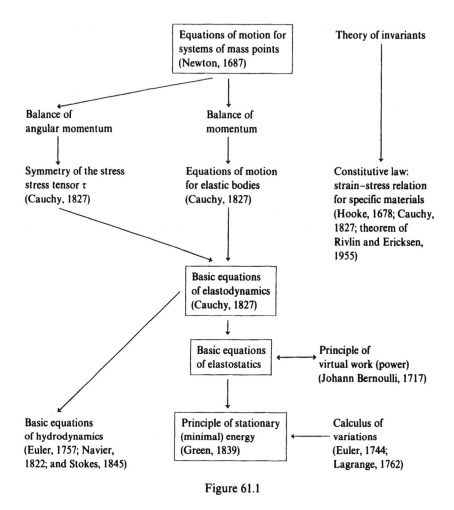

Figure 61.1

undeformed region. We thereby use the famous Piola transformation. It is therefore sufficient to solve the basic equations in the undeformed region.

The stress forces, which are effective in the deformed body, depend on the stress tensor τ. It is important that these forces can also be computed by using the reduced stress tensor σ. We emphasize, however, that one has to distinguish strictly between τ and σ, which is sometimes not clearly expressed in the literature.

As a consequence of general physical arguments τ is always symmetric, whereas the same need not be true for σ. We also introduce the symmetric second Piola–Kirchhoff tensor S in the undeformed region. The advantage of S is that it allows an especially simple formulation of the constitutive laws, as will be shown in Section 61.8.

In the stationary case, i.e., in the case of elastostatics, the basic equations lead to the principle of virtual work or, equivalently, to the principle of virtual power. This is discussed in Section 61.5.

Variational Approach

In order to reduce *difficulty* (ii) above we formulate, in Section 61.6, a class of *variational problems*

(P) $$\int_G L(x, u'(x))\,dx - \int_G Ku\,dx - \int_{\partial_2 G} Tu\,dO = \text{stationary!},$$

$$u = u_0 \quad \text{on } \partial_1 G,$$

which corresponds to the *principle of stationary potential energy*. In this case we speak of hyperelastic material. The corresponding Euler equations are

(E) $$\text{div } \sigma + K = 0 \quad \text{on } G,$$

$$u = u_0 \quad \text{on } \partial_1 G,$$

$$\sigma n = T \quad \text{on } \partial_2 G.$$

The important point is that (E) agrees with the basic equations of elastostatics. The function L represents the density of the elastic potential energy and is also called the stored energy function. The *advantage* of this variational approach is that from L, one immediately obtains the stress tensor σ and the constitutive law by using the process of differentiation, i.e.,

$$\sigma = L_{u'}.$$

Moreover, one finds

stability criteria;
duality relations between displacement u and the stress tensor σ; and
an elegant method for constructing approximation models.

In Section 61.7 we will show how, for a number of models, one obtains explicit expressions for the elastic energy, i.e., for L. We thereby reach our goal of obtaining a maximum of information from a minimum of model assumptions.

Note that not all basic equations (E) of elastostatics are Euler equations of a variational problem (P), i.e., elastic material need not be hyperelastic.

Theory of Invariants and General Constitutive Laws

In Section 61.8 we want to show how, in the framework of the exact theory, the theory of invariants can be used to obtain the general structure of constitutive laws and stored energy functions. We emphasize that

there exist serious restrictions on the form of the exact constitutive laws.

An important role in this connection is played by the axiom of material frame indifference, which, roughly speaking, postulates that constitutive laws are invariant under rotations.

On the other hand, many constitutive laws, which are used in the literature, do not satisfy these restrictions. Such constitutive laws only correspond to approximation models.

Approximation Models

In order to obtain approximation models, the following two options are available:

(a) One starts from the basic equations of elasticity (or the principle of virtual work) and neglects appropriate terms.
(b) One uses the variational approach and neglects appropriate terms in the variational problem, i.e., the stored energy function L is replaced with an appropriate approximation L_{approx}.

Throughout this volume we will stress the fact that method (b) is much simpler than method (a). In (a) one has to control approximations of a complex system of equations (or, in the case of the principle of virtual work, an integral identity with many components). In (b), on the other hand only approximations of a *single* function L need to be controlled. Furthermore, the function L has a significant physical meaning (density of elastic energy). Method (b) corresponds to a general strategy in theoretical physics:

Use variational principles via Lagrangian functions L and try to obtain important physical and mathematical information from the structure of L.

Let us note that method (a) can only be recommended if the basic equations (E) do not correspond to a variational principle. This, however, is an exceptional case.

General Existence Theory

In Chapters 61 and 62 our main goal will be the proof of general existence theorems. We thereby consider two important ways to approach nonlinear elasticity:

(I) *Local existence and uniqueness of smooth solutions via the implicit function theorem and the continuation method (Section 61.12).*
(II) *Compensated compactness and global existence for polyconvex material (Section 62.13).*

In connection with (I) we shall also explain how this method is related to stability and bifurcation, and, in addition, we will present an important approximation method.

The simple *key idea in* (I) is the following. Write the basic equations of elastostatics as an operator equation

(E) $$F(u) = (K, u_0),$$

where u is the unknown displacement of the elastic body, K denotes the given density of outer forces, and u_0 is the given displacement of the boundary of the body. It is important then that the linearization

(E_{lin}) $$F'(\bar{u})h = (\bar{K}, \bar{u}_0)$$

of (E), at a so-called *strongly stable* state of the deformation $\bar{u}$, corresponds to a linear strongly elliptic system. The general theory of such systems, which will be discussed in Section 61.11, implies that, with respect to suitable function spaces (Hölder spaces or Sobolev spaces), $F'(\bar{u})$ is a *bijective* operator. From the implicit function theorem (Theorem 4.B) we then immediately obtain that equation (E) has unique solutions u in a neighborhood of $\bar{u}$ if K and u_0 lie in a small neighborhood of $\bar{K}$ and $\bar{u}_0$, respectively. Thereby $F(\bar{u}) = (\bar{K}, \bar{u}_0)$.

If we start, for example, with a known strongly stable initial state $\bar{u}$, let's say the rest state $\bar{u} = 0$, then this solution of equation (E) can be continued as long as the continuation remains strongly stable. The discretization of this continuation procedure yields an important approximation method, whose convergence can easily be proved by using well-known methods for ordinary differential equations in B-spaces. This will be done in Section 61.16.

In Section 61.14, we shall show that the loss of stability of an equilibrium state can lead to bifurcation, i.e., to new equilibrium states.

The *basic idea of* (II) is the following. We begin by applying standard arguments of existence theory for variational problems from Chapter 38, i.e., we choose a minimal sequence and select a weakly convergent subsequence. The important fact thereby is that the functional in the variational problem (the potential energy of the elastic body) is *not* convex. Thus we have the usual difficulties in proving that the weak limit of the minimal sequence is a solution of our problem. This difficulty can be overcome by using weak limit properties of integral identities which typically contain terms of the form det A and adj A. In this connection, the skew symmetry of det A and adj A plays a decisive role. In order to apply this technique it is crucial to have the polyconvexity of the energy functional.

Existence Theory for Approximation Models

In (I) and (II) above the full basic equations of nonlinear elasticity are solved without using any additional approximation assumptions. We will, however, also consider several approximation models:

(i) Korn's inequality and generalized solutions in linear elastostatics (application of the main theorem on quadratic variational problems (Theorem 22.A)).

(ii) Classical solutions in linear elastostatics (strongly elliptic systems).
(iii) Generalized solutions in linear elastodynamics (application of the main theorem on linear second-order evolution equations (Theorem 24.A)).
(iv) Duality theory for a class of models with linear and nonlinear Hooke's law, e.g., linear elastic material and nonlinear Hencky material (application of the main theorem of duality theory for monotone potential operators (Theorem 51.B)).
(v) Existence theorem in linear quasi-statical plasticity (quadratic variational inequalities).
(vi) Existence and uniqueness theorem in linear statical plasticity via duality.

As we will show in Chapter 62, the advantage of duality theory is that it leads to effective error estimates for approximation methods (Ritz method) and that it allows a simple approach to statical plasticity theory.

Concrete Problems

In Chapters 63–66 the following topics will be discussed:

(α) Signorini problem for nonlinear material (application of the main theorem on elliptic variational inequalities (Theorem 54.A)).
(β) Buckling of rods and beams and bifurcation theory for variational inequalities.
(γ) Buckling of plates (application of the main theorem on pseudomonotone operators (Theorem 27.A) and of the main theorem of bifurcation theory for potential operators (Theorem 45.A)).
(δ) Quasi-dynamical plasticity theory (application of the main theorem on evolution variational inequalities of first order (Theorem 55.A)).

Invariant Formulation and Functional Analysis

Elastic properties of bodies do not depend on the choice of the coordinate system. Thus the theory is developed in an invariant way, i.e., coordinate-free. We thereby use methods of functional analysis in the space V_3. The stress tensor, for example, is regarded as a linear operator

$$\tau: V_3 \to V_3,$$

while its coordinate representation (τ_j^i) plays only a secondary role. Similarly as in vector calculus, this functional analytic approach simplifies the formulas. One works with geometrical objects rather than with coordinates. The coordinate formulas, however, can easily be obtained from these invariant formulas and will also be given.

Strain Tensor, Stress Tensor, and Constitutive Laws

The two most important notions in elasticity are:

(i) *strain tensor* $\mathscr{E}$; and
(ii) *stress tensor* τ.

As we will show in Section 61.2, $\mathscr{E}$ contains all the important information about the *geometry of deformations*. The stress tensor τ describes the *stress forces* acting in an elastic body. We note here that those two basic notions were already introduced by Cauchy (1827). In Sections 61.3 and 61.4 we will discuss the properties of τ. The strain tensor $\mathscr{E}$ depends in a nonlinear fashion on the displacement vector $u(x)$ of the elastic body, i.e.,

$$\mathscr{E}(x) = \tfrac{1}{2}(u'(x) + u'(x)^* + u'(x)^* u'(x)),$$

where

$$y = x + u(x)$$

describes the deformation of the elastic body. The nonlinear relation between u and $\mathscr{E}$ is responsible for the fact that elasticity theory and hydrodynamics are basically *nonlinear* theories. The *key* relation for the stress tensor τ is

$$\int_{\partial H'} \tau n' \, dO = \text{stress force acting on } H'$$

for all deformed subregions H'. Here, n' is the outer unit normal vector to $\partial H'$. For a homogeneous body, the constitutive law, i.e., the relation between strain and stress, is given by

$$S = B(\mathscr{E}),$$

where S denotes the second Piola–Kirchhoff tensor, i.e.,

$$\tau(y) = y'(x) S(x) y'(x)^* \det x'(y).$$

In the case of hyperelastic material, B is a potential operator, i.e.,

$$S = A'(\mathscr{E})$$

and for the stored energy function we obtain

$$L = A(\mathscr{E}).$$

These important relations show the connection between strain, stress, and elastic potential energy.

It is a remarkable fact that the tensors $\mathscr{E}$, τ, and S are *symmetric*, i.e., they are symmetric linear operators on the H-space V_3. This allows us to apply the *principal axis theorem* to these operators, i.e., these operators have three orthonormal eigenvectors with corresponding real eigenvalues in V_3. The eigenvalues of $(I + 2\mathscr{E}(x))^{1/2}$ are called the principal strains at the point x and the eigenvalues of $\tau(y)$ are called the principal stresses at the point y. The

61.1. Notations

We use again the notations of Section 58.1 and add several new ones. Let

$$L(V_3) \quad \text{and} \quad L_{\text{sym}}(V_3)$$

denote the space of linear and linear, symmetric operators

$$\gamma: V_3 \to V_3,$$

respectively. For $\gamma, \mu \in L(V_3)$ we define

$$\text{tr}\,\gamma = \gamma_i^i, \tag{1}$$

$$[\gamma, \mu] = \text{tr}(\gamma\mu) = \gamma_j^i \mu_i^j, \tag{2}$$

$$\det \gamma = \det(\gamma_j^i). \tag{3}$$

Thereby $\text{tr}\,\delta$ is called the *trace* of γ. As usual, the sum is taken over two equal upper and lower indices from 1 to 3. For example,

$$\text{tr}\,\gamma = \sum_{i=1}^{3} \gamma_i^i.$$

These definitions are independent of the coordinate system. This can be verified explicitly. More elegantly, the invariance is deduced from the general tensor calculus of Chapter 74, since no free indices occur in (1) and (2). For symmetric γ we have

$$\text{tr}\,\gamma = \lambda_1 + \lambda_2 + \lambda_3,$$

where the λ_i are the eigenvalues of γ.

On $L(V_3)$ we introduce a *scalar product* through

$$(\gamma|\mu) = [\gamma, \mu^*] \quad \text{for all} \quad \gamma, \mu \in L(V_3),$$

where the star denotes the adjoint operator on the H-space V_3. In Cartesian coordinates this gives

$$(\gamma|\mu) = \sum_{i,j=1}^{3} \gamma_j^i \mu_j^i.$$

As a special case, we obtain in $L_{\text{sym}}(V_3)$ the scalar product

$$(\gamma|\mu) = [\gamma, \mu] \quad \text{for all} \quad \gamma, \mu \in L_{\text{sym}}(V_3).$$

If, furthermore, $\gamma(\cdot)$ is a tensor field, i.e., $\gamma(x) \in L(V_3)$ for all x in a subset of V_3, then we define the *divergence* of γ,

$$\text{div}\,\gamma(x),$$

61.1. Notations

in an invariant way through the integral formula

$$\int_G \operatorname{div} \gamma \, dx = \int_{\partial G} \gamma n \, dO \tag{4}$$

with outer unit normal vector n. Using integration by parts we obtain that in Cartesian coordinates

$$\int_G (D^j \gamma^i_j) e_i \, dx = \int_{\partial G} (\gamma^i_j n^j) e_i \, dO.$$

This implies

$$\operatorname{div} \gamma(x) = D^j \gamma^i_j(x) e_i \tag{5}$$

with $x = \xi^i e_i$ and $D^j = D_j = \partial/\partial \xi^j$. For smooth $\gamma(\cdot)$ the invariance of $\operatorname{div} \gamma(x)$ follows from

$$\operatorname{div} \gamma(x) = \lim_{G \to x} \frac{1}{\operatorname{meas} G} \int_{\partial G} \gamma n \, dO.$$

This relation is independent of the coordinate system.

Let $A: V_3 \to V_3$ be a bijective linear operator. We then define the operator

$$\operatorname{adj} A = (\det A) A^{-1},$$

and observe that

$$\operatorname{adj}(AB) = \operatorname{adj} B \operatorname{adj} A, \qquad (\operatorname{adj} A)^* = \operatorname{adj} A^*.$$

In Section 61.2 we will consider the *Piola identity*

$$\operatorname{div} \operatorname{adj} y'(x)^* = 0$$

which plays a key role in nonlinear elasticity.

In Cartesian coordinates, the coordinates of $\operatorname{adj} A$ are the adjoint subdeterminants of the matrix corresponding to A^*. We use this property to define $\operatorname{adj} A$ for an arbitrary linear operator $A: V_3 \to V_3$. Then

$$\operatorname{adj} A: V_3 \to V_3$$

is a linear operator as well, and its definition is independent of the choice of the Cartesian coordinate system.

Given two vectors $a, b \in V_3$, we define the *dyadic* product

$$\sigma = a \circ b$$

as the uniquely determined linear operator $\sigma: V_3 \to V_3$ with

$$(a \circ b)x = a(bx) \qquad \text{for all} \quad x \in V_3.$$

In Cartesian coordinates this gives $a = a^i e_i$, $x = \xi^i e_i$, etc., and hence $\sigma x = \sigma^i_j \xi^j e_i = a^i b_j \xi^j e_i$, i.e.,

$$(a \circ b)^i_j = a^i b_j.$$

Moreover, we will use the following notation:

- u displacement vector;
- $\mathscr{E}$ strain tensor;
- γ linearized strain tensor;
- τ stress tensor (in the defomed region);
- σ first Piola–Kirchhoff stress tensor (reduced stress tensor in the undeformed region);
- S second Piola–Kirchhoff tensor (in the undeformed region);
- x position vector in the undeformed region;
- y position vector in the deformed region, $y = x + u(x)$;
- G undeformed region;
- G' deformed region (deformation of G);
- H undeformed subregion of G;
- H' deformation of H;
- F density of outer forces in the deformed region;
- K density of the outer forces with respect to the undeformed region;
- T density of the outer boundary forces with respect to the undeformed region;
- L stored energy function;
- U elastic potential energy;
- ρ mass density of the deformed body;
- ρ_0 mass density of the undeformed body.
- $\bar{\gamma}$ deviation of γ, $\bar{\gamma} = \gamma - 3^{-1}\operatorname{tr}\gamma I$.

61.2. Strain Tensor and the Geometry of Deformations

Our first goal is to introduce the strain tensor $\mathscr{E}$. We thereby keep in mind that

(i) $\mathscr{E}$ is invariant under rigid motions (rotations and translations); and
(ii) $\mathscr{E}$ contains all necessary information about the geometry of the deformation.

We shall use the following two steps.

(a) First, we consider the so-called stretch tensor E, which is defined for all orientation-preserving linear operators from V_3 to V_3.
(b) Second, we consider the linearizations of arbitrary deformations. The corresponding stretch tensor then leads to $\mathscr{E}$.

A *stretch tensor* thereby is a linear, symmetric, strongly positive operator $E\colon V_3 \to V_3$. Such a tensor has a simple intuitive interpretation. From the principal axis theorem it follows that there exists an orthonormal system of eigenvectors $\{\bar{e}_1, \bar{e}_2, \bar{e}_3\}$ with

$$E\bar{e}_i = (1 + \lambda_i)\bar{e}_i, \qquad i = 1, 2, 3. \tag{6}$$

61.2. Strain Tensor and the Geometry of Deformations

The positive eigenvalues $1 + \lambda_i$ of E are called *principal strains*. Thus E describes a transformation with a strain of

$$\Delta l_i / l_i = \lambda_i$$

in the direction of the $\bar{e}_i$-axis for all i. These axes are called the *principal axes of strain*.

We now consider, more generally, an arbitrary affine, orientation-preserving map of V_3. Such a map has the form

$$y = Ah + b, \quad h \in V_3, \tag{7}$$

where $A \in L(V_3)$ and $\det A > 0$. The fixed element $b \in V_3$ describes a translation. We want to represent transformation (7) as the product of a translation, a rotation, and a stretch tensor. This decomposition is of central importance in elasticity. If we consider physical quantities, which occur in connection with elastic deformations, then we expect that such quantities, for example, the elastic energy, do not depend on the rotations and translations of the body, but only on the strain described by the corresponding stretch tensor. As usual, we mean by a *rotation*, an operator $R \in L(V_3)$ with

$$R^*R = I \quad \text{and} \quad \det R = 1.$$

The following proposition contains the key to the strain tensor $\mathscr{E}$ below.

Proposition 61.1 (Normal Form). *If $A: V_3 \to V_3$ is a linear operator with $\det A > 0$, then it can be uniquely written as*

$$A = RE,$$

i.e., as a product of a stretch tensor E and a rotation R. It is

$$E = (A^*A)^{1/2}.$$

PROOF. The operator A^*A is symmetric and strongly positive. This follows from $(A^*A)^* = A^*A$ and

$$(x | A^*Ax) = (Ax | Ax) \geq 0$$

as well as from the fact that $\det A^*A = \det A^* \det A > 0$.

The uniqueness of E is a consequence of $E = R^{-1}A$ and

$$E^2 = E^*E = (R^{-1}A)^*(R^{-1}A) = A^*(R^{-1})^*R^{-1}A$$
$$= A^*A.$$

Hence we obtain $E = (A^*A)^{1/2}$.

The existence of the decomposition follows by letting

$$E = (A^*A)^{1/2} \quad \text{and} \quad R = AE^{-1}. \qquad \square$$

61.2a. Taylor Expansion and Strain Tensor

We now study the local behavior of a C^1-map

$$y = x + u(x), \quad x \in V_3 \tag{8}$$

from the space V_3 into itself. The Taylor expansion at the point x yields

$$y(x + h) = x + u(x) + h + u'(x)h + o(\|h\|), \quad h \to 0, \tag{9}$$

and thus the linearization of the map (8) at the point x has the form

$$y = b + Ah$$

with

$$b = x + u(x) \quad \text{and} \quad A = I + u'(x).$$

Proposition 61.1 then shows that the corresponding local stretching can be described by

$$E = (A^*A)^{1/2}$$

with

$$A^*A = I + u'(x) + u'(x)^* + u'(x)^*u'(x).$$

We set

$$A^*A = I + 2\mathscr{E}(x).$$

Definition 61.2. The linear operator $\mathscr{E}(x): V_3 \to V_3$ with

$$\mathscr{E}(x) = \tfrac{1}{2}(u'(x) + u'(x)^* + u'(x)^*u'(x))$$

is called the *strain tensor* of the transformation (8) at the point x.

The linear operator $\gamma(x): V_3 \to V_3$ with

$$\gamma(x) = \tfrac{1}{2}(u'(x) + u'(x)^*)$$

is called the *linearized strain tensor* at x.

It is clear that both $\mathscr{E}(x)$ and $\gamma(x)$ are *symmetric* linear operators on V_3.
In the future we shall call $\mathscr{E}$ and γ simply strain tensors. We note, however, that $\mathscr{E}$ and γ will always refer to the strain tensor and the linearized strain tensor, respectively. From

$$u'(x)h = (D_j u^i(x)h^j)e_i$$

we obtain in Cartesian coordinates the representation

$$\mathscr{E}^i_j = \tfrac{1}{2}(D_j u^i + D^i u_j + D^i u_k D_j u^k) \tag{10}$$

for the matrix $(\mathscr{E}^i_j)$ which corresponds to $\mathscr{E}(x)$ according to Section 58.1. As usual, we set $D^j = D_j = \partial/\partial \xi^j$ and $x = \xi^j e_j$, $u^k = u_k = u^k(x)$. The matrix which

61.2. Strain Tensor and the Geometry of Deformations

corresponds to $\gamma(x)$ satisfies

$$\gamma_j^i = \tfrac{1}{2}(D_j u^i + D^i u_j). \tag{11}$$

In order to give some intuitive meaning to $\gamma(x)$, we set

$$\gamma_- = \tfrac{1}{2}(u'(x) - u'(x)^*).$$

From (9) then follows that

$$y(x + h) = (x + u(x)) + (h + \gamma(x)h + \gamma_-(x)h) \\ + o(\|h\|), \quad h \to 0. \tag{12}$$

Passing to Cartesian coordinates we obtain

$$\gamma_-(x)h = \omega \times h \quad \text{for all} \quad h \in V_3$$

with the fixed vector

$$\omega = \tfrac{1}{2} \operatorname{curl} u(x).$$

Hence it follows from (12) that:

The linearization of the transformation $y = x + u(x)$ at the point x is the sum of

(i) *a translation about the vector $x + u(x)$;*
(ii) *the stretching $(I + \gamma(x))h$; and*
(iii) *the infinitesimal rotation $\gamma_-(x)h$.*

For the change in volume we obtain

$$\int_{H'} dy = \int_{H} J(x)\,dx = \int_{H} (1 + \operatorname{div} u)\,dx + o(\|u'\|),$$

where $J(x) = \det y'(x)$. Hence, in first-order approximation, $\operatorname{div} u(x)$ is the relative local change in volume. Roughly, one has that $\Delta V' = (1 + \operatorname{div} u(x))\Delta V$ and hence

$$(\Delta V' - \Delta V)/\Delta V = \operatorname{div} u(x).$$

The observations above illustrate the intuitive meaning of $\operatorname{curl} u(x)$ and $\operatorname{div} u(x)$, where $u(x)$ denotes the displacement vector.

61.2b. Geometrical Meaning of the Strain Tensor

Let H be an undeformed subregion, Σ an undeformed surface with unit normal vector n, and C and C_* two undeformed curves with parametrization

$$x = x(\alpha) \quad \text{and} \quad x_* = x_*(\alpha),$$

respectively. Let $s = s(\alpha)$ denote the arclength of C, and $\dot{s}$ the derivative with respect to α.

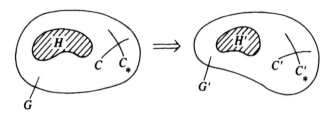

Figure 61.2

We then consider a C^1-deformation

$$y = x + u(x)$$

from the undeformed region G onto the deformed region G'. We set $J(x) = \det y'(x)$ and suppose that $J(x) > 0$ on G. Moreover, we set

$$A = y'(x)$$

and recall that $A = RE$, where R is a rotation, and

$$E^2 = A^*A = I + 2\mathscr{E}(x).$$

With H', Σ', C', C'_* we denote the deformations of the sets H, Σ, C, C_*, respectively (Fig. 61.2). The deformed curve C' then has the parametrization

$$y = x(\alpha) + u(x(\alpha)).$$

Obviously, we obtain $\dot{y} = \dot{x} + u'\dot{x}$ and hence

$$\dot{y} = A\dot{x}, \qquad \dot{y}_* = A\dot{x}_*.$$

By observing that $\dot{s}^2 = (\dot{x}|\dot{x})$ and $\dot{s}'^2 = (\dot{y}|\dot{y})$, we find the first two of the following four *basic formulas*:

(I) $\qquad (\dot{y}|\dot{y}_*) = (\dot{x}|\dot{x}_*) + 2(\dot{x}|\mathscr{E}(x)\dot{x}_*),$

(II) $\qquad \dot{s}'^2 = \dot{s}^2 + 2(\dot{x}|\mathscr{E}(x)\dot{x}),$

(III) $\qquad dV' = \det A \, dV = \sqrt{\det(I + 2\mathscr{E}(x))} \, dV,$

(IV) $\qquad dO' = |(\text{adj } A)^* n| \, dO = |\sqrt{\text{adj}(I + 2\mathscr{E}(x))} \, n| \, dO.$

The definition of the operator adj A has been given in Section 61.1. It leads to the following integral formulas for the *change* in arclength, volume, and surface under the considered deformation:

(II*) $\qquad s'(\beta) = \int_0^\beta \sqrt{(\dot{x}|\dot{x}) + 2(\dot{x}|\mathscr{E}(x)\dot{x})} \, d\alpha, \qquad x = x(\alpha),$

(III*) $\qquad \int_{H'} dV' = \int_H \sqrt{\det(I + 2\mathscr{E}(x))} \, dV,$

(IV*) $\qquad \int_{\Sigma'} dO' = \int_\Sigma |\sqrt{\text{adj}(I + 2\mathscr{E}(x))} \, n| \, dO.$

61.2. Strain Tensor and the Geometry of Deformations

In fact, (III) and (IV) represent a short-hand writing for (III*) and (IV*), respectively. A proof will be given below. The key to (IV*) is the so-called Piola identity.

Let us note that the formulas above show that the change in arclength, volume, and surface can be described exclusively in terms of A, $\det A$, and $\operatorname{adj} A$. This observation will play a fundamental role in the theory of polyconvex materials.

The following discussion should help to illustrate the intuitive idea behind these formulas. Let $\{\bar{e}_1, \bar{e}_2, \bar{e}_3\}$ be an orthonormal system of eigenvectors of E with corresponding eigenvalues s_1, s_2, s_3, which are called the principal strains at the point x. The $\bar{e}_i$-axes are called the principal axes of strain at x. If ε_i are the eigenvalues of the strain tensor $\mathscr{E}(x)$, then

$$s_i = \sqrt{1 + 2\varepsilon_i}, \quad i = 1, 2, 3.$$

In terms of s_i we obtain

$$dV' = s_1 s_2 s_3 \, dV,$$
$$dO' = \sqrt{n_1^2 s_2^2 s_3^2 + n_2^2 s_1^2 s_3^2 + n_3^2 s_1^2 s_2^2} \, dO,$$

where $n = \sum_{i=1}^{3} n_i \bar{e}_i$. If $\|\mathscr{E}(x)\|$ is small, i.e., if $\varepsilon_1, \varepsilon_2, \varepsilon_3$ are small, then

$$dV' = (1 + \varepsilon_1 + \varepsilon_2 + \varepsilon_3) dV + o(|\varepsilon|)$$
$$= (1 + \operatorname{tr} \mathscr{E}(x)) dV + o(|\varepsilon|), \quad \varepsilon \to 0,$$
$$dO' = (1 + n_1^2(\varepsilon_2 + \varepsilon_3) + n_2^2(\varepsilon_1 + \varepsilon_3) + n_3^2(\varepsilon_1 + \varepsilon_2)) dO + o(|\varepsilon|),$$

where $|\varepsilon| = |\varepsilon_1| + |\varepsilon_2| + |\varepsilon_3|$.

61.2c. Curve Deformation and the Strain Tensor

In order to illustrate the intuitive meaning of the components $\mathscr{E}_j^i$ of $\mathscr{E}(x)$, with respect to an arbitrary Cartesian coordinate system, we fix the point x and choose an orthonormal system of vectors $\{e_1, e_2, e_3\}$. We then consider the straight lines

$$x_i = x + \alpha e_i, \quad i = 1, 2, 3$$

through the point x. These straight lines are deformed into the curves

$$y_i = x + \alpha e_i + u(x + \alpha e_i), \quad i = 1, 2, 3$$

with tangent vectors

$$e_i' = e_i + u'(x) e_i, \quad i = 1, 2, 3$$

at the point $y = x + u(x)$. Denote the angle between e_i' and e_j' by φ_{ij} (Fig. 61.3).

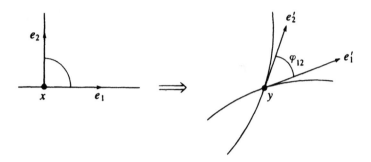

Figure 61.3

The components $\mathscr{E}^i_j$ of $\mathscr{E}(x)$ are given by

$$\mathscr{E}(x)e_j = \mathscr{E}^i_j e_i.$$

Setting $\mathscr{E}_{ij} = \mathscr{E}^i_j$ and $|\mathscr{E}| = \sum_{i,j=1}^{3} \mathscr{E}_{ij}$, we obtain

$$e'_i e'_j = \delta_{ij} + 2\mathscr{E}_{ij}, \qquad i, j = 1, 2, 3,$$

and hence

$$|e'_i| = \sqrt{1 + 2\mathscr{E}_{ii}},$$

$$\cos \varphi_{ij} = \frac{e'_i e'_j}{|e'_i||e'_j|} = \frac{2\mathscr{E}_{ij}}{\sqrt{1 + 2\mathscr{E}_{ii}}\sqrt{1 + 2\mathscr{E}_{jj}}}$$

for all $i \neq j$. Note here that $\cos \varphi_{ij} = \sin(\pi/2 - \varphi_{ij})$, so that, for small $|\mathscr{E}|$, the two *key* formulas

$$|e'_i| = 1 + \mathscr{E}_{ii} + o(|\mathscr{E}|), \qquad |\mathscr{E}| \to 0,$$

$$\frac{\pi}{2} - \varphi_{ij} = 2\mathscr{E}_{ij} + o(|\mathscr{E}|), \qquad i \neq j$$

are valid.

61.2d. Volume Deformation and the Strain Tensor

Using the well-known transformation rule for integrals, we find that

$$\int_{H'} dy = \int_{H} \det y'(x)\, dx.$$

We then observe that $\det A = \det y'(x)$ and that

$$\det(I + 2\mathscr{E}(x)) = \det A^* \det A = (\det A)^2.$$

61.2e. The Piola Identities

The following two so-called Piola identities hold true for the undeformed region G:

(P1) $$\operatorname{div}_x \operatorname{adj} y'(x)^* = 0,$$

(P2) $$\operatorname{div}_x \tau(y(x)) \operatorname{adj} y'(x)^* = J(x) \operatorname{div}_y \tau(y).$$

The index thereby indicates with respect to which variable div has to be applied. Suppose that $\tau: G' \to L(V_3)$ is C^1, i.e., $\tau(\cdot)$ is a C^1-tensor field on the deformed region G'. Differentiating the identity $y = y(x(y))$ with respect to y, we obtain the relation $I = y'(x)x'(y)$. This implies

$$\operatorname{adj} y'(x)^* = (\det y'(x)^*) y'(x)^{*-1} = J(x) x'(y)^*,$$

and thus we can write the Piola identities in the form

(P1*) $$\operatorname{div}_x J(x) x'(y(x))^* = 0,$$

(P2*) $$\operatorname{div}_x J(x) \tau(y(x)) x'(y(x))^* = J(x) \operatorname{div}_y \tau(y).$$

The transformation

$$\sigma(x) = \tau(y) \operatorname{adj} y'(x)^*$$
$$= J(x) \tau(y) x'(y)^*, \quad y = y(x)$$

is called the *Piola transformation*. It will play an important role in Section 61.5, where the reduced stress tensor will be introduced via the Piola transformation.

Let us now prove (P1) and (P2). Identity (P1) is precisely formula (12.14) of Part I. Recall that this formula was the key to our simple analytic approach to the mapping degree of Chapter 12. In Cartesian coordinates (P1) and (P1*) read as

$$\frac{\partial}{\partial \xi^i}\left(J \frac{\partial \xi^i}{\partial \eta^j}\right) = 0.$$

This implies (P2). In fact, we obtain

$$\frac{\partial}{\partial \xi^j}\left(J \tau_k^i \frac{\partial \xi^j}{\partial \eta^k}\right) = J \frac{\partial \tau_k^i}{\partial \xi^j} \frac{\partial \xi^j}{\partial \eta^k} = J \frac{\partial \tau_k^i}{\partial \eta^k}.$$

61.2f. Surface Deformation and the Strain Tensor

We use the Piola transformation $\sigma(x) = \tau(y(x)) \operatorname{adj} y'(x)^*$. Formula (P2) then implies

$$\int_{\partial H} \sigma n \, dO = \int_H \operatorname{div} \sigma \, dx = \int_H J(x) \operatorname{div}_y \tau(y(x)) \, dx$$
$$= \int_{H'} \operatorname{div}_y \tau(y) \, dy = \int_{\partial H'} \tau n' \, dO'.$$

Setting $\tau = I$, we obtain

$$\int_{\partial H} (\text{adj } y'(x)^*) n \, dO = \int_{\partial H'} n' \, dO'$$

and hence, taking norms,

$$\int_{\partial H} |(\text{adj } y'(x)^*) n| \, dO = \int_{\partial H'} dO',$$

i.e.,

$$dO' = |(\text{adj } y'(x)^*) n| \, dO.$$

By definition of $\mathscr{E}$, there exists a rotation R such that

$$y'(x) = RE, \qquad E = \sqrt{I + 2\mathscr{E}(x)}.$$

From $RR^* = I$ and $\det R = 1$ we obtain

$$\text{adj } R^* = R,$$

and hence

$$\text{adj } y'(x)^* = \text{adj}(E^* R^*) = R \, \text{adj } E = R(\text{adj } E^2)^{1/2}.$$

Note that $E^* = E$. Thus

$$dO' = |\sqrt{\text{adj } I + 2\mathscr{E}(x)}\, n| \, dO.$$

61.3. Basic Equations

In the general case, the motion of an elastic body is described by the following three equations.

(i) *Deformation*

$$y = x + u(x, t), \qquad (x, t) \in \bar{G} \times [0, t_0]. \tag{13}$$

(ii) *Equation of motion*

$$\rho(Q)\ddot{y} = \text{div } \tau(Q) + F(Q), \qquad Q = (y, t). \tag{14}$$

(iii) *Constitutive law*

$$\tau(Q) = \Phi(u(P), u_x(P), P), \qquad P = (x, t). \tag{15}$$

We are looking for

the displacement u.

In (14) div is applied with respect to y. Moreover, we have

$$\ddot{y} = u_{tt}(x, t).$$

We now discuss the basic equations.

61.3. Basic Equations

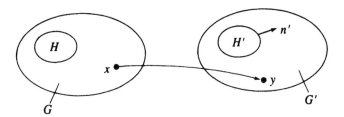

Figure 61.4

61.3a. Deformation

The deformation of the region G occurs in such a way that at time t

the point $x \in \bar{G}$ becomes the point y

(Fig. 61.4). The quantity $\ddot{y}$ is then the acceleration of the point y at time t. The subregion H of G becomes H' at time t. In order to obtain a regular deformation, we have to assume that formula (13) describes

an orientation preserving diffeomorphism from $\bar{G}$ onto $\bar{G}'$

for every t. Because of the complexity of the basic equations in nonlinear elasticity, one often shows just the existence of the displacement u without proving the diffeomorphism property.

We call G and G' the reference configuration and the deformed configuration, respectively. Moreover, the coordinates x in G and y in G' are called *Lagrangian* and *Eulerian* coordinates, respectively. In elasticity one frequently uses Lagrangian coordinates, while in hydrodynamics one usually works with Eulerian coordinates.

Differentiation of the identity $y = y(x(y), t)$ with respect to y yields

$$y_x(x, t) x_y(y, t) = I, \qquad \det y_x(x, t) \det x_y(x, t) = 1,$$

where $y = x + u(x, t)$. These formulas will constantly be used in the following.

61.3b. Stress Tensor

The stress tensor

$$\tau(Q): V_3 \to V_3$$

should be a linear, *symmetric* operator. The *physical* meaning of $\tau(Q)$ is that

$$\int_{\partial H'} \tau n' \, dO = \text{stress force acting on } H' \tag{16}$$

holds true, where n' is the outer unit normal vector to the deformed subregion H'. A typical property of stress forces is that they can be described by *surface* integrals, i.e., they represent surface forces. Here, dO denotes the surface element of $\partial H'$. Moreover, we have

$$\int_{H'} F \, dy = \text{outer force acting on } H', \tag{17}$$

$$\int_{H} \rho_0(x) \, dx = \text{mass of } H, \tag{18}$$

$$\int_{H'} \rho(y) \, dy = \text{mass of } H'. \tag{19}$$

Let τ_1, τ_2, τ_3 be the eigenvalues of the stress tensor τ at a fixed point y with corresponding orthonormal system of eigenvectors n_1, n_2, n_3. The decomposition

$$n' = \alpha^1 n_1 + \alpha^2 n_2 + \alpha^3 n_3 = \alpha^i n_i$$

of the outer unit normal vector n' at the point y yields

$$\tau n' = \alpha^1 \tau_1 n_1 + \alpha^2 \tau_2 n_2 + \alpha^3 \tau_3 n_3 = \alpha^i (\tau_i n_i).$$

Thus, we obtain

$$\int_{\partial H'} \alpha^i (\tau_i n_i) \, dO = \text{stress force acting on } H'.$$

This explains the intuitive meaning of the so-called *principal stresses* τ_1, τ_2, τ_3 and *principal axes of stress* n_1, n_2, n_3.

61.3c. Mass Conservation

We postulate mass conservation. This gives

$$\int_{H} \rho_0 \, dx = \int_{H'} \rho \, dy$$

for all subregions H. Letting

$$J(x, t) = \det y_x(x, t),$$

we obtain

$$\int_{H'} \rho \, dy = \int_{H} \rho J \, dx.$$

By contracting H into a single point we find that

$$\rho(y, t) J(x, t) = \rho_0(x)$$

with $y = y(x, t)$. This is a consequence of the mean value theorem of integral calculus. Hence, for a given mass density $\rho_0(x)$ of the undeformed body, we

61.3. Basic Equations

have to set

$$\rho(y,t) = \rho_0(x)J(x,t)^{-1} \tag{19*}$$

in the basic equation (14).

61.3d. Constitutive Law

The constitutive law (15) describes the connection between

deformation (strain) and stress.

Constitutive laws depend critically on the material under consideration. In a rigorous theory, the constitutive law cannot be prescribed in an arbitrary fashion but must satisfy serious restrictions. This will be discussed in Section 61.8. The general form (15) of the constitutive law corresponds to approximation models in elasticity.

61.3e. Typical Difficulties

In Cartesian coordinates, the equation of motion (14) is equal to

$$\rho(Q)u_{tt}^i(P) = D'^j \tau_j^i(Q) + F^i(Q), \quad i = 1, 2, 3 \tag{20}$$

with $y = \eta^i e_i$ and $D'^j = \partial/\partial \eta^j$.

The main difficulty with these basic equations is that the differentiation on the right-hand side of (20) is taken with respect to the unknown deformed region G'. Therefore (20) only apparently has a simple structure. Also, the number of constitutive laws, which are being used in the literature, is rather large. This is a consequence of the fact that these laws cannot be uniquely determined from general physical arguments.

The mathematical difficulties are caused by the physical reality that bodies may become plastic or even break under deformations.

61.3f. Initial and Boundary Conditions

In order to obtain the complete system of basic equations, we need to add initial and boundary conditions:

(a) The initial conditions are obtained by prescribing the initial values

$$u(x,0) \quad \text{and} \quad u_t(x,0) \quad \text{on } \bar{G},$$

i.e., by prescribing the initial position and the initial velocity.

(b) In order to formulate the boundary conditions, we decompose the boundary ∂G of the undeformed region into the two disjoint sets

$$\partial G = \partial_1 G \cup \partial_2 G$$

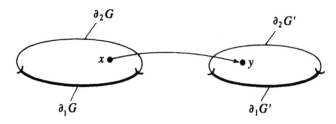

Figure 61.5

(Fig. 61.5). We then prescribe the displacement

$$u(x,t) \quad \text{on } \partial_1 G$$

and the stress forces

$$\tau n' \quad \text{on } \partial_2 G'$$

for all times t.

If the *displacement u* is known, then one obtains the *stress tensor* from the constitutive law (15) and the *stress forces*, which are acting on all parts of the deformed body, from the integral formula (16). These forces are of particular interest to the engineer, since they determine how the material in buildings, bridges, etc. is stressed.

61.3g. Elastostatics

If the displacement u does not depend on time, then we have the common *stationary problem*. In this case the general basic equations above become the basic equations of elastostatics. The term $\rho \ddot{y}$ disappears in the equation of motion (14) and time t does not explicitly appear in the basic equations. Here y denotes the position of the point x after the deformation.

61.4. Physical Motivation of the Basic Equations

Our observations are based on the following three points:

(i) Structure of the stress forces and the stress tensor.
(ii) Momentum.
(iii) Angular momentum.

We assume in the following that all functions, transformations, and regions are sufficiently regular.

61.4a. Stress Forces and Stress Tensor

Following Cauchy (1827) the main assumption of elasticity theory is that a deformation causes stress forces which can be described like this:

$$\int_{\partial H'} \bar{\tau}(y, t; n') \, dO = \text{stress force acting on } H'.$$

The heuristic idea behind this is that the stress force

$$\bar{\tau} \, \Delta O,$$

which acts from the domain $G' - H'$ onto H', is applied to the surface element ΔO of $\partial H'$, whereby this force depends on the outer unit normal vector n' of $\partial H'$. In the following we shall simply write n instead of n'. From Newton's principle of actio = reactio, we may, conversely, assume that the stress force

$$-\bar{\tau} \, \Delta O$$

acts from the domain H' onto the surface element ΔO of $G' - H'$. We therefore postulate

$$\bar{\tau}(y, t; -n) = -\bar{\tau}(y, t; n). \tag{21}$$

for all n. Below, we shall prove the linearity of $\bar{\tau}$ with respect to n, i.e., there exists a linear operator $\tau: V_3 \to V_3$ with

$$\bar{\tau}(y, t; n) = \tau(y, t)n \tag{22}$$

for all n. The operator $\tau(y, t)$ is called *stress tensor*. Moreover, we shall demonstrate the symmetry of the stress tensor, i.e.,

$$\tau(y, t) = \tau(y, t)^*. \tag{23}$$

All important information will follow from the integral form of the equations of motion, which we will now discuss.

61.4b. Momentum Balance and Angular Momentum Balance

In Section 58.7, i.e., in the context of point mechanics we derived the following important balance laws for momentum and angular momentum:

Time derivative of the total momentum = *total force*.
Time derivative of the total angular momentum = *total torque*.

This will be our key to the derivation of the basic equations. If we decompose the deformed region G' into small parts, then, after summation and passing to limits, the following assumptions make sense:

$$\frac{d}{dt} \int_{H'} \rho u_t \, dy = \int_{H'} F \, dy + \int_{\partial H'} \bar{\tau} \, dO, \tag{24}$$

$$\frac{d}{dt}\int_{H'} \rho u_t \times y\, dy = \int_{H'} F \times y\, dy + \int_{\partial H'} \vec{\tau} \times y\, dO. \tag{25}$$

Formulas (24) and (25) are the momentum balance and angular momentum balance, respectively. Note that $\dot{y} = u_t$. For the time derivatives we obtain

$$\frac{d}{dt}\int_{H'} \rho u_t\, dy = \frac{d}{dt}\int_H \rho_0 u_t\, dx = \int_H \rho_0 u_{tt}\, dx$$

$$= \int_{H'} \rho u_{tt}\, dy,$$

$$\frac{d}{dt}\int_{H'} \rho u_t \times y\, dy = \frac{d}{dt}\int_H \rho_0 u_t \times y\, dx$$

$$= \int_H \rho_0 u_{tt} \times y\, dx = \int_{H'} \rho u_{tt} \times y\, dy.$$

Moreover, observe that $\rho J = \rho_0$, $dy = J\, dx$, and $u_t \times \dot{y} = u_t \times u_t = 0$. From (22), which has yet to be proved and (4) it follows that

$$\int_{\partial H'} \vec{\tau}\, dO = \int_{\partial H'} \tau n\, dO = \int_{H'} \operatorname{div} \tau\, dy.$$

The momentum balance (24) gives

$$\int_{H'} (\rho u_{tt} - F - \operatorname{div} \tau)\, dy = 0,$$

and by contracting H into a point, we obtain

$$\rho u_{tt} - F - \operatorname{div} \tau = 0.$$

This is the equation of motion (14).

61.4c. Existence of the Stress Tensor

We prove (22). From the momentum balance (24) follows

$$\int_{H'} (\rho u_{tt} - F)\, dy = \int_{\partial H'} \vec{\tau}\, dO. \tag{26}$$

We choose H' as a tetrahedron with outer unit normal vectors

$$-e_1, -e_2, -e_3, \quad n = n^j e_j,$$

side surfaces S^1, S^2, S^3, S, and volume V. This implies that

$$S^j = Sn^j.$$

From (26) and the mean value theorem of integral calculus it follows that

$$V(\rho u_{tt} - F) = S\vec{\tau}(\cdot, n) + S^j \vec{\tau}(\cdot, -e_j).$$

61.4. Physical Motivation of the Basic Equations

More precisely, this relation holds for the components at suitable points of the tetrahedron. Contracting the tetrahedron with a similarity transformation into a single point, we obtain after dividing by S that

$$\bar{\tau}(\cdot, n) = -n^j \bar{\tau}(\cdot, -e_j) = n^j \bar{\tau}(\cdot, e_j).$$

This follows since $V/S \to 0$, and hence we obtain the linearity of $n \mapsto \bar{\tau}(\cdot, n)$. Thus, there exists a linear operator $\tau: V_3 \to V_3$ such that

$$\bar{\tau}(\cdot, n) = \tau n \quad \text{for all} \quad n \in V_3.$$

This is (22).

61.4d. Symmetry of the Stress Tensor

We shall prove (23), i.e., the symmetry of τ, by using the angular momentum balance and the key identity

$$\int_{H'} (\tau^* - \tau) \, dy = \int_{H'} \text{div} \, \tau \times y \, dy - \int_{\partial H'} \tau n \times y \, dO$$

for arbitrary smooth $\tau \in L(V_3)$. The latter equation follows from passing to Cartesian coordinates and using integration by parts. In fact, we have that

$$\int \varepsilon_{mij} \eta^i D^r \tau_r^j \, dy - \int \varepsilon_{mij} \eta^i \tau_r^j n^r \, dO$$

$$= \int \varepsilon_{mij} (\eta^i D^r \tau_r^j - D^r(\eta^i \tau_r^j)) \, dy$$

$$= \int \varepsilon_{mij} \tau_i^j \, dy.$$

In order to prove

$$\tau^* = \tau,$$

we need

$$\int_{H'} \text{div} \, \tau \times y \, dy = \int_{\partial H'} \tau n \times y \, dO$$

for all H'. But this follows from the angular momentum balance (25), i.e.,

$$\int_{H'} \rho u_{tt} \times y \, dy = \int_{H'} F \times y \, dy + \int_{\partial H'} \tau n \times y \, dO,$$

and from the equation of motion

$$\rho u_{tt} = \text{div} \, \tau + F.$$

Note that $\bar{\tau} = \tau n$.

61.5. Reduced Stress Tensor and the Principle of Virtual Power

The basic equations in Section 61.3 contain the stress tensor

$$\tau(y, t)$$

which corresponds to the *deformed* region G'. In order to avoid this complication, we introduce a new (reduced) stress tensor

$$\sigma(x, t)$$

in the *undeformed* region G, which has the following physical meaning:

$$\int_{\partial H} \sigma(x, t) n \, dO = \text{stress force acting on } H'. \tag{27}$$

On the other hand, we obtain the following relation

$$\int_{\partial H'} \tau(y, t) n' \, dO = \text{stress force acting on } H' \tag{28}$$

for τ. Note that, in (27), the integral is taken over the boundary of the undeformed subregion H with outer unit normal vector n, and, in (28), over the boundary of the deformed subregion H' with outer unit normal vector n'.

The fact that (28) can be transformed into (27), the linearity of τ implying the linearity of σ, is by *no* means trivial. As we will see, this follows from the *Piola identity*.

Moreover, we set

$$\int_H K(x, t) \, dx = \text{outer force acting on } H'. \tag{29}$$

In Section 61.3, we used

$$\int_{H'} F(y, t) \, dy = \text{outer force acting on } H'. \tag{30}$$

61.5a. Transformed Basic Equations

We shall see below that the basic equations for the deformed region of Section 61.3 imply the following new basic equations for the *undeformed* region which are easier to handle:

61.5. Reduced Stress Tensor and the Principle of Virtual Power

(i) *Deformation*

$$y = x + u(x,t), \qquad (x,t) \in \bar{G} \times [0, t_0]. \tag{31}$$

(ii) *Equation of motion*

$$\rho_0(x) u_{tt}(P) = \text{div } \sigma(P) + K(P), \qquad P = (x,t). \tag{32}$$

(iii) *Constitutive law*

$$\sigma(P) = \Omega(u(P), u_x(P), P). \tag{33}$$

We are looking for the

displacement u.

Then one obtains the deformation from (31) and the stress forces, which are effective in the deformed body, from (27) and (33). In equation (32), div is used with respect to the variable x.

Furthermore, it is necessary to add the following *initial* and *boundary conditions*. As in Section 61.3, we use the decomposition

$$\partial G = \partial_1 G \cup \partial_2 G, \qquad \partial_1 G \cap \partial_2 G = \emptyset$$

of the boundary ∂G for the undeformed region G (Fig. 61.5).

(a) We prescribe the initial values

$$u(x,0) \quad \text{and} \quad u_t(x,0) \quad \text{on } \bar{G},$$

i.e., the initial position and the initial velocity of the deformed body.

(b) We prescribe the displacement

$$u(x,t) \quad \text{on } \partial_1 G$$

and the stress forces

$$\sigma(x,t) n \quad \text{on } \partial_2 G$$

for all times t.

In Cartesian coordinates, the equation of motion (32) becomes

$$\rho_0(x) u_{tt}^i(P) = D^j \sigma_j^i(P) + K^i(P), \qquad i = 1, 2, 3 \tag{34}$$

with $x = \xi^j e_j$ and $D^j = \partial/\partial \xi^j$.

61.5b. Transformation of Forces

The fundamental relation between the forces is given by

$$K(x,t) = J(x,t) F(y,t)$$

with $y = y(x,t)$ and corresponding inverse $x = x(y,t)$. Recall that

$$J(x,t) = \det y_x(x,t),$$

and
$$\rho_0(x) = J(x,t)\rho(y,t).$$

If, for example, the gravitational force F is applied to the deformed body, then
$$F(y,t) = -\rho(y,t)ge_3.$$

This implies that
$$K(x) = -\rho_0(x)ge_3,$$

i.e., the transformation of F onto the undeformed region assumes a particularly simple form.

61.5c. The First Piola–Kirchhoff Tensor

By definition, the important relation between σ and τ is given by the Piola transformation
$$\sigma(x,t) = J(x,t)\tau(y,t)x_y(y,t)^*.$$

The linear map $\sigma(x,t): V_3 \to V_3$ is called the *reduced* stress tensor or first Piola–Kirchhoff tensor. Recall that, according to Section 58.1, the space V_3 is an H-space and the star denotes the adjoint operator. In the future τ and σ will simply be called stress tensors, but τ and σ will always refer to the stress tensor and the reduced stress tensor, respectively.

61.5d. The Second Piola–Kirchhoff Tensor

The second Piola–Kirchhoff tensor S is defined as
$$S(x,t) = J(x,t)x_y(y,t)\tau(y,t)x_y(y,t)^*.$$

Because of $(AB)^* = B^*A^*$ we have
$$S^* = S,$$

i.e., S is symmetric, while for σ this need not be true. In Cartesian coordinates we can write
$$\sigma_j^i = J\tau_k^i \frac{\partial \xi^j}{\partial \eta^k},$$

and
$$S_j^i = \sigma_j^k \frac{\partial \xi^i}{\partial \eta^k}.$$

61.5e. Proof of the Transformation Formulas

The key here is the Piola identity

$$\text{div}_x \sigma(x, t) = J(x, t) \text{div}_y \tau(y, t), \tag{35}$$

which was proved in Section 61.2e. From (29) and (30) follows for the transformation of forces that

$$\int_H K \, dx = \int_H FJ \, dx.$$

Contracting H into a single point, we find

$$K = FJ.$$

The basic equation (14) therefore implies that

$$\rho_0 u_{tt} - \text{div}\, \sigma - K = (\rho u_{tt} - \text{div}\, \tau - F)J = 0.$$

This is the transformed basic equation (32). Equality of (27) and (28) then follows from

$$\int_{H'} \text{div}\, \tau \, dy = \int_{H'} (\text{div}\, \tau) J \, dx = \int_H \text{div}\, \sigma \, dx.$$

According to (4), this implies that

$$\int_{\partial H'} \tau n' \, dO = \int_{\partial H} \sigma n \, dO.$$

61.5f. Basic Equations of Elastostatics

We now consider the stationary case. The displacement u in this case does not depend on time t, and the basic equations of Section 61.5a attain the form:

(i) *Deformation*

$$y = x + u(x), \qquad x \in \bar{G}.$$

(ii) *Equilibrium condition*

$$\text{div}\, \sigma(x) + K(x) = 0 \quad \text{on } G. \tag{36}$$

(iii) *Constitutive law*

$$\sigma(x) = \Omega(u(x), u'(x), x).$$

In order to complete this system we must add *boundary conditions*. We thereby use the decomposition

$$\partial G = \partial_1 G \cup \partial_2 G, \qquad \partial_1 G \cap \partial_2 G = \emptyset,$$

and prescribe the displacement

$$u = u_0 \quad \text{on } \partial_1 G,$$

and the stress forces

$$\sigma n = T \quad \text{on } \partial_2 G.$$

Generally, these basic equations represent a complex system of nonlinear differential equations for the unknown displacement u.

In Cartesian coordinates, equilibrium condition (36) becomes

$$D^j \sigma_j^i + K^i = 0 \quad \text{on } G, \quad i = 1, 2, 3. \tag{36*}$$

In Section 61.8 we will show that the constitutive law for a homogeneous body must have the natural form

$$S = B(\mathscr{E}),$$

i.e., the symmetric second Piola–Kirchhoff tensor S depends only on the symmetric strain tensor $\mathscr{E}$. This explains the role of S in nonlinear elasticity.

61.5g. Principle of Virtual Power in Elastostatics

Our goal is the transformation of the differential equation (36) into the integral identity

$$\int_G [\sigma, h'^*] \, dx - \int_G K h \, dx - \int_{\partial_2 G} T h \, dO = 0 \tag{37}$$

for all $h \in X(\partial_1 G)$. This represents the *generalized* form of the equilibrium condition (36).

Let $X(\partial_1 G)$ denote the set of all C^1-maps $h: \bar{G} \to V_3$ with

$$h = 0 \quad \text{on } \partial_1 G.$$

Multiplication of (36*) with h_i, and integration by parts, yields

$$\int_G \sigma_j^i D^j h_i \, dx - \int_G K^i h_i \, dx - \int_{\partial_2 G} T^i h_i \, dO = 0 \tag{37*}$$

for all $h \in X(\partial_1 G)$. This is (37).

In the literature the expression on the left-hand side of (37) is called virtual work. Therefore (37) is called the principle of virtual work. Transformation

$$y = x + u(x) + th(x) \quad \text{with} \quad h \in X(\partial_1 G) \tag{38}$$

describes a time-dependent perturbation of the deformation $y = x + u(x)$ with deformation velocity

$$\dot{y} = h.$$

Condition $h = 0$ on $\partial_1 G$ guarantees that all these perturbations satisfy the

61.5. Reduced Stress Tensor and the Principle of Virtual Power

same boundary condition on $\partial_1 G$ as does u. Comparison with Section 58.11 then shows that equation (36) with $h = \dot{y}$ can be regarded as a generalization of the *principle of virtual power* of point mechanics. The equivalence between the principle of virtual work and virtual power has been discussed in Section 58.3.

We now consider the special case where σ is symmetric. From

$$\gamma(x) = \tfrac{1}{2}(h'(x) + h'(x)^*),$$

and equation (37) we obtain the equation

$$\int_G [\sigma, \gamma] \, dx - \int_G Kh \, dx - \int_{\partial_2 G} Th \, dO = 0 \tag{39}$$

for all $h \in X(\partial_1 G)$. In fact, using $\gamma_j^i = \tfrac{1}{2}(D^i h_j + D^j h_i)$ and $\sigma_j^i = \sigma_i^j$ for all i, j, equation (37*) implies that

$$\int_G \sigma_j^i \gamma_i^j \, dx - \int_G K^i h_i \, dx - \int_{\partial_2 G} T^i h_i \, dO = 0 \tag{39*}$$

for all $h \in X(\partial_1 G)$. This is (39).

In Section 66.4 this version of the principle of virtual work will be used in the study of plastic materials.

61.5h. Basic Formulas for the Stress Tensors

For the convenience of the reader we now summarize the formulas which connect the stress tensor τ to the first and second Piola–Kirchhoff tensors σ and S, respectively.

For $y = x + u(x)$ we have

$$\sigma(x) = \tau(y) x'(y)^* \det y'(x),$$
$$S(x) = x'(y) \tau(y) x'(y)^* \det y'(x),$$
$$\sigma(x) = y'(x) S(x).$$

The inverse formulas are

$$\tau(y) = \sigma(x) y'(x)^* \det x'(y),$$
$$\tau(y) = y'(x) S(x) y'(x)^* \det x'(y).$$

Note that $\det x'(y) = (\det y'(x))^{-1}$.

The star denotes the adjoint operator on the H-space V_3.

If the displacement u depends on time t, then the same applies to the stress tensors τ, σ, and S.

61.6. A General Variational Principle (Hyperelasticity)

We want to show how a large class of problems in elastostatics can be obtained from a general variational principle. This variational approach to nonlinear elasticity is very important in the applications. At the same time it yields a general method to construct the stress tensor σ and to obtain reasonable constitutive laws. Moreover, we find reasonable approximation models. The main idea is to consider these cases where the equilibrium condition div σ + $K = 0$ of Section 61.5f is the Euler equation of a variational problem (principle of minimal or stationary potential energy).

Our starting point is the *variational problem*

$$\int_G L(x, u'(x))\, dx - \int_G Ku\, dx - \int_{\partial_2 G} Tu\, dO = \text{stationary!},$$

$$u = u_0 \quad \text{on } \partial_1 G, \tag{40}$$

where G is the undeformed region. Given K, T, and u_0, we are looking for the displacement u.

We call

$$U = \int_G L(x, u'(x))\, dx$$

the *elastic potential energy* and

$$W = \int_G Ku\, dx + \int_{\partial_2 G} Tu\, dO$$

the *work* of the outer forces. Relation (40) is known as the *principle of stationary potential energy* for elastic bodies. The function L, which represents the density of the elastic potential energy, is called the stored energy function.

Moreover, we define the stress tensor $\sigma(x) \in L(V_3)$ through

$$\sigma(x) = L_{u'}(x, u'(x)). \tag{41}$$

This identification has to be understood in the sense of the scalar product on $L(V_3)$, i.e.,

$$L_{u'}(\cdot)\gamma = [\sigma(\cdot), \gamma^*] \quad \text{for all } \gamma \in L(V_3).$$

Equation (41) is the required constitutive law. Our discussion below will show that this is a satisfactory definition of σ.

The deformation of G follows from

$$y = x + u(x).$$

According to (27) the stress forces, which act on the deformed body, are obtained from

$$\int_H \sigma n\, dO = \text{stress force acting on } H'.$$

61.6a. Euler Equations and the Equilibrium Condition

Below, we will show that
$$\sigma n = T \quad \text{on } \partial_2 G.$$
This explains the physical meaning of T. Moreover, we have
$$\int_H K\,dx = \text{outer force acting on } H'.$$

61.6a. Euler Equations and the Equilibrium Condition

In order to explain the relation with the basic equations of elastostatics of Section 61.5f, we consider the variational problem (40) with sufficiently smooth data.

(H1) G is a bounded region in $\mathbb{R}^3$ with sufficiently smooth boundary, i.e., $\partial G \in C^{0,1}$.

(H2) The boundary of G admits the decomposition
$$\partial G = \overline{\partial_1 G} \cup \overline{\partial_2 G},$$
where $\partial_1 G$ and $\partial_2 G$ are disjoint, open subsets of ∂G.

(H3) The stored energy function L is a C^2-map for all arguments, i.e., $L \in C^2(\overline{G} \times L(V_3))$.

(H4) The functions $K: \overline{G} \to V_3$ and $T: \overline{\partial_2 G} \to V_3$ are continuous.

Theorem 61.A (Equilibrium Condition). *Assume (H1) to (H4). Let $u: \overline{G} \to V_3$ be a C^2-map. Then the following three statements are equivalent:*

(a) u *is a solution of the variational problem* (40).
(b) u *satisfies the variational equation*
$$\int_G [\sigma, h'^*]\,dx - \int_G Kh\,dx - \int_{\partial_2 G} Th\,dx = 0 \tag{42}$$
for all $h \in X(\partial_1 G)$.
(c) u *satisfies the equilibrium condition*
$$\text{div } \sigma + K = 0 \quad \text{on } G,$$
$$\sigma n = T \quad \text{on } \partial_2 G. \tag{43}$$

Recall that $X(\partial_1 G)$ is the set of all C^1-maps $h: \overline{G} \to V_3$ with $h = 0$ on $\partial_1 G$.

Remark 61.3. In Cartesian coordinates, $u'(x)$ corresponds to the matrix $(D_j u^i(x))$. Equations (41), (42), (43) then become
$$\sigma^i_j = \frac{\partial L}{\partial D_j u^i}, \tag{41*}$$
$$\int_G \sigma^i_j D^j h_i\,dx - \int_G K^i h_i\,dx - \int_{\partial_2 G} T^i h_i\,dO = 0, \tag{42*}$$

$$D^j \sigma_j^i + K^i = 0 \quad \text{on } G, \quad i = 1, 2, 3,$$
$$\sigma_j^i n^j = T^i \quad \text{on } \partial_2 G.$$
(43*)

Moreover, we have

$$L_{u'u'} h'^2 = \frac{\partial^2 L}{\partial D_j u^i \, \partial D_r u^s} D_j h^i D_r h^s.$$

The sums are taken over two equal indices from 1 to 3.

PROOF OF THEOREM 61.A. We use the standard arguments of variational calculus as described in Section 18.3. We set

$$\varphi(t) = U(u + th) - W(u + th) \quad \text{for } h \in X(\partial_1 G).$$

Note that all $u + th$ are admitted as candidates in (40). This follows since $h = 0$ on $\partial_1 G$ and hence

$$u + th = u_0 \quad \text{on } \partial_1 G.$$

Formula (40) means that

$$\varphi'(0) = 0.$$

This gives (42*) and hence (42). Integration by parts in (42*) yields

$$-\varphi'(0) = \int_G (\text{div } \sigma + K) h \, dx + \int_{\partial_2 G} (T - \sigma n) h \, dO = 0$$

for all $h \in X(\partial_1 G)$. If we make the special choice $h_i \in C_0^\infty(G)$, then the variational lemma (Proposition 18.2) implies that

$$\text{div } \sigma + K = 0 \quad \text{on } G.$$

An analogous argument shows that

$$T - \sigma n = 0 \quad \text{on } \partial_2 G. \qquad \square$$

All this yields the following result:

(i) In the case of smooth solutions the Euler equation for the variational problem (40), i.e., the principle of stationary potential energy, is equivalent to the *equilibrium condition* (43).
(ii) Condition $\sigma n = T$ on $\partial_2 G$ is a so-called *natural boundary condition*, because it does not appear in the original variational problem (40), but is automatically obtained as a necessary solvability condition.
(iii) Equation (42) is called the variational equation or *generalized equation* for the Euler equation (43). It corresponds to the principle of virtual power which is also called the principle of virtual work (see Section 61.5g).

For C^2-maps $u: \bar{G} \to V_3$, relations (42) and (43) are equivalent. The importance of equation (42) is that it remains meaningful if the displacement u has fewer smoothness properties. In fact, one can find solutions of (42) which are

61.6b. Stable Solutions

Definition 61.4. The C^2-solution u of the variational problem (40) is called *strictly stable* if and only if

$$\delta^2 U(u; h) \equiv \int_G L_{u'u'}(x, u'(x))h'^2 \, dx > 0 \tag{44}$$

for all $h \in X(\partial_1 G)$ with $h \not\equiv 0$, i.e., the second variation of the potential energy is strictly positive.

We want to motivate this definition. Let u be an equilibrium position and consider a displacement

$$v = u + th \quad \text{with} \quad h \in X(\partial_1 G).$$

Moreover, let

$$\varphi(t) = U(u + th) - W(u + th).$$

The work done by the outer forces and the elastic forces during a displacement of the body from u to v is equal to

$$\Delta W = W(v) - W(u) \quad \text{and} \quad -\Delta U = U(u) - U(v),$$

respectively. According to the general stability principle of Section 58.12, a stable equilibrium position u must satisfy the relation

$$\Delta W - \Delta U < 0$$

for the entire work. This shows that

$$\varphi(t) > \varphi(0) \tag{45}$$

for all $h \in X(\partial_1 G)$ and all t in a neighborhood of zero. Condition (45) means that φ has a strict local minimum at $t = 0$. This is satisfied if

$$\varphi'(0) = 0 \tag{45a}$$

and

$$\varphi''(0) > 0. \tag{45b}$$

Now observe that (45a) is precisely the variational equation (42). Moreover, (45b) corresponds to the strict stability condition (44).

We will show in Section 61.12b that a *strongly stable* solution u of the variational problem (40) corresponds to a strict local minimum of the potential energy, i.e., u is a solution of the minimum problem

$$\int_G L \, dx - \int_G Ku \, dx - \int_{\partial_2 G} Tu \, dO = \min!,$$

$$u = u_0 \quad \text{on } \partial_1 G.$$

61.6c. Equilibrium of Torques

As we have seen, the Euler equation for the variational problem (40) yields the equilibrium condition

$$\text{div } \sigma + K = 0.$$

In integral form this becomes

(E*) $$\int_H K \, dx + \int_{\partial H} \sigma n \, dO = 0$$

for all subregions H. This condition states that the outer forces and the stress forces are in an equilibrium. Using the Piola transformation (35) we obtain

$$\text{div } \sigma(x) = J(x) \text{div}_y \tau(y), \qquad K(x) = J(x) F(y)$$

and hence

$$\text{div}_y \tau(y) + F(y) = 0.$$

In integral form, this becomes the equilibrium condition for the forces

(E) $$\int_{H'} F \, dy + \int_{\partial H'} \tau n' \, dO = 0$$

for all deformed subregions H'. The equilibrium condition for the torques reads as

(T) $$\int_{H'} F \times y \, dy + \int_{\partial H'} \tau n' \times y \, dO = 0$$

for all deformed subregions H'. According to Section 61.4d, the equilibrium condition for the forces div $\tau + F = 0$ implies the following:

The equilibrium condition for the torques (T) is equivalent to the symmetry of the stress tensor τ.

Condition (T) refers to the deformed body. In Problem 61.5 we show that this implies the natural condition

(T*) $$\int_H K \times y \, dx + \int_{\partial H} \sigma n \times y \, dO = 0$$

for all undeformed subregions H.

We emphasize that for general stored energy functions L in (40) condition (T) need not be satisfied. In the following, we consider two important cases where (T) is exactly or approximately satisfied.

Case 1: $L = A(\mathscr{E})$.

In this case, which is a physically important situation, the stored energy function L depends only on the strain tensor $\mathscr{E}$. We will show that then

61.6. A General Variational Principle (Hyperelasticity)

condition (T) is satisfied. To this end, we choose a Cartesian coordinate system and set

$$T_j^i = \frac{\partial A}{\partial \mathscr{E}_j^i}.$$

We have the symmetry condition

$$T_j^i = T_i^j \quad \text{for all } i, j$$

and (10) implies that

$$\sigma_j^i = \frac{\partial L}{\partial D_j u^i} = T_r^k \frac{\partial \mathscr{E}_r^k}{\partial D_j u^i}$$

$$= T_j^i + T_j^k D_k u^i = T_j^k D_k \eta^i$$

with $D_i = \partial/\partial \xi^i$. Note that $y = x + u(x)$. This yields

$$\sigma(x) = y'(x) T(x)$$

and hence $T(x) = x'(y)\sigma(x)$. From Section 61.5h it follows that T is precisely the second Piola–Kirchhoff tensor, i.e., $T = S$ and hence

$$S = A'(\mathscr{E}).$$

This identification has to be understood in the sense of the scalar product on $L(V_3)$, i.e.,

$$A'(\mathscr{E})\gamma = [S, \gamma^*] \quad \text{for all } \gamma \in L(V_3).$$

The symmetry of S implies the symmetry of the stress tensor

$$\tau(y) = y'(x) S(x) y'(x)^* \det x'(y),$$

i.e., condition (T) is satisfied.

Case 2: σ is symmetric.
This means that $\sigma_j^i = \sigma_i^j$, and hence that

$$\frac{\partial L}{\partial D_j u^i} = \frac{\partial L}{\partial D_i u^j} \quad \text{for all } i, j.$$

Using div $\sigma + K = 0$, the same computation as in Section 61.4d yields

$$(T_{\text{approx}}) \qquad \int_H K \times x \, dx + \int_{\partial H} \sigma n \times x \, dO = 0$$

for all undeformed subregions H. Thus it follows from $y = x + u(x)$ that for small displacements, i.e., for small $\|u(x)\|$, the torque condition (T*) is satisfied in first-order approximation.

This Case 2 often occurs in approximation models.

61.6d. General Strategy of Hyperelasticity

We summarize the basic ideas of hyperelasticity. The starting point is the variational problem

(P) $$\int_G L\,dx - \int_G Ku\,dx - \int_{\partial_2 G} Tu\,dO = \text{stationary!},$$

$$u = u_0 \quad \text{on } \partial_1 G.$$

We are given the functions

$L = L(x, u')$ (stored energy function),
$K = K(x)$ (density of outer forces),
$T = T(x)$ (density of boundary forces),
$u = u_0(x)$ (displacement of the boundary part $\partial_1 G$),

and are looking for the displacement

$$u = u(x).$$

(i) *Deformation.* If we have a solution $u = u(x)$ of (P), then the deformation of the region G has the form

$$y = x + u(x).$$

(ii) *Constitutive law.* The first Piola–Kirchhoff stress tensor is given by

$$\sigma(x) = L_{u'}(x, u'(x)).$$

From this we obtain the stress tensor

$$\tau(y) = \sigma(x) y'(x)^* \det x'(y).$$

In order to find exact models we need the symmetry of τ. From Section 61.6c, this symmetry condition is satisfied if

$$L = A(x, \mathscr{E}),$$

where

$$\mathscr{E}(x) = \tfrac{1}{2}(u'(x) + u'(x)^* + u'(x)^* u'(x))$$

is the strain tensor.

(iii) *Stress forces.* Let H be a subregion of G and let H' denote the corresponding deformed set. Then we have

$$\int_{\partial H} \sigma n\,dO = \text{stress force acting on } H'.$$

This force is equal to

$$\int_{\partial H'} \tau n\,dO.$$

61.6. A General Variational Principle (Hyperelasticity)

(iv) *Outer forces.* We also have

$$\int_H K\,dx = \text{outer force acting on } H'.$$

Setting $F(y) = K(x) \det x'(y)$, this force is equal to

$$\int_{H'} F\,dy.$$

(v) *Boundary forces.* Let B be a subregion of the boundary part $\partial_2 G$ and let B' denote the corresponding deformed set. We then have

$$\int_B T\,dO = \text{boundary force acting on } B'.$$

(vi) *Euler equations.* Sufficiently smooth solutions of the variational problem (P) satisfy

$$\operatorname{div}\sigma + K = 0 \quad \text{on } G,$$

$$\sigma n = T \quad \text{on } \partial_2 G.$$

This means

$$\int_H K\,dx + \int_{\partial H} \sigma n\,dO = 0,$$

i.e., the outer forces and the stress forces are in an equilibrium, and

$$\int_B T\,dO = \int_B \sigma n\,dO,$$

i.e., the outer boundary forces are equal to the corresponding stress forces.

(vii) *Stability.* The deformation $y = x + u(x)$ is strictly stable if and only if

$$\int_G L_{u'u'}(x, u'(x))h'^2\,dx > 0$$

for all nonzero C^1-maps $h\colon \bar{G} \to V_3$ with $h = 0$ on $\partial_1 G$.

If u is *strongly stable* in the sense of Definition 61.18, then u corresponds to a strict local minimum of the original variational problem (P).

61.6e. A General Strategy for Obtaining Approximation Models

The easiest procedure to get approximation models in elasticity is the following:

(a) We start with an exact model (P) in hyperelasticity.

(b) We replace the stored energy function L with an approximation function L_{approx}.
(c) We consider the corresponding variational problem (P_{approx}).

This gives us a clear idea of the approximation process. In the following chapters we will frequently use this strategy.

One advantage of hyperelasticity is that we can apply the methods of duality theory. This will be explained in Chapter 62. The basic idea is to consider the minimum problem

(P) $$E_{pot}(u) = \min!$$

for the potential energy together with the dual maximum problem

(P*) $$E_{dual}(\sigma) = \max!$$

for the dual energy, where E_{pot} and E_{dual} depend on the displacement u and the stress tensor σ, respectively.

In Section 62.9 we will apply this duality theory to quasi-statical plasticity.

61.7. Elastic Energy of the Cuboid and Constitutive Laws

To be able to apply the variational principle (P) of the previous section, we need physically meaningful expressions for the elastic potential energy

$$U = \int L \, dx.$$

We want to show how such expressions can be obtained from general observations. Our strategy is the following:

(i) We consider the stretching of a cuboid in a special coordinate system, using special stresses and Hooke's law.
(ii) The calculated elastic energy in this case will be formulated in an invariant way.
(iii) This invariant expression then is independent of the special situation.

In the case of an arbitrary body, U is obtained by using a decomposition of the body into cuboids and summation or integration.

61.7a. Deformation of a Cuboid

In Cartesian coordinates with basis vectors e_i we consider an axially parallel cuboid

$$C = \{x : |\xi_i| \leq l_i, i = 1, 2, 3\}.$$

61.7. Elastic Energy of the Cuboid and Constitutive Laws

Suppose the cuboid C is stretched by the transformation

$$\eta_i = (1 + \lambda_i)\xi_i, \quad i = 1, 2, 3.$$

We then obtain the displacements $u_i = \lambda_i \xi_i$ and

$$\gamma_{ij} = 2^{-1}(D_i u_j + D_j u_i) = \lambda_i \delta_{ij}.$$

The strain of C in the direction of the ξ_i-axis is equal to

$$\Delta l_i / l_i = \lambda_i.$$

Suppose the side of the cuboid with $\xi_i = \pm l_i$ has the surface S_i, and the tensile force $\pm K_i e_i$ with $K_i \geq 0$ acts on it. This corresponds to the stress

$$\sigma_i = K_i / S_i.$$

According to Hooke's law (60.1) and (60.2) the stress σ_1 leads to an extension

$$\Delta l_1 / l_1 = \sigma_1 / E$$

and a lateral contraction

$$\Delta l_i / l_i = -\mu \sigma_1 / E, \quad i = 2, 3.$$

Taking the lateral contractions into account, which are caused by σ_2 and σ_3, one obtains

$$\Delta l_1 / l_1 = E^{-1} \sigma_1 - \mu E^{-1}(\sigma_2 + \sigma_3)$$

together with analogous expressions for $\Delta l_i / l_i$. This yields

$$\Delta l_i / l_i = (1 + \mu) E^{-1} [\sigma_i - (1 + \mu)^{-1}(\sigma_1 + \sigma_2 + \sigma_3)] \tag{46}$$

for $i = 1, 2, 3$. Solving for σ_i, one finds

$$\sigma_i = \lambda(\gamma_{11} + \gamma_{22} + \gamma_{33}) + 2\kappa \gamma_{ii} \tag{47}$$

with the so-called Lamé constants

$$\lambda = E\mu / (1 + \mu)(1 - 2\mu), \quad \kappa = E/2(1 + \mu).$$

From Hooke's law (60.1) and (60.2), the material constants E and μ can then be experimentally determined. Table 61.1 contains several values. Recall that on the surface of the earth 1 kp is the force which is caused by 1 kg, i.e., 1 kp = 9.81 N. Table 61.1 shows that $\lambda > 0$ and $\kappa > 0$ are reasonable general assumptions.

Table 61.1.

	E (kp/cm^2)	μ
Steel	$2.18 \cdot 10^6$	0.29
Iron	$2.17 \cdot 10^6$	0.28
Copper	$1.23 \cdot 10^6$	0.35
Aluminium	$0.74 \cdot 10^6$	0.34

We now want to compute the elastic energy of the cuboid. During the time interval $0 \leq t \leq 1$, we stretch the cuboid by letting

$$\eta_i = (1 + t\lambda_i)\xi_i.$$

The work which thereby is done equals

$$U = \int_0^1 2K^i(t)\dot\eta_i(t)\,dt = K^i(1)\lambda_i l_i.$$

Note that $\xi_i = l_i$, i.e., $\dot\eta_i = \lambda_i l_i$ and

$$K^i(t) = tK^i(1),$$

since the stress forces, which act on all sides of the cuboid, depend linearly on the dilatation at time t, according to (47). Let V be the volume of the cuboid. Because of

$$K^i(1) = \sigma_i/S_i \quad \text{and} \quad V = 2l_i S_i$$

we find that

$$U = \sigma^i \gamma_{ii} V/2. \tag{48}$$

If we define $\sigma \in L(V_3)$ through $\sigma e_i = \sigma_i e_i$, then (47) and (48) may be written in the invariant form

$$\sigma = \lambda \operatorname{tr} \gamma I + 2\kappa \gamma, \tag{49}$$

$$U = (2^{-1}\lambda (\operatorname{tr}\gamma)^2 + \kappa[\gamma,\gamma])V. \tag{50}$$

We interpret U as the elastic energy which, by the stretching, is kept in the cuboid.

61.7b. The Yield Condition in Plasticity

Our next goal is the plasticity condition (51) below which is also called the yield condition. If we consider the extension of a wire, we observe the following behavior:

$$\sigma^2 < \sigma_0^2 \quad \text{no plasticity},$$
$$\sigma^2 \geq \sigma_0^2 \quad \text{plasticity occurs},$$

where σ is the stress of the wire. We want to generalize this condition to three-dimensional bodies, whereby nothing about the concrete microscopical structure of the bodies shall be used. The way to do this is by using a purely mathematical argument based on the invariants of the stress tensor σ.

We define the deviations $\bar\gamma, \bar\sigma$ of the tensors γ, σ through

$$\bar\gamma = \gamma - 3^{-1}\operatorname{tr}\gamma I, \qquad \bar\sigma = \sigma - 3^{-1}\operatorname{tr}\sigma I.$$

This gives
$$\text{tr}\,\bar{\gamma} = \text{tr}\,\bar{\sigma} = 0,$$
and in the special coordinate system used above, this yields
$$3[\bar{\gamma},\bar{\gamma}] = (\lambda_1 - \lambda_2)^2 + (\lambda_2 - \lambda_3)^2 + (\lambda_3 - \lambda_1)^2,$$
$$3[\bar{\sigma},\bar{\sigma}] = (\sigma_1 - \sigma_2)^2 + (\sigma_2 - \sigma_3)^2 + (\sigma_3 - \sigma_1)^2.$$
Hence, we obtain
$$[\bar{\gamma},\bar{\gamma}] = 0 \quad \text{iff} \quad \lambda_1 = \lambda_2 = \lambda_3,$$
i.e., the cuboid is stretched uniformly in all three axial directions. The greater $[\bar{\gamma},\bar{\gamma}]$ becomes, the more nonuniformly the cuboid is stretched.

Similarly, the greater $[\bar{\sigma},\bar{\sigma}]$ becomes, the more nonuniform the stresses in the cuboid become.

EXAMPLE 61.5 (The Plasticity Condition of von Mises (1913)). For the cuboid we write
$$\begin{aligned} [\bar{\sigma},\bar{\sigma}] &< \sigma_0^2 \quad \text{no plasticity,} \\ [\bar{\sigma},\bar{\sigma}] &\geq \sigma_0^2 \quad \text{plasticity occurs.} \end{aligned} \quad (51)$$

This criterium is based on the experimental result that a strongly nonuniform stress in the cuboid causes plasticity, whereas in the case of a strongly uniform stress, i.e., for strong stresses with $[\bar{\sigma},\bar{\sigma}] = 0$, no plasticity is observed.

Replacing $\bar{\sigma}$ in (51) with $\bar{\sigma}(x)$, we obtain the plasticity condition for general three-dimensional bodies which will be used in the following chapters.

The constant σ_0 depends on the material.

61.7c. The General Linear Model

EXAMPLE 61.6 (Linear Elasticity Theory). We consider a homogeneous and isotropic body. According to (50), a decomposition of the body into small cuboids suggests the following ansatz for the elastic energy:
$$U = \int_G 2^{-1}\lambda(\text{tr}\,\gamma)^2 + \kappa[\gamma,\gamma]\,dx. \quad (52)$$

Here γ depends on x. The energy expression (52) corresponds to the so-called linear elasticity theory. Since we used the linear Hooke's law to motivate (50) and (52), we have to assume that $\|u'(x)\|$ is small, i.e., that all derivatives $|D_j u_i|$ are small.

The constitutive law, which follows from (52) by using the general formula (41), namely $\sigma = L_{u'}$, is equal to Hooke's law (49).

61.7d. Nonlinear Models

EXAMPLE 61.7 (Nonlinear Elasticity Theory). A general ansatz for U is

$$U = \int_G A(x, \mathscr{E}(x)) \, dx.$$

For small $\|u'(x)\|$ the strain tensor $\mathscr{E}(x)$ of Section 61.2 can be approximately replaced with $\gamma(x)$.

EXAMPLE 61.8 (Nonlinear Hencky Material). We write (52) in the form

$$U = \int_G 2^{-1} k (\operatorname{tr} \gamma)^2 + \kappa \varphi([\bar{\gamma}, \bar{\gamma}]) \, dx \tag{53}$$

with $k = \lambda + 2\kappa/3$ and $\varphi(t) \equiv t$.

The corresponding nonlinear material behavior occurs for nonlinear functions φ in (53).

The idea for this ansatz is that for a nonuniform strain, i.e., for $[\bar{\gamma}, \bar{\gamma}] \neq 0$, the cuboid in Section 61.7a exhibits nonlinear behavior.

61.8. Theory of Invariants and the General Structure of Constitutive Laws and Stored Energy Functions

We now want to show that natural symmetry conditions restrict the possible structure of constitutive laws. The main results will be Theorems 61.B and 61.C below.

Let $\gamma \in L_{\text{sym}}(V_3)$, i.e.,

$$\gamma: V_3 \to V_3$$

is a symmetric linear operator. *Principal invariants* of γ are the coefficients of the characteristic equation

$$\det(\gamma - \lambda I) = 0. \tag{54}$$

Proposition 61.9. *Let $\gamma \in L_{\text{sym}}(V_3)$ and let $\lambda_1, \lambda_2, \lambda_3$ be the eigenvalues of γ. Then the principal invariants of γ are given by*

$$\lambda_1 + \lambda_2 + \lambda_3 = \operatorname{tr} \gamma, \qquad \lambda_1 \lambda_2 \lambda_3 = \det \gamma,$$
$$\lambda_1 \lambda_2 + \lambda_2 \lambda_3 + \lambda_3 \lambda_1 = \tfrac{1}{2}((\operatorname{tr} \gamma)^2 - [\gamma, \gamma]) = \operatorname{tr}(\operatorname{adj} \gamma). \tag{55}$$

PROOF. Note that $\det(\gamma - \lambda I)$, $\operatorname{tr} \gamma$, and $[\gamma, \gamma]$ are invariants of γ, i.e., they are independent of the chosen coordinate system. Let $\{e_1, e_2, e_3\}$ be an orthonormal system of eigenvectors which correspond to the eigenvalues $\lambda_1, \lambda_2, \lambda_3$. We then have

$$\gamma e_i = \lambda_i e_i, \quad i = 1, 2, 3. \tag{56}$$

If we choose a Cartesian coordinate system with axes e_1, e_2, e_3, we obtain

$$\gamma(\xi^i e_i) = \xi^i \gamma_i^j e_j, \quad \gamma_i^j = \lambda_i \delta_i^j,$$

and hence equation (54) assumes the form

$$(\lambda - \lambda_1)(\lambda - \lambda_2)(\lambda - \lambda_3) = 0. \qquad \square$$

61.8a. Isotropic Real Tensor Functions

Proposition 61.10. *Let $f: L_{\text{sym}}(V_3) \to \mathbb{R}$ be an isotropic function, i.e.,*

$$f(R^{-1}\gamma R) = f(\gamma) \tag{57}$$

for all rotations R and all $\gamma \in L_{\text{sym}}(V_3)$.

Then f is a function of the principal invariants, i.e., there exists a function $g: \mathbb{R}^3 \to \mathbb{R}$ such that

$$f(\gamma) = g(\operatorname{tr} \gamma, [\gamma, \gamma], \det \gamma). \tag{58}$$

PROOF. Let $\gamma, \sigma \in L_{\text{sym}}(V_3)$. It suffices to show that the relation

$$\sigma = R^{-1}\gamma R \tag{59}$$

holds precisely if the principal invariants of γ and σ are equal. Here, R denotes a rotation.

From (59) it follows that

$$\det(\sigma - \lambda I) = \det(\gamma - \lambda I).$$

Hence, the principal invariants equal each other.

If, conversely, the principal invariants are equal, then the characteristic equations are identical, and hence the same is true for the eigenvalues. It follows that there exists a rotation which transforms the corresponding eigenvectors of γ and σ into each other. This implies (59). $\qquad \square$

To explain the intuitive meaning of (57), consider the equation

$$y = \gamma x.$$

A rotation R sends x and y into $x_R = Rx$ and $y_R = Ry$, respectively. Hence, we obtain

$$y_R = \gamma_R x_R, \quad \text{where} \quad \gamma_R = R^{-1}\gamma R,$$

and condition (57) means that $f(\gamma_R) = f(\gamma)$.

61.8b. Isotropic Tensor Functions

Proposition 61.11 (Rivlin and Ericksen (1955)). *Let $T: L_{\text{sym}}(V_3) \to L_{\text{sym}}(V_3)$ be an isotropic tensor function, i.e.,*

$$R^{-1}T(\gamma)R = T(R^{-1}\gamma R) \tag{60}$$

for all rotations R and all $\gamma \in L_{\text{sym}}(V_3)$.
 Then there exist real functions $a, b, c: \mathbb{R}^3 \to \mathbb{R}$ such that

$$T(\gamma) = aI + b\gamma + c\gamma^2 \quad \text{for all} \quad \gamma \in L_{\text{sym}}(V_3), \tag{61}$$

where

$$a = a(\operatorname{tr}\gamma, [\gamma, \gamma], \det \gamma),$$

together with analogous expressions for b and c.

PROOF.

(I) Let $\{e_1, e_2, e_3\}$ be an orthonormal system of eigenvectors of γ with corresponding eigenvalues $\lambda_1, \lambda_2, \lambda_3$, i.e.,

$$\gamma e_i = \lambda_i e_i, \quad i = 1, 2, 3.$$

We show that this is also an orthonormal system of eigenvectors for $T(\gamma)$, i.e.,

$$T(\gamma) = t_i e_i, \quad i = 1, 2, 3.$$

We define a rotation by

$$Qe_1 = e_1, \quad Qe_2 = -e_2, \quad Qe_3 = -e_3.$$

From (60) and $Q^{-1}\gamma Q = \gamma$ it follows that

$$Q^{-1}T(\gamma)Q = T(\gamma).$$

In the language of matrices, this means

$$\begin{pmatrix} 1 & 0 & 0 \\ 0 & -1 & 0 \\ 0 & 0 & -1 \end{pmatrix} \begin{pmatrix} t_{11} & t_{12} & t_{13} \\ t_{21} & t_{22} & t_{23} \\ t_{31} & t_{32} & t_{33} \end{pmatrix} \begin{pmatrix} 1 & 0 & 0 \\ 0 & -1 & 0 \\ 0 & 0 & -1 \end{pmatrix}$$

$$= \begin{pmatrix} t_{11} & -t_{12} & -t_{13} \\ -t_{21} & t_{22} & t_{23} \\ -t_{31} & t_{32} & t_{33} \end{pmatrix}$$

and hence $t_{12} = t_{13} = t_{21} = t_{31} = 0$. Similarly, we obtain $t_{23} = t_{32} = 0$.

(II) According to (I), there exists a rotation R which diagonalizes both γ and $T(\gamma)$. From

$$R^{-1}T(\gamma)R = T(R^{-1}\gamma R)$$

with
$$R^{-1}T(\gamma)R = \begin{pmatrix} t_1 & 0 & 0 \\ 0 & t_2 & 0 \\ 0 & 0 & t_3 \end{pmatrix}, \qquad R^{-1}\gamma R = \begin{pmatrix} \lambda_1 & 0 & 0 \\ 0 & \lambda_2 & 0 \\ 0 & 0 & \lambda_3 \end{pmatrix}$$

it follows that
$$t_i = T_i(\lambda_1, \lambda_2, \lambda_3), \qquad i = 1, 2, 3$$

with fixed functions T_1, T_2, T_3 for all $\gamma \in L_{\text{sym}}(V_3)$.

(III) Any permutation of the λ_i leads to a corresponding permutation of the t_i. This follows from (60) by using rotations like
$$R = \begin{pmatrix} 0 & 1 & 0 \\ 1 & 0 & 0 \\ 0 & 0 & 1 \end{pmatrix}.$$

(IV) *Case 1:* $\lambda_1 \neq \lambda_2 \neq \lambda_3 \neq \lambda_1$. Since
$$\begin{vmatrix} 1 & \lambda_1 & \lambda_1^2 \\ 1 & \lambda_2 & \lambda_2^2 \\ 1 & \lambda_3 & \lambda_3^2 \end{vmatrix} = (\lambda_1 - \lambda_2)(\lambda_2 - \lambda_3)(\lambda_3 - \lambda_1) \neq 0,$$

the linear system
$$t_i = a + b\lambda_i + c\lambda_i^2, \qquad i = 1, 2, 3,$$

has a unique solution a, b, c. Thus, a is a function of $\lambda_1, \lambda_2, \lambda_3, t_1, t_2, t_3$, i.e., a function of $\lambda_1, \lambda_2, \lambda_3$, and hence a function of the principal invariants of γ, according to Proposition 61.10. This proves assertion (61) in Case 1.

Case 2: $\lambda_1 = \lambda_2, \lambda_2 \neq \lambda_3$. From (II) it follows that $t_1 = t_2$. We then consider the linear system
$$t_i = a + b\lambda_i, \qquad i = 2, 3.$$

Case 3: $\lambda_1 = \lambda_2 = \lambda_3$. In this case we obtain from (II) that $t_1 = t_2 = t_3$, and hence that $t_i = a, i = 1, 2, 3$. □

61.8c. Structure of the Constitutive Laws for Homogeneous Bodies

We consider the constitutive law
$$\tau(y) = \Phi(I + u'(x)) \tag{62}$$

for a homogeneous body, i.e., Φ does not depend on x. It is reasonable then to postulate
$$R^{-1}\Phi(R + Ru'(x))R = \Phi(I + u'(x)) \tag{63}$$

for all rotations R and all x. This condition is called the axiom of *material frame indifference*. In order to motivate (63), we consider the stress vector

$$\vec{\tau}(y) = \tau(y)n, \qquad y = x + u(x). \tag{64}$$

Using a rotation R, we pass from y to

$$y_R = Ry = Rx + Ru(x).$$

We set

$$\tau_R(y) = \Phi(R + Ru'(x)),$$

and observe that it is quite natural to require

$$R\vec{\tau}(y) = \tau_R(y)Rn,$$

which implies

$$\vec{\tau}(y) = (R^{-1}\tau_R(y)R)n. \tag{64*}$$

A comparison of (64) with (64*) yields (63).

Proposition 61.12. *Condition (63) is equivalent to*

$$S(x) = B(\mathscr{E}(x)), \tag{65}$$

where S is the second Piola–Kirchhoff stress tensor and $\mathscr{E}$ the strain tensor.

PROOF.

(I) (63) $\Rightarrow$ (65). We set $C = I + u'(x)$. From Proposition 61.1 it follows that

$$C = RU, \qquad U = (I + 2\mathscr{E}(x))^{1/2},$$

and Section 61.5h implies

$$CS(x)C^* = (\det C)\tau(y). \tag{66}$$

Noting $R^{-1} = R^*$, condition (63) then yields

$$\tau(y) = \Phi(C) = \Phi(RU) = R\Phi(U)R^{-1} = CU^{-1}\Phi(U)U^{-1}C^*.$$

This implies (65).

(II) (65) $\Rightarrow$ (63). Reverse the argument. □

Proposition 61.12 shows that the most general constitutive law for homogeneous bodies is given by (65).

61.8d. Structure of the Constitutive Laws for Homogeneous Isotropic Bodies

Let $y = x + u(x)$ denote a deformation of a homogeneous body. The body is called *isotropic* if and only if the following natural conditions

(i) $\qquad y(x) = y(Rx),$

(ii) $\qquad \tau(y) = \tau(Ry),$

(iii) $\qquad S(Rx) = B(\mathscr{E}(Rx))$

are satisfied for all points x and all rotations R.

Theorem 61.B (Rivlin–Ericksen Theorem (1955)). *The most general constitutive law for a homogeneous isotropic body is given by*

$$S = aI + b\mathscr{E} + c\mathscr{E}^2, \tag{67}$$

where

$$a = a(\operatorname{tr} \mathscr{E}, [\mathscr{E}, \mathscr{E}], \det \mathscr{E}),$$

and there exist analogous expressions for b and c. Here, S is the second Piola–Kirchhoff stress tensor and $\mathscr{E}$ is the strain tensor.

PROOF. Condition (i) implies

$$Rx + u(Rx) = x + u(x).$$

Differentiation with respect to x yields

$$R + u'(Rx)R = I + u'(x)$$

and hence

$$I + 2\mathscr{E}(Rx) = (I + u'(Rx))^*(I + u'(Rx)) = R(I + u'(x))^*(I + u'(x))R^{-1}.$$

This gives

$$\mathscr{E}(Rx) = R\mathscr{E}(x)R^{-1}. \tag{68}$$

Similarly, we obtain from (ii) and (66) that

$$S(Rx) = RS(x)R^{-1}.$$

Finally, condition (iii) shows that B is an isotropic tensor function. The assertion then follows from Proposition 61.11. □

61.8e. The Stored Energy Function

From Proposition 61.12, the most general constitutive law for homogeneous bodies is given by

$$S = B(\mathscr{E})$$

with $B: L_{\text{sym}}(V_3) \to L_{\text{sym}}(V_3)$. We then consider the special case where B is a potential operator with potential A, i.e.,

$$S = A'(\mathscr{E}). \tag{69}$$

Comparison with Section 61.6c shows that it makes sense to call $L = A(\mathscr{E})$ a stored energy function. Moreover, using (68) it is reasonable to postulate that the stored energy function of an isotropic body has the property

$$A(R^{-1}\mathscr{E}R) = A(\mathscr{E})$$

for all rotations R and all $\mathscr{E}$, i.e., A is isotropic.

Theorem 61.C. *The stored energy function of a homogeneous isotropic body has the form*

$$L = L(\operatorname{tr}\mathscr{E}, [\mathscr{E},\mathscr{E}], \det \mathscr{E}).$$

PROOF. This follows immediately from Proposition 61.10. □

61.8f. Special Constitutive Laws

EXAMPLE 61.13 (Saint Venant–Kirchhoff Material). We begin with the most general constitutive law for homogeneous isotropic bodies

$$S = aI + b\mathscr{E} + c\mathscr{E}^2,$$

where $a = a(\operatorname{tr}\mathscr{E}, [\mathscr{E},\mathscr{E}], \det \mathscr{E})$ and analogous expressions hold for b, c. We assume that $|\mathscr{E}|$ is small. Taylor's theorem then implies that

$$a = a_0 + a_1 \operatorname{tr}\mathscr{E} + o(|\mathscr{E}|), \qquad \mathscr{E} \to 0.$$

We postulate that $\mathscr{E} = 0$ yields $S = 0$, i.e., $a_0 = 0$. Setting $a_1 = \lambda$ and $b(0) = 2\kappa$, we obtain

$$S = \lambda \operatorname{tr}\mathscr{E}\, I + 2\kappa\mathscr{E} + o(|\mathscr{E}|), \qquad \mathscr{E} \to 0.$$

By definition, the linear term

$$S = \lambda \operatorname{tr}\mathscr{E}\, I + 2\kappa\mathscr{E} \tag{70}$$

corresponds to the Saint Venant–Kirchhoff material. Setting

$$A(\mathscr{E}) = \tfrac{1}{2}\lambda(\operatorname{tr}\mathscr{E})^2 + \kappa[\mathscr{E},\mathscr{E}], \tag{71}$$

we obtain

$$S = A'(\mathscr{E}).$$

Hence, $L = A(\mathscr{E})$ is the corresponding stored energy function.

If we replace the strain tensor $\mathscr{E}$ with the linearized strain tensor γ, then we obtain linear elastic materials. This shows that the constitutive law for linear elastic materials can be obtained from general considerations without using Hooke's law as in Section 61.7.

EXAMPLE 61.14 (Rubberlike Material of Ogden (1972)). Set $A = I + u'(x)$. Let $\lambda_1, \lambda_2, \lambda_3$ be the eigenvalues of $(A^*A)^{1/2}$, which are called the principal stretches of A. If μ_1, μ_2, μ_3 are the eigenvalues of $\mathscr{E}(x)$, then

$$\lambda_i = (1 + 2\mu_i)^{1/2} \quad \text{for all } i.$$

The stored energy function has the form

$$L = C(\lambda_1^p + \lambda_2^p + \lambda_3^p - 3) + D((\lambda_1\lambda_2)^q + (\lambda_2\lambda_3)^q + (\lambda_1\lambda_3)^q - 3) + f(\lambda_1\lambda_2\lambda_3),$$

where $p, q \geq 1$, $C, D > 0$, and $f:]0, \infty[\to \mathbb{R}$ is convex with

$$\lim_{d \to +0} f(d) = +\infty.$$

In terms of A this yields

$$L = C\operatorname{tr}(E^p - I) + D\operatorname{tr}(\operatorname{adj}(E^q) - I) + f(\det A), \tag{71*}$$

where $E = (A^*A)^{1/2}$ is the stretch tensor.

More generally, Ogden's material consists of sums of such expressions. Mooney–Rivlin material corresponds to the special case $p = q = 2$.

EXAMPLE 61.15 (Polyconvex Material of Ball (1977)). From Section 61.2b, the deformation of curves, surfaces, and volume elements depends on A, adj A, and det A where $A = I + u'(x)$. Thus, it is natural to assume that the stored energy function has the form

$$L = P(A, \operatorname{adj} A, \det A).$$

The function L is called *polyconvex* if and only if the function P is convex with respect to its three arguments, i.e., if

$$P: L(V_3) \times L(V_3) \times]0, \infty[\to \mathbb{R}$$

is convex.

From Problem 61.1, Ogden's stored energy function in Example 61.14 is polyconvex. In the special case of Mooney–Rivlin material, the polyconvexity of L follows easily from

$$L = C\operatorname{tr}(A^*A - I) + D\operatorname{tr}(\operatorname{adj}(A^*A) - I) + f(\det A)$$

(see Problem 61.1b).

61.9. Existence and Uniqueness in Linear Elastostatics (Generalized Solutions)

We now want to show that the main theorem about quadratic variational problems (Theorem 22.A) immediately leads to an existence and uniqueness theorem in linear elastostatics.

61.9a. The Classical Problem

The principle of minimal potential energy for linear material yields the following variational problem for the displacement vector u:

$$\int_G L(u'(x))\,dx - \int_G Ku\,dx = \min!, \qquad (72)$$

$$u = u_0 \quad \text{on } \partial G,$$

where

$$L = \tfrac{1}{2}\lambda(\operatorname{tr}\gamma)^2 + \kappa[\gamma,\gamma], \qquad \gamma = \tfrac{1}{2}(u'(x) + u'(x)^*).$$

Note that $\operatorname{tr}\gamma = \operatorname{div} u$. The corresponding Euler equations are

$$\operatorname{div}\sigma + K = 0, \qquad (73)$$

$$\sigma = L_{u'} = \lambda \operatorname{tr}\gamma\, I + 2\kappa\gamma.$$

In terms of the displacement vector u, we obtain the so-called Lamé equations

$$\kappa\,\Delta u + (\lambda + \kappa)\operatorname{grad}\operatorname{div} u + K = 0 \quad \text{in } G, \qquad (74)$$

$$u = u_0 \quad \text{on } \partial G.$$

61.9b. The Key Inequality

We set

$$a(u,v) = \int_G \lambda\operatorname{tr}\gamma(u)\operatorname{tr}\gamma(v) + 2\kappa[\gamma(u),\gamma(v)]\,dx,$$

$$b(u) = -\int_G Ku\,dx.$$

The key observation, in our existence proof, is the inequality

$$a(u,u) \geq 2\kappa \int_G [\gamma(u),\gamma(u)]\,dx \geq \kappa \int_G [u'(x), u'(x)]\,dx \qquad (75)$$

$$\text{for all } u \in C_0^\infty(G, V_3).$$

Here, $u \in C_0^\infty(G, V_3)$ means that the components of u belong to $C_0^\infty(G)$.

This is a simple special case of Korn's inequality of Chapter 62. In fact, for all $u_i \in C_0^\infty(G)$, integration by parts yields

$$\int_G D^i u_j D^j u_i\,dx = -\int_G (D^i D^j u_j) u_i\,dx = \int_G D^j u_j D^i u_i\,dx$$

$$= \int_G (\operatorname{div} u)^2\,dx \geq 0.$$

61.9. Existence and Uniqueness in Linear Elastostatics (Generalized Solutions)

Hence

$$\int_G [\gamma(u), \gamma(u)] \, dx = \int_G \tfrac{1}{4}(D_i u^j + D^j u_i)(D^i u_j + D_j u^i) \, dx$$

$$\geq \tfrac{1}{2} \int_G D_i u^j D^i u_j \, dx = \tfrac{1}{2} \int_G [u'(x), u'(x)] \, dx.$$

Furthermore, the Poincaré–Friedrichs inequality of Section 18.9 implies

$$\int_G [u'(x), u'(x)] \, dx \geq C \int_G u(x)^2 \, dx. \tag{76}$$

We set $X = \mathring{W}_2^1(G; V_3)$, i.e., X consists of all vector functions $u: G \to V_3$ where, in a fixed coordinate system, the components of u belong to $\mathring{W}_2^1(G)$. This definition is independent of the choice of the coordinate system. According to (76), the space X is a real H-space with scalar product

$$(u|v)_X = \int_G [u'(x), v'(x)] \, dx.$$

61.9c. The Generalized Problem

The generalized problem which belongs to the classical variational problem (72) is

$$\tfrac{1}{2} a(w + u_0, w + u_0) - b(w + u_0) = \min!, \quad w \in X. \tag{77}$$

The corresponding Euler equation

$$a(w + u_0, v) = b(v) \quad \text{for all} \quad v \in X \tag{78}$$

is the generalized equation of the classical Lamé equations (74). Our hypotheses are the following:

(H1) G is a bounded region in $\mathbb{R}^3$ with $\partial G \in C^{0,1}$.
(H2) We are given the density of outer forces $K \in L_2(G; V_3)$ and the boundary displacement $u_0 \in W_2^1(G; V_3)$.

Theorem 61.D (Main Theorem of Linear Elastostatics). *Assume* (H1), (H2). *Then both problems* (77) *and* (78) *have the same unique solution* $w \in X$. *The corresponding displacement vector is given by* $u = w + u_0$.

PROOF. This is an immediate consequence of Theorem 22.A. The strong positivity of $a(\cdot, \cdot)$ follows from Korn's inequality (75). □

Theorem 22.A also yields the convergence of the Ritz method, together with error estimates.

61.10. Existence and Uniqueness in Linear Elastodynamics (Generalized Solutions)

We use the same notation as in Section 61.9. The *classical problem* of linear elastodynamics is

$$\rho_0 u_{tt} = \operatorname{div} \sigma + K,$$

i.e.,

$$\begin{aligned}
\rho_0 u_{tt} &= \kappa \Delta u + (\lambda + \kappa) \operatorname{grad} \operatorname{div} u + K \quad \text{in} \quad G \times \,]0, \infty[, \\
u(x, t) &= u_0(x, t) \quad \text{on} \quad \partial G \times \,]0, \infty[, \\
u(x, 0) &= u_1(x) \quad \text{on } G, \\
u_t(x, 0) &= u_2(x) \quad \text{on } G.
\end{aligned} \tag{79}$$

For the sake of simplicity, we set $u_0 \equiv 0$. Letting $w = u - u_0$, the general case can always be reduced to this special case.

With $(u|v) = \int_G \rho_0 uv \, dx$, the *generalized problem* reads as follows:

$$\frac{d^2}{dt^2}(u|v) + a(u, v) = b(v) \quad \text{for all} \quad v \in C_0^\infty(G; V_3), \tag{80}$$

$$u(0) = u_1, \quad u'(0) = u_2.$$

We obtain (80) from (79) by multiplying with $v \in C_0^\infty(G; V_3)$ and integrating by parts.

Set $Q_T = G \times \,]0, T[$ and

$$X = \mathring{W}_2^1(G; V_3), \quad H = L_2(G; V_3).$$

Theorem 61.E (Main Theorem of Linear Elastodynamics). *Let G be a bounded region in $\mathbb{R}^3$, and $T > 0$. Suppose we are given the mass density $\rho_0 \in C(\bar{G})$, the density of outer forces $K \in L_2(Q_T)$, the initial position $u_1 \in X$, and the initial velocity $u_2 \in H$.*

Problem (80) then has a unique solution

$$u \in L_2(0, T; X), \quad u' \in L_2(0, T; H), \quad u'' \in L_2(0, T; X^*).$$

PROOF. This is an immediate consequence of Theorem 24.A. The key inequality (75) yields the strong positivity of $a(\cdot, \cdot)$. □

In Section 24.2 we also proved the convergence of the corresponding Galerkin method.

61.11. Strongly Elliptic Systems

The following results serve as a preparation for our general considerations in the following sections on nonlinear elasticity. The key observation is that the linearization of the equations of nonlinear elasticity at strongly stable solutions yields linear strongly elliptic systems. In order to be able to apply the implicit function theorem and the methods of bifurcation theory, we need information about the solutions of such linear strongly elliptic systems.

Consider the system

$$-a_{ijkm}D_iD_ju_k + a_{jkm}D_ju_k = K_m \quad \text{in } G$$
$$u_m = g_m \quad \text{on } \partial G, \quad m = 1,\ldots,M, \tag{81}$$

together with the homogeneous adjoint system

$$-D_iD_j(a_{ijkm}u_m^*) - D_j(a_{jkm}u_m^*) = 0 \quad \text{in } G$$
$$u_m^* = 0 \quad \text{on } \partial G, \quad m = 1,\ldots,M. \tag{81*}$$

We employ the notation

$$x = (\xi_1,\ldots,\xi_N), \quad D_i = \partial/\partial\xi_i, \quad u = (u_1,\ldots,u_M),$$

and sum over two equal indices, where $i, j = 1,\ldots,N$ and $k, m = 1,\ldots,M$. Moreover, let us make the following assumptions:

(H1) G is a bounded region in $\mathbb{R}^N$ with $\partial G \in C^\infty$.
(H2) $a_{ijkm}, a_{jkm} \in C^\infty(\bar{G})$ for all i, j, k, m.
(H3) The system is *strongly elliptic*, i.e., there is a constant $c > 0$ such that

$$a_{ijkm}(x)d_id_jv_kv_m \geq c|d|^2|v|^2$$

for all $d \in \mathbb{R}^N, v \in \mathbb{R}^M, x \in \bar{G}$.

We are now looking for solutions of (81) with the given functions

$$K_m \in W_2^{k-2}(G), \quad g_m \in W_2^{k-1/2}(\partial G)$$

for all m and $k = 2, 3,\ldots$. We expect solutions of the form

$$u_m \in W_2^k(G) \quad \text{for all } m.$$

Proposition 61.16(ii) below shows that uniqueness implies the strongest possible existence result.

Setting

$$Au = (K, g),$$

equation (81) defines an operator

$$A: W_2^k(G)^M \to W_2^{k-2}(G)^M \times W_2^{k-1/2}(\partial G)^M$$

for all $k = 2, 3,\ldots$. The spaces $W_2^{k-1/2}(\partial G)$ of boundary functions were introduced in A₂(48). The important point is that the embedding

$$W_2^k(G) \subseteq W_2^{k-1/2}(\partial G)$$

is continuous and *surjective*. Conversely, there exists a continuous extension operator

$$T: W_2^{k-1/2}(\partial G) \to W_2^k(G)$$

such that $Tu = u$ on ∂G. Using this fact one can reduce (81) to the homogeneous case.

Proposition 61.16. *Assume* (H1)–(H3). *Then*:

(i) *The operator A is Fredholm of index zero for all $k = 2, 3, \ldots$.*
    If $K \in C^\infty(\bar{G})^M$ and $g \in C^\infty(\partial G)^M$, then each solution u of (81) belongs to $C^\infty(\bar{G})^M$.
(ii) *If the homogeneous equation (81) has only the trivial solution $u = 0$, then A is a linear homeomorphism.*
(iii) *Let $K \in L_2(G)^M$ and $g = 0$ on ∂G. Then equation (81) has a solution if and only if*

$$\int_G \sum_{m=1}^M K_m u_m^* \, dx = 0$$

for all solutions u^* of the adjoint equation (81*).

The homogeneous equations (81) and (81*) have an equal finite number of linearly independent solutions.

PROOF. Use the same argument as in Chapter 22 for strongly elliptic equations including regularization. The key is Gårding's inequality

(G) $$\int_G u_m L_m u \, dx \geq c \int_G (D_k u_m)^2 \, dx - d \int_G u_m^2 \, dx$$

for all $u \in C_0^\infty(G)^M$ with constants $c, d > 0$, where $L_m u$ denotes the left-hand side of the first line in (81). Compare Browder (1954) and Nirenberg (1955). In (G), we sum over $k, m = 1, \ldots, M$. □

We are now interested in a substantially stronger result, i.e., we are looking for solutions

$$u_m \in C^{2,\alpha}(\bar{G}) \quad \text{for all } m, \quad 0 < \alpha < 1$$

for given functions

$$K_m \in C^\alpha(\bar{G}), \quad g_m \in C^{2,\alpha}(\partial G) \quad \text{for all } m.$$

The use of Hölder spaces will be essential. We note that the following proposition is wrong in case that $\alpha = 0$, which underlines the importance of Hölder spaces in the theory of elliptic differential equations.

Let us make the following assumptions:

(A1) G is a bounded region in $\mathbb{R}^N$ with $\partial G \in C^{2,\alpha}$ for fixed $0 < \alpha < 1$.
(A2) $a_{ijkm}, a_{jkm} \in C^\alpha(\bar{G})$ for all i, j, k, m.
(A3) The system (81) is strongly elliptic.

Now consider the operator

$$A: C^{2,\alpha}(\bar{G})^M \to C^\alpha(\bar{G})^M \times C^{2,\alpha}(\partial G)^M$$

defined through $Au = (K, g)$ by equation (81).

Proposition 61.17. *Assume* (A1)–(A3). *Then*:

(i) *The operator A is Fredholm of index zero.*
(ii) *If the homogeneous equation (81) has only the trivial solution $u = 0$, then A is a linear homeomorphism.*

PROOF. This is a profound and extremely sharp result in the theory of elliptic partial differential equations. A detailed proof would be rather long. We sketch here the main ideas for (ii).

(I) First, consider the case of C^∞-coefficients. By using Proposition 61.16, one obtains a solution $u \in W_2^2(G)^M$. The L_p-estimates in Agmon, Douglis, and Nirenberg (1959), Part II yield $u \in W_p^2(G)^M$ for all $p \geq 2$. The Sobolev embedding theorems then imply that $u \in C^{1,\alpha}(\bar{G})^M$ and the result follows from sharp Schauder a priori estimates in Agmon, Douglis, and Nirenberg (1959), Part II, Theorem 9.3. See also Morrey (1966, M), Chapter 6.

(II) Now suppose the coefficients are in $C^\alpha(\bar{G})$. Note that the $C^{2,\alpha}(\bar{G})$-a priori estimates in (I) depend only on an upper bound for the $C^\alpha(\bar{G})$-norms of the coefficients. Use the smoothing operator of Section 18.14 in order to approximate the C^α-coefficients by C^∞-coefficients and observe that the $C^\alpha(\bar{G})$-norms of all the approximating coefficients are uniformly bounded. This is a consequence of the properties of the smoothing operator. A compactness argument together with the $C^{2,\alpha}$-a priori estimates yields the final assertion. □

61.12. Local Existence and Uniqueness Theorem in Nonlinear Elasticity via the Implicit Function Theorem

61.12a. Basic Ideas

The basic idea in our approach to nonlinear elasticity is the following:

(i) In order to solve the basic equations of elastostatics, we *linearize* around so-called *strongly stable* solutions, which correspond to known strongly stable deformation states of the elastic body (e.g., the rest state $u = 0$).

(ii) It is important then that these linearizations correspond to linear *strongly elliptic systems* which have *unique* solutions as a consequence of the strong stability.
(iii) The Fredholm property of linear strongly elliptic systems implies that the linearization corresponds to a *bijective* operator. Thus the *implicit function theorem* (Theorem 4.B) can be applied. This leads to Theorem 61.F below.
(iv) A known strongly stable solution of the nonlinear basic equations of elastostatics may be continued via the implicit function theorem, as long as this continuation remains strongly stable, i.e., the linearization corresponds to a bijective operator.

The *discretization* of this continuation procedure yields an approximation method, the convergence of which is proved in Theorem 61.H below. This convergence proof uses standard arguments from the theory of ordinary differential equations in B-spaces.

In order to obtain classical solutions, we work in Hölder spaces. Using the results from Section 61.11, this approach also applies to Sobolev spaces, where it yields generalized solutions.

To be able to give a simple physical interpretation in terms of stability theory via potential energy, we use a variational approach. Our method, however, is also applicable to the general basic equations of nonlinear elastostatics, which do not correspond to variational problems. Let us call a deformation state u *admissible* if and only if the linearization of the basic equations at u yields a linear strongly elliptic system which has a unique solution, i.e., where the corresponding linear operator is bijective. Then, roughly speaking, the following general and very natural local result follows from the implicit function theorem.

In a sufficiently small neighborhood of a known admissible (e.g., strongly stable) deformation state of an elastic body, there exist uniquely determined new deformation states if we consider small changes in the outer forces and boundary displacements.

The connection with *bifurcation theory* is the following. If the known deformation state u is *not* admissible, then bifurcation may occur, i.e., small changes in the outer forces and boundary displacements may lead to several nonuniquely determined new deformation states of the elastic body. In this case, nature chooses the new state with the "greatest stability" (e.g., the lowest potential energy). Such bifurcation situations correspond, for example, to the buckling of rods, beams, plates, and shells.

In the case of the variational approach it is crucial that the linearization of the basic equation at u is equal to the Euler equation of the accessory quadratic variational problem, i.e., we have the situation of Figure 61.6. Recall that, in Section 29.12, we used accessory quadratic variational problems in order to obtain general sufficient criteria for minima (eigenvalue criteria).

61.12. Local Existence and Uniqueness Theorem in Nonlinear Elasticity

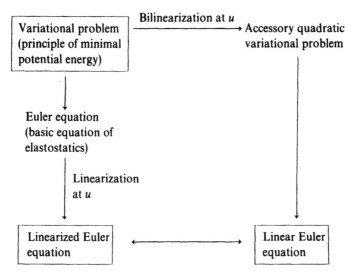

Figure 61.6

61.12b. Variational Problem and Strongly Stable States

We consider the principle of minimal potential energy

$$\int_G L(u'(x))\,dx - \int_G Ku\,dx = \min!, \tag{82}$$

$$u = g \quad \text{on } \partial G.$$

Let us assume that a fixed Cartesian coordinate system is given and that the sum is taken over two equal indices from 1 to 3. From Theorem 61.A, the Euler equations to (82) are

$$-a_{ijkm}D_iD_ju_k = K_m \quad \text{in } G,$$
$$u_m = g_m \quad \text{on } \partial G, \quad m = 1, 2, 3, \tag{83}$$

where

$$a_{ijkm} = \frac{\partial^2 L(u'(x))}{\partial D_i u_m\, \partial D_j u_k}.$$

Let L be C^∞. The elastic potential energy of the body is then given by

$$U = \int_G L(u'(x))\,dx.$$

An important role will be played by the second variation

$$\delta^2 U(u; h) = \int_G L_{u'u'}(u'(x))h'(x)^2\,dx$$

$$= \int_G a_{ijkm}D_i h_m D_j h_k\,dx$$

and by the so-called *accessory quadratic variational problem*

$$\tfrac{1}{2}\delta^2 U(u;h) - \int_G Kh\,dx = \min!, \tag{84}$$

$$h = g \quad \text{on } \partial G$$

for the unknown function h. Set

$$\mathring{C}^2(\bar{G})^3 = \{h \in C^2(\bar{G})^3 : h = 0 \text{ on } \partial G\}.$$

A necessary condition for a solution u of (82) to exist is that

$$\delta^2 U(u;h) \geq 0 \quad \text{for all} \quad h \in \mathring{C}^2(\bar{G})^3.$$

According to Section 18.17b, this leads to the Legendre–Hadamard condition

$$L_{u'u'}(u'(x))(v \circ d)^2 \geq 0 \quad \text{for all} \quad v, d \in V_3,$$

where $v \circ d$ denotes the dyadic product. In components,

$$L_{u'u'}(u'(x))(v \circ d)^2 = a_{ijkm} d_i d_j v_k v_m.$$

Definition 61.18. A function $u \in C^2(\bar{G})^3$ is called *strongly stable* with respect to U if and only if

$$\delta^2 U(u;h) > 0 \quad \text{for all nonzero} \quad h \in \mathring{W}_2^1(G)^3 \tag{85}$$

and the strong Legendre–Hadamard condition is valid, i.e.,

$$L_{u'u'}(u'(x))(v \circ d)^2 > 0 \tag{86}$$

for all nonzero $v, d \in V_3$ and all $x \in \bar{G}$.

From Gårding's inequality of Section 29.19 and Hestenes' theorem (Proposition 22.39), it follows that (85) implies the existence of a constant $C > 0$ such that

$$\delta^2 U(u;h) \geq C\|h\|^2 \quad \text{for all} \quad h \in \mathring{W}_2^1(G)^3, \tag{85*}$$

where $\|h\|$ denotes the norm on $\mathring{W}_2^1(G)^3$.

Using a simple continuity argument, it follows from condition (86) that there exists a constant $c > 0$ such that

$$L_{u'u'}(u'(x))(v \circ d)^2 \geq c|v|^2|d|^2 \quad \text{for all } v, d \in V_3 \tag{86*}$$

and all $x \in \bar{G}$, i.e., the system (83) is *strongly elliptic*.

According to (85*) and Theorem 29.L, each strongly stable $C^2(\bar{G})$-solution of the Euler equation (83) yields a strict local minimum of the original variational problem (82) with respect to the space $C^1(\bar{G})^3$.

61.12c. Local Continuation

In order to prove a local continuation theorem, we assume the following:

(H1) G is a bounded region in $\mathbb{R}^3$ with $\partial G \in C^{2,\alpha}$ for fixed $0 < \alpha < 1$.
(H2) The stored energy function $L: \mathbb{R}^9 \to \mathbb{R}$ is C^∞.
(H3) We know a strongly stable solution $\bar{u} \in C^{2,\alpha}(\bar{G})^3$ of (83) with corresponding density of outer forces $\bar{K}$ and boundary displacement $\bar{g}$.

Set

$$X = C^{2,\alpha}(\bar{G})^3, \qquad Y = C^\alpha(\bar{G})^3 \times C^{2,\alpha}(\partial G)^3.$$

Theorem 61.F (Local Existence and Uniqueness). *Assume* (H1)–(H3). *Then, there exist neighborhoods*

$$V(\bar{u}) \quad \text{in } X \quad \text{and} \quad W(\bar{K}, \bar{g}) \quad \text{in } Y$$

such that, for each $(K, g) \in W$, *equation* (83) *has a unique solution* $u \in V$.

PROOF.

(I) Equation (83) defines an operator $F: X \to Y$ by letting

$$F(u) = (K, g).$$

The linearized equation

$$F'(u)h = (K, g)$$

corresponds to the linearization of (83), i.e.,

$$-\left(\frac{\partial a_{ijkm}(u(x))}{\partial D_r u_s} D_i D_j u_k\right) D_r h_s - a_{ijkm}(u(x)) D_i D_j h_k = K_m \quad \text{in } G, \tag{87*}$$

$$h_m = g_m \quad \text{on } \partial G, \qquad m = 1, 2, 3,$$

whereby we are looking for h. Formula (87*) can be written as

$$-D_i(a_{ijkm}(u(x))D_j h_k) = K_m \quad \text{in } G$$

$$h_m = g_m \quad \text{on } \partial G, \qquad m = 1, 2, 3. \tag{87}$$

The *key* observation then is that (87) is precisely the Euler equation to the accessory variational problem (84).

(II) Let $h \in X$ be a solution of

$$F'(\bar{u})h = 0,$$

i.e., h is a solution of (87) with $u = \bar{u}$ and $K = 0$, $g = 0$. Integration by parts yields

$$\delta^2 U(\bar{u}; h) = \int_G a_{ijkm} D_i h_m D_j h_k \, dx$$

$$= -\int_G h_m D_i(a_{ijkm} D_j h_k) \, dx = 0$$

and hence $h = 0$, since $\bar{u}$ is strongly stable.

(III) Proposition 61.17 shows that $F'(\bar{u}): X \to Y$ is bijective. The assertion then follows from the implicit function theorem (Theorem 4.B). □

EXAMPLE 61.19. Let

$$L = M(\gamma(u)),$$

where $M: L_{\text{sym}}(V_3) \to \mathbb{R}$ is C^∞ and suppose that there exists a constant $c > 0$ such that

$$M''(\gamma)\mu^2 \geq c[\mu, \mu] \qquad \text{for all} \quad \mu, \gamma \in L_{\text{sym}}(V_3). \tag{88}$$

Then, each solution u of (83) is strongly stable.

Condition (88) is fulfilled for linear materials, i.e.,

$$L = \tfrac{1}{2}\lambda(\text{tr }\gamma)^2 + \kappa[\gamma, \gamma].$$

PROOF. In this case, we obtain

$$a_{ijkm} = \frac{\partial^2 M}{\partial \gamma_{im} \partial \gamma_{jk}}.$$

Korn's inequality (75), (76) implies

$$\int_G a_{ijkm} D_i h_m D_j h_k \, dx = \int_G M''(\gamma(u))\gamma(h)^2 \, dx$$

$$\geq c \int_G [\gamma(h), \gamma(h)] \, dx \geq C\|h\|^2 \qquad \text{for all} \quad h \in C_0^\infty(G)^3.$$

This is (85). Moreover, from

$$0 = a_{ijkm} d_i d_j v_k v_m \geq \frac{c}{4}(d_i v_m + v_i d_m)^2$$

we obtain $d_i v_m = -v_i d_m$ for all i, m, and hence $d = 0$ or $v = 0$. This is (86). □

EXAMPLE 61.20. Let

$$L = M(\mathscr{E}(u)),$$

where $M: L_{\text{sym}}(V_3) \to \mathbb{R}$ is C^∞ and suppose that there exists a constant $c > 0$

such that

$$M''(0)\mu^2 \geq c[\mu,\mu] \quad \text{for all} \quad \mu \in L_{\text{sym}}(V_3). \tag{89}$$

Then $u = 0$ is strongly stable. This is proved analogously as Example 61.19. Condition (89) is fulfilled for Saint Venant–Kirchhoff materials, i.e.,

$$L = \tfrac{1}{2}\lambda(\operatorname{tr}\mathscr{E})^2 + \kappa[\mathscr{E},\mathscr{E}].$$

Proposition 61.21. *Suppose we have the situation of Example 61.19 or 61.20. The trivial solution $u = 0$ (rest state) of the Euler equation (83) with $K = 0$ and $g = 0$ is then strongly stable. From Theorem 61.F follows that, for all sufficiently small smooth boundary displacements g and densities of outer forces K, there exists a unique classical solution u of (83).*

61.13. Existence and Uniqueness Theorem in Linear Elastostatics (Classical Solutions)

Theorem 61.G. *Choose $0 < \alpha < 1$. Let G be a bounded region in $\mathbb{R}^3$ with $\partial G \in C^{2,\alpha}$. Let*

$$L = \tfrac{1}{2}\lambda(\operatorname{tr}\gamma)^2 + \kappa[\gamma,\gamma].$$

Then, for each $K \in C^{\alpha}(\bar{G})^3$ and $g \in C^{2,\alpha}(\partial G)^3$, the Lamé equations (83) have a unique solution $u \in C^{2,\alpha}(\bar{G})^3$.

PROOF. This follows from Proposition 61.21 and the linearity of (83) in this special case. □

61.14. Stability and Bifurcation in Nonlinear Elasticity

We consider a homogeneous and isotropic body. From Section 61.8, the stored energy function must have the form

$$L = M(\mathscr{E}).$$

Suppose that $M: L_{\text{sym}}(V_3) \to \mathbb{R}$ is C^∞. We use the notation of Section 61.12. Let Σ be the set of all

$$u \in X, \quad (K, g) \in Y,$$

which are solutions of the basic equations of nonlinear elasticity (83). Our preceding results lead to the following clear picture. Let $P = (\bar{u}, \bar{K}, \bar{g})$.

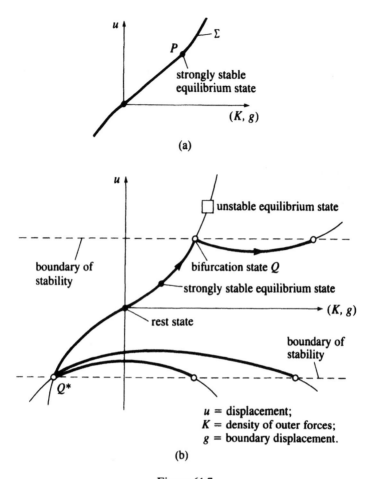

Figure 61.7

(i) The points P of Σ are *equilibrium states* of the elastic body. They correspond to critical points of the potential energy

$$E_{\text{pot}} = \int_G L\,dx - \int_G Ku\,dx.$$

(ii) If $P \in \Sigma$ and u is *strongly stable*, then, by Theorem 61.F, Σ behaves locally like a curve (Fig. 61.7(a)).
(iii) *Bifurcation* may only occur at points $Q \in \Sigma$ which are not strongly stable (Fig. 61.7(b) shows a subset of Σ).
(iv) By definition, the *boundary of stability* is given by all the states $u \in X$ which are "almost" strongly stable, i.e.,

$$\delta^2 U(u;h) \geq 0 \quad \text{for all} \quad h \in \mathring{W}_2^1(G)^3$$

and u is *not* strongly stable, i.e.,

$$\delta^2 U(u;h) = 0 \quad \text{for some nonzero} \quad h \in \mathring{W}_2^1(G)^3,$$

61.14. Stability and Bifurcation in Nonlinear Elasticity

or
$$L_{u'u'}(u'(x))(v \circ d)^2 = 0$$
for some $x \in \bar{G}$ and some nonzero $v, d \in V_3$.

(v) Points $P \in \Sigma$ with
$$\delta^2 U(u; h) < 0 \quad \text{for some nonzero} \quad h \in \mathring{W}_2^1(G)^3$$
correspond to unstable equilibrium states of the body. In this case, the potential energy E_{pot} has a critical point at u which is not a minimum.

Summarizing our observations we obtain the following basic principle of nonlinear elasticity theory:

Loss of strong stability can lead to bifurcation, i.e., to new equilibrium states.

Bifurcation, for example, can lead to the buckling of rods, beams, plates, and shells. Of special interest are such bifurcation points where precisely one new strongly stable branch occurs. Let us, for instance, look at point Q in Figure 61.7(b). Nature will follow the new strongly stable branch (e.g., a buckled state of a plate). At point Q^* in Figure 61.7(b), there occur two new strongly stable branches. In this case, nature will follow the strongly stable branch with the *lower* potential energy.

For the study of bifurcation problems one can use the main theorem of bifurcation theory for potential operators (Theorem 29.K). Applications to the theory of nonlinear plates will be considered in Chapter 65.

In Example 61.20 we have shown that the natural condition
$$M''(0)\mu^2 \geq c[\mu, \mu] \quad \text{for all} \quad \mu \in L_{\text{sym}}(V_3)$$
with $c > 0$ implies that the rest state $u = 0$, $K = 0$, $g = 0$ is strongly stable. Hence, in a neighborhood of the origin, Σ looks like a "curve" (Fig. 61.7(a)).

If we know $(u, K, g) \in \Sigma$, then all the interesting physical quantities can easily be computed. The second Piola–Kirchhoff tensor is obtained from
$$S = M'(\mathscr{E}),$$
and the stress tensor in the deformed region follows from
$$\tau(y) = (\det y'(x))^{-1} y'(x) S(x) y'(x)^*, \quad y = x + u(x).$$
According to Sections 61.3 and 61.5, the stress forces acting on a deformed subregion H' are given by
$$\int_{\partial H} \sigma n \, dO = \int_{\partial H'} \tau n' \, dO'$$
and the outer forces by
$$\int_H K \, dx = \int_{H'} K \det x'(y) \, dy.$$

The model, considered above, is an exact model in nonlinear elasticity, i.e., the basic equations in Section 61.3 are strictly satisfied.

Our approach also applies to approximation models, where $L = L(u'(x))$ or, more generally, $L = L(u'(x), x)$ (inhomogeneous bodies). We then compute the reduced stress tensor from

$$\sigma = L_{u'}$$

and the stress tensor in the deformed region from

$$\tau(y) = (\det y'(x))^{-1} \sigma(x) y'(x)^*.$$

Note, however, that in this general case, the symmetry of τ may be violated, i.e., the basic requirement that "total torque = 0" may not hold.

In several situations it is an open problem as to what the global structure of the set Σ of equilibrium states and the region of strong stability look like. Because of bifurcation, rupture, and plasticity in nature, we expect the global structure of Σ to be very complex.

In the special case of linear elasticity, Σ is a plane in $X \times Y$ and all points of Σ are strongly stable, according to Example 61.19. This shows again that linear elasticity is an unrealistic model for large displacements.

61.15. The Continuation Method in Nonlinear Elasticity and an Approximation Method

If one knows a strongly stable solution of the basic equation (83), for example, the rest state $u = 0$, $K = 0$, $g = 0$, then one can try to *continue* this solution. This can be done, for example, by replacing K and g in (83) with tK and tg, where $0 \le t \le 1$, i.e., we obtain an operator equation

$$F(u(t)) = (tK, tg), \qquad u(0) = 0. \tag{90}$$

Intuitively, this means that we follow the curve Σ in Figure 61.7(a) starting at the origin. As in the proof of Theorem 61.F, the implicit function theorem tells us that, at least for small t, there exists a unique solution curve

$$u = u(t).$$

The operator $F: X \to Y$ is C^∞ and bounded on bounded sets. By the implicit function theorem, the derivative $u'(t)$ exists. From (90) follows

$$F'(u(t))u'(t) = (K, g). \tag{91}$$

Moreover, we can continue $u = u(t)$ as long as $u(t)$ remains strongly stable, i.e., $F'(u(t))^{-1}$ exists on Y.

If we replace the differential quotient in (91) with the difference quotient, then we find the natural approximation method

$$F'(u^t)u^{t+\Delta t} = F'(u^t)u^t + \Delta t(K, g). \tag{92}$$

61.15. Continuation Method in Nonlinear Elasticity; An Approximation Method

Suppose that Δt, K, and g are given and that we have already computed u^t for fixed $t = 0, \Delta t, 2\Delta t, \ldots$. We are then looking for the new deformation state

$$u^{t+\Delta t}.$$

Our method starts with the rest state $u^t = 0$ if $t = 0$. Note that t is an index.

In the proof of Theorem 61.F we computed the F-derivative $F'(u)h$ in terms of a linear elliptic system. Therefore, we can translate (92) into a system of differential equations. According to (87), equation (92) corresponds to the system

$$-D_i(a_{ijkm}(u^t)D_j h_k) = K_m^t \quad \text{in } G, \quad m = 1, 2, 3,$$
$$h = u^t + \Delta t\, g \quad \text{on } G \tag{93}$$

for the unknown function

$$h = u^{t+\Delta t}$$

with the new density of the outer forces

$$K_m^t = \bar{K}_m^t + \Delta t\, K_m. \tag{94}$$

Here the old density of the outer forces $\bar{K}_m^t$ is given by the equilibrium condition

$$-D_i(a_{ijkm}(u^t)u^t) = \bar{K}_m^t. \tag{95}$$

Equation (93) represents a strongly elliptic system for determining $u^{t+\Delta t}$ if u^t is strongly stable.

Moreover, equation (93) for determining $h = u^{t+\Delta t}$ is the Euler equation for the accessory variational problem

$$\tfrac{1}{2}\delta^2 U(u^t; h) - \int_G K^t h\, dx = \min!,$$
$$h = u^t + \Delta t\, g \quad \text{on } \partial G. \tag{93_{var}}$$

Hence the easiest way to compute $u^{t+\Delta t}$ is to solve (93$_{var}$) approximately by a Ritz method, whereby error estimates are obtained from the duality theory of the following chapter.

61.15a. Physical Interpretation

This approximation method admits a very natural physical interpretation. We obtain the continuation $u^{t+\Delta t}$ from the known state u^t by using:

(a) *linear material corresponding to the elastic potential energy*

$$u \mapsto \tfrac{1}{2}\delta^2 U(u^t; u); \quad \text{and}$$

(b) *the new density of outer forces K^t.*

Note that formula (94) for K^t is very natural because (95) is exactly the equilibrium condition for the displacement u^t with respect to the linear material in (a).

More precisely, we set

$$\mathscr{L}^t(u') = \tfrac{1}{2} L_{u'u'}(u^t) u'^2.$$

Then we have

$$\tfrac{1}{2} \delta^2 U(u^t; u) = \int_G \mathscr{L}^t(u') \, dx,$$

i.e., $\mathscr{L}^t$ is the stored energy function to the linear material corresponding to u^t in (a). Furthermore, we introduce the stress tensor

$$\sigma^t(u) = \mathscr{L}^t_{u'}(u')$$

which corresponds to this linear material.

Approximation Method 61.22. *The method (93)–(95) corresponds to the equation*

$$\operatorname{div} \sigma^t(u^{t+\Delta t}) = K^t \quad \text{in } G,$$
$$u^{t+\Delta t} = u^t + \Delta t \, g \quad \text{on } \partial G$$
(93*)

with

$$K^t = \bar{K}^t + \Delta t \, K, \tag{94*}$$

where $\bar{K}^t$ is obtained from

$$\operatorname{div} \sigma^t(u^t) = \bar{K}^t. \tag{95*}$$

We start with $u^t = 0$ *and* $\bar{K}^t = 0$ *if* $t = 0$ *and compute successively*

$$u^{t+\Delta t} \quad \text{for} \quad t = 0, \Delta t, 2\Delta t, \ldots.$$

We are given the step length Δt, the density of outer forces K, and the boundary displacement g.

We call

$$\mu^t = \sigma^t(u^{t+\Delta t}) - \sigma^t(u^t)$$

the supplementary stress with respect to the state u^t. We then obtain the equilibrium condition

$$\operatorname{div} \mu^t = \Delta t \, K,$$

i.e., the supplementary stress is in equilibrium with the outer force corresponding to $\Delta t \, K$.

61.15b. The Singular Bifurcation Case

This method can be applied as long as u' remains strongly stable; it breaks down if u' loses its strong stability. The same is true for $u(t)$.

In the bifurcation case for the basic equation (83), i.e., for

$$F(u(t)) = (tK, tg),$$

we can apply the methods of bifurcation theory from Chapter 8. If the linearized system $F'(u(t))h = 0$ remains strongly elliptic, but loses its unique solvability, then $F'(u(t))$ is a Fredholm operator of index zero, according to Section 61.11, and hence we have to apply the Fredholm alternative from Proposition 61.16 in order to get the branching equations of Ljapunov–Schmidt. In concrete cases it may be hard to apply this method, because we need explicit expressions for the solutions of the linearized problem.

There exists, however, another method for handling the bifurcation case. It consists in an application of the main theorem of bifurcation theory for potential operators (Theorem 29.K). This theorem tells us that each nontrivial solution of the linearized problem corresponds to a bifurcation point and, roughly speaking, that the number of branches is at least equal to the number of the linearly independent nontrivial solutions of the linearized problem. This approach has been studied in detail in Section 29.20. We will come back to this in Chapter 65 in the context of buckling of plates.

61.16. Convergence of the Approximation Method

The following theorem proves the convergence of this approximation method. We construct the continuous curve $t \mapsto v_{\Delta t}(t)$ by letting

$$v_{\Delta t}(n \Delta t) = u^{n \Delta t}, \qquad n = 0, 1, \ldots, N$$

and using linear interpolation (Fig. 61.8(a)). Here $N \Delta t = T$.

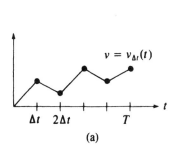 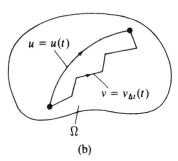

Figure 61.8

Let
$$X_\beta = C^{2,\beta}(\bar{G})^3, \qquad Y_\beta = C^\beta(\bar{G})^3 \times C^{2,\beta}(\partial G)^3, \qquad 0 < \beta < \alpha,$$
and recall that $X_\alpha = X$, $Y_\alpha = Y$.

Theorem 61.H. *Let $u = u(t)$, $0 \leq t \leq T$ be a solution curve of (90) such that $u(t)$ is strongly stable for all $t \in [0, T]$. Then*
$$\lim_{\Delta t \to 0} \max_{0 \leq t \leq T} \|v_{\Delta t}(t) - u(t)\|_{X_\beta} = 0,$$
i.e., $v_{\Delta t}$ converges to $u(\cdot)$ as $\Delta t \to 0$.

PROOF. The *key* formulas in this simple proof are
$$\begin{aligned} u(t) - u(n\,\Delta t) &= \int_{n\Delta t}^t F'(u(s))^{-1}(K,g)\,ds, \\ v_{\Delta t}(t) - v_{\Delta t}(n\,\Delta t) &= \int_{n\Delta t}^t F'(v_{\Delta t}(n\,\Delta t))^{-1}(K,g)\,ds \end{aligned} \qquad (96)$$
for all $t \in J$ where $J = [n\,\Delta t, (n+1)\,\Delta t]$.

(I) The inverse operator $F'(u(t))^{-1}$ exists for all $t \in [0, T]$. Using the continuity of inverse formation (Problem 1.7) and the compactness of $[0, T]$, there exists an open bounded neighborhood Ω of the solution curve in X such that
$$F'(u)^{-1} \quad \text{exists on } Y \text{ for all } \quad u \in \Omega \qquad \text{(Fig. 61.8(b))}.$$
The operator $F: X \to Y$ is C^2 and bounded on bounded sets. From Problem 1.7, the operator
$$u \mapsto F'(u)^{-1}$$
is bounded and Lipschitz continuous on Ω.

(II) Formula (96) implies that
$$\|u(t) - v_{\Delta t}(t)\| \leq \|u(n\,\Delta t) - v_{\Delta t}(n\,\Delta t)\|$$
$$+ \frac{C\,\Delta t}{2} \max_{s \in J} \|u(s) - v_{\Delta t}(n\,\Delta t)\|$$
for all $t \in J$. By using induction, $u(0) = v_{\Delta t}(0)$ and
$$C\,\Delta t + C^2(\Delta t)^2 + C^3(\Delta t)^3 + \cdots \leq \frac{C\,\Delta t}{1 - C\,\Delta t},$$
we obtain that, for Δt sufficiently small, $v_{\Delta t}$ remains in Ω.

(III) Convergence of a subsequence. Since
$$\sup_{u \in \Omega} \|F'(u)^{-1}\| < \infty,$$

equation (96) implies the equicontinuity of $\{v_{\Delta t}\}$. The embedding $X \subseteq X_\beta$ is compact. Hence the Arzelà–Ascoli theorem $A_1(24)$ tells us that there exists a subsequence $(v_{\Delta t})$ and a function $v(\cdot)$ such that

$$\max_{0 \le t \le T} \|v_{\Delta t}(t) - v(t)\|_{X_\beta} \to 0 \quad \text{as} \quad \Delta t \to 0. \tag{97}$$

The operator $F: X_\beta \to Y_\beta$ has the same properties as $F: X \to Y$ by Proposition 61.17. Letting $\Delta t \to 0$, we obtain from (96) that

$$v(t) = \int_0^t F'(v(s))^{-1}(K, g)\, ds.$$

Differentiation of this integral shows that

$$F'(v(t))v'(t) = (K, g), \quad v(0) = 0.$$

Thus, $u'(t)$ and $v'(t)$ are classical solutions of the same strongly elliptic differential equation and hence $u'(t) = v'(t)$. Integration yields $u(t) = v(t)$.

(IV) The same argument proves that each subsequence which satisfies (97) converges to $u(\cdot)$. The convergence principle in Proposition 10.13(1) shows that (97) is valid for the original sequence. □

Our approach to nonlinear elasticity in Sections 61.12–61.16 has been strongly influenced by the work of Beckert (1975), (1984, S).

PROBLEMS

61.1. *Polyconvex functions.* Let $A \in L(V_3)$, i.e., the operator $A: V_3 \to V_3$ is linear. The eigenvalues $\lambda_1, \lambda_2, \lambda_3$ of $(A^*A)^{1/2}$ are called the principal stretches of A.

61.1a. Show that the function $f: L(V_3) \to \mathbb{R}$,

$$f(A) = \operatorname{tr} A^*A = \lambda_1^2 + \lambda_2^2 + \lambda_3^2$$

is convex.
   Solution: $\delta^2 f(A) H^2 = 2 \operatorname{tr} H^*H \ge 0$ for all $H \in L(V_3)$.

61.1b. Show that the function $L = P(A, \operatorname{adj} A)$ with

$$L = C \operatorname{tr} A^*A + D \operatorname{tr} \operatorname{adj}(A^*A)$$
$$= C(\lambda_1^2 + \lambda_2^2 + \lambda_3^2) + D((\lambda_1\lambda_2)^2 + (\lambda_2\lambda_3)^2 + (\lambda_1\lambda_3)^2)$$

is polyconvex.
   Solution: Use $\operatorname{adj}(A^*A) = (\operatorname{adj} A)^*(\operatorname{adj} A)$ and Problem 61.1a.

61.1c. Define a function $f: L(V_3) \to \mathbb{R}$ through

$$f(A) = h(\lambda_1, \lambda_2, \lambda_3).$$

Suppose that $h: \mathbb{R}^3_+ \to \mathbb{R}$ is symmetric, convex, and nondecreasing in each argument. Show that f is convex.
   Hint: Let the operators $A, B: V_3 \to V_3$ be linear, *symmetric*, and positive. Let $\lambda_1 \ge \lambda_2 \ge \lambda_3$, $\mu_1 = \mu_2 = \mu_3$, and $\rho_1 \ge \rho_2 \ge \rho_3$ be the eigenvalues of A, B,

and $tA + sB$, respectively, where $0 \le t \le 1$, $s = 1 - t$. From the Courant maximum–minimum principle we obtain that

$$\rho_1 \le t\lambda_1 + s\mu_1, \qquad \rho_2 \le t\lambda_2 + s\mu_2, \qquad \rho_3 \le t\lambda_3 + s\mu_3$$

(see Riesz and Nagy (1978), p. 239). Hence

$$f(tA + sB) \le h(t\lambda_1 + s\mu_1, t\lambda_2 + s\mu_2, t\lambda_3 + s\mu_3)$$
$$\le th(\lambda_1, \lambda_2, \lambda_3) + sh(\mu_1, \mu_2, \mu_3)$$
$$= tf(A) + sf(B).$$

In the general case, this argument must be modified. See Ball (1977), p. 363.

61.2. *The theorem of Cauchy.* For a function $f: \prod_{i=1}^{N} V_3 \to \mathbb{R}$ the following two statements are equivalent.
  (i) f is isotropic, i.e.,

$$f(Rx_1, \ldots, Rx_N) = f(x_1, \ldots, x_N)$$

for all rotations and reflections R, and all $x_1, \ldots, x_N \in V_3$.
  (ii) f can be expressed as a function of the inner products $x_i x_j$, $i, j = 1, \ldots, N$.
Hint: See Truesdell and Noll (1965), p. 29.

61.3. *Elastic energy of a cuboid with linear state of stress.* Let $K = \sum_i K_i e_i$ be the density of the outer forces and assume that

$$K_2 = K_3 = 0$$

in Section 61.7a, i.e., the stress occurs only in the direction of the ξ_1-axis. We call this a linear state of stress. Compute, analogously to (50), the elastic energy U of the cuboid.
Solution: We obtain

$$U = \frac{EV}{2} \gamma_{11}^2,$$

because from (48) follows that $U = \sigma_1 \gamma_{11} V/2$ and Hooke's law yields $\sigma_1 = E\gamma_{11}$. Here, V is the volume of the cuboid.

61.4. *Elastic energy of a cuboid with planar state of stress.* Let

$$K_3 = 0$$

in Section 61.7a, i.e., no stress occurs in direction of the ξ_3-axis. Compute the elastic energy of the cuboid.
Solution: Analogous to Section 61.7a we obtain

$$\sigma_i = \bar{\lambda}(\gamma_{11} + \gamma_{22}) + 2\kappa\gamma_{ii}, \qquad i = 1, 2, \tag{98}$$

$$U = \frac{V}{2}(\bar{\lambda}(\gamma_{11} + \gamma_{22})^2 + 2\kappa(\gamma_{11}^2 + \gamma_{22}^2)) \tag{99}$$

with $\bar{\lambda} = \mu E/(1 - \mu^2)$. Note that here, other than in the general case of three-dimensional states of stresses of Section 61.7a, the quantity $\bar{\lambda}$ occurs instead of $\lambda = \mu E/(1 + \mu)(1 - 2\mu)$.

Because of $\sigma_3 = 0$, we obtain the relation

$$\gamma_{ii} = \frac{1+\mu}{E}\left(\sigma_i - \frac{\mu}{1+\mu}(\sigma_1 + \sigma_2)\right), \quad i = 1, 2.$$

This implies (98). Furthermore, it follows from (48) that

$$U = \tfrac{1}{2}(\gamma_{11}\sigma_1 + \gamma_{22}\sigma_2).$$

If we allow arbitrary rotations in the (ξ_1, ξ_2)-plane, then (99) takes the invariant form

$$U = \frac{V}{2}(\bar{\lambda}(\gamma_{11} + \gamma_{22})^2 + 2\gamma_j^i \gamma_i^j),$$

where the sum is taken over $i, j = 1, 2$.

61.5. **Equilibrium of torques in elastostatics.** We consider the basic equations of elastostatics in Section 61.5f. In this stationary case we obtain the torque condition

$$\int_{H'} F \times y \, dy + \int_{\partial H'} \tau n' \times y \, dO = 0 \tag{100}$$

for all deformed subregions H' from the general balance of angular momentum (25). This condition refers to the deformed body. In terms of the undeformed body we expect the torque condition

$$\int_{H} K \times y \, dx + \int_{\partial H} \sigma n \times y \, dO = 0 \tag{100*}$$

for all undeformed subregions H. Prove (100*).

Solution: Using the equilibrium condition $K^j + D^r \sigma_r^j = 0$ and integrating by parts we find the general identity

$$J \stackrel{\text{def}}{=} \int \varepsilon_{mij} \eta^i K^j \, dx + \int \varepsilon_{mij} \eta^i \sigma_r^j n^r \, dO$$

$$= \int \varepsilon_{mij}(\eta^i K^j + D^r(\eta^i \sigma_r^j)) \, dx$$

$$= \int \varepsilon_{mij} \sigma_r^j D^r \eta^i \, dx.$$

We now use the second Piola–Kirchhoff tensor S given by

$$\sigma(x) = y'(x) S(x).$$

According to Section 61.4d the torque condition (100) implies the symmetry of τ. Now S is also symmetric according to Section 61.5h. In Cartesian coordinates we have

$$\sigma_j^i = S_j^k D_k \eta^i$$

and $S_r^k = S_k^r$. Noting $\varepsilon_{mij} = -\varepsilon_{mji}$ we obtain

$$J = \int \varepsilon_{mij} S_r^k D_k \eta^j D^r \eta^i \, dx = 0.$$

This is (100*).

References to the Literature

Monographs (mathematical point of view): von Mises (1962), Nečas and Hlavaček (1981), Gurtin (1981), Ciarlet (1983), Marsden and Hughes (1983).

Monographs (physical point of view): Sommerfeld (1970) (classical standard work), Prager (1961), Landau and Lifšic (1962), Vol. 7, Solomon (1968), Washizu (1968), Lurje (1980), Grinčenko (1985, M), Vols. 1-6.

Handbook of Physics: Flügge (1956), Vol. VIa/1-4.

Rational Mechanics: Truesdell and Noll (1965, S), Truesdell (1977, M).

Thermomechanics: Ziegler (1983, M).

(See also References to the Literature to Chapter 62.)

CHAPTER 62

Monotone Potential Operators and a Class of Models with Nonlinear Hooke's Law, Duality and Plasticity, and Polyconvexity

There is no science, which did not develop from a knowledge of the phenomena; but in order to gain something from this knowledge, it is necessary to be a mathematician.

Daniel Bernoulli (1700–1782)

When solving variational problems with the Ritz method, it is important to estimate the quality of the approximation for the minimal values. The Ritz method yields upper bounds for those minimal values. In 1927, Erich Trefftz introduced a method—applicable to the Dirichlet and related problems—which allows approximations of the solution of a variational problem and at the same time yields lower bounds for the minimal value.

In the following the same goal will be reached by using a different and more fundamental approach. In general, we can assign to each minimization problem a dual maximization problem, whose maximal value is equal to the minimal value of the original problem.[1] The basic principle thereby corresponds to the Legendre transformation of point mechanics.

Kurt Otto Friedrichs (1929)

When studying any physical problem in applied mathematics, three essential stages are involved.

(i) *Modeling*: An appropriate mathematical model, based on the physics or the engineering of the situation, must be found. Usually, these models are given a priori by the physicists or the engineers themselves. However, mathematicians can also play an important role in this process, especially considering the increasing emphasis on nonlinear models of physical problems.
(ii) *Mathematical study of the problem*: A model usually involves a set of ordinary or partial differential equations or an (energy) functional to be minimized. One of the first tasks is to find a suitable function space in which to study the problem. Then comes the study of existence and uniqueness or nonuniqueness of solutions. An important feature of linear theories is the

---

[1] This is also called the principle of complementary energy.

existence of unique solutions depending continuously on the data (well-posed problems in the sense of Hadamard (1865–1963)). But with nonlinear problems, nonuniqueness is a prevalent phenomenon. For instance, bifurcation of solutions is of special interest.

(iii) *Numerical analysis of the model*: By this is meant the description of, and the mathematical analysis of, approximation schemes which *can* be run on a computer in a "reasonable" time to get "reasonably accurate" numbers.

Phillipe Ciarlet (1983)

Despite this recent progress, numerical analysis of nonlinear partial differential equations in three dimensions, as well as many other frontier questions, still await new mathematical methods.

Arthur Jaffe (1984)

Existence theorems under the assumption that the stored energy function L is *convex* with respect to u' have been given by several authors. Unfortunately, these results are only of mathematical interest, since convexity of L with respect to u' is unacceptable physically....

A wide variety of realistic models of nonlinear elastic materials satisfy the hypotheses of our existence theorem for *polyconvex* stored energy functions. In particular, these include the Mooney–Rivlin material and the Ogden material.

John Ball (1977)

62.1. Basic Ideas

In this chapter we consider:

(i) a class of approximation models with *convex* stored energy function. We investigate:
existence and uniqueness,
duality, and
approximation methods (Ritz method and Trefftz method, projection–iteration method, and gradient method);
(ii) a class of exact models with *polyconvex* stored energy functions (existence via compensated compactness).

Moreover, we consider:

(a) duality and statical plasticity;
(b) variational inequalities and quasi-statical plasticity.

A crucial analytic tool thereby is Korn's inequality, which will be proved later on by using an equivalent norm on $L_2(G)$ via negative norms.

We shall observe the following:

(α) Duality theory in elasticity is a special case of the general duality theory of Chapter 51 for monotone potential operators. One uses conjugate functionals and the abstract Young inequality (Theorem 51.B).
(β) The classical version of duality theory in elasticity is a special case of a general classical duality theory in the calculus of variations (the Friedrichs

62.1a. Convex Functionals and Monotone Operators

We begin by discussing the basic ideas of (i), and the convexity of the elastic potential energy will thereby be of central interest. Our starting point is the variational problem

$$\int_G M(\gamma(u))\,dx - \int_G Ku\,dx - \int_{\partial_2 G} Tu\,dO = \min!, \qquad (1)$$

$$u = u_0 \quad \text{on } \partial_1 G,$$

which corresponds to the principle of minimal potential energy. We set $v = u - u_0$ and write (1) as

$$F(v) - b(v) = \min!, \qquad v \in X \qquad (2)$$

with the Euler equation

$$F'(v) = b, \qquad v \in X, \qquad (3)$$

where X denotes a suitable function space (Sobolev space). It is important that M satisfies a so-called *strong stability condition*

$$M''(\gamma)\mu^2 \geq c[\mu,\mu] \qquad \text{for all} \quad \gamma, \mu \in L_{\text{sym}}(V_3)$$

with a constant $c > 0$. This condition guarantees that M is convex. Thus the energy functional F is *convex* and the derivative

$$F': X \to X^*$$

is a *strongly monotone* Lipschitz-continuous potential operator. The proof of this will be based on Korn's inequality. Hence the entire apparatus of the theory of monotone potential operators of Parts II and III is available for the Euler equation (3). Let us recall that the G-derivative of a convex functional is a monotone operator.

62.1b. Convexity and Duality

In applying the abstract duality theory of Part III to elasticity theory we obtain a picture of remarkable clarity with regard to both physical and mathematical aspects. Let us describe the basic idea. First, we write the original variational problem (1) in the form

(P) $\qquad E_{\text{pot}}(u) = \min!, \qquad u - u_0 \in X,$

where u denotes the *displacement* of the elastic body, where

$$E_{\text{pot}}(u) = \int_G M(\gamma(u))\,dx - \int_G Ku\,dx - \int_{\partial_2 G} Tu\,dO$$

is the potential energy and where

$$\gamma(u)(x) = \tfrac{1}{2}(u(x) + u'(x)^*)$$

is the linearized strain tensor, which will also be written simply as $\gamma(x)$. The constitutive law which corresponds to (P) is

(C) $\qquad\qquad\qquad \sigma = M'(\gamma)$

with inverse constitutive law

$$\gamma = M'^{-1}(\sigma).$$

The key formula of convex analysis $M^{*\prime} = M'^{-1}$ from Part III then yields

(C*) $\qquad\qquad\qquad \gamma = M^{*\prime}(\sigma),$

where M^* denotes the conjugate functional to M. Recall that σ is the stress tensor (first Piola–Kirchhoff tensor). We now define the so-called dual energy

$$E_{\text{dual}}(\sigma) = -\int_G M^*(\sigma)\,dx + \int_G [\sigma, \gamma(u_0)] - Ku_0\,dx - \int_{\partial_2 G} Tu_0\,dO$$

and consider the *dual variational* problem

(P*) $\qquad\qquad E_{\text{dual}}(\sigma) = \max!, \qquad \sigma \in \Sigma$

for the *stress tensor* σ. The precise definition of the set Σ will be given in Section 62.4. Roughly speaking, Σ is characterized by the relation

$$\int_G [\sigma, \gamma(h)]\,dx - \int_G Kh\,dx - \int_{\partial_2 G} Th\,dO = 0 \qquad \text{for all} \quad h \in X.$$

From Section 61.5g, this means that $\sigma \in \Sigma$ if and only if σ satisfies the principle of virtual work (power) for all "virtual displacements" $h \in X$. Intuitively, this says that $\sigma \in \Sigma$ if and only if the stress tensor σ is in an *equilibrium* with the outer forces K and T. In Section 62.4 we will show that for sufficiently smooth $\sigma \in \Sigma$, the simpler classical expression

$$E_{\text{dual}}(\sigma) = -\int_G M^*(\sigma)\,dx + \int_{\partial_1 G} (\sigma n)u_0\,dO$$

is valid.

Let us summarize our observations.

(α) The original variational problem (P) refers to the potential energy with respect to all possible displacements of the elastic body which satisfy the given boundary displacement.

(β) The dual variational problem refers to the dual energy with respect to all stress tensors which are in an equilibrium with the outer forces (volume force K and boundary force T).

Our main result is as follows.

The original variational problem (P) *and the dual variational problem* (P*) *have unique solutions u and σ, respectively, which are related by the constitutive law*

$$\sigma = M'(\gamma(u)),$$

i.e., σ *is precisely the stress tensor which is observed in the equilibrium state corresponding to the displacement u.*

Moreover, the extremal values of (P) *and* (P*) *are the same, i.e.,*

$$E_{pot}(u) = E_{dual}(\sigma).$$

In order to find an intuitive physical interpretation for the dual variational principle, let us introduce the stress energy

$$E_{stress}(\sigma) = \int_G M^*(\sigma)\,dx.$$

It is positive for all reasonable models. The dual problem then becomes

$$E_{dual}(\sigma) \equiv \int_{\partial_1 G} (\sigma n)u_0\,dO - E_{stress}(\sigma) = \max!, \qquad \sigma \in \Sigma.$$

This means that if we consider all stresses which are in an equilibrium with the outer forces, then, in an equilibrium state of the elastic body, the actual stress corresponds to a maximal difference between the work done by the boundary stress forces and the stress energy in the interior of the body.

The dual problem can also be written in the equivalent form

$$-E_{dual}(\sigma) = \min!, \qquad \sigma \in \Sigma.$$

This is called the principle of minimal complementary energy. We note the remarkable fact that the integrand M^* of E_{dual} corresponds to the inverse constitutive law (C*) above. Thus, we might say:

The dual variational problem refers to the inverse constitutive law of the elastic body.

In Section 51.5 we found that the abstract duality theory for monotone potential operators is based on the abstract Young inequality for conjugate functionals in the sense of convex analysis. In Example 51.4 we further showed that the classical Legendre transformation provides the motivation for conjugate functionals. This abstract duality theory is a generalization of very simple classical results. We will discuss this classical background, which was discovered by Friedrichs (1929), in Section 62.16. There it will be shown that the general duality theory of Friedrichs leads to a quite remarkable relation between elasticity theory and point mechanics. Table 62.1 illustrates how, by

Table 62.1

Point mechanics	Elasticity theory
Lagrangian function Hamiltonian function (Conjugate Lagrangian function)	Stored energy function M Conjugate stored energy function M^*
Position Velocity Momentum	Displacement u Strain tensor γ Stress tensor σ
Legendre transformation (between velocity and momentum)	Constitutive law $\sigma = M'(\gamma)$

applying the Legendre transformation, duality in elasticity can be viewed as a generalization of duality in point mechanics.

In applying the general theory of this chapter, we consider special stored energy functions M which correspond to:

(a) linear elasticity theory, and
(b) nonlinear Hencky material.

62.1c. Generalized Solutions

The solution u of the original variational problem (P) above with $u - u_0 \in X$, which corresponds to the classical problem

$$\int_G M(\gamma(u))\,dx - \int_G Ku\,dx - \int_{\partial_2 G} Tu\,dO = \min!$$

$$u = u_0 \quad \text{on } \partial_1 G,$$

is also a solution of the abstract Euler equation (3), as well as a *generalized* solution of the classical Euler equation

$$\operatorname{div} \sigma + K = 0 \quad \text{on } G, \qquad \sigma = M'(\gamma),$$

$$u = u_0 \quad \text{on } \partial_1 G,$$

$$\sigma n = T \quad \text{on } \partial_2 G,$$

which is a mixed boundary-value problem. On the boundary parts $\partial_1 G$ and $\partial_2 G$ we are given the displacement and the stress forces (boundary forces), respectively.

62.1d. Approximation Methods

Since the original variational problem (P) is convex, we can use the *Ritz method* (projection method) in order to obtain approximate solutions of (P).

62.1. Basic Ideas

The Ritz method, applied to the dual problem (P*), is the *Trefftz method*. In case of the variational problem (P) the Ritz method yields upper bounds for the minimal value of (P). If we use both the Ritz method and the Trefftz method, then we find two-sided error estimates for the minimal value of (P) and error estimates for the solution u of (P) in terms of the so-called energetic norm, i.e., the norm on the space X.

Moreover, since the operator F', which appears in the abstract Euler equation (3), is a strongly monotone and Lipschitz continuous potential operator, we can apply:

the projection–iteration method, and
the gradient method (method of steepest descent).

Let us summarize the advantages of duality in elasticity theory. By using duality, we obtain:

(a) approximate solutions for the displacements and stresses of the elastic body;
(b) two-sided error estimates for the minimal potential energy of the elastic body;
(c) error estimates for the displacements in terms of an energy norm (and error estimates for the stresses);
(d) a natural approach to quasi-statical plasticity by using the principle of maximal dual energy together with a plasticity inequality for the stress tensor as a side condition; and
(e) a natural approach to statical plasticity by using both

$$\text{the principle of maximal dual energy} \\ + \text{ the plasticity inequality for the stress tensor}$$

and

$$\text{the principle of minimal potential energy} \\ + \text{ the corresponding inequality for the strain tensor.}$$

In this way we obtain a rigorous justification of the empirical theory of Haar and von Kármán (1909).

62.1e. Linear Elasticity

In the case of linear elasticity we have that

$$M(\gamma) = \tfrac{1}{2}\lambda(\operatorname{tr}\gamma)^2 + \kappa[\gamma,\gamma],$$

whereby it is clear that if the parameters λ and κ are positive, then M is strongly stable. Equation (2) assumes the special form

$$\tfrac{1}{2}a(u,u) - b(u) = \min!, \qquad u \in X,$$

where $a: X \times X \to \mathbb{R}$ is a strongly positive, symmetric bilinear form. Here, the strong positivity, i.e., the fact that

$$a(v, v) \geq c\|v\|^2 \quad \text{for all} \quad v \in X$$

with fixed $c > 0$, follows from Korn's inequality. The existence and uniqueness result then is an immediate consequence of the main theorem on quadratic variational problems (Theorem 22.A).

In Section 61.9 we have used this approach for the boundary condition

(F) $\quad\quad u = u_0 \quad \text{on } \partial G \quad$ (first boundary-value problem).

In this chapter we consider the more general boundary conditions

(M) $\quad\quad \begin{aligned} u &= u_0 \quad \text{on } \partial_1 G \quad \text{(mixed boundary-value problem)} \\ \sigma n &= T \quad \text{on } \partial_2 G, \end{aligned}$

and

(S) $\quad\quad \sigma n = T \quad \text{on } \partial G \quad$ (second boundary-value problem).

For (F) and (M) we obtain unique solutions u.

Note that in (S) only the boundary forces are given. We expect that there will exist several deformation states u which satisfy this condition, and that there will be some equilibrium condition between the outer forces K and the boundary forces T. In fact, the displacement u is only determined up to infinitesimal rigid motions and a solution u exists if and only if the equilibrium condition

$$\int_G K \, dx + \int_{\partial G} T \, dO = 0$$

$$\int_G K \times x \, dx + \int_{\partial G} T \times x \, dO = 0$$

is satisfied. The first condition means that the total force is equal to zero. The second condition means that the total torque is equal to zero in the sense of (T_{approx}) of Section 61.6c.

62.1f. Convexity and Approximation Models

In Section 61.8 we have seen that, in the context of the exact theory, the stored energy function of a homogeneous body must have the form

$$L = M(\mathscr{E}).$$

If we replace the strain tensor

$$\mathscr{E}(x) = \tfrac{1}{2}(u'(x) + u'(x)^* + u'(x)^* u'(x))$$

with the linearized strain tensor

$$\gamma(x) = \tfrac{1}{2}(u'(x) + u'(x)^*),$$

then we obtain our model (1). Consequently, this model has only an *approximation character*.

62.1g. Polyconvexity

More general models are obtained from polyconvex stored energy functions

$$L = P(A, \operatorname{adj} A, \det A),$$

where $P: L(V_3) \times L(V_3) \times \,]0, \infty[\to \mathbb{R}$ is convex and

$$A = I + u'(x).$$

Hence

$$E \stackrel{\text{def}}{=} (A^*A)^{1/2} = (I + 2\mathscr{E}(x))^{1/2}.$$

In the special case of rubberlike Ogden material with

$$L = C \operatorname{tr} E^p + D \operatorname{tr} \operatorname{adj} E^r + f(\det A) + \text{const}$$

and $p, r \geq 1$, we obtain the *exact* stored energy function

$$L = M(\mathscr{E}).$$

In fact, because of $\det A = \det E$ and the definition of E, the quantity L in this case depends only on $\mathscr{E}$.

In Section 62.13 we prove a general existence theorem for polyconvex material including the Ogden material. The key thereby is the important recent method of *compensated compactness* whose basic idea we explain in Section 62.12.

Recall that in the preceding chapter we have used the implicit function theorem in order to obtain a rigorous *local* approach to nonlinear elasticity. The variational approach via polyconvexity is *global* in nature.

But it is clear that there remain many open problems in nonlinear elasticity.

As in the preceding chapter, we use an invariant, i.e., coordinate-free, approach. This simplifies formulas and a variety of arguments. The reader, who prefers coordinates, can easily rewrite everything by using the results of Section 61.1. The constitutive law

$$\sigma = M'(\gamma),$$

for example, becomes

$$\sigma_j^i = \frac{\partial M(\gamma)}{\partial \gamma_i^j}, \qquad i, j = 1, 2, 3.$$

Before studying this chapter one should again take a short look at Section

61.6 on the variational approach to elasticity, and at Section 21.2 on Sobolev spaces.

62.2. Notations

Let G be a bounded region in the three-dimensional vector space V_3. Recall that V_3 may be identified with $\mathbb{R}^3$. The sets $\partial_1 G$ and $\partial_2 G$ correspond to the boundary decomposition

$$\partial G = \overline{\partial_1 G} \cup \overline{\partial_2 G}, \qquad \partial_1 G \cap \partial_2 G = \varnothing,$$

where $\partial_1 G$ and $\partial_2 G$ are open subsets of the boundary ∂G.

The space

$$X = \mathring{W}_2^1(G, \partial_1 G; V_3),$$

which plays a *key* role in this chapter, has the structure of a Sobolev space (see Part II). More precisely, the space X consists of all displacements $v \colon G \to V_3$ with components

$$v^i \in W_2^1(G), \quad i = 1, 2, 3,$$
$$v^i = 0 \quad \text{on } \partial_1 G. \tag{4}$$

The boundary condition is understood in the generalized sense (see Section 21.2).

If relation (4) holds in a fixed coordinate system, then, as a consequence of the linearity of the coordinate transformations, it holds in *every* coordinate system. Recall that

$$(f|g)_{W_2^1(G)} = \int_G \sum_{j=1}^3 fg + D_j f D_j g \, dx.$$

As a scalar product in X we choose

$$(v|w)_X = \sum_{i=1}^3 (v^i|w^i)_{W_2^1(G)}.$$

This expression is independent of the choice of the Cartesian coordinate system. Because of

$$[v'(x), w'(x)^*] = D_j v^i D^j w_i$$

we obtain the invariant equation

$$(v|w)_X = \int_G vw \, dx + \int_G [v'(x), w'(x)^*] \, dx.$$

If we consider (4) without the boundary condition $v^i = 0$ on $\partial_1 G$ for all i, then we obtain the space

$$W_2^1(G, V_3)$$

62.2. Notations

instead of $\mathring{W}_2^1(G, \partial_1 G; V_3)$. The space $\mathring{W}_2^1(G, V_3)$ consists of all $v\colon G \to V_3$, whose components v^i belong to $\mathring{W}_2^1(G)$ for all i. In particular,

$$\mathring{W}_2^1(G, \partial G; V_3) = \mathring{W}_2^1(G, V_3).$$

Let us motivate the definition of the space X. Given $u_0 \in W_2^1(G, V_3)$ we will prove the existence of solutions

$$u = u_0 + v, \qquad v \in X$$

of equation (2). These solutions may be viewed as generalized solutions of the corresponding classical variational problem (1). The following two points are important in this regard:

(i) Because of (4), we have that

$$u = u_0 \quad \text{on } \partial_1 G,$$

i.e., the boundary condition in (1) is satisfied in the generalized sense.

(ii) The variational problem (1) contains first-order partial derivatives. Because of (4), u has such derivatives in the generalized sense.

Let

$$L_2(G, V_3)$$

denote the space of all vector functions $K\colon G \to V_3$ with

$$K^i \in L_2(G), \qquad i = 1, 2, 3.$$

Analogously, $L_2(\partial_j G, V_3)$ is defined.

Moreover, let

$$Y = L_2(G, L_{\text{sym}}(V_3))$$

be the space of all tensor functions $\gamma\colon G \to L_{\text{sym}}(V_3))$ with

$$\gamma_j^i \in L_2(G), \qquad i, j = 1, 2, 3.$$

As a scalar product in Y we choose

$$(\gamma \mid \mu)_Y = \int_G [\gamma(x), \mu(x)] \, dx.$$

The space Y will play an important role in duality theory.

The spaces X and Y are real H-spaces. This follows from the properties of Sobolev spaces and the fact that, in Section 61.1, we introduced the scalar product $[\gamma, \mu]$ on $L_{\text{sym}}(V_3)$. We set

$$|\mu| = [\mu, \mu]^{1/2} \qquad \text{for all} \quad \mu \in L_{\text{sym}}(V_3).$$

All spaces above are defined in an invariant way, i.e., the definitions do not depend on the choice of the coordinate system.

62.3. Principle of Minimal Potential Energy, Existence, and Uniqueness

We consider the variational problem for the displacement u of an elastic body

$$\int_G M(\gamma(u))\,dx - \int_G Ku\,dx - \int_{\partial_2 G} Tu\,dO = \min!, \tag{5}$$

$$u = u_0 \quad \text{on } \partial_1 G$$

with the linearized strain tensor

$$\gamma(u)(x) = \tfrac{1}{2}(u'(x) + u'(x)^*).$$

The key to our approach is the so-called strong stability condition

$$M''(\gamma)\mu^2 \geq c|\mu|^2 \quad \text{for all} \quad \gamma, \mu \in L_{\text{sym}}(V_3) \tag{6}$$

and fixed $c > 0$. We assume, in addition, the growth conditions

$$|M(\gamma)| \leq \text{const}\,|\gamma|^2,$$

$$|M'(\gamma)\mu| \leq \text{const}\,|\gamma||\mu|, \tag{7}$$

$$|M''(\gamma)\mu^2| \leq \text{const}\,|\mu|^2$$

for all $\gamma, \mu \in L_{\text{sym}}(V_3)$. More precisely, we make the following assumptions:

(H1) G is a bounded region in the three-dimensional vector space V_3 with sufficiently smooth boundary, i.e., $\partial G \in C^{0,1}$. Intuitively, G corresponds to the region of the undeformed body.
(H2) The boundary of G can be decomposed as $\partial G = \overline{\partial_1 G} \cup \overline{\partial_2 G}$ where $\partial_1 G$ and $\partial_2 G$ are disjoint, open subsets of ∂G, and $\partial_1 G$ is *nonempty*.
(H3) The map $M: L_{\text{sym}}(V_3) \to \mathbb{R}$ is C^2 and satisfies the strong stability condition (6) and the growth conditions (7). Let $M(0) = 0$ and $M'(0) = 0$.
(H4) On G the density of outer forces $K \in L_2(G, V_3)$ is given.
(H5) On the boundary part $\partial_1 G$, the boundary displacement u_0 is given. More precisely, let $u_0 \in W_2^1(G, V_3)$.
(H6) On the boundary part $\partial_2 G$, the density of the boundary stress forces $T \in L_2(\partial G_2, V_3)$ is given.

Theorem 62.A. *If* (H1) *to* (H6) *hold, then the variational problem* (5), *i.e., more precisely, the generalized problem*

$$\int_G M(\gamma(u))\,dx - \int_G Ku\,dx - \int_{\partial_2 G} Tu\,dO = \min!, \quad u - u_0 \in X$$

has a unique solution. This solution is strictly stable.

The proof will be given in Section 62.5.

If (H1) to (H6) hold with $\partial_1 G = \emptyset$, then $\partial_2 G = \partial G$. In this special case, the existence theorem for the so-called *second boundary-value problem* will be proved in Problem 62.6. It is a typical property of this *statically undetermined* problem that the outer forces K and T must satisfy an equilibrium condition and that the displacements are only determined up to translations and infinitesimal rotations, i.e., up to infinitesimal rigid motions. Intuitively, one would expect that the displacements are determined up to translation and rotations. The appearnace of infinitesimal rotations results from the fact that our model is *not* exact. This is because we replace the strain tensor $\mathscr{E}$ with the linearized strain tensor γ.

62.4. Principle of Maximal Dual Energy and Duality

Let us write the principle of minimal potential energy (5) in the form

(P) $\qquad E_{\text{pot}}(u) = \min!, \qquad u - u_0 \in X.$

In addition to this principle, we consider in this section the principle of maximal dual energy

(P*) $\qquad E_{\text{dual}}(\sigma) = \max!, \qquad \sigma \in \Sigma$

as an example of a dual variational problem. Theorem 62.B below then will show that the solution of (P*) is the stress tensor σ which corresponds to the solution u of (P).

In preparation we observe that, by passing to components, relation (61.41) yields the constitutive law

$$\sigma = M'(\gamma) \qquad (8)$$

for (5). This is to be understood in the sense of the scalar product on $L_{\text{sym}}(V_3)$, i.e.,

$$M'(\gamma)\mu = [\sigma, \mu] \qquad \text{for all} \quad \mu \in L_{\text{sym}}(V_3).$$

The important point thereby is the following. Because of the strong stability condition (6), it follows from Corollary 42.8 that the derivative M' is strongly monotone. Thus, the main theorem on monotone operators (Theorem 26.A) shows that equation (8) has a unique solution γ, namely

$$\gamma = M'^{-1}(\sigma). \qquad (9)$$

In addition, we will use the conjugate functional M^* to $M: L_{\text{sym}}(V_3) \to \mathbb{R}$. From Proposition 51.5 we obtain the following key formula of convex analysis

$$M'^{-1} = M^{*'}.$$

More precisely, we find that

$$M^*(\sigma) = \int_0^1 [\sigma, M'^{-1}(t\sigma)]\, dt$$

for all $\sigma \in L_{\text{sym}}(V_3)$. The inverse constitutive law (9) becomes

$$\gamma = M^{*\prime}(\sigma). \tag{9*}$$

This is a very natural condition:

In order to obtain the dual (inverse) constitutive law (9*) from the original constitutive law (8), we must replace the stored energy function M with its conjugate functional M^*.

We now define the potential energy of the elastic body

$$E_{\text{pot}}(u) = \int_G M(\gamma(u))\,dx - \int_G Ku\,dx - \int_{\partial_2 G} Tu\,dO,$$

and the corresponding dual energy

$$E_{\text{dual}}(\sigma) = -\int_G M^*(\sigma)\,dx + \int_G [\sigma, \gamma(u_0)] - Ku_0\,dx - \int_{\partial_2 G} Tu_0\,dO.$$

By definition, we have that $\sigma \in \Sigma$ if and only if $\sigma \in L_2(G, L_{\text{sym}}(V_3))$ and

$$\int_G [\sigma, \gamma(v)]\,dx = \int_G Kv\,dx + \int_{\partial_2 G} Tv\,dO$$

for all $v \in X$. According to Section 61.5, this relation can be regarded as a generalized form of the equilibrium condition

$$\text{div } \sigma + K = 0 \quad \text{on } G,$$

$$\sigma n = T \quad \text{on } \partial_2 G.$$

Roughly speaking, one minimizes in (P) over all possible displacements u, which satisfy the boundary condition $u = u_0$ on $\partial_1 G$. In the dual problem (P*), one maximizes over all possible stress tensors σ which are in an equilibrium with the outer forces K and T.

If $\sigma \in \Sigma$ and σ is sufficiently smooth, then we obtain from integration by parts the simpler expression

$$E_{\text{dual}}(\sigma) = -\int_G M^*(\sigma)\,dx + \int_{\partial_1 G} (\sigma n)u_0\,dO.$$

In Problem 62.9 we will show that

$$M^*(\sigma) = [\sigma, \gamma] - M(\gamma), \quad \sigma = M^{\prime -1}(\gamma)$$

for all $\sigma \in L_{\text{sym}}(V_3)$. Hence M and M^* correspond to the Lagrangian function and the Hamiltonian function in point mechanics, respectively. In Section 62.16 we shall prove that duality in elasticity theory is a special case of a general duality in the calculus of variations.

The following theorem contains a precise formulation of duality between displacements and stresses in elasticity.

Theorem 62.B (Duality). *We make the assumptions* (H1) *to* (H6) *of Theorem 62.A above.*

Problems (P) *and* (P*) *then have the unique solutions $\tilde{u}$ and $\tilde{\sigma}$, respectively, related through the constitutive law* (8), *i.e.,* $\tilde{\sigma} = M'(\gamma(\tilde{u}))$.

Furthermore, the two extremal values of (P) *and* (P*) *equal each other, i.e.,* $E_{\text{pot}}(\tilde{u}) = E_{\text{dual}}(\tilde{\sigma})$.

Corollary 62.1 (Error Estimates). *If u and σ satisfy the side conditions of* (P) *and* (P*), *respectively, then we obtain the two-sided estimate for the minimal potential energy*

$$E_{\text{dual}}(\sigma) \leq E_{\text{pot}}(\tilde{u}) \leq E_{\text{pot}}(u).$$

Moreover, for the displacement $\tilde{u}$, we find the estimate

$$\tfrac{1}{2}cc_1 \|u - \tilde{u}\|_X^2 \leq E_{\text{pot}}(u) - E_{\text{dual}}(\sigma).$$

The positive constants c and c_1 thereby appear in the strong stability condition (6) *above and in Korn's inequality* (11) *below.*

The advantage of these error estimates is that we can choose u and σ arbitrarily. Only the side conditions of (P) and (P*) need to be satisfied, respectively.

62.5. Proof of the Main Theorems

The proofs of:

Theorem 62.A, and
Theorem 62.B together with Corollary 62.1

are easily obtained from

Theorem 42.A (main theorem on free convex minimum problems), and
Theorem 51.B (main theorem of the duality theory for monotone potential operators).

The key is the important and nontrivial inequality of Korn (1907). This inequality, which will be proved in Section 62.15, is intimately related to the famous inequalities of Poincaré and Friedrichs of Part II. Thus, our proof consists of two parts:

(i) concrete analytic substance (Korn's inequality);
(ii) abstract functional-analytic substance (Theorems 42.A and 51.B).

62.5a. Proof of Theorem 62.A

We set $\gamma = Du$, i.e.,

$$Du(x) = \tfrac{1}{2}(u'(x) + u'(x)^*).$$

Furthermore, we set $v = u - u_0$. The variational problem (5) then becomes

$$F(v) - b(v) = \min!, \quad v \in X \qquad (10)$$

with

$$F(v) = \int_G M(Dv + Du_0)\,dx - \int_G Ku_0\,dx - \int_{\partial_2 G} Tu_0\,dO,$$

$$b(v) = \int_G Kv\,dx + \int_{\partial_2 G} Tv\,dO.$$

Important is the following result.

Lemma 62.2 (Functional Analytic Formulation of Korn's Inequality). *Through*

$$(v|w)_E = \int_G [Dv, Dw]\,dx$$

an equivalent scalar product is defined on the space X.

Letting $\|v\|_E = (v|v)^{1/2}$, this result states that there exist positive constants c_1 and c_2 such that

$$c_1 \|v\|_X^2 \leq \|v\|_E^2 \leq c_2 \|v\|_X^2 \quad \text{for all} \quad v \in X. \qquad (11)$$

Analogously as in Section 22.1 the norm $\|v\|_E$ is called an energy norm. The left-hand inequality in (11) is Korn's inequality which will be proved in Section 62.15 (see (57)). The right-hand inequality in (11) follows easily from the definition of $\|v\|_X$ in Section 62.1 and Hölder's inequality (see Problem 62.1).

Lemma 62.3. *The functional $b: X \to \mathbb{R}$ is linear and continuous.*

The simple proof follows from Hölder's inequality (see Problem 62.2).

We now study the functional F which represents the elastic potential energy. To do this we set

$$\varphi(t) = F(v + tw) \quad \text{for all} \quad t \in \mathbb{R} \text{ and fixed } v, w \in X.$$

We then have that

$$\varphi'(0) = \langle F'(v), w \rangle, \qquad \varphi''(0) = \delta^2 F(v; w).$$

Computation of $\varphi'(0)$ and $\varphi''(0)$ yields

$$\langle F'(v), w \rangle = \int_G M'(Dv + Du_0) Dw\,dx,$$

$$\delta^2 F(v; w) = \int_G M''(Dv + Du_0)(Dw)^2\,dx \quad \text{for all} \quad v, w \in X.$$

As a consequence of stability condition (6) we then obtain that

$$\delta^2 F(v; w) \geq c\|w\|_E^2 \geq cc_1 \|w\|_X^2 \quad \text{for all} \quad v, w \in X. \qquad (12)$$

62.5. Proof of the Main Theorems

Notice Korn's inequality (11). We thus obtain the key result:

The strong stability condition (6) for the stored energy function M implies that the second variation of the elastic potential energy F is uniformly strongly positive.

In particular, according to Corollary 42.8, inequality (12) implies the convexity of $F: X \to \mathbb{R}$.

Lemma 62.4. *The functional $F: X \to \mathbb{R}$ is C^1 and convex.*
The operator $F': X \to X^*$ is a strongly monotone potential operator. More precisely, we have that

$$\langle F'(v) - F'(w), v - w \rangle \geq \tfrac{1}{2}c\|v - w\|_E^2 \quad \text{for all} \quad v, w \in X. \tag{13}$$

PROOF. As in Section 42.7, the growth conditions (7) ensure that the differentiation of φ can be performed under the integral sign and that F, F' are continuous. This is a consequence of the majorant theorem on the differentiation of parameter dependent integrals $A_2(25)$.

Inequality (13) follows from Corollary 42.8. □

Theorem 42.A then implies that the minimum problem (10) is equivalent to the Euler equation

$$F'(u) - b = 0, \quad u \in X, \tag{14}$$

which has a unique solution u. Formula (12) implies that

$$\delta^2 F(u; w) > 0 \quad \text{for all} \quad w \in X - \{0\},$$

and this is the strict stability of u, according to Definition 61.4.

The proof of Theorem 62.A is complete.

In the context of approximation methods (projection–iteration methods and gradient methods) the following additional information is useful.

Lemma 62.5. *The operator $F': X \to X^*$ is Lipschitz continuous.*

PROOF. By passing to components, formula (7) implies that the second-order partial derivatives of M are bounded. Thus M' is Lipschitz continuous, i.e.,

$$|M'(\gamma)\tau - M'(\mu)\tau| \leq \text{const}|\gamma - \mu||\tau|$$

for all $\gamma, \mu, \tau \in L_{\text{sym}}(V_3)$. Hölder's inequality implies that

$$|\langle F'(u) - F'(v), w \rangle| \leq \int |M'(Du + Du_0)Dw - M'(Dv + Du_0)Dw|\, dx$$

$$\leq \text{const} \left(\int |Du - Dv|^2\, dx \right)^{1/2} \left(\int |Dw|^2\, dx \right)^{1/2}$$

$$= \text{const}\, \|u - v\|_E \|w\|_E \leq \text{const}\, \|u - v\|_X \|w\|_X$$

for all $u, v, w \in X$, and this means that

$$\|F'(u) - F'(v)\| \leq \text{const } \|u - v\|_X \quad \text{for all} \quad u, v \in X. \qquad \square$$

62.5b. Proof of Theorem 62.B

Let us show that we are in the situation of Theorem 51.B. To this end, recall that

$$X = \mathring{W}_2^1(G, \partial_1 G, V_3), \quad Y = L_2(G, L_{\text{sym}}(V_3)),$$

and that $\gamma = Du$, where

$$Du(x) = \tfrac{1}{2}(u'(x) + u'(x)^*).$$

The operator

$$D: X \to Y$$

is linear and continuous, and on X we impose the energy norm of Lemma 62.2 above. We then have that

$$\|Dv\|_Y = \|v\|_E \quad \text{for all} \quad v \in X.$$

Using the operator D we can write the original variational problem (10) in the form

(P) $$\qquad H(Dv) - b(v) = \min!, \qquad v \in X$$

with the functional

$$H(\gamma) = \int_G M(\gamma + Du_0)\, dx - b(u_0) \quad \text{for all} \quad \gamma \in Y.$$

According to Section 51.1, the corresponding dual problem reads as

(P*) $$\qquad -H^*(\sigma) = \max!, \qquad \sigma \in \Sigma$$

with

$$\Sigma = \{\sigma \in Y : (\sigma | Dv)_Y = b(v) \text{ for all } v \in X\}$$

and conjugate functional

$$H^*(\sigma) = \int_0^1 (\sigma | H'^{-1}(t\sigma))_Y\, dt - H(H'^{-1}(0)).$$

We identify the H-space Y with its dual space Y^*.

In order to be able to apply Theorem 51.B to (P*) we have to study the properties of H. We will show that the operator

$$H': Y \to Y^*$$

is strongly monotone and Lipschitz continuous. The main theorem on mono-

62.5. Proof of the Main Theorems

tone operators (Theorem 26.A) then implies that the inverse operator

$$H'^{-1}: Y^* \to Y$$

is also strongly monotone and Lipschitz continuous. Finally, Proposition 51.5 implies that

$$H^{*\prime} = H'^{-1}.$$

Case 1: We begin with the special case $u_0 = 0$.
For the G-derivative of the functional $H: Y \to \mathbb{R}$ we find

$$H'(\gamma)\mu = \int_G M'(\gamma)\mu\, dx \quad \text{for all} \quad \gamma, \mu \in Y,$$

and for the second variation, we obtain from stability condition (6) that

$$\delta^2 H(\gamma; \mu) = \int_G M''(\gamma)\mu^2\, dx \geq c\|\mu\|_Y^2 \quad \text{for all} \quad \gamma, \mu \in Y.$$

Thus, Corollary 42.8 shows that the operator $H': Y \to Y^*$ is strongly monotone. The same argument, as has been used for the operator F' in Section 62.5a then shows that H' is Lipschitz continuous.

If we set

$$\sigma = M'(\gamma)$$

in the sense of the scalar product on $L_{\text{sym}}(V_3)$, i.e.,

$$[\sigma, \mu] = M'(\gamma)\mu \quad \text{for all} \quad \mu \in L_{\text{sym}}(V_3),$$

we obtain

$$H'(\gamma)\mu = (\sigma|\mu)_Y \quad \text{for all} \quad \mu \in Y.$$

Thus, we have

$$H'(\gamma) = \sigma$$

with $\sigma(x) = M'(\gamma(x))$ on G. This gives

$$H^*(\sigma) = \int_0^1 \int_G [\sigma, M'^{-1}(t\sigma)]\, dx\, dt,$$

i.e., we obtain the natural formula

$$H^*(\sigma) = \int_G M^*(\sigma)\, dx.$$

Theorem 62.B and Corollary 62.1 are then an immediate consequence of Theorem 51.B.

Case 2: The general case, i.e., $u_0 \in W_2^1(G, V_3)$.
Analogous arguments apply. Note that now $H'(\gamma) = \sigma$ means that $\sigma =$

$M'(\gamma + Du_0)$, i.e.,

$$\gamma = M'^{-1}(\sigma) - Du_0,$$

and because of $M'^{-1}(0) = 0$ it follows that

$$H^*(\sigma) = \int_0^1 \int_G [\sigma, M'^{-1}(t\sigma) - Du_0] \, dx \, dt + b(u_0),$$

i.e.,

$$H^*(\sigma) = \int_G M^*(\sigma) - [\sigma, Du_0] \, dx + b(u_0).$$

Hence

$$H^*(\sigma) = -E_{\text{dual}}(\sigma).$$

Thus the proof of Theorem 62.B and Corollary 62.1 is complete.

62.6. Approximation Methods

For approximate solutions of our model we can use, for example, Ritz' method, Trefftz' method, projection–iteration methods, or gradient methods. Let us explain this.

62.6a. The Ritz Method

In Section 42.5 of Part III, we studied the Ritz method for free convex minimum problems. This method can be applied to the original variational problem (10), i.e.,

$$F(v) - b(v) = \min!, \quad v \in X. \tag{15}$$

The idea of the Ritz method is to replace the space $X = \mathring{W}_2^1(G, \partial_1 G; V_3)$ with a finite-dimensional subspace X_n, i.e., to consider the new variational problem

$$F(v_n) - b(v_n) = \min!, \quad v_n \in X_n. \tag{15_n}$$

For $n = 1, 2, \ldots$ we choose subspaces with

$$X_1 \subseteq X_2 \subseteq X_3 \subseteq \cdots \subseteq X \quad \text{and} \quad \overline{\bigcup_n X_n} = X.$$

The convergence in X of this method and error estimates are obtained from Theorem 42.A. Observe that, according to Lemma 62.4, the operator F': $X \to X^*$ is strongly monotone.

In order to obtain a handy formulation, let us write the original problem (15) in the form

(P) $\qquad E_{\text{pot}}(u) = \min!, \quad u - u_0 \in X.$

62.6. Approximation Methods

The Ritz method (15_n) then becomes

(P$_n$) $\qquad E_{\text{pot}}(u_n) = \min!, \qquad u_n - u_0 \in X_n$

with $n = 1, 2, \ldots$. All displacements u_n which satisfy the boundary condition

$$u_n = u_0 \quad \text{on } \partial_1 G$$

are admitted as candidates in (P$_n$). More precisely, we have to make the ansatz

$$u_n = u_0 + \sum_{j=1}^{n} d_j u^{(j)} \tag{16}$$

with the unknown real coefficients $d_1, \ldots, d_n$ and the fixed given displacements $u^{(j)} \in X$, $j = 1, 2, \ldots, n$, where

$$u^{(j)} = 0 \quad \text{on } \partial_1 G \qquad \text{for all } j.$$

In the Appendix to Part II we have shown, how such basis functions $u^{(j)}$ can be chosen in the form of finite elements, which may be, for example, piecewise linear functions, obtained from a triangulation of the region G by using linear interpolation. Problem (P$_n$) is of the form

$$f(d_1, \ldots, d_n) = \min!.$$

This leads to the generally nonlinear system of equations

$$\frac{\partial f(d_1, \ldots, d_n)}{\partial d_j} = 0, \qquad j = 1, 2, \ldots, n.$$

If u and u_n are solutions of (P) and (P$_n$), respectively, then, by taking (13) into account, Theorem 42.A yields the error estimate

$$\tfrac{1}{2} cc_1 \|u_n - u\|_X^2 \leq \tfrac{1}{2} c \|u_n - u\|_E^2 \leq E_{\text{pot}}(u_n) - E_{\text{pot}}(u).$$

In order to eliminate the unknown value $E_{\text{pot}}(u)$, we choose an arbitrary $\sigma_n \in \Sigma$. Theorem 62.B then implies that $E_{\text{dual}}(\sigma_n) \leq E_{\text{pot}}(u)$. Hence we obtain

$$\tfrac{1}{2} cc_1 \|u_n - u\|_X^2 \leq \tfrac{1}{2} c \|u_n - u\|_E^2 \leq E_{\text{pot}}(u_n) - E_{\text{dual}}(\sigma_n) \tag{17}$$

for arbitrary fixed $\sigma_n \in \Sigma$. The advantage of this *error estimate* is that on the right-hand side, only the known values $E_{\text{dual}}(\sigma_n)$ and $E_{\text{pot}}(u_n)$ appear. Moreover, the *energy* $E_{\text{pot}}(u)$ of the elastic body satisfies the estimate

$$E_{\text{dual}}(\sigma_n) \leq E_{\text{pot}}(u) \leq E_{\text{pot}}(u_n). \tag{17*}$$

Inequalities (17) and (17*) contain the two *key* estimates of the Ritz method.

62.6b. The Trefftz Method

Ritz' method for the dual problem

(P*) $\qquad E_{\text{dual}}(\sigma) = \max!, \qquad \sigma \in \Sigma$

is

(P$_n^*$) $\qquad E_{\text{dual}}(\sigma_n) = \max!, \qquad \sigma_n \in \Sigma_n,$

where

$$\Sigma_1 \subseteq \Sigma_2 \subseteq \cdots \subseteq \Sigma \quad \text{and} \quad \overline{\bigcup_n \Sigma_n} = \Sigma.$$

This method is also called the Trefftz method.

The convergence of this method is a consequence of Theorem 42.A. From Section 62.5b follows that problem (P*) can be written as

$$-H^*(\sigma) = \max!, \qquad \sigma \in \Sigma,$$

where the operator $H^{*\prime}: Y^* \to Y$ is strongly monotone. We note that Σ is obtained from a translation of a closed, linear subspace of the H-space Y.

Let us show that a combination of the Ritz method of Section 62.6a and the Trefftz method can be very useful in practical computations. Suppose u_n and σ_n are solutions of (P$_n$) and (P$_n^*$), respectively. From Theorem 62.B follows that $E_{\text{pot}}(u) = E_{\text{dual}}(\sigma)$ for the solutions u and σ of (P) and (P*), respectively. Our convergence proofs then show that

$$E_{\text{pot}}(u_n) \to E_{\text{pot}}(u) \quad \text{and} \quad E_{\text{dual}}(\sigma_n) \to E_{\text{dual}}(\sigma) \qquad \text{as } n \to \infty.$$

Thus, we obtain the *key* relation

$$E_{\text{pot}}(u_n) - E_{\text{dual}}(\sigma_n) \to 0 \qquad \text{as } n \to \infty.$$

A *good* choice of σ_n in the error estimates (17) and (17*) is therefore a solution σ_n of the Trefftz method (P$_n^*$).

These observations show the importance of duality theory for elasticity theory.

62.6c. Projection–Iteration Method

According to Section 62.5a and Theorem 42.A, the original variational problem (P) is equivalent to the operator equation

(E) $\qquad F'(v) - b = 0, \qquad v \in X$

with $u = v + u_0$. Moreover, as a consequence of Section 62.5a, the operator $F': X \to X^*$ is strongly monotone and Lipschitz continuous where X is a real, separable H-space. Hence, projection–iteration methods, which have been studied in Section 25.4 of Part II, can be applied to (E). The algorithm is

(E$_n^*$) $\qquad (u_{n+1}|u^{(j)})_E = (u_n|u^{(j)})_E - t\langle F'(u_n - u_0) - b, u^{(j)}\rangle$

for $j = 1, 2, \ldots, n$. We start with the known u_0 and compute successively the quantities u_{n+1} for $n = 0, 1, \ldots$ which are given by the ansatz (16).

The *advantage* of this projection–iteration method over the projection method (Ritz' method) (P_n), (E_n) is that here, at each step, we only have to solve *linear* systems of equations for the $d_1, \ldots, d_n$.

For practical computations one uses (E_n^*) with respect to a Cartesian coordinate system. This gives

$$(u|v)_E = \int_G \gamma_j^i(u)\gamma_i^j(v)\,dx,$$

$$\langle F'(u - u_0) - b, v\rangle = \int_G \sigma_j^i(u)\gamma_i^j(v)\,dx - \int_G K^i v_i\,dx - \int_{\partial_2 G} T^i v_i\,dO$$

with

$$\gamma_j^i(u) = \tfrac{1}{2}(D^i u_j + D^j u_i), \qquad \sigma_j^i(u) = \frac{\partial M(\gamma(u))}{\partial \gamma_i^j}.$$

62.6d. Gradient Method

We may also apply gradient methods to the original problem (P), (E) as well as to the Ritz equations (E_n). This method has been studied in Theorem 42.B. It is important, in this connection, that the operator $F'\colon X \to X^*$ is strongly monotone and Lipschitz continuous.

The Ritz problem (P_n) is a free convex optimization problem, and hence, in order to solve (P_n), the entire apparatus of convex optimization is available.

62.7. Applications to Linear Elasticity Theory

We consider a homogeneous, isotropic body. As a consequence of Example 61.6 we obtain the *stored energy function*

$$M(\gamma) = \tfrac{1}{2}\lambda(\operatorname{tr}\gamma)^2 + \kappa[\gamma, \gamma]$$

with the Lamé constants $\lambda, \kappa > 0$. If we set $\psi(t) = M(\gamma + t\mu)$ and compute $\psi'(0)$ and $\psi''(0)$, then we obtain that

$$M'(\gamma)\mu = \lambda \operatorname{tr}\gamma \operatorname{tr}\mu + 2\kappa[\gamma, \mu],$$
$$M''(\gamma)\mu^2 = 2M(\mu) \qquad \text{for all} \quad \gamma, \mu \in L_{\text{sym}}(V_3).$$

Thus the strong stability condition

$$M''(\gamma)\mu^2 \geq 2\kappa[\mu, \mu] \qquad \text{for all} \quad \gamma, \mu \in L_{\text{sym}}(V_3)$$

is trivially satisfied and all results of the previous sections can be applied. The constitutive law $\sigma = M'(\gamma)$ becomes

$$\sigma = \lambda \operatorname{tr}\gamma\, I + 2\kappa\gamma.$$

This is precisely *Hooke's law*. Note that $[\sigma, \mu] = M'(\gamma)\mu$ and $[I, \mu] = \operatorname{tr}\mu$.

The inverse constitutive law $\gamma = M'^{-1}(\sigma)$ is

$$\gamma = \lambda^* \operatorname{tr} \sigma I + 2\kappa^* \sigma$$

with the dual Lamé constants

$$\kappa^* = \frac{1}{\kappa}, \qquad \lambda^* = -\frac{\lambda}{(2\kappa + 3\lambda)\kappa}.$$

From Section 62.4 we thus obtain the *dual* stored energy function

$$M^*(\sigma) = \tfrac{1}{2}\lambda^*(\operatorname{tr}\sigma)^2 + \kappa^*[\sigma, \sigma].$$

62.8. Application to Nonlinear Hencky Material

As an important example for a stored energy function, leading to a nonlinear Hooke's law, we consider

$$M(\gamma) = \tfrac{1}{2}k(\operatorname{tr}\gamma)^2 + \kappa\varphi([\bar{\gamma},\bar{\gamma}])$$

with the material constant $k = \lambda + 2\kappa/3$ and the strain deviator

$$\bar{\gamma} = \gamma - \tfrac{1}{3}\operatorname{tr}\gamma\, I.$$

The expression for M has been physically motivated in Example 61.7. In the special case $\varphi(t) = t$, we obtain the stored energy function of linear elasticity theory, which has been considered in the previous section.

Proposition 62.6. *The function $M\colon L_{\mathrm{sym}}(V_3) \to \mathbb{R}$ satisfies the strong stability condition (6) and the growth conditions (7) if the material function φ satisfies the following properties:*

(i) $\varphi\colon \mathbb{R}_+ \to \mathbb{R}$ *is C^2 with $\varphi(0) = 0$.*
(ii) *There exist constants $a, b > 0$ and a natural number n such that the following inequalities hold on $\mathbb{R}_+$:*

$$a \le \varphi'(t) \le 1, \qquad -b \le \varphi''(t) \le 0,$$

$$\frac{1}{n} \le \varphi'(t) + 2\varphi''(t)t \le n.$$

This proposition shows that all the results from Sections 62.3–62.6 can be applied to this nonlinear model. Conditions (i) and (ii) are satisfied, for example, for concave functions φ which display an almost linear behavior in neighborhoods of $t = 0$ and $t = +\infty$ (Fig. 62.1).

The constitutive law $\sigma = M'(\gamma)$ is

$$\sigma = (k - \tfrac{2}{3}\kappa\varphi'(\Gamma))\operatorname{tr}\gamma\, I + 2\kappa\varphi'(\Gamma)\gamma$$

with $\Gamma = [\bar{\gamma},\bar{\gamma}]$.

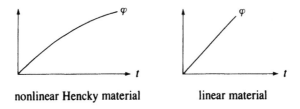

Figure 62.1

PROOF. Let $\gamma, \mu \in L_{\text{sym}}(V_3)$, and let $\psi(t) = M(\gamma + t\mu)$. Computation of the derivatives $\psi'(0)$ and $\psi''(0)$ implies

$$M'(\gamma)\mu = k \operatorname{tr}\gamma \operatorname{tr}\mu + 2\kappa\varphi'(\Gamma)[\bar{\gamma}, \bar{\mu}],$$
$$M''(\gamma)\mu^2 = k(\operatorname{tr}\mu)^2 + 4\kappa\varphi''(\Gamma)[\bar{\gamma}, \bar{\mu}]^2 + 2\kappa\varphi'(\Gamma)[\bar{\mu}, \bar{\mu}].$$

Schwarz' inequality

$$[\gamma, \mu]^2 \leq |\gamma|^2|\mu|^2 \quad \text{with} \quad |\gamma|^2 = [\gamma, \gamma]$$

then yields

$$M''(\gamma)\mu^2 \geq k(\operatorname{tr}\mu)^2 + 2\kappa\varphi'(\Gamma)|\bar{\mu}|^2 - 4\kappa|\varphi''(\Gamma)||\bar{\gamma}|^2|\bar{\mu}|^2$$
$$= k(\operatorname{tr}\mu)^2 + 2\kappa(\varphi'(\Gamma) + 2\varphi''(\Gamma)\Gamma)|\bar{\mu}|^2$$
$$\geq k(\operatorname{tr}\mu)^2 + \frac{2\kappa}{n}|\bar{\mu}|^2$$
$$= \left(k - \frac{2\kappa}{3n}\right)(\operatorname{tr}\mu)^2 + \frac{2\kappa}{n}|\mu|^2$$
$$\geq \frac{2\kappa}{n}|\mu|^2,$$

and this is the strong stability condition (6). The growth conditions (7) follow from Schwarz' inequality and the boundedness of φ' and φ''. □

62.9. The Constitutive Law for Quasi-Statical Plastic Material

We now want to generalize the one-dimensional quasi-statical plasticity model from Section 60.4 to three-dimensional bodies (Fig. 62.2(b)). According to Section 61.7 we assume the following plasticity condition of von Mises:

$$[\bar{\sigma}, \bar{\sigma}] < \sigma_0^2 \quad \text{no plasticity,}$$
$$[\bar{\sigma}, \bar{\sigma}] = \sigma_0^2 \quad \text{plasticity occurs,} \tag{18}$$

where σ_0 is a given positive number called the plasticity limit.

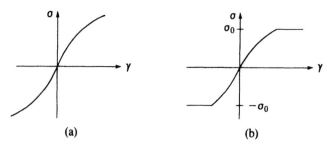

Figure 62.2

In what follows we consider the H-space $L_{\text{sym}}(V_3)$ with the scalar product $[\sigma, \mu]$. Set

$$C = \{\sigma \in L_{\text{sym}}(V_3): [\bar{\sigma}, \bar{\sigma}] \leq \sigma_0^2\}$$

and let χ denote the indicator function of C, i.e.,

$$\chi(\sigma) = \begin{cases} 0 & \text{if } \sigma \in C, \\ +\infty & \text{if } \sigma \notin C. \end{cases}$$

In Section 62.4 we considered the constitutive law $\sigma = M'(\gamma)$ with the inverse formula

$$\gamma = M^{*\prime}(\sigma) \tag{19}$$

(Fig. 62.2(a)). Here $\gamma, \sigma \in L_{\text{sym}}(V_3)$. Motivated by Section 60.4, we describe the corresponding quasi-statical plasticity by the *constitutive law*

$$\gamma \in M^{*\prime}(\sigma) + \partial \chi \tag{20}$$

(Fig. 62.2(b)). According to the definition of the subgradient in Section 47.3, equation (20) is equivalent to

$$\chi(\mu) - \chi(\sigma) \geq [\gamma - M^{*\prime}(\sigma), \mu - \sigma] \tag{20*}$$

for all $\mu \in L_{\text{sym}}(V_3)$. This, in turn, is equivalent to the following variational inequality

$$[\gamma - M^{*\prime}(\sigma), \mu - \sigma] \leq 0 \quad \text{for all} \quad \mu \in C. \tag{20**}$$

In particular, this implies

$$\gamma = M^{*\prime}(\sigma) \quad \text{if } [\bar{\sigma}, \bar{\sigma}] < \sigma_0^2.$$

In Cartesian coordinates we set $\sigma_{ij} = \sigma_i^j$, etc., and thus, we get

$$[\sigma, \mu] = \sum_{i,j=1}^{3} \sigma_{ij} \mu_{ij},$$

$$[\bar{\sigma}, \bar{\sigma}] = \sum_{i,j=1}^{3} (\sigma_{ij} - \tfrac{1}{3}(\sigma_{11} + \sigma_{22} + \sigma_{33}))^2,$$

$$C = \{(\sigma_{ij}) \in \mathbb{R}^9_{\text{sym}}: [\bar{\sigma}, \bar{\sigma}] \leq \sigma_0^2\},$$

where $\mathbb{R}^9_{\text{sym}}$ denotes the set of all real tuples (σ_{ij}) in $\mathbb{R}^9$ with $\sigma_{ij} = \sigma_{ji}$ for all i, j.

62.10. Principle of Maximal Dual Energy and the Existence Theorem for Linear Quasi-Statical Plasticity

As in Section 62.7 we consider linear material, i.e., we choose the stored energy function

$$M(\gamma) = \tfrac{1}{2}\lambda(\operatorname{tr}\gamma)^2 + \kappa[\gamma,\gamma]$$

with the Lamé constants $\lambda, \kappa > 0$, and hence

$$M^*(\sigma) = \tfrac{1}{2}\lambda^*(\operatorname{tr}\sigma)^2 + \kappa^*[\sigma,\sigma]$$

with the dual Lamé constants $\kappa^* = 1/\kappa$ and $\lambda^* = -\lambda/\kappa(2\kappa + 3\lambda)$. Moreover, we have the constitutive law

$$\gamma = \lambda^* \operatorname{tr}\sigma\, I + 2\kappa^*\sigma.$$

This implies

$$\bar{\gamma} = 2\kappa^*\bar{\sigma},$$

which will be of particular interest for the statical model in the following section.

Our goal is to generalize the variational inequality

$$\int_G [\gamma(u) - M^{*\prime}(\sigma), \mu - \sigma]\,dx \leq 0 \qquad \text{for all } \mu \in \Sigma^1 \tag{21}$$

in order to obtain a quasi-statical model in plasticity for which we are able to prove an existence theorem. This plasticity model corresponds to the constitutive law

$$\gamma = M^{*\prime}(\sigma)$$

in linear elasticity with the mixed boundary conditions

$$\begin{aligned} u &= u_0 & \text{on } \partial_1 G, \\ \sigma n &= T & \text{on } \partial_2 G. \end{aligned}$$

Note that (21) corresponds to (20**).

62.10a. The Smooth Case

We make the following assumptions:

(H1) Let G be a bounded region in the three-dimensional vector space V_3 with $\partial G \in C^1$. Suppose that there exists a decomposition of the boundary

$$\partial G = \overline{\partial_1 G} \cup \overline{\partial_2 G}, \qquad \partial_1 G \cap \partial_2 G = \emptyset,$$

with $\partial_1 G$ and $\partial_2 G$ sufficiently regular.

(H2) Suppose the two maps $K\colon \bar{G} \to V_3$ and $T\colon \overline{\partial_2 G} \to V_3$ are continuous.

(H3) Let Σ^1 denote the set of all C^1-maps $\sigma\colon \bar{G} \to L_{\text{sym}}(V_3)$ which satisfy the two equilibrium conditions

$$\operatorname{div} \sigma + K = 0 \quad \text{on } G,$$

$$\sigma n = T \quad \text{on } \partial_2 G,$$

and the plasticity condition

$$[\bar{\sigma}, \bar{\sigma}] \leq \sigma_0^2 \quad \text{on } G$$

for some fixed $\sigma_0 > 0$.

(H4) Let $u\colon \bar{G} \to V_3$ and $u_0\colon \overline{\partial_1 G} \to V_3$ be continuous maps with

$$u = u_0 \quad \text{on } \partial_1 G.$$

We consider the variational problem

$$-\int_G M^*(\sigma)\, dx + \int_{\partial_1 G} u_0(\sigma n)\, dO = \text{max!}, \qquad \sigma \in \Sigma^1. \tag{22}$$

This problem corresponds to the *principle of maximal dual energy* of Section 62.4 with the additional side condition $[\bar{\sigma}, \bar{\sigma}] \leq \sigma_0^2$. Hence, the formulation of (22) is very natural.

Proposition 62.7. *Assume* (H1) *to* (H4). *If σ is a solution of* (22), *then σ satisfies the variational inequality* (21).

PROOF. Let $\sigma \in \Sigma^1$ be a solution of (22). Let $F(\sigma)$ denote the left-hand side of (22) and set

$$\varphi(t) = F(\sigma + t(\mu - \sigma)), \qquad t \in \mathbb{R},$$

where $\mu \in \Sigma^1$. Then $\varphi'(0) \leq 0$, and this implies

$$-\int_G [M^{*\prime}(\sigma), \mu - \sigma]\, dx + \int_{\partial_1 G} u_0(\mu - \sigma)n\, dO \leq 0$$

for all $\mu \in \Sigma^1$. Integration by parts then yields

$$\int_G [\gamma(u), \mu - \sigma]\, dx = \int_{\partial_1 G} u_0(\mu - \sigma)n\, dO - \int_G u \operatorname{div}(\mu - \sigma)\, dx$$

for all $\mu \in \Sigma^1$ and hence we obtain (21). Note that $\operatorname{div} \sigma = \operatorname{div} \mu = -K$. □

This proposition shows that we must solve the maximum problem (22) in order to guarantee that the stress tensor σ and the given displacement vector u in (H4) satisfy our plasticity model (21).

62.10b. Existence of a Generalized Solution

In order to solve (22) in a suitable H-space, we need a generalized definition of the boundary integral

$$F_u(\sigma) = \int_{\partial_1 G} u(\sigma n)\, dO$$

in the case of weak assumptions on σ. To this end we set

$$\Sigma(\sigma_0) = \text{closure of } \Sigma^1 \text{ in the H-space } L_2(G, L_{\text{sym}}(V_3)).$$

Using $\sigma_\varepsilon = S_\varepsilon \sigma$, where S_ε is the smoothing operator from Section 18.14, the definition of the set Σ in Section 62.4 shows that

$$\Sigma(\sigma_0) = \{\sigma \in \Sigma : [\bar{\sigma}, \bar{\sigma}] \le \sigma_0^2 \text{ on } G\}.$$

Let us suppose the following:

(A1) G is a bounded region in V_3 with $\partial G \in C^{0,1}$, and $\partial_1 G$ and $\partial_2 G$ are open subsets of ∂G with

$$\partial G = \overline{\partial_1 G} \cup \overline{\partial_2 G}, \qquad \partial_1 G \cap \partial_2 G = \emptyset.$$

(A2) We are given the densities of the outer forces

$$K \in L_2(G, V_3), \qquad T \in L_2(\partial_2 G, V_3)$$

and a vector function $u_0 \in W_2^1(G, V_3)$, where u_0 on $\partial_1 G$ corresponds to the given boundary displacement on $\partial_1 G$.

For all C^1-maps $u: \bar{G} \to V_3$ and all $\sigma \in \Sigma^1$, we obtain from integrating by parts that

$$\int_G u \operatorname{div} \sigma\, dx = -\int_G [\sigma, \gamma(u)]\, dx + \int_{\partial_2 G} Tu\, dO + F_u(\sigma)$$

and hence

$$F_u(\sigma) = \int_G [\sigma, \gamma(u)]\, dx - \int_G Ku\, dx - \int_{\partial_2 G} Tu\, dO. \tag{23}$$

Obviously, this expression makes sense also in case that $u \in W_2^1(G, V_3)$ and $\sigma \in \Sigma$. In this general situation we define $F_u(\sigma)$ through (23).

The embedding $W_2^1(G) \subseteq W_2^{1/2}(\partial G)$ is continuous. Thus, for $u \in W_2^1(G, V_3)$, it follows that $u \in W_2^{1/2}(\partial G, V_3)$, and we can make the following more general assumption.

(A2*) Replace $T \in L_2(\partial_2 G, V_3)$ with

$$T \in W_2^{1/2}(\partial G, V_3)^*.$$

The map $T: W_2^{1/2}(\partial G, V_3) \to \mathbb{R}$ is a continuous linear functional and we can

define
$$\int_{\partial_2 G} Tu \, dO = T(u).$$

In place of the classical variational problem (22), we now consider the more general variational problem

$$-\int_G M^*(\sigma) \, dx + F_{u_0}(\sigma) = \max!, \qquad \sigma \in \Sigma(\sigma_0). \tag{24}$$

In case that $\sigma_0 = +\infty$, formula (24) is identical to the principle of maximal dual energy of Section 62.4.

Theorem 62.C (Linear Quasi-Statical Plasticity). *Under assumptions* (A1), (A2), (A2*), *the variational problem* (24) *has a unique solution.*

PROOF. If we write (24) as a minimum problem, our problem represents a strongly positive quadratic variational problem on the closed convex set $\Sigma(\sigma_0)$ (see Problem 62.8). The assertion then follows from Proposition 46.1. □

We interpret the solution σ of (24) as the stress tensor of our quasi-statical plasticity model.

62.11. Duality and the Existence Theorem for Linear Statical Plasticity

Quasi-statical models in plasticity theory have the disadvantage that the strain tensor γ, and hence the displacement u, are undetermined in case the plasticity limit σ_0 has been reached. This can already be seen in the one-dimensional situation, pictured in Figure 62.2(b). The constitutive law of quasi-statical models therefore represents a strong idealization. In this section we study a so-called statical model in plasticity which yields *unique* stresses and displacements. The idea is this:

Use the principles of minimal potential energy and maximal dual energy and add the plasticity condition $[\bar{\sigma}, \bar{\sigma}] \leq \sigma_0^2$ *as a side condition.*

Our assumptions are the following:

(A1) Conditions (H1) to (H6) of Section 62.3 are satisfied, and $\sigma_0 > 0$ is fixed.
(A2) We consider linear material, i.e., the stored energy function M has the same form as in Section 62.10. In particular, the inverse constitutive law $\gamma = M^{*\prime}(\sigma)$ implies $\bar{\gamma} = 2\kappa^* \bar{\sigma}$ and hence

$$[\bar{\gamma}, \bar{\gamma}] = 4\kappa^{*2}[\bar{\sigma}, \bar{\sigma}].$$

62.11. Duality and the Existence Theorem for Linear Statical Plasticity

This yields the following key relation in our approach: The stress tensor σ satisfies the plasticity inequality

$$[\bar{\sigma}, \bar{\sigma}] \leq \sigma_0^2$$

if and only if the strain tensor γ satisfies the inequality

$$[\bar{\gamma}, \bar{\gamma}] \leq 4\kappa^2 \sigma_0^2.$$

This observation motivates us to consider the following two variational problems

$$E_{\text{pot}}(u) = \min!, \qquad u \in X(\sigma_0), \tag{25}$$

$$E_{\text{dual}}(\sigma) = \max!, \qquad \sigma \in \Sigma(\sigma_0), \tag{25*}$$

where we set

$$X(\sigma_0) = \{u : u - u_0 \in X \text{ and } [\bar{\gamma}, \bar{\gamma}] \leq 4\kappa^2 \sigma_0^2 \text{ on } G\}$$

$$\Sigma(\sigma_0) = \{\sigma \in \Sigma : [\bar{\sigma}, \bar{\sigma}] \leq \sigma_0^2 \text{ on } G\}.$$

We observe that if $\sigma_0 = \infty$, then (25) and (25*) coincide with problems (P) and (P*) of Section 62.4, respectively.

Theorem 62.D (Linear Statical Plasticity). *Assume* (A1), (A2). *Then problem* (25), *as well as problem* (25*), *has a unique solution.*

PROOF. The sets $\Sigma(\sigma_0)$ and $X(\sigma_0)$ are closed and convex in the H-spaces $L_2(G, L_{\text{sym}}(V_3))$ and X, respectively (Problem 62.8). Both problem (25) and (25*) represent strongly positive quadratic variational problems, and the assertion therefore follows from Proposition 46.1. □

We interpret the solutions u and σ as displacement vector and reduced stress tensor, respectively.

This statical plasticity model is reasonable from both the physical and the mathematical point of view.

(a) Instead of the "unrealistic" and arbitrary constitutive law in the quasi-statical model, we start here with a general physical principle (minimization of energy).
(b) We obtain a unique displacement u.
(c) The unique stress tensor σ is the same as in the quasi-statical model, since problem (25*) is identical to problem (24).
(d) As in the quasi-statical model, the zone of plasticity is given by the set of all $x \in G$ with $[\bar{\sigma}(x), \bar{\sigma}(x)] = \sigma_0^2$.
(e) For implementation on computers, we can use the well-established methods of convex optimization.
(f) This method can also be applied to nonlinear material. For example, in Chapter 86, we consider plastic torsion of rods, consisting of nonlinear

Hencky material. There, we will also show the close relation of the torsion problem with stationary conservation laws in hydrodynamics.

Theorem 62.D represents a rigorous justification of the classical empirical theory of Haar and von Kármán (1909).

62.12. Compensated Compactness

There are four important methods to obtain *existence results* in nonlinear functional analysis and, for that matter, for nonlinear partial differential equations:

(i) Compactness.
(ii) Monotonicity.
(iii) Convex extremal problems.
(iv) Compensated compactness.

Prototypes for (i)–(iii) are the following:

(i) Leray–Schauder principle (Theorem 6.A) and the generalized Weierstrass theorem for extremal problems (Theorem 38.B).
(ii) Main theorem for monotone operators (Theorem 26.A).
(iii) Minimization of continuous convex functionals or, more generally, of weak sequentially lower semicontinuous functionals (Theorem 38.A).
   As we have seen in Chapter 42, there exists a close relation between (ii) and (iii).

The method of compensated compactness stands for the following:

The lack of (i)–(iii) *is compensated for by using additional information* (*side conditions*).

We will apply this method to two important problems.

(a) Existence theorem for polyconvex material in nonlinear elasticity (Section 62.13).
(b) Existence theorem for transonic flow (Part V).

In (a) we shall make essential use of properties of adj A and det A (Proposition 62.10).

In (b) we shall introduce an additional entropy condition. This will enable us to apply the embedding theorem of Murat (Proposition 62.11). The method of compensated compactness is especially transparent in case (b) of transonic flows. There, we have, roughly, the following situation:

(α) The corresponding variational problem looks like

$$F(u) = \min!, \quad u \in K. \qquad (26)$$

(β) If there exist subregions with a supersonic flow, then F loses its convexity.

62.12a. Properties of adj A

Subsequently, we use simple identities for determinants and the method of integration by parts. We make the following assumption:

(H) G is a bounded region in $\mathbb{R}^3$ with $\partial G \in C^{0,1}$. Let $2 \leq p < \infty$ and $q \leq r < \infty$ where $p^{-1} + q^{-1} = 1$. We choose a fixed Cartesian coordinate system. Let $A: V_3 \to V_3$ be a linear operator. Recall that the components of the linear operator

$$\operatorname{adj} A: V_3 \to V_3$$

are the adjoint subdeterminants of A^*, i.e.,

$$(\operatorname{adj} A)_{ij} = \tfrac{1}{2}\varepsilon_{ikl}\varepsilon_{jmn} A_{mk} A_{nl}, \tag{27}$$

where, as usual, $\varepsilon_{ikl} = \operatorname{sgn}\begin{pmatrix} 1 & 2 & 3 \\ i & k & l \end{pmatrix}$. We sum over two equal indices from 1 to 3.

Lemma 62.8. *Assume* (H). *If*

$$a \rightharpoonup b \quad \text{in } W_p^1(G)^3,$$
$$\operatorname{adj} a' \rightharpoonup c \quad \text{in } L_r(G)^9,$$

then $\operatorname{adj} b' = c$.

PROOF. Let $a \in W_p^1(G)^3$. Equation (27) yields

$$(\operatorname{adj} a')_{ij} = \tfrac{1}{2}\varepsilon_{ikl}\varepsilon_{jmn} D_k a_m D_l a_n. \tag{28}$$

(I) We begin by proving the key identity

$$\int_G (\operatorname{adj} a')_{ij}\varphi\, dx = -\frac{1}{2}\int_G \varepsilon_{ikl}\varepsilon_{jmn} a_m D_l a_n D_k \varphi\, dx \tag{29}$$

for all $\varphi \in C_0^\infty(G)$. The important fact is that no second-order derivatives appear on the right-hand side. In case that $a \in C^\infty(\bar{G})^3$, this follows from (28), employing integration by parts and the fact that

$$\varepsilon_{ikl}\varepsilon_{jmn} D_k D_l a_n = 0.$$

In the general case, we use that $C^\infty(\bar{G})$ is dense in $W_p^1(G)$ and that the embedding $W_p^1(G) \subseteq L_q(G)$ is continuous. Note that $2 \leq p < \infty$.

(II) The embedding $W_p^1(G) \subseteq L_q(G)$ is compact, and thus
$$a \to b \quad \text{in } L_q(G)^3.$$

From (29) follows
$$\int_G c_{ij} \varphi \, dx = -\frac{1}{2} \int_G \varepsilon_{ikl} \varepsilon_{jmn} b_m D_l b_n D_k \varphi \, dx$$
for all $\varphi \in C_0^\infty(G)$ and
$$\int_G (\operatorname{adj} b')_{ij} \varphi \, dx = \int_G c_{ij} \varphi \, dx$$
for all $\varphi \in C_0^\infty(G)$. Hence $\operatorname{adj} b' = c$. □

Lemma 62.9. *Assume* (H). *If* $a \in W_p^1(G)^3$ *and*
$$\operatorname{adj} a' \in L_r(G)^9, \tag{30}$$
then $\det a' \in L_1(G)$ *and*
$$\int_G (\det a') \varphi \, dx = -\int_G a_1 (\operatorname{adj} a')_{i1} D_i \varphi \, dx \tag{31}$$
for all $\varphi \in C_0^\infty(G)$.

This result is *typical* for the method of compensated compactness. In case that $a \in W_p^1(G)^3$, we merely know that
$$\det a' \colon G \to \mathbb{R}$$
is a measurable function. Together with the additional information (30), however, we obtain the much stronger result that $\det a' \in L_1(G)$. We set
$$b_i = (\operatorname{adj} a')_{i1}.$$
Equation (31) is the distributive form of the classical identity
$$\det a' = D_i(a_1 b_i).$$
This identity is a consequence of
$$\det a' = b_i D_i a_1 \tag{32}$$
and
$$D_i b_i = 0 \quad \text{for} \quad a \in C^2(\bar{G})^3. \tag{33}$$
Equation (33) follows from (28). The simple proof idea is to use integration by parts as well as density arguments in order to overcome a lack of smoothness.

PROOF. Let $a \in C^\infty(\bar{G})$. From (33) and integration by parts follows that
$$\int_G b_i D_i \psi \, dx = 0 \quad \text{for all} \quad \psi \in C_0^\infty(G).$$

62.12. Compensated Compactness

By using a density argument the same is true for $a \in W_p^1(G)^3$. Thus we obtain the key identity

$$\int_G b_i(\varphi D_i c)\, dx = -\int_G b_i(D_i \varphi) c\, dx \tag{34}$$

for all $\varphi \in C_0^\infty(G)$ and $c \in C^\infty(\bar{G})$. We then choose $c \in C^\infty(\bar{G})$ with

$$c \to a_1 \quad \text{in } W_p^1(G),$$

and from $b_i \in L_q(G)$ and (32) we obtain that

$$b_i D_i c \to \det a' \quad \text{in } L_1(G).$$

Formula (31) therefore follows from (34). □

Proposition 62.10. *Assume* (H). *If*

$$a \to b \quad \text{in } W_p^1(G)^3,$$
$$\text{adj } a' \to c \quad \text{in } L_r(G)^9,$$
$$\det a' \to d \quad \text{in } L_s(G), \quad 1 < s < \infty,$$

then $\text{adj } b' = c$ *and* $\det b' = d$.

This is the key to the proof of Theorem 62.E in the following section.

PROOF. Lemma 62.8 implies that $\text{adj } b' = c$. By passing to the limit in (31), we obtain

$$\int_G d\varphi\, dx = -\int_G b_1 (\text{adj } b')_{i1} D_i \varphi\, dx$$

for all $\varphi \in C_0^\infty(G)$, and hence $d = \det b'$, according to Lemma 62.9. Notice that our assumptions imply that

$$a \to b \quad \text{in } L_p(G)^3$$
$$\text{adj } a' \to c \quad \text{in } L_q(G)^9.$$

□

62.12b. Embedding Theorem of Murat

Recall that

$$C_0^\infty(G)_+ = \{\varphi \in C_0^\infty(G): \varphi \geq 0 \text{ on } G\}.$$

Proposition 62.11 (Murat (1981)). *Let G be a bounded region in $\mathbb{R}^N$, $N \geq 1$ with $\partial G \in C^{0,1}$.*

(a) *Suppose that*

$$F_n \to F \quad \text{in } W_2^1(G)^* \qquad \text{as } n \to \infty, \tag{35}$$

$$F_n(\varphi) \geq 0 \quad \text{on } C_0^\infty(G)_+ \qquad \text{for all } n. \tag{36}$$

Then

$$F_n \to F \quad \text{in } W_p^1(G)^* \quad \text{as } n \to \infty \quad \text{for all } 2 < p \leq \infty. \tag{37}$$

(b) *Assertion* (a) *remains true if we replace* $W_2^1(G)$ *and* $W_p^1(G)$ *with* $\mathring{W}_2^1(G)$ *and* $\mathring{W}_p^1(G)$, *respectively. Assumption* $\partial G \in C^{0,1}$ *drops out in this case.*

Observation (b) can be stated as follows. If G is a bounded region in $\mathbb{R}^N$, $N \geq 1$, then the embedding

$$\mathring{W}_2^1(G)^*_+ \subseteq \mathring{W}_p^1(G)^*$$

is compact for all $2 < p \leq \infty$. The key thereby is the *positivity* condition (36) which compensates for the lack of compactness of the embedding $\mathring{W}_2^1(G)^* \subseteq \mathring{W}_p^1(G)^*$.

Actually, Proposition 62.11(a) is stronger than the corresponding theorem in Murat (1981) and goes back to Feistauer, Mandel, and Nečas (1984). We need this stronger result in our approach to transonic flows of Chapter 86.

The proof of Proposition 62.11 makes essential use of interpolation theory and a priori estimates due to Agmon, Douglis, and Nirenberg (1959) and Meyers (1963). We also use standard properties of embeddings which may be found in Section 21.11.

Since the embedding $W_q^1(G) \subseteq W_p^1(G)$, $1 \leq p \leq q \leq \infty$ is continuous, the dual embedding

$$W_p^1(G)^* \subseteq W_q^1(G)^*, \quad 1 \leq p \leq q \leq \infty$$

is continuous as well. Hence it is sufficient to prove (37) for all p with $2 < p \leq q_0 < \infty$, where q_0 is fixed. In particular, we can assume that $p < \infty$.

PROOF OF (a) IN A SPECIAL CASE. We assume that $\partial G \in C^1$ and that (36) holds on $C^\infty(\bar{G})$.

In this special case we can give an "elementary" proof. The general case will be considered below.

(I) We set $\|\varphi\| = \max_{x \in \bar{G}} |\varphi(x)|$. The obvious inequality

$$-\|\varphi\| \leq \varphi \leq \|\varphi\| \quad \text{on } \bar{G} \quad \text{for all} \quad \varphi \in C^\infty(\bar{G})$$

and (36) imply

$$|F_n(\varphi)| \leq \|\varphi\| F_n(1) \quad \text{for all} \quad \varphi \in C^\infty(\bar{G}).$$

From (35) follows that $F_n(1) \to F(1)$ as $n \to \infty$ and hence

$$0 \leq F_n(1) \leq C \quad \text{for all } n.$$

Since $C^\infty(\bar{G})$ is dense in $C(\bar{G})$, we obtain the *key* relation

$$|F_n(\varphi)| \leq C\|\varphi\| \quad \text{for all} \quad \varphi \in C(\bar{G}).$$

The set $\{F_n\}$ is therefore bounded in $C(\bar{G})^*$.

The proof then follows from the standard arguments (II) to (IV) below.

62.12. Compensated Compactness

(II) The embedding

$$W_q^1(G) \subseteq C(\bar{G})$$

is compact for all $q > N$. This yields the compactness of the dual embedding

$$C(\bar{G})^* \subseteq W_q^1(G)^* \qquad \text{for all} \quad q > N.$$

Thus, there exists a subsequence $(F_{n'})$ with

$$F_{n'} \to \bar{F} \quad \text{in } W_q^1(G)^* \qquad \text{as} \quad n \to \infty.$$

For $q > 2$, the embedding $W_q^1(G) \subseteq W_2^1(G)$ is continuous. Hence the embedding $W_2^1(G)^* \subseteq W_q^1(G)^*$ is continuous as well. Hypothesis (35) therefore implies that

$$F_n \rightharpoonup F \quad \text{in } W_q^1(G)^* \qquad \text{as} \quad n \to \infty,$$

and thus, $\bar{F} = F$.

The convergence principle of Proposition 10.13(1) then yields the convergence of the original sequence, i.e.,

$$F_n \to F \quad \text{in } W_q^1(G)^* \qquad \text{as} \quad n \to \infty. \tag{38}$$

(III) If $p \geq q$, then (38) implies that

$$F_n \to F \quad \text{in } W_p^1(G)^* \qquad \text{as} \quad n \to \infty.$$

Observe the continuity of the embedding $W_p^1(G) \subseteq W_q^1(G)$ which implies the continuity of the embedding $W_q^1(G)^* \subseteq W_p^1(G)^*$.

(IV) In case that $2 < p < q$ and $q < \infty$, we use the convergence trick of interpolation theory (Section 21.17). Every p can be represented as

$$\frac{1}{p} = \frac{1-t}{q} + \frac{t}{2}, \qquad 0 < t < 1.$$

The well-known interpolation relation

$$W_p^1(G)^* = [W_q^1(G)^*, W_2^1(G)^*]_t$$

then yields

$$\|F\|_{1,p} \leq C \|F\|_{1,q}^{1-t} \|F\|_{1,2}^t \qquad \text{for all} \quad F \in W_2^1(G)^*, \tag{39}$$

where $\|F\|_{1,p}$ denotes the norm on $W_p^1(G)^*$. From assumption (35) follows that the sequence (F_n) is bounded in $W_2^1(G)^*$. Assertion (37) then follows from (38) and (39).

PROOF OF (b). Choose proper subregions

$$H \subset\subset A \subset\subset B \subset\subset G$$

with C^∞-boundaries and set $Du\, D\varphi = \sum_{i=1}^N D_i u\, D_i \varphi$. By hypothesis,

$$F_n \to F \quad \text{in } \mathring{W}_2^1(G)^* \quad \text{as} \quad n \to \infty, \tag{40}$$

$$F_n(\varphi) \geq 0 \quad \text{on } C_0^\infty(G)_+ \quad \text{for all } n. \tag{41}$$

(I) The *trick* in our proof is to represent F_n in the form

$$-\Delta u_n = F_n, \quad u_n \in \mathring{W}_2^1(G),$$

i.e.,

$$\int_G Du_n\, D\varphi\, dx = F_n(\varphi) \quad \text{for all} \quad \varphi \in C_0^\infty(G).$$

According to Theorem 25.I, the function u_n is uniquely determined by F_n, where

$$\|u_n\|_{\mathring{W}_2^1(G)} \leq \text{const } \|F_n\|_{\mathring{W}_2^1(G)^*} \quad \text{for all } n.$$

From (40) follows that the sequence (F_n) is bounded in $\mathring{W}_2^1(G)^*$, and hence that

$$(u_n) \text{ is bounded in } \mathring{W}_2^1(G). \tag{42}$$

Below we shall prove the following:

There exists a sequence of subregions $H_1 \subset H_2 \subset \cdots \subset G$ such that $\text{meas}(G - H_k) \to 0$ as $k \to \infty$, and there exists a subsequence of (u_n) such that

$$(u_{n'}) \text{ converges in } W_s^1(H) \quad \text{for all} \quad 1 < s < 2 \tag{43}$$

and all $H = H_k$, $k = 1, 2, \ldots$.

We show that (43) implies assertion (b). Let $p > 2$, and choose s with $p^{-1} + s^{-1} = 1$. We then have the splitting

$$F_n(\varphi) - F_m(\varphi) = \int_G (Du_n - Du_m) D\varphi\, dx$$

$$= \int_H (Du_n - Du_m) D\varphi\, dx + \int_{G-H} (Du_n - Du_m) D\varphi\, dx.$$

Observe that

$$\left|\int_H (Du_n - Du_m) D\varphi\, dx\right| \leq \|u_n - u_m\|_{W_s^1(H)} \|\varphi\|_{\mathring{W}_p^1(G)}$$

and

$$\left|\int_{G-H} (Du_n - Du_m) D\varphi\, dx\right|$$

$$\leq \left(\int_{G-H} |Du_n - Du_m|^2 dx\right)^{1/2} \left(\int_{G-H} |D\varphi|^p dx\right)^{1/p} \left(\int_{G-H} dx\right)^{(1/2)-(1/p)}$$

$$\leq (\text{meas } G - H)^{(1/2)-(1/p)} \|u_n - u_m\|_{\mathring{W}_2^1(G)} \|\varphi\|_{\mathring{W}_p^1(G)}.$$

62.12. Compensated Compactness

This implies

$$\|F_{n'} - F_{m'}\| < \varepsilon \quad \text{in } \mathring{W}_p^1(G)^* \quad \text{if } n', m' \geq n_0(\varepsilon).$$

Notice (42) and (43). Hence

$$F_{n'} \to \bar{F} \quad \text{in } \mathring{W}_p^1(G)^* \quad \text{as } n \to \infty.$$

From (40) follows that $\bar{F} = F$. The convergence principle in Proposition 10.13(1) then yields the assertion

$$F_n \to F \quad \text{in } \mathring{W}_p^1(G)^* \quad \text{as } n \to \infty.$$

(II) In order to finish our proof it only remains to show that (43) holds true.

(II-1) Choose $\bar{\varphi} \in C_0^\infty(G)$ with $\bar{\varphi} = 1$ on B. From

$$-\|\varphi\|\bar{\varphi} \leq \varphi \leq \|\varphi\|\bar{\varphi} \quad \text{on } B \quad \text{for all } \varphi \in C_0^\infty(B)$$

and hypothesis (41), we obtain the *key* estimate

$$|F_n(\varphi)| \leq C\|\varphi\| \quad \text{for all } \varphi \in C_0^\infty(B)$$

and all n. Since $C_0^\infty(B)$ is dense in $C(\bar{A})$, we find that

$$(F_n) \text{ is bounded in } C(\bar{A})^*.$$

Choose $q > \max\{N, 2\}$. The embedding $\mathring{W}_q^1(A) \subseteq C(\bar{A})$ is then compact and hence the embedding $C(\bar{A})^* \subseteq \mathring{W}_q^1(A)^*$ is also compact. Thus there exists a subsequence of (F_n) such that

$$(F_{n'}) \text{ converges in } \mathring{W}_q^1(A)^*. \tag{44}$$

(II-2) The interior L_p-estimates in Agmon, Douglis, and Nirenberg (1959) yield

$$\|u_n\|_{W_r^1(H)} \leq \text{const } \|F_n\|_{\mathring{W}_q^1(A)^*} \quad \text{for all } n,$$

where $r^{-1} + q^{-1} = 1$. Hence

$$(u_{n'}) \text{ converges in } W_r^1(H). \tag{45}$$

(II-3) We prove that

$$(u_{n'}) \text{ converges in } W_s^1(H) \quad \text{for all } 1 < s < 2. \tag{46}$$

Observe that $1 < r < 2$. If $1 < s \leq r$, then (46) is valid as a consequence of the continuity of the embedding $W_r^1(H) \subseteq W_s^1(H)$.

In case that $r < s < 2$, we choose $0 < t < 1$ with

$$\frac{1}{s} = \frac{1-t}{r} + \frac{t}{2}.$$

Interpolation yields

$$W_s^1(H) = [W_r^1(H), W_2^1(H)]_t,$$

i.e.,

$$\|u\|_{1,s} \leq \text{const } \|u\|_{1,r}^{1-t}\|u\|_{1,2}^{t} \quad \text{for all} \quad u \in W_2^1(H).$$

This, together with (42), (45) implies (46).

(II-4) Now choose a sequence $H_1 \subset H_2 \subset \cdots \subset G$. Formula (43) then follows from a diagonal procedure.

PROOF OF (a) IN THE GENERAL CASE. The key to the proof is the *orthogonal projection* (47), together with the *splitting argument* (49), (49*) below. In particular, we will use a result due to Meyers (1963).

(I) Let $\varphi \in W_p^1(G)$, $p > 2$. Then $\varphi \in W_2^1(G)$. From Section 22.4, there exists the orthogonal decomposition

$$\varphi = \varphi_1 + \varphi_2 \quad \text{where} \quad \varphi_2 \in \mathring{W}_2^1(G) \tag{47}$$

and $\varphi_1 \in C^\infty(G)$ with

$$\Delta\varphi_1 = 0 \quad \text{on } G, \quad \varphi_1 = \varphi \quad \text{on } \partial G. \tag{48}$$

A standard result in Hilbert space theory says that the mappings

$$\varphi \mapsto \varphi_i, \quad i = 1, 2 \tag{47*}$$

are orthogonal projections on the H-space $W_2^1(G)$, and hence are continuous on $W_2^1(G)$. We need, however, a much stronger result here. The paper of Meyers (1963) shows that there exists a real number $q_0 > 2$ such that the mappings (47*) are continuous from

$$W_r^1(G) \text{ to } W_r^1(G) \quad \text{if} \quad 2 \leq r \leq q_0.$$

In the following we assume that $2 < p \leq q_0$.

(II) We show that the set

$$K = \{\varphi_1 \colon \|\varphi\|_{W_p^1(G)} \leq 1\}$$

is relatively compact in $W_2^1(G)$. In fact, it follows from (48) that

$$\|\varphi_1\|_{W_2^1(G)} \leq \text{const } \|\varphi\|_{W_2^{1/2}(\partial G)} \quad \text{for all} \quad \varphi \in W_2^1(G),$$

according to Theorem 25.I. We then observe that the embedding $W_p^1(G) \subseteq W_p^{1-1/p}(\partial G)$ is continuous and that the embedding

$$W_p^{1-1/p}(\partial G) \subseteq W_2^{1/2}(\partial G)$$

is compact. Denoting the closure of K in $W_2^1(G)$ by $\bar{K}$, we find that $\bar{K}$ is compact.

(III) From hypothesis (35), the sequence (F_n) is bounded in $W_2^1(G)^*$, and

$$|F_n(\varphi) - F_n(\psi)| \leq \|F_n\| \, \|\varphi - \psi\|$$

for all n and all $\varphi, \psi \in W_2^1(G)$. This yields the equicontinuity of $\{F_n\}$ on $W_2^1(G)$.

(IV-1) From (35),
$$F_n(\varphi) \to F(\varphi) \quad \text{as} \quad n \to \infty \quad \text{for all} \quad \varphi \in \bar{K}.$$

According to Problem 19.14, the compactness of $\bar{K}$ and the equicontinuity of $\{F_n\}$ imply that this convergence is uniform, i.e.,

$$\sup_{\|\varphi\|_{1,p} \leq 1} |F_n(\varphi_1) - F(\varphi_1)| \to 0 \quad \text{as} \quad n \to \infty. \tag{49}$$

(IV-2) The embedding $W_2^1(G)^* \subseteq \mathring{W}_2^1(G)^*$ is continuous. From (35) follows that

$$F_n \rightharpoonup F \quad \text{in} \quad \mathring{W}_2^1(G)^* \quad \text{as} \quad n \to \infty.$$

Proposition 62.11(b) then shows that

$$F_n \to F \quad \text{in} \quad \mathring{W}_p^1(G)^* \quad \text{as} \quad n \to \infty,$$

and hence

$$\sup_{\|\varphi\|_{1,p} \leq 1} |F_n(\varphi_2) - F(\varphi_2)| \to 0 \quad \text{as} \quad n \to \infty. \tag{49*}$$

Notice the continuity of the map $\varphi \mapsto \varphi_2$ from $W_p^1(G)$ to $\mathring{W}_p^1(G)$. From (49), (49*) and $\varphi = \varphi_1 + \varphi_2$, we obtain the assertion

$$F_n \to F \quad \text{in} \quad W_p^1(G)^* \quad \text{as} \quad n \to \infty. \qquad \square$$

62.13. Existence Theorem for Polyconvex Material

We consider the principle of minimal potential energy

$$\int_G L(I + u'(x)) \, dx - \int_G Ku \, dx - \int_{\partial_2 G} Tu \, dO = \min!, \tag{50}$$

$$u = u_0 \quad \text{on} \quad \partial_1 G$$

in the case that the stored energy function has the form

$$L(A) = P(A, \text{adj } A, \det A) \tag{51}$$

for all $A \in L(V_3)_+$. We set

$$L(V_3)_+ = \{A \in L(V_3) : \det A > 0\},$$

and define the norm

$$|A| = (\operatorname{tr} A^* A)^{1/2} \quad \text{for all} \quad A \in L(V_3).$$

In Cartesian coordinates we have

$$|A| = \left(\sum_{i,j=1}^3 a_{ij}^2 \right)^{1/2}.$$

The connection between A and the displacement u is given by
$$A = I + u'(x).$$
We now choose a fixed Cartesian coordinate system and make the following assumptions:

(H1) G is a bounded region in $\mathbb{R}^3$ with $\partial G \in C^{0,1}$ and there exists a decomposition
$$\partial G = \overline{\partial_1 G} \cup \overline{\partial_2 G}, \qquad \partial_1 G \cap \partial_2 G = \emptyset, \qquad \partial_1 G \neq \emptyset,$$
where $\partial_1 G$ and $\partial_2 G$ are open in ∂G.

(H2) *Polyconvexity.* There exists a convex function
$$P: L(V_3) \times L(V_3) \times \,]0, \infty[\to \mathbb{R}$$
such that (51) is valid. Moreover, in a neighborhood of a fixed point, P is bounded from above.

(H3) *The limit* $\det A \to 0$. If
$$A_n \to A \quad \text{and} \quad B_n \to B \quad \text{in } L(V_3) \qquad \text{as } n \to \infty$$
and $d_n \to +0$, then
$$P(A_n, B_n, d_n) \to +\infty \qquad \text{as } n \to \infty.$$

(H4) *Coerciveness.* Assume that there exist constants $C > 0$ and D such that
$$P(A, \operatorname{adj} A, \det A) \geq C(|A|^p + |\operatorname{adj} A|^r + |\det A|^s) + D$$
for all $A \in L(V_3)_+$. Here $2 \leq p < \infty$, $q^{-1} + p^{-1} = 1$, $q \leq r < \infty$, and $1 < s < \infty$.

(H5) *Outer forces.* We are given
$$K \in L_q(G)^3, \qquad T \in L_q(\partial_2 G)^3.$$

(H6) *Boundary displacement.* Let V be the set of all
$$u \in W_p^1(G)^3$$
with the additional properties
$$\operatorname{adj}(I + u') \in L_r(G)^9, \qquad \det(I + u') \in L_s(G),$$
$$\det(I + u'(x)) > 0 \quad \text{a.e. on } G.$$

We are given $u_0 \in V$.

Our final problem looks like
$$F(u) = \min!, \qquad u \in U, \qquad (52)$$
where
$$U = \{u \in V : u = u_0 \text{ on } \partial_1 G\}$$

62.13. Existence Theorem for Polyconvex Material

and

$$F(u) = \int_G L\, dx - \int_G Ku\, dx - \int_{\partial_2 G} Tu\, dO.$$

Theorem 62.E (Ball (1977)). *Assume (H1)–(H6) with $F(u_0) < \infty$. Then there exists a solution of the variational problem (52).*

PROOF. We use standard arguments of the calculus of variations, together with the key result of Proposition 62.10.

(I) Preparations. Define

$$\bar{P}(A, B, d) = \begin{cases} P(A, B, d) & \text{if } d > 0, \\ +\infty & \text{if } d \leq 0. \end{cases}$$

The convex function P in (H2) is continuous according to Proposition 47.5. Hence the function

$$\bar{P}: L(V_3) \times L(V_3) \times \mathbb{R} \to]-\infty, +\infty]$$

is continuous according to (H3).

Furthermore, we set

$$X = W_p^1(G)^3 \times L_r(G)^9 \times L_s(G)$$

and

$$Hu(x) = (x + u(x), \operatorname{adj}(I + u'(x)), \det(I + u'(x))). \tag{53}$$

(I-1) On $W_p^1(G)$ we introduce the equivalent norm

$$\left(\int_G |Dv|^p\, dx + \int_{\partial_1 G} |v|^p\, dO \right)^{1/p},$$

and consequently, we get

$$\|u\|_{1,p} \leq \text{const}\, \|Du\|_p + \text{const} \qquad \text{for all} \quad u \in U. \tag{54}$$

Notice that $u = u_0$ on $\partial_1 G$.

(I-2) Theorem of Mazur. Let (v_n) be a sequence in an arbitrary B-space X with *weak* convergence

$$v_n \rightharpoonup v \quad \text{in } X \qquad \text{as} \quad n \to \infty.$$

Then there exist convex linear combinations

$$w_n = \sum_{i=1}^{N(n)} \lambda_{ni} v_i$$

such that $N(n) \to \infty$ as $n \to \infty$, and we have *strong* convergence

$$w_n \to v \quad \text{in } X \qquad \text{as} \quad n \to \infty.$$

The proof may be found in Yosida (1965, M).

(I-3) Recall that in case
$$f_n \to f \text{ in } L_p(G) \quad \text{as} \quad n \to \infty$$
and $1 < p < \infty$, there exists a subsequence such that
$$f_{n'}(x) \to f(x) \quad \text{a.e. on } G \quad \text{as} \quad n \to \infty.$$

(II) **Minimal sequence.** Set
$$\mu = \inf_{u \in U} F(u).$$
From (H4), we have
$$-\infty < F(u) \leq +\infty \quad \text{for all} \quad u \in U.$$
Since $F(u_0) < \infty$, the value μ is *finite*. We choose a sequence (u_n) in U such that
$$F(u_1) \geq F(u_2) \geq \cdots \geq F(u_n) \to \mu \quad \text{as} \quad n \to \infty.$$
Coerciveness (H4) and (54) then yield that
$$(Hu_n) \text{ is bounded in } X.$$

(III) **Weak convergence.** Since X is reflexive, there exists a subsequence, denoted again by (Hu_n), such that
$$Hu_n \rightharpoonup v \text{ in } X \quad \text{as} \quad n \to \infty.$$
From Proposition 62.10 follows
$$v = Hu.$$
This is the key to our proof.

(IV) **Strong convergence.** We set $v_n = Hu_n$ and choose (w_n) as in (I-2). By passing to a subsequence, if necessary, we may assume that
$$w_n \to Hu \text{ in } X \quad \text{as} \quad n \to \infty,$$
$$w_n(x) \to Hu(x) \text{ a.e. on } G \quad \text{as} \quad n \to \infty.$$

(V) **Lemma of Fatou.** The convexity of P yields
$$P(w_n) \leq \sum_{i=1}^{N(n)} \lambda_{ni} P(Hu_i).$$
From the coerciveness condition (H4) we have that
$$P(w_n) \geq \text{const} \quad \text{for all } n,$$
and thus the lemma of Fatou A_2(19c) yields
$$\int_G \bar{P}(u)\,dx = \int_G \lim_{n \to \infty} P(w_n)\,dx \leq \lim_{n \to \infty} \int_G \sum_{i=1}^{N(n)} \lambda_{ni} P(Hu_i)\,dx.$$

Hence

$$\bar{F}(u) \leq \lim_{n\to\infty} \sum_{i=1}^{N(n)} \lambda_{ni} F(u_i) \leq \lim_{n\to\infty} F(u_{N(n)}) = \mu,$$

i.e.,

$$\int_G \bar{P}(u)\,dx - \int_G Ku\,dx - \int_{\partial_2 G} Tu\,dO \leq \mu.$$

(VI) We show that $u \in U$. We have $\bar{P}(u) < \infty$. Hence the construction of $\bar{P}$ yields

$$\det(I + u'(x)) > 0 \quad \text{a.e. on } G.$$

Since $u_n = u_0$ on $\partial_1 G$ for all n, and since convex linear combinations of the u_n converge to u in $W_p^1(G)$, we also have that

$$u = u_0 \quad \text{on } \partial_1 G.$$

Observe the continuity of the embedding $W_p^1(G) \subseteq W_p^{1-1/p}(\partial G)$.
Consequently, $F(u) = \mu$ and $u \in U$, i.e., u is a solution of (52). □

62.14. Application to Rubberlike Material

We consider Ogden's material from Example 61.15, and wish to show that Theorem 62.E can be applied to such a material. The stored energy function has the form

$$L = P(A, \operatorname{adj} A, \det A)$$

with $A = I + u'(x)$ and

$$P = C \operatorname{tr} E^p + D \operatorname{tr} \operatorname{adj} E^r + f(\det A) + \text{const}, \tag{55}$$

where $E = (A^*A)^{1/2}$. In the special case of Mooney–Rivlin material $p = r = 2$ we have

$$P = C|A|^2 + D|\operatorname{adj} A|^2 + f(\det A) + \text{const}.$$

Let us assume that:

(A1) The function $f:]0, \infty[\to \mathbb{R}$ is convex and continuous with

$$f(d) \geq ad^s \quad \text{for all} \quad d > 0,$$

where $0 < a < \infty$ and $1 < s < \infty$. Moreover, $f(d) \to +\infty$ as $d \to +0$.
(A2) The constants in (55) satisfy

$$2 \leq p < \infty, \quad q^{-1} + p^{-1} = 1, \quad q \leq r < \infty, \quad C, D > 0.$$

Proposition 62.12. *Assume* (A1), (A2). *Then Theorem 62.E can be applied to Ogden's rubberlike material* (55).

PROOF. Let $\lambda_1, \lambda_2, \lambda_3$ be the eigenvalues of $(A^*A)^{1/2}$. The well-known inequality

$$\alpha(\lambda_1^2 + \lambda_2^2 + \lambda_3^2)^{1/2} \leq (\lambda_1^p + \lambda_2^p + \lambda_3^p)^{1/p}$$

with a constant $\alpha > 0$ implies that

$$\alpha^p |A|^p \leq \operatorname{tr} E^p.$$

Similarly, we obtain

$$\alpha^r |\operatorname{adj} A|^r \leq \operatorname{tr} \operatorname{adj} E^r.$$

Thus the coerciveness condition is satisfied. Polyconvexity follows from Example 61.15. □

62.15. Proof of Korn's Inequality

Our goal is to give a simple proof of Korn's inequality. This is the most important inequality in elasticity theory. Our approach follows Nečas and Hlaváček (1981). The key will be a standard result about equivalent norms on the Lebesgue space $L_2(G)$ in terms of negative norms, which correspond to norms on the dual Sobolev space $\mathring{W}_2^1(G)^*$ (Proposition 62.18).

We make the following assumptions:

(H1) G is a bounded region in V_3 with $\partial G \in C^{0,1}$.
(H2) We have the boundary decomposition

$$\partial G = \overline{\partial_1 G} \cup \overline{\partial_2 G}, \qquad \partial_1 G \cap \partial_2 G = \emptyset,$$

where $\partial_1 G$ and $\partial_2 G$ are open subsets of ∂G.

As in Section 62.2 we set

$$X = \mathring{W}_2^1(G, \partial_1 G; V_3).$$

Recall that this space corresponds to the boundary condition

$$u = 0 \quad \text{on } \partial_1 G.$$

Moreover, recall that the linearized strain tensor is given by

$$\gamma(u)(x) = \tfrac{1}{2}(u'(x) + u'(x)^*),$$

and define

$$X_0 = \{u \in X : \gamma(u) = 0 \text{ on } G\},$$

which is a linear, closed subspace of the H-space X, consisting of all displacements with vanishing linearized strain. Let $X_0^\perp$ denote the orthogonal complement to X_0, i.e.,

$$X = X_0 \oplus X_0^\perp.$$

62.15. Proof of Korn's Inequality

Theorem 62.F (Korn's Inequality). *Assume* (H1), (H2). *Then there exists a constant* $c_1 > 0$ *such that*

$$\int_G [\gamma(u), \gamma(u)] \, dx \geq c_1 \|u\|_X^2 \qquad \text{for all} \quad u \in X_0^\perp. \tag{56}$$

Corollary 62.13 (Special Korn's Inequality). *Assume* (H1), (H2) *with* $\partial_1 G \neq \emptyset$. *Then there exists a constant* $c_1 > 0$ *such that*

$$\int_G [\gamma(u), \gamma(u)] \, dx \geq c_1 \|u\|_X^2 \qquad \text{for all} \quad u \in X. \tag{57}$$

In Cartesian coordinates we have

$$\gamma_j^i(u) = \tfrac{1}{2}(D^i u_j + D^j u_i),$$

$$[\gamma(u), \gamma(u)] = \gamma_j^i \gamma_i^j, \tag{58}$$

$$\|u\|_X^2 = \int_G u^i u_i + D^i u_j D_i u^j \, dx.$$

Moreover, we introduce

$$|u|_X^2 = \int_G D^i u_j D_i u^j \, dx. \tag{59}$$

Corollary 62.14 (Equivalent Norm). *Assume* (H1), (H2) *with* $\partial_1 G \neq \emptyset$. *Then, on the H-space* X, *the norms* $\|u\|_X$ *and* $|u|_X$ *are equivalent, i.e., there are constants* $C, D > 0$ *such that*

$$C|u|_X \leq \|u\|_X \leq D|u|_X \qquad \text{for all} \quad u \in X.$$

Consequently, we can replace $\|u\|_X^2$ with $|u|_X^2$ in (57). For smooth functions, inequalities (56) and (57) go back to a fundamental paper of Korn (1907) on the existence of classical solutions in linear elasticity theory. We note that Korn's inequality is by no means trivial. Observe that, in contrast to the left-hand side in (57), the right-hand side in (57) contains only special products of the derivatives $D^i u_j$.

In the special case $\partial_1 G = \partial G$, we gave a simple proof of Korn's inequality (57) in Section 61.9b by using integration by parts and the inequality of Poincaré–Friedrichs. The general proof proceeds in several steps.

62.15a. The Structure of the Space X_0

Lemma 62.15. *Assume* (H1), (H2) *with* $\partial_1 G = \emptyset$. *Then* X_0 *consists of all infinitesimal rigid motions, i.e., we have* $u \in X_0$ *if and only if there exist vectors* a,

$b \in V_3$ such that
$$u(x) = a + b \times x \quad \text{on } G. \tag{60}$$

PROOF. See Problem 62.3. □

Recall that $X = W_2^1(G, V_3)$ if $\partial_1 G = \emptyset$, i.e.,
$$X_0 = \{u \in W_2^1(G, V_3): \gamma(u) = 0 \text{ on } G\}. \tag{61}$$
This lemma therefore states the very natural result that the linearized strain tensor vanishes identically precisely for all infinitesimal rigid motions.

Lemma 62.16. *Assume* (H1), (H2) *with* $\partial_1 G \neq \emptyset$. *Then* $X_0 = \{0\}$.

PROOF. See Problem 62.4. □

Recall that
$$X_0 = \{u \in W_2^1(G, V_3): \gamma(u) = 0 \text{ on } G \text{ and } u = 0 \text{ on } \partial_1 G\}. \tag{62}$$
This lemma states the following natural result. If $\partial_1 G$ is nonempty, then all infinitesimal rigid motions u with $u = 0$ on $\partial_1 G$ are trivial.

62.15b. Equivalent Norms on Sobolev Spaces

We now recall two important results on Sobolev spaces which were proved in Part II.

Proposition 62.17. *Assume* (H1), (H2) *with* $\partial_1 G \neq \emptyset$. *Then*
$$\left(\int_G D^i v\, D_j v\, dx + \int_{\partial_1 G} v^2\, dO \right)^{1/2}$$
and
$$\|v\|_{1,2} = \left(\int_G v^2 + D^j v\, D_j v\, dx \right)^{1/2}$$
are equivalent norms on the Sobolev space $W_2^1(G)$.

Let $f \in L_2(G)$. We set
$$F(v) = \int_G f v\, dx,$$
$$H(v) = -\int_G f D_j v\, dx.$$

Hölder's inequality implies that $F, H: \mathring{W}_2^1(G) \to \mathbb{R}$ are continuous, linear functionals, i.e., $F, H \in \mathring{W}_2^1(G)^*$. The corresponding norms of these functionals $\|F\|$ and $\|H\|$ are denoted by $\|f\|_{-1,2}$ and $\|D_j f\|_{-1,2}$, respectively, i.e., we obtain the so-called negative norms

$$\|f\|_{-1,2} = \sup_{v \in B} \left| \int_G fv \, dx \right|,$$

$$\|D_j f\|_{-1,2} = \sup_{v \in B} \left| \int_G f D_j v \, dx \right|,$$

with the closed unit ball B in $\mathring{W}_2^1(G)$, i.e.,

$$B = \{v \in \mathring{W}_2^1(G): \|v\|_{1,2} \leq 1\}.$$

Recall that the norm $\|v\|_2$ on $L_2(G)$, is given by

$$\|v\|_2 = \left(\int_G v^2 \, dx \right)^{1/2}.$$

Proposition 62.18. *Assume* (H1). *Then*

$$\|f\|_* = \|f\|_{-1,2} + \sum_{j=1}^{3} \|D_j f\|_{-1,2}$$

is an equivalent norm on $L_2(G)$, i.e., there are constants $C, D > 0$ such that

$$C\|f\|_2 \leq \|f\|_* \leq D\|f\|_2 \quad \text{for all} \quad f \in L_2(G).$$

PROOF. See Problem 21.9. □

62.15c. The Algebraic Key Relation

The following lemma shows that the second partial derivatives of the displacements can be expressed in terms of the first partial derivatives of the linearized strain tensor.

Lemma 62.19. *Set $\gamma^{st} = \gamma_t^s$. There exist real constants $A_{klm,rst}$ such that*

$$D_k D_l u_m = A_{klm,rst} D^r \gamma(u)^{st} \quad \text{for all} \quad u_m \in C_0^\infty(\mathbb{R}^3),$$

which satisfy the symmetry property $A_{kml,rst} = A_{klm,rts}$. This holds for all possible indices. The sum is taken over r, s, t from 1 to 3.

PROOF. Let U_m be the Fourier transform of u_m. We use the key property of Fourier transforms that differentiation is transformed into multiplication, i.e., $D_j u_m$ is transformed into $i\xi_j U_m$ and vice versa. Therefore, it is sufficient to prove

$$\xi_k \xi_l U_m = \tfrac{1}{2} A_{klm,rst} \xi_r (\xi_s U_t + \xi_t U_s).$$

This is a purely algebraic problem. We make the ansatz

$$A_{klm,rst}\xi_r = P_{st}(\xi_1, \xi_2, \xi_3)$$

with $P_{st} = P_{ts}$ for all t, s. Note that P_{st} depends also on k, l, m. For example, let $m = 1$. Hence it is sufficient to find homogeneous polynomials P_{st} of first degree which satisfy the relations

$$\xi_k\xi_l = P_{11}\xi_1 + P_{12}\xi_2 + P_{13}\xi_3,$$
$$0 = P_{12}\xi_1 + P_{22}\xi_2 + P_{23}\xi_3,$$
$$0 = P_{13}\xi_1 + P_{23}\xi_2 + P_{33}\xi_3.$$

First, we consider the special case $k = 1$, $l = 2$. In this case, our relations are valid if we set

$$P_{11} = \xi_2 \quad \text{and} \quad P_{ij} = 0 \quad \text{for all the other indices.}$$

For arbitrary k, l, m, we may use the same argument. □

62.15d. Coerciveness of Strains

Lemma 62.20. *Assume* (H1), (H2) *with* $\partial_1 G = \partial G$. *Then there exists a constant* $c > 0$ *such that*

$$\int_G [\gamma(u), \gamma(u)] + u^2\, dx \geq c\|u\|_X^2 \quad \text{for all} \quad u \in X.$$

Observe that, in this case, $X = W_2^1(G, V_3)$. In Cartesian coordinates the lemma states that

$$\int_G \gamma(u)_i^j \gamma(u)_j^i + u^i u_i\, dx \geq c \int_G D^i u_j D_i u^j + u^i u_i\, dx$$

for all $u_i \in W_2^1(G)$ and all i. The key to our proof will be Proposition 62.18.

PROOF. Since $C_0^\infty(\mathbb{R}^3)$ is dense in $W_2^1(G)$, it suffices to show that the assertion is valid for all $u_i \in C_0^\infty(\mathbb{R}^3)$. In the following, all positive constants are denoted by C.

(I) From Proposition 62.18 follows the key relation

$$\|D_l u_m\|_2 \leq C\left(\|D_l u_m\|_{-1,2} + \sum_k \|D_k D_l u_m\|_{-1,2}\right).$$

(II) Lemma 62.19 implies that

$$\|D_k D_l u_m\|_{-1,2} \leq C \sum_{r,s,t} \|D_r \gamma_t^s\|_{-1,2},$$

where $\gamma = \gamma(u)$. From Proposition 62.18 follows

$$\|D_r \gamma_t^s\|_{-1,2} \leq C\|\gamma_t^s\|_2$$

and hence

$$\|D_k D_l u_m\|_{-1,2} \leq C \sum_{s,t} \|\gamma_t^s\|_2,$$

$$\|D_l u_m\|_{-1,2} \leq C \|u_m\|_2.$$

Inequality (I) yields

$$\|D_l u_m\|_2 \leq C \left(\sum_{s,t} \|\gamma_t^s\|_2 + \|u_m\|_2 \right),$$

and this proves our assertion. □

62.15e. Proof of Theorem 62.F

In Section 21.2, we introduced a general proof strategy. In order to obtain equivalent norms (inequalities) we used compact embeddings. Here we want to apply this strategy. The key is the following well-known embedding theorem of Rellich:

$$\text{The embedding } W_2^1(G) \subseteq L_2(G) \text{ is compact.} \tag{63}$$

Suppose that Theorem 62.F is not true. Then there exists a sequence (u_n) in $X_0^\perp$ such that $\|u_n\|_X = 1$ and

$$\int_G [\gamma(u_n), \gamma(u_n)] \, dx < \frac{1}{n} \quad \text{for all } n. \tag{64}$$

Using (63), we find a subsequence, again denoted by (u_n), such that

$$u_n \to u \quad \text{in } L_2(G, V_3) \quad \text{as } n \to \infty.$$

From the coerciveness of strains (Lemma 62.20) and (64), it follows that (u_n) is a Cauchy sequence in the H-space X. Hence we obtain the stronger convergence

$$u_n \to u \quad \text{in } X \quad \text{as } n \to \infty. \tag{65}$$

Relation (64) yields $\gamma(u) = 0$ on G, i.e.,

$$u \in X_0.$$

From (65) and $u_n \in X_0^\perp$ for all n, we obtain

$$u \in X_0^\perp.$$

Since $X = X_0 \oplus X_0^\perp$, we get $u = 0$. At the same time, it follows from (65) and $\|u_n\|_X = 1$ for all n that $\|u\|_X = 1$. This contradicts $u = 0$.

The proof of Theorem 62.F is complete.

Corollary 62.13 follows from Theorem 62.F and Lemma 62.16. Corollary 62.14 is an immediate consequence of Proposition 62.17. Observe that $u = 0$ on $\partial_1 G$ for $u \in X$.

62.16. Legendre Transformation and the Strategy of the General Friedrichs Duality in the Calculus of Variations

In 1929, Friedrichs discovered a general method for obtaining dual problems in the calculus of variations. Special cases of this so-called Friedrichs duality are:

(i) the Trefftz duality (1927) for the Dirichlet problem (Section 62.17); and
(ii) the duality in elasticity theory between displacement and stress (Section 62.18).

In order to explain the basic idea of this method, we consider a simple, but typical example. In the following all functions are assumed to be sufficiently smooth.

62.16a. The Original Variational Problem

We consider the variational problem

(P) $\qquad F(u) \stackrel{\text{def}}{=} \int_G M(Du) - Ku \, dx - \int_{\partial_2 G} Tu \, dO = \text{stationary!},$

$$u = u_0 \quad \text{on } \partial_1 G.$$

Here, G is a bounded region in $\mathbb{R}^N$ with $\partial G \in C^{0,1}$ and the boundary decompositon

$$G = \overline{\partial_1 G} \cup \overline{\partial_2 G}, \qquad \partial_1 G \cap \partial_2 G = \emptyset,$$

where $\partial_1 G$ and $\partial_2 G$ are open subsets of ∂G. We are looking for a solution

$$u \colon \bar{G} \to \mathbb{R}$$

of (P). We set

$$Du = (D_1 u, \ldots, D_N u).$$

Moreover, for brevity, we write

$$\sigma n = \sum_{i=1}^N \sigma_i n_i, \qquad \text{div } \sigma = \sum_{i=1}^N D_i \sigma_i,$$

where $\sigma = (\sigma_1, \ldots, \sigma_N)$, $n = (n_1, \ldots, n_N)$, and

$$DM' = \sum_{i=1}^N D_i M_{D_i u}, \qquad nM' = \sum_{i=1}^N n_i M_{D_i u},$$

where $M_{D_i u} = \partial M / \partial D_i u$.

62.16. Legendre Transformation, Strategy of the General Friedrichs Duality

Lemma 62.21. *If u is a solution of (P), then u satisfies the Euler equation*

$$DM'(Du) + K = 0 \quad \text{on } G$$
$$u = u_0 \quad \text{on } \partial_1 G, \qquad nM' = T \quad \text{on } \partial_2 G, \tag{66}$$

where n is the outer unit normal vector at boundary points of G.

PROOF. We use the standard technique of the calculus of variations described in Section 18.3. To this end we set

$$\varphi(t) = F(u + t\,\delta u), \qquad t \in \mathbb{R},$$

where the function δu satisfies the boundary condition $\delta u = 0$ on $\partial_1 G$. Letting $\delta F = \varphi'(0)$, we obtain

$$\delta F = \int_G M'D\,\delta u - K\,\delta u\,dx - \int_{\partial_2 G} T\,\delta u\,dO.$$

If u is a solution of (P), then $\delta F = 0$. Integration by parts yields

$$\delta F = -\int_G (DM' + K)\,\delta u\,dx + \int_{\partial_2 G} (nM' - T)\,\delta u\,dO = 0$$

for all δu, and hence we obtain (66). □

62.16b. Additional Side Condition

The first idea of Friedrichs was to replace (P) with the equivalent problem

$$\int_G M(\gamma) - Ku\,dx - \int_{\partial_2 G} Tu\,dO = \text{stationary!},$$
$$u = u_0 \quad \text{on } \partial_1 G, \tag{67}$$
$$\gamma = Du \quad \text{on } G.$$

Here, we are looking for u and γ.

62.16c. The Free Problem

The second idea of Friedrichs was to set

$$\bar{F}(u, \gamma, \sigma, q) = \int_G M(\gamma) - Ku - \sigma(\gamma - Du)\,dx$$
$$- \int_{\partial_1 G} q(u - u_0)\,dO - \int_{\partial_2 G} Tu\,dO$$

and to consider the new problem

$$\bar{F}(u, \gamma, \sigma, q) = \text{stationary!}. \tag{68}$$

Here, the variation is taken over all u, γ, σ, q. The important point is that this is a free problem, i.e., no side conditions appear. We obtain $\bar{F}$ from the problem (67) with side conditions, by using our general strategy from Part III: Reduce problems with side conditions to free problems by adding terms which contain the side conditions and Lagrange multipliers. In (68), the role of the Lagrange multipliers is played by the functions

$$\sigma: \bar{G} \to \mathbb{R}^N \quad \text{and} \quad q: \partial_1 G \to \mathbb{R}.$$

Lemma 62.22. *If (u, γ, σ, q) is a solution of (68), then it satisfies the Euler equation*

$$\sigma = M'(\gamma) \quad \text{(Legendre transformation)},$$
$$\operatorname{div} \sigma + K = 0 \quad \text{on } G \quad \text{(equilibrium condition)},$$
$$\gamma = Du \quad \text{on } G, \tag{69}$$
$$\sigma n = q \quad \text{and} \quad u = u_0 \quad \text{on } \partial_1 G,$$
$$\sigma n = T \quad \text{on } \partial_2 G \quad \text{(equilibrium condition)}.$$

In terms of the components, the Legendre transformation becomes

$$\sigma_i = \frac{\partial M(\gamma)}{\partial \gamma_i}, \quad i = 1, \ldots, N.$$

PROOF. Similarly, as in the proof of Lemma 62.21, we obtain

$$\delta \bar{F} = \int_G M' \, \delta \gamma - K \, \delta u - \delta \sigma (\gamma - Du) - \sigma (\delta \gamma - D \, \delta u) \, dx$$
$$- \int_{\partial_1 G} \delta q (u - u_0) + q \, \delta u \, dO - \int_{\partial_2 G} T \, \delta u \, dO,$$

where $\delta u = 0$ on $\partial_1 G$. Integration by parts yields the *key* relation

$$\delta \bar{F} = \int_G (M' - \sigma) \, \delta \gamma - (K + \operatorname{div} \sigma) \, \delta u + (Du - \gamma) \, \delta \sigma \, dx$$
$$+ \int_{\partial_1 G} (\sigma n - q) \, \delta u - (u - u_0) \, \delta q \, dO + \int_{\partial_2 G} (\sigma n - T) \, \delta u \, dO. \tag{70}$$

From $\delta \bar{F} = 0$ we obtain (69). □

62.16d. The Dual Problem

We make the following crucial assumption:

(H) The Legendre transformation

$$\sigma = M'(\gamma)$$

62.16. Legendre Transformation, Strategy of the General Friedrichs Duality

can be solved for γ, i.e., the inverse transformation

$$\gamma = \dot{M}'^{-1}(\sigma)$$

exists.

In order to motivate the following definition of the function H, we consider the original problem for the special case $N = 1$. This problem corresponds to a problem in point mechanics. There, we have

- M Lagrangian function;[1]
- u position;
- γ velocity ($\gamma = Du$);
- σ momentum.

Similarly, as in Section 58.20, we define the Hamiltonian function

$$H(\sigma) = \sigma\gamma - M(\gamma), \qquad \gamma = M'^{-1}(\sigma),$$

where $\sigma\gamma = \sum_{i=1}^{N} \sigma_i \gamma_i$. Moreover, we set

$$\Phi(\sigma) = -\int_G H(\sigma)\,dx + \int_{\partial_1 G} (\sigma n) u_0 \, dO$$

and consider the *dual problem*

(P*) $\qquad\qquad \Phi(\sigma) = $ stationary!,

$$\operatorname{div} \sigma + K = 0 \quad \text{on } G,$$

$$\sigma n = T \quad \text{on } \partial_2 G.$$

Theorem 62.G (The Friedrichs Duality (1929)). *Suppose that we have a sufficiently smooth situation and that* (H) *holds true.*

If u is a solution of the original problem (P), *then $\sigma = M'(Du)$ is a solution of the dual problem* (P*), *and the corresponding extremal values are identical, i.e., $F(u) = \Phi(\sigma)$.*

PROOF. The trick is to compute $\bar{F}(u, \gamma, \sigma, q)$ for the special choice

$$\gamma = M'^{-1}(\sigma), \qquad q = \sigma n$$

and

$$\operatorname{div} \sigma + K = 0 \quad \text{on } G, \qquad \sigma n = T \quad \text{on } \partial_2 G.$$

---

[1] More precisely, the Lagrangian function is equal to $L = M(\gamma) - Ku$, but this is not important for the following.

This is motivated by (69). Integration by parts yields the *key* relation

$$\bar{F} = \int_G M(\gamma) - (K + \operatorname{div} \sigma)u - \sigma\gamma\, dx$$

$$+ \int_{\partial G} (\sigma n)u\, dO - \int_{\partial_1 G} \sigma n(u - u_0)\, dO - \int_{\partial_2 G} Tu\, dO$$

$$= \int_G M(\gamma) - \sigma\gamma\, dx + \int_{\partial_1 G} (\sigma n) u_0\, dO = \Phi(\sigma).$$

Now let $\gamma = Du$. According to (70), we then obtain

$$\delta\Phi = \delta\bar{F} = 0,$$

i.e., σ is a solution of (P*). Moreover, we have

$$F(u) = \bar{F}(u, Du, \sigma, \sigma n) = \Phi(\sigma). \qquad \square$$

62.17. Application to the Dirichlet Problem (Trefftz Duality)

We consider the Dirichlet problem

(P) $\quad \dfrac{1}{2}\displaystyle\int_G |Du|^2 - Ku\, dx - \int_{\partial_2 G} Tu\, dO = \text{stationary!},$

$$u = u_0 \quad \text{on } \partial_1 G.$$

This corresponds to the original problem in Section 62.16 with

$$M(Du) = \tfrac{1}{2}|Du|^2.$$

The Euler equation to (P) becomes

$$\Delta u + K = 0 \quad \text{on } G,$$

$$u = u_0 \quad \text{on } \partial_1 G, \qquad \frac{\partial u}{\partial n} = T \quad \text{on } \partial_2 G.$$

The Legendre transformation has the simple form

$$\sigma = M'(\gamma) = \gamma$$

with $\sigma = (\sigma_1, \ldots, \sigma_N)$. The Hamiltonian function is

$$H(\sigma) = |\sigma|^2 - \tfrac{1}{2}|\sigma|^2 = \tfrac{1}{2}|\sigma|^2,$$

i.e., $H(\sigma) = M(\sigma)$. Thus, the dual problem in Section 62.16 becomes

(P*) $\quad -\dfrac{1}{2}\displaystyle\int_G |\sigma|^2\, dx + \int_{\partial_1 G} (\sigma n)u_0\, dO = \text{stationary!},$

$$\operatorname{div} \sigma + K = 0 \quad \text{on } G, \qquad \sigma n = T \quad \text{on } \partial_2 G.$$

Theorem 62.G tells us that if u is a solution of the Dirichlet problem (P), then
$$\sigma = Du$$
is a solution of the dual problem (P*). In this case,
$$\operatorname{div} \sigma + K = 0 \quad \text{means} \quad \Delta u + K = 0.$$
This is the duality of Trefftz (1927), which came up in Section 51.6 in a more general functional-analytic context.

62.18. Application to Elasticity

If we apply the general strategy of Section 62.16 to the principle of stationary potential energy

(P) $$\int_G M(Du) - Ku \, dx - \int_{\partial_2 G} Tu \, dO = \text{stationary!},$$

$$u = u_0 \quad \text{on } \partial_1 G,$$

then we obtain precisely the duality between the displacement u and the stress tensor σ of Section 62.4.

In this case, $\gamma = Du$ is the linearized strain tensor, i.e.,
$$Du = \tfrac{1}{2}(u'(x) + u'(x)^*).$$

As in Section 62.16, we obtain the Legendre transformation
$$\sigma = M'(\gamma),$$
which is nothing other than the constitutive law. The Hamiltonian function becomes
$$H(\sigma) = [\sigma, \gamma] - M(\gamma), \qquad \gamma = M'^{-1}(\sigma),$$
and the dual problem becomes

(P*) $$-\int_G H(\sigma) \, dx + \int_{\partial_1 G} (\sigma n) u_0 \, dO = \text{stationary!},$$

$$\operatorname{div} \sigma + K = 0 \quad \text{on } G, \qquad \sigma n = T \quad \text{on } \partial_2 G.$$

If M satisfies the assumptions made in Section 62.3, then
$$H(\sigma) = M^*(\sigma),$$
according to Problem 62.9, i.e., the Hamiltonian function is the conjugate stored energy function.

PROBLEMS

62.1. *An inequality.* Set $Z = W_2^1(G, V_3)$. Show that
$$\int_G [\gamma(u), \gamma(u)] \, dx \leq \text{const} \, \|u\|_Z^2 \qquad \text{for all} \quad u \in Z.$$

Solution: In Cartesian coordinates we have
$$\gamma_j^i = \tfrac{1}{2}(D^i u_j + D^j u_i).$$

Hölder's inequality yields
$$\int \gamma_j^i \gamma_i^j \, dx \leq \text{const} \sum_{i,j,k,m} \int |D^i u_j| |D^k u_m| \, dx$$
$$\leq \text{const} \sum_{i,j,k,m} \left(\int |D^i u_j|^2 \, dx\right)^{1/2} \left(\int |D^k u_m|^2 \, dx\right)^{1/2}$$
$$\leq \text{const} \int \sum_{r,s} |D^r u_s|^2 + |u_s|^2 \, dx = \text{const} \, \|u\|_Z^2.$$

62.2. Proof of Lemma 62.3.
Solution: By Hölder's inequality
$$\left|\int_G K u \, dx\right|^2 \leq \left(\int |K||u| \, dx\right)^2 \leq \int K^2 \, dx \int u^2 \, dx$$
$$\leq \left(\int_G K^2 \, dx\right) \|u\|_X^2.$$

Moreover, Hölder's inequality and the continuous embedding $W_2^1(G) \subseteq L_2(\partial G)$ imply
$$\left|\int_{\partial_2 G} T u \, dO\right|^2 \leq \int_{\partial_2 G} T^2 \, dO \int_{\partial G} u^2 \, dO \leq c \left(\int_{\partial_2 G} T^2 \, dO\right) \|u\|_X^2.$$

62.3. Proof of Lemma 62.15.
Solution: Let $M(G)$ be the set of all $u \in W_2^1(G, V_3)$ which have the representation
$$u(x) = a + b \times x \quad \text{on } G$$
with $a, b \in V_3$, i.e., $M(G)$ is the set of all infinitesimal rigid motions on G.

(I) Smooth case. Let $u \in C^\infty(\bar{G}, V_3)$ with
$$\gamma(u) = 0 \quad \text{on } G.$$

According to Lemma 62.19, the second-order partial derivatives of all the components u_i vanish on G. Hence the u_i's are polynomials of first degree, i.e.,
$$u_i = a_i + \sum_{j=1}^{3} b_{ij} \xi_j, \quad i = 1, 2, 3.$$

Moreover, $\gamma_j^i = \tfrac{1}{2}(D^i u_j + D^j u_i) = 0$ implies that
$$b_{ij} = -b_{ji} \quad \text{for all } i, j.$$

Hence $u \in M(G)$.

(II) General case. Let $u \in W_2^1(G, V_3)$ with $\gamma(u) = 0$ on G. Choose a proper subregion $H \subset\subset G$. Let S_h be the smoothing operator from Section 18.14, and set
$$u_h = S_h u \quad \text{for all small } h > 0.$$

Then $u_h \in C^\infty(\bar{H}, V_3)$, and $\gamma(u) = 0$ on G implies

$$\gamma(u_h) = S_h\gamma(u) = 0 \quad \text{on } H \quad \text{for all small} \quad h > 0.$$

From (I) follows that $u_h \in M(H)$ for all small $h > 0$. Moreover, the properties of S_h yield

$$u_h \to u \quad \text{in } W_2^1(H, V_3) \quad \text{as} \quad h \to 0.$$

Since $M(H)$ is a *finite-dimensional* subspace of $W_2^1(H, V_3)$, it follows that $M(H)$ is closed and hence that $u \in M(H)$.

Because of the arbitrariness of H, we find that $u \in M(G)$.

62.4. *Proof of Lemma 62.16.*
Solution: Suppose that

$$a + b \times x = 0 \quad \text{on } \partial_1 G$$

with $\partial_1 G \neq \emptyset$. We have to show that $a = b = 0$.

If $b = 0$, then $a = 0$. So let $b \neq 0$. The coefficient matrix of the linear system

$$\begin{pmatrix} 0 & -b_3 & b_2 \\ b_3 & 0 & -b_1 \\ -b_2 & b_1 & 0 \end{pmatrix}$$

has rank $= 2$. Hence $\partial_1 G$ is contained in a line. But this contradicts the fact that $\partial_1 G$ is open in ∂G.

62.5. *Gradient method.* Consider the original variational problem, which corresponds to the principle of minimal potential energy, and prove the convergence of the gradient method for this problem.

Hint: Use Section 62.6d.

62.6. *Existence proof for the second boundary-value problem and the equilibrium condition.* Consider the original variational problem (5), namely,

(P) $\qquad \int_G M(\gamma(u))\,dx - \int_G Ku\,dx - \int_{\partial G} Tu\,dO = \min!,$

with assumption (H1) to (H6), as formulated in Section 62.3. Let $\partial_2 G = \partial G$, i.e., $\partial_1 G = \emptyset$. Then the boundary condition is

$$\sigma n = T \quad \text{on } \partial G.$$

In contrast to the first and mixed boundary-value problem, the boundary displacements are here completely unknown. We only know the boundary forces. We are looking for the displacements u of the elastic body.

Show: There exists a solution u if and only if the following two equilibrium conditions are satisfied:

$$\int_G K\,dx + \int_{\partial G} T\,dO = 0,$$

$$\int_G K \times x\,dx + \int_{\partial G} T \times x\,dO = 0,$$
(71)

i.e., the total outer force and the total outer torque vanish (in the sense of (T_{approx})

of Section 61.6c). Moreover, the solution u is uniquely determined up to infinitesimal rigid motions.

Solution: As in Section 62.5 we write the variational problem (P) in the form

$$F(u) - b(u) = \min!, \quad u \in X \tag{72}$$

with the Euler equation

$$\langle F'(u), v \rangle = b(v) \quad \text{for all} \quad v \in X. \tag{73}$$

The trick of the proof is to solve first the *modified* problem

$$F(u) - b(u) = \min!, \quad u \in X_0^\perp. \tag{74}$$

The space X_0 has been introduced in Section 62.15. According to Theorem 62.F (Korn's inequality), the energetic norm

$$\|u\|_E = \left(\int_G [D(u), D(u)] \, dx \right)^{1/2}, \quad Du \equiv \gamma(u)$$

is an equivalent norm on the closed subspace $X_0^\perp$ of $X = W_2^1(G, V_3)$. Analogously to Section 62.5a, we thus obtain that (74) has a unique solution u and

$$\langle F'(u), v \rangle = b(v) \quad \text{for all} \quad v \in X_0^\perp. \tag{75}$$

We want to show that u is also a solution of the original problem (72). Because of the strong stability condition (6) we obtain

$$\delta^2 F(u; v) = \int_G M''(Du)(Dv)^2 \, dx \geq 0 \quad \text{for all} \quad v \in X.$$

Hence $F: X \to \mathbb{R}$ is convex, according to Section 42.3, and problems (72) and (73), as well as problems (74) and (75), are equivalent. Important is then the fact that

$$F(u + w) = \int_G M(Du + Dw) \, dx = F(u) \quad \text{for all} \quad w \in X_0,$$

since $Dw = 0$ on X_0. This implies

$$\langle F'(u), w \rangle = 0 \quad \text{for all} \quad w \in X_0.$$

Since $X = X_0 \oplus X_0^\perp$ it follows that u is a solution of (73), and hence of (72), if and only if

$$b(w) = 0 \quad \text{for all} \quad w \in X_0. \tag{76}$$

Lemma 62.15 shows that X_0 consists precisely of all infinitesimal rigid motions

$$w(x) = a + b \times x \quad \text{on } G$$

with arbitrary vectors $a, b \in V_3$. Therefore, noting the definition of b in (10), equation (76) is equivalent to

$$a \left(\int_G K \, dx + \int_{\partial G} T \, dO \right) + b \left(\int_G x \times K \, dx + \int_{\partial G} x \times T \, dO \right) = 0$$

for all $a, b \in V_3$. This is the equilibrium condition (71).

Problems

62.7. *Existence and uniqueness of solutions for the mixed boundary-value problem in linear elasticity theory (main theorem of linear elastostatics).* Give a direct proof of Theorem 62.A in the special case of linear elasticity theory by using the main theorem about quadratic variational problems (Theorem 22.A). In this way we obtain an immediate generalization of Theorem 61.D of Section 61.9c (first boundary-value problem).

Solution: Set

$$\mathscr{L}(\gamma, \mu) = \lambda \operatorname{tr} \gamma \operatorname{tr} \mu + 2\kappa[\gamma, \mu]$$

for all $\gamma, \mu \in L_{\text{sym}}(V_3)$. The principle of minimal potential energy (5) then becomes the quadratic variational problem

$$\tfrac{1}{2}a(u, u) - b(u) = \min!, \qquad u - u_0 \in X, \tag{77}$$

where

$$a(u, v) = \int_G \mathscr{L}(\gamma(u), \gamma(v))\, dx,$$

$$b(u) = \int_G K u \, dx + \int_{\partial_2 G} T u \, dO.$$

If we set $v = u - u_0$, then we obtain

$$\tfrac{1}{2}a(v, v) - f(v) = \min!, \qquad v \in X \tag{78}$$

with $f(v) = b(v) - a(v, u_0)$. Obviously, we have

$$\mathscr{L}(\gamma, \gamma) \geq 2\kappa[\gamma, \gamma].$$

The *key* to the proof is Korn's inequality (57), i.e.,

$$a(v, v) = \int_G \mathscr{L}(\gamma(v), \gamma(v))\, dx \geq 2\kappa \int_G [\gamma(v), \gamma(v)]\, dx$$

$$\geq 2\kappa c_1 \|v\|_X^2 \qquad \text{for all} \quad v \in X.$$

Hence the bilinear, symmetric functional $a: X \times X \to \mathbb{R}$ is *strongly positive*. According to Theorem 22.A, equation (78) has a unique solution.

62.8. *Special closed sets in Sobolev spaces.* The following two simple results will be used frequently. Let G be a bounded region in $\mathbb{R}^N$ with $N \geq 1$ and choose a fixed number $c > 0$.

62.8a. Show that the set

$$S = \{u \in L_2(G): |u| \leq c \text{ a.e. on } G\}$$

is closed, bounded, and convex in the Lebesgue space $L_2(G)$.

As usual, "a.e." stands for "almost everywhere". Each function in $L_2(G)$ can be changed on a set of measure zero. In this sense we can write

$$S = \{u \in L_2(G): |u| \leq c \text{ on } G\}.$$

For the sake of brevity, we will use this notation in the future.

Solution: Obviously, S is convex and bounded.

We show that S is closed. Let (u_n) be a sequence in S with $u_n \to u$ in $L_2(G)$ as

$n \to \infty$. From $A_2(36)$, there exists a subsequence such that $u_{n'}(x) \to u(x)$ a.e. on G as $n \to \infty$. Hence $u \in S$.

62.8b. Assume $\partial G \in C^{0,1}$ and suppose that there exists a decomposition of the boundary

$$\partial G = \overline{\partial_1 G} \cup \overline{\partial_2 G}, \qquad \partial_1 G \cap \partial_2 G = \emptyset,$$

where $\partial_1 G$ and $\partial_2 G$ are open subsets of ∂G, and $\partial_1 G$ is nonempty. Choose a fixed function $u_0 \in W_2^1(G)$ and let T denote the set of all functions $u \in W_2^1(G)$ with

$$|Du| \leq c \quad \text{on } G \qquad \text{and} \qquad u = u_0 \quad \text{on } \partial_1 G,$$

where

$$|Du| = \left(\sum_{i=1}^{N} |Du_i|^2 \right)^{1/2}.$$

Show that T is a closed, bounded, and convex set in the Sobolev space $W_2^1(G)$.

Solution: The set T is closed according to Problem 62.8a. Obviously, T is convex.

On the space $W_2^1(G)$, an equivalent norm is given by

$$\left(\int_G |Du|^2 \, dx + \int_{\partial_1 G} |u|^2 \, dO \right)^{1/2}.$$

Thus, T is bounded. Note that $u = u_0$ on $\partial_1 G$ and $u_0 \in L_2(\partial G)$.

62.9. *The dual stored energy function M^*.* Set $Z = L_{\text{sym}}(V_3)$ and consider a stored energy function $M: Z \to \mathbb{R}$ as in Section 62.3. Show that

$$M^*(\sigma) = [\sigma, \gamma] - M(\gamma) \qquad \text{for all} \quad \sigma \in Z,$$

where $\gamma = M'^{-1}(\sigma)$.

Solution: The space Z is an H-space with scalar product $[\sigma, \mu]$. By Definition 51.1,

$$M^*(\sigma) = \sup_{\gamma \in Z} F(\gamma),$$

where $F(\gamma) = [\sigma, \gamma] - M(\gamma)$ for all $\gamma \in Z$. The strong stability condition (6) implies $F(\gamma) \to -\infty$ as $|\gamma| \to \infty$. Hence the concave function F has a maximal point. This implies

$$M^*(\sigma) = F(\gamma), \qquad F'(\gamma) = 0.$$

The assertion now follows from

$$F'(\gamma)\mu = [\sigma, \mu] - M'(\gamma)\mu \qquad \text{for all} \quad \mu \in Z,$$

i.e., $F'(\gamma) = 0$ if and only if $\sigma = M'(\gamma)$.

References to the Literature

Classical works: Korn (1907) and Lichtenstein (1924) (existence proofs in linear elasticity theory), Ritz (1908) and Trefftz (1927) (approximation methods), Friedrichs (1929) (duality).

Dual variational principles in linear elasticity: Gurtin (1972, S), Washizu (1968, M).
Introduction to the existence theory in elasticity theory: Nečas and Hlaváček (1981, M), Ciarlet (1983, L), Marsden and Hughes (1983, M).

(A) *Existence proofs in linear elasticity theory*
Differential equations: Fichera (1972, S) (handbook article), Duvaut and Lions (1972, M), Nečas and Hlaváček (1981, M).
Semigroups and elastodynamics: Marsden and Hughes (1983, M).
Potential theory and singular integral equations: Kupradze (1976, M).
Plane elasticity, complex function theory, and singular integral equations: Muskelišvili (1954, M) (classical monograph), Babuška (1960, M).
Piecewise homogeneous bodies and singular integral equations: Jentsch (1977, S) (three-dimensional problems), Maul (1976, S) (two-dimensional problems).
Korn's inequality: Duvaut and Lions (1972, M), Nečas and Hlaváček (1981, M).
Linear strongly elliptic systems: Browder (1954), Nirenberg (1955), Agmon, Douglis, and Nirenberg (1959), Part II, Morrey (1966, M).

(B) *Existence proofs in nonlinear elasticity theory*
Implicit function theorem: Stoppeli (1954), Ciarlet and Rabier (1980, L), Ciarlet (1983, L). Marsden and Hughes (1983, M).
Continuation method: Beckert (1975), (1977), (1982), (1984, S), Beyer (1979), Benkert (1987) (structure of strongly stable solutions).
Polyconvex material and compensated compactness: Morrey (1952) (quasi-convexity and lower semicontinuity of multiple integrals), Ball (1977) (fundamental paper), Ciarlet (1983, L), Nečas (1983, L), Evans (1986) (partial regularity).
Method of compensated compactness: Tartar (1979), (1983), Dacarogna (1982, L) (recommended as an introduction), Murat (1981) and Feistauer, Mandel, and Nečas (1984) (embedding theorem of Murat), Murat (1987, S).
Semigroups and elastodynamics: Hughes, Kato, and Marsden (1977), Marsden and Hughes (1983, M), Ebin (1986).
Variational methods for approximation models: Beju (1972), Oden and Reddy (1976, M), Langenbach (1976, M) (monotone potential operators).
Global analysis and nonlinear elasticity: Marsden and Hughes (1983, M).
Symmetry and bifurcation: Chillingworth, Marsden, and Wan (1983).
Uniqueness in nonlinear elasticity and approximation methods: John (1972), (1985) (collected works).
Propagation of waves and the life-span of solutions of the instationary equations of nonlinear elasticity: John (1977), (1983).

(C) *Approximation methods*
Ciarlet (1977, M) (finite elements), Oden and Reddy (1976, M), Oden (1980, M), Nečas and Hlaváček (1981, M).
The boundary integral method: Cf. the References to the Literature for Chapter 22.

(See also the References to the Literature for Chapters 63–66)

CHAPTER 63

Variational Inequalities and the Signorini Problem for Nonlinear Material

Signorini (1959) posed the problem, now bearing his name, that in simplest terms is to determine the displacements in a heavy, linearly elastic body resting on a rigid, frictionless horizontal plane. The essential difficulty of this problem is that the region of contact between the body and the plane is not known a priori. It is conceivable that the contact set could be especially complicated.

Fichera (1964) was the first to study the existence and uniqueness of this problem which is nonlinear because position fields, satisfying the governing equations, are subjected to a unilateral constraint that restricts their values lying in a half space.

Stuart Antman (1983)

With regard to the general nonlinear model of Section 62.3 we now consider boundary-value problems, for which the elastic body is supported on parts of its boundary. The boundary conditions thereby have the form of inequalities. In functional-analytic terms, this leads to convex variational problems on convex sets. The corresponding Euler equations are variational inequalities. We shall use Theorem 46.A of Part III in order to obtain a general existence and uniqueness theorem. Throughout, the same notation as in the previous chapter will be employed.

63.1. Existence and Uniqueness Theorem

Similarly, as in Section 62.3, we consider the variational problem

$$\int_G M(\gamma(u))\,dx - \int_G Ku\,dx - \int_{\partial_2 G} Tu\,dO = \min!,$$
$$u = 0 \quad \text{on } \partial_1 G,$$
$$un \le 0 \quad \text{on } \partial_3 G,$$

(1)

63.1. Existence and Uniqueness Theorem

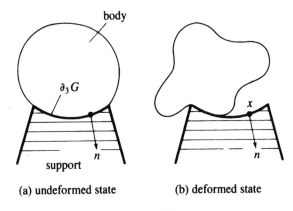

Figure 63.1

where n denotes the outer unit normal vector at the points of the boundary part $\partial_3 G$.

The boundary inequality "$un \leq 0$ on $\partial_3 G$" means that we are in the situation of Figure 63.1. The undeformed elastic body rests on a rigid support which comes from the boundary part $\partial_3 G$. The possible deformations of the body are restricted by this support, i.e., for points x of $\partial_3 G$ only those displacements u are possible which have an obtuse angle with n.

More precisely, we can write (1) in the form

$$F(u) - b(u) = \min!, \quad u \in C \tag{2}$$

with

$$C = \{u \in X : un \leq 0 \text{ on } \partial_3 G\}.$$

The H-space $X = \mathring{W}_2^1(G, \partial_1 G; V_3)$ has been defined in Section 62.2. Analogously to Section 62.3, our assumptions are the following:

(H1) G is a bounded region in $\mathbb{R}^3$ with sufficiently smooth boundary, i.e., $\partial G \in C^{0,1}$.
(H2) The boundary of G admits the decomposition

$$\partial G = \overline{\partial_1 G} \cup \overline{\partial_2 G} \cup \overline{\partial_3 G}$$

with pairwise disjoint, open subsets $\partial_j G$ of ∂G, $j = 1, 2, 3$. Moreover, suppose that $\partial_1 G$ and $\partial_3 G$ are nonempty, while $\partial_2 G$ may be empty.
(H3) The stored energy function M satisfies assumption (H3) of Section 62.3.
(H4) On G the density of the outer forces $K \in L_2(G, V_3)$ is prescribed, and on $\partial_2 G$ the density of the outer boundary stress forces $T \in L_2(\partial_2 G, V_3)$ is given.

In connection with our problem (2) we consider the variational inequality

$$\langle F'(u) - b, v - u \rangle \geq 0 \quad \text{for all} \quad v \in C. \tag{3}$$

We are looking for a solution $u \in C$. From Section 62.5 it follows that (3) has the explicit form

$$\int_G [\sigma, \gamma(v) - \gamma(u)] \, dx - \int_G K(v-u) \, dx - \int_{\partial_2 G} T(v-u) \, dO \geq 0, \quad (4)$$

where $\gamma(u)$ is the linearlized strain tensor which corresponds to the displacement u (see (62.5)), and

$$\sigma = M'(\gamma(u))$$

is the stress tensor (first Piola–Kirchhoff tensor).

Theorem 63.A. *The variational problem (2) has a unique solution, which coincides with the unique solution of the variational inequality (3).*

PROOF. We want to apply Theorem 46.A.

The set C is closed and convex in X. This is a consequence of Problem 62.8b.

From Section 62.5a it follows that b is a linear, continuous functional on X, and the operator $F': X \to X^*$ is strongly monotone and Lipschitz continuous. Consequently, the functional $F: X \to \mathbb{R}$ is strictly convex, continuous, and coercive, according to Section 42.3.

Theorem 46.A then yields the assertion. $\square$

63.2. Physical Motivation

Let u be a sufficiently smooth solution of problem (3). After integration by parts, the variational inequality (4) becomes

$$-\int_G (\operatorname{div} \sigma + K)(v-u) \, dx + \int_{\partial_2 G} (\sigma n - T)(v-u) \, dO$$

$$+ \int_{\partial_3 G} (\sigma n)(v-u) \, dO \geq 0 \quad (5)$$

for all $v \in C$.

We specialize v. First we choose $v \in C_0^\infty(G, V_3)$, i.e., all components of v, in a fixed Cartesian coordinate system, belong to $C_0^\infty(G)$. Then we have $v \in C$ and hence (5) holds for all $v \in C_0^\infty(G, V_3)$. This yields the equilibrium condition

$$\operatorname{div} \sigma + K = 0 \quad \text{on } G. \quad (6)$$

Next, we choose $v \in C^\infty(\bar{G}, V_3)$ with $v = 0$ on $\partial_1 G$, $v =$ arbitrary on $\partial_2 G$, and $v = u$ on $\partial_3 G$. Thus we have again that $v \in C$ and (5) implies

$$\sigma n = T \quad \text{on } \partial_2 G. \quad (7)$$

Therefore (5) is reduced to

$$\int_{\partial_3 G} (\sigma n)(v-u) \, dO \geq 0 \quad \text{for all} \quad v \in C. \quad (8)$$

Let us now analyze this relation. We set $T = \sigma n$ on $\partial_3 G$. Then T is the density of the outer boundary forces on $\partial_3 G$. Let t be a tangent vector at a fixed point x of $\partial_3 G$ and let n denote the outer unit normal vector at x. We decompose v at x as

$$v = v_t t + v_n n.$$

If we choose v with $v_n = u_n$ and $v_t =$ arbitrary, then it follows from (8) that

$$Tt = 0 \quad \text{on } \partial_3 G, \tag{9}$$

i.e., the tangent component of the outer boundary forces vanishes on $\partial_3 G$. Therefore no friction occurs on $\partial_3 G$. If we choose v with $v_t = u_t$ and $v_n \le 0$, then (8) implies the following relation at the points of $\partial_3 G$:

$$\begin{aligned} Tn &= 0 \quad \text{if} \quad un < 0, \\ Tn &\le 0 \quad \text{if} \quad un = 0, \end{aligned} \tag{10}$$

which means:

(i) If $un < 0$ at the point x of $\partial_3 G$, i.e., in case there exists no contact between the elastic body and the rigid support at x, then no boundary force occurs at x in the direction of n (Fig. 63.1(b)).
(ii) If $un = 0$ at x, i.e., in case there exists a contact between the body and the support at x, then the boundary force at x forms an obtuse angle with n.

These conditions are very natural. Finally, let us add the boundary condition

$$u = 0 \quad \text{on } \partial_1 G, \tag{11}$$

which is an immediate consequence of the fact that $u \in C$ and of the definition of C and X above.

This shows that our original problem (2), which was solved in Theorem 63.A, can be regarded as a generalized problem to the classical problem (6), (7), (9)–(11). It is called the Signorini problem. The boundary conditions (7) and (10) do not explicitly appear in the original problem (2). They are natural boundary conditions.

PROBLEMS

63.1. *Penalty technique for variational inequalities.* Consider the variational inequality

$$\langle F'(u) - b, v - u \rangle \ge 0 \quad \text{for all} \quad v \in C. \tag{12}$$

We are looking for a solution $u \in C$, and analogously to Section 46.7, we also study the penalty equation

$$F'(u_n) + \varepsilon_n^{-1} P'(u_n) = b, \quad u_n \in X. \tag{13}$$

The variational inequality corresponds to the variational problem

$$F(u) - b = \min!, \quad u \in C, \tag{12*}$$

and the operator equation (13) corresponds to the variational problem

$$F(u_n) + \varepsilon_n^{-1} P(u_n) - b = \min!, \quad u_n \in X. \tag{13*}$$

We assume:
(i) C is a nonempty, closed, convex set in the real H-space X.
(ii) The functionals $F, P: X \to \mathbb{R}$ are C^1. The operator $F': X \to X^*$ is strongly monotone and Lipschitz continuous.
(iii) The operator $P': X \to X^*$ is monotone and Lipschitz continuous. Moreover, there holds the key condition
$$P'(u) = 0 \Leftrightarrow u \in C.$$
(iv) We have that $b \in X^*$, and (ε_n) is a sequence of positive numbers with $\varepsilon_n \to 0$ as $n \to \infty$.

Prove the following:
(a) Problems (12) and (12*) as well as (13) and (13*) are equivalent.
(b) For each $n \in \mathbb{N}$, there exists a unique solution u_n of (13*), and the sequence (u_n) converges to the unique solution of (12*) as $n \to \infty$.

This result has the advantage that the convex variational problem (12*) and the variational inequality (12) can be approximately solved by using the family of free variational problems (13*).

Hint: Use analogous arguments as in Section 46.7. See Nečas and Hlaváček (1981, M), p. 298.

63.2. *Penalty technique for the Signorini problem.* Under the same assumptions as in Section 63.1, consider the variational problem

$$\int_G M(\gamma(u))\,dx - \int_G Ku\,dx - \int_{\partial_2 G} Tu\,dO + \varepsilon_n^{-1} P(u) = \min!, \qquad u \in X \quad (14)$$

with penalty functional

$$P(u) = \int_{\partial_3 G} [un]^+ \,dO.$$

Here, we set $[\varphi]^+ = \max(\varphi, 0)$. Since the assumptions of Problem 63.1 are satisfied, one obtains a method for solving the Signorini problem.

Give a physical interpretation of (14).

Solution: As in Section 63.2, one shows that formula (14) corresponds to the natural boundary condition

$$Tn = -\varepsilon_n^{-1}[un]^+ \quad \text{on } \partial_3 G.$$

This means that, in contrast to the original problem, the rigid support on $\partial_3 G$ is approximated by an elastic support. Because of

$$\varepsilon_n^{-1} \to +\infty \quad \text{as } n \to \infty,$$

this elastic support becomes more and more rigid as $n \to \infty$.

63.3. *General displacement on $\partial_1 G$.* Prove a statement, analogous to Theorem 63.A, by replacing the boundary condition in (1) with

$$u = u_0 \quad \text{on } \partial_1 G,$$
$$un \le 0 \quad \text{on } \partial_3 G,$$

where $u_0 \in W_2^1(G, V_3)$ with $u_0 n \le 0$ on $\partial_3 G$ is given.

Solution: As in Section 62.5 set $u = u_0 + v$ with $v \in C$.

63.4. *Second boundary-value problem.* Consider problem (1) with $\partial_1 G = \emptyset$, i.e.,

$$\partial G = \overline{\partial_2 G} \cup \overline{\partial_3 G}, \qquad \partial_2 G \cap \partial_3 G = \emptyset.$$

In this case, the body is supported on $\partial_3 G$, and the boundary stress forces are prescribed on $\partial_2 G$. Moreover, we use the stored energy function of linear elasticity, i.e.,

$$M(\gamma) = \tfrac{1}{2}\lambda(\operatorname{tr}\gamma)^2 + \kappa[\gamma,\gamma].$$

The original problem (1) then becomes

$$\int_G M(\gamma(u))\,dx - \int_G Ku\,dx - \int_{\partial_2 G} Tu\,dO = \min!, \qquad un \leq 0 \quad \text{on } \partial_3 G \quad (15)$$

with the corresponding functional-analytic formulation

$$F(u) - b(u) = \min!, \qquad u \in C \quad (16)$$

and the corresponding variational inequality

$$\langle F'(u) - b, v - u \rangle \geq 0 \qquad \text{for all} \quad v \in C. \quad (17)$$

Here we set $X = W_2^1(G, V_3)$ and

$$C = \{u \in X : un \leq 0 \text{ on } \partial_3 G\}.$$

Let R be the set of all infinitesimally small rigid motions, i.e., we have $u \in R$ if and only if

$$u(x) = a + b \times x \qquad \text{on } G$$

with $a, b \in V_3$. Analogously to Problem 62.6, we consider the following equilibrium condition

$$\int_G Ku\,dx + \int_{\partial_2 G} Tu\,dO \leq 0 \qquad \text{for all} \quad u \in R \cap C \quad (18)$$

for the outer forces.

Prove: The variational problem (16) is equivalent to the corresponding variational inequality (17).

Problem (16) has a solution if and only if the equilibrium condition (18) is satisfied. Two solutions of (16) differ by an infinitesimal rigid motion.

Solution: Apply Proposition 54.3, and note that

$$F(u) = 0, \qquad u \in X \Leftrightarrow u \in R,$$

according to Lemma 62.15. The semicoerciveness follows from Korn's inequality (Theorem 62.F).

See also Nečas and Hlaváček (1981, M), p. 301 for a penalty technique.

References to the Literature

Classical works: Signorini (1959), Fichera (1964).
Fichera (1972a, S, H), Kinderlehrer (1981), Nečas and Hlaváček (1981, M).

Signorini problem with friction: Haslinger (1983).
Signorini problem for polyconvex material: Ciarlet and Nečas (1984).
Variational inequalities in mechanics: Duvaut and Lions (1972, M), Kinderlehrer and Stampacchia (1981, M), Friedman (1982, M), Nečas and Hlavaček (1986, M).
Approximation methods: Glowinski, Lions, and Trémolières (1976, M).
Strong regularity results for the Signorini problem via pseudodifferential operators: Schumann (1987), (1988).

CHAPTER 64

Bifurcation for Variational Inequalities

> On the occasion of the problem of the rope curve, we encountered another equally outstanding problem. It concerns the bending of beams.
> Jacob Bernoulli (1691)

> The maximal load, which a column is capable of bearing, is indirectly proportional to the square of the height.
> Leonhard Euler (1744)

> In 1691, Jacob Bernoulli proposed the problem of a bent beam, elastic bar, or simply elastica. Bernoulli's problem was twofold: first derive the governing equations, then solve them.
> In mathematical practice today it is, unfortunately, often forgotten that to derive basic equations is as much a mathematician's duty as it is to study their properties.
> Clifford Ambrose Truesdell (1983)

> Bifurcation theory began with Euler's (1744) analysis of the planar equilibrium configurations of the elastica subjected solely to end forces.
> Stuart Antman (1983)

64.1. Basic Ideas

In Theorem 43.B we proved the existence of a bifurcation point for equations with potential operators. Thereby we used a maximum principle. Here we will apply the same proof technique to variational inequalities of the form

$$\lambda \langle F'(u), v - u \rangle \geq \langle G'(u), v - u \rangle \quad \text{for all} \quad v \in C. \tag{1}$$

We are looking for $u \in C$ with $u \neq 0$ and $\lambda \in \mathbb{R}$, i.e., we are looking for eigensolutions. Thereby C is a closed convex cone in a real H-space. As in Part I,

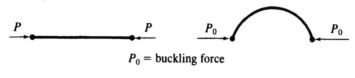

P_0 = buckling force

Figure 64.1

we call $(\lambda, 0)$ a *bifurcation point* of (1) if and only if there exists a sequence of solutions (λ_n, u_n) of (1) with $u_n \neq 0$ for all $n \in \mathbb{N}$ and

$$(\lambda_n, u_n) \to (\lambda, 0) \quad \text{as} \quad n \to \infty.$$

We then call λ a bifurcation value.

Problems of the form (1) often occur in elasticity theory. Let F be the elastic potential energy of a body and $PG(u)$ the work done by the outer forces depending on a parameter P, which can be changed, i.e., we increase, for example, the compressive forces, acting on a rod (Fig. 64.1). The variational principle, which will be motivated below by using stability arguments, is

$$F(u) - PG(u) = \min!, \quad u \in C. \tag{2}$$

If there are *no* restrictions imposed on the displacements u, such as in Figure 64.1, then we have $C = X$. The Euler equation to (2) is

$$F'(u) - PG'(u) = 0. \tag{3}$$

In Figure 64.1 one observes experimentally that no buckling occurs for small forces P, i.e., we have $u = 0$. If we increase the force P, then for a critical value P_0 we observe buckling. Here, P_0 is the smallest eigenvalue of (3) and is called the buckling force.

If, however, *restrictions* are imposed on the displacements through obstacles as in Figure 64.2, then, for a convex set C, we obtain from the minimum problem (2) the variational inequality (1) with $\lambda = P^{-1}$, according to Section 46.1. A concrete example is considered in Section 64.6. There we will show that, as expected, the buckling force in Figure 64.2 is greater than in Figure 64.1.

We now want to motivate the minimum problem (2). Let u be an equilibrium position of the elastic body. In order that u is stable, we require

$$[PG(v) - PG(u)] - [F(v) - F(u)] < 0 \tag{4}$$

Figure 64.2

for all admissible displacements $v \in C$ with $v \neq u$. As in Section 61.6b, the expression on the left-hand side in (4) is equal to the work needed to bring the body from state u to state v. Note that the elastic work is equal to the negative potential energy difference. Therefore, $[F(v) - F(u)]$ appears in (4) with a minus sign. Inequality (4) corresponds to the general stability principle of Section 58.12. From (4) follows (2). Our motivation for (2) is complete.

Besides the variational inequality

(V) $\qquad \lambda \langle F'(u), v - u \rangle \geq \langle G'(u), v - u \rangle \qquad$ for all $\quad v \in C$,

we also consider its linearization at the point $u = 0$, that is

(L) $\qquad \lambda \langle F''(0)u, v - u \rangle \geq \langle G''(0)u, v - u \rangle \qquad$ for all $\quad v \in C$.

This is a quadratic variational inequality. Our goal is the following important physical result:

(L) *has a greatest positive eigenvalue* λ_0, *and* λ_0 *is, at the same time, the greatest bifurcation value of* (V).

Physically, $P_0 = \lambda_0^{-1}$ corresponds to the buckling force. For rods, beams, and columns the buckling force was computed for the first time by Euler (1744). This will be discussed in (36).

64.2. Quadratic Variational Inequalities

We begin with the simplest case and consider

$$\lambda a(u, v - u) \geq b(u, v - u) \qquad \text{for all} \quad v \in C. \tag{5}$$

We are looking for $u \in C$ and $\lambda \in \mathbb{R}$. In addition, we consider the maximum problem

$$\lambda_0 = \max_{u \in \dot{C}} \frac{b(u, u)}{a(u, u)} \tag{6}$$

with $\dot{C} = C - \{0\}$. We assume:

(H1) C is a closed, convex cone in a real H-space X, i.e., from $u \in C$, $t \geq 0$ follows $tu \in C$, and the set C is closed and convex.
(H2) The bilinear forms $a, b: X \times X \to \mathbb{R}$ are symmetric. Moreover, a is strongly positive, and b is *compact*.
(H3) There exists an element $w \in C$ with $b(w, w) > 0$.

Proposition 64.1. *Assume* (H1)–(H3). *Then the maximum problem for the Rayleigh quotient* (6) *has a solution. The number* λ_0 *is the largest positive eigenvalue of the quadratic variational inequality* (5) *and, at the same time, a bifurcation value of* (5).

Corollary 64.2 (Solution Set). *For $u \neq 0$, problems (5) and (6) with $\lambda = \lambda_0$ are equivalent. The solution set of (5) with $\lambda = \lambda_0$ is a closed cone.*

These results have many applications in linear elasticity theory. An example is considered in Section 64.6. In the following two sections we generalize Proposition 64.1 to nonquadratic variational inequalities.

PROOF OF PROPOSITION 64.1. We set $v = 0$ in (5). It follows that $\lambda \leq \lambda_0$.

(I) The trick of the proof is to replace the original problem (6) with the new problem

$$b(u,u) = \max!, \quad u \in C,$$
$$a(u,u) \leq r, \tag{7}$$

where $r > 0$. This maximum problem for a weakly sequentially continuous functional over a closed, convex, bounded set has a solution u according to Section 38.3. Because of (H3) we get $u \neq 0$. For reasons of homogeneity we therefore have a solution of (7) with $a(u,u) = r$.

Hence $u/\sqrt{r}$ is a solution of (6). Thus we obtain $\lambda_0 = b(u,u)/a(u,u)$.

(II) We show that the solution u of (7) is also a solution of the variational inequality (5) with $\lambda = \lambda_0$. For $r \to 0$ it follows then that $(\lambda_0, 0)$ is a bifurcation point of (5).

For fixed $v \in C$ we set

$$w = (1 + \alpha)(u + \varepsilon(v - u))$$

and solve the quadratic equation $a(w,w) = r$ for α. This yields

$$\alpha(\varepsilon) = -\varepsilon \frac{a(u, v-u)}{a(u,u)} + o(\varepsilon), \quad \varepsilon \to 0.$$

We have $w \in C$. From (7) we thus obtain that $b(w,w) \leq b(u,u)$. This yields (5) with $\lambda = b(u,u)/a(u,u)$, hence $\lambda = \lambda_0$. □

The proof of Corollary 64.2 will be given in Problem 64.1.

64.3. Lagrange Multiplier Rule for Variational Inequalities

In addition to the maximum problem

$$G(u) = \max!, \quad u \in C,$$
$$F(u) = r, \tag{8}$$

with fixed $r \in \mathbb{R}$, we consider the variational inequality

$$\lambda \langle F'(u), v - u \rangle \geq \langle G'(u), v - u \rangle \quad \text{for all} \quad v \in C. \tag{9}$$

64.3. Lagrange Multiplier Rule for Variational Inequalities

We want to show that the solutions of (8) are also solutions of (9). Important is the nondegeneracy condition

$$\langle F'(u), u \rangle \neq 0. \tag{10}$$

We assume:

(H1) C is a closed, convex cone on the real B-space X.
(H2) $F, G: X \to \mathbb{R}$ are given, and u is a solution of the maximum problem (8) with (10).
(H3) The functional F is continuous at the point u and the F-derivatives $F'(u)$, $G'(u)$ exist.

Proposition 64.3 (Lagrange Multiplier Rule). *If* (H1) *to* (H3) *hold, then there exists a real number λ such that the solution u of the maximum problem (8) is also a solution of the variational inequality (9).*

If we choose $C = X$, then we obtain

$$\lambda F'(u) - G'(u) = 0.$$

This is the Lagrange multiplier rule of Section 43.2, whereby the eigenvalue λ is called a Lagrange multiplier.

Proposition 64.3 is the key to our approach, since it allows us to reduce eigenvalue problems for variational inequalities to maximum problems.

PROOF. From the proof of Proposition 43.6 there exist for each $v \in C$ two zero sequences $(\alpha_n), (\delta_n)$ with $\alpha_n > 0$ for all n such that $F(u_n) = r$ with

$$u_n = u + \alpha_n(\beta + \delta_n)u + \alpha_n v$$

and

$$\beta = -\langle F'(u), v \rangle / \langle F'(u), u \rangle.$$

Since the set C is a cone, we have

$$x \in C, \quad t > 0 \Rightarrow tx \in C \quad \text{and} \quad x, y \in C \Rightarrow x + y \in C.$$

Thus we obtain $u_n \in C$ for large n, i.e., u_n is an admissible element for the maximum problem (8). Consequently,

$$G(u_n) \leq G(u) \quad \text{for all} \quad n \geq n_0.$$

This means

$$G(u) + \alpha_n \langle G'(u), \beta u + v \rangle + o(\alpha_n) \leq G(u).$$

For $n \to \infty$ it follows that

$$\langle G'(u), \beta u + v \rangle \leq 0.$$

This is the variational inequality (9) with

$$\lambda = \langle G'(u), u \rangle / \langle F'(u), u \rangle. \qquad \square$$

64.4. Main Theorem

We study the eigenvalue problem for the variational inequality

$$\lambda \langle F'(u), v - u \rangle \geq \langle G'(u), v - u \rangle \qquad \text{for all} \quad v \in C \qquad (11)$$

on the real H-space X. We are looking for $u \in C$ with $u \neq 0$ and $\lambda \in \mathbb{R}$. Thereby (11) is a perturbation of the quadratic variational inequality

$$\lambda a(u, v - u) \geq b(u, v - u) \qquad \text{for all} \quad v \in C. \qquad (12)$$

Moreover we consider, as in Section 64.2, the maximum problem for the Rayleigh quotient

$$\lambda_0 = \max_{u \in \dot{C}} \frac{b(u, u)}{a(u, u)}. \qquad (13)$$

Our assumptions are:

(A1) The set C is a closed, convex cone in the real H-space X. The bilinear forms $a, b: X \times X \to \mathbb{R}$ are symmetric. Furthermore, a is strongly positive and b is compact.

There exists an element $w \in C$ with $b(w, w) > 0$.

(A2) The functionals $F, G: U(0) \subseteq X \to \mathbb{R}$ are F-differentiable with $F'(0) = G'(0) = 0$. Moreover, F is weakly sequentially lower semicontinuous, and G is weakly sequentially continuous.

(A3) For $u \to 0$ we have the approximation formulas

$$F(u) = \tfrac{1}{2} a(u, u) + o(\|u\|^2), \qquad F(u) = \tfrac{1}{2} \langle F'(u), u \rangle + o(\|u\|^2),$$
$$G(u) = \tfrac{1}{2} b(u, u) + o(\|u\|^2), \qquad G(u) = \tfrac{1}{2} \langle G'(u), u \rangle + o(\|u\|^2).$$

Theorem 64.A. *Assume* (A1) *to* (A3). *Then the variational problem* (13) *has a solution with* $\lambda_0 > 0$, *and* λ_0 *is the largest bifurcation value of* (11) *and the largest eigenvalue of the linearized problem* (12).

Corollary 64.4 (Linearization Principle). *Assume* (A2). *If the second-order F-derivatives* $F''(0), G''(0)$ *exist and if* $F(0) = G(0) = 0$, *then also* (A3) *is valid with*

$$a(u, v) = \langle F''(0)u, v \rangle, \qquad b(u, v) = \langle G''(0)u, v \rangle$$

for all $u, v \in X$.

If, in addition, (A1) *is satisfied for these a and b, then every positive bifurcation value of* (11) *is an eigenvalue of the linearized problem* (12).

Theorem 64.A goes back to Miersemann (1975).

In assumption (A1), we require the existence of a point $w \in C$ with $b(w, w) > 0$. If we replace this condition with

$$b(w, w) < 0 \qquad \text{for a} \quad w \in C,$$

then we can substitute G and b with $-G$ and $-b$, respectively. Multiplication with (-1) means that, in (11) and (12), the number λ is replaced by $-\lambda$ and "$\leq$" by "$\geq$". According to Theorem 64.A we also obtain an existence result for this new problem.

64.5. Proof of the Main Theorem

We prove Theorem 64.A and follow Zeidler (1976). For a proof of Corollary 64.4, see Problem 64.2.

In the following we assume that $0 < r \leq r_0$ with sufficiently small $r_0 > 0$. The proof idea is as follows:

(i) We apply the Lagrange multiplier rule from Section 64.3.
(ii) For this we have to solve the maximum problem

(M) $$\max_{u \in C \cap \partial M_r} G(u) = G(u_r)$$

over the set $C \cap \partial M_r$.

(iii) In order to solve (M) we consider the simpler maximum problem

(M*) $$\max_{u \in C \cap M_r} G(u) = G(u_r)$$

over the weakly sequentially compact set $C \cap M_r$. Then, we apply the so-called boundary trick, i.e., we show that the solution u_r of (M*) lies on the boundary ∂M_r. Consequently, u_r is also a solution of (M).

(I) Geometrical preparation. We introduce the new equivalent norm $|u| = a(u, u)^{1/2}$ on the H-space X. We define the ball $B_r = \{u \in X : |u| \leq r\}$ and the set

$$M_r = \left\{ u \in X : F(u) \leq \frac{r^2}{2}, |u| < 2r \right\}.$$

Because of (A3) we find that, for small r, the closed set M_r lies in a small open ball. Because of the continuity of F we find

$$\partial M_r = \left\{ u \in X : F(u) = \frac{r^2}{2}, |u| < 2r \right\}.$$

The functional F is weakly sequentially lower semicontinuous. Therefore the set M_r is weakly sequentially compact. According to (A3) there exist functions $\alpha = \alpha(r)$ and $\beta = \beta(r)$ with

$$B_\alpha \leq M_r \leq B_\beta \tag{14}$$

and $\alpha/r, \beta/r \to 1$ for $r \to 0$.

(II) We consider the three maximum problems

$$\max_{u \in C \cap B_1} b(u,u) = \lambda_0,$$

$$\max_{u \in C \cap M_r} b(u,u) = b(v_r, v_r),$$

$$\max_{u \in C \cap M_r} G(u) = G(u_r),$$

with corresponding solutions v_r and u_r. According to Section 38.3, there exist solutions to all these problems, because weakly sequentially continuous functionals are maximzed over weakly sequentially compact sets.

Important is now the condition

$$\lim_{r \to 0} \frac{\langle G'(u_r), u_r \rangle}{r^2} = \lambda_0 \qquad (15)$$

which will be proved at the end. From Section 64.2 follows that $\lambda_0 > 0$.

(III) Boundary trick. We prove that $u_r \in \partial M_r$. Thus we obtain

$$\max_{u \in C \cap \partial M_r} G(u) = G(u_r). \qquad (16)$$

Suppose we have $u_r \in \text{int}(M_r \cap C)$. Since C is a cone, we then also have

$$u_r + \varepsilon u_r \in M_r \cap C \text{ for small } \varepsilon > 0,$$

hence $G(u_r + \varepsilon u_r) \leq G(u_r)$. For $\varepsilon \to 0$ we obtain $\langle G'(u_r), u_r \rangle \leq 0$, which contradicts (15). Hence $u_r \in \partial M_r$.

(IV) Lagrange multipliers. From $u_r \in \partial M_r$ and (A3) follows $\langle F'(u_r), u_r \rangle / r^2 \to 1$ as $r \to 0$. Thus the nondegeneracy condition

$$\langle F'(u_r), u_r \rangle \neq 0$$

is satisfied. Since the proof in Section 64.3 has a purely local character, we may apply Proposition 64.3 to (16), according to (I), and obtain

$$\langle G'(u_r), v - u_r \rangle \leq \lambda_r \langle F'(u_r), v - u_r \rangle \qquad \text{for all} \quad v \in C. \qquad (17)$$

For $v = 0$ and $v = 2u_r$ this gives

$$\lambda_r = \frac{\langle G'(u_r), u_r \rangle r^{-2}}{\langle F'(u_r), u_r \rangle r^{-2}}. \qquad (18)$$

Thus we obtain $\lambda_r \to \lambda_0$ as $r \to 0$. Moreover, we have $u_r \to 0$ as $r \to 0$. Therefore $(\lambda_0, 0)$ is a bifurcation point of the original equation (11).

(V) We show that the number λ_0 is the largest bifurcation value of (11). Suppose that in (17) we have

$$u_r \to 0 \qquad \text{and} \qquad \lambda_r \to \lambda_1 \qquad \text{as} \quad r \to 0 \quad \text{with} \quad u_r \neq 0.$$

From (A3) follows

$$\lim_{r \to 0} \frac{\langle F'(u_r), u_r \rangle}{|u_r|^2} = 1$$

and furthermore,

$$\lim_{r \to 0} \frac{\langle G'(u_r), u_r \rangle}{|u_r|^2} = \lim_{r \to 0} \frac{b(u_r, u_r)}{|u_r|^2} \leq \lambda_0.$$

Equation (18) implies $\lambda_1 = \lim_{r \to 0} \lambda_r \leq \lambda_0$.

(VI) Proof of (15). For reasons of homogeneity we have

$$\max_{u \in C \cap B_r} b(u, u) = \lambda_0 r^2.$$

Because of (14) this implies

$$\lambda_0 \alpha^2 \leq b(v_r, v_r) \leq \lambda_0 \beta^2.$$

Thus we find

$$\lambda_0 = \lim_{r \to 0} \frac{b(v_r, v_r)}{r^2} = \lim_{r \to 0} \frac{2G(v_r)}{r^2}.$$

The second equation is a consequence of (A3). From $u_r \in C \cap M_r$ follows $|u_r| \leq \beta$, according to (14), hence

$$2G(v_r) \leq 2G(u_r) = b(u_r, u_r) + o(|u_r|^2)$$
$$\leq \lambda_0 \beta^2 (1 + o(1)), \quad \beta \to 0.$$

For $r \to 0$ we thus obtain that $G(u_r)/r^2 \to \lambda_0$, and (A3) implies (15).

The proof of Theorem 64.A is complete.

Let us note that this proof can greatly be simplified if we make the additional assumption that

$$G(u) > 0, \quad u \in C \Rightarrow \langle G'(u), u \rangle > 0.$$

64.6. Applications to the Bending of Rods and Beams

As in Section 64.3 we consider a rod or a beam of length l, which is clamped at both boundary points. Moreover, we assume that the left boundary point is fixed, and suppose that a compressive force of magnitude P acts on the right movable boundary point

$$x = l - \Delta l.$$

Moreover, we assume that obstacles exist as in Figure 64.3. For a mathematical description we use the general observations of Section 64.1.

Let $u(\xi)$ denote the deflection of the beam at the point ξ. The length of the buckled beam is

$$l_1 = \int_0^x \sqrt{1 + u'^2} \, d\xi.$$

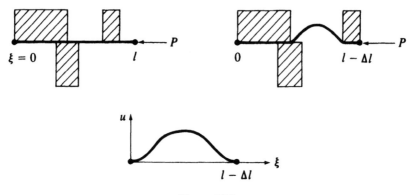

Figure 64.3

Since the beam is clamped at the boundary points we obtain

$$u(\xi) = u'(\xi) = 0 \quad \text{for} \quad \xi = 0, \ l - \Delta l. \tag{19}$$

The work of the compressive force is obtained from the product of force times length, that is

$$PG(u) = P\,\Delta l. \tag{20}$$

We work in a linearized theory, i.e., we make the following assumption:

$$|u|, |u'|, |u''| \text{ and } \Delta l \text{ are small, and we} \tag{21}$$
consider only lowest order terms,

In the next section we will show that, in the context of (21), the elastic potential energy of the beam satisfies:

$$F(u) = \frac{1}{2} EJ \int_0^{l-\Delta l} u''^2 \, d\xi. \tag{22}$$

Thereby E is the elasticity module of Section 60.1 and J will be given in (32).

Besides the deflection u we also have to determine the change in length Δl, i.e., in (19) we have a free boundary condition. In order to eliminate this difficulty, we replace (19) *approximately* with the boundary condition

$$u(\xi) = u'(\xi) = 0 \quad \text{for} \quad \xi = 0, l. \tag{23}$$

because of the smallness of Δl. This means that we are looking for a solution u on $[0, l]$, which is not equal to zero on $[l - \Delta l, l]$ as in the real situation, but only small. Moreover, we assume that $u = u(\xi)$ is the equation of the neutral fiber, which is not subject to length changes. Thus we have $l_1 = l$, hence

$$l = \int_0^{l-\Delta l} \sqrt{1 + u'^2} \, d\xi = \int_0^{l-\Delta l} 1 + \frac{1}{2} u'^2 + o(u'^2) \, d\xi$$

$$= l - \Delta l + \int_0^{l-\Delta l} \frac{1}{2} u'^2 + o(u'^2) \, d\xi.$$

64.6. Applications to the Bending of Rods and Beams

Together with (21) this yields the *final formulas*:

$$PG(u) = P\,\Delta l = \frac{1}{2}P\int_0^l u'^2\, d\xi,$$

$$F(u) = \frac{1}{2}EJ\int_0^l u''^2\, d\xi.$$

To take the obstacles in Figure 64.3 into account, we choose two subsets M and N of $[0, l]$ and assume

$$\begin{aligned} u(\xi) &\leq 0 \quad \text{for} \quad \xi \in M, \\ u(\xi) &\geq 0 \quad \text{for} \quad \xi \in N. \end{aligned} \qquad (24)$$

From the problem

$$F(u) - PG(u) = \min!,$$

which has been discussed in Section 64.1, we obtain the *final problem*:

$$\lambda a(u, u) - b(u, u) = \min!, \qquad u \in C \qquad (25)$$

with

$$X = \mathring{W}_2^2(0, l) \quad \text{and} \quad C = \{u \in X : (24) \text{ holds}\}.$$

Moreover, we have $\lambda = 1/P$ and

$$a(u, v) = EJ\int_0^l u''v''\, d\xi, \qquad b(u, v) = \int_0^l u'v'\, d\xi.$$

With X, the boundary conditions (23) are taken into account. From Section 64.1 we then obtain the following variational inequality

(P) $\qquad \lambda_0 a(u, v - u) \geq b(u, v - u) \qquad \text{for all} \quad v \in C.$

We are looking for $u \in C$ and $\lambda \in \mathbb{R}$. As in Section 64.2 we also consider

$$\lambda_0 = \max_{u \in C} \frac{b(u, u)}{a(u, u)}. \qquad (26)$$

Theorem 64.B. *Let $C \neq \{0\}$. Then (26) has a solution $\lambda_0 > 0$ and λ_0 is the largest eigenvalue of (P) and, at the same time, a bifurcation value of (P). The solution set of (P) with $\lambda = \lambda_0$ is a closed cone.*

PROOF. Because of the continuous embedding $\mathring{W}_2^2(0, l) \subseteq C^1[0, l]$ it follows that C is closed. Obviously, C is a convex cone.

From $A_2(54)$ follows that $a(u, u)^{1/2}$ is an equivalent norm on X, and hence a is strongly positive.

The bilinear form b is compact, because from

$$u_n \rightharpoonup u \quad \text{and} \quad v_n \rightharpoonup v \quad \text{in } X \quad \text{as} \quad n \to \infty$$

follows $u_n \to u$ and $v_n \to v$ in $W_2^1(0,l)$, since the embedding $\mathring{W}_2^2(0,l) \subseteq W_2^1(0,l)$ is compact, hence $b(u_n, v_n) \to b(u, v)$ as $n \to \infty$.

Proposition 64.1 yields the assertion. □

We now want to give a physical motivation of Theorem 64.B. Let $P_0 = 1/\lambda_0$. For

$$0 < P < P_0$$

there only exists the trivial solution $u = 0$ of (P), hence there occurs *no* buckling. For $P = P_0$ we obtain a nontrivial solution. Therefore P_0 is the *buckling force*. Without obstacles, we obtain $C = X$. Hence the eigenvalue λ_0 in (26) increases or remains equal if we pass from the obstacle case $C \subset X$ to the case without obstacles $X = C$. As expected, this implies the following natural result:

In the presence of obstacles the buckling force is at least as big as without obstacles.

Roughly speaking, obstacles stabilize the beam.

When no obstacles are present, then the Euler equation for the classical variational problem (25) is the well-known linearized rod (beam) equation

$$\begin{aligned} EJu^{(4)} &= -Pu'' \quad \text{on }]0,l[, \\ u(\xi) &= u'(\xi) = 0 \quad \text{for } \xi = 0, l. \end{aligned} \tag{27}$$

The buckling force without obstacles is the smallest positive eigenvalue of (27).

As in the case of all linear eigenvalue models of elasticity theory, our model also has the disadvantage that several solutions occur for the buckling force P_0, i.e., there exists no uniquely determined state of buckling (Fig. 64.4(a)). Experimentally, one observes the following fact for increasing forces $P > P_0$. The deflections are uniquely determined by the force P and they become larger and larger (Fig. 64.4(b)). Such bifurcation branches exist in more realistic nonlinear models.

Important, for our abstract model in Theorem 64.A, is the following:

(i) The elastic potential energy $u \mapsto a(u, u)$ is strongly positive.
(ii) The work $u \mapsto b(u, u)$ of the outer forces is compact.

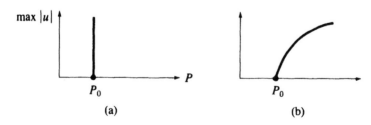

Figure 64.4

In the case of the beam, we have as a typical structure:

(S) *The elastic potential energy a(u, u) contains derivatives which are of higher order than those of the work b(u, u) of the outer forces.*

The proof of Proposition 64.5 shows that (S) implies (ii), while (i) follows from the fact that $a(u,u)^{1/2}$ is an equivalent norm on the Sobolev space $\mathring{W}_2^2(0, l)$. In fact, (S) occurs in many other situations in elasticity theory. For example, the following chapter will show that (S) occurs for plates. Thus we can also apply Theorem 64.A to plates with obstacles (see Section 65.8).

64.7. Physical Motivation for the Nonlinear Rod Equation

We consider here the same situation as in the previous section, but without obstacles. In order to obtain a better model, we choose the arclength s as a parameter. Suppose the equation for the neutral fiber of the rod or beam of length l is

$$\xi = \xi(s), \qquad \zeta = \zeta(s). \tag{28}$$

Since s is the arclength, we have $\xi'^2 + \zeta'^2 = 1$. Therefore it suffices to compute ζ (Fig. 64.5). The angle between the tangent to the curve and the ξ-axis is denoted by $\varphi(s)$. Thus we have

$$\xi'(s) = \cos \varphi(s), \qquad \zeta'(s) = \sin \varphi(s), \tag{29}$$

and

$$\varphi(s) = \arcsin \zeta'(s). \tag{30}$$

As is well known $\varphi'(s)$ is the curvature at the point s, and $r = 1/\varphi'$ is called the radius of curvature. Our goal is the variational problem

$$\int_0^l \frac{1}{2} A\varphi'^2 + P \cos \varphi \, ds - Pl = \min!,$$

$$\varphi(0) = \alpha, \qquad \varphi(l) = \beta \tag{31}$$

for fixed given angles α and β. The material constant A is called flexible

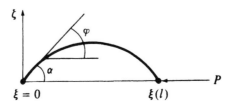

Figure 64.5

stiffness. If the rod lies symmetrically to the ξ-axis and has a constant cross section Q, then we have

$$A = EJ. \tag{32}$$

Thereby E is the elasticity module of Section 60.1 and

$$J = \int_Q \zeta^2 \, d\zeta \, d\eta.$$

Moreover, l is the length of the rod and P the outer force. For $P > 0$, we have a compressive force. If φ is expressed in terms of ζ, then we obtain the following variational problem which is equivalent to (31):

$$\int_0^l \frac{1}{2} \frac{A\zeta''^2}{(1-\zeta'^2)} + P\sqrt{1-\zeta'^2} \, ds - Pl = \min!,$$

$$\zeta(0) = \zeta(l) = 0, \qquad \varphi(0) = \alpha, \quad \varphi(l) = \beta. \tag{33}$$

Note that $\zeta'' = \varphi' \cos \varphi$. In Section 37.28j we have studied problems of type (33) by making use of catastrophe theory. As we shall see during the next section, the form (31) is better suited for the construction of explicit solutions.

Motivation of (31). The work PG of the outer force is equal to force times displacement, hence

$$PG = P(l - \xi(l)) = Pl - P\int_0^l \cos \varphi \, ds.$$

Moreover, we assume that the elastic energy F depends only on the curvature φ', i.e.,

$$F = \int_0^l L(\varphi') \, ds.$$

We require $L(0) = 0$ and $L(-\varphi') = L(\varphi')$, i.e., L does not depend on the sign of the curvature. The Taylor expansion $L(\varphi') = 2^{-1}A\varphi'^2 + o(\varphi'^2)$ yields approximately

$$F = \frac{1}{2}A \int_0^l \varphi'^2 \, ds = \frac{1}{2}A \int_0^l \zeta''^2 (1-\zeta'^2)^{-1} \, ds.$$

According to Section 64.1 we obtain (31) from

$$F - PG = \min!.$$

Motivation of (32). We want to compute the material constant A more accurately. Thereby we obtain another more general motivation for F. Consider Figure 64.6. Suppose the rod lies symmetrically to the ξ-axis. We first look at the (ξ, ζ)-plane. Since the upper fiber is stretched and the lower fiber shortened we make, with Jacob Bernoulli, the following assumptions:

(i) The ξ-axis remains unstretched, i.e., it corresponds to the so-called neutral fiber.

64.8. Explicit Solution of the Rod Equation

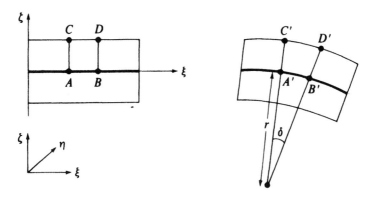

Figure 64.6

(ii) Straight lines, perpendicular to the ξ-axis, pass without stretching into straight lines, perpendicular to the neutral fiber.

At the point A' of Figure 64.6 the neutral fiber behaves locally like a circle of radius $r = 1/\varphi'$. For the distances we find

$$|AB| = |A'B'| = r\delta \quad \text{and} \quad |C'D'| = (r + \zeta(C))\delta.$$

For the dilatation γ of $\overrightarrow{CD}$ we therefore obtain

$$\gamma = (|C'D'| - |CD|)/|CD| = \zeta(C)/r,$$

hence $\gamma = \zeta(C)\varphi'$.

We now consider the three-dimensional case. Let $Q(\xi)$ be the cross section of the rod for fixed ξ. Suppose that all planes parallel to the (ξ, ζ)-plane behave analogously to the (ξ, ζ)-plane, whereby no dilatation occurs in the direction of the η-axis. We choose a small axially parallel cuboid. According to Problem 61.3, its elastic energy is equal to

$$\tfrac{1}{2}E\gamma^2 \, \Delta V = \tfrac{1}{2}E\zeta^2\varphi'^2 \, \Delta V.$$

Summation over the small cuboids yields the total elastic potential energy

$$F = \frac{1}{2}E \int_0^l J\varphi'^2 \, ds \quad \text{with} \quad J(s) = \int_{Q(s)} \zeta^2 \, d\zeta \, d\eta.$$

For a constant cross section we obtain $J = \text{const}$. This yields (32).

64.8. Explicit Solution of the Rod Equation

For a constant cross section of the rod the Euler equation for the variational problem (31) is

$$\varphi''(s) = -\mu^2 \sin \varphi(s) \tag{34}$$

with $\mu = \sqrt{P/EJ}$. As usual, let $sn(\cdot)$ denote the Jacobian elliptic sinus amplitudinis.

Proposition 64.5. *For each $k \in [0, 1]$ one obtains from*

$$\sin\frac{\varphi}{2} = k\,sn(\mu s, k) \tag{35}$$

a solution of (34) with $\varphi(0) = 0$ and $\varphi'(0) = 2k\mu$.

PROOF. After multiplication with φ' and integration it follows from (34) that

$$\varphi'^2 = 2\mu^2(\cos\varphi - \cos\varphi_0) = 4\mu^2\left(\sin^2\frac{\varphi_0}{2} - \sin^2\frac{\varphi}{2}\right).$$

Letting $k = \sin\varphi_0/2$ and $\sin\varphi/2 = k\sin\psi$ we obtain

$$\mu s = \int_0^\psi \frac{d\psi}{\sqrt{1 - k^2\sin^2\psi}}.$$

By reversing this elliptic integral we find $\psi = am(\mu s, k)$ and

$$k\sin\psi = k\sin am(\mu s, k) = k\,sn(\mu s, k).$$

This is (35). □

The bending of the rod was studied for the first time by Jacob Bernoulli in 1691. From 1696 on, he, and his younger brother Johann, worked as embittered rivals on a further development of variational calculus, after Johann Bernoulli had posed the famous problem of the brachystochrone. The variational principle (31) goes back to Johann's son, Daniel Bernoulli. In 1742, he communicated it in a letter to Euler. In a wonderful piece of work, Euler (1744) studied the differential equation (34) in great detail. The theory of elliptic functions was not known at this time. It was developed only later during the nineteenth century by Jacobi and Weierstrass. Nevertheless, using great computational skills, Euler was able to classify nine qualitatively different types of solutions. This classification of (35) may be found in Geckeler (1928), p. 189. Interestingly, the family of solutions of (35) also contains such strange rod curves as shown in Figure 64.7.

If we consider a pendulum as in Figure 64.8, then we again obtain the differential equation (34) with

$$\mu^2 = g/L$$

Figure 64.7

Problems

Figure 64.8

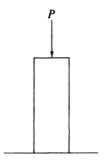

Figure 64.9

and g = gravitational acceleration, L = length of the pendulum, s = time. Using this observation one can determine solutions of (34) experimentally and give intuitive interpretations.

For small φ, equation (34) becomes approximately

$$\varphi'' + \mu^2 \varphi = 0, \qquad \varphi(0) = \varphi(l) = 0.$$

The smallest eigenvalue is $\mu = \pi/l$ with eigenfunction $\varphi = \sin \mu s$. Thus, for the buckling force we obtain *Euler's* (1744) *famous formula*:

$$P_0 = \frac{EJ\pi^2}{l^2}. \tag{36}$$

Euler considered a column as in Figure 64.9. For the critical force $P = P_0$ the column collapses. For a detailed discussion see Section 29.13.

PROBLEMS

64.1. *Proof of Corollary 64.2.*
   Solution: The proof of Proposition 64.1 shows that every solution of (6) is also a solution of (5) with $\lambda = \lambda_0$. Let, conversely, u be a solution of (5) with $\lambda = \lambda_0$. If we choose $v = 0$ and $v = 2u$, then it follows that $\lambda_0 = b(u, u)/a(u, u)$.
   Let S be the solution set of (5) with $\lambda = \lambda_0$. The continuity and homogeneity of a and b imply that S is a closed cone.

64.2. *Proof of Corollary 64.4.*
   Hint: See Zeidler (1976), pp. 44, 49.

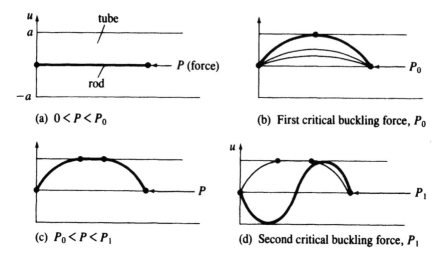

(a) $0 < P < P_0$

(b) First critical buckling force, P_0

(c) $P_0 < P < P_1$

(d) Second critical buckling force, P_1

The heavily drawn situations are observed in the experiment up to symmetry with respect to the P-axis.

Figure 64.10

64.3. Bifurcation for variational inequalities on convex sets, higher eigenvalues, and applications to beams and plates. See the papers of Miersemann in the References to the Literature to this chapter.

64.4. Bending of rods with obstacles and optimal control. Let us consider the situation of Figure 64.10, namely, a rod under the action of a compressive force P with the side condition

$$|u| \leq a$$

for the deflection u. According to Section 64.6, we obtain the following problem:

$$\frac{1}{2} \int_0^l Aw^2 - Pv^2 \, d\xi = \min!,$$

$$v = u', \quad w = v',$$

$$|u| \leq a,$$

$$u(0) = u(l) = 0.$$

(37)

This is a typical optimal control problem to which the Pontrjagin maximum principle applies (see Section 48.6). A careful study of this problem may be found in the paper of Phú (1987), where the author introduces a new method, which he calls the method of region analysis. In Phú (1987a) this method is also applied to the optimal control of hydro-electric power stations. Figure 64.10 shows some of the typical solutions which coincide with physical experiments. In those experiments, the compressive force is increased and one observes an abrupt change of the rod. More and more complex configurations occur.

A typical and interesting feature of the method of region analysis is the fact that the optimal solutions depend only on the structure of the problem, and not on the precise formulation of the problem. Thus, one only needs a *rough* modeling of the actual situation.

References to the Literature

Classical works: Bernoulli (1691), Euler (1744).
Modern development: Antman (1983, S, B, H).
Article about rods in the handbook of physics: Geckeler (1928, B, H), Antman (1972, B).
Bifurcation for rods: Antman (1977, S), Antman and Rosenfeld (1978, S), Reiss (1977, S), Marsden (1980), Antman and Kenney (1981).
Bifurcation for variational inequalities and its applications: Miersemann (1981, S) (recommended as an introduction), (1975), (1978), (1979), (1981), (1981a, b), Do (1975), (1976), (1977), Dias (1975), Dias and Hernandez (1975), Zeidler (1976), Kučera, Nečas, and Souček (1978), Kučera (1982), (1982a).
Explicit solutions for the rod: Euler (1744), Geckeler (1928), Pflüger (1965, M).
Rope and bifurcation: Dickey (1976, L).
Rotating rods, chains, and bifurcation: Stuart (1976, S), Reeken (1977) (1979), Antman and Nachman (1980).
Rod and catastrophe theory: Zeeman (1976), Golubitsky (1978, S), Poston and Stuart (1978, M), Gilmore (1981, M).
Mooney–Rivlin material and the bending of rods: Ball (1977).
Bending of rods, optimal control, and the method of region analysis: Phú (1987), (1987a) (optimal control of hydro-electric power stations).
Variational inequalities in mechanics: Panagiotopoulos (1985, M).
History: Antman (1983, S), Truesdell (1983, S).

CHAPTER 65

Pseudomonotone Operators, Bifurcation, and the von Kármán Plate Equations

The historical development of nonlinear plate and shell theories has not been as felicitous as that of rod theories up to the creation of Kirchhoff's theory. Kirchhoff (1824–1887) established a satisfactory linear theory of plates begun by Navier (1785–1836). A popular nonlinear plate theory was finally developed by von Kármán in 1910.

These equations are "derived" by introducing, in the standard procedure of mechanics, a number of geometric approximations that are roughly analogous to the replacement of $\sin \varphi$ in the rod equation

$$\varphi'' + \sin \varphi = 0$$

by its cubic approximation $\varphi - \varphi^3/6$. Moreover, the von Kármán's equations are based upon a certain linear stress–strain law.

Stuart Antman (1977)

The number of papers about plates is infinitely large, and gets larger and larger.

Mathematical folclore

65.1. Basic Ideas

In this chapter we consider a plate which is clamped at the boundary. Our method of proof, however, can also be applied to other boundary conditions. We use the following tools:

(I) Implicit function theorem (Theorem 4.B).
(P) Main theorem about pseudomonotone operators (Theorem 27.A).
(B) Main theorem of bifurcation theory for potential operators (Theorem 45.A).

The von Kármán plate equations, which will be formulated in Section 65.2

65.1. Basic Ideas

and motivated in Section 65.7, lead to the operator equation

$$w - pLw + Cw = f, \qquad w \in X \tag{1}$$

in the real H-space X. This equation has the following structure:

p = real parameter;
L = linear, symmetric, strongly continuous operator;
C = strongly continuous potential operator which is homogeneous of degree 3;
f = fixed element in X.

Roughly speaking, these quantities allow the physical interpretation:

$w \Rightarrow$ deflection of the plate perpendicular to the plate plane;
$f \Rightarrow$ outer volume forces;
$pL \Rightarrow$ outer boundary forces.

For $p = 0$, the outer boundary forces vanish, while for $|p| \to \infty$ they increase. Let us investigate two problems:

(i) *Existence.* For arbitrary f and small $|p|$, i.e., for arbitrary volume forces and small boundary forces, equation (1) has a solution.

For small $\|f\|$ and $|p|$, i.e., for small volume and boundary forces, the solution is unique.

Because of possible bifurcations, there exist no uniqueness results for arbitrary forces.

(ii) *Bifurcation.* Let $f = 0$, i.e., the volume forces vanish. Then equation (1) has the trivial solution $w = 0$, which corresponds to a state of the plate without buckling.

We are interested in buckled states which correspond to nontrivial solutions $w \neq 0$. We will show that, in mathematical terms, the bifurcation points $(p, 0) \in \mathbb{R} \times X$ of equation (1) with $f = 0$ are in a one-to-one correspondence with the characteristic numbers p of the linearized problem

$$w - pLw = 0, \qquad w \in X. \tag{2}$$

In physical terms, this means the following.

Precisely all these characteristic numbers of equation (2), for which buckling of the plate occurs, correspond to critical boundary forces. If one increases the boundary forces, beginning with forces equal to zero, i.e., if p increases from zero onwards, then buckling occurs for the first time at the smallest characteristic number

$$p_0 > 0$$

of equation (2).

The knowledge of p_0 is very important to engineers in order to determine the stability of plates. If

$$0 < p < p_0,$$

then engineers expect that the plate is stable for the boundary force corresponding to p. By using modifications at the plate, for example, by adding additional forces, they try to obtain the largest possible value for p_0.

We now explain the *basic ideas* for the proofs of (i) and (ii).

(a) It is clear that (ii) is an immediate consequence of (B).
(b) The uniqueness result in (i) follows from (I) or directly from the Banach fixed-point theorem.
(c) The general existence result in (i) is a consequence of (P), because the operator

$$A = I - pL + C$$

is a strongly continuous perturbation of the identity and hence A is pseudomonotone and bounded. It is important that we show the coerciveness of the operator A. We thereby make essential use of the special structure of the operator C. Namely, there exists a bilinear, bounded, and strongly continuous operator

$$B: X \times X \to X$$

such that

$$Cw = B(w, B(w, w)) \quad \text{for all} \quad w \in X \tag{3}$$

and

$$(Cw|w) = \|B(w, w)\|^2. \tag{4}$$

Thus we have

$$(Aw|w) = \|w\|^2 - p(Lw|w) + (Cw|w)$$
$$\geq \|w\|^2 (1 - |p| \|L\|)$$

and hence

$$\frac{(Aw|w)}{\|w\|} \to +\infty \quad \text{as} \quad \|w\| \to +\infty \tag{5}$$

for small $|p|$. This is the coerciveness of A for small $|p|$.

In Problem 65.2 we show, how, in contrast to (i), the existence of solutions for arbitrary boundary and volume forces can be proved.

In Section 65.7 we give a detailed mathematical and physical motivation for the von Kármán equations. In this connection, we use our general strategy for obtaining approximation models in elasticity, which has been described in Section 61.6e. The basic idea is as follows:

(a) We start with an exact stored energy function $L = A(\mathscr{E})$, where $\mathscr{E}$ is the strain tensor (Saint Venant–Kirchhoff material).

(b) We replace $\mathscr{E}$ with an approximation, where, roughly speaking, we assume that all effects *perpendicular* to the plate plane are small.
(c) We consider the principle of stationary potential energy with respect to the approximate stored energy function.
(d) We compute the corresponding Euler equations, which represent a nonlinear system of partial differential equations for the three displacements u_1, u_2, and u_3.
(e) We simplify the Euler equations by introducing a function U, called the Airy stress function. This function plays the role of a potential for u_1 and u_2. In this way we obtain the von Kármán plate equations for the two unknown functions $w = u_3$ and U.

For readers who are interested in practical applications, we remark that, for example, the construction of the Metro in Prague and the Prague Ice Palace was based on the Signorini problem and the von Kármán–Vlasov shell equations, respectively.

There still remain many open problems in shell theory. This applies to model building as well as to mathematical existence theory. For shells, the stress depends, in a more sensible way, on higher order derivatives of the displacements than this is the case for plates. This is the reason, why there exist, for example, equations of degree eight and higher, in contrast to the fourth-order elliptic equations for plates. Some deep work in shell theory has been done by John (1965), (1971), (1985) (collected works), who found a priori estimates for the stresses in shells. This way, he rigorously derived interior shell equations.

65.2. Notations

In this chapter we work in a fixed Cartesian coordinate system with coordinates ξ_1, ξ_2, ξ_3 and a corresponding system of orthonormal basis vectors e_1, e_2, e_3. Instead of the partial derivative

$$D_i w = \frac{\partial w}{\partial \xi_i},$$

we simply write

$$D_i w = w_{,i}.$$

This is a usual convention in elasticity theory. For example, we have

$$D_j w_k = w_{k,j} \quad \text{and} \quad D_i D_j w = w_{,ij}.$$

Let $\|\cdot\|_q$ and $\|\cdot\|_{m,q}$ denote the norms on $L_q(\Omega)$ and on the Sobolev space $W_q^m(\Omega)$, respectively.

65.3. The von Kármán Plate Equations

Let Ω be a bounded, simply connected region in the (ξ_1, ξ_2)-plane with sufficiently regular boundary, i.e., $\partial\Omega \in C^{0,1}$. The von Kármán plate equations are

$$\Delta^2 U = -2^{-1} E \varphi(w, w) \quad \text{on } \Omega, \tag{6}$$

$$\Delta^2 w = 2aD^{-1}(\varphi(w, U + pU_0) + K_3) \quad \text{on } \Omega, \tag{7}$$

with the boundary conditions

$$w = w_{,1} = w_{,2} = 0 \quad \text{on } \partial\Omega, \tag{8}$$

$$U = U_{,1} = U_{,2} = 0 \quad \text{on } \partial\Omega. \tag{9}$$

We set

$$\varphi(v, w) = v_{,11} w_{,22} + v_{,22} w_{,11} - 2 v_{,12} w_{,12}$$

and note that

$$D = 2Ea^3/3(1 - \mu^2),$$

$E =$ elasticity module, $\quad \mu =$ Poisson number,

$2a =$ thickness of the plate.

In order to simplify the notation we choose a physical system of units for which

$$2^{-1} E = 2aD^{-1} = 1.$$

We are looking for the real functions w and U.

In Section 65.7 we shall show that all displacements and stress forces of the plate are obtained from a solution w, U, i.e.,

> *all physically interesting quantities of the plate follow from a solution w, U of the von Kármán equation.*

Let us restrict ourselves here to the following remarks:

(i) *Displacements.* In the undeformed state, the plate is in the region

$$G = \Omega \times]-a, a[$$

of $\mathbb{R}^3$ (Fig. 65.1). Thus Ω lies in the medium plate plane. Moreover, w is

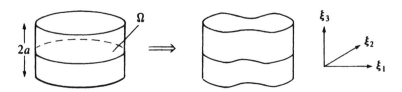

Figure 65.1

65.4. The Operator Equation

the displacement of Ω in the direction of the ξ_3-axis, i.e., $(\xi_1, \xi_2, 0)$ is deformed into $(\xi_1, \xi_2, w(\xi_1, \xi_2))$. The boundary condition (8) means that the plate is clamped at the boundary.

(ii) *Outer forces.* The volume force

$$\left(\int_G K_3 \, dx \right) e_3 = \left(2a \int_\Omega K_3(\xi_1, \xi_2) \, d\xi_1 \, d\xi_2 \right) e_3$$

and the boundary stress force

$$2a \int_{\partial \Omega} (T_1(s) e_1 + T_2(s) e_2) \, ds$$

act on the deformed plate, where s is the arclength of the boundary curve which corresponds to $\partial \Omega$. We orient this curve in such a way that Ω lies to the left of this curve.

The function pU_0 with the real parameter p, given on $\bar{\Omega}$, then describes the boundary stress forces as

$$p \frac{d}{ds} U_{0,2} = T_1, \quad -p \frac{d}{ds} U_{0,1} = T_2 \quad \text{on } \partial \Omega. \tag{10}$$

We thereby make the additional assumption that

$$\Delta^2 U_0 = 0 \quad \text{on } \Omega.$$

(iii) *Stress forces.* See Section 65.7. If one knows a solution w, U of the von Kármán equations, then the displacements in the direction of the ξ_i-axis with $i = 1, 2$ are determined only up to infinitesimally small rigid motions. This will be discussed in Section 65.7.

65.4. The Operator Equation

We now want to show that the generalized problem to the von Kármán equations leads to the operator equation

$$w - pLw + Cw = f, \quad w \in X \tag{11}$$

with $X = \mathring{W}_2^2(\Omega)$.

Roughly, the *idea* is:

(i) to eliminate the function U by solving equation (6) for U; and
(ii) inserting this expression into (7).

Classically, we thereby have to solve a boundary-value problem for the equation

$$\Delta^2 U = -\varphi(w, w) \quad \text{on } \Omega,$$

with w given.

Functional analytically, this can be done, very elegantly, by introducing on the Sobolev space $X = \mathring{W}_2^2(\Omega)$ the equivalent energetic scalar product

$$(v|w) = \int_\Omega \sum_{i,j=1}^2 v_{,ij} w_{,ij}\, dx$$

(see A_2(53b)). Integration by parts yields

$$(v|w) = \int_\Omega \Delta v\, \Delta w\, dx \qquad \text{for all}\quad v, w \in X,$$

and multiplying (6) and (7) with Z and z, respectively, we again apply integration by parts to obtain

$$(U|Z) = -b(w, w, Z),$$
$$(w|z) = b(w, U + pU_0, z) + F(z) \qquad \text{for all}\quad z, Z \in C_0^\infty(\Omega), \tag{12}$$

where

$$b(v, w, z) = \int_\Omega \varphi(v, w) z\, dx, \qquad F(z) = \int_\Omega K_3 z\, dx.$$

65.4a. The Generalized Problem

Definition 65.1. Set $X = \mathring{W}_2^2(\Omega)$. The generalized problem to the von Kármán plate equations (6)–(9) is the following. Suppose

$$p \in \mathbb{R}, \qquad K_3 \in L_2(\Omega), \qquad U_0 \in W_2^2(\Omega)$$

with $\Delta^2 U_0 = 0$ on Ω are given.
We are then looking for $U, w \in X$ such that equation (12) is satisfied.

Every classical solution of (6)–(9) satisfies (12), i.e., is a generalized solution. If we reverse integration by parts in (12), then every sufficiently smooth generalized solution is also a classical solution of (6)–(9). This is a consequence of the arbitrariness of z and Z.
We use the space X, since it follows from $w, U \in X$ that the boundary conditions (8), (9) are satisfied in the generalized sense.

65.4b. Elimination of the Airy Stress Function U via the Theorem of Riesz

Our next goal is to obtain the system

$$U = -B(w, w),$$
$$w = B(w, U) + pLw + f \tag{13}$$

65.4. The Operator Equation

from the generalized problem (12) by using the theorem of Riesz of Section 18.11. To this end we set

$$(B(v, w)|z) = b(v, w, z),$$
$$(Lw|z) = b(w, U_0, z), \tag{14}$$
$$(f|z) = F(z) \quad \text{for all} \quad v, w, z \in X.$$

For this to make sense we have to show that

$$|(b(v, w, z)| \leq \text{const} \|z\|,$$
$$|b(w, U_0, z)| \leq \text{const} \|z\|, \tag{15}$$
$$|F(z)| \leq \text{const} \|z\| \quad \text{for all} \quad z \in X$$

and arbitrarily fixed $v, w \in X$. In fact, if we fix $v, w \in X$ and suppose that the first inequality in (15) holds true, then

$$z \mapsto b(v, w, z)$$

is a linear, continuous functional on the H-space X. According to the theorem of Riesz, there exists an element $B(v, w) \in X$ such that the first line of (14) holds true. In a similar way we obtain the second and third line of (14) from the corresponding lines of (15).

The proof of (15) will be given below. From (12), (14) follows (13), since $C_0^\infty(\Omega)$ is dense in X.

65.4c. Properties of the Operator Equation

We now set
$$Cw = B(w, B(w, w)).$$

Then the system (13) becomes

$$w = -Cw + pLw + f, \quad w \in X.$$

This is the operator equation (11).

Proposition 65.2. *The generalized problem (12) to the von Kármán equations is equivalent to the operator equation (11), whereby the following holds:*

(a) $L: X \to X$ *is linear, symmetric, and strongly continuous.*
(b) $B: X \times X \to X$ *is bilinear, symmetric, and strongly continuous.*
(c) $C: X \to X$ *is a strongly continuous potential operator which satisfies the key property*
$$(Cw|w) = \|B(w, w)\|^2 \quad \text{for all} \quad w \in X.$$

PROOF. We use arguments analogously to Section 27.4. All constants will be denoted by c.

The point is that the integral, which corresponds to $b(v, w, z)$ in (12), contains three factors. In order to estimate this integral, we will use Hölder's inequality for three factors

$$\left| \int_\Omega efg\, dx \right| \leq \|e\|_2 \|f\|_4 \|g\|_4, \tag{16}$$

where $e \in L_2(\Omega)$, and $f, g \in L_4(\Omega)$, and also the following Sobolev embedding theorem:

$$\text{The embedding } \mathring{W}_2^2(\Omega) \subseteq W_4^1(\Omega) \text{ is compact.} \tag{17}$$

This is a consequence of our general results in $A_2(45)$.

In fact, (17) is the heart of the proof, since it implies the strong continuity of B.

(I) Preparations. Integration by parts yields the symmetry relation

$$\int_\Omega \varphi(v, w) U\, dx = \int_\Omega \varphi(U, w) v\, dx \tag{18}$$

for all $v, w, U \in C_0^\infty(\Omega)$. Furthermore, we have the important divergence representation

$$\varphi(v, w) = D_1 \psi(v, w) + D_2 \chi(v, w) \tag{19}$$

with $D_j = \partial/\partial \xi_j$ and

$$\psi(v, w) = v_{,1} w_{,22} - v_{,2} w_{,12},$$

$$\chi(v, w) = v_{,2} w_{,11} - v_{,1} w_{,12}.$$

Again, using integration by parts we find that

$$\int_\Omega \varphi(v, w) U\, dx = -\int_\Omega \psi(v, w) U_{,1} + \chi(v, w) U_{,2}\, dx \tag{20}$$

for all $v, w, U \in C_0^\infty(\Omega)$.

Note that relations (18) and (20) also hold true if, in each case, two functions only belong to $C^2(\bar{\Omega})$.

In the following it is important that ψ and χ in (20) only contain first-order derivatives of v.

Let $Y = W_4^1(\Omega)$ with norm

$$|v| \stackrel{\text{def}}{=} \|v\|_{1,4},$$

and recall that $\|v\|$ denotes the norm on $X = \mathring{W}_2^2(\Omega)$. From (16), (17), and (20) we obtain the *key estimate*

$$\left| \int_\Omega \varphi(v, w) U\, dx \right| \leq c \|v\|_{1,4} \|w\|_{2,2} \|U\|_{1,4} \tag{21}$$

for all $v, w, U \in \mathring{W}_2^2(\Omega)$. Notice that this is also true if $w \in W_2^2(\Omega)$.

65.4. The Operator Equation

In order to obtain (21), we first prove this relation for the space $C_0^\infty(\Omega)$. Then we use the density of $C_0^\infty(\Omega)$ in $\mathring{W}_2^2(\Omega)$. In the case $w \in W_2^2(\Omega)$, we need the density of $C^\infty(\bar{\Omega})$ in $W_2^2(\Omega)$.

(II) Proof of (14). From (21) follows that

$$|F(z)| \leq \|K_3\|_2 \|z\|_2 \leq c\|z\|,$$

$$|b(v, w, z)| \leq c|v| \|w\| \|z\|,$$

$$|b(w, U_0, z)| \leq c|w| \|z\| \quad \text{for all} \quad z \in X.$$

This is (14). Moreover, we obtain

$$\|Lw\| \leq c|w|,$$
$$\|B(v, w)\| \leq c|v| \|w\| \quad \text{for all} \quad v, w \in X. \tag{22}$$

(III) Proof of (a). Because of (17) the embedding $X \subseteq Y$ is compact. Consequently, the weak convergence

$$w_n \rightharpoonup w \quad \text{in } X \quad \text{as} \quad n \to \infty$$

implies the strong convergence $w_n \to w$ in Y, i.e.,

$$|w_n - w| \to 0 \quad \text{as} \quad n \to \infty.$$

From (22) follows

$$\|Lw_n - Lw\| \leq c|w_n - w| \to 0 \quad \text{as} \quad n \to \infty.$$

Therefore, the operator $L: X \to X$ is strongly continuous.

Equation (18) tells us that

$$b(v, U_0, w) = b(w, U_0, v) \quad \text{for all} \quad v, w \in X.$$

Thus, L is symmetric.

(IV) Proof of (b). Because of the symmetry of φ we have

$$b(v, w, z) = b(w, v, z) \quad \text{for all} \quad v, w, z \in X,$$

i.e., B is symmetric. The strong continuity of B follows from (22) by using a similar argument as in (III) for L. In this connection we note that there exists a splitting

$$\|B(e, f) - B(g, h)\| = \|B(e - g, f) + B(f - h, g)\|$$
$$\leq c(|e - g| \|f\| + |f - h| \|g\|),$$

and observe the boundedness of weakly convergent sequences.

(V) Proof of (c). The strong continuity of C follows from the strong continuity of B. Equation (18) implies

$$(B(v, w)|z) = (B(z, w)|v) \quad \text{for all} \quad v, w, z \in X.$$

Thus, according to Problem 65.1, it follows that $C: X \to X$ is a potential operator. Moreover, setting $z = B(v, v)$, we obtain

$$(Cv|v) = (B(z, v)|v) = (B(v, v)|z) = \|B(v, v)\|^2. \quad \square$$

65.5. Existence Theorem

Theorem 65.A (Existence). Set $X = \mathring{W}_2^2(\Omega)$. *The operator equation* (11), *which corresponds to the generalized problem for the von Kármán plate equations, has a solution for every $f \in X$ and every $p \in \mathbb{R}$ with sufficiently small $|p|$.*

Corollary 65.3 (Uniqueness). *If p is not a characteristic number of the linear operator L in* (11), *then there exists a constant $r_0 > 0$ such that the operator equation* (11) *has a unique solution $w \in X$ with*

$$\|w\| \le r_0$$

for every $f \in X$ with sufficiently small norm.

PROOF OF THEOREM 65.A. Let $A = I - pL + C$. The operator $A: X \to X$ is a strongly continuous perturbation of the identity. According to Section 27.2, A is pseudomonotone and bounded. From (5) follows that A is coercive.

The main theorem about pseudomonotone operators (Theorem 27.A) yields the existence result. □

PROOF OF CROLLARY 65.3. We write equation (11) in the form

$$F(w, p, f) = 0.$$

Because of $F_w(0, p, 0) = I - pL$, the assertion follows from the implicit function theorem (Theorem 4.B). □

The operator $L: X \to X$ is symmetric and compact. Therefore, the set of characteristic numbers p of L is nonempty, at most countable, and $\pm\infty$ are the only possible accumulation points.

The solutions in Theorem 65.A can be computed approximately by a Galerkin procedure. According to Theorem 27.A, a subsequence of the Galerkin sequence converges to a solution w of (11). Moreover, the entire Galerkin sequence converges to w if the solution w of (11) is unique.

The physical meaning of these results have already been discussed in Section 65.1. Recall that f and p correspond to the volume and boundary forces, respectively.

65.6. Bifurcation

We make the following assumptions:

(H) It is $K_3 = 0$, i.e., no outer volume forces are present. The function $U_0 \in W_2^2(\Omega)$ is given as a solution of the differential equation $\Delta^2 U_0 = 0$ on Ω.

65.6. Bifurcation

According to (10), the boundary values of pU_0 are related to the boundary stress forces.

The fundamental operator equation (11) then becomes

$$w = pLw - Cw, \quad w \in X \qquad (23)$$

with the linearization

$$w = pLw, \quad w \in X. \qquad (24)$$

Here, $p \in \mathbb{R}$.

Theorem 65.B (Bifurcation). *Assume* (H). *Then $(p, 0)$ is a bifurcation point of* (23) *if and only if p is a characteristic number of* (24).

Equation (24) *has at least one characteristic number p.*

Corollary 65.4 (Multiplicity). *If the multiplicity of the characteristic number $p_0 > 0$ of the linearized problem* (24) *is equal to $n \geq 1$, then the nonlinear problem* (23) *has at least n different nontrivial solutions (p, w) with*

$$|p - p_0| < r, \quad 0 < \max_{x \in \bar{\Omega}} |w(x)| \leq r$$

for every $r > 0$.

PROOF. From Proposition 8.2 we know that if $(p, 0)$ is a bifurcation point of (23), then p is a characteristic number of (24).

The converse follows from the main theorem of bifurcation theory for potential operators (see Theorem 45.A and Example 45.2).

Corollary 65.4 is a consequence of Theorem 45.A. Note that

$$\max_{x \in \bar{\Omega}} |w(x)| \leq \text{const} \|w\|_X,$$

which is implied by the continuity of the embedding $\mathring{W}_2^2(\Omega) \subseteq C(\bar{\Omega})$ and $X = \mathring{W}_2^2(\Omega)$. □

The important physical interpretation of these bifurcation results has already been given in Section 65.1. Note that because of $w \neq 0$, there exists a buckling of the plate in the direction of the ξ_3-axis for each bifurcation solution (p, w).

Since the operator C is analytic, one can also apply the apparatus of analytic bifurcation theory of Chapter 8 to equation (23). For this, explicitly known eigensolutions of the linearized equation (24) are needed. For example, in case that the region Ω is a disk or a rectangle, such eigensolutions are known.

In addition, we can use the results of topological bifurcation theory of Chapter 15 which concern the global behavior of the bifurcation branches in the case that the multiplicity of the characteristic number p of (24) is odd.

65.7. Physical Motivation of the Plate Equations

Our goal is to motivate the von Kármán plate equations by using only the natural model assumptions (H1)–(H4) below.

We set $G = \Omega \times]-a, a[$ and, as in Section 61.6, we start from the *principle of stationary potential energy*

$$\int_G A(\mathscr{E}) \, dx - \int_G Ku \, dx - \int_{\partial G} Tu \, dO = \text{stationary!} \qquad (25)$$

with boundary conditions

$$u_3 = u_{3,1} = u_{3,2} = 0 \quad \text{on } \partial\Omega \times]-a, a[. \qquad (26)$$

Here, the displacement of the plate is given by

$$u = u_1 e_1 + u_2 e_2 + u_3 e_3$$

(Fig. 65.1 in Section 65.3) and $\mathscr{E}$ is the strain tensor with components

$$\mathscr{E}_{ij} = \tfrac{1}{2}(u_{i,j} + u_{j,i} + u_{k,i}u_{k,j}), \quad i, j = 1, 2, 3.$$

The sum is taken over $k = 1, 2, 3$.

We now formulate a number of assumptions. Roughly speaking, we require the following:

All effects perpendicular to the plate plane, i.e., in the direction of the ξ_3-axis, are small.

(H1) In the undeformed state, the plate corresponds to the set $\bar{G}$, where $G = \Omega \times]-a, a[$. The thickness of the plate $2a$ is small compared with the diameter of Ω, where the region Ω lies in the (ξ_1, ξ_2)-plane. Suppose that

$$\mathscr{E}_{i3} = \mathscr{E}_{3j} = 0 \quad \text{for } i, j = 1, 2, 3,$$

i.e., there exist no extensions or contractions in the direction of the ξ_3-axis.

(H2) The density of the outer volume forces has the form $K = \sum_{i=1}^{3} K_i e_i$ with

$$K_1 = K_2 = 0, \qquad K_3 = K_3(\xi_1, \xi_2).$$

The density T of the outer boundary stress forces has the form

$$T_3 = 0 \quad \text{and} \quad T_i = T_i(\xi_1, \xi_2) \quad \text{for } i = 1, 2.$$

We assume that no boundary stress forces are acting on the covering surfaces of the plate, i.e.,

$$T_i = 0 \quad \text{on } \Omega \times \{\pm a\} \quad \text{for } i = 1, 2, 3.$$

In the following we sum from 1 to 2 over two equal indices.

(H3) For the stored energy function, which represents the elastic potential

65.7. Physical Motivation of the Plate Equations

energy of the plate, we make the ansatz

$$A = \tfrac{1}{2}\bar\lambda(\mathscr{E}_{11} + \mathscr{E}_{22})^2 + \kappa \mathscr{E}_{ij}\mathscr{E}_{ij}$$

with $\bar\lambda = E\mu/(1-\mu^2)$ and $\kappa = E/2(1+\mu)$.

This expression was obtained in Problem 61.4 by considering a cuboid, where the stress forces were not acting in the direction of the ξ_3-axis. In order to obtain a nonlinear model, we replace here the linearized strain tensor γ of Problem 61.4 with the strain tensor $\mathscr{E}$.

(H4) The displacements satisfy the following assumptions. The displacement u_3, i.e., the deflection of the plate in the ξ_3-direction, does not depend on ξ_3. We set $w = u_3$, hence

$$w = w(\xi_1, \xi_2).$$

We expand the displacements u_1 and u_2 with respect to ξ_3 and take only linear terms into account, i.e.,

$$u_i = \bar u_i(\xi_1, \xi_2) + \hat u_i(\xi_1, \xi_2)\xi_3, \qquad i = 1, 2.$$

In $\mathscr{E}_{ij}$ we only keep the quadratic terms with $k = 3$, i.e.,

$$\mathscr{E}_{ij} = \tfrac{1}{2}(u_{i,j} + u_{j,i} + w_{,i}w_{,j}), \qquad i = 1, 2, 3. \tag{27}$$

This is motivated by the assumption that the first-order derivatives of u_1 and u_2 are small compared with those of u_3.

We now study the purely mathematical consequences of these model assumptions.

Step 1: Computation of the displacements u_i and of the strain tensor $\mathscr{E}_{ij}$.

From $\mathscr{E}_{13} = \mathscr{E}_{23} = 0$ and (27) it follows that $u_{i,3} = -u_{3,i}$, $i = 1, 2$, hence $\hat u_i = -w_{,i}$. Consequently, for $i = 1, 2$, we have

$$u_i = \bar u_i - w_{,i}\xi_3, \tag{28}$$

$$\mathscr{E}_{ij} = \bar{\mathscr{E}}_{ij} - w_{,ij}\xi_3, \tag{29}$$

with

$$\bar{\mathscr{E}}_{ij} \stackrel{\text{def}}{=} \bar e_{ij} + \tfrac{1}{2}w_{,i}w_{,j}, \qquad \bar e_{ij} \stackrel{\text{def}}{=} \tfrac{1}{2}(\bar u_{i,j} + \bar u_{j,i}).$$

Step 2: Auxiliary quantities $\mu_{ij} = \partial A(\bar{\mathscr{E}})/\partial \bar{\mathscr{E}}_{ij}$.

We introduce these quantities in order to be able to express the first variation below in a simple form. The physical interpretation of μ_{ij} will be given at the end of this section. Explicitly, we have

$$\mu_{ij} = \bar\lambda(\bar{\mathscr{E}}_{11} + \bar{\mathscr{E}}_{22})\delta_{ij} + 2\kappa\bar{\mathscr{E}}_{ij}, \qquad i, j = 1, 2.$$

By solving this system of equations we obtain

$$\bar{\mathscr{E}}_{11} = E^{-1}(\mu_{11} - \mu\mu_{22}), \qquad \bar{\mathscr{E}}_{22} = E^{-1}(\mu_{22} - \mu\mu_{11}),$$
$$\bar{\mathscr{E}}_{12} = \bar{\mathscr{E}}_{21} = E^{-1}(1+\mu)\mu_{12}. \tag{30}$$

All quantities μ_{ij} depend only on ξ_1 and ξ_2.

Step 3: Computation of the first variation.
The original variational problem (25) now becomes

$$J \stackrel{\text{def}}{=} \int_\Omega \int_{-a}^{a} A(\mathscr{E}) \, dx - 2a \int_\Omega K_3 w \, d\xi_1 \, d\xi_2 \qquad (31)$$

$$- \int_{\partial\Omega} T_1 \bar{u}_1 + T_2 \bar{u}_2 \, ds = \text{stationary!}$$

with boundary condition

$$w = w_{,1} = w_{,2} = 0 \quad \text{on } \partial\Omega. \qquad (32)$$

We thereby replace all $\mathscr{E}_{ij}$ in $A(\mathscr{E})$ with the expressions in (29). The integrands then depend on $\bar{u}_1, \bar{u}_2, w$, and ξ_3.

According to our standard procedure for variational problems, we replace

$$\bar{u}_i \text{ and } w \quad \text{with} \quad \bar{u}_i + th_i \quad \text{and} \quad w + th,$$

respectively, where $h_i, h \in C^\infty(\bar{\Omega})$. For $w + th$ to be an admissible variation, we assume that

$$h = h_{,1} = h_{,2} = 0 \quad \text{on } \partial\Omega.$$

In this way, the function $w + th$ satisfies the boundary condition (32). The integral in (31) now depends on the real parameter t, i.e., we obtain $J = J(t)$. We set $\delta J = J'(0)$.

If $\bar{u}_1, \bar{u}_2, w$ is a solution of (31), (32), then

$$\delta J = 0.$$

After integration over $[-a, a]$ with respect to ξ_3 and integration by parts, we obtain the *key* relation

$$\delta J = \int_\Omega (A_i h_i + Ch) \, d\xi_1 \, d\xi_2 + \int_{\partial\Omega} B_i h_i \, ds = 0$$

with

$$A_i = -2a(\mu_{i1,1} + \mu_{i2,2}), \quad i = 1, 2,$$

$$B_i = 2a(\mu_{ij} n_j - T_i),$$

$$C = D \Delta^2 w - 2aK_3 + A_i w_{,i} - 2a\mu_{ij} w_{,ij}.$$

Here, $n = n_1 e_1 + n_2 e_2$ is the outer unit normal vector of $\partial\Omega$.

Step 4: The Euler equations.
Because of the arbitrariness of h_i and h, we obtain, in the usual way, from $\delta J = 0$, the Euler equations

$$A_i = 0 \quad \text{on } \Omega, \quad i = 1, 2, \qquad (33)$$

$$B_i = 0 \quad \text{on } \partial\Omega,$$

$$C = 0 \quad \text{on } \Omega. \qquad (34)$$

65.7. Physical Motivation of the Plate Equations

Step 5: Simplification of the Euler equation (33) by using Airy's stress function V.

If we choose a smooth function V with

$$\frac{d}{ds}V_{,2} = T_1 \quad \text{and} \quad -\frac{d}{ds}V_{,1} = T_2 \quad \text{on } \partial\Omega, \tag{35}$$

and if we set

$$\mu_{11} = V_{,22}, \quad \mu_{22} = V_{,11} \quad \text{on } \Omega,$$
$$\mu_{12} = -V_{,12} \quad \text{on } \Omega, \tag{36}$$

then μ_{ij} is a solution of (33).

However, in order to obtain the von Kármán equations in a rigorous way, we also need that the converse is true. This is sometimes overlooked in the engineering literature. In fact, if $\partial\Omega$ is sufficiently smooth, then every smooth solution μ_{ij} of (33) can be represented in the form (35), (36). This will be shown in Problem 65.5.

Step 6: Decomposition of $V = U + pU_0$.

Suppose that the boundary forces have the form

$$T_i = pT_i^0 \quad \text{on } \partial\Omega, \quad i = 1, 2$$

with the real parameter p. Let $p \neq 0$. We want to show that it is possible to find a uniquely determined decomposition of $V = U + pU_0$, where U is independent of the boundary forces.

First, assume that V is given with the boundary condition (35). By definition, U is the unique solution of the classical boundary-value problem

$$\Delta^2 U = -\Delta V \quad \text{on } \Omega,$$
$$U = U_{,1} = U_{,2} = 0 \quad \text{on } \partial\Omega.$$

Now set $pU_0 = V - U$. Then,

$$\Delta^2 U_0 = 0 \quad \text{on } \Omega, \tag{37}$$

and

$$pU_0 = V, \quad pU_{0,1} = V_{,1}, \quad pU_{0,2} = V_{,2} \quad \text{on } \partial\Omega. \tag{38}$$

Hence, the function pU_0 is uniquely determined by V. Moreover, (35) implies that

$$p\frac{d}{ds}U_{0,2} = T_1, \quad -p\frac{d}{ds}U_{0,1} = T_2 \quad \text{on } \partial\Omega. \tag{39}$$

Conversely, if U_0 satisfies (37) and (39) and if we choose a function U with the boundary condition

$$U = U_{,1} = U_{,2} = 0 \quad \text{on } \partial\Omega,$$

then (38) holds and hence $V = U + pU_0$ satisfies (35).

Step 7: The second von Kármán equation (7).
Inserting (36) into the second Euler equation (34), and using $V = U + pU_0$, we obtain (7).

Step 8: The first von Kármán equation (6).
It is important here that the relation

$$\bar{e}_{ij} = \tfrac{1}{2}(\bar{u}_{i,j} + \bar{u}_{j,i}) \tag{40}$$

implies the so-called compatibility condition

$$\bar{e}_{11,22} + \bar{e}_{22,11} - 2\bar{e}_{12,12} = 0. \tag{41}$$

Conversely, if all the $\bar{e}_{ij}$'s are given, then we obtain the u_i's from (40) in case the compatibility condition (41) is satisfied. This will be shown in Problem 65.4.

The second von Kármán equation is an equation for the two functions w and U. We therefore need another equation. Suppose we know w and U. From U we obtain μ_{ij} according to (36). From μ_{ij} we obtain $\bar{e}_{ij}$ according to (30). In order to determine the displacements $\bar{u}_1$ and $\bar{u}_2$, we have to solve equation (40). Thus we need the compatibility condition (41).

In fact, the first von Kármán equation is nothing other than (41). This follows, since, according to Step 1

$$\bar{e}_{ij} = \bar{\mathscr{E}}_{ij} - \tfrac{1}{2} w_{,i} w_{,j},$$

and from (30), we can express $\bar{\mathscr{E}}_{ij}$ in terms of μ_{ij}. Formula (41) then yields (6).

Step 9: Computation of the physically important quantities.
Suppose we have a solution w, U of the von Kármán equations (6)–(9). We then need to show how to compute the displacements and the stresses of the plate.

(i) *Displacements.* From (36) and (30) we find that

$$(w, U) \Rightarrow V = U + pU_0 \Rightarrow \mu_{ij} \Rightarrow \bar{\mathscr{E}}_{ij} \Rightarrow \bar{e}_{ij}.$$

The first von Kármán equation (6) guarantees that the compatibility condition is satisfied. Thus, we can use Problem 65.4 in order to obtain the displacements $\bar{u}_1$ and $\bar{u}_2$ from equation (40). Up to infinitesimally small rigid motions they are uniquely determined.

From (28) we obtain, in addition, the quantities

$$u_1, u_2, \quad \text{and} \quad u_3 = w,$$

i.e., the displacements are known.

(ii) *Stresses.* According to Section 61.6, we derive the first Piola–Kirchhoff stress tensor σ from the variational principle (25) by using

$$\sigma_j{}^i = \frac{\partial A}{\partial u_{i,j}} \quad \text{for} \quad i, j = 1, 2, 3.$$

If Ω_1 is a subregion of Ω, then, after the deformation, the region $H = \Omega_1 \times \,]-a, a[$ becomes H'. On the deformed region H', there acts the stress force

$$\int_{\partial H} \sigma n\, dO$$

with

$$\sigma n = \sum_{i,j=1}^{3} \sigma_j^i n^j e_i,$$

where $n = \sum_{j=1}^{3} n^j e_j$ is the outer unit normal vector of the boundary of H. This follows from Section 61.6.

In particular, we obtain that

$$\sigma_j^i = \mu_{ij} \quad \text{for} \quad \xi_3 = 0 \quad \text{and} \quad i, j = 1, 2.$$

This yields the desired physical interpretation of the quantities μ_{ij}.

Remark 65.5 (Alternate Approach). The von Kármán equations can also be obtained by showing that, under suitable assumptions, those equations represent the first term in an asymptotic expansion derived from the basic equations of three-dimensional nonlinear elasticity theory. This can be found in Ciarlet and Rabier (1980, M).

65.8. Principle of Stationary Potential Energy and Plates with Obstacles

Our goal is to formulate the principle of stationary potential energy in terms of the vertical displacement $w = u_3$. We start from the fundamental operator equation of Section 65.4, i.e., we consider again

$$w - pLw + Cw = f, \quad w \in X \tag{42}$$

with $X = \mathring{W}_2^2(\Omega)$.

Proposition 65.6 (Principle of Stationary Potential Energy for Plates). *Equation (42) is the Euler equation to the variational problem*

$$\frac{1}{2}\|w\|^2 + \frac{1}{4}\|U\|^2 - \frac{p}{2}(Lw|w) - (f|w) = \text{stationary!},$$

$$U = -B(w, w), \quad w \in X. \tag{43}$$

PROOF. According to Problem 65.1, the operator C has the potential $w \mapsto 4^{-1}\|B(w, w)\|^2$. □

From (14), we obtain $(Lw|w) = b(w, U_0, w)$ and $(f|w) = \int_\Omega K_3 w\, dx$. Hence the variational problem (43) corresponds to the classical problem

$$\int_\Omega \frac{1}{2}(\Delta w)^2 + \frac{1}{4}(\Delta U)^2 - \frac{p}{2}\varphi(w, U_0)w - K_3 w\, dx = \text{stationary!},$$

$$\Delta^2 U = -\varphi(w, w) \quad \text{on } \Omega,$$

$$w = w_{,1} = w_{,2} = 0 \quad \text{on } \partial\Omega, \qquad (43^*)$$

$$U = U_{,1} = U_{,2} = 0 \quad \text{on } \partial\Omega,$$

where

$$\varphi(w, v) = w_{,11} v_{,22} + v_{,11} w_{,22} - 2 w_{,12} v_{,12}.$$

The function U_0 satisfies

$$\Delta^2 U_0 = 0 \quad \text{on } \Omega,$$

$$p\frac{d}{ds} U_{0,2} = T_1, \qquad -p\frac{d}{ds} U_{0,1} = T_2 \quad \text{on } \partial\Omega.$$

As a consequence of Section 65.3, T_1 and T_2 correspond to the boundary forces and K_3 corresponds to the volume forces. Recall that we use a special physical system of units where $2^{-1}E = 2aD^{-1} = 1$.

The advantage of the variational problem (43) over the operator equation (42) is the fact that it allows a study of the more general problem of plates with obstacles. This is pictured in Figure 65.2. We proceed here in the same way as in the case of beams in Section 64.6. According to (43), the elastic potential energy of the plate is

$$F(w) = \tfrac{1}{2}\|w\|^2 + \tfrac{1}{4}\|U(w)\|^2$$

with $U(w) = -B(w, w)$, and the work of the outer forces is given by

$$pG(w) = \frac{p}{2}(Lw|w).$$

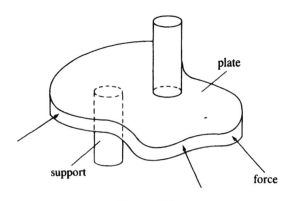

Figure 65.2

65.8. Principle of Stationary Potential Energy and Plates with Obstacles

Here we assume that there exist no outer volume forces, i.e., $K_3 = 0$, and hence $f = 0$. For the second F-derivatives we obtain

$$\langle F''(0)w, v \rangle = (w|v), \qquad \langle G''(0)w, v \rangle = (Lw|v).$$

Let C be the set of all $w \in X$ with

$$w(x) \geq 0 \quad \text{on } \Omega_1,$$
$$w(x) \leq 0 \quad \text{on } \Omega_2,$$

where Ω_1 and Ω_2 are subsets of Ω. This condition describes the obstacles. We then study the variational problem

$$F(w) - pG(w) = \min!, \qquad w \in C.$$

The corresponding variational inequality is

(V) $\qquad \lambda \langle F'(w), v - w \rangle \geq \langle G'(w), v - w \rangle \qquad \text{for all } v \in C$

with $\lambda = 1/p$.

We make the following assumptions:

(H1) Let Ω be a simply connected region in $\mathbb{R}^2$ with $\partial \Omega \in C^{0,1}$. Moreover, let Ω_1 and Ω_2 be subsets of Ω such that the set C is nontrivial, i.e., $C \neq \{0\}$.
(H2) There are no outer volume forces, i.e., $K_3 = 0$.
(H3) There exists a fixed vertical displacement $w \in X$ such that the work of the outer boundary forces is positive, i.e., $(Lw|w) > 0$.
(H4) We are given $U_0 \in W_2^2(\Omega)$ with $\Delta^2 U_0 = 0$ on Ω.

Theorem 65.C (Plates with Obstacles). *Assume* (H1)–(H4). *The maximum problem*

(M) $\qquad \lambda_0 = \max_{w \in \bar{C}} \dfrac{(Lw|w)}{(w|w)}$

then has a solution and $\lambda_0 > 0$ is the largest bifurcation value of the variational inequality (V).

PROOF. Recall that $X = \mathring{W}_2^2(\Omega)$. The embedding $X \subseteq C(\bar{\Omega})$ is continuous. Hence C is a closed, convex cone in X.

Proposition 65.2 shows that the convex functional $w \mapsto \|w\|^2$ is weakly sequentially lower continuous, and that $w \mapsto \|U(w)\|^2$ is weakly sequentially continuous. Moreover, it follows that $w \mapsto (Lw|w)$ is weakly sequentially continuous.

Hence, the assertion follows from Theorem 64.A and Corollary 64.4. □

In physical terms, this theorem guarantees that there exists a buckling force, which corresponds to the force parameter $p_0 = 1/\lambda_0$ of the boundary forces.

In (M), recall that

$$(w|w) = \int_\Omega (\Delta w)^2 \, dx,$$

$$(Lw|w) = \int_\Omega \varphi(w, U_0) w \, dx.$$

The boundary forces acting on the plate are given by

$$2a \int_{\partial\Omega} (T_1 e_1 + T_2 e_2) \, ds$$

with

$$T_1 = p \frac{d}{ds} U_{0,2}, \qquad T_2 = -p \frac{d}{ds} U_{0,1} \quad \text{on } \partial\Omega,$$

and arbitrary $p > 0$. Here s denotes the arclength of $\partial\Omega$. The buckling force corresponds to $p = p_0$. The function U_0 has been introduced above.

Remark 65.7 (Multiply Connected Plates). The existence result (Theorem 64.A) as well as the bifurcation result (Theorem 64.B) and Theorem 64.C remain valid if Ω is an arbitrary region, i.e., if Ω is *not* simply connected. In fact, the proofs of Theorems 64.A–64.C do not use the fact that Ω is simply connected. Thus these theorems describe properties of the vertical deflection w for arbitrary plates. In Step 9 of Section 65.7, however, we computed the horizontal deflections u_1 and u_2 of the plate by solving the equations

(E) $\qquad \frac{1}{2}(D_i u_j + D_j u_i) = \bar{e}_{ij} \quad \text{on } \Omega, \qquad i, j = 1, 2.$

In the case that Ω is simply connected, we can solve (E) for u_1 and u_2. This is because the integrability conditions, which correspond to the first von Kármán equation, are satisfied. If Ω is not simply connected, then a more careful discussion of equation (E) is required.

PROBLEMS

65.1. Special potential operators. Let $B: X \times X \to X$ be a bilinear, bounded, symmetric operator on the H-space X. Set $Cu = B(u, B(u, u))$ and assume that

$$(B(u, v)|w) = (B(w, v)|u) \quad \text{for all} \quad u, v, w \in X.$$

Prove: The map $C: X \to X$ is a potential operator.

Solution: According to Section 41.3 we begin with

$$F(u) = \int_0^1 (C(tu)|u) \, dt = \frac{1}{4}((Bu, u)|B(u, u)).$$

We find

$$F'(u)h = (B(h, u)|B(u, u)) = (Cu|h) \quad \text{for all} \quad h \in X,$$

and hence $F' = C$.

65.2. *Existence theorem for plates with general boundary forces via a reduction trick.* In order to explain the reduction trick we drop the condition

$$\Delta^2 U_0 = 0 \quad \text{on } \Omega,$$

which was used in Step 6 of Section 65.7. Thus we have to add the term $-p\Delta^2 U_0$ to the right-hand side of equation (6). Moreover, we replace U_0 with ρU_0, where the function ρ is chosen so that U_0 and ρU_0 have the same boundary values and normal derivatives on $\partial\Omega$. Then the corresponding boundary forces in (10) do not change.

The trick is to make a convienent choice of the function ρ. Analogously to Section 65.4, we then obtain the system

$$U = -B(w, w) + p\rho U_0,$$
$$w = pB(w, \rho U_0) + B(w, U) + f.$$

Let $h = (w, U)$. In the product space $\mathring{W}_2^2(\Omega) \times \mathring{W}_2^2(\Omega)$, we then obtain an operator equation

$$h - pL_\rho h + Qh = f_0$$

with $f_0 = (p\rho U_0, f)$ and the linear operator L_ρ. Furthermore, Q is homogeneous of degree two. From

$$(B(w,w)|U) = (B(w,U)|w)$$

it follows that $(Qh|h) = 0$. Let $A = I - pL_\rho + Q$. Then we find

$$\frac{(Ah|h)}{\|h\|} \geq \|h\|(1 - |p|\|L_\rho\|).$$

Suppose that $\rho = 1$. For small $|p|$ the operator A is then coercive. Moreover, A is a strongly continuous perturbation of the identity. Hence, A is pseudo-monotone. From Theorem 27.A we thus obtain a solution h of the equation

$$Ah = f_0$$

for every f_0.

Prove: After a suitable choice of ρ one can always assure that the operator norm $\|L_\rho\|$ becomes arbitrarily small. The key is here that the operator A with $\rho = \rho(p)$ is then coercive for arbitrary p, and one obtains an existence result for the von Kármán equations with arbitrary boundary forces.

Hint: See Nečas and Hlaváček (1981, M), p. 286.

65.3. *Integrability conditions.* Let G be a *simply connected* region in $\mathbb{R}^N$ with $N \geq 1$. Let $f_i \in C^1(G)$ for $i = 1, \ldots, N$. The differential equation

$$U_{,i} = f_i \quad \text{on } G, \quad i = 1, \ldots, N \qquad (44)$$

then has a C^2-solution U if and only if the so-called integrability conditions

$$f_{i,j} = f_{j,i} \quad \text{for all } i, j \qquad (45)$$

are satisfied.

Up to an additive constant, the solution U is uniquely determined.

In the case that (45) holds, we obtain a solution of (44) through

$$U(x) = U(x_0) + \int_{x_0}^{x} f\,dx, \qquad x \in G$$

for fixed $x_0 \in G$, whereby the line integral is path independent.
The necessity of (45) follows from $U_{,ij} = U_{,ji}$.
This classical result should be used in the following two problems.

65.4. *Computation of the displacements from the linearized strain tensor.* Let G be a simply connected region in $\mathbb{R}^N$. We consider the differential equation

$$\tfrac{1}{2}(u_{i,j} + u_{j,i}) = \gamma_{ij} \quad \text{on } G, \qquad i,j = 1, \ldots, N, \tag{46}$$

where the functions $\gamma_{ij} \in C^2(G)$ with $\gamma_{ij} = \gamma_{ji}$ for all i, j are given.

Prove: Equation (46) has a C^3-solution $u = (u_1, \ldots, u_N)$ if and only if the so-called *compatibility conditions*

$$(\gamma_{ij,k} - \gamma_{kj,i})_{,r} = (\gamma_{ir,k} - \gamma_{kr,i})_{,j} \tag{47}$$

are satisfied for all $i, j, k = 1, \ldots, N$.

The solution is uniquely determined up to infinitesimally small rigid motions, i.e., up to functions

$$u_i = a_i + \sum_{j=1}^{N} w_{ij}\xi_j, \qquad i = 1, \ldots, N \tag{48}$$

with arbitrary constants a_i and w_{ij}, where $w_{ij} = -w_{ji}$ for all i, j.

Solution: The trick is to define the functions

$$w_{ij} = \tfrac{1}{2}(u_{i,j} - u_{j,i}).$$

From (46) follows then that

$$u_{i,j} = \gamma_{ij} + w_{ij}, \tag{49}$$

$$w_{ik,j} = \gamma_{ij,k} - \gamma_{kj,i}. \tag{50}$$

(I) Necessity of (47). If u is a solution of (46), then (50) holds, and the compatibility condition (47) follows from the integrability condition

$$w_{ik,jr} = w_{ik,rj}.$$

(II) Sufficiency of (47). If (47) holds, then the integrability condition is satisfied for equation (50). From Problem 65.3 follows that (50) has a solution w_{ik}. By making a suitable choice of the arbitrary constants of the solutions, one can get that

$$w_{ik} = -w_{ki} \qquad \text{for all } k, i. \tag{51}$$

Equation (50) implies that the integrability condition is satisfied for equation (49). Thus (49) has a solution u, which, as a consequence of (51) also satisfies the original equation (46).

(III) Uniqueness. Let u be a solution of the original equation (46) with

$$\gamma_{ij} = 0 \qquad \text{for all } i, j.$$

From (50) follows that

$$w_{ik} = \text{const} \quad \text{for all } i, k,$$

and equation (49) yields (48). Substituting (48) into the original equation (46) with $\gamma_{ij} = 0$, we obtain that $w_{ij} + w_{ji} = 0$.

Conversely, the functions u_i in (48) are always solutions of the original equation (46) with $\gamma_{ij} = 0$.

65.5. *Equilibrium condition in plane elasticity theory and the Airy stress function V.* Let Ω be a bounded, simply connected region in $\mathbb{R}^2$ with smooth boundary, i.e., $\partial\Omega \in C^1$. For $i = 1, 2$ we consider the boundary-value problem

$$\sigma_{ij,j} = 0 \quad \text{on } \Omega, \tag{52}$$

$$\sigma_{ij} n_j = T_i \quad \text{on } \partial\Omega \tag{53}$$

with $\sigma_{12} = \sigma_{21}$. Over equal indices it is summed from 1 to 2. Let $n = (n_1, n_2)$ denote the outer unit normal vector to $\partial\Omega$. The system of equations corresponds to the equilibrium condition

$$\text{div } \sigma = 0 \quad \text{on } \Omega, \tag{52*}$$

$$\sigma n = T \quad \text{on } \partial\Omega \tag{53*}$$

for the stress tensor σ in the plane. Suppose $T_1, T_2 \in C^1(\partial\Omega)$ are given.

Prove: The $C^1(\bar\Omega)$-functions σ_{ij} are solutions of (52), (53) if and only if there exists a $C^3(\bar\Omega)$-function V with

$$\sigma_{11} = V_{,22}, \quad \sigma_{22} = V_{,11} \quad \text{on } \Omega,$$
$$\sigma_{12} = -V_{,12} \quad \text{on } \Omega, \tag{54}$$

$$\frac{d}{ds} V_{,2} = T_1, \quad -\frac{d}{ds} V_{,1} = T_2 \quad \text{on } \partial\Omega. \tag{55}$$

Here, s denotes the arclength on $\partial\Omega$, i.e.,

$$\frac{dU}{ds} = -n_2 U_{,1} + n_1 U_{,2}.$$

Solution: From (54), (55) one immediately obtains (52), (53).

Let, conversely, σ_{ij} be a solution of equation (52), (53). The trick is then to construct functions A and B through

$$A_{,1} = -\sigma_{12}, \quad A_{,2} = \sigma_{11},$$
$$B_{,1} = \sigma_{22}, \quad B_{,2} = -\sigma_{12}. \tag{56}$$

The integrability conditions for this system are satisfied as a consequence of the equilibrium condition (52). We construct the desired function V through

$$V_{,1} = B, \quad V_{,2} = A. \tag{57}$$

and note that, in this case, the integrability conditions are satisfied because of (56). Finally, we find that

$$(56), (57) \Rightarrow (54) \quad \text{and} \quad (53) \Rightarrow (55).$$

65.6.* *Bifurcation for rectangular plates and the general imperfection principle in bifurcation theory.* Consider a plate which is given by the rectangular region

$$\Omega = \,]0, a[\, \times \,]0, b[.$$

It is important that
(i) for noncritical ratios $a:b$ the smallest characteristic number $p_0 > 0$ of the linearized problem is simple; and
(ii) there exist critical ratios $a:b$ for which p_0 is double.

In case (i) Theorem 8.A can be used in order to find a unique bifurcating branch.

In case (ii) the branching equation has to be studied in detail. In order to obtain a generic situation one has to use the method of Theorem 8.F (generic point bifurcation) and methods of catastrophe theory.

The bifurcation diagrams are then rather complicated. This can be seen from Chow and Hale (1982, M), p. 284.

Let us emphasize the following general imperfection principle in bifurcation theory:

> *In order to obtain a complete and natural picture for the bifurcation of mechanical and more general physical systems, one has to add imperfections.*

Mathematically, this means that, in order to obtain a structurally stable situation in the sense of catastrophe theory, one has to add additional parameters (see Section 73.16).

In Section 37.28, this principle has already been discussed in connection with the bifurcation of beams. In this case, one needs to add a small additional vertical load. With regard to plates one has to add
(a) a small vertical load εK_3; and
(b) a small vertical displacement αw_0 in the case that no outer forces are present.

Hence, the complete (generic) bifurcation diagram depends on the following parameters:

a, b (side length of the plate), p (boundary force),

ε (small vertical load), α (imperfection of the rest state).

In this direction, study Chow and Hale (1982, M).

References to the Literature

Classical works: von Kármán (1910), Friedrichs and Stoker (1942) (nonlinear circular plates), Berger and Fife (1967).

Explicit computations: Dickey (1976, L).

Introduction: Vlasov (1964, M) and Pflüger (1965, M) (shells), Szillard (1974, M) (plates).

Modern introduction: Ciarlet and Rabier (1980, L), Djubek, Kodnár, and Škaloud (1983, M) (existence and numerical methods).

Handbook article about plates and shells: Naghdi (1972, B).

Existence theory: Vorovič (1955), Berger and Fife (1967), Langenbach (1976, M), Ciarlet and Rabier (1980, L), Nečas and Hlaváček (1981, M).

Bifurcation for plates, imperfections, and catastrophe theory: Chow, Hale, and Mallet-Paret (1975), Chow and Hale (1982, M).

Bifurcation for rectangular plates: Knightly and Sather (1974), List (1978), Recke (1978), Chow and Hale (1982, M) (recommended as an introduction).

Bifurcation for circular plates: Dickey (1976, L), Antman (1980) (global branches).

Bifurcation for shells: Sather (1976, S), Recke (1978), Knightly and Sather (1980).

Cones and positive eigensolutions for the plate equation: Miersemann (1979).

Buckling of plates with obstacles: Do (1975), (1977), Miersemann (1975), (1981, S), (1982).

Control of the stability of plates: Beckert (1972), Hofmann (1986) (general theory for the control of the smallest eigenvalue of self-adjoint compact operator in H-spaces by using a constructive approach via subgradient methods).

Derivation of the von Kármán equations from three-dimensional elasticity via asymptotic expansions: Ciarlet and Rabier (1980, L).

Numerical methods: Ciarlet (1977, M), Brezzi (1978), Djubeck, Kodnár, and Škaloud (1983, M), Bernadou and Boisserie (1982, M) (finite elements in thin shell theory).

A priori estimates for the stresses in shells and interior shell equations: John (1965), (1971), (1985, S) (fundamental papers).

Shells: Koiter (1972), (1980), Ramm (1982, M), Niordson (1985, M), Antman (1986).

CHAPTER 66

Convex Analysis, Maximal Monotone Operators, and Elasto-Viscoplastic Material with Linear Hardening and Hysteresis

It was shown by Moreau (1976) that for materials without hardening it is inevitable, in general, to use irreflexive spaces of L_∞-type or C-type. If some kind of hardening of the material is involved, then the situation is much simpler since satisfactory *a priori* estimates for the solutions of the problem are available under natural assumptions.

Konrad Gröger (1979)

In this chapter we generalize the results of Chapter 60 about the wire to three-dimensional bodies. Our goal is to clarify the following points.

(i) The *subgradient* of functionals, which was studied in Part III in the context of convex analysis, can be used to formulate *multivalued* constitutive laws which, in a mathematically elegant form, model plastic behavior and more generally elasto-viscoplastic behavior.

(ii) We consider slow deformation processes of the form

$$u = u(t),$$

where u and t describe displacement and time, respectively. The basic idea is to replace the classical strain–stress relation

$$\gamma = F(\sigma)$$

with the multivalued constitutive law

$$\dot\gamma(t) \in \partial F(\sigma(t)).$$

The dot stands for the time derivative, and ∂F is the subgradient of F.

In this way we obtain first-order evolution equations which contain multivalued *maximal monotone operators*. Hence we are able to apply the main theorem about these equations (Theorem 55.A).

Recall that the subgradient of convex lower semicontinuous functionals is maximal monotone (Theorem 47.F).

(iii) In order to model hardening effects which occur in physical experiments, it is convenient to introduce so-called *internal state variables*.

(iv) In this chapter we will show that the hardening operator B leads to a mathematical regularization of the problem. In the proof of the main theorem in Section 66.3, we shall make essential use of the fact that the operator PA is symmetric and positive. Note that $PA + B$ is symmetric and *strongly* positive. Hence, according to the main theorem on monotone operators (Theorem 26.A), the inverse operator $(PA + B)^{-1}$ exists, and can be used in order to introduce an equivalent norm. This greatly simplifies the original problem.

In summarizing let us say:

The study of plastic and viscoplastic material automatically leads to nonlinear problems.

The abstract model of Section 66.1 clearly shows that the heart of elasticity and plasticity theory has a *functional-analytic character*. Therefore the use of methods from functional analysis is natural.

In Section 66.4 we shall apply the abstract model to several concrete situations.

The reader should also keep the following in mind:

(a) In Chapter 62, as well as the present chapter, we shall use models of plastic behavior in which the *displacements* of the body are calculated. Such behavior is close to elastic behavior (e.g., the slow extension or contraction of wires).

(b) In Chapter 86 we consider viscoplastic liquids. We compute the slow *velocities* of the liquid. A typical example in this regard is the transport of extremely viscous liquids in pipe lines. This is one of the main problems in the chemical industry.

We should like to recommend that the reader look again at Chapters 60 and 61 before studying this chapter. In particular, we recommend Section 60.3, because the following abstract model is a generalization of the concrete situation of wires which was considered there.

66.1. Abstract Model for Slow Deformation Processes

Our functional-analytic model for the description of slow deformation processes for elasto-viscoplastic bodies with a linear hardening law will be described in abstract terms below. The physical meaning of the model will be discussed in Sections 66.2 and 66.4. We assume that all equations below are valid for almost all times $t \in [0, t_0]$ with fixed $t_0 > 0$.

A characteristic property of our model is the fact that the observable strain tensor γ and the corresponding observable stress tensor σ can be decomposed by (i) and (ii) below, into internal state variables e, p and q, r, respectively, where p is the viscoplastic part of the strain tensor γ. These quantities are not directly observable. Our abstract model consists of the following elements:

(i) *Deformation process*:
$$\gamma(t) = e(t) + p(t).$$

(ii) *Stress process*:
$$\sigma(t) = q(t) + r(t).$$

(iii) *Viscoplastic constitutive law*:
$$\dot{p}(t) \in \partial F(q(t)).$$

(iv) *Linear elastic constitutive law*:
$$\sigma(t) = Ae(t), \qquad r(t) = Bp(t).$$

(v) *Linear relation between displacement and (linearized) strain tensor γ*:
$$\gamma(t) = Du(t).$$

(vi) *Equilibrium condition for the stress tensor σ and the outer forces K*:
$$D^*\sigma(t) = K(t).$$

(vii) *Generalized inequality of Korn*:
$$\|Du\|_{\Gamma^*} \geq d\|u\|_U \qquad \text{for all} \quad u \in U \quad \text{and fixed} \quad d > 0.$$

(viii) *Initial conditions at time $t = 0$*:
$$u(0) = u_0, \qquad \gamma(0) = \gamma_0, \qquad p(0) = p_0,$$
$$\sigma(0) = \sigma_0, \qquad q(0) = q_0.$$

We assume that the initial values satisfy the following relations
$$D^*\sigma_0 = K(0), \qquad \sigma_0 = A(Du_0 - p_0) = q_0 + Bp_0.$$

(ix) *Spaces*: For all times $t \in [0, t_0]$, we have
$$u(t) \in U,$$
$$\gamma(t), e(t), p(t) \in \Gamma,$$
$$\sigma(t), q(t), r(t) \in \Gamma^*$$

with the following real, separable H-spaces:

U: H-space of the displacements u,

U^*: H-space of the outer forces K,

66.1. Abstract Model for Slow Deformation Processes

Γ: H-space of the strains γ,

Γ^*: H-space of the stresses σ.

As usual, the star denotes the dual space. Our Hilbert space model clearly shows that there exists a *duality* between displacement u and outer force K on the one side, and between strain γ and stress σ on the other side. We note that the generalized Korn inequality is very natural because it implies that

$$\gamma = 0 \Rightarrow Du = 0 \Rightarrow u = 0,$$

which means that the vanishing of strain implies the vanishing of displacement.

(x) *Operators*. We assume:

$F: \Gamma^* \to]-\infty, \infty]$ is convex, lower semicontinuous and $F \not\equiv +\infty$,

$A, B: \Gamma \to \Gamma^*$ are linear, continuous, symmetric, and strongly positive,

$D: U \to \Gamma$ is linear and continuous.

Suppose the function

$$t \mapsto K(t)$$

is given which describes the change in time for the outer force. We are looking for the change in time of

$$\gamma, e, p \quad \text{and} \quad \sigma, q, r.$$

If we eliminate e and r by using relations (i) to (vi), then we obtain the new *basic equations*

$$D^*(q + Bp) = K,$$
$$q + (A + B)p = ADu, \qquad (1)$$
$$\dot{p} \in \partial F(q).$$

Moreover, we find that

$$D^*\sigma_0 = K(0), \qquad q_0 + Bp_0 = \sigma_0, \qquad (2)$$

where

$$\sigma_0 = A(Du_0 - p_0).$$

Thus we are led to the following problem.

Problem 66.1. Assume that the outer force process

$$K \in W_2^1(0, t_0; U^*),$$

as well as the initial values of the displacement

$$u_0 \in U,$$

the plastic deformation

$$p_0 \in \Gamma,$$

and the viscoplastic stress

$$q_0 \in \Gamma^*,$$

with (2) and $F(q_0) < \infty$ are given.

We are looking for a process

$$(u, p, q) \in W_2^1(0, t_0; U \times \Gamma \times \Gamma^*),$$

which satisfies equation (1) and the initial condition

$$(u, p, q)(0) = (u_0, p_0, q_0).$$

The spaces $W_2^1(0, t_0; X)$ with a real, separable H-space X have been introduced in Section 55.3. Recall that

the map $f: [0, t_0] \to X$ belongs to $W_2^1(0, t_0; X)$

if and only if f is continuous and has a generalized derivative on $[0, t_0]$ with $\dot{f} \in L_2(0, t_0; X)$, i.e.,

$$\int_0^{t_0} \|\dot{f}(t)\|^2 \, dt < \infty.$$

If one knows a solution of Problem 66.1, then γ, e, and r are easily obtained from

$$\gamma = Du, \qquad e = \gamma - p, \qquad \sigma = Ae, \qquad r = \sigma - q.$$

From $q + (A + B)p = ADu$ we then obtain the simpler formula

$$r = Bp.$$

66.2. Physical Interpretation of the Abstract Model

An important concrete example will be considered in Section 66.4. Here we only discuss some general aspects of the model. As in Figure 66.1, our model can be interpreted as the parallel connection of a viscoplastic element F and an elastic element B, which are both subject to the plastic deformation p. Furthermore, p is related to the plastic stress q through

$$\dot{p} \in \partial F(q)$$

66.2. Physical Interpretation of the Abstract Model

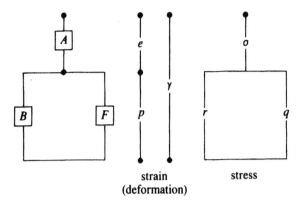

F = viscoplastic element;
A, B = elastic elements.

Figure 66.1

and to the Hooke stress r through

$$r = Bp.$$

These add up to the total stress

$$\sigma = q + r.$$

This total stress causes another elastic deformation

$$e = A^{-1}\sigma$$

so that the total deformation adds up to

$$\gamma = e + p.$$

There exists a remarkable analogy between Figure 66.1 and electrical circuits. In terms of electricity, we find that

$$\text{strain} \Rightarrow \text{voltage},$$

$$\text{stress} \Rightarrow \text{current},$$

$$\text{strain-stress relation} \Rightarrow \text{Ohm's law}.$$

In case of an electrical circuit, A, B, and F in Figure 66.1 correspond to switching elements.

As in the concrete case of the elastoplastic wire of Section 60.3, the operator B describes a *hardening* effect. The presence of the physically reasonable hardening operator B is very welcome from a mathematical point of view, since the operator

$$PA + B$$

is symmetric and strongly positive and hence has an inverse.

In order to motivate the constitutive law

$$\dot{p}(t) \in \partial F(q(t))$$

in (iii) of Section 66.1 we now consider several important special cases of the constitutive law

$$\dot{\gamma}(t) \in \partial F(\sigma(t))$$

between strain γ and stress σ. In this direction, compare the concrete examples of Section 60.2. Further concrete examples will be considered in Section 66.4.

EXAMPLE 66.2 (Plastic Behavior). Let C be a convex, closed, nonempty set in the real H-space Γ^*. Moreover, let χ be the indicator function of C, i.e.,

$$\chi(\sigma) = \begin{cases} 0 & \text{if } \sigma \in C, \\ +\infty & \text{if } \sigma \notin C. \end{cases}$$

If we set $F = \chi$, then the constitutive law

$$\dot{\gamma}(t) \in \partial \chi(\sigma) \tag{3}$$

describes plastic behavior. Explicitly, relation (3) means that $\sigma(t) \in C$ and

$$\langle \sigma(t), \dot{\gamma}(t) \rangle_U \geq \langle \sigma, \dot{\gamma}(t) \rangle_U \quad \text{for all} \quad \sigma \in C. \tag{4}$$

This condition admits a direct physical interpretation if we observe that

$$\langle \sigma, \gamma \rangle \quad \text{with} \quad \gamma = Du$$

is equal to the *virtual work* which corresponds to the displacement u under the action of the stress σ, without taking the outer force K into account. If we set $\delta \gamma = \dot{\gamma}(t) \Delta t$, then (4) assumes the form

$$\langle \sigma(t), \delta \gamma \rangle \geq \langle \sigma, \delta \gamma \rangle \quad \text{for all} \quad \sigma \in C.$$

Therefore, (4) can be regarded as the principle of the *maximal virtual plastic work*.

By definition, the total virtual work, caused by the displacement u, is equal to

$$\langle \sigma, \gamma \rangle - \langle K, u \rangle \quad \text{with} \quad \gamma = Du,$$

which corresponds to our observations in Section 61.5. The principle of *virtual work* is given by

$$\langle \sigma, Du \rangle - \langle K, u \rangle = 0 \quad \text{for all} \quad u \in U.$$

This definition is reasonable, because it implies that

$$\langle D^*\sigma - K, u \rangle = 0 \quad \text{for all} \quad u \in U,$$

i.e.,

$$D^*\sigma = K,$$

and this is the equilibrium condition (vi) of Section 66.1.

EXAMPLE 66.3 (Viscous Behavior). Let $F_1: \Gamma^* \to \mathbb{R}$ be a convex C^1-functional on the real H-space Γ^*. Then, by definition, the constitutive law $\dot{y}(t) \in \partial F_1(\sigma(t))$, i.e.,

$$\dot{y}(t) = F_1'(\sigma(t))$$

describes viscous behavior. In addition, we assume the existence of the inverse operator $F_1'^{-1}: \Gamma^* \to \Gamma$.

EXAMPLE 66.4 (Viscoplastic Behavior). By setting $F = \chi + F_1$ with χ and F_1 as in the two preceding examples, we obtain from Theorem 47.B that

$$\partial F = \partial \chi + \partial F_1.$$

The constitutive law $\dot{y}(t) \in \partial F(\sigma(t))$ then corresponds to

$$\dot{y}(t) \in (\partial \chi + F_1')(\sigma(t)).$$

This is a superposition of plastic and viscous behavior.

66.3. Existence and Uniqueness Theorem

Theorem 66.A (Gröger (1979)). *Problem 66.1 has a unique solution.*

After some elementary transformations, this theorem is an immediate consequence of the main theorem about first-order evolution equations with maximal monotone operators (Theorem 55.A). The main step in the following proof is (III).

PROOF.

(I) Operator properties. The operator

$$D^*AD \in L(U, U^*)$$

is linear, symmetric, and strongly positive, since

$$\langle D^*ADu, u \rangle = \langle ADu, Du \rangle \geq a\|Du\|^2 \geq ad^2\|u\|^2$$

for all $u \in U$ with $a, d > 0$. Consequently, the inverse operator

$$(D^*AD)^{-1} \in L(U^*, U)$$

exists (see Theorem 26.A). We define the operators $S \in L(U^*, \Gamma^*)$ and $Q, P \in L(\Gamma^*, \Gamma^*)$ through

$$S = AD(D^*AD)^{-1}, \quad Q = SD^*, \quad P = I - Q.$$

A simple computation shows that

$$P^2 = P, \quad Q^2 = Q, \tag{5}$$

i.e., P and Q are projection operators. Moreover, we have
$$PAD = 0, \quad D^*PA = 0, \quad D^*S = I,$$
$$QS = S, \quad PS = 0. \tag{6}$$
In addition, let us prove that
$$N(Q) = N(D^*), \quad R(Q) = R(AD). \tag{7}$$
Clearly, we have $N(D^*) \subseteq N(Q)$ and in order to prove $N(Q) \subseteq N(D^*)$ we consider the equation
$$Qg = AD(D^*AD^{-1})^{-1}D^*g = 0,$$
which yields $D^*g = 0$. According to assumption (vii) and (x) of Section 66.1 the inverse operators
$$A^{-1}: \Gamma^* \to \Gamma, \quad D^{-1}: R(D) \to U$$
exist and thus we find that $N(Q) = N(D^*)$.

Furthermore, we have the obvious relation $R(AD) \subseteq R(Q)$, and the equation $h = ADg$ implies that $g = (D^*AD)^{-1}D^*h$, and hence $h = Qh$. This shows that $R(AD) = R(Q)$.

(II) Equation
$$ADu = g, \quad u \in U$$
has a unique solution for every $g \in \Gamma^*$ with $Pg = 0$. In fact, since $g \in R(Q)$ and $R(Q) = R(AD)$, there exists a solution to this equation and from $ADu = 0$ it follows that $u = 0$. Hence the solution is unique and equal to $u = (D^*AD)^{-1}D^*g$.

(III) Main step. We show that the auxiliary equation
$$S\dot{K} \in \dot{q} + (PA + B)\partial F(q),$$
$$q(0) = q_0, \quad q \in W_2^1(0, t_0; \Gamma^*) \tag{8}$$
has a unique solution.

The operator
$$PA \in L(\Gamma, \Gamma^*)$$
is symmetric and positive (see Problem 66.2). Therefore the operator
$$PA + B \in L(\Gamma, \Gamma^*)$$
is symmetric and *strongly positive*, and hence the inverse operator
$$(PA + B)^{-1} \in L(\Gamma^*, \Gamma)$$
exists. This is the key to our proof. Recall that the operator B describes hardening effects of the material. Thus, the hardening operator B is responsible for the regularization of the operator PA.

66.3. Existence and Uniqueness Theorem

On the H-space Γ^*, we introduce the *equivalent scalar product*

$$(x|y)_1 = \langle x, (PA + B)^{-1}y \rangle \qquad \text{for all} \quad x, y \in \Gamma^*$$

and obtain that

$$(PA + B)\partial F(q) = \partial_1 F(q),$$

where $\partial_1 F$ denotes the subgradient with respect to $(\cdot|\cdot)_1$. This follows from the definition of the subgradient in Section 47.3. Therefore, (8) becomes the equivalent equation

$$S\dot{K} \in \dot{q} + \partial_1 F(q), \qquad (9)$$
$$q(0) = q_0, \qquad q \in W_2^1(0, t_0; \Gamma^*).$$

According to Theorem 55.A, equation (9) has a unique solution. Note that the subgradient

$$\partial_1 F \colon \Gamma^* \to 2^{\Gamma^*}$$

is a *maximal monotone operator*, according to Theorem 47.F.

(IV) Uniqueness for Problem 66.1. Let

$$(u, p, q) \in W_2^1(0, t_0; U \times \Gamma \times \Gamma^*)$$

be a solution of Problem 66.1, i.e.,

$$D^*(q + Bp) = K, \qquad q + (A + B)p = ADu, \qquad \dot{p} \in \partial F(q), \qquad (10)$$
$$(u, p, q)(0) = (u_0, p_0, q_0).$$

After multiplication with S and P we obtain

$$Q(q + Bp) = SK, \qquad Pq + P(A + B)p = 0.$$

Addition yields

$$q + (PA + B)p = SK. \qquad (11)$$

Differentiation with respect to time t implies

$$\dot{q} + (PA + B)\dot{p} = S\dot{K}, \qquad (12)$$

and $\dot{p} \in \partial F(q)$ yields (8). Together with (III) this shows that there exists at most one solution of Problem 66.1.

(V) Existence for Problem 66.1. By reversing the argument in (IV), we show that the solution q of (8) induces a solution (u, p, q) of Problem 66.1.

We begin by constructing p as a unique solution of equation (11). An application of the operator P to (11) yields

$$P(q + (A + B)p) = 0.$$

Using (II), we may thus construct u as the unique solution of the equation

$$ADu = q + (A + B)p. \qquad (13)$$

In addition, an application of the dual operator D^* to equation (11) yields

$$D^*(q + Bp) = K. \tag{14}$$

From (8) and (12) follows

$$\dot{p} \in \partial F(q), \tag{15}$$

and thus we obtain from (13) to (15) that (u, p, q) is a solution of the original equation (10), i.e., a solution of Problem 66.1. □

66.4. Applications

A first application of Theorem 66.A to wires and bars has been given in Section 60.3. Here we want to show how the abstract model of Section 66.1 can naturally be realized for three-dimensional, homogeneous, isotropic bodies. We use the same notations as in Sections 61.5 and 62.2.

Let G be the undeformed region in $\mathbb{R}^3$. We assume that G is bounded with $\partial G \in C^{0,1}$. The deformation of G is given by

$$y = x + u(x)$$

with

$$u = 0 \quad \text{on } \partial G_1,$$

where $\partial G = \overline{\partial_1 G} \cup \overline{\partial_2 G}$ with $\partial_1 G \neq \emptyset$ as in Section 62.2. Therefore, we choose $u \in U$ with

$$U = \mathring{W}_2^1(G, \partial_1 G; V_3).$$

The strain tensor is

$$\gamma = Du, \qquad Du = \tfrac{1}{2}(u'(x) + u'(x)^*).$$

If we set

$$\Gamma = L_2(G, L_{\text{sym}}(V_3)),$$

then $\gamma \in \Gamma$. Recall that Γ is an H-space with the scalar product

$$(\sigma|\gamma) = \int_G [\sigma(x), \gamma(x)]\, dx.$$

We identify Γ^* with Γ, i.e., we identify the elements $\sigma \in \Gamma$ with $\sigma \in \Gamma^*$ in the sense of

$$\langle \sigma, \gamma \rangle = (\sigma|\gamma) \qquad \text{for all} \quad \gamma \in \Gamma.$$

Hence we obtain the usual relation $\sigma(x) \in L_{\text{sym}}(V_3)$ for the stress tensor.

The classical Korn inequality tells us that $(Du|Dv)$ is an equivalent scalar

66.4. Applications

product on the H-space U (see Lemma 62.2). Thus we have

$$\|Du\| \geq d\|u\| \quad \text{for all} \quad u \in U \text{ and fixed } d > 0.$$

This coincides with the generalized Korn inequality of Section 66.1.

Let us now consider the outer forces $\bar{K}$ (volume forces) and T (boundary forces). Suppose that

$$\bar{K} \in L_2(G, V_3), \qquad T \in L_2(\partial_2 G, V_3).$$

Then

$$\langle K, u \rangle = \int_G \bar{K} u \, dx + \int_{\partial_2 G} T u \, dO \quad \text{for all} \quad u \in U$$

defines a linear, continuous functional $K \colon U \to \mathbb{R}$, i.e.,

$$K \in U^*.$$

This is the reason why the dual space U^* of Section 66.1 is called the space of outer forces.

According to equation (61.39), the principle of virtual work corresponds to

$$\langle \sigma, \gamma \rangle - \langle K, u \rangle = 0 \quad \text{for all} \quad u \in U, \tag{16}$$

where $\gamma = Du$. From Section 66.2 follows that this is equivalent to the equilibrium condition

$$D^* \sigma = K. \tag{16*}$$

In Section 61.5g we used integration by parts to see that (16), and hence (16*) corresponds to the classical equilibrium condition

$$\operatorname{div} \sigma + \bar{K} = 0 \quad \text{on } G,$$

$$\sigma n = T \quad \text{on } \partial_2 G.$$

To complete our model we will now be concerned with the realization of the constitutive laws of Section 66.1.

EXAMPLE 66.5 (Plastic Behavior). Let $\varphi \colon L_{\text{sym}}(V_3) \to \mathbb{R}$ be convex and continuous. We choose $\sigma_0 > 0$ such that the set

$$C = \{\sigma \in \Gamma^* \colon \varphi(\sigma(x)) \leq \sigma_0^2 \text{ on } G\}$$

is not empty. Given $\sigma \in C$ we find that $\sigma(x) \in L_{\text{sym}}(V_3)$ and that C is a closed, convex subset of Γ^* (see Problem 62.8).

Let χ denote the indicator function of C. From Example 66.2, it follows that the constitutive law

$$\dot{p}(t) \in \partial \chi(q(t))$$

models plastic behavior. Thus, if we choose

$$\varphi(\sigma) = [\bar{\sigma}, \bar{\sigma}]$$

with $\bar{\sigma} = \sigma - \frac{1}{3}\operatorname{tr}\sigma I$, then Example 61.5 shows that C corresponds to the plasticity condition of von Mises.

EXAMPLE 66.6 (Linear Elastic Behavior). According to Section 66.1, we need a constitutive law of the form

$$\sigma = Ae.$$

The idea is to use a linear Hooke's law

$$\sigma(x) = \lambda \operatorname{tr} e(x) I + 2\kappa e(x).$$

Thus we construct the operator $A: \Gamma \to \Gamma^*$ through

$$\langle Ae, \gamma \rangle = \int_G [\sigma(x), \gamma(x)] \, dx \qquad \text{for all} \quad e, \gamma \in \Gamma.$$

Dropping the argument x we find

$$[\sigma, \gamma] = \lambda \operatorname{tr} e \operatorname{tr} \gamma + 2\kappa [e, \gamma].$$

This implies

$$\langle Ae, \gamma \rangle = \langle A\gamma, e \rangle \qquad \text{for all} \quad e, \gamma \in \Gamma,$$

and

$$\langle Ae, e \rangle \geq 2\kappa \int_G [e, e] \, dx = 2\kappa \|e\|^2 \qquad \text{for all} \quad e \in \Gamma,$$

i.e., the operator A is linear, symmetric, and strongly positive.

In the same way one can choose the operator B for the constitutive law $r = Bp$.

Our model is now complete.

PROBLEMS

66.1. *Proof of (5) and (6).*
Hint: Elementary computation.

66.2. *Operator PA.* Show that the operator PA in the proof of Theorem 66.A is symmetric and positive.
Solution: The symmetry follows from

$$(QA)^* = A^*Q^* = AQ^* = ADS^* = (AD)(D^*AD)^{-1}D^*A^*$$
$$= SD^*A = QA,$$
$$(PA)^* = A^* - (QA)^* = A - QA = PA.$$

The positivity follows from

$$\langle PA\gamma, \gamma \rangle = \langle A\gamma, P^*\gamma \rangle = \langle A(P^*\gamma + Q^*\gamma), P^*\gamma \rangle$$
$$= \langle AP^*\gamma, P^*\gamma \rangle \geq 0 \qquad \text{for all} \quad \gamma \in \Gamma.$$

Note that we have $PAQ^* = P(QA)^* = PQA = 0$, since $PQ = 0$.

66.3.* *Application to plastic torsion.* In Chapter 86 the plastic torsion problem is considered by using a statical model. Formulate the torsion problem in terms of our abstract model of Section 66.1 and use Theorem 66.A to prove existence and uniqueness.

Hint: See Hünlich (1979). This reference also contains numerical results.

References to the Literature

Internal state variables: Nguyen (1973), Gröger (1979, S, B) (general abstract model), Hünlich (1979) (plastic torsion), Nečas and Hlaváček (1981, M).

Classical works in plasticity theory: See the References to the Literature for Chapter 60.

Textbooks on classical plasticity theory: Hill (1950, M), Prager and Hodge (1951, M, H), Prager (1955, M), Sokolovskii (1955, M), Kačanov (1969, M).

Handbook of Physics: Flügge (1956), Vol. VIa, Part 3, Geiringer (1972, S).

Explicit examples in plasticity theory: Nguyen (1982, L).

Modern introduction to plasticity theory: Duvaut and Lions (1972, M), Gröger (1979, S, B), Nečas and Hlaváček (1981, M), Temam (1983a, M).

Existence proofs: Duvaut and Lions (1972, M), Nguyen (1973), Halphen and Nguyen (1975), Moreau (1976), Gröger (1979), Miersemann (1980) (regularity of generalized solutions), Nečas and Hlaváček (1981, M), Temam (1983a, M), (1986).

General systems with hysteresis: Krasnoselskii and Pokrovskii (1983, M).

Classical plastic torsion: Prager and Hodge (1951, M), Geiringer (1972, S).

Modern plastic torsion: Duvaut and Lions (1972, M), Lanchon (1974), and Friedman (1982, M) (introduction).

Ting (1969), (1972, S), Gajewski (1970a), Brézis (1972), Langenbach (1976, M), Gerhardt (1976), Hünlich (1979), Glowinski (1980, L).

Approximation methods: Glowinski, Lions, and Trémolières (1976, M), Hünlich (1979), Glowinski (1980, L), (1984, M), Korneev and Lange (1984, L), Miyoshi (1985, M).

Plasticity, defects in crystals, and the methods of the modern gauge field theory: Kleinert (1987, P).

APPLICATIONS IN THERMODYNAMICS

>Imperfection of matter sows the seed of death.
>Thomas Mann

>The greatest joy of a thinking man is to have explored the explorable and just to admire the unexplorable.
>Johann Wolfgang von Goethe

Many processes in nature have a thermodynamical character. Among them we find cosmological, chemical, and biological processes. During the nineteenth century a great intellectual achievement was marked by the fact that it was possible to separate two fundamental concepts from an abundance of very diversified experimental data, namely: *energy and entropy*. The concept of entropy is closely related to the phenomenon of evolution. Through the second law of thermodynamics, time gets a direction. It is possible to explicitly distinguish between past and future, because actually all real processes in nature are irreversible, i.e., other than in classical mechanics, they cannot occur in the opposite time direction. The development of modern sciences is characterized in all subject areas (cosmology, biology, chemistry, physics) by an increased recognition of the role played by evolution, which leads from simple to complicated structures. One may think, for example, of the process of life. However, the particular mechanisms, leading to evolution, are only vaguely known, since a comprehensive mathematical–physical theory for the production of entropy is *not* available.

In Chapter 67 we will be concerned with phenomenological thermodynamics of the quasi-equilibrium and the equilibrium. A deeper understanding of thermodynamics, however, is only possible in the context of statistical physics. This will be discussed in Chapter 68.

The two fundamental physical quantities, which determine the specific character of thermodynamics, are *temperature T* and *entropy S*. In phenomenological thermodynamics, the existence of these quantities for thermodynamical equilibrium states and certain quasi-equilibrium states is postulated. Statistical physics shows that, in fact, T and S are of statistical nature and meaningful for *systems* of particles. The systems which occur in macroscopical physics are systems with a great number of particles, whereby statistical fluctuations are small. This is the reason why the statistical character of macrophysical phenomena is not that obvious. As in the case for chemical processes, the number of particles may vary. In statistical physics, temperature T is a parameter on which the probability w of the particular state z depends. The state z thereby is characterized through energy and number of particles. In case the number of particles varies, w also depends on the chemical potential μ, i.e.,

$$w = w(T, \mu).$$

From a mathematical point of view, *entropy S* is nothing other than *information*. The fundamental concept of information has been introduced in Section 43.11 of Part III. The basic idea is the following. Consider a system Σ having the possible states $z_1, \ldots, z_n$. Let w_i be the probability for the realization of state z_i. The corresponding entropy (information) is then defined by

$$S = -k \sum_{i=1}^{n} w_i \ln w_i.$$

In Section 43.11 we showed that this is a very natural definition. In fact, one can prove that entropy is uniquely determined by few very natural axioms (cf. Problem 43.7e). It is quite remarkable that the concept of entropy was discovered by physicists in the middle of the nineteenth century, while the mathematical concept of information was introduced only approximately one hundred years later. The constant k in the definition of S is the so-called Boltzmann constant. This universal constant plays a fundamental role in thermodynamics.

We emphasize, however, that not every arbitrary physical system with a great number of particles has a temperature and an entropy. Roughly speaking, the system must exhibit a regular probability behavior with respect to energy and number of particles. A necessary condition for this is that in a physically meaningful way one can speak of the concepts of mean energy and average number of particles.

In the following two chapters we shall see that the Lagrange multiplier rule of Part III plays an important role in phenomenological thermodynamics, as well as in statistical physics.

The following quotation, taken from the "Scientific Autobiography" of Max Planck (1858–1947), should give an impression of the historical development of the fundamental notion of entropy and its relations to probability, statistical physics, and quantum theory:

> My original decision to devote myself to science was a direct result of the discovery (which has never ceased to fill me with enthusiasm since my early

youth), of the comprehension of the far from obvious fact that the laws of human reasoning *coincide* with the laws governing the sequences of impressions we receive from the world about us; that, therefore, *pure* reasoning can enable man to gain an insight into the mechanism of the latter. In this connection, it is of paramount importance that the outside world is something independent from man, something *absolute*, and the quest for the laws which apply to this absolute appeared to me as the most sublime scientific pursuit in life.

These views were bolstered and furthered by the excellent instruction which I received, through many years, in the Maximilian Gymnasium in Munich from my mathematics teacher, Hermann Müller, a middle-aged man with a keen mind and a great sense of humor, a past master of the art of making his pupils visualize and understand the meaning of the laws of physics.

After my graduation from the gymnasium, I attended university, first in Munich for three years, then in Berlin for another year. It was in Berlin that my scientific horizon widened considerably under the guidance of Hermann Helmholtz (1821–1894) and Gustav Kirchhoff (1824–1887), whose pupils had every opportunity to follow their pioneering activities, known and watched all over the world. I must confess that the lectures of these men netted me no perceptible gain. It was obvious that Helmholtz never prepared his lectures properly. We had the unmistakable impression that the class bored him at least as much as it did us.

Kirchhoff was the very opposite. He would always deliver a carefully prepared lecture. Not a word too few, not one too many. But it would sound like a memorized text, dry and monotonous. We would admire him, but not what he was saying.

Under such circumstances, my only way to quench my thirst for advanced scientific knowledge was to do my own reading on subjects which interested me. One day, I happened to come across the treatises of Rudolf Clausius (1822–1888), whose lucid style and enlightening clarity of reasoning made an enormous impression on me.

Clausius (1865) deduced his proof of the Second Law of Thermodynamics from the hypothesis that

heat will not pass spontaneously from a hotter to a colder body.

But this hypothesis must be supplemented by a clarifying explanation. For it is meant to express not only that heat will not pass from a colder into a warmer body, but also that it is impossible to transmit, by any means, heat from a colder into a hotter body without there remaining in nature some change to serve as compensation.

In my endeavor to clarify this point as fully as possible, I discovered a way to express this hypothesis in a form which I considered to be simpler and more convenient, namely:

The process of heat conduction cannot be completely reversed by any means.

A process which in no manner can be completely reversed I called a "natural" one. The term for it in universal use today, is: "irreversible."

I found the meaning of the Second Law of Thermodynamics in the principle that:

In every natural process the sum of the entropies of the bodies involved increase.

I worked out these ideas in my doctoral dissertation at the University of Munich, which I completed in 1879 (at the age of 21 years). The effects of my dissertation on the physicists of those days was *nil*. Helmholtz probably did not even read my paper at all. Kirchhoff expressly disapproved of its contents.... Clausius did not answer my letters.

The universal acceptance of my thesis was ultimately brought about by considerations of an altogether different type of argument, unrelated to the arguments which I had adduced in support of it—namely, by the atomic theory, as represented by Ludwig Boltzmann (1844–1906).

Boltzmann (1872) succeeded in establishing, for a given gas in a given state, a function H, which has the property that its value constantly decreases with time (the famous H-theorem of Boltzmann). It suffices, therefore, to identify the negative value of this H with entropy, to arrive at the principle of the increase of entropy. This discovery demonstrated, at the same time, irreversibility to be a characteristic of the processes occurring in a gas....

My new universal *radiation formula* was submitted to the Berlin Physical Society, at the meeting on October 19, 1900. On the very day when I formulated this law by using purely formal arguments, I began to devote myself to the task of investing it with a true physical meaning. This quest led me to study the interrelation of *entropy S* and *probability W*. Since the entropy S is an additive magnitude, but the probability W is a multiplicative one, I simply postulated that

$$S = k \log W,$$

where k is a universal constant.[1]

I found that k represents the so-called *absolute gas constant*. It is, understandably, often called *Boltzmann's constant*. However, this calls for the comment that Boltzmann *never* introduced this constant.

Now, as for the magnitude W with respect to radiation in a cavity, I found that in order to interpret it as a probability, it was necessary to introduce a universal constant, which I called h. Since it had the dimension of action (energy × time), I gave it the name, *elementary quantum of action*. Thus the nature of entropy as a measure of probability, in the sense indicated by Boltzmann, was established in the domain of radiation, too.

While the significance of the quantum of action for the interrelation between entropy and probability was established, the part played by the new constant in general physical processes still remained an open question. I therefore tried immediately to weld the elementary quantum of action into the framework of classical theory. But in the face of all such attempts continued for a number of years, this constant showed itself to be obdurate. For it heralded the advent of something entirely unprecedented (namely, quantum physics), and was destined to remodel basically the physical outlook and thinking of man which, ever since Leibniz (1646–1716) and Newton (1643–1727) laid the groundwork for infinitesimal calculus, were founded on the assumption that all causal interactions in nature are continuous.

Max Planck (1945)

The Second Law of Thermodynamics is essentially different from the First Law (conservation of energy), since it deals with a question in no way touched upon by the First Law, namely, the direction in which a process takes place in nature. Not every change, which is consistent with the principle of the conservation of energy, also satisfies the additional conditions which the Second Law imposes on the processes which actually take place in nature.

---

[1] This is an excellent example which shows that fundamental relations in physics are often obtained by extremely simple, but ingenious ideas. Note that W is not a probability in the terminology of today, but a statistical weight. This will be explained in Problem 68.6.

The Second Law of Thermodynamics states that there exists in nature, for each system of bodies. a quantity which by all changes of the system either remains constant (in reversible processes) or increases in value (in irreversible processes). This fundamental quantity is called, following Clausius (1865), the entropy of the system.

Max Planck, *Treatise on Thermodynamics*, (1913)

CHAPTER 67
Phenomenological Thermodynamics of Quasi-Equilibrium and Equilibrium States

> Temperature, energy, and entropy are state variables, i.e., they are independent of the history of the body.
> The energy of the universe is constant (first law).
> The entropy of the universe tends towards a maximum (second law).
> Rudolf Clausius (1865)

> In the huge factory of natural processes, entropy plays the role of a president, because it prescribes the way in which the entire course of business takes place. The energy principle plays the role of the book-keeper by balancing debit and credit.
> Robert Emden (1938)

> Only through the second law a certain direction is given to the course of the world. This is not present in the mechanical world picture.
> Arnold Sommerfeld (1954)

The main objectives of phenomenological thermodynamics are:

(i) Characterization of thermodynamical quasi-equilibrium states (Gibbs' fundamental equation of Section 67.2).
(ii) Characterization of thermodynamical processes (laws of thermodynamics of Section 67.4).
(iii) Computation of thermodynamical equilibrium states from quasi-equilibrium states (extremal properties of thermodynamical potentials of Section 67.6).

For didactical reasons we begin with Gibbs' fundamental equation rather than with the three laws of thermodynamics. Later on we will discuss their interrelation. Especial emphasis is placed on *general principles*. Through this the reader might be able to overlook the entire phenomenological thermodynamics with its plentiful applications. The mathematical substance of phenomenological thermodynamics is contained in the solution of:

(a) equations for differential forms (Gibbs' fundamental equation); and
(b) extremal problems with side conditions (Lagrange's multiplier rule).

We consider the following applications:

(α) In connection with (a) we solve Gibbs' fundamental equation in Section 67.3 for a general thermodynamical system with two degrees of freedom. We thereby obtain the most general mathematical description for gases which are compatible with the laws of thermodynamics.

(β) In connection with (b) we consider:
Gibbs' phase rule (Section 67.7); Law of mass action (Section 67.8).

These are two famous general thermodynamical relations, which are frequently used in chemistry.

(γ) In Problems 67.3 and 67.4 we discuss the impossibility of the perpetuum mobile of the second kind as well as Carnot's cycle. During the nineteenth century, these two phenomena formed the experimental basis for the formulation of the second law.

The first and second law of thermodynamics appears in the physical literature in the form

$$dE = \delta Q + \delta W,$$
$$T\,dS \geq \delta Q. \tag{1}$$

One speaks of "infinitesimal small changes" of inner energy E, heat Q, work W, and entropy S. Here T denotes the absolute thermodynamical temperature measured in K (degree Kelvin). However, more precisely meant are the processes

$$\dot{E} = \dot{Q} + \dot{W},$$
$$T\dot{S} \geq \dot{Q}. \tag{2}$$

The dot means derivative with respect to time. Here

$$E(t), \quad T(t), \quad S(t)$$

corresponds to the value of E, T, S at time t. On the other hand,

$$Q(t) \quad \text{and} \quad W(t)$$

denote the heat and work (energy), respectively, which is added to the system *during the time interval* $[t_0, t]$, for a fixed initial time t_0. It is important to keep the different meanings of $E(t)$, $S(t)$, $T(t)$ on the one hand, and of $Q(t)$, $W(t)$ on the other hand, in mind. Note that Q and W are always used in connection with time-dependent processes. It doesn't make any sense to speak of "heat Q or work W at time t."

Since (1) often leads to misunderstandings, we will only use (2), but the relation between both notations is obvious.

We emphasize that, for a mathematical treatment of phenomenological thermodynamics, it is necessary to begin with time-dependent thermodynamical processes and to formulate the two laws of thermodynamics in form (2). Often, in the mathematically oriented literature, one finds equations

$$dE = \mathcal{Q} + \mathcal{W},$$

$$T\,dS = \mathcal{Q},$$

where $\mathcal{Q}$ and $\mathcal{W}$ are differential forms. As we shall show in Section 67.4d, this approach is a special case of (2). But, actually, (2) is more general and allows the study of changes from quasi-equilibrium states to equilibrium states. This is one of the main objectives in thermodynamics and will be discussed in Sections 67.6 to 67.8.

In order to treat thermodynamical problems effectively, one needs to change parameters, depending on the particular problem. In the language of manifolds this means chart change. In fact, the theory is independent of the particular parametrization. An invariant formulation is obtained by introducing a state space Z, which is a manifold, and by using differential forms on Z. For a first glimpse at this chapter we recommend that the reader should treat differentials in a naive fashion. Precise definitions of differentials and differential forms on manifolds may be found in Section 73.23. The theory of manifolds is the natural setting for phenomenological thermodynamics.

67.1. Thermodynamical States, Processes, and State Variables

For an understanding of thermodynamics it is important to give precise definitions of the three concepts mentioned in the heading. We begin with a verbal characterization. A precise mathematical formulation can be found in Section 67.4. We distinguish between:

(α) physical states; and
(β) thermodynamical states (quasi-equilibrium states).

As an example we consider a gas G in a container which consists of N molecules. In classical mechanics, the *physical state* of G at time t is characterized by position vectors $q_i(t)$ and momentum vectors $p_i(t)$, $i = 1, \ldots, N$. If one knows the forces, then according to Newton's mechanics, one can compute the motion for all times. If, in addition, each molecule is regarded as an oscillating and rotating system, then further position and momentum vectors are needed to characterize the complete classical state. Since N has an order of magnitude of 10^{23}, a great number of parameters is needed in order to describe the system. A mathematical treatment in this setting, which would lead to practical results, is impossible. The same is true for a quantum mechanical treatment. Schrödinger's wave function then depends on the N position vectors q_i. However, experience shows that the macroscopical behavior of many systems with a large number of particles can be described by substantially less parameters. Such states, for which the temperature plays an important role, are called thermodynamical *quasi-equilibrium states*.

EXAMPLE 67.1 (Homogeneous Gas). Gas states can completely be described by volume V, pressure P, and temperature T,

$$T = T(V, P) \quad \text{(equation of state)}.$$

EXAMPLE 67.2 (Inhomogeneous Gas). In order to understand the propagation of sound waves in the air, the situation of the previous example is not sufficient. Here states have to be described by the pressure distribution $P = P(x)$, density $\rho = \rho(x)$, and temperature distribution $T = T(x)$, together with the equation of state

$$T = T(\rho, P).$$

Sound waves are then time periodic fluctuations of these quantities.

In an abstract setting, quasi-equilibrium states are described by the elements z of a set Z, which will be called *state space*. In Example 67.1 the state space Z consists of all points

$$z = (V, P, T(V, P)) \quad \text{in } \mathbb{R}^3$$

with $T, V \in \mathbb{R}^3_+$, i.e., Z is a surface in $\mathbb{R}^3$. A state function is a function

$$\Omega = \Omega(z) \quad \text{on } Z.$$

Important examples are temperature T, inner energy E, and entropy S. In Example 67.1, the state functions can also be represented through

$$\Omega = \Omega(V, P).$$

Furthermore, the representation

$$\Omega = \Omega(V, T) \quad \text{or} \quad \Omega = \Omega(P, T)$$

is possible if the equation $T = T(V, P)$ can be solved for P or V. One thereby recognizes that states and state functions can be parametrized in very different ways. This fact is widely used in thermodynamics.

State changes are given by

$$z = z(t),$$

i.e., at each time t, the system is in a quasi-equilibrium state. State changes depend on the particular state space Z. In Example 67.1, we only allow state

67.1. Thermodynamical States, Processes, and State Variables

changes of the form

$$V = V(t), \quad P = P(t) \quad \text{and} \quad T = T(V(t), P(t)).$$

Such state changes can only be obtained in idealized form, by very slowly relaxing the gas or compressing it. In Example 67.2 a state change is described by

$$P = P(x,t), \quad \rho = \rho(x,t), \quad T = T(x,t), \quad T = T(\rho, P). \tag{3}$$

It is possible that at times $t = 0$ and $t = t_1$, i.e., at the beginning and the end of the process we have

$$P = \text{const}, \quad \rho = \text{const} \quad \text{and} \quad T = \text{const},$$

i.e., we are in the situation of Example 67.1, whereas during the process, we are in the more general situation (3). Such a process cannot be described in the state space of Example 67.1.

Strictly speaking, besides $z = z(t)$ for a process, one would also have to know the heat $Q(t)$ and other forms of energy $W(t)$, which during the time interval $[t_0, t]$ are added to the system.

The concept of quasi-equilibrium states is introduced as an idealization, since it greatly increases the number of possible applications. Most important is the concept of thermodynamical *equilibrium states*. It means the following. If a system is subject to time-independent outer conditions, then one expects that, after a long period of time, the system passes into a state in which it will remain and which is characterized by a minimal number of parameters. These states are called thermodynamical equilibrium states. A necessary condition for the thermodynamical equilibrium is that the system has a common temperature. Quasi-equilibrium states may become equilibrium states. We mentioned already earlier that quasi-equilibrium states are described by a smaller number of parameters than arbitrary physical states. If one passes from a quasi-equilibrium state to an equilibrium state, then the number of state parameters may further decrease. Example 67.1 shows a thermodynamical equilibrium, while in Example 67.2 we only have a quasi-equilibrium. Here temperature, pressure, and density compensations are still possible.

EXAMPLE 67.3 (Chemical Reactions). In a container of volume V, pressure P, and temperature T, we have N_i particles of the ith substance with $i = 1, \ldots, n$. Quasi-equilibriums are described by

$$z = (T, V, \mu, P(T, V, \mu), N(T, V, \mu))$$

with $\mu = (\mu_1, \ldots, \mu_n)$ and $N = (N_1, \ldots, N_n)$. Thereby μ_i is the chemical potential of the ith substance. If chemical reactions are possible, then the number of particles N_i may still change. The thermodynamical equilibrium then corresponds to the chemical equilibrium, which occurs after a long period of time. In Section 67.8 we will see how this chemical equilibrium can be computed (law of mass action). Thereby we use a general method for the computation of thermodynamical equilibriums, which is discussed in Section 67.7.

From the standpoint of statistical physics, thermodynamical equilibriums correspond to time-independent probability distributions of the entire system. In Example 67.1, this is the distribution of gas molecules onto possible velocities. This is a time-independent Gaussian normal distribution (Maxwell's velocity distribution of Section 68.3). Quasi-equilibriums occur after dividing a system into subsystems, and assigning probability distributions to each subsystem. In Example 67.2, the subsystems consist of small volume elements, whereby for each such element the situation of Example 67.1 is assumed.

An important advantage of thermodynamical equilibriums and quasi-equilibriums is the following. In these states the system completely forgets its history. This, for example, makes it possible to compute the state of our universe at a temperature of 10^{11} K, one hundredth of a second after the Big Bang (see Section 68.6). Without this "forgetting" we would be in a hopeless situation, since we do not know the initial conditions of the Big Bang.

67.2. Gibbs' Fundamental Equation

Gibbs' fundamental equation for quasi-equilibrium states is

$$dE = T\,dS - P\,dV + \sum_{i=1}^{r} \mu_i\,dN_i + \sum_{j=1}^{s} A_j\,da_j. \qquad (4)$$

The quantities have the following meaning: E is inner energy, T is absolute temperature, S is entropy, P is pressure, V is volume, μ_i is chemical potential, and N_i is number of particles of the ith substance, and A_j, a_j are further parameters. We write $N = (N_1, \ldots, N_r)$, $a = (a_1, \ldots, a_s)$. We are looking for functions

$$E = E(y), \quad S = S(y), \quad P = P(y), \quad \mu = \mu(y), \quad A = A(y)$$

with $y = (T, V, N, a)$ which satisfy (4). A quasi-equilibrium state is characterized by

$$z = (y, P(y), \mu(y), A(y)). \qquad (5)$$

The collection of all these z forms the state space Z. Explicitly, (4) is a system of first-order partial differential equations:

$$E_T = TS_T, \quad E_V = TS_V - P, \quad E_{N_i} = TS_{N_i} + \mu_i,$$

$$E_{a_j} = TS_{a_j} + A_j. \qquad (6)$$

In the following section we integrate (6) for an important special case. Every solution of (4) describes possible quasi-equilibrium states of thermodynamical systems. In Section 67.4c we will show how (4) can be derived from the laws of thermodynamics.

67.3. Applications to Gases and Liquids

In (5), the state space Z has been parametrized with y. Since $y \in \mathbb{R}^{2+r+s}$ the number $2 + r + s$ is called the number of degrees of freedom of the system. If P, μ, and A in (5) are C^1-functions on an open set, then Z is a C^1-manifold which lies in $\mathbb{R}^{3+2r+2s}$. This manifold is important itself, and not the way in which it is parametrized by choosing suitable parameters. In Section 67.5 we will show how equation (4) can be solved in a general form. We will use thermodynamical potentials.

The importance of the chemical potential μ will become clear from the following section and Sections 67.7 and 67.8.

67.3. Applications to Gases and Liquids

We now discuss the important question: Which quantities of a gas must be experimentally determined in order to compute its inner energy, its entropy, and its chemical potential? By integrating the fundamental equation (4) we want to show:

(A) Equation of state $P = p(T, \rho)$ and specific heat capacity $c = c(T, \rho)$ must be determined experimentally. Thereby ρ denotes the density of the gas.

In fact, it is our goal to give a complete description of the structure of *all* thermodynamical systems which solely depend on density ρ and temperature T. Our observations are applicable as well to liquids. The main result is contained in Theorem 67.A below.

Quasi-equilibrium states of the gas are described by

$$z = (T, V, N, P(T, V, N)),$$

where T is temperature, V is volume, P is pressure, and N is the number of particles (Fig. 67.1). In fact, this describes thermodynamical equilibrium states since, by experience, the number of independent parameters cannot be reduced. In statistical physics the use of the particle number N is appropriate. In our phenomenological discussion we want to eliminate the molecular structure of the gas, and replace N with the mass M of the gas. We then find $M = m \cdot N$, where m is the molecular mass and

$$\rho = M/V.$$

The fundamental equation (4) takes the form

$$dE = T\, dS - P\, dV + \mu\, dN \tag{7}$$

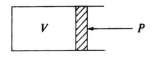

Figure 67.1

with a modified μ. We are looking for P, μ, E, S as C^2-functions of (T, V, M) for $T > 0$, $V > 0$, $M > 0$.

Consider the special process

$$T = T(t), \qquad V = V(t), \qquad M = M_0 = \text{const},$$

i.e., temperature and volume are changed while the gas mass remains constant. Let $W(t)$ and $Q(t)$ denote work and heat, respectively, which are added to the gas during the time interval $[t_0, t]$. We then have

$$\dot{W}(t) = -P(t)\dot{V}(t)$$

with $P(t) = P(T(t), V(t), M_0)$ (see Problem 67.1). The first law of thermodynamics yields, for the energy balance,

$$\dot{E}(t) = \dot{W}(t) + \dot{Q}(t) \tag{8}$$

with $E(t) = E(T(t), V(t), M_0)$ (see Section 67.4). Now we assume that also volume V remains constant, i.e., $V(t) = V_0 = \text{const}$ and we set

$$T(t) = T_0 + t.$$

Then

$$C(T_0, V_0, M_0) = \dot{Q}(0)$$

is called the heat capacity (added heat per temperature change). This quantity can be experimentally determined. From (8) with $\dot{W}(t) = 0$ it follows then that

$$E_T(T_0, V_0, M_0) = C(T_0, V_0, M_0). \tag{9}$$

We now want to separate the dependence on M. To this end, we set

$$E(T, V, M) = e(T, \rho)M, \qquad S(T, V, M) = s(T, \rho)M, \qquad P(T, V, M) = p(T, \rho). \tag{10}$$

We keep in mind that E and S are extensive quantities, i.e., homogeneous of degree one in V and M, while P is intensive, i.e., homogeneous of degree zero in V and M. Moreover, we set

$$C(T, V, M) = c(T, \rho)M, \qquad c(T, \rho) = e_T(T, \rho), \tag{11}$$

and call e, s, and c specific inner energy, specific entropy, and specific heat capacity, respectively.

We now assume that we know a C^2-solution of the basic equation (7). Because of

$$V = M/\rho \qquad \text{and} \qquad dV = dM/\rho - M\,d\rho/\rho^2$$

we obtain

$$(de - T\,ds - p\,d\rho/\rho^2)M + (e - Ts + p/\rho - \mu)\,dM = 0.$$

Since M is arbitrary it follows that

$$\mu = e - Ts + p/\rho \qquad \text{and} \qquad de = T\,ds + p\,d\rho/\rho^2.$$

67.3. Applications to Gases and Liquids

Because of $c = e_T$ we find
$$ds = c\,dT/T + (e_\rho/T - p/T\rho^2)\,d\rho,$$
hence
$$s_T = c/T \quad \text{and} \quad s_\rho = e_\rho/T - p/T\rho^2.$$

The integrability condition $s_{T\rho} = s_{\rho T}$ yields
$$e_\rho = p/\rho^2 - p_T T/\rho^2, \qquad e_T = c,$$

and the integrability condition $e_{\rho T} = e_{T\rho}$ implies
$$c_\rho = -p_{TT} T/\rho^2. \tag{12}$$

We set $q = (T, \rho)$. Integration yields

$$e(q) = e(q_0) + \int_{q_0}^{q} c\,dT + \rho^{-2}(p - p_T T)\,d\rho,$$
$$s(q) = s(q_0) + \int_{q_0}^{q} T^{-1}c\,dT - \rho^{-2} p_T\,d\rho, \tag{13}$$
$$\mu(q) = e - Ts + p/\rho.$$

Theorem 67.A (Main Theorem for Gases and Liquids). *We are given the C^1-function $c = c(T, \rho)$ and the C^2-function $P = p(T, \rho)$ for $T > 0, \rho > 0$ with (12). Then for fixed values $e(q_0)$ and $s(q_0)$ the functions e, s, and μ of $q = (T, \rho)$ are uniquely determined through (13) with $q = (T, \rho)$. These integrals are path independent.*

PROOF. Apply the above observations in opposite directions. Because of (12), the integrability conditions for the integrals in (13) are satisfied. □

For an important example, this theorem illustrates which kinds of information can be obtained from Gibbs' fundamental equation.

EXAMPLE 67.4 (Ideal Gas). For gases which have a temperature T close to the "room temperature" T_0 and small densities ρ close to ρ_0, the experiment shows in first-order approximation the following two results.

$$p = r\rho T \quad \text{(equation of state)}, \tag{14}$$
$$c = \alpha r/2 \quad \text{(specific heat capacity)}. \tag{15}$$

Here r is the gas constant and $\alpha = 3, 5, 6$ (one-atomic, two-atomic, n-atomic gas with $n \geq 3$). Thereby equation (12) is satisfied and from (13) we obtain for $e = e(T, \rho)$, $s = s(T, \rho)$, $\mu = \mu(T, \rho)$ the relations

$$e = cT + \text{const},$$
$$s = c\ln(T\rho^{1-\gamma}) + \text{const}, \qquad \gamma = 1 + r/c, \tag{16}$$
$$\mu = e - Ts + p/\rho = rT\ln\rho + \mu_0(T),$$

where $\mu_0(T)$ is defined by the last line. Note that (14), (15) and hence also (16) are only valid in a neighborhood of (T_0, ρ_0), i.e., *not* for low temperatures.

For a deeper understanding of (14) and (15) statistical physics is required. If we neglect oscillations and interactions of the gas molecules, then the inner energy of the N molecules is equal to the mean kinetic energy. The law of equipartition of statistical physics (see Problem 68.3) states that the mean kinetic energy per molecular degree of freedom is equal to $kT/2$. This implies

$$E = \beta N k T/2, \tag{17}$$

where k = Boltzmann constant and β = number of molecular degrees of freedom. Moreover, it follows from statistical physics that

$$PV = NkT.$$

In Section 68.3 this will be proved for one-atomic gases. The pressure, however, which is exerted onto the container wall is independent of the number of atoms of the molecule. Hence this relation is generally true. It follows that

$$p = NkT\rho/M.$$

A comparison with (14) yields

$$r = Nk/M.$$

This implies that $e = E/M = \beta r T/2$ and hence

$$c = e_T = \beta r/2.$$

We thus find that $\alpha = \beta$. This yields a direct physical interpretation of the values of α. Namely,

$$\alpha = \text{number of the molecular degrees of freedom}.$$

A one-atomic molecule has $\alpha = 3$ translational degrees of freedom. For a two-atomic or n-atomic molecule, $n \geq 3$, we find in addition two or three rotational degrees of freedom (see Section 58.13), i.e., we then have $\alpha = 5$ or $\alpha = 6$, respectively.

In fact, an n-atomic molecule may have further degrees of freedom, for example, those given by oscillations. The experimental values for α show that at a "room temperature" these additional degrees of freedom are of no significance. Physicists call these degrees of freedom "frozen." However, at higher temperatures they become important. At extremely low temperatures near $T = 0$, on the other hand, we encounter completely new physical effects, which can be understood only in the context of quantum theory. This will be discussed in Chapter 94 of Part V.

67.4. The Three Laws of Thermodynamics

We now explain the physical basis for the fundamental equation (4).

67.4a. First Law or the Theorem about Conservation of Energy

The key formula is

$$\dot{E} = \dot{Q} + \dot{W}. \tag{18}$$

More precisely, we start with a C^1-manifold Z, which we shall call state space. Concrete examples have already been given in Section 67.1. The points z in Z are called *quasi-equilibrium states* of the system or simply states. A *thermodynamical process* consists of a C^1-curve

$$z = z(t) \quad \text{in } Z$$

and the two C^1-functions

$$Q = Q(t) \quad \text{and} \quad W = W(t)$$

on $[t_0, t_1]$. For simplicity, we set $t_0 = 0$. We interpret:

$Q(t)$ = amount of heat which is added to the system during the time interval $[0, t]$.

$W(t)$ = work which is done at the system during the time interval $[0, t]$, i.e., if $W(t) > 0$, then the system gains energy.

More precisely, $W(t)$ represents all kinds of energy, except for heat energy. Obviously, we must have that

$$Q(0) = W(0) = 0.$$

Note that $Q(t)$ and $W(t)$ can be positive as well as negative. If, for example, $Q(t) < 0$, then this means negative supply of heat, i.e., during the time interval $[0, t]$ the system supplies the amount of heat $|Q(t)|$ to the exterior environment.

From a physical point of view, a thermodynamical process describes the fact that a system passes through quasi-equilibrium states, whereby heat and energy is added to the system.

The functions on Z are called *state functions*. The physical importance of these functions is that they only depend on the state of the system and *not* on the process, i.e., in particular, they do not depend on the history of the system.

The *first law of thermodynamics* states:

(i) *There exists a C^1-function*

$$E: Z \to \mathbb{R}$$

which is called inner energy.

(ii) *Only such processes occur, for which on $[0, t_1]$ equation (18) is satisfied. Thereby we set $E(t) = E(z(t))$.*

An important consequence of (i) is that the inner energy E is a state function. If we set

$$\Delta E = E(t_a) - E(t_b)$$

and, analogously, for Q and W, then by integrating we immediately obtain from (18) that

$$\Delta E = \Delta Q + \Delta W.$$

For a cycle with $z(t_a) = z(t_b)$ we have $\Delta E = 0$, hence

$$\Delta Q + \Delta W = 0.$$

The state space Z is called *regular* if and only if it admits a C^1-function

$$T: Z \to \mathbb{R}.$$

We call T the absolute thermodynamical temperature or simply *temperature*. From a physical point of view, every state of the system has a certain temperature in this case.

The assumption that a state function T exists is called the *zero-th law* of thermodynamics. We emphasize, however, that a restriction to regular state spaces is too stringent. For a description of temperature fields and a study of temperature compensation in systems, one needs nonregular state spaces. This is discussed in Example 67.6 in Section 67.6.

A thermodynamical process is called *closed* if and only if

$$Q(t) \equiv 0 \quad \text{and} \quad W(t) \equiv 0.$$

A system is called closed if and only if in the corresponding state space only closed thermodynamical processes are allowed. From a physical point of view those systems are insulated against supply of outer heat and outer energy.

A thermodynamical process is called *adiabatic* if and only if

$$Q(t) \equiv 0,$$

i.e., if during the entire process no heat is added to the system.

67.4b. Second Law or Entropy Theorem

The key formula is

$$T\dot{S} \geq \dot{Q}. \tag{19}$$

Precisely, the *second law of thermodynamics* postulates the following:

(i) *There exists a C^1-function*

$$S: Z \to \mathbb{R}$$

which is called entropy.
(ii) *For closed thermodynamical processes, i.e., for processes $z = z(t)$, $Q(t) \equiv 0$ and $W(t) \equiv 0$ on $[0, t_1]$ we have*

$$\dot{S}(t) \geq 0 \quad \text{on } [0, t_1].$$

Thereby we set $S(t) = S(z(t))$.

67.4. The Three Laws of Thermodynamics

(iii) *For thermodynamical processes on a regular state space, equation* (19) *is satisfied. Thereby we set* $T(t) = T(z(t))$.

These assumptions greatly restrict the number of thermodynamical processes which are possible in nature. An important consequence of (i) is that entropy S is a state function.

A process
$$z = z(t), \quad Q = Q(t), \quad W = W(t)$$
is called *reversible* if and only if the process can occur with time reversed, i.e., also the process
$$z = z(-t), \quad Q = Q(-t), \quad W = W(-t)$$
is possible. From (19) follows that
$$T(t)\dot{S}(t) \geq \dot{Q}(t) \quad \text{and} \quad -T(-t)\dot{S}(-t) \geq -\dot{Q}(-t)$$
for a reversible process, hence
$$T\dot{S} = \dot{Q}. \tag{20}$$

This is a necessary condition for reversible processes. Hence a process, for which (19) is satisfied on $[0, t_1]$ without having strict equality everywhere, is irreversible. All real processes in nature are irreversible. Reversible thermodynamical processes are idealizations. In order to describe the deviations of a process from reversibility we define S_e and S_i through
$$S_e(t) = S(0) + \int_0^t \dot{Q}(t)/T(t)\,dt, \tag{21}$$
and
$$S(t) = S_e(t) + S_i(t). \tag{22}$$

According to (21), the quantity S_e describes the change in entropy which is caused by outer supply of heat Q. The quantity S_i on the other hand contains that portion of the entropy change which is caused by inner processes of the system. From (19) and (21) we obtain
$$T\dot{S}_e = \dot{Q}, \quad \dot{S}_i \geq 0. \tag{23}$$

In a closed system, for example, we have that $\dot{Q}(t) \equiv 0$ for all occurring processes. This implies $\dot{S}_e(t) \equiv 0$.

To completely understand irreversible processes, we need explicit expressions for the entropy production $S_i(t)$.

The great importance of equation (19) is the fact that it relates entropy and heat. For reversible processes, the change in entropy can be determined from (20) by measuring the amount of heat supply. It is important then that "reversible" means intuitively that the process also can occur backwards in time. But, in experiments, this can only approximately be realized by using extremely slow processes.

67.4c. Gibbs' Fundamental Equation for Thermodynamical Quasi-Equilibrium States

Our objective here is to consider a basic mathematical model for which both the first and the second law are satisfied. In this model, which frequently is used in thermodynamics, states are interpreted as thermodynamical quasi-equilibrium states. One thereby obtains first-order partial differential equations for the inner energy E and the entropy S.

Our starting point is Gibbs' fundamental equation

$$dE = T\,dS - P\,dV + \sum_{i=1}^{r} \mu_i\,dN_i + \sum_{j=1}^{s} A_j\,da_j. \tag{24}$$

A possible physical interpretation for these quantities has already been given in Section 67.2. We set

$$y = (T, V, N, a),$$

where y varies on $Y = \mathring{\mathbb{R}}_+ \times X$ and X is an open set in $\mathbb{R}^{1+r+s}$. The state space Z is the set of all points

$$z = (y, P(y), \mu(y), A(y))$$

with $y \in Y$, where $\mu = (\mu_1, \ldots, \mu_r)$ and $A = (A_1, \ldots, A_s)$. Moreover, let

$$P, \mu_i, A_j \colon Y \to \mathbb{R}$$

be C^1-functions. We assume that there exist two C^1-functions

$$E, S \colon Y \to \mathbb{R}$$

on Y with (24).

Obviously, E and S can also be considered as C^1-functions on Z. We choose Z as above, since then all physical state parameters are explicitly contained. In fact, Z and Y are diffeomorphic and hence can in this sense be identified.

We now want to show that in this model both the first and second law of thermodynamics are satisfied. To this end we consider, analogously to (24), the two differential forms

$$W = -P\,dV + \sum_i \mu_i\,dN_i + \sum_j A_j\,da_j,$$

$$Q = T\,dS. \tag{25}$$

A thermodynamical process is now, by definition, given through

$$z = z(t)$$

and $W = W(t)$, $Q = Q(t)$ with

$$\dot{W} = -P\dot{V} + \sum_i \mu_i \dot{N}_i + \sum_j A_j \dot{a}_j,$$

$$\dot{Q} = T\dot{S},$$

67.4. The Three Laws of Thermodynamics

and $W(0) = Q(0) = 0$. Thereby one obtains

$$T(t), \quad V(t), \quad N(t) \quad \text{and} \quad a(t)$$

immediately from $z(t)$. Moreover, we set $P(t) = P(y(t))$. Analogously, we obtain

$$\mu(t), \quad A(t) \quad \text{and} \quad S(t).$$

All these processes are reversible. Obviously, we have

$$\dot{E} = \dot{Q} + \dot{W}, \qquad T\dot{S} = \dot{Q},$$

i.e., both the first and the second law are satisfied for all these processes.

In classical analysis, one works with differentials in a naive fashion. The theory of manifolds, on the other hand, leads to a precise concept of the differential and the differential form. This will be discussed in detail in Section 73.23. In this sense, C^k-differential forms on Z are identical with C^k-cotangent vector fields on Z. In this precise sense, W and Q are C^k-differential forms on Z with $k = 1$ and $k = 0$, respectively. Furthermore, we have

$$dV(\dot{z}(t)) = \dot{V}(t)$$

where $V(t) = V(z(t))$, etc., hence

$$\dot{W}(t) = W(\dot{z}(t)), \qquad \dot{Q}(t) = Q(\dot{z}(t)).$$

Physically, the states z are interpreted as thermodynamical quasi-equilibrium states. This model is motivated by the physical idea that it is possible to pass from one quasi-equilibrium state to another by using reversible processes. From

$$\dot{Q} = T\dot{S}$$

one obtains the heat function $Q = Q(t)$, which describes the amount of heat which has to be added to the system during the time interval $[0, t]$ to guarantee a reversible process. The assumption that reversible processes exist is an idealization.

This model also allows the study of more general thermodynamical processes, which need not be reversible. Thermodynamical processes are then given by

$$z = z(t), \qquad W = W(t) \quad \text{and} \quad Q = Q(t)$$

with

$$\dot{E} = \dot{W} + \dot{Q}, \qquad T\dot{S} \geq \dot{Q}.$$

Here we set $E(t) = E(y(t))$ and $S(t) = S(y(t))$. In this case $W(\cdot)$ and $Q(\cdot)$ are *not* computed from the differential forms W and Q as above. For example, one may choose $W(\cdot)$ with $\dot{W} = -P\dot{V}$, i.e., more precisely

$$\dot{W}(t) = -P(y(t))\dot{V}(t).$$

Here the outer work which is added to the system results solely from the volume change and corresponding pressure. This will be studied more precisely in Problem 67.1.

Equation (24) by itself does not uniquely determine all state functions E, S, P, μ, and A, but the importance of (24) lies in the fact that this equation yields conditions which *every* thermodynamical quasi-equilibrium system of this type must satisfy. Mathematical analysis then is used to determine which state functions have to be known in order to uniquely describe the system, i.e., one obtains the physically important information, which state functions have to be experimentally determined. We followed this path in Section 67.3 in the analysis of gases.

67.4d. Differential Forms

The model of Section 67.4c can be substantially generalized. For this, we assume:

(H1) Let a real, $(n + 1)$-dimensional C^1-manifold Z be given, which we call the state space. Thereby Z is C^1-diffeomorphic to $Y = \mathring{\mathbb{R}}_+ \times X$, where X is a real, n-dimensional C^1-manifold. In this sense, we can identify Z with Y. The elements of Y have the form

$$y = (T, x),$$

where $T > 0$ denotes the temperature and $x \in X$ the system parameters.

(H2) Assume that

$$E, S: Y \to \mathbb{R}$$

are C^1-functions and W is a C^1-differential form on Y such that on Y the following holds:

$$dE = \mathcal{Q} + W,$$
$$\mathcal{Q} = T \, dS. \tag{26}$$

If $(\xi_1, \ldots, \xi_n)$ are local coordinates of x, then

$$W = \sum_{i=1}^{n} W_i(T, x) \, d\xi_i,$$

holds, i.e., W does not depend on dT.

The thermodynamical processes and the thermodynamical interpretation are then obtained as in Section 67.4c. In particular, the reversible thermodynamical processes are obtained from $z = z(t)$, $W = W(t)$, and $Q = Q(t)$ with

$$\dot{W}(t) = W(\dot{z}(t)), \qquad \dot{Q}(t) = \mathcal{Q}(\dot{z}(t)).$$

This model can immediately be extended to Banach manifolds.

67.4e. Third Law

The key formulas are

$$\lim_{T \to 0} S(y) = 0, \qquad \lim_{T \to 0} S'(y) = 0 \qquad (27)$$

with $y = (T, x)$ and S' denotes the derivative with respect to y. More precisely, the *third law of thermodynamics* postulates the following:

(i) *The state space Z satisfies assumptions* (H1) *of Section 67.4d. In particular, Z is regular.*
(ii) *For the C^1-function entropy $S: Z \to \mathbb{R}$, relation (27) holds for all x.*

From a physical point of view, this represents an assumption on the behavior of the entropy in a neighborhood of the absolute temperature zero $T = 0$. In Example 67.4, relation (27) is not satisfied for ideal gases. In fact, formulas (14)–(16) are only valid for temperatures which are not too low. Especially (14), i.e., the equation of state

$$p = r\rho T$$

is not satisfied for ideal gases in a neighborhood of $T = 0$. In these regions of extremely low temperatures, the methods of classical physics do not suffice. Methods of quantum statistics are required. The third law is closely related to the behavior of quantum systems in the ground state.

67.5. Change of Variables, Legendre Transformation, and Thermodynamical Potentials

In thermodynamics, in order to obtain a convenient description for various different processes, it is important to change the independent variables. In Section 67.3, for example, in connection with the study of gases we chose T, V, and N as the independent variables. The pressure was then obtained from

$$P = P(T, V, N).$$

For processes with $P = \text{const}$, however, it is more convenient to choose T, P, N as independent variables. We then have

$$V = V(T, P, N).$$

All variable changes in thermodynamics are obtained easily from Legendre transformations, which have been discussed in Section 58.16. There we also gave conditions under which the Legendre transformations are possible. In order not to obscure the simple idea of this method, we will not explicitly state these conditions below. Our starting point is Gibbs' fundamental equation

$$dE = T\,dS - P\,dV + \mu\,dN + A\,da. \qquad (28)$$

Here we simply write

$$\mu \, dN = \sum_{i=1}^{r} \mu_i \, dN_i \quad \text{and} \quad A \, da = \sum_{j=1}^{s} A_j \, da_j.$$

Furthermore, E, S, P, μ, A are functions of (T, V, N, a).

Inner energy E. A look at (28) shows that it is more convenient to choose as independent variables S, V, N, a. This yields

$$E = E(S, V, N, a)$$

and

$$E_S = T, \quad E_V = -P, \quad E_N = \mu, \quad E_a = A. \tag{29}$$

For a reversible process we have

$$T\dot{S} = \dot{Q}.$$

Processes with $\dot{Q}(t) \equiv 0$ are called adiabatic. Such processes are heat isolated. For reversible, adiabatic processes we therefore have

$$S = \text{const.}$$

The general solution of the fundamental equation (28) is obtained by prescribing an arbitrary C^1-function $E = E(S, V, N, a)$. The remaining quantities then follow from (29). It is assumed thereby that it is possible to uniquely pass to the variables (S, V, N, a).

Entropy S. From (28) follows

$$dS = T^{-1}(dE + P\, dV - \mu \, dN - A\, da). \tag{30}$$

The natural (canonical) independent variables for S are therefore (E, V, N, a). From

$$S = S(E, V, N, a)$$

we obtain

$$S_E = 1/T, \quad S_V = P/T, \quad S_N = -\mu/T, \quad S_a = -A/T.$$

Free energy F. From (28) and $T\, dS = d(TS) - S\, dT$ follows

$$dF = -S\, dT - P\, dV + \mu \, dN + A\, da \tag{31}$$

with $F = E - ST$. For $F = F(T, V, N, a)$ we find

$$F_T = -S, \quad F_V = -P, \quad F_N = \mu, \quad F_a = A.$$

In order to understand the meaning of the free energy, we consider a reversible process. The two laws $\dot{E} = \dot{Q} + \dot{W}$ and $T\dot{S} = \dot{Q}$ imply

$$\dot{F} = -S\dot{T} + \dot{W}.$$

Table 67.1

Thermodynamical potential	Total differential	Canonical independent variables	Meaning of derivatives
Inner energy E	$dE = T\,dS - P\,dV + \mu\,dN + A\,da$	$E(S, V, N, a)$	$E_S = T, E_V = -P,$ $E_N = \mu, E_a = A$
Entropy S	$T\,dS = dE + P\,dV - \mu\,dN - A\,da$	$S(E, V, N, a)$	$TS_E = 1, TS_V = P,$ $TS_N = -\mu, TS_a = -A$
Free energy $F = E - TS$	$dF = -S\,dT - P\,dV + \mu\,dN + A\,da$	$F(T, V, N, a)$	$F_T = -S, F_V = -P,$ $F_N = \mu, F_a = A$
Free enthalpy $G = F + PV$	$dG = -S\,dT - V\,dP + \mu\,dN + A\,da$	$G(T, P, N, a)$	$G_T = -S, G_P = -V,$ $G_N = \mu, G_a = A$
Enthalpy $H = E + PV$	$dH = T\,dS - V\,dP + \mu\,dN + A\,da$	$H(S, P, N, a)$	$H_S = T, H_P = -V,$ $H_N = \mu, H_a = A$
Statistical potential $\Omega = F - \mu N$	$d\Omega = -S\,dT - P\,dV - N\,d\mu + A\,da$	$\Omega(T, V, \mu, a)$	$\Omega_T = -S, \Omega_V = -P,$ $\Omega_\mu = -N, \Omega_a = A$

For isothermal processes, i.e.,

$$T(t) \equiv \text{const.},$$

we find that $\dot{F} = \dot{W}$ and hence

$$\Delta F = \Delta W.$$

This means, for reversible, isothermal processes, the change of the free energy is equal to the work added to the system. The free energy will be applied during the following chapter in connection with Planck's radiation law and the study of the universe immediately after the Big Bang.

Table 67.1 gives a survey about further important thermodynamical potentials, which can be obtained analogously. During the following discussion we will frequently use statements from Table 67.1.

67.6. Extremal Principles for the Computation of Thermodynamical Equilibrium States

We want to show how equilibrium states can be computed from quasi-equilibrium states. The idea thereby is the following.

(i) The first and second law imply that, depending on the process, one can choose suitable thermodynamical potentials which exhibit a favorable growth behavior with respect to time (e.g., $\dot{S} \geq 0$ or $\dot{F} \leq 0$).

(ii) We use this behavior in order to define equilibrium states and then to compute them.

Principle 67.5 (Entropy). *Assume the system is closed, i.e., no energy is added to the system neither in the form of heat nor in the form of work. We then have*

$$\dot{S} \geq 0$$

for all processes which occur in the system. For a reversible process we have $\dot{S} = 0$, i.e., $S = $ const.

PROOF. This follows from the second law. □

EXAMPLE 67.6 (Temperature and Pressure Compensation). We consider a system Σ which consists of n subsystems Σ_j. We assume that each Σ_j is in a quasi-equilibrium state, which is parametrized by

$$y_j = (E_j, V_j, N_j).$$

According to Section 67.5, we obtain from

$$S_j = S_j(y_j)$$

that

$$\frac{\partial S_j}{\partial E_j} = \frac{1}{T_j}, \qquad \frac{\partial S_j}{\partial V_j} = \frac{P_j}{T_j}, \qquad \frac{\partial S_j}{\partial N_j} = -\frac{\mu_j}{T_j}.$$

A process in Σ is described by $y_j = y_j(t)$ and

$$T_j = T_j(y_j(t)), \qquad P_j = P_j(y_j(t)), \qquad \mu_j = \mu_j(y_j(t))$$

for all j. The physicist imagines that the subsystems Σ_j are separated by walls and that those are slowly removed. The systems are assumed to be statistically independent from each other, i.e., we neglect interactions (e.g., water and steam). We then find entropy S and inner energy E of Σ by adding the corresponding quantities of Σ_j. Moreover, we assume that Σ is a closed system. It follows then that

$$E = \text{const}, \qquad V = \text{const}, \qquad N = \text{const}.$$

According to Principle 67.5, only processes with $\dot{S} \geq 0$ may occur in Σ. We expect that these processes tend towards a thermodynamical equilibrium as $t \to \infty$. Therefore, it makes sense to compute these equilibrium states by using the following extremal principle:

$$S = \Sigma_j S_j = \max!,$$
$$E = \Sigma_j E_j = \text{const}, \qquad V = \Sigma_j V_j = \text{const}, \qquad N = \Sigma_j N_j = \text{const}. \tag{32}$$

Assume that all functions S_j are C^2. According to Section 43.10 of Part III we solve this problem by using Lagrange's multiplier rule. For this, we choose

the Lagrange function
$$L = S - \alpha E - \beta V - \gamma N.$$
The necessary solvability condition for (32) is
$$L'(y_0) = 0,$$
i.e., all partial derivatives of L with respect to E_j, V_j, N_j are equal to zero. This gives
$$1/T_j = \alpha, \qquad P_j/T_j = \beta, \qquad -\mu_j/T_j = \gamma \qquad (33)$$
for all j. Thus, in a thermodynamical equilibrium, all temperatures T_j, all pressures P_j, and all chemical potentials μ_j must equal each other. The values for the multipliery α, β, γ are obtained by inserting (33) into the side conditions of equation (32). In order to verify that the solution $y_0 = (y_1^0, y_2^0, \ldots)$, computed this way, does actually yield a maximum, we have, according to Section 43.10, to show that there exists a number $c > 0$ with
$$S''(y_0)(\Delta y)^2 \leq c|\Delta y|^2 \qquad (34)$$
and $\Delta y = y - y_0$ for all y, which satisfy the side conditions in (32). Thereby $S''(y_0)(\Delta y)^2$ is a quadratic form with the second-order partial derivatives of S at the point y_0 as coefficient matrix. If (34) holds, then physicists speak of a stable thermodynamical equilibrium. They symbolize this by
$$\delta^2 S < 0.$$
Observe the following fact. According to the general result of Section 43.10, we obtain equation (34) where S is replaced with L. But since the side conditions are linear, we find that $L''(y) = S''(y)$.

EXAMPLE 67.7 (Chemical Reactions). The following observations will be used later on. We consider, for example, the chemical reaction
$$\alpha_1 SO_2 + \alpha_2 O_2 \rightleftharpoons \alpha_3 SO_3 \qquad (35)$$
with $\alpha_1 = 2, \alpha_2 = 1, \alpha_3 = 2$. This means that two SO_2-molecules are combined with one O_2-molecule to form two SO_3-molecules, or the opposite reaction occurs. Let $N_1(t), N_2(t), N_3(t)$ be the number of SO_2, O_2, SO_3-molecules at time t, respectively. We then have
$$\dot{N}_1(t) = \alpha_1 \beta, \qquad \dot{N}_2(t) = \alpha_2 \beta, \qquad \dot{N}_3(t) = -\alpha_3 \beta,$$
where β is called the *reaction velocity*. As a more convenient mathematical notation for chemical reactions we will use in the future:
$$\sum_{i=1}^{n} \gamma_i [N_i] = 0. \qquad (36)$$
In case (35) we have $n = 3$ and $\gamma_1 = \alpha_1, \gamma_2 = \alpha_2, \gamma_3 = -\alpha_3$. For general reactions of the form (36) we have
$$\dot{N}_i(t) = \gamma_i \beta \qquad \text{for all } i.$$

In the following chapter we will also consider reactions of elementary particles of the form (36).

Principle 67.8 (Free Energy). *Assume that no work is done at a given system and that its temperature is constant. We then have*

$$\dot{F} \leq 0$$

for all processes which occur in the system. For reversible processes we obtain $\dot{F} = 0$, i.e., F = const.

PROOF. The first and second laws imply that $\dot{E} = \dot{Q}$ and $T\dot{S} \geq \dot{Q}$. For $F = E - ST$ we obtain $\dot{F} = \dot{E} - \dot{S}T - S\dot{T} = \dot{E} - \dot{S}T \leq 0$. □

EXAMPLE 67.9 (Photon Gas). Consider a gas which consists of N photons in a container with a fixed temperature T and fixed volume V. The photons are emitted from the walls into the container. Each N corresponds to a quasi-equilibrium state. The free energy is

$$F = F(T, V, N).$$

According to Principle 67.8, it makes sense to assume that for fixed T and V the free energy F has a minimum at the thermodynamical equilibrium with repect to N. This implies

$$\mu = F_N(T, V, N) = 0.$$

In Section 68.4 Planck's famous radiation law will be derived from this relation. In Section 68.5 we will show that Planck's law implies the Stefan–Boltzmann law for the radiation of black bodies.

EXAMPLE 67.10 (Chemical Reactions). For fixed temperature T and fixed volume V, we consider n substances in a container. We have

$$F = F(T, V, N_1, \ldots, N_n).$$

Assume that the chemical reaction (36) occurs inside the container. Thus we have

$$\dot{N}_i = \gamma_i \beta.$$

Since N_i depends on the time, we cannot use assumption F = min! for the definition of thermodynamical equilibrium, but instead have to use the process

$$F(t) = F(T, V, N_1(t), \ldots, N_n(t)).$$

This is sometimes overlooked in the physical literature. According to Principle 67.8, we can characterize the equilibrium states through the condition

$$\dot{F}(t) = 0.$$

This yields $\sum_i (\partial F/\partial N_i)\dot{N}_i = 0$, and hence

$$\sum_i \gamma_i \mu_i(T, V, N_1, \ldots, N_n) = 0. \qquad (37)$$

67.7. Gibbs' Phase Rule

Principle 67.11 (Free Enthalpy). *Let $G = F + PV$. Suppose pressure and temperature of the system are constant, and assume that besides mechanical work through volume change no further work is done at the system. It follows then that*

$$\dot{G} \leq 0$$

for all processes which occur in the system. For reversible processes we have $\dot{G} = 0$, i.e., $G = $ const.

PROOF. The first and second laws imply $\dot{E} = \dot{Q} - P\dot{V}$ and $T\dot{S} \geq \dot{Q}$. For $G = E - TS + PV$ we obtain

$$\dot{G} = \dot{E} - \dot{T}S - T\dot{S} + \dot{P}V + P\dot{V} = \dot{E} - T\dot{S} + P\dot{V} = \dot{Q} - T\dot{S} \leq 0. \quad \square$$

Applications will be given in the following two sections.

Principle 67.12 (Enthalpy). *Let $H = E + PV$. Suppose the pressure is constant, the system is heat isolated, and assume that besides mechanical work through volume changes no further work is done at the system. It follows then that*

$$\dot{H} = 0$$

for all processes which occur in the system, i.e., $H = $ const.

PROOF. The first law implies that $\dot{E} = -P\dot{V}$. For $H = E + PV$ we thus obtain $\dot{H} = \dot{E} + P\dot{V} = 0$. $\quad \square$

The principle of enthalpy conservation, for example, plays an important role in connection with liquefaction of air.

67.7. Gibbs' Phase Rule

The phase rule states:

Suppose a system of K substances is given which can be in Φ phases (gaseous, liquid, solid, etc.). The system has temperature T and pressure P. No chemical reactions occur. If the system is in a thermodynamical equilibrium, then it has

$$f = 2 + K - \Phi$$

degrees of freedom, i.e., the possible values of T, P, and the phase concentrations depend on f parameters.

Before proving this, we consider, as an example, a system which consists of water and steam. Here we have $K = 1$ and $\Phi = 2$. The system has one degree of freedom. We are free to choose T and obtain

$$P = P(T).$$

If water, ice, and steam are in a thermodynamical equilibrium, then we have

$K = 1$ and $\Phi = 3$. The system has zero degrees of freedom, i.e., T and P are fixed. This gives the triple point

$$T = 0.008 \,°C, \qquad P = 0.006 \text{ kp/cm}^2.$$

We now prove the phase rule. Let $N_{k\varphi}$ be the number of particles of the kth substance in the φth phase. According to Principle 67.11 we compute the equilibrium by using the following extremal principle:

$$G(T, P, N) = \min!,$$
$$N_k = \sum_\varphi N_{k\varphi} = \text{const}, \qquad k = 1, \ldots, K. \tag{38}$$

Let N be the tuple of all $N_{k\varphi}$. We set

$$L = G - \sum_k \alpha_k N_k,$$

and apply Lagrange's multiplier rule of Section 43.10. This gives

$$\partial G / \partial N_{k\varphi} = \alpha_k,$$

hence $\mu_{k\varphi} = \alpha_k$, i.e.,

$$\mu_{k1} = \mu_{k2} = \cdots = \mu_{k\varphi}, \qquad k = 1, \ldots, K. \tag{39}$$

We now assume that for fixed φ the quantity $\mu_{k\varphi}$ only depends on P, T, and the concentrations in the φth phase. These concentrations are

$$c_{r\varphi} = N_{r\varphi} \Big/ \sum_k N_{k\varphi}, \qquad r = 1, \ldots, K.$$

Since the sum of all these concentrations is equal to one, $\mu_{k\varphi}$ only depends on $2 + (K - 1)$ variables. Relation (39) therefore contains $K(\Phi - 1)$ equations for $2 + \Phi(K - 1)$ variables. Thus in the regular case the solution of equation (39) depends on

$$f = 2 + \Phi(K - 1) - K(\Phi - 1) = 2 - \Phi + K$$

parameters.

67.8. Applications to the Law of Mass Action

As in Example 67.7 we consider a chemical reaction

$$\sum_{i=1}^n \gamma_i [N_i] = 0 \tag{40}$$

between n substances. Let $c_i = N_i / \sum_i N_i$ be the concentration of the ith substance. Then

$$\prod_{i=1}^n c_i^{\gamma_i} = K(T, P) \tag{41}$$

is satisfied where K is a constant, depending on temperature T and pressure P. Strictly speaking, (41) is valid only for ideal gases, but, approximately, this relation can also be applied to real gases and liquids.

We will now derive (41). From (40) follows that

$$\dot{N}_i = \gamma_i \beta.$$

We fix T and P and consider the free enthalpy

$$G = G(P, T, N_1, \ldots, N_n).$$

Analogous to Example 67.10 and, according to Principle 67.11, we now use the condition

$$\dot{G} = 0$$

to determine the equilibrium. This yields $\sum_i (\partial G/\partial N_i) \dot{N}_i = 0$, and hence

$$\sum_i \gamma_i \mu_i = 0. \tag{42}$$

The chemical potential of an ideal gas satisfies

$$\mu_i = kT \ln P_i + \mu_{0i}(T). \tag{43}$$

This follows from Example 67.4 by simply changing the notation. In this example we used the mass $M = mN$ instead of N. Hence μ in (16) has to be divided by m. Note also that we use $p = NkT\rho/M$ in (16). The quantity P_i in (43) denotes the pressure of the ith gas. We assume that the gases do not interact with each other. From Example 67.4 it follows then that

$$PV = NkT \quad \text{and} \quad P_i V = N_i kT.$$

This implies $P_i = Pc_i$, and hence (41) follows from $\exp \sum_i \gamma_i \mu_i = 1$ and (43).

PROBLEMS

67.1. *Power during volume increase.* Assume a gas in a bounded region G of $\mathbb{R}^3$ is under pressure P. We increase the volume V by transforming G via

$$y(t) = x + u(t)$$

into $G(t)$. Let $W(t)$ denote the work done during the time interval $[0, t]$. Compute the power $\dot{W}(0)$.

Solution: An application of the transformation formula for volume integrals and integration by parts yields

$$V(t) = \int_{G(t)} dy = \int_G dx + t \int_{\partial G} n\dot{u}(0) \, dO + o(t), \quad t \to 0$$

with outer unit normal vector n. The force $-Pn \, \Delta O$ acts on the surface element ΔO. Hence the power is equal to

$$\dot{W}(0) = -\int_{\partial G} Pn\dot{u}(0) \, dO = -P\dot{V}(0).$$

This observation becomes completely elementary if one looks at the situation of Figure 67.1. There one obtains

$$W(t) = P(V(0) - V(t))$$

for constant P. Note that pressure = force/surface and work = force · distance.

67.2. Adiabatic process for ideal gases. What is the relation between pressure P and density ρ of ideal gases for reversible, adiabatic processes.

Solution: Principle 67.5 implies that S = const. From (16) it follows

$$T\rho^{-r/c} = \text{const},$$

and $P = r\rho T$ implies

$$P\rho^{-1-r/c} = \text{const}.$$

67.3. Perpetuum mobile of the second kind. Show: According to the second law, it is impossible to construct a periodically working machine which, during one cycle, emits energy in the form of mechanical work whereby it only absorbes heat from precisely one heat reservoir.

Solution: Let t_1 be the period. Since the entropy S only depends on the state, we find that

$$S(0) = S(t_1).$$

Integration of the second law $\dot{S} \geq \dot{Q}/T$ over $[0, t_1]$ yields

$$\int_0^{t_1} \dot{Q}\, dt/T \leq 0.$$

Because of $\dot{Q} > 0$ and $T > 0$ this is a contradiction.

67.4. Carnot's cycle. We consider a periodically working machine with period t_4. Let $T(t)$ be the temperature at time t, and $Q(t)$ the heat added to the machine during the time interval $[0, t]$. We consider four intervals $0 < t_1 < t_2 < t_3 < t_4$ and set $Q_i = Q(t_i) - Q(t_{i-1})$, where $t_0 = 0$.
(i) For $[0, t_1]$ let $Q_1 < 0$ and $T(t) \geq T_1$ (e.g., heat emission into the surrounding region).
(ii) For $[t_1, t_2]$ and $[t_3, t_4]$ let $Q(t) \equiv 0$.
(iii) For $[t_2, t_3]$ let $Q_3 > 0$ and $T(t) \leq T_3$ (heat absorbtion). Assume that $T_1 < T_3$.
Let A denote the work done by the machine during one run through $[0, t_4]$. The ratio

$$\eta = A/Q_3$$

between work done and absorbed heat is called the efficiency. Compute η.

Solution: As in Problem 67.3 the important point is that $S(0) = S(t_4)$ for the entropy S. From the second law $\dot{S} \geq \dot{Q}/T$ follows the inequality

$$0 \geq \frac{Q_1}{T_1} + \frac{Q_3}{T_3} \tag{44}$$

by integration over $[0, t_4]$. The first law implies $\dot{E} = \dot{Q} + \dot{W}$. By integration it follows from $E(0) = E(t_4)$ that

$$0 = Q_1 + Q_3 + W(t_4),$$

and hence $A = Q_1 + Q_3$ and $\eta = 1 - |Q_1|/Q_3$. From (44) follows

$$\eta \leq 1 - T_1/T_3. \tag{45}$$

In the case of reversible processes and $T = $ const in (i) and (iii), we have equality in (44). It follows, then that

$$\eta = 1 - T_1/T_3.$$

This is the ideal efficiency of a machine which has the form (i) to (iii).

The important fact about (45) is that, because of $\eta < 1$, heat cannot be completely transformed into work. This was Carnot's fundamental technological discovery of 1824 which, during the nineteenth century, led to the phenomenological formulation of the second law by Clausius.

67.5. *Further concrete problems.* Study the numerous examples in Sommerfeld (1954, M), Vol. 5 and Landau and Lifšic (1962, M), Vol. 5.

References to the Literature

Classical work: Clausius (1865), Boltzmann (1871), (1872), Planck (1913, M).
Sommerfeld (1954, M), Vol. 5, Landau and Lifšic (1962, M), Vols. 5, 9, 10, Glansdorff and Prigogine (1971, M).
Mathematical theory of entropy: Martin (1981, S).
Conceptual analysis of the laws of thermodynamics: Serrin (1979), (1983, S), (1986, P), Owen (1984, M).
Thermomechanics: Ziegler (1983, M).
(See also the References to the Literature to Chapters 68 and 86.)

CHAPTER 68

Statistical Physics

> The true logic in this world lies in probability theory.
> James Clerk Maxwell (1831–1879)

> In 1866, Ludwig Boltzmann began his scientific career with an attempt to give a purely mechanical explanation of the second law of thermodynamics. He gradually regognized the need to introduce statistical concepts in order to understand irreversibility and the second law.
> M. J. Klein (1973)

> Don't trust any statistics that you didn't falsify yourself.
> Folclore

During the study of the Big Bang in Section 58.15 we already made essential use of Planck's radiation law. In order to find this law, Planck formulated his famous hypothesis about the quantization of energy for the harmonic oscillator. This was the hour of birth of quantum theory. Planck's radiation law implies the Stefan–Boltzmann radiation law, which will be used during the following chapter in the discussion of Carleman's radiation problem. In the present chapter we want to show how these important physical laws can be derived from general principles of statistical physics. The development of statistical physics is mainly connected with the names of Maxwell (1831–1879), Boltzmann (1844–1906), Gibbs (1839–1903), Planck (1858–1947), and Einstein (1879–1955).

In classical statistical physics, it was assumed that, for example, the motion of molecules in a gas is governed by the laws of classical mechanics. The statistical point of view was only an auxiliary tool to mask complicated mechanical systems. But as we already saw in Section 59.12, the concept of probability has far deeper roots. In fact, in the microcosmos, the laws of classical mechanics have to be replaced with the laws of quantum mechanics. These laws have *primarily* a probability theoretical character.

We begin our discussion of statistical physics with the basic equations which, in Section 43.12 of Part III, have been obtained from the principle of maximal entropy (information) by using Lagange's multiplier rule. This implies the following special cases:

(i) Bose statistics and Fermi statistics which, other than classical statistics, also describe correctly the behavior of thermodynamical systems at low temperatures.
(ii) Classical statistics and quasi-classical statistics.

As applications we consider:

(a) Ideal gases (equation of state, Maxwell's velocity distribution).
(b) Photon gas, black-body radiation. Planck's radiation law, and the Stefan–Boltzmann radiation law.
(c) Cosmos at the temperature of 10^{11} K at time $t = 10^{-2}$ s after the Big Bang.
(d) Basic equation of star models and the maximal Chandrasekhar mass of white dwarf stars.
(e) Law of equipartition.

In Part V we will consider problems in modern quantum statistics and irreversible thermodynamics. In fact, there are still many open problems in this area.

68.1. Basic Equations of Statistical Physics

The following equations stand at the beginning of statistical physics:

$$w_r = \frac{e^{(\mu N_r - E_r)/kT}}{\sum_r e^{(\mu N_r - E_r)/kT}}, \tag{1}$$

$$E = \sum_r w_r E_r, \qquad N = \sum_r w_r N_r, \tag{2}$$

$$S = -k \sum_r w_r \ln w_r, \tag{3}$$

$$F = E - ST. \tag{4}$$

We consider a system which might be in the different states Z_r. To each Z_r belongs the energy E_r and the number of particles N_r. We are looking for the *probability* w_r that the system is in the state Z_r.

68.1a. Motivation

We determine w_r from the assumption that under the side condition (2) the entropy becomes maximal, i.e., we solve the problem

$$S = \max!,$$
$$\sum_r w_r E_r = E, \quad \sum_r w_r N_r = N, \quad \sum_r w_r = 1$$

for fixed E and N. In Section 43.12 of Part III we used the Lagrange multiplier rule to show that the solution w_r of this problem is given by (1). The Lagrange multipliers T and μ were obtained from (1), (2) as functions of E (mean energy) and N (mean particle number). More precisely, we find from the equations

$$E = \sum_r w_r(T,\mu) E_r,$$
$$N = \sum_r w_r(T,\mu) N_r,$$

that

$$T = T(E,N), \quad \mu = \mu(E,N).$$

By making a comparison with phenomenological thermodynamics, we will show below that it is useful to identify T with the *absolute temperature* and μ with the *chemical potential*.

68.1b. Mean Value and Dispersion of Arbitrary Physical Quantities

Consider an arbitrary physical quantity $\mathscr{A}$ and assume that $\mathscr{A}$ has the value A_r in state Z_r for all r. We then define the corresponding *mean value* as

$$A = \sum_r w_r A_r \tag{5}$$

and the *dispersion* as

$$(\Delta A)^2 = \sum_r w_r (A_r - A)^2. \tag{6}$$

Those two values are important in linking the theory to physical measurements. Roughly speaking, the mean value A is the value expected to be measured in physical experiments. Moreover, we would assume that the difference $\mathscr{A} - A$ between the actual measurement value $\mathscr{A}$ and the mean value A is small if the dispersion $(\Delta A)^2$ is small. The precise mathematical formulation is given in the Chebyshev inequality (C) below. For this we define

$p(|\mathscr{A} - A| \leq \varepsilon) =$ probability for measuring some value $\mathscr{A}$ with $|\mathscr{A} - A| \leq \varepsilon$.

Chebyshev's inequality tells us that

(C) $$p(|\mathscr{A} - A| \leq \varepsilon) \geq 1 - \frac{(\Delta A)^2}{\varepsilon^2}.$$

68.1. Basic Equations of Statistical Physics

for all $\varepsilon > 0$. In particular, letting $\varepsilon = 4\Delta A$, we obtain that

$$p(|\mathscr{A} - A| \leq 4\Delta A) \geq \tfrac{15}{16}.$$

Thus, roughly speaking, in most cases, the measurement value $\mathscr{A}$ lies in the interval $[A - 4\Delta A, A + 4\Delta A]$. If $\Delta A = 0$, we find

$$p(\mathscr{A} = A) = 1.$$

To prove (C), let

$$\chi_r = \begin{cases} 0 & \text{for } |A_r - A| \leq \varepsilon, \\ 1 & \text{for } |A_r - A| > \varepsilon. \end{cases}$$

This yields

$$p(|\mathscr{A} - A| > \varepsilon) = \sum_r \chi_r w_r.$$

From

$$(\Delta A)^2 = \sum_r (A_r - A)^2 w_r \geq \sum_r \varepsilon^2 \chi_r w_r$$

follows that

$$p(|\mathscr{A} - A| \leq \varepsilon) = 1 - p(|\mathscr{A} - A| > \varepsilon)$$
$$\geq 1 - (\Delta A)^2/\varepsilon^2,$$

and this is (C).

In Example 68.6 below we will show that for a system with a great number N of independent particles the dispersion $(\Delta E)^2$ of energy is extremely small. We shall derive the crucial formula

$$\Delta E/E \sim 1/\sqrt{N},$$

where E is the mean value of the energy. As usual we have $N \sim 10^{23}$ (the number of molecules in a gas of reasonable size). Using the same argument, analogous formulas for the dispersion $(\Delta A)^2$ of other physical quantities can be obtained. This is the reason for the surprising fact that in macroscopical physics the measurement values seem to be constant without any fluctuations.

68.1c. Computation of All Interesting Physical Quantities

The following proposition provides the key for computations in statistical physics.

Proposition 68.1. *All important thermodynamical quantities can be computed from the function*

$$\Omega(\mu, T) = -kT \ln \sum_r e^{(\mu N_r - E_r)/kT}.$$

We have

$$S = -\Omega_T, \quad N = -\Omega_\mu,$$
$$E = -T^2(\Omega/T)_T + \mu N = F + ST, \quad F = \Omega + \mu N, \qquad (7)$$
$$w_r = e^{(\Omega + \mu N_r - E_r)/kT}.$$

PROOF. This follows easily from (1) to (3). □

If Ω depends also on the volume V, then we define the pressure

$$P = -\Omega_V. \qquad (8)$$

The number $k = 1.380 \cdot 10^{-23}$ J/K is a universal constant (Boltzmann constant). If we compare (7), (8) with the last line of Table 67.1 of Section 67.5, then it makes sense to identify T with the absolute temperature. Moreover, we may identify: S as entropy, F as free energy, E as inner energy, μ as chemical potential, and Ω as statistical potential.

68.1d. The Role of Entropy

The concept of entropy is more general than that of temperature. Actually one can always compute the entropy S from (3) if one knows arbitrary probabilities w_r for the states Z_r with

$$\sum_r w_r = 1.$$

The probabilities w_r in (1) are characterized by the assumption of maximal entropy. Thereby one obtains the temperature T as a Lagrange multiplier. Physically, (1) corresponds to quasi-equilibrium states of thermodynamical systems in the sense of Section 67.1 with a common temperature.

68.1e. The Role of the Partition Function

Remark 68.2 (Partition Function and the Counting of Different States). It is very important that the so-called *partition function*

$$\sum_r e^{(\mu N_r - E_r)/kT}$$

in Proposition 68.1 is to be taken over all *different* states Z_r, which the system may occupy. In particular, different states may have the same energy. The following examples will show that there exist many different ways to count "different" states. This is the reason for the fact that there exist different forms of statistical physics. In fact, the classification of "different states" is a physical problem. The main difference, for example, between classical statistical physics and quantum statistics lies in the different classification of states.

68.1. Basic Equations of Statistical Physics

(i) *Prototype of classical statistical physics.* If, for instance, we have two particles A and B, with possible energies α, β, then the various states of Table 68.1 are obtained. Thereby A/α means that A has the energy α, etc. This is the classical way of counting states, where a strict distinction between particle A and particle B is exercised. In Table 68.1 we obtain four states Z_r, $r = 1,\ldots, 4$ with corresponding energy E_r and particle number N_r.

Table 68.1

r	Z_r	E_r	N_r
1	A/α, B/β	$\alpha + \beta$	2
2	A/β, B/α	$\alpha + \beta$	2
3	A/α, B/α	2α	2
4	A/β, B/β	2β	2

(ii) *Prototype of Bose statistics.* In quantum theory there exists the important principle of *indistinguishability* for the particles, i.e., one cannot distinguish between particles A and B. This leads to Table 68.2.

Table 68.2

r	Z_r	E_r	N_r
1	$\begin{Bmatrix} A/\alpha,\ B/\beta \\ A/\beta,\ B/\alpha \end{Bmatrix}$	$\alpha + \beta$	2
2	A/α, B/α	2α	2
3	A/β, B/β	2β	2

State Z_1, for example, can be described by saying that one particle has energy α and one particle has energy β, but we cannot say which one has which energy. In Table 68.2 we therefore only have three different states Z_r, $r = 1, 2, 3$ in contrast to Table 68.1. Note, however, that the situation of Table 68.2 only applies to particles with integer spin (e.g., photons).

(iii) *Prototype of Fermi statistics.* Particles in quantum theory with half-numberly spin (e.g., electrons) have the property that *no* two particles can be in the same state. This yields Table 68.3, where only one state exists.

Table 68.3

r	Z_r	E_r	N_r
1	$\begin{Bmatrix} A/\alpha,\ B/\beta \\ A/\beta,\ B/\alpha \end{Bmatrix}$	$\alpha + \beta$	2

We therefore see that it is important to first clarify what we mean by "states," before starting with statistics. Different classifications of states may lead to entirely different results in statistical physics.

Remark 68.3 (Pure Energy Statistics). If the number of particles is *not* subject to statistical fluctuations, i.e., we have $N_r = N$ for all r, then

$$w_r = e^{-E_r/kT} \Big/ \sum_r e^{-E_r/kT}.$$

This formula follows as in Section 68.1a by omitting the side condition $\sum_r w_r N_r = N$. This way one obtains all quantities (7), (8) by formally letting $N = 0$ in Ω. Only $N = -\Omega_\mu$ is no longer valid, but instead N is fixed.

68.2. Bose and Fermi Statistics

EXAMPLE 68.4 (Bose Statistics). Here the state Z_r with $r = 0, 1, \ldots$ is characterized by the fact that it contains

$$N_r = r \quad \text{particles with energy} \quad E_r = r\varepsilon_n$$

for fixed $\varepsilon_n > 0$ and fixed n. For $\mu < 0$ one obtains the geometrical series

$$\Omega_n = -kT \ln \sum_{r=0}^{\infty} e^{(r\mu - r\varepsilon_n)/kT}$$

$$= kT \ln(1 - e^{(\mu - \varepsilon_n)/kT}). \tag{9}$$

From (7) one immediately obtains

$$N_n = -\partial \Omega_n / \partial \mu = 1/(e^{(\varepsilon_n - \mu)/kT} - 1), \qquad E_n = N_n \varepsilon_n$$

for the average number of particles N_n and the mean energy E_n, respectively.

We now assume, in addition, that the particles may assume different energies ε_n and that for different n the states are statistically independent. The average number of particles is then defined by

$$N = \sum_n N_n.$$

Also we assume

$$E = \sum_n E_n, \qquad S = \sum_n S_n, \qquad F = \sum_n F_n.$$

This is obtained by letting

$$\Omega = \sum_n \Omega_n$$

and applying the formulas of Proposition 68.1 to Ω. In particular, we have

$$E = \sum_n N_n \varepsilon_n.$$

EXAMPLE 68.5 (Fermi Statistics). We assume that in the previous example only $r = 0, 1$ is possible, i.e., only one particle of energy ε_n is possible for each state. Then we obtain

$$\Omega_n = -kT \ln \sum_{r=0}^{1} e^{(r\mu - r\varepsilon_n)/kT}$$
$$= -kT \ln(1 + e^{(\mu - \varepsilon_n)/kT}).$$

As in Example 68.4 we then obtain the following formulas with "+":

$$N_n = 1/(e^{(\varepsilon_n - \mu)/kT} \pm 1), \qquad E_n = N_n \varepsilon_n,$$
$$-\Omega = \pm kT \sum_n \ln(1 \pm e^{(\mu - \varepsilon_n)/kT}). \tag{10}$$

The sign "$-$" corresponds to the Bose statistics in Example 68.4.

In the case of elementary particles, the particles with half-numberly spin are governed by the Fermi statistics (e.g., electrons, positrons, protons, neutrons, and neutrinos) and particles with integer spin are governed by the Bose statistics (e.g., photons). This fact can be proved in the context of axiomatic quantum field theory (see Streader and Wightman (1964, M)).

68.3. Applications to Ideal Gases

By an ideal gas we mean a system which consists of particles that do not interact. Hence, one obtains the statistics for this system by applying the Bose and Fermi statistics to all possible energy states ε_n of a particle. Let q and p denote the position and momentum vector of the particle, respectively. We assume for the energy of the particle that

$$\varepsilon = \varepsilon(p, q) + Q_s, \qquad s = 1, 2, \ldots \tag{11}$$

holds. Thereby $\varepsilon(p, q)$ describes the kinetic energy of the particle and its potential energy $U(q)$ in an outer field. In the context of classical mechanics we have

$$\varepsilon(p, q) = p^2/2m + U(q) \tag{12}$$

with m = mass of the particle. In the special theory of relativity we have

$$\varepsilon(p, q) = \sqrt{m_0^2 c^4 + c^2 p^2} + U(q) \tag{13}$$

with m_0 = rest mass of the particle and c = velocity of light (see Section 75.11). If the particle is a molecule, then Q_s are its quantum mechanical energy states which can be computed from the Schrödinger equation. Classically, ε is the total mechanical energy of the molecule (kinetic energy plus potential energy in a given outer field and in the inner field of electrostatic attraction). The quantities Q_s then correspond to rotational and oscillation energies.

The statistical potential for (11) is

$$-\Omega = \pm kTg \sum_s \int \ln(1 \pm e^{(\mu - \varepsilon(p,q) - Q_s)/kT}) \frac{dp\,dq}{h^3} \quad (14)$$

with Planck's constant $h = 6.625 \cdot 10^{-34}$ J s. We obtain (14) from (10) by formally replacing the summation Σ_n over the states with an integration over $dp\,dq$ and a summation over s and the spin states of the particle. We assume the particle has g spin states, and hence the factor g appears. The factor $1/h^3$ yields the correct quantum mechanical description. It corresponds to the hypothesis that the phase space is quantized, i.e., if G is a region in the (p, q)-phase space, then

$$\int_G dp\,dq/h^3$$

is the number of corresponding states (see Section 59.21). In (14) we choose "+" and "−" for the Fermi and Bose statistics, respectively. The computation of Ω in (14) is difficult in the general case and only approximations are possible. For large molecules it is already very difficult to numerically solve the Schrödinger equation in order to determine Q_s. However, if one knows Ω, then one can use Proposition 68.1 to compute all interesting physical quantities E, F, S, and pressure P as functions of μ, T, and V.

For the following computations we use the classical *integral formulas* which are given in Problem 68.5.

EXAMPLE 68.6 (Ideal One-Atomic Gas). We consider an ideal gas which consists of N atoms of mass m in the volume V. The particle density is equal to $\rho = N/V$. Moreover, we set

$$\alpha = e^{\mu/kT}.$$

In this case we have

$$\varepsilon = p^2/2m \quad \text{and} \quad g = 1, \quad Q_s = 0.$$

We use the Bose statistics. Because of $\int dq = V$ we obtain

$$\Omega(\mu, T, V) = kTVh^{-3} \int \ln(1 - e^{(\mu - \varepsilon(p))/kT})\,dp. \quad (15)$$

This implies

$$N = -\Omega_\mu = \int n(p)\,dp \quad (16)$$

with

$$n(p) = V/h^3 (e^{(\varepsilon(p) - \mu)/kT} - 1)$$
$$= \alpha V e^{-\varepsilon(p)/kT}/h^3 (1 - \alpha e^{-\varepsilon(p)/kT}) \quad (17)$$
$$= \alpha V h^{-3} e^{-p^2/2mkT} + O(\alpha^2), \quad \alpha \to 0,$$

68.3. Applications to Ideal Gases

as well as

$$E = \int \varepsilon(p) n(p) \, dp \tag{18}$$

and

$$P = -\Omega_V, \qquad F = \Omega + \mu N = E - ST.$$

In this way one can compute all important quantities as functions of the parameters μ, T, V. Integration by parts in (16) yields the important relation

$$\Omega = -2E/3. \tag{19}$$

Since Ω depends linearly on V, we have $\Omega = -PV$, and hence

$$PV = 2E/3. \tag{20}$$

The number $n(p)$ describes the distribution of the particles onto the momentum vectors and because of

$$p = mv$$

also onto the velocity vectors v. The number of particles with $p \in G$ is equal to

$$\int_G n(p) \, dp.$$

This can be seen by evaluating the partition function above only over momentums with $p \in G$. The integration in (15) then is to be taken over G. The number $n(p)$ depends on the chemical potential μ. From (16) this can be determined in the form

$$\mu = \mu(N, V, T).$$

We obtain

$$Nh^3/V = \alpha \int e^{-p^2/2mkT} \, dp + O(\alpha^2), \qquad \alpha \to 0, \tag{21}$$

where $\alpha = e^{\mu/kT}$.

We now want to perform the computation for fixed T and small particle densities ρ. As expected we will obtain the quantities for ideal one-atomic gases of phenomenological thermodynamics of Example 67.4. Note, however, that there we had $\rho = M/V$ with $M = mN$, and that the chemical potentials of Example 67.4 and here are in the ratio of $m:1$. We essentially use

$$\int_{-\infty}^{\infty} e^{-\beta p^2} \, dp = (\pi/\beta)^{3/2}$$

and the formulas which are obtained by differentiation with respect to β, for example,

$$\int_{-\infty}^{\infty} p^2 e^{-\beta p^2} \, dp = 3\pi^{3/2}/2\beta^{5/2}.$$

From (21) follows $\rho = O(\alpha)$, $\alpha \to 0$. The implicit function theorem (Theorem 4.B) implies that for small ρ equation (21) can be solved for α. This gives

$$\alpha = e^{\mu/kT} = \rho h^3 (2\pi kmT)^{-3/2} + O(\rho^2), \qquad \rho \to 0. \tag{22}$$

Using this, we find for the *chemical potential*

$$\mu = kT \ln[\rho h^3 (2\pi kmT)^{-3/2} + O(\rho^2)], \qquad \rho \to 0. \tag{23}$$

From (17) and (22) we obtain

$$n(p) = Nw(p) + O(\rho^2), \qquad \rho \to 0$$

with

$$w(p) = (2\pi mkT)^{-3/2} e^{-p^2/2mkT}.$$

This is a Gauss distribution with $\int w(p)\,dp = 1$. The number of particles with momentum $p \in G$ is equal to

$$N \int_G w(p)\,dp + O(\rho^2), \qquad \rho \to 0.$$

This is the so-called Maxwell momentum distribution from which, with $p = mv$, *Maxwell's velocity distribution* follows. From (18) we obtain for the energy

$$E = N \int \varepsilon(p) w(p)\,dp + O(\rho^2) = 3NkT/2 + O(\rho^2), \qquad \rho \to 0.$$

Neglecting the terms of order $O(\rho^2)$ we obtain from (20) the equation of state

$$PV = NkT.$$

Moreover, we find for the free energy

$$\begin{aligned} F &= \Omega + \mu N = -2E/3 + \mu N \\ &= NkT + NkT \ln \rho h^3 (2\pi kmT)^{-3/2} \end{aligned} \tag{24}$$

and the entropy

$$S = T^{-1}(E - F) = Nk(\tfrac{5}{2} - \ln \rho h^3 (2\pi kmT)^{-3/2}).$$

Finally, we consider the important question of how large the energy fluctuations are. For a single atom the mean energy is equal to

$$\bar{\varepsilon} = \int \varepsilon(p) w(p)\,dp = 3kT/2$$

and the dispersion is equal to

$$\begin{aligned} (\Delta \varepsilon)^2 &= \int (\varepsilon(p) - \bar{\varepsilon})^2 w(p)\,dp \\ &= \int \varepsilon(p)^2 w(p)\,dp - \bar{\varepsilon}^2 = 3k^2 T^2/2. \end{aligned}$$

68.3. Applications to Ideal Gases

The total energy is equal to the sum of the energies of the single atoms. Since the atoms are assumed to be *independent*, the dispersions are added in the usual way, i.e.,

$$(\Delta E)^2 = N(\Delta \varepsilon)^2 = 2E^2/3N.$$

The relative energy fluctuation is therefore equal to

$$\Delta E/E = 1/\sqrt{1.5N}.$$

Hence, for the usual number of particles $N \sim 10^{20}$, one obtains the small quantity $\Delta E/E \sim 10^{-10}$. Physically, such small energy fluctuations occur if the total energy of the gas is not fixed, i.e., if the gas is in contact with a big heat reservoir (e.g., the earth). This way an exchange of energy is possible.

EXAMPLE 68.7 (Relativistic Particles). According to the special theory of relativity, the energy of a free particle is equal to

$$\varepsilon(p) = \sqrt{m_0^2 c^4 + c^2 p^2}$$

with rest mass m_0 and velocity of light c. We assume that $\varepsilon(p) \gg m_0 c^2$, i.e., the energy is substantially larger than the rest energy. Then we can let approximately

$$\varepsilon(p) = c|p|.$$

This relation is exactly true for photons. From (14) we then obtain

$$-\Omega = \pm kTVg \int \ln(1 \pm e^{(\mu - c|p|)/kT}) \frac{dp}{h^3}.$$

We integrate over $\mathbb{R}^3$, i.e., we assume that all states are maximally occupied. For the number of particles we obtain

$$N = -\Omega_\mu = \int n(p)\, dp$$

with $n(p) = Vg/h^3 (e^{(c|p| - \mu)/kT} \pm 1)$. The energy is equal to

$$E = \int c|p| n(p)\, dp.$$

We now derive the fundamental relation

$$P = E/3V,$$

i.e., the pressure P is equal to one-third of the energy density. We have

$$\Omega = \mp 4\pi g kTVh^{-3}c^{-3} \int_0^\infty \varepsilon^2 \ln(1 \pm e^{(\mu - \varepsilon)/kT})\, d\varepsilon.$$

Integration by parts yields

$$\Omega = -\frac{4\pi Vg}{3h^3 c^3} \int_0^\infty \frac{\varepsilon^3\, d\varepsilon}{e^{(\varepsilon - \mu)/kT} \pm 1} = -\frac{E}{3}.$$

The assertion now follows from $\Omega = -PV$.

For the special case $\mu = 0$, we obtain the *number of particles*

$$N = \int n(p)\,dp, \qquad n(p) = \frac{Vg}{h^3(e^{c|p|/kT} \pm 1)}$$

and the *energy* $E = \int c|p|n(p)\,dp$, i.e.,

$$E = \begin{cases} gaT^4V/2, & \text{Bose statistics,} \\ 7gaT^4V/16, & \text{Fermi statistics,} \end{cases} \tag{25}$$

with $a = 8\pi^5 k^4 T^4/15c^3 h^3$. To compute (25), one can use the integral formula (38). Applications of the fundamental formula (25) will be given in the following three sections.

If in a gas the states are occupied only up to a maximal limiting momentum, then all integrals $\int \ldots dp$ are only to be evaluated over a ball of radius p_0. This will be the case in the study of white dwarf stars in Section 68.8. Integration by parts then gives

$$PV = E/3 + o(p_0), \qquad p_0 \to +\infty.$$

Thus $PV = E/3$ is an ideal limiting case.

By using the energy-momentum tensor one can motivate that $PV < E/3$ must always be satisfied for macroscopial bodies (e.g., gas or liquids). This can be found in Weinberg (1972, M), Chapter 2, (2.10.24).

68.4. Planck's Radiation Law

We consider a container with volume V and assume that its walls constantly emit and absorb photons. We are interested in the photon gas in the container which, for a fixed temperature T, is in a thermodynamical equilibrium.

Photons have no rest mass, hence its energy is equal to

$$\varepsilon(p) = c|p|,$$

according to Example 68.7. Furthermore, we have the quantum formula

$$\varepsilon(p) = h\omega/2\pi$$

with angular frequency ω and wave length $\lambda = 2\pi c/\omega$, i.e.,

$$\varepsilon(p) = ch/\lambda.$$

The photon has the spin positions ± 1, so that $g = 2$, and it satisfies the Bose statistics. Thus we can apply Example 68.7. The free energy has the form

$$F = F(T, V, N).$$

For a fixed volume V and a fixed temperature T, we are still free to choose the number of particles N. We determine N from the assumption that F is minimal, i.e., $F_N = 0$. This gives

$$\mu = F_N(T, V, N) = 0$$

(see Example 67.9). From Example 68.7, we find for the number of particles

$$N = \int n(p)\,dp, \qquad n(p) = 2V/(e^{c|p|/kT} - 1)h^3$$

and for the energy

$$E = \int c|p|n(p)\,dp.$$

From Example 68.7, we obtain

$$\Omega = -2kTVh^{-3}\int \ln(1 - e^{c|p|/kT})\,dp = -E/3.$$

In order to obtain the energy distribution with respect to the wave length λ, we choose spherical coordinates $dp = p^2\,d|p|\,dO$. With $|p| = h/\lambda$ we obtain from (25)

$$E = \int_0^\infty 8\pi hcV\,d\lambda/\lambda^5(e^{hc/kT\lambda} - 1) = aT^4V. \qquad (26)$$

This is *Planck's famous radiation formula* for a container which is filled with radiation which is in a thermodynamical equilibrium. Equation (26) determines, at the same time, how the total energy E is distributed onto the different wave lengths λ. We now compute the remaining thermodynamical quantities. Because of $\mu = 0$ we have $F = \Omega$, hence

$$F = -E/3, \qquad S = (E - F)/T = 4aT^3V/3.$$

Moreover, we have $P = -\Omega_V = aT^4/3$, and hence

$$P = E/3V.$$

68.5. Stefan–Boltzmann Radiation Law for Black Bodies

Consider a body with temperature T and surface measure A (e.g., a stove or a star). The energy E_A which is emitted from the surface, during the time interval $[0, t]$, is equal to

$$E_A = caT^4At/4 = \sigma T^4At, \qquad (27)$$

where

$$\sigma = 2\pi^5/15c^2h^3 = 5.7\cdot 10^{-8}\ \text{W/m}^2\ \text{K}^4.$$

This is the so-called *Stefan–Boltzmann radiation law*. More precisely, the distribution of the emitted energy onto the wave lengths λ is given by the formula

$$E_A = 2\pi hc^2 tA \int_0^\infty \frac{d\lambda}{\lambda^5(e^{hc/kT\lambda} - 1)}, \qquad (27*)$$

whereby an essential assumption is that the body is a so-called *black radiator* (*black body*), i.e., it absorbs all incoming radiation. Stars, for example, can approximately be regarded as black bodies. Let us motivate (27) and (27*).

(I) We begin by considering a container C, filled with radiation (photons) in a thermodynamical equilibrium. Let T denote the temperature of C. We then look at a small boundary part of C with surface measure ΔA (Fig. 68.1(a)). Let $E_{\Delta A}$ be the radiation energy which arrives at ΔA during the time interval $[0, t]$. As in Figure 68.1(b) we consider an arbitrarily small volume element ΔV of C with energy $\rho \Delta V$. According to (26) the energy density in C is equal to

$$\rho = aT^4 = \int_0^\infty 8\pi hc \, d\lambda / \lambda^5 (e^{hc/kT\lambda} - 1).$$

Let r be the distance between ΔA and ΔV. For symmetry reasons we assume that after time $\tau = r/c$ the energy in ΔV has been distributed by radiation onto the surface of a ball of radius r. Thus on ΔA we have the energy

$$\frac{\Delta A \cos \vartheta}{4\pi r^2} \rho \Delta V$$

(see Fig. 68.1(b)). By choosing spherical coordinates we find

$$\Delta V = r^2 \sin \vartheta \, d\vartheta \, d\varphi \, dr.$$

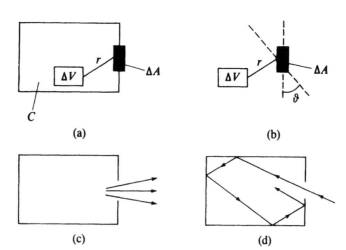

Figure 68.1

During the time interval $[0, t]$ all volume elements ΔV with $0 \leq r \leq ct$ radiate energy onto ΔA, where c is the velocity of light. Summing all these energies we obtain

$$E_{\Delta A} = \int \frac{\Delta A \cos \vartheta}{4\pi r^2} \rho \, dV$$

$$= \frac{\rho \Delta A}{4\pi} \int_0^{\pi/2} \cos \vartheta \sin \vartheta \, d\vartheta \int_0^{2\pi} d\varphi \int_0^{ct} dr$$

$$= ct\rho \, \Delta A/4.$$

This yields (27) and (27*) with A replaced by ΔA.

(II) We now consider the situation of Figure 68.1(c), i.e., we look at the container C of (I) with a slot of small surface measure ΔA. According to (I), this slot emits the energy $E_{\Delta A}$ during the time interval $[0, t]$. The crucial hypothesis of physicists then is that this slot behaves like a black radiator. In fact, the slot absorbs almost completely all incoming radiation (see Fig. 68.1(d)).

(III) Next consider any arbitrary black radiator of surface measure A. We regard its small surface elements ΔA as black radiators which emit the same energy as the slot in (II). Summation over ΔA yields

$$E_A = \sum E_{\Delta A} = ct\rho A/4.$$

This gives (27) and (27*). Our motivation is complete.

In Problem 68.4 Wien's displacement law, which is a consequence of (27*), will be discussed.

A deeper understanding of radiation processes, however, is only possible by using quantum electrodynamics.

68.6. The Cosmos at a Temperature of 10^{11} K

In Section 58.15 we made essential use of the fact that at 10^{11} K the cosmos has an energy density which is equal to 9/2 times the corresponding energy density of photons. We want to show that this is a consequence of (25). To this end we make the following two assumptions:

(i) At 10^{11} K the universe is in a thermodynamical equilibrium. The critical temperatures in Table 58.3 show that at 10^{11} K essentially only photons, electrons, positrons, and neutrinos exist.
(ii) The chemical potentials of these particles are equal to zero. This will be motivated in Problem 68.1.

According to (25), the energy density of the universe at $T = 10^{11}$ K is then equal to

$$(1 + 2 \cdot \tfrac{7}{8} + 4 \cdot \tfrac{7}{16})aT^4 = \tfrac{9}{2}aT^4.$$

Note that the photons satisfy the Bose statistics, and that the electrons, positrons, and neutrinos satisfy the Fermi statistics. Moreover, Table 58.3 implies for photons $g = 2$ holds (two spin positions). For electrons and positrons we also have $g = 2$, while for the four neutrino species we have $g = 1$.

68.7. Basic Equation for Star Models

For a radially symmetric star the *basic equation* is

$$P'(r) = -GM(r)\rho(r)/r^2,$$
$$P = p(\rho, T). \tag{28}$$

Thereby r is the distance from one point to the center of the star. Moreover, we have, P as pressure, ρ as density, T as temperature, G as the gravitational constant, and $M(r)$ is the mass of a ball of radius r around the center of a star, i.e.,

$$M(r) = \int_0^r 4\pi\rho(r)r^2\,dr.$$

We motivate (28). If we consider a small volume element ΔV at a distance r from the center, then ΔV is affected by the gravitational force. According to Problem 58.5, the effect is the same as if the entire mass $M(r)$ were located at the center. Hence, this force is equal to

$$-(GM(r)\rho(r)\Delta V/r^2)e_r.$$

Thereby e_r is a unit radial vector. More precisely, the force density is equal to

$$K = -(GM(r)\rho(r)/r^2)e_r.$$

Pressure and gravitational force must be in an equilibrium. We assume that only pressure forces are effective, i.e., the stress tensor has the diagonal form $\sigma_j^i = -P\delta_j^i$ (see Section 70.2). From the equilibrium condition

$$\operatorname{div}\sigma + K = 0$$

of Section 61.5 it follows that $(\operatorname{div}\sigma)e_r = -Ke_r$. This is (28).

68.8. Maximal Chandrasekhar Mass of White Dwarf Stars

White dwarfs have a very high density. The best-known white dwarf is the Sirius companion. Based on the irregular motion of Sirius, Bessel (1784–1846) predicted this companion, and in 1862 it was discovered by Clark. Only in 1915 it was found by using spectroscopical methods that both stars have the

68.8. Maximal Chandrasekhar Mass of White Dwarf Stars

same effective surface temperature of 10,000 K. Since the companion has a 10^4 times smaller luminosity than Sirius, and its mass is equal to the mass of the sun, it has to be very small. Its radius is about one hundredth of the radius of the sun and its density is approximately 1,000 kg/cm³. At these densities, the electron shell of the atoms has been crushed. The electrons move freely. The pressure of this electron gas is in an equilibrium with the gravitational force.

For the maximal mass of such a star one obtains the famous *formula of Chandrasekhar*

$$M_{crit} = \mu^{-2}(3\pi)^{1/2}\alpha_G^{-3/2} m_N = \mu^{-2}\, 5.87\, M_{sun}. \tag{29}$$

Thereby we have: G as gravitational constant, m_N as average mass of a nucleon (proton or neutron), and M_{sun} as mass of the sun. Moreover,

$$\alpha_G = \frac{Gm_N^2}{\hbar c} = 6 \cdot 10^{-39}$$

is the so-called fine structure constant of gravitation.

The key number μ in (29) is the number of nucleons, which belongs to one electron. For a hydrogen star we have $\mu = 2$, hence $M_{crit} = 1.5 M_{sun}$. For iron we have $\mu = 56/26$, hence

$$M_{crit} = 1.26 M_{sun}.$$

This value is important, since for elements in the periodic system, which are heavier than iron, the nuclear synthesis does not produce energy but consumes energy. Therefore, iron represents an especially important limiting case. We find that a white dwarf can have maximally about 1.2 sun masses, whereby this value still depends a little bit on the particular model.

This maximal mass was determined by Chandrasekhar, in 1933, in his Cambridge dissertation. In Section 76.16, we explain why this mass is so important for the final state of a star. In 1983, Chandrasekhar and Fowler received the Nobel prize for their contributions towards an understanding of the structure of stars.

We now motivate (29).

Step 1. Equation of state.

A white dwarf has a very high density. We assume that, in principle, the heavy nucleons are motionless, and hence have no kinetic energy. Therefore they do not contribute to the pressure P. Thus P is only caused by the freely moving electrons. According to the Pauli principle, two electrons cannot be in the same quantum mechanical state. Thus in one cell of the phase space (space-momentum space) with volume h^3 we can have only two electrons with different spin positions described by the spin quantum number S_z. Because of the huge matter density we assume that all possible states are occupied, i.e., in every cell there are exactly two electrons. Let V be the volume of the star and p the momentum vector of an electron. The number N of electrons in the phase volume $V \int dp$ is therefore equal to

$$N = 2Vh^{-3} \int dp = \int n(p)\,dp.$$

We have to integrate over a ball of radius p_0. Thus we have

$$N = \frac{8\pi V}{h^3} \int_0^{p_0} p^2\,d|p| = \frac{8\pi V p_0^3}{3h^3}.$$

For the limiting momentum we find

$$p_0 = (3h^3 N/8\pi V)^{1/3}.$$

The density of the star is

$$\rho = (Nm_e + N\mu m_N)/V$$

with electron mass m_e, nucleon mass m_N, and number of nucleons μ per electron. Because of $m_N/m_e = 1{,}836$ we set approximately $\rho = N\mu m_N/V$. This gives

$$p_0 = \left(\frac{3h^3 \rho}{8\pi \mu m_N}\right)^{1/3}.$$

Moreover, we assume that the electrons have a very high velocity v, which lies in the neighborhood of the velocity c of light. For the relativistic energy of the electrons we find

$$\varepsilon = m_e c^2/\sqrt{1 - v^2/c^2} = \sqrt{m_e^2 c^4 + p^2 c^2}$$

(see Section 75.11). Therefore we may set approximately $\varepsilon = |p|c$. Hence the energy of the electrons is given by

$$E = \int n(p) c |p|\,dp = \frac{8\pi c V}{h^3} \int_0^{p_0} |p|^3\,d|p|$$

$$= \frac{8\pi c V p_0^4}{4h^3} = \frac{hc}{8\pi^3}\left(\frac{3\pi^2 \rho}{m_N \mu}\right)^{4/3} V.$$

Motivated by Example 68.7, we assume the limiting relation $PV = E/3$ between pressure P and energy E. This implies the important *equation of state*

$$P = C\rho^\gamma,$$

$$C = \frac{hc}{24\pi^3}\left(\frac{3\pi^2}{m_N \mu}\right)^{4/3}, \qquad \gamma = 4/3. \tag{30}$$

Step 2: Application of the basic equation (28).
Differentiation of (28) yields

$$\frac{d}{dr}\left(\frac{r^2}{\rho(r)}\frac{dP(r)}{dr}\right) = -4\pi G r^2 \rho(r). \tag{31}$$

68.8. Maximal Chandrasekhar Mass of White Dwarf Stars

We set $\rho_0 = \rho(0)$ and introduce new coordinates ξ and η through

$$r = \xi \left(\frac{C\gamma}{4\pi G(\gamma - 1)}\right)^{1/2} \rho_0^{(\gamma-2)/2}, \qquad \rho = \rho_0 \eta^{1/(\gamma-1)}.$$

Then we have

$$P = C\rho_0^\gamma \eta^{\gamma/(\gamma-1)}.$$

From (30) and (31) we obtain the differential equation

$$\eta'' + \frac{2}{\xi}\eta' + \eta^{1/(\gamma-1)} = 0 \tag{32}$$

with initial conditions

$$\eta(0) = 1, \qquad \eta'(0) = 0. \tag{33}$$

The first condition follows from $\rho(0) = \rho_0$. The second condition is $\rho'(0) = 0$. In fact, $\eta'(0) = 0$ is a necessary consequence of the Taylor expansion for η at the point $\xi = 0$ and (32).

The solutions $\eta = \eta(\xi)$ of (32) with (33) are called Emden's functions. They may be found in Chandrasekhar (1939, M). For $\gamma = 4/3$ and $\xi > 0$ this function has a first zero at $\xi_0 = 6.90$. Moreover, we have $-\xi_0^2 \eta'(\xi_0) = 2.02$. The differential equation (32) becomes

$$\xi^{-2}(\xi^2 \eta')' + \eta^{1/(\gamma-1)} = 0.$$

It follows therefore that

$$\int_0^{\xi_0} \xi^2 \eta(\xi)^{1/(\gamma-1)} d\xi = -\xi_0^2 \eta'(\xi_0) = 2.02 \tag{34}$$

holds. On the boundary of the star we must have the density $\rho = 0$, hence $\xi = \xi_0$. For the radius $R = r(\xi_0)$ of the star we obtain

$$R = \xi_0 \left(\frac{C\gamma}{4\pi G(\gamma - 1)}\right)^{1/2} \rho_0^{(\gamma-2)/2}.$$

Because of (34) we find for the mass of the star

$$M = \int_0^R 4\pi r^2 \rho \, dr = 4\pi \left(\frac{C\gamma}{4\pi G(\gamma - 1)}\right)^{3/2} \cdot 2.02.$$

This is the required formula (29).

In our model we have used the maximally possible pressure for the electron gas which would be present if all electrons had the maximal kinetic energy, i.e., would, practically, move with the velocity of light. The actual pressure therefore is smaller. It follows that for $M > M_{\text{crit}}$ the pressure in the interior of the star cannot be in an equilibrium with the gravitational force. A gravitational collapse occurs which results in a much denser neutron star or at around $M > 2M_{\text{crit}}$ in a black hole with extremely large density. This will be discussed in Section 76.16.

Let us now briefly consider neutron stars. In this case all protons p and all electrons e^- are changed into neutrons n and neutrinos ν. This is a consequence of the reaction $p + e^- \to n + \nu$. For the computation of neutron stars, one therefore has to deal with a neutron gas instead of an electron gas. In this direction, compare Weinberg (1972, M), Chapter 11, §4.

PROBLEMS

68.1. *Chemical potential of elementary particles in the cosmos at high temperatures.* Motivate assumption (ii) in Section 68.6 by using general symmetry arguments.

Solution: Use the free energy

$$F = F(T, V, N_1, \ldots, N_k).$$

Thereby T is the temperature of the cosmos, V is the volume of the cosmos, and N_i is the number of particles of the ith particle species. According to Table 67.1 of Section 67.5, we find

$$\mu_i = \partial F/\partial N_i$$

for the chemical potential of the ith particle species. Let, for example, N_1, N_2, N_3 be the number of photons γ, of particles p and of the corresponding antiparticles p^*, respectively. According to Example 67.10, the annihilation reaction

$$p + p^* \to \gamma$$

yields $\mu_2 + \mu_3 = \mu_1$. Motivated by Section 68.4 we assume that

$$\mu_1 = 0.$$

Thus we have $\mu_2 = -\mu_3$. Furthermore, we may assume that $\mu_i(T, V, N_1, \ldots, N_k)$ changes its sign, when particles and antiparticles are interchanged. As above, all reactions between elementary particles correspond to relations between the chemical potentials. However, we do not consider all these concrete reactions in detail, but assume in summary that all μ_i can be uniquely determined from the conserved quantities for elementary particles, presently known:

Baryon number N_B (number of nucleons and hyperons minus number of the corresponding antiparticles),

lepton number N_L (number of electrons, neutrinos, and muons minus number of the corresponding antiparticles), and

charge number N_Q.

The numbers N_B, N_L and N_Q change their sign, when particles and antiparticles are interchanged. Thus, according to our assumption above, all μ_i are odd functions of N_B, N_L, and N_Q. In the cosmos, however, we have approximately

$$N_B = N_L = N_Q = 0.$$

Thus we have

$$\mu_i = 0 \quad \text{for all } i.$$

Further details may be found in Weinberg (1972, M), Chapter 15.

68.2. *Quasi-classical statistics.* We consider a gas of N particles with space coordinates $q = (q_1, \ldots, q_{3N})$ and momentum coordinates $p = (p_1, \ldots, p_{3N})$. Let the energy be given by

$$E = E(q, p).$$

Determine the statistical potential Ω, the entropy S, and the mean value $\bar{A}$ of a quantity $A(q, p)$.

Solution: From Proposition 68.1 and Remark 68.3 we obtain

$$\Omega = -kT \ln \int e^{-E(q,p)/kT} \, dq \, dp / h^{3N} N!.$$

For the calculation of the partition function, observe that the (q, p)-phase space is decomposed into cells of volume h^{3N}. Moreover, because of the indistinguishability of the particles, one has to identify all points of the phase space which are obtained from one another by a permutation of particles. This is the reason why the factor $N!$ occurs. In *classical* statistics, the factor $h^{3N} N!$ does *not* occur in Ω. For the entropy we find

$$S = -\Omega_T.$$

The probability density $w(q, p)$ satisfies the normalization condition

$$\int w(q, p) \, dq \, dp = 1.$$

Similarly to $w_r = e^{(\Omega - E_r)/kT}$, in the discrete case of Proposition 68.1, we obtain in the continuous case

$$w(q, p) = e^{-E(q,p)/kT} \bigg/ \int e^{-E(q,p)/kT} \, dq \, dp$$

and

$$\bar{A} = \int A(q, p) w(q, p) \, dq \, dp. \tag{35}$$

The complete analogy with the discrete case is obtained if everywhere $dq \, dp$ is replaced with $dq \, dp / h^{3N} N!$. This, however, does not change $\bar{A}$.

68.3. *Classical law of equipartition.* We choose

$$A(p) = (2m)^{-1}(p_1^2 + \cdots + p_{3N}^2),$$

i.e., $A(p)$ is equal to the kinetic energy of the gas. Moreover, let

$$E(q, p) = A(p) + U(q)$$

with potential energy U. Prove that

$$\bar{A} = 3N(kT/2),$$

i.e., to each of the $3N$ degrees of freedom corresponds the mean kinetic energy $kT/2$.

Solution: Use (35) and the formula for the integration by parts

$$\int_{-\infty}^{+\infty} \alpha x^2 e^{-\alpha x^2/\beta} \, dx = \frac{\beta}{2} \int_{-\infty}^{\infty} e^{-\alpha x^2/\beta} \, dx \tag{36}$$

for $\alpha, \beta > 0$. Implicitly it is assumed that U is given in such a form that $\bar{A}$ exists.

Generalization. If the gas is described by $q = (q_1, \ldots, q_f)$, $p = (p_1, \ldots, p_f)$ and if

$$E(q, p) = A(p) + U(q),$$

where A is a positive definite quadratic form, then one can use an orthogonal transformation from A into a sum of squares and (36) in order to obtain the relation

$$\bar{A} = fkT/2.$$

In the same way one can prove the following. If $E = E(q, p)$ is a positive definite quadratic form in the variables (q, p), then the mean energy satisfies

$$\bar{E} = fkT.$$

Such a situation occurs for oscillating systems. In particular, we have $\bar{E} = kT$ for an harmonic oscillator with $f = 1$ and

$$E(q, p) = \frac{p^2}{2m} + \frac{\omega^2 m q^2}{2}.$$

According to Problem 58.11, a general oscillating system with f degrees of freedom and in normal coordinates has the energy

$$E(q, p) = \sum_{i=1}^{f} \frac{p_i^2}{2} + \frac{\omega_i^2}{2} q_i^2,$$

where $p = \dot{q}$.

Thus we obtain the following important result:

The mean energy of an oscillating system with f degrees of freedom is given by
$$\bar{E} = fkT.$$

Experience shows that quasi-classical statistics and hence the law of equipartition is only valid for sufficiently high temperatures. For low temperatures one has to apply the quantum statistics of Section 68.2.

68.4. Wien's displacement law and an iteration scheme. Determine the wave length λ_{max} for which, according to the Stefan–Boltzmann law, a body (black radiator) emits the most energy.

Solution: According to Section 68.5, one finds for the energy, which during the time interval $[0, t]$ is emitted from a surface A,

$$E_A = 4^{-1} c A t \int_0^\infty \rho_\lambda \, d\lambda, \qquad \rho_\lambda = \frac{8\pi h c}{\lambda^5 (e^{hc/kT\lambda} - 1)}.$$

For λ_{max} the value of ρ_λ is maximal, i.e., $d\rho_\lambda/d\lambda = 0$. If we set $x = hc/kT\lambda$, then we obtain

$$x = 5e^{-x}(e^x - 1).$$

The corresponding iteration scheme $x_{n+1} = f(x_n)$ with $x_0 = 5$ converges very rapidly to $x = 4.965$. This implies

$$\lambda_{\max} T = hc/kx = 2.9 \cdot 10^{-3} \text{ mK}. \tag{37}$$

In Section 76.17 we will use this law in connection with the vaporization of black holes.

68.5. Classical integrals. Prove the following formulas:

$$\int_0^\infty \frac{x^{a-1}}{e^x + 1} dx = \begin{cases} (1 - 2^{1-a})\Gamma(a)\zeta(a) & \text{if } a > 0, a \neq 1, \\ \ln 2 & \text{if } a = 1, \\ \dfrac{2^{2n-1} - 1}{2^n} \pi^{2n} B_n & \text{if } a = 2n, \end{cases} \tag{38}$$

$$\int_0^\infty \frac{x^{a-1} dx}{e^x - 1} = \begin{cases} \Gamma(a)\zeta(a) & \text{if } a > 1, \\ \dfrac{(2\pi)^{2n} B_n}{4n} & \text{if } a = 2n, \end{cases}$$

where $n = 1, 2, \ldots$. Moreover, ζ is the Riemannian ζ-function, and B_n are the Bernoulli numbers with $B_1 = \frac{1}{6}$, $B_2 = \frac{1}{30}$, and $B_3 = \frac{1}{42}$.

Hint: See Landau and Lifšic (1962), Vol. V, §58.

68.6. Connection between the general entropy definition and Planck's classical definition. Consider n different particles $P_1, \ldots, P_n$ which may assume the energy values $E_1, \ldots, E_R$. The key question in classical statistical physics is: "How, at a given temperature T, are these particles distributed onto the energy values?

68.6a. Modern approach via information theory. How can this problem be solved in the context of our theory?

Solution: Consider one particle and assign the possible states $Z_r = (E_r, N_r)$ with $r = 1, \ldots, R$ to it. Thereby E_r denotes the energy of the state, and $N_r = 1$ for all r, the number of particles. Let w_r be the probability that state Z_r is realized. The corresponding entropy is

$$S = -k \sum_{r=1}^R w_r \ln w_r. \tag{39}$$

In order to determine w_r we solve the problem

$$S = \max!,$$

$$\sum_r w_r = 1, \quad \sum_r N_r w_r = N, \quad \sum_r E_r w_r = E, \tag{40}$$

with $N = 1$, according to Section 68.1a. The Lagrange multiplier rule yields the solution (1), i.e.,

$$w_r = e^{(\mu N_r - E_r)/kT} \Big/ \sum_r e^{(\mu N_r - E_r)/kT}.$$

From $N = 1$ we obtain $\mu = 0$ for the chemical potential. If n particles are given, then we find a distribution of

$$n_r = w_r n = \frac{n e^{-E_r/kT}}{\sum_r e^{-E_r/kT}}, \tag{41}$$

Table 68.4.

E_1	E_2	n_1	n_2	$W(n_1,n_2)$ Statistical weight	$w(n_1,n_1)$ Probability
P_1P_2	—	2	0	1	1/4
—	P_1P_2	0	2	1	1/4
P_1	P_2	1	1	2	1/2
P_2	P_1	1	1	2	1/2

i.e., exactly n_r particles have the energy E_r at temperature T. This is the famous formula of classical statistical physics.

68.6b. *Classical derivation of* (41). In contrast to the procedure above, we now start from n different particles $P_1, \ldots, P_n$ instead of just from one. Boltzmann's basic idea then was to consider all possible distributions of $P_1, \ldots, P_n$ onto the energies. Table 68.4 shows this for two particles P_1, P_2 and two possible energy values E_1, E_2. Thereby n_r is the number of particles with energy E_r. Let $w(n_1, \ldots, n_R)$ denote the probability that exactly n_r particles have the energy E_r for all r. In order to compute this probability in a simple way, we observe the generalized binomial formula

$$(a_1 + \cdots + a_R)^R = \sum_{n_1 + \cdots + n_R = n} W(n_1, \ldots, n_R) a_1^{n_1} \ldots a_R^{n_R}$$

with

$$W(n_1, \ldots, n_R) = \frac{n!}{n_1! \ldots n_R!}.$$

One easily sees that the number of favorable cases for our problem equals $W(n_1, \ldots, n_R)$. By setting all a_r equal to 1, we obtain

$$\sum W(n_1, \ldots, n_R) = R^R,$$

and hence

$$w(n_1, \ldots, n_R) = W(n_1, \ldots, n_R)/R^R.$$

The quantities $W(n_1, \ldots, n_R)$ are called *statistical weights*. Following the classical example of Max Planck we set

$$S_p = k \ln W(n_1, \ldots, n_R)$$

and compute the required distribution from the problem:

$$S_p = \max!,$$

$$\sum_{r=1}^{R} n_r = n, \qquad nE = \sum_{r=1}^{R} n_r E_r. \tag{42}$$

To clearly realize the connection with our approach, we introduce the numbers

$$w_r = n_r/n$$

and use Stirling's approximation formula

$$n! \sim \left(\frac{n}{e}\right)^n$$

for large n. In case that also all n_r are large, it follows that

$$S_p \sim -k\sum_r w_r \ln w_r.$$

Using this approximation formula, we obtain problem (40) from (42), which has been computed above. Consequently, in the context of this approximation, we find in expression (41) a solution of (42).

References to the Literature

Classical works: Boltzmann (1873) (Boltzmann equation), (1909) (collected works), Gibbs (1902, M), Planck (1913, M).
History of statistical physics: Sommerfeld (1954, M), Vol. 5, Cohen and Thirring (1973, P).
Introduction: Kittel (1969, M).
Classical textbooks: Planck (1913), Sommerfeld (1954), Vol. 5, Landau and Lifšic (1962), Vols. 5, 9, 10 (much material), Huang (1963).
Modern presentation: Ruelle (1978, S), Wehrl (1978, S), Martin (1981, S), Balian (1982, L), Thirring (1983, M), Vol. 4, Kubo (1983, M), (1985, M).
Infinite-dimensional systems in classical statistical mechanics: Petrina (1983, S).
Thermodynamics for nonequilibriums and the Boltzmann equation: Sommerfeld (1954, M), Vol. 5 (classical methods), Landau and Lifšic (1962), Vol. 10 (standard work), Groot and Mazur (1962, M), Glansdorff and Prigogine (1971, M), Cohen and Thirring (1973, P), Cercignani (1975, M), Greenberg (1986, M), Röpke (1987).
Methods of nonstandard analysis, stochastic processes in mathematical physics, and the Boltzmann equation: Albeverio (1986, M).
Synergetics and self-organization in nonequilibrium systems: Glansdorff and Prigogine (1971, M), Nicolis and Prigogine (1977, M), Haken (1977, M), (1983, M), Ebeling and Feistel (1982, M), Ebeling and Klimontowitsch (1985, L).
Rational mechanics and gas kinetics: Truesdell and Muncaster (1980, M).
Modern quantum statistics: Thirring (1983, M), Vol. 4 (introduction), von Neumann (1932, M) (classical work), Ruelle (1969, M), (1978, S), Bratteli and Robinson (1979, M) (von Neumann algebras), Bogoljubov (1980, M), (1984, M), Alberti and Uhlmann (1981, L), Fick and Sauermann (1982, M), Simon (1987, M).
Solvable models of quantum statistics: Dubin (1974, M), Baxter (1982, M), Sinai (1982, M), Simon (1987, M).
Critical phenomena, self-similarity, and phase transition: Wilson (1982, S, B) (recommended as an introduction), Ma (1982, M), Kubo, Toda, and Saitô (1983, M), Fröhlich (1983, M).
Spin glass theory: Mezard (1987, P).

CHAPTER 69

Continuation with Respect to a Parameter and a Radiation Problem of Carleman

We solve this problem with Carleman (1921) by using a method which was first employed by Ed. Le Roy and which, with a striking success, was used by S. Bernstein around 1900. It consists in a stepwise continuation of the solution in dependence of a suitable parameter .... It is of great importance to find *a priori* bounds for the solution of a differential equation, which contains a parameter, as well as to find these bounds for all partial derivatives up to a certain degree.

Leon Lichtenstein (1931)

In Chapter 6 we studied the important method of continuation with respect to a parameter. In the present chapter we discuss a nontrivial physical application. We want to show how, by using existence theorems for linear problems and *a priori* estimates, one can find existence results for nonlinear problems. The proof technique used here may also be applied to many other problems.

In the first two sections we discuss the basic equations for the heat conduction.

69.1. Conservation Laws

Many basic equations in physics have the form of conservation laws

$$\rho_t + \operatorname{div} \vec{j} = p. \tag{1}$$

We want to explain in which sense (1) represents a conservation law. We interpret ρ as the density of the chemical substance S, $\vec{j}$ as the mass current density vector, and p as the mass production per volume and time. If G is a bounded region in $\mathbb{R}^3$, then, by definition, we have

$$\int_G \rho(x,t)\,dx = \text{mass in } G \text{ at time } t.$$

Moreover, we set

$J(t) = $ mass of S which during the time interval $[0,t]$ flows out of G,

$M(t) = $ mass of S which during the time interval $[0,t]$ is produced in G as a consequence of chemical reactions.

By definition, we have

$$\dot{J} = \int_{\partial G} \vec{j}n\,dO,$$

$$\dot{M} = \int_G p\,dx,$$

with outer unit normal vector n. Obviously, there exists the relation

$$\frac{\partial}{\partial t}\int_G \rho\,dx = \dot{M} - \dot{J}.$$

Integration by parts yields

$$\int_G (\rho_t + \operatorname{div}\vec{j} - p)\,dx = 0. \tag{2}$$

We assume that all functions and regions are sufficiently regular. By contracting G into a single point x we obtain (1) from (2), according to the mean value theorem of integral calculus.

For the special case $p \equiv 0$, relation (1) describes mass conservation.

Moreover, if

$$p(x,t) > 0 \quad \text{or} \quad p(x,t) < 0,$$

then there exists a *source* (production of mass) or a *sink* (annihilation of mass) at the point x at time t (production of mass). This may have its cause, for example, in chemical reactions.

69.2. Basic Equations of Heat Conduction

Let $T = T(x,t)$ be the temperature at the point $x \in \mathbb{R}^3$ at time t. We replace the mass of the previous section with the amount of heat. Then ρ is the heat density and p the heat production per volume and time.

If now μ is the mass density, we find

$$\rho_t = c\mu T_t,$$

because the temperature change ΔT in a small volume element ΔV causes the change in heat

$$\Delta Q = c\mu \Delta V \Delta T$$

with specific heat $c = c(T)$. Note that $\Delta Q/\Delta V \Delta t \to \rho_t$. Thus we obtain from (1) the following basic equations of heat conduction.

(i) Heat balance

$$c\mu T_t + \text{div}\,\vec{j} = p. \tag{3}$$

(ii) Constitutive law

$$\vec{j} = \Psi(T, T_x). \tag{4}$$

EXAMPLE 69.1 (Fourier's Law). For an isotropic body one often uses

$$\vec{j} = -\kappa\,\text{grad}\,T, \tag{5}$$

where the so-called heat conductivity number κ may still depend on the position x; note that κ is positive. For a homogeneous body we have, in addition, that $\kappa(x) = \text{const}$. Table 69.1 contains several values for κ. Note that there 1 kcal = 4,185 J.

By inserting (5) into (3) we obtain the *basic equation*

$$c\mu T_t - \text{div}(\kappa\,\text{grad}\,T) = p. \tag{6}$$

For $\kappa = \text{const}$ it is

$$c\mu T_t - \kappa \Delta T = p, \tag{7}$$

where Δ is the Laplace operator.

One still has to add boundary and initial conditions.

(i) The *initial condition* consists in prescribing $T(x, 0)$, i.e., the temperature in the region G at time $t = 0$.
(ii) As *boundary condition* one can use the temperature T or the normal component $\vec{j}n$ of the heat flow density vector on ∂G.

Table 69.1

Material	Heat conductivity κ	Heat conduction
Silver	0.1 kcal/ms K	good
Iron	0.017	-
Lead	0.008	
H_2O	$1.43 \cdot 10^{-4}$	average
Rubber	$0.3 \cdot 10^{-4}$	
CO_2	$3 \cdot 10^{-6}$	bad
N_2	$6 \cdot 10^{-6}$	

69.3. Existence and Uniqueness for a Heat Conduction Problem

According to (5) we have

$$\vec{j}n = -\kappa \partial T/\partial n.$$

Also, the boundary condition

$$\partial T/\partial n = f(T) \quad \text{on } \partial G \tag{8}$$

is possible, i.e., the heat flow on the boundary depends on the temperature.

In Section 37.7 we have seen that it is also meaningful to prescribe inequalities for $\vec{j}n$. For example, $\vec{j}n \leq 0$ means that no heat flows out in the direction of the outer unit normal vector. Such a condition makes sense if the outer temperature is greater than the temperature of the body.

In conclusion, we want to motivate Fourier's law (5). From experience we assume that the heat flow depends only on the temperature difference. This makes the ansatz $\vec{j} = \Psi(T_x)$ with $\Psi(0) = 0$ plausible. Taylor expansion yields

$$\vec{j} = \Psi'(0)T_x + \cdots.$$

Since no direction is preferred for an isotropic body, we must have that

$$\Psi'(0) = -\kappa I.$$

This is (5). The positive sign of κ in (5) corresponds to the fact that heat always flows in the direction of falling temperature.

69.3. Existence and Uniqueness for a Heat Conduction Problem

We consider a homogeneous body in a bounded region Ω of $\mathbb{R}^3$. We assume that there is a heat source at the point $x_0 \in \Omega$, and moreover, that the body emits heat. We investigate if in this situation a stationary temperature state exists for the body.

Let $T(x)$ be the temperature at the point x. We obtain the following boundary-value problem:

$$\begin{aligned} \Delta T &= 0 \quad \text{in } \Omega - \{x_0\}, \\ T(x) &= u(x) + \frac{a}{|x - x_0|} \quad \text{in } \Omega - \{x_0\}, \\ T &> 0 \quad \text{in } \bar{\Omega} - \{x_0\}, \\ \frac{\partial T}{\partial n} &= -kT^4 \quad \text{on } \partial \Omega. \end{aligned} \tag{9}$$

By a classical solution we mean $u \in C^2(\Omega) \cap C^1(\bar{\Omega})$. Our assumptions are:

(H) Ω is a bounded region in $\mathbb{R}^3$ with smooth boundary, i.e., more precisely, let $\partial \Omega \in C^{2,\mu}$ with fixed $\mu \in]0,1[$. Moreover, let a and k be positive constants.

Theorem 69.A (Carleman (1921)). *If (H) holds, then (9) has a unique classical solution.*

The proof will be given in the following section. At first, however, we will motivate the problem physically.

Because of the homogeneity of the body we obtain from (5) for the heat flow density vector $\vec{j} = -\kappa \operatorname{grad} T$. The singularity $a/|x - x_0|$ yields the portion

$$\vec{j} = \kappa a(x - x_0)/|x - x_0|^3.$$

Therefore during one unit of time, e.g., a second, the amount of heat

$$\int \vec{j} n \, dO = 4\pi\kappa a$$

flows through the surface of a ball of radius R and center x_0. Thus the singularity models the heat source.

The boundary condition in (9) means that during one unit of time the amount of heat

$$\int_{\partial\Omega} \vec{j} n \, dO = -\int_{\partial\Omega} \kappa \frac{\partial T}{\partial n} dO = k\kappa \int_{\partial\Omega} T^4 \, dO$$

is emitted. This expression corresponds to the Stefan–Boltzmann law of Section 68.5.

69.4. Proof of Theorem 69.A

We use the method of continuation with respect to a parameter. Thereby we apply, without proof, a number of theorems of classical potential theory. The analytical key result is contained in Lemma 69.4. In particular, we will, repeatedly, make use of the maximum principle in the form of Problem 7.2. In this section we assume (H) of the preceeding section and choose

$$X = C^\mu(\partial\Omega) \quad \text{for fixed} \quad \mu \in \,]0, 1[.$$

Then X is a B-space with norm

$$\|f\| = \max_{x \in \partial\Omega} |f(x)| + H_\mu(f).$$

The Hölder constant $H_\mu(f)$ is the smallest number c with

$$|f(x) - f(y)| \le c|x - y|^\mu \quad \text{for all} \quad x, y \in \partial\Omega.$$

Lemma 69.2. *For (9) there exists at most one classical solution.*

The proof will be given in Problem 69.1.
We now discuss the existence proof. In order to eliminate the singularity in

69.4. Proof of Theorem 69.A

(9) we consider, for fixed x_0, Green's function $x \mapsto G(x, x_0)$ for the Laplace operator, i.e.,

$$\Delta G = 0 \quad \text{in } \Omega - \{x_0\},$$
$$G(x, x_0) = v(x) + a/|x - x_0| \quad \text{in } \Omega - \{x_0\}, \tag{10}$$
$$G = 0 \quad \text{on } \partial\Omega,$$

with $v \in C^2(\Omega) \cap C^1(\bar{\Omega})$. The maximum principle implies

$$G > 0 \quad \text{in } \Omega - \{x_0\}, \tag{11}$$
$$\partial G/\partial n < 0 \quad \text{on } \partial\Omega. \tag{12}$$

We now set

$$T = G + w$$

and consider, instead of (9), the problem

$$\Delta w = 0 \quad \text{in } \Omega,$$
$$w > 0 \quad \text{in } \Omega, \tag{13}$$
$$-\partial w/\partial n = \partial G/\partial n + (1 - t)w + tkw^4 \quad \text{on } \partial\Omega,$$

with parameter $t \in [0, 1]$. If, for $t = 1$, we can find a classical solution $w \in C^2(\Omega) \cap C^1(\bar{\Omega})$ of (13), then we obtain a classical solution of (9) from $T = G + w$.

Lemma 69.3. *For fixed $t \in [0, 1]$ there exists at most one classical solution of (13).*

This is proved in complete analogy to Lemma 69.2.

Lemma 69.4. *Let $b, f \in X$ with $0 < \alpha \leq b(x) \leq \beta$ on $\partial\Omega$. Then the third boundary-value problem*

$$\Delta w = 0 \quad \text{in } \Omega,$$
$$\partial w/\partial n + bw = f \quad \text{on } \partial\Omega \tag{14}$$

has exactly one classical solution w. If we set

$$T(b)f = w \quad \text{on } \partial\Omega,$$

then $T(b): X \to X$ is a linear and continuous operator, i.e.,

$$\|T(b)\|_X \leq \delta \|f\|_X \quad \text{for all } f \in X. \tag{15}$$

Thereby it is important that the constant δ does not depend on b, but only on the bounds α and β.

This classical result of potential theory will be essentially used in the following.

Corollary 69.5. *From $f > 0$ on $\partial\Omega$ follows $T(b)f > 0$ on $\partial\Omega$.*

PROOF. Let x be the point for which w assumes its minimum on $\partial\Omega$. The maximum principle implies that $\partial w(x)/\partial n \leq 0$. From (14) it follows then that $w(x) > 0$. □

Lemma 69.6 (*A Priori* Bounds). *There exist positive numbers α, β, γ, independent of t, such that the classical solution of (13) satisfies:*

$$0 < \alpha \leq w(x) \leq \beta \quad \text{on } \partial\Omega, \tag{16}$$

$$\|w\|_X \leq \gamma. \tag{17}$$

PROOF.

(I) For fixed $t \in [0, 1]$ let w be a classical solution of (13). Moreover, let $x \in \partial\Omega$ be a point for which w achieves its minimum on $\partial\Omega$. According to the maximum principle, we have $\partial w(x)/\partial n \leq 0$. The boundary condition in (13) implies

$$(1 - t)w(x) + tkw(x)^4 \geq -\partial G(x)/\partial n \geq \delta > 0$$

on $\partial\Omega$. Thereby δ denotes the minimal value of $-\partial G/\partial n$ on $\partial\Omega$. For fixed $t \in [0, 1]$ the function

$$\varphi_t(\xi) = (1 - t)\xi + tk\xi^4$$

is strictly monotone increasing on $\mathbb{R}_+$. Therefore there exists a unique $\alpha(t) > 0$ with $\varphi_t(\alpha(t)) = \delta$. This implies

$$w(x) \geq \alpha(t) > 0 \quad \text{for all} \quad x \in \partial\Omega.$$

Since the function $\alpha(\cdot)$ is continuous, there exists a number α with

$$\alpha(t) \geq \alpha > 0 \quad \text{for all} \quad t \in [0, 1].$$

(II) By replacing the minimum with the maximum one obtains β in an analogous fashion.

(III) A well-known result in potential theory states that if

$$\Delta w = 0 \quad \text{on } \Omega$$

and $w \in C^2(\Omega) \cap C^1(\bar{\Omega})$, then

$$|w(x) - w(y)| \leq c\left(\max_{\partial\Omega} |\partial w/\partial n|\right) |x - y|^\mu$$

is satisfied for all $x, y \in \partial\Omega$. This estimate, which may be found in Lichtenstein (1930), p. 327, is obtained in the following way. One can use a potential of the simple layer with continuous density ρ in order to represent the solution of the second boundary-value problem. Thereby ρ is obtained as a solution of a Fredholm integral equation. Then one

69.4. Proof of Theorem 69.A

uses the properties of the potential of the simple layer (see Smirnov (1956), Vol. IV, Section 206 and Günter (1957), Chapter 2, §2).

Because of $\alpha \leq w \leq \beta$ on $\partial\Omega$ we obtain from the boundary condition in (13) that the Hölder constant of w is bounded. This implies (17). □

Using these classical tools, it is then easy to give the following existence proof.

Lemma 69.7. *For $t = 0$ equation (13) has a classical solution.*

PROOF. This follows from Lemma 69.4, because (12) and Corollary 69.5 imply that $w > 0$ on $\partial\Omega$, and hence also $w > 0$ on $\bar{\Omega}$, according to the maximum principle. □

Our *goal* is to continue this solution for $t = 0$ up to the parameter value $t = 1$. For fixed t let w_t be a classical solution of (13), where t denotes an index in w_t. For $t + \tau$ we are looking for a solution of (13) which has the form

$$w = w_t + v.$$

This yields

$$\begin{aligned} \Delta v &= 0 \quad \text{in } \Omega, \\ \partial v/\partial n + bv &= f \quad \text{on } \partial\Omega, \end{aligned} \quad (18)$$

with

$$b = 1 - t - \tau + 4k(t + \tau)w_t^3,$$
$$f = \tau(w_t - kw_t^4) - k(t + \tau)(6w_t^2 v^2 + 4w_t v^3 + v^4).$$

The following properties of f are important:

All terms of f, either contain the small parameter τ, or are of higher order than one in v. $\quad (19)$

Lemma 69.8. *There exists a number τ_0 independent of t such that (18) has a classical solution v for every τ with $0 < \tau \leq \tau_0$ and $t + \tau \leq 1$.*

PROOF. According to Lemma 69.4 we write (18) as an operator equation

$$v = T(b)f(v), \qquad v \in X. \quad (20)$$

Because of the *a priori* estimate (16) and Lemma 69.4 we have $\|T(b)\| \leq \text{const}$ uniformly for all t. Banach's fixed-point theorem (Theorem 1.A) applied to the ball $\{v \in X : \|v\| \leq r\}$ with sufficiently small r and (19) yield a solution of (20) (see Problem 69.2).

Because of the *a priori* estimate (17) we can choose r and τ independently of t. Because of (16) we can also assure that $w_t + v > 0$ on $\partial\Omega$, hence $w_t + v > 0$ on $\bar{\Omega}$, according to the maximum principle. □

Lemma 69.8 shows that if we start with the solution at $t = 0$, then after finitely many steps we obtain a solution for $t = 1$.

PROBLEMS

69.1. *Proof of Lemma 69.2.*
Solution: Suppose T_1 and T_2 are solutions of (9). We set $T = T_1 - T_2$. Integration by parts yields

$$0 = -\int_\Omega T \Delta T \, dx = \int_\Omega (\operatorname{grad} T)^2 \, dx - \int_{\partial\Omega} T \, \partial T/\partial n \, dO.$$

Because of the monotonicity of $T \mapsto T^4$ on $\mathbb{R}_+$ we obtain

$$-T \, \partial T/\partial n = k(T_1^4 - T_2^4)(T_1 - T_2) \geq 0. \tag{21}$$

This implies that $\operatorname{grad} T = 0$ on $\bar{\Omega}$, hence $T = \text{const}$. From (21) follows then that $T \equiv 0$.

69.2. *Application of Banach's fixed-point theorem.* Verify explicitly that Banach's fixed-point theorem can, in fact, be applied to (20).
Solution: From (19) it follows that (20) has the precise structure of (H3) and (H4) in Section 8.12 of Part I.

69.3. *Generalizations.* Prove Theorem 69.A for the more general boundary condition

$$-\partial T/\partial n = F(T) \quad \text{on } \partial\Omega,$$

where $F: \mathbb{R} \to \mathbb{R}$ is C^1 with $F(0) \leq 0$, $F'(T) > 0$ for all $T > 0$ and $F(T) \to +\infty$ as $T \to +\infty$.
Hint: Use arguments analogous to Section 69.4. See Lichtenstein (1931), p. 54.

References to the Literature

Classical works: Carleman (1921), Lichtenstein (1931, M).
Classical potential theory: Kellogg (1929, M) and Günter (1957, M) (standard works), Smirnov (1956, M), Vol IV (introduction), Lichtenstein (1921, S, B, H) (classical survey article), (1929, M), (1930).
Approximation methods for heat conduction problems: Aziz and Na (1984, M), Shih (1984, M), Cebeci (1984, M).
Boundary value problems in kinetic theory: Greenberg (1986, M).

APPLICATIONS IN HYDRODYNAMICS

Although I envy a great generality with regard to the nature of fluids and the forces that are being applied to their particles, I have no fear of the reproaches often leveled with good reasons at those who have undertaken to generalize the researches of others. Often a great generality causes more confusion than it does illuminate, and sometimes it leads to such voluminous computations that it becomes extremely difficult to derive any consequences, even in the simplest cases. If generalizations have this disadvantage, then we should retreat from them and restrict ourselves to the special cases.

The generality that I embrace, far from dazzling our lights, will reveal to us rather the veritable laws of Nature in all their brilliance, and in them we shall find even stronger reason to admire her beauty and her simplicity. It will be an important lesson to learn that some principles, till now believed bound to some special cases, are of greater breadth. Finally, these researches will demand calculations scarcely any more troublesome, and it will be easy to apply them to all special cases we might set up.

Leonhard Euler, *General Principles of the State of Equilibrium of Fluides* (1755)

CHAPTER 70

Basic Equations of Hydrodynamics

The tensor surface $n(\tau n) = -1$ of hydrostatic pressure is a sphere. This law was probably discovered by Pascal (1624–1662).

The first two papers of Euler (1707–1783) about equilibrium and motion of fluids appeared in 1755...

Bernoulli's equation represents the most important theorem in hydrodynamics. It was discovered in 1738 by Daniel Bernoulli (1700–1782), even before Euler's equations were known, by using an argument which can be regarded as an early version of the energy conservation law ....

Today we consider the equation of Navier (1822) and Stokes (1845) as basic for the representation of all properties of fluids. In the technology of the nineteenth century, however, it was the conviction that there exists a deep gap between mathematical–physical hydrodynamics and technological hydraulic. This gap was closed by the profound experimental and theoretical work of the British engineer and physicist Osborne Reynolds (1842–1912). He worked with a colored fluid thread in a glass tube. For a small diameter d and a small velocity V the thread moved on a straight line (laminar flow). For large d or large V, irregular side motions of the thread occurred (turbulent flow). Only by applying the aspects of a similarity law, was Reynolds able to obtain some systematics.

Arnold Sommerfeld (1944)

As we shall see, the basic equations of hydrodynamics for liquids and gases are obtained by modifying the basic equations of elastodynamics.

The main difference is that, in hydrodynamics, it is not the motion of the single volume element that is studied, but instead, the motion of the fluid particles, which at various times are at a fixed point y. Instead of an individual description, such as in elasticity theory, one uses an anonymous description in hydrodynamics.

In order to obtain the equations of gas dynamics as a special case, we also consider thermodynamical effects for the basic equations. In Section 70.5, the basic equations follow from the so-called transport theorem.

70.1. Basic Equations

The motions of a liquid or a gas are described by a velocity field

$$v = v(y, t).$$

Thereby $v(y, t)$ is the velocity vector at the point y at time t. The basic equations which will physically be motivated in Section 70.5, are as follows:

(i) *Momentum balance (equations of motion)*

$$(\rho v)_t + \operatorname{div} \rho v \circ v = K - \operatorname{grad} p + \operatorname{div} \tilde{\tau}. \tag{1}$$

(ii) *Mass balance (continuity equation)*

$$\rho_t + \operatorname{div} \rho v = 0. \tag{2}$$

(iii) *Entropy balance*

$$(s\rho)_t + \operatorname{div} s\rho v = ([\tilde{\tau}, Dv] - \operatorname{div} q)/T. \tag{3}$$

(iv) *Thermodynamical equations*

$$p = \prod(\rho, T), \qquad s = \sum(\rho, T). \tag{4}$$

(v) *Constitutive laws*

$$\tilde{\tau} = \Phi(Dv, \rho, T), \qquad q = \Psi(\operatorname{grad} T, \rho, T),$$
$$[\tilde{\tau}, Dv] \geq 0, \qquad q \operatorname{grad} T \leq 0. \tag{5}$$

The quantities above, which all depend on position y and time t, have the following meaning:

- ρ density;
- p pressure;
- T temperature;
- $\tilde{\tau}$ stress tensor for the inner friction;
- q heat flow density vector;
- s specific entropy density.

The outer force which is exerted onto the flow region H is

$$\int_H K \, dy$$

and the entropy of H is

$$\int_H \rho s \, dy.$$

70.1. Basic Equations

The stress tensor τ of Section 61.3 has the form

$$\tau = -pI + \tilde{\tau}. \tag{6}$$

Thus, according to Section 61.3, the stress force

$$\int_{\partial H} \tau n \, dO = -\int_{\partial H} pn \, dO + \int_{\partial H} \tilde{\tau} n \, dO \tag{7}$$

acts on H. Here n is the outer unit normal vector to ∂H. If the inner friction vanishes, i.e., $\tilde{\tau} = 0$, then (7) corresponds to the well-known fact that, in a fluid, the pressure force $-pn \Delta O$ is applied to the surface element ΔO in direction of the inner normal vector $-n$ (Pascal's law).

Moreover, we set

$$Dv(y) = 2^{-1}(v'(y) + v'(y)^*),$$

i.e., Dv is constructed analogously to the strain tensor $\gamma = Du$. The constitutive law relates the inner friction $\tilde{\tau}$ with Dv, i.e., with the first-order derivatives of the velocities with respect to space. As in the elasticity theory of Section 61.4 the stress tensor τ is assumed to be symmetric, i.e., $\tau \in L_{\text{sym}}(V_3)$. Hence it follows that

$$\tilde{\tau} \in L_{\text{sym}}(V_3).$$

For $a, b \in V_3$ we define the dyadic product $\sigma = a \circ b$ as the uniquely determined linear operator $a \circ b \colon V_3 \to V_3$ with

$$(a \circ b)x = a(bx) \quad \text{for all} \quad x \in V_3.$$

In Cartesian coordinates we have $\sigma_j^i = a^i b_j$, so that the equation of motion (1) takes the form

$$(\rho v^i)_t + D^j(\rho v_j v^i) = K^i - D^i p + D^j \tilde{\tau}_j^i. \tag{8}$$

According to Section 69.1, all basic equations (i) to (iii) have the form of conservation laws. Thus the right-hand sides in (i) and (iii) describe the production of momentum and entropy per volume and time.

Initial and boundary conditions have to be added to the basic equations. As initial conditions, one prescribes at time $t = 0$ velocity $v(y, 0)$, density $\rho(y, 0)$, and pressure $p(y, 0)$ in the flow region G. If G is not in motion, then the normal component of the velocity vector must vanish on the boundary ∂G, i.e.,

$$vn = 0 \quad \text{on } \partial G.$$

For viscous fluids, the fluid sticks to the boundary. Thus we must have

$$v = 0 \quad \text{on } \partial G.$$

Further possible boundary conditions will be discussed in the following chapter.

In order to be able to conveniently transform the basic equations, we define in an invariant way

$$\Delta v = 2 \operatorname{div} Dv - \operatorname{grad} \operatorname{div} v, \tag{9}$$

$$(v \operatorname{grad})v = \operatorname{div}(v \circ v) - v \operatorname{div} v. \tag{10}$$

In Cartesian coordinates this gives

$$\Delta v = (\Delta v^j)e_j, \qquad (v\,\text{grad})v = (v^i D_i v^j)e_j.$$

The continuity equation allows us to simplify conservation laws, because from (2) it follows that for an arbitrary real function f:

$$(\rho f)_t + \text{div}\, f\rho v = \rho_t f + \rho f_t + f \text{div}\, \rho v + \rho v\, \text{grad}\, f$$
$$= \rho f_t + \rho v\, \text{grad}\, f. \qquad (11)$$

In particular, we obtain for the equation of motion (1):

$$(\rho v)_t + \text{div}\, \rho v \circ v = \rho v_t + \rho(v\,\text{grad})v. \qquad (12)$$

Moreover, we have

$$\text{curl}\,\text{curl}\,v = \text{grad}\,\text{div}\,v - \Delta v.$$

All expressions in these formulas are defined in an invariant way. Therefore it suffices to verify these relations in Cartesian coordinates. This strategy will be used throughout this chapter.

70.2. Linear Constitutive Law for the Friction Tensor

Based on general symmetry arguments we want to formulate the constitutive law for a fluid. In addition to the universal pressure, we observe forces of friction in a fluid, which are caused by the fact that the particles of fluid move with different velocities. We therefore assume in (6) that the stress tensor can be written in the form

$$\tau = -pI + \tilde{\tau},$$

whereby the friction tensor $\tilde{\tau}$ depends on the space derivatives of v. More precisely, we assume that

$$\tilde{\tau} = \tilde{\tau}(Dv)$$

in analogy to the constitutive $\tau = \tau(Du)$ of elasticity theory. Note that $v = u_t$. According to Proposition 61.11, the most general rotation invariant map between the symmetric tensors $\tilde{\tau}$ and Dv has the form

$$\tilde{\tau} = 2\eta Dv + (\eta' - 2\eta)(\text{tr}\, Dv)I + \eta''(Dv)^2 \qquad (13)$$

with

$$\eta = \eta(\text{tr}\, Dv, [Dv, Dv], \det Dv)$$

and analogous expressions for η' and η''. In first-order approximation we want to obtain a *linear* law in (13). If η and η' are expanded in a Taylor series, then, in first-order approximation, one may assume that η and η' are constants.

70.2. Linear Constitutive Law for the Friction Tensor

Here η is called the *viscosity* and $\eta/\rho = \nu$ the *kinematic viscosity*. Consequently, we set $\eta'' = 0$. This gives

$$\text{div }\tilde{\tau} = \eta \Delta v + (\eta' - \eta)\,\text{grad div } v \qquad (14)$$
$$= -\eta\,\text{curl curl } v + \eta'\,\text{grad div } v.$$

According to (7), the stress force

$$\int_{\partial H} \tau n \, dO = -\int_{\partial H} pn \, dO + \int_H \text{div }\tilde{\tau}\, dy \qquad (15)$$

is applied to a region H of the fluid. As we shall see in Section 70.5, curl v and div v are a measure for the change in time of the rotational motion of a fluid particle, and the change in time of the volume, respectively. Thus, (14) shows that η and η' describe the so-called friction of vorticity and friction of compression, respectively. For incompressible fluids, i.e., div $v = 0$, the quantity η' does not occur. But experience also shows that for compressible fluids η' can be neglected compared with η.

EXAMPLE 70.1 (Parallel Flow). In Cartesian coordinates we consider the parallel flow

$$v = \varphi(\xi_2)e_1$$

(Fig. 70.1), whereby the velocity in direction of the ξ_2-axis varies. This implies tr $Dv = 0$ in (13). Let H be a cuboid parallel to the coordinate axes as shown in Fig. 70.1. By (13), this cuboid is subject to the stress force

$$\int_{\partial H} \tau n \, dO = -\int_{\partial H} pn \, dO + 2\eta \int_{\partial H} (Dv)n \, dO.$$

In particular, the force

$$\tau e_2 \, \Delta O = -pe_2 \, \Delta O + \eta \varphi' e_1 \, \Delta O$$

is applied to this side of the cuboid which has the outer normal vector e_2. As expected, the force of friction in Figure 70.1 is effective in direction e_1. The larger the derivative of the speed φ' is, the larger the force of friction becomes.

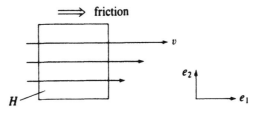

Figure 70.1

70.3. Applications to Viscous and Inviscid Fluids

EXAMPLE 70.2 (Viscous Fluids). We neglect temperature effects. The basic equations for viscous fluids then follow from the basic equations of Section 70.1 and the constitutive law (13) by using (12) and (14). They are called *Navier–Stokes equations*.

(i) Equation of motion

$$\rho v_t + \rho(v\,\text{grad})v - \eta \Delta v + (\eta - \eta')\,\text{grad div}\, v = K - \text{grad}\, p. \qquad (16)$$

(ii) Continuity equation

$$\rho_t + \text{div}\, \rho v = 0. \qquad (17)$$

(iii) Density-pressure relation

$$\rho = \rho(p). \qquad (18)$$

From (14) it follows that (i) also has the form

$$\rho v_t + \rho(v\,\text{grad})v + \eta\,\text{curl curl}\, v - \eta'\,\text{grad div}\, v = K - \text{grad}\, p. \qquad (19)$$

EXAMPLE 70.3 (Special Cases). If $\rho = \text{const}$, then the fluid is called *incompressible*. From (ii) it follows immediately that $\text{div}\, v = 0$.

If $\text{curl}\, v \equiv 0$, then the flow is called *irrotational*. If C is a closed curve, then

$$\Gamma = \int_C v\, dy$$

is called the *circulation* of C. If a given particle of fluid moves on a closed orbit which corresponds to C, then we have $\Gamma \neq 0$, since $v\dot{y}$ is proportional to v^2 for a parametrization $y = y(t)$ of C. This is not possible in an irrotational flow, because in this case, Stokes' theorem implies that

$$\Gamma = \iint \text{curl}\, v\, df = 0.$$

The motion is called *stationary* if the density ρ and the velocity v do not depend on time.

Proposition 70.4. *For a solution of the basic equations (16)–(18) in a connected flow region, the Bernoulli equation*

$$\frac{1}{2}v^2 + U + \int_{p_0}^{p} \frac{dp}{\rho(p)} = \text{const} \qquad (20)$$

is valid if the flow is stationary and irrotational, $\eta' = 0$ holds, and the outer forces possess a potential, i.e., $K = -\rho\,\text{grad}\, U$.

All functions are assumed to be sufficiently smooth. Equation (20) corresponds to the energy conservation law.

PROOF. Let L denote the left-hand side in (20). Because of

$$(v \, \text{grad})v = \text{grad} \frac{v^2}{2} - v \times \text{curl} \, v \qquad (21)$$

the equation of motion (19) assumes the form

$$v_t + \text{grad} \, L = v \times \text{curl} \, v - \frac{\eta}{\rho} \text{curl curl} \, v + \frac{\eta'}{\rho} \text{grad div} \, v.$$

From $v_t = 0$, curl $v = 0$ and $\eta' = 0$ follows grad $L = 0$, and hence $L = $ const. □

EXAMPLE 70.5 (Inviscid Fluid). A fluid is called *inviscid* if and only if the inner friction vanishes, i.e., $\eta = \eta' = 0$. In this case the basic equations (16)–(18) are called *Euler equations*.

Inviscid fluids are also called *ideal* fluids.

70.4. Tube Flows, Similarity, and Turbulence

To understand a number of important phenomena of the Navier–Stokes equations, we consider the simple case of a tube flow.

EXAMPLE 70.6 (Law of Hagen 1839 and Poiseuille 1840). As in Example 70.1 we consider a tube of length l, with radius R and pressure difference

$$\Delta p = p_1 - p_2 > 0.$$

We set $r = (\xi_2^2 + \xi_3^2)^{1/2}$, i.e., r is the distance from the tube axis. One easily verifies that for $\eta' = 0$, a solution of the stationary equations of motion (16)–(18) is given by:

$$v = \frac{\Delta p}{4\eta l}(R^2 - r^2)e_1, \qquad (22)$$

$$p = p_1 - \xi_1 \Delta p / l, \qquad \rho = \text{const}.$$

This is the typical parabolic velocity profile, shown in Figure 70.2. The mass

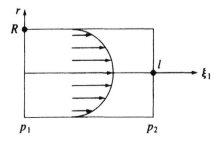

Figure 70.2

M, which during the time Δt flows through the tube, is equal to

$$M = \Delta t \int_0^R \rho |v| 2\pi r \, dr = \frac{\rho}{8\eta l} \pi R^4 \Delta p. \tag{23}$$

From this formula the viscosity η can experimentally be determined. Also, the expression kinematic viscosity for η/ρ becomes clear. Table 70.1 contains several values.

Table 70.1

Substance	Viscosity η	Kinematic viscosity η/ρ
Air	$1.8 \cdot 10^{-4}$ g/s cm	0.15 cm²/s
Water	0.01	0.01
Glycerine	15	12

The number

$$\text{Re} = \rho V R/\eta$$

is called *Reynolds number*, where V is the mean velocity.

The following fact is now especially important. Expression (22) describes, a mathematically correct solution of the Navier–Stokes equations, which is valid for all values of Δp, l, ρ, η, and R. In experiments, however, this solution is observed only for small Reynolds numbers. One speaks of a *laminar* flow. For large Reynolds numbers the solution (22) becomes unstable and in experiments one observes turbulent behavior, which for $\text{Re} \to +\infty$ becomes more and more chaotic. For example,

$$M \sim \Delta p,$$

as in (23), is no longer valid in a turbulent flow, but only approximately

$$M \sim (\Delta p)^{1/2}.$$

The great achievement of Reynolds in 1885 was his discovery that the change from laminar flow into turbulence does not depend on the single values of R, V, η, and ρ, but only on the number Re. This is Reynold's similarity law. We have

$$\text{Re} \begin{cases} < \text{Re}_{\text{crit}} & \text{laminar flow,} \\ > \text{Re}_{\text{crit}} & \text{turbulent flow,} \end{cases} \tag{24}$$

with $\text{Re}_{\text{crit}} = 1{,}150$. But Re_{crit} also depends on the particular way the fluid flows into the tube. For a very good smoothing of the muzzle we may assume values of 20,000 for Re_{crit}.

We want to show how, by general dimension arguments, one is led to the Reynolds number Re. We assume that the sudden change of laminar flow into turbulence depends on a dimensionless number, since the physical phenomenon cannot depend on the system of units. Our ansatz is $\text{Re} = f(\rho, \eta, R, V)$.

More precisely, let
$$Re = \rho^\alpha \eta^\beta R^\gamma V^\delta.$$
This gives the dimension $m^{-3\alpha-\beta+\gamma+\delta} \, kg^{\alpha+\beta} \, s^{-\beta-\delta}$. In order that this expression becomes dimensionless, we must have $\beta = -\alpha$, $\delta = \alpha$, $\gamma = \alpha$, hence $Re = (\rho R V/\eta)^\alpha$. The value of α, however, is unimportant.

70.5. Physical Motivation of the Basic Equations

We use the same notations as in elasticity theory of Section 61.3. As a consequence of the deformation

$$y = x + u(x, t), \tag{25}$$

the point x at time t becomes the point y, and the region H becomes $H' = H(t)$. According to Figure 70.3, the point x denotes the individual particles of fluid. We may assume that $u(x, 0) = 0$ for all x. Thus the velocity of the individual particle of fluid, which at time t is at the point y, is equal to

$$v(y, t) = u_t(x, t).$$

From Section 61.2 it follows that, in first-order approximation, $\alpha = \operatorname{div} u$ describes the relative change in volume and $\omega = 2^{-1} \operatorname{curl} u$ the vector of the infinitesimal rotation of the volume elements. This implies

$$\alpha_t = \operatorname{div} v \quad \text{and} \quad \omega_t = 2^{-1} \operatorname{curl} v.$$

The following formula is important for obtaining the basic equations

$$\frac{d}{dt}\left(\int_{H(t)} h \, dy\right) = \int_{H(t)} (h_t + \operatorname{div} hv) \, dy. \tag{26}$$

Proposition 70.7 (Transport Theorem). *Let H denote a bounded region in $\mathbb{R}^3$ which after applying the C^2-diffeomorphism (25) becomes $H(t)$ for every $t \in [0, t_0]$. More precisely, let (25) be a C^2-diffeomorphism from $\overline{H}$ onto $\overline{H(t)}$. Let $h: \mathbb{R}^3 \times [0, t_0] \to \mathbb{R}$ be C^1. Then (26) holds for every $t \in [0, t_0]$.*

The proof will be given in Problem 70.1. In the following we assume that all functions are sufficiently smooth.

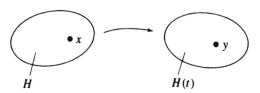

Figure 70.3

Mass balance. Conservation of mass means $(d/dt)\int_{H(t)} \rho\, dy = 0$. From (26) follows

$$\int_{H(t)} \rho_t + \operatorname{div} \rho v\, dy = 0.$$

Contracting $H(t)$ into one point, we obtain $\rho_t + \operatorname{div} \rho v = 0$. This is the continuity equation (2).

Momentum balance. Similarly as in Section 58.7, the time derivative of the total momentum of $H(t)$ is equal to the effective force on $H(t)$. This implies

$$\frac{d}{dt}\int_{H(t)} \rho v\, dy = \int_{H(t)} K\, dy + \int_{\partial H(t)} \tau n\, dO. \tag{27}$$

From (26) and $\int \tau n\, dO = \int \operatorname{div} \tau\, dy$ follows then

$$\int_{H(t)} [(\rho v)_t + \operatorname{div} \rho v \circ v - K - \operatorname{div} \tau]\, dy = 0.$$

This immediately implies the equation of motion (1).

Kinetic energy balance. The kinetic energy of $H(t)$ is

$$E_{\mathrm{kin}}(t) = \int_{H(t)} 2^{-1}\rho v^2\, dy.$$

If one multiplies the equation of motion (1) with v, then one obtains the following relation from the continuity equation (2):

$$(\rho v^2)_t + \operatorname{div}(\rho v^2)v = 2(Kv + v\operatorname{div} \tau).$$

Equation (26) yields

$$\dot{E}_{\mathrm{kin}}(t) = \int_{H(t)} (vK + v\operatorname{div} \tau)\, dy.$$

Thermodynamical system. We consider the fluid in $H(t)$ as a thermodynamical system with mass M, inner energy E, and entropy S. We assume that these quantities can be represented by specific densities of the form

$$M = \int \rho\, dy, \qquad E = \int \rho e\, dy, \qquad S = \int \rho s\, dy.$$

The area of integration is $H(t)$. We postulate that for e, p, and s the same relations are valid as in Section 67.3. This means

$$de = T\, ds + p\, d\rho/\rho^2,$$
$$p = p(\rho, T), \qquad e = e(\rho, T), \qquad s = s(\rho, T). \tag{28}$$

First law of thermodynamics. Let $W(t)$ be the work done at $H(t)$ during the time interval $[0, t]$. In point mechanics, the time derivative of the work is equal

70.5. Physical Motivation of the Basic Equations

to the power of all forces. Analogously, we set

$$\dot{W}(t) = \int_{H(t)} Kv \, dy + \int_{\partial H(t)} (\tau n) v \, dO.$$

Moreover, let $Q(t)$ be the amount of heat which during the time interval $[0, t]$ flows into $H(t)$. If q is the heat flow density vector, then it follows from Section 69.1 that

$$\dot{Q}(t) = -\int_{\partial H(t)} qn \, dO.$$

The total energy of $H(t)$ is $E + E_{\text{kin}}$. Thus the first law of thermodynamics implies

$$\dot{E} + \dot{E}_{\text{kin}} = \dot{W} + \dot{Q}.$$

According to (26) this means

$$\int_{H(t)} (\rho e)_t + \operatorname{div} \rho e v + v \operatorname{div} \tau \, dy = \int_{\partial H(t)} (\tau n) v - qn \, dO.$$

Passing to components and integration by parts yields

$$\int_{H(t)} (\rho e)_t + \operatorname{div} \rho e v - [\tau, Dv] + \operatorname{div} q \, dy = 0,$$

hence

$$(\rho e)_t + \operatorname{div} \rho e v - [\tau, Dv] + \operatorname{div} q = 0. \tag{29}$$

We want to replace e with s. From (28) follows

$$\dot{e} = T\dot{s} + p\dot{\rho}/\rho^2.$$

Because of $e = e(y(x, t), t)$ we obtain

$$\dot{e} = v \operatorname{grad} e + e_t.$$

This implies

$$\rho(e_t + v \operatorname{grad} e) = \rho T(s_t + v \operatorname{grad} s) + \frac{p}{\rho}(\rho_t + v \operatorname{grad} \rho). \tag{30}$$

The continuity equation together with (11) and (29) yields

$$(\rho s)_t + \operatorname{div} \rho s v = ([\tilde{\tau}, Dv] - \operatorname{div} q)/T. \tag{31}$$

Note that $\tau = -pI + \tilde{\tau}$ and $[-pI, Dv] = -p \operatorname{div} v$. Equation (31) is the same as the basic equation (3).

Second law of thermodynamics. Motivated by the second law

$$\dot{S} \geq \dot{Q}/T,$$

and noticing the dependence of the temperature on (y,t), we postulate the inequality

$$\frac{d}{dt}\int_{H(t)} \rho s\, dy \geq -\int_{\partial H(t)} qn/T\, dO.$$

From (26), (31), and (61.4) this implies

$$\int \left(T^{-1}([\tilde{\tau}, Dv] - \operatorname{div} q) + \operatorname{div}\frac{q}{T}\right) dy \geq 0$$

and hence $T^{-1}[\tilde{\tau}, Dv] - T^{-2}q\operatorname{grad} T \geq 0$. This inequality is always satisfied if $[\tilde{\tau}, Dv] \geq 0$ and $q\operatorname{grad} T \leq 0$.

This motivates the basic equations of Section 70.1.

70.6. Applications to Gas Dynamics

The basic equations of Section 70.1 are not only valid for liquids but also for gases.

EXAMPLE 70.8 (Inviscid Ideal Gas without Heat Conduction). We neglect inner friction and heat conduction, i.e., $\tilde{\tau} = 0$ and $q = 0$.

The basic equations (1)–(5) then take the form:

(i) Equation of motion

$$\rho v_t + \rho(v\operatorname{grad})v = K - \operatorname{grad} p.$$

(ii) Continuity equation

$$\rho_t + \operatorname{div}\rho v = 0.$$

(iii) Entropy balance

$$s_t + v\operatorname{grad} s = 0.$$

(iv) Thermodynamical equations for ideal gases

$$p = r\rho T, \qquad s = c\ln T\rho^{-r/c}.$$

Relations (i)–(iii) follow from Section 70.1 and (iv) has been considered in Example 67.4.

Problems in gas dynamics (e.g., shock waves and transonic flow) will be considered in Part V. We will see that a *generalized* form of the basic equations is needed in order to obtain the important Rankine–Hugoniot *jump conditions* for shock waves.

Problems

PROBLEMS

70.1. *Proof of Proposition 70.7.* Let $y = x + u(x,t)$ and $D_j = \partial/\partial \xi_j$. Observe that $v = y_t$. The transformation yields

$$\int_{H(t)} h\, dy = \int_H hJ\, dx$$

with $J(x,t) = \det y_x(x,t)$. If A_{ij} denotes the adjoint subdeterminant to $D_j \eta_i$ in J, then the well-known expansion formula

$$\delta_{ik} J = A_{ij} D_j \eta_k$$

is valid. The sum is taken over equal indices from 1 to 3. According to the product rule, the derivative of a determinant is obtained by differentiating the ith line and then adding all these determinants, i.e.,

$$J(x,t)_t = (D_j \eta_i)_t A_{ij} = (D_j v_i) A_{ij}$$

$$= \frac{\partial v_i}{\partial \eta_k}(D_j \eta_k) A_{ij} = \frac{\partial v_i}{\partial \eta_i} J$$

$$= J \operatorname{div}_y v.$$

This implies

$$\frac{d}{dt} \int_{H(t)} h(y,t)\, dt = \frac{d}{dt} \int_H h(y(x,t),t) J(x,t)\, dx$$

$$= \int_H (v \operatorname{grad}_y h + h_t) J + hJ \operatorname{div}_y v\, dx$$

$$= \int_H (h_t + \operatorname{div}_y hv) J\, dx$$

$$= \int_{H(t)} (h_t + \operatorname{div}_y hv)\, dy.$$

70.2. *The basic equations in hydrostatics of Newton (1687).*

70.2a. Formulate these equations for constant temperature T.

Solution: We use the basic equations of hydrodynamics (1) to (5). In the special case of hydrostatics, we have $v \equiv 0$, i.e., the velocity of the fluid particles is equal to zero. Hence, the inner friction vanishes, i.e., $\tilde{\tau} \equiv 0$. From (1) to (5), we obtain the following basic equations of hydrostatics:

(i) *Equilibrium condition*

$$K = \operatorname{grad} p. \tag{32}$$

(ii) *Density-pressure relation*

$$\rho = \rho(p, T). \tag{33}$$

Here, p is pressure, and ρ is density. The outer force, which is exerted onto an arbitrary flow region H, is equal to

$$\int_H K\,dy.$$

In a simply connected region, equation (32) determines the pressure p up to a constant. Thus, we may describe the pressure p_0 at a fixed point y_0, i.e.,

$$p(y_0) = p_0.$$

70.2b. Give a physical interpretation of (32).
Solution: Integration by parts yields

$$\int_H K\,dy = \int_{\partial H} pn\,dO.$$

This means that, for each flow region H, the outer forces onto H and the pressure onto the boundary ∂H are in an equilibrium.

70.2c. Formulate the basic equations of hydrostatics in case the outer force possesses a potential U, i.e.,

$$K = -\rho\,\text{grad}\,U.$$

Solution: Instead of $\rho = \rho(p, T)$ we write briefly $\rho = \rho(p)$. From (32) we obtain

$$\text{grad}\left(U + \int_{p_0}^{p} \frac{dp}{\rho(p)}\right) = 0.$$

In a simply connected region, this yields the following basic equations:
(i) *Energy conservation*

$$U + \int_{p_0}^{p} \frac{dp}{\rho(p)} = \text{const} = U(y_0). \tag{34}$$

(ii) *Density-pressure relation*

$$\rho = \rho(p). \tag{35}$$

Here, we are given the potential $U = U(y)$ and the pressure $p(y_0) = p_0$ at a fixed point y_0. From (34) we obtain the pressure $p = p(y)$, and from (35) we obtain the density $\rho = \rho(p(y))$ at y.

If the fluid is incompressible, then

$$\rho = \text{const}, \tag{36}$$

and equation (34) yields

$$\rho U + p = \text{const}. \tag{37}$$

Further important material can be found in the problem sections of Chapters 71, 72, and 86.

References to the Literature

Classical works: Newton (1687) (hydrostatics), Bernoulli (1738) (Bernoulli's law—conservation of energy), Euler (1755) (fundamental paper—equation of motions), Navier (1822) and Stokes (1845) (Navier–Stokes equations), Darcy (1856) (basic equations of filtration theory), Helmholtz (1858) (motion of vortices), Kelvin (1869) (conservation of circulation), Riemann (1860) (integration of the equations of gas dynamics in special cases), Rankine (1870) and Hugoniot (1889) (jump conditions in gas dynamics), Reynolds (1885) (turbulence and similarity), Prandtl (1904) (boundary layer equation), Lichtenstein (1929, M) (existence theorems for inviscid fluids via potential theory). (See also "classical works" in the References to the Literature for Chapters 71 and 72.)

Article in the handbook of physics: Serrin (1959).

Classical monographs: Lamb (1924, H), Courant and Friedrichs (1948, H) (gas dynamics), Kotschin (1954), Jacob (1959), Milne-Thomson (1960).

Standard works from the physical point of view: Sommerfeld (1954, M), Vol. 2, Landau and Lifšic (1962, M), Vol. 6, Lighthill (1986, M).

Mathematical point of view: Hughes and Marsden (1976, L), and Chorin and Marsden (1979, L) (recommended as introductions), Batchelor (1967, M), Meyer (1971, M), Shinbrot (1973, M), Schreier (1982, M) (compressible fluids), Ockendon and Taylor (1983, M) (inviscid fluids), Chipot (1984) (porous media), Pipkin (1986) (visco-elasticity).

Free boundary problems and variational inequalities: Friedman (1982, M).

Capillarity: Finn (1985, M).

Global analysis and hydrodynamics: Arnold (1966), Ebin, Fischer, and Marsden (1972), Marsden (1972, L).

Gas dynamics and shock waves (mathematical point of view): Courant and Friedrichs (1948, M), Bers (1958, M), Roždestvenskii and Janenko (1978, M), Morawetz (1981, L), Smoller (1983, M) (recommended as an introduction), Cole (1986, M) (transonic aerodynamics).

Gas dynamics and shock waves (physical point of view): Landau and Lifšic (1962, M), Vol. 6, Guderley (1957, M), Becker (1965, M), Sauer (1960, M), (1966, M), Oswatitsch (1976, M) (standard work).

Recent results in hydrodynamics and plasma physics: Marsden (1984, P).

Open questions in the dynamics of liquids and gases: Smoller (1983a), Majda (1984, M), Marsden (1984, P).

Numerical methods: Cf. the References to the Literature for Chapter 72.

(Cf. also the References to the Literature for Chapters 71, 72, and 86.)

CHAPTER 71

Bifurcation and Permanent Gravitational Waves

In one of his last papers, Lord Rayleigh, in 1917, computed approximate solutions up to order 6. He also showed by means of a numerical example that the relative error is not greater than $2.5 \cdot 10^{-6}$. We give here a rigorous existence proof for permanent gravitational waves of infinite depth which is also constructive.

Tullio Levi-Civita (1925)

I have tried to avoid long numerical computations, thereby following Riemann's postulate that proofs should be given through ideas and not voluminous computations.

David Hilbert, *Report on Number Theory* (1897)

In this chapter we study the existence of nontrivial water waves in a channel of finite depth. As shown in Figure 71.1 we find that, in addition to the trivial parallel flow, there occur nontrivial wave motions at certain critical velocities c. Such waves were studied during the nineteenth century by British hydrodynamicists such as Airy, Stokes, Kelvin, and Rayleigh. They solved the linearized problems and calculated nonlinear approximations up to order 6. No convergence proofs, however, were given. The linearized theory of Airy (1845) shows that for

$$c^2 = \frac{g\lambda}{2\pi} \tanh \frac{2\pi h}{\lambda} \tag{1}$$

nontrivial waves occur with:

- c propagation velocity of the wave;
- λ wave length;
- h average channel depth;
- g gravitational acceleration.

71. Bifurcation and Permanent Gravitational Waves

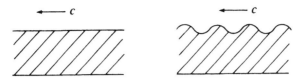

Figure 71.1

Relations between propagation velocity and wave length, of which formula (1) is an example, are called dispersion relations, and one of the main goals in the theory of permanent water waves is to obtain such dispersion relations for various situations.

For about 20 years Levi-Civita (1873–1941) worked on a rigorous solution for the complete nonlinear problem. In 1925, he found a very complicated existence proof for channels of infinite depth. Analogously to Section 71.6, he calculated the solutions as a power series with respect to a small parameter s. The main difficulty, which he overcame by performing voluminous computations, was to show the convergence of the formal solution by using a majorant method.

In our functional-analytic approach this difficulty is avoided, since existence and analyticity of solutions follows from the main theorem of analytic bifurcation theory (Theorem 8.A). In addition, we obtain uniqueness results which Levi-Civita did not have. Our approach shows that, by using the abstract methods of functional analysis, the proofs can be given through ideas following Hilbert's requirement cited at beginning of this chapter.

Independently of Levi-Civita, the Russian mathematician Nekrasov (1883–1957) also worked on this wave problem. In 1921, he gave the first existence proof for channels of infinite depth by using a nonlinear integral equation and a complicated majorant method. Using the method of Levi-Civita, Struik, in 1926, proved the existence of permanent gravitational waves in channels of finite depth. This proof was even more complicated than the proof of Levi-Civita for channels of infinite depth.

The functional-analytic approach of this section goes back to the author. The monograph by Zeidler (1968) contains detailed historical remarks as well as applications of this method to various classical wave problems. There, we presented, for the first time, a number of new existence proofs. The key is Theorem 8.A by the author. Further applications may be found in Zeidler (1971), (1972a, b), (1973), (1977, S), and Beyer and Zeidler (1979).

From a physical point of view our main result in this chapter is as follows. We are given:

h average channel depth;
ρ_0 constant density of the liquid;
λ wave length;
p_0 constant barometric pressure on the surface of the liquid.

We find that in a neighborhood of the critical velocity c, given by (1), there occur waves which satisfy:

(i) Wave surface
$$y_0(x) = \frac{\lambda}{2\pi} s \cos \frac{2\pi x}{\lambda} + O(s^2), \qquad s \to 0.$$

(ii) Average velocity c at the channel bottom
$$c^2 = \frac{g\lambda}{2\pi} \tanh \frac{2\pi h}{\lambda} + \varepsilon_2 s^2 + O(s^4).$$

(iii) Channel bottom
$$y = -h.$$

(iv) Flux
$$\psi_0 = hc.$$

(v) Velocity vector $v = ae_1 + be_2$ of the fluid particles at the wave surface
$$a = c + O(s), \qquad b = -sc \sin \frac{2\pi x}{\lambda} + O(s^2), \qquad s \to 0$$

and in the flow region
$$a = c + O(s), \qquad b = O(s), \qquad s \to 0.$$

(vi) Pressure in the flow region
$$p = p_0 + O(s), \qquad s \to 0.$$

Thereby s is a small real parameter and $\varepsilon_2 > 0$. What this means is that the nontrivial waves appear supercritical, i.e., above the classical critical velocity (1), which is obtained from (ii) with $s = 0$. According to the formal stability theory of Section 8.7, such solutions are stable, see Figure 8.6. The trivial parallel flow of Figure 71.1 corresponds to $s = 0$. Roughly speaking, the waves above are the only ones that occur in a neighborhood of the critical velocity (1). More precisely, this will be stated in Theorem 71.B of Section 71.4.

We use a Cartesian (x, y)-coordinate system in which the wave surface seems to be at rest. In such a system the fluid particles move from left to right with a mean velocity of c. That this is a very natural result can be seen as follows. If we choose another coordinate system in which the wave surface travels from right to left with velocity c (Fig. 71.1), then the fluid particles have a mean velocity of zero, i.e., they just perform small oscillations.

Our present problem is a *free boundary-value problem* which exhibits the typical difficulty that, a priori, the form of the wave surface (i) is not known. In order to overcome this obstacle, we introduce a conformal mapping. The main idea is this:

(a) The complex flow potential is used, to obtain a conformal map between a suitable portion of the flow region and the circular ring.

(b) Thereby the original free boundary-value problem reduces to a nonlinear boundary-value problem for an analytic function on the circular ring.
(c) Using $C^{2,\alpha}$-Hölder spaces, we transform problem (b) into an operator equation on a B-space.
(d) We then apply the main theorem of analytic bifurcation theory (Theorem 8.A) to this operator equation. The following relations are valid:

$$\text{trivial solution} \Rightarrow \text{trivial parallel flow},$$
$$\text{bifurcation point} \Rightarrow \text{critical velocity (1)},$$
$$\text{bifurcation branch} \Rightarrow \text{nontrivial permanent waves}.$$

During the last few years, various other free boundary-value problems have been studied using the theory of variational inequalities. A discussion of this may be found in Friedman (1982, M). In Part V we will consider an elegant variational approach to problems with permanent waves.

71.1. Physical Problem and Complex Velocity

Let us now consider a water wave of wave length λ which propagates with a constant velocity in a channel of finite depth. This problem is most easily studied in a Cartesian (x, y)-coordinate system, in which the fluid surface seems to be at rest, i.e., where the surface is described by a λ-periodic even function

$$y = y_0(x), \tag{2}$$

which satisfies the normalization condition

$$\int_{-\lambda/2}^{\lambda/2} y_0(x)\, dx = 0. \tag{3}$$

As the equation of the channel bottom we choose

$$y = -h_0 \tag{4}$$

(Fig. 71.2). Let e_1, e_2 denote the orthonormal basis vectors in the (x, y)-system.

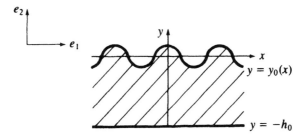

Figure 71.2

Suppose the velocity vector has the form

$$v = ae_1 + be_2 \tag{5}$$

with velocity components $a = a(x, y)$ and $b = b(x, y)$, i.e., we are interested in a plane stationary velocity field. Moreover, we assume that the components a and b have period λ with respect to x, and satisfy the following symmetry conditions

$$a(-x, y) = a(x, y), \qquad b(-x, y) = -b(x, y). \tag{6}$$

At the channel bottom we have

$$b(x, -h_0) = 0. \tag{7}$$

We define the *average velocity* at the channel bottom as

$$c = \frac{1}{\lambda} \int_{-\lambda/2}^{\lambda/2} a(x, -h_0)\, dx. \tag{8}$$

According to our assumption above, the surface of the wave in the (x, y)-systems seems to be at rest. However, our existence proof below shows that the particles of fluid move with an average velocity of c from left to right.

By definition, the flux is equal to

$$\psi_0 = \int_{-h_0}^{y_0(x)} a(0, y)\, dy. \tag{9}$$

The quantity

$$h = \frac{\psi_0}{c}$$

is called the average channel depth. Our existence proof below yields

$$h_0 = h + O(s^2), \qquad s \to 0.$$

Since we consider an inviscid, incompressible, irrotational, stationary flow with constant density ρ_0, formulas (70.16) and (70.19) yield the following Euler equations for the flow region:

$$\operatorname{grad}(\tfrac{1}{2}\rho_0 v^2 + \rho_0 U + p) = 0, \tag{10}$$

$$\operatorname{div} v = 0, \tag{11}$$

$$\operatorname{curl} v = 0. \tag{12}$$

Thereby p is the pressure and

$$K = -\rho_0 \operatorname{grad} U$$

is the density of the outer forces. The effective outer force in our wave problem is the gravitational force, i.e., $K = -\rho_0 g e_2$ and hence

$$U = gy.$$

71.1. Physical Problem and Complex Velocity

Theorem 71.A. *Consider the basic equations (10)–(12) for a planar, irrotational flow of an incompressible inviscid fluid in a region G. These equations are equivalent to the following conditions:*

(a) *The so-called complex velocity function*

$$V(z) = a(z) - ib(z)$$

with $z = x + iy$ is holomorphic in G, where a and b are the components of the velocity vector, i.e., $v = ae_1 + be_2$.
(b) *There exists a constant B_0 such that the Bernoulli equation*

$$\tfrac{1}{2}\rho_0 v^2 + \rho_0 U + p = B_0 \tag{13}$$

is valid in G.

PROOF. Equations (11) and (12) are equivalent to the Cauchy–Riemann equations

$$a_x = -b_y, \qquad a_y = b_x.$$

Hence $a - ib$ is holomorphic in G. □

Recall that the Bernoulli equation (13) describes conservation of energy. Theorem 71.A allows us to apply methods of complex function theory to planar hydrodynamics.

In the case of our wave problem, the following boundary conditions are valid:

$$\tfrac{1}{2}\rho_0 v^2 + \rho_0 g y_0 + p_0 = B_0 \quad \text{along} \quad y = y_0(x), \tag{14}$$

where p_0 is the outer atmospheric pressure, and

$$y_0' a - b = 0 \quad \text{along} \quad y = y_0(x), \tag{15}$$

$$b = 0 \quad \text{along} \quad y = -h_0. \tag{16}$$

Conditions (15) and (16) state that the wave surface and the channel bottom are streamlines, i.e., the velocity vector is tangential. This leads us to the following problem:

(i) *We are looking for a holomorphic function $V = a - ib$ in the flow region*

$$G = \{(x, y) \in \mathbb{R}^2 : -h_0 < y < y_0(x), -\infty < x < \infty\},$$

whereby $y = y_0(x)$ is unknown.
(ii) *The boundary conditions (14)–(16) are to be satisfied whereby the Bernoulli constant B_0 is unknown.*

If a solution is found, then the velocity is known. Furthermore, the pressure p in the flow region follows from equation (13) with $U = gy$.

71.2. Complex Flow Potential and Free Boundary-Value Problem

Let us simplify our problem, and introduce the holomorphic function

$$W(z) = \varphi + i\psi \tag{17}$$

via the integral

$$W(z) = \int_{iy_0(0)}^{z} V(w)\,dw.$$

Since the flow region is simply connected, this so-called complex flow potential is uniquely determined in this region. From

$$W'(z) = \varphi_x - i\varphi_y \tag{18}$$

we obtain

$$a = \varphi_x, \qquad b = \varphi_y, \tag{19}$$

with

$$\varphi(0, y_0(0)) = 0. \tag{20}$$

Moreover, the Cauchy–Riemann equations

$$\varphi_x = \psi_y, \qquad \varphi_y = -\psi_x \tag{21}$$

are valid in the flow region with

$$\psi(0, y_0(0)) = 0. \tag{22}$$

The components φ and ψ of the complex flow potential admit a simple interpretation. From (19) we have

$$v = \operatorname{grad} \varphi, \tag{23}$$

i.e., φ is the negative velocity potential. By definition, the curve $C: x = x(t)$, $y = y(t)$ is a streamline if and only if the velocity vector is tangential along C, i.e.,

$$-x'(t)\varphi_y(x(t), y(t)) + y'(t)\varphi_x(x(t), y(t)) = 0 \quad \text{along } C.$$

From (21) follows that

$$\frac{d}{dt}\psi(x(t), y(t)) = -x'\varphi_y + y'\varphi_x.$$

Hence C is a streamline if and only if

$$\psi = \text{const} \quad \text{along } C. \tag{23*}$$

From Section 71.1, the wave surface $y = y_0(x)$ is a streamline, i.e.,

$$\psi = 0 \quad \text{along } y = y_0(x)$$

71.2. Complex Flow Potential and Free Boundary-Value Problem

and the channel bottom is a streamline as well, i.e.,

$$\psi = -\psi_0 \quad \text{along} \quad y = -h_0.$$

The values of the constants follow from (9), (22), and $a = \psi_y$.
From (6), together with (19), (21), we obtain

$$\varphi(-x, y) = -\varphi(x, y), \qquad \psi(-x, y) = \psi(x, y), \tag{24}$$

and the λ-periodicity of a and b with respect to x implies the same for φ_x and ψ_x. This yields

$$\psi(x + \lambda, y) = \psi(x, y), \tag{25}$$
$$\varphi(x + \lambda, y) = \varphi(x, y) + c\lambda.$$

Notice that ψ is even with respect to x, and observe (8) with $a = \varphi_x$. Furthermore, it follows from (24) and (25) that

$$\varphi\left(\pm\frac{\lambda}{2}, y\right) = \pm\frac{c\lambda}{2}. \tag{26}$$

This leads to the following problem.

Free Boundary-Value Problem 71.1. We are looking for a flow region

$$G = \{(x, y) : -h_0 < y < y_0(x), -\infty < x < \infty\}$$

and a function ψ with

$$\Delta\psi = 0 \quad \text{on } G$$

which satisfies the boundary conditions

$$\psi = 0 \qquad \text{for} \quad y = y_0(x),$$
$$\psi = -\psi_0 \qquad \text{for} \quad y = -h_0,$$

and

$$\tfrac{1}{2}\rho_0(\psi_x^2 + \psi_y^2) + \rho_0 g y_0 + p_0 = B_0 \tag{27}$$

for $y = y_0(x)$. Moreover, let y_0 and ψ be even and λ-periodic with respect to x. Finally, let $\psi_0 > 0$ and for y_0 assume the normalization condition (3).

Since the boundary $y = y_0(x)$ is unknown and hence has to be determined together with ψ, one speaks of a free boundary-value problem. In particular, we are searching for a function $y_0(x) \not\equiv 0$.

To further simplify our problem let us introduce the function

$$\omega = \vartheta + i\tau.$$

We make the ansatz

$$W'(z) = \eta c e^{-i\omega},$$

where the constant η has still to be specified. This means that
$$a = \eta c e^\tau \cos \vartheta, \quad b = \eta c e^\tau \sin \vartheta,$$
and hence ϑ is the angle between the velocity vector v and the x-axis and
$$|v| = \eta c e^\tau.$$
For this to make sense, we assume that no stationary points exist in the flow region, i.e., $v \neq 0$ in G. Hence $W'(z) \neq 0$ in G. The function ω together with W is holomorphic in G. Since the maps a and b are λ-periodic in x, it follows that ϑ and τ have the same property. The symmetry of a and b implies
$$\vartheta(-x, y) = -\vartheta(x, y), \quad \tau(-x, y) = \tau(x, y).$$
Moreover, we find that at the channel bottom
$$\vartheta = 0 \quad \text{for} \quad y = -h_0,$$
and from (27) we obtain
$$\tfrac{1}{2}\rho_0 \eta^2 c^2 e^{2\tau} + \rho_0 g y_0 + p_0 = B_0 \tag{28}$$
for $y = y_0(x)$. The relation between η and τ is given in (31) below.

71.3. Transformed Boundary-Value Problem for the Circular Ring

In order to avoid the difficulty of a free boundary $y = y_0(x)$, we transform our problem onto the circular ring.

As shown in Figure 71.3, only the portion G_λ of the flow region is considered, where $-\lambda/2 < x < \lambda/2$. We then use the two mappings
$$W = W(z)$$
and
$$\zeta = \exp \frac{2\pi W}{ic\lambda}.$$

This way, the region G_λ is mapped conformally onto a circular ring, cut along the negative real axis. We transform the function $\omega = \vartheta + i\tau$ into the ζ-plane and denote this transformed function again by $\omega = \vartheta + i\tau$. The transformed function ω is then holomorphic on the open circular ring, cut along the negative, real axis.

In the ζ-plane we introduce polar coordinates ρ and σ through
$$\zeta = \rho e^{i\sigma}.$$
Because of $\vartheta(\rho, \pm\pi) = 0$ and
$$\vartheta(\rho, -\sigma) = -\vartheta(\rho, \sigma), \quad \tau(\rho, -\sigma) = \tau(\rho, \sigma),$$

71.3. Transformed Boundary-Value Problem for the Circular Ring

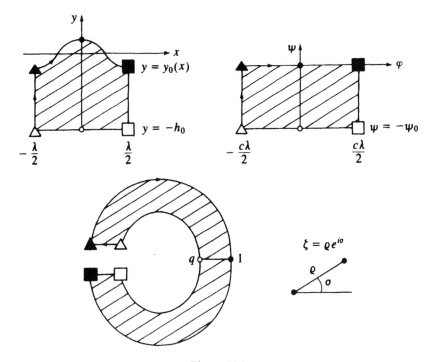

Figure 71.3

we can use Schwarz's reflection principle to continue the function ω analytically onto the open circular ring

$$A_q \stackrel{\text{def}}{=} \{\zeta \in \mathbb{C} : q < |\zeta| < 1\},$$

where $q = e^{-2\pi\psi_0/c\lambda}$, i.e.,

$$q = e^{-2\pi h/\lambda}.$$

Furthermore, we set

$$S_r \stackrel{\text{def}}{=} \{\zeta \in \mathbb{C} : |\zeta| = r\}.$$

Note that the boundary of A_q consists of the two circles S_1 and S_q (Fig. 71.4). From

$$dz = \frac{dW}{W'(z)} = \frac{i\lambda}{2\pi\eta\zeta} e^{i\omega(\zeta)} d\zeta$$

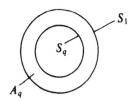

Figure 71.4

we obtain the back transformation

$$z = \frac{i\lambda}{2\pi\eta} \int_1^\zeta e^{i\omega} \frac{d\zeta}{\zeta} + iy_0(0). \tag{29}$$

This integral is to be evaluated along a curve in the circular ring, cut along the negative real axis. The free surface $y = y_0(x)$ corresponds to the unit circle $\zeta = e^{i\sigma}$, and hence has a parameter representation

$$x = -\frac{\lambda}{2\pi\eta} \int_0^\sigma e^{-\tau(\sigma)} \cos \vartheta(\sigma) \, d\sigma,$$

$$y_0 = -\frac{\lambda}{2\pi\eta} \int_0^\sigma e^{-\tau(\sigma)} \sin \vartheta(\sigma) \, d\sigma + y_0(0).$$

Here we write $f(\sigma)$ instead of $f(1,\sigma)$. The boundary relation (28) then becomes

$$\frac{1}{2}\rho_0\eta^2 c^2 e^{2\tau} - \frac{\rho g\lambda}{2\pi\eta} \int_0^\sigma e^{-\tau} \sin \vartheta \, d\sigma + p_0 = B_0$$

on S_1. Differentiation with respect to σ yields

$$\tau_\sigma = \mu e^{-3\tau} \sin \vartheta \quad \text{on } S_1 \tag{30}$$

with

$$\mu = \frac{g\lambda}{2\pi c^2 \eta^3}.$$

Under the back transformation the point $\zeta = qe^{\pm i\pi}$ becomes $z = \mp \lambda/2 - ih_0$. Hence, it follows from (29) that

$$\eta = \frac{1}{2\pi} \int_{-\pi}^{\pi} e^{-\tau(q,\sigma)} \, d\sigma. \tag{31}$$

In summarizing, the free boundary-value problem of Section 71.2 has been transformed into the following boundary-value problem for the circular ring.

Boundary-Value Problem 71.2. We are looking for functions

$$\vartheta, \tau \in C^2(\bar{A}_q)$$

which satisfy the Cauchy–Riemann differential equations

$$\rho\vartheta_\rho = \tau_\sigma, \qquad \rho\tau_\rho = -\vartheta_\sigma \quad \text{on } A_q \tag{32}$$

with boundary conditions

$$\tau_\sigma = \mu e^{-3\tau} \sin \vartheta \quad \text{on } S_1,$$
$$\vartheta = 0 \quad \text{on } S_q. \tag{33}$$

Furthermore, we assume the symmetry conditions

$$\vartheta(\rho, -\sigma) = -\vartheta(\rho,\sigma), \qquad \tau(\rho,-\sigma) = \tau(\rho,\sigma) \quad \text{on } \bar{A}_q \tag{34}$$

and the normalization condition

$$\int_{-\pi}^{\pi} \tau(q,\sigma)\,d\sigma = 0. \tag{35}$$

The trivial solution $\vartheta = \tau = 0$ corresponds to a parallel flow. We are interested, however, in parameters μ, for which nontrivial solutions exist. This is a typical bifurcation problem.

71.4. Existence and Uniqueness of the Bifurcation Branch

Our goal is to formulate the Boundary-Value Problem 71.2 as an operator equation.

For this, we let X be the set of all function pairs

$$(\vartheta, \tau) \in C^2(\bar{A}_q) \times C^2(\bar{A}_q),$$

which satisfy the following additional properties:

(i) The Cauchy–Riemann differential equations (32) are valid on the open circular ring A_q.
(ii) The boundary condition $\vartheta = 0$ is valid on S_q.
(iii) We assume the symmetry condition (34) and the normalization condition (35).

Obviously, X is a closed, linear subspace of the B-space $C^2(\bar{A}_q) \times C^2(\bar{A}_q)$. Hence X is a B-space.

The Boundary-Value Problem 71.2 can now be formulated in the following form. We are looking for an element $(\vartheta, \tau) \in X$ which satisfies

$$\tau_\sigma = \mu e^{-3\tau} \sin\vartheta \quad \text{on } S_1. \tag{36}$$

With regard to the final operator equation (38) below, let us now consider the boundary condition

$$\tau_\sigma = f \quad \text{on } S_1 \tag{37}$$

and formulate the solution of the corresponding boundary-value problem in functional-analytical terms. For this, we let Y be the B-space of all C^2-functions

$$f: S_1 \to \mathbb{R}$$

which are odd with respect to σ, i.e., $f(-\sigma) = -f(\sigma)$.

Lemma 71.3. *For each $f \in Y$ there exists a unique $(\vartheta, \tau) \in X$ which satisfies (37). If we set*

$$(\vartheta, \tau) = Rf,$$

then the operator $R: Y \to X$ is linear and compact.

This functional-analytical result contains the solution of the following classical boundary-value problem:

$$\rho\vartheta_\rho = \tau_\sigma, \qquad \rho\tau_\rho = -\vartheta_\sigma \quad \text{on } A_q, \tag{37*}$$

$$\tau_\sigma = f \quad \text{on } S_1,$$

$$\vartheta = 0 \quad \text{on } S_q,$$

$$\vartheta(\rho, -\sigma) = -\vartheta(\rho, \sigma), \qquad \tau(\rho, -\sigma) = \tau(\rho, \sigma) \quad \text{on } \bar{A}_q,$$

$$\int_{-\pi}^{\pi} \tau(q, \sigma)\, d\sigma = 0.$$

Let $f \in Y$ be given.

PROOF. Integration of equation (37) with respect to σ yields the boundary values of τ on S_1, i.e.,

$$\tau(1, \sigma) = \tau(1, -\pi) + \int_{-\pi}^{\sigma} f(\sigma)\, d\sigma.$$

Because of $f(-\sigma) = -f(\sigma)$ we obtain $\tau(1, \pi) = \tau(1, -\pi)$, i.e., for given $\tau(1, -\pi)$ the function τ is uniquely determined on S_1.

We then solve the boundary-value problem (37*), where the condition "$\tau_\sigma = f$ on S_1" is replaced with

$$\tau(1, \sigma) = a + \int_{-\pi}^{\sigma} f(\sigma)\, d\sigma \quad \text{on } S_1.$$

According to Problems 71.1 and 71.2, this classical boundary-value problem has a unique solution. Notice that the normalization condition, i.e., the last line in (37*), uniquely determines the constant a.

The *Schauder estimates* (46) of Problem 71.1 imply that the solution operator

$$\tau(1, \sigma) \mapsto (\vartheta, \tau)$$

is a continuous linear operator

from $\quad C^{2,\alpha}(S_1) \quad$ to $\quad C^{2,\alpha}(\bar{A}_q) \times C^{2,\alpha}(\bar{A}_q)$

for all $0 < \alpha < 1$. Thus it is a compact linear operator

from $\quad C^3(S_1) \quad$ to $\quad C^2(\bar{A}_q) \times C^2(\bar{A}_q),$

since the embeddings $C^3(S_1) \subseteq C^{2,\alpha}(S_1)$ and $C^{2,\alpha}(\bar{A}_q) \subseteq C^2(\bar{A}_q)$ are compact. Consequently, the solution operator

$$f(\sigma) \mapsto (\vartheta, \tau)$$

is a compact linear operator

from $\quad C^2(S_1) \quad$ to $\quad C^2(\bar{A}_q) \times C^2(\bar{A}_q),$

i.e., the operator $R: Y \to X$ is compact. □

71.4. Existence and Uniqueness of the Bifurcation Branch

Using Lemma 71.3, the Boundary-Value Problem 71.2 can be reduced to the following equivalent problem.

Operator Equation 71.4. We are looking for $(\vartheta, \tau) \in X$ and $\mu \in \mathbb{R}$ such that

$$(\vartheta, \tau) = \mu R(e^{-3\tau} \sin \vartheta) \tag{38}$$

is satisfied.

On $\mathbb{R} \times X$, this equation has the trivial solution $(\mu, 0)$, which corresponds to a parallel flow. We set

$$\mu_m = m(1 + q^{2m})(1 - q^{2m})^{-1}.$$

Theorem 71.B. *A point $(\mu, 0)$ is a bifurcation point of the operator equation (38) in $\mathbb{R} \times X$ if and only if*

$$\mu = \mu_m, \qquad m = 1, 2, \ldots.$$

In a sufficiently small neighborhood of such a point in $\mathbb{R} \times X$, all bifurcation solutions lie on a curve which can be analytically parametrized by a small parameter s.

Corollary 71.5. *On S_1 the bifurcation solution satisfies*

$$\vartheta = s \sin m\sigma + O(s^2), \qquad s \to 0,$$
$$\mu = \mu_m + O(s^2).$$

In Section 71.6 we shall describe a method which allows us to explicitly compute the solution up to any given order of s. For small $|s|$, the series for μ, ϑ, and τ then converge in the space $\mathbb{R} \times X$. In particular, ϑ and τ converge uniformly on the closed circular ring $\bar{A}_q$, and the bifurcation branches have the form depicted in Figure 71.5.

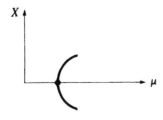

Figure 71.5

71.5. Proof of Theorem 71.B

We apply Theorem 8.A. The key step in the proof is the verification of the bifurcation condition of Theorem 8.A. The main tools we use are Fourier series.

Let $f: S_1 \to \mathbb{R}$ be a continuous function. For $k = 1, 2, \ldots$, we denote the Fourier coefficients of f by

$$a_k(f) = \frac{1}{\pi} \int_{-\pi}^{\pi} f(\sigma) \cos k\sigma \, d\sigma,$$

$$b_k(f) = \frac{1}{\pi} \int_{-\pi}^{\pi} f(\sigma) \sin k\sigma \, d\sigma.$$

Lemma 71.6. *If $f: S_1 \to \mathbb{R}$ is C^n, $n \geq 2$, then*

$$|a_k| + |b_k| \leq \mathrm{const} \cdot k^{-n}, \qquad k = 1, 2, \ldots.$$

The simple proof will be given in Problem 71.3.

Lemma 71.7. *For every $(\vartheta, \tau) \in X$ there exists the following Fourier expansion on $\bar{A}_q$:*

$$\vartheta(\rho, \sigma) = \sum_{k=1}^{\infty} b_k \frac{\rho^k - q^{2k}\rho^{-k}}{1 - q^{2k}} \sin k\sigma,$$

$$\tau(\rho, \sigma) = -\sum_{k=1}^{\infty} b_k \frac{\rho^k + q^{2k}\rho^{-k}}{1 - q^{2k}} \cos k\sigma, \qquad (39)$$

where $b_k = b_k(\vartheta(1, \sigma))$.

PROOF. Since the function $\omega = \vartheta + i\tau$ is holomorphic on the circular ring A_q, we make the ansatz

$$\omega = \sum_k \alpha_k \zeta^k + \beta_k \zeta^{-k}.$$

By using the classical root convergence criterium for series, we immediately observe that the series (39) on A_q are arbitrarily often differentiable and satisfy the Cauchy–Riemann differential equations (32). On S_1 and S_q the convergence of (39) follows from Lemma 71.6. Because of

$$\int_{-\pi}^{\pi} \tau(q, \sigma) \, d\sigma = 0,$$

we obtain from Cauchy's integral theorem that

$$\oint \omega \, d\zeta = 0$$

71.5. Proof of Theorem 71.B

for all circles $|\zeta| = r$ with $q \leq r \leq 1$. Hence no constant term appears in the expansion of τ. □

Lemma 71.8. *Let* $m = 1, 2, \ldots$ *be fixed. If* $f: S_1 \to \mathbb{R}$ *is* C^2 *and odd with respect to* σ, *then there exists an element* $(\vartheta, \tau) \in X$ *with*

$$\tau_\sigma = \mu_m \vartheta + f \quad \text{on } S_1, \tag{40}$$

if and only if the solvability condition

$$b_m(f) = 0 \tag{40*}$$

is satisfied. We find that

$$b_k(\vartheta(1, \sigma)) = \frac{b_k(f)}{\mu_k - \mu_m} \quad \text{for all} \quad k \neq m, \tag{41}$$

and $b_m(\vartheta(1, \sigma))$ *is arbitrary. Moreover,* ϑ *and* τ *on* $\bar{A}_q$ *are obtained from* (39).

PROOF. Formulas (39) and (40) imply that

$$\mu_k b_k = \mu_m b_k + b_k(f). \qquad \square$$

We are now ready to prove Theorem 71.B. Condition (40*) will play a key role. Set $w = (\vartheta, \tau)$ and write the fundamental operator equation (38) in the form

$$F(\mu, w) = 0, \quad w \in X. \tag{42}$$

Because of the linearization

$$e^{-3\tau} \sin \vartheta = \vartheta + \cdots,$$

the partial F-derivative $F_w(\mu, 0) \colon X \to X$ is equal to

$$F_w(\mu, 0)w = w - \mu R\vartheta \quad \text{for all} \quad w \in X.$$

The operator $F_w(\mu, 0)$ is therefore a compact perturbation of the identity and hence a Fredholm operator of index zero.

The eigenvalue equation

$$F_w(\mu, 0)w = 0, \quad \mu \in \mathbb{R}, \quad w \in X \tag{42*}$$

corresponds to (40) with $f = 0$ and μ in place of μ_m. According to Lemma 71.8, the eigensolutions of (42*) correspond to

$$\mu = \mu_m, \quad b_m = \text{arbitrary}, \quad b_k = 0 \quad \text{for all} \quad k \neq m,$$

and according to (39) we have

$$\vartheta_m = \frac{\rho^m - q^{2m}\rho^{-m}}{1 - q^{2m}} \sin m\sigma,$$

$$\tau_m = \frac{\rho^m + q^{2m}\rho^{-m}}{q^{2m} - 1} \cos m\sigma.$$

Thus the linearized eigenvalue problem (42*) has the eigensolutions

$$\mu = \mu_m, \qquad w_m = (\vartheta_m, \tau_m), \qquad m = 1, 2, \ldots.$$

We now verify the decisive *bifurcation condition* of Theorem 8.A. We set $\mu = \mu_m + \varepsilon$, and have to show that

$$F_w(\mu_m, 0)w = -F_{\varepsilon w}(\mu_m, 0)w_m, \qquad w \in X \qquad (42^{**})$$

has *no* solution. This equation is equivalent to $(\vartheta, \tau) \in X$ and

$$\tau_\sigma = \mu_m \vartheta + \sin m\sigma \quad \text{on } S_1.$$

Lemma 71.8 implies that this latter equation has *no* solution.
Theorem 71.B and Corollary 71.5 thus follow directly from Theorem 8.A. □

71.6. Explicit Construction of the Solution

We want to show that the method described in Corollary 8.25 can be applied to our present case, i.e., we want to explicitly compute the solution. We make the following ansatz

$$\begin{aligned} \vartheta &= s\vartheta_m + s^2\vartheta_{m,2} + s^3\vartheta_{m,3} + \cdots, \\ \tau &= s\tau_m + s^2\tau_{m,2} + s^3\tau_{m,3} + \cdots, \\ \mu &= \mu_m + s\varepsilon_1 + s^2\varepsilon_2 + \cdots, \end{aligned} \qquad (43)$$

where s is a small real parameter. According to Theorem 8.A we have

$$b_m(\vartheta) = s \quad \text{on } S_1.$$

This yields the *first key relation*

$$b_m(\vartheta_{m,k}) = 0 \quad \text{on } S_1 \qquad \text{for all } k. \qquad (43^*)$$

Note that $b_m(\vartheta_m) = 1$.
Our goal is to successively compute

$$\vartheta_{m,k}, \tau_{m,k} \quad \text{and} \quad \varepsilon_{k-1} \quad \text{for } k = 2, 3, \ldots,$$

by using the solvability condition (40*).

(i) The functions ϑ_m and τ_m are already known. Suppose we know $\vartheta_{m,r}, \tau_{m,r}$, and $\varepsilon_{m,r-1}$ for all r with $2 \le r \le k-1$.
(ii) We insert (43) into the Boundary-Value Problem 71.2, and comparing the terms with s^k we find that

$$\frac{d\tau_{m,k}}{d\sigma} = \mu_m \vartheta_{m,k} + \varepsilon_{k-1}\vartheta_m + f_k \quad \text{on } S_1,$$

where f_k depends only on already computed quantities. The solvability condition (40*) of Lemma 71.8 then yields the *second key relation*

$$b_m(\varepsilon_{k-1}\vartheta_m + f_k) = 0 \quad \text{on } S_1,$$

which shows that

$$\varepsilon_{k-1} = -b_m(f_k).$$

We then compute $\vartheta_{m,k}$ and $\tau_{m,k}$ from Lemma 71.8, and observe that these functions are uniquely determined by condition (43*).

Since products of sine and cosine functions can always be transformed into appropriate sums, it follows that f_k is a finite Fourier series. Hence it follows from Lemma 71.8 that $\vartheta_{m,k}$ and $\tau_{m,k}$ can explicitly be computed. The functions correspond to finite sums of type (39).

Using the back transformation (29) from the circular ring, cut along the negative real axis, to the flow plane, we obtain the basic formulas (i)–(vi) from the introduction of this chapter.

Recall that the operator equation depends on A_q, i.e., the solution depends on the parameter

$$q = e^{-2\pi h/\lambda}.$$

We therefore can prescribe the wave length λ and the average channel depth h.

Remark 71.9. Note that $h_0 = h + O(s^2)$, $s \to 0$. Hence we obtain (i) and (iii) from the introduction of this chapter by applying a small translation in the direction of the y-axis.

PROBLEMS

71.1.* *First mixed boundary-value problem for analytic functions on the circular ring.* We use the notations of Section 71.3 and consider the boundary-value problem for the Cauchy–Riemann differential equations

$$\rho\vartheta_\rho = \tau_\sigma, \quad \rho\tau_\rho = -\vartheta_\sigma \quad \text{on } A_q,$$

$$\tau = f \quad \text{on } S_1, \tag{44}$$

$$\vartheta = g \quad \text{on } S_q,$$

where $0 < q < 1$ (see Fig. 71.4). For fixed $n = 0, 1, \ldots$ and $0 < \alpha < 1$ we assume that the functions

$$f: S_1 \to \mathbb{R} \quad \text{and} \quad g: S_q \to \mathbb{R} \quad \text{are } C^{n,\alpha}.$$

Prove that (44) has a unique classical solution (ϑ, τ). Moreover, this solution satisfies

$$\|\tau(q,\sigma)\|_{n,\alpha} + \|\vartheta(1,\sigma)\|_{n,\alpha} \leq c(\|f\|_{n,\alpha} + \|g\|_{n,\alpha}). \tag{45}$$

Hint: Use Fourier series to construct integral representations and apply the classical theorem of Fatou–Privalov about singular integrals. See Zeidler (1968, M), p. 63.

If $n = 2$, then similarly as in Section 6.3, one obtains the Schauder estimates

$$\|\tau\| + \|\vartheta\| \leq c_1(\|f\|_{2,\alpha} + \|g\|_{2,\alpha}), \tag{46}$$

where $\|\cdot\|$ denotes the norm in $C^{2,\alpha}(\bar{A}_q)$.

71.2. *Symmetry properties of the solutions.* Prove the following result for the equations (44): If f is even and g is odd with respect to σ, then the solution τ is even and ϑ is odd with respect to σ on $\bar{A}_q$.

Solution: Set $\tau_1(\rho,\sigma) = \tau(\rho,-\sigma)$ and $\vartheta_1(\rho,\sigma) = -\vartheta(\rho,-\sigma)$, and observe that in addition to ϑ, τ, also ϑ_1, τ_1 solves (44). Uniqueness of solutions of (44) then implies that $\tau = \tau_1$ and $\vartheta = \vartheta_1$.

71.3. *Proof of Lemma 71.6.*

Solution: Let $f \in C^1(S_1)$. Integration by parts yields

$$\pi a_k = \int_{-\pi}^{\pi} f(\sigma) \cos k\sigma \, d\sigma$$

$$= -\frac{1}{k} \int_{-\pi}^{\pi} f'(\sigma) \sin k\sigma \, d\sigma,$$

which implies

$$|a_k| \leq \frac{2}{k} \max_{\sigma \in S_1} |f'(\sigma)|.$$

A repeated application of integration by parts then yields

$$|a_k| \leq \frac{\text{const}}{k^n} \max_{\sigma \in S_1} |f^{(n)}(\sigma)|$$

for $f \in C^n(S_1)$ and $n \geq 2$.

71.4. *Explicit computation of the solution.* Use the method of Section 71.6 to compute the solution (i)–(vi) from the introduction of this chapter up to order 2.

Hint: See Zeidler (1968, M), p. 104.

71.5. *Channels with infinite depth.* Show that by making this assumption one obtains a problem analogous to the Boundary-Value Problem 71.2 on the unit circle and solve it.

Hint: See Zeidler (1968, M), p. 89.

71.5.* *Sinusoidal waves, cnoidal, and solitary waves.* During the nineteenth century the following types of permanent water waves were discovered in channels of finite depth h by using first-order approximations. Rigorous existence proofs appeared in this century. Let s and a denote small parameters.

(a) Sinusoidal waves (Airy, 1845)

$$y_0(x) = \frac{\lambda}{2\pi} s \cdot \sin \frac{2\pi x}{\lambda}, \qquad c^2 = \frac{g\lambda}{2\pi} \tanh \frac{2\pi h}{\lambda}.$$

(b) Cnoidal waves (Korteweg and de Vries, 1895)

$$y_0(x) = 3a^2hk^2 \cdot cn^2\left(\frac{3ax}{2h}, k\right), \qquad 0 < k < 1,$$

$$c^2 = gh, \qquad \lambda = \frac{4K(k)}{3a}h,$$

where "cn" denotes the Jacobian elliptic function "amplitude cosine".

(c) Solitary waves (Boussinesq, 1871, Rayleigh, 1876)

$$y_0(x) = 3a^2h \cdot \operatorname{sech}^2\left(\frac{3ax}{2h}\right) \qquad \text{(see Fig. 71.6)},$$

$$c^2 = gh, \qquad \lambda = \infty.$$

(d) Sinusoidal waves under the influence of surface tension β (Kelvin, 1871)

$$y_0(x) = \frac{\lambda}{2\pi} s \cdot \sin\frac{2\pi x}{\lambda}, \qquad c^2 = \left(\frac{g\lambda}{2\pi} + \frac{2\pi\beta}{\lambda\rho_0}\right)\tanh\frac{2\pi h}{\lambda}.$$

(e) Cnoidal waves under the influence of surface tension β (Korteweg and de Vries, 1895)

$$y_0(x) = 3a^2h\left(1 - \frac{3\beta}{g\rho_0 h^2}\right)k^2 \cdot cn^2\left(\frac{3ax}{2h}, k\right),$$

$$c^2 = gh, \qquad \lambda = \frac{4K(k)}{3a}h, \qquad 0 < k < 1.$$

A detailed discussion may be found in Zeidler (1968, M), (1971), where also rigorous existence proofs are given which clarify the following relation between sinusoidal waves, cnoidal waves, and solitary waves.

(i) Using conformal mappings we obtain a nonlinear boundary-value problem for the circular ring

$$A_q = \{\zeta \in \mathbb{C}: q < |\zeta| < 1\}, \qquad q = e^{-2\pi h/\lambda}.$$

This boundary-value problem can be reduced to an operator equation

$$F(x, p, q) = 0 \tag{47}$$

for functions on the unit circle $\{\zeta \in \mathbb{C}: |\zeta| = 1\}$, where p is an eigenvalue parameter.

(ii) We fix q and arrive, as in Section 71.5, at a bifurcation problem. The first-order approximation of (47) corresponds to a linear equation which

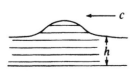

Figure 71.6

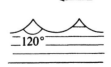

Figure 71.7

has the solution (a). The main theorem of analytic bifurcation theory (Theorem 8.A) then yields a bifurcation branch, parametrized by the small parameter s. This leads to sinusoidal waves of finite depth h.

(iii) Passing to the limit $q \to 0$, i.e., $h \to \infty$ we obtain sinusoidal waves of infinite depth, for which

$$c^2 = \frac{g\lambda}{2\pi}$$

is valid.

(iv) Now comes the important point. Consider the limit $q \to 1$, i.e., $h/\lambda \to 0$. A nontrivial asymptotic expansion of (47) for small $(1 - q)$ shows that the first-order approximation of (47) with $p = (1 - q)^{-1}$ leads to a *nonlinear* equation which has the solution (b) or (e). The small parameter a in (b) and (e) is related to $(1 - q)$.

The existence of cnoidal waves therefore corresponds to a singular bifurcation phenomenon.

(v) The limit $k \to 1$ with $K(k) \to +\infty$ yields solitary waves.

In summarizing, we find that cnoidal waves correspond to

$$h/\lambda \to 0,$$

where the channel depth h is small compared to the wave length λ (shallow water theory). Letting

$$\lambda \to +\infty$$

we arrive at solitary waves.

A direct existence proof for solitary waves may be found in Friedrichs and Hyers (1954) and Amick and Toland (1981).

71.6.* *Sinusoidal waves and solitary waves of maximal height.* Around 1880, Stokes conjectured that such waves have a sharp crest with an angle of 120° (Fig. 71.7). This conjecture was proved by Amick, Fraenkel, and Toland (1982), who used methods of global bifurcation theory.

71.7. *Coupled unharmonic oscillators, the Korteweg–de Vries equation, and solitons.* We study here the most elementary nonlinear oscillating system, and show how it leads to the Korteweg–de Vries equation (KdV equation). This explains why this equation appears in many branches of physics. Originally, Korteweg and de Vries derived their equation in 1895 in order to approximately describe long water waves in channels with small depth (shallow water theory). They found cnoidal waves and solitary waves. We emphasize, however, that the KdV equation is only a nonlinear first-order *approximation* for the many nonlinear models in physics.

We first consider a nonlinear spring which has the position coordinate

$$\xi(t) = h + \eta(t)$$

at time t, where $\eta(0) = 0$, and $K(\eta)$ is the spring force (Fig. 71.8). A Taylor expansion yields

$$K = -k(\eta + a\eta^2) + O(\eta^3), \qquad \eta \to 0, \qquad (48)$$

Problems

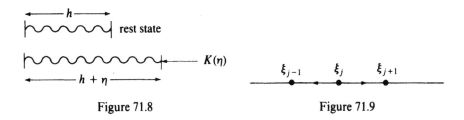

Figure 71.8   Figure 71.9

and the equation of motion is

$$m\ddot{\eta} = -k(\eta + a\eta^2). \tag{49}$$

If $a = 0$, then the general solution of (49) has the form

$$\eta = C \sin \omega t, \quad \omega = \sqrt{k/m}.$$

Next we consider an infinite number of points with identical mass m and position coordinates

$$\xi_j(t) = jh + \eta_j(t), \quad j = 0, \pm 1, \pm 2, \ldots$$

at time t, where $\eta_j(0) = 0$ for all j (Fig. 71.9). Suppose these points are connected to their closest neighbors via nonlinear springs under the force law (48). We then obtain the equation of motion

$$m\ddot{\eta}_j = K_{j-1} - K_{j+1}, \quad j = 0, \pm 1, \pm 2, \ldots \tag{50}$$

with forces

$$K_{j-1} = -k(\eta_j - \eta_{j-1}) - ka(\eta_j - \eta_{j-1})^2,$$
$$K_{j+1} = -k(\eta_{j+1} - \eta_j) - ka(\eta_{j+1} - \eta_j)^2.$$

In 1955, Fermi, Pasta, and Ulam used numerical experiments in the study of equation (50). To their great surprise they did not observe a trend towards equipartition of energy among the possible degrees of freedom at a given time.

Let us now consider a continuous version of (50), where

$$\zeta(t) = \xi + \eta(\xi, t)$$

is the position of the point $\xi \in \mathbb{R}$ at time t with $\zeta(0) = \xi$. We set

$$\eta(jh, t) = \eta_j(t),$$

and apply a Taylor expansion to (50). It follows that

$$\eta_{tt} = \omega^2 h^2 \left(\eta_{\xi\xi} + 2ah\eta_\xi \eta_{\xi\xi} + \frac{h^2}{12} \eta_{\xi\xi\xi\xi} \right) + O(h^5), \quad h \to 0,$$

where $\omega^2 = k/m$. We are looking for traveling waves and make the ansatz

$$\eta(\xi, t) = U(\xi - h\omega t, t).$$

Setting $u = U_\xi$, we find that

$$u_t + \omega h \left(2ahuu_\xi + \frac{h^2}{12} u_{\xi\xi\xi} \right) = 0.$$

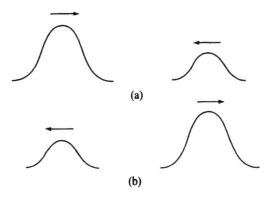

Figure 71.10

Finally, we obtain the classical Korteweg–de Vries equation (KdV equation)

$$u_t + 6uu_\zeta + u_{\zeta\zeta\zeta} = 0, \tag{51}$$

by rescaling t, ζ, and u. In 1895, Korteweg and de Vries discovered that equation (51) has a family of solutions of the form of solitary waves

$$u = 2k^2 \operatorname{sech}^2 k(\zeta - 4k^2 t - \zeta_0)$$

with velocity $c = 4k^2$ and amplitude $2k^2$ (Fig. 71.6), where k and ζ_0 are constants. Such solitary waves are also refered to as solitons.

Let us summarize some important modern results on the KdV equation.

(i) *Stability of solitons.* In Figure 71.10 two solitons interact. Kruskal and Zabusky, in 1963, made the remarkable discovery that this interaction does not effect the shape and velocity of the solitons. This stability behavior of solitons was of great interest to physicists, because solitons behave like elementary particles.

(ii) *Inverse scattering theory and the solution of the initial-value problem for the KdV equation.* In 1967, Gardner, Greene, Kruskal, and Miura discovered a new and important method in mathematical physics to explicitly solve nonlinear evolution equations. We explain here the basic idea.

Suppose u is a solution of the *nonlinear* Korteweg–de Vries equation (51). Then the scattering data of the *linear* Schrödinger equation

$$-\psi_{\zeta\zeta} + u\psi = \lambda\psi \tag{52}$$

are known. Conversely, if the initial values $u_0(\zeta)$ of equation (51) are given, then the scattering data of (52) are known for *all* times t. Using inverse scattering theory, one determines the potential $u = u(\zeta, t)$ of (52) from the known scattering data. This potential u is the solution of the initial-value problem for (51) (see Problem 30.7).

(iii) *Lax pairs.* The basic connection between the Korteweg–de Vries equation (51) and the Schrödinger equation (52) was discovered by Lax (1968) in terms of functional analysis.

Introducing the two operators

$$L(t)v(\xi) = (-D^2 + u(\xi,t))v(\xi), \qquad D = \partial/\partial\xi,$$

$$A(t)v(\xi) = (4D^3 - 3(uD + Du))v(\xi),$$

equations (51) and (52) can be written as

$$u_t = LA - AL \qquad (51^*)$$

and

$$L(t)\psi = \lambda(t)\psi, \qquad (52^*)$$

respectively, where $A^* = -A$ for a suitable specification of the domain of definition $D(A) \subseteq L_2(-\infty, \infty)$. Hence, the operator

$$U(t) = e^{B(t)} \quad \text{with} \quad B(t) = \int_0^t A(s)\,ds$$

is unitary for all t.

Let u be a solution of (51), i.e., u satisfies (51*). Then the relation

$$\frac{d}{dt}U(t)L(t)U(t)^{-1} = e^B(AL - LA + L')e^{-B} = 0$$

is valid because $L' = u_t$ and (51*). This leads to the key observation that

$$U(t)L(t)U(t)^{-1} = L(0) \qquad \text{for all } t.$$

Consequently, the eigenvalues in (52*) do not depend on t. This implies the following two facts.

(a) The eigenvalues in (52*) depend only on the initial values $u(x, 0)$. The inverse scattering method in (ii) above is closely related to this observation.

(b) Since the λ's are time-independent, they are conserved quantities for the solutions of (51). Thus the KdV equation has an infinite number of conserved quantities.

We call (L, A) a Lax pair.

(iv) *The KdV equation as a completely integrable infinite-dimensional Hamiltonian system*. This was discussed in Problem 40.9. Physicists like to reduce their problems to Hamiltonian systems in order to get a maximal insight. In this regard the KdV equation has favorable properties.

For details, one may consult Novikov et al. (1980, M). Moreover, we recommend Lax (1968), Lamb (1980, M), Bullough and Caudrey (1980, P) (applications to physics), Ablowitz and Segur (1981, M), Calogero and Degasperis (1982, M), Rajaraman (1982, M) (solitons, instantons, and elementary particles), Faddeev and Tahktadjan (1986, M) (infinite-dimensional Hamiltonian systems, solitons, the Riemann–Hilbert problem, and inverse scattering theory), Knörrer (1986, S) (solitons and algebraic geometry). A collection of important papers on solitons and instantons can be found in Rebbi (1984).

71.8. *The dead water problem, cavities, and free boundaries.*

71.8a. Physical problem. We consider the situation of Figure 71.11. Suppose that a fixed obstacle (arc) AB is immersed in a planar stationary irrotational flow of an inviscid, incompressible fluid without outer forces. Behind the obstacle there occurs a finite or infinite cavity. If the obstacle AB is regarded as a "ship," then, behind the obstacle, there occurs a so-called wake or dead water region. The problem is to compute the flow outside the cavity and the boundary of the cavity which corresponds to the unknown "free streamline Γ."

Formulate the corresponding mathematical problem.

Solution: As in Section 71.2 we use the complex flow potential

$$W = \varphi + i\psi,$$

where the velocity field is given by

$$v = \text{grad } \varphi.$$

Then the problem is the following:

$$\Delta \psi = 0 \quad \text{in the flow region}, \tag{53}$$

$$\psi = \text{const} \quad \text{along } AB \text{ and } \Gamma, \tag{54}$$

$$|\text{grad } \psi| = \text{const} \quad \text{along } \Gamma. \tag{55}$$

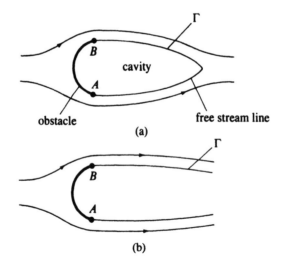

Figure 71.11

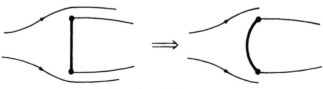

deformation of the obstacle

Figure 71.12

Equation (54) tells us that the obstacle and Γ correspond to a streamline. Moreover, equation (55) is a consequence of Bernoulli's law

$$\tfrac{1}{2}\rho v^2 + p = \text{const} \tag{56}$$

in the flow region and $p = \text{const}$ along Γ. Observe that $\varphi_x = \psi_y$, $\varphi_y = -\psi_x$. If we have a solution ψ of (53)–(55), then equation (56) yields the pressure in the flow region.

71.8b.* *Leray–Schauder mapping degree and existence theorems.* Explicit solutions for special obstacles were obtained by Helmholtz and Kirchhoff around 1870 (see Jacob (1959, M)). The first general topological existence proof via mapping degree was given in a famous paper by Leray (1935). We recommend for study Serrin (1952). The main idea of this paper is as follows.
(i) Using a conformal mapping, the original problem is reduced to a nonlinear integro-differential equation which is due to Villat (1911).
(ii) This equation is solved by applying the mapping degree. In order to compute the mapping degree, which is equal to one, the given obstacle AB is continuously deformed into a straight segment (Fig. 71.12).

71.8c.* *Variational approach, the decisive trick of indicator functions, and existence theorems.* The method of conformal mappings can only be applied to planar problems. The advantage of the modern variational approach to free boundary problems is that one can also treat three-dimensional problems. Study Friedman (1982, M), Chapter 3, in this connection, where axially symmetric jets and cavities are considered (Fig. 71.13). The basic trick is contained in the following result:

Let G be a region in $\mathbb{R}^N$ with a sufficiently smooth boundary. Let $\partial_1 G$ be a nonempty open subset of ∂G. We are given a function ψ_0 with

$$\psi_0 \geq 0 \quad \text{on } G, \qquad \psi_0 \in L_1(G), \qquad \text{grad } \psi_0 \in L_2(G),$$

and consider the variational problem

$$\int_G |\text{grad } \psi|^2 + \chi_{\{\psi > 0\}} \, dx = \min!, \tag{57}$$

$$\psi = \psi_0 \quad \text{on } \partial_1 G, \qquad \psi \in L_1(G), \qquad \text{grad } \psi \in L_2(G).$$

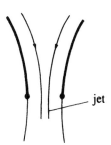

Figure 71.13

Here we set

$$\chi_{\{\psi>0\}}(x) = \begin{cases} 0 & \text{if } \psi(x) > 0, \\ +\infty & \text{if } \psi(x) \leq 0, \end{cases}$$

i.e., $\chi_{\{\psi>0\}}$ is the indicator function of the set $\{\psi > 0\} = \{x \in G: \psi(x) > 0\}$.
Show: If ψ is a local minimum of (57), then

$$\Delta \psi = 0 \quad \text{on } \partial\{\psi > 0\},$$

$$\lim_{\varepsilon \to +0} \int_{\partial\{\psi > \varepsilon\}} (|\operatorname{grad} \psi|^2 - 1) hn \, dO = 0$$

for all $h \in C_0^\infty(G)^N$, where n denotes the outer unit normal vector.

Hence, if ψ and the free boundary $\partial\{\psi > 0\}$ are sufficiently smooth, then we obtain the key condition

$$|\operatorname{grad} \psi| = 1 \quad \text{on } \partial\{\psi > 0\}. \tag{58}$$

The important point is that the appearence of the indicator function in the variational problem (57) guarantees the free boundary condition (58) at least in a generalized sense.

71.9. *Flow around a body.*

71.9a. *Planar flow and the Kutta–Jukovski formula.* We consider a simply connected region G in $\mathbb{R}^2$ with the boundary curve $\partial G: z = z(t)$ oriented as in Figure 71.14. We regard G as a rigid body (e.g., a ship). Moreover, we consider a planar flow in the exterior of the body $\mathbb{R}^2 - \bar{G}$ which satisfies the following properties.

(i) The flow is irrotational, incompressible, and inviscid with density ρ_0 and velocity vector $v = ae_1 + be_2$.
(ii) The flow approaches the velocity $-c_0 e_1$ at infinity.
(iii) The circulation of the flow around the body is Γ, i.e.,

$$\int_{\partial G} a \, dx + b \, dy = \Gamma. \tag{59}$$

(iv) There are no outer forces.

Let F be the force acting onto the body. Prove:

$$F = \rho_0 c_0 \Gamma e_2. \tag{60}$$

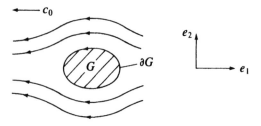

Figure 71.14

This formula contains the so-called d'Alembert paradox: the body experiences no drag, i.e., no force in the direction of $-e_1$. Think, for example, of a ship G, which moves in a fluid resting at infinity. Let $c_0 e_1$ be the velocity of the ship. We expect that there is a friction force acting on the ship in the direction of $-e_1$. This force is caused by the viscosity. If we change the system of reference, then we obtain the situation of Figure 71.14. Hence, the d'Alembert paradox appears because we assume that the fluid is inviscid.

Solution: We identify vectors $\alpha e_1 + \beta e_2$ with points $\alpha + i\beta$ in the complex plane. Let $z = x + iy$. From Theorem 71.A we have the following problem.

(a) We are looking for a complex velocity function $V = a - ib$ which is holomorphic in $\mathbb{R}^2 - \bar{G}$ and $V(\infty) = -c_0$.
(b) We have the boundary condition

$$\operatorname{Im} V(z(t))z'(t) = 0 \quad \text{along } \partial G. \tag{61}$$

This condition means that the velocity vector is tangential along the surface of the rigid body.

If we know V, then the pressure p can be computed from

$$\tfrac{1}{2}\rho_0 |V|^2 + p = B_0 \quad \text{in } \mathbb{R}^2 - \bar{G}, \tag{62}$$

where B_0 is a constant. Since $V(\infty) = -c_0$, the function V has the Laurent expansion

$$V(z) = -c_0 + \frac{c_1}{z} + \frac{c_2}{z^2} + \cdots$$

at infinity. Condition (61) implies

$$\Gamma = \int_{\partial G} V\, dz = \oint V\, dz = 2\pi i c_1 \tag{63}$$

and hence

$$c_1 = \Gamma/2\pi i.$$

The force exerted on the body is equal to

$$F = -\int_{\partial G} pn\, ds,$$

where n is the outer unit normal vector to ∂G. This yields

$$F = i \int_{\partial G} pz'(t)\, dt = i \int_{\partial G} \left(B_0 - \frac{\rho_0}{2}(a^2 + b^2)\right) dz.$$

Because of $\int_{\partial G} dz = 0$ and since (61) implies $b\, dx = a\, dy$, we obtain

$$\bar{F} = \frac{i\rho_0}{2} \int_{\partial G} V^2\, dz.$$

Since

$$V^2 = -\frac{2c_0 c_1}{z} + c_0^2 + O\left(\frac{1}{z^2}\right),$$

we have
$$\bar{F} = 2\pi c_0 c_1 \rho_0 = -ic_0 \Gamma \rho_0,$$
and this is (60).

71.9b. Boundary-value problem. Reduce the problem (i)–(iv) above to a boundary-value problem for the Laplace equation.

Solution: Let $0 \in G$ and set
$$V = -c_0 + \frac{\Gamma}{2\pi i z} + V_1.$$

Equation (63) implies that $\int_{\partial G} V_1 \, dz = 0$. Hence there exists a function $W_1 = \varphi_1 + i\psi_1$ with
$$W_1' = V_1 \quad \text{in } \mathbb{R}^2 - \bar{G},$$
where W_1 is holomorphic in $\mathbb{R}^2 - \bar{G}$. From (61) we obtain
$$\operatorname{Im}\left(W_1'(z(t))z'(t) + \frac{\Gamma}{2\pi i} \frac{z'(t)}{z(t)} - c_0 z'(t) \right) = 0.$$

This yields
$$\Delta \psi_1 = 0 \quad \text{in } \mathbb{R}^2 - \bar{G},$$
$$\psi_1 - \frac{\Gamma}{2\pi} \ln|z(t)| - c_0 y(t) = \text{const} \quad \text{on } \partial G.$$

Moreover, we have $W_1' = O(1/z^2)$ at $z = \infty$.

71.9c. Planar subsonic flow around a body and quasi-conformal mappings. Study Bers (1954).

71.9d. Three-dimensional subsonic flow around a body and Leray–Schauder theory. Study Finn and Gilbarg (1957).

71.9e. Flow around a gas bubble. Study this free boundary-value problem in Zeidler (1968, M).

71.9f. Optimal design of aircraft. Study Morawetz (1982, S).

71.10. Equilibrium forms of rotating fluids. In a Cartesian (ξ', η', ζ')-inertial system we consider a homogeneous incompressible fluid of constant temperature which rotates counterclockwise around the ζ-axis with constant angular velocity ω. Suppose the fluid is in a hydrostatic equilibrium with respect to a Cartesian (ξ, η, ζ)-system Σ which rotates together with the fluid. Moreover, assume that the fluid occupies the simply connected region Ω in Σ (Fig. 71.15). Let p_0 be the constant outer pressure on $\partial \Omega$. Determine the equation for the unknown boundary $\partial \Omega$.

Solution: The density of the outer forces in Σ is
$$K = -\rho(\omega \times (\omega \times x)) + \int_\Omega \frac{\rho G(y-x)\,dy}{|y-x|^3},$$

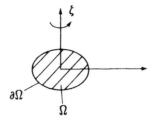

Figure 71.15

which corresponds to the centrifugal force (Section 58.6) and the gravitational force. Here ρ is the constant density and $\omega = \omega e_3$. Observe that the volume elements ΔV of the fluid are resting in Σ, and $K \Delta V$ is the corresponding force onto ΔV. Moreover, we have $K = -\rho \operatorname{grad} U$ with the potential

$$U(x) = -\frac{\omega^2}{2}(\xi^2 + \eta^2) - \int_\Omega \frac{G\, dy}{|y - x|}.$$

From (70.37) we obtain $\rho U + p = \operatorname{const}$ on $\bar{\Omega}$. Hence we arrive at the basic equation

$$\rho U(x) + p_0 = \operatorname{const} \quad \text{on } \partial\Omega. \tag{64}$$

This is the desired equation for the unknown boundary $\partial\Omega$.

Historical Remarks

The classical problem of equilibrium forms of rotating fluids (see Problem 71.10) attracted the interest of physicists and mathematicians, since it models rotating stars. At the beginning of the twentieth century, this problem played an important role in the development of bifurcation theory and the theory of nonlinear integral equations. The first explicit solutions for equation (64) were discovered by Maclaurin (1698–1746) in the form of ellipsoids. In 1834, Jacobi (1804–1851) found that, for critical angular velocities, there bifurcate families of new ellipsoids from the ellipsoids of Maclaurin. The following general problem arose.

(P) Suppose we are given a one-parameter family of equilibrium forms. Find the critical values of angular velocity for which new equilibrium forms bifurcate from the known solutions.

In 1885, Poincaré (1854–1912) studied this problem in detail. The first rigorous existence proof, however, was given by Ljapunov (1906). He thereby created the bifurcation theory for nonlinear integral equations. Many results can be found in the classical monograph by Lichtenstein (1933), who himself made important contributions to this problem.

A variational approach to rotating fluids may be found in Friedman (1982, M).

References to the Literature

Classical works: Nekrasov (1921), Levi-Civita (1925), Lavrentjev (1946), and Friedrichs and Hyers (1954) (solitary waves).
Monograph about permanent waves: Zeidler (1968, B, H).
Permanent waves: Zeidler (1977, S), (1971), (1972a, b), (1973), Beyer and Zeidler (1979), Beale (1979), Turner (1981), Amick and Toland (1981), Amick, Fraenkel, and Toland (1982) (proof of the Stokes conjecture).
The blowing-up lemma and applications of bifurcation theory to capillary-gravity waves: Zeidler (1968, M), Jones and Toland (1986).
Homoclinic bifurcation of dynamical systems and permanent waves: Kirchgässner (1988).
Survey about wave problems in physics: Whitham (1974, M, B) (standard work), Stoker (1957, M) (water waves), Lighthill (1978, M) (1986), LeBlond (1978, M) (waves in the ocean), Friedlander (1981) (geophysics), Debnath (1985, P) (nonlinear waves), Brekhovskikh (1985, M), Ghil (1987, M) (atmospheric dynamics, dynamo theory, and climate dynamics), Washington (1987, M) (climate modelling).
Dead water problems and cavities: Leray (1935) (classical work), Serrin (1952), Birkhoff and Zarantonello (1957, M), Hilbig (1964), (1982), Socolescu (1977).
Variational approach to jets and cavities: Friedman (1987, M) (recommended as an introduction), Alt and Caffarelli (1981).
Equilibrium forms of rotating fluids: Lichtenstein (1933, M) (classical monograph), Chandrasekhar (1969, M), Lebovitz (1977, S), Friedman (1982, M).
Rotating stars: Tassoul (1978, M).
Solitons: Novikov (1980, M) and Ablowitz and Segur (1981, M, H) (recommended as an introduction), Lax (1968), Bullough and Caudrey (1980, P) (applications in physics), Lamb (1980, M), Calogero and Degasperis (1982, M), Drazin (1983, M), Rajaraman (1983, M), (solitons, instantons, and elementary particles), Davydov (1984, M) (solitons in molecular systems), Faddeev and Takhtadjan (1986, M) (infinite-dimensional Hamiltonian systems, solitons, the Riemann–Hilbert problem, and inverse scattering theory), Knörrer (1986, S) (solitons and algebraic geometry).
Collection of classical articles on solitons: Rebbi (1984).
The initial-value problem for the Korteweg–de Vries equation: Bona and Smith (1975), Kato (1983).
The initial-value problem for water waves: Shinbrot (1976).

CHAPTER 72

Viscous Fluids and the Navier–Stokes Equations

For nonlinear equations, such as the Navier–Stokes equations, it is known that a regular solution for the nonstationary problem need not exist for all times $t \geq 0$. At some finite time the solution may go to infinity or loose its regularity. Even if the solution exists for all $t \geq 0$, it need not converge towards the solution of the stationary problem as $t \to +\infty$, when the boundary conditions and the forces converge towards a stationary situation.
Olga Aleksandrovna Ladyženskaja (1970)

For many years now, the Navier–Stokes equations have attracted the attention of engineers and mathematicians. The reason lies in the great number of interesting and difficult problems which are connected with them and which lead to important applications. Many of these problems are still unsolved. Beginning with the work of Jean Leray, during the 1930s, a number of deeper results were obtained for individual solutions of these equations. But from a physical point of view, the study of only individual solutions is not always justified. For large Reynolds numbers, i.e., roughly speaking, for large velocities or small viscosities, the flow becomes turbulent. Therefore, it is reasonable to look for a statistical description, analogously to the kinetic gas theory.
Mark Iosifovič Višik and Andrei Vladimirovič Fursikov (1980)

The distinguishing feature of a turbulent flow is that its velocity field appears to be random and varies unpredictably. The flow does, however, satisfy the Navier–Stokes differential equations, which are not random. This contrast is the source of much of what is interesting in turbulence theory....

If the Reynolds number Re is small, then the Navier–Stokes equations have a unique solution which can be observed in nature. When Re is large, even when a solution can be obtained, it is not observed in nature. The reason lies in the fact that the solutions are unstable; very small perturbations, too small to be measured by the experimenter, can be amplified and induce large changes in the flow. Note that this fact makes uniqueness theorems rather meaningless for a person trying to describe physical reality, since the possible uniqueness rests on an assumed uniqueness of the data, which cannot be assured in any meaningful sense....

The main problem of turbulence theory is to isolate statistical properties of solutions at the Navier–Stokes equations which are independent of the precise statistical properties of random data, if this is possible, and then use the knowledge thus acquired to construct reasonable predictive procedures for specific problems. One should keep in mind that a practical person is usually interested only in mean properties of a small number of functionals of the flow (e.g., lift and drag in the case of a flow past a wing), and these could conceivably be obtained even when the details of the flow are unknown....
Alexandre Joel Chorin (1975)

The recent improvement of our understanding of the nature of turbulence has three different roots. The first is the injection of new mathematical ideas from the theory of dynamical systems (strange attractors). The second is the availability of powerful computers which permit, among other things, experimental mathematics on dynamical systems and numerical simulation of hydrodynamic equations. The third is the improvement of experimental techniques, in particular, Doppler measurements of velocities by use of a laser beam, and then numerical Fourier analysis of the time series obtained....

Extensive computer studies of low-dimensional dynamical systems have shown that sensitive dependence on initial data is quite common, but mostly appears in systems for which we have no good mathematical theory.
David Ruelle (1983)

72.1. Basic Ideas

In this chapter we apply, step-by-step, the following functional-analytical results:

(i) Leray–Schauder principle (Theorem 6.A).
(ii) Main theorem about pseudomonotone operators (Theorem 27.A).
(iii) Main theorem about first-order evolution equations (Theorem 23.A).
(iv) Implicit function theorem (Theorem 4.B).
(v) Main theorem of analytic bifurcation theory (Theorem 8.A).

According to Section 70.3, the Navier–Stokes equations for an incompressible fluid are

$$\rho v_t + \rho (v \operatorname{grad}) v - \eta \, \Delta v = K - \operatorname{grad} p \quad \text{on } G,$$

$$\operatorname{div} v = 0 \quad \text{on } G, \qquad (1)$$

$$v = 0 \quad \text{on } \partial G,$$

where:

$v(x, t)$ velocity vector at point x and time t;
$p(x, t)$ pressure;
ρ constant density of the fluid;
$\int_G K \, dx$ outer total force acting onto the fluid;
η viscosity (a positive constant).

72.1. Basic Ideas

We choose a physical system of units such that $\rho = 1$. The fluid is located in the fixed spatial region G. More precisely, we assume:

G is a bounded region in $\mathbb{R}^3$ with sufficiently regular boundary, i.e., $\partial G \in C^{0,1}$. (2)

Let n be the outer unit normal vector to ∂G. The boundary condition "$v = 0$ on ∂G" means that the viscous fluid sticks to the boundary.

We consider equation (1) for time t in a suitable interval and add the initial condition

$$v(x,0) = v_0(x) \quad \text{on } G.$$

We are looking for velocity v and pressure p.

In the stationary case, v and p do not depend on t. We then have $v_t = 0$ in (1).

As we will show below, the solution of equation (1) can be greatly simplified by eliminating the incompressibility condition $\operatorname{div} v = 0$ through an orthogonal projection in an H-space. In this way pressure p can be eliminated as well, i.e., we obtain an operator equation for the velocity v.

72.1a. Stationary Case

The functional-analytical treatment of equation (1) leads to the operator equation

$$\eta v - Qv = f, \quad v \in V \qquad (3)$$

in the H-space V, where f depends only on the outer force K. The operator Q is strongly continuous and hence compact. Moreover, Q is homogeneous of degree two. The following relation

$$(Qv|v) = 0 \quad \text{for all } v \in V, \qquad (4)$$

is essential and follows from the incompressibility condition $\operatorname{div} v = 0$. It implies the central coerciveness relation

$$\lim_{\|v\| \to \infty} \frac{(\eta v - Qv|v)}{\|v\|} = \lim_{\|v\| \to \infty} \eta \|v\| = +\infty. \qquad (5)$$

According to Section 27.2 the operator $\eta I - Q$ is pseudomonotone because it is a strongly continuous perturbation of the operator ηI. Thus the main theorem about pseudomonotone operators (Theorem 27.A) guarantees the existence of a solution v for every $f \in V$. If this solution v is unique, then the Galerkin sequence converges towards v. If it is not unique, then, at least, a subsequence of the Galerkin sequence converges towards a solution of the original problem.

Physically, this means that, for all outer forces K, the stationary problem has a solution. We shall prove uniqueness for the case where $\|f\|/\eta^2$ is small, i.e., for small

$$\frac{\rho^2}{\eta^4} \int_G K^2 \, dx,$$

that is, the velocity v is uniquely determined, and, up to a constant, the pressure p is also uniquely determined. Clearly, more cannot be expected for p, since as with p also $p + \text{const}$ is a solution of (1).

An alternative existence proof for (3) can be obtained by using the Leray–Schauder principle (Theorem 6.A). Let us write (3) in the form

$$v = \frac{\tau}{\eta}(f + Qv), \qquad v \in V, \quad 0 \le \tau \le 1. \tag{6}$$

If v is a solution of (6), then we obtain from (4) that

$$(v|v) \le |(f|v)|/\eta \le \|f\| \|v\|/\eta,$$

which yields the crucial *a priori* estimate

$$\|v\| \le \|f\|/\eta.$$

Hence, Theorem 6.A implies that equation (3) has a solution for every $f \in V$. We thereby use the fact that Q is compact.

In the usual formulation of the generalized problem to equation (1), it follows from $\text{div } v = 0$ that, after an integration by parts, pressure p drops out, so that p does not appear in the operator equation (3). On the other hand pressure p is obtained from (3), by applying an important orthogonal decomposition of the H-space $L_2(G, V_3)$, see below. This decomposition corresponds to the classical result of Stokes (1849), that every smooth vector field on G can be written as

$$\begin{aligned} v &= \text{curl } A + \text{grad } U \quad \text{on } G, \\ n \, \text{curl } A &= 0 \quad \text{on } \partial G. \end{aligned} \tag{7}$$

In classical physics, the vector field A is called the vector potential and the function U is called the potential. Decomposition (7) greatly simplifies the integration of the Maxwell equations in electrodynamics. If we set $v_1 = \text{curl } A$ and $v_2 = \text{grad } U$, then we obtain

$$\begin{aligned} v &= v_1 + v_2 \quad \text{on } G, \\ \text{div } v_1 &= 0, \qquad \text{curl } v_2 = 0, \end{aligned} \tag{8}$$

i.e., the vector field v can be decomposed into a source free and irrotational vector field. If

$$nv_1 = 0 \quad \text{on } \partial G.$$

then the decomposition (8) is unique.

72.1b. Nonstationary Problem

The simple idea for our existence proof is to use the general trick for regular semilinear equations, which was introduced in Section 8.12. For this we write the original equation (1) in the form

$$v'(t) + Av(t) + Bv(t) = f(t) \quad \text{for} \quad t \geq 0,$$
$$v(0) = v_0, \tag{9}$$

with the corresponding linearized problem

$$v'(t) + Av(t) = f(t) \quad \text{for} \quad t \geq 0,$$
$$v(t) = v_0. \tag{10}$$

Let the unique solution of (10) be denoted by $v = S(f, v_0)$. Formula (9) then becomes

$$v = S(f - Bv, v_0). \tag{11}$$

Equation (11) can be solved for small $\|f\|$ and $\|v_0\|$ by using the implicit function theorem or the Banach fixed-point theorem.

Physically, this means that for small outer forces $|K|$ and small initial velocities $|v_0|$ there exist unique solutions for all times. We do not expect to have turbulence in this case. The monographs, cited in the References to the Literature to this chapter, contain a number of stronger existence results. Here, however, we restrict ourselves to situations which require only a minimal amount of mathematical tools. For example, for arbitrary forces K and initial velocities v_0, one can prove existence of weak generalized solutions of (1) by assuming less regularity, but without obtaining uniqueness. On the other hand, one can prove uniqueness by assuming more regularity, but without obtaining existence. This strange situation might be a consequence of turbulence phenomena.

72.1c. Approximation Methods

To engineers it is important to have effective numerical methods, which may be found in Temam (1977, M), Girault and Raviart (1981, L), (1986, M), Thomasset (1981, M), Glowinski (1984, L), and Murman (1985, P) (supercomputers). They consist, for example, of Galerkin methods with finite elements as basis functions and difference methods. A typical difficulty which arises is that finite elements are needed which satisfy

$$\text{div } v = 0 \quad \text{on } G.$$

One trick to avoid this problem is to consider div $v = 0$ as a side condition and to apply optimization methods for problems with side conditions. In this direction, algorithms of Uzawa and Arrow–Hurewicz are very practicable (see Problems 50.2 and 50.3 and Temam (1977, M), Girault and Raviart (1981, L)). Another trick to avoid div $v = 0$, is to assume that the density has a weak compressibility, i.e., a pressure dependence of the form $\rho = \rho_0 + \varepsilon p$. This gives

$$p_t + \varepsilon \operatorname{div} v = 0$$

and one can study the limiting process $\varepsilon \to 0$. This is contained in Temam (1977, M), where one can also find the method of fractional steps. The idea thereby is to discretize the time derivative and to use, at each time step, only a certain portion of the spatial part of equation (1).

72.1d. Bifurcation

As two important examples for bifurcation phenomena we consider:

(a) the Taylor problem (viscous fluid between two rotating cylinders); and
(b) the Bénard problem (heated viscous fluid between two horizontal plates).

To both problems we apply the main theorem of analytic bifurcation theory (Theorem 8.A). We obtain existence and uniqueness results for the bifurcation branch and also derive effective methods for its construction. The important point thereby is that the so-called bifurcation condition be satisfied. In order to guarantee simple zero solutions of the linearized problem, the following trick is used. We consider the corresponding operator equations in spaces of functions with suitable symmetry properties.

72.1e. Hydrodynamic Similarity

In order to write the basic equation (1) dimensionless, we set

$$v = V\bar{v}, \qquad x = R\bar{x}, \qquad t = T\bar{t}, \qquad p = P\bar{p}, \qquad K = k\bar{K}, \qquad (12)$$

with

$$k = \rho V^2 / R, \qquad P = \rho V^2,$$

where the overbarred quantities $\bar{v}, \bar{x}$, etc. are dimensionless. The quantities V, R, etc., on the other hand have respective dimensions and numerical values, which depend on the specific problem. For example, V may be chosen as the average velocity of the fluid, and R as the diameter of G. Moreover, we set

$$G = R \cdot \Omega,$$

i.e., the flow region G is obtained from the region Ω by multiplying all coordinates with R. If we set $u = \bar{v}$, then the original Navier–Stokes equations

(1) assume the following *dimensionless form*

$$\alpha u_t + (u\,\text{grad})u - \frac{1}{\text{Re}}\Delta u = \bar{K} - \text{grad}\,\bar{p} \quad \text{on } \Omega,$$

$$\text{div } u = 0 \quad \text{on } \Omega, \tag{13}$$

$$u = 0 \quad \text{on } \partial\Omega,$$

where

$$\text{Re} = \frac{\rho VR}{\eta}$$

is the crucial Reynold's number and $\alpha = R/VT$. The space derivatives in (13) are taken with respect to $\bar{x}$. Thus we find the following similarity principle:

If one knows one solution of the dimensionless equation (13), then one obtains a family of solutions of the original Navier–Stokes equations (1) from (12) which are similar in the sense of these transformation formulas.

This shows that only the dimensionless numbers Re and α are of significance to (1) and not the specific values of density ρ, viscosity η, etc.

This principle is being used by engineers in order to simulate the flight of aircraft in wind tunnels.

72.2. Notations

We will use the same notations as in Section 62.2 and set

$$X = \mathring{W}_2^1(G, V_3), \qquad Y = L_2(G, V_3)$$

equipped with the scalar products

$$(v|w)_Y = \int_G vw\,dx,$$

$$(v|w)_X = \int_G (vw + [v', w'^*])\,dx.$$

In Cartesian coordinates this gives

$$[v'(x), w'(x)^*] = D_j v^i(x) D^j w_i(x).$$

On the real H-space X we introduce the equivalent energetic scalar product

$$(v|w) = \int_G [v'(x), w'(x)^*]\,dx.$$

In order to give a functional-analytical formulation of the side condition div $v = 0$ on G and $v = 0$ on ∂G, we let

$$C_0^\infty(G, \text{div})$$

be the set of all vector fields $v: G \to V_3$, whose components belong to $C_0^\infty(G)$, and which satisfy div $v = 0$ on G. Moreover, we define

$$H = \text{closure of } C_0^\infty(G, \text{div}) \text{ in } Y,$$
$$V = \text{closure of } C_0^\infty(G, \text{div}) \text{ in } X.$$

We denote the norm on V by

$$\|v\| = (v|v)^{1/2}.$$

As for any H-space, there exists the *orthogonal decomposition*

$$Y = H \oplus H^\perp, \qquad (14)$$

where H is a closed subspace of the H-space Y.

Lemma 72.1. *Under assumption* (2) *we have that*

$$H^\perp = \{v \in Y: v = \text{grad } p \text{ for some } p \in W_2^1(G)\},$$

and, up to a constant, the function p is uniquely determined by $v \in H^\perp$.

The proof will be given in Problem 72.1. The intuitive meaning was discussed already in connection with the classical formula (7).

Throughout the following we assume that components of vectors correspond to a fixed Cartesian coordinate system.

72.3. Generalized Stationary Problem

We multiply the stationary Navier–Stokes equations (1) with $w \in C_0^\infty(G, \text{div})$ and use integration by parts to obtain

$$\eta(v|w) - b(v, v, w) = c(w) \qquad \text{for all} \quad w \in C_0^\infty(G, \text{div}), \qquad (15)$$

where

$$b(u, v, w) = \int_G uw'v \, dx,$$

$$c(w) = \int_G Kw \, dx,$$

or in component notation

$$b(u, v, w) = \int_G u_i v^j D_j w^i \, dx. \qquad (16)$$

72.3. Generalized Stationary Problem

Notice that because of $\operatorname{div} v = \operatorname{div} w = 0$, the integration by parts in the derivation of equation (15) yields

$$\int v \operatorname{grad} p \, dx = -\int p \operatorname{div} v \, dx = 0$$

and

$$\int w(v \operatorname{grad}) v \, dx = \int w^i(v^j D_j) v_i \, dx$$
$$= -\int v_i v^j D_j w^i \, dx = -b(v, v, w).$$

Definition 72.2. Suppose the region G satisfies assumption (2). The *generalized problem* to the stationary Navier–Stokes equations (1) is this.

Given the density of the outer forces $K \in L_2(G, V_3)$, we are looking for the velocity $v \in V$ of the fluid so that equation (15) is satisfied.

72.3a. Computation of the Pressure

The derivation of (15) shows that every classical solution of (1) is a generalized solution. Conversely, if v is a sufficiently smooth generalized solution of (1), then we set

$$g = \eta \Delta v - (v \operatorname{grad}) v + K.$$

The components of g are denoted by g_i. From (15) and integration by parts it follows that

$$(g | w)_Y = 0 \quad \text{for all} \quad w \in C_0^\infty(G, \operatorname{div}).$$

This implies $g \in H^\perp$. Lemma 72.1 then shows that

$$g = \operatorname{grad} p, \quad p \in W_2^1(G),$$

and the pressure p is uniquely determined up to an additive constant. Therefore v, p is a solution of the original stationary equation (1).

72.3b. The Pressure in the Weak Sense

This argument can be refined in the case that the generalized solution v is not sufficiently smooth, and instead only $v \in V$, i.e.,

$$v_i \in \mathring{W}_2^1(G), \quad i = 1, 2, 3$$

is valid, such as in Definition 72.2. We thereby use the calculus of distributions, which is summarized in $A_2(62)$, and consider the components v_i of v as

distributions, i.e.,
$$v_i \in \mathscr{D}'(G), \quad i = 1, 2, 3.$$
Since distributions are arbitrarily often differentiable, it follows that
$$\Delta v_i \in \mathscr{D}'(G), \quad i = 1, 2, 3.$$
Let $u = (v\,\mathrm{grad})v$, i.e., $u_i = v^j D_j v_i$. Thus, $u_i \in L_1(G)$ and hence
$$u_i \in \mathscr{D}'(G), \quad i = 1, 2, 3.$$
From $K_i \in L_2(G)$, we find $K_i \in \mathscr{D}'(G)$, and consequently,
$$g_i \in \mathscr{D}'(G), \quad i = 1, 2, 3.$$
Explicitly, we have
$$g_i(\varphi) = \int_G (\eta v_i \Delta\varphi - (v^j D_j v_i)\varphi + K_i \varphi)\,dx$$
for all $\varphi \in C_0^\infty(G)$. We define
$$\bar{g}(w) = \sum_{i=1}^{3} g_i(w_i),$$
and integration by parts yields
$$\bar{g}(w) = -\eta(v|w) + b(v, v, w) + c(w)$$
for all $w_i \in C_0^\infty(G)$. From the generalized problem (15) it then follows that
$$\bar{g}(w) = 0 \quad \text{for all} \quad w \in C_0^\infty(G, \mathrm{div}).$$
Now to the point. The proof of Lemma 72.4 below implies
$$|g_i(\varphi)| \leq \mathrm{const}\,\|\varphi\|_{1,2} \quad \text{for all} \quad \varphi \in C_0^\infty(G).$$
This means that $g_i \in W_2^{-1}(G)$. Therefore, Problem 72.2 shows that there exists a function $p \in L_2(G)$ with
$$g = \mathrm{grad}\,p \quad \text{on } G$$
in the sense of distributions. Up to a constant, p is uniquely determined.

Hence v, p is a solution of the original stationary equation (1) in the sense of distributions, and Definition 72.2 therefore makes sense.

72.3c. Operator Properties

Lemma 72.3. *It is $c \in V^*$.*

PROOF. We have
$$|c(w)| = |(K|w)_Y| \leq \|K\|_Y \|w\|_Y$$
$$\leq \mathrm{const}\,\|K\|_Y \|w\|. \quad \square$$

72.3. Generalized Stationary Problem

In order to study $b(u, v, w)$, we make essential use of the following Sobolev embedding theorem:

$$\text{The embedding} \quad W_2^1(G) \subseteq L_4(G) \quad \text{is compact,} \tag{17}$$

see $A_2(45)$. We choose a fixed Cartesian coordinate system and denote the norm on $L_4(G, V_3)$ by

$$|v| = \sum_{i=1}^{3} \|v_i\|_4.$$

From (17) follows that

$$\text{the embedding} \quad V \subseteq L_4(G) \quad \text{is compact.} \tag{17*}$$

In particular, this implies

$$|v| \leq \text{const} \, \|v\| \quad \text{for all} \quad v \in V.$$

Recall that $\|v\|$ denotes the norm on V and $\|f\|_p$ is the norm of $f \in L_p(G)$.

Lemma 72.4. *For all $u, v, w \in V$ we have*

$$|b(u, v, w)| \leq \text{const} \, |u| \, |v| \, \|w\| \tag{18}$$
$$\leq \text{const} \, \|u\| \, \|v\| \, \|w\|,$$

$$b(v, v, v) = 0. \tag{19}$$

PROOF. The Hölder inequality for three factors (65.16) and (16) imply

$$|b(u, v, w)| \leq \|u_i\|_4 \|v^j\|_4 \|D_j w^i\|_2$$
$$\leq \text{const} \, |u| \, |v| \, \|w\|.$$

This is (18).

Integration by parts and $\text{div} \, v = 0$ yield

$$2b(v, v, v) = \int 2v^j v_i D_j v^i \, dx = \int v \, \text{grad} \, v^2 \, dx$$
$$= -\int v^2 \, \text{div} \, v \, dx = 0$$

for all $v \in C_0^\infty(G, \text{div})$. Since this set is dense in V, we obtain (19) from (18). □

72.3d. Equivalent Operator Equation

Our goal is to show that the generalized stationary problem (15) is equivalent to the operator equation

$$\eta v - Qv = f, \quad v \in V. \tag{20}$$

To this end, we set

$$(f|w) = c(w),$$
$$(B(u,v)|w) = b(u,v,w) \quad \text{for all} \quad w \in V \tag{21}$$

and fixed $u, v \in V$. In order to justify (21), we use the theorem of Riesz of Section 18.11. In fact, from Lemma 72.3, we have that $c: V \to \mathbb{R}$ is a linear continuous functional and hence there exists an element $f \in V$ with (21). In the same way, Lemma 72.4 guarantees the existence of a $B(u,v) \in V$ with (21).

Moreover, we set

$$Qv = B(v,v) \quad \text{for all} \quad v \in V.$$

Obviously, equation (15) is equivalent to (20).

Lemma 72.5. *The operator $B: V \times V \to V$ is bilinear, symmetric, bounded, and strongly continuous.*

PROOF. The operator B is bounded. This follows from (18) and hence

$$\|B(u,v)\| \leq \text{const}\,|u|\,|v| \leq \text{const}\,\|u\|\,\|v\|$$

for all $u, v \in V$.

We show that B is strongly continuous. Let

$$u_n \rightharpoonup u \quad \text{and} \quad v_n \rightharpoonup v \quad \text{in } V \quad \text{as } n \to \infty.$$

Then the sequences (u_n) and (v_n) are bounded in V. Because of the compact embedding $V \subseteq L_4(G)$ we have that $|u_n - u| \to 0$ and $|v_n - v| \to 0$ as $n \to \infty$. Hence

$$\|B(u_n, v_n) - B(u,v)\| = \|B(u_n - u, v_n) + B(u, v_n - v)\|$$
$$\leq |u_n - u|\,|v_n| + |u|\,|v_n - n| \to 0 \quad \text{as } n \to \infty.$$

The remaining claims are obvious. □

72.4. Existence and Uniqueness Theorem for Stationary Flows

Theorem 72.A. *Assume condition (2) for a fixed flow region G. The generalized stationary problem (15) then has a solution for every $K \in L_2(G, V_3)$.*

This solution is unique if

$$\frac{\rho^2}{\eta^4} \int_G K^2\,dx$$

is sufficiently small.

PROOF.

(I) Existence. From Lemma 72.5 it follows that the operator $Q: V \to V$ is strongly continuous, and (19) implies that

$$(Qv|v) = b(v, v, v) = 0 \quad \text{for all} \quad v \in V.$$

Thus the existence result follows from Section 72.1.

(II) Uniqueness. From (6) it follows that

$$\|v\| \leq \|f\|/\eta$$

for every solution $v \in V$ of equation (20). For two solutions u and v we therefore have

$$\eta\|u - v\| = \|Qu - Qv\| = \|B(u - v, u) + B(v, u - v)\|$$
$$\leq \text{const}\, \|u - v\|(\|u\| + \|v\|)$$
$$\leq \text{const}\, \|u - v\| \|f\|/\eta,$$

which shows that $\|u - v\| = 0$ if $\|f\|/\eta^2$ is sufficiently small. This gives the uniqueness of the solution.

From Lemma 72.3 follows that

$$\|f\|^2 \leq \text{const} \int_G K^2 \, dx.$$

In case we do not choose $\rho = 1$, such as in Section 72.1, we have to replace η, K with η/ρ, K/ρ. □

If we introduce dimensionless quantities, as in (13), then we obtain unique solutions of (13) if the dimensionless number

$$(\text{Re})^4 \int_\Omega \bar{K}^2 \, dx$$

is sufficiently small for fixed Ω.

72.5. Generalized Nonstationary Problem

In order to understand this section we need some tools which were presented in Chapter 23 in connection with linear evolution equations. In particular, we need the spaces

$$L = L_2(0, T; V) \quad \text{and} \quad W = W_2^1(0, T; V, H).$$

Recall that $v \in W$ means that

$$v \in L_2(0, T; V) \quad \text{and} \quad v' \in L_2(0, T; V^*).$$

The spaces V and H have been introduced in Section 72.2.

Moreover,

$$\text{``}V \subseteq H \subseteq V^*\text{''}$$

is an evolution triple in the sense of Section 23.4. Let $[0, T]$ be a fixed, but otherwise arbitrary time interval with $0 < T < \infty$.

In order to obtain the generalized problem, we multiply equation (1) with $w \in C_0^\infty(G, \text{div})$. After integrating by parts we then obtain

$$\frac{d}{dt}(v|w)_H + \eta(v|w) - b(v, v, w) = (K|w)_Y \quad \text{for all} \quad w \in V,$$

$$v(0) = v_0. \tag{22}$$

In contrast to the stationary case, v and K depend here on time t.

Definition 72.6. Assume condition (2) for the region G. The *generalized problem to the nonstationary Navier–Stokes equations* (1) is the following.

Given the initial velocity $v_0 \in H$ and the density of outer forces $K \in L_2(0, T; Y)$ with $Y = L_2(G, V_3)$, we are looking for

$$v \in W$$

such that equation (22) is valid on the time interval $]0, T[$.

The time derivative on $]0, T[$ in (22) is understood in the generalized sense.

As in Section 72.3, we see that this is a meaningful generalization of the original problem (1).

By using (21) we can write (22) in the form

$$\frac{d}{dt}(v|w)_H + \eta(v|w) - (B(v, v)|w) = (f|w) \quad \text{for all} \quad w \in V,$$

$$v(0) = v_0, \tag{23}$$

where v and K depend on t. Recall that $(\cdot|\cdot)$ denotes the scalar product on V.

72.5a. The Linearized Problem

The linearized problem to (23) is

$$\frac{d}{dt}(v|w)_H + \eta(v|w) = (g|w) \quad \text{for all} \quad w \in V,$$

$$v(0) = v_0. \tag{24}$$

According to the main theorem about linear evolution equations of first order (Theorem 23.A), this equation has a unique solution $v \in W$ for every $v_0 \in H$ and $g \in L$. We set

$$v = S(g, v_0).$$

72.5. Generalized Nonstationary Problem

Lemma 72.7. *The solution operator $S: L \times H \to W$ is linear and continuous.*

PROOF. From Theorem 23.A follows

$$\|v\|_W \leq c(\|g\|_L + \|v_0\|_H). \tag{25}$$

□

72.5b. Operator Properties

Letting $g = f - B(v, v)$ we can write (23) equivalently as

$$v = S(f - B(v, v), v_0), \qquad v \in W. \tag{26}$$

Observe that v depends on t. We need to show that f and B are correctly defined, i.e., that $f \in L$ and $B(v, v) \in L$ for all $v \in W$.

Lemma 72.8. *It is $f \in L$.*

PROOF. By assumption we have $K \in L_2(0, T; Y)$, i.e., $K(t) \in Y$ for all $t \in]0, T[$. From

$$(f(t)|w) = (K(t)|w)_Y \qquad \text{for all} \quad w \in V$$

it follows that $\|f(t)\|_V \leq \text{const} \|K(t)\|_Y$, according to Lemma 72.3, and hence

$$\|f\|_L^2 = \int_0^T \|f(t)\|_V^2 \, dt \leq \text{const} \int_0^T \|K(t)\|_Y^2 \, dt. \qquad \square$$

Lemma 72.9. *The bilinear, bounded operator $B: V \times V \to V$ can naturally be extended to a bilinear, bounded operator $B: W \times W \to L$.*

PROOF. All constants will be denoted by c. Set

$$Z = \text{closure of } C_0^\infty(G, \text{div}) \quad \text{in } L_4(G, V_3).$$

Since the embedding $\mathring{W}_2^1(G) \subseteq L_4(G)$ is continuous, it follows that the embedding $V \subseteq Z$ is continuous as well. Moreover, V is dense in Z. Therefore

$$\text{``}V \subseteq Z \subseteq V^*\text{''}$$

is an evolution triple. From Proposition 23.23 it then follows that the embedding

$$W \subseteq C([0, T], Z)$$

is continuous. Note that the definition of $W = W_2^1(0, T; V, H)$ depends only on V and V^*, and not on the concrete form of H, i.e., $W_2^1(0, T; V, H) = W_2^1(0, T; V, Z)$. We have the important estimate

$$\max_{0 \leq t \leq T} |w(t)| \leq c \|w\|_W \qquad \text{for all} \quad w \in W. \tag{27}$$

Recall that $|\cdot|$ and $\|\cdot\|$ denote the norm on $Z \subseteq L_4(G, V_3)$ and V, respectively. Let

$$z(t) = B(u(t), v(t)).$$

Then for all $u, v \in W$ we have

$$\|z\|_L^2 = \int_0^T \|B(u(t), v(t))\|^2 \, dt$$

$$\leq \int_0^T c |u(t)|^2 |v(t)|^2 \, dt$$

$$\leq c \left(\max_t |u(t)| \right)^2 \left(\max_t |v(t)| \right)^2$$

$$\leq c \|u\|_W^2 \|v\|_W^2. \qquad \square$$

72.6. Existence and Uniqueness Theorem for Nonstationary Flows

Theorem 72.B. *Assume condition* (2) *for a fixed flow region* G. *Suppose that we are given the initial velocities* $v_0 \in H$ *and the density of the outer forces* $K \in L_2(0, T; Y)$.

The generalized nonstationary problem (22) *then has at most one solution. If the norms of* v_0 *and* K *are sufficiently small, i.e., if*

$$\|v_0\|_H^2 + \int_0^T \int_G K(x, t)^2 \, dx \, dt \leq r^2 \tag{28}$$

for fixed small $r > 0$, *then there exists a unique solution.*

PROOF.

(I) Existence. In case that the norms $\|v_0\|_H$ and $\|f\|_L$ are sufficiently small, we apply the implicit function theorem (Theorem 4.B) to equation (26). Observe that, because of Lemma 72.9, the right-hand side of (26) is analytic.

(II) Uniqueness. Assume that $u, v \in W$ are solutions of the generalized problem (26), and set $w = u - v$. Then, according to (27), there exists a number $R > 0$ with

$$|u(t)|, |v(t)| \leq R \qquad \text{for all} \quad t \in [0, T].$$

From (26) follows

$$w = S(B(w, u) + B(v, w), 0).$$

This yields

$$|w(T)|^2 \leq c\|w\|_W^2 \leq c\|S\|^2 \int_0^T (|w||u| + |w||v|)^2 \, dt$$

$$\leq 4cR^2\|S\|^2 \int_0^T |w(t)|^2 \, dt.$$

The function w is a generalized solution on each time interval $[0, s]$ with $s \leq T$, which belongs to the space $W_2^1(0, s; V, H)$. Therefore, in the last estimate, we may replace T with s. Note that, according to Theorem 23.A, the constant c in (25) does not depend on the subinterval $[0, s]$ of $[0, T]$. Consequently, the norm $\|S\|$ of the solution operator is uniformly bounded with respect to all subintervals $[0, s]$ of $[0, T]$. Gronwall's lemma of Section 3.5 then implies that

$$w(s) = 0 \quad \text{for all} \quad s \in [0, T]. \qquad \square$$

72.7. Taylor Problem and Bifurcation

In this and the following section we need results about B-spaces of Hölder-continuous functions which have been discussed in Section 6.2. We use arguments which can also be applied to bifurcation problems for more general elliptic systems of partial differential equations.

72.7a. The Physical Problem

As in Figure 72.1, we consider a viscous fluid between two concentric cylinders, whereby the outer cylinder is at rest and the inner cylinder rotates counterclockwise around the z-axis with angular velocity ω. Let the cylinder radia be r_1 and r_2 with $r_1 < r_2$. Important is the Reynolds number Re. We set

$$\lambda = \text{Re} = \rho \omega r_1^2 / \eta.$$

In experiments one observes a critical number λ_0 with the following properties.

(i) For $\lambda < \lambda_0$, i.e., for small angular velocities ω there exists an axisymmetric flow which does not depend on the z-coordinate. This is the so-called Couette flow.
(ii) For $\lambda = \lambda_0$, so-called Taylor vortices occur, which are periodic in z (Fig. 72.2).
(iii) If the angular velocity ω gets larger and larger, i.e., for increasing λ, one obtains more and more complicated flow pictures until at a certain $\lambda = \lambda_{\text{crit}}$ turbulence occurs.

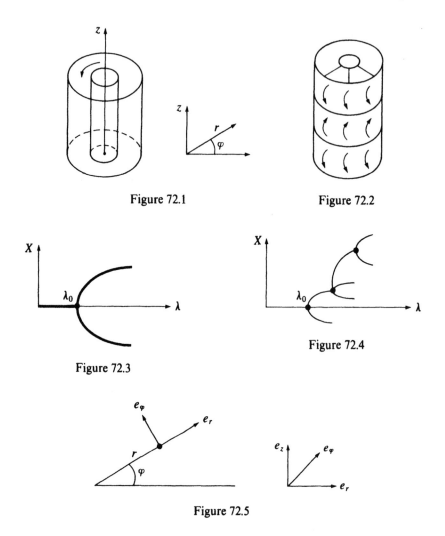

Figure 72.1

Figure 72.2

Figure 72.3

Figure 72.4

Figure 72.5

Our goal is to treat this Taylor problem as a bifurcation problem in a neighborhood of the first bifurcation point λ_0 (Fig. 72.3). Up until today, however, it has not been possible to follow the solution branch as $\lambda \to +\infty$. One expects successive secondary bifurcations as shown in Figure 72.4.

72.7b. The Mathematical Problem

As in formula (12) we introduce dimensionless quantities u, $\bar{x}$, and $\bar{p}$ through

$$v = \omega r_1 u, \qquad x = r_1 \bar{x}, \qquad p = \rho \omega^2 r_1^2 \bar{p},$$

and describe $\bar{x}$ with respect to the cylinder coordinates r, φ, z and the corresponding orthonormal basis vectors e_r, e_φ, e_z (Fig. 72.5). Moreover, we let

$$R = r_2/r_1.$$

72.7c. Bifurcation from the Couette Flow

By using the dimensionless stationary Navier–Stokes equations (13), our problem can then be stated as follows.

We are looking for a velocity field u and a pressure function $\bar{p}$ which satisfies

$$-\frac{1}{\lambda}\Delta u + (u\,\mathrm{grad})u = -\mathrm{grad}\,\bar{p},$$
$$\mathrm{div}\,u = 0 \qquad (29)$$

for all arguments $(r,z) \in [1,R] \times \mathbb{R}$ and the boundary conditions

$$u = 0 \quad \text{for} \quad r = 1,$$
$$u = e_\varphi \quad \text{for} \quad r = R. \qquad (30)$$

More precisely, we are looking for axisymmetric solutions, which have the period P with respect to z.

72.7c. Bifurcation from the Couette Flow

By writing (29) in cylindrical coordinates, we immediately observe that the so-called Couette flow

$$u_0(r) = (1 - R^2)^{-1}(r - R^2 r^{-1})e_\varphi,$$
$$\bar{p}_0(r) = u_0^2 \ln r + \mathrm{const}$$

is a solution of (29), (30). In order to find solutions which bifurcate from this solution, i.e., which correspond to the Taylor vortices, we set

$$u = u_0 + w, \qquad \bar{p} = \bar{p}_0 + q$$

with

$$w = U e_r + V e_\varphi + W e_z.$$

Transformation of (29) and (30) onto cylindrical coordinates yields

$$-\lambda^{-1} A_1 U + U_r + W U_z - r^{-1} V^2 - 2r^{-1} V V_0 + q_r = 0,$$
$$-\lambda^{-1} A_1 V + U V_r + W V_z + r^{-1} U V + r^{-1}(V_0 + r V_0')U = 0, \qquad (31)$$
$$-\lambda^{-1}(A_1 W + r^{-2} W) + U W_r + W W_z + q_z = 0,$$

with $V_0 = |u_0|$ and also the elliptic differential operator

$$A_1 U = U_{zz} + U_{rr} + r^{-1} U_r - r^{-2} U.$$

From $\mathrm{div}\,v = 0$ we obtain

$$(rU)_r + (rW)_z = 0, \qquad (32)$$

and the boundary conditions are

$$U = V = W = 0 \quad \text{for} \quad r = 1 \text{ and } r = R. \qquad (33)$$

72.7d. The Stream Function

In order to simplify equation (31) we use the so-called stream function ψ, which can be uniquely determined from the equation

$$(r\psi)_z = rU, \qquad (r\psi)_r = -rW$$

with $(r, z) \in [1, R] \times \mathbb{R}$ and the normalization condition $\psi(1, 0) = 0$. Because of (32), the integrability conditions are satisfied (see Problem 65.3). In addition, we assume the symmetry condition

$$\psi(r, -z) = -\psi(r, z), \qquad V(r, -z) = V(r, z). \tag{34}$$

Our next goal is to reduce this problem to a corresponding one for the functions ψ and V. In order to eliminate the pressure q in (31), we differentiate the first and third equation in (31) with respect to z and r, respectively, and subtract the results. If we add the second equation of (31), we obtain

$$\begin{aligned} A_1^2 \psi + \lambda(aV_z + N_1(\psi, V)) &= 0, \\ A_1 V + \lambda(b\psi_z + N_2(\psi, V)) &= 0, \end{aligned} \tag{35}$$

with symmetry conditions (34) and boundary conditions

$$\psi = \psi_r = V = 0 \quad \text{for} \quad r = 1 \quad \text{and} \quad r = R. \tag{36}$$

We thereby have

$$a(r) = 2(R^2 - r^2)/(R^2 - 1)r^2, \qquad b = 2/(R^2 - 1),$$

and

$$rN_1 = ((r\psi)_r A_1 \psi)_z - r(\psi_z A_1 \psi)_r + 2VV_z,$$
$$rN_2 = (r\psi)_r V_z - (rV)_r \psi_z.$$

The concrete form of the nonlinear terms N_1 and N_2 is not important for the existence proof. We will only need the fact that they are bilinear with respect to ψ and V, and that N_1 and N_2 contain derivatives up to order 3 and 1, respectively. Notice, furthermore, that a and b are positive on $]1, R[$.

We are looking for functions

$$\psi = \psi(r, z) \quad \text{and} \quad V = V(r, z)$$

with $(r, z) \in [1, R] \times \mathbb{R}$, which have period P with respect to z. If we know a solution of (35), then we can derive pressure q from equation (31), because (35) contains the integrability conditions for q.

Our problem is therefore reduced to the question of finding nontrivial solutions for the boundary-value problem (34)–(36).

72.7e. The Operator Equation

For a functional-analytic formulation of this problem we set

$$x = (\psi, V),$$

72.7. Taylor Problem and Bifurcation

and write the differential equation (35) in the form

$$Ax + \lambda(Bx + Nx) = 0, \quad x \in X \tag{37}$$

with linear operators

$$Ax = (A_1^2 \psi, A_1 V), \quad Bx = (aV_z, b\psi_z)$$

and the nonlinear operator

$$Nx = (N_1(x), N_2(x)).$$

In order to obtain the operators

$$A, B, N \colon X \to Y,$$

we define the real B-spaces X and Y in the following way. Set

$$Q = [1, R] \times [0, P],$$

and let $C_{\text{per}}^{k,\alpha}(Q)$ denote the B-space of all $C^{k,\alpha}(Q)$-functions, which can be extended to $C^{k,\alpha}$-functions on $[1, R] \times \mathbb{R}$ and have the period P with respect to the z-coordinate. Obviously, $C_{\text{per}}^{k,\alpha}(Q)$ is a closed linear subspace of the B-space $C^{k,\alpha}(Q)$ (see Section 6.2).

We now define X as the B-space of all function pairs

$$(\psi, V) \in C_{\text{per}}^{4,\alpha}(Q) \times C_{\text{per}}^{2,\alpha}(Q)$$

with $0 < \alpha < 1$, where in addition ψ and V satisfy the symmetry conditions (34) on Q as well as the boundary conditions (36).

Moreover, we define Y as the B-space of all function pairs

$$(f, g) \in C_{\text{per}}^{\alpha}(Q) \times C_{\text{per}}^{\alpha}(Q),$$

which satisfy the symmetry conditions

$$f(r, -z) = -f(r, z), \quad g(r, -z) = g(r, z) \quad \text{on } Q.$$

Note that the spaces X and Y depend on the period P.

72.7f. Main Result

Theorem 72.C (Taylor Problem). *One can choose the period $P > 0$ in such a way that there exists a positive number λ_0 which has the following properties.*

(i) *For $0 < \lambda < \lambda_0$, the operator equation (37) has only the trivial solution $x = 0$.*

(ii) *$(\lambda_0, 0)$ is a bifurcation point of (37). More precisely, there exists a unique bifurcation branch in a neighborhood of $(\lambda_0, 0)$ in $\mathbb{R} \times X$. This branch depends analytically on the small real parameter s, i.e.,*

$$\begin{aligned} x &= sx_1 + s^2 x_2 + \cdots, \\ \lambda &= \lambda_0 + \varepsilon_1 s + \varepsilon_2 s^2 + \cdots. \end{aligned} \tag{38}$$

In the following section we will show that Theorem 72.C is a direct consequence of the main theorem of analytic bifurcation theory (Theorem 8.A). Corollary 8.25 contains an effective method for a successive computation of the coefficients in (38). Letting $C = A + \lambda_0 B$ we repeatedly have to solve the equation

$$Cx = y, \quad x \in X \tag{39}$$

in order to determine (x_{j+1}, ε_j) for $j = 1, 2, \ldots$. Moreover, we will compute (x_1, λ_0), with $Cx_1 = 0$. Equation (39) means that we have to solve a linear elliptic system. By using a Fourier expansion this can be reduced to a boundary-value problem for fourth-order ordinary differential equations, see (47) below.

Numerical computations of Kirchgässner and Sorger (1969) have shown that

$$\varepsilon_1 = 0 \quad \text{and} \quad \varepsilon_2 > 0.$$

The bifurcation branch has therefore the structure shown in Figure 72.3. The result coincides with the physical experiments described in Section 72.7a. The trivial solution $x = 0$ corresponds to the Couette flow, and for Reynolds numbers $\lambda > \lambda_0$ in a neighborhood of λ_0, the bifurcation solutions yield the Taylor vortices. Further numerical studies in connection with the Taylor problem may be found in Meyer and Keller (1980) and Frank and Meyer (1981).

72.8. Proof of Theorem 72.C

In order to apply Theorem 8.A, we need the following basic results.

(i) The linearized operator $A + \lambda B$ is Fredholm of index zero.
(ii) The linearized equation $(A + \lambda B)x = 0$ has a simple solution.
(iii) The generic bifurcation condition is satisfied.

In connection with (ii) and (iii), we use the spectral theory of linear integral equations with positive kernels. This is related to Chapter 7. Furthermore, we apply Fourier series, and we force condition (ii) by using a suitable period P (trick of changing the periodicity). The analytical key to the following proof is contained in Lemmas 72.12 and 72.14.

Lemma 72.10. *For all* $(\psi, V), (\varphi, W) \in X$ *we have*

$$-\int_Q (A_1 V) Wr \, dr \, dz = \int_Q (V_r W_r + V_z W_z + r^2 VW) r \, dr \, dz$$

$$= -\int_Q (A_1 W) Vr \, dr \, dz, \tag{40}$$

72.8. Proof of Theorem 72.C

$$\int_Q \psi(A_1^2\varphi) r\, dr\, dz = \int_Q (A_1\psi)(A_1\varphi) r\, dr\, dz.$$

PROOF. Use integration by parts. □

Lemma 72.11. *The operators $A, B, N: X \to Y$ have the following properties: A is linear and continuous, B is linear and compact, and N is analytic with $Nx = O(\|x\|^2)$ as $x \to 0$.*

PROOF. According to Section 6.2, this follows from the compact embedding

$$C^{k+1,\alpha}(Q) \subseteq C^{k,\alpha}(Q)$$

and estimates of the form

$$\|fg\|_\alpha \leq \text{const} \|f\|_\alpha \|g\|_\alpha,$$

where $\|\cdot\|_\alpha$ denotes the C^α-norm. Notice that B, other than A, contains only low-order derivatives. □

Lemma 72.12. *Equation*

$$Ax = y, \qquad x \in X \tag{41}$$

has a unique solution for every $y \in Y$.

PROOF. This statement follows from standard results about linear elliptic differential equations. Explicitly, equation (41) means that

$$A_1^2 \psi = f, \qquad A_1 V = g, \qquad (\psi, V) \in X \tag{42}$$

with $(f, g) \in Y$.

(I) Uniqueness. From $Ax = 0$ and (40) we obtain at once that $x = 0$.
(II) Existence. If f and g are expanded in Fourier series as in (47) below, and if $f, g \in C^\infty_{\text{per}}(Q)$, then it is easy to obtain a solution (ψ, V) in the form of a Fourier series. This classical solution is also a generalized solution, i.e.,

$$\int_Q \psi(A_1^2 \varphi) r\, dr\, dz = \int_Q f\varphi r\, dr\, dz,$$

$$\int_Q V(A_1 \varphi) r\, dr\, dz = \int_Q g\varphi r\, dr\, dz$$

for all $\varphi \in C_0^\infty(Q)$. From (40) and the Hölder inequality we derive the estimates

$$\|\psi\| \leq \text{const} \|f\|, \qquad \|V\| \leq \text{const} \|g\|$$

in $L_2(Q)$-norms.

For $(f, g) \in Y$, i.e., for $f, g \in C^\alpha_{\text{per}}(Q)$ we then obtain generalized solutions (ψ, V) by applying these estimates and a simple approximation argument.

According to the regularity results in Agmon, Douglis, and Nirenberg (1959), these generalized solutions (ψ, V) have classical derivatives up to the natural order prescribed in (42), i.e., $\psi \in C^{4,\alpha}(Q)$ and $V \in C^{2,\alpha}(Q)$. This implies that $(\psi, V) \in X$. □

Lemma 72.13. *For every $\lambda \in \mathbb{R}$, the operator $A + \lambda B: X \to Y$ is Fredholm of index zero.*

PROOF. Lemma 72.12 implies that A is Fredholm of index zero. According to Section 8.4, the compact perturbation $A + \lambda B$ then has the same property. □

Lemma 72.14. *There exists a period P and a number $\lambda_0 > 0$ such that the equation*

$$Ax + \lambda Bx = 0, \qquad x \in X \tag{43}$$

has only the trivial solution $x = 0$ for every $\lambda \in \,]0, \lambda_0[$, and precisely one linearly independent nontrivial solution $x_1 = (\psi_1, V_1)$ for $\lambda = \lambda_0$ with

$$\psi_1 = \alpha(r) \sin \frac{2\pi z}{P}, \qquad V_1 = \beta(r) \cos \frac{2\pi z}{P},$$

where $\alpha > 0$ and $\beta > 0$ on $\,]1, R[$. Equation

$$Ax + \lambda_0 Bx = Bx_1, \qquad x \in X \tag{44}$$

has no solution.

This key lemma will be proved at the end of this section.

PROOF OF THEOREM 72.C. We set $\lambda = \lambda_0 + \varepsilon$. The operator equation (37) then takes the form

$$F(\varepsilon, x) = 0, \qquad (\varepsilon, x) \in \mathbb{R} \times X$$

with

$$F(0,0) = 0, \qquad F_x(0,0) = A + \lambda_0 B, \qquad F_{\varepsilon x}(0,0) = B.$$

We set $C = F_x(0,0)$ and choose an element $x_1^* \in Y^*$ with $x_1^* \neq 0$ and $C^* x_1^* = 0$. Such an x_1^* exists according to Lemmas 72.13 and 72.14 (see Section 8.4). This implies that

$$\langle x_1^*, Bx_1 \rangle \neq 0, \tag{45}$$

since otherwise, equation (44) would have a solution.

Observe that (45) is precisely the generic bifurcation condition of Theorem 8.A. Thus, Theorem 72.C follows from Theorem 8.A. □

72.8. Proof of Theorem 72.C

PROOF OF LEMMA 72.14. We make essential use of the positivity of certain Green's functions and the spectral theory of linear integral operators with positive kernels.

(I) Equivalent homogeneous integral equation. The operator equation (43) is equivalent to the differential equation

$$A_1^2 \psi + \lambda a V_z = 0, \qquad A_1 V + \lambda b \psi_z = 0 \qquad (46)$$

with $(\psi, V) \in X$. According to the regularity theory for linear elliptic differential equations, every such solution is arbitrarily often differentiable in $[1, R] \times \mathbb{R}$. Recall that, in the definition of the space X of the previous section, periodicity conditions with respect to z, symmetry conditions (34), and boundary conditions (36) were included. As a consequence of these conditions, the solutions of (46) have Fourier expansions of the form

$$\psi(r, z) = \sum_{n=1}^{\infty} \alpha_n(r) \sin \frac{2\pi n z}{P},$$

$$V(r, z) = \beta_0(r) + \sum_{n=1}^{\infty} \beta_n(r) \cos \frac{2\pi n z}{P}. \qquad (47)$$

From (46) it follows that

$$L^2 \alpha_n = \lambda a k \beta_n \quad \text{and} \quad L \beta_n = \lambda b k \alpha_n \quad \text{on }]1, R[,$$

$$\alpha_n(r) = \alpha'_n(r) = \beta_n(r) = 0 \quad \text{for } r = 1 \text{ and } r = R \qquad (48)$$

for fixed $n \in \mathbb{N}$ with $k = 2\pi n / P$ and

$$L\alpha = -\alpha'' - r^{-1} \alpha' + r^{-2} + k^2.$$

Moreover, we have

$$L\beta_0 = 0 \quad \text{on }]1, R[\quad \text{and} \quad \beta_0(1) = \beta_0(R) = 0.$$

This implies $\beta_0 = 0$, since integration by parts yields

$$\int_1^R \beta_0 (L\beta_0) r \, dr \geq \int_1^R \beta_0^2 r^{-1} \, dr.$$

Let G and H be the respective Green's function for L and L^2 with boundary conditions as in (48). Then equation (48) is equivalent to the system of integral equations

$$\alpha_n(r) = \lambda k \int_1^R H(r, s) a(s) \beta_n(s) \, ds,$$

$$\beta_n(s) = \lambda k \int_1^R G(s, t) b(t) \alpha_n(t) \, dt. \qquad (49)$$

This implies

$$\alpha_n(r) = \mu \int_1^R K(r,t)\alpha_n(t)\,dt \tag{50}$$

with $\mu = \lambda^2$ and

$$K(r,t) = k^2 \int_1^R H(r,s)G(s,t)a(s)b(t)\,ds.$$

From Problem 72.4 it follows the crucial fact that

$$H \text{ and } G \text{ are positive} \quad \text{on }]1,R[\times]1,R[.$$

The same property is valid for K. Problem 72.5 implies that the eigenvalue problem (50) has a simple characteristic number $\mu_0(n) > 0$ of smallest absolute value with an eigenfunction

$$\alpha_{n,0}(r) > 0 \quad \text{on }]1,R[.$$

If we set $\lambda_0(n) = \sqrt{\mu_0(n)}$ and

$$\beta_{n,0}(s) = \lambda_0(n)k \int_1^R G(s,t)b(t)\alpha_{n,0}(t)\,dt,$$

then $\lambda_0(n)$, $\alpha_{n,0}$, $\beta_{n,0}$ is a solution of (49). Hence we derived the following result:

For fixed $n \in \mathbb{N}$, the number $\lambda_0(n) > 0$ is the eigenvalue of smallest absolute value of (48). It is simple, and the corresponding eigenfunctions $\alpha_{n,0}$ and $\beta_{n,0}$ are positive on $]1,R[$.

(II) The inhomogeneous integral equation

$$\alpha(r) = \mu_0(n) \int_1^R K(r,t)\alpha(t)\,dt + \gamma(r) \tag{51}$$

with either $\gamma(r) > 0$ or $\gamma(r) < 0$ on $]1,R[$ has no solution.

To see this, let α^* be the solution of the adjoint integral equation to (50) with $\mu = \mu_0(n)$. Problem 72.5 shows that $\alpha^*(r) > 0$ on $]1,R[$. According to the classical theory, (51) has a solution if and only if

$$\int_1^R \alpha^*(r)\gamma(r)\,dr = 0.$$

This condition, however, cannot be satisfied because of the positivity of the integrand.

(III) Trick of changing the periodicity. It is possible that

$$\lambda_0(n) = \lambda_0(m) \quad \text{for} \quad n \neq m.$$

Thus $\lambda_0(n)$ need not be necessarily a simple eigenvalue of the original equation (43).

However, this can be guaranteed by modifying the space X. Recall that X depends on the period P. Our idea is to change P in such a way that certain eigenfunctions do not lie in X.

From Problem 72.6 it follows that

$$\lambda_0(n) \geq \frac{8\pi^2 n^2}{P^2} \left(\max_{1 \leq r \leq R} (a(r) + b(r)) \right)^{-1}, \tag{52}$$

and this implies that $\lambda_0(n) \to +\infty$ as $n \to \infty$. Let m be the largest natural number with

$$\lambda_0(m) \geq \min_{n \geq 1} \lambda_0(n),$$

and set

$$\lambda_0 = \lambda_0(m), \qquad \alpha = \alpha_{m,0}, \qquad \beta = \beta_{m,0}.$$

This is the required simple solution of Lemma 72.14 if, in the construction of X,

we replace period P with P/m.

Note that the function $\sin(2\pi n z/P)$ for $n = 1, 2, \ldots, m-1$ does *not* have period P/m. Thus the corresponding eigenfunctions are *not* in the modified space X.

(IV) Proof of (44). If x is a solution of equation (44), then we expand this solution in a Fourier series of the form (47). As in (I) we obtain an integral equation for α_1 of the form (51), which, according to (II), has no solution. This proves Lemma 72.14. □

The proof of Theorem 72.C is now complete.

72.9. Bénard Problem and Bifurcation

The Taylor vortices were experimentally discovered by Taylor in 1923. But already in 1901, Bénard had found another bifurcation phenomenon for viscous fluids. In order to explain this phenomenon, we consider here a viscous fluid between two plates, as shown in Figure 72.6, where the temperature T_0 of the lower plate and the temperature T_1 of the upper plate satisfy the condition

$$T_0 > T_1.$$

If $T_0 - T_1$ is sufficiently small, then the fluid is at rest. If the temperature difference is increased, then, at a critical value, so-called Bénard cells appear in the fluid. These cells have a hexagonal structure.

In experiments a pan with silicon oil is heated with hot water from below. The fluid flow is made visible through small equally distributed aluminum

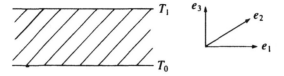

Figure 72.6

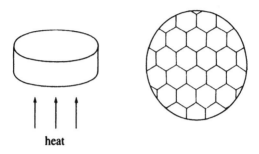

Figure 72.7

pieces. After reaching the critical temperature difference, hexagonal cells appear in the pan, which are shown from above in Figure 72.7.

Bénard cells correspond to a bifurcation phenomenon. Physically, they arise by combining the gravitational force and the heat convection flow. During the past years, physicists, chemists, and biologists have shown a great deal of interest in these Bénard cells, because one observes the formation of a complicated structure. This is a process which frequently occurs in the evolution of life.

72.9a. Physical Problem

We begin with the so-called Boussinesq approximation for the Navier–Stokes equations with temperature effects. This is

$$\rho_0 v_t + \rho_0(v \, \text{grad})v + \text{grad} \, p = K + \eta \, \Delta v,$$
$$T_t + v \, \text{grad} \, T = \delta \, \Delta T, \qquad (53)$$
$$\text{div} \, v = 0,$$

with the density relation

$$\rho = \rho_0 - \alpha \rho_0 (T - T_0). \qquad (54)$$

The physical motivation for these equations will be given in Section 72.10. The quantities there have the following meaning:

ρ_0 constant average density;
ρ density;

72.9. Bénard Problem and Bifurcation

v velocity vector;
p pressure;
T temperature;
T_0 constant average temperature;
η viscosity;
$v = \eta/\rho_0$ kinematic viscosity;
κ heat conductivity;
c specific heat capacity;
$\delta = \kappa/\rho_0 c$ heat diffusion coefficient;
$\int_\Omega K \, dV$ outer force which is applied to the region Ω.

The positive constant α in (54) must be determined experimentally. We now consider the situation of the Bénard problem. There K corresponds to the gravitational force, i.e.,

$$K = -\rho g e_3, \tag{55}$$

where g denotes the gravitational acceleration. We consider a stationary problem, i.e., the time derivatives v_t and T_t in (53) are identically zero. We choose a Cartesian (x, y, z)-system. The equations for the lower and upper plate in Figure 72.6 are given by $z = 0$ and $z = h$, respectively. Hence we require

$$T = T_0 \quad \text{for} \quad z = 0 \quad \text{and} \quad T = T_1 \quad \text{for} \quad z = h.$$

Proposition 72.15. *The stationary problem to (53), (54) has the solution*

$$\begin{aligned} v^* &= 0, \quad T^* = T_0 - \beta z, \quad p^* = p_0 - g\rho_0(z + \tfrac{1}{2}\alpha\beta z^2), \\ \rho^* &= \rho_0(1 + \alpha\beta z), \end{aligned} \tag{56}$$

where $\beta = (T_0 - T_1)/h$.

PROOF. Explicit computation. □

The trivial solution (56) corresponds to a fluid at rest, where temperature and pressure depend on the position. In order to give a mathematical description of the Bénard effect, we search for solutions which are perturbations of (56), i.e., we choose the following ansatz

$$v = v^* + \bar{v}, \quad T = T^* + \bar{T}, \quad p = p^* + \bar{p},$$
$$\rho = \rho^* + \bar{\rho}.$$

Moreover, we pass to dimensionless quantities through

$$\xi_1 = \frac{x}{2h}, \quad \xi_2 = \frac{y}{2h}, \quad \xi_3 = \frac{z-1}{2h},$$

$$\bar{v}_i = \frac{2v}{h} w_i, \qquad i = 1, 2, 3,$$

$$\bar{T} = \frac{v\beta h}{2\delta R} w_4, \qquad \bar{p} = \frac{4\rho_0 v^2}{h^2} w_5.$$

The stationary problem (53), (54) then yields the following key equations

$$\Delta w_i + w_4 \delta_{i3} - D_i w_5 = N_i(w), \qquad i = 1, 2, 3,$$
$$\Delta w_4 + R w_3 = N_4(w), \tag{57}$$

with divergence condition

$$\sum_{i=1}^{3} D_i w_i = 0, \tag{58}$$

where

$$N_i(w) = \sum_{j=1}^{3} w_j D_j w_i, \qquad i = 1, 2, 3,$$

$$N_4(w) = P \sum_{j=1}^{3} w_j D_j w_4.$$

The two dimensionless numbers

$$R = \frac{g\alpha\beta}{v\delta}\left(\frac{h}{2}\right)^4, \qquad P = \frac{v}{\delta}$$

are called the Rayleigh number R and Prandtl number P. In order for Bénard cells to occur, the value of R is essential, while P, in the following, is a fixed but arbitrary value. In experiments Bénard cells appear at the critical value $R = 1{,}700 \pm 50$. Mathematical analysis shows that $R_{\text{crit}} = 1{,}708$ (see Problem 72.8).

72.9b. Mathematical Problem

Equations (57), (58) have the trivial solution $w_i = 0$ for all i. We are looking for nontrivial solutions. In order to present the key idea in the proof of the existence of bifurcation solutions as clearly as possible, we begin with a simple situation, i.e., we consider solutions which are periodic in ξ_1 and ξ_2. This leads to rectangular Bénard cells, which are not observed in the experiment. Thus our proof is somewhat academic. But the case of hexagonal cells can be treated in a similar way (see Problem 72.7). Furthermore, we emphasize the functional-analytic aspects of the existence proof. The same methods, which are used in the following, can also be applied to bifurcation problems for general elliptic systems. We consider the strip region

$$G = \{x \in \mathbb{R}^3 : -1 < \xi_3 < 1\}, \qquad x = (\xi_1, \xi_2, \xi_3)$$

72.9. Bénard Problem and Bifurcation

and the periodicity cuboid

$$Q = \{x \in \mathbb{R}^3 : 0 \leq \xi_1 \leq p_1, 0 \leq \xi_2 \leq p_2, -1 \leq \xi_3 \leq 1\}$$

where $p_1 > 0$ and $p_2 > 0$.

Problem 72.16. We are looking for five real functions w_i and a real number $R > 0$ which satisfy the following conditions:

(i) Differential equation. Equations (57) and (58) are valid on G.
(ii) Boundary condition.

$$w_i = 0 \quad \text{on } \partial G, \quad i = 1, 2, 3, 4. \tag{59}$$

(iii) Periodicity condition. For all i and all $x \in G$ we have

$$w_i(\xi_1 + p_1, \xi_2 + p_2, \xi_3) = w_i(\xi_1, \xi_2, \xi_3). \tag{60}$$

(iv) Normalization condition for the pressure w_5.

$$\int_0^{p_1} \int_0^{p_2} w_5(\xi_1, \xi_2, 0) \, d\xi_1 \, d\xi_2 = 0. \tag{61}$$

(v) Symmetry condition.

$$\begin{aligned} w_1 \in A_1, S_2, \quad & w_2 \in S_1, A_2, \\ w_3, w_4, w_5 \in S_1, S_2. & \end{aligned} \tag{62}$$

Here A_j and S_j mean antisymmetry and symmetry with respect to the variable ξ_j, respectively.

72.9c. The Operator Equation

Our goal is to transform this boundary-value problem into an operator equation

$$B_R w = N(w), \quad w \in X, \quad R > 0. \tag{63}$$

Thereby (63) corresponds to the system (57). The spaces X and Y are constructed in such a way that the operators $B_R, N : X \to Y$ are continuous.

By definition, the B-space X consists of all function tuples $w = (w_1, \ldots, w_5)$ which satisfy the following properties:

(a) $w_i \in C^{2,\alpha}(Q)$ for $i = 1, 2, 3, 4$ with fixed $\alpha \in \,]0, 1[$.
(b) $w_5 \in C^{1,\alpha}(Q)$.
(c) Relations (58)–(62) are valid.

The norm on X is naturally given by

$$\|w\| = \sum_{i=1}^{4} \|w_i\|_{2,\alpha} + \|w_5\|_{1,\alpha}.$$

Moreover, we introduce on X the scalar product:

$$(w|u) = \int_Q R_0 \sum_{i=1}^{3} w_i u_i + w_4 u_4 \, dx.$$

Furthermore, let Y be the B-space of all tuples $f = (f_1, \ldots, f_4)$ with $f_i \in C^\alpha(Q)$ such that the periodicity condition (60) is satisfied for all i. The norm on Y is given by

$$\|f\| = \sum_{i=1}^{4} \|f_i\|_\alpha.$$

72.9d. The Linearized Equation

We consider the inhomogeneous linear equation

$$B_{R_0} w = f, \quad w \in X, \tag{64}$$

which corresponds to the linearization of (63). The following condition is important.

(H) For fixed $R_0 > 0$, equation (64) with $f = 0$ has precisely one linearly independent eigensolution $\bar{w} \in X$. We normalize this solution through $(\bar{w}|\bar{w}) = 1$.

Lemma 72.17. *Assume* (H). *Let* $f \in Y$ *be given. Equation* (64) *then has a solution if and only if*

$$(\bar{w}|f) = 0. \tag{65}$$

The necessity of the solvability condition (65) follows immediately from (57) after multiplication with $\bar{w}_i$ and integration by parts over Q.

A proof of the sufficiency of (65), based on deep results about linear elliptic systems, may be found in Fife (1970). There, Fourier transformations and the Riesz–Schauder theory are used. It is important that the so-called complementing condition of Agmon, Douglis, and Nirenberg (1959) can be verified. It guarantees that the necessary *a priori* estimates of type $C^{k,\alpha}$ are satisfied.

72.9e. Existence and Uniqueness of the Bifurcation Branch

Theorem 72.D. *Assume* (H). *Then there exist positive numbers* $s_0, r_0, \varepsilon_0 > 0$ *such that for given real s with* $0 < |s| \leq s_0$ *there is a unique solution* $w \in X, R > 0$ *of the given problem* (63) *with*

$$\|w\| \leq r_0, \quad (\bar{w}|w) = s, \quad |R - R_0| \leq \varepsilon_0.$$

72.9. Bénard Problem and Bifurcation

This solution is analytic with respect to s, i.e.,

$$w = s\bar{w} + O(s^2), \qquad R = R_0 + \sum_{k=1}^{\infty} \varepsilon_k s^k. \tag{66}$$

The series for w converges absolutely in the B-space X.

Remark 72.18. The expansion coefficients in (66) can be determined from the procedure in Corollary 8.25. At each step, one has to solve a linear elliptic system of the form (64).

Corollary 72.19. *The periods p_1 and p_2 can be chosen in such a way that the smallest positive eigenvalue $R = R_0$ of equation $B_R w = 0$, $w \in X$, satisfies condition (H), and hence Theorem 72.D can be applied to this R_0.*

PROOF OF THEOREM 72.D. We follow Zeidler (1972) and apply the main theorem of analytic bifurcation theory (Theorem 8.A).

For this, we set $R = R_0 + \varepsilon$ and the original operator equation (63) becomes

$$B_{R_0} w = \varepsilon L w + N(w), \qquad w \in X \tag{67}$$

with

$$(Lw)_i = -w_3 \delta_{i4}, \qquad i = 1, 2, 3, 4.$$

We write (67) in the form

$$F(\varepsilon, w) = 0, \qquad w \in X,$$

which implies that

$$F_w(0, 0) = B_{R_0}.$$

From Lemma 72.17 it follows that the operator $F_w(0,0): X \to Y$ is Fredholm of index zero. We construct a linear continuous functional $w^* \in X^*$ by setting

$$\langle w^*, w \rangle = (\bar{w}|w) \qquad \text{for all} \quad w \in X.$$

Lemma 72.17 implies

$$\langle w^*, F_w(0,0)w \rangle = 0 \qquad \text{for all} \quad w \in X,$$

i.e., $F_w(0,0)^* w^* = 0$.

The key bifurcation condition of Theorem 8.A is

$$\langle w^*, F_{\varepsilon w}(0,0)\bar{w} \rangle \neq 0, \tag{68}$$

and explicitly, this means $(\bar{w}|L\bar{w}) \neq 0$, i.e.,

$$-\int_Q \bar{w}_3 \bar{w}_4 \, dx \neq 0.$$

Let us prove this. The equation

$$B_{R_0} \bar{w} = 0$$

corresponds to the homogeneous problem (57), (58) with $R = R_0$, and integration by parts yields the key relation

$$\sum_{i=1}^{4} \int_Q (\operatorname{grad} \bar{w}_i)^2 \, dx = -\sum_{i=1}^{4} \int_Q \bar{w}_i \Delta \bar{w}_i \, dx$$

$$= (1 + R_0) \int_Q \bar{w}_3 \bar{w}_4 \, dx.$$

Now suppose that

$$\int_Q \bar{w}_3 \bar{w}_4 \, dx = 0.$$

We obtain $\operatorname{grad} \bar{w}_i = 0$ for $i = 1, \ldots, 4$, and together with the boundary condition $\bar{w}_i = 0$ on ∂G it follows that

$$\bar{w}_i = 0 \quad \text{on } G \quad \text{for } i = 1, \ldots, 4.$$

The homogeneous equations (57) for $\bar{w}_i$ and the normalization condition (61) then imply

$$\bar{w}_5 = 0 \quad \text{on } G,$$

and hence $\bar{w} = 0$. This contradicts $(\bar{w}|\bar{w}) = 1$. Consequently, the generic bifurcation condition (68) is satisfied.

Theorem 72.D therefore follows from Theorem 8.A. □

The proof of Corollary 72.19 may be found in Rabinowitz (1968). It is analogous to the proof of Lemma 72.14.

72.10. Physical Motivation of the Boussinesq Approximation

As in Section 70.5, the Navier–Stokes equations for a viscous fluid, including thermodynamical effects, have the following form. We assume that no outer heat sources are present. In this connection, observe (70.12) to (70.16).

(i) Momentum balance

$$\rho v_t + \rho (v \operatorname{grad}) v - \eta \Delta v + (\eta - \eta') \operatorname{grad} \operatorname{div} v = K - \operatorname{grad} p. \quad (69)$$

(ii) Mass balance

$$\rho_t + \operatorname{div} \rho v = 0. \quad (70)$$

(iii) Balance of inner energy

$$\rho e_t + \rho v \operatorname{grad} e = [\tilde{\tau} - pI, Dv] - \operatorname{div} q \quad (71)$$

with the tensor of inner friction

$$\tilde{\tau} = 2\eta \, Dv + (\eta' - 2\eta)(\operatorname{tr} Dv) I.$$

(iv) Thermodynamical equation

$$\rho = \rho(T), \quad e = e(T). \tag{72}$$

(v) Fourier law for the heat flow

$$q = -\kappa \operatorname{grad} T. \tag{73}$$

Typical for the Boussinesq approximation is the fact that the temperature difference

$$T - T_0$$

is regarded as small. Hence, in the Taylor expansion for the density ρ and specific inner energy density e, only the linear terms are considered, i.e.,

$$\rho = \rho_0 - \rho_0 \alpha (T - T_0),$$
$$e = e_0 + c(T - T_0).$$

Relation (67.11) shows that $c = e_T$ is the specific heat capacity.

In (69)–(71), we only consider the zero-th approximation of the density, i.e., we set $\rho = \rho_0 = \text{const}$. This implies

$$\operatorname{div} v = 0 \tag{74}$$

and

$$\rho_0 v_t + \rho_0 (v \operatorname{grad}) v - K + \operatorname{grad} p - \eta \Delta v = 0. \tag{75}$$

The temperature dependence of the density is considered only for the gravitational force, i.e., we set

$$K = -\rho g e_3. \tag{76}$$

Moreover, in the balance of inner energy (71), we neglect the inner friction, i.e., we set $\tilde{\tau} = 0$. Thus we obtain

$$T_t + v \operatorname{grad} T = \delta \Delta T \tag{77}$$

with $\delta = \kappa/\rho_0 c$.

Equations (74)–(77) correspond to the basic equations (53).

72.11. The Kolmogorov 5/3-Law for Energy Dissipation in Turbulent Flows

We shall use a very rough argument in order to obtain information about turbulent flows. Our notations are:

- η viscosity;
- ρ density;
- ε total rate of energy dissipation;
- λ diameter of eddies;
- k wave number of eddies ($k \stackrel{\text{def}}{=} 2\pi/\lambda$).

It is a typical property of turbulent flows that there exist eddies of different diameters λ, where

$$\lambda_{min} \leq \lambda \leq \lambda_{max}.$$

One may think, for example, of clouds in the air or of nebulas in astronomy. One finds that the large eddies tend to break down into smaller eddies. This way energy from large eddies flows to smaller eddies. Moreover, physicists assume that the energy of the smallest eddies with $\lambda = \lambda_{min}$ is transformed into heat by friction (energy dissipation). Viscosity is of significance only for small eddies. We define

$$\varepsilon = \frac{\text{loss of kinetic energy by dissipation}}{\text{mass} \cdot \text{time}}.$$

This is a very important physical quantity. Note that ε can be measured in experiments; it is equal to the produced heat. We are interested here in the distribution of ε with respect to λ. To this end, we make the ansatz

$$\varepsilon = \int_{\lambda_{min}}^{\lambda_{max}} \mu(\lambda)\, d\lambda, \tag{78}$$

i.e.,

$$\int_{\lambda_{min}}^{\lambda_0} \mu(\lambda)\, d\lambda$$

is the loss of kinetic energy per mass and time, caused by all eddies of diameter λ with $\lambda_{min} \leq \lambda \leq \lambda_0$.

Kolmogorov Law 72.20. *We have*

$$\mu(\lambda) = C(\lambda/\lambda_{min})\frac{\eta}{\rho}\varepsilon^{2/3}\lambda^{-7/3}, \tag{79}$$

where C is dimensionless.

If λ is near λ_{min}, then we can replace C with a constant. Frequently, one uses the so-called wave number of eddies $k = 2\pi/\lambda$ and makes the ansatz

$$\varepsilon = \frac{\eta}{\rho} \int_{k_{min}}^{k_{max}} E(k)k^2\, dk.$$

Then, for large k near k_{max}, one obtains

$$E(k) = \text{const} \cdot \varepsilon^{2/3} k^{-5/3}. \tag{80}$$

This is called the *Kolmogorov 5/3-law*.

Let us motivate (79). It is natural to assume that $\mu(\lambda)$ depends only on η, ρ, ε, and λ. We therefore make the ansatz

$$\mu = C\eta^a \rho^b \varepsilon^c \lambda^d.$$

The dimensions are

$$[\eta] = \text{kg/ms}, \quad [\rho] = \text{kg/m}^3, \quad [\varepsilon] = \text{m}^2/\text{s}^3, \quad [\lambda] = \text{m},$$
$$[\mu] = [\varepsilon/\lambda] = \text{m/s}^3,$$

and comparison of dimensions yields

$$b = -a, \quad c = 1 - a/3, \quad d = -1 - 4a/3.$$

Finally, let us motivate that $a = 1$. The energy dissipation in the region G is equal to the time derivative of the kinetic energy in the region G, i.e.,

$$\frac{d}{dt}\int_G \tfrac{1}{2}\rho v^2 \, dx = \int_G \rho v v_t \, dx$$

if the density ρ is constant. The Navier–Stokes equations yield $\rho v_t = \eta \, \Delta v + \cdots$. Hence it is natural to postulate that the energy dissipation depends on η^a with $a = 1$.

Formula (80) is obtained the same way.

Finally, we want to apply a typical argument used by physicists in order to get some information about the magnitude of $\lambda_{\min}$. We make the ansatz

$$\lambda_{\min} = \text{const} \cdot \eta^a \rho^b \varepsilon^c.$$

Comparison of dimensions yields $a = -b = \tfrac{3}{4}$, $c = -\tfrac{1}{4}$. Physical experience shows that dimensionless constants are not too small and not too large. Hence, we assume that

$$\lambda_{\min} \sim \left(\frac{\eta^3}{\rho^3 \varepsilon}\right)^4.$$

72.12. Velocity in Turbulent Flows

In a turbulent flow, the velocity field $v = v(x, t)$ is assumed to be a stochastic process. What can be measured in experiments are the mean velocity vector

$$\overline{v(x, t)}$$

at the point x and time t, and the dispersions

$$(\Delta v_i)^2 = \overline{u_i(x, t)^2}, \quad i = 1, 2, 3$$

for the velocity components v_i. Here we set $u_i = v_i - \bar{v}_i$. The bar denotes expectation values. The mean density of kinetic energy is equal to

$$\tfrac{1}{2}\overline{\rho v(x, t)^2}.$$

In order to describe the correlation between the velocity vectors at different points, one uses the correlation coefficients

$$C_{ij}(x, x + h; t) = \frac{\overline{u_i(x, t) u_j(x + h, t)}}{\Delta v_i(x, t) \Delta v_j(x + h, t)}, \quad i, j = 1, 2, 3.$$

We have $0 \le C_{ij} \le 1$. The larger C_{ij} is, the larger the correlation between $v_i(x,t)$ and $v_j(x+h,t)$ becomes. If $v_i(x,t)$ and $v_j(x+h,t)$ are independent of each other, then $C_{ij} = 0$.

As an important example, let us consider a one-dimensional flow, i.e., $v = v_1(\xi)e_1$ and $x = \xi e_1 + \cdots$.

Kolmogorov Law 72.21. *If λ is not too large and not too close to a critical distance $\lambda_{\min}$, then*

$$\overline{(v_1(\xi + \lambda) - v_1(\xi))^2} = \text{const} \cdot (\varepsilon\lambda)^{2/3}. \tag{81}$$

Here ε is the total rate of energy dissipation as used in the previous section.

In order to motivate (81) let $\lambda_{\min}$ be the smallest eddy size. Physicists assume that the viscosity η does not play any role if the eddy size satisfies $\lambda \gg \lambda_{\min}$. Hence we make the ansatz const $\cdot \rho^a \varepsilon^b \lambda^c$. Comparison of dimensions yields (81).

Problems

72.1.* *Functional analytic decomposition of vector fields.* Prove Lemma 72.1.

Hint: A short and elegant proof can be given by using the deep result of de Rham of Problem 72.2. See Temam (1977, M), p. 15 and p. 19. A direct proof, which uses the solution of the second boundary-value problem for the Laplace equation, and further material can be found in Girault and Raviart (1981, L), p. 31.

72.2.* *Pressure as a mathematical distribution.* Show:
  (i) Let G be an open set in $\mathbb{R}^3$ and $g = (g_1, g_2, g_3)$, where the g_i are distributions, i.e., elements of $\mathscr{D}'(G)$ (see $A_2(64)$). Then

$$g = \text{grad } p \tag{82}$$

has a solution $p \in \mathscr{D}'(G)$ if and only if

$$\sum_{i=1}^{3} g_i(v_i) = 0 \tag{83}$$

holds for all $v_i \in C_0^\infty(G)$, $i = 1, 2, 3$ with div $v = 0$.
  (ii) If the set G satisfies the regularity assumption (2) and if $g_i \in W_2^{-1}(G)$ for all i, then equation (82) has a solution $p \in L_2(G)$ if and only if (83) is satisfied. Moreover, up to an additive constant, p is uniquely determined. As usual, $W_2^{-1}(G)$ denotes the dual space to the Sobolev space $\mathring{W}_2^1(G)$.

Hint: See Temam (1977, M), p. 14. Statement (i), which, in an analogous form, holds in $\mathbb{R}^n$, is a special case of a general theorem of de Rham about differential forms with distributions as coefficients (see de Rham (1960, M), p. 114). An elementary proof of (ii) can be found in Temam (1977, M), p. 19.

Let $g_i \in L_2(G)$ for all i. Then the necessity of condition (83) follows by using integration by parts. In fact, equation (82) implies

$$\int_G gv\,dx = \int_G (\text{grad } p)v\,dx = -\int_G p\,\text{div } v\,dx = 0$$

for all $v_i \in C_0^\infty(G)$, $i = 1, 2, 3$ with div $v = 0$.

72.3.* *Very weak solutions for the nonstationary problem.* As in Chapter 30, use the Galerkin method to show that the generalized nonstationary problem for the Navier–Stokes equations of Section 72.5 has a solution

$$v \in L_2(0, T; V)$$

for every

$$v_0 \in H, \qquad K \in L_2(0, T; Y)$$

with $Y = L_2(G, V_3)$. In contrast to Theorem 72.B, the uniqueness for these very weak solutions has not yet been verified.

Hint: See Temam (1977, M), p. 282. There one also finds uniqueness results for weakly regular solutions. However, the existence in this case remains open.

72.4.* *Positivity of Green's functions.* We set

$$Lu = -Au'' + Bu' + C$$

and consider the two boundary-value problems

$$Lu = f \quad \text{on } [a, b], \qquad u(a) = u(b) = 0,$$

and

$$L^2 u = f \quad \text{on } [a, b], \qquad u(a) = u'(a) = u(b) = u'(b) = 0,$$

where $-\infty < a < b < \infty$. For the coefficients we assume

$$A, B, C \in C^4[a, b] \qquad \text{and} \qquad A, C > 0 \quad \text{on } [a, b].$$

Prove: Green's functions for L and L^2 are positive on $]a, b[\times]a, b[$.

Hint: Use analogous arguments as in Problem 7.2g. See Kirchgässner (1961), p. 18. More general results may be found in Karlin (1967, M), p. 534, in connection with the theory of oscillating kernels of Krein and Gantmacher.

72.5.* *Integral equations with positive kernels.* Let $K: [a, b] \times [a, b] \to \mathbb{R}$ be continuous and positive on $]a, b[\times]a, b[$ for $-\infty < a < b < \infty$.

Show: The integral equation

$$u(r) = \mu \int_a^b K(r, s) u(s)\,ds$$

has a simple characteristic number $\mu_0 > 0$ of smallest absolute value. The corresponding eigenfunction is positive on $]a, b[$.

Hint: This is a sharpening of Example 7.30. See Jentzsch (1912), p. 248.

72.6. *Proof of (52).*

Solution: We set

$$\lambda = \lambda_0(n), \qquad \alpha = \alpha_{n,0}, \qquad \beta = \beta_{n,0}.$$

From

$$L^2\alpha = \lambda k a \beta \quad \text{and} \quad L\beta = \lambda k b \alpha$$

follows, after integration by parts over $[1, R]$, that

$$\Gamma \stackrel{\text{def}}{=} \lambda k \int_1^R (a+b)\alpha\beta r\, dr$$

$$= \int_1^R (\alpha L^2 \alpha + \beta L \beta) r\, dr$$

$$\geq \int_1^R (k^2\beta^2 + k^4\alpha^2) r\, dr \stackrel{\text{def}}{=} \Gamma_1.$$

Because of the positivity of a, b, α, β, we have $\Gamma > 0$. From $2\alpha\beta \leq k\alpha^2 + \beta^2/k$ follows

$$\Gamma \leq \frac{\lambda \Gamma_1}{2k^2} \max_{1\leq r \leq R} (a(r) + b(r)).$$

Since $\Gamma_1 \leq \Gamma$ we thus obtain (52).

72.7.* *Bénard problem with hexagonal cells.* Give an existence proof for this case.

Hint: See Judovič (1967). The proof is analogous to the proof of Section 72.9, but the treatment of the linearized problem is now more complicated.

The following question is important: Why does nature choose hexagonal cells? The answer must have something to do with stability analysis. Up to now there exists no general stability analysis for the complete nonlinear problem. Partial results can be found in Sattinger (1977), where group theory and bifurcation theory are combined. In this direction, see also Sattinger (1979, L), (1980, S), Knightly and Sather (1985).

72.8.* *Weak stability analysis for the Bénard problem.* Starting with the time-dependent equations (53) find solutions of the linearized problem in the form of two-dimensional waves.

Show: There exists a critical Rayleigh number $R_{\text{crit}} = 1{,}708$ with the following property. For $R < R_{\text{crit}}$ these waves are stable, while for $R > R_{\text{crit}}$ they become unstable.

Hint: See Chandrasekhar (1961, M) and Ebeling (1976, M). This result is in good correspondance with experiment. Bénard cells occur for $R = 1{,}700 \pm 50$.

Many concrete stability results can be found in Joseph (1976, M).

72.9. *Boundary layers and singular perturbation problems.*

72.9a. *A mathematical model.* We consider the differential equation

$$\varepsilon y'' + y' = b \tag{84}$$

with the small parameter $\varepsilon > 0$. The general solution is

$$y = bx + C + De^{-x/\varepsilon}. \tag{85}$$

If $\varepsilon = 0$, then the general solution of (84) is

$$y = bx + C. \tag{86}$$

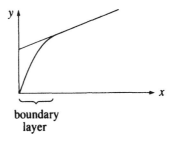

boundary layer

Figure 72.8

The *first key* result is that the solutions (85) and (86) differ essentially only in a thin boundary layer near the boundary point $x = 0$. For example, if we add the boundary conditions

$$y(0) = 0, \qquad y(1) = 1 \tag{87}$$

to equation (84), then we obtain the unique solution

$$y_\varepsilon = bx + (1-b)\frac{1 - e^{-x/\varepsilon}}{1 - e^{-1/\varepsilon}} \tag{88}$$

(Fig. 72.8). Problem (84), (87) is called a singular perturbation problem, since the term "$\varepsilon y''$" with highest derivative disappears as $\varepsilon \to 0$, i.e., the type of the problem changes drastically. One may regard "$\varepsilon y''$" as a small viscosity term. In fact, singular perturbation problems occur very frequently in the natural sciences, and the existence of boundary layers is typical for such problems.

72.9b. Rescaling. If we set

$$\xi = x/\varepsilon,$$

then we obtain

$$y_\varepsilon = (1-b)(1 - e^{-\xi}) + O(\varepsilon), \qquad \varepsilon \to 0.$$

This yields the *second key* result: The behavior of the solution in the boundary layer ($0 \leq \xi \leq 1$) can be described best by a suitable rescaling.

In the following we describe a general method to obtain the correct rescaling by just using the original equation. We make the ansatz

$$\xi = x/\varepsilon^k.$$

From (84) follows

$$\varepsilon^{1-2k}\frac{d^2 y}{d\xi^2} + \varepsilon^{-k}\frac{dy}{d\xi} = b.$$

We now postulate that $\varepsilon^{1-2k} = \varepsilon^{-k}$. Hence $k = 1$ and

$$\frac{d^2 y}{d\xi^2} + \frac{dy}{d\xi} = \varepsilon b.$$

This is a *regular* perturbation problem.

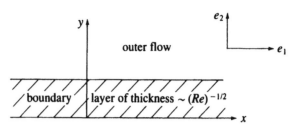

Figure 72.9

72.9c. *The Prandtl boundary-layer equation* (1904). We want to apply the preceding argument to the Navier–Stokes equations and consider a planar incompressible viscous flow in the upper half-plane $G = \{(x, y) \in \mathbb{R}^2 : y > 0\}$ with velocity vector

$$v = ae_1 + be_2 \quad \text{(Fig. 72.9)}.$$

Suppose that x, y, a, b are dimensionless. According to (13) the dimensionless Navier–Stokes equations are

$$\alpha a_t + aa_x + ba_y = -p_x + \varepsilon \Delta a,$$
$$\alpha b_t + ab_x + bb_y = -p_y + \varepsilon \Delta b, \tag{89}$$
$$a_x + b_y = 0$$

on G with the boundary conditions

$$a = b = 0 \quad \text{if} \quad y = 0. \tag{90}$$

Here we set

$$\varepsilon = 1/\text{Re}.$$

We assume that the Reynolds number Re is *large*, i.e., ε is small. If we set $\varepsilon = 0$ in (89), then we obtain the Euler equations for inviscid fluids. But in the case of an inviscid fluid, we only have the boundary condition

$$b = 0 \quad \text{if} \quad y = 0. \tag{91}$$

This contradicts (90). In 1904, Prandtl proposed the following model for large Reynolds numbers.
(i) Outside a thin boundary layer of thickness $\varepsilon^{1/2} = (\text{Re})^{-1/2}$ one uses equation (89) with $\varepsilon = 0$.
(ii) Inside the boundary layer one uses the so-called Prandtl boundary layer equations

$$\alpha a_t + aa_x + ba_y = -p_x + \varepsilon a_{yy},$$
$$p_y = 0, \tag{92}$$
$$a_x + b_y = 0,$$

with the boundary condition

$$a = b = 0 \quad \text{if} \quad y = 0.$$

Motivate this model.

Solution: The idea is that the essential effects occur in the direction of the y-axis. Therefore we use the following rescaling

$$X = x, \quad Y = y/\varepsilon^k, \quad A = a, \quad B = b/\varepsilon^m.$$

Formula (89) implies

$$A_t + AA_X + \varepsilon^{m-k}BA_Y = -p_X + \varepsilon A_{XX} + \varepsilon^{1-2k}A_{YY},$$

$$\varepsilon^{m+k}B_t + \varepsilon^{m+k}AB_X + \varepsilon^{2m}BB_Y = -p_Y + \varepsilon^{m+k+1}B_{XX} + \varepsilon^{m-k+1}B_{YY}, \quad (93)$$

$$A_X + \varepsilon^{m-k}B_Y = 0.$$

We assume that the Y-derivatives of A are essential. From the first line of (93), a natural choice is $\varepsilon^{m-k} = \varepsilon^{1-2k} = 1$. Hence

$$m = k = \tfrac{1}{2}.$$

Letting $\varepsilon = 0$ in (93) and going back to the original variables we obtain (92).

Observe that it is very important to use dimensionless quantities and the rescaling argument. Otherwise one will not find a clear motivation for the Prandtl boundary-layer equation.

72.9d.* *Existence and uniqueness theorems for the Prandtl boundary layer equations.* Study Oleinik (1968). Information about the qualitative behavior of the solutions may be found in Nickel (1958), (1963), where differential inequalities play an important role. See also Walter (1964, M).

72.10.* *Generic finiteness of the solution set.* We consider the stationary Navier–Stokes equations

$$\rho(v\,\text{grad})v - \eta\,\Delta v = K - \text{grad}\,p \quad \text{in } G,$$

$$\text{div}\,v = 0 \quad \text{in } G,$$

$$v = 0 \quad \text{on } \partial G.$$

Study Foias and Temam (1977). There it is shown that, for every fixed ρ and η, the solution set is finite for "almost all" forces K. The proof is based on the Smale theorem (Theorem 4.K).

72.11. *Partial regularity of the solutions to the nonstationary problem.*

72.11a. *Hausdorff measure.* Let A be a subset of $\mathbb{R}^n$. By definition, the m-dimensional Hausdorff measure

$$H^m(A) = \inf t$$

of A is the infimum of the set of all t such that $0 \leq t \leq \infty$ and for every $\varepsilon > 0$ there exists a countable covering C of the set A with

$$\sum_{S \in C} \left(\frac{\text{diam}(S)}{2}\right)^m \leq t$$

and $\text{diam}(S) < \varepsilon$ for all $S \in C$.

If m is an integer and V_m denotes the volume of the m-dimensional unit ball, then $V_m H^m(A)$ agrees with the surface area of smooth m-dimensional surfaces A. See Federer (1969, M).

72.11b.** *Navier–Stokes equations.* Let $v = v(x, t)$ be a solution of the nonstationary Navier–Stokes equations. A point (x, t) is called singular if and only if v is not essentially bounded in any neighborhood of (x, t) in the sense of the space L_∞. Study Caffarelli, Kohn, and Nirenberg (1982). There it is shown that for "suitable weak" solutions v of the nonstationary Navier–Stokes equations on an open set in space–time, the associated set of singular points has a one-dimensional Hausdorff measure zero.

This shows that singular points are rare. For example, they cannot form a regular curve.

72.12.* *Statistical solutions of the Navier–Stokes equations.* Let S be the set of all solutions $v = v(x, t)$ of the nonstationary Navier–Stokes equations. Let S_0 be the set of all possible initial values $v_0 = v_0(x)$, where $v_0(x) = v_0(x, t)$. Suppose we are given a probability measure p_0 on S_0, i.e.,

$$p_0(A) = \text{probability of } v_0 \in A.$$

We are looking for a probability measure p on S, i.e.,

$$p(B) = \text{probability of } v \in B,$$

where p and p_0 are compatible, i.e.,

$$p(v: v_0 \in A) = p_0(A).$$

We call p a statistical solution of the Navier–Stokes equations. The existence of statistical solutions and their properties are studied in detail in Višik and Fursikov (1980, M). The main idea is to use a Galerkin method and a theorem of Prohorov on weak compactness of sets in measure spaces, where the measures are defined over metric spaces.

72.13. *Turbulence, Feigenbaum bifurcation, chaos, and universality theory.* See Problem 17.13.

72.14. *Turbulence and strange attractors.*

72.14a. *The Hénon attractor.* Let $a = 1.4$ and $b = 0.3$. Compute

$$\begin{aligned} x_{n+1} &= y_n + 1 - ax_n^2, \\ y_{n+1} &= bx_n \end{aligned} \tag{94}$$

for $n = 0, 1, \ldots, 10^5$ on a computer. For a bad choice of the initial values x_0, y_0, the sequence will tend to infinity. For a good choice of x_0, y_0, the sequence will rapidly approach the Hénon attractor (Figure 72.10 shows

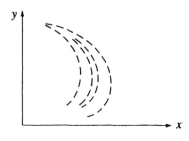

Figure 72.10

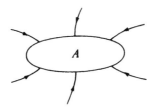

Figure 72.11

this schematically). One will observe that the sequence behaves in a strange way.

Roughly speaking, an attractor A for a dynamical system is a set which attracts the trajectories in a neighborhood of A (Fig. 72.11). Such an attractor is called *strange* if the trajectories depend *sensitively* on the initial data.

If one chooses $a = 1.3$ and $b = 0.3$, then the strange attractor disappears and an attractor of period 7 appears.

72.14b. *The Lorenz attractor and meteorology.* Compute the trajectory of the initial-value problem

$$\dot{x} = a(y - x), \qquad \dot{y} = bx - y - xz, \qquad \dot{z} = xy - c, \tag{95}$$
$$x(0) = y(0) = z(0) = 0$$

with $a = 10$, $b = 28$, $c = 8/3$. In a Cartesian (x, y, z)-system, the trajectory makes one loop to the right, then a few loops to the left, then to the right, and so on in an irregular fashion. In fact, the trajectory approaches a strange attractor which lies in a neighborhood of the origin. See Guckenheimer and Holmes (1983), p. 92.

This system was considered by Lorenz (1963) as a simple Galerkin system for the heat convection in a two-dimensional horizontal fluid layer heated from below. The components of (95) correspond to the Fourier coefficients of velocity and temperature. Lorenz was interested in this system as a model of the heat convection in the atmosphere of the earth. Thus we expect to have turbulence for critical data and the Lorenz attractor reflects this. Since the trajectories depend sensitively on the initial data, this provides some theoretical excuse for the unreliability of weather forecasting.

72.14c. *Continuous frequency spectrum, strange attractors, and turbulence.* Let $f(t)$ be a time-dependent physical quantity (e.g., the velocity component at a fixed point in a fluid). Physicists analyze such quantities by using the Fourier integral

$$f(t) = \int_{-\infty}^{\infty} g(\omega) e^{-i\omega t} \, d\omega. \tag{96}$$

This is the superposition of special periodic motions

$$g(\omega) e^{-i\omega t},$$

where ω is the angular frequency and $|g(\omega)|$ is the amplitude. If the frequency diagram has only a finite number of sharp peaks such as in Figure 72.12(a),

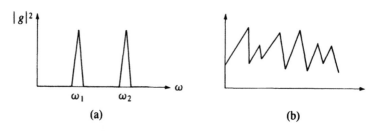

Figure 72.12

then this corresponds to a special quasi-periodic behavior, i.e.,

$$f(t) = \sum_{k=1}^{N} g_k e^{-i\omega_k t}. \tag{97}$$

By definition, a general quasi-periodic behavior is described by

$$f(t) = \sum_{n_1,\ldots,n_N} g_{n_1\ldots n_N} e^{-i(n_1\omega_1 + \ldots + n_N\omega_N)t},$$

where the sum is taken over all integers $n_1, \ldots, n_N$. This corresponds to sharp peaks at $n_1\omega_1 + \cdots + n_N\omega_N$ in the frequency diagram. If $g(\omega) \neq 0$ for a continuum of angular frequencies ω, then one speaks of a continuous spectrum (Fig. 72.12(b)).

The classical turbulence theory, due to Landau and Hopf, was based on the following idea. There exists a sequence of critical Reynolds numbers

$$\text{Re}_1 < \text{Re}_2 < \text{Re}_3 \ldots.$$

The velocity vector has the quasi-periodic structure

$$v(x,t) = \sum_{n_1,\ldots,n_N} v_{n_1\ldots n_N}(x) e^{-i(n_1\omega_1 + \ldots + n_N\omega_N)t},$$

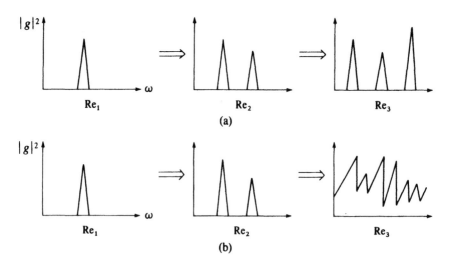

Figure 72.13

where the number of angular frequencies $\omega_1, \ldots, \omega_N$ increases at the critical Reynolds numbers. Those critical Reynolds numbers Re_k converge to a limit Re_{turb}, at which turbulence occurs. Turbulence, in this picture, corresponds to quasi-periodic motions with a very large number of basic frequencies, which simulates a continuous frequency spectrum. Schematically this corresponds to Figure 72.13(a).

In 1971, Ruelle and Takens proposed a completely different approach to turbulence. They assumed that strange attractors are responsible in many cases for the occurrence of turbulence. Mathematically, it is possible that a quasi-periodic motion breaks down and a strange attractor occurs. Strange attractors correspond to a continuous frequency spectrum. Ruelle and Takens therefore predicted a situation as shown in Figure 72.13(b), where a continuous spectrum appears very rapidly. Perhaps this can be verified by using delicate physical experiments (see Swinney and Gollub (1978)).

References to the Literature

Classical work on the existence theory for inviscid fluids: Lichtenstein (1929, M). Recent existence proofs: Majda (1984, L), Kato and Lai (1984), DiPerna and Majda (1987).

Basic papers on the existence theory for viscous fluids: Leray (1934a), Hopf (1951).

Classical works on the existence theory for viscous fluids: Odquist (1930), Leray (1933), (1934), (1934a), Hopf (1950) (Burgers equation and modeling of turbulence), (1951) (initial-value problem for the Navier–Stokes equations), (1952) (statistical hydrodynamics), Ladyženskaja (1959), (1970, M), Finn (1959), (1961), (1965, S), Serrin (1963), Fujita and Kato (1964), Judovič (1966) and Velte (1966) (Taylor problem), Judovič (1967) and Rabinowitz (1968) (Bénard problem), Ladyženskaja and Solonnikov (1977) and Solonnikov (1984) (unbounded regions).

Introduction to viscous flow from the physical point of view: Prandtl (1949, M), Landau and Lifšic (1962, M), Vol. VI (standard work).

Introduction to the Navier–Stokes equations from the mathematical point of view: Temam (1977, M).

Monographs: Ladyženskaja (1970), Temam (1977), Višik and Fursikov (1980) (statistical solutions), Girault and Raviart (1981), (1986), Telionis (1981), von Wahl (1985).

Numerical methods: Temam (1977, M), Chorin (1973), (1977), (1978), (1982), Girault and Raviart (1981, L), (1986, M), Thomasset (1981, M), Glowinski (1981, L), (1983, S), (1984, M), Fortin and Glowinski (1983), Holt (1984, M), Peyret (1985, M), Sod (1985), Vols. 1, 2, Chavent (1986) (finite elements and reservoir simulation).

Numerical methods on supercomputers: Murman (1985, P).

Numerical weather prediction: Haltiner and Williams (1980, M).

Recent trends: Temam (1983, S), Berkeley (1983, P), (1986, P), Ruelle (1983, S), Constantin, Foias, and Temam (1985) (turbulence and the dimension of attractors), Ladyženskaja (1986).

Boundary layers in mathematics and singular perturbation theory: Ljusternik and Višik (1957) (classical work), Trenogin (1970, S), Kervorkian and Cole (1981, M), Goering (1983, S) (see also the References to the Literature for Chapter 79).

Rheology: Reiner (1958, S) (handbook article), Fredrickson (1964, M), Wilkinson (1960, M) and Showalter (1978, M) (non-Newtonian fluids), Duvaut and Lions (1972, M) and Naumann (1982) (existence theorems).

Stability for concrete problems in fluid dynamics: Chandrasekhar (1961, M), Joseph (1976, M).

Stability and bifurcation, Taylor problem, and Bénard problem: Kirchgässner (1975, S) and Sattinger (1980, S) (general survey), Serrin (1959), (1959a), Judovič (1966), (1966a), (1967), Velte (1966), Rabinowitz (1968), Kirchgässner and Sorger (1969), Joseph and Sattinger (1972), Zeidler (1972), Kirchgässner and Kielhöfer (1973), Kirchgässner (1975a), Sattinger (1977), (1979, L), (1980, S), Knightly and Sather (1985).

Free boundary-value problems for the Navier–Stokes equations: Pukhnačov (1972), Socolescu (1980), Solonnikov (1983).

Genericity and structure of solutions to the Navier–Stokes equations: Foias and Temam (1977).

Partial regularity of solutions to the nonstationary Navier–Stokes equations: Scheffer (1980), Caffarelli, Kohn, and Nirenberg (1982).

Unbounded regions, infinite channels, and tubes: Ladyženskaja and Solonnikov (1977), (1980), Amick (1978), Solonnikov (1984).

Existence, regularity, and decay of solutions: Heywood (1980, S).

Asymptotic behavior of the kinetic energy of viscous fluids in external regions: Galdi and Maremonti (1986).

Existence theory for the Euler equations for incompressible and compressible inviscid fluids: Majda (1984, L) and Kato and Lai (1984) (especially recommended), Kato (1967), (1972), Temam (1979), Schochet (1986), Di Perna and Majda (1987).

Applications of the methods of global analysis: Arnold (1966), Ebin and Marsden (1970), Marsden (1974, L).

Boundary layer equation: Prandtl (1904) (classical work), Garabedian (1960, M), Schlichting (1960, M), Oleinik (1968) (existence proofs), Nickel (1958), (1963), Walter (1964, M) (differential inequalities), Pukhnačov (1975, M).

Asymptotic methods in fluid dynamics: Zeytounian (1987, M).

Statistical solutions in hydrodynamics: Foias (1973), Višik and Fursikov (1980, M).

Survey on turbulence: Lin and Reid (1963) (handbook article), Frost and Moulden (1977, M) (handbook), Bernard and Ratiu (1977, P), Berkeley (1983, P).

Monographs on turbulence: Chorin (1975, L) (introductory), Batchelor (1982, M), Dwoyer (1985, M).

Kolmogorov flow: Obuhov (1983, S).

Turbulence and universality theory: Feigenbaum (1980, S), Berkeley (1983, P), Vul, Sinai, and Chanin (1984, S).

Turbulence and self-organization: Ebeling and Klimontowitsch (1985, L).

Turbulence, chaos, and strange attractors: Ruelle (1980, S), (1983, S) (introductory), Ruelle and Takens (1971), Bothe (1982) (topological structure of attractors), Sparrow (1982, M) (Lorenz equation), Guckenheimer and Holmes (1983, M) (recommended as an introduction to strange attractors), Bergé, Pomeau, and Vidal (1984, M).

Visual representations: Abraham (1983, M), Peitgen and Richter (1985, M).

Estimates for the dimension of attractors: Ladyženskaja (1982), Babin and Višik (1983, S), Constantin, Foias, and Temam (1985, S), Constantin and Foias (1985) (Kaplan–Yorke formulas).

MANIFOLDS AND THEIR APPLICATIONS

Some four years ago, I observed that a certain number of most significant theorems and constructions of modern mathematics have undergone the following evolution, and that one might even talk of a principle. Viewed historically, at first one knew certain natural objects (e.g., spaces); then certain "abstract objects" were discovered, or one was forced to introduce them. Finally, with considerable effort and brilliance of mind, it was proved that these objects were "simply" subspaces of the well-known spaces, and that in some (favorable) cases they were indeed isomorphic with "classical" objects. The more natural the spaces were, the more difficult it was to prove the corresponding embedding theorems.

I realized that the evolution principle of modern mathematics which I had observed was an exact illustration of the famous parable of the cave from the seventh book of Plato's "Politea." The following correspondences were found:

(a) Shadow's on the cave's walls are classical mathematical objects: e.g., planes in the Euclidean space, algebraic projective varieties, etc.
(b) Ideas are "abstract objects," e.g., Riemann spaces, Hodge manifolds.
(c) The dramatic "descent" into the cave is the corresponding embedding theorem.

And who is the prisoner at first fettered, then released, and dragged (by force) into the sunlight, and finally descending again into the cave? It is mathematics itself as a whole, for it is different researchers belonging to different generations of mathematicians who have accomplished the ascent, the creation of awakening of a great mathematical idea, e.g., the Riemann surface, and an embedding or uniformization theorem which is often, several decades later, proved by quite different mathematicians.

The well-known *bon mot* that "European philosophy" is only a footnote to Plato is perhaps true, but I should venture the much truer one: modern mathematics is only a footnote to Riemann.

Krysztof Maurin (1982)

CHAPTER 73

Banach Manifolds

The categories of differentiable manifolds and vector bundles provide a useful context for the mathematics needed in mechanics, especially the new topological and qualitative results.
                               Ralph Abraham and Jerrold Marsden (1978)

Too often in the physical sciences, the space of states is postulated to be a linear space when the basic problem is essentially nonlinear; this confuses the mathematical development.
                                                        Steve Smale (1980)

The proof that ... is left as a masochistic exercise for the reader. Rest assured that we will never have to do this sort of abstract nonsense.
                                                     Michael Spivak (1979)

Typical examples of manifolds are sufficiently smooth curves and surfaces in $\mathbb{R}^n$ which have a tangent space (tangent line, tangent plane) at each point. Manifolds will always be manifolds *without* boundary. One may think, for example, of the surface of a ball. Manifolds with boundary, such as the ball itself, will be considered in Section 73.19.

The concept of manifolds is one of the most important concepts in mathematical physics, perhaps the most important one. The reason for this is that a description of scientific phenomena uses the process of measurements, i.e., phenomena are locally described by parameters. In astronomy, for instance, we use space and time coordinates. The choice of these coordinates is somewhat arbitrary. *A priori*, it is not clear, for example, how we have to set our clocks. We may adjust them to atomic oscillations or to the daily rotations of the earth. Different observers will generally use different systems of reference. Thus it is important to have a way of comparing the results of measurements. Locally, a manifold looks like $\mathbb{R}^n$ or more generally like a B-space. For

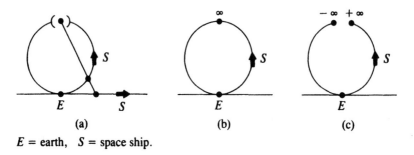

E = earth, S = space ship.

Figure 73.1

a local description, we allow different B-spaces (coordinate or parameter spaces). It is then important to have a transformation rule for these different coordinates. In the simplest case, one may use for the plane, for example, Cartesian or oblique coordinates. From a mathematical or physical point of view, those properties of a manifold are most important which are independent of the choice of the local coordinates. Moreover, it should be noted that besides local properties, manifolds also have global properties. Consider, for example, a space flight in $\mathbb{R}^1$. A space ship, which starts at some point E in $\mathbb{R}^1$, and flies on a straight line, can never, like the flying dutchman in Wagner's opera, reach its goal. It will travel through space $\mathbb{R}^1$ forever (Fig. 73.1(a)). Now, let us use the one-point compactification of $\mathbb{R}^1$, i.e., we add an element ∞ to $\mathbb{R}^1$ such that the stereographic projection maps $\mathbb{R}^1 \cup \{\infty\}$ homeomorphically onto the unit circle S^1 (Fig. 73.1(b)). Our space ship may now return to its starting point. If, however, we use the two-point compactification of $\mathbb{R}^1$, then our space ship may reach the end of the world $+\infty$ (Fig. 73.1(c)). Note that the sets $\mathbb{R}^1$, $\mathbb{R}^1 \cup \{\infty\}$ and $\mathbb{R}^1 \cup \{+\infty\} \cup \{-\infty\}$ have the same local structure in points $x \in \mathbb{R}^1$. But, as we have seen above, the differences in the global structure may have far-reaching physical consequences. In fact, today we do not really understand the global structure of our universe, i.e., the global structure of the corresponding four-dimensional space–time manifold. We will discuss two cosmological models in Chapter 76.

The last seven chapters of this volume and the beginning chapters of the following Part V deal with various aspects of the theory of manifolds and its plentiful applications. Several important features will be discussed more carefully at the beginning of each of these chapters. In this chapter, we will give a number of basic definitions with emphasis on a detailed motivation. The following three chapters contain applications to the theory of surfaces and the special and general theory of relativity. We then give applications to fixed-point theory, the theory of stability for dynamical systems, electrodynamics, mechanics (symplectic geometry), as well as to quantum theory in connection with the theory of Lie groups and Lie algebras. In many textbooks, the definition of tangent bundles is presented, after the more general concept of vector bundles has been introduced. Tangent bundles are then special cases of vector bundles. From experience, I know that this procedure, because of its abstractness, causes difficulties for the student. For didactical reasons, we therefore choose the following presentation. We begin with a definition of tangent and cotangent bundles, and explain the simple connection with abstract bundles in Section 73.7. We show that the concept of tangent bundles arises naturally, when introducing higher-order F-derivatives and vector fields on Banach manifolds. In Section 73.21, we demonstrate the usefulness of tangent bundles in the proof of the fundamental Whitney embedding theorem. Then, after the reader has become somewhat familiar with the concept of bundles, we introduce vector bundles in Section 73.22. We choose a geometric definition which can be generalized to fiber bundles in Part V. Also in Part V, we try to explain the great importance of vector and fiber bundles for global analysis and mathematical physics.

73.1. Local Normal Forms for Nonlinear Double Splitting Maps

Consider a map

$$f: U(x_0) \subseteq X \to Y,$$

which satisfies the following conditions:

(H1) X and Y are B-spaces over $\mathbb{K}$ and f is C^k, $k \geq 1$.
(H2) f is double splitting at x_0, i.e., the null space $N(f'(x_0))$ splits X and the range $R(f'(x_0))$ splits Y (cf. $A_1(22i)$).

We let $N = N(f'(x_0))$ and $R = R(f'(x_0))$. For a Fredholm operator $f'(x_0)$, condition (H2) is always satisfied. In this case, N and $R^\perp$ are finite dimensional. It follows from (H2) that X and Y have the topological direct sum decompositions

$$X = N \oplus N^\perp \quad \text{and} \quad Y = R \oplus R^\perp.$$

Let $U_N(0)$ and $U_R(0)$ denote neighborhoods of 0 in N and R.
In the following we will use the decomposition $Y = R \oplus R^\perp$ to construct the local normal form

$$f(\varphi(n,r)) = f(x_0) + r + g(n,r) \tag{1}$$

for all $n \in U_N(0)$, $r \in U_R(0)$ such that the following conditions are satisfied: $\varphi(0,0) = x_0$ and

$$r \in R, \quad g(n,r) \in R^\perp,$$
$$g(0,0) = 0, \quad g'(0,0) = 0. \tag{2}$$

Let $P: N \times R \to R$, $P(n,r) = r$, denote the natural projection. It follows then from (1) that, after a suitable coordinate change $x = \varphi(n, r)$, f behaves locally, in first-order approximation, like $f(x_0) + P(n, r)$, i.e., locally, in first-order approximation, double splitting maps are translations of projections.

If f is a submersion at x_0, i.e., $f'(x_0)$ is surjective, then the normal form becomes particularly simple, since in this case $R^\perp = \{0\}$ and (1) is satisfied with $g \equiv 0$. Here, f actually equals $f(x_0) + P(n, r)$ in a neighborhood of x_0. In Section 78.9, normal form (1) will be a main tool in proving the Sard–Smale theorem.

Another normal form that we consider is

$$f(\psi(x)) = f(x_0) + f'(x_0)(x - x_0) + a(x) \quad \text{on } V(x_0) \tag{3}$$

with $a(x_0) = 0$, $a'(x_0) = 0$ and

$$a(x) \in R^\perp \quad \text{on } V(x_0).$$

This is a variant of Taylor's theorem, and if f is a submersion at x_0, then $a \equiv 0$ because of $R^\perp = \{0\}$. Let $\psi(x_0) = x_0$.

Proposition 73.1 (Local Normal Forms). *Let* (H1), (H2) *be satisfied. It follows that*:

(i) *There exists a neighborhood $W(x_0)$ of x_0 in X and a C^k-diffeomorphism $\varphi: U_N(0) \times U_R(0) \to W(x_0)$ such that normal form* (1), (2) *is satisfied.*
(ii) *There exist neighborhoods $V(x_0)$ and $W(x_0)$ of x_0 in X and a C^k-diffeomorphism $\psi: V(x_0) \to W(x_0)$ such that normal form* (3) *is satisfied.*

PROOF. (i) Without loss of generality, let $x_0 = 0$ and $f(x_0) = 0$. The proof idea is to apply the inverse mapping theorem to

$$F(x) = (x_1, f_1(x))$$

and to let $\varphi = F^{-1}$.

(I) The splittings

$$X = N \oplus N^\perp \quad \text{and} \quad Y = R \oplus R^\perp$$

yield the decompositions

$$x = x_1 + x_2 \quad \text{and} \quad f(x) = f_1(x) + f_2(x).$$

Since $f(0) = 0$ and $f'(0)h = f_1'(0)h + f_2'(0)h$ with $f'(0)h \in R$ for all $h \in X$, we obtain that

$$f_1(0) = f_2(0) = 0 \quad \text{and} \quad f_2'(0) = 0.$$

(II) The map $F: U(0) \subseteq X \to N \times R$, as defined above, is C^k with $F(0) = 0$ and

$$F'(0)h = (h_1, f_1'(0)h) = (h_1, f'(0)h).$$

Since $f'(0): N^\perp \to R$ is bijective, it follows that $F'(0): X \to N \times R$ is bijective, and the inverse mapping theorem (Theorem 4.F) implies that F is a local C^k-diffeomorphism at $x_0 = 0$.

(III) Letting $\varphi = F^{-1}$, we get $\varphi(n, r) = x$ for $n = x_1$ and $r = f_1(x)$. Thus

$$f(\varphi(n, r)) = f_1(x) + f_2(x) = r + f_2(\varphi(n, r)).$$

This is (1) with $g(n, r) = f_2(\varphi(n, r))$.

Finally, we obtain $g(0, 0) = 0$ from $f_2(0) = 0$ and $\varphi(0, 0) = 0$ and

$$g'(0, 0) = f_2'(0)\varphi'(0, 0) = 0$$

from $f_2'(0) = 0$.

(ii) For $X = N \oplus N^\perp$ let $Q: X \to N$ be the natural projection and

$$A: N^\perp \to R$$

the restriction of $f'(x_0)$ to $N^\perp$. The operator Q is linear and continuous, and A is a linear homeomorphism. For each $x \in X$ there exists a unique decomposition

$$x = x_0 + x_1 + x_2 \quad \text{with} \quad x_1 \in N \quad \text{and} \quad x_2 \in N^\perp,$$

thus, there is a unique pair $(n, r) \in N \times R$ with $x = x_0 + n + A^{-1}r$, and

$$n = Q(x - x_0), \quad r = A(x - x_0 - n) = f'(x_0)(x - x_0).$$

The map $x \mapsto (n, r)$, constructed this way, is a C^∞-diffeomorphism from X onto $N \times R$. Letting $\psi(x) = \varphi(n, r)$, we obtain

$$f(\psi(x)) = f(\varphi(n, r)) = f(x_0) + r + g(n, r)$$
$$= f(x_0) + f'(x_0)(x - x_0) + a(x)$$

with $a(x) = g(n, r)$, and this is (3). □

73.2. Banach Manifolds

It is important for the reader of this chapter to get a good geometric understanding of the concept of manifolds. One may think of the surface of the earth as a standard example of a nontrivial manifold. Because of its curvature, it cannot be described with just one chart, but instead a geographical atlas is needed. In such an atlas, the same city T may appear in different charts with different local coordinates. Thus we need a rule which allows us to pass from one local coordinate system to another. During the following formal definitions of charts, atlas, ..., one may keep the example of the surface of the earth in mind. Recall that, according to $A_1(8)$, all topological spaces are assumed to be separated.

Definition 73.2. Let M be a topological space. A *chart* (U, φ) in M is a pair

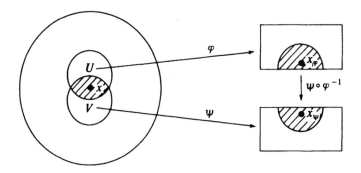

Figure 73.2

where the set U is open in M and $\varphi: U \to U_\varphi$ is a homeomorphism onto an open subset U_φ of a B-space X_φ. We call φ a *chart map*.

The B-space X_φ is called chart space and U_φ is called chart image. For $x \in U$, we call

$$x_\varphi = \varphi(x)$$

the representative of x in the chart (U, φ) or the local coordinate of x in the local coordinate system φ (Fig. 73.2). The point $x \in M$ may have different local coordinates $x_\varphi = \varphi(x)$ and $x_\psi = \psi(x)$ for two different charts (U, φ) and (V, ψ). The transformation rules between them are

$$x_\varphi = \varphi(\psi^{-1}(x_\psi)) \quad \text{and} \quad x_\psi = \psi(\varphi^{-1}(x_\varphi)). \tag{4}$$

Definition 73.3. Let M be a topological space. Two charts (U, φ) and (V, ψ) are called C^k-*compatible* if and only if $U \cap V = \emptyset$ or $\varphi \circ \psi^{-1}$ and $\psi \circ \varphi^{-1}$ are C^k, $k \geq 0$.

If both maps are analytic, the charts are called analytic compatible.

Figure 73.2 shows the two bijective maps

$$\varphi \circ \psi^{-1}: \psi(U \cap V) \to \varphi(U \cap V)$$

and

$$\psi \circ \varphi^{-1}: \varphi(U \cap V) \to \psi(U \cap V).$$

Let $k \geq 1$. If $\varphi \circ \psi^{-1}$ is C^k, then the inverse mapping theorem (Theorem 4.F) makes it into a C^k-diffeomorphism, i.e., $\psi \circ \varphi^{-1}$ is C^k as well. If the chart spaces are complex, then the C^k-maps $\varphi \circ \psi^{-1}$ and $\psi \circ \varphi^{-1}$ are automatically analytic.

Definition 73.4. Let M be a topological space. A C^k-*atlas* for M, $0 \leq k \leq \infty$, is a collection of charts $(U_\alpha, \varphi_\alpha)$ (α ranging in some indexing set), which satisfies the following conditions:

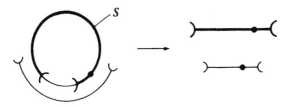

Figure 73.3

(i) The U_α cover M.
(ii) Any two charts are C^k-compatible.
(iii) All chart spaces X_α are B-spaces over $\mathbb{K}$.

M is said to be a C^k-*Banach manifold* if and only if there exists a C^k-atlas for M. Keeping the geographical atlas of the earth in mind, it might be useful to add new charts to the atlas. We call a chart in M, which is C^k-compatible with all atlas charts, an *admissible* chart. In particular, all atlas charts are admissible. The collection of all admissible charts forms a new atlas, which is called the *maximal atlas* for M.

Usually, we will have different chart spaces. If, however, all chart spaces are equal to a fixed B-space X, then M is called a C^k-Banach manifold modeled on X. A Banach manifold is called real or complex if all chart spaces are real or complex. Often we will simply use manifold instead of Banach manifold. If all chart spaces have the same dimension d, $0 \le d \le \infty$, then $d = \dim M$ is called the *dimension* of the manifold M. In this case, M is said to have a dimension. If all atlas charts are analytic compatible, then M is called an analytic manifold. Finally, C^0-manifolds are also called topological manifolds.

EXAMPLE 73.5. Obviously, each open set M in a B-space X is a C^∞-manifold which is also analytic. A chart (U, φ) is obtained with $U = M$ and $\varphi = $ id, i.e., the identity on U. In particular, each open set in $\mathbb{R}^n$ and $\mathbb{C}^n$, is an n-dimensional real or complex analytic manifold, respectively.

EXAMPLE 73.6. The boundary S of a disk in $\mathbb{R}^2$ is a one-dimensional, real C^∞-manifold, which is also analytic. One can choose two charts (U, φ) and (V, ψ) as in Figure 73.3, i.e., U and V are two open curves which cover S. Similarly, the boundary of a ball in $\mathbb{R}^3$ (e.g., the surface of the earth) is a two-dimensional real C^∞-manifold which is also analytic. A detailed proof follows in Section 73.11.

73.3. Strategy of the Theory of Manifolds

So far, it is a little disturbing that we describe manifolds with different atlases. Thinking of the surface of the earth, for example, we have the feeling that its structure does not depend on the form of the specific atlas. We therefore make the following definition.

Definition 73.7. Two C^k-atlases for M are called *equivalent* if and only if all charts are C^k-compatible, $k \geq 0$. Two C^k-Banach manifolds M and N are said to have the same differentiable structure if and only if the following two conditions are satisfied:

(i) The topological spaces M and N are equal.
(ii) The atlases are equivalent.

In Example 73.5, for instance, one can get an equivalent atlas for M by choosing all pairs (U, φ) with U open in M and φ equal to the identity on U.

Obviously, two C^k-Banach manifolds have the same differentiable structure if and only if (i) is satisfied and all admissible charts in M coincide with those in N, i.e., the corresponding maximal atlases are equal. In the 1950s, Milnor made the surprising observation that the seven-dimensional unit sphere S^7, with topology induced by $\mathbb{R}^8$, has more than one differentiable structure, i.e., there exist C^k-atlases, $k \geq 1$, which are not equivalent.

Using the methods of modern gauge field theory, Donaldson (1983) proved the deep result that only certain four-dimensional topological manifolds are smoothable, i.e., they are also differentiable manifolds. Gompf (1985) showed that the four-dimensional Euclidean space $\mathbb{R}^4$ is exotic, i.e., the set $\mathbb{R}^4$ possesses infinitely many different differentiable structures.

Banach manifolds look locally like B-spaces, but not necessarily globally; the important properties of manifolds are global. We propose the following strategy for a study of manifolds:

(a) One works locally in charts, i.e., in local coordinate systems where the calculus of B-spaces is available;
(b) One applies only such concepts which are invariant, i.e., independent of the particular choice of atlas charts and which remain unchanged in equivalent atlases.

This means that we only allow concepts which are independent of the choice of admissible charts.

We explain this procedure in the following two sections which deal with C^k-maps and tangent vectors.

It is an important task in many applications (tangent bundles, jet spaces, etc.) to transform sets M, which are not originally equipped with a topology, into manifolds. The method here is to use *abstract charts* (U, φ) where

$$\varphi: U \to U_\varphi$$

is not a homeomorphism, but a bijection onto an open set of a B-space. It follows from Problem 73.2 that the presence of an abstract atlas implies, in a natural way, the existence of a topology for M. With this atlas, M becomes a manifold.

We say that a manifold M has a *countable basis* if and only if there exists a countable collection $\{U_\alpha\}$ of open sets such that each open set in M is the union of some U_α's. If M has an equivalent atlas which contains countably many charts and all chart spaces are separable (e.g., finite dimensional), then M has a countable basis. Every B-space is regular since each of its points has a neighborhood basis of closed sets. Therefore, every Banach manifold is regular. The metrization theorem of Section 13.10 implies that a Banach manifold with a countable basis is metrizable, i.e., there exists a metric which induces the given topology. This metric need not be unique.

73.4. Diffeomorphisms

Recall that a map $g: U(x) \subseteq X \to Y$ is said to be C^r at a point x if and only if it is C^r in a neighborhood of x.

Definition 73.8. Let M and N be C^k-Banach manifolds with chart spaces over $\mathbb{K}$, $k \geq 1$. Then $f: M \to N$ is called C^r, $r \leq k$ if and only if f is C^r at each point $x \in M$ in fixed admissible charts.

Moreover, f is called a C^r-*diffeomorphism* if and only if f is bijective and f, f^{-1} are both C^r.

We discuss this definition. Let (U, φ) and (V, ψ) be charts in M and N, respectively, with $x \in U$ and $f(x) \in V$. Figure 73.4 shows the map

$$\bar{f} = \psi \circ f \circ \varphi^{-1},$$

which is well defined in a sufficiently small neighborhood of $\varphi(x)$. This map is assumed to be C^r in the usual sense and will be called a *representative* of f. The chain rule implies that the representatives remain C^r after passing to other admissible charts. Thus Definition 73.8 is invariant in the sense of Section 73.3, i.e., independent of the choice of the representatives.

The role that is played by topological spaces and homeomorphisms in topology is played by manifolds and diffeomorphisms in the study of mani-

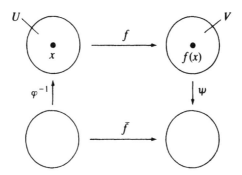

Figure 73.4

folds. For $x \in M$, the map $f: M \to N$ is called a *local C^r-diffeomorphism* at the point x if and only if f maps an open neighborhood $U(x)$ of x C^r-diffeomorphically onto an open neighborhood $V(f(x))$ of $f(x)$.

Proposition 73.9 (Structure of One-Dimensional Manifolds). *A connected, one-dimensional, real C^∞-manifold with a countable basis is C^∞-diffeomorphic either to $\mathbb{R}$ (open curve) or to the unit circle S^1 (closed curve).*

The proof of this important structure theorem may be found in the appendix of Milnor (1965, M). The Whitney embedding theorem implies that any such manifold can be embedded in $\mathbb{R}^3$.

73.5. Tangent Space

For surfaces in $\mathbb{R}^3$, one has an intuitive understanding of tangent vectors and tangent planes (tangent spaces) (Fig. 73.5). The importance of tangent spaces is the following. For $x \in M$, the space TM_x is a linear space which gives a first-order approximation of the manifold M in a neighborhood of x. This allows us to define a derivative

$$f'(x): TM_x \to TN_{f(x)}$$

at the point x for a map $f: M \to N$ by linearization. Thus the concept of F-derivative carries over to Banach manifolds. For general manifolds it is not possible to define tangent vectors in some surrounding space. However, we can use the concept of equivalent curves. A collection of equivalent curves through a point x is defined in such a way that by existence of a surrounding space Σ, all curves of the collection would have the same tangent vector in Σ with respect to x (Fig. 73.6).

Definition 73.10. Let M be a C^k-manifold, $k \geq 1$, and $x \in M$. Two C^1-curves in M, which pass through the point x, are called *equivalent* at the point x if and only if the representatives have the same tangent vector at x in some fixed admissible chart.

A *tangent vector v to M at x* consists of all C^1-curves which are equivalent at x to a fixed C^1-curve.

We will now discuss these definitions in some detail (see Fig. 73.7). A C^1-curve

$$x = x(t)$$

in M through x is a C^1-map $x(\cdot): U(t_0) \subseteq \mathbb{R} \to M$ such that $x(t_0) = x$ at some fixed t_0. The representative of this curve in the chart (U, φ) is

$$x_\varphi = x_\varphi(t) \qquad \text{with} \qquad x_\varphi(t) = \varphi(x(t)).$$

In the corresponding chart space, this curve has a "concrete" tangent vector

73.5. Tangent Space

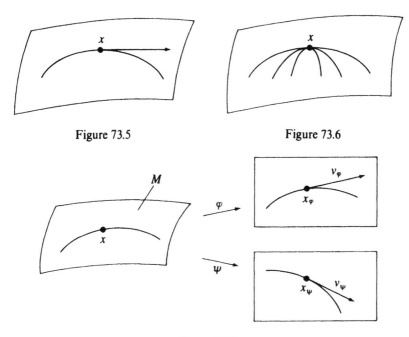

Figure 73.5

Figure 73.6

Figure 73.7

at x,
$$v_\varphi = x'_\varphi(t_0).$$

The "abstract" tangent vector of Definition 73.10 to the curve $x = x(t)$ at the point x is denoted by v or more suggestively by $x'(t_0)$. Then v_φ is called the *representative* or local coordinate of v.

Let (V, ψ) be another chart with $x \in U \cap V$. Then

$$x_\psi = F(x_\varphi) \tag{5}$$

with $F = \psi \circ \varphi^{-1}$ and $x_\psi(t) = F(x_\varphi(t))$. Differentiation with respect to t at the point t_0 gives

$$v_\psi = F'(x_\varphi)v_\varphi, \tag{6}$$

which is the important transformation rule for the local coordinates of the tangent vectors. The same argument shows that the equivalence of Definition 73.10 is invariant, i.e., independent of the choice of the admissible chart. Also, it is easily seen to be an equivalence relation.

Definition 73.11. Let M be a C^k-Banach manifold. *The tangent space TM_x to M at the point x is the set of all tangent vectors at x.* Here, $k \geq 1$.

Proposition 73.12. *The tangent space TM_x is a topological vector space which is linear homeomorphic to each chart space X_φ at the point x.*
In particular, we have $\dim TM_x = \dim X_\varphi$.

PROOF. We choose a chart (U, φ) in M with $x \in U$. Then there exists a one-to-one correspondence between the elements $v \in TM_x$ and the representatives $v_\varphi \in X_\varphi$. We define the linear operations in such a way that

$$v + w \quad \text{and} \quad \alpha v$$

correspond to

$$v_\varphi + w_\varphi \quad \text{and} \quad \alpha v_\varphi,$$

i.e., we use representatives. Because of (6), this definition is independent of the choice of the chart. This makes TM_x into a linear space.

Moreover, the map F of (5) is a C^k-diffeomorphism. Hence with Theorem 4.F, the map $F'(x_\varphi): X_\varphi \to X_\psi$ is a linear homeomorphism between the chart spaces. Thus all chart spaces are linear homeomorphic at x. Now, there exists a linear bijective map $b: TM_x \to X_\varphi$ and by using its inverse b^{-1} we can carry the topology of X_φ onto TM_x. Also, X_ψ induces the same topology on TM_x, so that TM_x becomes a topological vector space (see A_1(22o)). □

Since the norms $\|v_\psi\|$ and $\|v_\varphi\|$ in (6) might be different, i.e., they might depend on the choice of the charts, it is not always possible to make TM_x into a B-space in an invariant fashion. If M is open in a B-space X, then TM_x can be identified with X for all $x \in M$.

73.6. Tangent Map

Let $f: U(x) \subseteq X \to Y$ be C^1, and X and Y be B-spaces. In Chapter 4 we gave the definition for the F-derivative $f'(x): X \to Y$ by using the process of linearization. Now we will see how this important concept carries over to Banach manifolds.

Proposition 73.13. *Let $f: M \to N$ be C^1, and M and N C^1-Banach manifolds with chart spaces over $\mathbb{K}$. Then there exists a linear, continuous map $f'(x): TM_x \to TN_{f(x)}$ at each point $x \in M$, given by (7) below.*

PROOF. Let $v \in TM_x$ and choose a C^1-curve

$$x = x(t)$$

passing through a fixed $x \in M$ such that this curve belongs to the tangent vector v at x. The map f takes this curve into the curve

$$y = y(t)$$

on N such that $y(t) = f(x(t))$. Let w denote the tangent vector to $y = y(t)$ at $f(x)$, and define $f'(x)$ through

$$w = f'(x)v. \tag{7}$$

73.7. Higher-Order Derivatives and the Tangent Bundle

Passing to charts, we obtain that f takes equivalent curves at x into equivalent curves at $f(x)$. Thus (7) is well defined, i.e., independent of the curve $x(t)$, which belongs to v.

Linearity and continuity of $f'(x)$ follow as in (8) below by passing to representatives. □

Definition 73.14. The map $f'(x)\colon TM_x \to TN_{f(x)}$ is called the *tangent map* of $f\colon M \to N$ at the point x. Another way to write $f'(x)$ is $T_x f$.

In Part III we also used $Tf(x)$ for $T_x f$, but in this Part IV this might cause confusion with $Tf(x, v)$ of Section 73.7. Sometimes, also $Df(x)$ or $df(x)$ is used for $f'(x)$ in the literature.

Let $\bar{f}$ be the representative of f in the admissible charts (U, φ) and (V, ψ) with $x \in U$ and $f(x) \in V$, as pictured in Figure 73.4. The local coordinates v_φ and w_ψ are related as in (7) through

$$w_\psi = \bar{f}'(x_\varphi) v_\varphi. \tag{8}$$

This follows immediately from the definition of the local coordinates. Therefore,

$$\bar{f}'(x_\varphi)\colon X_\varphi \to X_\psi$$

is the representative of

$$f'(x)\colon TM_x \to TN_{f(x)},$$

i.e., in local coordinates $f'(x)$ is the usual F-derivative. The chain rule of Section 4.3 implies the following general result.

Proposition 73.15 (Chain Rule). *Let M, N, and P be C^1-Banach manifolds with chart spaces over $\mathbb{K}$, and let $f\colon M \to N$ and $g\colon N \to P$ be C^1. Then we have*

$$(g \circ f)'(x) = g'(f(x)) \circ f'(x) \tag{9}$$

for all $x \in M$.

Formula (9) can be rewritten as

$$T_x(g \circ f) = T_{f(x)} g \circ T_x f, \tag{10}$$

i.e., linearization and composition commute.

73.7. Higher-Order Derivatives and the Tangent Bundle

Let $f\colon U \subseteq X \to Y$ be C^2, U open in X, and X, Y B-spaces over $\mathbb{K}$. Then $f'(x)\colon X \to Y$ exists for all $x \in U$, i.e., $f'(x) \in L(X, Y)$. This induces a map

$$f'\colon U \to L(X, Y).$$

Thus the second-order derivative can be defined through $f''(x) = g'(x)$, where $g = f'$.

This procedure carries over to manifolds after making the following modifications. Let

$$f: M \to N$$

be C^2, where M and N are open sets in the B-spaces X and Y. Let $TM = M \times X$ and $TN = N \times Y$ and

$$Tf(x, v) = (f(x), f'(x)v) \qquad \text{for all} \quad x \in M, \quad v \in X.$$

This defines a map

$$Tf: TM \to TN.$$

If we write T^2 instead of TT, we obtain

$$T^2 f: T^2 M \to T^2 N$$

as a substitute for the second-order derivative. For arbitrary manifolds M this leads to the concept of the tangent bundle TM. The proof of the Whitney embedding theorem in Section 73.21 is a good illustration for the usefulness of TM.

Definition 73.16. Let M be a C^k-Banach manifold, $k \geq 1$. The *tangent bundle* TM is the set of all pairs

$$(x, v) \quad \text{with} \quad x \in M, \quad v \in TM_x.$$

The map $\pi: TM \to M$, where $\pi(x, v) = x$ is called the natural projection.

Intuitively, TM is obtained by attaching the tangent space TM_x at each point $x \in M$, neglecting the fact that different tangent spaces may intersect (Fig. 73.8). If M is a surface in $\mathbb{R}^3$, then TM will be a four-dimensional manifold, since each component of a pair $(x, v) \in TM$ is determined by two coordinates. Indeed, we need two coordinates to describe the point x on M, and, moreover, we need two coordinates to describe the tangent vector v in the tangent plane at x.

Proposition 73.17. *Let M be a C^k-manifold, $k \geq 1$. The tangent bundle TM is a C^{k-1}-Banach manifold. If M has dimension d, then TM has dimension $2d$.*

Figure 73.8   Figure 73.9

73.7. Higher-Order Derivatives and the Tangent Bundle

PROOF. The simple proof idea is as follows. Choose (x_φ, v_φ) as the local coordinates for (x, v) and then transform them according to (5) and (6).

More precisely, we assign a chart (TU, φ_T) in TM to each chart (U, φ) in M, with

$$TU = \{(x, v) : x \in U, v \in TM_x\}$$

and

$$\varphi_T(x, v) = (x_\varphi, v_\varphi).$$

It follows from Problem 73.2 that, in a natural way, TM can be given a separated topology. This makes TM with the atlas above into a Banach manifold. □

Since (x_φ, v_φ) is an element of $U_\varphi \times X_\varphi$, the space TM looks locally like the product $U_\varphi \times X_\varphi$ with chart space $X_\varphi \times X_\varphi$. Globally, however, TM need not have a product structure.

Definition 73.18. Let $f: M \to N$ be C^k, and M and N C^k-Banach manifolds with chart spaces over $\mathbb{K}$, $k \geq 1$. The *tangent map* $Tf: TM \to TN$ is defined through

$$Tf(x, v) = (f(x), f'(x)v) \qquad \text{for all} \quad (x, v) \in TM.$$

Passing to local coordinates, we find that Tf is C^{k-1} since the local representative of Tf is

$$(x_\varphi, v_\varphi) \mapsto (\bar{f}(x_\varphi), \bar{f}'(x_\varphi)v_\varphi).$$

Through iteration we obtain a C^{k-2}-map $T^2f: T^2M \to T^2N$ for $k \geq 2$, etc., where $T^2f = T(Tf)$ and $T^2M = T(TM)$. Thus, using the tangent bundle, we obtain a substitute for higher-order derivatives of f. In Problem 73.6 we prove the following very convenient *chain rule* for higher-order derivatives

$$T^r(g \circ f) = T^r g \circ T^r f.$$

As another application of tangent bundles, we now look at vector fields on manifolds. One may think of a vector field v on M as having a tangent vector v_x attached at each point $x \in M$ (Fig. 73.9).

Definition 73.19. A C^k-*vector field* on a Banach manifold M is a C^k-section of the tangent bundle TM, i.e., a C^k-map $v: M \to TM$ with

$$\pi(v(x)) = x \qquad \text{for all} \quad x \in M.$$

More precisely, we have $v(x) = (x, v_x)$ for $x \in M$ and $v_x \in TM_x$. Such a map is C^k if and only if for each point $x \in M$ there exists an admissible chart (U, φ) in M such that the map $x_\varphi \mapsto v_{\varphi(x)}$ from U_φ into X_φ is C^k. Here $v_{\varphi(x)}$ denotes the representative of v_x for φ.

Definition 73.19 makes sense if TM is a C^k-manifold, i.e., M is a C^{k+1}-manifold, $k \geq 0$.

The concept of sections can be defined for general bundles.

Definition 73.20. A *bundle* is a triple (π, B, M) where B and M are topological spaces and $\pi: B \to M$ is a surjective, continuous map.

B is called bundle space, M is called basis space and π is called natural projection. The set $F_x = \pi^{-1}(x)$ is called a *fiber* over $x \in M$. A *section* is a continuous map

$$s: M \to B \quad \text{with} \quad \pi(s(x)) = x \quad \text{for all} \quad x \in M,$$

i.e., $s(x) \in F_x$.

In many examples of bundles, the fibers are either linear spaces (e.g., vector bundles) or groups (e.g., principal fiber bundles).

EXAMPLE 73.21. A standard example of a bundle (π, B, M) is obtained by considering the product

$$B = M \times Y,$$

where M and Y are topological spaces and

$$\pi: M \times Y \to M$$

is the natural projection, $\pi(x, y) = x$. Here, the fiber $\pi^{-1}(x)$ equals $\{x\} \times Y$ and is homeomorphic to Y. Figure 73.10 shows a section.

A *bundle morphism* (f, g) between two bundles (π_j, B_j, M_j), $j = 1, 2$ is a pair of continuous maps such that the following diagram commutes:

$$\begin{array}{ccc} B_1 & \xrightarrow{f} & B_2 \\ \pi_1 \downarrow & & \downarrow \pi_2 \\ M_1 & \xrightarrow{g} & M_2 \end{array}$$

Thus, f and g are compatible with the fiber structure, i.e., f maps the fiber F_x into the fiber $F_{g(x)}$. If f and g are homeomorphisms, then (f, g) is called a *bundle isomorphism*.

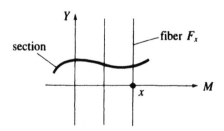

Figure 73.10

If, in the above definitions, topological spaces, continuous maps, and homeomorphisms are replaced by C^k-Banach manifolds, C^k-maps, and C^k-diffeomorphisms, then we will speak of C^k-bundles, C^k-bundle morphisms, and C^k-bundle isomorphisms, respectively.

The tangent bundle TM of a C^k-Banach manifold, $k \geq 1$, corresponds to the C^{k-1}-bundle (π, TM, M). The fiber over x is the tangent space TM_x. Similarly, the cotangent bundle TM^*, which will be defined below, corresponds to the C^{k-1}-bundle (π, TM^*, M). The fiber over x is TM_x^*, i.e., is equal to the dual space of the tangent space TM_x.

73.8. Cotangent Bundle

The dual space X^* of all linear, continuous functionals on a B-space X plays an important role in mathematics. A generalization to manifolds uses the concept of cotangent bundles.

Definition 73.22. Let M be a C^k-Banach manifold, $k \geq 1$. The *cotangent bundle* TM^* consists of all pairs (x, w^*) with $x \in M$ and $w \in TM_x^*$.

Each tangent space TM_x is a topological vector space. Thus, the dual space TM_x^* is defined as the set of all linear, continuous functionals on TM_x. Intuitively, the dual space TM_x^* is attached at each point $x \in M$. Let $\pi: TM^* \to M$ be the natural projection.

An element $w^* \in TM_x^*$ is called a *cotangent vector* to M at x, or simply a covector. We call TM_x^* the *cotangent space* of M at x.

Proposition 73.23. *The cotangent bundle TM^* is a C^{k-1}-Banach manifold. If M has dimension d, then TM^* has dimension $2d$.*

PROOF. Let (U, φ) be a chart in M. The local coordinates for $x \in M$ and $v \in TM_x$ are $x_\varphi \in U_\varphi$ and $v_\varphi \in X_\varphi$. For $w^* \in TM_x^*$ we define a functional $w_\varphi^* \in X_\varphi^*$ through

$$w_\varphi^*(v_\varphi) \stackrel{\text{def}}{=} w^*(v) \qquad \text{for all} \quad v \in TM_x. \tag{11}$$

The map $w^* \mapsto w_\varphi^*$ between TM_x^* and X_φ^* is a linear bijection. The pair (x_φ, w_φ^*) is considered as a local coordinate for $(x, w^*) \in TM^*$, and w_φ^* is called a representative of w^*.

More precisely, we use the concept of abstract atlases of Problem 73.2, which means that we assign an abstract chart (TU^*, φ_{T^*}) in TM^* to each chart (U, φ) in M with

$$TU^* = \{(x, w^*): x \in U, w^* \in TM_x^*\}$$

and

$$\varphi_{T^*}(x, w^*) = (x_\varphi, w_\varphi^*).$$

With this abstract atlas, TM^* becomes a C^{k-1}-Banach manifold. This can be seen by passing to another chart (V, ψ) in M. Then

$$x_\psi = F(x_\varphi) \tag{12}$$

with $F = \psi \circ \varphi^{-1}$ and

$$v_\psi = F'(x_\varphi) v_\varphi.$$

Formula (11) implies that $w_\psi^*(v_\psi) = w_\varphi^*(v_\varphi)$, so that

$$\langle w_\psi^*, F'(x_\varphi) v_\varphi \rangle = \langle w_\varphi^*, v_\varphi \rangle \qquad \text{for all} \quad v_\varphi \in X_\varphi.$$

From this we obtain the following transformation rule for the representatives

$$F'(x_\varphi)^* w_\psi^* = w_\varphi^*. \tag{13}$$

$\square$

Since $(x_\varphi, w_\varphi^*) \in U_\varphi \times X_\varphi^*$, the cotangent bundle TM^* looks locally like the product $U_\varphi \times X_\varphi^*$ with chart space $X_\varphi \times X_\varphi^*$. Globally, however, TM^* need not have a product structure.

Definition 73.24. A C^k-*covector field* on a Banach manifold M is a C^k-section of the cotangent bundle, i.e., a C^k-map $w^*: M \to TM^*$ with

$$\pi(w^*(x)) = x \qquad \text{for all} \quad x \in M.$$

More precisely, we have $w^*(x) = (x, w_x^*)$ with $x \in M$ and $w_x^* \in TM_x^*$. Such a map is C^k if and only if for each $x \in M$ there exists an admissible chart (U, φ) in M such that the map

$$x_\varphi \mapsto w_{\varphi(x)}^*$$

from U_φ into X_φ^* is C^k. Here $w_{\varphi(x)}^*$ denotes the representative of w_x^* for φ.

Definition 73.24 makes sense, if TM^* is a C^k-manifold, i.e., M is a C^{k+1}-manifold, $k \geq 0$.

73.9. Global Solutions of Differential Equations on Manifolds and Flows

Consider the differential equation

$$x'(t) = v(x(t), t), \qquad x(0) = x_0 \tag{14}$$

on a manifold M, which satisfies the following conditions:

(H1) M is a C^{r+1}-Banach manifold, $r \geq 1$.
(H2) v is a time-dependent vector field on M, i.e., there exists a C^r-map

$$V: M \times \mathbb{R} \to TM \qquad \text{with} \qquad V(x, t) = (x, v(x, t)).$$

73.9. Global Solutions of Differential Equations on Manifolds and Flows

Figure 73.11

Condition (H2) is satisfied if for each $x \in M$ there exists an admissible chart (U, φ) in M with $x \in U$ such that the representative

$$(x_\varphi, t) \mapsto v_\varphi(x_\varphi, t)$$

is a C^r-map from $U_\varphi \times \mathbb{R}$ into X_φ.

Moreover, condition (H2) assigns a tangent vector $v(x(t), t)$ to each point $x(t) \in M$ at time t. Thus a solution of (14) is a curve $x = x(t)$ in M such that the tangent vector $x'(t)$ at the point $x(t)$ is equal to the given vector $v(x(t), t)$ of the vector field. A simple model for this differential equation is as follows. Consider a moving liquid on M and let $x = x(t)$ be the trajectory of a particle of fluid. Suppose at time $t = 0$ it is at the point x_0. At time t it will be at the point $x(t)$ with velocity $x'(t) = v(x(t), t)$. One may think, for example, of an ocean current (Fig. 73.11). Writing $F_t x_0 = x(t)$ and $f(x_0, t) = F_t x_0$, we obtain the concept of flows for time-independent velocity fields $v = v(x)$.

Definition 73.25. Let M be a C^r-Banach manifold, $r \geq 0$. A *flow* on M is a map

$$(t, x) \mapsto f(t, x)$$

from $\mathbb{R} \times M$ into M such that the following conditions are satisfied for $F_t x = f(t, x)$:

(i) $F_0 = I$ (identity on M).
(ii) $F_{t+s} = F_t F_s$ for all $t, s \in \mathbb{R}$.

If f is C^r, then f is called a C^r-*flow*.

If f is only defined for $t \geq 0$ and (ii) holds for all $t, s \geq 0$, then f is called a *semiflow*.

If f is defined only locally, i.e., only for (t, x) in a neighborhood of some point $(0, x_0)$ and (i), (ii) hold locally, then f is called a *local flow*. The notation for flows will be either f or $\{F_t\}$.

A C^r-flow $\{F_t\}$ has an inverse $F_t^{-1} = F_{-t}$ and therefore $F_t: M \to M$ is a C^r-diffeomorphism for all $t \in \mathbb{R}$. For this reason, a flow is also called a one-parameter group of diffeomorphisms or simply a one-parameter group. Similarly for semigroups.

Theorem 73.A (Existence Theorem for Differential Equations on Manifolds). *Assume* (H1) *and* (H2). *For each initial value* $x_0 \in M$ *there exists a maximal open* interval J in $\mathbb{R}$ with $0 \in J$, on which (14) has a solution $x = x(t)$. This solution is unique.

We set $f(x_0, t) = x(t)$. Then f is C^r on its domain of definition and assumes values in M. For the autonomous equation

$$x'(t) = v(x(t)),$$

f is a local C^r-flow.

Corollary 73.26 (Flows on Compact Manifolds). *Assume* (H1) *and* (H2), *and let M be a compact manifold. Then we have $J = \mathbb{R}$ in the above theorem, i.e., for each initial value $x_0 \in M$ there exists a unique solution of* (14) *which is defined for all $t \in \mathbb{R}$. If* (14) *is an autonomous equation, then f is a C^r-flow.*

PROOF OF THEOREM 73.A. Working in charts, we obtain a differential equation for the representatives analogous to (14), where the right-hand side is of class C^r. Theorem 4.D implies that, locally, there exists a unique solution. Now consider all open intervals J_α in $\mathbb{R}$ with $0 \in J_\alpha$ such that (14) has a solution in J_α. This solution is unique with maximal interval of existence equal to $J = \bigcup_\alpha J_\alpha$.

Theorem 4.D implies that f is C^r, and the proof below shows that f is a local flow. □

PROOF OF COROLLARY 73.26. Let M be compact. Suppose that $J \neq \mathbb{R}$, i.e., J may be equal to $J = \,]t_1, t_2[$ with $t_2 < \infty$. It follows then from the compactness of M that $x(t) \to x$ as $t \to t_2 - 0$. Thus, for the initial-value problem in x, we can continue the solution beyond t_2, which is a contradiction.

We next show that f is a flow. Let $x = x(t)$ be a solution of

$$x'(t) = v(x(t)), \qquad x(0) = x_0.$$

Let $y(t) = x(t + s)$ for some fixed s. Then $y = y(t)$ is a solution of

$$y'(t) = v(y(t)), \qquad y(0) = x(s).$$

Therefore, we have $y(t) = F_{t+s} x_0$ and

$$y(t) = F_t x(s) = F_t F_s x_0,$$

and hence $F_{t+s} = F_t F_s$. □

An important fact about the concept of flows and semiflows is that it allows us to describe scientific processes which satisfy the principle of causality. Let $\{F_t\}$ be a semiflow. The points $x \in M$ represent the states of a system. Suppose the system is in x_0 at time $t = 0$ and in $F_t x_0$ at time t. Formula

$$F_{t+s} x_0 = F_t(F_s x_0) \tag{15}$$

shows that the final state of the process depends only on its initial state and the time period in between. Each intermediate state $F_s x_0$ can be viewed as a

73.9. Global Solutions of Differential Equations on Manifolds and Flows

new initial state, which leads to the final state $F_t(F_s x_0)$ after time t. From (15), this is the same as the final state $F_{t+s} x_0$, corresponding to the initial state x_0, after time $t + s$. Thus, formula (15) represents a principle of causality. Semiflows are only defined for $t \geq 0$, i.e., only for future times. They describe irreversible processes such as diffusion and heat conduction, while reversible processes like wave motions are described by flows.

We will now indicate why it is often easier to work with flows and semiflows, rather than with the differential equation itself.

EXAMPLE 73.27. A standard example of a flow is the linear flow generated by the differential equation

$$x'(t) = Ax(t), \qquad x(0) = x_0, \tag{16}$$

where $A: X \to X$ is a linear, continuous operator on a B-space X. The solution has the form $x(t) = e^{tA} x_0$, and

$$F_t = e^{tA}$$

describes the flow on X. The multiplication rule for the exponential function $e^{(t+s)A} = e^{tA} e^{sA}$ is equivalent to the group property (15).

There are, however, many time-dependent scientific processes (16) where A is not a bounded operator. A typical case is:

(H) $A: D(A) \subseteq X \to X$ is a linear, self-adjoint operator on an H-space X.

The proofs of the following results may be found in Chapter 19 of Part II.

EXAMPLE 73.28 (Parabolic Case). Assume (H) with $(Ax|x) \leq 0$ for all $x \in D(A)$. Then

$$F_t = e^{tA}$$

defines a semiflow on X, and $x(t) = F_t x_0$ is the unique solution of (16) which passes through $x_0 \in D(A)$ and is defined for all $t \geq 0$.

Thus we have the important situation that $F_t x_0$ is defined for all $x_0 \in X$ and $t \geq 0$, whereas the corresponding differential equation might not have a solution for each $x_0 \in X$, if $D(A) \neq X$. Furthermore, $F_t: X \to X$ is linear and continuous for each $t \geq 0$.

As a typical application, we consider the parabolic boundary initial-value problem

$$x_t = \Delta x,$$

$$x = x_0 \qquad \text{for} \quad t = 0 \quad \text{(initial condition)}, \tag{17}$$

$$x = 0 \quad \text{on } \partial G \qquad \text{for} \quad t \geq 0 \quad \text{(boundary condition)},$$

for the sought function $x: \mathbb{R}_+ \times G \to \mathbb{R}$. Here G is a bounded region in $\mathbb{R}^n$. We may think of $x(t, p)$ as a temperature at the point $p \in G$ at time t.

Let $X = L_2(G)$. The boundary condition suggests to choose $C_0^\infty(G)$ as the domain for the Laplacian Δx. The Laplacian is a symmetric operator which, according to Section 19.11, can be extended to a self-adjoint operator A: $D(A) \subseteq X \to X$ with $(Ax|x) \leq 0$ for all $x \in D(A)$. Formula (16) is a functional-analytic version of (17) and

$$x(t) = e^{tA}x_0$$

is a generalized solution of (17) through $x_0 \in X$.

EXAMPLE 73.29 (Schrödinger Case). Consider the differential equation

$$x'(t) = -iAx(t), \quad x(0) = x_0. \tag{18}$$

Here A is the energy operator of the quantum system. Assume (H) and let X be a complex H-space. Then $F_t = e^{-iAt}$ defines a flow on X. For $x_0 \in D(A)$,

$$x(t) = e^{-iAt}x_0$$

is the unique solution of (18) defined for all $t \in \mathbb{R}$.

The Schrödinger equation of quantum mechanics has precisely the form (18) if we set $\hbar = 1$.

73.10. Linearization Principle for Maps

Consider the following question:

(L) *Which map behaves, locally, after a suitable coordinate change, like its linearization?*

In this section we show that for etale mappings, submersions, immersions, and subimmersions, (L) has a positive answer. The following section contains the corresponding global results. We make the following general assumption.

(H) Let $f: M \to N$ be C^k, $k \geq 1$, where M and N are C^k-Banach manifolds with chart spaces over $\mathbb{K}$.

The splitting of linear subspaces, which has been used already in Part I, will play an important role. The linearization of f at the point x is the map

$$f'(x): TM_x \to TN_{f(x)}.$$

Let Y be a linear subspace of TM_x. The tangent space TM_x is a topological vector space. Thus, it follows from A_1 (22l) that the splitting of Y is well defined. The tangent space TM_x and the corresponding chart spaces X_φ are linear homeomorphic. The invariance of the splitting under linear homeomorphisms implies that Y splits TM_x if and only if there exists a chart (U, φ) in M with $x \in U$ such that the representative Y_φ of Y splits the chart space X_φ. Thus, Y

splits TM_x, if one of the following three conditions is satisfied:

(i) $\dim M < \infty$.
(ii) $\dim Y < \infty$ or $\operatorname{codim} Y < \infty$ and Y is closed.
(iii) Y is closed, and some chart space X_φ at the point x is an H-space.

Note $A_1(22l)$. Analogous results hold for TN_y.

Definition 73.30. Assume (H).

(a) f is called an *etale mapping* at x if and only if $f'(x)$: $TM_x \to TN_{f(x)}$ is bijective.
(b) f is called a *submersion* at x if and only if $f'(x)$ is onto and the null space $N(f'(x))$ splits TM_x.
(c) f is called an *immersion* at x if and only if $f'(x)$ is injective, and the range $R(f'(x))$ splits $TN_{f(x)}$.
(d) f is called a *subimmersions* at x_0 if and only if rank $f'(x)$ is constant in a neighborhood of x_0.

Since rank $f'(x) = \dim R(f'(x))$, this provides a natural classification of maps between manifolds according to the behavior of the linearizations. Properties (a)–(d) are satisfied if they are satisfied for a representative of f in local charts. If M and N are finite-dimensional then the splitting is automatic. The following Theorems 73.B–73.F contain important applications of the definitions above, which have already been studied in the context of B-spaces in Chapter 4. For the convenience of the reader, we begin with a number of definitions which often occur in connection with manifolds.

We begin with a global version of Definition 73.30. Let (H) be satisfied. A map $f: M \to N$ which is a *submersion* at every point $x \in M$ is simply called a submersion, and analogous definitions are given, for etale mappings, immersions, and subimmersions. A map $f: M \to N$ is called *closed* if it takes closed sets into closed sets, and it is called *proper* if the preimage of compact sets is compact.

EXAMPLE 73.31. Let $f: U(x_0) \subseteq \mathbb{K}^n \to \mathbb{K}^m$ be C^1, $\mathbb{K} = \mathbb{R}, \mathbb{C}$. Then we have $f'(x_0) = (\partial f_i(x)/\partial \xi_j)$ with $i = 1, \ldots, m$ and $j = 1, \ldots, n$.

(i) If $n \leq m$ and rank $f'(x_0) = n$, then f is an immersion at x_0.
(ii) If $n \geq m$ and rank $f'(x_0) = m$, then f is a submersion at x_0.
(iii) If rank $f'(x)$ is constant in a neighborhood of x_0, then f is a subimmersion at x_0.

The matrix $f'(x_0)$ in (i) and (ii) has a nonvanishing subdeterminant of maximal size. Continuity implies that it is nonvanishing in a neighborhood of x_0, and thus (iii) follows from (i) as well as from (ii). Therefore, we have in finite-dimensional spaces: every submersion and immersion at x_0 is also a subimmersion at x_0.

The map $f: \mathbb{R}^2 \to \mathbb{R}^1$ with $f(x) = \xi_1^2 - \xi_2^2$ is not a subimmersion at $x = (0, 0)$, and hence is not an immersion and submersion since rank $f'(x) = 0$ (or $= 1$) for $x = 0$ (or $x \neq 0$). Note that $f'(x) = (2\xi_1, -2\xi_2)$.

Definition 73.32. Assume (H) for $f: M \to N$.

(a) A point $x \in M$ is called a *regular* point of f if and only if f is a submersion at x. Otherwise x is called a *singular* point.
(b) A point $y \in N$ is caled a *regular value* of f if and only if the set $f^{-1}(y)$ is empty or consists only of regular points. Otherwise y is called a *singular value*, i.e., $f^{-1}(y)$ contains at least one singular point.

Instead of singular (or regular) point, one sometimes says critical or degenerate (or noncritical or nondegenerate) point.

Definition 73.33. Let (H) be satisfied. A map $f: M \to N$ is called a *Fredholm operator* at x if and only if the linearization $f'(x): TM_x \to TN_{f(x)}$ is a Fredholm operator, i.e., we have dim $N(f'(x)) < \infty$ and codim $R(f'(x)) < \infty$. We define the index as

$$\text{ind } f'(x) = \dim N(f'(x)) - \text{codim } R(f'(x)).$$

Moreover, if $f: M \to N$ is a Fredholm operator at all $x \in M$, then f is simply called a Fredholm operator. If ind $f'(x)$ is constant on M, then this number is called the index of f. We write ind f.

A map $f: M \to N$ is a Fredholm operator at x if and only if the representatives of f in local charts are Fredholm operators at the corresponding points. The invariance of the index of Fredholm operators on B-spaces under perturbations, which has been discussed in Section 8.4, implies that ind $f'(x)$ is locally constant, i.e., ind $f'(x)$ is constant on M, if M is connected.

Theorem 73.B (Diffeomorphisms). *Assume* (H) *for the C^k-map $f: M \to N$, $k \geq 1$. Then f is a local C^k-diffeomorphism at x if and only if f is an etale mapping at x.*

f is a C^k-diffeomorphism if and only if f is a bijective etale mapping.

PROOF. Working in local charts, the theorem follows from the inverse mapping theorem (Theorem 4.F) applied to the representatives. □

Theorem 73.B is the linearization principle for diffeomorphisms. We now look at the local behavior of f in the case that $f'(x)$ is not bijective, i.e., we consider subimmersions, submersions, and immersions at x. The corresponding global results will be discussed in the following section.

Proposition 73.34 (Local Subimmersion Theorem). *Let X and Y be finite-dimensional B-spaces over $\mathbb{K}$, and $f: U(x_0) \subseteq X \to Y$ is a C^k-subimmersion at*

73.10. Linearization Principle for Maps

x_0, $k \geq 1$. Then there exist local C^k-diffeomorphisms φ and ψ with $\varphi(x_0) = 0$ and $\psi(0) = f(x_0)$ such that the following diagram commutes:

$$\begin{array}{ccc} U(x_0) \subseteq X & \xrightarrow{f} & Y \\ \varphi \downarrow & & \uparrow \psi \\ U(0) \subseteq X & \xrightarrow{f'(x_0)} & Y \end{array}.$$

If f is analytic, then also φ and ψ are analytic.

Corollary 73.35. *The same conclusions hold if, in the above diagram, $f'(x_0)$ is replaced by g with*

$$g(\xi_1, \ldots, \xi_n) = (\xi_1, \ldots, \xi_r, 0, \ldots, 0).$$

Here, we identify X and Y with $\mathbb{K}^n$ and $\mathbb{K}^m$ and note that $r = \operatorname{rank} f'(x_0)$.

PROOF. In Problem 4.4b we proved this proposition for g (rank theorem). Furthermore, we know from linear algebra that there exist linear isomorphisms A and B with $g = Af'(x_0)B$ (normal form for matrices of rank r). This proves the proposition in the case of $f'(x_0)$. □

Thus, after introducing new coordinates, f looks, locally at x_0, like its derivative $f'(x_0)$. In the case of submersions and immersions, this important linearization result can be generalized to infinite-dimensional B-spaces. For simplicity, let $f(x_0) = 0$ and $x_0 = 0$. This can always be obtained by using translations.

Proposition 73.36 (Local Submersion Theorem). *Let X and Y be B-spaces over $\mathbb{K}$ and $f\colon U(x_0) \subseteq X \to Y$ a C^k-submersion at x_0, $k \geq 1$, with $f(x_0) = 0$. Then there exists a local C^k-diffeomorphism φ with $\varphi(x_0) = 0$ and $\varphi'(x_0) = I$ such that the following diagram commutes:*

$$\begin{array}{ccc} U(x_0) \subseteq X & \xrightarrow{f} & Y \\ \varphi \downarrow & \nearrow{f'(x_0)} & \\ U(0) \subseteq X & & \end{array}.$$

If f is analytic, then also φ is analytic.

PROOF. This is a consequence of results of Section 73.1, but we give here a direct proof. Let $N = N(f'(x_0))$. Since f is a submersion at x_0, N splits the B-space X. Thus, there exists a projection operator $P\colon X \to N$. Let $P^\perp = I - P$ and $N^\perp = P^\perp(X)$. Then we obtain that $X = N \oplus N^\perp$ and also that $f'(x_0)\colon N^\perp \to Y$ is bijective. Let its inverse be denoted by $A\colon Y \to N^\perp$ and define

$$\varphi(x) = Px + Af(x). \tag{19}$$

Then we have $\varphi'(x_0) = P + Af'(x_0) = P + P^\perp = I$. The inverse mapping theorem (Theorem 4.F) implies that φ is a local C^k-diffeomorphism at x_0. Multiplication of both sides of (19) with $f'(x_0)$ gives $f'(x_0)(\varphi(x)) = f(x)$. □

Moreover, we obtain that the following diagram commutes:

$$\begin{array}{ccc} U(x_0) \subseteq X & \xrightarrow{f} & Y \\ \varphi \downarrow & & \uparrow f'(x_0) \\ U(0) \subseteq X & \xrightarrow{P^\perp} & N^\perp \end{array}$$

This, and the diagram of Proposition 73.36, show that, after a suitable coordinate change, f looks, locally at x_0, like $f'(x_0)$ or like the projection operator $P^\perp$.

Proposition 73.37 (Local Immersion Theorem). *Let X and Y be B-spaces over $\mathbb{K}$, and $f: U(0) \subseteq X \to Y$ a C^k-immersion, $k \geq 1$, at $x_0 = 0$. Then there exists a local C^k-diffeomorphism φ with $\varphi(0) = f(0)$ and $\varphi'(0) = I$ such that the following diagram commutes:*

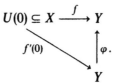

If f is analytic, then also φ is analytic.

PROOF. Without loss of generality, let $f(0) = 0$. Since f is an immersion at $x_0 = 0$, it follows that $R(f'(0))$ splits the B-space Y. Let $P: Y \to R(f'(0))$ be a projection operator and define

$$\varphi(y) = f(f'(0)^{-1}Py) + (I - P)y. \tag{20}$$

It follows from $Pf'(0)x = f'(0)x$ that $\varphi(f'(0)x) = f(x)$, i.e., the above diagram commutes. Moreover, we have

$$\varphi'(0)h = f'(0)f'(0)^{-1}Ph + (I - P)h = h,$$

and hence $\varphi'(0) = I$. The inverse mapping theorem (Theorem 4.F) implies that φ is a local C^k-diffeomorphism at $x_0 = 0$. □

73.11. Two Principles for Constructing Manifolds

We are looking for general and easy to apply criteria to decide whether or not a given set is a Banach manifold. We will discuss the following two principles.

73.11. Two Principles for Constructing Manifolds

(P1) Implicit construction by solving an equation $f(x) = y$ for some fixed y (Theorems 73.C and 73.D).
(P2) Explicit construction by using an embedding (Theorem 73.E).

Let us explain the basic ideas. Principles (P1) and (P2) correspond to (i) and (ii) below, respectively. In $\mathbb{R}^n$, $n \geq 2$, there are two ways to construct a curve C.

(i) The curve C is defined as the set of all points $x \in \mathbb{R}^n$, which satisfy the $(n - 1)$ equations

$$f_i(x) = 0, \quad i = 1, \ldots, n - 1. \tag{21}$$

(ii) The curve C is defined by using a parameter equation $x = x(t)$.

In both cases we will not always obtain manifolds. In (ii), for example, the curve C may intersect itself. In (i), the rank of $f'(x)$ plays an important role. More precisely, we obtain from Section 4.18:

If $f: \mathbb{R}^n \to \mathbb{R}^{n-1}$ is C^k, $k \geq 1$, and 0 is a regular value of f, i.e., the rank of the Jacobian matrix $f'(x) = (\partial f_i(x)/\partial \xi_j)$ is equal to $n - 1$ for all x which satisfy (21), then the solution set of (21) is a one-dimensional C^k-manifold.

In many cases (i) is more convenient than (ii). For example, a circle is defined through $\xi^2 + \eta^2 = 1$. Similarly for a sphere. In this case, parameter equations are more elaborate.

We now generalize these two principles. Submersions (regular values) play an important role in (i) and immersions in (ii). As in Theorem 73.B above, we use the process of linearization, i.e., the behavior of the nonlinear map is described by corresponding properties of the derivative.

As we saw in Section 4.16, formula (21) can be used to construct manifolds even if the rank of $f'(x)$ is not equal to $n - 1$, i.e., is not maximal. More precisely, we have:

If $f: \mathbb{R}^n \to \mathbb{R}^{n-1}$ is C^k, $k \geq 1$, and f is a subimmersion on $f^{-1}(0)$ of rank r, then the solution set of (21) is a $(n - r)$-dimensional C^k-manifold.

Subimmersion on $f^{-1}(0)$ means that in the neighborhood of each point $x \in \mathbb{R}^n$, which satisfies (21), the rank of $f'(x)$ is equal to r. A generalization of this yields the subimmersion theorem below, which is valid only in the finite-dimensional case.

We need the concept of a submanifold. As a motivation, consider a curve C on the surface of the earth M, with no endpoints (open or closed curve without self-intersections). One may think of a circle of longitude or a circle of latitude or a river without source and mouth. We call C a submanifold of M if for each $r \in C$ there exists a chart in M, for which C looks like a straight line (Fig. 73.12). Naturally, we do not require that such a chart is contained in our atlas. We only assume that such a chart exists and that it is compatible with our atlas, i.e., is an admissible chart in the sense of Section 73.2. Since M is a C^∞-manifold, and a chart change uses C^∞-maps, a one-dimensional

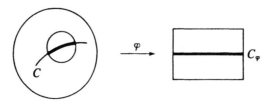

Figure 73.12

submanifold C must be a C^∞-curve. According to the general definition below, points and open subsets of the surface of the earth are zero-dimensional and two-dimensional submanifolds. Also, our definition implies that submanifolds of Banach manifolds look, locally, in suitable admissible charts, like closed linear subspaces of the chart spaces. In order to obtain more convenient methods for constructing submanifolds, we make the additional assumptions that these subspaces split the corresponding chart spaces. This condition is automatically satisfied for finite-dimensional manifolds.

Definition 73.38. Let M be a C^k-Banach manifold, $k \geq 0$. A subset S of M is called a *submanifold* of M if and only if for each point $x \in S$ there exists an admissible chart (U, φ) in M with $x \in U$ such that the following hold:

(i) The chart space X_φ contains a linear, closed subspace Y_φ which splits X_φ.
(ii) The chart image $\varphi(U \cap S)$ is an open set in Y_φ.

Proposition 73.39. *Every submanifold S of a C^k-Banach manifold is itself a C^k-Banach manifold.*

PROOF. Choose as atlas charts of S the restrictions of the special charts (U, φ) of Definition 73.38. That is, the charts of S are $(U \cap S, \varphi_{|U \cap S})$ with chart spaces Y_φ. Here, $\varphi_{|U \cap S}$ denotes the restriction of φ onto $U \cap S$. □

For the following three global theorems about submersions (regular values), immersions, and subimmersions we make the assumption:

(H) The map $f \colon M \to N$ is C^k, $k \geq 1$, with M and N C^k-Banach manifolds with chart spaces over $\mathbb{K}$.

Theorem 73.C (Preimage Theorem). *Assume* (H) *for the map $f \colon M \to N$. If y is a regular value of f, then $S = f^{-1}(y)$ is a submanifold of M, i.e., in particular, a C^k-Banach manifold.*

The tangent space TS_x is equal to $N(f'(x))$ for all $x \in S$. If M and N have a dimension, then codim $S = \dim N$.

73.11. Two Principles for Constructing Manifolds

In Section 78.8 we prove the important Sard–Smale theorem. It states that, for Fredholm operators f, most points $y \in N$ are regular values of f.

Corollary 73.40. *If $f: M \to N$ is a C^k-submersion, $k \geq 1$, such that (H) is satisfied, then each image point $y \in f(M)$ is a regular value of f.*

The codimension codim S of a submanifold S of M is the codimension of TS_x in TM_x, if this number is independent of $x \in S$. Recall that M and N have a dimension, if dim TM_x and dim TN_y are independent of $x \in M$ and $y \in N$. We then have dim M = dim TM_x and dim N = dim TN_y. Therefore, we get

$$\dim S + \operatorname{codim} S = \dim M.$$

Theorem 73.C then implies

$$\dim S + \dim N = \dim M. \tag{22}$$

Thus, Theorem 73.C generalizes the following well-known result of linear algebra. If $M = \mathbb{R}^m$, $N = \mathbb{R}^n$, $y = 0$, and if f is linear, surjective, then S is the solution set of the system of linear equations $f(x) = 0$. This is a linear subspace of M with dim $S = m - n$. Similarly, Theorems 73.D and 73.E below are important generalizations of known results about systems of linear equations.

Corollary 73.40 immediately follows from Definition 73.32.

PROOF OF THEOREM 73.C. It suffices to study the local problem. Thus we look at S in a neighborhood of some given point x.

(I) Let M and N be B-spaces. The local submersion theorem (Proposition 73.36) implies that there exists an admissible chart (U, φ) in M with $\varphi(x) = 0$ and $\varphi'(x) = I$ such that

$$f(\varphi^{-1}(h)) = f'(x)h + y$$

for all h in a neighborhood of 0. Thus the solution set for the equation

$$f(z) = y$$

in a neighborhood of $z = x$ corresponds to the solution set of the equation

$$f'(x)h = 0$$

in a neighborhood of $h = 0$. Thus in this chart, S looks, locally, like $N(f'(x))$. Since f is a submersion at x, the null space $N(f'(x))$ splits the B-space $TM_x = M$.

We now compute the tangent space TS_x. Let $x = x(t)$ be a C^1-curve in S with $x(0) = x$. It follows from $f(x(t)) = 0$ that $f'(x)x'(0) = 0$, and hence that $TS_x \subseteq N(f'(x))$. Conversely, let $f'(x)v = 0$. If we choose $x(t) = \varphi^{-1}(tv)$, then this curve lies in S and $x'(0) = v$. Therefore, we have $TS_x = N(f'(x))$.

For a linear operator L we always have codim $N(L)$ = dim $R(L)$. Since

$f'(x)\colon TM_x \to TN_{f(x)}$ is surjective, it follows that codim $TS_x = \dim TN_{f(x)}$. For the special case at hand we have $TM_x = M$ and $TN_{f(x)} = N$.

(II) Let M and N be Banach manifolds. Then, locally, they look like B-spaces. Thus we obtain Theorem 73.C by applying (I) to the charts of M and N and to the corresponding representatives of f. □

EXAMPLE 73.41. Let $f\colon X \to \mathbb{R}$ be C^k, $k \geq 1$, X a B-space and $f'(x) \neq 0$ for all solutions x of the equation

$$f(x) = 0.$$

The solution set S of this equation is a submanifold of X and thus a C^k-Banach manifold. For the tangent space we have $TS_x = N(f'(x))$ and codim $TS_x = 1$ for all $x \in S$.

In the special case that X is a H-space, and if we choose $f(x) = (x|x) - r^2$, then S is equal to the sphere of radius $r > 0$. Consequently, S is a C^∞-Banach manifold with codim $S = 1$. For the tangent space we have

$$TS_x = N(f'(x)) = \{h \in X\colon (x|h) = 0\}.$$

PROOF. Since $f'(x) \neq 0$, it follows that $f'(x)\colon X \to \mathbb{R}$ is surjective. Moreover, we have codim $N(f'(x)) = 1$, because if h_0 satisfies $f'(x)h_0 = 1$, then every $h \in X$ has a unique representation $h = h_1 + (f'(x)h)h_0$ with $f'(x)h_1 = 0$. Theorem 73.C then yields the assertion. □

Theorem 73.D (Subimmersions). *Let $f\colon M \to N$ be C^k, $k \geq 1$, and assume* (H). *Suppose that all chart spaces of M and N are finite-dimensional. If f is a subimmersion at every point of $S = f^{-1}(y)$ for some fixed $y \in N$, then S is a submanifold of M, i.e., in particular, a C^k-Banach manifold.*

The tangent space TS_x is equal to $N(f'(x))$. If $\dim M = m$, $\dim N = n$, and rank $f'(x) = r$ for all $x \in S$, then $\dim S = m - r$.

PROOF. One applies the local subimmersion theorem (Proposition 73.34) and follows the proof of Theorem 73.C. □

In view of Theorem 73.E below, we consider the following important question.

(Q) *Which continuous maps have a continuous inverse?*

COUNTEREXAMPLE 73.42. Consider the following continuous, injective map $f\colon \mathbb{R} \to \mathbb{R}^2$ of Figure 73.13. The behavior of the image $f(\mathbb{R})$ at the point $f(0)$ plays an important role. We have $f(x) \to f(0)$ as $x \to 0$ and $x \to +\infty$. Thus f^{-1} is not continuous at the point $f(0)$.

If f is sufficiently smooth, then the above map $f\colon \mathbb{R} \to \mathbb{R}^2$ is also an example of an injective C^k-immersion, where $f(\mathbb{R})$ is not a submanifold of $\mathbb{R}^2$ because

73.11. Two Principles for Constructing Manifolds

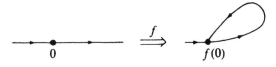

Figure 73.13

there exists no chart in $\mathbb{R}^2$, for which $f(\mathbb{R})$ looks like a straight line in a neighborhood of $f(0)$. Intuitively, $f(\mathbb{R})$ has no tangent line at $f(0)$.

Definition 73.43. Let M and N be topological spaces. Then $f: M \to N$ is called a *homeomorphic embedding* (short: e-homeomorphism) if and only if f is a homeomorphism onto its image $f(M)$, i.e., f is injective and continuous and f^{-1} is continuous as well.

If M and N are C^k-Banach manifolds, $k \geq 1$, then $f: M \to N$ is called an *embedding* if and only if f is an e-homeomorphism and an immersion.

Proposition 73.44. *If $f: M \to N$ is injective and continuous, and M and N are topological spaces, then f is an e-homeomorphism if one of the following three conditions is satisfied:*

(i) *f is closed.*
(ii) *f is proper, and N satisfies the first axiom of countability, i.e., every point in N has a countable neighborhood basis.*
(iii) *M is compact.*

Here, (ii) and (iii) are special cases of (i).

Corollary 73.45. *If $f: M \to N$ is an injective C^k-immersion, and M and N are C^k-Banach manifolds, $k \geq 1$, then f is an embedding if M is compact or f is closed or proper.*

PROOF. (i) If f is closed, then $f(M)$ is closed and f^{-1} is continuous. This follows because the closedness of K in M implies the closedness of $f(K)$ (see $A_1(11)$).

(ii) An analogous argument, as used in the proof of Proposition 4.44, shows that f is closed.

(iii) If K is a closed subset of M, then K and $f(K)$ are compact, i.e., f is closed. □

Since every Banach manifold looks locally like a B-space, it follows that every point has a countable neighborhood basis. Thus, Corollary 73.45 is an immediate consequence of Proposition 73.44.

Theorem 73.E (Embeddings). *Let $f: M \to N$ be a C^k-embedding, $k \geq 1$, and let (H) be satisfied. Then $S = f(M)$ is a submanifold of N, i.e., in particular, a C^k-Banach manifold.*

The tangent space TS_x is equal to $R(f'(x))$. If M has a dimension, then $\dim S = \dim M$.

The following proof is analogous to the proof of Theorem 73.C. Here, we use the local immersion theorem instead of the local submersion theorem in Theorem 73.C. It follows from (I) below that the difficulties of Counterexample 73.42 at $f(0)$ of Figure 73.13 will not occur.

PROOF.

(I) The map f is an e-homeomorphism. Thus it follows that if U is an open neighborhood of x in M, then $f(U)$ is an open neighborhood of $f(x)$ in $f(M)$. Hence there exists an open set V in N with $f(U) = V \cap f(M)$ (see the definition of the subspace topology in $A_1(9)$).

(II) Let M and N be B-spaces and $x \in M$. The local immersion theorem (Proposition 73.37) shows that, in suitable coordinates, f looks like $f'(x)$ in a neighborhood of x. Thus there exists an open neighborhood U of x such that $f(U)$ looks like $R(f'(x))$ in an admissible chart for N. Hence, $f(M)$ is a submanifold of N.

Looking at curves through $f(x)$, one finds that $TS_x = R(f'(x))$. This follows similarly as in the proof of Theorem 73.C. Since $f'(x): TM_x \to R(f'(x))$ is bijective, we obtain that $\dim TM_x = \dim R(f'(x))$.

(III) If M and N are Banach manifolds, then (II) can be applied to the charts. □

73.12. Construction of Diffeomorphisms and the Generalized Morse Lemma

The following important example illustrates the method of continuation of a parameter. It allows us to construct diffeomorphisms by solving ordinary differential equations on B-spaces, which yield normal forms. Consider the Taylor expansion

$$f(x_0 + h) = f(x_0) + f'(x_0)h + f''(x_0)h^2/2 + \cdots.$$

Let $f'(x_0) = 0$. Then f is not a submersion at x_0 and Proposition 73.36 *cannot* be used to construct the local normal form $f(\varphi(y)) = f(x_0) + f'(x_0)y$ in a neighborhood of zero. Thus the following question arrises. Is there a coordinate transformation $x = \varphi(y)$ with $x_0 = \varphi(0)$ for maps $f: U(x_0) \subseteq X \to \mathbb{R}$ with $f'(x_0) = 0$ such that

$$f(\varphi(y)) = f(x_0) + f''(x_0)y^2/2 \tag{23}$$

is satisfied for all y in a neighborhood of zero? In this case f will behave like the quadratic part of the Taylor expansion in a neighborhood of x_0.

73.12. Construction of Diffeomorphisms and the Generalized Morse Lemma

Let X be a B-space over $\mathbb{K}$, and let $Q: X \times X \to \mathbb{K}$ be a bounded, symmetric bilinear form. We define the operator A through

$$(Ax)y = Q(x, y).$$

It follows that $\|(Ax)y\| \le q\|x\|\|y\|$ and hence $\|Ax\| \le q\|x\|$. Thus, $Ax \in X^*$, and the linear operator $A: X \to X^*$ is continuous.

Definition 73.46. The bilinear form Q is called *nondegenerate* (or *weakly nondegenerate*) if and only if $A: X \to X^*$ is bijective (or injective).

For example, let $X = \mathbb{R}^n$, and let

$$Q(x, y) = \sum_{i,j=1}^{n} a_{ij}\xi_i\eta_j$$

with a real, symmetric matrix (a_{ij}). Then $Q: X \times X \to \mathbb{R}$ is a bounded, symmetric bilinear form. If we identify X with X^*, then $A: X \to X$ corresponds to the matrix (a_{ij}), i.e., $y = Ax$ corresponds to

$$\eta_i = \sum_{j=1}^{n} a_{ij}\xi_j.$$

Moreover, Q is nondegenerate if and only if A^{-1} exists, i.e., $\det(a_{ij}) \ne 0$.

If $\dim X < \infty$, then the notions of nondegenerate and weakly nondegenerate coincide.

Let X be a B-space. It follows from Definition 73.32 that a map $f: U(x_0) \subseteq X \to \mathbb{R}$ has a *singular point* at x_0 if and only if $f'(x_0) = 0$. Such a point is called *nondegenerate* if and only if the bilinear form $(h, k) \mapsto f''(x_0)hk$ is nondegenerate.

Theorem 73.F (Generalized Morse Lemma of Palais (1969)). *Let X be a real B-space, and let $f: U(x_0) \subseteq X \to \mathbb{R}$ be a C^{k+2}-map, $k \ge 1$, which has a nondegenerate singular point at x_0.*

Then there exists a local C^k-diffeomorphism $\varphi: U(0) \subseteq X \to X$ with $\varphi(0) = x_0$ such that (23) holds in a neighborhood of zero.

PROOF. Without loss of generality, let $x_0 = 0$ and $f(0) = 0$. The proof idea is as follows. Let $g(x) = f''(0)x^2/2$ and use the homotopy

$$H(x, t) = tf(x) + (1 - t)g(x).$$

Then solve the linear equation

$$H_x(x, t)h + H_t(x, t) = 0 \qquad (24)$$

with respect to h. This allows us to solve the ordinary differential equation

$$\varphi_t = h(\varphi, t), \qquad \varphi(0, x) = x. \qquad (25)$$

Letting $z = (\varphi(t, x), t)$, we obtain

$$\frac{\partial}{\partial t} H(z) = H_x(z)h(z) + H_t(z) = 0,$$

and hence

$$H(\varphi(1, x), 1) = H(\varphi(0, x), 0) = H(x, 0)$$

and therefore

$$f(\varphi(1, x)) = g(x).$$

Hence, the map $x \mapsto \varphi(1, x)$ is our desired local diffeomorphism at the origin. We now justify our formal considerations.

(I) Solution of (24). It follows from Taylor's theorem (Theorem 4.A) with integral remainder that

$$f(x) = \int_0^1 (1 - \tau) f''(\tau x) x^2 \, d\tau$$

and

$$f'(x) = \int_0^1 f''(\tau x) x \, d\tau.$$

Moreover, we have

$$H_x(x, t) = tf'(x) + (1 - t)g'(x) = B(x, t)x$$

with

$$B(x, t) = f''(0) + t \int_0^1 (f''(\tau x) - f''(0)) d\tau,$$

and

$$H_t(x, t) = f(x) - g(x) = C(x)x^2$$

with

$$C(x) = \int_0^1 [(1 - \tau) f''(\tau x) - 2^{-1} f''(0)] \, d\tau$$

and $C(0) = 0$. It follows then from $f'(x) \in X^*$ and $f''(x) \in L(X, X^*)$ that $B(x, t), C(x) \in L(X, X^*)$.

Since the bilinear form $(h, k) \mapsto f''(0)hk$ is nondegenerate, we obtain from the open mapping theorem $A_1(36)$ that $f''(0)^{-1} \in L(X^*, X)$. Thus $B(x, t)^{-1}$ exists for all $t \in [0, 1]$ and all x in a fixed neighborhood of zero. This is a consequence of the continuity of the formation of inverse mappings (Problem 1.7). Letting

$$h(x, t) = -B(x, t)^{-1} C(x) x,$$

we obtain that $Bxh = Bhx = -Cx^2$, i.e., h is a solution of (24).

(II) Since $C(0) = 0$, we have $h(0, t) = 0$ and $h_x(0, t) = 0$ for all t.

(III) Solution of (25). It follows from Theorem 3.A that for each x, equation (25) has a unique solution in $[0, 1]$, which lies in a sufficiently small neighborhood of the origin in X. Note that, because of (II), the Lipschitz constant of h can be chosen arbitrarily small. From Theorem 4.D (Dependence on Data) it follows that φ is C^k, since h is C^k.

If we pose the initial-value problem at $t = 1$ backwards with respect to t, then we obtain from uniqueness that $x \mapsto \varphi(1, x)$ is a local diffeomorphism at $x = 0$. □

A general criteria, which allows the transformation of functionals to not necessarily quadratic normal forms, will be proved in Problem 73.8a. From this, one obtains a nice version of Theorem 73.F for variational integrals.

73.13. Transversality

Consider the following two questions.

(i) When is the intersection of manifolds a manifold?
(ii) When is the preimage of a manifold a manifold?

Interestingly, the geometric concept of transversality provides answers to both of these questions. Transversality is certainly one of the most important concepts of modern mathematics. It is frequently used to express that a qualitative phenomenon is natural or *nondegenerate*. As a first illustration, consider Figure 73.14. In Figure 73.14(a) the two curves in $\mathbb{R}^2$ intersect transversally but in Figure 73.14(b) they intersect nontransversally. There they only touch each other. It is convenient to speak of transversal intersection at some point y if both curves do not intersect there. To get an impression of how transversality can be used to answer question (i), note that in Figure 73.14(a) the intersection of the two curves M and N forms a manifold (a point or the empty set). In the case of nontransversal intersection, however, the intersection $M \cap N$ need not be a manifold. In Figure 73.15, $M \cap N$ is not a manifold because of the two endpoints P and Q.

Let us make the following three assumptions:

(H1) M, N, and Y are C^k-Banach manifolds with chart spaces over $\mathbb{K}$, $k \geq 1$, and N is a submanifold of Y.
(H2) Besides (H1), M is also a submanifold of Y.
(H3) Besides (H1), the map $f: M \to Y$ is C^k.

Definition 73.47. If L is a linear space, then *two* linear subspaces A and B of L are called *transversal* at 0 if and only if

$$A + B = L, \qquad (26)$$

i.e., A and B span L.

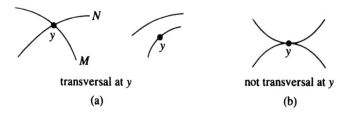

transversal at y (a)   not transversal at y (b)

Figure 73.14

Figure 73.15

This is the key idea for the general definition of transversality below. Note that this definition depends on L. In $L = \mathbb{R}^3$, for example, two straight lines can never intersect transversally. Besides transversality between linear subspaces, we also consider transversality between linear maps and linear subspaces.

The linear map $f: L \to L$ is called *transversal* to B at 0 if and only if the image $f(L)$ and B are transversal at 0, i.e.,

$$f(L) + B = L. \tag{27}$$

These definitions can be generalized to B-spaces and Banach manifolds by

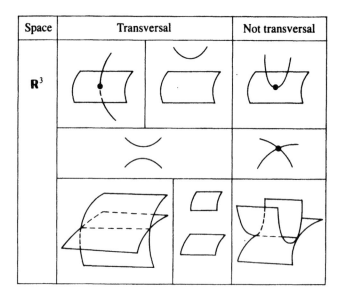

Figure 73.16

using linearization, i.e., by considering

$$TM_y + TN_y = TY_y \tag{28}$$

and

$$R(f'(x)) + TN_y = TY_y. \tag{29}$$

For technical reasons, we also need the following conditions about splitting:

The intersection $TM_y \cap TN_y$ splits TM_y. (28a)

The preimage $f'(x)^{-1}(TN_y)$ splits TM_x for all x with $f(x) = y$. (29a)

Note that the linearization of $f: M \to Y$ at the point x is equal to $f'(x)$: $TM_x \to TY_y$ with $y = f(x)$. Moreover, we have $N \subseteq Y$. For finite-dimensional manifolds, (28a) and (29a) are automatically satisfied.

Definition 73.48. If (H2) holds, then M and N are called *transversal* in Y if and only if (28) and (28a) are satisfied for all points y in the intersection $M \cap N$. We write $M \pitchfork N$ mod Y.

If (H3) holds, then f is called *transversal* to N in Y if and only if (29) and (29a) are satisfied for all points y in the intersection $f(M) \cap N$. We write $f \pitchfork N$ mod Y.

Thus for $M \cap N = \emptyset$ and $f(M) \cap N = \emptyset$, we always have transversality. In Figure 73.14(a) the two curves are transversal in $\mathbb{R}^2$, because they do not intersect at all or the tangent lines at the points of intersection span $\mathbb{R}^2$. If two curves touch each other as in Figure 73.14(b), then they are not transversal. Figure 73.16 shows several examples of transversal intersections in $\mathbb{R}^3$.

EXAMPLE 73.49. If (H3) holds for the map $f: M \to Y$, then $y \in Y$ is a regular value of f iff f is transversal to $N = \{y\}$ in Y. Compare Definition 73.32 and note that $TN_y = \{0\}$.

In Part I we saw that regular values play an important role in fixed-point theory. The example, at hand, shows the connection with transversality.

Theorem 73.G (Transversality). *If* (H3) *holds and the map* $f: M \to Y$ *is transversal to* N *in* Y, *then* $f^{-1}(N)$ *is a submanifold of* M.

If M, N, *and* Y *have dimensions, then* $f^{-1}(N)$ *has the same codimension in* M *as* N *in* Y.

Corollary 73.50. *If* (H2) *holds and the submanifolds* M *and* N *are transversal in* Y, *then the intersection* $M \cap N$ *is a submanifold of* Y.

As shown in Example 73.49, this theorem is an immediate generalization of the preimage theorem (Theorem 73.C).

PROOF OF THEOREM 73.G.

(I) We first consider the special case where M and Y are B-spaces, and N is a linear subspace of Y which splits Y. Let $P: Y \to N^\perp$ be a projection operator onto $N^\perp$ with corresponding topological direct sum decomposition $Y = N \oplus N^\perp$. The proof idea is as follows. Consider the map

$$g: X \xrightarrow{f} N \oplus N^\perp \xrightarrow{P} N^\perp,$$

instead of f, i.e., $g = Pf$. Then $f^{-1}(N) = g^{-1}(0)$. Using the preimage theorem (Theorem 73.C), it suffices to show that 0 is a regular value of g. Note that P is linear and continuous.

In fact, $g'(x) = Pf'(x)$. Let $g(x) = 0$, i.e., $f(x) \in N$. Transversality of f and N in Y means that

$$R(f'(x)) + N = Y$$

and $f'(x)^{-1}(N)$ splits M. The first condition implies that $PR(f'(x)) = N^\perp$, i.e., $g'(x)$ is surjective. The second condition shows that the null space $N(g'(x))$ splits M.

(II) Theorem 73.C implies that codim $g^{-1}(0) = \dim N^\perp$. Thus we have

$$\text{codim } f^{-1}(N) = \dim N^\perp = \text{codim } N.$$

(III) If M, N, and Y are manifolds, then Theorem 73.G follows immediately from (I) by using charts. □

PROOF OF COROLLARY 73.50. Let $f: M \to Y$ be the trivial embedding for $M \subseteq Y$. Then $M \pitchfork N$ mod Y is equivalent to $f \pitchfork N$ mod Y. Because of $f^{-1}(N) = M \cap N$, Corollary 73.50 is a special case of Theorem 73.G. □

73.14. Taylor Expansions and Jets

Let $f: U(x_0) \subseteq X \to Y$ be C^k, $k \geq 1$, and X and Y B-spaces over $\mathbb{K}$. Define

$$j_x f(h) = f(x) + \sum_{r=1}^{k} f^{(r)}(x) h^r / r!,$$

i.e., $j_x f$ is equal to the Taylor expansion of f at the point x up to the kth derivative. Moreover, we define the k-jet coordinate of f at the point x as

$$J^k f(x) = (x, f(x), f'(x), \ldots, f^{(k)}(x)),$$

and let

$$P^k f(x) = (f'(x), \ldots, f^{(k)}(x)).$$

We now give the global generalizations of $J^k f$ to manifolds.

In preparation we study the change of jet coordinates after coordinate transformations. Let $L^n_{\text{sym}}(X, Y)$ be the B-space of all n-linear, symmetric, and

73.14. Taylor Expansions and Jets

bounded maps $A: X \times \cdots \times X \to Y$. Also let

$$P^k(X, Y) = \prod_{r=1}^{k} L^r_{sym}(X, Y).$$

Then $P^k f(x) \in P^k(X, Y)$ and

$$J^k f(x) \in X \times Y \times P^k(X, Y).$$

Define

$$g = \psi \circ f \circ \varphi,$$

where φ and ψ are C^k. More precisely, we choose $\varphi: U(z) \subseteq X \to X$ with $\varphi(z) = x$ and $\psi: U(f(x)) \subseteq Y \to Y$. Then, we obtain the *key relation*

$$P^k g(z) = B P^k f(x), \tag{30}$$

where $B: P^k(X, Y) \to P^k(X, Y)$ is linear and continuous. This follows from the chain rule. If φ and ψ are local diffeomorphisms, then, conversely, $P^k f(x)$ can be expressed in terms of $P^k g(z)$. Thus B is bijective and from the open mapping theorem $A_1(36)$ it follows that B is a linear homeomorphism. We now assume:

(H) M and N are C^∞-Banach manifolds with chart spaces over $\mathbb{K}$.

Two C^∞-maps $f, g: U(x) \subseteq M \to N$ are called *k-equivalent* at a point $x \in M$ if and only if

$$J^k \bar{f}(\bar{x}) = J^k \bar{g}(\bar{x}) \tag{31}$$

holds for the representatives $\bar{f}, \bar{g}$, and $\bar{x}$ of f, g, and x in fixed admissible charts. This means that the k-jet coordinates of the representatives coincide at the point $\bar{x}$. From (30) it follows that this definition is invariant, i.e., independent of the choice of the charts.

Definition 73.51 (Jets). If (H) holds, then $J^k f(x)$ denotes the set of all C^∞-maps $g: U(x) \subseteq M \to N$ which are k-equivalent to the C^∞-map $f: U(x) \subseteq M \to N$ at the point $x \in M$. Moreover, let $J^k(X, Y)$ be the set of all possible $J^k f(x)$ where f and x vary.

We call $J^k f(x)$ a *k-jet*.

Proposition 73.52. *The set $J^k(M, N)$ is a C^∞-Banach manifold.*

PROOF. Let (U, φ) and (V, ψ) be charts in M and N with $x \in U$ and $f(x) \in V$ and chart spaces X_φ and X_ψ. Let $J^k f(x)$ have the local coordinates $J^k \bar{f}(\bar{x})$, where $\bar{f}$ and $\bar{x}$ are the representatives of f and x. For a chart change, we obtain the transformation rule for

$$J^k \bar{f}(\bar{x}) = (\bar{x}, \bar{f}(\bar{x}), \ldots, \bar{f}^{(k)}(\bar{x})),$$

analogously to (30). This follows from an application of the chain rule to the representatives.

In order to obtain a manifold structure, we use local coordinates and the method of abstract atlases of Problem 73.2. We choose (W_{UV}, χ) as abstract charts in $J^k(M, N)$. Here, W_{UV} is the set of all $J^k f(x)$ with $x \in U$ and $f(x) \in V$. Moreover, we have

$$\chi(J^k f(x)) = J^k \overline{f}(\overline{x}).$$

Since

$$J^k \overline{f}(\overline{x}) \in U_\varphi \times V_\psi \times P^k(X_\varphi, Y_\psi),$$

it follows that the set on the right-hand side is the chart image. Note that for each element h in this set, there exists a C^∞-map $f: U(x) \subseteq M \to N$ with $J^k \overline{f}(\overline{x}) = h$. To see this choose as the representative of f a polynomial. The chart space is equal to $X_\varphi \times Y_\psi \times P^k(X_\varphi, Y_\psi)$.

The collection of all charts (W_{UV}, χ), where U and V vary, then forms an abstract C^∞-atlas. This induces a topology on $J^k(M, N)$ and gives us a manifold structure (Problem 73.2). □

In this way we obtain a map

$$J^k f: M \to J^k(M, N)$$

for every C^∞-map $f: M \to N$ which is called the k-jet extension of f. This map is also of class C^∞.

EXAMPLE 73.53 ($J^k(M, N)$). For every C^∞-map $f: \mathbb{R} \to \mathbb{R}$ we have

$$J^k f(x) = (x, f(x), f'(x), \ldots, f^{(k)}(x)) \in \mathbb{R}^{2+k}. \tag{32}$$

Actually, $J^k f(x)$ in the above definition is a collection of maps. But since all of them are characterized by the same k-jet coordinate at x, we may identify $J^k f(x)$ with the k-jet coordinate as in (32). Thus we obtain

$$J^k(\mathbb{R}, \mathbb{R}) = \mathbb{R}^{2+k}.$$

Note that for each $h \in \mathbb{R}^{2+k}$ there exists a C^∞-map $f: U(x) \subseteq \mathbb{R} \to \mathbb{R}$ with $J^k f(x) = h$.

If M and N are open in $\mathbb{R}$, then we obtain, analogously, for the C^∞-map $f: M \to N$ that

$$J^k f(x) = (x, f(x), f'(x), \ldots, f^{(k)}(x)) \in M \times N \times \mathbb{R}^k. \tag{33}$$

Thus we have

$$J^k(M, N) = M \times N \times \mathbb{R}^k,$$

and (33) describes the map $J^k f: M \to J^k(M, N)$.

For C^∞-maps $f: \mathbb{R}^n \to \mathbb{R}$ we have

$$J^k f(x) = (x, f(x), D^\alpha f(x))_{1 \leq |\alpha| \leq k}, \tag{34}$$

i.e., $J^k f(x)$ contains all of the partial derivatives up to order k. Derivatives,

73.14. Taylor Expansions and Jets

which are different in the order of differentiation, are only accounted for once. For example, only the first of the following two derivatives is considered; $\partial^2 f(x)/\partial \xi_1 \partial \xi_2$ and $\partial^2 f(x)/\partial \xi_2 \partial \xi_1$. If M and N are open in $\mathbb{R}^n$ and $\mathbb{R}$, respectively, then

$$J^k(M, N) = M \times N \times \mathbb{R}^r.$$

Here, $2 + r$ is the number of components of (34).

EXAMPLE 73.54 (Transversality and Nondegeneracy). The transversality of k-jets can effectively be used to describe the nondegeneracy of higher-order derivatives. For example, let $f: \mathbb{R} \to \mathbb{R}$ be C^∞, and $g = J^k f$, $k \geq 1$. It follows then from (32) that

$$g'(x)h = (h, f'(x)h, \ldots, f^{(k+1)}(x)h)$$

for all $h \in \mathbb{R}$. Let

$$S = \{(\xi_1, \ldots, \xi_{k+2}) \in \mathbb{R}^{k+2}: \xi_{k+2} = 0\}.$$

This implies the following statement:

From $f^{(k)}(x) = 0$ follows $f^{(k+1)}(x) \neq 0$ if and only if

$$J^k f \pitchfork S \quad \mod J^k(\mathbb{R}, \mathbb{R}). \tag{35}$$

For $k = 1$, the transversality condition

$$J^1 f \pitchfork S \quad \mod J^1(\mathbb{R}, \mathbb{R})$$

means precisely that f has only nondegenerate singular points.

PROOF. The condition $J^k f(x) \cap S \neq \emptyset$ is equivalent to $f^{(k)}(x) = 0$. Moreover, $f^{(k+1)}(x) \neq 0$ is equivalent to

$$g'(x)(\mathbb{R}) + S = \mathbb{R}^{k+2}. \qquad \square$$

Condition (35), which appears somewhat complicated, is chosen in view of the important transversality theorem of Thom (Proposition 73.57).

The definition of jets allows an elegant introduction of the so-called Whitney topology, which is the main topology in modern differential topology and especially catastrophe theory.

Definition 73.55 (C^∞-Whitney Topology). Let M and N be C^∞-Banach manifolds with chart spaces over $\mathbb{K}$, and let $C^\infty(M, N)$ denote the set of all C^∞-maps $f: M \to N$.

Let $k = 0, 1, \ldots$, be a nonnegative integer and U an arbitrary open set in $J^k(M, N)$. Define

$$O_{k,U} = \{f \in C^\infty(M, N): J^k f(M) \subseteq U\}.$$

A set S in $C^\infty(M, N)$ is called ∞-open if and only if it is the union of sets of the form $O_{k,U}$. The ∞-open sets define a topology on $C^\infty(M, N)$ which is called the C^∞-*Whitney topology*.

This definition is equivalent to the definition of $A_1(67)$ for a more special situation. Roughly speaking, two maps $f: M \to N$ and $g: M \to N$ are considered close in the C^∞-Whitney topology if all local Taylor expansions for the representatives are sufficiently close. This means that all jet coordinates are sufficiently close. The following example shows this more precisely.

EXAMPLE 73.56. Let $f_n, f \in C^\infty(\mathbb{R}^m, \mathbb{R})$ for all n. Then the statement that $f_n \to f$ as $n \to \infty$ in the C^∞-Whitney topology is equivalent to the following two conditions:

(i) For each $k = 0, 1, \ldots$ there exists a compact set K in $\mathbb{R}^n$ such that $D^\alpha f_n \rightrightarrows D^\alpha f$ on K for all α with $0 \leq |\alpha| \leq k$ (uniform convergence of the derivatives up to order k).
(ii) $f_n = f$ outside of K except for finitely many n.

The proof, which is left as an exercise for the reader, can be found in Golubitsky and Guillemin (1977, M), Chapter 2, §3. We also suggest to show that Definition 73.55 actually gives a topology for $C^\infty(M, N)$ (see Problem 73.3). As a typical application of the Whitney topology, we mention the important transversality theorem of Thom. Let $C^\infty(M, N)$ be equipped with the C^∞-Whitney topology.

Proposition 73.57. *Let M and N be finite-dimensional real C^∞-manifolds with countable bases. Also, let S be a submanifold of $J^k(M, N)$ for some fixed $k = 0, 1, \ldots$. Then, the set*

$$\{f \in C^\infty(M, N): J^k f \pitchfork S \bmod J^k(M, N)\}$$

is residual and dense in $C^\infty(M, N)$. Moreover, this set is open in $C^\infty(M, N)$ if S is closed.

Let Y be a closed submanifold of N. Then, the set

$$\{f \in C^\infty(M, N): f \pitchfork Y \bmod N\}$$

is open and dense in $C^\infty(M, N)$.

A proof may be found in Golubitsky and Guillemin (1977, M), Theorem 4.9. Recall that a set is called residual precisely if it is the intersection of countably many open and dense sets. Since $C^\infty(M, N)$ with the C^∞-Whitney topology is a Baire space, all residual sets are dense (see $A_1(64)$). In (35), we saw that transversality conditions of the form $J^k f \pitchfork S \bmod J^k(M, N)$ include nondegeneracy conditions for higher-order derivatives. Thus, Proposition 73.57 is a precise formulation of the following general principle:

(G) *Nondegeneracy is generic, i.e., holds in most cases.*

73.15. Equivalence of Maps

This section should prepare the reader for the next following one. We begin with the transformation rule

$$\psi[f(\varphi(u))] = g(u), \tag{36}$$

which takes the map f into g. If f and g are real functions, for example, then (36) follows from

$$f(x) = y$$

and a coordinate change in x and y, i.e., $x = \varphi(u)$ and $v = \psi(y)$. This gives

$$g(u) = v.$$

Formula (36) defines an equivalence between maps which allows a transformation of f into a normal form g.

Definition 73.58. Two C^k-maps $f\colon X \to Y$ and $g\colon U \to V$ between C^k-Banach manifolds are called C^k-equivalent if and only if there exist C^k-diffeomorphisms φ and ψ such that the following diagram commutes:

$$\begin{array}{ccc} U & \xrightarrow{g} & V \\ \varphi \downarrow & & \uparrow \psi \\ X & \xrightarrow{f} & Y \end{array} \tag{37}$$

i.e., $g = \psi \circ f \circ \varphi$. If, in addition, $\psi = \mathrm{id}$, $Y = V$ (or $\varphi = \mathrm{id}$, $X = U$) we speak of *right equivalence* (or left equivalence).

This definition has the following local form. A map f is C^k-equivalent at a point x_0 to a map g at a point u_0 if and only if φ and ψ in (37) are local C^k-diffeomorphisms at u_0 and $f(x_0)$, with $\varphi(u_0) = x_0$, respectively, and $\psi(f(x_0)) = g(u_0)$. In this case f and g need only be defined in a neighborhood of x_0 and u_0.

Similarly, local right and left equivalence are defined. For example, the C^k-right equivalence of f at x_0 to g at u_0 is expressed through

$$f(\varphi(u)) = g(u) \tag{38}$$

for all u in a neighborhood of u_0. More precisely, we have that

$$f\colon U(x_0) \subseteq X \to Y \quad \text{and} \quad g\colon U(u_0) \subseteq U \to Y$$

are C^k, and $\varphi\colon V(u_0) \subseteq U \to X$ is a local C^k-diffeomorphism at u_0 with $\varphi(u_0) = x_0$.

In this language we can formulate the results of Section 73.10 quite elegantly.

EXAMPLE 73.59 (Linearization of Maps). Let $f\colon U(x) \subseteq X \to Y$ be C^k, $k \geq 1$, and X and Y B-spaces over $\mathbb{K}$. Let $g = j_x^1 f$, i.e.,

$$g(u) = f(x) + f'(x)u.$$

If f is a submersion (or immersion) at x, then f is C^k-right equivalent at x to g at 0 (or C^k-left equivalent).

If $X = \mathbb{K}^n$ and $Y = \mathbb{K}^m$, and if f is a subimmersion at x, then f is C^k-equivalent at x to g at 0. Moreover, if rank $f'(x) = r$, then f is C^k-equivalent at x to $h: X \to Y$ at 0 with

$$h(\xi_1, \ldots, \xi_n) = (\xi_1, \ldots, \xi_r, 0, \ldots, 0).$$

Thus the following question arises. Under which conditions is f equivalent to $j_x^k f$, $k \geq 2$, if some of the above assumptions are not satisfied (multilinearization)? As of yet, no general theory is available. The important special case of real functions is discussed in the following section.

A general multilinearization principle for systems of real equations will be considered in Problem 73.9 (the blowing-up lemma). In fact, there are two *basic tools* in modern bifurcation theory, namely,

(i) the implicit function theorem, and
(ii) the blowing-up lemma (the generalized implicit function theorem).

Experience shows that (i) and (ii) are enough in order to handle many cases occuring in applications of bifurcation theory to concrete problems in the natural sciences.

73.16. Multilinearization of Maps, Normal Forms, and Catastrophe Theory

> At the tenth bifurcation
> my true love gave to me:
> Ten dual parabolics
> Nine parabolic umbilics
> Eight hyperbolic umbilics
> Seven elliptic umbilics
> Six dual butterflys
> Five butterflies;
> Four swallowtails
> Three dual cusps
> Two standard cusps
> and a fold catastrophe.
>
> Folclore
>
> Universal unfoldings generalize analytic continuation.
>
> René Thom

Let us begin with the following two questions, which are of major importance for a qualitative understanding of many scientific phenomena.

73.16. Multilinearization of Maps, Normal Forms, and Catastrophe Theory

(Q1) *Principal part problem.* When does the Taylor expansion up to some order k

$$f(x) + f'(x)u + \cdots + f^{(k)}(x)u^k/k!$$

provide enough information to understand the local behavior of a function f at x?

(Q2) *Normal form problem.* What are the local normal forms for parameter families of functions?

In this section we generalize the results of Section 37.28. We discuss a number of deeper results without proof. In particular, we generalize the linearization procedure of Section 73.15 and the Morse lemma of Section 73.12. The central ideas are

transversality,
structural stability,
genericity.

Also, we need the concept of k-determination of functions and universal unfoldings. Methodically, it is quite interesting that purely algebraic operations between polynomials give information about normal forms which are obtained from diffeomorphisms (multilinearization). This is a substantial generalization of classical results about solving analytic equations by using the Weierstrass preparation theorem and Newton diagrams. This has been discussed in Problem 8.3. For proofs, we recommend Golubitsky (1978, S) and Poston and Stewart (1978, M). Many interesting applications are contained in Poston and Stewart (1978, M) and Gilmore (1981, M). A number of deeper results about normal forms of singularities may be found in Arnold (1985).

The so-called catastrophe theory, which was initiated by René Thom in the 1960s, tries to answer questions (Q1) and (Q2) above. The classification of mathematical objects, by using normal forms, is one of the main objectives in mathematics. From a scientific point of view, this aims at getting a survey of all possible qualitative structures. For example, the classification of Lie algebras, which exists in part, gives a good picture of all possible symmetries in our world. Symmetry is one of the main tools of physicists in classifying elementary particles and their processes. Unfortunately, at present, many classification problems seem to be too difficult. This, for example, applies to the classification of topological spaces and homeomorphisms as well as to manifolds and diffeomorphisms.

We emphasize in the following such methods, which actually allow a computation of normal forms. This might help scientists and engineers to get acquainted with mathematical tools which allow a qualitative description of phenomena.

(C) *Convention.* In this section we only consider real C^∞-functions

$$f, g \colon U(0) \subseteq \mathbb{R}^n \to \mathbb{R} \quad \text{with} \quad f(0) = g(0) = 0,$$

and parameter families, i.e., real C^∞-functions

$$F, G: U(0) \subseteq \mathbb{R}^n \times \mathbb{R}^d \to \mathbb{R}$$

of the form $F = F(x, p)$ and $G = G(y, q)$ with $F(0, 0) = G(0, 0) = 0$.

We think of x and y as independent variables (state variables) and of p and q as parameters (outer control parameters). The dimension d of the parameter space may depend on F and G. Let n be fixed. Local C^∞-maps are always maps which are defined in a neighborhood of zero and map zero into zero. If such a map is a local C^∞-diffeomorphism at 0, then it is simply called a local C^∞-diffeomorphisms. Furthermore, we write j^k for j_x^k with $x = 0$, i.e.,

$$j^k f(u) = f'(0)u + f''(0)u^2/2! + \cdots + f^{(k)}(0)u^k/k!.$$

The main transformation rule is

$$g(\varphi(u)) = j^k f(u) \tag{39}$$

in a neighborhood of zero.

Definition 73.60 (k-Determination). The function $f: U(0) \subseteq \mathbb{R}^n \to \mathbb{R}$ is called *k-determined* in $\mathbb{R}$ if and only if for each function $g: U(0) \subseteq \mathbb{R}^n \to \mathbb{R}$ with

$$j^k g = j^k f$$

there exists a local C^∞-diffeomorphism φ in $\mathbb{R}^n$ which satisfies (39), i.e., all g are right equivalent to $j^k f$ at 0.

f is called *strongly k-determined* if, in addition, $\varphi'(0) = I$.

This is a satisfactory definition for the principal part problem. Namely, if f is k-determined, then all functions, which differ from f only in terms of order higher than k, behave qualitatively like $j^k f$ in a neighborhood of zero. Roughly, this means: the Taylor expansion of f up to order k completely determines f and its perturbations with terms of order higher than k.

In science and engineering, mathematical expressions are often simplified by neglecting higher-order terms. In order not to loose important qualitative information, one has to secure that the kth order approximation is also k-determined. Consider, for example

$$f(\xi, \eta) = a\xi^2 + 2b\xi\eta + c\eta^2 + O_3.$$

With O_3 we denote terms of at least order 3. According to the following example, f is 2-determined if $ac - b^2 \neq 0$. In this case, all the terms O_3 can be neglected. If, however, $ac - b^2 = 0$, we also need to consider higher-order terms, and this in such a way that f is k-determined with $k \geq 3$.

EXAMPLE 73.61. Let $f: U(0) \subseteq \mathbb{R}^n \to \mathbb{R}$.

(a) If $f'(0) \neq 0$, i.e., not every first-order partial derivative of f vanishes at 0, then f is 1-determined in $\mathbb{R}^n$.

73.16. Multilinearization of Maps, Normal Forms, and Catastrophe Theory

(b) If $f'(0) = 0$, and the matrix $f''(0)$ for the second-order partial derivatives of f at 0 is invertible, then f is 2-determined in $\mathbb{R}^n$.
(c) If $n = 1$, and if $f'(0) = \cdots = f^{(k-1)}(0) = 0$ and $f^{(k)}(0) \neq 0$, $k \geq 2$ then f is k-determined on $\mathbb{R}^1$.
(d) $\xi^2 \eta$ is not k-determined in $\mathbb{R}^2$ for arbitrary k.

In (a), the point 0 is a regular point of f. In (b), the point 0 is a nondegenerate singular point of f.

PROOF. (a), (b) This is another formulation of Example 73.59 and the Morse lemma of Section 73.12.

(c) We have $f(x) = x^k(a + O(x))$ with $a \neq 0$. Let $g(x) = x^k(a + O(x))$. We solve

$$g(\varphi(u)) = au^k,$$

using the following ansatz: $\varphi(u) = u\psi(u)$. Then ψ is obtained from the inverse function theorem.

(d) The solution sets of $\xi^2\eta = 0$ and $\xi^2\eta + \eta^{2r+1} = 0$ with $r \geq 2$ have a completely different structure (two straight lines and one straight line). □

The case of k-determination, $k \geq 3$, will be discussed below.

In science, perturbations are often described by parameters, i.e., one considers parameter families in the sence of (C) above. In what follows the important transformation rule is

$$G(y, q) = F(x(y, q), p(q)) + c(q), \tag{40}$$

i.e., the parameters p and q are separated from the state variables x and y.

Definition 73.62 (Unfoldings). Consider parameter families F and G in the sense of (C) above.

(a) F induces G if and only if (40) holds, where x, p, and c are local C^∞-maps and $y \mapsto x(y, q)$ is a local C^∞-diffeomorphism for all q in a neighborhood of 0.

If, in addition, p is a local C^∞-diffeomorphism, then F and G are called *equivalent*.

(b) F is called an *unfolding* or deformation of f if and only if $F(x, 0) = f(x)$ in a neighborhood of 0.
(c) An unfolding F of f is called *versal* if and only if F induces every unfolding G of f.
(d) Every versal unfolding with a minimal number of parameters is called *universal*.

Roughly speaking, universal unfoldings provide a survey of all perturbations of f with finitely many parameters.

Note that the equivalence of parameter families F and G is more special than the equivalence of maps of Section 73.15. If F and G are equivalent, then the parameter spaces of F and G have the same dimension. In the following, we present a number of criteria for the numerical computation of normal forms. Some examples are given in Problem 73.4. For

$$P_0^k \subseteq P_{\text{hom}}^k \subseteq P^k,$$

$k \geq 1$, we use the following notations:

- P^k set of all polynomials of degree $\leq k$ with respect to the real variables $\xi_1 \ldots \xi_n$;
- P_0^k polynomials in P^k which vanish at the origin;
- P_{hom}^k homogeneous polynomials of degree k.

We also write $P^k(\mathbb{R}^n)$, etc. For the criteria below, the following expressions are important:

(I) $j^{k+1}[Q_1 j^k(D_1 f + D_1 Q) + \cdots + Q_n j^k(D_n f + D_n Q)]$,
(II) $j^{k+1}[Q_1 j^{k-1}(D_1 f) + \cdots + Q_n j^{k-1}(D_n f)]$,
(III) $j^k[Q_1 j^k(D_1 f) + \cdots + Q_n j^k(D_n f)]$,

where $D_i = \partial/\partial \xi_i$. Also, recall condition (C) above.

(C1) (*Criterium for k-determination*). *The map $f: U(0) \subseteq \mathbb{R}^n \to \mathbb{R}$ is k-determined in $\mathbb{R}^n$ if and only if the following holds: If $Q \in P_{\text{hom}}^{k+1}$, then every $P \in P_{\text{hom}}^{k+1}$ can be expressed through (I) with $Q_i \in P^r$, $r \geq 1$.*

If f is k-determined in $\mathbb{R}^n$, then every $P \in P_{\text{hom}}^{k+1}$ can be expressed through (II) with $Q_i \in P^r$, $r \geq 1$.

(C2) (*Criterium for strong k-determination*). *The map f is strongly k-determined in $\mathbb{R}^n$ if and only if every $P \in P_{\text{hom}}^{k+1}$ can be expressed through (II) with $Q_i \in P^r$, $r \geq 2$.*

(C3) (*Construction of universal unfoldings*). *Let f be k-determined in $\mathbb{R}^n$ with $f'(0) = 0$. If one chooses a cobasis $\{v_1, \ldots, v_r\}$, then*

$$F(x, p) = f(x) + p_1 v_1(x) + \cdots + p_r v_r(x)$$

is a universal unfolding of f.

We explain the concept of cobasis for a k-determined function f in $\mathbb{R}^n$. Let ideal(f') be the set of all polynomials in P_0^k which can be expressed through (III) with $Q_i \in P^k$. Here, P_0^k is considered as a linear space. Then, dim P_0^k is equal to the maximal number of linear independent polynomials in P_0^k. The *codimension* of f is defined as

$$\text{codim } f = \dim P_0^k - \dim(\text{ideal}(f')).$$

Let $r = \text{codim } f$. Then $\{v_1, \ldots, v_r\}$ is a *cobasis* if and only if

$$P_0^k = \text{ideal}(f') \oplus \text{span}\{v_1, \ldots, v_r\}.$$

EXAMPLE 73.63. The function $f: \mathbb{R} \to \mathbb{R}$ with $f(x) = x^k$ is k-determined in $\mathbb{R}$.

Let $k \geq 3$. Then

$$P_0^k(\mathbb{R}) = \text{span}\{x,\ldots,x^k\},$$

$$\text{ideal}(f') = \text{span}\{x^{k-1}, x^k\}.$$

Consequently, $\{x,\ldots,x^{k-2}\}$ is a cobasis, i.e., codim $f = k - 2$. From (C3) we obtain a universal unfolding F of f through

$$F(x, p) = x^k + p_1 x^1 + \cdots + p_{k-2} x^{k-2}.$$

For $k = 2$, we have $F = f$.

The following result is the *main theorem in catastrophe theory*.

(C4) (Local normal forms for parameter families). *A parameter family with at most five parameters and n state variables is "in general" versal. If this is the case, then it is induced by one of the universal unfoldings of Table 73.1. These universal unfoldings are "structurally stable."*

A precise definition for the expressions "in general" and "structurally stable" will be given below in (C6) and (C7). The normal forms of Table 73.1 with codim $f \geq 1$, are called catastrophies. For codim $f \leq 4$, the catastrophies are called the seven elementary catastrophies. In concrete applications, one uses Table 73.1 in the following way.

(i) Given a parameter family $G = G(y, q)$, let q be equal to zero and consider $g(y) = G(y, 0)$. If one can show, by using (C1), that g is k-determined with codim $g \leq 5$, then G is induced by one of the normal forms F of Table 73.1 with codim $f =$ codim g, and f of Table 73.1 is right equivalent to g.

(ii) If, conversely, g is given and k-determined with codim $g \leq 5$, then one can construct from (C3) a universal unfolding G of g which is equivalent to a unversal unfolding F of Table 73.1 with codim $f =$ codim g.

In Table 73.1, we have used, for reasons of simplicity, a, b, c, d, e instead of $p_1, \ldots, p_5$. Letting a, b, c, d, e in F be equal to zero, one obtains f. For codim $f \geq 1$, all these f have a degenerate singular point at $x = 0$, and codim f measures the degree of degeneracy. The intuitive meaning of codim f is as follows: If $f: U(0) \subseteq \mathbb{R}^n \to \mathbb{R}$ is k-determined with codim $f = r$, then all sufficiently small perturbations of f have no more than $r + 1$ singular points in a small neighborhood of zero.

Table 73.1 gives a survey of the main structure of parameter families in a neighborhood of zero. The importance of this for science will become clear in the following section.

(C5) (Transversality criterium for versal unfoldings). *Let f be k-determined and let $F = F(x, p)$ be an unfolding of f with $x = (\xi_1, \ldots, \xi_n)$ and $p = (p_1, \ldots, p_r)$. Let*

$$v_i(x) = j^k(\partial F(x, 0)/\partial p_i).$$

Then F is versal if and only if span$\{v_1, \ldots, v_r\}$ *and ideal* (f') *are transversal in* $P_0^k(\mathbb{R}^n)$, *i.e., they span* $P_0^k(\mathbb{R}^n)$.

Table 73.1

Name	Normal form F (universal unfolding of f)	f is obtained by letting $a = b = c = d = e = 0$ codim f
Regular point	ξ_1	0
Nondegenerate singular point	$\xi_1^2 + \cdots + \xi_r^2 - \xi_{r+1}^2 - \cdots - \xi_n^2$, $\quad 0 \leq r \leq n$	0
Fold	$\xi_1^3 + a\xi + A$	1
Cusp	$\pm(\xi^4 + a\xi^2 + b\xi) + A$	2
Swallow tail	$\xi^5 + a\xi^3 + b\xi^2 + c\xi + A$	3
Elliptic umbilics	$\xi^2\eta - \eta^3 + a\xi^2 + b\eta + c\xi + B$	3
Hyperbolic umbilic	$\xi^2\eta + \eta^3 + a\xi^2 + b\eta + c\xi + B$	3
Parabolic umbilic	$\pm(\xi^2\eta + \eta^4 + a\eta^2 + b\xi^2 + c\eta + d\xi) + B$	4
Butterfly	$\pm(\xi^6 + a\xi^4 + b\xi^3 + c\xi^2 + d\xi) + A$	4
Wigwam	$\xi^7 + a\xi^5 + b\xi^4 + c\xi^3 + d\xi^2 + e\xi + A$	5
Second elliptic umbilic	$\xi^2\eta - \eta^5 + a\eta^3 + b\eta^2 + c\xi^2 + d\eta + e\xi + B$	5
Second hyperbolic umbilic	$\xi^2\eta + \eta^5 + a\eta^3 + b\eta^2 + c\xi^2 + d\eta + e\xi + B$	5
Symbolic umbilic	$\pm(\xi^3 + \eta^4 + a\xi\eta^2 + b\eta^2 + c\xi\eta + d\eta + e\xi) + B$	5

notations

$$\xi = \xi_1, \quad \eta = \xi_2$$
$$A = \xi_2^2 + \cdots + \xi_r^2 - \xi_{r+1}^2 - \cdots - \xi_n^2, \quad 1 \leq r \leq n$$
$$B = \xi_3^2 + \cdots + \xi_r^2 - \xi_{r+1}^2 - \cdots - \xi_n^2, \quad 2 \leq r \leq n$$

a, b, c, d, e are real parameters

Definition 73.64 (Structural Stability). Roughly speaking, a parameter family F is called structurally stable if every $F + P$ is equivalent to F for a sufficiently small perturbation P. More precisely,

$$F: U(0) \subseteq \mathbb{R}^n \times \mathbb{R}^d \to \mathbb{R}$$

is called *locally structurally stable* if and only if there exists a neighborhood U of 0 in $\mathbb{R}^n \times \mathbb{R}^d$ such that all parameter families

$$F + P: U(0) \subseteq \mathbb{R}^n \times \mathbb{R}^d \to \mathbb{R}$$

are equivalent to F for $F + P$ in a neighborhood of F in $C^\infty(U, \mathbb{R})$. Here, $C^\infty(U, \mathbb{R})$ is endowed with the C^∞-Whitney topology.

(C6) (Structural stability). *Every locally structurally stable parameter family is versal. All universal unfoldings F of Table 73.1 are locally structurally stable.*

We now make the expression "in general" of (C4) more precise.

(C7) (Genericity). *In $C^\infty(\mathbb{R}^n \times \mathbb{R}^d, \mathbb{R})$, $d \leq 5$, there exists an open and dense set in the C^∞-Whitney topology such that each function $H = H(x, p)$ in this set has the following property. If (x_0, p_0) is fixed and*

$$G(x, p) = H(x_0 - x, p_0 - p) - H(x_0, p_0),$$

then G is locally structurally stable and a versal unfolding of $g(x) = G(x, 0)$. Moreover, G is induced by a universal unfolding F of Table 73.1.

Unfortunately, for $d \geq 6$, we no longer get such nice results. Structural stability is an important basic concept for describing scientific phenomena. We are convinced that the essential features in nature are structurally stable. Roughly, Proposition (C7) states that most parameter families H in $C^\infty(\mathbb{R}^n \times \mathbb{R}^d, \mathbb{R})$ behave reasonably.

73.17. Applications to Natural Sciences

In Section 37.28 of Part III we already gave some simple applications to the van der Waals gas (phase transitions) and bifurcations in connection with the bending of beams. Now, let us make some principal remarks. We consider:

(i) Useless models.
(ii) Incomplete models.
(iii) Structurally stable perturbations of models.

We begin with (i). Suppose a scientist or engineer models a phenomenon with a function $\xi^2\eta$ by making a number of simplifying assumptions. This model is not very useful since, from Example 73.61, there exists no k such that $\xi^2\eta$

is k-determined. Arbitrarily small perturbations, i.e., perturbations of arbitrarily high order, may change the qualitative picture of $\xi^2\eta$ completely.

For (ii), we consider $\xi^3 + \xi\eta^3$. It follows from Problem 73.4 that this function is 4-determined, but not 3-determined. This means that this model is incomplete, because 4-determination implies stability against perturbations of order ≥ 5, while perturbations of order 4 may influence the qualitative picture. Thus, the criteria (C1) and (C2) are quite useful in applications and we suggest doing Problem 73.4 in order to get some practice with this.

We now discuss (iii). Each phenomenon in nature or engineering is subject to outer perturbations which often are described by outer parameters $p = (p_1, \ldots, p_r)$, i.e., we consider equations of the form $y = F(x, p)$. Two important questions arise.

(a) How does a measurement curve change during the repetition of measurements, which are subject to *stable* perturbations?
(b) At least, how many parameters are needed to describe these stable perturbations caused by outer influences?

The answer to (a) is: F must be a versal unfolding of f with $f(x) = F(x, 0)$. The answer to (b) is: F must be a universal unfolding of f.

EXAMPLE 73.65. Consider $f: U(0) \subseteq \mathbb{R} \to \mathbb{R}$ with

$$f(x) = ax^k + O(x^{k+1}) \quad \text{and} \quad a \neq 0, \quad k \geq 3.$$

From Examples 73.61 and 73.63 it follows that f is k-determined with codim $f = k - 2$ and f has the universal unfolding

$$F(x, p) = f(x) + p_1 x + p_2 x^2 + \cdots + p_{k-2} x^{k-2}.$$

Each parameter family $G = G(y, q)$ with $G(y, 0) = f(y)$ is induced by F, i.e., we have

$$G(y, q) = F(x(y, q)), p(q)) + c(q)$$

with the properties of Definition 73.62. If $k = 2$, then $F(x) = f(x)$, i.e., the universal unfolding F contains no parameters. Let $k = 3$. Then

$$F(x, p) = f(x) + px.$$

Figure 73.17 shows the perturbations F of $f(x) = x^3$ for a fixed p. According to Table 73.1, $F(x, p) = x^3 + px$ is the first elementary catastrophe (fold).

In many applications $F = F(x, p)$ is thought of as energy or thermodynamic potential, or as a generating function of a dynamical system

$$x'(t) = F_x(x(t), p). \tag{41}$$

Another field of applications is bifurcation theory. For this, we recommend Golubitsky and Schaeffer (1979), (1984, M). In connection with energy questions, we consider the following example. Let $F = F(x, p)$ be the potential energy of a mechanical system, which is described by the state variable $x \in \mathbb{R}^n$

73.17. Applications to Natural Sciences

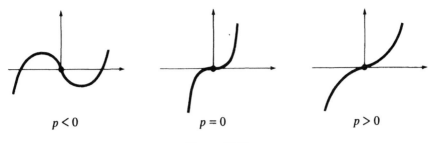

Figure 73.17

and outer parameters $p \in \mathbb{R}^d$. The equilibrium states (x, p) of this system are obtained from $F(x, p) = \min!$, i.e., from

$$F_x(x, p) = 0. \tag{42}$$

Any reasonable energy function should be structurally stable, i.e., according to (C6), F should be versal. In case we are interested in the minimal number of outer parameters which influence a structurally stable system, it is useful to assume that F is universal. To be more concrete, let $n = 1$ and

$$F(x, 0) = ax^k + O(x^{k+1})$$

with $x \in \mathbb{R}$, $k \geq 2$ and $a > 0$. From Example 73.63 it follows that the minimal number of parameters is equal to $k - 2$. Assume, for instance, that $k = 4$. Then $x \mapsto F(x, 0)$ has a degenerate minimum at $x = 0$ and a universal unfolding

$$F(x, p_1, p_2) = ax^4 + O(x^5) - p_1 x^2 - p_2 x.$$

This corresponds to the second elementary catastrophe (cusp) of Table 73.1. The equilibrium states (x, p_1, p_2), i.e., the solutions of (42) follow from

$$4ax^3 + O(x^4) - 2p_1 x - p_2 = 0.$$

Figure 73.18 shows the diagrams for $p_2 = 0$ and $p_2 \gtrless 0$. Here, $p_2 = 0$ corresponds to the broken line. Since in reality, one always deals with perturbations, a diagram with $p_2 = 0$ cannot be expected in experiments, but only the perturbed diagrams with $p_2 \gtrless 0$. This is actually the case.

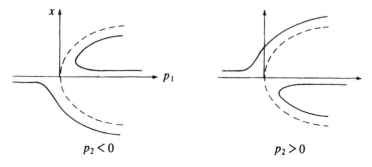

Figure 73.18

73.18. Orientation

In a finite-dimensional B-space X, an orientation is determined by specifying a basis $\{b_1,\ldots,b_n\}$. A linear map

$$L: X \to X$$

is called *orientation preserving* if and only if $\det L > 0$. Note that the definition of $\det L$ is invariant, i.e., independent of the basis in X (see Section 4.16). Two bases $\{b_1,\ldots,b_n\}$ and $\{c_1,\ldots,c_n\}$ in X are called *orientation equivalent* if and only if there exists a linear, orientation preserving map $L: X \to X$ with

$$Lb_j = c_j \quad \text{for all } j.$$

The corresponding equivalence classes of bases are called *orientations* of X. In X, there exist exactly two orientations with representatives $\{b_1, b_2, \ldots, b_n\}$ and $\{-b_1, b_2, \ldots, b_n\}$. Figure 73.19 shows the representatives for both orientations of $\mathbb{R}^2$. A C^1-map $f: U \subseteq X \to X$, where U is open in X, is called orientation preserving if and only if every derivative has this property, i.e., $\det f'(x) > 0$ for all $x \in U$.

We now extend this concept to manifolds. Let us consider the surface of the earth M. Intuitively, an orientation of M is determined by specifying two small curved coordinate axes $\{a, b\}$ at each point of M. This should be done globally, without causing contradictions. To this end we require that, for each fixed chart, the chart images of $\{a, b\}$ have the same orientation, and that this orientation is preserved under chart changes (see Fig. 73.20). Depending on the choice of the geographical atlas, this might not be possible; however, we can always find an equivalent atlas with this property. Proceeding like this, we obtain exactly two orientations for the surface of the earth (Fig. 73.21). On the other hand, there exists no orientation for the well-known Moebius band. This is a surface in $\mathbb{R}^3$ which is obtained by joining two opposite sides of a rectangle as in Figure 73.22, i.e., A and A' are identified. It is told that Moebius (1790–1868) discovered this surface while watching his wife sewing a garter. Now, if one moves a coordinate cross along BOB' in Figure 73.22, the orientation at B' is different from the orientation at B. But since both points on the surface correspond to the same point, the Moebius band cannot be given an orientation. Let us make these heuristic arguments more precise. Assume:

(H) Let M be a real, n-dimensional C^k-manifold, $k \geq 1$.

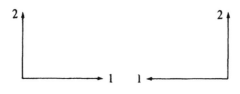

Figure 73.19

73.18. Orientation

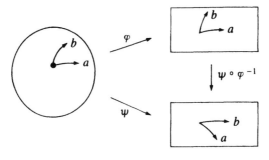

Figure 73.20

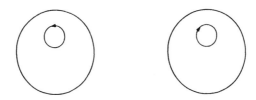

Figure 73.21

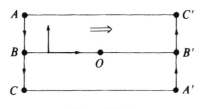

Figure 73.22

Two admissible charts (U, φ) and (V, ψ) in M are called *orientation compatible* if and only if $U \cap V$ is empty or the map

$$\psi \circ \varphi^{-1} \colon \varphi(U \cap V) \to \psi(U \cap V),$$

describing the chart change, is orientation preserving (Fig. 73.20). An equivalent atlas of M is called *oriented* if and only if all of its charts are orientation compatible.

Definition 73.66. A manifold M is called *orientable* if and only if M has an equivalent atlas which is oriented.

In order to formalize the concept of orientation, we say that two oriented, equivalent atlases for M have the same orientation if and only if all charts in the two atlases are orientation compatible. The corresponding equivalence classes of oriented atlases are called orientations of M. If M is connected and orientable, then M admits exactly two orientations. We say that M is oriented if we specify an oriented, equivalent atlas for M. All admissible charts in M, which are orientation compatible with this atlas, are called admissible oriented charts.

73.19. Manifolds with Boundary

According to our previous definition of manifolds, the open disk in $\mathbb{R}^2$ is a two-dimensional manifold, while the closed disk is not a manifold at all, since a neighborhood of a boundary point x of M is not homeomorphic to an open set in $\mathbb{R}^2$. Such a neighborhood, however, is homeomorphic to a relatively open set of the half space $H\mathbb{R}^2$ (Fig. 73.23). This leads to the following definition of manifolds with boundary. Let

$$H\mathbb{R}^n = \{(\xi_1, \ldots, \xi_n) \in \mathbb{R}^n : \xi_1 \leq 0\}.$$

A set A in $H\mathbb{R}^n$ is called relatively open if and only if there exists an open set B in $\mathbb{R}^n$ such that $A = B \cap H\mathbb{R}^n$.

Recall that from Definition 4.22 a map $f: A \to \mathbb{R}^m$ is C^k, $k \geq 1$, if and only if f is C^k at every inner point of A, and for every boundary point of A there exists an open neighborhood in $\mathbb{R}^n$ such that f can be extended to a C^k-map on this neighborhood.

The definition of a real, n-dimensional C^k-*manifold M with boundary* is then analogous to the definition of manifolds of Section 73.2. The only difference is that the chart images U_φ are no longer open sets in $\mathbb{R}^n$, but relatively open sets in $H\mathbb{R}^n$. A point $x \in M$ is called a *boundary point* or *inner point* in M if and only if there exists a chart (U, φ) in M, such that $\varphi(x)$ is a boundary point or inner point of the chart half space $H\mathbb{R}^n$. In the case of disks and balls, this corresponds to our intuition. If the boundary is empty, then we obtain a manifold in the previous sense.

EXAMPLE 73.67. Let M be a subset of $\mathbb{R}^n$. If each point in M has a neighborhood which is C^k-diffeomorphic to a relatively open subset of $H\mathbb{R}^n$, then M is a real, n-dimensional manifold with boundary.

In particular, the finite interval $[a, b[$ is a real one-dimensional C^∞-manifold with boundary point a. The closed ball in $\mathbb{R}^n$ is a real, n-dimensional C^∞-manifold with boundary.

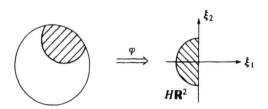

Figure 73.23

73.19. Manifolds with Boundary

Proposition 73.68. *Every connected, real, one-dimensional C^∞-manifold with nonempty boundary and countable basis is C^∞-diffeomorphic to the unit interval $[0, 1]$ or $[0, 1[$.*

For a proof see Milnor (1965, M), Appendix.

The *orientation* of manifolds with boundary is defined analogously to the orientation of manifolds without boundary in Section 73.18.

Let e_i denote the ith unit vector in $\mathbb{R}^n$, i.e., $e_1 = (1, 0, \ldots, 0)$, etc. Let $N = H\mathbb{R}^n$. We say that N and ∂N are coherently oriented if and only if the orientations of N and ∂N are given either by

$$(e_1, \ldots, e_n) \quad \text{and} \quad (e_2, \ldots, e_n)$$

or by

$$(-e_1, e_2, \ldots, e_n) \quad \text{and} \quad (-e_2, e_3, \ldots, e_n),$$

respectively (Fig. 73.24). This definition coincides with the following more general situation.

Proposition 73.69 (Coherent Orientation). *Let M be a real, oriented, n-dimensional C^k-manifold with boundary ∂M, $k \geq 1$. Then the oriented atlas for M induces in a natural way an oriented atlas for ∂M.*

This way an orientation is induced on ∂M. The corresponding orientations of M and ∂M are called *coherent*.

PROOF. The basic idea is contained in Figure 73.24. Let us consider a fixed boundary point $x \in \partial M$. In two different charts, a neighborhood of x in M is described by the local coordinates

$$(\xi_1, \ldots, \xi_n), \quad \xi_1 \leq 0$$

and

$$(\eta_1, \ldots, \eta_n), \quad \eta_1 \leq 0.$$

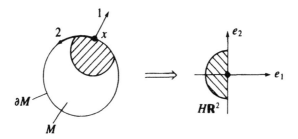

Figure 73.24

Correspondingly, a neighborhood of x in ∂M is described by the local coordinates

$$(0, \xi_2, \ldots, \xi_n) \quad \text{or} \quad (0, \eta_2, \ldots, \eta_n).$$

Let

$$\eta_j = \eta_j(\xi_1, \ldots, \xi_n), \quad j = 1, \ldots, n$$

be the coordinate change on M. Since M is oriented, we have

$$\frac{\partial(\eta_1, \ldots, \eta_n)}{\partial(\xi_1, \ldots, \xi_n)} > 0.$$

The corresponding coordinate change on ∂M is described by

$$0 = \eta_1(0, \xi_2, \ldots, \xi_n),$$
$$\eta_j = \eta_j(0, \xi_2, \ldots, \xi_n), \quad j = 2, \ldots, n.$$

Thus, on ∂M, we obtain

$$0 < \frac{\partial(\eta_1, \ldots, \eta_n)}{\partial(\xi_1, \ldots, \xi_n)} = \begin{vmatrix} D_1\eta_1 & D_1\eta_2 & \cdots & D_1\eta_n \\ 0 & D_2\eta_2 & \cdots & D_2\eta_n \\ \vdots & \vdots & & \vdots \\ 0 & D_n\eta_2 & \cdots & D_n\eta_n \end{vmatrix}$$

$$= D_1\eta_1 \frac{\partial(\eta_2, \ldots, \eta_n)}{\partial(\xi_2, \ldots, \xi_n)},$$

where $D_i = \partial/\partial\xi_i$. Since $D_1\eta_1 > 0$, this implies

$$\frac{\partial(\eta_2, \ldots, \eta_n)}{\partial(\xi_2, \ldots, \xi_n)} > 0,$$

i.e., ∂M is oriented. □

Figure 73.25 shows the two coherent orientations of surfaces in $\mathbb{R}^3$, if these surfaces are connected submanifolds of $\mathbb{R}^3$ with boundary. The coherent orientation will play an important role in the formulation of the generalized integral theorem of Stokes of Section 74.24.

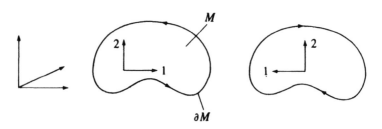

Figure 73.25

For a convenient formulation of this theorem we need the concept of a submanifold M with boundary. One may think of M as an open set in $\mathbb{R}^3$ or a closed ball, and of S as a sufficiently regular surface or curve with or without boundary which lies in M.

Let M be a real, n-dimensional C^k-manifold with boundary, $k \geq 0$. Moreover, let S be a subset of M. Then S is called an m-dimensional C^k-*submanifold of M with boundary* if and only if the following holds: For each point $x \in M$ there exists an admissible chart in M such that M looks locally like a relatively open set in $H\mathbb{R}^n$ and S looks locally like a relatively open set in $H\mathbb{R}^m$. Here

$$H\mathbb{R}^n = \{(\xi_1, \ldots, \xi_n) \in \mathbb{R}^n : \xi_1 \leq 0\}.$$

$$H\mathbb{R}^m = \{(\xi_1, \ldots, \xi_n) \in H\mathbb{R}^n : \xi_i = 0 \text{ for all } i > m\}.$$

More precisely, for each $x \in M$ there exists an admissible chart (U, φ) in M with $x \in U$ such that $\varphi(U)$ and $\varphi(U \cap S)$ are the corresponding relatively open sets in $H\mathbb{R}^n$ and $H\mathbb{R}^m$.

From these admissible charts in M, we obtain charts in S. Thus, S becomes a real, m-dimensional C^k-manifold with boundary.

Analogously, one can define Banach manifolds with boundary. The only difference with the Banach manifolds of Section 73.2 is that here the chart images are no longer open sets in B-spaces X, but relatively open sets in half spaces HX of B-spaces, i.e.,

$$HX = \{x \in X : f(x) \leq 0\},$$

where $f: X \to \mathbb{R}$ is linear and continuous.

Manifolds mean always manifolds in the sense of Section 73.2, i.e., manifolds without boundary. For manifolds with boundary we will always add "with boundary."

73.20. Sard's Theorem

Theorem 73.H (Sard (1942)). *Let M and N be real, finite-dimensional C^∞-manifolds with countable basis. If $f: M \to N$ is C^k with*

$$k > \max(0, \dim M - \dim N),$$

then the set of singular values of f has measure zero in N. The set of regular values of f is residual and dense in N.

This important classical theorem states that almost all values of f are regular. A proof may be found in Abraham and Robbin (1967, M), §15. The proof for C^∞-functions f is much easier and is contained in Lang (1972, M), p. 173. The important generalization to Banach manifolds is proved in Section 78.8 (Theorem of Sard–Smale). In Chapter 78 we show how to use Sard's theorem for an elegant and constructive approach to fixed-point theory and mapping degree. This also leads to effective, numerical programs for computers.

73.21. Whitney's Embedding Theorem

Theorem 73.I (Whitney (1936)). *Let M be a real, n-dimensional C^∞-manifold with a countable basis. Then there exists a C^∞-embedding $f: M \to \mathbb{R}^{2n+1}$.*

This fundamental structure theorem states that M can be realized as a submanifold $f(M)$ in $\mathbb{R}^{2n+1}$. A proof may be found in Golubitsky and Guillemin (1973, M), Chapter 2, §5. We will give here a simple proof for a special case which illustrates the geometric meaning of this theorem, and shows how the space $\mathbb{R}^{2n+1}$ comes up. Here, the tangent bundle TM plays a central role. The proof should also convince the reader of the usefulness of the abstract construction TM. Our additional assumption is:

(S) M is a compact submanifold of $\mathbb{R}^m$, $m \geq n$.

For $m \leq 2n + 1$, Theorem 73.I is trivially true. Thus, only the case $m > 2n + 1$ is of interest. In order to get a good intuitive understanding, we transform the tangent space TM_x into the origin of $\mathbb{R}^m$. Then TM_x is a linear subspace of $\mathbb{R}^m$. For the tangent bundle we have

$$TM = \{(x, v) \in M \times \mathbb{R}^m : v \in TM_x\}.$$

Then TM is a real, $2n$-dimensional C^∞-manifold in $\mathbb{R}^{2m}$. Roughly speaking, the space $\mathbb{R}^{2n+1}$ appears because $\dim TM < 2n + 1$. Let us make this precise. The key in the proof is Sard's theorem of the previous section.

PROOF IN CASE OF (S).

(I) Geometrical argument. Below, we prove the following: If $g: M \to \mathbb{R}^k$ is an injective immersion with $k > 2n + 1$, then there exists a nonvanishing vector $a \in \mathbb{R}^k$ such that

$$\pi \circ g: M \to a^\perp$$

is an injective immersion. Here $a^\perp$ denotes the orthogonal complement of a in $\mathbb{R}^k$ and $\pi: \mathbb{R}^k \to a^\perp$ is the corresponding orthogonal projection.

(II) Let $m > 2n + 1$. Since $a^\perp$ is isomorphic to $\mathbb{R}^{k-1}$, we are able to construct an injective C^∞-immersion $f: M \to \mathbb{R}^{2n+1}$ by succesively applying (I) to the trivial embedding $g: M \to \mathbb{R}^m$.

Because of the compactness of M, f is proper, i.e., from Corollary 73.45, an embedding. This proves the assertion.

(III) Proof of (I). We define the following two maps

$$G: TM \to \mathbb{R}^k, \qquad G(x, v) \stackrel{\text{def}}{=} (T_x g)(v),$$

and
$$H: M \times M \times \mathbb{R} \to \mathbb{R}^k, \qquad H(x, y, t) \stackrel{\text{def}}{=} t(g(x) - g(y)).$$

The preimage spaces of G and H have dimensions $2n$ and $2n + 1$ and the image spaces have dimensions $k > 2n + 1$. This is essential. According to *Sard's theorem* of Section 73.20, the set of singular values of G and H has measure zero in $\mathbb{R}^k$. Since the dimensions are lowered, each image point is a singular value. Therefore, there exists an $a \in \mathbb{R}^k$ which is not an image point of G and H. Since $G(0) = H(0) = 0$, we have $a \neq 0$.

The map $\pi \circ g: M \to a^\perp$ is injective. This is true because $\pi(g(x)) = \pi(g(y))$ implies that $g(x) - g(y) = ta$ for $t \in \mathbb{R}$. If $x \neq y$, then $t \neq 0$ since g is injective. Thus $H(x, y, 1/t) = a$, which contradicts the choice of a.

Moreover, $\pi \circ g$ is an immersion. Suppose, there exists a $v \neq 0$ with $T_x(\pi \circ g)(v) = 0$. The chain rule implies that $(\pi \circ T_x g)(v) = 0$, i.e., $(T_x g)(v) = ta$ for some $t \in \mathbb{R}$. Since g is an immersion, it follows that $t \neq 0$. Consequently, $G(x, v/t) = a$, which contradicts the choice of a. □

73.22. Vector Bundles

In Definition 73.20 the abstract notion of bundles was introduced. Vector bundles may be characterized in the following two ways.

(i) Vector bundles are bundles where the fibers are linear spaces, and which behave locally like products of open sets with B-spaces.
(ii) Vector bundles are obtained if a B-space F_x is attached at each point of a topological space or manifold M. Here, we *neglect* that different spaces F_x may intersect.

In (ii), we may also think of a vector bundle as a family $\{F_x\}$ of vector spaces which continuously depends on a parameter x, where x runs through M. Here M is called a basis space and F_x is called a fiber. The bundle space B is defined as
$$B = \{(x, t): x \in M, t \in F_x\}.$$
Also, we let $\pi(x, t) = x$. This induces a triple (π, B, M).

The reader should convince himself that the following definition is very natural, as is the case with every important definition in mathematics.

Let X and Y be B-spaces. As usual, $L(X, Y)$ denotes the B-space of linear, continuous maps from X into Y.

Definition 73.70. A *vector bundle* (π, B, M) is a triple which satisfies the following conditions:

(V1) *Bundle property.* The map $\pi: B \to M$ is continuous and surjective where B and M are topological spaces.

(V2) *Linear fibers.* For each $x \in M$ the fiber $F_x = \pi^{-1}(x)$ is a linear space over $\mathbb{K}$.

(V3) *Fiber preserving local trivializations.* There exists a covering $\{U_i\}$ of the basis space M with open sets, and for every U_i there exist homeomorphisms

$$\tau_i: \pi^{-1}(U_i) \to U_i \times Y_i,$$

where Y_i is a B-space over $\mathbb{K}$. These *trivializations* τ_i are fiber preserving, i.e., we always have

$$\tau_i(F_x) = \{x\} \times Y_i,$$

and the τ_i define linear maps from F_x onto Y_i.

The B-space Y_i is called a *typical fiber*.

(V4) *Change of trivializations.* Let $x \in U_i \cap U_j$. The following diagram

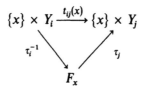

commutes and defines the so-called *transition functions*

$$t_{ij}(x): Y_i \to Y_j$$

between the typical fibers where we require that

$$t_{ij}: U_i \cap U_j \to L(Y_i, Y_j)$$

is continuous.

Definition 73.71. Let $k \geq 0$. If one replaces, in the above definition, topological spaces, continuous maps, and homeomorphisms with C^k-Banach manifolds, C^k-maps, and C^k-diffeomorphisms, then we will speak of C^k-*vector bundles*.

A *section* of a vector bundle (π, B, M) is a continuous map $s: M \to B$ with

$$\pi(s(x)) = x \qquad \text{for all} \quad x \in M.$$

This is equivalent to saying that $s(x) \in F_x$ for all $x \in M$.

If we are given a C^k-vector bundle and if, in addition, s is C^k, then we speak of a C^k-*section*.

Remark 73.72 (Meaning of (V4)). Condition (V3) implies that τ_i defines a homeomorphism from F_x onto $\{x\} \times Y_i$. Therefore $t_{ij}(x): Y_i \to Y_j$ is always a linear homeomorphism.

In the finite-dimensional case, i.e., if all fibers are finite dimensional, we identify Y_i with $\mathbb{K}^{n_i}$. Then, all transition functions $t_{ij}(x)$ are invertible matrices. In this case, the definition of vector bundles can be simplified because (V4) is then an obvious consequence of (V1)–(V3).

73.22. Vector Bundles

Instead of τ_i, one often calls (U_i, τ_i) a trivialization or bundle chart.

The definition of vector bundles can easily be generalized. In Part V, for example, we consider the concept of fiber bundles which plays an important role in modern physics. Roughly, in this case, the typical fibers Y_i are equal to a topological space or a manifold Y, and the transition functions $t_{ij}(x)$ are elements of a transformation group on Y.

EXAMPLE 73.73 (Standard Example of a Vector Bundle). The simplest example of a vector bundle (π, B, M) is the product $B = M \times Y$, where M is a topological space and Y a B-space. We let $\pi(x, y) = x$.

As a covering $\{U_i\}$, we choose the trivial covering of M by M. The corresponding trivialization is then the identity $\tau: M \times Y \to M \times Y$.

A section $s: M \to M \times Y$ has the form $s(x, y) = (x, f(x))$, where $f: M \to Y$ is continuous.

If M is a C^k-manifold, $k \geq 0$, then we obtain in this way a C^k-vector bundle, and s is a C^k-section if $f: M \to Y$ is C^k.

General vector bundles are obtained from Definition 73.70 by gluing such products $U_i \times Y_i$ together. For this gluing together, one uses the transition functions $t_{ij}(x)$.

EXAMPLE 73.74 (Cylinder as a Typical Geometrical Example of a Vector Bundle). Let B be the surface of a cylinder and let M be the cylinder equator (Fig. 73.26(a)). Let $\pi: B \to M$ denote the orthogonal projection onto M. Then (π, B, M) is a C^∞-vector bundle. The fibers F_x generate the surface of the cylinder.

A C^k-section is a C^k-curve $s: M \to B$ with $s(x) \in F_x$ for all $x \in M$ (Fig. 73.26(b)).

EXAMPLE 73.75 (Normal Bundle of a Curve). Let M be the boundary of a disk (Fig. 73.27(a)). If we draw the normal N_x at every point $x \in M$, and neglect that different normals may intersect, then we obtain a vector bundle with fibers $F_x = N_x$ which has the same structure as the surface of the cylinder of Example 73.74.

More generally, if M is a sufficiently smooth curve in $\mathbb{R}^2$, then we obtain in a similar way the normal bundle (π, B, M) of M (Fig. 73.27(b)). Precisely, we proceed like this. Let N_x be the oriented normal at x. The bundle space B consists of all pairs (x, P) with $x \in M$ and $P \in N_x$. The trivialization

$$\tau: B \to M \times \mathbb{R}$$

is defined through $\tau(x, P) = (x, p)$, where p is the distance between P and x with the correct sign.

EXAMPLE 73.76 (Tangent Bundle as a Vector Bundle). Let M be a C^k-Banach manifold, $k \geq 1$. Then (π, TM, M) is a C^{k-1}-vector bundle.

PROOF. The bundle space $B = TM$ consists of all pairs (x, v) with $x \in M$ and $v \in TM_x$. The map $\pi: TM \to M$ is defined through $\pi(x, v) = x$. The fiber $F_x = \pi^{-1}(x)$ is the tangent space TM_x at x.

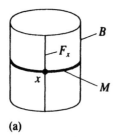 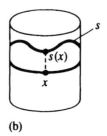

Figure 73.26

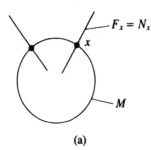 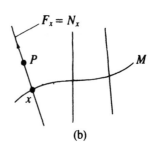

Figure 73.27

Let $\{(U_i, \varphi_i)\}$ be a collection of charts in M. We let $\varphi = \varphi_i$ for a fixed i. The local trivialization

$$\tau_i: \pi^{-1}(U_i) \to U_i \times X_\varphi$$

is defined through

$$\tau_i(x, v) = (x, v_\varphi).$$

Here v_φ is the coordinate of the tangent vector $v \in TM_x$ with respect to φ. Thus the typical fiber X_φ is the chart space. To see that this actually yields a vector bundle, we have to look at the change of the local trivializations. Let (U_j, φ_j) be another chart with $x \in U_i \cap U_j$ and let $\psi = \varphi_j$. From (6), we obtain

$$v_\psi = t_{ij}(x) v_\varphi$$

with

$$t_{ij}(x) = F'(\varphi(x)) \quad \text{and} \quad F = \psi \circ \varphi^{-1}.$$

The map F, which describes the chart change, is a C^k-map between the chart spaces X_φ and X_ψ. Thus F' is a C^{k-1}-map from X_φ into $L(X_\varphi, X_\psi)$. The smoothness of maps in M is defined with respect to the corresponding chart spaces. Therefore $x \mapsto t_{ij}(x)$ is a C^{k-1}-map from U_i into $L(X_\varphi, X_\psi)$. □

EXAMPLE 73.77 (Cotangent Bundle as Vector Bundle). In a similar way as in the previous example, one can prove that (π, TM^*, M) is a C^{k-1}-vector bundle, if M is a C^k-Banach manifold, $k \geq 1$.

73.22. Vector Bundles

Often, in mathematics, one has the following situation:

(i) One is given certain objects (e.g., groups, topological spaces, C^k-manifolds, etc.).
(ii) For these objects there exist maps which are called morphisms (e.g., group homomorphisms, continuous maps, C^k-maps, etc.).
(iii) Bijective maps b between these objects are called isomorphisms if b and b^{-1} are morphisms (e.g., group isomorphisms, homeomorphisms, C^k-diffeomorphisms, etc.).

In this way, one immediately recognizes common properties between different mathematical structures. Morphisms are always defined in such a way that the corresponding isomorphisms preserve in some intuitive sense the structure of these objects. Here isomorphic objects are considered to be essentially equal. Classes of objects, together with their morphisms, form categories (category of groups, category of topological spaces, category of C^k-manifolds, etc.). In category theory, one then studies general properties of these categories, i.e., general properties of mathematical structures. In order to obtain a category of vector bundles, we need to define the corresponding morphisms for our objects "vector bundles." This will be done in the following definition.

Definition 73.78. Let $V_j = (\pi_j, B_j, M_j)$, $j = 1, 2$, be two vector bundles. A *morphism* from V_1 to V_2 is a pair (f, g) of maps which satisfies the following two properties.

(M1) *Fiber preserving property*. The diagram

$$\begin{array}{ccc} B_1 & \xrightarrow{f} & B_2 \\ \pi_1 \downarrow & & \downarrow \pi_2 \\ M_1 & \xrightarrow{g} & M_2 \end{array}$$

commutes where f and g are continuous, i.e.,

$$f(F_x) \subseteq F_{g(x)} \qquad \text{for all} \quad x \in M_1.$$

This way, we obtain a linear map from F_x into $F_{g(x)}$.

(M2) *Mapping of typical fibers*. For each $x \in M_1$, there exist trivializations for V_j:

$$\tau_j \colon \pi_j^{-1}(U_j) \to U_j \times Y_j, \qquad j = 1, 2$$

with $x \in U_1$, $g(x) \in U_2$ and $g(U_1) \subseteq U_2$, such that the map

$$\bar{f}(x) \colon Y_1 \to Y_2,$$

which is naturally defined through the following commuting diagram

$$\begin{array}{ccc} \{x\} \times Y_1 & \xrightarrow{\bar{f}(x)} & \{g(x)\} \times Y_2 \\ {\scriptstyle \tau_1^{-1}}\downarrow & & \uparrow{\scriptstyle \tau_2} \\ F_x & \xrightarrow{f} & F_{g(x)} \end{array}$$

has the property that $\bar{f} \colon U_1 \to L(Y_1, Y_2)$ is continuous.

If V_1 and V_2 are C^k-vector bundles, $k \geq 0$, then we obtain C^k-*morphisms* if "continuous" is everywhere replaced by "C^k-map."

Two vector bundles V_1 and V_2 are called *isomorphic* if and only if there exists a morphism (f, g) from V_1 to V_2 such that f and g in condition (M1) are bijective, and (f^{-1}, g^{-1}) is a morphism from V_2 to V_1.

Analogously, C^k-isomorphisms are defined.

Remark 73.79 (Meaning of (M2)). One may think of $\bar{f}$ as the representative of f with respect to the typical fibers. Since τ_1 and τ_2 are homeomorphisms, condition (M1) implies that the map $\bar{f}(x) \colon Y_1 \to Y_2$ is always linear and continuous, i.e., $\bar{f}(x) \in L(Y_1, Y_2)$.

In the finite-dimensional case, i.e., if all fibers of V_1 and V_2 are finite dimensional, one can simplify this definition because (M2) is then a consequence of (M1), so that (M2) is not needed.

In mathematical physics, the concept of vector bundles arises naturally if at each point x of a manifold M a linear space F is attached (tangent spaces for the description of vector fields, tensor spaces for the description of tensor fields, spinor spaces in relativistic quantum theory, etc.). The concept of vector bundles allows a global description of these linear objects on M.

Conversely, one can use the algebraic structure of the collection of all vector bundles on M to describe topological properties of M. This is done in K-theory. Important is the fact that it is possible to form a linear algebra for vector bundles by reducing the algebraic operations between vector bundles to the corresponding operations between typical fibers. K-theory—a generalized cohomology theory—plays a central role in the formulation and proof of the Atiyah–Singer index theorem, which describes a fundamental connection between the topology on M and the index of differential and integral operators on M. This will be discussed in Part V.

Vector bundles also play a fundamental role in gauge theories. Model equations for elementary particles are the Yang–Mills equations. These are intimately related to the connection in principal fiber bundles. The solutions of these complicated nonlinear partial differential equations can explicitely be given. This uses the Penrose transformation, which reduces the problem to a study of analytic vector bundles on the three-dimensional complex projective space $\mathbb{C}P^3$. In this connection, it is important to master the theory of these bundles in algebraic geometry. For this, we recommend Atiyah (1979, L).

73.23. Differentials and Derivations on Finite-Dimensional Manifolds

In this last section of the chapter on Banach manifolds we will give several definitions and results that are used later on in the differential calculus for finite-dimensional manifolds. In order to help the reader with the study of the literature, we try to incorporate various different notations used by different authors. Our general assumption is:

(H1) M and N are C^1-Banach manifolds with chart spaces over $\mathbb{K}$.

Let $f: U(x) \subseteq M \to N$ be C^1 on an open neighborhood of x. As in Section 73.6

$$f'(x): TM_x \to TN_x$$

denotes the tangent map of f at x.

Definition 73.80. Similarly as in Section 4.2, we define the *differential* of f at the point x in the direction of v through

$$df(x; v) = f'(x)v.$$

This definition is valid for all tangent vectors v at x, i.e., for all $v \in TM_x$. Another notation for $f'(x)$ is $df(x)$ or df_x, and $df(x)$ or df_x is called the *differential* of f at x.

Thus "tangent map $f'(x)$" and "differential df_x" are synonymous concepts. We have

$$df_x[v] = df(x)[v] = df(x; v) = f'(x)v$$

for all $v \in TM_x$.

For the special case $N = \mathbb{K}$ we have that $df_x: TM_x \to \mathbb{K}$ is a linear, continuous functional on TM_x, i.e., $df_x \in TM_x^*$. Consequently, df_x is a cotangent vector to M at x. Moreover,

$$\langle df_x, v \rangle = \langle df(x), v \rangle = f'(x)v \tag{43}$$

for all $v \in TM_x$. Instead of (43), one finds in the literature also

$$v(f) \stackrel{\text{def}}{=} f'(x)v. \tag{44}$$

As in classical analysis, $f'(x)v$ is also called a *directional derivative* of f at the point x in the direction of v.

We now want to show that, with the definition of df_x, the differential calculus of $\mathbb{R}^n$ carries over to finite-dimensional manifolds. Formally, the same rules apply. For this, we specialize assumption (H1).

(H2) Let M be an n-dimensional C^1-manifold with chart spaces $\mathbb{K}$.

Fix a point $x \in M$ and let (U, φ) be a chart in M with $x \in U$. The elements of the chart space $\mathbb{K}^n$ have the form $u = (u^1, \ldots, u^n)$, i.e.,

$$u = u^i e_i,$$

where e_i is the unit vector in the direction of the u^i-coordinate. Here, and in the following, we use Einstein's summation convention, i.e., we sum from 1 to n over equal upper and lower indices.

Definition 73.81. If (H2) holds, then we let b_i denote the tangent vector to M at x which corresponds to the vector e_i.

For a C^1-curve $x = x(t)$ in M, we previously agreed to denote the tangent vector at $x(t)$ with $x'(t)$ or $\dot{x}(t)$. The representative of $x = x(t)$ and $\dot{x}(t)$ in the u-chart is $u = u(t)$ and

$$\dot{u}(t) = \dot{u}^i(t) e_i,$$

respectively. If we choose the u^i-coordinate line through the point $u = \varphi(x)$, then we have $t = u^i$ and $\dot{u} = e_i$. Consequently, b_i corresponds to the u^i-coordinate line through $u = \varphi(x)$. In accordance with our convention for $x'(t)$, we write $b_i = \partial x(u)/\partial u^i$ with $u = \varphi(x)$ or in short notation

$$b_i = \frac{\partial x}{\partial u^i}, \qquad i = 1, \ldots, n. \tag{45}$$

If M lies in $\mathbb{K}^{n+1}$, then (45) actually represents a geometric vector in $\mathbb{K}^{n+1}$ which corresponds to b_i. Figure 73.28 shows this for the case $n = 2$. In general, however, b_i cannot be represented by a vector in some surrounding space; one has to stick with the corresponding curves in M.

Since $\{e_1, \ldots, e_n\}$ is a basis for the chart space, it follows that $\{b_1, \ldots, b_n\}$ is a basis for the tangent space TM_x. Consequently, every tangent vector $v \in TM_x$ can be written as

$$v = v^i b_i \tag{46}$$

with $v^i \in \mathbb{K}$ for all i. The representative of v in the chart space to (U, φ) is

$$\bar{v} = v^i e_i.$$

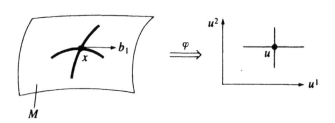

Figure 73.28

73.23. Differentials and Derivations on Finite-Dimensional Manifolds

In order to obtain a basis $\{b^1,\ldots,b^n\}$ for the dual space TM_x^*, we define linear, continuous functionals b^j on TM_x through
$$b^j[v] = v^j, \quad j = 1,\ldots,n.$$
Thus $b^j \in TM_x^*$ and
$$b^j[b_i] = \delta_i^j, \quad i,j = 1,\ldots,n.$$
We therefore obtain for each cotangent vector $w \in TM_x^*$
$$w[v] = w[v^i b_i] = w_i b^i[v]$$
with $w_i = w[b_i]$. This implies that
$$w = w_i b^i. \tag{47}$$
Hence $\{b^1,\ldots,b^n\}$ is in fact a basis for TM_x^*. The representative $\bar{w}$ of $w \in TM_x^*$ with respect to the chart (U,φ) is obtained from $\bar{w}[\bar{v}] = w[v]$, i.e.,
$$\bar{w} = w_i e^i.$$
Here, e^j is defined through $e^j[v^i e_i] = v^j$, i.e., e^j is a linear, continuous functional on the chart space $\mathbb{K}^n$.

Definition 73.82. If (H2) holds, then $\{b_1,\ldots,b_n\}$ is called a *natural basis* for the tangent space TM_x with respect to the chart (U,φ). Moreover, $\{b^1,\ldots,b^n\}$ is called a natural basis for the cotangent space TM_x^* with respect to (U,φ).

The numbers v^i and w_i in (46) and (47) are called *natural coordinates* of $v \in TM_x$ and $w \in TM_x^*$.

Recall that $b_i = \partial x/\partial u^i$. Below we will show that $b^i = du^i$. In the literature, sometimes also $b_i = \partial/\partial u^i$ is used.

Proposition 73.83 (C^k-Fields). *Assume* (H2), *and let M be a C^k-manifold, $k \geq 1$. Then a vector field*
$$x \mapsto (x, v(x))$$
[or covector field $x \mapsto (x, w(x))$] on M is C^k if and only if for each $x \in M$ there exists a chart (U,φ) with $x \in U$ such that the natural coordinate functions
$$u \mapsto v^i(u), \quad i = 1,\ldots,n$$
[or $u \mapsto w_i(u)$] are C^k-functions on the chart image $\varphi(U)$, i.e., all v^i and w_i are C^k-functions on M.

PROOF. This follows by using the representations of the representatives $\bar{v}$ and $\bar{w}$ in the chart space. □

We now study how the b_i and b^i are transformed under chart changes, i.e., we pass from the u^i-coordinates to the new $u^{i'}$-coordinates and denote the new basis vectors by $b_{i'}$ and $b^{i'}$.

We let
$$A_i^{i'} = \frac{\partial u^{i'}}{\partial u^i}(u) \quad \text{and} \quad A_{i'}^i = \frac{\partial u^i}{\partial u^{i'}}(u'),$$
where u and u' are the corresponding chart images of $x \in M$.

Proposition 73.84. *From assumption* (H2) *we obtain under chart changes that*
$$b_{i'} = A_{i'}^i b_i, \tag{48}$$
$$b^{i'} = A_i^{i'} b^i. \tag{49}$$
The natural coordinates v^i and w_i are transformed in the same way as b^i and b_i.

In Part V, these key relations will enable us to pass from invariant tensor calculus on manifolds to coordinate representations.

PROOF. Ad(48). Consider the $u^{i'}$-coordinate line through u'. Under chart changes, this curve becomes
$$u^i = u^i(u^{1'}, \ldots, u^{n'}), \quad i = 1, \ldots, n,$$
where $u^{i'}$ varies, and all other $u^{j'}$ are fixed. The tangent vector to this curve at u' is
$$\frac{\partial u^i}{\partial u^{i'}}(u') e_i.$$

But this is the representative of $b_{i'}$ in the u-chart. Hence, formula (48) follows. Reversing the roles of u^i and $u^{i'}$ formula (48) implies that
$$b_i = A_i^{i'} b_{i'}.$$
From $v = v^i b_i = v^{i'} b_{i'}$ we then obtain that
$$v^{i'} = A_i^{i'} v^i.$$

Ad(49). This relation follows immediately from $b^{i'}(v) = v^{i'}$ and $b^i(v) = v^i$. Finally, it follows from $w = w_i b^i = w_{i'} b^{i'}$ and (49) that
$$w_{i'} = A_{i'}^i w_i. \qquad \square$$

Proposition 73.85. *If* (H2) *holds, then*
$$b^i = du_x^i.$$

PROOF. Consider the map $f: U \subseteq M \to \mathbb{K}$, defined through $f(x) = u^i$, i.e., f assigns to each point $x \in M$ the coordinate u^i in the chart space. It follows that $df_x = du_x^i$. According to Section 73.6, one computes $f'(x)v$ by passing to representatives, i.e.,
$$f'(x)v = \bar{f}'(u)\bar{v} \quad \text{with} \quad u = \varphi(x).$$

73.23. Differentials and Derivations on Finite-Dimensional Manifolds

Thus we obtain
$$df_x[b_j] = f'(x)b_j = \bar{f}'(u)e_j. \tag{50}$$

But this is the directional derivative of $\bar{f}$ at u in the direction of e_j, i.e.,
$$df_x[b_j] = \frac{\partial u^i}{\partial u^j} = \delta^i_j$$

for all j. This means that $df_x = b^i$. □

Instead of du^i_x, one often uses the shorter notation du^i, i.e.,
$$b^i = du^i, \quad i = 1, \ldots, n. \tag{51}$$

We will see below that this formula is the key for the differential calculus on n-dimensional manifolds. It is worth emphasizing that du^i is a well-defined object, not just a formal symbol. It provides a rigorous justification for the old formal differential calculus which goes back to Leibniz (1646–1716), and supports our general experience that every sucessful formal calculus admits a rigorous justification. During this century, for instance, it was possible to develop a mathematical framework for the Heaviside calculus and Dirac's delta function which were earlier introduced in a formal way by physicists. This was done, by using the Laplace transformation and the theory of distributions. The *differential* du^i is a linear, continuous *functional* on the tangent space TM_x. For an arbitrary tangent vector $v = v^i b_i$ we have
$$du^j[v] = v^j, \quad i, j = 1, \ldots, n. \tag{52}$$

The transformation rules (48) and (49) now assume the following form
$$\frac{\partial x}{\partial u^{i'}} = \frac{\partial u^i}{\partial u^{i'}} \frac{\partial x}{\partial u^i}, \tag{53}$$

$$du^{i'} = \frac{\partial u^{i'}}{\partial u^i} du^i, \tag{54}$$

which correspond to the classical formulas.

Corollary 73.86. *If* (H2) *holds, then one obtains precisely all cotangent vectors $w \in TM_x^*$ through*
$$w = w_i du^i, \tag{55}$$

with arbitrary $w_i \in \mathbb{K}$.

PROOF. This follows from $w = w_i b^i$ and (51). □

Because of formula (55), cotangent vectors are also called forms or *differential forms*. We will see in Part V that formula (55) allows a very elegant formulation of the canonical formalism of classical mechanics for the general situation that the state space is a manifold.

In order to derive other useful consequences from the key formula (55) we make the following assumption.

(H3) Let $f, g: U(x) \subseteq M \to \mathbb{K}$ be two C^1-functions defined on an open neighborhood of x, and assume (H2) for M.

Definition 73.87. Assume (H3). Let

$$\frac{\partial f}{\partial u^i}(x) \stackrel{\text{def}}{=} df_x[b_i], \qquad i = 1, \ldots, n.$$

By (44), this is the directional derivative of f at x in the direction of b_i.

Corollary 73.88. *If $\bar{f}$ denotes the chart representative of f, i.e., $\bar{f}(u) = f(x)$ with $u = \varphi(x)$, then*

$$\frac{\partial f}{\partial u^i}(x) = \frac{\partial \bar{f}}{\partial u^i}(u), \qquad i = 1, \ldots, n.$$

PROOF. This follows from (50). □

Proposition 73.89 (Product Formula). *Let $f, g: U(x) \subseteq M \to \mathbb{K}$ be two C^1-functions defined on an open neighborhood of x. Furthermore, let M be a C^1-Banach manifold with chart spaces over $\mathbb{K}$. Then*

$$d(fg)_x = (df_x)g(x) + f(x)\, dg_x. \tag{56}$$

In short notation, one writes

$$d(fg) = (df)g + f\, dg.$$

PROOF. We have

$$(fg)'(x)v = (f'(x)v)g(x) + f(x)g'(x)v.$$

This follows from an application of the product rule of Section 4.3 to the chart representatives. □

In conclusion, we consider the concept of derivation.

Definition 73.90. Let M be a C^∞-Banach manifold with chart spaces over $\mathbb{K}$. We fix a point $x \in M$. Let C_x^∞ denote the collection of all C^∞-functions $f: U(x) \subseteq M \to \mathbb{K}$, defined on an open neighborhood of x, where $U(x)$ may depend on f. A *derivation* at x is a map $D: C_x^\infty \to \mathbb{K}$ with

$$D(\alpha f + \beta g) = \alpha D(f) + \beta D(g),$$
$$D(fg) = D(f)g(x) + f(x)D(g)$$

for all $\alpha, \beta \in \mathbb{K}$ and $f, g \in C_x^\infty$.

73.23. Differentials and Derivations on Finite-Dimensional Manifolds

If we fix a tangent vector $v \in TM_x$, then the directional derivative at x in the direction of v

$$D(f) = df_x[v]$$

is obviously a derivation at x. This follows from the product rule (56). The following proposition shows that all derivations on finite-dimensional C^∞-manifolds are obtained in this way. Recall the notation

$$df_x[v] = v(f).$$

Thus, we can also write $D(f) = v(f)$.

Proposition 73.91. *Let M be an n-dimensional real or complex C^∞-manifold. Then each tangent vector $v \in TM_x$, i.e., $v = v^i b_i$ generates a derivation at x through*

$$v(f) = v^i \frac{\partial f}{\partial u^i}(x). \tag{57}$$

Every derivation at x is generated like this. In particular, we have

$$b_i(f) = \frac{\partial f}{\partial u^i}(x). \tag{58}$$

PROOF. Let $v \in TM_x$. Formulas (57) and (58) follow from

$$v(f) = df_x[v^i b_i] = v^i df_x[b_i] = v^i \frac{\partial f}{\partial u^i}(x).$$

Conversely, let D be a derivation at x. For $f = g = 1$ we obtain $D(1) = 2D(1)$ and thus $D(1) = 0$. In the same way we obtain $D(f) = 0$ for every constant function f. Let $u_0 = \varphi(x)$ and let $\bar{f}$ be the chart representative of f. Letting $D(\bar{f}) = D(f)$ we obtain a derivation for the chart space $\mathbb{K}^n$. Taylor's theorem of Section 4.6, with integral remainder, gives the following representation for the C^∞-function $\bar{f}: U(x_0) \subseteq \mathbb{K}^n \to \mathbb{K}$,

$$\bar{f}(u) = \bar{f}(u_0) + (u^i - u_0^i) g_i(u),$$

in a neighborhood of u_0 where the g_i are C^∞ with $g_i(u_0) = \partial \bar{f}(u_0)/\partial u^i$ for all i. Because of $D(\text{const}) = 0$ we obtain

$$D(\bar{f}) = D(u^i - u_0^i) g_i(u_0).$$

Letting $w^i = D(u^i - u_0^i)$, it follows that

$$D(f) = w^i \frac{\partial f}{\partial u^i}(x).$$

Recall Corollary 73.88 in this connection. □

Proposition 73.91 yields a one-to-one correspondence between tangent vectors and derivations on finite-dimensional C^∞-manifolds. In the literature, the notion of derivations is often used to define tangent vectors. Starting from (57), one writes in suggestive form

$$v = v^i \frac{\partial}{\partial u^i}.$$

A comparison with $v = v^i b_i$ yields a correspondence between b_i and $\partial/\partial u^i$ which formally coincides with (58).

For infinite-dimensional Banach manifolds, Proposition 73.91 is not true. In the general case, the geometric definition of Section 73.5 is most appropriate for the development of a corresponding theory.

Problems

73.1.* *Proof of Proposition 73.9.*
Hint: See Milnor (1965, M).

73.2. *Abstract atlases.* In Section 73.2 we gave a definition of a manifold M, using the fact that M is a topological space. Another possible way to construct Banach manifolds is the following. Let M be an arbitrary set. An *abstract C^k-atlas* for M is a set of pairs $(U_\alpha, \varphi_\alpha)$, which will be called abstract charts and satisfy the following properties:
  (i) The sets U_α cover M.
  (ii) Every φ_α maps U_α bijectively onto an open set $\varphi_\alpha(U_\alpha)$ in a B-space X_α over $\mathbb{K}$.
  (iii) If $U_\alpha \cap U_\beta$ is not empty, then

  $$\varphi_\beta \circ \varphi_\alpha^{-1} : \varphi_\alpha(U_\alpha \cap U_\beta) \to \varphi_\beta(U_\alpha \cap U_\beta)$$

  is a C^k-map defined on the open set $\varphi_\alpha(U_\alpha \cap U_\beta)$.
  Prove: If an abstract C^k-atlas belongs to M, $k \geq 0$, then a topology may be introduced on M, such that M becomes a C^k-Banach manifold in the sense of Section 73.2, whenever this topology is separated.
  Note that in Section 73.2, as throughout the book, topological spaces are always assumed to be separated (see $A_1(7)$). But even if the topology on M is not separated one still can develop the theory. We then speak of nonseparated Banach manifolds.
  Solution: A set U in M is called open if and only if for each $x \in U$ there exists a chart $(U_\alpha, \varphi_\alpha)$ with $x \in U_\alpha \cap U$ such that $\varphi_\alpha(U_\alpha \cap U)$ is open.

73.3. *Whitney topology.* Prove that Definition 73.55 actually gives a topology on $C^\infty(M, N)$.
Hint: See Golubitsky and Guillemin (1973, M), Chapter 2, §3. There, one also finds a proof of Example 73.56.

73.4. *k-determination.* Using (C1) and (C2) of Section 73.16, prove the following statements in $\mathbb{R}^2$, and by using (C3) construct the corresponding universal unfoldings.
  (i) $a\xi + b\eta$ is strongly 1-determined for $a^2 + b^2 \neq 0$ and not 1-determined for $a^2 + b^2 = 0$.
  (ii) $a\xi^2 + 2b\xi\eta + c\eta^2$ is strongly 2-determined for $ac - b^2 \neq 0$ and not 2-determined for $ac - b^2 = 0$.

(iii) $a\xi^2\eta + b\eta^3$ is strongly 3-determined for $ab \neq 0$ and not 3-determined for $ab = 0$.
(iv) $\xi^2\eta$ is not k-determined.
(v) $3\xi^2 + 2\xi^3 - \xi\eta^2$ is not k-determined for $k = 1, 2, 3$, but strongly 4-determined.
(vi) $\xi^3 + \xi\eta^3$ is not 3-determined and not strongly 4-determined.
(vii) $\xi^5 + \eta^5$ is not 5-determined, but strongly 6-determined.

Hint: See Poston and Stewart (1978, M).

73.5. *Normal forms for functions of two real variables.* In Table 73.2 we give normal forms for k-forms which are obtained from linear, bijective transformations of the independent variables.

Table 73.2

		Normal form	
k	k-form	Strongly k-determined	Not k-determined
1	$a\xi + b\eta$	η	0
2	$a\xi^2 + b\xi\eta + c\eta^2$	$\pm(\xi^2 + \eta^2)$, $\xi^2 - \eta^2$	$\pm\eta^2, 0$
3	$a\xi^3 + b\xi^2\eta + c\xi\eta^2 + d\eta^3$	$\xi^2\eta \pm \eta^3$	$\xi^2\eta, \eta^3, 0$

Hint: See Poston and Stewart (1978, M), Chapter 2. If f is a cubic form with $f \not\equiv 0$, then one can determine the type of the normal form by solving the equation $f(\xi, \eta) = 0$ and studying the qualitative behavior of the solution set (three lines, one simple line, one line and a double line, one triple line). For quadratic forms the sign of f plays a role.

73.6 *General chain rule.* Let $f: M \to N$ and $g: N \to P$ be C^r, $r \geq 1$, where M, N, and P are C^r-Banach manifolds. Prove:

$$T^r(g \circ f) = T^r g \circ T^r f. \tag{59}$$

Solution: It is sufficient to use local coordinates. The chain rule implies that

$$Tf(x, v) = (f(x), f'(x)v),$$

$$T(g \circ f)(x, v) = (g(f(x)), g'(f(x))f'(x)v),$$

$$Tg(f(x), f'(x)v) = (g(f(x)), g'(f(x))f'(x)v).$$

This gives (59) for $r = 1$. Moreover, we have

$$T^2(g \circ f) = T(Tg \circ Tf) = T^2 g \circ T^2 f.$$

73.7. *Whitney's embedding theorem.* Study the proofs in Guillemin and Pollack (1974, M), Hirsch (1976, M), and Golubitsky and Guillemin (1973, M) in this order.

73.8. *Generalized Morse lemma.* Our goal is to generalize Theorem 73.F of Section 73.12 which is important for applications in partial differential equations. The key to this and further applications is the result of Problem 73.8a.

73.8a. *General criterium for right equivalence of Golubitsky and Marsden (1983).* We study the following important question. When is $f + g$ C^m-right equivalent to

f at 0?, i.e., letting $h = f + g$ we try to find a map ψ with

$$h(\psi(x)) = f(x)$$

in a neighborhood of 0. Here, ψ is a local C^m-diffeomorphism at 0 with $\psi(0) = 0$. The following equations are important:

$$g(x) = f'(x)A(x) \quad \text{on } U(0), \tag{60}$$

$$g'(x) = f'(x)B(x) \quad \text{on } U(0), \tag{61}$$

$$A(0) = 0, \quad B(0) = 0. \tag{62}$$

Our assumptions are:
(i) X is a real B-space. The maps $f, g: U(0) \subseteq X \to \mathbb{R}$ are C^m, $m \geq 1$ and $f(0) = g(0) = 0$.
(ii) There exist C^m-maps $A: U(0) \to X$ and $B: U(0) \to L(X,X)$ which satisfy (60)–(62).

Prove similarly as in Section 73.12: The map $f + g$ is C^m-right equivalent at 0 to f at 0.

Solution:
(I) Let

$$H(x,t) = f(x) + tg(x)$$

and

$$C(x,t) = -(I + tB(x))^{-1}A(x).$$

Moreover, let V be a sufficiently small neighborhood of 0 in X. For each fixed $x \in V$ we solve the ordinary differential equation

$$\varphi_t(x,t) = C(\varphi(x,t),t), \qquad \varphi(x,0) = x. \tag{63}$$

It is important that this solution is defined for all $t \in [0,1]$. To see this note that C is defined on $V \times [0,1]$ for sufficiently small V, since $B(0) = 0$. Furthermore, (63) has for $x = 0$ the solution $\varphi(0,t) \equiv 0$, since $C(0,t) \equiv 0$. From the dependence of solutions on the initial data (Theorem 4.D) it follows that V can be chosen so small that (63) has a solution on $[0,1]$ for all $x \in V$.

(II) We let $\psi(x) = \varphi(x,1)$. Then ψ is a local C^m-diffeomorphism at 0 with $\psi(0) = 0$. This follows from uniqueness of solutions of (63) and from Theorem 4.D. Note that the map C is C^m.

(III) Let $P = (x,t)$. After a short computation it follows from (63) and (60), (61) that

$$\frac{\partial}{\partial t} H(\varphi(P),t) = f'(\varphi(P))\varphi_t(P) + g(\varphi(P)) + tg'(\varphi(P))\varphi_t(P) = 0.$$

Therefore $H(\varphi(x,t),t) = \text{const}$, and hence

$$H(\varphi(x,1),1) = H(\varphi(x,0),0).$$

This yields

$$f(\psi(x)) + g(\psi(x)) = f(x).$$

73.8b. Morse-Tromba lemma. Let X be a real B-space with a scalar product $(\cdot|\cdot)$, i.e., the map $(x, y) \mapsto (x|y)$ from $X \times X$ into $\mathbb{R}$ is bilinear, symmetric, strictly positive, and continuous. Also we assume:

(i) The map $h: U(0) \subseteq X \to \mathbb{R}$ is C^k, $k \geq 3$, and $h(0) = 0$, $h'(0) = 0$.

(ii) There exists a linear, continuous, and bijective map $T: X \to X$ with

$$h''(0)xy = (Tx|y) \qquad \text{for all} \quad x, y \in X.$$

(iii) There exists a C^{k-1}-map $H: U(0) \subseteq X \to X$ with

$$h'(x)y = (H(x)|y)$$

for all $x \in U(0)$ and $y \in X$. We let

$$f(x) = 2^{-1}h''(0)x^2 = 2^{-1}(Tx|x).$$

Prove with Problem 73.8a that h is C^{k-2}-right equivalent to f.

Discussion. The lemma states that h can be transformed into a normal form f. The assumptions here are weaker than in Theorem 73.F and can often be verified in applications to variational problems. This also is in contrast to Theorem 73.F. Compare Tromba (1976), (1983) and Marsden, Buchner, and Schecter (1983) for a more thorough discussion.

If we have a continuous embedding $X \subseteq Y$, and if Y is a H-space, then one can choose the scalar product on Y as the scalar product on X. In variational problems this situation occurs, for example, if $X = W_2^m(G)$, $m \geq 1$, and $Y = L_2(G)$. Assumptions (ii) and (iii) mean that the first- and second-order differential of h can be expressed in terms of the scalar product.

Solution:

(I) We prove (61). Define g through

$$h = f + g.$$

Then g is C^k with $g(0) = 0$, $g'(0) = 0$, and $g''(0) = 0$. Moreover, we have

$$g'(x)y = (G(x)|y)$$

with $G = H - T$. Note that

$$f'(x)z = h''(0)xz = (Tx|z), \tag{64}$$

where G is C^{k-1} with $G(0) = 0$ and $G'(0) = 0$. Consequently,

$$g'(x)y = \left(\int_0^1 G'(tx)x\,dt \,\Big|\, y\right).$$

From $g''(x)zy = (G'(x)z|y)$ and the symmetry of $g''(x)$, it follows that $(G'(x)z|y) = (G'(x)y|z)$, and thus

$$g'(x)y = \left(\int_0^1 G'(tx)y\,dt \,\Big|\, x\right).$$

Because of $(Tx|z) = (x|Tz)$, (ii) implies that

$$g'(x)y = \left(Tx \,\Big|\, T^{-1}\int_0^1 G'(tx)y\,dt\right).$$

Thus
$$g'(x)y = f'(x)B(x)y \qquad (65)$$
with
$$B(x)y = T^{-1}\int_0^1 G'(tx)y\, dt.$$

This is formula (61).

(II) Relation (60) follows immediately from (65) and (64), since
$$g(x) = \int_0^1 g'(\tau x)x\, d\tau = f'(x)A(x)$$
with
$$A(x) = \int_0^1 \tau B(\tau x)x\, d\tau.$$

Now apply Problem 73.8a with $m = k - 2$.

73.8c. *Splitting lemma and infinite-dimensional catastrophe theory.* In Problem 73.8b we assumed that $T: X \to X$ is bijective, i.e., $h''(0)$ is nondegenerate. In the degenerate case, i.e., if $T: X \to X$ is a Fredholm operator of index zero, one may find a normal form in Golubitsky and Marsden (1983). Using this result, catastrophy theory in $\mathbb{R}^n$ can be extended to B-spaces. In particular, there exists a map g for h such that the critical points of h in a neighborhood of 0 correspond to those of g in the finite-dimensional null space $N(T)$.

73.9.* *The multilinearization principle and the fundamental blowing-up lemma.* In order to study the structure of the solution set of the equation
$$f(x) = 0 \qquad (66)$$
in a neighborhood of $x = 0$ in $\mathbb{R}^N$, we consider the simpler multilinearized equation
$$g(x) = 0, \qquad (66^*)$$
where $g = (g_1, \ldots, g_M)$, and g_i corresponds to certain Taylor polynomials of f_i, i.e., more precisely,
$$g_i(x) = \sum_{\langle \alpha | \beta \rangle = \gamma_i} D^\alpha f_i(0) \frac{x^\alpha}{\alpha!}$$
for $i = 1, \ldots, M$. Here, $\alpha = (\alpha_1, \ldots, \alpha_N)$ denotes a tuple of non-negative integers, and
$$D^\alpha = D_1^{\alpha_1}\ldots D_N^{\alpha_N}, \qquad x^\alpha = \xi_1^{\alpha_1}\ldots \xi_N^{\alpha_N}$$
$$\alpha! = \alpha_1!\alpha_2!\ldots \alpha_N!$$
with $x = (\xi_1, \ldots, \xi_N)$ and $D_i = \partial/\partial \xi_i$.

Moreover, we set
$$\langle \alpha | \beta \rangle = \sum_{j=1}^N \alpha_j \beta_j.$$

We assume:

(H1) The map $f: \mathbb{R}^N \to \mathbb{R}^M$ is C^∞, where N and M are fixed positive integers with $N \geq M + 1$.

We are given fixed tuples $\beta = (\beta_1, \ldots, \beta_N)$ and $\gamma = (\gamma_1, \ldots, \gamma_N)$ of positive integers.

(H2) *Vanishing derivatives of f*. We have $f(0) = 0$ and

$$D^\alpha f_i(0) = 0, \quad i = 1, \ldots, M$$

for all α with $\langle \alpha | \beta \rangle \leq \gamma_i - 1$.

(H3) *Nondegeneracy*. If x is a nonzero solution of the multilinearized equation (66*), then x is a *regular* point of g, i.e., rank $g'(x) = M$.

Prove: If (H1)–(H3) hold, then, in a sufficiently small neighborhood of the point $x = 0$, the solution set of the original equation (66) and the solution set of the multilinearized equation (66*) have the same structure. More precisely, we have the following result:

(a) There exists a homeomorphism

$$h: f^{-1}(0) \cap U \to g^{-1}(0) \cap V$$

with $h(0) = 0$, where U and V are fixed, sufficiently small open neighborhoods of the point $x = 0$ in $\mathbb{R}^N$.

(b) The restricted map

$$h: (f^{-1}(0) \cap U) - \{0\} \to (g^{-1}(0) \cap V) - \{0\}$$

is C^∞, and the local solution set $(f^{-1}(0) \cap U) - \{0\}$ of the original equation (66) is a C^∞-submanifold of $\mathbb{R}^N$.

Hint: This result generalizes the blowing-up lemma of Problem 8.22. Use a similar argument as in the proof of Problem 8.22. Cf. Buchner, Marsden, and Schecter (1983a) and Jones and Toland (1986). The latter paper contains an interesting application to the bifurcation of capillary-gravity water waves.

In fact, there are two basic results in modern bifurcation theory, namely, the implicit function theorem and the blowing-up lemma, which can be viewed as a generalization of the implicit function theorem and the Morse lemma as well.

73.10. *A simple example.* We set $N = 2$, $M = 1$, $x = (\xi, \eta)$, and

$$f(x) = \xi^3 - \xi^2 \eta + \text{terms of order} \geq 4.$$

In this case, it is quite natural to set

$$g(x) = \xi^3 - \xi^2 \eta.$$

Letting $\beta = (1, 1)$ and $\gamma = 3$, we may apply the blowing-up lemma of Problem 73.9 to the equation $f(x) = 0$. Note that

$$\langle \alpha | \beta \rangle = \gamma \quad (\text{resp.} \leq \gamma - 1)$$

means that $\alpha_1 + \alpha_2 = 3$ (resp. ≤ 2).

Moreover, note that

$$g'(x) = (3\xi^2 - 2\xi\eta, -\xi^2),$$

and hence rank $g'(x) = 1$ if $g(x) = 0$ and $x \neq 0$.

The blowing-up lemma of Problem 73.9 tells us that, in a sufficiently small neighborhood of $x = 0$, the solution set of the equation

$$f(x) = 0$$

looks like the solution set of the multilinearized equation

$$\xi^3 - \xi^2\eta = 0.$$

References to the Literature

Introduction: Guillemin and Pollack (1974, M) and Marsden, Abraham, and Ratiu (1983, M).

Standard reference: Lang (1972, M).

Transversality and dynamical systems: Abraham and Robbin (1967, M).

Finite-dimensional manifolds: Guillemin and Pollack (1974, M) (introduction), Warner (1971, M), Golubitsky and Guillemin (1973, M), Hirsch (1976, M), and Westenholz (1981, M).

Applications to mathematical physics: Westenholz (1981, M) (introduction), Marsden (1974, L), (1980, L); Abraham and Marsden (1978, M), and Choquet-Bruhat (1982, M).

Catastrophe theory: Golubitsky (1978, S), Poston and Stewart (1978, M), Gilmore (1981, M), and Arnold (1985) (see also the References to the Literature to Section 37.28).

Catastrophe theory and bifurcation theory: Golubitsky and Schaeffer (1979), (1984, M) and Chow and Hale (1982, M).

Generalized Morse lemma: Golubitsky and Marsden (1983) (recommended as an introduction); Tromba (1976), (1983), and Buchner, Marsden, and Schecter (1983).

Blowing-up lemma: Buchner, Marsden, and Schecter (1983a), Jones and Toland (1986).

Infinite-dimensional catastrophe theory: Golubitsky and Marsden (1983).

Vector bundles: Lang (1972, M), Osborn (1982, M), Vols. 1–3, and Marsden, Abraham, and Ratiu (1983, M).

History of manifolds: Scholz (1980).

Differential forms: Cf. the References to the Literature for Chapter 82.

CHAPTER 74

Classical Surface Theory, the Theorema Egregium of Gauss, and Differential Geometry on Manifolds

> The curvature K of a surface depends only on the coefficients g_{ij} of the first fundamental form and their first- and second-order derivatives. Therefore K is an intrinsic property of the surface.
>
> Theorema egregium of Gauss (1827)
>
> His spirit lifted the deepest secrets of numbers, space, and nature; he measured the orbits of the planets, the form and the forces of the earth; in his mind he carried the mathematical science of a coming century.
>
> Under the picture of Carl Friedrich Gauss (1777–1855)
in the German Museum of Munich
>
> Classical differential geometry books are filled with monstrosities of long equations with many upper and lower indices. The modern revolt against the classical point of view has been so complete in certain quarters that some mathematicians will give a three-page proof that avoids coordinates in preference to a three-line proof that uses them.
>
> Michael Spivak (1970)
>
> Tensor calculus is an application of the chain rule.
>
> Mathematical folclore

In this and the following two chapters we consider three central applications of the theory of manifolds:

(i) Classical surface theory of Gauss.
(ii) Riemannian and affine connected manifolds.
(iii) Einstein's general theory of relativity (1916).

One should note that (ii) is a consequent development of (i), and (iii) is based on (ii). In fact, (i)–(iii) represent extraordinary achievements of mankind. We like to stress this line of thought in mathematics and physics. Furthermore, we want to emphasize the relation between differential geometry on Banach manifolds and its intuitive classical roots.

This chapter contains results which form the hard core of differential geometry. It is organized as follows:

(a) Tensor calculus in $\mathbb{R}^n$ and covariant differentiation (Sections 74.1–74.8).
(b) Applications to classical surface theory (Sections 74.9–74.16).
(c) Generalization to manifolds (Sections 74.17–74.21).
(d) Further development of the calculus in $\mathbb{R}^n$ and on manifolds (alternating differential forms, Lie derivatives; Sections 74.22–74.25).

The surface theory of Gauss was strongly influenced by Gauss' practical work as a surveyor. Under great physical pains he worked from 1821 to 1825 as a land surveyor in the kingdom of Hannover in the northern part of Germany. It almost led to his physical exhaustion. In 1822, he submitted his prize memoir "General solution of the problem of mapping parts of a given surface onto another given surface in such a way that image and preimage become similar in their smallest parts," to the Royal Society of Sciences in Copenhagen for which he received the official prize. What was the importance of his work?

The mapping of surfaces onto one another, which satisfy certain given properties, is a basic problem in cartography; in particular, the reproduction of parts of the surface of the earth in plane geographical charts. It is impossible, for example, to map parts of the surface of the earth onto the plane and preserve the length. This will be an easy consequence of the theorema egregium of Section 74.14. Thus one has to look for other mappings. Of great practical use are the conformal maps, i.e., angle preserving maps. Angle preservation of geographical charts is important in navigation, i.e., in determining routes of ships on charts. As we will see in Section 74.14, conformal maps are also similar in the small. Special cases of conformal maps from the surface of the earth onto the plane are stereographic projections (Fig. 74.1), which were already known to the Greeks, and the projection of Mercator (1512–1594) is still being used in the cartography of today. Gauss succeeded in finding a procedure to determine all conformal maps in the small for analytic surfaces. The study of conformal maps in the large began with the dissertation of Bernhard Riemann (1826–1866), which was written in 1851. It contains a development of the theory of complex function theory and the famous Riemannian mapping theorem. When writing his prize memoir, Gauss had apparently already worked on a more general surface theory, because he added

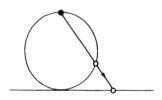

Figure 74.1

74. Classical Surface Theory, Theorema Egregium of Gauss

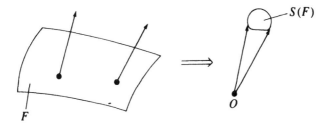

Figure 74.2

the following Latin saying to his title page. "Ab his via sternitur ad maiora" (From here the path to something more important is prepared).

The development of the general surface theory, however, was difficult, though the basic ideas were already known to Gauss since 1816. On February 19, 1826 he wrote to Olbers: "I hardly know any period of my life, where I earned so little real gain for truly exhausting work, as during this winter. I found many, many beautiful things, but my work on other things has been unsuccessful for months." Finally, on October 8, 1827, Gauss presented the general surface theory. The title of his paper was "Disquisitiones generales circa superficies curvas" (Investigations about curved surfaces). The most important result of this masterpiece in the mathematical literature is the theorema egregium—the beautiful theorem. Gauss begins with the following definition of curvature at a surface point. He considers a piece of the surface F with surface measure $m(F)$ and draws the unit normal vectors of F at the origin. This gives the spherical picture $S(F)$ of F on the unit sphere (Fig. 74.2). The absolute value of the Gaussian curvature $|K(P)|$ at the point $P \in F$ satisfies by definition

$$|K(P)| = \lim_{F \to P} m(S(F))/m(F). \qquad (1)$$

If F is the surface of a ball of radius r, then obviously $|K(P)| = 1/r^2$ for all points P of F. The above definition of $|K(P)|$ depends on the embedding of F in $\mathbb{R}^3$. The fundamental result of the Theorema egregium of Gauss is that $K(P)$ can be computed independently of the embedding in $\mathbb{R}^3$, i.e., the curvature of a surface is *independent* of the surrounding space. It may be determined solely by measurements on the surface itself. We shall try to explain why this is not only an important mathematical statement, but also has a fundamental impact on our physical view of the world.

In his famous habilitation talk of 1854 "On the hypotheses which form the fundaments of geometry" Riemann extended the Gaussian surface theory to n-dimensional manifolds for which a differential of the arclength ds and hence a metric is defined (Riemannian manifolds). Here the Gaussian curvature $K(P)$ is replaced with the Riemannian curvature tensor which is defined without reference to the surrounding space. The ingenious idea of Einstein, in his general theory of relativity (1916), was that the Riemannian curvature tensor of our four-dimensional space-time universe E_4 is determined by its masses, and the force of gravity arises from the fact that the orbits of the planets correspond to geodesics in E_4. In this way, Einstein gave a geometrical explanation for the gravitational force, i.e., gravity was reduced to curvature. Until the end of his life in 1955, Einstein, unsuccessfully, tried to find a unified theory for all physical interactions using the concept of geometrization. Today we have some hope that this program might be realized in the context of gauge field theories. The idea is that the connection in principal fiber bundles induces a curvature which causes the four fundamental interactions: strong, weak, electromagnetic, and gravitative. Such a unified theory would include the microcosmos (elementary particles) as well as the macrocosmos (cosmology).

If one looks at the history of the concept of manifolds, then during the last one hundred years since Riemann's habilitation talk, five important mathematical innovations have emerged, which are also relevant from a physical point of view. We shall call a property an intrinsic property of a manifold if it can be defined independently of the surrounding space and independently of the particular choice of charts:

(i) The existence of tangent vectors is an intrinsic property.
(ii) Curvature is an intrinsic property.
(iii) The possibility of parallel transport of objects is an intrinsic property. This allows a generalization of the concept of straight lines (geodesics).
(iv) Given a connection (Christoffel symbols), one can define parallel transport and curvature. Moreover, one obtains an invariant differential calculus on manifolds (covariant differentiation).
(v) For every metric (Riemannian manifold) there exists a connection. The notion of connection, however, can be explained independently of the existence of a metric (affine connected spaces, principal fiber bundles, vector bundles).

Point (i) has already been discussed in the previous chapter. In a more general form we shall discuss (ii)–(v) in Part V for Banach manifolds. In order to prepare the reader for the general theory, we give the standard examples for (ii)–(v) in this chapter, and applications to the special and general theory of relativity in the next following chapters. From (i)–(v) it is clear that the concept of a connection is of central importance in differential geometry.

In Part V we shall develop a tensor calculus and a differential geometry for Banach manifolds which generalizes the classical vector analysis. This elegant calculus is introduced in an invariant way, i.e., coordinate free. This is very natural, since in infinite-dimensional B-spaces there are no coordinates to work with. For a better understanding, however, it might be useful to be aware of the classical tensor calculus with coordinates which will be introduced in this chapter and applied to the general theory of relativity in Chapter 76. This classical calculus has the disadvantage that many indices appear. Its advantage, on the other hand, is that because of the index principle of Section 74.5 *it works on its own*. This is one of the basic requirements of Leibniz

(1646–1716) on a good calculus. This way, one gets many important hints as to which theorems to expect in the general calculus on Banach manifolds. Also, the definitions of the general calculus get an intuitive interpretation. It is really useful to master both calculi, in order to be able to apply the most advantageous in a concrete situation. One should *not* underestimate the power of the classical calculus.

In this chapter we will use the following strategy.

(a) We develop the tensor calculus in $\mathbb{R}^n$ (covariant differentiation, alternating differential forms, Lie derivatives).
(b) This calculus carries over to finite-dimensional manifolds without further thought (see Section 74.17).
(c) The heart of classical tensor calculus is the *index principle* of Section 74.5. It allows us to write an equation in such a way that we know its form in $\mathbb{R}^n$ for any arbitrary coordinate system. At the same time we obtain equations on manifolds which are chart independent, i.e., which have a geometrical meaning. Such notation is called covariant.

In mathematical physics, (c) is of particular importance because the physicist is interested in transforming his equations into other systems of reference.

In surface theory we proceed as follows:

(i) The behavior of surfaces is studied in a particular coordinate system. This yields, for example, in Section 74.11 a very simple and intuitive analytic definition of the Gaussian curvature.
(ii) Using the index principle, we write these equations in covariant from for arbitrary coordinate systems, i.e., geometrically invariant.
(iii) The basic equations of surface theory are the so-called fundamental equations, i.e., the equations for the change in the natural trihedral of the surface.
(iv) The integrability conditions, i.e., the necessary solvability conditions for (iii) yield the curvature tensor and the Theorema egregium.
(v) The solution of (iii) yields the main theorem of surface theory: Locally, a surface is uniquely determined by the first and second Gaussian fundamental form, except for motions in $\mathbb{R}^3$.
(vi) The generalization of surface theory and covariant differentiation to $\mathbb{R}^n$ leads naturally to Riemannian and affine connected manifolds.

It might be helpful for the reader to be aware of the classical origin of the curvature tensor as well as to get some intuitive understanding. The curvature tensor is the key to the general theory of relativity of Chapter 76.

The main analytic tool, which we use in this chapter, is the theorem of Frobenius (1849–1917) of Chapter 4. We will use it in the following way:

(α) Main theorem of surface theory.
(β) Main theorem for Riemannian manifolds: Such a manifold is locally flat if and only if the Riemannian curvature tensor is identically zero.

We note that (α) and (β) correspond to the following general strategy in differential geometry:

(S1) One finds differential equations for structures.
(S2) The necessary solvability conditions follow from the integrability conditions, i.e., from differentiating and equating the mixed derivatives. For example, it immediately follows from $u_x = a$ and $u_y = b$ that $a_y = b_x$ because $u_{xy} = u_{yx}$.
(S3) One shows that the integrability conditions are also sufficient. Here, the theorem of Frobenius or the theorem of Poincaré of Section 74.23 play an important role.

In order to shed some light onto the historical background of the general theory of relativity, we consider in Section 74.21 models for non-Euclidean geometries. In the problems to this chapter we explain important connections between topology and analysis (Theorem of Gauss–Bonnet–Chern, theorem of de Rham, duality theorem of Poincaré, index theorems of Poincaré–Hopf and Morse, theorem of Adams on vector fields). In Part V we will study the connection between these subjects and the Atiyah–Singer index theorem. In this chapter we only assume some knowledge of elementary differential and integral calculus and the theorem of Frobenius of Chapter 4. Also, we need the concepts of manifolds and tangent spaces of the previous chapter.

For readers who want to get acquainted with the life of Gauss, we recommend the Gauss biographies of Worbs (1955), Wussing (1974), and Bühler (1981). Furthermore, we suggest taking a look at the collected works of Gauss (1863). One will be fascinated by both, the depth of thought and the clarity and simplicity of the language. We conclude these introductory remarks with several citations from and about Gauss, which may help bring about a better understanding of the scientist and human being Gauss.

> Science should be the friend of applications, not its slave.
>
> Gauss

> I thank you, highly honored Sir, in the name of mankind, for presenting us with a picture of the highest intellectual power and force together with an inspiring and never ending warmth of feelings.
>
> Alexander von Humboldt to Gauss (1853)

> It is not the knowledge but the learning, not the possessing but the earning, not the being there but the getting there, which gives us the greatest pleasure.
>
> Gauss to Bolyai

> It is quite extraordinary how much the young mathematicians here in Berlin and, as I hear, in all parts of Germany adore Gauss. For them, he is the incorporation of mathematical perfection.
>
> Niels Henrik Abel (1825)

> By explanations the scientist means nothing else than a reduction to very few and simple basic rules, which cannot be reduced any further, but which allow a complete deduction of the phenomena.
>
> Gauss in *Electromagnetism and Magnetometer*

> My theories are important to me, but infinitely much more—the truth.
>
> Gauss

> Pauca sed matura (few, but ripe).
>
> Inscription of Gauss' seal

> "Princeps mathematicorum"—Prince of the mathematicians—Gauss is called on a commemorative coin, which the king of Hannover ordered after his death in his honor.
>
> Erich Worbs (1955)

> Since three days, the almost, for this world too heavenly angel, is my bride. I am unboundedly happy.
>
> The twenty-seven-year-old Gauss in a letter to Bolyai (1804)

> You were kind enough to invite me for a visit after my wife was well again. She is well now. Yesterday evening, at 8 o'clock I closed her angelic eyes in which I have found heaven for the last five years. Heaven give me the strength to bear this blow. Grant me a few weeks, dear Olbers to gather new strength in the arms of your friendship—strength for a life which is only valuable because it belongs to my three small children.
>
> The thirty-two-year-old Gauss in a letter to Olbers (1809)

> Sartorius von Waltershausen reports that Gauss once said there were questions of infinitely higher value than the mathematical ones, namely, those about our relation to god, our determination, and our future. Only, he concluded, their solutions lie far beyond our comprehension, and completely outside the field of science.
>
> Erich Worbs (1955)

74.1. Basic Ideas of Tensor Calculus

The main goal of tensor calculus is to write equations in such a way that they might be identified in arbitrary coordinate systems. This is very important in mathematical physics. For example, the Poisson equation $\Delta f = h$ in a fixed Cartesian coordinate system of $\mathbb{R}^3$ with coordinates (u^1, u^2, u^3) has the form

$$\sum_{i=1}^{3} D_i^2 f = h, \qquad (2)$$

where $D_i = \partial/\partial u^i$. This form is preserved if one passes to another Cartesian coordinate system. However, it is *not* preserved if one uses arbitrary curved coordinates such as the spherical coordinates. In an arbitrary coordinate system, (2) takes the form

$$g^{ij} \nabla_i \nabla_j f = h. \qquad (3)$$

The sum is taken over equal upper and lower indices from 1 to 3. The quantities g^{ij} and the symbol ∇_i for covariant differentiation are defined in such a way that, for Cartesian coordinates, equation (3) coincides with equation (2). Tensor calculus has the great advantage that one immediately sees whether or not an equation holds, independently of the particular choice of the coordinate system. According to the *index principle* of Section 74.5, one only needs to check if the index picture is right, i.e., if the free indices are equal in all terms; an index is called free if it is not summed. In the special case (3), for instance, the index picture is right, because the number of the free indices is equal to 0 for both sides of the equation.

Einstein's Summation Convention 74.1. In this chapter, we always sum from 1 to n over equal upper and lower indices, unless the contrary is explicitly stated. For example, we have

$$t^{ij}s_j = \sum_{j=1}^{n} t^{ij}s_j.$$

Smoothness Convention 74.2. In this chapter, all functions and manifolds are of class C^∞, unless the contrary is explicitly stated.

This last convention is only made for convenience. Actually, many statements are true under much weaker smoothness assumptions. Often C^k-functions suffice with $k = 1, 2$.

In order to present the key ideas of tensor calculus as clearly as possible, we begin with the most important relations. Later on, we will give a more detailed exposition. Tensor calculus is based on the following two well-known transformation rules for partial derivatives and differentials

$$D_{i'}f = A_{i'}^i D_i f, \tag{4}$$

$$du^{i'} = A_i^{i'} du^i, \tag{5}$$

where $D_i = \partial/\partial u^i$ and $A_j^i = \partial u^i/\partial u^j$. Here u^i and $u^{i'}$ denote different coordinate systems (see (10)). Also, we essentially invoke the identity

$$\delta_{j'}^{i'} = A_i^{i'} A_{j'}^j \delta_j^i, \tag{6}$$

where δ_j^i is the Kronecker symbol, i.e., $\delta_j^i = 1$ for $i = j$ and $\delta_j^i = 0$ for $i \neq j$. In fact, the chain rule implies

$$\frac{\partial u^{i'}}{\partial u^i} \frac{\partial u^j}{\partial u^{j'}} \delta_j^i = \begin{cases} 1 & \text{for } i' = j', \\ 0 & \text{for } i' \neq j'. \end{cases}$$

The development of differential calculus makes essential use of the following formulas:

$$\nabla_i f = D_i f,$$

$$\nabla_i t^j = D_i t^j + \Gamma_{is}^j t^s, \tag{7}$$

$$\nabla_i t_j = D_i t_j - \Gamma_{ij}^s t_s,$$

where Γ_{is}^j are the so-called Christoffel symbols. They also play an important role in the general theory of manifolds in connection with the definition of

74.2. Covariant and Contravariant Tensors

parallel transport and the geodesics. Of central importance are:
(i) The defining equation (23) for the Christoffel symbols.
(ii) The differential equation (30) for the parallel transport.
(iii) The differential equation (35) for straight lines in arbitrary coordinate systems in $\mathbb{R}^n$, which immediately yields the differential equation for geodesics on manifolds.

In Sections 74.2–74.8 we develop the calculus for $\mathbb{R}^n$. This then immediately yields the more general calculus for real n-dimensional manifolds.

74.2. Covariant and Contravariant Tensors

Let us make the following conventions for Sections 74.2–74.8.

(A) Let G be an open, nonempty subset of $\mathbb{R}^n$. Let x denote the points in G as well as the corresponding radius vectors (Fig. 74.3). A coordinate system for G is a C^∞-diffeomorphism $u = u(x)$, which maps G onto an open set in $\mathbb{R}^n$. Its inverse is denoted by $x = x(u)$. This way, one assigns the coordinates $u = (u^1, \ldots, u^n)$ to a point x. The curve $x = x(u)$ in G, which is obtained for variable u^i and fixed u^j for all $j \neq i$, is called u^i-coordinate line. Since $x = x(u)$ is a diffeomorphism, the inverse mapping theorem of Section 4.13 implies that the vectors

$$b_i = \frac{\partial x(u)}{\partial u^i} \tag{8}$$

are linearly independent. We call $\{b_i\}$ the natural basis at the point $x(u)$. Note that this basis depends on the choice of the point x as well as the coordinate system. Obviously, b is a tangent vector for the u^i-coordinate line at the point x (Fig. 74.3). If we choose a Cartesian coordinate system in $\mathbb{R}^n$, i.e., $x = u^i e_i$, and the vectors $e_1, \ldots, e_n$ form a positively oriented coordinate system, then $b_i = e_i$. In the physical literature one often uses in place of b_i the unit vectors $b_i/|b_i|$. This, however, destroys much of the elegance of tensor calculus.

EXAMPLE 74.3 (Polar Coordinates). We choose a fixed, positively oriented orthonormal system $\{e_1, e_2\}$ in $\mathbb{R}^2$ and define $x = \xi^1 e_1 + \xi^2 e_2$. Through

$$\xi^1 = r \cos \varphi, \qquad \xi^2 = r \sin \varphi, \tag{9}$$

we introduce polar coordinates in the usual way (Fig. 74.4). Here we have $u^1 = r, u^2 = \varphi$. The u^i-coordinate lines are rays for $i = 1$ and circles for $i = 2$. Moreover, $b_1 = \partial x/\partial r$ and $b_2 = \partial x/\partial \varphi$, hence

$$b_1 = \cos \varphi \, e_1 + \sin \varphi \, e_2, \qquad b_2 = -r \sin \varphi \, e_1 + r \cos \varphi \, e_2.$$

In order to precisely obtain situation (A), we let r and φ vary in the corresponding regions $0 < r < \infty$ and $-\pi < \varphi < \pi$.

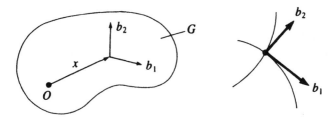

Figure 74.3

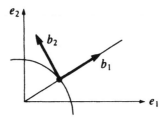

Figure 74.4

Again we consider situation (A). If $v = v(x)$ is another coordinate system in G, then the corresponding coordinates are denoted for convenience by

$$v = (u^{1'}, \ldots, u^{n'}).$$

From $u = u(x)$ and $v = v(x)$ we obtain the map $u = u(v)$ and its inverse $v = v(u)$ by eliminating x. The corresponding partial derivatives

$$A_i^{i'}(x) = \frac{\partial u^{i'}}{\partial u^i}(u(x)),$$

$$A_{i'}^{i}(x) = \frac{\partial u^{i}}{\partial u^{i'}}(v(x))$$

(10)

are called the transformation coefficients. They will essentially be used in the definition of tensors. For simplicity in notation we will skip the argument x, i.e., we simply write $A_{i'}^{i}$, t^i instead of $A_{i'}^{i}(x)$, $t^i(x)$, etc. The tensorial transformation rules (11)–(13), below, therefore actually depend on x. The following definition is fundamental.

Definition 74.4 (Tensors). Assume (A), and let all indices run from 1 to n.

A *scalar field* f on G is a function $f: G \to \mathbb{R}$, which assigns a real number to each point $x \in G$, independently of the coordinate system in G.

A *covariant* tensor field t_i on G is an assignment of a tuple $\{t_i\}$ of real numbers to each point $x \in G$, depending on the coordinate system in G. Passing to another coordinate system, these numbers are transformed like the *deriva-*

74.2. Covariant and Contravariant Tensors

tives $D_i f$ in (4), i.e.,

$$t_{i'} = A_{i'}^i t_i. \tag{11}$$

Analogously a *contravariant* tensor field t^i is defined. Instead of (11), we require that the t^i are transformed like the *differentials* du^i in (5), i.e.,

$$t^{i'} = A_i^{i'} t^i. \tag{12}$$

An *r-fold covariant* and *s-fold contravariant* tensor field

$$t_{i_1 \ldots i_r}^{j_1 \ldots j_s}$$

on G is an assignment of a tuple $\{t_{i_1 \ldots i_r}^{j_1 \ldots j_s}\}$ of real numbers to each point $x \in G$, depending on the coordinate system in G. Passing to another coordinate system, these numbers are transformed like the corresponding product

$$t_{i_1} t_{i_2} \ldots t_{i_r} t^{j_1} \ldots t^{j_s}$$

according to (11) and (12). The number $r + s$ is called the *degree* of the tensor field. For example,

$$t_{i'} t^{j'} = A_{i'}^i A_j^{j'} t_i t^j,$$

hence

$$t_{i'}^{j'} = A_{i'}^i A_j^{j'} t_i^j. \tag{13}$$

A one-fold or two-fold covariant tensor field is also called a simple or twice covariant tensor field, etc.

Two tensor fields are called of the same *type* if and only if they have the same number of corresponding upper and lower indices. For instance, t_{jk}^i and s_{jk}^i are of third degree and of the same type, while t_{jk}^i and s_i^{jk} are of third degree but not of the same type.

According to our Smoothness Convention 74.2, let $t_{\ldots}^{\ldots}$ denote a C^∞-function on G with fixed indices. For simplicity, we will often speak of tensors instead of tensor fields.

Remark. Our definition of tensors naturally arises if one studies physical processes of measurement. These depend on the choice of a coordinate system. Here we need rules which transform one coordinate system or system of reference into another. It is, for example, an important task in physics to decide how physical observables—such as space, time, velocity, electric and magnetic field strengths—change when passing to another system of reference. Einstein's special theory of relativity, for example, shows that the transformation rules of classical mechanics for space, time, and velocity are no longer valid for high velocities, i.e., velocities which are close to the velocity of light. Scalar fields will play an important role, since at some fixed point they have the same numerical value for each coordinate system. Such values are also called invariants.

EXAMPLE 74.5. Let $f: G \to \mathbb{R}$ be a function. Then, from (4), the partial derivatives $t_i = D_i f$ form the standard example of a covariant tensor field.

EXAMPLE 74.6. If $w = w(x)$ is a vector field on G, which is decomposed at the point x with respect to the natural basis $\{b_i\}$, then

$$w(x) = w^i b_i,$$

and the components w^i provide the standard example of a contravariant tensor field in G. In fact, it follows from (8) and the chain rule that

$$b_{i'} = \frac{\partial x}{\partial u^{i'}} = \frac{\partial x}{\partial u^i} \frac{\partial u^i}{\partial u^{i'}},$$

thus

$$b_{i'} = A^i_{i'} b_i. \tag{14}$$

This yields

$$w = w^i b_i = w^i A^{i'}_i b_{i'}.$$

Moreover, we have the representation $w = w^{i'} b_{i'}$ for the new components. Since the $b_{i'}$ are linearly independent, we obtain $w^{i'} = A^{i'}_i w^i$.

EXAMPLE 74.7 (Unit Tensor). Let $i, j = 1, \ldots, n$. We assign to each point $x \in G$ and every u^i-coordinate system, the numbers

$$\delta^j_i = \begin{cases} 1 & \text{for } i = j, \\ 0 & \text{for } i \neq j. \end{cases}$$

Then δ^j_i is a simple covariant and simple contravariant tensor. This remarkable fact immediately follows from (13) and (6).

As shown in the following example, there exists no covariant tensor field g_{ij} which is equal to the Kronecker symbol in each coordinate system. Thus in tensor calculus we will *not* use δ_{ij} for the Kronecker symbol.

The tensor δ^i_j is called the *unit tensor* because it is the unit element for the tensor multiplication of Section 74.3.

EXAMPLE 74.8 (Metric Tensor). Set $b_i b_j = \langle b_i | b_j \rangle$. The quantities

$$g_{ij} = b_i b_j \tag{15}$$

form a twice covariant tensor field, which is called the metric tensor field. In fact, it follows immediately from (14) that $g_{i'j'} = A^i_{i'} A^j_{j'} g_{ij}$.

We have $g_{ij} = g_{ji}$. The linear independence of the b_i implies that $\det(g_{ij}) \neq 0$. In Cartesian coordinates we have $b_i = e_i$, i.e., g_{ij} is equal to the Kronecker symbol. This property, however, is generally lost in arbitrary coordinate systems.

EXAMPLE 74.9 (Inverse Metric Tensor). Let g^{ij} denote the elements of the inverse matrix to (g_{ij}) i.e.,

$$g_{is} g^{sj} = \delta^j_i \tag{16}$$

74.3. Algebraic Tensor Operations

in an arbitrary coordinate system. Then g^{ij} is a twice contravariant tensor field and, moreover,

$$g^{ij} = g^{ji}. \tag{17}$$

PROOF. If we write $g_{i'j'} = A_{i'}^i A_{j'}^j g_{ij}$ as a matrix equation

$$G' = A^T G A,$$

we obtain $G'^{-1} = A^{-1} G^{-1} (A^{-1})^T$. From (6) it follows then that $g^{i'j'} = A_i^{i'} A_j^{j'} g^{ij}$. $\square$

EXAMPLE 74.10 (Natural Basis b_i and Natural Dual Basis b^i). In Definition 74.4 we gave a definition of tensors as number tuples which satisfy a certain transformation behavior. It should be noted that various other objects exhibit this same transformation behavior. For instance, we know already that the vectors b_i of the natural basis at some point x behave like the components of a simple covariant tensor field under coordinate changes, i.e.,

$$b_{i'} = A_{i'}^i b_i. \tag{18}$$

Let $V_n = \text{span}\{b_1, \ldots, b_n\}$. For $i = 1, \ldots, n$ we define linear functionals $b^i \in V_n^*$ through $b^i(b_j) = \delta_j^i$. Then:

$$b^{i'} = A_i^{i'} b^i, \tag{19}$$

i.e., the linear functionals b^i are transformed like the components of a simple contravariant tensor field. In fact, we have $b^i(v^j b_j) = v^i$ and

$$b^{i'}(v^j b_j) = b^{i'}(v^{j'} b_{j'}) = v^{i'} = A_i^{i'} v^i = A_i^{i'} b^i(v^j b_j).$$

74.3. Algebraic Tensor Operations

The following propositions are immediate consequences of the linearity of the tensor definition and identity (6).

Proposition 74.11. *The sum of tensor fields of equal type is a tensor field of the same type.*

The product of tensor fields is again a tensor field where the type is given by the indices.

If a tensor field vanishes at some point with respect to a fixed coordinate system, then the same is true for every coordinate system.

PROOF. For example, it immediately follows from

$$t^{i'} = A_i^{i'} t^i \quad \text{and} \quad s^{i'} = A_i^{i'} s^i$$

that

$$t^{i'} + s^{i'} = A_i^{i'}(t^i + s^i),$$

$$t^{i'} s^{j'} = A_i^{i'} A_j^{j'}(t^i s^j),$$

i.e., $t^i + s^i$ or $t^i s^j$ is a simple contravariant or twice contravariant tensor field. Moreover, from $t^i(x) = 0$, it always follows that $t^{i'}(x) = 0$. □

An index is called *free* if it is not summed.

Proposition 74.12 (Contraction). *If one sums over one upper and lower index of a tensor field, then one obtains a tensor field whose type is given by the free indices.*

This operation is called a contraction.

PROOF. This is an immediate consequence of (6). If, for instance,

$$t^{i'}_{j'} = A^{i'}_i A^j_{j'} t^i_j,$$

then it follows at once from (6) that

$$t^{i'}_{i'} = \delta^j_i t^i_j = t^i_i,$$

i.e., $t^i_i = t^1_1 + \cdots + t^n_n$ is a scalar field. □

The contraction t^i_{ikm} with respect to i, for example, makes the tensor field t^i_{jkm} into a twice covariant tensor field. The free indices are here k and m.

Proposition 74.13. *Permutations of upper (or lower) indices of a tensor field do not change the type of the tensor field.*

PROOF. For example, it follows immediately from $t_{i'j'} = A^i_{i'} A^j_{j'} t_{ij}$ and $s_{ij} = t_{ji}$ that

$$s_{i'j'} = A^i_{i'} A^j_{j'} s_{ij}.$$
□

A tensor field $t_{i_1 \ldots i_r}$ is called *symmetric* (or *antisymmetric*) if and only if $t_{\ldots}$ remains unchanged under permutations of the indices (or is multiplied with the sign of the permutation). Let, for example, t_{ij} be a tensor field. Then one obtains from Propositions 74.11 and 74.13 a symmetric and antisymmetric tensor field through $s_{ij} = t_{ij} + t_{ji}$ and $a_{ij} = t_{ij} - t_{ji}$, respectively. Analogously, $t_{i_1 \ldots i_r}$ can be symmetrized and antisymmetrized. Here we use the notations

$$\text{Alt } t_{i_1} \ldots t_{i_r} = \frac{1}{r!} \sum_\pi (\text{sgn } \pi) t_{\pi(i_1 \ldots i_r)},$$

and

$$\text{Sym } t_{i_1} \ldots t_{i_r} = \frac{1}{r!} \sum_\pi t_{\pi(i_1 \ldots i_r)}.$$

74.4. Covariant Differentiation

One sums over all permutations π of the indices. For instance, $\text{Alt } t_{ij} = (t_{ij} - t_{ji})/2$ and $\text{Sym } t_{ij} = (t_{ij} + t_{ji})/2$.

Definition 74.14. The index picture for an equation is called *right* if and only if the free indices are the same for all additive terms. Permutations are allowed.

Principle of the Index Picture 74.15. All tensor operations above automatically yield the right form, if one only uses terms where the index picture is right.

EXAMPLE 74.16. If t_{ij}, s_{ij} and w^k are tensor fields, then

$$c_{ij} = t_{ij} + s_{ij}, \qquad c_{ij} = t_{ij} + s_{ji},$$
$$c_{ijkm} = t_{ij} s_{km}, \qquad c_i = t_{ij} w^j,$$

are also tensor fields. On the other hand, the index picture in $t_{ij} + w^k$ is not right since the additive terms t_{ij} and w^k contain different free indices. In fact, $t_{ij} + w^k$ is not a tensor field.

74.4. Covariant Differentiation

If t_j is a tensor field, then the partial derivatives $D_i t_j$ do not form a tensor field. In order to make them into a tensor field we need to add additional terms.

Definition 74.17. We define *covariant differentiation* of a scalar field f or a simple covariant tensor field t_j and a simple contravariant tensor field t^j through

$$\nabla_i f = D_i f, \tag{20}$$
$$\nabla_i t_j = D_i t_j - \Gamma_{ij}^s t_s, \tag{21}$$
$$\nabla_i t^j = D_i t^j + \Gamma_{is}^j t^s. \tag{22}$$

The Γ_{ij}^k are the so-called *Christoffel symbols* which are defined through

$$\Gamma_{ij}^k = g^{km} \Gamma_{m,ij},$$
$$\Gamma_{m,ij} = \tfrac{1}{2}(D_i g_{mj} + D_j g_{mi} - D_m g_{ij}). \tag{23}$$

The intuitive meaning of these symbols is explained in Problem 74.4. Because of $g_{ij} = g_{ji}$ we have

$$\Gamma_{m,ij} = \Gamma_{m,ji} \quad \text{and} \quad \Gamma_{ij}^k = \Gamma_{ji}^k \quad \text{for all } i, j.$$

The definition of ∇_i is very simple. This is clear if one looks at the summation index s and the index picture in (21) and (22). The Christoffel symbols were introduced in 1869 by Elwin Bruno Christoffel (1829–1900).

Proposition 74.18. *The covariant differentiation* (20)–(22) *defines tensor fields where the type is given by the index picture.*

PROOF. From the transformation rules

$$g_{i'j'} = A^i_{i'} A^j_{j'} g_{ij}, \qquad g^{i'j'} = A^{i'}_i A^{j'}_j g^{ij},$$

we obtain

$$\Gamma^{k'}_{i'j'} = A^i_{i'} A^j_{j'} A^{k'}_k \Gamma^k_{ij} + (D_{i'} A^s_{j'}) A^{k'}_s \qquad (24)$$

by differentiation. This is the central formula. It shows that the Γ^k_{ij} are *not* transformed tensorial but, instead, an additional term appears. From (24) and the transformation formulas for t_j and t^j, a straightforward computation yields the tensorial transformation formulas for (21) and (22). We have, for example,

$$\nabla_{i'} t_{j'} = A^i_{i'} A^j_{j'} \nabla_i t_j. \qquad (25)$$
□

A geometric motivation for the covariant differentiation follows in Section 74.6.

The definition of covariant differentiation for general tensor fields uses the following rule.

(i) Besides the given tensor field $t^{j_1 \ldots j_s}_{i_1 \ldots i_r}$, one considers the corresponding product $t_{i_1} \ldots t_{i_r} t^{j_1} \ldots t^{j_s}$.
(ii) One computes the covariant derivative of the product by formally applying the product rule and (21), (22).
(iii) The covariant derivative of the given tensor field in (i) is defined analogously to (ii).

We obtain, for example,

$$\nabla_i t_j t^k = (\nabla_i t_j) t^k + t_j \nabla_i t^k$$
$$= D_i(t_j t^k) - \Gamma^s_{ij} t_s t^k + \Gamma^k_{is} t_j t^s.$$

Thus we define

$$\nabla_i t^k_j = D_i t^k_j - \Gamma^s_{ij} t^k_s + \Gamma^k_{is} t^s_j.$$

Analogously, one obtains

$$\nabla_k t_{ij} = D_k t_{ij} - \Gamma^s_{ki} t_{sj} - \Gamma^s_{kj} t_{is}.$$

Besides the partial derivative, we obtain a term with $\Gamma^{\ldots}_{\ldots}$ for each index of $t^{\ldots}_{\ldots}$ which is formed analogously to (21) and (22). For some practise we recommend Problem 74.2.

This definition guarantees that Proposition 74.18, as well as the sum and product rule, hold for the covariant differentiation of general tensor fields.

Definition 74.19 (Absolute Differential). If τ is a real parameter, then the *absolute τ-derivative* along the curve $u^i = u^i(\tau)$ is defined through

$$\frac{D}{d\tau}(t^{j_1\ldots j_s}_{i_1\ldots i_r}) = \frac{du^k}{d\tau}\nabla_k t^{j_1\ldots j_s}_{i_1\ldots i_r}.$$

Analogously, the *absolute differential* is defined through

$$Dt^{j_1\ldots j_s}_{i_1\ldots i_r} = du^k \nabla_k t^{j_1\ldots j_s}_{i_1\ldots i_r}.$$

These two expressions are transformed in the same way as $t^{j_1\ldots j_s}_{i_1\ldots i_r}$, because from (5), $du^i/d\tau$ and du^i are transformed like a simple contravariant tensor field.

74.5. Index Principle of Mathematical Physics

In order that equations hold in arbitrary coordinate systems, one only has to write them as equations for tensor fields. If, for instance, f and h are scalar fields, i.e., functions, then

$$\nabla_i \nabla_j g^{ij} f = h \tag{26}$$

is an equation for tensor fields because the left- and right-hand sides consist of scalar fields. In Cartesian coordinates one has the following special situation:

(i) The quantities g_{ij} and g^{ij} are equal to the Kronecker symbol, and for $g = \det(g_{ij})$ we have $g = 1$. The Christoffel symbols are identical zero.
(ii) The covariant derivative ∇_i becomes the partial derivative D_i.
(iii) $D/d\tau$ becomes $d/d\tau$.
(iv) The vectors b_i of the natural basis form a positively oriented orthonormal system, i.e., $b_i = e_i$.

Thus (26) becomes in Cartesian coordinates

$$\sum_{i=1}^n D_i^2 f = h. \tag{27}$$

This is the Poisson equation. Hence, (26) is a version of (27) in arbitrary coordinate systems. This represents a general method in mathematical physics. We will formulate a suggestive version.

Index Principle 74.20. Let an equation (E) be given in Cartesian coordinates. One may think, for example, of (27). In order to obtain (E), in a form which holds in arbitrary coordinate systems, write (E) as an equation of tensor fields, i.e., replace the partial derivative D_i with ∇_i and assure that the index picture in the sense of Definition 74.14 is right.

An inverse index principle is formulated in Problem 74.3i.

EXAMPLE 74.21. Instead of
$$D_i A_j \pm D_j A_i = 0$$
one writes
$$\nabla_i A_j \pm \nabla_j A_i = 0.$$

EXAMPLE 74.22. Instead of
$$\text{div } w = D_i w^i$$
with $w = w^i e_i$, one writes $w = w^i b_i$ and
$$\text{div } w = \nabla_i w^i. \tag{28}$$

EXAMPLE 74.23. The equation
$$w = \text{grad } f$$
becomes $w^i = D_i f$ in Cartesian coordinates. The correct tensor form is $w^i = g^{ij} \nabla_j f$. Thus for a scalar field f we obtain
$$\text{grad } f = (g^{ij} \nabla_j f) b_i. \tag{29}$$

Formulas (28) and (29) also follow immediately from the Index Principle 74.20, if b_i is treated formally as a covariant tensor field. The form of curl w in $\mathbb{R}^3$ is given in Example 74.30 using pseudotensors.

In Problem 74.3 we consider equations in elasticity theory as well as the Navier–Stokes equations, and we will give a number of other useful applications of the index principle.

74.6. Parallel Transport and Motivation for Covariant Differentiation

In this section, we consider the differential equation
$$\dot{v}^k + \Gamma_{ij}^k \dot{u}^i v^j = 0 \tag{30}$$
for the parallel transport of a contravariant tensor field v^k along the curve $u^i = u^i(\tau)$. The dot means differentiation with respect to the real curve parameter τ, i.e., $\dot{v}^k = \dot{u}^i D_i v^k$. Equation (30), which is of fundamental importance for all of differential geometry, can also be written as
$$\dot{u}^i \nabla_i v^k = 0, \tag{31}$$
or especially short and elegant as
$$\frac{Dv^k}{d\tau} = 0. \tag{32}$$

Theorem 74.A (Parallel Transport). *Let situation* (A) *of Section* 74.2 *be given, and let C be a curve in the open set G, denoted by* $x = x(\tau)$ *or in coordinates by* $u^i = u^i(\tau)$, $i = 1, \ldots, n$.

The vector field $v = v^k b_k$ *along the curve C is constant if and only if equation* (30) *holds along C.*

PROOF. Being constant in Cartesian coordinates is equivalent to $\dot{v}^k = 0$ along C. Moreover, (32) is a correct tensor equation which, in Cartesian coordinates, becomes $dv^k/d\tau = 0$ and hence $\dot{v}^k = 0$. □

In Problem 74.4, we give a direct but somewhat elaborate proof of this theorem, without using tensor calculus. This may help illustrate the geometric meaning of the Christoffel symbols. According to (31), the importance of covariant differentiation is that it allows a simple description of parallel transport. Analogously to (32), we define the parallel transport of arbitrary tensor fields along C through

$$\frac{D}{d\tau}(t^{j_1 \ldots j_s}_{i_1 \ldots i_r}) = 0. \tag{33}$$

EXAMPLE 74.24. The differential equation for a straight line $u^k = u^k(\tau)$, $k = 1, \ldots, n$ becomes, in arbitrary coordinate systems,

$$\frac{D}{d\tau}\left(\frac{du^k}{d\tau}\right) = 0. \tag{34}$$

This is equivalent to

$$\ddot{u}^k + \Gamma^k_{ij} \dot{u}^i \dot{u}^j = 0. \tag{35}$$

PROOF. Equation (34) is a correct tensor equation because it has the same transformation behavior as (32). In Cartesian coordinates, all the Γ^k_{ij} are equal to zero, i.e., (34) and (35) become $\ddot{u}^k = 0$. □

The great importance of formulas (30) and (35) is that they allow us to define parallel transport and generalized straight lines (geodesics) in the same way for Riemannian manifolds and more general manifolds with affine connection. This will be discussed in Sections 74.18 and 74.19.

74.7. Pseudotensors and a Duality Principle

We assume again situation (A) of Section 74.2. Let $D = \det(A^i_{i'})$ denote the Jacobian determinant. For tensor fields t^i, t_{ij}, etc., we have the transformation rules

$$t^{i'} = \alpha A^{i'}_i t^i, \qquad t_{i'j'} = \alpha A^i_{i'} A^j_{j'} t_{ij} \tag{36}$$

with $\alpha = 1$. If (36) holds with $\alpha = \text{sgn } D$, then we speak of pseudotensor fields. Here D is chosen at the corresponding point.

Definition 74.25. A pseudotensor field $t_{i_1 \ldots i_r}^{j_1 \ldots j_s}$ is transformed like the corresponding tensor field with the exception that the transformation rule is multiplied with $\alpha = \text{sgn } D$ such as in (36).

If $\text{sgn } D = -1$, then the natural basis $\{b_1, \ldots, b_n\}$ has another orientation than $\{b_{1'}, \ldots, b_{n'}\}$. Therefore pseudotensors often occur in connection with orientations.

EXAMPLE 74.26 (Standard Example of a Pseudoscalar). If, as usual, we specify a positive orientation of $\mathbb{R}^n$, and let $s = \pm 1$ denote the orientation of $\{b_1, \ldots, b_n\}$, then s is a pseudoscalar field, since under coordinate changes we have $s' = (\text{sgn } D)s$.

EXAMPLE 74.27 (Standard Example of a Pseudotensor Field). Let $g = \det(g_{ij})$ and let $\varepsilon_{i_1 \ldots i_n}$ denote the sign of the permutation $\binom{1 \ldots n}{i_1 \ldots i_n}$. In particular, for two equal indices, $\varepsilon \ldots$ is equal to zero. For $n = 2$, for example, we obtain $\varepsilon_{12} = -\varepsilon_{21} = 1$ and $\varepsilon_{11} = \varepsilon_{22} = 0$. Then

$$E_{i_1 \ldots i_n} = |g|^{1/2} \varepsilon_{i_1 \ldots i_n},$$
$$E^{i_1 \ldots i_n} = |g|^{-1/2} \varepsilon_{i_1 \ldots i_n},$$

are pseudotensor fields where the type is given by the index picture of E.

PROOF. From $g_{i'j'} = A_{i'}^i A_{j'}^j g_{ij}$ follows $g' = D^2 g$ according to the multiplication rule for determinants. Therefore

$$|g'|^{1/2} = D \, \text{sgn } D |g|^{1/2}.$$

Furthermore, it follows from the definition of determinants that

$$\varepsilon_{i'_1 \ldots i'_n} D = A_{i'_1}^{i_1} \ldots A_{i'_n}^{i_n} \varepsilon_{i_1 \ldots i_n}.$$

This implies

$$E_{i'_1 \ldots i'_n} = \text{sgn } D A_{i'_1}^{i_1} \ldots A_{i'_n}^{i_n} E_{i_1 \ldots i_n}.$$

Analogous arguments apply to $E^{i_1 \ldots i_n}$. □

Proposition 74.28. *The sum of pseudotensor fields of the same type is again a pseudotensor field of this type.*

The product of a pseudotensor fields with a tensor field (or pseudotensor field) is again a pseudotensor field (or tensor field).

PROOF. As in Section 74.3 this follows immediately from the definition. □

74.7. Pseudotensors and a Duality Principle

Thus the Index Principle 74.20 remains valid for pseudotensors. One only has to make sure that only pseudotensors (or only tensors) appear in each additive term.

EXAMPLE 74.29 (Duality Principle). The pseudotensors of Example 74.27 can be used to transform tensors into *dual pseudotensors*. Let, for example, $n = 3$. Then we obtain the pseudotensor field

$$w^i = E^{ijk} t_{jk}, \tag{37}$$

from the tensor field t_{ik}.

If we let $s = 1$ (or $= -1$), if $\{b_1, b_2, b_3\}$ is a right-hand (or left-hand) system, then s is a pseudoscalar and

$$v^i = s E^{ijk} t_{jk} \tag{38}$$

is a contravariant tensor field. More generally, tensor fields $t_{\dots}$ and $t^{\dots}$ generate dual pseudotensor fields through

$$\begin{aligned} p^{i_1 \cdots i_k} &= E^{i_1 \cdots i_n} t_{i_{k+1} \cdots i_n}, \\ p_{i_1 \cdots i_k} &= E_{i_1 \cdots i_n} t^{i_{k+1} \cdots i_n} \end{aligned} \tag{39}$$

with type given by the index picture.

EXAMPLE 74.30 (curl w in $\mathbb{R}^3$). For Cartesian coordinates in $\mathbb{R}^3$, equation

$$\text{curl } w = v$$

is equivalent to $v = v^i e_i$ and $v^i = \varepsilon_{ijk} D_j w^k$, where the sum is taken over j and k. According to the index principle and (38), the correct tensor expression is

$$v^i = E^{ijk} \nabla_j g_{kr} w^r.$$

We therefore obtain in arbitrary coordinates

$$\text{curl } w = s(E^{ijk} \nabla_j w_k) b_i \tag{40}$$

with $w = w^k b_k$ and $w_k = g_{kr} w^r$.

Because of the orientation factor s, curl w is called an axial vector in the physical literature. From Section 74.5 we have for an arbitrary coordinate system in $\mathbb{R}^3$ that

$$\text{grad } f = (\nabla^i f) b_i, \tag{41}$$

$$\text{div } w = \nabla_i w^i \tag{42}$$

with $\nabla^i = g^{ij} \nabla_j$. Since no factor s appears in (41), grad f is called a polar vector in the physical literature.

EXAMPLE 74.31 (Vector Product). Analogously to (40), one obtains for the vector product in arbitrary coordinates

$$v \times w = s(E^{ijk} v_j w_k) b_i. \tag{43}$$

Therefore $v \times w$ is called an axial vector. The scalar product is discussed in Problem 74.3a.

The elegance with which one obtains formulas (40)–(43), without any computations, already illustrates the usefulness of tensor calculus in classical vector analysis.

EXAMPLE 74.32 (Curl of a Vector Field). In $\mathbb{R}^n$ the tensor field

$$t_{ij} = \nabla_i w_j - \nabla_j w_i$$

is called the Curl of w_i. Since the Christoffel symbols cancel out, we also have

$$t_{ij} = D_i w_j - D_j w_i$$

for an arbitrary coordinate system.

The fact that the Christoffel symbols cancel by alternations and that partial derivatives D_i can be used instead of ∇_i is the *key* to the fundamental calculus of alternating differential forms of Section 74.24.

The dual pseudotensor to t_{ij} is given by

$$p^{i_1 \cdots i_{n-2}} = E^{i_1 \cdots i_{n-2} ij} t_{ij}.$$

In $\mathbb{R}^3$, i.e., in the special case $n = 3$, we obtain

$$\operatorname{curl} w = (sp^k)b_k = (sE^{kij}t_{ij})b_k$$

from (40). This formula shows that in $\mathbb{R}^3$ there exists a close relation between the tensor field "Curl" and the vector field "curl." In the general case of $\mathbb{R}^n$, $n > 3$, however, only the tensor field "Curl," i.e., t_{ij} is available.

74.8. Tensor Densities

If g_{ij} is a tensor field on $\mathbb{R}^n$ and h_{ij} is a pseudotensor field, then

$$g_{i'j'} = A_{i'}^i A_{j'}^j g_{ij}, \qquad h_{i'j'} = (\operatorname{sgn} D) A_{i'}^i A_{j'}^j h_{ij}$$

with $A_{i'}^i = \partial u^i / \partial u^{i'}$ and $D = \det(A_{i'}^i)$. Letting $g = \det(g_{ij})$ and $h = \det(h_{ij})$, we obtain from the multiplication rule for determinants that

$$g' = D^2 g, \qquad h' = (\operatorname{sgn} D)^n D^2 h.$$

We call $t_{i_1 \cdots i_r}^{j_1 \cdots j_s}$ a *tensor density* (or pseudotensor density) of weight w if and only if $t_{i_1 \cdots i_r}^{j_1 \cdots j_s}$ is transformed like a tensor field (or pseudotensor field), however, with an *additional* multiplication factor

$$|D|^w \quad (\text{or } \operatorname{sgn} D |D|^w).$$

For a tensor density t_j^i, for example, of weight w, we have the formula

$$t_{j'}^{i'} = |D|^w A_i^{i'} A_{j'}^j t_j^i.$$

It follows from above that g is a scalar density of weight 2, and h is a scalar density (or pseudodensity) of weight 2 for n odd (or n even). Because of

$$\sqrt{|g'|} = |D|\sqrt{|g|},$$

$\sqrt{|g|}$ is a scalar density of weight 1. This density plays the central role in Riemannian geometry in connection with computations of volumina.

If a is a scalar density of weight 1, then the transformation rule for integrals implies that

$$J = \int a\,du^1 \ldots du^n = \int a|D|\,du^{1'} \ldots du^{n'}$$
$$= \int a'\,du^{1'} \ldots du^{n'} = J',$$

i.e., J is a scalar. Analogously to Section 74.3, one can prove the following:

(i) The product of two tensor densities (or pseudotensor densities) is a tensor density where the weights are added.
(ii) The product of a tensor density with a pseudotensor density is a pseudotensor density where again the weights are added.
(iii) Contractions of densities preserve the weights.

74.9. The Two Fundamental Forms of Gauss of Classical Surface Theory

In the following sections we consider the classical local surface theory of Gauss as an elegant application of tensor calculus. It also provides the standard example for a differential geometry on general manifolds. If one specializes the results of Section 74.19 about n-dimensional Riemannian manifolds to $n = 2$, one also obtains a global surface theory.

Our starting point is as follows.

(B) Let x be the radius vector in $\mathbb{R}^3$ and G a fixed region in $\mathbb{R}^2$, i.e., in the (u^1, u^2)-plane. A surface S in $\mathbb{R}^3$ is given by the equation

$$x = x(u), \quad u = (u^1, u^2) \tag{44}$$

on G (Fig. 74.5), where $x\colon G \to \mathbb{R}^3$ is a bijective C^∞-map, which is also an

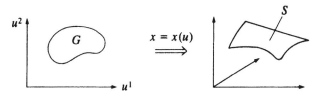

Figure 74.5

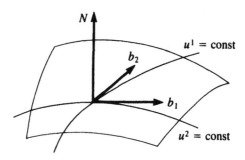

Figure 74.6

immersion. In Cartesian coordinates for $\mathbb{R}^3$ we have $x = \xi^i e_i$, and (44) becomes

$$\xi^i = \xi^i(u), \qquad i = 1, 2, 3.$$

Then $x(\cdot)$ is as C^∞-immersion if and only if all functions $\xi^i(\cdot)$ are C^∞ and

$$\operatorname{rang}(\partial \xi^i / \partial u^j) = 2 \tag{45}$$

holds on G for $i = 1, 2, 3$ and $j = 1, 2$. The *natural basis* $\{b_1, b_2\}$ of the tangent plane at the surface point $x(u)$ is defined as

$$b_i = \frac{\partial x}{\partial u^i}, \tag{46}$$

i.e., b_i is a tangent vector to the u^i-coordinate line (Fig. 74.6). Because of (45), b_1 and b_2 are linearly independent. The unit normal vector N at $x(u)$ is defined through

$$N = b_1 \times b_2 / |b_1 \times b_2|. \tag{47}$$

A study of the trihedral b_1, b_2, N is essential for classical surface theory.

An *admissible* coordinate system for the surface is any C^∞-diffeomorphism $v = v(u)$ which maps G onto a region of $\mathbb{R}^2$. We write $v = (u^{1'}, u^{2'})$. In the following let u^1, u^2 and $u^{1'}, u^{2'}$ denote arbitrary admissible coordinate systems.

Principle of Coordinate Independence 74.33. We consider only such properties of the surface which are independent of the choice of the admissible coordinate system.

As we shall see, this can effectively be done with the aid of tensor calculus on G. Using the vector calculus for $\mathbb{R}^3$, we also make sure that the properties under consideration are invariant under motions of the surface in $\mathbb{R}^3$. In the following, a coordinate system will always be an admissible coordinate system.

Convention 74.34. Up to Section 74.16 all indices run from 1 to 2.

74.9. The Two Fundamental Forms of Gauss of Classical Surface Theory

The following definition is basic.

Definition 74.35 (Fundamental Forms). If $x = x(\tau)$ is a curve on the surface, which will also be denoted as $u^i = u^i(\tau)$, $i = 1, 2$, then, for an arbitrary coordinate system, the following two expressions represent the first and second fundamental form:

$$\dot{x}^2 = g_{ij}\dot{u}^i\dot{u}^j \tag{48}$$

and

$$-\dot{x}\dot{N} = h_{ij}\dot{u}^i\dot{u}^j. \tag{49}$$

The dot means derivative with respect to τ. In short notation one also writes

$$ds^2 = g_{ij}\,du^i\,du^j, \qquad ds^2 \equiv (dx)^2, \tag{48a}$$

$$-dx\,dN = h_{ij}\,du^i\,du^j. \tag{49a}$$

We let $b_{ij} = D_i b_j$ with $D_i = \partial/\partial u^i$, and hence $b_{ij} = D_i D_j x$. Moreover, we let $N_i = D_i N$.

Proposition 74.36. *For an arbitrary coordinate system we have*

$$g_{ij} = b_i b_j, \tag{50}$$

$$h_{ij} = Nb_{ij} = -b_i N_j. \tag{51}$$

Moreover, g_{ij} or h_{ij} is a symmetric tensor field or symmetric pseudotensor field on G, respectively, where the type is given by the index picture.

Corollary 74.37. *Let $g = \det(g_{ij})$ and $h = \det(h_{ij})$. Then $g_{11}, g_{22}, g > 0$ and*

$$g' = D^2 g, \qquad h' = D^2 h \tag{52}$$

with $D = \det(\partial u^i/\partial u^{i'})$. Consequently, h/g is a scalar field on G.

PROOF. It follows from $\dot{x} = b_i \dot{u}^i$ that

$$\dot{x}^2 = b_i b_j \dot{u}^i \dot{u}^j = g_{ij} \dot{u}^i \dot{u}^j.$$

Since $\dot{u}^i$ and $\dot{u}^j$ are arbitrary, we obtain (50).

From $\dot{N} = N_j \dot{u}^j$ it follows immediately that

$$-\dot{x}\dot{N} = -b_i N_j \dot{u}^i \dot{u}^j,$$

hence $h_{ij} = -b_i N_j$. Differentiating $Nb_j = 0$ gives $N_i b_j + Nb_{ij} = 0$, and hence (51).

As usual let $A^i_{i'} = \partial u^i/\partial u^{i'}$, hence $D = \det(A^i_{i'})$. Differentiation gives

$$\dot{u}^{i'} = A^{i'}_i \dot{u}^i, \qquad b_{i'} = A^i_{i'} b_i. \tag{53}$$

Therefore $g_{ij} = b_i b_j$ is a twice covariant tensor field. Since b_1, b_2, N is always a right-hand system, $-\dot{x}\dot{N}$ changes its sign if we pass to a $u^{i'}$-coordinate system with change in the orientation, i.e., sgn $D = -1$. Thus $-\dot{x}\dot{N}$ is a pseudoscalar, i.e.,

$$h_{ij}\dot{u}^i\dot{u}^j = \text{sgn } D h_{i'j'}\dot{u}^{i'}\dot{u}^{j'}.$$

Using the first equation in (53) and noting that the $\dot{u}^i$, $\dot{u}^j$ can be chosen arbitrarily, we obtain

$$h_{ij} = \text{sgn } D A_i^{i'} A_j^{j'} h_{i'j'},$$

i.e., h_{ij} is a twice covariant pseudotensor field. □

PROOF OF COROLLARY 74.37. The vector equation $|b_1 \times b_2|^2 = \det(b_i b_j)$ implies that $\det g > 0$. Equation (52) has been proved in Section 74.8. Therefore we have $h'/g' = h/g$. □

As we shall see, Corollary 74.37 plays an important role in the investigation of the Gaussian curvature.

Definition 74.38. Using the quantities g_{ij} we define the Christoffel symbols Γ_{ij}^k and the covariant differentiation ∇_i as in Section 74.4.

It is important that an application of ∇_i to the tensor fields yields again a tensor field. This can be proved as in Proposition 74.18. There we only needed the fact that the g_{ij} are transformed like a twice covariant tensor field. Hence, the entire tensor calculus is available. We will make essential use of this in the following.

74.10. Metric Properties of Surfaces

We want to show that a knowledge of the first fundamental form, i.e., a knowledge of the metric tensor field g_{ij}, suffices to determine the metric properties of the surface.

Let H be a subregion of G in the (u^1, u^2)-plane. We then define the *surface measure* m of the corresponding surface area through

$$m = \int_H \sqrt{g}\, du^1\, du^2. \tag{54}$$

If $x = x(\tau)$, $\tau_1 \leq \tau \leq \tau_2$ is a curve on the surface, then we define the *arclength* s of this curve between the points $x(\tau_1)$ and $x(\tau)$ through $s(\tau) = \int_{\tau_1}^\tau |\dot{x}(\tau)|\, d\tau$, i.e.,

$$s(\tau) = \int_{\tau_1}^\tau \sqrt{g_{ij}\dot{u}^i\dot{u}^j}\, d\tau. \tag{55}$$

More precisely, we have to write $g_{ij}(u(\tau))\dot{u}^i(\tau)\dot{u}^j(\tau)$. Differentiation with re-

74.10. Metric Properties of Surfaces

spect to τ gives

$$\dot{s}^2 = g_{ij}\dot{u}^i\dot{u}^j.$$

This motivates the notation

$$ds^2 = g_{ij}du^i\,du^j$$

for the first fundamental form.

If $x = y(\tau)$ is another curve on the surface with coordinate representation $u^i = v^i(\tau)$ and if $x = x(\tau)$ and $x = y(\tau)$ intersect for fixed τ, then the angle φ between the two curves satisfies

$$\dot{x}\dot{y} = |\dot{x}||\dot{y}|\cos\varphi.$$

Note that $\dot{x}$ and $\dot{y}$ are the tangent vectors at the point of intersection. Because of $\dot{x} = b_i\dot{u}^i$ and $\dot{y} = b_i\dot{v}^i$ we have

$$\cos\varphi = \frac{g_{ij}\dot{u}^i\dot{v}^j}{\sqrt{g_{ij}\dot{u}^i\dot{u}^j}\sqrt{g_{ij}\dot{v}^i\dot{v}^j}}. \tag{56}$$

More precisely, we must write $g_{ij}(u(\tau))$, $\dot{u}^i(\tau)$, $\dot{v}^i(\tau)$.

Proposition 74.39. *Expressions (54)–(56) do not depend on the particular choice of the coordinate system.*

The advantage of formulas (54)–(56) is that they remain valid for Riemannian manifolds. This will be shown in Section 74.19. The following proof is given in view of this generalization.

PROOF. The transformation formula for integrals yields the invariance of (54), since

$$m = \int \sqrt{g}\,du^1\,du^2 = \int \sqrt{g}|D|\,du^{1'}\,du^{2'} = \int \sqrt{g'}\,du^{1'}\,du^{2'} = m'.$$

According to (53), $\dot{u}^i$ is transformed like a contravariant tensor field. Consequently, $g_{ij}\dot{u}^i\dot{u}^j$ is a scalar, and hence s and $\cos\varphi$ in (55) and (56) are scalars. □

A motivation for (56) has already been given. A motivation for (55) is provided by the approximation formula $\Delta s = |\Delta x|$, i.e.,

$$\Delta s = \left|\frac{\Delta s}{\Delta \tau}\right|\Delta\tau,$$

together with summation and passing to limits.

We now want to motivate (54). To do this, we consider, as in Figure 74.7, a small rectangle R which is parallel to the coordinate axes in the (u^1, u^2)-plane and has the area $\Delta u^1\,\Delta u^2$. The Taylor expansion of $x = x(u)$ shows that in

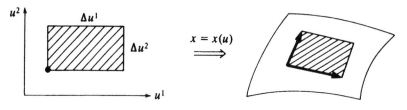

Figure 74.7

first-order approximation R becomes a parallelogram which is spanned by $(\partial x/\partial u^1)\Delta u^1$ and $(\partial x/\partial u^2)\Delta u^2$, i.e., by the vectors $b_1 \Delta u^1$ and $b_2 \Delta u^2$ in the tangent plane. The area of this parallelogram is

$$\Delta m = |b_1 \times b_2|\Delta u^1 \Delta u^2.$$

We have $|b_1 \times b_2|^2 = \det(b_i b_j)$. Consequently,

$$\Delta m = \sqrt{g}\,\Delta u^1 \Delta u^2.$$

Summation and passing to limits gives (54).

74.11. Curvature Properties of Surfaces

We will use the first and second fundamental form to define the curvature of surfaces. Our starting point is the following simple situation.

EXAMPLE 74.40. As in Figure 74.8 we consider a surface

$$\zeta = \zeta(\xi, \eta)$$

which passes through the origin where it has the (ξ, η)-plane as the tangent plane, i.e.,

$$\zeta(0,0) = \zeta_\xi(0,0) = \zeta_\eta(0,0) = 0.$$

The Taylor expansion in a neighborhood of zero is therefore

$$\zeta = a\xi^2 + 2b\xi\eta + c\eta^2 + O_3.$$

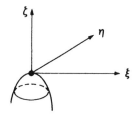

Figure 74.8

74.11. Curvature Properties of Surfaces

Using a rotation around the ζ-axis one can always secure that

$$\zeta = 2^{-1}(\alpha\xi^2 + \beta\eta^2) + O_3. \tag{57}$$

In mathematics, properties of curvatures are always given by the quadratic terms of the Taylor expansion. We define the *Gaussian curvature* K and the *mean curvature* H of the surface at the origin very simply through

$$K = \alpha\beta, \qquad H = (\alpha + \beta)/2. \tag{58}$$

Moreover, we call $R = 1/\alpha$ and $r = 1/\beta$ the *principal curvature radii*, and the ξ-direction and η-direction the *principal curvature directions*. For $\alpha = \beta$ every direction is by definition a principal curvature direction.

Using the tensor calculus for K and H, we are now looking for expressions which are valid for all coordinate systems. To this end we let e_1, e_2, e_3 be the coordinate unit vectors in Figure 74.8.

Then the surface equation takes the form

$$x = \xi e_1 + \eta e_2 + \zeta(\xi, \eta)e_3.$$

Consequently, we have at the origin

$$b_1 = x_\xi = e_1, \qquad b_2 = x_\eta = e_2$$

and $N = e_1 \times e_2 = e_3$. Therefore $g_{ij}(0,0) = e_i e_j$ and

$$h_{ij}(0,0) = N(D_i D_j x) = D_i D_j \zeta(0,0).$$

This gives at the origin

$$\begin{aligned} g_{11} = g_{22} = 1, \qquad g_{12} = g_{21} = 0, \\ h_{11} = \alpha, \qquad h_{22} = \beta, \qquad h_{12} = h_{21} = 0. \end{aligned} \tag{59}$$

For the inverse matrix (g^{ij}) of (g_{ij}) we obtain at the origin $g^{11} = g^{22} = 1$ and $g^{12} = g^{21} = 0$. This gives

$$K = h/g, \qquad H = 2^{-1} g^{ij} h_{ij}. \tag{60}$$

Definition 74.41. The Gaussian curvature K and the mean curvature H at some surface point is defined through (60).

This definition is independent of the choice of the coordinate system. More precisely, H changes its sign under changes in the orientation, because Proposition 74.36 implies that h_{ij} is a pseudotensor, and hence H is a pseudoscalar, and Corollary 74.37 shows that K is a scalar.

Next we want to show that this definition of the Gaussian curvature allows a pleasant geometric interpretation of the sign of K. For this we first show how to compute α and β in an invariant way.

At a fixed surface point P one can always, as in Figure 74.9, introduce a local (ξ, η, ζ)-coordinate system which gives the situation of Example 74.40, where P corresponds to the origin. The (ξ, η)-plane is equal to the tangent

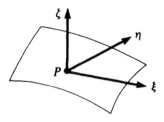

Figure 74.9

plane at P, and the ζ-axis points in the direction of the normal vector N. Then (57) holds locally. Knowing K and H, one obtains α and β as the zeros of the quadratic equation $(\lambda - \alpha)(\lambda - \beta)$, i.e., with (58) as the zeros of

$$\lambda^2 - 2H\lambda + K = 0. \tag{61}$$

From (60) it follows that H changes its sign under changes in the orientation. Thus the same holds true for α and β, i.e., α and β are pseudoscalars. Furthermore, α and β are the *zeros* of the equation

$$\det(\lambda g_{ij} - h_{ij}) = 0. \tag{62}$$

Without any computation this can be seen as follows. If λ is a pseudoscalar, then $t_{ij} = \lambda g_{ij} - h_{ij}$ is a pseudotensor. According to Section 74.8, the left-hand side of (62) is multiplied by D^2 under coordinate changes. Thus the solutions of (62) are independent of the coordinate system. In the special case (59), however, formula (62) has the solutions $\lambda = \alpha, \beta$.

From (57), one now immediately obtains a geometric interpretation for the sign of the Gaussian curvature.

(i) If $K > 0$ at the point P, then $\alpha\beta > 0$. Thus, in a neighborhood of P, the surface lies on one side of the tangent plane at P.
(ii) If $K < 0$ at the point P, then $\alpha\beta < 0$. Thus, in a neighborhood of P, the surface lies on both sides of the tangent plane at P.
(iii) If $K = 0$ and $H = 0$ at the point P, then $\alpha = \beta = 0$. Thus, in a neighborhood of P, the surface behaves like a plane, except for terms of at least order 3.
(iv) If $K = 0$ and $H \neq 0$ at the point P, then $\alpha = 0$ and $\beta \neq 0$ or $\beta = 0$ and $\alpha \neq 0$. Thus, in a neighborhood of P, the surface behaves like a cylinder, except for terms of at least order 3.

EXAMPLE 74.42. In the case when the surface is a plane, no quadratic terms appear in the Taylor expansion, i.e., $\alpha = \beta = 0$ and $K = H = 0$. For a sphere of radius R we have

$$\zeta = R - \sqrt{R^2 - \xi^2 - \eta^2} = 2^{-1}R^{-1}(\xi^2 + \eta^2) + O_3,$$

and hence $\alpha = \beta = H = 1/R$ and $K = 1/R^2$.

If we choose the surface equation for the sphere as $\zeta = \sqrt{R^2 - \xi^2 - \eta^2} - R$, then we obtain $\alpha = \beta = H = -1/R$. This shows that in contrast to K, the signs of α, β, and H have no absolute geometrical meaning.

So far we have used the following convenient strategy.

(S) We choose a special coordinate system in which the problem at hand is most easily treated. Then, by using the index principle of tensor calculus, we obtain results in a form which are valid for arbitrary coordinate systems.

Let us consider another example.

EXAMPLE 74.43 (Lines of Curvature). A line of curvature is a curve

$$u^i = u^i(\tau), \quad i = 1, 2,$$

on the surface where at each point the tangent direction is a principal curvature direction. In Example 74.40, we find at the origin $\dot{\xi} = 0$ or $\dot{\eta} = 0$ for the lines of curvature, or more precisely

$$\dot{\xi}\dot{\eta}(\alpha - \beta) = 0,$$

because for $\alpha = \beta$ every direction is a principal curvature direction. Following the index principle, we write this as

$$E^{ij}g_{ir}h_{js}\dot{u}^r\dot{u}^s = 0, \qquad (63)$$

where $E^{ij} = |g|^{-1/2}\varepsilon_{ij}$. In fact, (59) and (63) imply that $\dot{u}^1\dot{u}^2(\alpha - \beta) = 0$. Equation (63) holds for arbitrary coordinate systems. Thus for $\tau_1 \leq \tau \leq \tau_2$, equation (63) is the differential equation for lines of curvature. More precisely, we would have to write $g_{ir}(u(\tau))$ and $h_{js}(u(\tau))$. Finally, (63) is equivalent to

$$\begin{vmatrix} \dot{u}^1\dot{u}^1 & -\dot{u}^1\dot{u}^2 & \dot{u}^2\dot{u}^2 \\ g_{11} & g_{12} & g_{22} \\ h_{11} & h_{12} & h_{22} \end{vmatrix} = 0.$$

74.12. Fundamental Equations and the Main Theorem of Classical Surface Theory

We take a look at the central problem of surface theory. From a local point of view, which data suffice to uniquely determine a surface? We begin with the so-called fundamental equations of Gauss and Weingarten:

$$D_i b_j = h_{ij} N + \Gamma^k_{ij} b_k \quad \text{(Gauss, 1827)}, \qquad (64a)$$

$$D_i N = -h^j_i b_j \quad \text{(Weingarten, 1861)}. \qquad (64b)$$

These equations describe the change in the local trihedral b_1, b_2, N of the surface. We have $D_i = \partial/\partial u^i$ and

$$h_i^j = g^{js}h_{si}.$$

The Christoffel symbols Γ_{ij}^k, which depend on g_{ij} and their first-order derivatives, have already been defined in (23). Recall $\Gamma_{ij}^k = \Gamma_{ji}^k$. The fundamental equations are the most important equations in surface theory.

By *integrability conditions* for (64) we mean all the equations for g_{ij} and h_{ij}, which follow from (64) by using

$$D_k D_i b_j = D_i D_k b_j \quad \text{and} \quad D_k D_i N = D_i D_k N. \tag{65}$$

There are exactly three equations, which have the explicit form (70), (71) below.

Proposition 74.44. *If assumption* (B) *of Section 74.9 holds, then* (64) *is true for all surface points and coordinate systems.*

PROOF. Ad(64b). Differentiation of $N^2 = 1$ yields $2ND_i N = 0$. Hence

$$D_i N = c_i^j b_j,$$

where the numbers c_i^j remain to be determined. It follows from (51) that $h_{si} = -b_s D_i N$, and hence

$$-h_{si} = c_i^j b_s b_j = c_i^j g_{sj}.$$

From $g^{ms} g_{sj} = \delta_j^m$ follows $c_i^m = -g^{ms} h_{si}$. This is (64b).

Ad(64a). We determine the coefficients α and β in the decomposition

$$D_i b_j = \alpha_{ij}^k b_k + \beta_{ij} N. \tag{66}$$

From (51), multiplication with N gives $\beta_{ij} = N D_i b_j = h_{ij}$.
Furthermore, $g_{sj} = b_s b_j$ implies that

$$D_i g_{sj} = b_j(D_i b_s) + b_s D_i b_j.$$

Interchanging the indices and summation gives

$$b_s D_i b_j = 2^{-1}(D_i g_{sj} + D_j g_{si} - D_s g_{ij}).$$

Thus from (23), $b_s D_i b_j = \Gamma_{s, ij}$. Multiplication of (66) with b_s gives

$$\Gamma_{s, ij} = \alpha_{ij}^k g_{sk}, \quad \text{and thus} \quad \alpha_{ij}^m = g^{ms} \Gamma_{s, ij}.$$

From (23), this means $\alpha_{ij}^m = \Gamma_{ij}^m$. □

Theorem 74.B (Main Theorem of Surface Theory of Bonnet 1867). *Choose the first and second fundamental form in such a way that the integrability conditions* (65) *are satisfied. Then, locally, except for motions, there exists exactly one surface which has these two fundamental forms.*

74.12. Fundamental Equations and the Main Theorem of Classical Surface Theory

Remark. More precisely, this theorem states the following. We are looking for a surface $x = x(u)$ in the sense of (B) of Section 74.9, where u varies in a neighborhood $U(u_0)$ of u_0. To this end we choose real C^∞-functions g_{ij} and h_{ij}, $i, j = 1, 2$ on $U(u_0)$ which satisfy

$$g_{ij} = g_{ji}, \quad h_{ij} = h_{ji}, \quad g_{ii} > 0, \quad \det(g_{ij}) > 0$$

and the integrability conditions (65). Moreover, we choose a point $x(u_0)$ and basis vectors $b_i^0 = D_i x(u_0)$ at $x(u_0)$ with

$$b_i^0 b_j^0 = g_{ij}(u_0) \quad \text{for all } i, j.$$

Also, we define

$$N^0 = b_1^0 \times b_2^0 / |b_1^0 \times b_2^0|.$$

Then there exists precisely one C^∞-surface $x = x(u)$ on a sufficiently small u_0-neighborhood whose fundamental form coincides with g_{ij} and h_{ij}.

As our proof shows, it suffices for the g_{ij} to be C^{k+1} and for the h_{ij} to be C^k with $k \geq 1$. Then $x = x(u)$ is of class C^{k+2}.

This immediately implies Theorem 74.B, because b_1^0, b_2^0, N^0 and $x(u_0)$ are, except for motions in $\mathbb{R}^3$, uniquely determined through $g_{ij}(u_0)$.

In the following proof we make essential use of the theorem of Frobenius of Section 4.12. From this, the initial-value problem

$$D_i w_j = f_{ij}(w, u), \quad w(u_0) = w_0 \tag{67}$$

with $w = (w_1, \ldots, w_m)$ and $i = 1, 2, j = 1, \ldots, m$, has a unique C^{r+1}-solution, $r \geq 1$, in a sufficiently small u_0-neighborhood, if the f_{ij} are real C^r-functions which satisfy the integrability conditions which follow from $D_k D_i w_j = D_i D_k w_j$.

PROOF.

(I) Construction of b_i and N. Consider the initial-value problem

$$b_i(u_0) = b_i^0, \quad N(u_0) = N^0$$

for the fundamental equations (64). Passing to Cartesian coordinates, this is a problem of type (67) and has a unique local solution, because the integrability conditions are satisfied by (65).

(II) Construction of $x = x(u)$. Consider the initial-value problem

$$D_i x = b_i, \quad x(u_0) = x_0,$$

which again is of type (67). The integrability conditions $D_j b_i = D_i b_j$ are satisfied since the right-hand side of (64a) is symmetric. Thus there exists a unique local solution $x = x(u)$.

(III) Uniqueness. Since each surface $x = x(u)$ satisfies the fundamental equations (64), uniqueness follows from (I) and (II).

(IV) Existence. We have to show that the given functions g_{ij} and h_{ij} correspond to the first and second fundamental form of $x = x(u)$.

Let b_i, N be the solutions of (64). We let

$$\alpha = N^2, \quad \beta_i = Nb_i \quad \text{and} \quad \gamma_{ij} = b_i b_j.$$

From (64) together with the product rule for D_k, and from the definition of ∇_k in (21), we obtain the system

$$\nabla_k \alpha = -2h_k^r \beta_r,$$
$$\nabla_k \beta_j = -h_k^r \gamma_{rj} + h_{jk} \alpha, \quad (68)$$
$$\nabla_k \gamma_{ij} = h_{ik} \beta_j + h_{jk} \beta_i,$$

and $\alpha(u_0) = 1$, $\beta_i(u_0) = 0$, $\gamma_{ij}(u_0) = g_{ij}(u_0)$.

From the lemma of Ricci it follows that

$$\nabla_k g_{ij} = 0$$

(see Problem 74.2d). Thus

$$\alpha = 1, \quad \beta = 0, \quad \gamma_{ij} = g_{ij}$$

is a solution of (68). This system is of type (67). Hence the solution is locally unique. Therefore

$$N^2 = 1, \quad Nb_i = 0, \quad b_i b_j = g_{ij},$$

i.e., g_{ij} corresponds to the first fundamental form, and we have $N = (b_1 \times b_2)/|b_1 \times b_2|$. From (64b) follows

$$b_k D_j N = -h_j^i b_i b_k = -g^{im} h_{mj} g_{ik} = -h_{kj},$$

i.e., h_{kj} corresponds to the second fundamental form. □

74.13. Curvature Tensor and the Theorema Egregium

We define the *curvature tensor* R^i_{jkm} for a surface through

$$R^i_{jkm} = D_k \Gamma^i_{mj} - D_m \Gamma^i_{kj} + \Gamma^i_{ks} \Gamma^s_{mj} - \Gamma^i_{ms} \Gamma^s_{kj}, \quad (69)$$
$$R_{ijkm} = g_{is} R^s_{jkm}.$$

In Section 74.18 we demonstrate how the R^i_{jkm} arise in a natural way in connection with the commutability of the covariant differentiation. Here we show how the curvature tensor of surface theory follows automatically from the integrability condition of the fundamental equations. Note that sometimes in the literature R^i_{jkm} is used with the opposite sign. Our definition is chosen so that the general equation (82) below appears in a symmetric form.

74.13. Curvature Tensor and the Theorema Egregium

Proposition 74.45. *The integrability conditions for the fundamental equations are*

$$\nabla_i h_{jk} = \nabla_k h_{ij}, \tag{70}$$

$$R_{ijkm} = h_{ik} h_{jm} - h_{im} h_{jk}. \tag{71}$$

A proof is given below. Equation (70) is called the equation of Mainardi–Codazzi and (71) is called the theorema egregium of Gauss. The tensor property of R_{ijkm} is a consequence of (71), because the right-hand side is a tensor. From $h_{ij} = h_{ji}$ and (71) follows

$$\begin{aligned} R_{ijkm} &= R_{kmij}, \\ R_{ijkm} &= -R_{jikm} = -R_{ijmk}. \end{aligned} \tag{72}$$

Thus R_{2112} is the only essential component and

$$R_{1212} = h_{11} h_{22} - (h_{12})^2 = h = Kg.$$

Thus (70) and (71) only contain three essential equations. Moreover, we obtain the following fundamental theorem.

Theorem 74.C (Theorema Egregium of Gauss (1827)). *The curvature K can be expressed through the metric tensor g_{ij} and its first- and second-order derivatives. More precisely, it is $K = R_{1212}/g$.*

This theorem shows that K is an intrinsic property of the surface, because g_{ij} is determined solely by measurements of length on the surface. For example, the curvature of the surface of the earth may be determined just by land surveying on the earth. We do not need the surrounding space $\mathbb{R}^3$. The theorema egregium was the starting point for Riemann's work (1854) on a theory of curved n-dimensional manifolds, which he developed without using a surrounding space. Riemann's point of view, in turn, was the key to Einstein's general theory of relativity (1916). Gauss discovered his theorem after tedious computations without using the curvature tensor. With the aid of covariant differentiation, our proof now becomes very short.

PROOF OF PROPOSITION 74.45. Letting

$$\nabla_i b_j \stackrel{\text{def}}{=} D_i b_j - \Gamma_{ij}^s b_s$$

similarly as in Section 74.4, we can write the fundamental equations (64) in the elegant form

$$\nabla_i b_j = h_{ij} N, \qquad D_i N = -h_i^j b_j. \tag{73}$$

(I) The equation $D_k D_i b_j = D_i D_k b_j$ is equivalent to

$$\nabla_k \nabla_i b_j - \nabla_i \nabla_k b_j = R_{jik}^m b_m. \tag{74}$$

Here we define $\nabla_k \nabla_i b_j$ in the same way as $\nabla_k t_{ij}$ of Section 74.4. Note that

$\nabla_k \nabla_i b_j = D_k D_i b_j + \cdots$. From (73) we obtain

$$\nabla_k \nabla_i b_j - \nabla_i \nabla_k b_j = (\nabla_k h_{ij} - \nabla_i h_{kj})N + (h_{jk} h_i^m - h_{ji} h_k^m) b_m. \tag{75}$$

Comparison of (74) with (75) gives (70) and (71).
(II) A short computation shows that

$$D_k D_i N = D_i D_k N$$

is equivalent to (70). This follows from (73) and the lemma of Ricci $\nabla_k g_{ij} = 0$ (Problem 74.2d). □

74.14. Surface Maps

The following considerations are important for applications in cartography. Let

$$x = x(u)$$

be a surface S which satisfies assumption (B) of Section 74.9. Consider a bijective map

$$\tilde{x} = \tilde{x}(x)$$

from S onto another surface $\tilde{S}$. Then $\tilde{S}$ can be written in the form

$$x = \tilde{x}(u)$$

and again we assume (B) for $\tilde{S}$. The map is called length preserving (or area preserving) if and only if the length of curves (or the surface area of subsets) is preserved. The preservation of angles means that the angle of intersection between curves is preserved. The metric tensor of S and $\tilde{S}$ is denoted by g_{ij} and $\tilde{g}_{ij}$. Recall that $g = \det(g_{ij})$.

Proposition 74.46 (Characterization of Maps). *The map is length preserving (or area preserving) if and only if*

$$g_{ij}(u) = \tilde{g}_{ij}(u)$$

(or $g(u) = \tilde{g}(u)$) for all u and $i, j = 1, 2$.

The map is angle preserving if and only if for each u there exists a number $c(u) > 0$ with

$$g_{ij}(u) = c(u)\tilde{g}_{ij}(u)$$

for all u and $i, j = 1, 2$.

PROOF. This follows easily from (54) to (56). The preservation of area, for example, means that

$$\int_H \sqrt{g}\, du = \int_H \sqrt{\tilde{g}}\, du$$

for all subregions H. This is equivalent to $\sqrt{g} = \sqrt{\tilde{g}}$.

The necessity of this condition for the preservation of angles follows from

(56). One uses coordinates for which $g_{ij}(u)$ is equal to the Kronecker symbol for fixed u, and then looks at special curves. □

Proposition 74.46 immediately yields the following results.

(i) Every length preserving map is also area preserving and angle preserving.
(ii) Every length preserving map preserves the Gaussian curvature K at every point. This follows from the theorema egregium (Theorem 74.C).
(iii) There is no length preserving map which maps parts of the sphere onto the plane. This follows, because K is different for the sphere and the plane.
(vi) For angle preserving maps, i.e., conformal maps, we have

$$ds^2 = c(u) d\tilde{s}^2$$

at each point u. Thus in first-order approximation, the lengths are multiplied by a fixed number in a small neighborhood of u. Therefore conformal maps are also called similar in the small. A geographical chart, which is made from a conformal map, therefore gives, in first-order approximation, a precise picture of a sufficiently small section of the earth in a fixed scale ratio.

74.15. Parallel Transport on Surfaces According to Levi-Civita

Let a curve C of the form $u^i = u^i(\tau)$, $i = 1, 2$ with curve parameter τ be given on a surface. Moreover, let $v = v^k b_k$ be a vector field along C. Analogously to (30), we say that v is parallel along C if and only if $Dv^k/d\tau = 0$, i.e.,

$$\dot{v}^k + \Gamma_{ij}^k \dot{u}^i v^j = 0 \tag{76}$$

along C. The dot means differentiation with respect to τ. This parallel transport allows the following very simple interpretation.

Proposition 74.47. *A vector field v is parallel along C if and only if $\dot{v}$ is always perpendicular to the tangent plane.*

Corollary 74.48. *Let v be parallel along the curve C. Consider v at a fixed point on the curve with parameter τ_0 and move $v(\tau_0)$ parallel along C in $\mathbb{R}^3$. This vector field is denoted by $w = \alpha^j b_j + \beta N$. We obtain*

$$\dot{v}^j(\tau_0) = \dot{\alpha}^j(\tau_0).$$

The last statement means that

$$\Delta v^i = \Delta \alpha^i + o(\Delta \tau) \quad \text{as } \Delta \tau \to 0.$$

One also says briefly: The parallel transport of v along the curve C is obtained through an infinitesimal parallel transport in $\mathbb{R}^3$ and a projection onto the tangent plane.

PROOF. From $\dot{v} = \dot{v}^j b_j + v^j \dot{b}_j$ and $\dot{b}_j = D_i b_j \dot{u}^i$ and the fundamental equations (64) it follows that

$$\dot{v} = (\dot{v}^k + \Gamma^k_{ij} \dot{u}^i v^j) b_k + (\ldots) N.$$

Hence, the proposition follows from (76). □

PROOF OF COROLLARY 74.48. From $\dot{w} = 0$ follows

$$0 = \dot{\alpha}^j b_j + \alpha^j D_i b_j \dot{u}^i + \dot{\beta} N + \beta \dot{N}.$$

Consider this equation for τ_0. It follows that $\beta(\tau_0) = 0$. The fundamental equations (64) yield

$$0 = (\dot{\alpha}^k + \Gamma^k_{ij} \dot{u}^i \alpha^j) b_k + (\ldots) N,$$

hence

$$\dot{\alpha}^k + \Gamma^k_{ij} \dot{u}^i \alpha^j = 0.$$

From $\alpha^j(\tau_0) = v^j(\tau_0)$ and (76) it follows that $\dot{\alpha}^k(\tau_0) = \dot{v}^k(\tau_0)$. □

Using parallel transport one can give a simple interpretation for covariant differentiation on a surface. For this, consider a fixed point on the curve C with parameter τ. Similarly as in Section 74.4, we define

$$\frac{Dt^k}{d\tau} = \dot{u}^i \nabla_i t^k = \dot{t}^k + \Gamma^k_{ij} \dot{u}^i t^j.$$

Let $t = t^i b_i$ and let $v = v^i b_i$ denote the vector field which is obtained from $t(\tau_0)$ by a parallel transport. Then $v(\tau_0) = t(\tau_0)$ and (76) implies

$$\frac{Dt^k(\tau_0)}{d\tau} = \dot{u}^i \nabla_i t^k = \dot{t}^k(\tau_0) - \dot{v}^k(\tau_0).$$

This then gives an intuitive interpretation of $D/d\tau$ and ∇_i on surfaces.

Since (76) depends on the curve $u^i(\tau)$, the parallel transport on surfaces usually depends on the path. More precisely, we obtain for a simply connected surface: The parallel transport is path independent if and only if the curvature tensor is identically zero. From the theorema egregium of Section 74.13 this is equivalent to $K \equiv 0$ (see Problem 74.7).

74.16. Geodesics on Surfaces and a Variational Principle

Section 74.6 provides the motivation for the following definition. A curve C on a surface with parameter representation $u^i = u^i(s)$ is called a *geodesic* if

74.16. Geodesics on Surfaces and a Variational Principle

and only if $D\dot{u}^i/ds = 0$, i.e.,

$$\ddot{u}^k + \Gamma^k_{ij}\dot{u}^i\dot{u}^j = 0, \qquad k = 1, 2 \tag{77}$$

along C. Here s denotes the arclength and the dot means derivative with respect to s. From Section 74.6, this definition generalizes the concept of straight lines in a plane. According to Section 74.15, the geometrical meaning of (77) is the following. The tangent vector field $t = \dot{u}^i b_i$ along C is obtained through a parallel transport.

Now we want to show that (77) is closely related to the variational problem of finding the shortest connecting line between two points on the surface.

$$\int_{s_1}^{s_2} ds = \min!. \tag{78}$$

Because of $\dot{s} = \sqrt{g_{ij}\dot{u}^i\dot{u}^j}$ we rewrite (78) in the form

$$\int_{\tau_1}^{\tau_2} \sqrt{g_{ij}\dot{u}^i\dot{u}^j}\, d\tau = \min!, \tag{79}$$

$$u^i(\tau_1) = u^i_1, \qquad u^i(\tau_2) = u^i_2.$$

One is looking for the shortest line, connecting two points P_1 and P_2 on the surface with corresponding coordinates u^i_1 and u^i_2 (Fig. 74.10).

Proposition 74.49. *If arclength is chosen as parameter, then every solution of the variational problem (79) satisfies (77).*

PROOF. Let $L = \sqrt{G}$ with $G = g_{ij}\dot{u}^i\dot{u}^j$. From Section 58.18 it follows that the Euler differential equations for (79) are equal to

$$\frac{d}{d\tau}\frac{\partial L}{\partial \dot{u}^k} - \frac{\partial L}{\partial u^k} = 0.$$

This means

$$0 = \frac{d}{d\tau}\frac{g_{kj}\dot{u}^j}{\sqrt{G}} - \frac{1}{2\sqrt{G}}\left(\frac{\partial}{\partial u^k}g_{ij}\right)\dot{u}^i\dot{u}^j$$

$$= \frac{1}{2G\sqrt{G}}\left[-\frac{dG}{d\tau}g_{kj}\dot{u}^j + 2G\frac{d}{d\tau}(g_{kj}\dot{u}^j) - G\left(\frac{\partial}{\partial u^k}g_{ij}\right)\dot{u}^i\dot{u}^j\right].$$

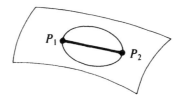

Figure 74.10

Letting $\tau = s$, gives $G = 1$ along the extremal. Using (23), this immediately implies (77). □

74.17. Tensor Calculus on Manifolds

We want to show that *without further thought*, the tensor calculus of $\mathbb{R}^n$ carries over to manifolds. Consider the following situation:

(C) M is a real, n-dimensional C^∞-manifold.

The indices in the following four sections run from 1 to n.

Definition 74.50. A tensor field

$$t^{j_1 \ldots j_s}_{i_1 \ldots i_r}$$

on M is an assignment of a tuple $\{t^{j_1 \ldots j_s}_{i_1 \ldots i_r}\}$ of real numbers to each point x of M in every chart that belongs to x, which is transformed under chart changes like a corresponding tensor field in $\mathbb{R}^n$.

Consequently, $t^{j_1 \ldots j_s}_{i_1 \ldots i_r}$ depends on the point x and on the charts.

Analogously, pseudotensor fields are defined. According to our Smoothness Convention 74.2, we assume that $t^{\ldots}_{\ldots}$ with fixed indices is of class C^∞ with respect to the corresponding chart coordinates.

We illustrate this definition with a simple example. Consider a fixed atlas for M. Let $u = u(x)$ and $v = v(x)$ denote two chart maps which assign coordinates $u = (u^1, \ldots, u^n)$ and $v = (u^{1'}, \ldots, u^{n'})$ to the point x. As before, we let

$$A^{i'}_i = \frac{\partial u^{i'}(u(x))}{\partial u^i} \quad \text{and} \quad A^i_{i'} = \frac{\partial u^i(v(x))}{\partial u^{i'}}.$$

Now, for t^i, we assume that

$$t^{i'} = A^{i'}_i t^i$$

under chart changes from the u-chart to the v-chart.

In a natural way, each tensor field can be extended to all admissible charts. This follows from the chain rule. Consider, for example, t^i. Let $w = w(x)$ be an admissible chart with $w = (u^{1''}, \ldots, u^{n''})$. We define

$$t^{i''} = A^{i''}_i t^i.$$

The chain rule implies

$$t^{i''} = A^{i''}_{i'} A^{i'}_i t^i = A^{i''}_{i'} t^{i'}.$$

Therefore it does not matter if we use $u = u(x)$ or $v = v(x)$ for the extension. Analogously, one shows that the tensorial transformation rule applies to changes in the admissible charts. We always extend tensor fields to the maximal atlas, i.e., to all admissible charts. Two tensor fields on M are called equal if and only if their extensions to the maximal atlas are identical.

Obviously, the algebraic tensor calculus of Section 74.3 applies to manifolds.

Let b_i denote the tangent vector at x with respect to the chart $u = u(x)$, which corresponds in the u-chart to the unit vector e_i in u^i-direction. Then every tangent vector $t \in TM_x$ can be written in the form

$$t = t^i b_i.$$

Let $\{b_1, \ldots, b_n\}$ denote the natural basis of the tangent space TM_x and t^i the natural coordinates of t in the u-chart. Moreover, we define linear, continuous functionals on TM_x through

$$b^j(t) = t^j.$$

From Section 73.23 it follows that

$$t^* = t_i b^i$$

for every $t^* \in TM_x^*$. We call $\{b^1, \ldots, b^n\}$ the natural basis of TM_x^* and t_i the natural coordinates of t^* in the u-chart. From Proposition 73.84, b_i and t_i are transformed like a simple covariant tensor under chart changes. Furthermore, b^i and t^i are transformed like a simple contravariant tensor. Finally, Section 73.23 implies that $b_i = \partial x/\partial u^i$ and $b^i = du^i$.

74.18. Affine Connected Manifolds

We now want to extend covariant differentiation to manifolds. To this end we need the Christoffel symbols.

Definition 74.51. Assume (C) of Section 74.17. The manifold M is called *affine connected* if and only if there exists a tuple $\{\Gamma_{ij}^k\}$ of real numbers for each point $x \in M$ in every chart that belongs to x, which is transformed under chart changes like:

$$\Gamma_{i'j'}^{k'} = A_k^{k'} A_{i'}^i A_{j'}^j \Gamma_{ij}^k + (D_{i'} A_{j'}^s) A_s^{k'}. \tag{80}$$

Consequently, Γ_{ij}^k depends on the point x and on the charts. Here, $D_{i'} = \partial/\partial u^{i'}$.

According to our Smoothness Convention 74.2, we assume that all the Γ_{ij}^k are C^∞-functions in chart coordinates.

As in $\mathbb{R}^n$ and surface theory we now use these Γ_{ij}^k to introduce the following concepts:

(i) Covariant differentiation ∇_i.
(ii) Absolute differentiation $D/d\tau$.
(iii) Parallel transport.
(iv) Affine geodesics as generalized straight lines.

(v) Curvature tensor R^i_{jkm}.
(vi) Torsion tensor $T^k_{ij} \stackrel{\text{def}}{=} \Gamma^k_{ij} - \Gamma^k_{ji}$.

We define ∇_i as in Section 74.4. For example, we set

$$\nabla_i t^j = D_i t^j + \Gamma^k_{is} t^s.$$

Direct computation shows that formula (80) implies the following: If

$$t^{j_1 \ldots j_s}_{i_1 \ldots i_r}$$

is a tensor field on M, then

$$\nabla_i t^{j_1 \ldots j_s}_{i_1 \ldots i_r}$$

is a tensor field on M where the type is given by the index picture.

Let C be a curve $x = x(\tau)$ on M, which in charts has the local representation $u^i = u^i(\tau)$. We define the *absolute derivative* along C as

$$Dt^{\cdots}_{\cdots}/d\tau = \dot{u}^i \nabla_i t^{\cdots}_{\cdots}.$$

A tensor field $t^{\cdots}_{\cdots}$ is called *parallel* along C if and only if

$$Dt^{\cdots}_{\cdots}/d\tau = 0 \quad \text{along } C.$$

The curve C is called an *affine geodesic* if and only if $D\dot{u}^i/d\tau = 0$ along C with respect to a suitable parameter τ, i.e.,

$$\ddot{u}^k + \Gamma^k_{ij} \dot{u}^i \dot{u}^j = 0. \tag{81}$$

The *curvature tensor* R^i_{jkm} is defined as in (69), using the Γ^k_{ij}. A direct computation shows

$$\nabla_k \nabla_m v^i - \nabla_m \nabla_k v^i = R^i_{jkm} v^j - T^j_{km} \nabla_j v^i. \tag{82}$$

Thus, analytically, one obtains the R^i_{jkm} and T^j_{km} in a natural way if one studies the commutability of the covariant differentiation. In $\mathbb{R}^n$ we have $\Gamma^k_{ij} \equiv 0$, and hence $R^i_{jkm} \equiv 0$ and $T^j_{km} \equiv 0$. From (80), T^k_{ij} is a tensor. Since the left-hand side of (82) is a tensor, the same is true for $R^i_{jkm} v^j$. Furthermore, since v^j may be chosen arbitrarily, the inverse index principle implies that R^i_{jkm} is a tensor as well (see Problem 74.3i).

The geometric meaning of R^i_{jkm} and T^k_{ij} is the subject of Theorem 74.D of Section 74.20 and Problem 74.7. A consequence is the following two results for simply connected manifolds M:

(i) The parallel transport in M is path independent if and only if $R^i_{jkm} = 0$ on M.
(ii) M is locally flat if and only if $R^i_{jkm} = 0$ and $T^k_{ij} = 0$ on M.

In the next section, we will introduce so-called Riemannian manifolds.

Each Riemannian manifold is affine connected.

74.19. Riemannian Manifolds

For affine connected manifolds, the concept of curve length is not available. For this we need Riemannian manifolds. The definition of Riemannian manifolds is based on the following formula

$$s_{\pm}(\tau) = \int_{\tau_1}^{\tau} \sqrt{\pm g_{ij}\dot{u}^i\dot{u}^j} \, d\tau \tag{83}$$

for the arclength. Thus we need to know the g_{ij}. From (83) it follows that

$$\dot{s}_{\pm}^2 = \pm g_{ij}\dot{u}^i\dot{u}^j.$$

In short notation, we write

$$ds_{\pm}^2 = \pm g_{ij} \, du^i \, du^j.$$

We set $g = \det(g_{ij})$ and $s = s_+$.

Definition 74.52. Assume (C) of Section 74.17. The manifold M is called a *Riemannian manifold* if and only if there exists a symmetric tensor field g_{ij} on M with $g \neq 0$. If $g > 0$ on M, then we speak of a proper Riemannian manifold.

Riemannian manifolds, which are not proper, are called pseudo-Riemannian manifolds. They occur in the general theory of relativity. One should note that many authors speak of Riemannian manifolds which in our terminology are proper. From convention 74.2 it follows that in chart coordinates g_{ij} is of class C^∞. We call g_{ij} the *metric tensor field*.

Using the g_{ij} we obtain as in $\mathbb{R}^n$ the following concepts:

(i) Curve length.
(ii) Angle between curves.
(iii) Volume.
(iv) The Christoffel symbols Γ_{ij}^k.

If C is a curve in M, which has the local parameter representation $u^i = u^i(\tau)$ in charts, we define the *arclength* through (83). This integral has the following meaning. We decompose the curve into parts so that each part lies in a single chart. As in (83) we integrate over these parts and then sum the results. Since $g_{ij}\dot{u}^i\dot{u}^j$ is a scalar, the result does not depend on the choice of the charts. Because of the additivity of the integral, the result also does not depend on the particular choice of the decomposition. In (83), we choose the sign "+" (or "−"), if $g_{ij}\dot{u}^i\dot{u}^j \geq 0$ (or ≤ 0) along C.

If H is a region in M with a compact closure, then the *volume* of H is defined through

$$m(H) = \int_H \sqrt{|g|} \, du^1 \ldots du^n. \tag{84}$$

This integral is to be understood in the sense of (83), i.e., it follows from a sufficiently small decomposition of H. Such decompositions can conveniently be described by using a partition of unity $\sum_\alpha \varphi_\alpha = 1$ in H (see A_1(12h)). We then let

$$m(H) = \sum_\alpha \int_H \varphi_\alpha \sqrt{|g|}\, du^1 \ldots du^n.$$

From compactness of H, we may assume that each support of φ_α lies in one chart and that the sum $\sum_\alpha$ is finite. Each subintegral is then evaluated in the corresponding chart.

As in Section 74.8, we obtain that the volume $m(H)$ does *not* depend on the choice of the chart coordinates $(u^1, \ldots, u^n)$.

This definition can be extended to arbitrary regions in M if M has a countable basis. In this case, M is paracompact as a locally compact space with a countable basis. Consequently, there exist partitions of unity in M. The sum $\sum_\alpha$, defining $m(H)$ might be an infinite series, so that, in addition, one has to assume convergence. Because of the additivity of the integral in charts, this definition is independent of the particular decomposition. Analogously, we define

$$J = \int_H f \sqrt{|g|}\, du^1 \ldots du^n$$

for a continuous $f: H \to \mathbb{R}$ as

$$J = \sum_\alpha \int_H \varphi_\alpha f \sqrt{|g|}\, du^1 \ldots du^n,$$

where we assume convergence of the corresponding series for $|f|$ if H is not compact. The general construction in A_2(75) shows that $m(H)$ defines a measure m on M. The corresponding Lebesgue integral $\int_H f\, dm$ coincides with J for a continuous f.

The *angle* φ between two curves is defined through (56).

Using the g_{ij}, the Christoffel symbols Γ^k_{ij} can be defined as in (23). A straightforward computation shows that the transformation rule for the g_{ij} implies the transformation rule (80) for the Γ^k_{ij}, which therefore determines an *affine connection*. Hence all the definitions and concepts of the previous section are available for Riemannian manifolds.

In the case of a Riemannian manifold, it follows from the symmetry of g_{ij} and from the definition of Γ^k_{ij} in (23), i.e.,

$$\Gamma^k_{ij} = \tfrac{1}{2} g^{km}(D_i g_{mj} + D_j g_{mi} - D_m g_{ij})$$

with $D_i = \partial/\partial u^i$, that

$$\Gamma^k_{ij} = \Gamma^k_{ji} \quad \text{for all } i, j,$$

and hence
$$T_{ij}^k = 0 \quad \text{for all } i, j,$$
i.e., the *torsion tensor* of a Riemannian manifold vanishes.

Let C be a C^∞-curve in M with $g_{ij}\dot{u}^i\dot{u}^j > 0$ (or < 0) along C. Then C is called a *geodesic* if and only if (81) holds with respect to the special parameter $\tau = s_+$ (or $\tau = s_-$). This definition is more special than the definition of affine geodesics of Section 74.18. The notation $s_\pm$ has been introduced in (83). In the following it is convenient to use s_α with $\alpha = \pm 1$ instead of $s_\pm$.

Proposition 74.53. *Consider the variational problem*
$$\int_{\tau_1}^{\tau_2} \sqrt{\alpha g_{ij} \dot{u}^i \dot{u}^j} \, d\tau = \text{stationary!},$$
$$u^i(\tau_1) = u_1^i, \quad u^i(\tau_2) = u_2^i, \quad i = 1, \ldots, n$$
with given starting point u_1^i and endpoint u_2^i as well as some fixed $\alpha = \pm 1$. Then every solution curve C with $\dot{s}_\alpha > 0$ along C is a geodesic.

This follows analogously as in the proof of Proposition 74.49. Note that Γ_{ij}^k remains unchanged when passing from g_{ij} to $-g_{ij}$. An important application of Proposition 74.53 is discussed in Section 76.2 in connection with the general theory of relativity.

74.20. Main Theorem About Riemannian Manifolds and the Geometric Meaning of the Curvature Tensor

We assume:

(R) Let M be a real, n-dimensional Riemannian C^∞-manifold with metric C^∞-tensor field g_{ij}.

The *Morse-index* m for the metric at the point $x_0 \in M$ is the Morse-index of the quadratic form
$$ds^2 = g_{ij} \, du^i \, du^j,$$
i.e., the number of negative eigenvalues of the matrix $(g_{ij}(x_0))$. According to the classical law of inertia of Sylvester for quadratic forms, m is invariant under coordinate transformations. Using a linear coordinate transformation, one can assume that

(N) $$ds^2 = \sum_{i=1}^n \varepsilon_i (du^i)^2$$

at the point x_0, where

$$\varepsilon_i = \begin{cases} -1 & \text{for } i = 1, \ldots, m, \\ 1 & \text{for } i > m. \end{cases}$$

We call $(\varepsilon_1, \ldots, \varepsilon_n)$ the signature type of the metric at x_0. For a proper Riemannian manifold we have $\varepsilon_i = 1$ for all i. For the general theory of relativity of Chapter 76, the signature type is $(-1, -1, -1, 1)$, where u^1, u^2, u^3 are the space coordinates and u^4 is the time coordinate.

M is called *locally flat* at x_0 if and only if there exists a coordinate system such that (N) holds in a neighborhood of x_0.

Theorem 74.D (Riemann (1861)). *If (R) holds, then M is locally flat at x_0 if and only if the curvature tensor R^i_{jkm} vanishes in a neighborhood of x_0.*

This theorem shows that the R^i_{jkm} are a measure for the deviation of a locally flat metric. Similarly as in the proof of the main theorem of surface theory of Section 74.12, this proof is an application of the theorem of Frobenius, i.e., we solve systems of explicit first-order partial differential equations by checking the integrability conditions. Also, as in Section 74.12, we use the lemma of Ricci. The key to the proof is (I).

PROOF. Since R^i_{jkm} contains the first- and second-order derivatives of g_{ij}, we immediately obtain from $g_{i'j'} = \text{const}$ that $R^{i'}_{j'k'm'} = 0$. The tensor property implies that $R^i_{jkm} = 0$ is a necessary condition.

To prove the sufficiency of $R^i_{jkm} = 0$, we choose a fixed u^i-system. Our goal is to construct a $u^{i'}$-system, in which $g^{i'j'}$ is constant in a neighborhood of x_0. Then also $g_{i'j'}$ is locally constant, and by using a linear transformation, we obtain that (N) holds locally.

(I) The system

$$\nabla_k h_j = 0$$

has a unique solution in a neighborhood of x_0 for any given $h_j(x_0)$ because this system is equivalent to

$$D_k h_j = \Gamma^s_{kj} h_s,$$

and the integrability conditions $D_r D_k h_j = D_k D_r h_j$ are satisfied because of $R^i_{jkm} = 0$.

(II) The function $\varphi = g^{ij} h_i h_j$ is locally constant because the product rule for ∇_k implies that

$$D_k \varphi \equiv \nabla_k \varphi = 0.$$

Note that from the lemma of Ricci,

$$\nabla_k g^{ij} = 0$$

(see Problem 74.2e). Furthermore, note that $\nabla_k h_i = 0$, according to (I).

(III) For h_j in (I), there always exists locally a function u' with
$$D_j u' = h_j,$$
since because of $\Gamma^s_{kj} = \Gamma^s_{jk}$ the integrability conditions $D_k D_j u' = D_j D_k u'$ are satisfied.

(IV) Now we construct the new $u^{i'}$-system through h_j^i with
$$\nabla_k h_j^i = 0 \quad \text{and} \quad D_j u^{i'} = h_j^i,$$
where $\det(h_j^i(x_0)) \neq 0$. Then $g^{i'j'} = g^{ij} D_i u^{i'} D_j u^{j'}$ which, similarly to (II), is locally constant. $\square$

This proof also provides us with a simple analytic approach for the definition of the curvature tensor using the integrability condition in (I). This, in principle, is the same way as was used by Riemann to obtain an analytic expression for the curvature tensor. In his prize memoir for the Academy of Paris, Riemann (1861) studied the problem of locally transforming the equation for the heat conduction equation
$$D_i(g^{ij} D_j f) = f_t$$
by a coordinate transformation into the normal form
$$\sum_{i=1}^{n} \varepsilon_i D_{i'}^2 f = f_t$$
which corresponds to a homogeneous body. One easily verifies that the coefficients g^{ij} are transformed like tensors under coordinate changes. Thus the local vanishing of the R^i_{jkm} is necessary and sufficient for the existence of such a local normal form.

Riemann's work of 1861 may be regarded as the hour of birth for the Riemannian curvature tensor. In fact, this tensor was already implicitly contained in Riemann's habilitation talk (1854) on the hypotheses of geometry. There he generalized the Gaussian surface curvature. This is discussed more thoroughly in Spivak (1979, M), Vol. 2, Chapter 4. This volume also contains seven variants of the proof of Theorem 74.D. It also gives an introduction to the various ways of developing a calculus for differential geometry. They all use the integrability condition, but often in a very implicit form. Bernhard Riemann died in 1866 at the age of 40. His collected works only fill one volume. But his ideas, revealing deep connections between analysis, topology, and geometry, profoundly influenced the mathematics and physics of our century. He was, for example, the first mathematician who discovered global analytic results in connection with his study of complex analytic functions.

74.21. Applications to Non-Euclidean Geometry

One of the most interesting developments in mathematics begins with the parallel axiom of Euclid around 325 B.C., and leads up to Einstein's general theory of relativity and the gauge field theories of our days. Generally, one is concerned with the question: What is geometry, and what is the role played by geometry in understanding the structure of our world?

In his "Elements," in which Euclid gave an axiomatic definition of geometry he postulated for arbitrary points P and straight lines g in the plane:

(P) If P is not on g, then there exists *exactly one* straight line through P which does not intersect g.

Actually, Euclid used another formulation, but for our purposes, version (P) is most convenient. It is contained in the standard book on the axiomatic foundations of geometry of Hilbert (1903), which was published in 1977 in its twelfth edition. Historically, the following question was important: Is (P) a consequence of the other axioms of Euclid or is (P) independent? Today we know that (P) is independent and that there exist elliptic and hyperbolic non-Euclidean geometries where (P) is replaced with (P_{ellip}) and (P_{hyp}):

(P_{ellip}) If P is not on g, then there exists *no* straight line through P which does not intersect g.

(P_{hyp}) If P is not on g, then there are infinitely many straight lines through P which do not intersect g.

In this section we want to show that there exists a natural connection between elliptic, Euclidean, and hyperbolic geometries, which is obtained by using spherical trigonometry and by passing to imaginary spherical radii, i.e., a change from constant Gaussian curvature, $K > 0$ (elliptic), to $K = 0$ (Euclidean), and $K < 0$ (hyperbolic).

During the first half of the nineteenth century, Gauss (1817), and also Janos Bolyai and Lobačevskii around 1830, independently, came to the conclusion that there exist hyperbolic geometries. Gauss, however, did not publish his results because he feared the verdict of small-minded philosophers and mathematicians of his time. Non-Euclidean geometries only became generally accepted after Beltrami (1868), Klein (1871), and Poincaré (1882) constructed simple models for these geometries.

EXAMPLE 74.54 (Elliptic Non-Euclidean Geometry). We choose $M_{\text{ellip}}(R)$ to be equal to the upper half of the surface of a ball of radius R, including the equator where antipodal points of the equator are identified. The "straight lines" on M_{ellip} are the great circles, i.e., the curves of intersection between M_{ellip} and planes passing through the center of the ball. As a "plane" we choose M_{ellip}. Identification of the antipodal points of the equator implies that two straight lines on M_{ellip} intersect at exactly one point and that only one straight line passes through two different points. Thus (P_{ellip}) holds. The geometry on M_{ellip} is the usual spherical geometry. In particular, all straight lines have length πR and the area of the plane is $2\pi R^2$. The Gaussian curvature of the sphere is $K = 1/R^2$.

Interestingly, in the 2000-year-long history of the parallel axiom, no one

74.21. Applications to Non-Euclidean Geometry

came up with the idea of using this simple model for a proof of the independence of (P). The reason for this was probably that it was implicitly assumed that straight lines are of infinite length. This is true for the hyperbolic geometry.

EXAMPLE 74.55 (Hyperbolic Non-Euclidean Geometry According to Poincaré (1882)). Let

$$M_{hyp} = \{(\xi, \eta) \in \mathbb{R}^2 : \eta > 0\}$$

and choose as metric for the upper half plane

$$ds^2 = (d\xi^2 + d\eta^2)/\eta^2. \tag{85}$$

Then M_{hyp} is a two-dimensional, proper Riemannian manifold. The hyperbolic geometry is the corresponding Riemannian geometry. We write

$$ds^2 = g_{ij} du^i du^j$$

with $\xi = u^1$, $\eta = u^2$, and $g_{11} = g_{22} = 1/\eta^2$, $g_{12} = g_{21} = 0$. Measurements of length, area, and angle on M_{hyp} are performed according to formulas (54)–(56). From (56) it follows, in particular, that the measurement of angles on M_{hyp} coincides with the Euclidean measurement of angles. This is one of the advantages of the Poincaré model. The area of the plane M_{hyp} is infinite since

$$\int \sqrt{g}\, du^1\, du^2 = \int_{-\infty}^{\infty} \int_0^{\infty} d\xi\, d\eta/\eta^2 = \infty.$$

As the Christoffel symbols we obtain

$$\Gamma_{11}^2 = -\Gamma_{22}^2 = 1/\eta \quad \text{and} \quad \Gamma_{2i}^i = -1/\eta.$$

All the other Γ_{ij}^k are identically zero. Thus $K = R_{1212}/g = -1$, i.e., M_{hyp} has constant negative Gaussian curvature. The "straight lines" in this geometry are the geodesics. They are most easily computed from the variational problem $\int ds = \min!$, i.e., from

$$\int_{\eta_1}^{\eta_2} \frac{\sqrt{\xi'^2 + 1}}{\eta}\, d\eta = \min!,$$

$$\xi(\eta_i) = \xi_i, \quad i = 1, 2. \tag{86}$$

The Euler equation is here $dL_{\xi'}/d\eta = 0$, i.e.,

$$L_{\xi'} = \xi'/\eta\sqrt{\xi'^2 + 1} = \text{const},$$

hence

$$(\xi - C)^2 + \eta^2 = D^2,$$

where C and D are constants. Therefore, all "straight lines" on M_{hyp} are Euclidean circles with center on the ξ-axis. From Figure 74.11 we obtain (P_{hyp}), because if one chooses a "straight line" g in the open, upper half-plane M_{hyp} and a point P on M_{hyp} which is not on g, then there exist infinitely many "straight lines" which do not intersect g. The boundary points of the upper

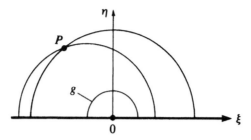

Figure 74.11

half plane, i.e., the points on the ξ-axis play the role of infinitely distant points on M_{hyp}. The "straight lines" on M_{hyp} are of infinite length because

$$\int ds = \int_{\eta_1}^{\eta_2} \frac{\sqrt{\xi'^2 + 1}}{\eta} d\eta \geq \int_{\eta_1}^{\eta_2} \frac{d\eta}{\eta}$$
$$= \ln \eta_2 - \ln \eta_1 \to \infty \quad \text{as} \quad \eta_1 \to 0.$$

The geometry on M_{hyp} admits a simple physical interpretation. The variational problem (86) corresponds to Fermat's principle of Example 37.4 with index of refraction equal to $n(\xi, \eta) = c/\eta$, i.e., we think of the upper half plane as filled with a medium of index with refraction equal to n. Then (86) states that light tries to get from one point to another in the shortest possible time. The geodesics are the light rays (half circles) as shown in Figure 74.11. The geodesic distance between two points $\int ds$ is the time t which is needed by the light to get from one point to another.

EXAMPLE 74.56 (Natural Connection Between Elliptic, Euclidean, and Hyperbolic Geometry). As in Example 74.54, we introduce geodesic polar coordinates φ, ρ on $M_{\text{ellip}}(R)$. Here φ is the geographical length and ρ is the distance between the corresponding point on the sphere and the North pole. Then

$$ds^2 = d\rho^2 + R^2 \sin^2(\rho/R) d\varphi^2. \tag{87}$$

Note that $\rho = R\vartheta$ where ϑ is the geographic latitude measured from the North pole.

On M_{hyp} we choose $\varphi = \xi$ as geodesic polar coordinates and

$\rho = $ distance from $(0, 0)$ in the metric for M_{hyp}.

After some computations this yields

$$ds^2 = d\rho^2 + \sinh^2 \rho \, d\varphi^2. \tag{88}$$

In usual polar coordinates for the Euclidean plane we get

$$ds^2 = d\rho^2 + \rho^2 d\varphi^2. \tag{89}$$

Now note the remarkable fact that, for $R = i$, equation (88) follows from (87) and (89) follows from (87) as $R \to \infty$. This is a special case of the following

74.21. Applications to Non-Euclidean Geometry

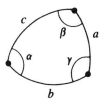

Figure 74.12

interesting result:

(L) If one has the formulas for the spherical geometry for a sphere of radius R, then for $R = i$ one gets the corresponding formulas for M_{hyp} and for $R \to \infty$ the formulas for the Euclidean geometry.

Therefore, M_{hyp} is often called a sphere with imaginary radius $R = i$. From

$$K = 1/R^2$$

on $M_{ellip}(R)$, for example, one obtains $K = 0$ as $R \to \infty$ and $K = -1$ on M_{hyp}. This coincides with Example 74.55. As another example we consider a triangle on $M_{ellip}(R)$ with angles α, β, γ and lengths a, b, c of the opposite sides (compare Fig. 74.12). Then

$$\alpha + \beta + \gamma = \pi + S/R^2,$$

where S is the area of the triangle. This implies

$$\alpha + \beta + \gamma = \pi$$

for the Euclidean geometry and

$$\alpha + \beta + \gamma = \pi - S$$

on M_{hyp}. In what follows, we list a number of basic formulas which, respectively, hold for the elliptic geometry on $M_{ellip}(R)$ with $R = 1$, the Euclidean geometry, and the hyperbolic geometry on M_{hyp}. All these results are obtained from (L). Note that the formulas for $M_{ellip}(R)$ follow from the formulas for $M_{ellip}(1)$ below, by dividing the quantities L, ρ, a, b, c (or S) by R (or R^2).

Circumference (ρ = radius of the circle):

$L = 2\pi \sin \rho$ (elliptic),

$L = 2\pi \rho$ (Euclidean),

$L = 2\pi \sinh \rho$ (hyperbolic).

Area of the circle:

$S = 2\pi(1 - \cos \rho),$

$S = \pi \rho^2,$

$S = 2\pi(\cosh \rho - 1).$

Pythagorean theorem:
$$\cos a = \cos b \cos c,$$
$$a^2 = b^2 + c^2,$$
$$\cosh a = \cosh b \cosh c.$$

Law of sines:
$$\sin \alpha : \sin \beta : \sin \gamma = \sin a : \sin b : \sin c,$$
$$\sin \alpha : \sin \beta : \sin \gamma = a : b : c,$$
$$\sin \alpha : \sin \beta : \sin \gamma = \sinh a : \sinh b : \sinh c.$$

Law of cosines:
$$\cos \alpha = \frac{\cos a - \cos b \cos c}{\sin b \sin c},$$
$$\cos \alpha = \frac{b^2 + c^2 - a^2}{2bc},$$
$$\cos \alpha = \frac{\cosh a - \cosh b \cosh c}{\sinh b \sinh c}.$$

Angular sum in a triangle:
$$\alpha + \beta + \gamma = \pi + S > \pi \quad \text{(elliptic)},$$
$$\alpha + \beta + \gamma = \pi \quad \text{(Euclidean)},$$
$$\alpha + \beta + \gamma = \pi - S \quad \text{(hyperbolic)}.$$

The last three formulas for the angular sum in a triangle are special cases of the important theorem of Gauss–Bonnet (cf. Problem 74.11). It might be the most important theorem in global differential geometry. Intuitively, it states that living beings in two dimensions are able to determine the geometry they live in solely on the basis of measurements at the circle or the triangle.

EXAMPLE 74.57 (Local Realization of the Hyperbolic Geometry on a Pseudosphere). The pseudosphere S_{pseu} is obtained through rotations of the curve $\zeta = \zeta(\xi)$ about the ζ-axis in a Cartesian (ξ, η, ζ)-coordinate system. Introducing polar coordinates φ, r we obtain

$$\xi = r \cos \varphi, \quad \eta = r \sin \varphi, \quad \zeta = \pm \int_1^r r^{-1} \sqrt{1 - r^2} \, dr$$

(Fig. 74.13). Naturally, the curves $\varphi = \text{const}$ and $r = \text{const}$ are called meridians and circles of latitude. The element of arc is $ds^2 = r^{-2} dr^2 + r^2 d\varphi^2$. Letting $\xi = \varphi, \eta = 1/r$, we obtain

$$ds^2 = (d\xi^2 + d\eta^2)/\eta^2.$$

74.21. Applications to Non-Euclidean Geometry

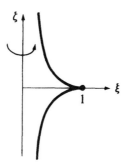

Figure 74.13

This is the metric (85) for M_{hyp}. Thus, the geometry of M_{hyp} with $\eta > 1$ can be realized on the northern half of S_{pseu} (Fig. 74.11).

Hilbert (1901), however, showed that the complete Riemannian manifold M_{hyp} cannot be realized as a surface in $\mathbb{R}^3$. Note that S_{pseu} is not a manifold, since at the equator points $r = 1$ there exist no tangent planes. Nevertheless, the Whitney embedding theorem (1936) of Section 73.21 implies that M_{hyp} can be embedded as a manifold in $\mathbb{R}^{2n+1}$, $n = 2$. One should note, though, that the embedded manifold need not have the same metric as M_{hyp}. However, the embedding theorem of Nash (1956), shows that every real, n-dimensional, proper Riemannian C^k-manifold M with countable basis, $3 \le k \le \infty$, can isometrically be embedded in $\mathbb{R}^m$ for m sufficiently large. This embedding is of class C^k. It suffices to choose $m = 2^{-1}n(n+1)(3n+11)$. If M is compact, one can choose $m = 2^{-1}n(n+5) + 3$. For M_{hyp} we have $n = 2$. In proving this, Nash solved a complicated system of nonlinear partial differential equations by using the hard implicit function theorem (see Problem 5.9). Embedding theorems for Riemannian manifolds are discussed in detail in Gromov and Rohlin (1970, S, B) and Gromov (1986, M).

An extremely simple and very elegant proof of the Nash embedding theorem can be found in Günther (1989).

EXAMPLE 74.58 (Model of Beltrami). In 1868, Beltrami published the first concrete model for a hyperbolic geometry. On the open unit disk

$$M = \{(u, v) \in \mathbb{R}^2 : u^2 + v^2 < 1\}$$

he introduced the metric

$$ds^2 = \frac{R^2}{(1 - u^2 - v^2)^2} [(1 - v^2) du^2 + 2uv \, du \, dv + (1 - u^2) dv^2].$$

For the Gaussian curvature, one finds that $K = -1/R^2$. The "straight lines," i.e., the geodesics in M are segments of straight lines. From Figure 74.14 we obtain (P_{hyp}).

This model played a famous role in the development of abstract manifolds, because it was an important example of an abstract manifold which could not be realized as a surface in $\mathbb{R}^3$. However, a local realization on the pseudosphere

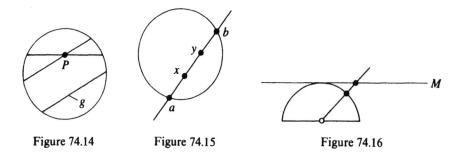

Figure 74.14 Figure 74.15 Figure 74.16

with equator radius R is possible. All this led to the realization of abstract manifolds as useful mathematical tools.

EXAMPLE 74.59 (Non-Euclidean Models of Felix Klein). In 1871, Felix Klein showed that elliptic and hyperbolic models may be constructed in the context of projective geometry. A model for the hyperbolic, non-Euclidean geometry is obtained by choosing the open unit disk as the plane M. The "straight lines" are segments of straight lines. From Figure 74.14 we obtain (P_{hyp}). As distance between two points x and y, one chooses

$$d(x, y) = \ln \frac{|a - x||b - y|}{|a - b||x - y|}.$$

This is the logarithm of the double ratio of the four points a, x, y, b where a and b lie on the boundary of the unit disk (Fig. 74.15). All points on the boundary of the unit disk are the infinitely distant points in this geometry.

The model for the elliptic geometry of Felix Klein is equivalent to M_{ellip} of Example 74.54. More precisely, we choose the tangent plane M at the north pole of M_{ellip} and project M_{ellip} from the center of the ball onto M (Fig. 74.16). We add infinitely distant points to M which correspond to the equator points of M_{ellip}, where antipodal points are identified. This way, we obtain M_∞ and by projecting this geometry we map M_{ellip} onto M_∞.

These models are carefully discussed in the standard book of Klein (1928, M, H).

We conclude this section with a general remark. In his famous work "Criticism of pure reasoning," Immanuel Kant in 1781 states that Euclidean geometry is "thought necessary," because it is immanent in the thoughts of every person. After Gauss in 1817 recognized that there exist non-Euclidean geometries, it was clear that physical laws determine the structure of space which therefore would have to be tested experimentally. This program then was realized by Einstein, 1916, as we shall see in Chapter 76. Interestingly, Euclidean geometry was revived in a modified form at the beginning of this century, since the geometry of infinite-dimensional H-spaces is Euclidean. Around 1925 this geometry then became the basis of quantum theory. In Part II we saw already that boundary-value problems for elliptic partial differential equations fit into this geometric concept. It is quite remarkable that the

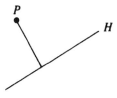

Figure 74.17

famous Dirichlet problem of the nineteenth century can be solved very easily by using the following fact: In an H-space one can construct the normal perpendicular to a closed hyperplane H from each point $P \notin H$ (Fig. 74.17). This implies the theorem of Riesz and in turn the existence theorem for quadratic variational problems (Section 18.11d). Some special applications of this existence theorem are the existence theorems for the basic problems in linear elasticity theory of Chapter 61. In Part V, we shall see that classical mechanics and statistical physics in phase space can best be understood in the context of symplectic geometry. Also, as has been mentioned already several times, the goal of the gauge field theories for elementary particles is the reduction of the fundamental physical interactions to the curvature of fiber bundles. All this illustrates the current trend towards a geometrization of physics.

The most general geometric properties are topological properties. Those are properties which remain unchanged under homeomorphisms and hence satisfy a strong form of structural stability. Deep mathematical results are those which show a connection between analytic and topological properties. In Problem 74.11 we consider, as an important example in this connection, the global theorem of Gauss–Bonnet, which states that the Gaussian total curvature of a closed surface is a topological property.

74.22. Strategy for a Further Development of the Differential and Integral Calculus on Manifolds

In the last sections of this chapter we consider several important supplements to tensor calculus, which will be discussed more thoroughly in Part V in the context of Banach manifolds:

(i) Alternating differentiation of alternating tensors (applications to Cartan's calculus of differential forms).
(ii) Lie derivative as a generalized directional derivative (applications to Lie algebras of vector fields and Lie groups).

The importance of (i) and (ii) is that one does not need the Christoffel symbols for the definition, i.e., these two derivative concepts exist for *arbitrary* manifolds. We use the following strategy:

(S) Using covariant differentiation, we look for expressions which do not explicitly contain the Γ_{ij}^k, such as, e.g., (90) below. This yields tensor fields which can be formed on general manifolds without Γ_{ij}^k, i.e., without affine connection.

Moreover, by integrating differential forms, we obtain a general method to formulate integral theorems on oriented, finite-dimensional manifolds. This includes classical volume, surface, and curve integrals.

74.23. Alternating Differentiation of Alternating Tensors

First we consider the situation in $\mathbb{R}^n$, i.e., we assume (A) of Section 74.2. In addition we require:

(T) The tensor field $t_{i_1 \ldots i_r}$ is antisymmetric.

Proposition 74.60. *If* (T) *holds, then*

$$\text{Alt } \nabla_i t_{i_1 \ldots i_r} = \text{Alt } D_i t_{i_1 \ldots i_r}, \tag{90}$$

PROOF. The terms with Γ_{ij}^k cancel out. For example,

$$\nabla_i t_j - \nabla_j t_i = D_i t_j - D_j t_i - \Gamma_{ij}^s t_s + \Gamma_{ji}^s t_s$$
$$= D_i t_j - D_j t_i. \qquad \square$$

Definition 74.61. If (T) holds, then we define alternating differentiation through

$$d_i t_{i_1 \ldots i_r} = \text{Alt } D_i t_{i_1 \ldots i_r}. \tag{91}$$

It is important that $d_i t_{i_1 \ldots i_r}$ is again an antisymmetric tensor field, because the left-hand side of (90) is a tensor, hence also the right-hand side. As in Section 74.17, we immediately are able to extend the operation d_i to arbitrary manifolds by using charts.

74.24. Applications to the Calculus of Alternating Differential Forms

This very elegant and supple calculus, which was created by Ellie Cartan (1869–1961), will be discussed more thoroughly in Part V, where we consider this calculus for Banach manifolds. Here we want to show how easily the

74.24. Applications to the Calculus of Alternating Differential Forms

invariance properties of this calculus for finite-dimensional manifolds are deduced from (91) and the substitution rule for integrals.

74.24a. Basic Ideas

The following four observations form the basis of Cartan's calculus.

(C1) *Substitution rule.* Let S be a bounded region in $\mathbb{R}^2$. Consider the integral $J = \int_S f\,du\,dv$. We also write

$$J = \int_S f\,du \wedge dv.$$

For the product symbol "$\wedge$" we use the formal rule

$$a \wedge b = -b \wedge a$$

and hence $a \wedge a = 0$. Under changes of variables $u = u(t,s)$, $v = v(t,s)$, we obtain

$$du \wedge dv = (u_t\,dt + u_s\,ds) \wedge (v_t\,dt + v_s\,ds) = (u_t v_s - u_s v_t)\,dt \wedge ds.$$

This gives

$$J = \int f \frac{\partial(u,v)}{\partial(t,s)}\,dt\,ds,$$

which is the correct substitution rule if the functional determinant is positive. Under changes in the orientation, J changes its sign.

(C2) *Theorem of Gauss.* In order to obtain the well-known formula for the integration by parts

$$\int_{\partial S} f\,du + g\,dv = \int_S (g_u - f_v)\,du\,dv$$

in a more elegant form

$$\int_{\partial S} \omega = \int_S d\omega, \qquad (92)$$

we set

$$\omega = f\,du + g\,dv,$$

and compute $d\omega$ according to

$$d\omega = df \wedge du + dg \wedge dv.$$

This gives

$$d\omega = (f_u\,du + f_v\,dv) \wedge du + (g_u\,du + g_v\,dv) \wedge dv$$
$$= (g_u - f_v)\,du \wedge dv.$$

(C3) *Theorem of Stokes.* We choose S as a surface in $\mathbb{R}^3$ and let
$$\omega = a\,du + b\,dv + c\,dw.$$
This gives
$$d\omega = da \wedge du + db \wedge dv + dc \wedge dw$$
$$= \alpha\,dv \wedge dw + \beta\,dw + dv + \gamma\,du \wedge dv$$
with
$$\alpha = c_v - b_w$$
and analogous formulas for β and γ using cyclic permutations. Setting
$$\vec{\omega} = ae_1 + be_2 + ce_3,$$
we obtain
$$\operatorname{curl}\vec{\omega} = \alpha e_1 + \beta e_2 + \gamma e_3.$$
Moreover, we find
$$\int_S d\omega = \int\!\!\int \left[\alpha\frac{\partial(v,w)}{\partial(t,s)} + \beta\frac{\partial(w,v)}{\partial(t,s)} + \gamma\frac{\partial(u,v)}{\partial(t,s)}\right] dt\,ds$$
$$= \int_S N\,\operatorname{curl}\vec{\omega}\,dm,$$
where N is the unit normal vector for the surface and m is the surface measure. Note that from Section 74.10
$$N = (x_t \times x_s)/|x_t \times x_s|$$
with $x = ue_1 + ve_2 + we_3$ and $dm = |x_t \times x_s|\,dt\,ds$. Thus (92) corresponds to the theorem of Stokes
$$\int_{\partial S} \vec{\omega}\,dx = \int_S N\,\operatorname{curl}\vec{\omega}\,dm.$$
If S is a bounded region in $\mathbb{R}^3$ and
$$\omega = a\,dv \wedge dw + b\,dw \wedge du + c\,du \wedge dv,$$
then
$$d\omega = da \wedge dv \wedge dw + db \wedge dw \wedge du + dc \wedge du \wedge dv$$
$$= (a_u + b_v + c_w)\,du \wedge dv \wedge dw$$
and
$$d\,d\omega = d(a_u + b_v + c_w) \wedge du \wedge dv \wedge dw = 0$$
since $du \wedge du = 0$, etc. Then (92) is equivalent to the theorem of Gauss

74.24. Applications to the Calculus of Alternating Differential Forms

in $\mathbb{R}^3$:

$$\int_{\partial S} N\vec{\omega}\, dm = \int_S \operatorname{div} \vec{\omega}\, dx,$$

where N denotes the outer unit normal vector of the boundary ∂S.

For all these formulas we assume that S and ∂S are coherently oriented (see Section 73.18).

(C4) *Integrability conditions.* Let G be a region in $\mathbb{R}^3$. Let

$$\omega = a\, du + b\, dv + c\, dw$$

and

$$\Omega = A\, dv \wedge dw + B\, dw \wedge du + C\, du \wedge dv.$$

Moreover, let $\vec{\omega} = ae_1 + be_2 + ce_3$ and $\vec{\Omega} = Ae_1 + Be_2 + Ce_3$. Given an Ω, we look for an ω such that

$$d\omega = \Omega \quad \text{on } G.$$

This equation is equivalent to

$$\operatorname{curl} \vec{\omega} = \vec{\Omega} \quad \text{on } G.$$

Because of $d\, d\omega = 0$ we immediately find the necessary solvability condition

$$d\Omega = 0, \quad \text{i.e.,} \quad \operatorname{div} \vec{\Omega} = 0 \quad \text{on } G.$$

If G is a convex or, more generally, a smoothly contractible region, then this condition is also sufficient for the existence of a solution ω. This follows from the theorem of Poincaré of Section 74.24c below.

This way many formulas from vector analysis can quickly be derived in this calculus. We shall also see how this calculus is generalized to classical vector analysis on manifolds. It is a very important tool in modern mathematical physics.

74.24b. Calculus in $\mathbb{R}^n$

Assume (A) of Section 74.2 and consider expressions of the form

$$\omega = t_{i_1 \ldots i_r}\, du^{i_1} \wedge \cdots \wedge du^{i_r}, \tag{93}$$

where the $t_{i_1 \ldots i_r}$ are antisymmetric in all indices. Thus instead of $\omega = a\, du^1 \wedge du^2$, for example, we write

$$\omega = t_{ij}\, du^i \wedge du^j = t_{12}\, du^1 \wedge du^2 + t_{21}\, du^2 \wedge du^1$$

with $t_{ij} = -t_{ji}$ and $t_{12} = a/2$. We formally use the antisymmetric product symbol $\wedge$ as above, i.e., $a \wedge b \wedge c = -b \wedge a \wedge c$, etc. For didactical reasons, this formal point of view is very convenient. A rigorous approach will be considered in Part V by defining $a \wedge b$, etc., via antisymmetric multilinear forms. However, this rigorous approach in Part V, which is also valid on infinite-dimensional Banach manifolds, will not change the rules of the calculus considered below.

We call ω an r-differential form or simply an r-form. Functions are called 0-forms. Making the change in variables

$$du^i = A^i_{i'}\, du^{i'},$$

we obtain

$$\omega = t_{i_1 \ldots i_r} A^{i_1}_{i'_1} \ldots A^{i_r}_{i'_r}\, du^{i'_1} \wedge \cdots \wedge du^{i'_r}$$
$$= t_{i'_1 \ldots i'_r}\, du^{i'_1} \wedge \cdots \wedge du^{i'_r}$$

with

$$t_{i'_1 \ldots i'_r} = A^{i_1}_{i'_1} \ldots A^{i_r}_{i'_r} t_{i_1 \ldots i_r}.$$

This yields the important result that the coefficients $t_{i_1 \ldots i_r}$ form an antisymmetric tensor. The derivative $d\omega$ is defined through

$$d\omega = d_i t_{i_1 \ldots i_r}\, du^i \wedge du^{i_1} \wedge \cdots \wedge du^{i_r}.$$

Because of the tensor property of $d_i t_{\ldots}$ of Section 74.23, this definition does *not* depend on the choice of the coordinate system. From (91) and the fact that $a \wedge b = -b \wedge a$ we obtain

$$d\omega = D_i t_{i_1 \ldots i_r}\, du^i \wedge du^{i_1} \wedge \cdots \wedge du^{i_r}.$$

This yields the *key* formula

$$d\omega = dt_{i_1 \ldots i_r} \wedge du^{i_1} \wedge \cdots \wedge du^{i_r}. \tag{D}$$

This formula is easily remembered and coincides with the result for $d\omega$ of Section 74.24a. From $D_i D_j = D_j D_i$ we immediately obtain the integrability condition

$$d\,d\omega = 0.$$

74.24c. Calculus on Manifolds

In this section we assume:

(H1) M is a real, n-dimensional C^∞-manifold.
(H2) S is an r-dimensional, compact, oriented submanifold of M with boundary ∂S where S and ∂S are coherently oriented if $\partial S \neq \emptyset$.

For $r = 1, 2$ one may think in (H2) of curves and surfaces S in $M = \mathbb{R}^3$.

Because of (H1) we can define ω on M as in (93) if $t_{i_1 \ldots i_r}$ is an antisymmetric C^∞-tensor field on M in the sense of Section 74.17. Moreover, we may define

74.24. Applications to the Calculus of Alternating Differential Forms

the derivative $d\omega$ as above. Because of the tensor property of $t_{...}$ and $d_i t_{...}$ these two definitions are correct, i.e., chart independent.

The definition of integrals

$$J = \int_S \omega$$

for r-forms with $r = \dim S$ follows automatically from this calculus. Consider, for example, a three-dimensional manifold M with local coordinates u, v, w and

$$\omega = a\, du \wedge dv + b\, dv \wedge dw,$$

where $u = u(t, s)$, $v = v(t, s)$, $w = w(t, s)$ is a local chart representation of S. Using transformations we obtain

$$\omega = \left(a \frac{\partial(u, v)}{\partial(t, s)} + b \frac{\partial(v, w)}{\partial(t, s)} \right) dt \wedge ds,$$

as in Section 74.24a. In computing J, we replace $dt \wedge ds$ with $dt\, ds$. Thereby, we obtain the subintegral of J over the local chart. As in Section 74.17 we find J by taking the sum of the local subintegrals using a partition of unity.

Again, this definition is correct, i.e., chart independent. This follows from the fact that this calculus respects the substitution rule. For example, we have

$$J_1 = \int du \wedge dv = \int \frac{\partial(u, v)}{\partial(t, s)} dt\, ds,$$

$$J_2 = \int du \wedge dv = \int \frac{\partial(u, v)}{\partial(\tau, \sigma)} d\tau\, d\sigma.$$

Because of the multiplication rule for functional determinants

$$\frac{\partial(u, v)}{\partial(t, s)} \frac{\partial(t, s)}{\partial(\tau, \sigma)} = \frac{\partial(u, v)}{\partial(\tau, \sigma)},$$

we obtain $J_1 = J_2$ if $\partial(t, s)/\partial(\tau, \sigma) > 0$. But since S is oriented we always have positive functional determinants for the chart changes.

Assuming (H1), (H2), we now state three fundamental results which we will prove in Part V. A region G in M is called *smooth contractible* if and only if it can smoothly be contracted into a point $x_0 \in G$, i.e., more precisely, there exists a C^∞-map $H: G \times [0, 1] \to G$ with $H(x, 0) = x$ and $H(x, 1) = x_0$ for all $x \in G$ (Fig. 74.18). For example, every convex region in $\mathbb{R}^n$ has this property.

(I) *Integrability condition*: It is $d\, d\omega = 0$.
(P) *Theorem of Poincaré*. Let G be a smooth contractible region in M. Then equation

$$d\omega = \Omega$$

has a solution ω in G if and only if $d\Omega = 0$ in G.

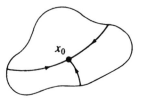

Figure 74.18

(S) *General theorem of Stokes*: It is

$$\int_{\partial S} \omega = \int_S d\omega.$$

This last theorem is one of the most important theorems in analysis. For the special case that S is an open interval in $\mathbb{R}$, theorem (S) reads:

$$\int_a^b f'(u)\,du = f(b) - f(a).$$

This is the fundamental theorem of differential and integral calculus. In fact, (S) follows from this quite easily by passing to quaders in the chart spaces and using a partition of unity.

74.24d. Pull-Back and Mapping Degree

If $f\colon M \to N$ is a map and ω a form on N, then the pull-back

$$f^*\omega \quad \text{on } M$$

is naturally obtained by mapping the N-coordinates in ω via f onto the M-coordinates. For example, it immediately follows from

$$\omega = a\,du \wedge dv$$

and $(u, v) = f(t, s)$ that

$$f^*\omega = a(u_t\,dt + u_s\,ds) \wedge (v_t\,dt + v_s\,ds)$$
$$= a(u_t v_s - u_s v_t)\,dt \wedge ds.$$

Let M and N be real, n-dimensional, compact, connected, and oriented C^∞-manifolds. One may think of closed curves in $\mathbb{R}^3$. Moreover, let ω be an n-form on N with $\int_N \omega \neq 0$. Then one can assign a number $\deg f$ to each C^∞-map $f\colon M \to N$ such that

$$\int_M f^*\omega = \deg f \int_N \omega.$$

It should be noted that this number does not depend on ω. This will be discussed more thoroughly in Part V. We call $\deg f$ the mapping degree of f. In gauge field theories, physicists call $\deg f$ a topological charge, since $\deg f$ is an integer and hence invariant under deformations, i.e., homotopies.

74.24e. Inner Product

If
$$\omega = t_{i_1 \ldots i_r} du^{i_1} \wedge \cdots \wedge du^{i_r}$$
is an r-form on the manifold M and $v = v^i b_i$ is a vector field on M, then we define the inner product through
$$i_v \omega = \frac{r!}{(r-1)!} v^{i_1} t_{i_1 \ldots i_r} du^{i_2} \wedge \cdots \wedge du^{i_r}.$$
Because of the tensor property of $v^{i_1} t_{i_1 \ldots}$ this definition is coordinate independent.

74.24f. De Rham Cohomology

An r-form ω with
$$d\omega = 0$$
is called an r-cocycle. If
$$\omega = d\alpha,$$
then ω is called an r-coboundary of α. The choice of these names is motivated by the duality between homology and cohomology, which will be discussed in Part V. Two r-forms ω_1 and ω_2 are called *cohomologous* if and only if
$$\omega_1 - \omega_2 = d\alpha,$$
i.e., if they differ in an r-coboundary. We write
$$\omega_1 \equiv \omega_2 \mod \alpha.$$
All forms which are cohomologous to ω_1 form the cohomology class of ω_1. All r-forms on a manifold M can be added and multiplied with real numbers. If one identifies cohomologous forms under these operations, one obtains a real linear space $H^r(M), r = 0, 1, \ldots$. It is called the rth de Rham cohomology group. It is already nontrivial to show that for the n-sphere S^n:
$$H^r(S^n) = \begin{cases} \mathbb{R} & \text{for } r = 0, n, \\ 0 & \text{otherwise,} \end{cases}$$
holds. In the language of linear algebra, $H^r(M)$ is the factor space of the

r-cocycles with respect to the r-coboundaries. The elements of $H^r(M)$ are the cohomology classes. This very formal definition provides us with an excellent tool to describe deep connections between analysis and topology. According to the theorem of de Rham, which will be discussed in Problem 74.11e, $H^r(M)$ is a topological invariant of M. Cohomology theory will be discussed in greater detail in Part V. The theorem of Poincaré may now be stated in the following way.

(P) If G is a smooth contractible region in M, then $H^r(G) = 0$ for all r.

Actually, $H^r(G) = 0$ means that all r-cocycles Ω are also r-coboundaries, i.e., from $d\Omega = 0$ follows the existence of an ω with $\Omega = d\omega$. This is a special case of the important phenomenon that global analytic existence theorems may be expressed in terms of a vanishing of cohomology groups.

74.24g. Duality of Forms and the Decomposition Theorem of Hodge

Let M be a real, n-dimensional, oriented Riemannian C^∞-manifold. Then the metric tensor g_{ij} and hence $g = \det(g_{ij})$ is defined. Therefore also the pseudotensor

$$E_{i_1 \ldots i_r} = \sqrt{|g|}\, \varepsilon_{i_1 \ldots i_r}$$

of Section 74.7 is defined on M. Given an r-form

$$\omega = \frac{1}{r!} t_{i_1 \ldots i_r}\, du^{i_1} \wedge \cdots \wedge du^{i_r}$$

we define a dual $(n-r)$-form $*\omega$ through

$$*\omega = \frac{1}{(n-r)!}\, *t_{i_{r+1} \ldots i_n}\, du^{i_{r+1}} \wedge \cdots \wedge du^{i_n}$$

with

$$*t_{i_{r+1} \ldots i_n} = \frac{1}{r!} t^{i_1 \ldots i_r} E_{i_1 \ldots i_n}.$$

Here $t^{\cdots}$ is obtained from $t_{\ldots}$ by lifting the indices with the aid of g^{ij}, i.e.,

$$t^{i_1 \ldots i_r} = g^{i_1 j_1} \ldots g^{i_r j_r} t_{j_1 \ldots j_r}.$$

Since $*t_{\ldots}$ is a pseudotensor, $*\omega$ is transformed like a pseudoscalar. Since M is oriented, a chart change uses positive functional determinants. Consequently, $*\omega$ behaves like a scalar. Thus the definition of $*\omega$ is chart independent.

Moreover, we define

$$\delta\omega = (-1)^{r(n-r)+r}(\operatorname{sgn} g) * d * \omega$$

and $\delta f = 0$ for functions f. Having this we define the Laplace operator for

forms as
$$\Delta\omega = -(d\,\delta\omega + \delta\,d\omega).$$

For functions, this is the classical Laplace operator $\Delta f = g^{ij}\nabla_i\nabla_j f$. Note that in the literature one often uses the definition $\Delta = d\delta + \delta d$, which does not coincide with the classical definition.

Now we assume in addition that M is a compact, proper Riemannian manifold. For r-forms ω and α we define a scalar product
$$\langle \omega, \alpha \rangle = \int_M \omega \wedge *\alpha.$$

Using a coordinate representation one easily verifies the following properties

(i) $\langle \omega, \omega \rangle \geq 0$, and $\langle \omega, \omega \rangle = 0 \Leftrightarrow \omega = 0$;
(ii) $\langle d\omega, \beta \rangle = \langle \omega, \delta\beta \rangle$, i.e., δ is the adjoint operator to d;
(iii) $\langle \Delta\omega, \beta \rangle = \langle \omega, \Delta\beta \rangle$ and $\langle -\Delta\omega, \omega \rangle \geq 0$.

Furthermore, one has the following fundamental theorems of Hodge.

(T1) *Decomposition theorem.* Every r-form ω admits a decomposition
$$\omega = d\alpha + \delta\beta + \gamma$$
with $\Delta\gamma = 0$. Here $d\alpha$, $\delta\beta$, and γ are uniquely determined through ω and pairwise orthogonal with respect to $\langle \cdot, \cdot \rangle$.

(T2) *Existence theorem.* Equation $\Delta\omega = \Omega$ on M has a solution ω if and only if $\langle \Omega, \gamma \rangle = 0$ for all γ with $\Delta\gamma = 0$ on M.

Here M is a real, compact, n-dimensional, oriented, proper Riemannian C^∞-manifold and all forms have C^∞-coefficients. A proof may be found in Warner (1974, M), p. 223. There one also finds a functional-analytic theory for elliptic operators on M, which is similar to the Sobolev space theory of Chapter 22. In Part V we show how d and δ can be used to obtain elegant versions for Maxwell's equations of electrodynamics and the equations of gauge field theories.

74.25. Lie Derivative

First we consider $\mathbb{R}^n$, i.e., we assume situation (A) of Section 74.2. Let a^i be a tensor field. We want to generalize the concept of directional derivative $a^s D_s f$ for a function f to tensor fields $t^{\ldots}_{\ldots}$. One possibility is $a^s \nabla_s t^{\ldots}_{\ldots}$. This definition, however, depends on the Christoffel symbols. Another important possibility is the *Lie derivative* $L_a t^{\ldots}_{\ldots}$.

Definition 74.62. Let $L_a f = a^s D_s f$ and
$$L_a t^i = a^s D_s t^i - t^s D_s a^i. \tag{94}$$
Another notation is $[a, t]^i$.

The definition of L_a for arbitrary tensor fields is then a consequence of the product rule which we assume to be valid. From
$$L_a(t^i t_i) = (L_a t^i) t_i + t^i L_a t_i$$
and $L_a(t^i t_i) = a^s D_s(t^i t_i)$ for all t^i we obtain the definition
$$L_a t_i = a^s D_s t_i + t_s D_i a^s. \tag{95}$$
The symmetry of definitions (94) and (95) becomes clear if one looks at the summation index and the index picture. Moreover, the product rule implies that
$$L_a(t^i t_j) = (L_a t^i) t_j + t^i L_a t_j,$$
and hence we define
$$L_a(t^i t_j) = a^s D_s(t^i t_j) - t_j t^s D_s a^i + t^i t_s D_j a^s,$$
and similarly
$$L_a t_j^i = a^s D_s t_j^i - t_j^s D_s a^i + t_s^i D_j a^s.$$
Analogously, one proceeds for arbitrary tensor fields. The general rule is very simple:

One writes $L_a t_{\cdots}^{\cdots} = a^s D_s t_{\cdots}^{\cdots}$ and adds a term of the form (94) or (95) to each index of $t_{\cdots}^{\cdots}$ and makes sure that the index picture is right.

Proposition 74.63. *The Lie derivative of a tensor field is again a tensor field of the same type.*

PROOF. If f is a function, i.e., a scalar field, then it follows from tensor calculus that $L_a f = a^s D_s f$ is again a scalar field. The tensor property of $L_a t^i$ and $L_a t_i$ is an immediate consequence of
$$L_a t^i = a^s \nabla_s t^i - t^s \nabla_s a^i, \qquad L_a t_i = a^s \nabla_s t_i + t_i \nabla_s a^s.$$
The tensor property of the other Lie derivatives follows then from our construction by applying the product rule. □

EXAMPLE 74.64 (Motivation for the Lie Derivative). If one tries to generalize the directional derivative to tensor fields t^i, one observes that $a^s D_s t^i$ is not a tensor field. On the other hand, $a^s \nabla_s t^i$ and $a^s \nabla_s t^i - t^s \nabla_s a^i$ are tensor fields. The first expression contains Γ_{ij}^k and can therefore not be used for a general theory on manifolds. The second expression does not contain any Γ_{ij}^k and is equal to $L_a t^i$.

We shall give another motivation which also admits a physical interpretation. Consider a point $u = (u^1, \ldots, u^n)$ together with a neighboring point
$$v^i = u^i + \varepsilon a^i(u).$$

74.25. Lie Derivative

At the point v we perform the coordinate transformation

$$v^{i'} = v^i - \varepsilon a^i(u),$$

i.e., $v^{i'} = u^i$. The transformation of $t^i(v)$ yields

$$t^{i'}(v) = \frac{\partial v^{i'}}{\partial v^j} t^j(v).$$

Explicitly, we have

$$t^{i'}(u + \varepsilon a) - t^i(u) = \varepsilon L_a t^i(u) + o(\varepsilon), \qquad \varepsilon \to 0. \tag{96}$$

PROOF. This follows from

$$\frac{\partial v^{i'}}{\partial v^j} = \delta^i_j - \varepsilon \frac{\partial a^i(u)}{\partial u^k} \frac{\partial u^k}{\partial v^j}, \qquad \frac{\partial u^k}{\partial v^j} = \delta^k_j + O(\varepsilon),$$

and $t^i(u + \varepsilon a) = t^i(u) + \varepsilon a^s D_s t^i(u) + o(\varepsilon)$, $\varepsilon \to 0$. □

A formula, analogous to (96), is obtained if one replaces t^i with an arbitrary tensor field $t^{...}_{...}$. Equation (96) provides the interpretation of L_a. Physically, it means the following. An observer travels from a point u, where he measures the field $t^i(u)$, to a neighboring point v. There he measures the field $t^i(v)$ which he transforms back into his old coordinates at the point u. This gives $t^{i'}(u + \varepsilon a)$. The difference

$$t^{i'}(u + \varepsilon a) - t^i(u)$$

is equal to $\varepsilon L_a t^i(u)$ except for terms $o(\varepsilon)$, i.e.,

$$L_a t^i(u) = \lim_{\varepsilon \to 0} \frac{t^{i'}(u + \varepsilon a) - t^i(u)}{\varepsilon}.$$

This interpretation motivates the following definition.

Definition 74.65. A tensor field $t^{...}_{...}$ is *constant in the sense of Lie* along the vector field $a = a^i b_i$ if and only if the directional Lie derivative with respect to a is identically zero, i.e., $L_a t^{...}_{...} \equiv 0$.

If one constructs a flow by using a and the differential equation

$$\dot{u}^i = a^i(u(\tau)),$$

whose tangent vectors are equal to a, then one says the tensor field $t^{...}_{...}$ is invariant with respect to this flow if and only if

$$L_a t^{...}_{...} \equiv 0.$$

Here, a denotes the tangent vector field of the flow. One may think of a water flow. Then a is the velocity field and $t^{...}_{...}$ is another physical field, e.g., in the case of t^i a force field or magnetic field.

The Lie derivative therefore plays an important role in the study of symmetries of tensor fields.

Suppose M is a finite-dimensional manifold. Then, similarly as in Section 74.17, one can define the Lie derivative for tensor fields $t_{...}^{...}$ on M as above by working in charts. In Definition 74.65, $\{b_i\}$ is the natural basis of the tangent spaces to M.

In Part V we shall give a definition of Lie derivatives for Banach manifolds using the concept of flows, which is equivalent to the above definition.

74.26. Applications to Lie Algebras of Vector Fields and Lie Groups

We want to show that Lie derivatives lead naturally to Lie algebras.

Definition 74.66. A Lie algebra X over $\mathbb{K}$ is a linear space over $\mathbb{K}$, where, in addition, a product $[v, w]$ is defined with

(i) $[v, w] = -[w, v]$;
(ii) $[v, [w, z]] + [w, [z, v]] + [z, [v, w]] = 0$.

More precisely, $(v, w) \mapsto [v, w]$ is a bilinear map from $X \times X$ into X with (i), (ii) for all $v, w, z \in X$. The dimension of the Lie algebra X is equal to the dimension of the linear space X.

EXAMPLE 74.67. The standard example of a real Lie algebra is $LGL(n, \mathbb{R})$, i.e., the set of all real $(n \times n)$-matrices v with the usual addition and bracket product

$$[v, w] = vw - wv.$$

The connection with the group $GL(n, \mathbb{R})$ of all real, regular $(n \times n)$-matrices is the following. For each $v \in LGL(n, \mathbb{R})$ and $t \in \mathbb{R}$ we have

$$e^{tv} = I + tv + o(t), \quad t \to 0$$

and from e^{tv} one obtains all elements of $GL(n, \mathbb{R})$ in a neighborhood of I. Furthermore,

$$e^{tv} e^{tw} e^{-tv} e^{-tw} = I + t^2 [v, w] + o(t^2), \quad t \to 0.$$

Therefore, the Lie algebra $LGL(n, \mathbb{R})$ is obtained from the Lie group $GL(n, \mathbb{R})$ by linearization. The commutator of the group thereby becomes the product $[u, w]$ of the Lie algebra. This was the ingenious idea of Sophus Lie (1842–1899), when he reduced the study of Lie groups to Lie algebras. This will be discussed in greater detail in Part V, in connection with applications to the theory of elementary particles (theory of quarks). Lie algebras are a basic tool in modern physics to explain quantum effects with symmetries.

74.26. Applications to Lie Algebras of Vector Fields and Lie Groups

EXAMPLE 74.68 (Lie Algebra of Vector Fields). Let M be a real, n-dimensional C^∞-manifold. If $v = v^i b_i$ and $a = a^i b_i$ are vector fields on M, then we define the Lie derivative of v in the direction of a through

$$L_a v = (L_a v^i) b_i.$$

We also write $[a, v] = L_a v$. Thus

$$[a, v] = (a^s D_s v^i - v^s D_s a^i) b_i.$$

This makes the collection of all C^∞-vector fields on M into a real *Lie algebra*. The Lie derivative of a covector field $v = v_i b^i$ is defined analogously through

$$L_a v = (L_a v_i) b^i,$$

where $\{b^i\}$ is the natural basis in the cotangent spaces of M (see Section 74.17). Analogously, we define the Lie derivative of the r-form

$$\omega = t_{i_1 \ldots i_r} du^{i_1} \wedge \cdots \wedge du^{i_r},$$

through

$$L_a \omega = (L_a t_{i_1 \ldots i_r}) du^{i_1} \wedge \cdots \wedge du^{i_r}.$$

We have

$$L_a \omega = i_a d\omega + d(i_a \omega).$$

This follows easily from a coordinate representation. As an exercise we recommend a proof of this.

Definition 74.69. A group G is a set on which a multiplication is defined, i.e., to each pair (g, h) of elements in G a unique element in G, gh is assigned, such that the following three conditions are satisfied:

(i) Associativity: $(fg)h = f(gh)$ for all $f, g, h \in G$.
(ii) Existence of a unit element: There exists an element e in G with $ge = eg = g$ for all $g \in G$.
(iii) Existence of an inverse element: For each $g \in G$ there exists an element $h \in G$ with $gh = e$.

For h in (iii) one writes g^{-1}. One can easily show that e and g^{-1} are uniquely determined. Besides $gg^{-1} = e$ one also has $g^{-1}g = e$.

Roughly speaking, a Lie group G is a group which can locally be parametrized, and the group operations depend smoothly on the parameters. This way, one can study G by using the methods of the theory of manifolds.

Definition 74.70. A *Lie group* G is a real, n-dimensional C^∞-manifold, whose elements form a group and the map

$$(h, g) \mapsto h^{-1} g$$

is a C^∞-map from $G \times G$ into G

If we choose $g = e$ and $h = f^{-1}$, then also the map $h \mapsto h^{-1}$ from G into G and the map $(f, g) \mapsto fg$ from $G \times G$ into G are of class C^∞, i.e., taking the inverse and multiplication are of class C^∞.

EXAMPLE 74.71. The standard example of a Lie group is $GL(n, \mathbb{R})$ of Example 74.67. All real $(n \times n)$-matrices g with $\|g - I\| < \varepsilon$ belong to $GL(n, \mathbb{R})$ for sufficiently small $\varepsilon > 0$, because the process of taking the inverse is continuous (see Problem 1.7). Letting $v = \ln g$, we obtain $g = e^v$ (see A_1(60b)). Thus $GL(n, \mathbb{R})$ can be parametrized by v in a neighborhood $U(I)$ of the unit element I. For an arbitrary group element g_0 we choose $g_0 U(I)$ for a parametrization. Thereby, $GL(n, \mathbb{R})$ naturally becomes an n^2-dimensional Lie group. Note that the set of $(n \times n)$-parameter matrices v can be identified with a zero neighborhood of $\mathbb{R}^{n^2}$.

EXAMPLE 74.72 (Lie Algebra of a Lie Group). Let G be a Lie group. Through

$$T_g x = gx$$

we obtain for each fixed $g \in G$ a C^∞-diffeomorphism $T_g: G \to G$. If $y = T_g x$, then the derivative

$$T_g': TG_x \to TG_y$$

maps the corresponding tangent spaces onto each other. Let v be a vector field on G. Then v is called *left invariant* if and only if

$$T_g' v = v \quad \text{for all} \quad g \in G,$$

i.e., $T_g'(v(x)) = v(gx)$ for all $x, g \in G$.

From the definition of $[a, v]$ in Example 74.68, the following important relation

$$T_g'[a, v] = [T_g' a, T_g' v]$$

is obtained by linearization. Thus, for left invariant vector fields a, v we have that $T_g'[a, v] = [a, v]$, i.e., also $[a, v]$ is left invariant.

Therefore all left invariant C^∞-vector fields on a Lie group G form a real Lie algebra with respect to

$$[a, v] = L_a v,$$

which is called the *Lie algebra LG* of G.

In Part V, we will compute a number of Lie algebras in view of physical applications. In particular, we will give a simple proof of the fact that the algebra $LGL(n, \mathbb{R})$ of Example 74.67 is isomorphic to the Lie algebra of $GL(n, \mathbb{R})$. The strategy for the theory of Lie groups consists in reducing the study of Lie groups to the much simpler object of Lie algebras. An important and deep structure theorem is the following: Every finite-dimensional, real Lie algebra is isomorphic to the Lie algebra of a Lie group (see Dieudonné (1975, M), Vol. 5, 21.23).

PROBLEMS

74.1. *Tensors.* How is the tensor field t^i_{jk} transformed?
Solution: From Definition 74.4 it follows that

$$t^{i'}_{j'k'} = A^{i'}_i A^j_{j'} A^k_{k'} t^i_{jk}.$$

How is $t^{j_1\cdots j_s}_{i_1\cdots i_r}$ transformed in case it is a tensor density (or pseudotensor density) of weight w?
Solution:

$$t^{j_1\cdots j_s}_{i_1\cdots i_r} = \alpha A^{j_1}_{j'_1}\cdots A^{j_s}_{j'_s} A^{i'_1}_{i_1}\cdots A^{i'_r}_{i_r} t^{j'_1\cdots j'_s}_{i'_1\cdots i'_r} \tag{97}$$

with $\alpha = D^w$ (or $\alpha = (\operatorname{sgn} D)D^w$). For $w = 0$ we obtain tensor fields (or pseudotensor fields).

74.2. *Covariant differentiation.*

74.2a. Compute $\nabla_i t^j_{km}$.
Solution:

$$\nabla_i t^j_{km} = D_i t^j_{km} + \Gamma^j_{is} t^s_{km} - \Gamma^s_{ik} t^j_{sm} - \Gamma^s_{im} t^j_{ks}.$$

Observe the summation index and the index picture.

74.2b. Prove that covariant differentiation and contraction commute. For $\nabla_i t^j_{jk}$, for instance, it does not matter whether ∇_i is applied before or after summing over j.
Hint: Use the definition of ∇_i and observe the different signs in (21) and (22).

74.2c. Prove that $\nabla_k \delta^i_j = 0$.
Solution:

$$\nabla_k \delta^i_j = D_k \delta^i_j + \Gamma^i_{ks} \delta^s_j - \Gamma^s_{kj} \delta^i_s = 0.$$

74.2d. *Lemma of Ricci.* Prove that $\nabla_k g_{ij} = 0$.
Solution: This follows from

$$\nabla_k g_{ij} = D_k g_{ij} - \Gamma^s_{ki} g_{sj} - \Gamma^s_{kj} g_{is}$$

and (23).

74.2e. Prove that $\nabla_k g^{ij} = 0$.
Solution: From $\delta^i_r = g^{is} g_{sr}$ and the product rule, it follows that

$$0 = \nabla_k g^{is} g_{sr} = g_{sr} \nabla_k g^{is} + g^{is} \nabla_k g_{sr} = g_{sr} \nabla_k g^{is}.$$

Multiplication with g^{rj} gives $0 = \delta^j_s \nabla_k g^{is}$.

74.2f. Verify (24) and (25).

74.3. *Index principle.* The following set of problems shows how naturally the index principle of Section 74.5 occurs.

74.3a. *Scalar product.* What is the value of the scalar product vw for vectors v, w for an arbitrary coordinate system of $\mathbb{R}^n$?
Solution: Let $v = v^i b_i$ and $w = w^i b_i$. It follows that

$$vw = g_{ij} v^i w^j = v^i w_i$$

with $w_i = g_{ij} w^j$.

74.3b. *Lifting and lowering of indices.* This uses the g_{ij} and g^{ij}. For example, one obtains

$$t^i = g^{ij}t_j, \qquad t^{ij} = g^{ir}g^{js}t_{rs}.$$

Prove that $w_i = g_{ij}w^j$ implies $w^j = g^{ji}w_i$.
Solution: From $g^{ki}g_{ij} = \delta^k_j$ it follows that $g^{ki}w_i = w^k$.

74.3c. *Cartesian coordinate system.* Why is the position of upper and lower indices not important for Cartesian coordinates?
Solution: Instead of v_i we can write v^i with $v^i = g^{ij}v_j$. This follows from $v_i = v^i$, because in Cartesian coordinates g^{ij} is equal to the Kronecker symbol.

74.3d. *Invariants for quadratic forms.* Let $a_{ij}v^iv^j$ be a quadratic form in $\mathbb{R}^n$ with $a_{ij} = a_{ji}$. Prove:
  (i) The number trace $a = \sum_i a_{ii}$ has the same value for all Cartesian coordinate systems. What is the value of trace a for an arbitrary coordinate system?
  (ii) For all $\lambda \in \mathbb{C}$, $\det(\lambda\delta_{ij} - a_{ij})$ has the same value for every Cartesian coordinate system. Therefore the solutions of the secular equation

$$\det(\lambda\delta_{ij} - a_{ij}) = 0,$$

i.e., the eigenvalues of (a_{ij}) are equal for all Cartesian coordinate systems. What is the form of this equation for an arbitrary coordinate system?
  (iii) $\operatorname{sgn}\det(a_{ij})$ is an invariant, i.e., a scalar.
  (iv) κ_+, κ_-, and κ_0, i.e., the number of eigenvalues of (a_{ij}) which are >0, <0, and $=0$ are invariant. Recall that κ_- is the Morse index of (a_{ij}).

Solution: Ad(i). Letting $a^i_j = g^{ik}a_{kj}$ we obtain that trace $a = a^i_i$ in Cartesian coordinates. For an arbitrary coordinate system this is a scalar.

Ad(ii). We let $c_{ij} = \lambda g_{ij} - a_{ij}$. From Section 74.7, $c = \det(c_{ij})$ is a scalar of weight 2, i.e., $c' = D^2 c$. For transformations between Cartesian coordinates we have $D = \det(A^i_j) = 1$. Hence, for any such transformation, c is a scalar. Furthermore, in Cartesian coordinates we have $g_{ij} = \delta_{ij}$. The general form of the secular equation is $\det c = 0$, i.e.,

$$\det(\lambda g_{ij} - a_{ij}) = 0.$$

Ad(iii). For $a = \det(a_{ij})$ we have $a' = D^2 a$, and hence $\operatorname{sgn} a' = \operatorname{sgn} a$.
Ad(iv). Using a rotation of the coordinate system one can always write $Q = a_{ij}v^iv^j$ as a sum of squares, i.e.,

$$Q = \sum_i \lambda_i v^{i'} v^{i'}.$$

Thereby the λ_i are the eigenvalues of (a_{ij}). From the law of inertia of Sylvester it follows that the number of squares with positive or negative sign is independent of the coordinate system.

74.3e. What is the form of $v \operatorname{grad} f$ for an arbitrary coordinate system of $\mathbb{R}^n$?
Solution: Letting $v = v^i b_i$ one obtains that $v \operatorname{grad} f = v^i \nabla_i f$.

74.3f. *The Navier–Stokes equations.* What is the form of

$$\rho v_t + \rho(v\,\mathrm{grad})v - \eta\,\Delta v = K - \mathrm{grad}\,p,$$

$$\mathrm{div}\,v = 0$$

in component notation for an arbitrary coordinate system of $\mathbb{R}^3$?

Solution: We let $v = v^i b_i$, $K = K^i b_i$. Then

$$\rho v_t^k + \rho v^i \nabla_i v^k - \eta g^{ij}\nabla_i \nabla_j v^k = K^k - \nabla^k p,$$

$$\nabla_i v^i = 0,$$

with $\nabla^k = g^{km}\nabla_m$.

74.3g. *Basic equation of elastodynamics.* What is

$$\rho u_{tt} = \mathrm{div}\,\tau + F$$

in component notation for an arbitrary coordinate system of $\mathbb{R}^3$?

Solution: In Cartesian coordinates we have $\mathrm{div}\,\tau = \sum_j D_j \tau_j^i e_i$, and hence $\mathrm{div}\,\tau = \nabla^j \tau_j^i b_i$ and

$$\rho u_{tt}^i = \nabla^j \tau_j^i + F^i,$$

where $u = u^i b_i$, $F = F^i b_i$, and the equation $u = \tau F$ corresponds to $u^i = \tau_j^i F^j$.

74.3h. *Construction of tensors.* Let G be a region in $\mathbb{R}^n$ and assume situation (A) of Section 74.2. In a fixed u-coordinate system of G we assign real numbers t^i to each point $x \in G$. For an arbitrary u'-system we then define

$$t^{i'} \stackrel{\mathrm{def}}{=} A_i^{i'} t^i.$$

Prove that through this a tensor field is obtained. Analogously, arbitrary tensor fields, pseudotensor fields, tensor densities, and pseudotensor densities are defined.

Solution: By definition we have

$$t^{i''} = A_i^{i''} t^i.$$

Now we must show that

$$t^{i''} = A_{i'}^{i''} t^{i'}.$$

In fact, $A_{i'}^{i''} t^{i'} = A_{i'}^{i''} A_i^{i'} t^i = A_i^{i''} t^i = t^{i''}$.

74.3i. *Inverse index principle.*

Prove: If for all tensor fields a^i,

$$a^i \alpha_{ij}$$

is transformed like a tensor field t_j, then also α_{ij} is a tensor field.

More generally: If for all tensor fields $a^{\cdots}_{\cdots}$, the quantity $a^{\cdots}_{\cdots}\alpha^{\cdots}_{\cdots}$ is transformed like a tensor field, where the type is given by the index picture, then $\alpha^{\cdots}_{\cdots}$ is also a tensor field where the type is given by the index picture.

This principle helps avoid tedious computations in proving the tensor property of $\alpha^{\cdots}_{\cdots}$. In (82), for example, the tensor property of the curvature tensor came up automatically.

Solution: From $a^{i'}\alpha_{i'j'} = A^j_{j'}(a^i\alpha_{ij})$ it follows that

$$A^{i'}_i a^i \alpha_{i'j'} = A^s_{j'} a^i \alpha_{is}.$$

Because of $A^{j'}_j A^s_{j'} = \delta^s_j$ this implies

$$a^i(A^{i'}_i A^{j'}_j \alpha_{i'j'}) = a^i \alpha_{ij}.$$

Since a^i is an arbitrary tensor field, and Problem 74.3h admits, for any fixed coordinate system, the computation of tensor fields with arbitrary numerical values, it follows that

$$\alpha_{ij} = A^{i'}_i A^{j'}_j \alpha_{i'j'},$$

i.e., α_{ij} is a tensor field.

74.4. Parallel transport and intuitive meaning of the Christoffel symbols. Give a direct proof of Theorem 74.A without using tensor calculus.

Solution: Let x be the radius vector in $\mathbb{R}^n$. We define vectors $b_{ij} \stackrel{\text{def}}{=} D_i b_j = D_i D_j x$ and numbers B^k_{ij} using the vector decomposition

$$b_{ij} = B^m_{ij} b_m. \tag{98}$$

(I) *Parallel transport.* Let C be the curve $x = x(\tau)$, i.e., $u^i = u^i(\tau)$. If the vector field $v = v^i b_i$ is constant along C, then differentiation with respect to τ gives:

$$0 = \dot{v}^i b_i + v^i \dot{b}_i = \dot{v}^m b_m + v^i b_{ij} \dot{u}^j.$$

From (98) we obtain $\dot{v}^m + B^m_{ij} v^i \dot{u}^j = 0$. This is equal to (30) if we can show that

$$B^m_{ij} = \Gamma^m_{ij}. \tag{99}$$

(II) *Proof of (99).* From

$$b_k b_r = g_{kr} \tag{100}$$

and (98) it follows that $b_{ij} b_r = B^m_{ij} g_{mr}$. Multiplication with g^{rk} gives

$$B^k_{ij} = g^{rk} b_{ij} b_r. \tag{101}$$

Differentiation of (100), i.e., of $b_r b_j = g_{rj}$ gives

$$b_{ir} b_j + b_r b_{ij} = D_i g_{rj}. \tag{102}$$

Interchanging the indices and summing yields

$$b_{ij} b_r = 2^{-1}(D_i g_{rj} + D_j g_{ir} - D_r g_{ij}) = \Gamma_{r,ij}.$$

(III) *Interpretation of the Christoffel symbols Γ^k_{ij}.* It follows from (98) that

$$D_i b_j = \Gamma^k_{ij} b_k,$$

i.e., Γ^k_{ij} describes the change in the natural basis $\{b_j\}$.

74.5. *Mean curvature H, Gaussian curvature K, and minimal surfaces.* For the surface area of minimal surfaces,

$$\int_G \sqrt{g}\, du^1\, du^2 = \min! \tag{103}$$

holds for any given boundary curve. Prove that a smooth solution of this variational problem satisfies $H = 0$ and $K \leq 0$ for all surface points.

Solution: To obtain the Euler equation for (103) is a purely local problem. Hence we consider the special local coordinate system of Example 74.40 with local surface representation

$$\zeta = 2^{-1}(\alpha\xi^2 + \beta\eta^2) + O_3.$$

The Euler equations to

$$\int \sqrt{1 + \zeta_\xi^2 + \zeta_\eta^2}\, d\xi\, d\eta = \min!$$

are

$$\frac{\partial}{\partial \xi} \frac{\zeta_\xi}{\sqrt{1 + \zeta_\xi^2 + \zeta_\eta^2}} + \frac{\partial}{\partial \eta} \frac{\zeta_\eta}{\sqrt{1 + \zeta_\xi^2 + \zeta_\eta^2}} = 0.$$

Because of $\zeta_\xi/\sqrt{1 + \zeta_\xi^2 + \zeta_\eta^2} = \alpha\xi + O_2$, this implies $\alpha + \beta = 0$ at the point $\xi = \eta = 0$, hence $H = 0$ and $K = \alpha\beta \leq 0$.

74.6. *Geometric meaning of the Gaussian curvature K.* We want to show that the analytic definition of K of Section 74.11 is the same as the original geometric definition of Gauss.

Let P be a surface point and Σ a small piece of the surface with $P \in \Sigma$ and surface area $m(\Sigma) = \int \sqrt{g}\, du^1\, du^2$. The spherical image $\tilde{\Sigma}$ of Σ is the piece of the surface obtained by drawing the unit normal vector N of Σ at the origin. Prove:

$$|K(P)| = \lim_{\text{diam } \Sigma \to 0} m(\tilde{\Sigma})/m(\Sigma).$$

Solution: The mean value theorem of integral calculus implies that $|K| = \sqrt{\tilde{g}}/\sqrt{g}$ at P. We have $\sqrt{g} = |b_1 \times b_2|$ and, analogously, $\sqrt{\tilde{g}} = |N_1 \times N_2|$ with $b_i = D_i x$ and $N_i = D_i N$. For computational reasons we choose the special coordinate system of Example 74.40. Then $P = (0, 0)$ and

$$x = \xi e_1 + \eta e_2 + \zeta e_3, \qquad \zeta = 2^{-1}(\alpha\xi^2 + \beta\eta^2) + O_3,$$

$$N = (x_\xi \times x_\eta)/|x_\xi \times x_\eta| = (e_3 - \zeta_\xi e_1 - \zeta_\eta e_2)/\sqrt{1 + \zeta_\xi^2 + \zeta_\eta^2}$$

$$= e_3 - \alpha\xi e_1 - \beta\eta e_2 + O_2.$$

At $(0,0)$ we have

$$\sqrt{g} = |x_\xi \times x_\eta| = 1 \quad \text{and} \quad \sqrt{\tilde{g}} = |N_\xi \times N_\eta| = \alpha\beta.$$

This means that $\sqrt{\tilde{g}}/\sqrt{g} = \alpha\beta = |K|$.

74.7.* *Geometric meaning of the curvature tensor R^i_{jkm} and the torsion tensor T^k_{ij}.* Let G be a simply connected region in an affine connected, real, n-dimensional C^∞-manifold M. Prove:

(i) The parallel transport in G is path independent if and only if $R^i_{jkm} = 0$ in G.

(ii) M is locally flat if and only if $R^i_{jkm} = 0$ and $T^k_{ij} = 0$ in M.

M is called *locally flat* if and only if for each point x_0 there exists a neighborhood $U(x_0)$ such that $\Gamma_{ij}^k = 0$ in $U(x_0)$ for an admissible chart, i.e., for suitable coordinates. For Riemannian manifolds this coincides with the definition of Section 74.20, because $D_i g_{jk}$ can be expressed as a linear combination of the $\Gamma_{\cdot\cdot}^{\cdot}$.

Hint: See Raschewski (1959, M), §106. For (i) use the curves $u^i = u^i(\tau, p)$ which depend on an additional parameter p, and differentiate the differential equation

$$\dot{v}^k + \Gamma_{ij}^k \dot{u}^i v^j = 0$$

for the parallel transport of v^i with respect to p. Then $\partial \dot{v}^k / \partial p \equiv 0$ means that the parallel transport is locally path independent.

74.8. *Non-Euclidean geometry.* Give an explicit proof of the trigonometric formulas for M_{hyp} of Example 74.56.

Hint: See Baule (1956, M), Vol. 7, §20. Use the geodesics on M_{hyp} (circular arcs) and the fact that goniometry on M_{hyp} is the same as Euclidean goniometry.

74.9. *Geodesics and classical dynamics.* In order to obtain fundamental properties for mechanical systems, (i) and (ii) below are of great importance. Consider, for instance, a mechanical system $q = q(t)$ with coordinates $q = (q^1, \ldots, q^n)$. Let kinetic and potential energy be equal to

$$T(q, \dot{q}) = g_{ij}(q) \dot{q}^i \dot{q}^j$$

and $U(q)$, respectively, with $g_{ij} = g_{ji}$. The Lagrangian is $L = T - U$. As usual, generalized momentum and force are defined as the partial derivatives $p_i = L_{\dot{q}^i}$ and $P_i = -U_{q^i}$. Moreover, we let $P^i = g^{ij} P_j$. Prove:

(i) The Lagrangian equations of motion $(d/dt)L_{\dot{q}^i} - L_{q^i} = 0$ can be written in the elegant form

$$\frac{D\dot{q}^k}{dt} = P^k, \qquad k = 1, \ldots, n, \qquad (104)$$

i.e.,

$$\ddot{q}^k + \Gamma_{ij}^k \dot{q}^i \dot{q}^j = P^k.$$

Here Γ_{ij}^k follows from the metric tensor g_{ij}. Formula (104) explicitly shows the tensorial character of the Lagrangian equations. It follows from (104) that in the case of vanishing forces, i.e., $P_i = 0$, the orbits are affine geodesics. The following construction of Jacobi yields geodesics for arbitrary forces.

(ii) We introduce a new metric

$$d\tilde{s}^2 = \tilde{g}_{ij} \, dq^i \, dq^j,$$

with $\tilde{g}_{ij} = 2(E - U) g_{ij}$ for fixed $E \in \mathbb{R}$ and assume that $E - U > 0$. The geodesics for this metric are exactly the orbits with total energy $E = U + T$.

According to Section 74.19, these geodesics are obtained from $D\dot{q}^k/d\tilde{s} = 0$, i.e.,

$$\ddot{q}^k + \tilde{\Gamma}_{ij}^k \dot{q}^i \dot{q}^j = 0.$$

The dot means derivative with respect to $\tilde{s}$. The corresponding variational principle

$$\int d\tilde{s} = \text{stationary!}$$

is called the principle of stationary action of Jacobi. Because of the energy theorem $E = U + T$, we have that $E - U > 0$ is equivalent to $T > 0$.
Hint: See Laugwitz (1960, M), 14.2.

The results remain unchanged if the state space M is a manifold, i.e., $q \in M$. Important applications of (i), (ii) in mechanics and especially in astronomy may be found in Abraham and Marsden (1978, M).

74.10. *Geodesic curvature of curves.* The curvature of a curve $x = x(s)$ in $\mathbb{R}^3$, parametrized with arclength s, is by definition equal to $\kappa = |\ddot{x}|$. If this curve lies on a surface, then we define the geodesic curvature as

$$\kappa_g = \ddot{x}(N \times \dot{x}).$$

We let $t = N \times \dot{x}$. It follows that t is a unit vector which, in the tangent plane to the surface, is perpendicular to the tangent $\dot{x}$ of the curve, and is obtained from $\dot{x}$ by using a rotation of 90° in the positive direction with respect to b_1, b_2. Prove that

$$\ddot{x} = \kappa_N N + \kappa_g t,$$

hence $\kappa^2 = \kappa_N^2 + \kappa_g^2$ and

$$\kappa_g t = b_k \frac{D\dot{u}^k}{ds} = b_k(\ddot{u}^k + \Gamma^k_{ij}\dot{u}^i\dot{u}^j).$$

The last equation shows that κ_g can be obtained solely through measurements on the surface. Under changes in the orientation, t changes its sign. Thus κ_g is a pseudoscalar. From Section 74.16 the curve $x = x(s)$, i.e., $u^i = u^i(s)$ is a geodesic if and only if

$$\kappa_g \equiv 0.$$

Thus κ_g is a measure for the deviation of the curve from a geodesic.
Hint: See Laugwitz (1960), II.3.5.

74.11. *The theorem of Gauss–Bonnet–Chern and connections between topology and analysis.* We shall try to explain, step by step, probably the most important theorem of global differential geometry—the theorem of Gauss–Bonnet–Chern. It relates topology, geometry, and analysis. The deepest-known generalization of this theorem, which also shows many other fundamental connections between topology and analysis, is the Atiyah–Singer index theorem. This will be discussed in Part V. In the following we assume that all manifolds, curves, functions, and vector fields are of class C^∞.

For introductory reading, we recommend Kreyszig (1957, M) and Guillemin and Pollack (1974, M). We also recommend Sulanke and Wintgen (1972, M) and Spivak (1979, M, B), Vol. 5. Recall that a quantity of a topological space is called a topological invariant if and only if it is preserved under homeomorphisms. Of particular interest are topological invariants in the

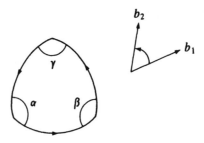

Figure 74.19

form of numbers, such as the Euler characteristic. This will be discussed below.

74.11a. *Theorem of Gauss* (1827) *on the angular sum in a triangle.* Besides the theorema egregium, the following important theorem is also contained in the "Disquisitiones generales circa superficies curvas" of Gauss. Let T be a triangle on a surface of $\mathbb{R}^3$ with angles α, β, γ and sides which are geodesics (Fig. 74.19). Then

$$\int_T K \, dm = \alpha + \beta + \gamma - \pi, \tag{105a}$$

where $dm = \sqrt{g} \, du^1 \, du^2$ is the surface measure. On the unit sphere we have $K = 1$, and hence $\alpha + \beta + \gamma - \pi$ equal to the surface area of the triangle. A triangle is always meant to have a simply connected region as its interior.

74.11b.* *Theorem of Bonnet* (1848). If the triangle sides of the previous problem are arbitrary curves, then

$$\int_T K \, dm + \oint_{\partial T} \kappa_g \, ds = \alpha + \beta + \gamma - \pi. \tag{105b}$$

Here, we move along the boundary ∂T in the positive direction with respect to b_1, b_2 (Fig. 74.19). If ∂T is a smooth curve without corners, we let $\alpha = \beta = \gamma = 0$.

The geodesic curvature κ_g has been defined in Problem 74.10. For geodesics we have $\kappa_g \equiv 0$. Hence (105a) is a special case of (105b). Prove (105b).

Hint: Equation (105b) follows directly from an application of the integral theorem of Gauss in the parameter plane, using a suitable orthogonal coordinate system (geodesic polar coordinates; see Kreyszig (1957, M), p. 206). An elegant proof, which uses the calculus of differential forms of Section 74.24, may be found in Blaschke (1950), §44, §46.

74.11c. *Global theorem of Gauss–Bonnet.* A closed, oriented surface M in $\mathbb{R}^3$, i.e., a real, two-dimensional, compact, and oriented manifold, is always homeomorphic to a sphere with p handles (Fig. 74.20). The number p, which is a topological invariant, is called the genus of M. For a sphere (or a torus) we have $p = 0$ (or $p = 1$). The number $\chi(M) = 2(1 - p)$ is called the Euler characteristic of M and is a fundamental topological invariant. Triangula-

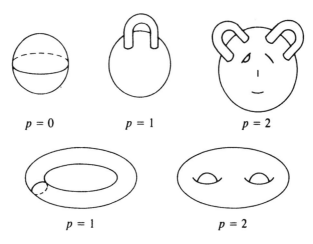

Figure 74.20

tion of M gives

$$\chi(M) = n_0 - n_1 + n_2, \tag{106}$$

where n_0, n_1, n_2 are the numbers of corners, edges, and surfaces. Equation (106) contains Euler's polyeder theorem which states that $n_0 - n_1 - n_2$ is independent of the triangulation. Prove that

$$\int_M K\,dm = m(S^k)\chi(M)/2, \tag{107}$$

where $k = 2$. Thereby $m(S^k)$ is the surface measure of the k-dimensional unit sphere S^k. In (107) this gives $m(S^k)\chi(M)/2 = 4\pi(1 - p)$. In the special case of the sphere (or the torus) with $p = 0$ (or $p = 1$) we have

$$\int_M K\,dm = 4\pi \quad (\text{or } = 0),$$

i.e., the total curvature of the torus is equal to zero.

This theorem, which is called the global theorem of Gauss–Bonnet, was formulated explicitly for the first time by Dyck (1885), who was a student of Felix Klein.

Solution: Let $p = 0$ and hence $\chi(M) = 2$. Then the topological type of M is the sphere S^2. We decompose M into four triangles which are naturally given on S^2 by the equator and one meridian. Summation of (105b) yields

$$\int_M K\,dm = 4\cdot 2\pi - 4\pi = 4\pi.$$

Analogously, one argues in the case of a torus, $p = 1$, and for $p > 1$. See Blaschke (1950, M), §47.

74.11d. *Meaning of the global theorem of Gauss–Bonnet.* Equation (107) is of principal value. On the left-hand side we have an analytic expression containing

K which is not a topological invariant, while the expression on the right-hand side is a topological invariant. In fact, (107) represents a fundamental connection between analysis and topology. Topological invariants are the most robust geometrical invariants. They remain unchanged under all homeomorphisms, i.e., roughly speaking, under all gum transformations. Analytic expressions like (107), which are topological invariants, might therefore be of particular importance for a study of the structure of our world. Actually, many physicists and mathematicians are now convinced that topological invariants play an important role in the theory of elementary particles. The problem, therefore, is to find these topological invariants in the form of analytic expressions and symmetries.

74.11e. *The theorem of de Rham and the duality theorem of Poincaré.* Because of the great importance of equation (107), many efforts have been made to generalize it as far as possible. Below we give two central results. Let us first remark that important topological invariants may be assigned to each topological space M—its singular homology groups $H_q(M, \mathbb{R})$ with real coefficients where $q = 0, 1, \ldots$. Actually, $H_q(M, \mathbb{R})$ is a real, linear space. The numbers

$$b_q = \dim H_q(M, \mathbb{R})$$

are called *Betti numbers*. If M is arcwise connected, then $H_0(M, \mathbb{R}) = \mathbb{R}$, and hence $b_0 = 1$. For the sphere S^n, one has

$$H_q(S^n, M) = \begin{cases} \mathbb{R} & \text{for } q = 0, n, \\ 0 & \text{otherwise,} \end{cases}$$

i.e., $b_0 = b_n = 1$ and $b_j = 0$ otherwise. The simple construction of $H_q(M, \mathbb{R})$ as well as the intuitive meaning of $H_q(S^n, \mathbb{R})$ and b_q will be explained in Part V. The Euler characteristic of M is defined similarly to (106) through

$$\chi(M) = \sum_{q=0}^{\infty} (-1)^q \dim H_q(M, \mathbb{R}),$$

provided the sum and all summands are finite. This is true, for example, for compact, finite-dimensional, real manifolds.

For arbitrary finite-dimensional, real manifolds with countable basis one has the following fundamental theorem of de Rham

$$H^q(M) \cong H_q(M, \mathbb{R})^*, \tag{108}$$

i.e., the de Rham cohomology groups $H^q(M)$ of Section 74.24 are isomorphic to the dual spaces of $H_q(M, \mathbb{R})$ (see Warner (1971, M)).

Let M be a compact, n-dimensional, real manifold. Then, $\dim H_q(M, \mathbb{R}) < \infty$ for all q. From (108) we immediately obtain that $H^q(M) \cong H_q(M, \mathbb{R})$ and $b_q = \dim H^q(M)$ as well as

$$\chi(M) = \sum_{q=0}^{\infty} (-1)^q \dim H^q(M).$$

Furthermore, the duality theorem of Poincaré states that

$$H^q(M) = H^{n-q}(M), \qquad q = 0, 1, \ldots, n,$$

hence $H_q(M, \mathbb{R}) = H_{n-q}(M, \mathbb{R})$ and $b_q = b_{n-q}$. The geometrical meaning of

Problems   689

this duality will be explained in Part V. The special case of the sphere yields

$$H^q(S^n) = \begin{cases} \mathbb{R} & \text{for } q = 0, n, \\ 0 & \text{otherwise.} \end{cases}$$

Moreover, we have $\chi(S^n) = 1 + (-1)^n$.

The importance of (108) is that $H^q(M)$ is defined purely analytically with differential forms, while the right-hand side of (108) is a topological invariant. Hence (108) represents a fundamental connection between analysis and topology, which essentially was already discovered by Poincaré at the end of the nineteenth century.

74.11f.* *Generalization of the global theorem of Gauss–Bonnet to $\mathbb{R}^{2n+1}$.* Let M be a real, $2n$-dimensional, compact, and oriented manifold in $\mathbb{R}^{2n+1}$. By definition, the Gauss map

$$g: M \to S^{2n}$$

assigns to each point x the outer unit normal vector $g(x)$. We define the Gaussian curvature through

$$K(x) = \det g'(x).$$

From Problem 74.6 it follows that for $n = 1$ this definition coincides with the definition of Section 74.11. For $n \geq 1$, we obtain (107) with

$$k = 2n.$$

Study the differential topological proof contained in Guillemin and Pollack (1974, M), p. 196. This very transparent proof makes essential use of the mapping degree on manifolds and the index theorem of Poincaré–Hopf (see Problem 74.11j). If k is odd, then (107) is wrong, because in this case $\chi(M) = 0$, whereas $\int_M K\, dm \neq 0$ holds for a sphere.

74.11g.* *The theorem of Chern* (1944). Formula (107) has the disadvantage that M has to be embedded in $\mathbb{R}^m$ and K is not an intrinsic property of M. Chern succeeded in finding a differential form γ for real, $2n$-dimensional, compact oriented Riemannian manifolds such that

$$\int_M \gamma = \chi(M). \tag{109}$$

Study this proof in Chern (1944), (1959, L) and in Sulanke and Wintgen (1972, M). The differential form γ is called the Euler class of M. It can explicitly be given through

$$\gamma = \frac{(-1)^n}{(4\pi)^n n!} \operatorname{sgn}\begin{pmatrix} 1 \ldots 2n \\ i_1 \ldots i_{2n} \end{pmatrix} \Omega_{i_2}^{i_1} \wedge \cdots \wedge \Omega_{i_{2n}}^{i_{2n-1}},$$

where Ω_j^i are the so-called curvature forms, i.e.,

$$\Omega_j^i = \tfrac{1}{2} R_{jkm}^i \, du^k \wedge du^m.$$

The differential form γ is a $2n$-cocycle, i.e., $d\gamma = 0$. Hence γ lies in one of the de Rham cohomology classes of $H^{2n}(M)$. From the theorem of Stokes it follows that one can replace γ in (109) with an arbitrary form of this class. Thus the integral in (109) depends only on the cohomology class.

74.11h. *The idea of characteristic classes.* Equation (109) shows that topological invariants may be constructed from differential forms which depend on the curvature. The theory of characteristic classes provides a systematic approach for such a construction. The important point is that vector bundles are considered as principal fiber bundles. This will be explained in Part V. Roughly speaking, characteristic classes measure the twisting of such bundles over M. In this context we recommend Spivak (1979, M), Vol. 5. The characteristic classes of Chern, Pontrjagin, Stiefel–Whitney, and Todd provide us with some deep insights into the properties of manifolds and their bundles. They also play a central role in the formulation of the Atiyah–Singer index theorem; compare Shanahan (1978, L). Characteristic classes are a wonderful tool to describe fundamental connections between analysis and topology; the standard types of which have already been found during the nineteenth century by Gauss, Riemann, and Poincaré.

By reading Gilkey (1984, M), Shanahan (1978, L), and Choquet-Bruhat (1981, M) simultaneously one soon discovers the interrelation between the following topics: pseudodifferential operators and the Atiyah–Singer index theorem for elliptic operators and elliptic complexes, de Rham cohomology, decomposition theorem of Hodge for differential forms, theorem of Gauss–Bonnet–Chern, index theorem of Poincaré-Hopf, Dolbeaut's cohomology and the theorem of Riemann–Roch–Hirzebruch. In Warner (1974, M) one finds the connection between de Rahm cohomology and the cohomology of sheaves discussed. The usefulness of cohomology of sheaves for the solution of fundamental problems in complex function theory, i.e., for the construction of analytic functions from their zeros or poles (Cousin's problems), may be seen from Hörmander (1967, M).

74.11i. *Connections in principal fiber bundles and gauge theories.* This problem will be treated in Part V. Here we just describe the basic idea, as it is closely related to the subject at hand. The standard type of a principal fiber bundle is the frame bundle, which was studied by Ellie Cartan (1869–1961) in connection with his fundamental investigations in differential geometry. His idea was to study the geometry on manifolds by using moving frames (method of moving frames). Thereby he employed his calculus of differential forms. A frame in $\mathbb{R}^n$ consists of n linearly independent vectors which are attached at one point (see Figure 74.21 with $n = 2$). On a manifold M there exists a frame of n linear independent tangent vectors $e_1, \ldots, e_n$ in the tangent space of a fixed point x (Fig. 74.22). The frame bundle $F(M)$ of M consists of all possible tuples

$$(x, e_1, \ldots, e_n).$$

If we represent e_j in the natural basis $\{b_i\}$, i.e.,

$$e_j = c_j^k b_k,$$

then (c_j^i) is a regular matrix. As coordinates of $(x, e_1, \ldots, e_n)$ we choose the coordinates

$$(u^i, c_j^k) \quad \text{with} \quad i, j, k = 1, \ldots, n.$$

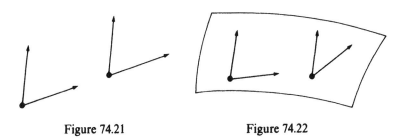

Figure 74.21   Figure 74.22

This way, $F(M)$ becomes a manifold and, as we shall see in Part V, even a principal fiber bundle. It is then important that for each principal fiber bundle, one can introduce a connection by choosing an appropriate 1-differential form ω, which admits the definition of parallel transport and covariant differentiation D. The 2-differential form Ω, which describes the curvature, is given by

$$\Omega = D\omega. \tag{110}$$

This equation is the basic equation of modern gauge field theories. Here ω represents a potential and Ω a field. Roughly, (110) establishes the relation

field = derivative of the potential.

As we shall see in Part V, a standard example of (110) is provided by classical electrodynamics. In this case ω corresponds to the four potential and Ω to the electric and magnetic field.

74.11j. *Index theorem of Poincaré–Hopf.* Let M be a real, finite-dimensional, compact, and oriented manifold. Moreover, let v be a vector field on M with finitely many zeros x_j. Then for the index sum we obtain

$$\sum_j \text{ind}(v, x_j) = \chi(M). \tag{111}$$

Here

$$\text{ind}(v, x_j) \stackrel{\text{def}}{=} \deg(\bar{v}, \bar{x}_j),$$

where $\bar{v}$ is the representative of v in a local chart and $\bar{v}(\bar{x}_j) = 0$. The zero index $\deg(\bar{v}, \bar{x}_j)$ has been defined in Section 12.3.

A proof may be found in Guillemin and Pollack (1974, M), p. 134. Therein, the index theorem is used to prove the global theorem of Gauss–Bonnet. Sulanke and Wintgen (1972, M), p. 236, on the other hand, first prove the global theorem of Gauss–Bonnet–Chern and then deduce (111) in a simple fashion. Hence, the theorem of Gauss–Bonnet and (111) are equivalent.

For the sphere S^2 in $\mathbb{R}^3$ we have $\chi = 2$. Thus (111) implies that there exists no nonvanishing vector field on S^2. This special case of (111) has been proved in Example 13.4 by using the mapping degree (hedgehog theorem). Note that vector fields on manifolds are, by definition, tangent vector fields.

74.11k. *Index theorem of Morse.* We choose M as in Problem 74.11j with $\dim M = n$ and assume that the function $f: M \to \mathbb{R}$ has only finitely many zeros x_j which are all nondegenerate. Prove

$$\sum_{i=0}^{n}(-1)^i M_i = \chi(M), \tag{112}$$

where M_i is the number of points x_j with Morse index i.

Solution: Consider the vector field v on M with $v(x) = f'(x)$. Let $f'(x_j) = 0$ and i the Morse index of x_j. In a local chart, the matrix $f''(x_j)$ has exactly i negative eigenvalues. Thus

$$\deg(f', x_j) = (-1)^i.$$

Hence the theorem follows from (111).

In many cases, this index theorem can be used to compute the Euler characteristic $\chi(M)$ quite easily. Let, for example, M be equal to the sphere S^n in $\mathbb{R}^{n+1}$. The function $\varphi(x) = \zeta_{n+1}$ has a maximum at the North pole with Morse index 0 and a minimum at the South pole with Morse index n. Consequently, $\chi = 1 + (-1)^n$.

74.111.** *Theorem of Adams for vector fields on spheres.* Let $n \geq 2$, and let the number of linearly independent tangent vector fields on the $(n-1)$-dimensional sphere S^{n-1} be equal to $\rho(n-1)$. Then $\rho(n-1)$ is obtained in the following way. Let $a, b, c, d \geq 0$ denote integers with $0 \leq c \leq 3$. We write n as the product of a power of 2 and an odd number

$$n = (2a(n) + 1)2^{b(n)}.$$

Dividing b by 4 gives the representation

$$b(n) = c(n) + 4d(n).$$

Finally, we set

$$\rho(n-1) = 2^{c(n)} + 8d(n) - 1.$$

For the special case of the circle S^1 and the sphere S^2 in $\mathbb{R}^3$ we obtain $\rho(1) = 1$ and $\rho(2) = 0$. An m-dimensional manifold is called parallelizable if and only if there exist m linearly independent vector fields on it. From the above result, exactly the spheres S^1, S^3 and S^7 are parallelizable.

Let k be even. Since each tangent vector field on S^k has a zero (Example 13.4), there is *no* linearly independent vector field on S^k, i.e., $\rho(k) = 0$. In fact, in this case we have $b(k+1) = 0$, and hence $c(k+1) = d(k+1) = 0$. Thus, $\rho(k) = 0$.

The proof of this fundamental topological theorem may be found in Adams (1962) and Schwartz (1968, L), p. 159. The proof makes essential use of K-theory, which will be discussed in Part V.

References to the Literature

Classical works: Euclid (325 B.C.) ("Elements"), Gauss (1827) (surface theory), Riemann (1854) (Riemannian manifolds), Beltrami (1868) (construction of a two-dimensional Riemannian manifold with negative curvature, with non-Euclidean geometry), Klein (1871) (models for non-Euclidean geometries in the context of projective geometry), Klein (1872) (Erlanger program—group-theoretical classification of the geometries), Hilbert (1903, M) (axiomatic foundation for the general geometry), Ricci and Levi-Civita (1901) (covariant differentiation), Einstein (1916) (applications of the calculus of covariant differentiation to the general theory of relativity), Levi-Civita (1917) (parallel transport).

History of non-Euclidean geometry and the concept of manifolds: Klein (1926, M), (1928, M), Scholz (1980, M).

Collected works which contain important contributions in the development of geometry: Gauss (1863), Riemann (1892), Klein (1921), Poincaré (1928), Hilbert (1932), Lie (1934), E. Cartan (1952), Einstein (1960).

Gauss biographies: Worbs (1955, M), Wussing (1974, M), Bühler (1981, M).

Classical surface theory: Kreyszig (1957, M, B, R), Laugwitz (1960, M).

Surface deformation in the large: Efimov (1957, M).

Introduction to classical tensor calculus: Schouten (1954, M) (standard work), Raschewski (1959, M), Zeidler (1979, S) (connection between vector analysis, tensor analysis, differential forms, and differential geometry).

Classical differential geometry: Blaschke (1923, M), (1950, M, H) (applications of differential forms).

Modern differential geometry: Spivak (1979, M, H, B), Vols. 1–5 (recommended as a comprehensive introduction), Helgason (1962, M), Kobayashi and Nomizu (1963, M), Sternberg (1964, M), Sulanke and Wintgen (1972, M), Greub, Halperin, and Vanstone (1972, M), Choquet-Bruhat (1982, M).

Riemannian geometry: Klingenberg (1983, M), Besse (1987, M).

Isometric embedding of Riemannian manifolds in the $\mathbb{R}^n$: Nash (1956) (fundamental work), Schwartz (1969, L), Gromov and Rohlin (1970, S, B), Gromov (1986, M).

Non-Euclidean geometry: Klein (1928, M).

Theorem of Gauss–Bonnet–Chern. Classical works: Gauss (1827), Bonnet (1848), Dyck (1885), Chern (1944). Introduction: Kreyszig (1957, M), Guillemin and Pollack (1974, M), Sulanke and Wintgen (1972, M). Connection with the theory of characteristic classes: Spivak (1979, M), Vol. 5 (recommended as an introduction), Chern (1959, L), Schwartz (1968, L), Greub, Halperin, and Vanstone (1972, M), Vol. 2 (general theory).

Differential forms: Maurin (1976, M), Vol. 2 and Westenholz (1981, M) (introduction), Kähler (1934, M) (applications to systems of partial differential equations), Hodge (1952, M), Cartan (1955, M), de Rahm (1960, M), Greub, Halperin, and Vanstone (1972, M), Vol.s. 1–3, Marsden (1974, L), Abraham and Marsden (1978, M) (applications to mechanics).

Lie groups: Choquet-Bruhat (1982) (introduction), Chevalley (1946, M), Montgomery and Zippin (1955, M), Pontrjagin (1966, M), Warner (1974, M), Dieudonné (1975, M), Vols. 4 and 5.

Modern differential geometry and its applications to mathematical physics: Dubrovin, Novikov, and Fomenko (1979, M), Westenholz (1981, M), Choquet-Bruhat (1982, M), Curtis (1986, M).

CHAPTER 75

Special Theory of Relativity

He who understands geometry may understand anything in this world.
Galileo Galilei (1564–1642)

The ways of people to the laws of nature are no less admirable than the laws themselves.
Johannes Kepler (1571–1630)

It is known that Maxwell's electrodynamics, when applied to moving bodies, leads to asymmetries which do not appear to be inherent in the phenomena. Take, e.g., the electrodynamic interaction between a magnet and a conductor. The observable phenomenon here depends only on the relative motion of the conductor and the magnet, whereas the customary view draws a sharp distinction between the two cases in which either the one or the other of these bodies is in motion.
Albert Einstein (1905)
(Beginning of his paper on the special theory of relativity)

If the energy of a body changes by ΔE, then its mass changes by

$$\Delta m = \Delta E/c^2.$$

Here c is the velocity of light.
Albert Einstein (1905a)

If Einstein's theory of relativity proves correct, which I expect, then he will be celebrated as the Kopernikus of the twentieth century.
Max Planck (1909)

Henceforth space by itself and time by itself are doomed to fade away into mere shadows and only a kind of union of the two will preserve an independent reality.
Hermann Minkowski (1909)

At the moment I am only concerned with the gravitational problem and I hope to overcome all the difficulties with the help of a local friend and mathematician (Marcel Grossmann). But it is true that, never in my life, I have worked so hard,

75. Special Theory of Relativity

and I am filled with a great respect for mathematics. In its more subtle parts, I have regarded it, in my simplicity, as pure luxury.

Albert Einstein in a letter of October 1912

We set

$$R_{ij} = \kappa(T_{ij} - 2^{-1}g_{ij}T).$$

This completes the general theory of relativity as a logical structure. The postulate of relativity in its most general form, which makes the space–time coordinates meaningless parameters, leads necessarily to a certain form of gravitational theory which explains the motion of the Perihelion.

Anyone, who really has grasped the general theory of relativity, will be captured by its beauty. It is a triumph of the general differential calculus, which was created by Gauss, Riemann, Christoffel, Ricci, and Levi-Civita.

Albert Einstein (1915)

The development of the general theory of relativity appears to me to be the greatest achievement of scientific thought over the laws of nature, an admirable unification of philosophical depth, physical intuition, and mathematical skills.

Max Born (1957)

In this and the following chapters we shall discuss the basic ideas of the general theory of relativity, explain its connection with the theory of manifolds, and give applications in the form of three interesting problems:

(i) Motion of the Perihelion of Mercury.
(ii) Big Bang and the expansion of the universe.
(iii) Black holes.

The present Chapter 75 on the special theory of relativity and Chapter 76 on the general theory of relativity form a unity. We therefore give problems and references to the literature at the end of Chapter 76. In connection with (i) we want to mention that this phenomenon is discussed in the physical literature only in first-order approximation, and the method used, is usually not clearly motivated. The first step of an iteration scheme is used. The difficulty, however, is that one has to solve an equation which has several solutions. This difficulty is avoided by formally choosing a solution which is physically meaningful. In Section 76.9 we present a consistent method which also uniquely determines all higher-order approximations and we shall prove the convergence of this method. We employ the same bifurcation methods which have been used in Section 8.12 in studying nonlinear oscillations. This method may also be applied to many other problems in celestial mechanics.

Einstein's theory of relativity has been developed in two fundamental papers, which appeared during the years 1905 (special theory of relativity) and 1916 (general theory of relativity). The special theory of relativity begins with the *principle of relativity*:

(R) *All physical processes have the same form for all inertial systems.*

This principle will be discussed more precisely in Section 75.2. Because the velocity of light is constant, a change of space and time between inertial systems is given by Lorentz transformations. In addition, all physical laws have to be given in such a way that they assume the same mathematical form for all inertial systems. This can be achieved by formulating these laws as geometrical laws for Minkowski's uncurved four-dimensional space–time manifold M_4.

The general theory of relativity represents an extension of Newton's theory of gravity to arbitrary systems of reference. Newton's gravitational force is replaced with the *curvature* of Einstein's four-dimensional space–time manifold E_4, which is caused by the mass distribution. Conversely, the curvature effects the mass motion, which occurs on geodesics in E_4. The following ingenious geometrization concept forms the basis for the general theory of relativity. It represents the deepest known connection between physics and mathematics:

(G) *Physical interactions can be reduced to geometrical properties.*

In the introduction to the previous chapter we already mentioned that at present many efforts are being made to use this geometrization principle to find a unified theory for all interactions. This will be discussed in Part V. The experimental results, which led Einstein to (R) and (G), were the fact that the velocity of light is constant (Michelson experiment) and the equivalence of gravitational and inert mass. Later on this will be further explained.

For the reader who wants to get acquainted with the modern development in the theory of relativity, we recommend Carelli (1979), Hawking and Israel (1979), Held (1980), and Schmutzer (1983). These are four voluminous conference reports, which appeared on the occasion of the centennial of Albert Einstein's (1879–1955) birthday. Furthermore, we recommend the collection of survey articles by Dewitt and Stora (1984) and the monographs by Einstein (1953) and (1965) on the theory of relativity and Einstein's "Weltbild," as well as the collected works of Einstein (1960). A standard work on the theory of relativity is Misner, Thorne, and Wheeler (1973). Finally, we recommend the conference report Ruffini (1987).

The following citations should illustrate the long historic development which led to the geometrization of physics in the context of the general theory of relativity.

> Geometry is the knowledge of what eternally exists.
> Plato (427–347 B.C.)

> Every process in nature will occur in the shortest possible way.
> Leonardo da Vinci (1452–1519)

> All human knowledge begins with intuition, thence passes to concepts, and ends with ideas.
> Immanuel Kant in his "Criticism of pure reasoning," 1781.
> (This citation was chosen by Hilbert (1903) as the motto of his *Foundations of Geometry*.)

> In humbleness, we have to admit that if "number" is a product of our imagina-

tion, "space" has a reality outside of our imagination, to which *a priori* we cannot assign its laws.

Carl Friedrich Gauss (1777–1855) in a letter to Bessel

Riemann, 1854, presented three topics for his inaugural lecture. Gauss, in recollection of his own struggle with Euclid's parallel axiom, chose—in breaking with tradition—the third one: "On the hypotheses, which form the basis of geometry." In his lecture, Riemann presented the fundamentals of a geometry for the n-dimensional curved metric space (Riemannian geometry). This must have made an extremely deep impression on Gauss, who at that time was already very weak. Later, on his way home, he spoke with unusual excitement to Wilhelm Weber about the depth of the presentation.

Erich Worbs in his Gauss biography (1955)

Every geometry is a theory about invariants of a transformation group.

Felix Klein (1872) (Erlanger program)

In physics there exists no concept, which *a priori* is necessary or justified. A concept only becomes justified through its clear and unique correspondence with events or physical experience. Newton's concepts of absolute simultaneity, absolute velocity, and absolute acceleration were abandoned in the theory of relativity, because a unique connection with the world of experience appeared to be impossible. The same applies to the concepts of the plane, straight line, etc., upon which Euclidean geometry is based. Every physical concept must be given a definition such that in a concrete situation, the validity or nonvalidity of this concept can principally be determined.

Albert Einstein (1920)

Formerly it was believed that if all things vanish from this world, space and time would remain, but according to the theory of relativity, space and time vanish together with all things.

Albert Einstein (1921)

"Every little boy in the streets of our mathematics-Göttingen knows more about four-dimensional geometry than Einstein," wrote David Hilbert with excusable exaggeration. "But in spite of this," Hilbert added, "Einstein has completed the work, not the mathematicians ...."

When the world was amazed with Einstein's theory of relativity, Minkowski said: "To me this came as a great surprise, because as a student, Einstein had been a lazy duck. Never has he been interested in mathematics."

Timothy Ferris (1977)

Only the genius Riemann, lonesome and unrecognized in the middle of the previous century, found the way towards a new conception of space, whereby space looses its stiffness, and gains the ability to participate in physical events.

Einstein (1953)

The following citations might help the reader to understand Einstein as a human being.

I saw Einstein for the first time in Berlin in 1921, when I was wandering through the streets, trying everything to enrol at the university where Planck, Laue, and Einstein taught. I felt miserable, since I didn't know anyone. I was as lonesome as one could possibly be in a great and hostile city. For weeks I waited for the chance to meet some influential people, only to find out, how little they cared, whether or not I would become a student at Berlin University. In my desperation I called Einstein, and to my greatest surprise I was invited to his house.

Kindness is a difficult thing to handle amidst all this coldness and hostility. Einstein welcomed me with a smile and offered me a cigarette, spoke to me like to one of his own, and took everything with a childlike confidence. This short discussion was an important event in my life. Instead of thinking about his ingenuity and his achievements in the area of physics, I thought then, as well as later, about his great kindness, his loud laughter, the shining of his eyes, and—the awkwardness with which, on a table covered with all sorts of papers, he looked for a piece of paper—about the mixture of great kindness and great remoteness.

Leopold Infeld (1969)

Not the person counts, but the work for the community.

Einstein (1929)

Few people are able to express opinions which differ from the common prejudices; most people are not even able to form them.

Einstein (1955)

Why do people always babble about my theory of relativity? I have done other useful things, maybe even better ones. But the audience does not take any notice of this.

Einstein (1955)

[In this connection we mention the paper (1905b) on the photoeffect in which Planck's quantum hypothesis has been used to predict the existence and properties of photons. Also, one may think of his paper (1905c) on the quantitative theory of Brownian motion, which later on was extended by Norbert Wiener to the theory of stochastic processes. In 1915, Einstein and de Haas observed experimentally an effect for ferromagnetica, which ten years later found its explanation through the discovery of the electron spin.]

Einstein is completely right that empiricism without bold ideas leads to nothing. A master is able to find the right mixture between both.

Max Born (1957)

One thing I have learned in a long life: that all our science, measured against reality, is primitive and childlike—and yet it is the most precious we have.

Einstein (1955)

History will tell that the best citizens in every country, the best defenders of honor, were always those, who, by risking their positions, their names, or even their lifes, spoke out against the errors and stupidities of their fellow-men.

Romain Rolland (1866–1944)

All students in Germany, all students in the entire world, should be brought here to see how horrible the war really is.

Einstein in 1922, when visiting the battlefields of Verdun (France)

The political apathy of people during times of peace is a sign of their later willingness to be massacred. Because today they are not willing to support disarmament, they will be forced tomorrow to loose their blood.

Einstein (1928)

> Dictatorship brings the muzzle and with it comes the lethargy. Science can only flourish in an athmosphere of free speech.
> Einstein (1929)

> The ideas and methods of the past did not prevent the wars; the ideas of the future must make them impossible.
> Einstein to the *New York Times* (1946)

> Society is in a crises which, in its full consequences, has not yet been recognized by those having the power to decide between good and bad. The released atomic force has changed everything except our way of thinking, and unprepared we slip into another catastrophy.
> Einstein (1955)

> It is the high determination of people to serve rather than to rule or to be supreme over others in any other form.
> Einstein (1955)

75.1. Notations

In this chapter $y = \xi e_1 + \eta e_2 + \zeta e_3$ denotes a position vector in a Cartesian coordinate system with point coordinates (ξ, η, ζ). Moreover, we let t denote time. Let $u = (u^1, u^2, u^3, u^4)$ with

$$u^1 = \xi, \quad u^2 = \eta, \quad u^3 = \zeta, \quad u^4 = ct,$$

where c is the velocity of light, i.e., $c = 299{,}793$ km/s. In the general theory of relativity, $u = (u^1, u^2, u^3, u^4)$ are arbitrary coordinates, where u^1, u^2, u^3 are space coordinates and u^4 is a timelike coordinate. This will be discussed more precisely in Section 76.2. Equal upper and lower Latin (or Greek) indices are always summed from 1 to 4 (or 1 to 3). A dot as in $\dot\xi$ means derivative with respect to time, i.e., $d\xi/dt$. On the other hand, the prime in ξ' does not stand for a derivative, but instead refers to the system Σ'.

75.2. Inertial Systems and the Postulates of the Special Theory of Relativity

At the beginning of his paper (1905) "On the electrodynamics of moving bodies," which represents the foundation of the special theory of relativity, Einstein formulated the following two postulates in a somewhat modified form. The concept of inertial system will be explained below:

(R) *Einstein's principle of relativity.* All inertial systems are physically equivalent, i.e., physical processes are the same in all inertial systems when initial and boundary conditions are the same.

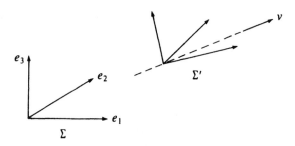

Figure 75.1

(C) *Constant velocity of light.* In every inertial system, light travels with the same constant velocity c in every direction.

We shall see in Section 75.5 and the following sections how these apparently very simple postulates lead to a fundamental revision of the classical concepts of time and physics. Also, we postulate:

(T) *Translation principle.* There exists an inertial system. If Σ is an inertial system, then also each Cartesian coordinate system Σ', which is obtained from Σ by a *constant translatory motion*, is an inertial system.

Recall that we mean by a translatory motion that Σ' is not rotated compared with Σ. By a *constant* translatory motion of Σ' we mean a constant motion of Σ' with respect to Σ with constant velocity vector v (Fig. 75.1).

In order to understand these postulates, we need a definition of inertial systems. A formal, mathematical definition will be given in Section 75.8. Here we shall give a heuristic description to illustrate the physical meaning.

(I) A Cartesian coordinate system Σ is an *inertial system* precisely if there exists a system time t for it such that each mass point, which is far enough away from other masses and shielded against fields (e.g., light pressure), remains at rest or moves rectilinearly with constant velocity.

75.3. Space and Time Measurements in Inertial Systems

A mathematical axiom system uses terms which are not explained any further. For example, Hilbert (1903), in his famous "Foundations of Geometry," used the concepts "points, straight lines, and planes" in this way. A fundamental difference in the way mathematicians and physicists work, is that a physicist does not have this luxury. He needs to know how physical concepts are connected to his experiments and how he can measure quantities such as space, time and momentum. A physicist is in the situation of a man who has to fight his way through a dark labyrinth with many dead ends. Permanently, hypotheses have to be made, which can only approximately be verified by experiments, and a great number of experiments have to be compared in order to find contradictions among those hypotheses. The theory of relativity and quantum theory both showed the importance of the process of measurement for physical theories.

Let us examine how a physicist can use (C), (T), and (I), to find an inertial system which then allows him to introduce meaningful processes of measurements for space and time. In connection with measurements of time, we always think of atomic clocks when speaking of clocks. Thereby we have Cesium-133 atoms in mind. Passing from one energy level to another, in these atoms, generates microwaves with 9,192,631,770 oscillations per second. The error of these clocks is around 10^{-9} seconds a day. Modern experimentalists use hydrogen masers instead.

Consider a physicist P in a space ship S, which is located at a far distance from stars and which flies without rocket propulsion. We expect that S is an inertial system. In order to verify this experimentally, P uses an atomic clock and drops an object without accelerating it. If P observes rest or a rectilinear motion with constant velocity, he concludes that S is an inertial system. More precisely, S may be chosen as the origin of an inertial system Σ. We introduce the following measurements of space and time in Σ. Note that we are dealing with cosmic distances.

(i) Time differences. All observers Q in Σ agree via radio to measure time differences with atomic clocks that are built the same way.
(ii) System time. In order to synchronize all clocks, P at his local time t sends a light signal (radar signal or laser) to Q which is reflected there and returns to P after time Δt. Then P informs the observer Q via radio that the light signal has been at Q at the system time $t + \Delta t/2$.
(iii) Measurement of distances. Because of (C), P measures the distance between Q and himself as $c \cdot \Delta t/2$. Thereby Euclidean geometry and a rectilinear motion of light are assumed.

Using a known inertial system Σ and the translation principle (T), one can now determine whether or not other systems Σ' in the universe are also inertial systems. Astronomical experience shows that the system Σ_{sun} is a good approximation of an inertial system. The origin of Σ_{sun} is the center of gravity of our solar system, which lies within the sun. The axes of Σ_{sun} point towards fixed stars which can be chosen arbitrarily. We only require a Cartesian coordinate system.

Recall that, by definition, the axes of a Cartesian coordinate system Σ are always positively oriented, i.e., are as in Figure 75.1. Our physicist P in his space ship knows this positive orientation even without his light system. He only needs the first three fingers of his right hand. But how can he communicate this positive orientation to distant creatures which might have no right hand? The answer to this is an experiment which was performed by Mrs. Wu and her co-workers in 1957, and which showed the violation of parity

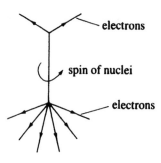

Figure 75.2

(asymmetry of space reflections) for weak interactions. The effect is that an β-decay of $^{60}_{27}Co$ kernels results in the emission of 30% more electrons antiparallel to the spin direction than parallel (Fig. 75.2).

75.4. Connection with Newtonian Mechanics

In classical mechanics, Newton assumed the existence of an absolute space and an absolute time. This allows the following very precise definition of an inertial system.

In the absolute space Σ_a we have the equation of motion

$$m_a \frac{d^2 y_a}{dt_a^2} = K_a, \tag{1}$$

where Σ_a stands for a fixed Cartesian coordinate system which is at rest in the absolute space, and where clocks show the absolute world time t_a. A particle is called force-free if and only if $K_a \equiv 0$. A Cartesian coordinate system Σ is an *inertial system* precisely if it is at rest in Σ_a or obtained from a translation. It follows from classical mechanics that the change from Σ_a to Σ is given by

$$y = y_a - tw + b, \quad t = t_a, \quad m = m_a \tag{2}$$

with fixed velocity vector w and fixed vector b. This is shown in Section 58.6. At time $t_a = 0$, the origin $y_a = 0$ of Σ_a has the coordinate $y = b$ in Σ. From (1) and (2) we obtain as the equation of motion in Σ

$$m \frac{d^2 y}{dt^2} = K_a, \tag{3}$$

which has the same form as (1). In Section 58.6 we also saw that systems, which are not inertial systems, i.e., which move accelerated in Σ_a, satisfy equation

$$m \frac{d^2 y}{dt^2} = K_a + A, \tag{4}$$

75.4. Connection with Newtonian Mechanics

where A is the force induced by the acceleration. From $K_a \equiv 0$ and (3) we immediately obtain that $y = y(0) + \dot{y}(0)t$. This implies:

(I_{class}) A Cartesian system Σ with world time t as system time is an inertial system if and only if every force-free particle remains at rest or moves rectilinearly with constant velocity.

Note that force-free refers to Σ_a and not to the forces $K = K_a + A$, which are observed in the system of reference Σ. As we shall see shortly, (I_{class}) cannot be used in the theory of relativity, since there Σ_a cannot be used to give a precise definition of the concept force-free. This is the reason we used formulation (I) in Section 75.2.

If Σ and Σ' are two inertial systems, then (2) yields the Galilei transformation

$$y' = y - tv + y_0, \qquad t' = t, \quad m' = m \tag{5}$$

with $v = w' - w$ and $y_0 = b' - b$. Furthermore, in Σ' we obtain the equation of motion

$$m \frac{d^2 y'}{dt^2} = K_a. \tag{6}$$

Comparison of (3) and (6) gives the classical principle of relativity:

(R_{class}) The *mechanical* processes are the same for all inertial systems when the initial conditions are the same.

If Σ' is obtained from the system Σ by a translation which may be accelerated, then we find that $y'(t) = y(t) - a(t)$ (Fig. 75.3). Differentiation with respect to t gives the following addition theorem for velocities

$$\dot{y}' = \dot{y} - \dot{a}. \tag{7}$$

In the general case, Section 58.6 implies that the transformation rule for motions $y = y(t)$ and $y' = y'(t)$ in Σ and Σ' is given by

$$\dot{y}' = \dot{y} - \dot{a} - (\omega \times y'). \tag{8}$$

Here $\dot{a}$ and ω are the translational and the rotational velocities of Σ' with respect to Σ. All quantities in (8) depend on t.

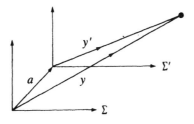

Figure 75.3

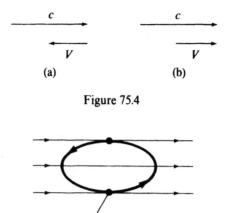

Figure 75.4

Figure 75.5

The physicists were confronted with the following important question.

(Q) *How can the absolute space Σ_a be determined experimentally?*

The following answers were proposed at the end of the last century.

(i) From the classical principle of relativity (R_{class}) it is clear that the absolute space cannot be determined by mechanical experiments. One may, however, try to use light.
(ii) Consider, as in Figure 75.4, two bodies (e.g., two cars) with velocities c and V. The classical addition theorem implies the relative velocities $c + V$ and $c - V$ in Figure 75.4(a) and (b), respectively.
(iii) Assume that (ii) can be applied to light with velocity c. It follows then that light cannot travel with the same velocity in *each* direction for every inertial system. There is at most one such inertial system.
(iv) Now assume that there actually exists an inertial system Σ_c for which the velocity of light is constant. We set $\Sigma_a = \Sigma_c$, i.e., we identify the absolute space with Σ_c.
(v) Since the earth moves in the inertial system Σ_{sun} of Section 75.2 on an elliptic orbit, i.e., accelerated, no system Σ_{labor} which is firmly connected to the earth can be an inertial system. In particular, we have $\Sigma_{labor} \neq \Sigma_a$. If we consider the orbit of the earth such as in Figure 75.5, then transformation formula (8) immediately implies that the velocity of light in Σ_{labor} cannot be constant for the whole year.
(vi) During the years from 1881 to 1887, Michelson performed very precise interference experiments. To his great surprise, he observed that the velocity of light in Σ_{labor} remained constant. This Michelson experiment was the experimental starting point for the special theory of relativity.

Let us describe this experiment. As in Figure 75.6 we consider a light source S which emits monochromatic light onto a semipermeable plate P. After

75.4. Connection with Newtonian Mechanics

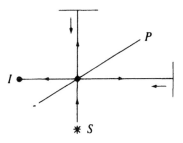

Figure 75.6

reflection, we obtain interference bands in I. Rotating the whole device about different angles α, one expects changes in the interference pictures, because the axes of the system travel for different α with different velocities relative to the absolute space. In spite of a very high accuracy in his measurements, Michelson did not observe any such changes.

In order to resolve this contradiction we make, with Einstein (1905), the following observations.

(a) Since (iv) leads to a contradiction we assume that light travels in every inertial system with a constant velocity. This is requirement (C) of Section 75.2.
(b) Furthermore, we take the more radical but logically very satisfactory point of view that no particular inertial system can be distinguished by means of physical experiments. This leads to the principle of relativity (R) of Section 75.2. In this way, one eliminates the special role of mechanics which is expressed in (R_{class}). The idea behind (R) is the unity of all physical phenomena. Since, according to (R), absolute space cannot be determined through experiments, we consider the concept of absolute space as physically meaningless. Therefore, the concepts of absolute rest and absolute velocity also lose their meaning.

In the following section we study the mathematical implications of (C) and (R). This leads to a revision of classical mechanics. In particular, transformation formula (5) and hence the addition theorem for the velocities does not hold any more. We will see in Part V that other than classical mechanics, Maxwell's theory of electrodynamics need not be changed if space and time, electric and magnetic field strengths, and charge densities and currents are transformed relativistically between inertial systems. The finding of these transformation rules was one of the main goals of Einstein's classical paper (1905).

The considerations above show a phenomenon which can be observed more generally in the history of physics. New theories are developed when experimental results can no longer be explained by means of the old theories. Under certain conditions, however, the old theory is contained as an approximation in the new theory. As we shall see, classical mechanics is a special case of relativistic mechanics if the velocities are small compared with the velocity of light, i.e., belong to our everyday experience.

75.5. Special Lorentz Transformation

Consider two Cartesian coordinate systems Σ and Σ' with corresponding space coordinates $y = (\xi, \eta, \zeta)$ and $y' = (\xi', \eta', \zeta')$. Assume also that Σ and Σ' are two inertial systems with corresponding system times t and t'. Furthermore, Σ' is obtained from system Σ by a constant translatory motion with velocity v. Using a fixed rotation of Σ and Σ' and a translation of the coordinates y' and t', one can always get the following more simple situation:

(S) At time $t = 0$, the two inertial systems Σ and Σ' coincide, and we have $t' = 0$. Moreover, $v = Ve_1$, i.e., the translation is performed for $V > 0$ along the ξ-axis, and for $V < 0$ in the opposite direction (Fig. 75.7).

The coincidence of Σ and Σ' means that both *origins* are equal at time $t = t' = 0$, and the corresponding coordinate axes have the same direction.

Postulate 75.1. The change from Σ to Σ' is given by the *special Lorentz transformation*

$$\xi' = \frac{\xi - Vt}{\sqrt{1 - V^2/c^2}}, \qquad t' = \frac{t - V\xi/c^2}{\sqrt{1 - V^2/c^2}}, \qquad \eta' = \eta, \quad \zeta' = \zeta, \qquad (9)$$

where c is the velocity of light and $|V| < c$.

The inverse transformation of (9) is

$$\xi = \frac{\xi' + Vt'}{\sqrt{1 - V^2/c^2}}, \qquad t = \frac{t' + V\xi'/c^2}{\sqrt{1 - V^2/c^2}}.$$

If the translational velocity V is very small compared to the velocity of light c, then we obtain from (9) in first-order approximation the classical Galilei

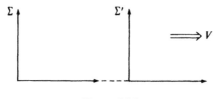

Figure 75.7

75.5. Special Lorentz Transformation

transformation

$$\xi' = \xi - Vt, \qquad t' = t.$$

As a *motivation*, we will show that the following two natural conditions immediately lead to (9).

(i) The change from Σ to Σ' is given by a linear regular transormation, i.e.,

$$\xi' = \alpha\xi + \beta t, \qquad t' = \gamma\xi + \delta t, \qquad \eta' = \eta, \quad \zeta' = \zeta. \tag{10}$$

(ii) We assume Einstein's two postulates (C) and (R) of Section 75.2.

From assumption (C) that the velocity of light is constant, it follows that $\xi = ct$ becomes $\xi' = ct'$. This yields the following central condition

$$c\alpha + \beta = c^2\gamma + c\delta.$$

For $\xi' = 0$ and t' arbitrary, we require $\xi = Vt$ for all t, i.e.,

$$\alpha V + \beta = 0.$$

The inverse transformation of (10) is

$$\xi = \mu(\delta\xi' - \beta t'), \qquad t = \mu(\alpha t' - \gamma\xi'), \qquad \mu^{-1} = \alpha\delta - \beta\gamma. \tag{11}$$

For $\xi = 0$ and t arbitrary, we require $\xi' = -Vt'$ for all t', i.e.,

$$\delta V + \beta = 0.$$

This implies

$$\beta = -\alpha V, \qquad \delta = \alpha, \qquad \gamma = -\alpha V/c^2, \qquad \mu^{-1} = \alpha^2(1 - V^2/c^2).$$

Assume that $V \geq 0$. For $t = 0$ and $\xi > 0$ we require $\xi' > 0$, and hence $\alpha > 0$. The free parameter α is then determined from the principle of relativity (R). This means that Σ and Σ' are *equivalent*. We therefore require in (11) that the coefficient of ξ' in the first equation is equal to the corresponding coefficient α in (10), i.e., $\mu\delta = \alpha$. This gives $\mu = 1$, and hence $\alpha = 1/\sqrt{1 - V^2/c^2}$. At the same time, we obtain $V < c$.

For $V \leq 0$ one uses an analogous argument. This motivates (9).

The following invariance relation

$$c^2 t^2 - \xi^2 = c^2 t'^2 - \xi'^2 \qquad \text{for all} \quad t, \xi \in \mathbb{R} \tag{12}$$

is the key for the geometrical interpretation of the special theory of relativity of Section 75.8. There we will use (12) to define, with Minkowski (1909), a Riemannian metric for the four-dimensional space–time manifold.

Proposition 75.2. *Formula* (12) *holds for all the special Lorentz transformations* (9). *Conversely, every linear transformation* (10) *which satisfies* (12) *is a special Lorentz transformation* (9), *except for reflections of the variables ξ' and t'.*

PROOF.

(I) It follows from (10) and (12) that

$$c^2t^2 - \xi^2 = c^2t^2(\delta^2 - \beta^2/c^2) - \xi^2(\alpha^2 - c^2\gamma^2)$$
$$+ 2\xi t(c^2\gamma\delta - \alpha\beta),$$

hence

$$\alpha^2 - c^2\gamma^2 = 1, \qquad \delta^2 - \beta^2/c^2 = 1, \tag{13}$$
$$c^2\gamma\delta = \alpha\beta. \tag{14}$$

We want to prove (9). From (13) we have $\delta^2, \alpha^2 \geq 1$. Using a reflection of the variables in (10), we can always assure that $\alpha, \delta \geq 1$. Let us assume that $\beta \geq 0$. From (14) it follows then that $\gamma \geq 0$. Therefore, there are numbers $0 \leq \rho, \sigma < \infty$ with

$$\alpha = \cosh \rho, \qquad \delta = \cosh \sigma.$$

From (13) it follows that

$$c\gamma = \sinh \rho, \qquad \beta = c \sinh \sigma.$$

Equation (14) implies that $\rho = \tanh \sigma$, hence $\rho = \sigma$. Now we choose a number $V \geq 0$ with $V^2/c^2 = 1 - 1/\alpha^2$. This implies (9). Analogous arguments are used for $\beta \leq 0$.

(II) Conversely, (12) is an immediate consequence of (9). □

75.6. Length Contraction, Time Dilatation, and Addition Theorem for Velocities

We consider in this section two inertial systems Σ and Σ' for which situation (S) of the previous section applies. Let P and P' denote observers in Σ and Σ', respectively.

EXAMPLE 75.3 (Length Contraction). Consider a rod of length l_0 which is at rest on the ξ-axis in Σ. Then P' in Σ' measures the length

$$l' = l_0\sqrt{1 - V^2/c^2}. \tag{15}$$

To see this, let ξ_1 and $\xi_2 = \xi_1 + l$ be the coordinates at both endpoints of the rod in Σ. From (9) we obtain in system Σ' as equations of motion for these endpoints

$$\xi'_j = \alpha(\xi_j - Vt), \qquad t'_j = \alpha(t - V\xi_j/c^2)$$

with variable t. Here $j = 1, 2$ and $\alpha = 1/\sqrt{1 - V^2/c^2}$. Solving for ξ_j and t gives

$$\xi_j = \alpha(\xi'_j + Vt'_j), \qquad t = \alpha(t'_j + V\xi'_j/c^2). \tag{16}$$

75.6. Length Contraction, Time Dilatation, and Addition Theorem

It is important then that P' measures the length $l' = \xi_2' - \xi_1'$ not at the same t time, but instead the same t' time. Letting $t_1' = t_2'$ in (16) we obtain from substraction $l_0 = \xi_2 - \xi_1 = \alpha l'$. This is (15).

Now, consider a cuboid of volume Q_0 in the inertial system Σ which is parallel to the axes and at rest. Because of (15) and $\eta' = \eta$, $\zeta' = \zeta$, one observes the volume

$$Q' = Q_0\sqrt{1 - V^2/c^2} \qquad (17)$$

in the inertial system Σ'. One easily sees that this formula is also true for an arbitrary cuboid in Σ which is at rest. By integration, (17) can then be extended to arbitrary volumina.

Formulas (15) and (17) are not in contradiction to the principle of relativity, because Σ and Σ' are not equivalent: the rod is at rest in Σ, while it moves in Σ'. In Σ', one observes the following velocities for the endpoints of the rod

$$\frac{d\xi_j'}{dt_j'} = \frac{d\xi_j'}{dt}\frac{dt}{dt_j'} = -V, \qquad j = 1, 2.$$

This implies: If a body moves in an inertial system rectilinearly with constant velocity V, then all lengths in this inertial system in the direction of motion are shortened by a factor $\sqrt{1 - V^2/c^2}$, compared to lengths in the system at rest.

EXAMPLE 75.4 (Time Dilatation). In an inertial system Σ two events take place at the same place $x = (\xi, \eta, \zeta)$ and at different times t_1 and $t_2 = t_1 + \Delta t$. If $\Delta t'$ is the time which passes between the two events for an observer in the inertial system Σ', then

$$\Delta t' = \Delta t/\sqrt{1 - V^2/c^2}, \qquad (18)$$

i.e., for P' the two events appear dilated compared to an observer P in Σ.

To show this we note that for P' the two events have coordinates (ξ_j', t_j'), $j = 1, 2$ with

$$t_j' = \alpha(t_j - V\xi/c^2), \qquad \xi_j' = \alpha(\xi - Vt_j)$$

Substraction yields (18).

This time dilatation does not contradict the principle of relativity, since for P the two events take place at the same place, while under the assumption $\Delta t \neq 0$ and $V \neq 0$ this is not the case for P'. Therefore, the two inertial systems Σ and Σ' are not equivalent.

EXAMPLE 75.5 (Relativistic Addition Theorem for Velocities). Suppose the motion of a point object is given by the equation

$$y = y(t) \quad \text{in the inertial system } \Sigma.$$

This motion corresponds to

$$y' = y'(t') \quad \text{in the inertial system } \Sigma'.$$

Thus one observes the velocities

$$w = dy(t)/dt \quad \text{and} \quad w' = dy'(t')/dt'$$

in Σ and Σ', respectively. The velocity components satisfy

$$w'_1 = (w_1 - V)/\gamma, \quad w'_2 = \beta w_2/\gamma, \quad w'_3 = \beta w_3/\gamma \qquad (19)$$

with $\beta = \sqrt{1 - V^2/c^2}$ and $\gamma = 1 - w_1 V/c^2$.

PROOF. From $\eta'(t') = \eta(t)$, $\zeta'(t') = \zeta(t)$, and

$$\xi' = \alpha(\xi(t) - Vt), \quad t' = \alpha(t - V\xi(t)/c^2)$$

it follows by differentiation

$$\frac{d\xi'}{dt'} = \frac{d\xi'}{dt} \bigg/ \frac{dt'}{dt} = (\dot{\xi} - V)/(1 - V\dot{\xi}/c^2),$$

$$\frac{d\eta'}{dt'} = \frac{d\eta'}{dt} \bigg/ \frac{dt'}{dt} = \dot{\eta}/\alpha(1 - V\dot{\xi}/c^2). \qquad \square$$

Equations (19) imply that

$$c^2 - w'^2 = \frac{(c^2 - w^2)(1 - V^2/c^2)}{(1 - w_1 V/c^2)^2}.$$

Hence, because of $|V| < c$ we have the following results:

(i) From $|w| < c$ follows $|w'| < c$, i.e., under light velocity remains under light velocity.
(ii) From $|w| = c$ follows $|w'| = c$, i.e., velocity of light remains velocity of light.
(iii) For over light velocities with $|w_1| > c$ in an inertial system Σ there always exists a $V > 0$, i.e., an inertial system Σ' for which transformation (19) becomes singular, i.e., $|w'|$ becomes infinitely large. $\square$

In Section 75.9 we will see that it is meaningful to require that physical effects cannot travel with over light velocities.

75.7. Lorentz Group and Poincaré Group

For all $V \in \mathbb{R}$ with $|V| < c$ we define the matrices

$$L(V) = \begin{pmatrix} \alpha & 0 & 0 & -\alpha V/c \\ 0 & 1 & 0 & 0 \\ 0 & 0 & 1 & 0 \\ -\alpha V/c & 0 & 0 & \alpha \end{pmatrix}, \quad R = \begin{pmatrix} \varepsilon_1 & 0 & 0 & 0 \\ 0 & \varepsilon_2 & 0 & 0 \\ 0 & 0 & \varepsilon_3 & 0 \\ 0 & 0 & 0 & \varepsilon_4 \end{pmatrix},$$

75.7. Lorentz Group and Poincaré Group

$$T = \begin{pmatrix} t_{11} & t_{12} & t_{13} & 0 \\ t_{21} & t_{22} & t_{23} & 0 \\ t_{31} & t_{32} & t_{33} & 0 \\ 0 & 0 & 0 & 1 \end{pmatrix}, \quad u = \begin{pmatrix} \xi \\ \eta \\ \zeta \\ ct \end{pmatrix},$$

with $\alpha = 1/\sqrt{1 - V^2/c^2}$ and $\varepsilon_i = \pm 1$. Moreover, let T be an orthogonal matrix with $\det T = 1$. Assume the same for T_1 and T_2. The special Lorentz transformation (9) can now be written in the short form

$$u' = L(V)u.$$

Definition 75.6. Precisely all maps of the form

$$u' = \Lambda u + a \tag{20}$$

with $\Lambda = RT_1 L(V) T_2$ are called *Poincaré transformations*. For $a = 0$ and $\varepsilon_4 = 1$ we speak of a Lorentz transformation. These Lorentz transformations are called *proper* if and only if R is equal to the unit matrix.

A proper Lorentz transformation consists of a spatial rotation T_2, a special Lorentz transformation $L(V)$, and a spatial rotation T_1. If, in addition, spatial reflections are allowed, we obtain Lorentz transformations. If, moreover, translations of the space and time coordinates and reflections of the time coordinates, i.e., $\varepsilon_4 = -1$, are allowed, we obtain Poincaré transformations. Because of $\det L(V) = \det T = 1$ we have: A Lorentz transformation is proper if and only if $\det \Lambda = 1$.

A Poincaré transformation with $a = 0$ is a Lorentz transformation precisely if $\partial t'/\partial t > 0$. Hence for a Lorentz transformation, the direction of time is preserved.

EXAMPLE 75.7. Consider two Cartesian coordinate systems Σ and Σ' with corresponding system times t and t'. Both systems are inertial systems, and Σ' is obtained from Σ by a constant translatory motion with velocity vector v. For $t = 0$, Σ and Σ' have the same origin. By using a rotation of Σ and Σ', the situation may be reduced to the more simple situation (S) of Section 75.5, for which the ξ-axis and the ξ'-axis have the same direction as the vector v. This implies

$$u' = T_1 L(V) T_2 u$$

with $V = |v|$. For a suitable choice of Σ and Σ', every proper Lorentz transformation can be obtained like this. Without assuming that both systems Σ and Σ' have the same origin at time $t = 0$, we obtain the more general equation

$$u' = \Lambda u + a$$

with fixed a and $\Lambda = T_1 L(V) T_2$.

Every matrix T_2 can be written as the product of rotations D_j about the jth space axis, $j = 1, 2, 3$. Because of $L(V)D_3 = D_3 L(V)$ we may always assume that T_2 is only the product of rotations D_1 and D_2. Thus the Lorentz transformations $T_1 L(V) T_2$ depend on $3 + 1 + 2 = 6$ parameters, and hence the Poincaré transformations depend on $6 + 4 = 10$ parameters.

Proposition 75.8. *The collection of all Lorentz transformations (or Poincaré transformations) forms a group, which is called the Lorentz group L (or Poincaré group P). The proper Lorentz transformations form a subgroup of L.*

PROOF. An elementary calculation shows that

$$L(V_1)L(V_2) = L(V_3) \quad \text{with} \quad V_3 = \frac{V_1 + V_2}{1 + V_1 V_2/c^2}$$

for $|V_j| < c$ and $j = 1, 2, 3$. Because of $L(0) = I$ and $L(V)L(-V) = I$, all matrices $L(V)$ form a group. Since all matrices T form a group as well, it follows that all Lorentz transformations form a group.

From $u' = \Lambda u + a$ and $u'' = \Lambda' u' + a'$ it follows that

$$u'' = \Lambda'\Lambda u + \Lambda' a + a'.$$

Therefore all Poincaré transformations form a group as well. □

It can easily be shown that L is a six-dimensional Lie group and P a ten-dimensional Lie group. In relativistic quantum field theories the invariance under the Poincaré group makes it possible to assign quantities like energy, momentum, rest mass, spin (angular momentum), and parity to elementary particles. Today, one expects that there exists PCT-invariance for all interactions between elementary particles. This means: If an elementary particle process π is possible in nature, then the same is true for π_{PCT}, where π_{PCT} is obtained from π by a spatial reflection P, a change to antiparticles C, and a time reflection T (reversing all velocities). Under certain assumptions a precise mathematical proof can be given for the PCT-invariance (see Streater and Wightman (1964, M)).

Proposition 75.9. *Let*

$$Q(u) = c^2 t^2 - \xi^2 - \eta^2 - \zeta^2$$

with $u = (\xi, \eta, \zeta, ct)$. *Then Q is invariant under every Poincaré transformation, i.e.,* $Q(u_1 - u_2) = Q(u_1' - u_2')$.

PROOF. This follows immediately from Proposition 75.2 and the fact that Q is invariant under spatial rotations and reflections of the coordinates. □

The converse of Proposition 75.9 is discussed in Problem 76.1.

75.8. Space–Time Manifold of Minkowski

In order to understand the geometrical meaning of the special theory of relativity and, later in Section 76.1, the difference between the special and the general theory of relativity, we construct with Minkowski (1909) a four-dimensional space–time manifold M_4 which will also be called Minkowski space. We use the notations of Section 75.1.

To every proper Lorentz transformation Λ and every $a \in \mathbb{R}^4$ we assign a sample of $\mathbb{R}^4$ which will be denoted by $\mathbb{R}^4(\Lambda, a)$. By definition, the change between points $u' \in \mathbb{R}^4(\Lambda', a')$ and $u \in \mathbb{R}^4(I, 0)$ is given by

$$u' = \Lambda' u + a'. \tag{21*}$$

Thus, for the change between the corresponding elements $u' \in \mathbb{R}^4(\Lambda', a')$ and $u'' \in \mathbb{R}^4(\Lambda'', a'')$, we have the formula

$$u'' = \bar{\Lambda} u' + \bar{a} \tag{21}$$

with the proper Lorentz transformation

$$\bar{\Lambda} = \Lambda'' \Lambda'^{-1} \quad \text{and} \quad \bar{a} = a'' - \Lambda'' \Lambda'^{-1} a'.$$

We consider $\mathbb{R}^4(\Lambda, a)$ as chart spaces and define the chart change through (21). According to the general construction of Problem 73.2, we thereby obtain a C^∞-manifold M_4, whose points $x = (u)$ consist of tuples with $u \in \mathbb{R}^4(\Lambda, a)$. Through (21) the elements u of the tuple are naturally connected to each other.

The physical interpretation is as follows. We think of Example 75.7. The chart spaces $\mathbb{R}^4(\Lambda, a)$ correspond to all possible inertial systems. Therefore, precisely the $\mathbb{R}^4(\Lambda, a)$ are called *inertial charts*. A point

$$x = (u) \quad \text{in } M_4$$

is called an event with coordinates $u = (u^1, u^2, u^3, u^4)$ in the inertial system for $\mathbb{R}^4(\Lambda, a)$. Here u^1, u^2, u^3 are spatial Cartesian coordinates and $u^4 = ct$, where t denotes the time and c the velocity of light. The transformation between the coordinates u' and u'' of the event x in the inertial systems for $\mathbb{R}^4(\Lambda', a')$ and $\mathbb{R}^4(\Lambda'', a'')$, respectively, is given by (21).

In order to introduce a metric for M_4, we set

$$ds^2 = g_{ij} du^i du^j \tag{22}$$

with

$$g_{44} = 1, \quad g_{11} = g_{22} = g_{33} = -1, \quad g_{ij} = 0 \quad \text{for } i \neq j \tag{23}$$

in $\mathbb{R}^4(I, 0)$. In $\mathbb{R}^4(\Lambda', a')$ we define g'_{ij} by transforming g_{ij} as a tensor with respect to the corresponding coordinate transformation (21*), $u' = \Lambda' u + a'$, i.e.,

$$g_{i'j'} = \frac{\partial u^i}{\partial u^{i'}} \frac{\partial u^j}{\partial u^{j'}} g_{ij}.$$

According to the general construction of Problem 74.3h, we thereby obtain a symmetric tensor field g_{ij} on M_4. Since $\det \Lambda = 1$ for all Λ, it follows that $\det \bar{\Lambda} = 1$ in (21). Thus M_4 is an oriented Riemannian C^∞-manifold and we can apply the tensor analysis and Riemannian geometry of Section 74.19. From (23) it follows that $R^i_{jkm} = 0$, i.e., the curvature tensor is identically zero on M_4. The following proposition shows that the geometry on M_4 is particularly simple. In mathematical terms it contains the physical fact that all inertial systems are equivalent.

Proposition 75.10. *For every inertial chart* $\mathbb{R}^4(\Lambda, a)$ *the metric tensor* g_{ij} *has the simple form* (23).

PROOF. If (21*) holds, then the right-hand side of (22) is transformed as $g_{ij}(u_1^i - u_2^i)(u_1^j - u_2^j)$. Proposition 75.9 then yields the desired result. □

Remark 75.11. In every inertial system, i.e., in every inertial chart, we have the following relations

$$ds^2 = c^2 dt^2 - d\xi^2 - d\eta^2 - d\zeta^2$$

and $\Gamma_{ij}^k = 0$. This implies that $\nabla_i = D_i$, i.e., the covariant derivative coincides with the classical derivative. Note, however, that g_{ij} need not have the form (23) for every possible admissible chart. For example, such charts may correspond to curved space coordinates. Also, there exist coordinates for which the distinction between space and time coordinates is lost. One may think, for instance, of $v^4 = u^1 + u^4$, i.e., $v^4 = \xi + ct$ and $v^i = u^i$ for $i = 1, 2, 3$.

Strategy 75.12. The goal of relativistic physics is to formulate all physical laws in such a way that they have the same form for every inertial system. This is Einstein's principle (R) of relativity of Section 75.2. Mathematically, this program can be realized by using only geometrical properties of M_4, i.e., properties which are independent of the choice of inertial charts. For example, tensor equations on M_4 satisfy this condition.

In the following sections we will use this geometrical method to formulate a relativistic mechanics. In Part V it will be used for a formulation of electrodynamics. A special role is played by the scalars on M_4. These are physical quantities which have the same value for all inertial systems. Examples are charge, rest mass, and entropy. As we shall see in Section 75.11, mass and energy are no scalars. This is one of the fundamental results of the theory of relativity.

75.9. Causality and Maximal Signal Velocity

Consider two events $x_1 \in M_4$ and $x_2 \in M_4$ with corresponding coordinates u_1 and u_2 in a fixed inertial chart, i.e., in a fixed inertial system Σ. Recall that

$$u = (u^1, u^2, u^3, u^4) = (\xi, \eta, \zeta, ct).$$

75.9. Causality and Maximal Signal Velocity

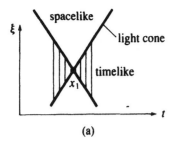
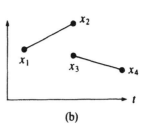

Figure 75.8

We define the square of the distance between the two events as

$$d = g_{ij}(u_1^i - u_2^i)(u_1^j - u_2^j)$$
$$= c^2(t_1 - t_2)^2 - (\xi_1 - \xi_2)^2 - (\eta_1 - \eta_2)^2 - (\zeta_1 - \zeta_2)^2. \tag{24}$$

Because of Proposition 75.9 this definition is independent of the inertial chart. The quotient $w = $ distance/time, i.e.,

$$w = \sqrt{(\xi_1 - \xi_2)^2 + (\eta_1 - \eta_2)^2 + (\zeta_1 - \zeta_2)^2}/|t_1 - t_2|$$

is called the signal velocity with respect to x_1 and x_2 in the inertial system Σ. For $t_1 = t_2$ we set $w = \infty$. If at point (ξ_1, η_1, ζ_1) and time t_1 one transmits a signal which travels rectilinearly with constant velocity w, then it reaches point (ξ_2, η_2, ζ_2) at time t_2. We say briefly that the signal connects the two events x_1 and x_2 with each other.

Definition 75.13. An event x_2 is called *timelike* (or *spacelike*, *lightlike*) with respect to x_1 if and only if $d > 0$ (or $d < 0$, $d = 0$). For a fixed x_1 we call the set of all $x_2 \in M_4$ with $d = 0$ the *light cone* with respect to x_1.

In Figure 75.8(a), which corresponds to $\eta = \zeta = 0$, the light cone is formed by the two straight lines which pass through the point x_1 and have slope $\pm c$. If we take the straight line through the point x_1 and parallel to the t-axis as the axis of the light cone, then exactly the inner and outer points of the light cone are timelike and spacelike with respect to x_1. Precisely the points x_2 on the light cone are lightlike with respect to x_1. The line segments in Figure 75.8(b) correspond to signals which travel rectilinearly and which connect the two events x_1 and x_2 (or x_3 and x_4). The constant signal velocity w is equal to the absolute value of the slope of the line segments.

Proposition 75.14. *An event x_2 is timelike (or spacelike, lightlike) with respect to x_1 if and only if the corresponding signal velocity w satisfies $w < c$ (or $w > c$, $w = c$) in every inertial system.*

PROOF. This follows immediately from the definition of d and w. □

Proposition 75.15. *An event x_2 is timelike with respect to x_1 if and only if there* exists an inertial system, in which x_1 and x_2 take place at the same point and at different times or the same time.

The event x_2 is *spacelike* with respect to x_1 if and only if there exists an inertial system, in which x_1 and x_2 take place at different points and at the same time.

PROOF. Let $x_1 \neq x_2$. We write $x_j \cong (\xi_j, \eta_j, \zeta_j, ct_j)$ in Σ if and only if x_j has these coordinates in Σ. Using a translation and a spatial rotation we can always find an inertial system Σ with $x_2 \cong (0,0,0,0)$ and $x_1 \cong (\xi_1, 0, 0, t_1)$ in Σ. Thus we have $d = c^2 t_1^2 - \xi_1^2$. Using a special Lorentz transformation,

$$\xi_j' = \alpha(\xi_j - Vt_j), \qquad t_j' = \alpha(t_j - V\xi_j/c^2), \qquad j = 1, 2,$$

we pass from Σ to Σ'. Here we have $\alpha = 1/\sqrt{1 - V^2/c^2}$. For $d > 0$ we can find a real number V with $|V| < c$ such that $\xi_2' = 0$. In system Σ' we therefore have $x_1 \cong (0, 0, 0, t_1')$ and $x_2 \cong (0, 0, 0, 0)$ with $t_1' \neq 0$.

For $d < 0$ one can always find a real number V with $|V| < c$ such that $t_1' = 0$. In Σ' this implies $x_1 \cong (\xi_1', 0, 0, 0)$ and $x_2 \cong (0, 0, 0, 0)$ with $\xi_1' \neq 0$. □

Now we formulate the following two general laws of nature as postulates:

(P) *Principle of causality.* If an event x_2 is spacelike with respect to x_1, then there exists no causal connection between x_1 and x_2.
(S) *Maximal signal velocity.* In an inertial system, physical signals can travel at most with the velocity of light.

We shall motivate (P) and (S). From Proposition 75.15, (P) is equivalent to the fact that there exists no causal connection between two events in an inertial system if they take place at different points and at the same time, i.e., physical signals cannot travel with infinite velocity. Suppose (S) does not hold. From Proposition 75.14 it follows then that there exist two events x_1 and x_2, where x_2 is *spacelike* with respect to x_1 and the event x_2 is effected by x_1 via a signal with a velocity *greater* than that of light. This contradicts (P).

EXAMPLE 75.16 (The Catastrophe in the Center of the Milky Way). We want to show that Newton's theory of gravity is not compatible with (S). For this reason, Einstein replaced it with his theory of gravity—the general theory of relativity. Our Milky Way consists of approximately 10^{11} stars. The sun is located at the boundary of the Milky Way and at a distance of approximately 30,000 light years from its center C. It rotates about C with a velocity of 268 km/s, i.e., ten times as fast as the earth rotates about the sun. Suppose there occurs a huge explosion at C which drastically changes the mass m_C of C. According to Newton's theory of gravity we obtain

$$\ddot{y} = \frac{Gm_C(y_C - y)}{|y_C - y|^3}$$

for the motion $y = y(t)$ of the sun. Thus the change of m_C causes an immediate change in the orbit of the sun. But, according to (S) this will be noticed only 30,000 years later. This is a contradiction.

At the end of this section, we try to find an expression for d which holds in arbitrary admissible charts of M_4. For this, we choose a line segment

$$u = u_1 + p(u_2 - u_1), \qquad 0 \le p \le 1,$$

which connects x_1 and x_2, i.e., u_1 and u_2 in a fixed inertial chart (Fig. 75.8(b)). This gives $d = \int ds^2$, i.e.,

$$d = \int_0^1 g_{ij}(u(p)) \dot{u}^i \dot{u}^j \, dp.$$

The dot means derivative with respect to p. The integrand can be transformed to arbitrary admissible charts. Since the integrand is a scalar, d remains unchanged.

75.10. Proper Time

Every curve $x = x(p)$ in M_4 will be called a world line. This is a set of events parametrized by a real parameter p. For an inertial chart, $x = x(p)$ corresponds to the coordinate representation $u = u(p)$, i.e.,

$$y = y(p), \qquad t = t(p) \tag{25}$$

with $y = (\xi, \eta, \zeta)$. The motion $y = y(t)$ of a body or signal in an inertial system is given by (25) with $t = p$. The arclength s of the world line between the points $x(p_1)$ and $x(p)$ is defined in the usual way through $s = \int ds$, i.e.,

$$s = \int_{p_1}^{p} \sqrt{g_{ij}(u(p)) \dot{u}^i(p) \dot{u}^j(p)} \, dp.$$

Here we assume that

$$g_{ij} \dot{u}^i \dot{u}^j \ge 0$$

along the world line. As the following observation shows, this condition is always satisfied if the world line corresponds to a motion of a mass point in an inertial system with under light velocity ($g_{ij}\dot{u}^i\dot{u}^j > 0$) or a motion of a light ray ($g_{ij}\dot{u}^i\dot{u}^j = 0$).

In an inertial chart we have

$$s = \int_{p_1}^{p} \sqrt{c^2 \dot{t}^2 - \dot{y}^2} \, dp.$$

The dot means derivative with respect to p. We now consider the motion $y = y(t)$ of a body in an inertial system with under light velocity, i.e., $|\dot{y}| < c$.

We choose $p = t$ and set

$$\tau = s/c.$$

Here, τ is called the *proper* time. It follows that

$$\tau = \int_{t_1}^{t} \sqrt{1 - \dot{y}^2/c^2}\, dt. \tag{26}$$

In Section 75.11 we shall prove the following:

> In an inertial system a particle can travel only with a velocity smaller than that of light.
>
> (27)

Postulate 75.17. If a clock in an inertial system Σ moves along an orbit $y = y(t)$, then τ is the *time* which has passed on the clock between the instants t_1 and t of the system time of Σ.

Therefore, the moving clock in Σ is always slow with respect to the system time of Σ. In order to motivate this postulate, we set $t_1 = 0$ and consider the following special motion $\xi = Vt$, $\eta = \zeta = 0$ of the clock in Σ with constant velocity V. Moreover, we choose a second inertial system Σ' in which the clock is at rest at the origin (Fig. 75.9). The special Lorentz transformation gives

$$t' = \frac{t - V\xi/c^2}{\sqrt{1 - V^2/c^2}} = \sqrt{1 - V^2/c^2}\, t = \tau.$$

Hence, in this case, the proper time τ of the clock coincides with the system time t' in the rest system Σ' of the clock. For arbitrary moving clocks, one considers small time intervals and momentary rest systems. This implies

$$\Delta\tau = \sqrt{1 - \dot{y}(t)^2/c^2}\, \Delta t.$$

Summation and passing to limits gives (26).

EXAMPLE 75.18 (The Twin Paradox). Suppose at time $t = 0$ and at the origin P_0 of an inertial system Σ the twins T_1 and T_2 are born. Shortly thereafter, T_2 is brought to a spaceship and begins a journey through the universe while T_1 remains at P_0. After several years, T_2 returns to T_1. Both are surprised to find

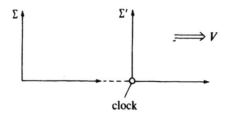

Figure 75.9

that T_2 is much younger than T_1. This fact can be easily explained if one assumes that the biological clock of T_j shows the proper time τ_j. The motion of T_j is $y = y_j(t)$ with $y_1(t) \equiv 0$. With $t_1 = 0$, it follows immediately from (26) that

$$\tau_2 < t = \tau_1.$$

The faster T_2 travels, the smaller τ_2 will be compared with τ_1.

In the limiting case of the velocity of light $|\dot{y}_2| = c$ one has $\tau_2 = 0$, i.e., T_2 remains young forever. Because of (27), however, this limiting case is impossible.

The twin paradox has caused many misunderstandings and controversies. The usual mistake made is that the effect is not calculated according to formula (26), instead one argues verbally. Furthermore, one often finds the false statement that the twin paradox can only be understood in the context of the general theory of relativity.

75.11. The Free Particle and the Mass–Energy Equivalence

> That sometimes, in his speculations, he went too far, such as, for example, in his hypothesis of the light-quantum, should not be held too much against him.
>
> Max Planck, in 1913, while recommending Einstein for the membership of the Prussian Academy. (In 1921, Einstein won the Nobel prize for his results about light-quanta of 1905.)

Already, the seemingly trivial case of a free particle leads to substantial revisions of Newton's mechanics, and also leads to Einstein's fundamental light quantum hypothesis. We present the following observations in a form which can be extended to the general theory of relativity in the following chapter.

In classical mechanics a free particle is described by the variational principle

$$\int_{t_1}^{t_2} L\, dt = \text{stationary!},$$

$$y(t_1) = y_1, \qquad y(t_2) = y_2, \tag{28}$$

with $L = m_0 \dot{y}^2/2 + \text{const}$, where t_1, t_2, y_1, and y_2 are given. The corresponding Euler equation

$$\frac{d}{dt} L_{\dot{y}} - L_y = 0 \tag{29}$$

yields the equation of motion $m_0 \ddot{y} = 0$. The momentum is

$$p = L_{\dot{y}}$$

and hence $p = m_0 \dot{y}$. The Hamiltonian function is

$$H = p\dot{y} - L,$$

i.e., $H = p^2/2m_0$. Here H is equal to the energy E of the particle. According to Section 58.23, the Hamilton–Jacobi equation is obtained from

$$E = H(p)$$

by replacing E with $-\partial S/\partial t$ and p with $\partial S/\partial y = \text{grad } S$. This gives

$$S_t + S_y^2/2m_0 = 0.$$

This Hamiltonian formalism has the advantage that it applies to every Lagrangian function L. Before using this fact we illustrate the difference between covariance and form invariance. Consider the vector equation

$$ab = 0 \qquad (30)$$

in $\mathbb{R}^3$. For an arbitrary coordinate system we obtain in component notation

$$g^{ij} a_i b_j = 0. \qquad (31)$$

For every Cartesian coordinate system, g^{ij} is equal to the Kronecker symbol, i.e., one has

$$\sum_{i=1}^{3} a_i b_i = 0. \qquad (32)$$

Here (30) is a geometrical equation which is formulated independently of a coordinate system, and (31) is a covariant equation. Contrary to this, (32) is called form-invariant with respect to a Cartesian coordinate system. Note also that, because of the changing coefficients g^{ij}, the form of (31) depends on the choice of the particular curved coordinate system, while (32) is independent of the choice of the Cartesian coordinate system.

In order to find a relativistic equation for the free particle, we require the following.

(i) The variational principle has a geometrical meaning for the space–time manifold M_4. Thereby we guarantee the covariance of the Euler equations.
(ii) In view of the principle of relativity (R) of Section 75.2, we assume that the equations of motion are form-invariant under changes in the inertial systems.
(iii) For velocities which are small compared with the velocity of light c, we assume that L coincides in first-order approximation with the classical expression (correspondence principle).

Assumption (ii) will follow from (i), because Proposition 75.10 implies that g_{ij} has the same numerical values for every inertial system. In order to obtain (i) we begin with

$$-m_0 c \int ds = \text{stationary!}. \qquad (33)$$

75.11. The Free Particle and the Mass–Energy Equivalence

From Section 75.10 this gives the relativistic Lagrangian function

$$L = -m_0 c^2 \sqrt{1 - \dot{y}^2/c^2}.$$

Because of

$$L = \frac{m_0 \dot{y}^2}{2} - m_0 c^2 + o\left(\frac{\dot{y}^2}{c^2}\right), \quad \frac{\dot{y}^2}{c^2} \to 0,$$

(iii) is satisfied. Now we immediately obtain the relativistic momentum

$$p = L_{\dot{y}} = m\dot{y} \tag{34}$$

with relativistic mass

$$m = m_0/\sqrt{1 - \dot{y}^2/c^2}. \tag{35}$$

The Hamiltonian function is

$$H = p\dot{y} - L = mc^2, \quad \text{i.e.,} \quad H^2 = m_0^2 c^4 + c^2 p^2.$$

For the *energy* $E = H$ of the particle we therefore obtain

$$E = mc^2, \tag{36}$$

$$E^2 = m_0^2 c^4 + c^2 p^2. \tag{37}$$

This implies for the Hamilton–Jacobi equation

$$S_t^2 = m_0^2 c^4 + c^2 S_y^2. \tag{38}$$

A comparison of (34) with the classical momentum definition shows that the expression relativistic mass for m is meaningful. For $\dot{y} = 0$ we have $m = m_0$, and we call m_0 the rest mass of the particle. Equation (36) is of fundamental importance. Note that (36) becomes

$$E = m_0 c^2$$

for a particle at rest. This is Einstein's famous formula, stating the *equivalence between mass and energy*. The energy production of all stars is based upon (36). For example, during the synthesis of helium from hydrogen in the sun, mass is transformed into energy. Formula (36) is a triumph for the mental ability of man; in a frightening way it also allows the self-destruction of mankind by atomic bombs.

Using (35), the equation of motion (29) now becomes

$$\frac{d}{dt}(m\dot{y}) = 0. \tag{39}$$

According to (35) this is only meaningful for $|\dot{y}| < c$. Therefore the inert mass m of the free particle increases whenever the absolute velocity $|\dot{y}|$ increases. For $|\dot{y}| \to c$, m and E become infinitely large. This means that a free particle with rest mass $m_0 > 0$ can never reach the velocity of light.

This is different for particles with $m_0 = 0$. For such particles, our theory is not applicable. The only formulas which are also meaningful for $m_0 = 0$ are the energy equation (37) and the Hamilton–Jacobi equation (38). In his paper (1905b), Einstein made the hypothesis that light consists of quanta of energy

$$E = h\nu = hc/\lambda. \tag{40a}$$

Here ν is the frequency of light, λ the wave length, and $h = 6.6 \cdot 10^{-34}$ Ws² Planck's quantum of action. Since a photon propagates with the velocity of light we conclude from (35) that its rest mass m_0 is equal to zero. From (37) we obtain for its momentum $|p| = E/c$, i.e.,

$$|p| = h\nu/c = h/\lambda. \tag{40b}$$

The two formulas (40a) and (40b) have been reaffirmed by many experiments and now form the basis of quantum electrodynamics, i.e., for the quantum field theory for electrons, positrons, and photons. In Section 68.4 we already saw how from these formulas, together with the Bose statistics, Planck's famous radiation law follows. In fact, Einstein (1905b) introduced his light quantum hypothesis in order to give a different derivation of Planck's radiation law. In his time, the light quantum hypothesis was a very bold and radical hypothesis. Before Einstein, light was always regarded as an electromagnetic wave in the context of Maxwell's theory. In fact, the wave picture explains numerous phenomena. In Chapter 59, we already discussed this dualism between wave and particle. At the present time, the quantum concept is the dominating idea in physics.

Formula (40a) can be motivated in the following way: From experience we know that the energy of light depends on the frequency ν. If we assume the proportionality $E = A\nu$, then A must have the dimension of an action, i.e., $A = \text{energy} \times \text{time}$. Letting $A = ah$, the Bose statistics of Section 68.4 yields Planck's radiation law for $a = 1$.

The equations for the free particle, formulated so far, do not explicitly exhibit covariance and form invariance. We now give a formulation for the equations of motion for which covariance and form invariance can be explicitly recognized. The motion of a particle corresponds to a world line $x = x(\sigma)$ with arclength

$$s = \int_{\sigma_0}^{\sigma} \sqrt{g_{ij}\dot{u}^i\dot{u}^j}\, d\sigma.$$

We assume that $g_{ij}\dot{u}^i\dot{u}^j > 0$ along the world line. According to Section 75.10, this implies under light velocity for every inertial system. As parameter σ we choose the proper time $\tau = s/c$. From (33), the world line for this motion corresponds to a geodesic. Proposition 74.53 then implies for the *equations of motion* that

$$m_0 c^2 \frac{D}{ds}\left(\frac{du^i}{ds}\right) = 0,$$

i.e.,

$$\frac{Dp^i}{d\tau} = 0, \qquad i = 1, 2, 3, 4 \tag{41}$$

75.12. Energy Momentum Tensor and Relativistic Conservation Laws   723

with the so-called four momentum

$$p^i = m_0 \frac{du^i}{d\tau} \tag{42}$$

and the four velocity $du^i/d\tau$. Equation (41) holds in every admissible chart of M_4. In an inertial chart, we have $\nabla_i = D_i$ and therefore $D/d\tau = d/d\tau$. Because of

$$\tau = \int_{t_0}^{t} \sqrt{1 - \dot{y}^2/c^2}\, dt$$

we explicitly have for an inertial system

$$p^\alpha = m\frac{du^\alpha}{dt}, \qquad p^4 = mc = E/c, \qquad \alpha = 1, 2, 3.$$

From (42) it follows that p^i is transformed like a tensor under chart changes; in particular, this is true for the Lorentz transformations, which describe the change between inertial systems. Therefore, in relativistic mechanics, energy E and momentum vector $p = p^\alpha e_\alpha$ form a unity.

The covariant form of the Hamilton–Jacobi equation (38) is

$$g^{ij} D_i S D_j S = m_0^2 c^2. \tag{43}$$

Note that $(g^{ij}) = (g_{ij})^{-1}$. Therefore (43) coincides with (38) for every inertial system.

In the general theory of relativity, (41) becomes the equation of motion for a particle in a gravitational field. Only the metric g_{ij} changes.

75.12. Energy Momentum Tensor and Relativistic Conservation Laws for Fields

The following purely mathematical ideas are of central importance for all relativistic field theories, e.g., fluids, the cosmos, electromagnetic fields, quantum fields, and elementary particles. We generalize here the statements about conserved quantities of Section 69.1 to relativistic fields. Our goal is to develop a formalism whereby quantities like charge, energy, momentum, angular momentum, and stress forces can be assigned to general fields. Important applications to relativistic ideal fluids will be given in the following section and to electromagnetic fields and quantum fields in Part V. There, we will also use the Noether theorem to derive the tensor fields T^i (currents) and T^{ij} (energy momentum tensor) from relativistically invariant variational principles.

We make the following assumptions:

(H1) T^i is a C^1-tensor field on M_4 with

$$\nabla_i T^i \equiv 0. \tag{44}$$

Consider an arbitrary, fixed inertial system, i.e., an inertial chart of M_4.

Then we have $V_i = D_i$ and $u = (y, ct)$. Let G be a bounded region in $\mathbb{R}^3$ and let

$$Q(G, t) = c^{-1} \int_G T^4(y, ct) \, dy.$$

As in Section 69.1, integration of (44) over G and an application of the integral theorem of Gauss gives

$$\frac{d}{dt} Q(G, t) = -\int_{\partial G} T^\alpha n_\alpha \, dO, \tag{45}$$

where $n = \sum_{\alpha=1}^{3} n_\alpha e_\alpha$ is the outer unit normal vector of ∂G. We say that T^i has a *compact* spacelike support if and only if there exists an inertial system and a bounded region $G_1 \subseteq \mathbb{R}^3$ therein such that

$$T^i(u) = 0 \quad \text{for all} \quad u = (y, ct) \quad \text{with} \quad y \notin G_1 \quad \text{for all } i.$$

In this case we let

$$Q(t) \stackrel{\text{def}}{=} Q(\mathbb{R}^3, t).$$

From (45) it follows that

$$\frac{dQ(t)}{dt} = 0 \quad \text{for all} \quad t \in \mathbb{R}, \tag{46}$$

i.e., Q is a *conserved quantity*. Since for every fixed t, Lorentz transformations take compact spatial sets into compact spatial sets, it follows from (45) that (46) is true for every inertial system.

Proposition 75.19. *If* (H1) *holds and if T^i has a compact spacelike support, then Q is a conserved quantity for every inertial system and Q is a scalar on M_4.*

The proof, which follows easily from Stokes' integral theorem for differential forms, will be given in Problem 76.3. In Part V, Q will represent an electrical charge or a chargelike quantum number in high-energy physics, for instance, the baryon number. We now assume:

(H2) T^{ij} is a symmetric C^1-tensor field on M_4 with

$$\nabla_i T^{ij} \equiv 0, \quad j = 1, 2, 3, 4.$$

We define

$$p^i(G, t) = c^{-1} \int_G T^{i4}(y, ct) \, dy.$$

As above we obtain the conservation law

$$\frac{d}{dt} p^i(G, t) = -\int_{\partial G} T^{i\alpha} n_\alpha \, dO. \tag{47}$$

75.12. Energy Momentum Tensor and Relativistic Conservation Laws

Proposition 75.20. *If* (H2) *holds and if* T^{ij} *has a compact spacelike support, then all the* p^i *are conserved quantities for every inertial system and* p^i *is a tensor on* M^4.

The proof will be given in Problem 76.3. In Part V, we will show how the energy–momentum tensor T^{ij} can be derived from a variational principle. In all field theories one requires that T^{44} can be physically explained as an *energy density*. Then p^i corresponds to the four momentum vector of Section 75.11, i.e., we interpret

$$p = p^\alpha e_\alpha$$

as the momentum vector and

$$E = cp^4$$

as the energy of the field. As an important example we consider in Section 75.13 a relativistic ideal fluid. We now compare (47) with classical mechanics. Letting $\sigma^{ij} = -T^{ij}$ for $i, j = 1, 2, 3$, we can write (47) for $i = 1, 2, 3$ as

$$\frac{dp}{dt} = \int_{\partial G} \sigma n \, dO.$$

The left-hand side represents a force (time derivative of the momentum vector). Therefore σ corresponds to the classical stress tensor of Section 61.3. This allows the following interpretation of T^{ij} for an inertial system:

$-T^{\alpha\beta}$ = component of the stress tensor for $\alpha, \beta = 1, 2, 3$,

$c^{-1} T^{\alpha 4}$ = component of the momentum density for $\alpha = 1, 2, 3$, (48)

T^{44} = energy density.

Moreover, we have $T^{ij} = T^{ji}$. In addition, a comparison of (47) with Section 69.1 gives

$\mathbf{p}_\alpha = T^{\alpha\beta} e_\beta$ current density vector for the αth momentum component,

$\mathbf{p}_4 = cT^{4\beta} e_\beta$ energy current density vector.

The index β is summed from 1 to 3.

We now study the *angular momentum* of the field. For this we set

$$S^{ijk} = u^i T^{jk} - u^j T^{ik}$$

for every inertial chart. Then S^{ijk} is a tensor on M_4, and

$$\nabla_k S^{ijk} = 0 \qquad (49)$$

holds for every inertial chart. This follows immediately from (H2) with $\nabla_i = D_i$ and $T^{ij} = T^{ji}$. As in Section 74.17, S^{ijk} can be extended to every admissible chart of M_4. Then (49) holds there, too. As above we define

$$M^{ij}(G, t) = c^{-1} \int_G S^{ij4}(y, ct) \, dy.$$

We have $M^{ij}(G,t) = -M^{ji}(G,t)$. By definition the angular momentum vector $J(G,t)$ of the field for an inertial system is

$$J(G,t) = M^{23}e_1 + M^{31}e_2 + M^{12}e_3,$$

where, precisely, we would have to write $M^{23}(G,t)$, etc. This definition is correct, because if we define the momentum *density* vector as

$$p_d = c^{-1}T^{\alpha 4}e_\alpha$$

and the angular momentum *density* vector as

$$J_d = y \times p_d,$$

then we obtain for the *momentum vector* of the field in the region G

$$p(G,t) = \int_G p_d(y,t)\,dy$$

and for the *angular momentum vector*

$$J(G,t) = \int_G J_d(y,t)\,dy.$$

For the *energy* of the field in the region G it follows that

$$E(G,t) = \int_G T^{44}(y,t)\,dy.$$

From (49) we obtain the important relation

$$\frac{d}{dt}M^{ij}(G,t) = -\int_{\partial G} S^{ij\alpha}n_\alpha\,dO.$$

Similarly to Section 69.1, we therefore call the vectors

$$J_1 = S^{23\alpha}e_\alpha, \qquad J_2 = S^{31\alpha}e_\alpha, \qquad J_3 = S^{12\alpha}e_\alpha$$

current density vectors of the corresponding angular momentum components J_1, J_2, J_3. We let $M^{ij} = M^{ij}(\mathbb{R}^3, t)$.

Corollary 75.21. *Under the assumptions of Proposition 75.20, all the M^{ij} are conserved quantities for every inertial system and M^{ij} is a tensor on M_4.*

The proof will be given in Problem 76.3.

75.13. Applications to Relativistic Ideal Fluids

It follows from Section 70.3 that the equation of motion and the equation of continuity for a classical *ideal* fluid in an inertial system are given by

$$D_4 c\rho w^\alpha + D_\beta(\rho w^\alpha w^\beta + \delta^{\alpha\beta}P) = 0, \qquad \alpha = 1, 2, 3, \tag{50}$$

$$D_4 \rho c^2 + D_\beta c\rho w^\beta = 0, \tag{51}$$

75.13. Applications to Relativistic Ideal Fluids

with $D_4 = \partial/\partial(ct)$. According to our convention in Section 75.1, β is summed from 1 to 3. Here $w = w^\beta e_\beta$ is the velocity vector, ρ the density, and P the pressure. The momentum density vector is $p_d = \rho w$. By letting

$$T_{\text{class}} = \begin{pmatrix} T^{\alpha\beta}_{\text{class}} & c\rho w \\ c\rho w & \rho c^2 \end{pmatrix}$$

with

$$T^{\alpha\beta}_{\text{class}} = \rho w^\alpha w^\beta + P\delta^{\alpha\beta}, \qquad \alpha, \beta = 1, 2, 3, \tag{52}$$

equations (50) and (51) can be written in the short form

$$D_i T^{ij}_{\text{class}} = 0. \tag{53}$$

Since equation (53) describes conservation of mass and momentum, we call T^{ij}_{class} the classical mass-momentum tensor. Nothing, however, has been said about the tensor property. It can be shown that T^{ij}_{class} is a tensor under classical Galileian transformations. We are now looking for a relativistic generalization.

Postulate 75.22. Let a tensor field v^i be given on M_4 such that $v^i v_i = c^2$. The basic equations of a relativistic ideal fluid are

$$\nabla_i T^{ij} = 0 \tag{54}$$

with energy–momentum tensor

$$T^{ij} = c^{-2}(\varepsilon + P)v^i v^j - Pg^{ij}. \tag{55}$$

Here ε and P are functions on M_4 (scalar fields). We have to add the constitutive law

$$\varepsilon = \varepsilon(P, \ldots),$$

where the dots stand for further variables like mass density, entropy density, etc.

The motion of a fluid particle $x = x(\tau)$ on M_4 with proper time τ follows from

$$\frac{du^i}{d\tau}(\tau) = v^i(u(\tau)) \tag{56}$$

for an arbitrary admissible chart. In an inertial system Σ, we get

$$v^\alpha = \frac{w^\alpha}{\sqrt{1 - w^2/c^2}}, \qquad v^4 = \frac{c}{\sqrt{1 - w^2/c^2}},$$

where w denotes the classical velocity vector of the particle.

We now want to motivate this postulate, whereby the following two points are important:

(i) T^{ij} is a tensor which, similarly as the corresponding classical expression, is quadratic in the velocities.
(ii) The component T^{44} is equal to the energy density.

First, we consider a fluid which is at rest in an inertial system Σ. The motion $y = y(t)$ of an arbitrary fluid particle is given then by $y(t) \equiv 0$, i.e.,

$$u^\alpha = 0 \quad \text{for} \quad \alpha = 1, 2, 3, \quad u^4 = ct$$

and t is equal to the proper time τ. From (56) it follows that

$$v^\alpha = 0 \quad \text{for} \quad \alpha = 1, 2, 3 \quad \text{and} \quad v^4 = c.$$

According to (48) it follows that:

$$-T^{\alpha\beta} = \text{component of the stress tensor for } \alpha, \beta = 1, 2, 3,$$

$$T^{44} = \text{energy density}.$$

The key trick is then to consider T^{ij} in the rest system Σ. Section 70.1 implies that

$$(T^{ij}) = \begin{pmatrix} P & 0 & 0 & 0 \\ 0 & P & 0 & 0 \\ 0 & 0 & P & 0 \\ 0 & 0 & 0 & \varepsilon \end{pmatrix}$$

in Σ, which can be rewritten as

$$T^{ij} = c^{-2}(\varepsilon + P)v^i v^j - Pg^{ij}, \quad i, j = 1, 2, 3, 4. \tag{57}$$

Notice that

$$(g^{ij}) = (g_{ij}) = \begin{pmatrix} -1 & 0 & 0 & 0 \\ 0 & -1 & 0 & 0 \\ 0 & 0 & -1 & 0 \\ 0 & 0 & 0 & 1 \end{pmatrix}$$

in Σ, and that the right-hand side in (57) is a tensor. Thus we have found an expression for the T^{ij} which is valid in *every* system of reference. Our discussion shows that the scalars P and ε have the following physical meaning:

P = pressure of the fluid in the rest system,

ε = energy density of the fluid in the rest system.

This motivates (55).

We have made use of the following general *strategy* which clearly illustrates the advantage of tensor calculus in physics:

(a) First, we consider a special coordinate system in which the physical problem assumes a simple form. This leads to certain equations.

75.13. Applications to Relativistic Ideal Fluids

(b) Second, we write these equations as tensor equations, which are then valid in *every* system of reference.

Let us now consider an arbitrary inertial system Σ. Then $\nabla_i = D_i$. The relativistic equation (54) with $j = 4$ is

$$\frac{\partial}{\partial t}\left(\frac{\varepsilon}{1 - w^2/c^2}\right) + \operatorname{div}\left(\frac{(\varepsilon + P)w}{1 - w^2/c^2}\right) = 0. \tag{58}$$

This describes energy conservation. In fact, it follows from (58) that the total energy

$$E = \int_{R^3} T^{44}\, dy = \int_{R^3} \frac{\varepsilon(y, t)}{1 - w(y, t)^2/c^2}\, dy$$

is a conserved quantity if the motion of the fluid is restricted to a bounded spatial region.

Generally, the law of mass conservation (51) is not valid in relativistic physics.

In the following chapter we will use relativistic fluids as a basis for cosmological models of the universe and for models of stars.

CHAPTER 76

General Theory of Relativity

> The general laws of nature are to be expressed in equations which are valid for all coordinate systems.
>
> Albert Einstein (1916)

> The fact that elementary particle physicists, astrophysicists, and cosmologists have become interested in the same questions is one of the most significant developments in physics within the last ten years.
>
> Alan H. Guth and Paul J. Steinhardt
> (*Scientific American*, July 1984)

76.1. Basic Equations of the General Theory of Relativity

We use the notations of Section 75.1. The *basic equations* of the general theory of relativity which determine the metric tensors g_{ij} of Einstein's four-dimensional space–time manifold E_4 are

$$R^{ij} - \tfrac{1}{2} g^{ij} R = \kappa T^{ij} \tag{1}$$

with the universal constant

$$\kappa = 8\pi G/c^4 = 2.07 \cdot 10^{-43}\,\text{N}^{-1}. \tag{2}$$

Here G is the gravitational constant.

The equations of motion for a *mass particle* are

$$\frac{D}{ds}\left(\frac{du^k}{ds}\right) = 0. \tag{3}$$

76.1. Basic Equations of the General Theory of Relativity

The equations of motion for *light rays* (photons) are

$$\frac{D}{d\sigma}\left(\frac{du^k}{d\sigma}\right) = 0, \tag{4}$$

$$\frac{ds}{d\sigma} = 0. \tag{5}$$

Equations (3) and (4) can be written more explicitly as

$$\ddot{u}^k + \Gamma^k_{ij}\dot{u}^i\dot{u}^j = 0. \tag{6}$$

The dot means derivative with respect to arclength s or parameter σ. In Theorem 76.A of Section 76.2 we state a unified variational principle for (3) and (4), (5).

These equations are motivated as follows. The masses which appear in the energy–momentum tensor T^{ij} effect the metric g_{ij} of E_4. In contrast to classical mechanics, there exists no force of gravity but the gravitational effect is caused by the metric. Mass particles move on geodesics, and light rays correspond to affine geodesic lines (4) with side condition (5), i.e., to zero lines. In general, T^{ij} also depends on g_{ij}.

We now explain these equations. Our starting point is a four-dimensional real Riemannian C^∞-manifold E_4 with metric

$$ds^2 = g_{ij}\,du^i\,du^j.$$

As before, let (g^{ij}) denote the inverse matrix to (g_{ij}). Let $(-1, -1, -1, 1)$ be the signature type of the metric tensor g_{ij} in the sense of Section 74.20. Thus, according to the law of inertia of Sylvester, we assume that at each point of E_4:

$$g_{44} > 0, \quad \begin{vmatrix} g_{33} & g_{34} \\ g_{43} & g_{44} \end{vmatrix} < 0, \quad \begin{vmatrix} g_{22} & g_{23} & g_{24} \\ g_{32} & g_{33} & g_{34} \\ g_{42} & g_{43} & g_{44} \end{vmatrix} > 0, \quad g < 0$$

with $g = \det(g_{ij})$. In physical models, the u^1, u^2, u^3 (and u^4) represent space (and time) coordinates. As in Chapter 74 we use the Christoffel symbols for g_{ij},

$$\Gamma^k_{ij} = \tfrac{1}{2}g^{ks}(D_i g_{sj} + D_j g_{is} - D_s g_{ij}) \tag{7}$$

with $D_i = \partial/\partial u^i$. To describe the curvature of E_4 we also use the curvature tensor

$$R^i_{jkm} = D_k \Gamma^i_{jm} - D_m \Gamma^i_{jk} + \Gamma^i_{sk}\Gamma^s_{jm} - \Gamma^i_{sm}\Gamma^s_{jk}. \tag{8}$$

Moreover, we define the *Ricci tensor* R_{jm} and the *scalar curvature* R by

$$R_{jm} = R^k_{jkm}, \quad R = g^{jm}R_{jm}, \\ R^{rs} = g^{rj}g^{sm}R_{jm}. \tag{9}$$

The energy–momentum tensor T^{ij} depends on the concrete physical situation.

For an ideal fluid, for example, we obtain from Section 75.13:

$$T^{ij} = c^{-2}(P + \varepsilon)v^i v^j - Pg^{ij}. \tag{10}$$

Here v^i is the four-dimensional velocity field, P the pressure, and ε the rest energy density of the fluid.

Lowering the indices by using the g_{ij}, we obtain the following equation which is equivalent to the basic equation (1):

$$R_{ij} - \tfrac{1}{2}g_{ij}R = \kappa T_{ij}, \tag{11}$$

where $T_{ij} = g_{ir}g_{js}T^{rs}$. An application of g^{ij} yields $R - 2R = \kappa T$ with $T = g^{ij}T_{ij}$. Thus (1) is equivalent to

$$R_{ij} = \kappa(T_{ij} - \tfrac{1}{2}g_{ij}T). \tag{12}$$

In the special case of absence of matter, electromagnetic radiation, and other outer fields, we obtain Einstein's equations for the metric tensor of the vacuum

$$R_{ij} = 0. \tag{13}$$

The main difference between the special and the general theory of relativity is the following. In the former case we use Minkowski's four-dimensional space–time manifold M_4 which has been introduced in Section 75.8. There exist coordinates (inertial charts) which correspond to inertial systems such that the metric tensor assumes the following special form

$$g_{11} = g_{22} = g_{33} = -1, \quad g_{44} = 1, \quad g_{ij} = 0 \quad \text{for } i \neq j.$$

Hence M_4 has zero curvature, i.e.,

$$\Gamma^k_{ij} = 0, \quad R^i_{jkm} = R^{ij} = R_{ij} = R = 0.$$

The four-dimensional space–time manifold E_4 of the general theory of relativity, on the other hand, has a nonzero curvature whenever matter is present. This curvature is responsible for the gravitational effect between the masses.

76.2. Motivation of the Basic Equations and the Variational Principle for the Motion of Light and Matter

We motivate (3). The motion of a mass particle corresponds to the world line $x = x(\sigma)$ with arclength

$$s = \int_{\sigma_0}^{\sigma_1} \sqrt{g_{ij}(u(\sigma))\dot{u}^i(\sigma)\dot{u}^j(\sigma)}\, d\sigma.$$

Here we assume $g_{ij}\dot{u}^i\dot{u}^j > 0$ along the world line. As proper time we define

$$\tau = s/c.$$

76.2. Motivation of the Basic Equations and the Variational Principle

By definition, this is the time shown by an atomic clock which moves together with the particle. Similarly as in Section 75.11, the variational principle to determine the orbit is given by $-m_0 c \int ds =$ stationary!, i.e.,

$$-m_0 c \int_{\sigma_0}^{\sigma_1} \sqrt{g_{ij} \dot{u}^i \dot{u}^j} \, d\sigma = \text{stationary!}, \tag{14}$$

$$u(\sigma_0) = u_0, \quad u(\sigma_1) = u_1$$

with σ_0, σ_1, u_0, and u_1 fixed. From Proposition 74.53 it follows that the Euler equations are equal to (3) if s is introduced as a curve parameter.

We motivate (4), (5). In the special theory of relativity we have

$$ds^2 = c^2 \, dt^2 - (dy)^2$$

for an inertial system. A photon moves rectilinearly with velocity c, i.e., $y(t) = vt + y_0$ with $v^2 = c^2$. This can also be written as

$$u^\alpha = \sigma v^\alpha + y_0^\alpha, \quad u^4 = c\sigma.$$

It follows that

$$\ddot{u}^i = 0 \quad \text{and} \quad \left(\frac{ds}{d\sigma}\right)^2 = c^2 - v^2 = 0,$$

which implies (4), (5). Hence in the general case, this system is a natural generalization of the situation in the special theory of relativity.

Theorem 76.A (General Variational Principle for Mass Particles and Photons). *The Euler equations for the variational problem $\int ds^2 =$ stationary!, i.e.,*

$$\int_{\sigma_0}^{\sigma_1} L \, d\sigma = \text{stationary!},$$

$$u(\sigma_0) = u_0, \quad u(\sigma_1) = u_1 \tag{15a}$$

with $L = g_{ij} \dot{u}^i \dot{u}^j$ are the equations of motion

$$\ddot{u}^k + \Gamma_{ij}^k \dot{u}^i \dot{u}^j = 0. \tag{15b}$$

PROOF. This follows from $(d/d\sigma) L_{\dot{u}^i} - L_{u^i} = 0$ after a short computation. □

For photons we have, in addition, that $\dot{s} = 0$. For mass particles we must have $\dot{s} > 0$. In this case, we choose σ equal to the proper time τ. We obtain $\dot{s}^2 = c^2$ because of $ds = c \, d\tau$. This yields the following additional conditions

$$L = 0 \quad \text{(light)},$$

$$L = c^2, \quad \sigma = \tau \quad \text{(matter particle)}.$$

In view of Section 75.9 we assume that physical signals can only travel on world lines $u = u(\sigma)$ with $\dot{s} \geq 0$.

The variational problem (15a) also provides a convenient method of computing the Christoffel symbols Γ_{ij}^k from (15b).

We motivate Einstein's basic equation (1). Thereby we use the heuristic principle of the greatest possible simplicity of a theory. In order to derive the field equations (1) for the vacuum we assume the variational principle

$$\int_H L_1 \, dm = \text{stationary!},$$

where H is a region in E_4 and $dm = \sqrt{|g|} \, du$ is the volume element in E_4. In order to obtain covariant equations, the integral has to be invariant, i.e., L_1 has to be a scalar. Since the curvature of E_4 should play an important role and R is the "simplest" scalar that can be formed from R^i_{jkm}, we assume $L_1 = R$. This way we obtain the *fundamental variational problem* of Hilbert

$$J \equiv \int_H L \, du = \text{stationary!} \tag{V}$$

with

$$L = R\sqrt{|g|}.$$

Since L depends on g_{ij} and the first- and second-order derivatives, we assume furthermore that all g_{ij} and $D_k g_{ij}$ remain fixed on the boundary ∂H.

In Problem 76.7 we prove

$$\delta J = -\int_H (R^{ij} - \tfrac{1}{2} g^{ij} R) \, \delta g_{ij} \, dm.$$

It follows then from $\delta J = 0$ that

$$R^{ij} - \tfrac{1}{2} g^{ij} R = 0. \tag{16*}$$

This is the basic equation (1) with $T^{ij} = 0$. In the case of matter fields we assume that the effect is described by a matter tensor S^{ij}, and we replace equation (16*) with

$$R^{ij} - \tfrac{1}{2} g^{ij} = S^{ij}. \tag{16}$$

In the case of an ideal fluid, T^{ij} in (10) is the simplest possible candidate. For dimensional reasons we need to assume that

$$S^{ij} = \kappa T^{ij}.$$

In Section 76.5, the constant κ will be exactly determined by comparison with the expansion of the universe, which has been obtained in Section 58.15 in the context of Newton's theory. The same value for κ also follows from approximation (19) below. Problem 76.5 implies that we have the identity

$$\nabla_i (R^{ij} - \tfrac{1}{2} g^{ij} R) = 0.$$

Hence for every solution of (1) we have

$$\nabla_i T^{ij} = 0. \tag{17}$$

76.2. Motivation of the Basic Equations and the Variational Principle

In the case of the ideal fluid (10), these are precisely the relativistic equations of motion for the fluid which have been discussed in Section 75.13.

In the general case we make the following assumptions on the T^{ij}:

(i) $T^{ij} = T^{ji}$ and T^{44} has the dimension of an energy density.
(ii) Equation (17) is physically meaningful for solutions of (1).

In Part V we will show how the T^{ij} for electromagnetic fields and other fields can be derived from a variational principle.

We now want to compare the field equations (1) with Newton's theory. We begin with the metric

$$ds^2 = c^2(1 + 2U/c^2)\,dt^2 - (1 - 2U/c^2)(dy)^2, \tag{18}$$

where t denotes the time and Cartesian coordinates are used as space coordinates. If ρ is the mass density, then $T^{44} = \rho c^2$ is the energy density. A straightforward computation from basic equation (1) yields for $i = j = 4$

$$\Delta U = 4\pi G\rho, \tag{19}$$

if $1/c$ is regarded as small and higher-order terms are neglected (see Problem 76.8). Equation (19), however, is the classical equation for Newton's gravitational potential

$$U(y) = -\int_H \frac{G\rho(y')}{|y - y'|}\,dy'.$$

For further motivation we might mention that the variational principle (14) for the case (18) becomes

$$\int_{t_0}^{t_1} L\,dt = \text{stationary!},$$

$$y(t_0) = y_0, \qquad y(t_1) = y_1,$$

with the Lagrangian

$$L = -m_0 c^2 \sqrt{(1 + 2U/c^2) - (1 - 2U/c^2)(\dot{y}^2/c^2)}$$

$$= \frac{m_0}{2}\dot{y}^2 - m_0 U - m_0 c^2 + \cdots .$$

The dots stand for higher-order derivatives with respect to $1/c$. The corresponding Euler equation $(d/dt)L_{\dot{y}} - L_y = 0$, however, coincides with the classical equation of motion

$$m_0 \ddot{y} = -m_0 U_y.$$

We note that the particular form of the factor by $(dy)^2$ in (18) cannot be determined from this variational argument, but instead may be determined from (19). Formula (18) is called a quasi-classical approximation of the general theory of relativity. This is well supported by experiments (e.g., the red shift of light).

The experimental basis for Einstein, in his general theory of relativity, was the equivalence between gravitational mass M and inert mass m. Let us explain this. According to Newton, the motion of a mass point in the gravitational field of the sun, for example, is given by the equation

$$m\ddot{y} = -GMM_{sun}y/|y|^3. \tag{20}$$

For $m = M$ the motion $y = y(t)$ is only affected by M_{sun} i.e., analogously to the electric field, we may speak of a gravitational field. The equality $M = m$, implicitly assumed by Newton, was confirmed experimentally by Eötvös, 1909, with an accuracy of 10^{-9} (today 10^{-11}). The field character of the gravitational force enables us to write the equations of motion (3) in a form where the mass of the particles does not appear.

An important role in the final formulation of the general theory of relativity, which Einstein worked on for about 10 years, was played by the principle of equivalence between gravitational and accelerational fields. This principle states that, locally, the gravitational effect can be eliminated by passing to an accelerated system. As an example, we consider an elevator Σ which moves downward with exactly the acceleration of gravity. A physicist in Σ, who drops a stone, will observe that the stone remains at rest in Σ. In mathematical terms, this local equivalence principle means the following. At each fixed point $x \in E_4$ one can introduce *locally* a coordinate system with

$$g_{ij} = \varepsilon_i \delta_{ij}, \qquad D_k g_{ij} = 0, \qquad \Gamma_{ij}^k = 0 \qquad \text{at } x,$$

where $\varepsilon_4 = 1$, $\varepsilon_i = -1$. This will be proved in Problem 76.4. At the point x, the equation of motion (6) then takes the form $\ddot{u}^k = 0$. Globally, however, the gravitational field can in general not be turned off by passing to an accelerated system. This follows already from Newton's mechanics, since the gravitational field of the sun and an accelerational field at infinity satisfy different boundary conditions.

The local equivalence principle played an important role for Einstein, since it led him to the idea of using the tensor calculus of Riemann, Ricci, and Levi-Civita. This point has caused many controversies because of the formal character of the requirement for the covariance of the equations. In fact, the hard core of general relativity is Einstein's idea that gravitational effects are caused by the *geometry* of the manifold E_4. Then a formulation of the theory which only uses invariant concepts of the manifold E_4 automatically leads to covariant equations by passing to charts, i.e., systems of reference.

76.3. Friedman Solution for the Closed Cosmological Model

In Section 74.21 we considered two-dimensional Riemannian manifolds with constant positive and negative curvature together with the corresponding

76.3. Friedman Solution for the Closed Cosmological Model

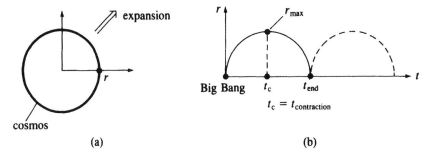

Figure 76.1

elliptic and hyperbolic non-Euclidean geometries. In the following we will show that the corresponding three-dimensional spaces can be used as models for our cosmos.

In the closed cosmological model, the cosmos consists of the surface S_r^3 of a ball of radius r in $\mathbb{R}^4$, i.e.,

$$\xi_1^2 + \xi_2^2 + \xi_3^2 + \xi_4^2 = r^2. \tag{21}$$

The radius r depends on time t, and $r = r(ct)$ can be determined from the basic equations

$$R_{ij} - \tfrac{1}{2}g_{ij}R = \kappa T_{ij} \tag{22}$$

of the general theory of relativity. Figure 76.1(a) shows the corresponding one-dimensional case. Therein we live on the boundary of a disk with radius r where r depends on t. More precisely, r will have the form of Figure 76.1(b). In the two-dimensional analog we live on the surface of a ball. The open cosmological model, on the other hand, which will be discussed in the following section, corresponds to Figure 76.2 as a two-dimensional analog. Therein we live on the surface of a cylinder, but the metric is not induced by the usual metric of $\mathbb{R}^3$, but instead is given by (39) below. For both models the cosmos is unbounded. The volume of the cosmos, on the other hand, is finite for the closed model but infinite for the open model.

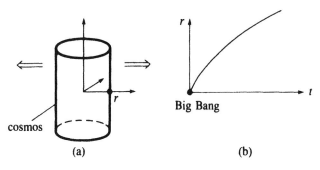

Figure 76.2

In S_r^3 we introduce the usual spherical coordinates:

$$\xi_4 = r\cos\psi, \qquad \xi_3 = r\sin\psi\cos\vartheta,$$
$$\xi_2 = r\sin\psi\sin\vartheta\cos\varphi, \qquad \xi_1 = r\sin\psi\sin\vartheta\sin\varphi, \qquad (23)$$

where $0 \le \varphi < 2\pi$ and $0 \le \vartheta, \psi \le \pi$. From $dl^2 = \sum_{i=1}^{4} d\xi_i^2$ we obtain

$$dl^2 = r^2[d\psi^2 + \sin^2\psi(\sin^2\vartheta\, d\varphi^2 + d\vartheta^2)]. \qquad (24)$$

The volume of the cosmos is $V = \int \sqrt{|g|}\, du$ with $u = (\varphi, \vartheta, \psi)$ and hence

$$V = \int_0^{2\pi}\int_0^{\pi}\int_0^{\pi} r^3 \sin^2\psi \sin\vartheta\, d\varphi\, d\vartheta\, d\psi = 2\pi^2 r^3. \qquad (25)$$

The greatest possible distance on S_r^3 is the distance between the North and the South pole, i.e.,

$$l = \int_0^{\pi} r\, d\psi = r\pi.$$

Note that analogously to the two-dimensional sphere S_r^2 in $\mathbb{R}^3$, we need at least two charts to describe the manifold S_r^3. The spherical coordinates are chart coordinates only if one removes North and South pole. We might consider everything in chart coordinates, but for convenience we shall use spherical coordinates. This causes no difficulties. For example, (25) can be obtained by passing to limits. Note that the choice for the position of the North pole is arbitrary. The equivalence of all points on S_r^3 corresponds to the following, today generally accepted cosmological principle:

(C) *All points in the cosmos are equivalent.*

As *metric* for the four-dimensional space–time manifold E_4 we choose

$$ds^2 = c^2\, dt^2 - dl^2 \qquad (26)$$

with $t > 0$. An atomic clock A which is located at a fixed point in the cosmos corresponds to the world line $\varphi = \varphi_0$, $\vartheta = \vartheta_0$, $\psi = \psi_0$, $t = $ arbitrary > 0. According to Section 76.2 the time parameter t corresponds to the proper time

$$\tau = c^{-1}\int ds = \int_0^t dt = t.$$

Therefore we may think of t as the world time. Time $t = 0$ corresponds to the Big Bang.

More precisely, we let $\mathbb{R}_> = \{t \in \mathbb{R}: t > 0\}$ and choose the four-dimensional space–time manifold E_4 in the form

$$E_4 = S_1^3 \times \mathbb{R}_>.$$

A point on E_4 is described then by a point on S_1^3, i.e., by the spherical

76.3. Friedman Solution for the Closed Cosmological Model

coordinates

$$(\varphi, \vartheta, \psi)$$

and by the world time t. We shall see below that $r = r(ct)$. Thus at time t a point $(\varphi, \vartheta, \psi, t)$ on E_4 corresponds to a point $(\varphi, \vartheta, \psi)$ in the cosmos $S^3_{r(ct)}$ at time t. In fact, we will see that the metric, obtained by solving Einstein's field equations (1), may have singularities for certain t-values. In the case of the matter cosmos, for example, we obtain that $r(ct_{end}) = 0$, and hence $\det g_{ij}(t_{end}) = 0$. If we want to rigorously work with Riemannian manifolds, we would have to exclude these singularities by modifying E_4. In the case of the matter cosmos, one can choose:

$$E_4 = S_1^3 \times\,]0, t_{end}[.$$

As energy–momentum tensor for the cosmos we choose the corresponding tensor

$$T^{ij} = -Pg^{ij} + c^{-2}(P + \varepsilon(P))v^i v^j \tag{27}$$

for an ideal fluid. This procedure is motivated as follows:

(i) For a study of the cosmos in the large, local properties are not important.
(ii) Astronomical data are not in conflict with the assumption that in the mean the mass is equally distributed in the cosmos.
(iii) There are no forces of friction between the masses in the cosmos.

One should not be irritated by the notion of a "fluid." As we saw in Section 75.13, formula (27) is the simplest ansatz for the motion of masses which satisfies (iii).

We neglect the individual motion of the galaxies, and assume that at every point in the cosmos, the mass is at rest with respect to our fixed system of reference. Thus we have

$$v^i = \frac{du^i}{d\tau} = \begin{cases} 0 & \text{for } i = 1, 2, 3, \\ c & \text{for } i = 4. \end{cases}$$

The following two differential equations of Friedman

$$3(\dot{r}^2 + \delta) = \kappa \varepsilon r^2, \tag{28}$$

$$\dot{\varepsilon} = -3(\varepsilon + P)\dot{r}/r, \tag{29}$$

are important for determining $r = r(ct)$ and the state equation

$$\varepsilon = \varepsilon(P) \tag{30}$$

between pressure P and rest energy density ε of the cosmos. The dot means derivative with respect to ct.

Theorem 76.B (Friedman (1922)). *Under the above assumptions, Einstein's equations (22) for the metric (26) are equivalent to (28) and (29) with $\delta = 1$ if we assume that $r \neq 0$ and $\varepsilon + P \neq 0$.*

PROOF: In the following, Greek indices run from 1 to 3. The nonvanishing components of $T_{ij} = g_{ir}g_{js}T^{rs}$ are

$$T_{\alpha\beta} = -Pg_{\alpha\beta}, \qquad T_{44} = \varepsilon.$$

We use a straightforward computation from (9) to obtain the Ricci tensor, and from (26) we find, for the nonvanishing components,

$$R_{\alpha\beta} = -[\ddot{r}r + 2(\dot{r}^2 + \delta)]g_{\alpha\beta}/r^2,$$

$$R_{44} = -3\ddot{r}/r.$$

We have

$$R = g^{ij}R_{ij} = g^{\alpha\beta}R_{\alpha\beta} + R_{44}$$

with $g^{\alpha\beta}g_{\alpha\beta} = 3$, and hence

$$R = -[6\ddot{r}r + 6(\dot{r}^2 + \delta)]/r^2. \tag{31}$$

Thus the field equations (22) at once yield the following system

$$\frac{2\ddot{r}}{r} + \frac{\dot{r}^2 + \delta}{r^2} = -\kappa P, \tag{32}$$

$$3(\dot{r}^2 + \delta)/r^2 = \kappa\varepsilon. \tag{33}$$

Differentiation of (33) with respect to ct yields (28) and (29). Conversely, formulas (32), (33) follow from (28), (29) after differentiating (28). □

Corollary 76.1. *The curvature scalar R of S_r^3 is equal to $6/r^2$.*

PROOF: This follows most easily from (31) with $r = $ const. Because of $ds^2 = -dl^2$ for $t = $ const, a sign change occurs in (31). □

Let us consider two important special cases.

EXAMPLE 76.2 (Closed Radiation Cosmos). According to Planck's radiation law of Section 68.4, we find the relation $\varepsilon = 3P$ for a space only filled with radiation. Formula (29) implies that

$$\varepsilon r^4 = \text{const} = C \tag{34}$$

and from Planck's radiation law follows that $\varepsilon = \text{const} \cdot T^4$. Consequently, the temperature T satisfies

$$Tr = \text{const.} \tag{35}$$

This relation has already essentially been used in Section 58.15. Letting $\xi = r^2$, the solution of (28) becomes:

$$r^2 = -c^2t^2 + 2ct\sqrt{\kappa C/3} \tag{36}$$

EXAMPLE 76.3 (Closed Matter Cosmos). If matter is dominant, then we may approximately set $P = 0$ and $\varepsilon = \rho c^2$ where ρ is the mass density. It follows from (29) that

$$\varepsilon r^3 = \text{const.}$$

Integration implies the conservation of the total mass M. More precisely, we have $\rho = M/V$, and hence

$$M = 2\pi^2 r^3 \rho = \text{const.} \tag{37}$$

If we introduce the new variable $\eta = \int dct/r(ct)$, the solution of (28) becomes:

$$r = \frac{\kappa M c^2}{12\pi^2}(1 - \cos \eta),$$

$$t = \frac{\kappa M c}{12\pi^2}(\eta - \sin \eta). \tag{38}$$

In both examples we have $r = 0$ at $t = 0$ (Big Bang). Moreover, r in (38) is time-dependent as shown in Figure 76.1(b), i.e., at a time $t_{\text{contraction}}$ the expansion of the cosmos changes into a contraction, and at time t_{end} the cosmos has collapsed into a single point. We find $\eta_{\text{end}} = 2\pi$, and thus

$$t_{\text{end}} = \kappa M c / 6\pi$$

and $t_{\text{contraction}} = t_{\text{end}}/2$.

76.4. Friedman Solution for the Open Cosmological Model

In the open model the cosmos is equal to the set $P_r^3 = S_r^2 \times \mathbb{R}_>$ with metric

$$dl^2 = r^2[d\psi^2 + \sinh^2 \psi(\sin^2 \vartheta \, d\varphi^2 + d\vartheta^2)]. \tag{39}$$

A point of P_r^3 is given by the coordinates $(\varphi, \vartheta, \psi)$ with $0 \leq \varphi < 2\pi$, $0 \leq \vartheta \leq \pi$ (spherical coordinates on S_r^2) and $0 < \psi < \infty$.

As four-dimensional space–time manifold E_4 we now choose

$$E_4 = P_1^3 \times \mathbb{R}_>.$$

If we know $r = r(ct)$, then a point $(\varphi, \vartheta, \psi, t)$ in E_4 corresponds to a point $(\varphi, \vartheta, \psi)$ in the cosmos $P_{r(ct)}^3$ at time t. We can then make analogous observations as in Section 76.3. But it is much easier to reduce this case to the case of the closed cosmological model by replacing r of Section 76.3 with ir and ψ with $i\psi$. For the volume of the cosmos we now find

$$V = \int_0^{2\pi} \int_0^{\pi} \int_0^{\infty} r^3 \sinh^2 \psi \sin \vartheta \, d\varphi \, d\vartheta \, d\psi = \infty.$$

Theorem 76.B holds with $\delta = -1$. From (28), (29) with $\delta = -1$ we obtain the following special cases.

EXAMPLE 76.4 (Open Radiation Cosmos). For $\varepsilon = 3P$ we have $\varepsilon r^4 = \text{const} = C$. Furthermore, $Tr = \text{const}$ and
$$r^2 = c^2 t^2 + 2ct\sqrt{\kappa C/3}.$$

EXAMPLE 76.5 (Open Matter Cosmos). For $P = 0$ and $\varepsilon = \rho c^2$ we have $\varepsilon r^3 = \text{const} = C$ and
$$r = \frac{\kappa C}{6}(\cosh \eta - 1), \qquad t = \frac{\kappa C}{6c}(\sinh \eta - \eta).$$

Here r behaves as shown in Figure 76.2(b). For the spatial curvature scalar R of the cosmos P_r^3 we obtain $-6/r^2$.

76.5. Big Bang, Red Shift, and Expansion of the Universe

In connection with Section 58.15 we consider the following questions:

(i) How can the red shift of galaxies (Hubble effect) be explained in the context of cosmological models in the general theory of relativity?
(ii) How can we decide whether the open or the closed model is the right model for our cosmos?
(iii) How can we determine the age of our universe?

We start with (i) and consider a galaxy as in Figure 76.3 which emits light signals at times t_G and $t_G + \Delta t_G$. Here on earth, we receive these light signals at times t_E and $t_E + \Delta t_E$. We begin with the closed cosmological model and, without loss of generality, assume that the galaxy is at the North pole of S_r^3. The motion of the photons is described by the affine geodesics equation (4). By computing the Γ_{ij}^k one shows that
$$\varphi = \text{const}, \qquad \vartheta = \text{const}, \qquad \psi = \psi(t)$$

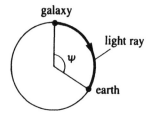

Figure 76.3

76.5. Big Bang, Red Shift, and Expansion of the Universe

are solutions of (4). This also can be motivated by symmetry arguments without any computations. The additional condition $ds^2 = c^2 dt^2 - r^2 d\psi^2 = 0$ implies that

$$\psi(t) = \int_{t_0}^{t} \frac{c\, dt}{r(ct)}.$$

The coordinate of the galaxy is $\psi_G = 0$. Let ψ_E be the coordinate of the earth. It follows then that

$$\psi_E = \int_{t_G}^{t_E} \frac{c\, dt}{r(ct)} = \int_{t_G + \Delta t_G}^{t_E + \Delta t_E} \frac{c\, dt}{r(ct)}.$$

This implies in first-order approximation

$$\frac{\Delta t_E}{r(ct_E)} = \frac{\Delta t_G}{r(ct_G)}.$$

By definition, the frequency v of light is equal to the number of oscillations per unit of time, i.e.,

$$\frac{v_E}{v_G} = \frac{\Delta t_G}{\Delta t_E} = \frac{r(ct_G)}{r(ct_E)}.$$

Let $t_E - t_G$ be small. Using a Taylor expansion, it follows that

$$r(ct_G) = r(ct_E) + c\dot{r}(ct_E)(t_G - t_E) + o(|t_G - t_E|).$$

We define the *Hubble constant*

$$H(t_E) = \frac{c\dot{r}(ct_E)}{r(ct_E)}.$$

Hence

$$\frac{v_E}{v_G} = 1 - H(t_E)(t_E - t_G) + o(t_E - t_G).$$

Because of $v\lambda = c$, we obtain for the wave length λ of the light:

$$\frac{\lambda_E - \lambda_G}{\lambda_G} = H(t_E)(t_E - t_G) + o(t_E - t_G). \tag{40}$$

Here, λ_G denotes the wave length of the light emitted from the galaxy, and λ_E denotes the wave length of the light received at the earth. According to (40), we obtain $\lambda_E > \lambda_G$ for $H(t_E) > 0$, i.e., we observe a *red shift* at the earth. This is *Hubble's law*:

For $H(t_E) > 0$, i.e., for an expanding universe and for small travel times of light $t_E - t_G$, the relative red shift is in first-order approximation proportional to the travel time.

In a contracting universe, we have $\dot{r}(ct_E) < 0$, i.e., $H(t_E) < 0$. This implies $\lambda_E < \lambda_G$. In this case, we observe a blue shift at the earth.

If the red shift (40) is interpreted in terms of a classical Doppler effect, then we obtain from Problem 58.6 that $V = HR$, where R is the distance and V the escape velocity of the galaxy. For the open cosmological model one uses analogous arguments.

We now consider (ii). For the matter cosmos with $P = 0$, $\varepsilon = \rho c^2$, which is the case of our cosmos right now, we obtain from (28):

$$3\delta c^2 = r^2(\kappa \rho c^4 - 3H^2). \tag{41}$$

This gives:

$\kappa \rho c^4 > 3H^2$: closed cosmological model ($\delta = 1$),

$\kappa \rho c^4 < 3H^2$: open cosmological model ($\delta = -1$).

Thus the critical density is

$$\rho_{\text{crit}} = 3H^2/\kappa c^4.$$

A comparison with (58.72) shows that

$$\kappa = 8\pi G/c^4.$$

This determines the universal constant κ which appears in Einstein's field equation (1). Interestingly, the same value for κ follows from approximation (19). In Section 58.15 we already mentioned that the physical data, presently known, admit no definite conclusion as to which model is the correct model of our cosmos.

Finally, we answer (iii). From (32) and (33), it follows that

$$6\ddot{r} = -\kappa(\varepsilon + 3P)r. \tag{42}$$

This equation is independent of the state equation $\varepsilon = \varepsilon(P)$. Since, at present, we observe a red shift, we have $H > 0$, and hence $\dot{r}(ct) > 0$. From (42) the curve $r = r(ct)$ is concave. Figure 76.4 shows that $r(ct)/ct \geq \dot{r}(ct)$. Therefore,

$$t \leq r(ct)/c\dot{r}(ct) = 1/H(t).$$

This estimate has also been obtained in Section 58.15 by using another derivation. It implies a maximal age of our universe of $20 \cdot 10^9$ years.

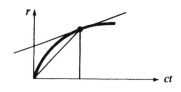

Figure 76.4

76.6. The Future of Our Cosmos

76.6a. The Future in the Closed Cosmological Model

In Section 58.15 we already saw that our cosmos is a matter cosmos. Consider the closed matter cosmos of Example 76.3.

EXAMPLE 76.6. The model of the closed matter cosmos is uniquely determined by the mass density ρ and the Hubble constant $H > 0$ at the present time. We define

$$\rho_{crit} = 3H^2/\kappa c^4, \qquad r_{crit} = c/H, \qquad \rho_{rel} = \rho/\rho_{crit},$$
$$M_{crit} = 2\pi^2 r_{crit}^3 \rho_{crit}, \qquad t_{crit} = \pi/H.$$

This implies

$$M_{crit} = 6\pi^2/\kappa cH, \qquad t_{crit} = \kappa M_{crit} c/6\pi.$$

The present radius of curvature of the cosmos is

$$r = r_{crit}(\rho_{rel} - 1)^{-1/2}.$$

From Section 76.3 it follows then that, at the present time, the maximal distance in the cosmos is equal to πr. The mass of the cosmos is $M = 2\pi^2 r^3 \rho$, hence

$$M = M_{crit}\rho_{rel}(\rho_{rel} - 1)^{-3/2}.$$

The age of the cosmos t is

$$t = \frac{\sin\eta(\eta - \sin\eta)}{H(1 - \cos\eta)^2}, \qquad \cos(\eta/2) = \sqrt{1/\rho_{rel}}.$$

Because of $0 < \eta < \pi$ we obtain

$$0 < t \leq 0.7/H.$$

At time $t_{contraction}$ the cosmos begins to collapse towards a single point. This process is completed at time t_{end}. We have

$$t_{end} = t_{crit}\rho_{rel}(\rho_{rel} - 1)^{-3/2}, \qquad t_{contraction} = t_{end}/2.$$

PROOF: The formula for r follows immediately from (41). Formulas (38) imply that

$$r = \frac{\kappa M c^2}{12\pi^2}(1 - \cos\eta), \qquad ct = \frac{\kappa M c^2}{12\pi^2}(\eta - \sin\eta). \tag{43}$$

Because of $M = 2\pi^2 r^3 \rho$ we obtain $r = \kappa r^3 \rho c^2(1 - \cos\eta)/6$, and hence

$$r = \sqrt{6/\rho\kappa c^2(1 - \cos\eta)}. \tag{44}$$

From $H = c\dot{r}(ct)/r(ct)$ together with (43) and the chain rule it follows that

$$rH = c(\sin \eta)/(1 - \cos \eta). \tag{45}$$

Equations (44) and (45) imply that $\cos(\eta/2) = \sqrt{\rho_{\text{crit}}/\rho}$ and from (43) and (45) we obtain the formula for t. □

The presently known data roughly yield $H = 1/20 \cdot 10^9$ years. This allows us to estimate the age t of the cosmos as

$$t \le 13 \cdot 10^9 \text{ years}$$

as well as

$$\rho_{\text{crit}} = 5 \cdot 10^{-27} \text{ kg m}^{-3}$$

and

$$r_{\text{crit}} = 20 \cdot 10^9 \text{ light years}, \qquad M_{\text{crit}} = 5 \cdot 10^{53} \text{ kg},$$

$$t_{\text{crit}} = 60 \cdot 10^9 \text{ years}.$$

The present mass density is estimated as 1 nucleon/m^3, i.e., $\rho_{\text{rel}} = 0.3$. This value, which favors the open cosmological model, however, is very uncertain. Also, ρ_{rel} may increase in the presence of dark masses between the galaxies and the expected neutrino mass of 20 eV/c^2. The proton has an approximate mass of 10^9 eV/c^2.

If we assume $\rho_{\text{rel}} = 1.3$, for instance, then we obtain $\eta = 1$ and a reasonable age of the universe of $t = 13 \cdot 10^9$ years. The radius of curvature r, the mass M of the cosmos, and the end time t_{end} of the cosmos are

$$r = 34 \cdot 10^9 \text{ light years},$$

$$M = 3 \cdot 10^{54} \text{ kg},$$

$$t_{\text{end}} = 400 \cdot 10^9 \text{ years}.$$

In this case, the cosmos begins to collapse at $200 \cdot 10^9$ years after the Big Bang. At first, only astronomers will notice a sudden blue shift in the spectrum of the galaxies. Slowly the sky will get brighter, and all living creatures will go blind and start to sweat until the inferno breaks loose. The cosmos becomes a gas ball which, after getting hotter and hotter, collapses into one point. According to Figure 76.1(b) there exists the theoretical possibility that in the end, i.e., after $400 \cdot 10^9$ years, there occurs another Big Bang. Astronomers on this earth, however, will have no chance to observe the beginning of the contracting phase, because already after $8 \cdot 10^9$ years, the radius of our sun has increased by a factor of 100 and its luminosity by a factor of 2,000, and thus all life on our planet will have been destroyed.

The model above describes a somewhat simplified situation, since in the beginning the cosmos was a radiation cosmos and only later on became a matter cosmos. This has been discussed in Section 58.15.

76.6b. The Future in the Open Cosmological Model

If the present ideas about the theory of elementary particles are correct, then we will have a gloomy future in the open cosmological model.

Similarly as in Example 76.6, we find as the radius of curvature

$$r = r_{\text{crit}}(1 - \rho_{\text{rel}})^{-1/2}$$

and as the age of the universe

$$t = \frac{\sinh \eta (\sinh \eta - \eta)}{H \cosh(\eta - 1)^2}, \qquad \cosh(\eta/2) = \sqrt{1/\rho_{\text{rel}}}.$$

Because of $0 < \eta < \infty$ we have

$$2/3H \leq t \leq 1/H.$$

This means $13 \cdot 10^9$ years $\leq t \leq 20 \cdot 10^9$ years. For $\rho_{\text{rel}} = 0.3$ we obtain $t = 14 \cdot 10^9$ years for the age of the universe and $r = 24 \cdot 10^9$ light years for the curvature radius of the universe.

In contrast to the closed model there occurs no collapse in the open model. In connection with a unified theory for elementary particles (SU(5)-theory or modifications), a decay of protons is predicted. If this is true, then after 10^{40} years, nucleons will no longer exist, but only black holes may exist. In Section 76.17 we explain why quantum effects may cause the vaporization of these black holes. After approximately 10^{150} years all black holes will have vaporized. In the dark and empty cosmos only a few low energy photons and neutrinos remain whose energy becomes weaker and weaker. This is a ghostly situation.

Although one might be sceptical with regard to some details of these models, most physicists are convinced that one of these two models is qualitatively the correct model for our universe. The numerical values differ in the literature, because the data for H and ρ are so uncertain.

76.7. The Very Early Cosmos

In this section we briefly mention a number of bold hypothetical ideas in modern physics which concern the very early state of our cosmos. We consider:

(a) Grand unification theory (GUT) and phase transitions.
(b) Inflationary universe.
(c) Quantum cosmology.
(d) Supersymmetry and supergravity.
(e) Superstring theory and the unification of all fundamental forces.

For further reading one may consult the References to the Literature at the end of this chapter.

76.7a. Grand Unification Theory and Phase Transitions

Phase transitions belong to the most interesting physical phenomena. The simplest phase transition, which occurs in our every-day life, is the transition

$$\text{steam} \Rightarrow \text{water} \Rightarrow \text{ice}$$

under decreasing temperatures. It is known that such phase transitions can take place with delay. For a certain amount of time the system is then in a metastable equilibrium state. For example, water may exist below the freezing point 0° Celsius. In this case, however, only a small impulse is needed to transform the supercool water into ice.

We now turn to the cosmos. In addition to the gravitational force, we presently observe three fundamental interactions:

(i) Strong interaction.
(ii) Weak interaction.
(iii) Electromagnetic interaction.

In the context of the SU(5)-standard model, (i)–(iii) can be given a unified description. This theory is called "grand unification theory," or briefly (GUT). It is a quantum field theory which has the character of a gauge field theory.

Table 76.1

Mean energy per degree of freedom of a particle $E = kT$	Temperature T of the cosmos	Time after the Big Bang	Interactions
$> 10^{19}$ GeV	$> 10^{32}$ K	$< 10^{-44}$ s	Quantum cosmology (strong coupling between gravitational and quantum effects; super-symmetry and supergravity)
10^{19} GeV (Planck energy)	10^{32} K (Planck temperature)	10^{-44} s (Planck time)	Only gravitation and a unified interaction Ω for all elementary particles exists
10^{15} GeV	10^{28} K	10^{-35} s	First phase transition (Ω splits into strong interaction Ω_s and electroweak interation Ω_{ew}; there exist gravitational, strong, and electroweak interaction
10^3 GeV	10^{16} K	10^{-12} s	Second phase transition (Ω_{ew} splits into weak and electro-magnetic interaction); there exist gravitational, strong, weak, and electromagnetic interaction
10 MeV	10^{11} K	10^{-2} s	See Section 58.15e

76.7. The Very Early Cosmos

All gauge field theories have the following mathematical mechanism in common: From the requirement of local gauge symmetry for the theory it follows that physical fields exist which cause the interactions. This will be discussed in Chapter 96 of Part V. A quantization of these fields yields the particles which are responsible for those interactions. Table 76.1 shows some of the typical phenomena. In the temperature region 10^{32} K $> T > 10^{28}$ K of the cosmos, only one common interaction exists besides gravitation. The cooling-off process of the cosmos then leads to the interactions (i)–(iii).

The mean energy per degree of freedom of a particle in the hot cosmos of temperature T is equal to $E = kT$. This follows from the equipartition law of statistical physics (see Problem 68.3). The energy density in the early cosmos can be calculated as

$$\varepsilon = \frac{\pi^2}{30} \frac{k^4}{c^3 \hbar^3} T^4 (N_B + \tfrac{7}{8} N_F).$$

Here N_B is the number of boson species where a particle with s spin positions is counted s-times. Analogously, N_F corresponds to the fermions. If, for example, only photons exist, then we have $N_B = 2$ and $N_F = 0$. The formula above for ε is a direct consequence of formula (68.25) for ideal Boson and Fermi gases with chemical potential $\mu = 0$. In Problem 68.1 we used symmetry arguments to motivate the fact that $\mu = 0$ is valid for the early cosmos.

76.7b. Inflationary Universe

In studying the Friedman model in Theorem 76.B we saw that the expansion of the universe $r = r(ct)$ depends mainly on the equation of state

$$\varepsilon = \varepsilon(P)$$

between pressure P and energy density ε. In the hypothetical model of the inflationary universe one assumes that the first phase transition in Table 76.1, which occurred 10^{-35} s after the Big Bang, took place with delay, i.e., during approximately the time interval $\Delta = [10^{-35}$ s, 10^{-20} s$]$ the cosmos was, as a consequence of quantum effects, in a metastable position. If one calculates this state from quantum field theoretical models, then the Friedman model implies that during the time interval Δ an enormous expansion of the universe in the order of magnitude of 10^{50} took place. Thereafter the development continued in a regular fashion. If this model is correct, then the radius of curvature of our present universe is much, much bigger than what has been calculated in Section 76.6. This inflationary model resolves a number of problems.

(α) *Critical density.* Because of the huge radius of curvature r, the relative mass density ρ_{rel} lies with great accuracy in the neighborhood of 1 according to Section 76.6. Consequently, we have that

$$|\rho - \rho_{crit}| = \text{very small}.$$

This way one can explain why the present mass density of the cosmos ρ lies in a neighborhood of ρ_{crit}.

(β) *Isotropy of the 3 K-radiation.* The 3 K-radiation, which was discovered in 1964, is with great precision isotropic. This, however, is difficult to understand, because parts of this radiation come from regions of the cosmos which are so far apart that no causal connection can exist between them. Note that physical effects can propagate with at most the velocity of light. In the inflationary model, this difficulty does not exist, since then the radiation which is presently observed comes from a small region of the early cosmos.

(γ) *Magnetic monopoles.* For superconductors of type II one obtains vortex filaments in case that the gauge symmetry is broken. The first phase transition 10^{-35} s after the Big Bang corresponds to a breaking of the SU(5)-gauge symmetry. Therefore physicists expect that at this time magnetic monopoles appeared with a mass at least 10^{16} times the proton masses, i.e., 10^{16} GeV/c^2. Without using the inflationary model, one obtains a monopole density for the cosmos, which is too large and implies that only some 10^4 years after the Big Bang the cosmos collapses. The inflationary model yields a significant decrease of the monopole density.

It should be noted, however, that the inflationary model has also several weak points, so that at the present time it can only be considered as a hypothesis. At any rate, it is interesting that the high-energy processes in the early universe might have had a significant influence on our present universe.

76.7c. Quantum Cosmology

We now discuss the first line of Table 76.1. Today the following universal constants are known:

- G gravitational constant (theory of gravity, general theory of relativity);
- k Boltzmann constant (thermodynamics and statistical physics);
- c velocity of light (Maxwell's theory of electromagnetism, theory of relativity);
- $\hbar$ Planck's action quantum (quantum theory);
- e_P elementary charge of the proton.

Numerical values may be found in Table 2 of the Appendix. All these constants are connected with fundamental physical theories. It is now very remarkable that any other physical dimension such as length, time, mass, temperature, and charge can be derived from them. The fact that these universal constants are so fundamental suggests basing, in a unique way, a natural system of units on them. Table 76.2 contains these natural units. In order to obtain the elementary length l_P (Planck length), for example, we make the ansatz

$$l_P = G^\alpha k^\beta c^\gamma \hbar^\delta e_P^\varepsilon.$$

76.7. The Very Early Cosmos

Table 76.2

Elementary units	
Planck length	$l_P = \sqrt{\hbar G/c^3} = 1.6 \cdot 10^{-35}$ m
Planck time	$t_P = l_P/c = 5.4 \cdot 10^{-44}$ s
Planck energy	$E_P = \hbar/t_P = 1.22 \cdot 10^{19}$ GeV
Planck mass	$m_P = E_P/c^2 = 1.22 \cdot 10^{19}$ GeV/c^2
	$= 1.3 \cdot 10^{19}$ proton masses
Planck temperature	$T_P = E_P/k = 1.4 \cdot 10^{32}$ K
Elementary charge (charge of the proton)	$e_P = 1.6 \cdot 10^{-19}$ As

Comparison of the dimensions (see Table 2 of the Appendix) implies *in a unique way* that $\beta = \varepsilon = 0$, $\alpha = \delta = \frac{1}{2}$ and $\gamma = -\frac{3}{2}$. Analogously, one obtains the remaining quantities of Table 76.2. Many physicists, using their experience with quantum theory, expect that beyond the elementary units of Table 76.2, completely new physical effects may occur. For example, there exists the hypothesis that lengths which are less than l_P cannot be measured. Since the velocity of light c is the largest possible velocity with which physical effects can propagate, it follows that also times which are less than $t_P = l_P/c$ cannot be measured. The particles in Table 76.1 have the mean energy $E_P = 10^{19}$ GeV per degree of freedom for the cosmic temperature $T_P = 10^{32}$ K. One hypothesis is that, for temperatures greater than T_P, a new kind of physics is required whereby quantum theory and the general theory of relativity are related in a significant way. This is quantum cosmology. Characteristic for quantum theory is the fact that physical states can only be realized with certain probabilities. Therefore one expects that the space–time metric (gravitational field) and the matter fields in quantum cosmology can only be realized with appropriate probabilities. This leads to the strange situation that the cosmos as a whole has a stochastic character and may change at random (see Hawking (1984, S)).

76.7d. Supersymmetry and Supergravity

The principle of *supersymmetry* states that between *bosons* (elementary particles with integer spin) and *fermions* (elementary particles with half-numberly spin), there exists a *complete* symmetry, i.e., each boson corresponds to a fermion and vice versa. It is then extremely important that the requirement of *local* supersymmetry for the theory leads to the *existence* of a tensor field g_{ij} which may be regarded as a metric of a space–time manifold. This field corresponds to *gravitation*. In a suggestive way this reads:

$$\text{local supersymmetry} \Rightarrow \text{gravitation}.$$

In the context of quantum field theory, all interactions are described by particles (quantization of fields). For example, it is expected that the gravitational field corresponds to the *graviton*. This is a boson of spin two. According to the supersymmetry, this boson must correspond to a fermion which is called a *gravitino*. Because of the existence of two gravitational particles, one speaks of *supergravity*.

In the present cosmos *no* supersymmetry is observed. But, the hypothesis exists that supersymmetry has been realized in the very early cosmos. Through the cooling-off process of the cosmos and phase transitions, this symmetry has been broken.

It is very interesting that a mathematical description of supersymmetry theories requires a *new differential calculus* for graded anticommutative algebras. One obtains supermanifolds, super-Lie groups, super-Lie algebras, etc. (see Leites (1980, S), Wess and Bagger (1983, M), Regge (1984, S), Pressley (1986, M), and West (1986, M)).

An elegant "supersymmetric" proof of the famous Atiyah–Singer index theorem can be found in Simon (1986, M).

76.7e. Superstring Theory and the Unification of All Fundamental Interactions in the Universe

The basic idea of string theory is the following:

Replace particles with strings.

We first want to explain the simple core of this idea in terms of the theory of special relativity. The crucial generalization to higher-dimensional curved spaces will be considered later on.

We use the same notation as in Section 75.1, i.e., we set

$$u = (u^1, u^2, u^3, u^4) = (\xi, \eta, \zeta, ct),$$

where ξ, η, ζ denote Cartesian coordinates, t denotes time, and c denotes the velocity of light. Moreover, we set

$$g_{11} = g_{22} = g_{33} = -1, \qquad g_{44} = 1,$$

and $g_{ij} = 0$ if $i \neq j$.

(i) *Particles and world lines.* In the theory of special relativity, the motion of a free particle is described by an equation of the form

$$u = u(t), \qquad t_1 \leq t \leq t_2,$$

where $u^4(t) \equiv ct$. This corresponds to a curve in the four-dimensional space–time manifold, which is called a world line. According to Section 75.11, the fundamental *variational principle* for the motion of the free particle $u = u(t)$

76.7. The Very Early Cosmos

reads as follows:

$$\int_{t_1}^{t_2} L(u_t(t))\,dt = \text{stationary!}, \tag{V}$$

$$u(t) = \text{fixed for } t = t_1, t_2,$$

where the Lagrangian L is given by

$$L(u_t) = -m_0 c (g_{ij} u_t^i u_t^j)^{1/2}$$

with $u_t = du/dt$. Here, m_0 is the so-called rest mass of the free particle.

(ii) *Strings and world sheets.* In contrast to a free particle, the motion of a string is described by an equation of the form

$$u = u(t,a), \qquad t_1 \leq t \leq t_2, \qquad a_1 \leq a \leq a_2,$$

where $u^4(t,a) \equiv ct$. This corresponds to a two-dimensional surface in the four-dimensional space–time manifold, which is called a world sheet. For fixed time t_0, the shape of the string in the usual three-dimensional space is given by

$$x = x(t_0, a), \qquad a_1 \leq a \leq a_2,$$

where $x = (u^1, u^2, u^3)$. Instead of (V), we now desribe the motion of the string $u = u(t,a)$ by the following *variational principle*:

$$\int_\Omega L(u_t(t,a), u_a(t,a))\,dt\,da = \text{stationary!}, \tag{V*}$$

$$u = \text{fixed on } \partial\Omega,$$

where Ω is a bounded region in $\mathbb{R}^2$. For example, we can choose $\Omega = \,]t_1, t_2[\, \times\,]a_1, a_2[$. The Lagrangian L is given by

$$L(u_t, u_a) = -T_0 c \begin{vmatrix} g_{ij} u_t^i u_t^j & g_{ij} u_t^i u_a^j \\ g_{ij} u_t^i u_a^j & g_{ij} u_a^i u_a^j \end{vmatrix}^{1/2}$$

with $u_t = \partial u/\partial t$ and $u_a = \partial u/\partial a$. Here, T_0 is the so-called rest tension of the string, and $|\cdot|$ denotes the absolute value of the determinant.

Note that the variational problem (V*) is formulated in an *invariant* way. In fact, if we change the coordinates u into u', then L remains invariant. If we change the parameters (t,a) into (t',a'), then

$$L(u_t, u_a) = \left| \frac{\partial(t',a')}{\partial(t,a)} \right| L(u_{t'}, u_{a'}),$$

and hence the integral $\int_\Omega L\,dt\,da$ remains invariant in (V*).

The Euler equations for the variational problems (V) and (V*) are the *equations of motion* for the free particle and the string, respectively.

We now want to discuss some possible generalizations. First, in the context of the theory of general relativity, we have to replace the special metric tensor g_{ij} with a general metric tensor of signature $(-1, -1, -1, 1)$.

In the more interesting context of higher-dimensional physics, we set

$$u = (u^1, \ldots, u^4; u^5, \ldots, u^d),$$

where u^1, u^2, u^3, u^4 correspond to the usual space–time coordinates, and in (V*), we choose g_{ij} as the metric tensor of a curved d-dimensional Riemannian manifold. Physicists regard $(u^5, \ldots, u^d)$ as additional degrees of freedom of space and time. However, these degrees of freedom are invisible to us, because they play only a role below the Planck length $l_P = 1.6 \cdot 10^{-35}$ m.

In Part V we will discuss in detail the following basic strategy of modern physics:

(a) For each Lagrangian L, it is possible to construct a quantum field theory by using the Feynman integral.
(b) Global symmetries of L yield conservation laws (e.g., conservation of energy and charge).
(c) Local symmetries (i.e., gauge symmetries) yield additional particles which are responsible for the interactions. (e.g., the photon is responsible for the interaction between electrons and positrons in quantum electrodynamics).

Roughly speaking, this strategy, applied to modifications of the variational principle (V*), leads to superstring theory. The following quotations refer to a fascinating recent development in theoretical physics.

> String theory is right now the *hot topic* of theoretical physics. String theory is a new view of what the fundamental constituents of nature are. According to this picture, the fundamental constituents of nature are not, in fact, particles or even fields, but are instead *little strings*, little elementary rubber bands that go zipping around, each in its own state of vibration. In these theories what we call a particle is just a string in a particular state of vibration, and what we call a *reaction* among particles, is just the *collision* of two or more strings, each in its own state of vibration, forming a single joined string which then later breaks up, forming several independent strings, each again in its own mode of vibration.
>
> It seems like a strange notion for physicists to have come to after all these years of talking about particles and fields, and it would take too long to explain why we think this is not an unreasonable picture of nature, but perhaps I can summarize it in one sentence:
>
> *String theories incorporate gravitation.*
>
> In fact, not only do they incorporate it, you *cannot* have a string theory *without* gravitation.
>
> The graviton, the quantum of gravitational radiation, the particle which is transmitted when a gravitational force is exerted between two masses, is just the lowest mode of vibration of a fundamental closed string (closed meaning that it is a loop). Not only do they incorporate and necessitate gravitation, but these string theories for the first time allow a description of gravitation on a microscopic quantum level which is *free* of mathematical inconsistencies.
>
> All other descriptions of gravity broke down mathematically, gave nonsensical results when carried to very *small* distances or very *high* energies. String theory is our first chance at a reasonable theory of gravity which extends from

76.7. The Very Early Cosmos

the very large down to the very small and as such, it is natural that we are all agog over it.

String theory itself has focused the attention of physicists on *branches of mathematics* that most of us weren't fortunate enough to have learned when we were students. You can easily see that a string (just think of a little bit of cord) traveling through space, sweeps out a two-dimensional surface. A very convenient (and, in fact, perhaps even more fundamental than talking about strings) description of string theory is to say that it is the *theory* of these *two-dimensional surfaces*.

The theory of two-dimensional surfaces is remarkably beautiful. There are ways of *classifying* all possible two-dimensional surfaces according to their *topology*, the number of handles on them and the number of boundaries, which simply don't exist in any higher dimension. The theory of two-dimensional surfaces is a branch of mathematics that, when you get into it, is one of the loveliest things you can learn. It was developed in the nineteenth century, again, I believe starting with *Riemann*, and further developed by mathematicians working in the late nineteenth century motivated by problems in complex analysis, and then continuing in the twentieth century. There are mathematicians who have spent their *whole* lives working on this theory of two-dimensional surfaces, who have never heard of string theory (or at least not until very recently). Yet when the physicists started to figure out how to solve the dynamical problems of strings, and they realized what they had to do was to perform sums over all possible two-dimensional surfaces in order to add up all the ways that reactions could occur, they found the mathematics just ready for their use, developed over the past 100 years.

String theory involves another branch of mathematics which goes back to group theory. The equations which govern these surfaces have a very large group of symmetries, known as the *conformal group*. One description of these symmetries is in terms of an algebraic structure (the Lie algebra) representing all the possible group transformations, which is actually *infinite-dimensional*. Mathematicians have been doing a lot of work developing the theory of these infinite-dimensional algebraic structures which underlie symmetry groups, again without any clear motivation in terms of physics, and certainly without knowing anything about string theory. Yet when the physicists started to work on it, there it was.

Speaking quite personally, I have found it exhilarating at my stage of life to have to go back to school and learn all this wonderful mathematics. Some of us physicists have enjoyed our conversations with mathematicians, in which we beg them to explain things to us *in terms we can understand*. At the same time the mathematicians are pleased and somewhat bemused that we are paying attention to them after all these years. The mathematics department of the University of Texas at Austin now allows physicists to use one of their lounges— which would have been unlikely in previous years.

Unfortunately, I must admit that there is *no* experimental evidence yet for string theory, and so, if theoretical physicists are spending *more* time talking to the mathematicians, they are spending *less* time talking to the experimentalists, which is *not* good.

Steve Weinberg (1986)

It appears likely that superstring theories unite gravity and quantum mechanics in a consistent manner. This is achieved by a modification of general relativity at *short* distances so that Einstein's theory emerges as a long-distance approximation. Furthermore, the quantum consistency of superstring theories provides very stringent restrictions on the possible unifying Yang–Mills gauge groups.

As a result, gravity is unified with the outer forces and particles in an almost unique manner. The only possible unifying groups are

$$SO(32) \quad \text{and} \quad E_8 \times E_8.$$

Here, SO(32) is a large orthogonal group while E_8 is the largest exceptional Lie group. The dimensionality of space–time is also required to take a special (or "critical") value

$$d = 10$$

in order to obtain a consistent superstring quantum theory. Clearly, in order to have any chance of describing the observed physics of an (approximately) four-dimensional world, *six* dimensions must turn out to be *curled up* (or "compactified") to a *very small size* (below the Planck space–time unit $1.6 \cdot 10^{-35}$ m s).

M. Green (1986)

Perhaps the most important news for this conference, so far as astro-particle physics is concerned, is the news of the emergence of a String Theory of Everything (TOE)—a theory which will embrace cosmology, all forces of nature, including gravitation and all matter. A field theory of closed strings, of the size of Planck loops (10^{-35} m s)—which naturally arises from excitations of closed strings is possibly *finite* to all loop orders in this formalism (i.e., the typical singularities of quantum field theory are renormalizable). If this statement is born out by future work, we shall have the *first quantum theory of gravity*: something which has eluded us all so far. For cosmologists, this will mean that we shall have, at last, a credible radiative extension of Einstein's equations—admittedly of us only when the Universe was very tiny in size, but of great conceptual significance nonetheless.

I cannot forbear from repeating a remark due to Chris Isham at Imperial College. Chris said, when he started research, he went to quantum gravity; his hope was to discover the origin of Planck's quantum of action h within the context of general coordinate transformations—Planck as a part of Einstein. With quantized strings one is succeeding, but in the *opposite* direction—*Einstein's theory appears to be emerging from a small part of quantum theory!*

Abdus Salam (1986)

76.8. Schwarzschild Solution

I have read your paper with the greatest interest. I did not expect that the exact solution of the problem can be formulated so easily. The analytic treatment of the problem seems to be brilliant.

Einstein in a letter to Schwarzschild on January 9, 1916.
(The astronomer Karl Schwarzschild wrote this paper on his death-bed).

In the following sections we try to explain some of the interesting physical consequences of the so-called Schwarzschild solution:

$$ds^2 = c^2(1 - r_s/r)\,dt^2 - r^2(d\vartheta^2 + \sin^2\vartheta\,d\varphi^2) - (1 - r_s/r)^{-1}\,dr^2. \quad (46)$$

Here

$$r_s = 2GM/c^2$$

76.8. Schwarzschild Solution

is the so-called *Schwarzschild radius*. For the sun (or the earth) we have $r_s = 3$ km (or 1 cm). As four-dimensional space–time manifold we choose

$$E_4 = \{(y,t) \in \mathbb{R}^4 : |y| > r_s\}.$$

Here r, ϑ, φ are spherical coordinates in $\mathbb{R}^3$.

Theorem 76.C (Schwarzschild (1916)). *The metric tensor on E_4, which belongs to (46), is a solution of Einstein's equation for the vacuum.*

The proof is a straightforward calculation, which is the subject of Problem 76.9, and also the more general and important theorem of Birkhoff (1923) is proved there. This theorem states that for $0 < r < r_s$ and $r_s < r$, (46) is the only spherically symmetric solution of Einstein's equation for the vacuum $R_{ij} = 0$. In Section 76.13 we shall see that for $0 < r_s < r$, the metric (46) corresponds to a black hole at rest.

In order to understand the physical meaning of (46) we present a simple approximation computation which implies (46). We combine the special theory of relativity with Newton's law of gravity and use the local equivalence principle. Let Σ be a Cartesian coordinate system, with the sun of mass M at the origin. Let Σ' be a box with rest mass m_0, which comes from infinity and falls radially towards the sun (Fig. 76.5). As a consequence of the free fall, an observer in Σ' will not observe any gravitation. He therefore treats Σ' as an inertial system and chooses the metric

$$ds^2 = c^2 \, dt'^2 - d\xi'^2 - d\eta'^2 - d\zeta'^2.$$

Σ' has velocity v at time t', which can be computed from the energy conservation law

$$E_{\text{kin}} + E_{\text{pot}} = \text{const} = C.$$

Newton's theory gives $E_{\text{pot}} = -GmM/r$. Moreover, we have $E = mc^2$ with rest energy $m_0 c^2$, and thus $E_{\text{kin}} = (m - m_0)c^2$. Also, we have $C = 0$, since in the beginning the box is at rest at infinity. This gives

$$(m - m_0)c^2 - GmM/r = 0.$$

Because of $m = m_0/\sqrt{1 - V^2/c^2}$ we obtain

$$\sqrt{1 - V^2/c^2} = 1 - r_s/2r, \qquad 1 - V^2/c^2 = 1 - r_s/r + \cdots.$$

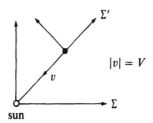

Figure 76.5

For the transformation from Σ' to Σ we use the formulas of the special theory of relativity. This yields

$d\xi' = dr/\sqrt{1 - V^2/c^2}$ (length contraction),

$dt' = dt\sqrt{1 - V^2/c^2}$ (time dilatation),

$d\eta' = r\,d\vartheta, \qquad d\zeta' = r\sin\vartheta\,d\varphi$ (invariance of transversal lengths).

This implies (46).

Therefore we can interpret (46) as the metric which is induced by a primary body of mass M. Furthermore, r, φ, ϑ are spherical coordinates in Σ. The proper time τ, which is shown by a clock at rest, is

$$\tau = \tau_0 + \int_{t_0}^{t} (1 - r_s/r)^{1/2}\,dt.$$

The equation for a radial light ray is $ds^2 = 0$. This yields $\rho = c(\tau - \tau_0)$ with

$$\rho = \int_{r_0}^{r} (1 - r_s/r)^{-1/2}\,dr$$

and $r_0 > r_s$. As in Section 75.3, one may perform measurements of lengths by using the travel time of light rays. Here one obtains ρ instead of r. Only ρ and τ have a physical meaning. If r_s/r is very small, then τ and ρ can be replaced by t and r. This is true for reasonable physical experiments, because on the surface of the sun or the surface of the earth we have approximately $r_s/r = 5 \cdot 10^{-7}$, if the primary body is the sun or the earth.

76.9. Applications to the Motion of the Perihelion of Mercury

> The last month was one of the most exciting and exhausting periods in my life, but also one of the most successful.... I recognized that my previous field equations of gravity were completely wrong. The Christoffel symbols have to be regarded as the natural expression for the components of the gravitational field.... The beautiful thing I experienced, was that not only Newton's theory could be obtained in first-order approximation, but also the motion of the Perihelion of Mercury in second-order approximation. For the deflection of light at the sun one finds a value twice as large as before.
>
> Einstein in a letter to Sommerfeld on November 28, 1915

During the nineteenth century many very precise computations of the orbits in our planetary system were performed using the methods of perturbation theory. Leverrier (1811–1877), who also predicted the orbit of Neptune, found a rotation of the Perihelion of Mercury of 43″ per century, which could not be explained by Newton's theory of gravity. The solution is a consequence of the general theory of relativity, as will be shown in the following. In Section 58.9,

76.9. Applications to the Motion of the Perihelion of Mercury

we saw that classical mechanics yields planetary orbits

$$r^{-1} = q(1 + \varepsilon \cos \varphi) \tag{47}$$

with $q = 1/a(1 - \varepsilon^2)$, where a is the great half axis and ε the eccentricity, i.e., εa is the distance of the sun from the center of the ellipse. The sun is located at a focus. The surface velocity S of each planet is constant, i.e.,

$$2^{-1}r^2\dot{\varphi} = \text{const} = S. \tag{48}$$

Moreover, we have $q = GM/4S^2$ with the gravitational constant G and mass of the sun equal to M. From (47) and (48) one obtains $\varphi = \varphi(t)$, i.e., the planetary motion in time.

The equations of motion for the general theory of relativity are

$$\ddot{u}^k + \Gamma^k_{ij}\dot{u}^i\dot{u}^j = 0. \tag{49}$$

In this section, the dot means derivative with respect to the proper time τ of the orbital motion. The Christoffel symbols Γ^k_{ij} belong to the Schwarzschild metric (46). We are looking for orbits $r = r(\varphi)$, which for $\varphi = 0$, coincide with the classical Kepler ellipse, i.e.,

$$r(0)^{-1} = q(1 + \varepsilon). \tag{50}$$

In the literature one finds approximate solutions for (49). Sometimes dubious arguments are used and the accuracy of the approximate solution is not clear. We prove here that the solution can be written as a convergent series of the small parameter $\eta = r_s q$. Also, we present a simple procedure which admits expansions up to any given order. We use the same methods as have been used in Chapter 8 for nonlinear oscillations. As we shall see in step (VI) of the following proof, the solution cannot be obtained by a direct iteration, but instead one uses a procedure which is related to the branching equation of Ljapunov.

As we shall show in (III) of the following proof, it is possible to describe explicitly the motion of planets $r^{-1} = E(\varphi)$ in the general theory of relativity by an elliptic function E. However, our method of proof via a convergent iteration method is also applicable to many other problems in celestial mechanics, where an explicit solution is not available. We assume:

(H) Let the mass of the sun M be given, as well as a number ε with $0 < \varepsilon < 1$ and a number $S \neq 0$. With this we construct $q = GM/4S^2$. Because of $r_s = 2GM/c^2$ we also have $q = r_s c^2/8S^2$. Furthermore, we let $\eta = r_s q$ and $a = 1/q(1 - \varepsilon^2)$.

Theorem 76.D. *If* (H) *holds, then there exists an* $\eta_0 > 0$ *such that for every* η *with* $0 < |\eta| \leq \eta_0$ *there exists a unique orbit* $r = r(\varphi)$ *for* (49), (50). *This is a periodic orbit in* φ *and has the form*

$$r^{-1} = q(1 + \varepsilon \cos \alpha\varphi) + O(\eta^2), \qquad \alpha = 1 - \frac{\Delta\varphi}{2\pi} + O(\eta^2) \tag{51}$$

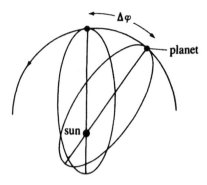

Figure 76.6

with $q = r_s^{-1}\eta$ and

$$\Delta\varphi = 3\pi\eta = 3\pi r_s/a(1 - \varepsilon^2).$$

Here, $O(\eta^2)$ denotes a convergent power series in η. The motion in time $\varphi = \varphi(\tau)$ can be uniquely determined from

$$2^{-1}r^2\dot{\varphi} = S \quad \text{and} \quad \varphi(0) = 0.$$

Remark. For the sun we have $r_s = 3$ km. In the case of Mercury (and Pluto) we find $\varepsilon^2 = 0.04$ (and $\varepsilon^2 = 0.06$). For all other planets ε^2 is significantly smaller. For the earth we find $\varepsilon^2 = 0.003$. Hence it follows for all planets that $q \approx 1/a$ and $\eta \approx r_s/a$. In the case of Mercury, which is the closest planet to the sun, we obtain $a = 60 \cdot 10^6$ km i.e., $\eta = 5 \cdot 10^{-8}$ assumes its largest value. Thus the terms which differ from the classical Kepler ellipses are very small. Orbit (51) has the period

$$2\pi/\alpha = 2\pi + \Delta\varphi + O(\eta^2).$$

Hence we find a rotation of the Perihelion of $\Delta\varphi$ per revolution (Fig. 76.6). Table 76.3 contains several values. Perturbations, caused by other planets, have already been taken into account for the observed values below. The proof will show that the orbital motion is given by

$$dt/d\tau = 1 + \zeta$$

where ζ is very small. Thus, except for a very small error, the proper time τ for all planets is equal to t, i.e., all planets have approximately the same proper time.

Table 76.3

Motion of the Perihelion in angular seconds per century	Mercury	Venus	Earth
Theory	43.03	8.6	3.8
Observation	43.11 ± 0.45	8.4 ± 4.8	5.0 ± 1.2

76.9. Applications to the Motion of the Perihelion of Mercury

We present a proof, which for $C = 0$, in addition, yields the deflection of light at the sun (see Section 76.10).

PROOF. The main idea is to reduce the problem to a second-order differential equation. This is done by differentiating. The linearized problem (57), however, can only be solved under suitable side conditions on the solution and the right-hand side. These side conditions yield an operator equation (59), from which the orbit and its *unknown* period can be determined. This operator equation can be solved by using the implicit function theorem. Equation (57) below need not have a unique solution. Carelessness with this leads to some inconsistent approximation procedures in the literature.

Our proof also will show that every solution of the original problem also must be a solution of the operator equation (59). Thus uniqueness of the solution of (59) implies the uniqueness of the solution of the original equation.

(I) Variational problem. The starting point is the equations of motion (49). As (15) shows, those are the Euler equations for the variational problem

$$\int g_{ij} \dot{u}^i \dot{u}^j \, d\tau = \text{stationary!}.$$

Thus we begin with

$$\int_{\tau_0}^{\tau_1} L \, d\tau = \text{stationary},$$

$$u(\tau_0) = u_0, \qquad u(\tau_1) = u_1, \tag{52}$$

where $L = g_{ij} \dot{u}^i \dot{u}^j$. From (46) it follows that

$$L = (1 - r_s/r)c^2 \dot{t}^2 - r^2(\dot{\vartheta}^2 + \dot{\varphi}^2 \sin^2 \vartheta) - (1 - r_s/r)^{-1} \dot{r}^2.$$

Because of $ds = c \, d\tau$ we have

$$L = Cc^2 \tag{53}$$

with $C = 1$. We assume that $\vartheta(0) = \pi/2$ and $\dot{\vartheta}(0) = 0$. This can always be achieved by rotating the coordinate system and by passing to $\vartheta - D\tau$. The Euler equation

$$\frac{d}{d\tau} L_{\dot{\vartheta}} - L_\vartheta = 0$$

is linear in $\cos \vartheta$, $\dot{\vartheta}$, and $\ddot{\vartheta}$. Hence $\vartheta(\tau) = \text{const} = \pi/2$ is a solution. The Euler equations

$$\frac{d}{d\tau} L_{\dot{\varphi}} - L_\varphi = 0 \quad \text{and} \quad \frac{d}{d\tau} L_{\dot{t}} - L_t = 0$$

yield

$$2^{-1} r^2 \dot{\varphi} = \text{const} = S,$$

$$(1 - r_s/r) \dot{t} = \text{const} = T. \tag{54}$$

Inserting the last equation into (53) gives an analog to the classical energy theorem.

(II) **First approximation.** We let $u = 1/r$, $r = r(\varphi)$, and $u' = du/d\varphi$. From (53) and (54) we obtain

$$4S^2 u'^2 + 4S^2 u^2 - T^2 + Cc^2(1 - r_s u) = 4S^2 r_s u^3. \tag{55}$$

We choose

$$T = \sqrt{C(1 - r_s q) + 4(1 + \varepsilon^2) S^2 q^2/c^2}.$$

By using $u = \eta v/r_s$ we pass to dimensionless quantities. From (55) it follows that

$$v'^2 = 2Cv - v^2 + \varepsilon^2 - 1 + \eta v^3, \qquad v(0) = 1 + \varepsilon. \tag{56}$$

For $\eta = 0$ we obtain the first-order approximation

$$v = C + \varepsilon \cos \varphi.$$

Because of $C = 1$ these are precisely the Kepler ellipses.

For the planets we find

$$4S^2/r^2 = r^2 \dot{\varphi}^2 \qquad \text{and} \qquad q = 1/a(1 - \varepsilon^2).$$

Hence $2Sq$ is approximately equal to the mean orbit velocity, which is very small compared to the velocity of light c. For the earth we find $4S^2 q^2/c^2 = 10^{-8}$. Thus T is almost equal to one. Hence (54) shows that the proper time τ is almost equal to t.

(III) **Elliptic functions.** From (56) it follows by integration that $\varphi = \varphi(v)$. The inverse function $v = v(\varphi)$ gives $r = r(\varphi)$. Since the polynomial in (56) has three real zeros for small $\eta \neq 0$, it follows that $v = v(\varphi)$ is an elliptic function with one real and one purely imaginary period.

(IV) **Decomposition.** Therefore the solution $v = v(\varphi)$ of (56) is periodic. Let the period be denoted by $2\pi/\alpha$. It is important then for the following approximation method that v can uniquely be decomposed as

$$v = C + \varepsilon \cos \alpha \varphi + H - H_0 \cos \alpha \varphi$$

with

$$\int_0^{2\pi/\alpha} H \cos \alpha \varphi \, d\varphi = 0 \qquad \text{and} \qquad H_0 = H(0).$$

This uses Fourier series and the fact that $v(0) = 1 + \varepsilon$.

(V) **Study of the central linear differential equation.** We set

$$a_k(f) = \pi^{-1} \int_0^{2\pi} f(\varphi) \cos k\varphi \, d\varphi.$$

This is the Fourier coefficient of f by $\cos k\varphi$ for $k \geq 1$. Now we consider the *key equation*

$$h'' + h = f. \tag{57}$$

76.9. Applications to the Motion of the Perihelion of Mercury

Let X be the B-space of all even, 2π-periodic C^4-functions $f: \mathbb{R} \to \mathbb{R}$ with

$$a_1(f) = 0.$$

Then $|a_k(f)| \leq \text{const}/k^4$. Thus we can differentiate the Fourier series of f twice with respect to φ. Comparison of the coefficients of the Fourier series shows that (57) has a unique solution $h \in X$ for each $f \in X$. We set

$$h = Af.$$

Then the solution operator $A: X \to X$ of (57) is linear and continuous (see Section 8.13).

It is important here that through the construction of X, i.e., through the side conditions

$$a_1(f) = 0 \quad \text{and} \quad a_1(h) = 0$$

we force (57) to have a *unique* solution. Without the side condition $a_1(h) = 0$, equation (57) will not have a unique solution, since $h = \cos \varphi$ is a solution of the homogeneous equation (57). Neglecting this fact leads to inconsistent approximation methods in the literature.

Moreover, one should note that the condition $a_1(f) = 0$ is a necessary solvability condition for (57), because from $h \in X$ it follows that

$$a_1(h'' + h) = 0.$$

Observe the Fourier expansion of h and also that $a_1(h) = 0$. For the function

$$f = a_2(f) \cos 2\varphi + a_3(f) \cos 3\varphi + \cdots$$

one obtains as a representation of the solution $h = Af$ of (57):

$$h = a_2(h) \cos 2\varphi + a_3(h) \cos 3\varphi + \cdots$$

with

$$a_k(h) = a_k(f)/(1 - k^2), \quad k = 2, 3, \ldots. \tag{57a}$$

(VI) *Equivalent operator equation.* Differentiation of (56) gives the *key equation*

$$v'' + v = C + 3\eta v^2/2. \tag{57b}$$

In order to determine the unknown period $2\pi\alpha$, we set $w(\varphi) = v(\varphi/\alpha)$. This gives

$$\alpha^2 w'' + w = C + 3\eta w^2/2.$$

We choose

$$\alpha^{-2} = 1 + 2\beta,$$

and analogously to (IV) we set

$$w = C + \varepsilon \cos \varphi + h - h_0 \cos \varphi$$

with $a_1(h) = 0$ and $h_0 = h(0)$. This yields

$$h'' + h = f, \qquad h \in X$$

with

$$f = g - 2\beta\varepsilon\cos\varphi.$$

The *trick* is to compute the unknown number β from

$$a_1(f) = 0.$$

According to (V) this condition is a necessary solvability condition. Hence we obtain the equivalent system

$$\begin{aligned} h'' + h &= g - a_1(g)\cos\varphi, \qquad h \in X, \\ \beta &= a_1(g)/2\varepsilon \end{aligned} \tag{58}$$

with

$$g = \tfrac{3}{2}\eta(1 + 2\beta)(C + \varepsilon\cos\varphi + h - h_0\cos\varphi)^2 - 2\beta(h - h_0\cos\varphi)$$

and the small parameter η. From (V) it follows that for $(h, \beta) \in X \times \mathbb{R}$, equation (58) is an operator equation of the form

$$\begin{aligned} h &= A(g - a_1(g)\cos\varphi), \\ \beta &= a_1(g)/2\varepsilon. \end{aligned} \tag{59}$$

(VII) Existence and uniqueness proof. According to the implicit function theorem of Section 4.7, equation (59) has a unique solution $(h, \beta) \in X \times \mathbb{R}$ for every η: $|\eta| \leq \eta_0$ which depends analytically on η.

(VIII) Computation of the solution. Since the solution is analytic, its coefficients of expansion can be determined by using an ansatz and comparison of coefficients in (58). For each step, one has to solve an equation of the form (57) where f is a finite Fourier series. Then also h is a finite Fourier series, which can be determined by (57a).

In order to compute the first-order approximation, we note that only that part of g will be needed which is obtained by letting $\beta = 0$ and $h = 0$. This gives

$$g = \tfrac{3}{2}\eta C^2 + 3\eta\varepsilon C \cos\varphi + \tfrac{3}{4}\eta\varepsilon^2(1 + \cos 2\varphi).$$

It follows from (58) that

$$\beta = 3\eta C/2 + O(\eta^2)$$

and hence $\alpha = 1 - 3\eta/2 + O(\eta^2)$ and

$$\begin{aligned} w &= C + \frac{3\eta\varepsilon^2}{4} + \frac{3\eta C^2}{2} - \frac{\eta\varepsilon^2}{4}\cos 2\varphi \\ &\quad + \left(\varepsilon - \tfrac{3}{2}\eta C^2 - \frac{\eta\varepsilon^2}{2}\right)\cos\varphi. \end{aligned} \tag{60}$$

$\square$

Equation (58) can also be solved by using an iteration scheme. On the right-hand side one replaces h, β with h_n, β_n and on the left-hand side with h_{n+1}, β_{n+1}, where $h_0 = \beta_0 = 0$. According to Section 4.7 this iteration scheme is convergent and differs significantly from the inconsistent scheme

$$v''_{n+1} + v_{n+1} = C + 3\eta v_n^2/2,$$

which can often be found in the literature in connection with (57b).

76.10. Deflection of Light in the Gravitational Field of the Sun

We want to show that in a neighborhood of the boundary of the sun a light ray behaves as pictured in Figure 76.7 with

$$\delta = r_s/d + o(r_s/d). \tag{61}$$

Hence, a light ray is deflected by 2δ. For the boundary of the sun we have $d = 0.686 \cdot 10^6$ km. Thus we obtain $2\delta = 8.7 \cdot 10^{-7}$, i.e., $2\delta = 1.75''$. This result admits an experimental test by means of the photographic registration of stars during a total eclipse of the sun. The observed values, however, vary between $1.4''$ and $2.7''$. For quasars (distant radio sources) the coincidence with the theory reaches 10%.

Now we prove (61). The motion of a photon in the Schwarzschild metric is given by equation (49), where the dot no longer means derivative with respect to τ, but instead with respect to σ. Also, we assume that $\dot{s} = 0$. Thus once again we can use the variational problem (52), where, however, (53) has to be replaced with $L = 0$. Hence, by letting $C = 0$ and $\varepsilon = 1$, the arguments in the proof of Theorem 76.D apply. From (60) we find the solution

$$r^{-1} = r_s^{-1}\eta\left[\left(1 - \frac{\eta}{2}\right)\cos\alpha\varphi + \frac{3\eta}{4} - \frac{\eta}{4}\cos 2\alpha\varphi\right] + O(\eta^3),$$
$$\alpha = 1 + O(\eta^2). \tag{62}$$

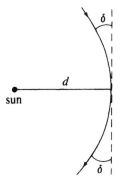

Figure 76.7

In order to find the asymptotes, we look for the angles $\gamma_\pm$ with $r \to \infty$ for $\varphi \to \gamma_\pm$. For $\varphi = \pi/2 + \delta$ we have

$$\cos \varphi = -\sin \delta = -\delta + O(\delta^3)$$

and

$$\cos 2\varphi = -\cos 2\delta = -1 + O(\delta^2).$$

By letting the right-hand side of (62) be equal zero we obtain

$$\delta - \eta + O_2(\delta, \eta) = 0.$$

This means $\gamma_\pm = \pm(\pi/2 + \eta + O(\eta^2))$.

The parameter η can be determined from $r = d$ and $\varphi = 0$. This implies

$$\eta = r_s/d.$$

For the boundary of the sun we have $d = 0.7 \cdot 10^6$ km and $\eta = 4 \cdot 10^{-6}$. This shows that in fact the parameter η is small.

76.11. Red Shift in the Gravitational Field

We consider again the Schwarzschild metric

$$ds^2 = c^2(1 - r_s/r) \, dt^2 - r^2(d\vartheta^2 + \sin^2 \vartheta \, d\varphi^2) - (1 - r_s/r)^{-1} \, dr^2.$$

For the radial motion of a photon we find the following equation

$$t = t_0 + c^{-1} \int_{r_0}^{r_1} (1 - r_s/r)^{-1} \, dr$$

from $ds = 0$. Consider two points P_0 and P_1 which lie on a radial ray through the origin with r-coordinates r_0 and r_1 with $r_j > r_s$ for $j = 1, 2$. Two signals which are emitted at P_0 at a timely distance Δt, reach P_1 with the same timely distance Δt. We now use atomic clocks in P_j. They show the proper time τ. According to Section 76.8 we have

$$\Delta \tau_j = (1 - r_s/r_j)^{1/2} \Delta t.$$

The frequency of light ν_j which is observed in P_j satisfies $\nu_1/\nu_0 = \Delta \tau_0/\Delta \tau_1$. For the wave length λ_j we then obtain the relative change

$$\frac{\lambda_1 - \lambda_0}{\lambda_0} = \sqrt{\frac{1 - r_s/r_0}{1 - r_s/r_1}} - 1 \approx \frac{1}{2}\left(\frac{r_s}{r_0} - \frac{r_s}{r_1}\right)$$

because of $\lambda = c/\nu$. The last value is true for small r_s/r_j. For $r_1 > r_0 > r_s$ we have $\lambda_1 > \lambda_0$, i.e., in P_1 we observe a *red shift* with respect to P_0.

EXAMPLE 76.7 (Experiment On the Earth). Consider a γ-source on the surface of the earth and a receiver (iron absorber) in a tower of height 22.5 m. Because of

$$r_s = 8.84 \cdot 10^{-3} \text{ m}, \qquad r_0 = 6.37 \cdot 10^6 \text{ m} \quad \text{and} \quad r_1 = r_0 + 22.5$$

we obtain

$$\Delta\lambda/\lambda = 2.5 \cdot 10^{-15}.$$

In 1960, this experiment had been performed by Pound and Repka at Harvard University with a brilliant confirmation of the theoretical result. These very small changes in the wave lengths can be measured by using the Mössbauer effect (recoilless absorbtion of γ-quanta).

76.12. Virtual Singularities, Continuation of Space–Time Manifolds, and the Kruskal Solution

Until now we only considered the Schwarzschild solution

$$ds^2 = c^2(1 - r_s/r)\,dt^2 - r^2(d\vartheta^2 + \sin^2\vartheta\,d\varphi^2) - (1 - r_s/r)^{-1}\,dr^2 \qquad (63)$$

for $r > r_s$. Here r is a space and t a time variable. In fact, the metric (63) is also a solution of Einstein's equations for the vacuum for $0 < r < r_s$ (see Problem 76.9). Since now the coefficients by dt^2 and dr^2 change their sign in (63), it follows that according to our definition of Section 76.1, r becomes a time and t a space variable.

(Q1) Does physics end at $r = r_s$?

We want to show that this is not the case and begin our discussion with a simple example which illustrates the key phenomenon. Let $M = \mathbb{R}$ and $M_\pm = \mathbb{R}_\pm$ with metric

$$ds^2 = d\xi^2 \quad \text{on } M.$$

Now on $M_\pm$ we introduce new coordinates through $\xi = \pm 2\sqrt{t}$, $0 < t < \infty$. This gives

$$ds^2 = dt^2/t,$$

i.e., for $t = 0$ we obtain a singularity. This shows that by choosing unfortunate chart coordinates one may pretend to have singularities. As an indication that, in fact, we have a virtual singularity at $t = 0$, we compute the arclength

$$s = \int_0^t dt/\sqrt{t} = 2\sqrt{t} \quad \text{for} \quad t > 0.$$

This is finite. In the following section we will see that a space ship in free fall in the metric (63) only needs a finite proper time to reach the boundary $r = r_s$.

This leads to the following additional question:

(Q2) Is it possible to extend the two disjoint space–time manifolds $S_>$ and $S_<$, which correspond to (63) for $r > r_s$ and $0 < r < r_s$ to a space–time manifold containing both?

As a matter of fact we now construct such an extension, which at the same time is maximal. To this end, we introduce the Kruskal transformation by letting

$$cv_\pm = ct \pm (r + r_s \ln|1 - r/r_s|)$$

and

$$z = \tfrac{1}{2}(e^{cv_+/2r_s} + \eta e^{-cv_-/2r_s}),$$

$$w = \tfrac{1}{2}(e^{cv_+/2r_s} - \eta e^{-cv_-/2r_s}),$$

with $\eta = 1$ for $r > r_s$ and $\eta = -1$ for $0 < r < r_s$. This implies

$$w^2 - z^2 = (1 - r/r_s)e^{r/r_s}$$

$$\frac{w}{z} = \begin{cases} \tanh ct/2r_s & \text{for } |w/z| < 1, \\ \coth ct/2r_s & \text{for } |w/z| > 1. \end{cases} \tag{64}$$

This way we obtain from (63) the *Kruskal metric*

$$ds^2 = \frac{4r_s}{r} e^{-r/r_s}(dw^2 - dz^2) - r^2(d\vartheta^2 + \sin^2\vartheta\, d\varphi^2) \tag{65}$$

with $r = r(w, t)$. In order to understand this metric, we consider the hyperbola $w^2 - z^2 = 1$ which corresponds to $r = 0$, and the two diagonals $w = \pm z$ which correspond to $t = \pm\infty$. As shown in Figure 76.8, the hyperbola and the two diagonals form the boundaries of open sets, which will be denoted as follows:

(i) K_W universe with a black hole;
K_{out} outer universe in K_W;
K_B black hole in K_W;
(ii) K_W^* universe with a white hole;
K_{out}^* outer universe in K_W^*;
K_B^* white hole in K_W^*.
(iii) K Kruskal universe (universe with a black–white dipole hole);
I unreal universe.

For example, I is equal to the closed set, shaded in Figure 76.8, which is bounded by the two branches of the hyperbola. Moreover, we have $K = \mathbb{R}^2 - I$.

The physical meaning of these suggestive notations will become clear during the following sections. The variables (r, t) vary in the form $r > 0$ and $-\infty < t < \infty$. The variables (z, w) vary in K. The coordinate lines $r = \text{const}$ in K are hyperbolas, while the coordinate lines $t = \text{const}$ correspond to straight lines

76.12. Virtual Singularities, Continuation of Space–Time Manifolds

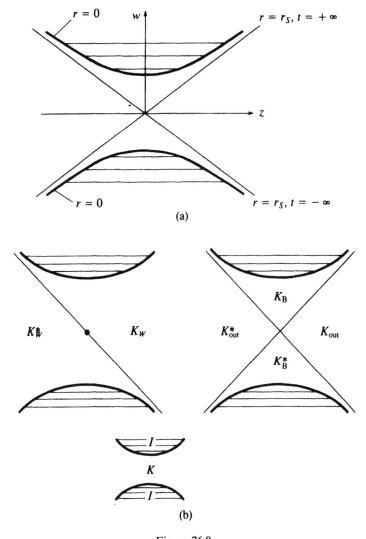

Figure 76.8

through the origin. In Figure 76.9(a) we indicate the direction of increasing r- and t-values. We now construct the set

$$E_4(K) = S^2 \times K.$$

Analogously, we let $E_4(K_{\text{out}}) = S^2 \times K_{\text{out}}$, etc. The points of $E_4(K)$ correspond to the coordinates $(\varphi, \vartheta, z, w)$, where φ, ϑ are spherical coordinates in S^2 and (z, w) lies in K. Here, S^2 denotes the boundary of the unit ball in $\mathbb{R}^3$.

Theorem 76.E (Kruskal (1960)). *With the Kruskal metric* (65), *the set* $E_4(K)$ *becomes a Riemannian C^∞-manifold of signature type* $(-1, -1, -1, 1)$. *On $E_4(K)$ the metric tensor satisfies Einstein's field equations for the vacuum.*

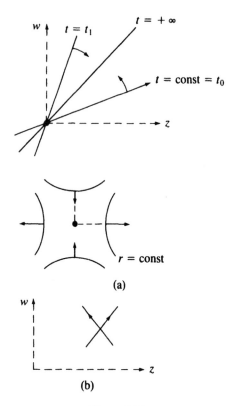

Figure 76.9

PROOF. The components g_{ij} of the metric tensor in (65) are C^∞-functions on $E_4(K)$ with $\det g_{ij} \neq 0$. The virtual singularities for $\vartheta = 0$ and $\vartheta = \pi$ only appear in spherical coordinates. They disappear in the usual way if one uses chart coordinates for S^2. For $0 < r < r_s$ and $r > r_s$ equation (65) follows from the Schwarzschild solution by using a coordinate transformation. Thus, in this region, (65) satisfies Einstein's equation

$$R_{ij} - \tfrac{1}{2} g_{ij} R = 0.$$

By continuity, g_{ij} is also a solution for $r = r_s$, and hence a solution on K_W. Using a reflection, one obtains K_W^* from K_W, where g_{ij} is equal in corresponding points. Therefore, g_{ij} satisfies $R_{ij} - \tfrac{1}{2} g_{ij} R = 0$ on K. □

Hence (65) is the desired extension of (63), where K_{out} and K_B in Figure 76.8(b) correspond to the Schwarzschild solutions for $r > r_s$ and $0 < r < r_s$. More precisely, we have $S_> = E_4(K_{\text{out}})$ and $S_< = E_4(K_B)$.

All radial light rays in $E_4(K)$, i.e., all light rays with $\varphi = \text{const}$ and $\vartheta = \text{const}$ satisfy $ds = 0$, so that

$$w = \pm z + \text{const},$$
$$\varphi = \text{const}, \quad \vartheta = \text{const} \tag{66}$$

holds. Hence all radial light rays in Fig. 76.8(a) are straight lines with slope ± 1. One easily shows that (66) is a solution of the equations of motion, which are the Euler equations of the variational problem (15).

The main reason for introducing the coordinates w and z is the simple form which the equation assumes for radial light rays. In the following we will see how this simplifies the treatment of the qualitative behavior of black holes.

Definition 76.8. Radial light rays are oriented in such a way that the future corresponds to increasing values of w (Fig. 76.9(b)).

According to Figure 76.9(a) and Figure 76.8, this corresponds to the following increasing values:

$$t \quad \text{in } K_{\text{out}} \quad \text{and} \quad -r \quad \text{in } K_B$$
$$t^* \quad \text{in } K_{\text{out}}^* \quad \text{and} \quad -r^* \quad \text{in } K_B^*,$$

where $t^* = -t$ and $r^* = -r$. This corresponds to the fact that the space coordinate for the outer universe K_{out} becomes a time coordinate inside the black hole K_B.

One can prove that the Kruskal metric cannot be extended beyond K, since on the boundary ∂K which is formed by the hyperbolas $r = 0$, there exists a proper singularity whose physical meaning we now explain.

76.13. Black Holes and the Sinking of a Space Ship

> After a short while, I was obsessed with the whirlpool. In spite of the sacrifice it would mean, I felt a strong desire to explore its depths; and that, what hurt the most, was the fact that I would never be able to tell my old comrades about the secrets that I would see.
> Edgar Allan Poe: *Down into the Maelström.*
>
> Many astrophysicists agree that Cyg X-1 is a black hole.
> Walter Sullivan (1979)

Intuitively, one means by a black hole a region in the cosmos which has such a strong gravitational force that neither light nor matter can leave it. Like a moloch, a black hole swallows all surrounding matter, which thereby is heated and emits X-rays as a cry of death. Therefore one expects that the X-ray source Cyg X-1 in the constellation of the swan is a black hole. These days, one could read in some newspapers that astronomers at Caltech in Pasadena discovered heated matter which tumbles into a black hole at the center of our universe. They studied photographs made with a large radio telescope in the desert of Socorro (New Mexico). This black hole apparently has a mass of 200 to $2 \cdot 10^6$ times the mass of the sun. One also expects that very bright quasars, which are located at huge distances, contain black holes. This will be discussed in Section

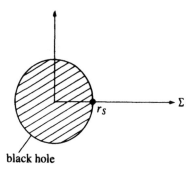

black hole

Figure 76.10

76.16. There we also briefly explain how black holes may occur at the end of the development of a star. For this reason one expects numerous black holes throughout the cosmos.

We now use the Schwarzschild metric and the Kruskal transformation to describe a mathematical model of a black hole, which is located at the origin, in a Cartesian coordinate system Σ and which has a radius of r_s (Fig. 76.10). We choose spherical coordinates r, φ, ϑ and use the Schwarzschild metric (63). More precisely, we extend this metric to the Kruscal metric on $E_4(K_W)$. In order to illustrate this situation, we choose a fixed plane with $\varphi = $ const and $\vartheta = $ const. In the (z, w)-coordinate system, the space–time manifold corresponds to the nonshaded open set in Figure 76.11(a). It consists of the

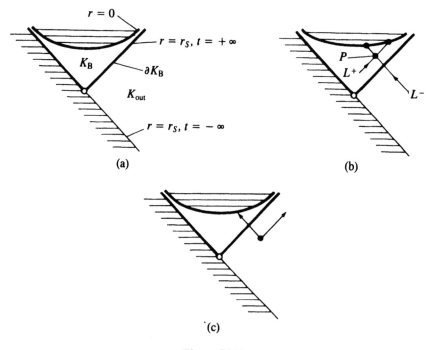

Figure 76.11

following open sets:

K_{out}: (exterior universe in Σ with $r > r_s$, $-\infty < t < \infty$),

K_B: (black hole in Σ with $0 < r < r_s$, $-\infty < t < \infty$),

and the diagonal $w = z, z \geq 0$, which corresponds to the boundary ∂K_B of the black hole ($r = r_s$ and "$t = +\infty$").

In order to determine the *mass M* of a black hole, we assume the same relation as in Section 76.8., i.e.,

$$M = r_s c^2 / 2G.$$

For $r_s = 3$ km (or $r_s = 1$ cm) this is the mass of the sun (or the mass of the earth), and implies an extraordinarily high mass density.

EXAMPLE 76.9 (Light Trap). According to Section 76.12, the radial light rays are the straight lines with slope ± 1 in the (z, w)-diagram. Important for a qualitative understanding of the following considerations is the fact that according to the previous section, increasing w-values for the light rays correspond to increasing time. In all figures below, the arrows of the light rays correspond to increasing time.

Exactly two radial light rays L^+ and L^- pass through a point P in the black hole K_B (Fig. 76.11(b)). The ray L^+, however, remains in K_B, while L^- corresponds to a ray which falls into K_B. Both rays reach the hyperbola in Figure 76.11(b), i.e., reach the singularity $r = 0$. We therefore obtain the important result that no radial light ray can leave the black hole K_B.

If a pair of photons is located on the boundary of the black hole, then, as in Figure 76.11(c), one of the two photons may disappear into the black hole while the other is radiated into the universe. In Section 76.17 we will use this effect in order to derive the formula for the vaporization of black holes.

EXAMPLE 76.10 (The Sinking of a Space Ship). From the earth, $r = r_E$, which is at rest in the system Σ, a space ship A takes off in a radial direction towards the black hole (Fig. 76.12). In order to save energy, the crew shuts off the engines, so that A falls freely. Then the following holds:

(i) The spaceship A only needs a finite proper time in order to reach the boundary $r = r_s$ of the black hole.
(ii) Observers on the earth find that, for the proper time on the earth, the space ship A never reaches the boundary $r = r_s$, i.e., it takes an infinite amount of time.
(iii) If space ship A has reached the boundary $r = r_s$, then it is in a hopeless situation. After proper time

$$\tau \leq \pi r_s / c$$

space ship A will have crashed at the singularity $r = 0$, even after the most forceful rocket stopping. If M is the mass of the black hole, then we have

$$r_s = 3(M/M_{sun}) \text{ km} \quad \text{and} \quad \tau \leq 3.2(M/M_{sun}) \cdot 10^{-5} \text{ s}.$$

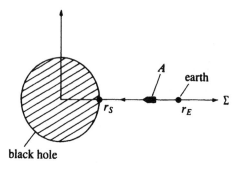

Figure 76.12

We prove (i) and (ii). From the variational principle of Theorem 76.A of Section 76.2 and the Schwarzschild metric (63) we obtain the equations of motion

$$(1 - r_s/r)\dot{t} = \text{const} = T,$$
$$(1 - r_s/r)c^2\dot{t}^2 - (1 - r_s/r)^{-1}\dot{r}^2 = c^2$$

(see (53), (54)). We choose $\dot{t}(0) > 1$. This implies $T > (1 - r_s/r_E)$ and

$$cT^2 \geq \dot{r}^2 = c^2 T^2 - c^2(1 - r_s/r) \geq \text{const} > 0$$

for $r_s \leq r \leq r_E$. The proper time needed by space ship A to travel from r_E to r_s is

$$\Delta\tau = \int_{r_E}^{r_s} dr/\dot{r} < \infty.$$

The t-time needed is

$$\Delta t = \int_{r_E}^{r_s+\varepsilon} \dot{t}\, dr/\dot{r} \geq \text{const} \int_{r_E}^{r_s+\varepsilon} \dot{t}\, dr \to \infty \quad \text{as} \quad \varepsilon \to 0$$

because of $\dot{t} = T/(1 - r_s/r)$. According to (63) with $r = r_E$, $\vartheta = \varphi = \text{const}$, the proper time which is measured on the earth is proportional to Δt.

Now we prove (iii). The transformation

$$cv = ct + r + r_s \ln|1 - r_s/r|$$

implies for the Schwarzschild metric

$$ds^2 = (1 - r_s/r)c^2\, dv^2 - 2c\, dr\, dv$$
$$- r^2(\sin^2\vartheta\, d\varphi^2 + d\vartheta^2).$$

This is a nonsingular metric for $r > 0$ (Eddington metric). For an arbitrary motion $r = r(\tau)$ and $v = v(\tau)$ we have $\dot{s}^2 = c^2$, and hence

$$(1 - r_s/r)c^2\dot{v}^2 - 2c\dot{r}\dot{v} - r^2((\sin^2\vartheta)\dot{\varphi}^2 + \dot{\vartheta}^2) = c^2.$$

For $r \leq r_s$ we therefore cannot have $\dot{r} = 0$. It follows that $\dot{r} < 0$ for $r \leq r_s$,

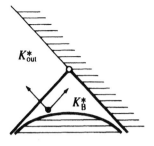

Figure 76.13

because for $r = r_s$ we have $\dot{r} < 0$ since space ship A travels radially towards the boundary. It follows from the Schwarzschild metric that

$$c^2(1 - r_s/r)\dot{t}^2 - (1 - r_s/r)^{-1}\dot{r}^2 - r^2((\sin^2 \vartheta)\dot{\varphi}^2 + \dot{\vartheta}^2) = c^2.$$

A look at the signs shows that $-(1 - r_s/r)^{-1}\dot{r}^2 \geq c^2$. From $\dot{r} < 0$ it follows that

$$\tau_{max} = -\int_{r_s}^0 d\tau/\dot{r} \leq c^{-1}\int_0^{r_s}[(r_s/r) - 1]^{-1/2}\,dr = \pi r_s/c.$$

This is (iii). Note that $\varphi = \vartheta =$ const.

Other important properties of black holes will be considered in Section 76.16.

76.14. White Holes

Consider the same Cartesian coordinate system Σ as in Section 76.13. We replace, however, time t with $t^* = -t$. Then the black hole K_B in Figure 76.11 becomes a white hole K_B^* with the same mass. Because of time reverse, the light rays which fall into the black hole now correspond to light rays which emerge from the white hole. The space–time manifold is now $E_4(K_W^*)$. In the (z, w)-diagram this corresponds to the nonshaded open set of Figure 76.13. This figure shows that light rays which emerge from the white hole K_B^* may leave it into the exterior universe K_{out}^*.

76.15. Black–White Dipole Holes and Dual Creatures Without Radio Contact to Us

> My suspicion is that the universe is not only queerer than we suppose, but queerer that we can suppose.
>
> J. Haldane

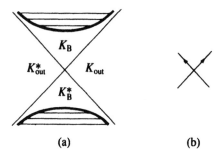

Figure 76.14

We now discuss the whole Kruskal universe. In the (z, w)-diagram this corresponds to the nonshaded open set of Figure 76.14(a). This universe consists of the two exterior universes K_{out} and K_{out}^*, a black hole K_B, and a white hole K_B^*. As in Figure 76.14(b) radial light rays correspond to straight lines with slope ± 1 and future direction is shown by the arrow. Therefore radial light rays cannot stay in the white hole, but instead are emitted into both, K_{out}^* and K_{out}. All light rays in the black hole K_B remain captured. Suppose that we live in K_{out}, then we cannot send radial light rays to K_{out}^* nor can we receive them from there.

We now investigate arbitrary physical signals. For the propagation of physical signals we have $\dot{s} \geq 0$. It follows from the Kruskal metric (65) that $\dot{w}^2 \geq \dot{z}^2$ in $E_4(K)$. This means $|dw/dz| \geq 1$. Consequently, the motion of matter and light in the Kruskal universe K (Fig. 76.14) is described by curves of the form

$$w = w(z) \quad \text{with} \quad |w'(z)| \geq 1.$$

For this reason K_{out} and K_{out}^* cannot be causally influenced by each other. Neither can we visit the dual creatures in K_{out}^* nor can we send radio signals to them.

During the last years, singularities of solutions of Einstein's equations have been intensely investigated. Roughly, the result was that singularities are not the exception but are instead quite common. This might lead to some surprises in astrophysics. In this direction we recommend Hawking and Ellis (1973, M), Tipler, Clarke and Ellis (1980, S), and Seifert (1983, S).

76.16. Death of a Star

> Thinking creatures on a distant star, who are able to identify parts of our television program, will be surprised that the occupants of planet Earth are mainly interested in the quality of detergents and similarly important things, or have been. Even after millions of years we may create a ridiculous impression.
> Harald Fritzsch (1983)

76.16. Death of a Star

Among the most interesting physical problems we encounter the study of star models. The stimulus of this problem is that a comprehensive knowledge of many physical disciplines is needed: Thermodynamics and statistical physics, the general theory of relativity, elementary particle physics, plasma physics, etc. One of the difficulties is that in part, matter is exposed to extreme conditions. In Section 68.7 the classical basic equation for star models has been motivated and used to calculate the critical Chandrasekhar mass for white dwarf stars. In Problem 76.13 we give the relativistic basic equation for star models, which contains the classical equation as a limiting case. In Problem 76.14 we consider the gravitational collapse. In the context of star models we recommend the monographs of Chandrasekhar (1939), Zeldovič and Novikov (1971), and Weinberg (1972).

A very good survey about the developments of stars is gained from complicated computer simulations. We only make some general remarks.

(i) *Hertzsprung–Russel diagram.* For astronomers, the two most important quantities for a star are its absolute temperature T and its luminosity L. The luminosity is the energy which the star emits per second. For the sun one sets $L = 1$. This corresponds to $4 \cdot 10^{26}$ W per second. In Figure 76.15 the so-called Hertzsprung–Russel diagram for L and T is depicted in a very schematic way. Most stars belong to the so-called main sequence.

(ii) *History of the sun.* This history is shown in Figure 76.15. The sun originated about five milliard years ago. At this time our region of the universe was dark and bitter cold. There was only a huge cloud of interstellar dust with as many atoms in 2,500 km^3 as are today in 1 cm^3 of air. One day this cloud exploded. There are two hypotheses for this. The reason may have been that a spiral arm of our Milky Way had been moving through this cloud or it may have been the shock wave of a supernova. Together with the sun the planets

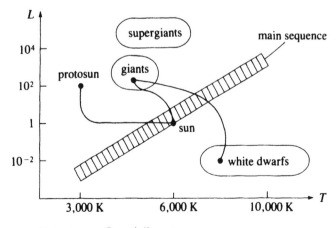

Hertzsprung–Russel diagram

Figure 76.15

appeared. The comets are probably the remains of the material which has been left over during the creation of the planets.

The sun produces its energy through nuclear fusion, i.e., by the burning of hydrogen into helium. During the next five milliard years the sun will not change very much. At the end of this epoch its luminosity L and its diameter d will have doubled, and already this leads to serious climatical difficulties on the earth. After eight milliard years L (and d) will have increased by a factor 2,000 (and 100). The sun becomes a red giant. At this time no life on this earth is possible any longer. The reason for the increase of the sun is the following. If large parts of hydrogen are burned, the central region becomes unstable. The equilibrium between radiation and gravitation has been disturbed. The boundary of the sun expands. The central region contracts and increases its temperature from $15 \cdot 10^6$ K to 10^8 K. Thereby the Salpeter process can take place where, among other things, helium is burned into beryllium under the production of energy. After the nuclear fusion has come to an end, the sun contracts, due to a gravitational collapse. This way a white dwarf occurs with a very high density of approximately 1,000 kg/cm^3. After cooling off the white dwarf slowly becomes a black dwarf. This process is finished after approximately 10^5 milliard years. One might question, however, if the universe will become that old.

(iii) *The critical Chandrasekhar mass* $M_{\text{crit}} = 1.2\, M_{\text{sun}}$. All stars with a mass $M < M_{\text{crit}}$ take a similar development as the sun.

If $M > M_{\text{crit}}$, then a huge supernova explosion may occur, whereby great parts of the masses of the star are thrown into the universe. If the remaining mass is smaller than twice the mass of the sun, then a neutron star occurs which consists of very densely packed neutrons. On an average such a neutron star has a diameter of only 30 km; its density, however, is 10^9 times larger than that of a white dwarf.

If, after a supernova explosion, the mass is bigger than twice the mass of the sun, then a black hole occurs due to the gravitational collapse. One estimates that 10^{11} suns and 10^6 black holes exist in our Milky Way. Furthermore, one expects that there exist at least 10^{11} galaxies.

(iv) *Pulsars*. These were discovered in 1967 in Cambridge (England) by the graduate student Jocelyn Bell. They are very fast oscillating radio sources with a period between 0.03 s and 4 s. One expects that they are very fast rotating neutron stars which have a strong magnetic field, not parallel to the rotational axis which therefore rotates as well. The best-known pulsar is located in the Crab Nebula. There, a supernova explosion occurred A.D. 1054 which had been observed by Chinese astronomers. The X-ray emission of Cyg X-1 cannot be explained by a pulsar, and seems to come from a very small source region. One expects that Cyg X-1 has a black hole as a binary star companion, analogously to Sirenius, which has a white dwarf as a companion.

(v) *Quasars and the red limit of the cosmos*. One of the most interesting astronomical objects are the quasars. The first quasar was detected in 1960 by Allan Sandage at Caltech (Mount Wilson Observatories), using the modern methods of radio astronomy. These quasars have a very strange spectrum, which at the beginning could not be explained at all. At first one thought that quasars were related to stars, and therefore called them quasi-stellar objects (quasars). Today we know that this was in error. Only in 1963 was it discovered at Caltech that the spectrum of quasars could be explained by enormous red shifts. Today approximately 2,000 quasars are known. Some of them are 10^3 times brighter than our galaxy, and the radiation comes from a region 10^5 times smaller than the diameter of our galaxy. Enormous energies are emitted. The nucleus of NGC 4151, for example, is not bigger than the interior of our solar system. However, it emits X-ray radiation in the order of magnitude of one milliard suns in the entire electromagnetic spectrum.

Today, one expects that quasars are young galaxies from the early history of our cosmos where huge explosions take place. It is possible that the centers of quasars contain black holes which absorb huge amounts of matter. But the physical mechanism of quasar explosions is not yet well understood. Also remarkable are the enormous red shifts. The light of the most distant quasars therefore takes approximately 15 milliard years to reach us. This is about the age of the universe. If, according to the Doppler effect of Problem 58.6, one computes the classical escape velocities from the red shift, one finds that the escape velocities for quasars are close to the velocity of light. If the age of the universe is t years, then we can observe only objects from which light has traveled for at most t years. It seems that for quasars this limit has almost been reached. Because of the red shift involved, this limit is also called "red limit." Timothy Ferris has written a very interesting book about this. In connection with the exciting history of modern astronomy and cosmology, we recommend this book as well as the book of Sullivan (1979).

All numerical data about our cosmos should be treated with some caution, because for very distant objects, astronomers can only measure the red shift in the spectrum. All statements about the corresponding distances in light years have a great error margin; this is because the Hubble constant is only approximately known and also, we derived Hubble's law (40) only under the assumption that the travel time of light, i.e., the distance was sufficiently small.

(vi) *Black holes*. In the context of the general theory of relativity, black holes are generally described by the Kerr–Newman solution:

$$ds^2 = [1 - B^{-1}(r_s r - Q^2)]c^2 dt^2$$
$$- [r^2 + a^2 + B^{-1}(r_s r - Q^2)a^2 \sin^2 \vartheta] \sin^2 \vartheta \, d\varphi^2$$
$$- B \, d\vartheta^2 - BD^{-1} dr^2 + B^{-1}[(r_s r - Q^2) 2a \sin^2 \vartheta] c \, dt \, d\varphi,$$

where $a^2 + Q^2 \leq r_s^2/4$ and

$$D = r^2 - r_s r + a^2 + Q^2,$$
$$B = r^2 + a^2 \cos^2 \vartheta.$$

This is a solution of Einstein's equation for the vacuum, which becomes the

Schwarzschild solution for $a = Q = 0$. This solution can be interpreted as a charged, rotating black hole with mass $M = r_s c^2/2G$, charge Q, and angular momentum $J = Ma$. The black hole rotates around the axis $\vartheta = 0$.

One expects that black holes, which represent the final state of a star, are of precisely this form, i.e., from the many parameters which describe a star, only the three parameters mass M, charge Q, and angular momentum J remain observable for the outer world. It is also expected that black holes are very stable objects.

As literature on this object we recommend the fundamental papers of Hawking (1972), (1973), (1975) and the standard work Chandrasekhar (1983).

76.17. Vaporization of Black Holes

We want to show that a black hole at rest cannot exist forever, but instead has a lifetime of

$$t = 10^{10} (M_0/M_{\text{crit}})^3 \text{ years,} \qquad (67)$$

where M_0 is the mass of the hole at the time of its creation measured in kilograms and $M_{\text{crit}} = 10^{10}$ kg. This critical mass corresponds to the Schwarzschild radius of a hole which is $r_s = 1.5 \cdot 10^{-7}$ cm. Today only such miniholes may vaporize, when they appeared 10^{10} years ago, i.e., the time of the Big Bang. Furthermore, a black hole of mass M has the temperature

$$T = 2 \cdot 10^{14}(M_{\text{crit}}/M) \text{ K.}$$

In Example 76.9 we already saw that photons may not escape from a black hole. Now we want to look at the very interesting quantum effect which causes the "boundary" of the black hole to radiate into the universe whereby the hole vaporizes, i.e., the hole decreases its mass and increases its temperature (Hawking effect).

The vacuum of quantum field theory, i.e., the ground state of quantum fields is nonempty. It contains virtual pairs of particles which can normally not be observed directly. Instead one observes indirect effects, e.g., on the spectrum of the hydrogen atom (Lamb shift). Through the process of fluctuations a virtual pair of photons $\{P_+, P_-\}$ may appear in the real world. The energy is given by $E_\pm = \pm h\nu$, where ν is the frequency. For states with finite lifetime we have the energy–time uncertainty relation

$$\Delta E \, \Delta t \sim h/2\pi.$$

Hence, the pair of photons has the time $\Delta t \sim 1/2\pi\nu$ to move apart, before annihilation occurs. If, however, $\{P_+, P_-\}$ appears on the boundary of a black hole, then according to Example 76.9, the photon P_- with negative energy may vanish into the black hole, whereas P_+ is radiated into the universe.

Since P_- cannot leave the black hole, the annihilation between P_- and P_+ is impossible. Therefore, the photon P_+ remains in our real world, whereas

P_- remains in the black hole. In fact, with P_-, the black hole gains the negative energy $-E$ of P_-, i.e., the black hole looses the energy E, and hence it looses the mass $m = E/c^2$, according to Einstein's fundamental energy–mass relation. Note that the total energy of this process is equal to zero, because the sum of the energies of the two photons is equal to zero. Thus, conservation of energy is not violated.

We now assume the radiation is black, i.e., it satisfies Planck's radiation law. According to the Stefan–Boltzmann law of Section 68.5, the energy emitted from the hole is equal to

$$E_A = \sigma T^4 A \cdot t \tag{68}$$

with $\sigma = 2\pi^5 k^4 / 15 c^2 h^3$. As a simple model computation, we consider uncharged, black holes which are at rest. The surface is then equal to $A = 4\pi r_s^2$ and the mass of the hole is given by

$$M = c^2 r_s / 2G.$$

We measure M in kilograms. In order to assign a temperature T to the black hole we note that, according to Wien's displacement law (68.37),

$$\lambda_{\max} T = hc/5k,$$

where $\lambda_{\max}$ is the wave length for which the maximum energy is emitted. We now assume that $\lambda_{\max}$ has the same order of magnitude as the radius of the black hole r_s. Therefore, we define $T = hc/5kr_s$, hence

$$T = hc^3 / 10 kGM.$$

This means that $T \sim (10^{24}/M)$ K if the mass is measured in kilograms. From (68) and $A = 4\pi r_s^2$ we obtain

$$E_A = (hc^6 / 5M^2 G^2) t.$$

Hence, approximately $(10^{38}/M^2)$ watt are emitted per second. Letting

$$E_A = Mc^2,$$

we find as time during which the entire mass of the hole is emitted:

$$t = 5M^3 G^2 / hc^4.$$

This yields assertion (67).

PROBLEMS

76.1. *Characterization of the Poincaré group.* Let $Q(u) = c^2 t^2 - \xi^2 - \eta^2 - \zeta^2$ with $u = (\xi, \eta, \zeta, ct)$.

Prove: A linear transformation $u' = \Lambda u + a$ is precisely a Poincaré transformation if Q remains invariant, i.e.,

$$Q(u_1 - u_2) = Q(u_1' - u_2') \qquad \text{for all} \quad u_1, u_2 \in \mathbb{R}^4. \tag{69}$$

Hint: Elementary matrix computations. See Neumark (1964, M), §7.

Many presentations of the theory of relativity begin with the invariance assumption (69). It is often said that (69) follows immediately from the fact that the velocity of light is constant, i.e., from (C) of Section 75.2. Actually, only

$$Q(u_1 - u_2) = 0 \iff Q(u'_1 - u'_2) = 0$$

follows immediately from (C). A complete derivation of (69) from (C) which, however, is quite long, may be found in Fock (1960, M), §8.

76.2.* *Characterization of inertial systems.* Let $u' = D(u)$ be a C^∞-diffeomorphism from $\mathbb{R}^4$ onto $\mathbb{R}^4$ with

$$c^2 dt^2 - d\xi^2 - d\eta^2 - d\zeta^2 = c^2 dt'^2 - d\xi'^2 - d\eta'^2 - d\zeta'^2. \qquad (70)$$

Prove that D is affine, i.e., $Du = \Lambda u + a$. From Problem 76.1, D is then a Poincaré transformation.

By relation (70) is meant to insert $u' = D(u)$ and consider the system of partial differential equations on $\mathbb{R}^4$, which is obtained by comparison of the coefficients of the differentials.

Hint: See Fock (1960, M), §8.

In order to understand this result we consider the metric $ds^2 = g_{ij} du^i du^j$ on M_4 of Section 75.8. Let C be an admissible chart in M_4 with $\mathbb{R}^4$ as chart space. We call C constant if and only if g_{ij} on C has the constant form (75.23). From above follows: Except for space and time reflections, precisely the inertial charts $\mathbb{R}^4(\Lambda, a)$ in M_4 are constant. If, therefore one considers g_{ij} in an $\mathbb{R}^4$-chart, then, except for uninteresting space and time reflections, one can determine whether or not one is in an inertial chart, i.e., physically speaking, in an inertial system.

76.3. *Conservation laws and differential forms.*

76.3a. *Proof of Proposition 75.19.*

Solution: Already in Section 75.12 we showed that $Q(t)$ is a conserved quantity. We now show that Q is a scalar on M_4, i.e., Q remains invariant under proper Lorentz transformations (change to other inertial systems).

For simplicity, we consider the special Lorentz transformation

$$t' = \alpha(t - V\xi/c^2), \qquad \xi' = \alpha(\xi - Vt)$$

with $\alpha = \sqrt{1 - V^2/c^2}$. From this follows the general case, since every proper Lorentz transformation can be obtained from such a special transformation and rotations. The key for this is the differential form on M_4:

$$\omega = \frac{1}{3!} T^s E_{sabc} du^a \wedge du^b \wedge du^c.$$

The pseudotensor E_{sabc} has been introduced in Example 74.27. For arbitrary coordinate transformations, ω behaves like a pseudoscalar. Since a chart change on M_4, by definition, uses a proper Lorentz transformation which has a positive Jacobian, ω behaves like a scalar under chart changes, i.e., ω is actually a differential form on M_4.

It follows that

$$cQ(\gamma) = \int_{\mathbb{R}^3} T^4(y, c\gamma) dy = \int_{t=\gamma} \omega.$$

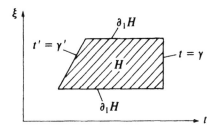

Figure 76.16

In an inertial chart it follows from $D_i T^i = 0$ that $d\omega = 0$. The theorem of Stokes of Section 74.24 implies

$$\int_{\partial H} \omega = \int_H d\omega = 0. \tag{71}$$

We assume that the region H is bounded by the two hyperplanes $t = \gamma$ and $t' = \gamma'$ as well as by other parts of the boundary $\partial_1 H$ as shown in Figure 76.16. Since T^i has a compact spacelike support, we can choose γ and γ' in such a way that $T^i = 0$ on $\partial_1 H$. Because of the coherent orientation of ∂H and H it follows from (71) that

$$\int_{t=\gamma} \omega = \int_{t'=\gamma'} \omega = 0,$$

and thus $Q'(\gamma') = Q(\gamma)$. Since Q and Q' are conserved quantities, it follows that $Q' = Q$, i.e., Q is a scalar.

Under space reflections (or time reflections), T^4 remains unchanged as the fourth component of a simple contravariant tensor (or changes its sign). It follows then immediately from the integral representation of $Q(t)$ above that the charge Q remains unchanged under space reflections, while it changes its sign under time reflections.

Passing from particles to antiparticles means a change in charge. This is why physicists say that antiparticles are nothing other than particles with time reverse. This can be formulated precisely in the context of quantum field theory.

76.3b. *Proof of Proposition* 75.20.
Solution: We now choose

$$\omega^i = \frac{1}{3!} T^{is} E_{sabc} \, du^a \wedge du^b \wedge du^c.$$

This implies

$$cp^i(\gamma) = \int_{\mathbb{R}^3} T^{i4}(y, c\gamma) \, dy = \int_{t=\gamma} \omega^i.$$

We have to show that the conserved properties p^i are transformed like simple contravariant tensors under proper Lorentz transformations. We use the same arguments as in Problem 76.3a and note that ω^i is transformed like a simple contravariant tensor under proper Lorentz transformations.

76.3c. *Proof of Corollary 75.21.*
Solution: We now choose

$$\omega^{ij} = \frac{1}{3!} S^{ijs} E_{sabc} du^a \wedge du^b \wedge du^c.$$

76.4. *Locally geodesic coordinates.* Let $x \in E_4$ be a fixed point. Prove that in every neighborhood of x one can introduce new coordinates such that at x

$$g_{ij} = \varepsilon_i \delta_{ij}, \tag{72}$$

$$\Gamma^k_{ij} = 0, \quad D_k g_{ij} = 0 \tag{73}$$

holds with $\varepsilon_4 = 1$ and $\varepsilon_i = -1$ for $i \neq 4$.
Solution: Let $\Gamma_{i,jk} = g_{is}\Gamma^s_{jk}$. Formula (7) implies

$$D_k g_{ij} = \Gamma_{i,jk} + \Gamma_{j,ik} \tag{74}$$

As in Section 74.20 it follows that $g_{i'j'} = \varepsilon_i \delta_{i'j'}$ at the point x after a linear coordinate transformation. For simplicity in notation let $u^{i'}(x) = 0$. We choose new coordinates

$$u^k = u^{k'} + \tfrac{1}{2}\Gamma^{k'}_{i'j'}(x) u^{i'} u^{j'}. \tag{75}$$

Transformation rule (74.24) implies that $\Gamma^k_{ij} = 0$. This immediately yields (72) and (73). Note that $\partial u^k(x)/\partial u^{k'} = \delta^k_{k'}$.

76.5. *Tensor identities.* Prove the following identities by direct computation, which can greatly be simplified by using locally geodesic coordinates. For mnemotechnical reasons we use commas:

$$R_{ij,km} = -R_{ji,km} = -R_{ij,mk}, \tag{76}$$

$$R_{ij,km} = R_{km,ij}, \tag{77}$$

$$R_{i,jkm} + R_{i,kmj} + R_{i,mjk} = 0, \tag{78}$$

$$\nabla_s R_{ij,km} + \nabla_i R_{js,km} + \nabla_j R_{si,km} = 0, \tag{79}$$

$$\nabla_i (R^{ij} - \tfrac{1}{2} g^{ij} R) = 0. \tag{80}$$

For an n-dimensional Riemannian manifold it follows that R_{ijkm} has $a(n) = n^2(n^2 - 1)/12$ independent components. For example, $a(2) = 1$, $a(3) = 6$, and $a(4) = 20$. Formula (79) is known as the Bianchi identity.

76.6. *Divergence.* Prove

$$\nabla_i t^i = \gamma^{-1} D_i(\gamma t^i) \tag{81}$$

on E_4 with $\gamma = \sqrt{-g}$.
Solution: This follows from

$$\nabla_i t^i = D_i t^i + \Gamma^i_{si} t^s = D_i t^i + \tfrac{1}{2}(g^{ij} D_s g_{ij}) t^s$$

and the differentiation rule for determinants $D_s g = g g^{ij} D_s g_{ij}$.

76.7. *The variational principle of Hilbert (1915).* As in Section 76.2, we let

$$J = \int_H R\gamma \, du$$

Problems

with $\gamma = \sqrt{-g}$. Prove

$$\delta J = \int_H (\tfrac{1}{2} g^{km} R - R^{km}) \gamma \delta g_{km} \, du, \qquad (82)$$

where

$$\delta g_{ij} = 0 \quad \text{and} \quad D_k(\delta g_{ij}) = 0 \quad \text{on } \partial H.$$

As usual, we have

$$\delta J = \varphi'(0)$$

with $\varphi(\varepsilon) = J(g_{ij} + \varepsilon h_{ij})$ and $\delta g_{ij} \stackrel{\text{def}}{=} h_{ij}$. Analogously, δR_{ij}, $\delta \Gamma^k_{ij}$, etc. are defined.

Solution: We have

$$\delta(R\gamma) \equiv \delta(g^{ij} R_{ij} \gamma) = g^{ij} \gamma \delta R_{ij} + R_{ij} \delta(g^{ij} \gamma).$$

(I) It is important that δR_{ij} does not contribute to δJ. To see this, let

$$w^s \stackrel{\text{def}}{=} g^{ij} \delta \Gamma^s_{ij} - g^{sj} \delta \Gamma^i_{ij}.$$

From the definition of R_{ij} in (9) it follows that

$$g^{ij} \delta R_{ij} = \gamma^{-1} D_s(\gamma w^s). \qquad (83)$$

Initially, this formula only holds for a locally geodesic coordinate system, because there we have $D_s \gamma = 0$ (Problem 76.4). The right-hand side of (83), however, is a tensor, and thus (83) holds for every coordinate system. In order to recognize the tensor property, we observe that $\delta \Gamma^k_{ij}$ is a tensor, since the second-order derivatives cancel out in the transformation rule (74.24) when passing from Γ^k_{ij} to $\delta \Gamma^k_{ij}$. Hence w^s is a tensor, and from (81) it follows that $g^{ij} \delta R_{ij} = \nabla_s w^s$.

Integration by parts yields

$$\int_H \gamma g^{ij} \delta R_{ij} \, du = \int_H D_s w^s \, du = 0.$$

Note that $\delta \Gamma^k_{ij} = 0$ on ∂H, and hence $w^s = 0$ on ∂H.

(II) Differentiation of $g = \det(g_{ij})$ gives

$$\delta \gamma = -\tfrac{1}{2} \gamma^{-1} \delta g = \tfrac{1}{2} \gamma g^{ij} \delta g_{ij}.$$

From $g^{ij} g_{jk} = \delta^i_k$ follows that

$$\delta g^{ij} g_{jk} + g^{ij} \delta g_{jk} = 0.$$

This implies

$$\delta(g^{ij} \gamma) = g^{ij} \delta \gamma + \gamma \delta g^{ij} = \gamma(\tfrac{1}{2} g^{ij} g^{km} - g^{ik} g^{jm}) \delta g_{km}.$$

The desired formula (82) then follows immediately from

$$\delta J = \int_H R_{ij} \delta(g^{ij} \gamma) \, du.$$

76.8. *Connection with classical Newtonian mechanics.* Prove that the Poisson equation (19) follows from metric (18) and Einstein's equation $R^{ij} - \frac{1}{2}g^{ij}R = \kappa T^{ij}$ with $T^{44} = \rho c^2$ if higher-order terms in $1/c$ are neglected.

Hint: Longwinding elementary computation. See Fock (1960, M), §55.

76.9. *The theorem of Birkhoff* (1923). We want to show that a radially symmetric gravitational field, which satisfies Einstein's equations $R_{ij} = 0$ for the vacuum, coincides locally with the Schwarzschild solution if some natural regularity conditions are satisfied.

76.9a. *Special case.* Consider the metric g_{ij} for $\eta = \pm 1$, i.e.,

$$ds^2 = \eta e^{a(r,t)}c^2\,dt^2 - \eta e^{b(r,t)}\,dr^2 - r^2(d\vartheta^2 + \sin^2\vartheta\,d\varphi^2), \qquad (84)$$

where a and b are C^2-functions.

Prove: If $R_{ij} = 0$ holds for a region G with $r > 0$ for all points in G, then (84) coincides with the Schwarzschild solution in G.

Solution: Let $a' = a_r$ and $\dot{a} = a_t$. The nonvanishing components of R_{ij} are

$$R_{14} = -\dot{b}/r,$$

$$4R_{11} = 2a'' + a'^2 - a'b' - (4b'/r) - e^{b-a}(2\ddot{b} + \dot{b}^2 - \dot{a}\dot{b}),$$

$$4R_{44} = -e^{a-b}(2a'' + a'^2 - a'b' + 4a'/r) + 2\ddot{b} + \dot{b}^2 - \dot{a}\dot{b},$$

$$-R_{22} = 1 - \eta e^{-b}(1 + r(a' - b')/2), \qquad R_{33} = (\sin^2\vartheta)R_{22}.$$

From $R_{14} = 0$ it follows that b does not depend on t. From $R_{22} = 0$ it follows that a' does not depend on t, i.e.,

$$a = \alpha(r) + \beta(t).$$

After the transformation $e^{\beta/2}\,dt = dt'$ we can assume in (84) that a is independent of t and $a(r_0) = b(r_0)$ for a fixed r_0. From $R_{11} = 0$ and $R_{44} = 0$ it follows that $(a' + b')/r = 0$, and hence $a = -b$. The substitution $f = \eta e^{-b}$ and $R_{22} = 0$ with $a' = -b'$ yields

$$rf' + f = 1$$

with the general solution

$$f = \eta e^{-b} = \eta e^a = 1 - r_s/r.$$

Here $r_s > 0$ is the constant of integration. For $\eta = -1$ (or $= 1$) we have $0 < r < r_s$ (or $r > r_s$). For these a and b we have in fact that $R_{11} = R_{44} = 0$.

76.9b. *General case.* Prove that

$$ds^2 = A\,dT^2 - B^2(d\vartheta^2 + \sin^2\vartheta\,d\varphi^2) \\ - C\,dR^2 + 2D\,dR\,dT \qquad (85)$$

can be locally transformed into (84). Thereby we assume that locally A, B, C, D are C^2-functions in T and R with

$$B > 0, \qquad B_R \neq 0, \qquad AC - D^2 < 0, \qquad A \neq 0. \qquad (86)$$

Solution: Let $r = B(T, R)$ and solve locally with respect to R. After change in notation, we can assume that (85) and (86) hold with $R = r$ and $B^2 = r^2$.

We let $T = \varphi(t, r)$ with

$$\varphi \stackrel{\text{def}}{=} -\int A^{-1}D\,dr + \psi(t),$$

where we choose ψ in such a way that locally $\varphi_t \neq 0$ is satisfied. Then $T = \varphi(t, r)$ can be solved locally with respect to t. Because of

$$A\varphi_r + D = 0 \quad \text{and} \quad dT = \varphi_t\,dt + \varphi_r\,dr,$$

we obtain

$$ds^2 = \alpha\,dt^2 - r^2(d\vartheta^2 + \sin^2\vartheta\,d\varphi^2) - \gamma\,dr^2$$

with $\alpha = A\varphi_t^2$ and $\gamma = A\varphi_r^2 - C$. Therefore, we have locally that $\alpha \neq 0$ and $\gamma = (D^2 - AC)/A$, and hence $\alpha\gamma < 0$.

76.10.** *The initial-value problem for Einstein's equations and causality.*

76.10a. *Causal structure.* Let E_4 be a four-dimensional space–time manifold of Section 76.1. A curve $x = x(\sigma)$ in E_4 is called timelike (or spacelike, lightlike) at a point $x(\sigma_0)$ if and only if the following holds in local coordinates:

$$g_{ij}(x(\sigma_0))\dot{u}^i(\sigma_0)\dot{u}^j(\sigma_0) > 0 \quad (\text{or } < 0, = 0).$$

Timelike curves are oriented in the direction of increasing proper time τ, i.e., increasing arclength s (future orientation).

We say the point x_1 causally effects x_2 if and only if x_1 and x_2 lie on a timelike curve where x_2 follows x_1.

By a spacelike 3-surface we mean a three-dimensional submanifold S of E_4 which is locally given by an equation

$$\varphi(u) = 0$$

with a C^∞-function φ and $g^{ij}D_i\varphi D_j\varphi < 0$ on S. If we choose M_4 with

$$ds^2 = c^2\,dt^2 - d\xi^2 - d\eta^2 - d\zeta^2,$$

then equation $t = 0$ describes a spacelike 3-surface S, where S is the set of all events which take place at time $t = 0$. The surface S can then be identified with $\mathbb{R}^3$.

76.10b. *Equivalent metrics.* Two metrics, i.e., two tensor fields g_{ij} and $g_{i'j'}$ on a C^∞-manifold M are called equivalent if and only if there exists a C^∞-diffeomorphism $f: M \to M$ which takes g_{ij} into $g_{i'j'}$. That is, by setting locally $u' = f(u)$, we obtain

$$g_{i'j'}(u') = A_{i'}^i A_{j'}^j g_{ij}(u) \quad \text{with} \quad A_{i'}^i = \partial u^i(u')/\partial u^{i'}.$$

76.10c. *Initial-value problem and Banach's fixed-point theorem.* Intuitively, one means by an initial-value problem that the gravitational field, i.e., g_{ij} is uniquely determined for all future times if it is known at an initial time $t = 0$. Moreover, because of the principle of causality one expects that $g_{ij}(x_2)$ only depends on all events x_1 with $t = 0$, which causally effect x_2 in the sense of Problem 76.10a.

Under suitable assumptions one can prove such existence and uniqueness theorems for Einstein's field equations. In place of the 3-surface $t = 0$ one then has a more general spacelike 3-surface. The main difficulty is that besides g_{ij} every other equivalent metric is also a solution of Einstein's equations. In order to avoid this nonuniqueness one assigns four side conditions for the components of the metric tensor g_{ij} (harmonic coordinates). These additional differential equations are called gauge conditions. The diffeomorphisms f of Problem 76.10b correspond to the so-called gauge transformations.

Study Hawking and Ellis (1973, M), Chapter 7 (hyperbolic second-order differential equations) and Fischer and Marsden (1972) (quasi-linear symmetric hyperbolic first-order system). In this last paper, the system of differential equations corresponds to an operator equation of the form

$$F(u)u = h. \tag{87}$$

Such equations are solved by studying the solutions $v = Tu$ for the linear equation

$$F(u)v = h \tag{88}$$

for some fixed u. Then (87) corresponds to the fixed-point equation

$$u = Tu,$$

which in the present case can be solved by Banach's fixed-point theorem. The main difficulty is the solution of (88). Under the assumption that u satisfies certain smoothness conditions one has to show that the solution v is sufficiently smooth. Otherwise, one cannot find a set in a Sobolev space which under T is mapped into itself. In fact, system (88) is a symmetric, linear, hyperbolic system. Since u appears in the coefficients one can obtain the desired strong regularity results for v by using the theory of semigroups.

A simpler approach to quasi-linear symmetric hyperbolic systems will be considered in Chapter 83 of Part V.

Furthermore, for the problems discussed here, we recommend the survey article of Choquet-Bruhat and Yorke (1980, S) and Marsden (1983). Existence proofs for global solutions of the initial-value problem, which correspond to small initial data, can be found in Christodoulou and Klainerman (1988, M).

76.11.** *Perturbation series and bifurcation of solutions of Einstein's equations for the vacuum.* An important problem is the following: Under which conditions is it possible to obtain from a known gravitational field g_{ij}^0 a new gravitational field by using the series expansion

$$g_{ij} = g_{ij}^0 + \varepsilon h_{ij} + O(\varepsilon^2), \qquad \varepsilon \to 0. \tag{89}$$

If we write Einstein's equations for the vacuum in the form

$$E(g) = 0, \tag{90}$$

then we have $E(g^0) = 0$ and for h_{ij} the linearized equation

$$E'(g^0)h = 0 \tag{91}$$

must be satisfied. Let M denote the set of all solutions of (90). The answer to our question, which was found only recently, is roughly like this.

(i) If g^0 has no symmetries, then M is a manifold in a neighborhood of g^0, i.e., for each h with (91) there exists an expansion (89), for which h is a solution of (90) for small $|\varepsilon|$. This means that precisely all h with (91) are tangent vectors to M at the point g^0.

(ii) If g^0 has symmetries, then g^0 is not a regular point of M. The series (89) are then solutions of (90) if and only if h is a solution of (91) and certain relations, which only depend on $E''(g^0)h^2$ and the symmetries, are satisfied. (The so-called Taub conserved quantities must vanish.)

Symmetries are described by the Lie group of the isometries of g^0. Study Marsden (1980, S) and Arms, Marsden, and Moncrief (1982). All explicitly known solutions of (90) have symmetries.

76.12.** *The twistor program of Penrose (1977).* In the following, a plane or straight line is always understood as a two-dimensional or one-dimensional linear subspace.

76.12a. *Twistors.* By a twistor we mean a point $T = (T_0, T_1, T_2, T_3)$ of $\mathbb{C}^4$. We set

$$\langle T, S \rangle = T_0 \bar{S}_0 + T_1 \bar{S}_1 - T_2 \bar{S}_2 - T_3 \bar{S}_3.$$

The twistor space $\mathbb{T}$ is the space $(\mathbb{C}^4, \langle \cdot, \cdot \rangle)$, i.e., $\mathbb{C}^4$ equipped with $\langle \cdot, \cdot \rangle$. A twistor T is called positive (or negative, isotropic) if and only if $\langle T, T \rangle > 0$ (or < 0, $= 0$). The geometry of these twistors is called twistor geometry.

Let $T \neq 0$. The set $T_p \stackrel{\text{def}}{=} \{\rho T : \rho \in \mathbb{C}, \rho \neq 0\}$ is called the projective twistor of T. The space of all projective twistors is the three-dimensional, projective, complex space $\mathbb{CP}^3$ and is denoted by $\mathbb{PT}$.

76.12b. *Penrose transformation.* Consider the space $\mathbb{R}^4$ with points (ξ, η, ζ, ct), where ξ, η, ζ are Cartesian space coordinates and t is the time coordinate for an inertial system. The corresponding set of the complex space–time points $p = (\xi, \eta, \zeta, ct)$ are denoted by $\mathbb{C}^4_{\text{space-time}}$. Our goal is to realize all p as planes in $\mathbb{T}$. This will be done in two steps.

(i) Spinor realization. We assign to each p a complex (2×2)-matrix X which is defined through

$$X = \begin{pmatrix} ct + \xi & \eta + i\zeta \\ \eta - i\zeta & ct - \xi \end{pmatrix}.$$

Conversely, to every X there corresponds a unique p.

(ii) To each X we assign the plane

$$\begin{pmatrix} T_0 \\ T_1 \end{pmatrix} = iX \begin{pmatrix} T_2 \\ T_3 \end{pmatrix} \tag{92}$$

in the twistor space $\mathbb{T}$. To different X correspond different planes. The planes in $\mathbb{T}$, which cannot be expressed in the form (92), are called ideal space–time points.

Since the set of all planes in $\mathbb{T}$ forms a compact manifold with respect to Plücker's coordinates (Grassmannian manifold), we obtain a compactification of $\mathbb{C}^4_{\text{space-time}}$ from this transformation. One may also think of (92) as a straight line in the projective twistor space $\mathbb{PT}$. This interpretation is often preferred, because in algebraic geometry one usually works with homogeneous coordinates, i.e., with projective structures.

The basic idea of the twistor program is to transform relativistic fields and their differential equations into objects of algebraic geometry by using the Penrose transformation (vector bundles, etc.). Thereby the differential equations vanish. One may think of this as a further development of the Fourier transformation. Thereby differential equations are transformed into algebraic equations. The twistor program has been very successful in solving the complicated, nonlinear Yang–Mills equations of gauge field theory. Here a reduction to a study of vector bundles was possible (see Atiyah (1979, L)).

We recommend Wells (1979, S) and Penrose and Rindler (1984, M), as well as the literature listed at the end of this chapter under the headline "twistor program."

76.13. *Basic equation for relativistic star models.* We now generalize the basic equation for the classical star models of Section 68.7. For this we use the metric

$$ds^2 = A(r)c^2\,dt^2 - B(r)\,dr^2 - r^2(\sin^2\vartheta\,d\varphi^2 + d\vartheta^2)$$

in the interior of the radially symmetric star and solve Einstein's equations

$$R^{ij} - \tfrac{1}{2}g^{ij}R = \kappa T^{ij},$$

therein by choosing T^{ij} as the energy–momentum tensor of an ideal fluid with pressure $P = P(r)$, mass density $\rho = \rho(r)$, and energy density $\varepsilon = \rho c^2$. Moreover, let

$$M(r) = \int_0^r 4\pi r^2 \rho(r)\,dr$$

be the mass of the ball of radius r. Prove that Einstein's equation implies the following equation:

$$-r^2 P' = GM\rho\left(1 + \frac{P}{c^2\rho}\right)\left(1 + \frac{4\pi r^2 P}{c^2 M}\right)\left(1 - \frac{2GM}{c^2 r}\right)^{-1}.$$

From this basic equation for star models one obtains the classical basic equation $-r^2 P' = GM\rho$ as $c \to \infty$.

Hint: See Weinberg (1972, M), Chapter 11, §1.

76.14. *Gravitational collapse.* We want to explain the physically interesting fact that a star with no pressure left inside can collapse into a point during a finite amount of time t_{collapse}.

76.14a. *Classical theory.* Derive this fact classically.

Hint: As in Figure 76.17 consider a point P with mass m on the surface of a ball of radius R. Let the mass of this ball be equal to M. During the gravitational collapse the motion of P is given by $r = r(t)$. For $t = 0$ we have

$$\dot{r}(0) = 0, \qquad r(0) = R.$$

The point P is only affected by the gravitational force of the ball, i.e.,

$$m\ddot{r} = -GMm/r^2.$$

Note the important fact that according to Problem 58.5 the ball affects P in

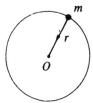

Figure 76.17

the same way as if its entire mass would be located at its center. Multiplication with $\dot{r}$ yields the energy theorem

$$2^{-1}\dot{r}^2 - MG/r = \text{const} = -MG/R.$$

Thus we obtain as collapse time

$$t_{\text{collapse}} = \int dt = \int_0^R dr/\dot{r} = \pi(R^3/8MG)^{1/2}.$$

76.14b.* *Relativistic theory.* Prove that the same formula for t_{collapse} can be obtained if, as in Section 76.3, one chooses the metric of the closed cosmological model for the interior of the star with energy–momentum tensor of an ideal fluid with vanishing pressure.

Hint: See Weinberg (1972, M), Chapter 11, §9.

76.15.** *The famous positive-energy theorem in general relativity.* Roughly speaking, this theorem says the following. For a nontrivial isolated physical system in general relativity, the total energy, including contributions from matter and gravitation, is positive.

The precise result can be found in Schoen and Yau (1979). We also recommend the survey article Choquet-Bruhat (1984).

76.16. *An unsolvable problem.* Compute the Riemannian invariant density

$$K(m) = \sqrt{-g}\,\text{sgn}\begin{pmatrix}\alpha_1\ldots\alpha_{2m}\\ \beta_1\ldots\beta_{2m}\end{pmatrix} R^{\beta_1\beta_2}_{\alpha_1\alpha_2} R^{\beta_3\beta_4}_{\alpha_3\alpha_4}\ldots R^{\beta_{2m-1}\beta_{2m}}_{\alpha_{2m-1}\alpha_{2m}}$$

for $m = 5$, where we sum over two equal indices from 1 to $2m$. This expression is closely related to the theorem of Gauss–Bonnet–Chern in (74.109) for $2m$-dimensional manifolds.

There are 3.6 million terms which would take 5 years to compute by hand. A computer needs 6 hours. Computer methods for numerical and algebraic solutions of Einstein's equations may be found in d'Inverno (1983).

References to the Literature

Classical works: Einstein (1905), (1905a) (special theory of relativity), Minkowski (1909) (geometrical formulation of the special theory of relativity—Minkowski space), Einstein (1915), (1916) (general theory of relativity).
Classical monograph: Einstein (1953).
Einstein's "Weltbild": Einstein (1965, M), Melcher (1979, M, B).

Collected works: Einstein (1960).

Einstein biographies: Frank (1949, M), Infeld (1969, M), Dukas and Hoffmann (1972, M), (1979, M).

Reminiscences of Einstein: Born and Infeld (1967, M).

Modern development of the general theory of relativity: Held (1980, P), Vols. 1, 2 (especially recommended), Carelli (1979, P), Hawking and Israel (1979, P), Schmutzer (1983, P), DeWitt and Stora (1984, P), Ruffini (1987, P), Vols. 1, 2.

Introduction: Landau and Lifšic (1962, M), Vol. 2, Stephani (1977, M), Sexl and Urbantke (1983, M) (especially recommended), Straumann (1984, M).

Standard works about the general theory of relativity and cosmology: Weinberg (1972, M, B), Misner, Thorne, and Wheeler (1973, M), Hawking and Ellis (1973, M).

Monographs: Weyl (1923, M) (classical monograph), Sommerfeld (1954, M), Vol. 3, Fock (1960, M), Schmutzer (1968, M), Sachs and Wu (1977, M), Rindler (1977, M), Dixon (1978, M), Triebel (1981, S), Wald (1984, M).

Lorentzian geometry: Beem and Ehrlich (1981, M), O'Neill (1983, M).

Experimental tests for the theory of relativity: Reasenberg and Shapiro (1983, S).

Computer methods for the general theory of relativity: d'Inverno (1983).

Initial-value problem for Einstein's equations: Choquet-Bruhat and Yorke (1980, S) (recommended as an introduction), Dionne (1962), Fischer and Marsden (1972) (quasi-linear symmetric hyperbolic system), Hawking and Ellis (1973, M), Hughes, Kato, and Marsden (1977), Friedrich (1981), Marsden (1983, S, B).

Global solutions for the initial-value problem: Christodoulou and Klainerman (1988, M).

Regular and singular points in the space of solutions for Einstein's equations and the Einstein–Yang–Mills equations: Marsden (1980, S), Fischer, Marsden, and Moncrief (1980), Arms, Marsden, and Moncrief (1982).

Positive-energy theorems: Schoen and Yau (1979), Choquet-Bruhat (1984, S).

Einstein manifolds: Besse (1987, M).

Group theory and general relativity: Carmeli (1977, M).

Black holes: Sexl (1981, M), Sexl and Urbantke (1983, M) (introduction), Chandrasekhar (1983, M) (standard work), Hawking (1972), (1973), (1975) (basic laws), Hawking and Israel (1979, P), Kaufmann (1979, M) (elementary introduction), Lotze (1980, S), Miller and Sciama (1980) (gravitational collapse), Seifert (1983, S), Novikov and Frolov (1986, M).

History of black holes: Sullivan (1979, M).

Singularities of gravitational fields: Hawking and Ellis (1973, M) (standard work), Penrose (1972, L), (1979, S), Hawking and Israel (1979, P), Tipler, Clarke, and Ellis (1980, S), Seifert (1983, S), O'Neill (1983, M).

Representation of all known solutions of Einstein's field equations: Kramer (1980, M).

Gravitational waves: Weinberg (1972, M) (mathematical theory), Weber (1980, S) (experiments), Ruffini (1987, P).

Quantization of the gravitational theory: DeWitt (1984, L) and Hawking (1984, S) (introduction), Misner, Thorne, and Wheeler (1973, M), Grib (1980, M), Marlow (1980, M), Birell and Davies (1982, M) (introduction), Narlikar and Padmanabhan (1982) (quantum cosmology via Feynman integrals), Novikov and Frolov (1986, M).

Textbooks on quantum field theory and modern particle physics: Lee (1981), Frampton (1987).

Elementary particle physics and cosmology—toward a theory of the universe: Weinberg (1983, P), (1986a, P), Kounas (1984, P), Audouze (1985, M), Haber (1987, P), Hincliffe (1987, P).

Magnetic monopoles: Craigie (1986, L) (standard work), Lee (1981, M), Rajaraman (1982, M).

Supersymmetry, superspace, supergravity, and unified theory of all interactions:

Mohapatra (1986, M) and West (1986, M) (recommended as an introduction), Hawking and Roček (1981, P), Wess and Bagger (1983, L), Gates (1983, L), Nieuwenhuizen (1984, S), Salam and Sezgin (1986, P), Haber (1987, P).

Supersymmetric Yang–Mills equations: Manin (1985).

Superstring theory and unified theory of all interactions: Green, Schwarz, and Witten (1986, M) (standard work), Green and Gross (1986, P), Witten (1986, S) and Green (1987, S) (fundamental survey articles), Haber (1987, P).

Topological tools for superstring theory: Witten (1986), Green, Schwarz, and Witten (1987, M), Segré (1987, S).

Loop groups: Pressley (1986, M).

Super-Lie groups, super-Lie algebras, supermanifolds: Leites (1980, S) (fundamental survey article), Gates (1983, L), Wess and Bagger (1983, L), Regge (1984, S), West (1986, M).

A new supersymmetric proof of the Atiyah–Singer index theorem: Simon (1986, M).

Penrose twistor program: Penrose (1977), Atiyah (1979, L) (applications to gauge field theory), Wells (1979, S), Penrose and Ward (1980, S), Manin (1982) (Yang–Mills–Dirac equations as Cauchy–Riemann equations in twistor space) (1984, M), Penrose and Rindler (1984, M), Vols. 1, 2.

Collection of problems for the general theory of relativity with solutions: Lightman (1975, M).

Elementary star models: Sexl (1981, M).

Relativistic astrophysics, star models and cosmology: Liang and Sachs (1980, S), Sexl and Urbantke (1983, M) (introduction), Weinberg (1972, M) (standard work), Chandrasekhar (1939, M) (classical exposition about star models), Zeldovič and Novikov (1971, M), (1971a) (astrophysics), Peebles (1980, M), Longair (1974, P) (cosmology), Hoyle and Wickramsinghe (1979, M) (life cloud), Ambarzumjan (1980, M) (cosmonogy).

Script for the development of the universe after the Big Bang: Weinberg (1972, M), (1977, M), Singh (1983) (unified theory of elementary particles, magnetic monopoles), Setti (1984, P), Haber (1987, P), Hincliffe (1987, P).

Inflationary universe: Guth (1981), Linde (1984, S), Hawking (1984, S), Börner (1985, S), Abbott and Pi (1985) (collection of important papers).

Historical survey about various cosmological models: Treder (1975, M).

History of modern astronomy and cosmology: Ferris (1977, M), Sullivan (1979, M).

History of astronomy: Moore (1977, M), Herrmann (1978, M).

Search for extraterrestial intelligent creatures: Sagan (1968, M), (1973, M), (1980a), Breuer (1978, M).

Encyclopedia of astronomy and space (1976).

Popular exposition of astronomy and cosmology: Ferris (1977, M), Sullivan (1979, M), Silk (1980, M), Sagan (1980, M), (1980a, M), (1982, M), Harrison (1981, M), Unsöld (1981, M), Fritzsch (1983, M), Trefil (1984, M), Field (1986, M), Kippenhahn (1987, M).

CHAPTER 77

Simplicial Methods, Fixed-Point Theory, and Mathematical Economics

> The highest praise for a mathematician is that his theorem, his proof, his theory is considered beautiful. Every mathematician polishes his proofs until they assume the most elegant form. After its first appearance, an important theorem will be proved in many different ways, and the most elegant proof, which usually is also the most simple one, will then be used in monographs and textbooks.
>
> Krysztof Maurin (1981)

> The purpose of this note is to give a short proof of Brouwer's fixed-point theorem. We first prove a lemma which contains the combinatorical key argument of a new and elegant proof of E. Sperner for the invariance of the dimension number. From this we deduce a combinatorical–topological theorem from which the above-mentioned fixed-point theorem, as well as Sperner's proof, follows.
>
> B. Knaster, C. Kuratowski, and S. Mazurkiewicz (1929)

> It has been typical for the study of nonlinear phenomena in analysis that they must be analyzed with one technique to obtain quantitative information (numerical values of the solution), and by another to obtain qualitative information (existence, uniqueness, multiplicity, stability, bifurcation). In a sense, this paper is devoted to a unified approach, i.e., for a number of topological methods which have been used in the past only to obtain qualitative insight, we will show how to exploit them and at the same time also provide numerical knowledge up to implementable and tested procedures.
>
> Heinz-Otto Peitgen and Michael Prüfer (1979)

> As our brief remarks below, concerning the history of fixed-point theory as it pertains to constructive methods, may indicate, there seems to have been a blockage concerning the development of algorithms for Brouwer's fixed-point theorem (1912). We suspect that we were not alone in our reaction, upon learning of the Scarf (1967) simplicial algorithm or the Kellogg, Li, and Yorke (1976) homotopy algorithm—approximately, "Yes! ... But of course!." The origin of the blockage is perhaps due to an interface between analysis or topology and computing, which will in time vanish, as the example of Brouwer's fixed-point theorem allows us to hope.
>
> Eugene Allgower and Kurt Georg (1980)

77. Simplicial Methods, Fixed-Point Theory, and Mathematical Economics

In the following two chapters, we present two ways of introducing fixed-point theory which have been developed intensely during the last years and provide an effective method for solving nonlinear equations on computers:

(i) Simplicial methods (Chapter 77).
(ii) Homotopy methods (Chapter 78).

In (i) we use triangulations and suitable labelings. The starting point is the lemma of Sperner of Section 77.1. In (ii), curves, i.e., one-dimensional manifolds are being followed. Thereby differential topological methods, especially Sard's theorem are employed. The disadvantage of (i) over (ii), in view of computer usage, is that with an increasing number of variables in the nonlinear equations, the necessary storage place increases very rapidly.

The content of this chapter is pictured schematically in Figure 77.1. In Chapter 2, we began the discussion of fixed-point theory with the negative retract principle. This way we obtained a quick proof of Brouwer's fixed-point theorem. In Chapter 9 we then used Brouwer's fixed-point theorem and

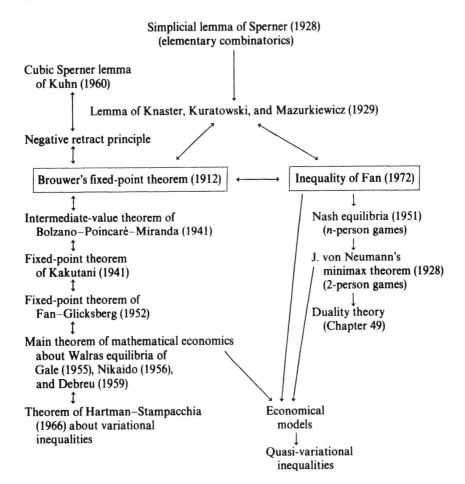

Figure 77.1

the partition of unity to prove existence theorems for variational inequalities. From this we derived fixed-point theorems for multivalued maps and gave applications to the minimax theorem. The negative retract principle is very intuitive, but, in fact, represents a deep topological result. Thus the presentation of Brouwer's fixed-point theorem in Chapter 2 is certainly of great geometrical clarity and simplicity, but actually does not represent an elementary approach. The goal of this chapter is to present an entirely elementary introduction to fixed-point theory, which can also be used as a first lecture about fixed-point theory. If one restricts attention to $\mathbb{R}^n$, then students only need to be familiar with elementary geometrical facts and the concept of continuous functions. We begin with an elementary combinatorial result—the so-called lemma of Sperner. This then implies the central lemma of Knaster–Kuratowski–Mazurkiewicz, which in turn implies Brouwer's fixed-point theorem. Furthermore, a simple generalization of the classical lemma of Knaster–Kuratowski–Mazurkiewicz in Section 77.4 yields the inequality of Fan. As can be seen from Figure 77.1 this inequality is of central importance for fixed-point theory, game theory, and mathematical economics. As several applications we consider:

Existence of Nash equilibrium points (main theorem of n-person games);

Saddle points and minimax theorem (main theorem of 2-person zero sum games);

Theorem of Hartman–Stampacchia about variational inequalities.

In Section 77.8 we derive the fixed-point theorem of Kakutani for multivalued maps on $\mathbb{R}^n$ from Brouwer's fixed-point theorem by using a simple approximation argument. The finite intersection property then immediately yields a generalization to locally convex spaces—the fixed-point theorem of Fan–Glicksberg. Figure 77.2 shows important consequences of this fixed-point theorem, which already have been proved in Section 9.3. Moreover, in Section 77.10 this fixed-point theorem is used to obtain the main theorem of mathematical economics, which can also be viewed as a general existence result for quasi-variational inequalities. Figure 77.3 shows important applications of Brouwer's fixed-point theorem to the theory of monotone operators of Part II.

At the same time, Figure 77.1 shows that very many different results are equivalent to Brouwer's fixed-point theorem. This will be proved in Section

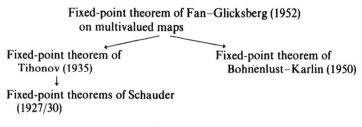

Figure 77.2

77.1. Lemma of Sperner

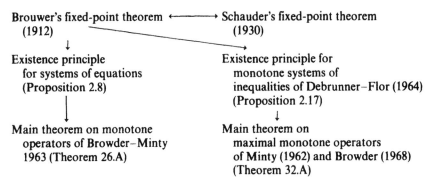

Figure 77.3

77.13. It follows that Brouwer's fixed-point theorem can be viewed as one of the possible formulations of a general mathematical existence principle, which we shall call *Brouwer's principle*. Another such fundamental existence principle is the principle of Hahn–Banach of Chapter 39. The annual numbers shown in Figures 77.1–77.3 illustrate that only during a long process the various applications of Brouwer's principle have been recognized. The many equivalent formulations of Brouwer's principle are the reason for the many different, but actually equivalent, approaches to fixed-point theory, game theory, and mathematical economics.

In Section 77.12, we show that Brouwer's fixed-point theorem is equivalent to an obvious generalization of the classical intermediate-value theorem of Bolzano to $\mathbb{R}^n$ (intermediate-value theorem of Bolzano–Poincaré–Miranda). This shows that Brouwer's principle is nothing other than a generalization of the intermediate-value theorem of Bolzano (1817).

The connection between simplicial methods and the mapping degree, as well as numerical methods, will be discussed in the problem section.

Recall that the notion of topological vector spaces and locally convex spaces has been introduced in $A_1(22)$ and $A_1(40)$, respectively. For example, the Euclidean space $\mathbb{R}^n$ and each B-space is a locally convex space and hence a topological vector space.

77.1. Lemma of Sperner

As clearly as possible we try to present the simple and ingenious proof idea for Brouwer's fixed-point theorem, which goes back to Knaster, Kuratowski, and Mazurkiewicz. In order to avoid clumsy notation, we restrict ourselves during the next three sections to $\mathbb{R}^2$. The reader then can easily extend everything to $\mathbb{R}^n$.

Let M be a closed triangle in $\mathbb{R}^2$ with vertices P_0, P_1, P_2. The r-dimensional

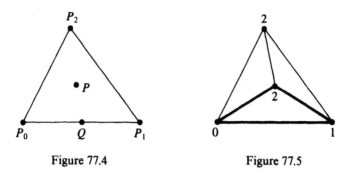

Figure 77.4   Figure 77.5

sides of M are the vertices P_0, P_1, P_2 for $r = 0$, the sides $\overline{P_0P_1}, \overline{P_1P_2}, \overline{P_2P_0}$ for $r = 1$, and the triangle itself for $r = 2$.

By definition the base of a point in M is the side of lowest dimension on which the point lies. In Figure 77.4 the base of P and Q is the triangle and the side $\overline{P_0P_1}$, respectively.

Lemma 77.1 (Sperner (1928)). *Let the triangle M be triangulated by subtriangles. Instead of P_0, P_1, P_2 we simply denote the vertices of M by 0, 1, 2, and assign to each knot point of the triangulation a number which belongs to some vertex of its base in M. Then there exists a Sperner simplex, i.e., a subtriangle having vertices with numbers 0, 1, 2 (Fig. 77.5).*

PROOF: A side of a subtriangle T is called distinguished if and only if it carries the numbers 0, 1. We have exactly the following two possibilities:

(i) T has precisely one distinguished side (Sperner simplex).
(ii) T has precisely two or no distinguished sides (no Sperner simplex).

But since the distinguished sides occur twice in the interior and in odd number on the boundary, the total number is odd. Therefore there exists a Sperner simplex. □

77.2. Lemma of Knaster, Kuratowski, and Mazurkiewicz

Lemma 77.2 (Knaster, Kuratowski, and Mazurkiewicz (1929)). *Suppose a closed triangle M of $\mathbb{R}^2$ with vertices P_0, P_1, P_2 is covered with three closed sets C_0, C_1, C_2 with*

$$\operatorname{co}\{P_i, P_j, P_k\} \subseteq C_i \cup C_j \cup C_k \tag{1}$$

for all possible index combinations, where index repetitions are allowed. Then, the intersection $C_0 \cap C_1 \cap C_2$ is not empty.

PROOF: For $k = 1, 2, \ldots$ consider a sequence of triangulations of M, where the maximal diameters of the subtriangles tend to zero as $k \to \infty$. For every triangulation there exists a Sperner simplex with vertices $P_0^{(k)}, P_1^{(k)}, P_2^{(k)}$ which carry the numbers 0, 1, 2. The construction of the labeling implies that

$$P_i^{(k)} \in \operatorname{co}\{P_i, \ldots\}$$

and from (1) it follows that

$$P_i^{(k)} \in C_i, \qquad i = 0, 1, 2. \tag{2}$$

We now choose a convergent subsequence with $P_i^{(k')} \to Q_i$ as $k \to \infty$. Since the diameters of the subtriangles tend to zero, we obtain $Q_0 = Q_1 = Q_2$. From (2) the claim $Q_0 \in C_i$ follows for all i. □

77.3. Elementary Proof of Brouwer's Fixed-Point Theorem

By an n-dimensional, closed simplex we mean the convex hull of $n + 1$ points $P_0, \ldots, P_n$ in $\mathbb{R}^n$ which do not all lie in a hyperplane.

Theorem 77.A (Brouwer (1912)). *Every continuous map $f: M \to M$ of an n-dimensional, closed simplex M into itself has a fixed point.*

PROOF: We apply Lemma 77.2. Let $n = 2$, i.e., M is a triangle in $\mathbb{R}^2$. For $n > 2$ one can use analogous arguments. Every point P in M has the representation

$$P = \lambda_0(P) P_0 + \lambda_1(P) P_1 + \lambda_2(P) P_2 \tag{3}$$

with barycentrical coordinates λ_i. Thereby we have

$$0 \leq \lambda_0, \lambda_1, \lambda_2 \leq 1, \qquad \lambda_0 + \lambda_1 + \lambda_2 = 1. \tag{4}$$

We set

$$C_i = \{P \in M : \lambda_i(f(P)) \leq \lambda_i(P)\}.$$

These sets satisfy the assumptions of Lemma 77.2, since the continuity of f and $\lambda_i(\cdot)$ implies that C_i is closed, and (1) follows because of $f(M) \subseteq M$ and (4). This is easily obtained by checking all possible cases. From $\lambda_0(P_0) = 1$, for example, it follows that $P_0 \in C_0$.

According to Lemma 77.2 there exists a point $P \in M$ with $P \in C_0 \cap C_1 \cap C_2$. This is the required fixed point, because from

$$\lambda_i(f(P)) \leq \lambda_i(P), \qquad i = 0, 1, 2 \tag{5}$$

and (4) we obtain equality in (5), and thus $f(P) = P$. □

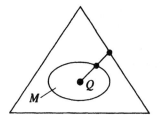

Figure 77.6

Corollary 77.3. *The assertion of Theorem 77.A remains valid if M is a compact, convex, nonempty set in $\mathbb{R}^n$.*

PROOF. Either M is a point or M lies on an m-dimensional plane and has an interior point Q there. In this last case, M is homeomorphic to a closed simplex as shown in Figure 77.6. The assertion then follows from Theorem 77.A. □

77.4. Generalized Lemma of Knaster, Kuratowski, and Mazurkiewicz

Proposition 77.4. *It is*

$$\bigcap_{x \in X} F(x) \neq \emptyset \tag{6}$$

if the following assumptions are satisfied:

(i) X is a nonempty set in a topological vector space E.
(ii) $F(x)$ is a closed, nonempty set in E for all $x \in X$.
(iii) $F(x_0)$ is compact for a fixed $x_0 \in X$.
(iv) For every finite subset $\{x_1, \ldots, x_n\}$ of X we have

$$\operatorname{co}\{x_1, \ldots, x_n\} \subseteq \bigcup_{i=1}^{n} F(x_i).$$

PROOF:

(I) If X is finite, then the claim follows similarly as in the proof of Lemma 77.2.
(II) Let X be infinite. Suppose (6) is not true. Because of (iii) and the finite intersection property $A_1(12g)$ there exists a tuple $\{x_1, \ldots, x_m\}$ with

$$\bigcap_{i=1}^{m} F(x_i) = \emptyset.$$

This contradicts (I). □

77.5. Inequality of Fan

Theorem 77.B (Fan (1972)). *It is*

$$\min_{y \in X} \sup_{x \in X} f(x, y) \leq \sup_{x \in X} f(x, x) \qquad (7)$$

if the following three assumptions are satisfied:

(a) *The function $f: X \times X \to \mathbb{R}$ is given where X is a compact, convex, and nonempty set in a topological vector space.*
(b) *f is quasi-concave in the first argument, i.e., $x \mapsto f(x, y)$ is quasi-concave on X for every fixed $y \in X$.*
(c) *f is lower semicontinuous in the second argument, i.e., $y \mapsto f(x, y)$ is lower semicontinuous on X for every fixed $x \in X$.*

The concepts of lower semicontinuity and quasi-convexity have been defined in Section 9.5. Moreover, g is upper semicontinuous and quasi-concave if and only if $-g$ is lower semicontinuous and quasi-convex, respectively. Often the following special case of (7) is used. If $f(x, x) \leq 0$ for all $x \in X$, then there exists a $y \in X$ with

$$f(x, y) \leq 0 \quad \text{for all} \quad x \in X, \qquad (8)$$

i.e., one obtains a solution y of this generalized variational inequality.

PROOF. We apply Proposition 77.4. For this we set $m = \sup_{x \in X} f(x, x)$ and define

$$F(x) = \{y \in X : f(x, y) \leq m\}.$$

Because of (c), the set $F(x)$ is closed and because of (a) it is compact. Moreover, we have

$$\mathrm{co}\{x_1, \ldots, x_n\} \subseteq \bigcup_{i=1}^{n} F(x_i).$$

Otherwise, there exists a convex linear combination

$$c = \sum_{i=1}^{n} \alpha_i x_1 \quad \text{with} \quad c \notin \bigcup_{i=1}^{n} F(x_i).$$

That is, $f(x_i, c) > m$ for all i. From (b) it follows that $f(c, c) > m$, which contradicts the definition of m.

From Proposition 77.4 there exists a $y \in \bigcap_x F(x)$. This is (7). □

In the following two sections we consider applications of this inequality.

77.6. Main Theorem for n-Person Games of Nash and the Minimax Theorem

Before reading this section, we recommend a look at Section 9.7 on game theory. There we have been restricted to only two players. Now we consider n players $P_1, \ldots, P_n$. Suppose, each player P_i has a strategy set K_i available. We assume:

(H) K_i is a compact, convex, nonempty set in a topological vector space E_i for all $i = 1, \ldots, n$.

The real number $f_i(p_1, \ldots, p_n)$ represents the loss of P_i if each player P_j chooses the strategy $p_j \in K_j$.

EXAMPLE 77.5. Let $E_i = \mathbb{R}^{n_i}$ and
$$K_i = \{p_i \in E_i: 0 \le p_{i,k} \le 1, k = 1, \ldots, n_i\}.$$
Similarly as in Section 9.7, the $p_{i,k}$ can be viewed as the probability with which player P_i makes his kth decision.

The main problem of game theory is to find useful definitions of equilibrium points and to prove their existence. Roughly speaking, equilibrium points correspond to a behavior most favorable to all players.

Definition 77.6. The point $(q_1, \ldots, q_n)$ with $q_j \in K_j$ for all j is called a *Nash equilibrium point* of the game if and only if for all $i = 1, \ldots, n$:
$$f_i(q_1, \ldots, q_n) = \min_{p_i \in K_i} f_i(q_1, \ldots, q_{i-1}, p_i, q_{i+1}, \ldots, q_n).$$

This definition has a simple interpretation. In a Nash equilibrium point no player has a reason to change his strategy if the other players keep their strategy (loss minimization).

Theorem 77.C (Nash (1951)). *A Nash equilibrium point exists if in addition to (H) the following two assumptions are satisfied:*

(i) *All maps $f_i: K_1 \times \cdots \times K_n \to \mathbb{R}$ are continuous.*
(ii) *All maps $p \mapsto f_i(p)$ are convex if we fix an arbitrary p_i in $p = (p_1, \ldots, p_n)$.*

PROOF: We let $X = K_1 \times \cdots \times K_n$ and
$$f(p, q) = \sum_{i=1}^{n} f_i(q) - f_i(p_1, \ldots, p_{i-1}, q_i, p_{i+1}, \ldots, p_n)$$
for all $p, q \in X$. Then f is concave in the first argument and continuous in the second argument. We have $f(p, p) = 0$. According to the inequality of Fan (8) there exists a $q \in X$ with
$$f(p, q) \le 0 \quad \text{for all} \quad p \in X.$$

If, for any fixed j, we choose a p with $p_i = q_i$ for all $i \neq j$, then we obtain
$$f_j(q) - f_j(p) \leq 0 \quad \text{for all } j. \qquad \square$$

In Section 9.6 we obtained the minimax theorem of John von Neumann by using a fixed-point theorem for multivalued maps. Now we will see that this theorem is a special case of Theorem 77.C.

EXAMPLE 77.7. We choose $n = 2$ in Theorem 77.C and suppose that $f_1 = -f_2$. Moreover, we let $f = f_1$. The Nash equilibrium point $q = (q_1, q_2)$ then implies that
$$\max_{p_2 \in K_2} f(q_1, p_2) = f(q_1, q_2) = \min_{p_1 \in K_1} f(p_1, q_2),$$
i.e., f has a saddle point q. Corollary 9.16 implies that
$$f(q_1, q_2) = \min_{p_1 \in K_1} \max_{p_2 \in K_2} f(p_1, p_2) = \max_{p_2 \in K_2} \min_{p_1 \in K_1} f(p_1, p_2).$$
This is a variant of the minimax theorem (Theorem 9.D). As we already saw in Section 9.7, this theorem describes the zero sum games for two persons. Because of $f_1 = -f_2$ the loss of player P_1 is equal to the gain of player P_2.

In Chapter 49 the existence of saddle points was our starting point for the discussion of duality theory.

77.7. Applications to the Theorem of Hartman–Stampacchia for Variational Inequalities

Consider the variational inequality
$$\langle Ay, x - y \rangle \geq 0 \quad \text{for all} \quad x \in K. \qquad (9)$$

Proposition 77.8 (Hartman–Stampacchia (1966)). *If K is a compact, convex, nonempty set in the B-space X and $A: K \to X^*$ is continuous, then (9) has a solution $y \in K$.*

PROOF. Apply the inequality of Fan (8) to $f(x, y) = -\langle Ay, x - y \rangle$, $x, y \in K$. $\square$

Corollary 77.9. *Instead of a B-space X we can also choose a locally convex space X, when X^* is equipped with the strong* topology.*

77.8. Fixed-Point Theorem of Kakutani

In Chapter 9 we obtained the fixed-point theorems of Kakutani and Fan–Glicksberg from Corollary 77.9. Here we want to prove these fixed-point theorems in a direct and intuitive fashion by using Brouwer's fixed-point theorem.

Proposition 77.10 (Kakutani (1941)). *The multivalued map $T: M \to 2^M$ has a fixed point x^*, i.e., $x^* \in T(x^*)$ if the following three conditions are satisfied:*

(i) M is a compact, convex, nonempty set in $\mathbb{R}^n$.
(ii) The set $T(x)$ is convex and nonempty for every $x \in M$.
(iii) The graph $G(T) = \{(x, T(x)): x \in M\}$ is closed in $M \times M$.

PROOF. The proof idea is to approximate T with a single-valued map for which Brouwer's fixed-point theorem can be applied and then passing to limits.

(I) Let $n = 2$, and M a triangle in $\mathbb{R}^2$. For $k = 1, 2, \ldots$ we choose a sequence of triangulations of M, for which the maximal diameter tends to zero as $k \to \infty$. For each triangulation we construct maps $T_k: M \to M$ by assigning to each knot point x a point $T_k(x)$ in $T(x)$, and by then extending T_k linearly to the subtriangles. According to Brouwer's fixed-point theorem (Theorem 77.A) each T_k has a fixed point x_k. Suppose it lies in the triangle with vertices $P_0^{(k)}$, $P_1^{(k)}$, $P_2^{(k)}$. From the definition of T_k it follows that

$$(P_i^{(k)}, T_k(P_i^{(k)})) \in G(T), \qquad i = 0, 1, 2. \tag{10}$$

Then there exist convergent subsequences, not separately denoted, with

$$P_i^{(k)} \to Q_i, \qquad T_k(P_i^{(k)}) \to S_i \quad \text{as} \quad k \to \infty.$$

Since the triangle diameters tend to zero as $k \to \infty$, we obtain $Q_0 = Q_1 = Q_2$, i.e., $x_k \to Q_0$ as $k \to \infty$. Moreover, we have

$$x_k \in \text{co}\{P_0^{(k)}, P_1^{(k)}, P_2^{(k)}\}.$$

By using the representation of the convex hull through convex linear combinations, we immediately obtain from $x_k = T_k x_k$ that

$$x_k \in \text{co}\{T_k(P_0^{(k)}), T_k(P_1^{(k)}), T_k(P_2^{(k)})\}.$$

By choosing convergent subsequences of the barycentrical coordinates as $k \to \infty$, we obtain the relation

$$Q_0 \in \text{co}\{S_0, S_1, S_2\}.$$

Furthermore, it follows from (10) and (iii) that $(Q_0, S_i) \in G(T)$, i.e.,

$$S_i \in T(Q_0) \qquad \text{for all } i.$$

Because of the convexity of $T(Q_0)$ we have $Q_0 \in T(Q_0)$, i.e., Q_0 is the required fixed point.

(II) For a simplex in $\mathbb{R}^n$ one uses analogous arguments.
(III) In the general case, one can choose a simplex S as in Corollary 77.3, which is mapped homeomorphically onto M by

$$h: S \to M.$$

Then one constructs approximations $T_k: S \to M$ of $T \circ h: S \to 2^M$ and applies Brouwer's fixed-point theorem to

$$h^{-1} \circ T_k: S \to S. \qquad \square$$

77.9. Fixed-Point Theorem of Fan–Glicksberg

Theorem 77.D (Fan (1952), Glicksberg (1952)). *The multivalued map $T: M \to 2^M$ has a fixed point if the following three conditions are satisfied:*

(i) *M is a compact, convex, nonempty set in the locally convex space X.*
(ii) *$T(x)$ is convex, closed, and nonempty for every $x \in M$.*
(iii) *T is upper semicontinuous.*

We shall prove this generalization of the fixed-point theorem of Kakutani by using this fixed-point theorem as well as the following lemma, which will be proved in Problem 77.3.

Lemma 77.11. *If C is a closed set in X, then also the following two sets*

$$Q = \{x \in M: x \in T(x) + C\},$$
$$P = \{(x, y): x \in M, y \in T(x) + C\}$$

are closed.

PROOF OF THEOREM 77.D.

(I) A convex set U in X is called absolute convex if $x \in U$, $|\alpha| \leq 1$ always implies that $\alpha x \in U$. Let B be a neighborhood basis of zero in X which consists of open, absolute convex sets. For every $U \in B$ we define the set

$$S_U = \{x \in M: x \in T(x) + \bar{U}\}.$$

According to Lemma 77.11 each of these sets is closed. Moreover, we show below in (II) that S_U is nonempty for every $U \in B$. From the choice of B it follows that the intersection of finitely many S_U is nonempty. Since M is compact, the finite intersection property A_1(12g) implies that there exists an x with $x \in S_U$ for all $U \in B$. This means that $x \in T(x)$, and hence x is the required fixed point.

(II) We show that $S_U \neq \emptyset$. Since M is compact, there exist finitely many points $x_1, \ldots, x_k \in M$ such that the open sets $x_1 + U, \ldots, x_k + U$ form a

covering of M. We set $K = \text{co}\{x_1, \ldots, x_k\}$ and define
$$T_U(x) = (T(x) + \bar{U}) \cap K.$$
Because of $T(x) \subseteq M$ and $U = -U$ it follows that $T_U(x)$ is nonempty, convex, and closed. The set K lies in a finite-dimensional subspace of X which can be identified with $\mathbb{R}^n$. According to Lemma 77.11 the map $T_U: K \to 2^K$ has a closed graph. The fixed-point theorem of Kakutani of Section 77.8 implies the existence of a point x with $x \in T_U(x)$, i.e., $S_U \neq \emptyset$. □

77.10. Applications to the Main Theorem of Mathematical Economics About Walras Equilibria and Quasi-Variational Inequalities

In order to explain the basic economical idea we consider a simple situation.

Standard Model 77.12 (Reasonable Price System). We define $P = \{p \in \mathbb{R}^n : 0 \leq p_i \leq 1 \text{ for all } i\}$ and let Q denote a compact, convex, nonempty set in $\mathbb{R}^n$.

Suppose there are n goods $G_1, \ldots, G_n$. Let p_i be the price of G_i with $0 \leq p_i \leq 1$. We assign a set $S(p) \subseteq Q$ to each price vector $p \in P$. Thereby
$$s \in S(p) \quad \text{with} \quad s = (s_1, \ldots, s_n)$$
means that the difference between supply and demand for G_i is equal to s_i. The number
$$\langle p | s \rangle = \sum_{i=1}^{n} p_i s_i$$
is therefore equal to the difference between the value of the goods which are supplied to the market and the value of the goods which are demanded by the market. Thus
$$\langle \bar{p} | \bar{s} \rangle = 0, \quad \bar{s} \in S(\bar{p})$$
is the mathematical formulation of the situation "supply equals demand." This ideal situation cannot always be realized. We therefore will be satisfied with the weaker result (12) below. First of all, it is reasonable to assume that
$$\langle p | s \rangle \geq 0 \quad \text{for all} \quad p \in P, \quad s \in S(p). \tag{11}$$
This is the so-called law of Walras. Roughly, it means that we only consider economical situations with a supply excess.

We call $(\bar{p}, \bar{s})$ with $\bar{s} \in S(\bar{p})$ a *Walras equilibrium* if and only if the following holds:
$$0 \leq \langle \bar{p} | \bar{s} \rangle \leq \langle p | \bar{s} \rangle \quad \text{for all} \quad p \in P. \tag{12}$$

77.10. Applications to the Main Theorem of Mathematical Economics

Roughly, this means that the value difference between supply and demand becomes minimal. The vector $\bar{p}$ is called equilibrium price system.

The fundamental problem of mathematical economics is to find conditions for the supply excess map S which guarantee the existence of a Walras equilibrium.

For the special case that

$$X = Y = \mathbb{R}^n, \qquad P = \text{unit cube}$$

and $f(p, s) = \langle p | s \rangle$ the following general assumptions (H1)–(H4) precisely correspond to our model. If we write (12) in the form

$$\langle p - \bar{p} | s \rangle \geq 0 \quad \text{for all} \quad p \in P,$$
$$\bar{s} \in S(\bar{p}), \tag{13}$$

then this is a so-called quasi-variational inequality. Therefore Theorem 77.E below represents a general existence theorem for quasi-variational inequalities. We assume that:

(H1) X and Y are locally convex spaces with compact, convex, nonempty subsets $P \subseteq X$ and $Q \subseteq Y$. Let a continuous function $f: P \times Q \to \mathbb{R}$, an upper semicontinuous map $S: P \to 2^Q$, and a real number c be given.
(H2) The map $f(\cdot, s)$ is quasi-convex on P for every fixed $s \in Q$.
(H3) $S(p)$ is convex, closed, and nonempty for every fixed $p \in P$.
(H4) $f(p, s) \geq c$ for all $p \in P$, $s \in S(p)$ (Walras law).

Theorem 77.E (Main Theorem of Mathematical Economics of Gale (1955), Nikaido (1956), and Debreu (1959)). *If (H1)–(H4) hold, then there exists a Walras equilibrium, i.e., there exists an element $\bar{p} \in P$ and an element $\bar{s} \in S(\bar{p})$ such that*

$$c \leq f(\bar{p}, \bar{s}) \leq f(p, \bar{s}) \quad \text{for all} \quad p \in P.$$

PROOF. We use the fixed-point theorem of Fan–Glicksberg (Theorem 77.D). For this we define a map $R: Q \to 2^P$ through

$$R(s) = \{p \in P : f(p, s) \leq f(q, s) \text{ for all } q \in P\}.$$

The set $R(s)$ is nonempty, because the continuous function $f(\cdot, s)$ assumes its minimum on the compact set P. Moreover, $R(s)$ is convex and closed. From the continuity of f it follows that the graph

$$G(R) = \{(s, p) : s \in Q, p \in R(s)\}$$

is a closed subset of the compact set $Q \times P$. From Problem 77.4 it follows that R is upper semicontinuous.

We now let $M = P \times Q$ and define the map $T: M \to 2^M$ through

$$T(p, s) = R(s) \times S(p).$$

Theorem 77.D implies the existence of a fixed point $(\bar{p}, \bar{s}) \in T(\bar{p}, \bar{s})$. This means that $\bar{s} \in S(\bar{p})$ and

$$f(\bar{p}, \bar{s}) \leq f(p, \bar{s}) \quad \text{for all} \quad p \in P. \qquad \square$$

Leon Walras, through his basic work (1874), is regarded as the creator of mathematical economics. However, only during the 1950s did his ideas appear in a precise mathematical form following the work of Arrow and Debreu (1954), Gale (1955), and Nikaido (1956). Thereby John von Neumann's minimax theorem of game theory (1928) and the fixed-point theorem of Kakutani (1941) played an important role.

77.11. Negative Retract Principle

Proposition 77.13. *Let B be a closed ball in $\mathbb{R}^n$, $n \geq 1$. Then the boundary ∂B is not a retract of B.*

PROOF. If $r: B \to \partial B$ is a retraction, then $-r: B \to B$ is fixed-point free, which contradicts Brouwer's fixed-point theorem. $\qquad \square$

77.12. Intermediate-Value Theorem of Bolzano–Poincaré–Miranda

Consider the equation

$$f(x) = 0, \quad x \in C. \qquad (14)$$

If $C = [a, b]$ is a compact interval, and the function $f: C \to \mathbb{R}$ is continuous with

$$f(a) \leq 0 \quad \text{and} \quad f(b) \geq 0, \qquad (15)$$

then equation (14) has a solution (Fig. 77.7(a)). We want to extend this

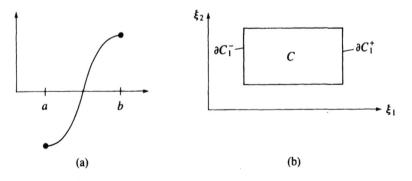

Figure 77.7

77.12. Intermediate-Value Theorem of Bolzano–Poincaré–Miranda

well-known intermediate-value theorem of Bolzano to the $\mathbb{R}^n$. To this end we consider the compact cuboid

$$C = \{x \in \mathbb{R}^n : a_i \leq \xi_i \leq b_i \text{ for all } i\},$$

where $x = (\xi_1, \ldots, \xi_n)$. Furthermore, let ∂C_i^- and ∂C_i^+ denote the side of C with $\xi_i = a_i$ and $\xi_i = b_i$ (Fig 77.7(b)). In place of (15), we have the very natural condition

$$\pm f_i(x) \geq 0 \quad \text{on } \partial C_i^\pm \qquad \text{for } i = 1, \ldots, n. \tag{16}$$

The following proposition is used in modern computer algorithms to determine cuboids C, for which (14) has a solution. Thereby one gets a rough idea about the position of the zero. This then can be improved by special algorithms.

Proposition 77.14 (Miranda (1941)). *Let C be a compact cuboid in $\mathbb{R}^n$ and $f: C \to \mathbb{R}$ a continuous function which satisfies (16). Then f has a zero in C.*

PROOF. We use Brouwer's fixed-point theorem. Let C be the standard cube, i.e., $a_i = -1$ and $b_i = 1$ for all i. The general case can always be reduced by an affine transformation to this special situation.

(I) First, we assume that (16) holds in the stronger form

$$\pm f_i(x) > 0 \quad \text{on } \partial C_i^\pm \qquad \text{for } i = 1, \ldots, n. \tag{16a}$$

Our simple trick is to let

$$F_i(x) = \xi_i - \varepsilon_i f_i(x).$$

Because of (16a) there exist numbers $\varepsilon_i > 0$ with

$$-1 \leq F_i(x) \leq 1 \quad \text{on } C \qquad \text{for all } i.$$

Thus we obtain $F(C) \subseteq C$. Brouwer's fixed-point theorem implies that F has a fixed point in C, i.e., f has a zero in C.

(II) If (16) holds, but not (16a), then we pass from f to

$$g(x) = f(x) + \varepsilon x, \qquad \varepsilon > 0.$$

We then can apply (I) to g. It follows that the equation

$$f(x_\varepsilon) + \varepsilon x_\varepsilon = 0, \qquad x_\varepsilon \in C \tag{17}$$

has a solution for each $\varepsilon > 0$. Because of the compactness of C there exists a convergent subsequence $x_\varepsilon \to x$ as $\varepsilon \to 0$. From (17) it follows that $f(x) = 0$. □

It is very interesting that this intermediate-value theorem (Proposition 77.14) is equivalent to Brouwer's fixed-point theorem. This clearly illustrates the analytical nature of Brouwer's fixed-point theorem.

The proof above shows that Proposition 77.14 is a consequence of Brouwer's fixed-point theorem. We prove the converse. Let C be the standard cube and $F: C \to C$ continuous, i.e.,

$$-1 \leq F_i(x) \leq 1 \quad \text{on } C \quad \text{for all } i.$$

We set $f(x) = x - F(x)$. With this (16) holds. From Proposition 77.14 it follows that f has a zero in C, i.e., F has a fixed point in C.

The general version of Brouwer's fixed-point theorem follows from this special case as in the proof of Corollary 77.3.

77.13. Equivalent Statements to Brouwer's Fixed-Point Theorem

The following statements are equivalent:

(i) Lemma of Knaster–Kuratowski–Mazurkiewicz.
(ii) Brouwer's fixed-point theorem.
(iii) Negative retract principle.
(iv) Inequality of Fan.
(v) Theorem of Hartman–Stampacchia.
(vi) Fixed-point theorem of Kakutani.
(vii) Fixed-point theorem of Fan–Glicksberg.
(viii) Main theorem of mathematical economy (Theorem 77.E).
(ix) Intermediate-value theorem of Bolzano–Poincaré–Miranda.
(x) Cubic Sperner lemma of Kuhn.

The equivalences are pictured in Figure 77.1. In this chapter we proved that

(i) $\Rightarrow$ (ii) $\Rightarrow$ (iii), (ii) $\Rightarrow$ (iv) $\Rightarrow$ (v), (ii) $\Rightarrow$ (vi) $\Leftrightarrow$ (vii) $\Rightarrow$ (viii)

and (ii) $\Leftrightarrow$ (ix).

In Section 2.3 we showed that (iii) $\Rightarrow$ (ii). Thus it remains to show that (v) $\Rightarrow$ (ii) $\Rightarrow$ (i) and (viii) $\Rightarrow$ (ii). This will be done in Problems 77.5 to 77.7.

The equivalence between the cubic Sperner lemma and Brouwer's fixed-point theorem will be discussed in Problem 77.10.

PROBLEMS

77.1. *Straightforward generalizations to the $\mathbb{R}^n$.* Generalize Lemmas 77.1 and 77.2 to the $\mathbb{R}^n$.
Hint: See Knaster, Kuratowski, and Mazurkiewicz (1929).

77.2. *Inequality of Fan and an economical model.* Consider Problem 9.1 and prove the existence of a reasonable price system. Do not use the fixed-point theorem of Kakutani, but instead give a shorter proof by using the inequality of Fan.
Solution: Let

$$X = \{p \in \mathbb{R}^n : p \geq 0, \sum_{i=1}^n p_i = 1\}$$

and

$$F_j(p) = \sum_{i=1}^{n} D_{ij}(p), \qquad f(q,p) = \langle q | F(p) \rangle - \langle q, a \rangle.$$

Assumption (9.32) states that $f(p,p) = 0$ for all $p \in X$. From the inequality of Fan (8) it follows that there exists an element $p^* \in X$ with $f(q, p^*) \leq 0$ for all $q \in X$. Thus we obtain that $F(p^*) \leq a$. This is assertion (9.33).

77.3. *Proof of Lemma* 77.11.

Solution: Let $P = \{(x, y): x \in M, y \in T(x) + C\}$. We show that the complement $(M \times X) - P$ is open. Let $(x_0, y_0) \notin P$, i.e., $x_0 \in M$ and $y_0 \notin T(x) + C$. Then there exists a neighborhood U of zero with

$$(y_0 + U) \cap (T(x_0) + C + U) = \emptyset.$$

Since T is upper semicontinuous, it follows from Proposition 9.5 that there exists a neighborhood $V(x_0)$ of x_0 in M with

$$x \in V(x_0) \Rightarrow T(x) \subseteq T(x_0) + U,$$

and hence

$$x \in V(x_0) \Rightarrow (y_0 + U) \cap (T(x) + C) = \emptyset.$$

Therefore a neighborhood of (x_0, y_0) in $M \times X$ does not belong to P.
Analogous arguments can be used for Q.

77.4. *Criterium for upper semicontinuity*. The spaces X and Y are locally convex with compact subsets $M \subseteq X$ and $N \subseteq Y$. Suppose the images $T(x)$ of the map

$$T: M \to 2^N$$

are closed for each $x \in M$.

Prove: T is upper semicontinuous if and only if the graph of T is closed.

Solution: The graph $G(T)$ is equal to the set $\{(x, y): x \in M, y \in T(x)\}$.

(I) Let T be upper semicontinuous. We show that $Z = (M \times N) - G$ is open. Then $G(T)$ is closed. Let $(x, y) \in Z$. Because of the compactness of $T(x)$ and since $y \notin T(x)$ there exist open sets V and W with

$$y \in V, \qquad T(x) \subseteq W, \qquad V \cap W = \emptyset.$$

Since T is upper semicontinuous, it follows from Proposition 9.5 that there exists an open set U with $x \in U$ and $T(U \cap M) \subseteq W$ and thus $U \times V \subseteq Z$. This means that $(x, y) \in \text{int } Z$.

(II) Let $G(T)$ be closed. If T is not upper semicontinuous, then Proposition 9.5 implies that there exists a point x and an open set W with $T(x) \subseteq W$ such that for all neighborhoods $U(x)$ the image $T(U(x) \cap M)$ does not completely lie in W. Hence, for every $U(x)$ there exists a point $(x_U, y_U) \in G(T)$ with $y_U \notin W$. Because of the compactness of $G(T)$ it follows from $A_1(17f)$ that there exists a convergent M–S-subsequence $(x_{U'}, y_{U'}) \to (x, y)$. Thus we find that $(x, y) \in G(T)$, and hence $y \in T(x)$. Because of $y_{U'} \notin W$ we obtain $y \notin W$. This gives the required contradiction $y \notin T(x)$.

77.5. From the theorem of Hartman–Stampacchia follows Brouwer's fixed-point theorem.

Solution: Let $T: K \subseteq \mathbb{R}^n \to K$ be continuous on the closed ball K. From Proposition 77.8 with $A = I - T$ there exists an element $y \in K$ with

$$\langle y - Ty | x - y \rangle \geq 0 \quad \text{for all} \quad x \in K.$$

For $x = Ty$ we obtain $y - Ty = 0$.

77.6. *From Brouwer's fixed-point theorem follows the lemma of Knaster–Kuratowski–Mazurkiewicz.*

Solution: We have to prove the statement of Proposition 77.4 for finite X with $\dim E < \infty$. Let, for example, $X = \{x_1, x_2\}$ and

$$\operatorname{co}\{x_1, x_2\} \subseteq F(x_1) \cup F(x_2) \quad \text{and} \quad x_i \in F(x_i) \quad \text{for} \quad i = 1, 2.$$

We have to show that $F(x_1) \cap F(x_2) \neq \emptyset$.

Suppose $F(x_1) \cap F(x_2) = \emptyset$. We let $K = \operatorname{co}\{x_1, x_2\}$, $d_i = \operatorname{dist}(x, F(x_i))$ and

$$f(x) = (d_1(x) x_1 + d_2(x) x_2)/(d_1(x) + d_2(x))$$

for all $x \in K$. The divisor is not equal to zero. Brouwer's fixed-point theorem implies that there exists an element $x \in K$ with

$$f(x) = x.$$

Because of $K \subseteq F(x_1) \cup F(x_2)$ we may assume that $x \in F(x_1)$, and hence $d_1(x) = 0$. From $x = f(x)$ it follows that $x = x_2$ and hence $x \in F(x_2)$. This is a contradiction to $F(x_1) \cap F(x_2) = \emptyset$.

77.7. *From Theorem 77.E follows Brouwer's fixed-point theorem.* Let $S: P \to P$ be a continuous map from the closed ball P into itself. For $r > 0$ we define

$$M_r = \{p \in P: |p - Sp| \leq r\}.$$

We prove that $M_r \neq \emptyset$. Otherwise, we have $|p - Sp| \geq r$ for all $p \in P$. If we choose

$$f(p, s) = |p - s|,$$

then Theorem 77.E implies that there exists an element $\bar{p} \in P$ with

$$r \leq f(p, S\bar{p}) \quad \text{for all} \quad p \in P.$$

Especially for $p = S\bar{p}$ we find $f(p, S\bar{p}) = 0$, which contradicts $r > 0$.

The intersection of finitely many sets M_r is always nonempty. From the finite intersection property $A_1(12g)$ there exists an element $p \in \bigcap_r M_r$. This is the required fixed point of S.

77.8. *Simplicial algorithm to determine fixed points.* We restrict ourselves to a description of the basic idea which is closely related to the proof of Brouwer's fixed-point theorem of Section 77.3.

77.8a. *Homotopy.* Let the map $f: \mathbb{R}^n \to \mathbb{R}^n$ be i-compact, i.e., f is continuous and $f(\mathbb{R}^n)$ is relatively compact. We consider the homotopy

$$H(x, t) = (1 - t) x_0 + t f(x).$$

By definition, the fixed-point set $\operatorname{Fix}(H)$ is equal to the set of all points

$(x, t) \in \mathbb{R}^n \times [0, 1]$ with

$$H(x, t) = x.$$

We look for a fixed point x_1 of f, i.e., $(x_1, 1) \in \text{Fix}(H)$. Let $x = (\xi_1, \ldots, \xi_n)$.

Show: Fix(H) contains a component C which connects the point $(x_0, 0)$ with a point $(x_1, 1) \in \text{Fix}(H)$ (Figure 77.8). Note that C need not be a curve.

Solution: Choose an open ball G which contains the image $H(\mathbb{R}^n \times [0, 1])$ and apply Theorem 14.C of Part I (Global Leray–Schauder principle).

77.8b. *Goal.* As in Figure 77.8 we want to construct a sequence of simplices which approximate C and lead us to a fixed point x_1 of f. The essential steps are:
  (i) Integer labeling.
  (ii) Door-in/door-out principle.
  (iii) Creation of new simplices by using quasi-reflections.

77.8c. *Labeling.* We assign an integer $i = 0, 1, \ldots, n$ to a point $(x, t) \in \mathbb{R}^n \times [0, 1]$, where i is the greatest number with the property

$$\xi_j \geq H_j(x, t) \quad \text{for } j = 1, \ldots, i.$$

An n-side of an $(n + 1)$-simplex in $\mathbb{R}^{n+1}$ is called regular or completely labeled if and only if its vertices carry the numbers $0, \ldots, n$. An $(n + 1)$-simplex is called regular if and only if it contains exactly two regular sides.

Prove: Every $(n + 1)$-simplex in $\mathbb{R}^n \times [0, 1]$ is either regular or contains no regular side.

Hint: Elementary arguments. See Allgower and Georg (1980), p. 34.

77.8d. *Quasi-reflection.* Let $\sigma(x_0, \ldots, x_{n+1})$ denote the closed $(n + 1)$-simplex with vertices $x_0, \ldots, x_{n+1}$. By a quasi-reflection with respect to x_k we mean the creation of a new simplex where x_k is replaced with

$$\bar{x}_k = x_{k_-} + x_{k_+} - x_k.$$

Here, k_- and k_+ is the left and right neighbor of k in Figure 77.9. For example, Figure 77.10 shows the quasi-reflection of $\sigma(x_0, x_1, x_2)$ with respect to x_0.

77.8e.* *Algorithms.* Let the map f satisfy the assumptions of Problem 77.8a. Then f has a fixed point.

Formulate the following steps as an algorithm and show that for any given $\varepsilon > 0$ after finitely many steps one finds a point $(x, 1)$ with

$$|f(x) - x| < \varepsilon.$$

Moreover, any subsequence generated by this algorithm converges to a fixed point of f.

  (i) One constructs a regular $(n + 1)$-start simplex σ such that $(x_0, 0)$ lies on a regular side S_0 of σ. One enters the simplex σ through S_0 and leaves it through the second regular side S_1 (Door-in/door-out principle).
  (ii) One performs a quasi-reflection of σ with respect to the vertex which does not lie on the exit side S_1. Thereby one obtains the simplex σ_1 which one enters through S_1 and leaves through the second regular side, etc. (See Figure 77.11 and Figure 77.8.)

Hint: See Allgower and Georg (1980), p. 37. There, and in Peitgen and Prüfer (1979), one may find further material together with numerical examples.

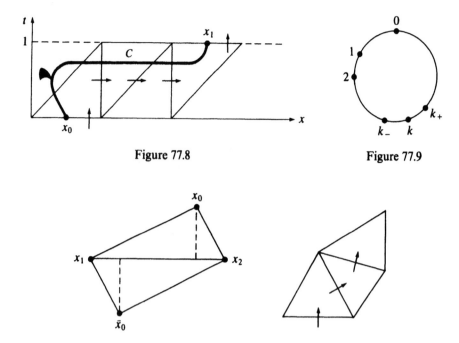

Figure 77.8   Figure 77.9

Figure 77.10   Figure 77.11

77.9.* *Sperner simplices and mapping degree.* Let $f: P \subseteq \mathbb{R}^n \to \mathbb{R}^n$ be a continuous map from the n-dimensional homogeneous polyeder P with $f(x) \neq 0$ on ∂P. Prove the following formula about the mapping degree:

$$\deg(f, \operatorname{int} P) = \sum_\sigma \operatorname{orientation}(\sigma). \tag{18}$$

The sum is taken over all Sperner simplices σ of P with respect to f. We explain this. The polyeder P is the union of finitely many closed n-simplices, whereby two of them always have exactly one $(n-1)$-side in common or are disjoint (Fig. 77.12). Therefore P possesses a natural triangulation. Let $\sigma = \operatorname{co}\{x_0, \ldots, x_n\}$ be a simplex of P. In $\mathbb{R}^n$ we specify a standard simplex $\sigma_0 = \operatorname{co}\{e_0, \ldots, e_n\}$ with $e_0 = 0$ and $e_i = (0, \ldots, 1, \ldots, 0)$, where 1 occurs at the ith position. Using a linear transformation L we map σ bijectively onto a translation of σ_0 with $Lx_i = e_i + \operatorname{const}$ for all i, and define

$$\operatorname{orientation}(\sigma) = \operatorname{sgn} \det L.$$

In Figure 77.13 we have $\operatorname{orientation}(\sigma_\pm) = \pm 1$.

We assign a number $i = 0, \ldots, n$ to every point $x \in P$. This is the smallest number i with

$$f_{i+1}(x) \leq 0.$$

Thereby let $i = n$ if $f_j(x) > 0$ for all j. Then σ is called a Sperner simplex with respect to f if and only if the vertices carry all numbers 0 to n. By orientation (σ) we mean the orientation of σ, which occurs if one runs through the vertices from 0 to n.

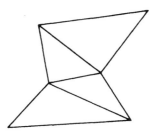

Figure 77.12

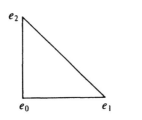 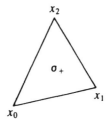 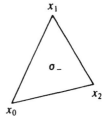

Figure 77.13

For the triangulation we require that to every $(n-1)$-simplex τ on ∂P there exists a number $i(\tau)$ such that all points of τ do not have the number $i(\tau)$. This condition is always satisfied for a sufficiently fine triangulation.

Hint: See Prüfer and Siegberg (1979). In Kliesch (1984), (1988), one can find a general approach which includes many different formulas of type (18). The coincidence with the classical mapping degree is shown there by using the axiomatic approach of Chapter 12.

The importance of (18) and similar formulas is that $\deg(f, \text{int } P) \neq 0$ implies the existence of a zero of f in P, and a computation of $\deg(f, \text{int } P)$ only requires finitely many values of f.

77.10.* *Cubic Sperner lemma of Kuhn* (1960).

77.10a. *Notations.* For the points $x, y \in \mathbb{R}^n$ we write $x = (\xi_1, \ldots, \xi_n)$ and $y = (\eta_1, \ldots, \eta_n)$. As usual, $x \leq y$ is equivalent to $\xi_i \leq \eta_i$ for all i, and $x < y$ means $x \leq y$ and $x \neq y$. Moreover, let $e \in \mathbb{R}^n$ denote the point $(1, \ldots, 1)$. Let $\mathbb{Z}$ be the set of all integers. For $p = 1, 2, \ldots$ we define the discrete cube

$$C_p = \{x \in \mathbb{Z}^n : 0 \leq \xi_i \leq p \text{ for all } i\}.$$

77.10b. *Lemma.* Let $y = f(x)$ be a map $f : C_p \to C_1$ with the property

$$\xi_i = 0 \Rightarrow \eta_i = 0,$$

$$\xi_i = p \Rightarrow \eta_i = 1.$$

Show: There exist points $x_0, x_1, \ldots, x_m \in C_p$ with

$$x_0 < x_1 < \cdots < x_m < x_0 + e, \qquad 0 \leq m \leq n$$

and

$$e \leq \sum_{r=0}^{m} f(x_r) \leq me.$$

Hint: See Kuhn (1960).

77.10c. Equivalence.
Show: The cubic Sperner lemma of Problem 77.10b is equivalent to Brouwer's fixed-point theorem.
Hint: See Kuhn (1960) and Riedrich (1976, L), p. 59.
The importance of this is that Brouwer's fixed-point theorem is shown to be equivalent to a purely combinatorical statement.

References to the Literature

Classical works: Sperner (1928), Knaster, Kuratowski, and Mazurkiewicz (1929), Scarf (1967), Fan (1972).

Survey about statements equivalent to Brouwer's existence principle: Gwinner (1981).

Applications to variational inequalities: Mosco (1976, S).

Inequality of Fan, multivalued maps, and mathematical economics: Aubin (1979, M), (1979a, M).

Survey about the numerical application of simplicial methods: Todd (1976, S), Allgower and Georg (1980, S, B, H), Peitgen and Prüfer (1979, S) (bifurcation), Schilling (1986, M).

Simpicial methods and mapping degree: Prüfer and Siegberg (1979), Kliesch (1983), (1984), (1988).

Mathematical economics: Karlin (1959, M, H), Nikaido (1968, M), (1970, M), Aubin (1979a, M), Ramanathan (1983, L).

Classical works on mathematical economics: Walras (1874, M), von Neumann (1928) (foundation of game theory), Arrow and Debreu (1954), Gale (1955), Nikaido (1956), Debreu (1959, M).

Topological methods in Walrasian economics: Dierker (1974, M).

Methods of global analysis in mathematical economics: Smale (1982, S).

Equilibrium in incomplete markets: Duffie and Shafer (1985), Husseini, Lasry, and Magill (1987) (generalization of the Brouwer fixed-point theorem to Grassmanians of k-dimensional subspaces of $\mathbb{R}^n$).

Handbook of mathematical economics: Arrow and Intrilligator (1982).

(See also the References to the Literature to Chapter 9.)

CHAPTER 78

Homotopy Methods and One-Dimensional Manifolds

The idea of considering the measure of the set of critical values of one function or of several functions is due to Marston Morse (1939).
Arthur Sard (1942)

The purpose of this note is to introduce a nonlinear version of Fredholm operators, and to prove that in this context Sard's theorem (1942) holds if zero measure is replaced by first category. Strictly speaking, our result is a generalization of a theorem of Brown (1935), an earlier special case of Sard's theorem.
Steve Smale (1965)

We illustrate that most existence theorems using degree theory are in principle relatively constructive.
Shui-Nee Chow, John Mallet-Paret, and James A. Yorke (1978)

The term continuation method derives from a familiar class of numerical methods dating back at least to Lahaye (1935), and also known as embedding methods. It is important to emphasize a distinction between classical embedding methods and the present continuation methods. The classical methods require that the homotopy parameter shall vary monotonically and the effort to follow a homotopy curve is abandoned when a critical point of the homotopy parameter, i.e., a turning point, is encountered. In contrast, the present continuation methods have faith and proceed beyond such critical points.
Eugene Allgower and Kurt Georg (1980)

In this chapter, Sard's theorem (Proposition 4.55 of Part I) plays a central role. Before reading this chapter, one should look again at this theorem as well as Definition 4.52 about regular values. For didactical reasons, we use a parametrized version of Sard's theorem already in Section 78.2, and present the proof afterwards in Section 78.7. The definition of the fixed-point index and the mapping degree of Section 78.6, however, only requires Sard's theorem and not the parametrized version. Sard's theorem is one of the most important theorems in modern mathematics. It gives a precise formulation of the following philosophy: Most situations in nature are generic, i.e., not degenerate.

As another central tool in this chapter we use the structure theorem about one-dimensional manifolds (Proposition 73.9). Roughly, this theorem states that one-dimensional manifolds behave like reasonable curves.

The results of Section 78.2 about regular solution curves admit a development of fixed-point theory which is characterized by great geometrical clarity and intuition. Surprisingly, this approach has only been developed during the last years.

78.1. Basic Idea

The simple basic idea of homotopy methods is the following. In order to solve the equation

$$F(x) = 0, \quad x \in \mathbb{R}^n, \tag{1}$$

for a map $F: \mathbb{R}^n \to \mathbb{R}^n$, we consider the more general equation

$$H(x, t) = 0, \quad t \in [0, 1], \quad x \in \mathbb{R}^n \tag{2}$$

with $H(x, 1) = F(x)$. Suppose (2) has a solution x_0 for $t = 0$ which is easy to determine. We look for a curve

$$x = x(s), \quad t = t(s) \tag{3}$$

with arclength s such that for $s = 0$ this curve begins at $x = x_0$, $t = 0$ and passes through a point $x = x_1$, $t = 1$ (Fig. 78.1). The classical continuation methods are designed to find solutions of (2) of the form $x = x(t)$. Such procedures, however, break down at a turning point T of Figure 78.1, and one can only follow curves such as in Figure 78.2. With (3), however, one can follow any curve whatever.

78.2. Regular Solution Curves

The important question is: When does the solution set of the equation

$$H(x, t) = 0, \quad (x, t) \in \mathbb{R}^{n+1} \tag{4}$$

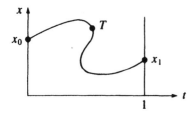
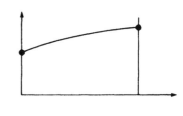

Figure 78.1   Figure 78.2

78.2. Regular Solution Curves

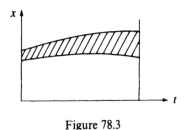

Figure 78.3

consist of reasonable curves which can be followed by numerical algorithms? Figure 78.3 shows that this need not always be the case.

Our assumption is:

(H) The map $H: \mathbb{R}^{n+1} \to \mathbb{R}^n$ is C^∞.

Definition 78.1. Assume (H). By a *regular solution curve* C of (4) we mean a set of solution points which is C^∞-diffeomorphic to the straight line $\mathbb{R}^1$ or the unit circle S^1, and each point $(x, t) \in C$ is a regular point of H. The solution set of (4) is called *regular* if and only if all its components are regular solution curves.

By an *index* of the solution point (x, t) of (4) we mean sgn det $H_x(x, t)$.

Proposition 78.2 (Regularity Criterium). *If* (H) *holds and zero is a regular value of* H, *then the solution set of* (4) *is regular.*

PROOF. From the preimage theorem (Theorem 4.J) it follows that the solution set is a one-dimensional manifold. The structure theorem for connected, one-dimensional manifolds (Proposition 73.9) yields the assertion. □

EXAMPLE 78.3 (Perturbation of H). If (H) holds, then for every $\varepsilon > 0$ there exists a $p \in \mathbb{R}^n$ with $|p| < \varepsilon$ such that the solution set of the equation

$$H(x, t) - p = 0, \quad (x, t) \in \mathbb{R}^{n+1}$$

is regular.

PROOF. From Sard's theorem we can choose p as a regular value of H, i.e., zero is a regular value of $H - p$. □

EXAMPLE 78.4 (Parametrization). Consider $H = H(x, t, p)$ with parameter $p \in \mathbb{R}^r$. Let $H: \mathbb{R}^{n+1} \times \mathbb{R}^r \to \mathbb{R}^n$ be C^∞. If zero is a regular value of H, then equation

$$H(x, t, p) = 0, \quad (x, t) \in \mathbb{R}^{n+1}$$

has a regular solution set for almost every parameter $p \in \mathbb{R}^r$.

As usual, "for almost all p" means "for all p except for a set of measure zero in $\mathbb{R}^r$."

PROOF. According to the parametrized version of Sard's theorem of Section 78.7, zero is a regular value of $H(\cdot, p)$ for almost all p. □

A common homotopy is

$$H(x, t, p) = (1 - t)(x - p) + tF(x), \qquad p \in \mathbb{R}^n.$$

For $t = 0$ we find that $x = p$ is the unique solution of

$$H(x, t, p) = 0, \qquad (x, t) \in \mathbb{R}^{n+1}. \tag{5}$$

EXAMPLE 78.5. If $F: \mathbb{R}^n \to \mathbb{R}^n$ is C^∞ with zero as a regular value, then equation (5) has a regular solution set for almost all starting values $p \in \mathbb{R}^n$.

PROOF. Zero is a regular value of H, because the linearization satisfies

$$H'(x, t, p) = (H_x(x, t, p), H_t(x, t, p), H_p(x, t, p))$$
$$= ((1 - t)I + tF'(x), \ldots, (t - 1)I).$$

It follows that $R(H'(x, t, p)) = \mathbb{R}^n$ for all (x, t, p). For $t = 1$ and $t \neq 1$ this is guaranteed by the first term $F'(x)$ and the last term $(t - 1)I$, respectively. □

Example 78.5 is of great practical use. If one picks an arbitrary starting value $x_0 = p$, then with probability one there passes a regular solution curve through it. According to Sard's theorem, any arbitrarily small perturbation of F will make zero into a regular value of the perturbed F. On the computer F is not exactly known. Thus on computers one always expects regular solution curves. Experience shows that this is the case as a rule. In summary we remark: Sard's theorem and its parametrized version guarantee that equation (4) can be treated in a naive way, i.e., the generic situation of regular solution curves may be assumed.

The great advantage of regular solution curves is that they cannot intersect themselves and cannot accumulate at one point. More precisely, we have the following result, which will frequently be used.

Proposition 78.6 (Door-In/Door-Out Principle). *Let Z be a bounded region in $\mathbb{R}^{n+1}$ and let the solution set of (4) be regular.*

If C is a regular solution curve of (4) which enters Z at the boundary point $(x, t) \in \partial Z$, then C cannot remain in the region Z.

If C passes through a point (x, t_0), whose index sgn det $H_x(x, t_0)$ is not equal to zero, then C has the form $x = x(t)$ in a neighborhood of (x, t_0).

The last statement is often used to assure that C enters Z (see Figure 78.9 of Section 78.5).

PROOF. If C remains in Z forever, then compactness of Z implies that C has an accumulation point $(x, t) \in \bar{Z}$. Continuity of H implies that $H(x, t) = 0$, i.e.,

78.3. Turning Point Principle and Bifurcation Principle

For simplicity in notation, we let $y = (x, t)$. From

$$H(y(s)) = 0 \tag{6}$$

we obtain, by differentiation with respect to s, the equation

$$H_x(y(s))x'(s) + H_t(y(s))t'(s) = 0. \tag{7}$$

Instead of (6), we use

$$H(y(s)) = 0,$$
$$\langle y'(s) | y'(s) \rangle = 1. \tag{8}$$

The last equation shows that s is the arclength. We set

$$A(s) \stackrel{\text{def}}{=} \begin{pmatrix} H_y(y(s)) \\ y'(s) \end{pmatrix}.$$

From (7) and (8) we obtain the matrix equation

$$A(s) \begin{pmatrix} x'(s) & I \\ t'(s) & 0 \end{pmatrix} = \begin{pmatrix} 0 & H_x(y(s)) \\ 1 & x'(s) \end{pmatrix}.$$

By passing to determinants we obtain the important relation

$$t'(s) \det A(s) = \det H_x(y(s)). \tag{9}$$

In the following let $H: \mathbb{R}^{n+1} \to \mathbb{R}^n$ be C^1. Moreover, let $s \mapsto y(s)$ be a solution curve C which satisfies (8), and let S denote the solution set of the equation $H(y) = 0$, $y \in \mathbb{R}^{n+1}$.

(L) *Local uniqueness.* If $\det A(s_0) \neq 0$, then S has no bifurcation point at $y(s_0)$ (Fig. 78.4).

We prove this. We may, eventually, after relabeling the variables, assume that $t'(s) \neq 0$. From (9) it follows that $\det H_x(y(s_0)) \neq 0$. The implicit function theorem (Theorem 4.B) then implies (L).

(C) *Constancy principle.* If C is a regular solution curve, then $\det A(s) \neq 0$ along C, i.e., the sign of this determinant remains constant as a result of continuity.

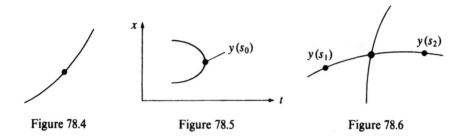

Figure 78.4   Figure 78.5   Figure 78.6

This follows from rank $H'(y(s)) = n$ and $H'(y(s))y'(s) = 0$ with $y'(s) \neq 0$, i.e., rank $A(s) = n + 1$. Note that $H' = H_y$.

(T) *Turning point principle.* If $\det A(s) \neq 0$ along C, and if the index jumps at s_0, i.e., $\det H_x(y(s))$ changes its sign at s_0, then $y(s_0)$ is a turning point of C (Fig. 78.5). This follows from (9).

(B) *Bifurcation principle.* Let $t'(s) \neq 0$ on $[s_1, s_2]$. If the index jumps, i.e., $\det H_x(y(s))$ has different signs for s_1 and s_2, then for the solution set S there exists a bifurcation point on C which lies between the curve points $y(s_1)$ and $y(s_2)$ (Fig. 78.6).

In particular, $y(s_0)$ is a bifurcation point if $\det A(s)$ changes its sign at s_0 and $t'(s_0) \neq 0$.

The first statement follows from the index jump principle (Proposition 15.1). For the second statement note that according to (9), the index $\det H_x(y(s))$ changes its sign at s_0.

78.4. Curve Following Algorithm

From a numerical point of view it is important to have algorithms available which follow the solution curve $s \mapsto y(s)$ of equation (8) very closely. Let us describe here the basic idea for such an algorithm (Fig. 78.7).

We start at a curve point $y(s)$ for fixed s and want to compute a new curve point $y(s_1)$.

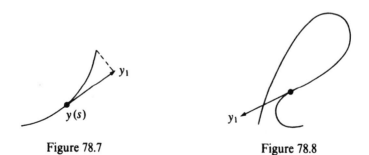

Figure 78.7   Figure 78.8

(i) *Predictor step.* We determine a solution h of
$$H_y(y(s))h = 0, \quad \langle h|h \rangle = 1.$$
We choose the sign of h such that $\det A(s) > 0$ if we replace $y'(s)$ in $A(s)$ with h. This implies that the point
$$y_1 = y(s) + \tau h$$
lies on the tangent line to the curve for fixed $\tau > 0$.

(ii) *Corrector step.* We choose y_1 as a starting point for a Newton-like method:
$$H_y(y_1)(y_{k+1} - y_k) = H(y_k), \quad k = 1, 2, \ldots,$$
$$\langle h|y_{k+1} - y_k \rangle = 0.$$
Under favorable conditions the sequence (y_k), computed this way, then converges to a curve point. In dangerous situations like in Figure 78.8, one eventually may have to choose a smaller parameter τ. This is discussed more thoroughly in Allgower and Georg (1980, S), p. 42. In order to avoid the computation of the inverse matrix "$H_y(y_1)^{-1}$," one can use a quasi-Newton procedure, where "$H_y(y_1)^{-1}$" is approximated by matrices which only depend on the values of H. An effective algorithm may be found in Georg (1981).

78.5. Constructive Leray–Schauder Principle

With a typical example we shall try to explain the applicability of the regular solution curves of Section 78.2 to constructive fixed-point theory. We assume:

(H1) The map $F: \mathbb{R}^n \to \mathbb{R}^n$ is C^∞ with zero as a regular value. We set
$$H(x, t) = (1 - t)(x - x_0) + tF(x).$$

(H2) There exists a bounded region G in $\mathbb{R}^n$ and a point $x_0 \in G$ such that equation $H(x, t) = 0$ has no solution in $\partial G \times [0, 1]$.

Proposition 78.7. *If (H1), (H2) hold, then equation $F(x) = 0$, $x \in G$ has a solution x_1.*

If $F: \bar{G} \to \mathbb{R}^n$ is only continuous and (H2) is satisfied, then $F(x) = 0$, $x \in G$ has a solution as well.

PROOF.

(I) Generic case. Assume (H1), (H2). We set $Z = G \times \,]0, 1[\,$ and denote the covering surfaces and the lateral surface of the cylinder by $Z_k = G \times \{k\}$ and $M = \partial G \times [0, 1]$. Here we have $k = 0, 1$ (Fig. 78.9(a)). If x_0 is only perturbed slightly so that (H2) remains valid, then we may assume from

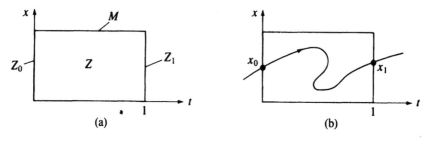

Figure 78.9

Example 78.5 that a regular solution curve C of the equation

$$(1 - t)(x - x_0) + tF(x) = 0, \quad (x, t) \in \mathbb{R}^{n+1} \tag{10}$$

passes through $(x_0, 0)$.

According to the door-in/door-out principle (Proposition 78.6), C enters the region Z through Z_0. Since $(x_0, 0)$ is the only solution of (10) for $t = 0$ and (H2) is satisfied, C cannot leave the region Z through either Z_0 or M. Hence C exits through Z_1 (Fig. 78.9(b)).

(II) Approximation argument for the weaker assumption. If $F: \bar{G} \to \mathbb{R}^n$ is continuous, then we uniformly approximate F on $\bar{G}$ with C^∞-functions $F_k: \mathbb{R}^n \to \mathbb{R}^n$ according to the Weierstrass approximation theorem, i.e., the components of F_k are polynomials and $F_k \rightrightarrows F$ on $\bar{G}$ as $k \to \infty$. By passing from F_k to $F_k - y_k$ we can assume that zero is a regular value of $F_k - y_k$ and $y_k \to 0$ as $k \to \infty$. Then $F_k - y_k$ satisfies assumptions (H1), (H2). We obtain a solution

$$F_k(x_k) = 0, \quad x_k \in G.$$

Since the set $\bar{G}$ is compact, there exists a convergent subsequence $x_{k'} \to x$. Thus we obtain $F(x) = 0$, $x \in \bar{G}$ as $k \to \infty$. But from (H2) it follows that $x \in G$. □

78.6. Constructive Approach for the Fixed-Point Index and the Mapping Degree

In Chapter 12 we gave an elementary introduction to the fixed-point index and the mapping degree. The only tools we used were the integral theorem of Gauss and the substitution rule for multiple integrals. Here we present an approach which is of great geometrical intuition, but uses some advanced methods. The simple geometric idea is contained in Figure 78.10.

If one uses this approach in lectures, one might choose the following presentation.

(a) One introduces the following concepts: Manifold in $\mathbb{R}^n$, regular value (Definition 4.52), preimage theorem (Theorem 4.J), Sard's theorem (Prop-

78.6. Constructive Approach for the Fixed-Point Index and the Mapping Degree

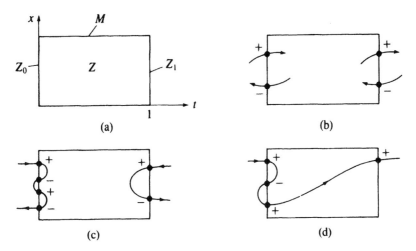

Figure 78.10

osition 4.55), and the structure theorem about one-dimensional manifolds (Proposition 73.9).

One might present the last two results at first without proof, in order not to exhaust the audience with long proofs.

(b) One discusses Sections 78.1 and 78.2.
(c) One formulates the axioms for the fixed-point index as in Sections 12.2 and 12.3 and adds the following observations.

As in Chapter 12 let $V_0(G, \mathbb{R}^n)$ denote the set of all functions $f: \bar{G} \to \mathbb{R}^n$ which satisfies the following properties:

(i) The set G is open and bounded in $\mathbb{R}^n$.
(ii) f is continuous on $\bar{G}$ and is C^1 on G.
(iii) f has at most finitely many fixed points, all of which are regular and do not lie on the boundary ∂G.

Let $V(G, \mathbb{R}^n)$ denote the set of all continuous functions $f: \bar{G} \to \mathbb{R}^n$ which have no fixed points on ∂G. We set $C^\infty(\mathbb{R}^n) = C^\infty(\mathbb{R}^n, \mathbb{R}^n)$.

Definition 78.8 (Fixed-Point Index $i(f, G)$). For maps $f \in V_0(G, \mathbb{R}^n) \cap C^\infty(\mathbb{R}^n)$ we let

$$i(f, G) = \sum_{j=1}^m \operatorname{sgn} \det F'(x_j),$$

where $F(x) = x - f(x)$, and $x_1, \ldots, x_m$ are precisely all fixed points of f on G. If f has no fixed points on G, then let $i(f, G) = 0$.

For $f \in V(G, \mathbb{R}^n)$ and $G \neq \emptyset$ we choose a map

$$\tilde{f} \in V_0(G, \mathbb{R}^n) \cap C^\infty(\mathbb{R}^n),$$

which has the approximation property
$$\sup_{x \in \partial G} \|f(x) - \bar{f}(x)\| \leq 3^{-1} \inf_{x \in \partial G} \|f(x) - x\|, \tag{11}$$

and let
$$i(f, G) = i(\bar{f}, G). \tag{12}$$

For $G = \emptyset$ let $i(f, G) = 0$.

The mapping degree is defined as
$$\deg(F, G, y) = i(f + y, G).$$

In this definition two points have to be justified: There always exists an $\bar{f}$ with (11), and $i(f, G)$ in (12) is independent of the choice of $\bar{f}$.

Proposition 78.9 (Existence of the Fixed-Point Index). *For every map $f \in V(G, \mathbb{R}^n)$ and all $V(G, \mathbb{R}^n)$ with arbitrary $n \in \mathbb{N}$ there exists precisely one fixed-point index which satisfies axioms (A1) to (A4) of Section 12.3. It is given by Definition 78.8.*

PROOF. The key formula is (14) below. The simple geometric proof idea can be obtained from Figure 78.10.

(I) Generic case. Let $f_0, f_1 \in V_0(G, \mathbb{R}^n) \cap C^\infty(\mathbb{R}^n)$ with $G \neq \emptyset$ and
$$\sup_{x \in \partial G} \|F_0(x) - F_1(x)\| < \inf_{x \in \partial G} \|F_0(x)\|, \tag{13}$$
where $F_j(x) = x - f_j(x)$. We want to show that
$$i(f_0, G) = i(f_1, G). \tag{14}$$

For this we let $Z = G \times \,]0, 1[$ and $Z_j = G \times \{j\}$ and also $M = \partial G \times [0, 1]$ (Fig. 78.10(a)). The main trick is to construct the homotopy
$$H(x, t) = (1 - t)F_0(x) + tF_1(x) - y$$
with $y \in \mathbb{R}^n$. Because of $F_0(x) \neq 0$ on ∂G and (13) we have
$$H(x, t) \neq 0 \quad \text{on } M \tag{15}$$
for small $\|y\|$. According to Sard's theorem, we can choose y such that zero is a regular value of H. From Proposition 78.2 it follows that the solution set of the equation
$$H(x, t) = 0, \quad (x, t) \in \mathbb{R}^{n+1} \tag{16}$$
is regular.

(I-1) In order to simplify the notations, let us first assume that $y = 0$. Let C be a regular solution curve of (16) which intersects Z_k for $k = 0$ or $k = 1$. According to the door-in/door-out principle (Proposition 78.6) the curve C enters the region Z. Because of (15), the set M contains no points of C. Thus we have precisely two possibilities for C.

78.6. Constructive Approach for the Fixed-Point Index and the Mapping Degree

(i) C does not reach the other side Z_m with $m \neq k$ and leaves Z through Z_k (Fig. 78.10(c)).

(ii) C reaches the other side Z_m (Fig. 78.10(d)).

We want to count the points of intersection, whereby we attach a sign to each of them. For this we define the *intersection number*

$$j(x, k) = \operatorname{sgn} \det H_x(x, k)$$

of the point of intersection (x, k) of the curve C with Z_k. Let s be the arclength of C. We choose an orientation of C such that $\det A(s) > 0$ along C. This is possible according to the constancy principle (C) of Section 78.3. From (9) it follows that

$$\operatorname{sgn} t'(s) = \operatorname{sgn} \det H_x(x(s), t(s)).$$

Thus j has the sign as shown in Figure 78.10(b). The definition of the fixed-point index implies that

$$i(f_k, G) = \sum_x j(x, k). \tag{17}$$

One sums over all solution points in the set Z_k, i.e., over all solutions x of $F_k(x) = 0$, $x \in G$. Through formula (17) the fixed point index is given a very intuitive geometrical interpretation.

Now (14) follows immediately from Figures 78.10(c) and 78.10(d):

(a) Suppose the curve C is of type (i). Then there exists an even number of intersection points on Z_k and the intersection numbers have pairwise different sign, i.e., they do not contribute to (17). Hence $i(f_0, G) = i(f_1, G) = 0$ (Fig. 78.10(c)).

(b) Suppose the curve C is of type (ii). Then there exists the same number of intersection points on Z_0 and Z_1 with equal sign in case an even number of intersection points with pairwise different sign is not accounted for. Hence $i(f_0, G) = i(f_1, G)$ (Fig. 78.10(d)).

(I-2) Now consider the case $y \neq 0$. As above, we obtain that

$$i(f_0 - y, G) = i(f_1 - y, G).$$

Note that if $f_k \in V_0(G, \mathbb{R}^n)$, then also $f_k - y \in V_0(G, \mathbb{R}^n)$ for sufficiently small $\|y\|$.

In fact, if $F_k(x) = 0$, $x \in G$ with $F_k(x) = x - f_k(x)$, then

$$\det F_k'(x) \neq 0.$$

Thus F_k is a local C^∞-diffeomorphism at x, and hence $F_k - y$ and F_k have the same finite number of zeros on G with equal sign of the Jacobian for sufficiently small $\|y\|$.

This argument also shows that

$$i(f_k - y, G) = i(f_k, G), \qquad k = 0, 1, \text{ i.e., (14) holds.}$$

(II) Existence of approximations. Let $f \in V(G, \mathbb{R}^n)$. On $\bar{G}$ the function f can then arbitrarily close be approximated with an element $g \in V(G, \mathbb{R}^n)$, whereby the components of g are polynomials. Sard's theorem implies that we can choose an element $y \in \mathbb{R}^n$ with arbitrarily small $\|y\|$ such that zero is a regular value of $\bar{f} = g - y$. From the inverse function theorem (Theorem 4.F) it follows that $\bar{f} \in V_0(G, \mathbb{R}^n)$.

(III) Fixed-point index for $f \in V(G, \mathbb{R}^n)$. If we choose $\bar{f} = f_0, f_1$ in $V_0(G, \mathbb{R}^n)$ with (11), then we obtain (13). Moreover, it follows from (14) that

$$i(f_0, G) = i(f_1, G).$$

This justifies Definition 78.8. Note that $i(f, G) = i(\bar{f}, G)$.

(IV) As in Section 12.6 one shows that axioms (A1) to (A4) are satisfied and also that the fixed-point index is unique. □

Now we can extend this fixed-point to B-spaces in analogy to Section 12.7. This provides all the tools needed to develop the theory of the fixed-point index and the mapping degree of Chapters 13–17 of Part I.

78.7. Parametrized Version of Sard's Theorem

Proposition 78.10. *Let the map*

$$H: M \times P \subseteq \mathbb{R}^m \times \mathbb{R}^r \to \mathbb{R}^n$$

be $C^k, k > \max(0, m - n)$, *where M and P are open sets in $\mathbb{R}^m$ and $\mathbb{R}^r$, respectively.*

If y is a regular value of H, then y is also a regular value of $H(\cdot, p)$ for almost all parameter values $p \in P$.

The significance of this theorem has already been discussed in Section 4.19. An important generalization to Banach manifolds may be found in Section 78.10.

PROOF. Let $H = H(x, p)$ with $x \in M$, $p \in P$. From the preimage theorem (Theorem 4.J) it follows that $H^{-1}(0)$ is a C^k-manifold in $\mathbb{R}^m \times \mathbb{R}^r$. Let $H^{-1}(0) \neq \emptyset$. Then we have

$$\dim H^{-1}(0) = m + r - n.$$

Let

$$\pi: H^{-1}(0) \to P, \qquad \pi(x, p) = p$$

denote the natural projection. Also, let $TH^{-1}(0)$ denote the tangent space to $H^{-1}(0)$ at the fixed point (x, p). According to Sard's theorem almost all $p \in P$ are regular values of π. Note that

$$\dim H^{-1}(0) - \dim P = m - n$$

and $k > \max(0, m - n)$. Let p be such a regular value with $\pi^{-1}(p) \neq \emptyset$. Then the linearization

$$T\pi(x, p): TH^{-1}(0) \to \mathbb{R}^r$$

is surjective. This is equivalent to saying that for each $q \in \mathbb{R}^r$ there exists an $x(q) \in \mathbb{R}^m$ with

$$H'(x, p)(x(q), q) = 0. \tag{18}$$

Note that $TH^{-1}(0)$ is given by an equation of the form (18). Equation (18) means

$$H_x(x, p)x(q) + H_p(x, p)q = 0 \quad \text{for all} \quad q \in \mathbb{R}^r. \tag{19}$$

Since zero is a regular value of H, we obtain that $R(H'(x, p)) = \mathbb{R}^n$. For the Jacobian matrix we get

$$H'(x, p) = (H_x(x, p), H_p(x, p)).$$

The linear dependence relation (19) shows that

$$\operatorname{rank} H'(x, p) = \operatorname{rank} H_x(x, p)$$

and hence $R(H_x(x, p)) = \mathbb{R}^n$. Thus, zero is a regular value of $x \mapsto H(x, p)$. □

78.8. Theorem of Sard–Smale

In this section we extend Sard's theorem to Banach manifolds. This provides us with a basic tool to generalize differential topological theorems on finite-dimensional manifolds to Banach manifolds. An important example in this direction will be considered in Section 78.10.

Recall that a set is called meager or of the first Baire category if and only if it is the union of at most countably many nowhere dense sets.

Definition 78.11. A set in a topological space is called *residual* (or *massive*) if and only if it can be represented as a countable intersection of open, dense subsets.

In particular, every open and dense subset is residual.

Theorem 78.A (Smale (1965)). *Let M and N be C^∞-Banach manifolds with chart spaces over $\mathbb{K}$, where M has a countable basis. If*

$$f: M \to N$$

is a C^k-Fredholm map with

$$k > \max(\operatorname{ind} f'(x), 0) \quad \textit{for all} \quad x \in M,$$

then the set of singular values of f is meager, and the set of regular values is residual.

A topological space is a Baire space if and only if all its residual sets are dense. According to $A_1(66)$, every B-space and every Banach manifold is a Baire space. Since subsets of meager sets are meager themselves and N behaves locally like a B-space, the topological results of $A_1(65)$ and $A_1(66)$ imply moreover: The set of regular values of f is dense in N and not meager, i.e., of the second Baire category.

For the following statement, we do not need that M has a countable basis. Instead, we assume that f is proper.

Corollary 78.12. *Let M and N be C^∞-Banach manifolds with chart spaces over $\mathbb{K}$. If $f: M \to N$ is a proper C^k-Fredholm map with $k > \max(\text{ind } f'(x), 0)$ for all $x \in M$, then the set of regular values of f is open and dense in N.*

Note that, for connected M, the index ind $f'(x)$ is constant on M. The simplest example of a C^∞-Banach manifold with chart spaces over $\mathbb{K}$ are the open sets M in a B-space X over $\mathbb{K}$. If X is separable, then M has a countable basis.

78.9. Proof of Theorem 78.A

By using the local normal form of Section 73.1 we transform Sard's theorem into infinite dimensions. Since Banach manifolds behave locally like B-spaces, we begin by studying the local properties. The global properties can then be easily deduced. Our assumption is:

(H) X and Y are B-spaces over $\mathbb{K}$ and $f: V(x_0) \subseteq X \to Y$ is a C^k-Fredholm map with $k > \max(\text{ind } f'(x_0), 0)$. Here, $V(x_0)$ is a neighborhood of x_0.

Lemma 78.13. *If (H) holds, then there exists an open neighborhood $W(x_0)$ in X such that the regular values of the restriction $f|_{W(x_0)}$ are dense in Y.*

PROOF. We make essential use of the normal form (73.1). Let $N = N(f'(x_0))$ and $R = R(f'(x_0))$. We choose topological direct sums

$$X = N \oplus N^\perp \quad \text{and} \quad Y = R \oplus R^\perp.$$

From Proposition 73.1 there exists a C^k-diffeomorphism

$$\varphi: U(0) \subseteq N \times R \to W(x_0),$$

such that the relation

$$h(u, v) = f(x_0) + v + g(u, v) \quad \text{on } U(0) \tag{20}$$

holds for $h(u, v) \overset{\text{def}}{=} f(\varphi(u, v))$, with $u \in N$, $v \in R$, and $g(u, v) \in R^\perp$ on $U(0)$.

The dimensions of N and $R^\perp$ are finite, because $f'(x_0)$ is a Fredholm

78.9. Proof of Theorem 78.A

operator. Since regular values are invariant under diffeomorphisms it suffices to show that the regular values of h are dense in Y.

Let $y \in Y$. We decompose

$$y = f(x_0) + y_1 + y_2 \quad \text{with} \quad y_1 \in R, \quad y_2 \in R^\perp$$

and let $\psi(u) = g(u, y_1)$. Then $\psi: V(0) \subseteq N \to R^\perp$ is C^k. Letting

$$m = \dim N - \dim R^\perp,$$

we obtain $m = \operatorname{ind} f'(x_0)$ and $k > \max(m, 0)$ from (H).

(I) According to Sard's theorem, the regular values of ψ are dense in $R^\perp$.

(II) We show: If y_2 is a regular value of ψ, then y is a regular value of h. Indeed, from $h(u, v) = y$ it follows that $v = y_1$ and $\psi(u) = y_2$. Moreover, we have

$$h'(u,v)(\bar u, \bar v) = \bar v + \psi'(u)\bar u + g_v(u, y_1)(0, \bar v).$$

Therefore, the surjectivity of $\psi'(u): N \to R^\perp$ implies the surjectivity of

$$h'(u,v): N \times R \to Y.$$

From (I) and (II) it follows that the regular values of h are dense in Y. □

Lemma 78.14. *If* (H) *holds, then* f *is locally closed, i.e.,* f *maps closed sets in a sufficiently small neighborhood of* x_0 *onto closed sets.*

PROOF. It suffices to show that h is closed. Let

$$h(u_n, v_n) \to w \quad \text{as} \quad n \to \infty,$$

where $u_n \in N$ and $v_n \in R$ for all n. We decompose

$$w = f(x_0) + w_1 + w_2, \quad w_1 \in R, w_2 \in R^\perp.$$

From (20) it follows that $v_n \to w_2$ as $n \to \infty$. Since $\dim N < \infty$ we have that $u_n \to u$, eventually after passing to a subsequence. This implies that $h(u, v) = w$. □

Lemma 78.15. *If* (H) *holds and* x_0 *is a regular value of* f, *then there exists a neighborhood of* x_0 *which contains only regular points of* f.

PROOF. This follows from normal form (20) with $R = Y$ and $g \equiv 0$. Note that

$$h'(u,v)(\bar u, \bar v) = \bar v \quad \text{for all} \quad \bar v \in Y$$

and all $(u,v) \in U(0)$, i.e., $h'(u,v)$ is surjective for these points (u,v). □

Corollary 78.16. *If* (H) *holds, then there exists an open neighborhood* $U(x_0)$ *in* X *such that the set of singular values of the restriction* $f|_{U(x_0)}$ *is closed in* Y *and the set of regular values is open and dense in* Y. *Thus the set of singular values is nowhere dense in* Y.

This follows immediately from the previous lemmata. Note that according to Lemma 78.15 the set of regular points is open and hence the complement of the set of regular points is closed.

PROOF OF THEOREM 78.A. For each point $x \in M$ we choose an open neighborhood $U(x)$ such that Corollary 78.16 holds in charts. Since M has a countable basis, it follows that M is Lindelöf, i.e., at most countably many $U(x)$ cover M. One observes then that y is a singular (regular) value of f if and only if y is a singular (regular) value for one of the (each of the) restrictions $f|_{U(x)}$. □

PROOF OF COROLLARY 78.12. The map

$$f: M \to N$$

is proper. Therefore $f^{-1}(y)$ is compact and f is closed (proof as in Proposition 4.44).

(I) The set $\text{Reg}(f)$ of regular values of f is open in N, because Lemma 78.15 implies that the set S of singular points of f is closed in M, and hence $f(S)$ is closed and $\text{Reg}(f) = N - f(S)$ is open.
(II) Let U be an open neighborhood of $f^{-1}(y)$. Then there exists a neighborhood $V(y)$ with $f^{-1}(V(y)) \subseteq U$. Otherwise there exists a convergent sequence $f(x_n) \to y$ as $n \to \infty$ with $x_n \notin U$ for all n. Since f is proper, we may assume that, eventually after passing to an M–S-subsequence, $x_n \to x$. This yields the desired contradiction $x \notin U$ and $f(x) = y$.
(III) The set $\text{Reg}(f)$ is dense in N. In order to prove this, we choose the neighborhood $V(y)$ so small that it lies in a chart of y. It suffices to show that $\text{Reg}(f)$ is dense in $V(y)$.

Let C and C_U be the set of singular values of $f: M \to N$ and $f|_U$ in $V(y)$, respectively. We choose $U(x)$ as in the proof of Theorem 78.A, that is, $f(U(x)) \subseteq V(y)$ for all $x \in f^{-1}(y)$. Let U be the union of finitely many $U(x)$, which already cover the compact set $f^{-1}(y)$. It follows then, as in the proof of Theorem 78.A, that C_U is meager in $V(y)$. From (II) it follows that $C = C_U$, eventually after decreasing $V(y)$.

Therefore C is meager in $V(y)$, and hence $V(y) - C$ is dense in $V(y)$ (see $A_1(65)$), i.e., $\text{Reg}(f)$ is dense in $V(y)$. □

78.10. Parametrized Version of the Theorem of Sard–Smale

Consider the operator equation

$$H(x, p) = z, \quad x \in G, \tag{21}$$

78.10. Parametrized Version of the Theorem of Sard–Smale

which depends on the parameter $p \in P$. Fix a point $z \in Z$. Our goal is to find conditions under which the solutions have a natural and favorable behavior for "most" parameter values p. As an application of the results of this section we will give conditions for which (21) has only finitely many solutions for "most" p. This will be done in the following section. It need not be emphasized that results of this type are of great mathematical and scientific interest.

Our assumptions are:

(H1) G, P, and Z are nonempty, metrizable C^∞-Banach manifolds with chart spaces over $\mathbb{K}$.

This condition is satisfied, for example, if G, P, and Z are open and nonempty sets in B-spaces over $\mathbb{K}$.

(H2) The C^k-map $H: G \times P \to Z$ with $k \geq 1$ has z as a *regular value*.
(H3) For each parameter $p \in P$, the map $H(\cdot, p): G \to Z$ is a Fredholm map, where

$$\text{ind } H_x(x, p) < k$$

for every solution $(x, p) \in G \times P$ of (21).

If G and Z are open sets in B-spaces, then in the usual way $H_x(x, p)$ denotes the partial F-derivative. In the general case, $H_x(x, p)$ is the tangent map of $H(\cdot, p): G \to Z$ at the point x.

(H4) Weak properness. The convergence $p_n \to p$ on P as $n \to \infty$ and

$$H(x_n, p_n) = z \quad \text{for all } n$$

implies the existence of a convergent subsequence $x_{n'} \to x$ as $n \to \infty$ with $x \in G$.

Let p be fixed. Recall that (H3) implies that a solution of (21) is regular if and only if the linearization $H_x(x, p): TG_x \to TZ_z$ is surjective. For the special case that Z is a B-space and G is an open set in the B-space X, we have $TG_x = X$ and $TZ_z = Z$.

Theorem 78.B (Parametrized Version of the Theorem of Sard–Smale). *If (H1)–(H4) hold, then there exists an open, dense subset P_0 of P such that z is a regular value of $H(\cdot, p)$ for each parameter $p \in P_0$.*

Corollary 78.17. *Fix an element $p \in P_0$. If there exists a number $n \geq 0$ with*

$$\text{ind } H_x(x, p) = n$$

for all solutions x of (21), then the solution set of (21) consists of an n-dimensional C^k-Banach manifold or the solution set is empty.

We prove Theorem 78.B in Section 78.12. Corollary 78.17 follows immediately from Theorem 78.B and the preimage theorem (Theorem 73.C).

78.11. Main Theorem About Generic Finiteness of the Solution Set

Theorem 78.C (Main theorem). *If* (H1)–(H4) *of the previous section hold with* $k = 1$, *and if*

$$\text{ind } H_x(x, p) = 0 \tag{22}$$

for all solutions $(x, p) \in G \times P$ *of equation* (21), *then there exists an open dense subset* P_0 *of* P *such that* (21) *has at most finitely many solutions for each fixed parameter* $p \in P_0$.

In addition, all these solutions are regular.

PROOF. We use Theorem 78.B. Let S be the set of solutions of

$$H(x, p) = z, \quad x \in G$$

for fixed $p \in P_0$. From (H4) it follows that S is compact. Theorem 78.B implies the surjectivity of $H_x(x, p)$: $TG_x \to TZ_z$ for all $x \in S$. Because of (22) this map is even bijective. The inverse mapping theorem (Theorem 73.B) implies that S consists of isolated points and from compactness it follows that it consists of at most finitely many points. □

An important version of Theorem 78.B, for which the compactness condition (H4) is not needed, will be considered in Problem 78.3.

78.12. Proof of Theorem 78.B

We make essential use of a simple result about linear operators. The starting point is the linear equation

$$Ax + Bp = z_0, \quad x \in X, \quad p \in Y \tag{23}$$

for a given $z_0 \in Z$. We set

$$D = \{(x, p) \in X \times Y : Ax + Bp = 0\}$$

and define the projection operator $Q: D \to Y$ as

$$Q(x, p) = p.$$

(L1) X, Y, and Z are B-spaces over $\mathbb{K}$.
(L2) The operators $A: X \to Z$ and $B: Y \to Z$ are linear and continuous, where A is a Fredholm operator.
(L3) Equation (23) has a solution for every $z_0 \in Z$.

Lemma 78.18. *Under the assumptions* (L1)–(L3) *we have:*

(i) *The operator* $Q: D \to Y$ *is Fredholm with* $\text{ind } Q = \text{ind } A$.
(ii) Q *is surjective if and only if* A *is surjective.*

78.12. Proof of Theorem 78.B

PROOF. We have

$$Q(x,p) = 0 \Leftrightarrow p = 0, Ax = 0.$$

This implies $\dim N(Q) = \dim N(A)$. We are done if we can show that

$$\operatorname{codim} R(Q) = \operatorname{codim} R(A).$$

From the definition of Q it follows that

$$R(Q) = B^{-1}(R(A)).$$

We choose linear subspaces Y_0 and Z_0 of Y and Z which induce the direct (algebraic) sum decompositions

$$Y = N(B) \oplus Y_0, \qquad Z = R(A) \oplus Z_0$$

(see $A_1(22k)$). Let the operator $B_0 \colon Y_0 \to R(B)$ be the restriction of B onto Y_0. Then B_0 is bijective. Thus we have

$$B^{-1}(R(A)) = N(B) \oplus B_0^{-1}(R(A)).$$

Because of (L3) we have $Z_0 \subseteq R(B)$, and hence

$$Y_0 = B_0^{-1}(R(B)) = B_0^{-1}(R(A)) \oplus B_0^{-1}(Z_0)).$$

This gives

$$Y = N(B) \oplus B_0^{-1}(R(A)) \oplus B_0^{-1}(Z_0)$$
$$= R(Q) + B_0^{-1}(Z_0).$$

Therefore

$$\operatorname{codim} R(A) = \dim Z_0 = \dim B_0^{-1}(Z_0)$$
$$= \operatorname{codim} R(Q). \qquad \square$$

Now we prove Theorem 78.B. We proceed as in the finite-dimensional case of Section 78.7. Instead of Sard's theorem we apply the theorem of Sard–Smale to the projection operator π.

For simplicity in notation, we assume that G and P are open, nonempty sets in the B-spaces X and Y, respectively, and that Z is a B-space. The proof of the general case is analogous, since Banach manifolds look locally like open sets in B-spaces.

Step 1: Solution manifold $M = H^{-1}(z)$.
Let M denote the set of all $(x,p) \in G \times P$ with

$$H(x,p) = z.$$

Because of (H2) and the preimage theorem (Theorem 73.C) it follows that M is a C^k-manifold. The tangent space TM_u at the point $u = (x_0, p_0)$ precisely consists of all points $(x,p) \in X \times Y$ with $H'(u)(x,p) = 0$, i.e,

$$H_x(u)x + H_p(u)p = 0. \qquad (24)$$

From (H2) it follows that the corresponding inhomogeneous equation $H'(u)(x, p) = z_0$, i.e.,

$$H_x(u)x + H_p(u)p = z_0, \quad (x, p) \in X \times Y$$

has a solution for each $z_0 \in Z$.

Step 2: Nonlinear projection operator $\pi: M \to P$.
We define

$$\pi(x, p) = p. \tag{25}$$

Moreover, for fixed u we let $D = TM_u$ and define the linear projection operator $Q: D \to Y$ through

$$Q(x, p) = p. \tag{26}$$

Then we have

$$Q = \pi'(u). \tag{27}$$

This follows because for each tangent vector $(x, p) \in D$ there exists a curve

$$t \mapsto (x(t), p(t)) \quad \text{on } M$$

with $x(0) = x_0$, $p(0) = p_0$ and $x'(0) = x$, $p'(0) = p$. If we insert this curve into equation (25), then we obtain (27) by differentiation.

Step 3: Application of the theorem of Sard–Smale to π.
From (H4) it follows that the operator

$$\pi: M \to P$$

is proper. To see this let P_1 be a compact set in P. If $\{(x_n, p_n)\}$ is a sequence in $\pi^{-1}(P_1)$, then

$$H(x_n, p_n) = z.$$

Since P_1 is compact there exists a convergent subsequence $p_{n'} \to p$ with $p \in P_1$. From (H4) follows the existence of a convergent subsequence $x_{n''} \to x$ with $x \in G$. Therefore $H(x, p) = z$, i.e., $(x, p) \in M$. Hence $\pi^{-1}(P_1)$ is compact. Note that because of the metrizability of G and P, compactness can be characterized through sequences (see A_1(21c)).

Moreover, it follows from Lemma 78.18 and (24) that the operator $Q: D \to Y$ is Fredholm with $\text{ind } Q = \text{ind } H_x(u)$. Thus we obtain from (27) that $\pi: M \to P$ is a C^k-Fredholm operator with

$$\text{ind } \pi'(u) = \text{ind } H_x(u).$$

According to Corollary 78.12 there exists an open, dense subset P_0 of P such that each $p_0 \in P_0$ is a regular value of π.

Let $p_0 \in P_0$. Then

$$\pi'(u): TM_u \to Y$$

is surjective for every $u = (x_0, p_0)$ in M, i.e., $Q: D \to Y$ is surjective. Lemma 78.18 shows that also

$$H_x(u): X \to Z$$

is surjective. Therefore z is a regular value of $H(\cdot, p_0)$. This proves Theorem 78.B.

PROBLEMS

78.1. *Numerical construction of bifurcation solutions using the perturbation trick.* Consider the situation of Section 78.3. Let $H: \mathbb{R}^{n+1} \to \mathbb{R}^n$ be a C^∞-map and let $s \mapsto y(s)$ be a solution curve C of the equation

$$H(y(s)) = 0. \tag{28}$$

Assume that det $A(s)$ changes its sign at s_0 and that $t'(s_0) \neq 0$. Then $y(s_0)$ is a bifurcation point of (28) (Fig. 78.11(a)). Besides (28) we study the perturbed problem

$$H^*(y^*(s), p) = 0 \tag{28*}$$

with $H^*(y, p) \stackrel{\text{def}}{=} H(y) + pf(y)$, where $p \in \mathbb{R}^n$ is fixed and $f: \mathbb{R}^{n+1} \to \mathbb{R}$ is a C^∞-map with

$$f(y) > 0$$

in a small open neighborhood V of the bifurcation point and $f = 0$ outside of V. The following is important:

$$H \text{ and } H^* \text{ coincide outside of } V. \tag{29}$$

Prove that if the bifurcation situation is sufficiently regular, then one can find a regular solution curve C^* of (28*) which runs into the other bifurcation branch of (28) outside of V (Fig. 78.11(b)).

By using the curve following algorithm of Section 78.4, one effectively can compute bifurcation branches. Numerical results and a well-written algorithm may be found in Georg (1981).

Solution: Zero is a regular value of $(y, p) \mapsto H^*(y, p)$ on $V \times \mathbb{R}^n$. This follows from

$$H^{*'}(y, p) = (H_y(y), f(y)I),$$

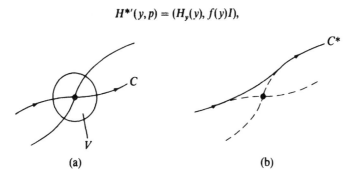

Figure 78.11

and hence $R(H^{*\prime}(y,p)) = \mathbb{R}^n$. From Example 78.4 it follows that for $\varepsilon > 0$ there exists a $p \in \mathbb{R}^n$ with $|p| < \varepsilon$ such that (28*) has a regular solution set. We choose a regular solution curve C^* of (28*) which intersects C in some small neighborhood of the bifurcation point. Then $\det A(s)$ changes its sign along C. But according to (C) of Section 78.3, the sign of $\det A^*(s)$ along C^* remains constant. From (29) it follows that A and A^* coincide outside of V. Thus C^* cannot run into C, but has to run outside of V into the other bifurcation branch.

78.2. Construction of a second solution with the p-trick. Let $F\colon \mathbb{R}^n \to \mathbb{R}^n$ be a C^∞-map which has the following properties:
(i) F has the zero x_0.
(ii) Zero is a regular value of F.
(iii) There exists an open neighborhood $U(x_0)$ and points $v, p \in \mathbb{R}^n$ which satisfy $\langle v|p\rangle > 0$ and $\langle v|F(x)\rangle > 0$ on $\partial U(x_0)$.

Use $H(x, t, p) = (1 - t)p + tF(x)$ to give a constructive proof that F has another zero x_1. It is important here that the zero indices of x_0 and x_1 are different, i.e.,

$$\operatorname{sgn} \det F'(x_0) \operatorname{sgn} \det F'(x_1) = -1.$$

Solution: Let $Z = U(x_0) \times {]}0, 1{[}$. We set $P = (x, t, p)$. Because of

$$H'(P) = (H_x(P), H_t(P), H_p(P)) = (tF'(x), \ldots, (1-t)I)$$

and (ii) it follows that zero is a regular value of H. Example 78.4 shows that, eventually after a small change of p, we may assume that a regular solution curve C of the equation

$$(1 - t)p + tF(x) = 0, \qquad (x, t) \in \mathbb{R}^{n+1}$$

passes through x_0. Because of (i), the curve C enters Z at the point $(x_0, 1)$. From (iii) it follows that $p \neq 0$. Therefore, C cannot leave the region Z on the side $Z_0 = U(x_0) \times \{0\}$. Also from (iii), it follows that C has no common points with $M = \partial U(x_0) \times [0, 1]$. Therefore C has to leave the region Z at a point $(x_1, 1)$ (Fig. 78.12). The statement about the indices follows from the different sign of intersection numbers (see the proof in Section 78.6).

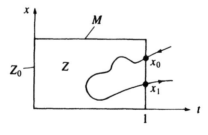

Fig. 78.12

78.3. Another form of the parametrized version of the theorem of Sard–Smale. As in Section 78.10, we consider the equation

$$H(x, p) = z, \qquad x \in G \qquad (30)$$

for fixed $z \in Z$ and assume:

(H1) G, P, and Z are nonempty C^∞-Banach manifolds with chart spaces over $\mathbb{K}$, where G and P have a countable basis.

This condition is satisfied, for example, if G and P are nonempty, open sets in separable B-spaces over $\mathbb{K}$, and Z is a B-space over $\mathbb{K}$.

(H2) The C^k-map $H: G \times P \to Z$, $k \geq 1$, has z as a regular value.

(H3) For each parameter $p \in P$, the map $H(\cdot, p): G \to Z$ is Fredholm with ind $H_x(x, p) < k$ for every solution $(x, p) \in G \times P$ of (30).

Prove: There exists a residual subset P_0 of P such that z is a regular value of $H(\cdot, p)$ for all $p \in P_0$.

Solution: Use similar arguments as in Section 78.12. But instead of Corollary 78.12, use Theorem 78.A of Section 78.8.

Further variants and generalizations may be found in Abraham and Robbin (1967, M), p. 48 (transversal density theorem).

References to the Literature

Classical works: Sard (1942), Smale (1965).

Parametrized version of Sard's theorem and fixed-point theory: Chow, Mallet-Paret, and Yorke (1978).

Theorem of Sard–Smale and Fredholm maps: Abraham and Robbin (1967, M), Tromba (1976), (1978), Borisovič (1977, S, B, H).

Numerical methods: Garcia, Zangwill (1983, M) (introduction), Allgower and Georg (1980, S, H, B), (1980a), (1988, M), Georg (1981), (1981a), Eaves (1982, P), Rheinboldt (1986, M).

(See also the References to the Literature to Chapter 6.)

CHAPTER 79

Dynamical Stability and Bifurcation in B-Spaces

> In case the differential equation can be integrated, the problem of stability presents no difficulty. It is important, however, to find methods which solve the stability problem independently of the integration.
>
> Alexander Mihailovič Ljapunov (1892)

> On passing through $\mu = 0$ let us now assume that none of the characteristic exponents vanishes, but a conjugate pair crosses the imaginary axis. This situation commonly occurs in nonconservative mechanical systems, for example, in hydrodynamics. The following theorem asserts that, with this hypotheses, there is always a periodic solution in the neighborhood of the equilibrium point. In the literature, I have not come across this bifurcation problem. However, I scarcely think that there is anything essentially new in the above theorem. The methods have been developed by Poincaré perhaps 50 years ago, and belong today to the classical conceptual structure of the theory of periodic solutions.[1]
>
> Eberhard Hopf (1942)

> Without the presence of stable phenomena, the world would pass into a state of complete chaos and its apparent structures would dissolve. There is no question that the discovery of interrelations in natural processes and its scientific description requires the existence of stable phenomena.
>
> Herbert Beckert (1977)

Because of its great importance for science and numerical analysis, stability questions have been discussed already in a number of chapters of this volume and the three previous ones. In the present chapter we examine the following two important principles:

(L) *Linearization principle.* The nonlinear differential equation has locally the same stability properties as the linearized differential equation.

---

[1] Hopf was not aware of the papers of the Russian mathematicians Andronov and Bautin from the years 1930 to 1941. Instead of Poincaré–Andronov–Hopf bifurcation, we simply speak of Hopf bifurcation.

(B) *Bifurcation principle.* Loss of stability of an equilibrium point leads to bifurcation.

Counterexamples show that (L) and (B) are not generally true. But the criteria for (L) and (B) which will be discussed below are widely applicable. In connection with (B) we discuss two important cases:

Simple curve bifurcation (Section 79.8).
Hopf bifurcation (Section 79.9).

As in Chapter 8 the central ingredients are transversality conditions (generic bifurcation conditions). In the following two Chapters 80 and 81 of Part V we continue the stability discussions and examine integral manifolds, i.e., manifolds which consist of trajectories (stable and unstable manifolds, center manifolds) as well as the method of Ljapunov functions.

In this chapter we shall use some results in spectral theory which can be found in $A_1(56)$ to $A_1(60)$.

The investigation of bifurcation problems for dynamical systems will be continued in Chapter 80 by using the theory of stable and unstable manifolds.

79.1. Asymptotic Stability and Instability of Equilibrium Points

We examine the differential equation

$$x' = F(x, t). \tag{1}$$

We are looking for a function $x = x(t)$ defined for all times $t \geq t_0$, where $x(t)$ lies in a B-space X. Equation (1) is called *autonomous* if and only if F does not depend on time t.

The point $x_0 \in X$ is called an *equilibrium point* of (1) if and only if

$$F(x_0, t) = 0 \quad \text{for all times} \quad t \geq t_0.$$

Then $x(t) = x_0$ is a solution of (1) for all $t \geq t_0$ and is called a stationary solution, because the system modeled by (1) remains in the same position x_0 for all times $t \geq t_0$. Equilibrium points are also called stationary or singular points. The following definition goes back to the fundamental paper of Ljapunov (1892).

Definition 79.1. Let x_0 be an equilibrium point of (1) for $t \geq t_0$. The point x_0 is called *stable* if and only if small perturbations of the initial condition $x(t_0) = x_0$ lead to solutions which remain in the neighborhood of x_0 for all times $t \geq t_0$. More precisely, for each $\varepsilon > 0$ there exists a $\delta(\varepsilon) > 0$ such that the initial-value problem $x(t_0) = a$ for (1) has a unique solution $x = x(t)$ for each $a \in X$ with

$$\|a - x_0\| < \delta(\varepsilon),$$

which exists for all times $t \geq t_0$. Moreover,
$$\|x(t) - x_0\| < \varepsilon \qquad \text{for all} \quad t \geq t_0.$$
If, in addition, there exists a $\delta_0 > 0$ such that
$$\|x(t_0) - x_0\| < \delta_0$$
implies
$$\lim_{t \to +\infty} x(t) = x_0,$$
then x_0 is called *asymptotically stable*.

The point x_0 is called *unstable* if and only if x_0 is not stable.

Note that the concept of stability implies the existence and uniqueness of the solution. Now, we assume more generally that $y = y_0(t)$ is a solution of the differential equation
$$y' = G(y, t) \tag{2}$$
for $t \geq t_0$. We set $y = y_0 + x$ and obtain
$$x' = F(x, t) \tag{3}$$
with $F(x, t) = G(y_0(t) + x, t) - y_0'(t)$. This implies
$$F(0, t) = 0 \qquad \text{for all} \quad t \geq t_0,$$
i.e., $x_0 = 0$ is an equilibrium point of (3).

Definition 79.2. The solution $y = y_0(t)$ of (2) is called stable, asymptotically stable, unstable if and only if the equilibrium point $x_0 = 0$ of (3) has the corresponding property.

We begin with the autonomous differential equation
$$x' = F(x) \tag{4}$$
for $t \geq t_0$. Let $\sigma(F'(x_0))$ denote the spectrum of $F'(x_0)$. If the B-space X is real, the spectrum corresponds to the complexification of $F'(x_0)$ in $X_\mathbb{C}$ (see A_1(23h)).

Proposition 79.3. Let $F: U(x_0) \subseteq X \to X$ be C^k in a B-space X over $\mathbb{K} = \mathbb{R}, \mathbb{C}$ with $F(x_0) = 0$. Then:

(a) If $\operatorname{Re} \lambda < 0$ for all $\lambda \in \sigma(F'(x_0))$ and if $k = 1$, then the equilibrium point x_0 of (4) is asymptotically stable.
(b) If $\operatorname{Re} \lambda > 0$ for some $\lambda \in \sigma(F'(x_0))$ and if $k = 2$, then x_0 is unstable.

This proposition contains the *linearization principle* of stability theory, i.e., the stability properties of the differential equation at x_0 depend only on the spectrum of the linearization $F'(x_0)$. Letting
$$F(x) = F'(x_0)(x - x_0) + r(x),$$

from Taylor's theorem of Section 4.6 it follows that:

$$r(x) = o(\|x - x_0\|) \quad \text{as} \quad x \to x_0 \quad \text{for} \quad k = 1,$$
$$r(x) = O(\|x - x_0\|^2) \quad \text{as} \quad x \to x_0 \quad \text{for} \quad k = 2.$$

Therefore Proposition 79.3 is a special case of the following Theorem 79.A. Instead of (4) we consider the time-dependent equation

$$x' = A(x - x_0) + f(x, t) \tag{5}$$

for all $t \geq t_0$ with time-independent linear principal part A.

(H1) The operator $A: X \to X$ defined on a B-space X over $\mathbb{K}$ is linear and continuous.
(H2) The map $f: U(x_0) \times [t_0, \infty[\to X$ with $U(x_0) \subseteq X$ is C^1 with $f(x_0, t) = 0$ for all $t \geq t_0$.
(H3) There exist numbers $r, \gamma > 0$ such that one of the following two smallness conditons is satisfied:

$$\lim_{x \to x_0} \|f(x, t)\|/\|x - x_0\| = 0 \quad \text{uniformly for all} \quad t \geq t_0 \tag{6}$$

or stronger

$$\|f(x, t)\| \leq \gamma \|x - x_0\|^q, \quad q > 1 \tag{7}$$

for all x, t with $\|x - x_0\| \leq r$ and $t \geq t_0$.

Theorem 79.A (Ljapunov's Main Theorem of Stability Theory in B-Spaces). *Assume* (H1) *and* (H2).

(a) *If* $\operatorname{Re} \lambda < 0$ *for all* $\lambda \in \sigma(A)$ *and if* (6) *holds, then the equilibrium point* x_0 *of equation* (5) *is asymptotically stable.*
(b) *If* $\operatorname{Re} \lambda > 0$ *for some* $\lambda \in \sigma(A)$ *and if* (7) *holds, then* x_0 *is unstable.*

Ljapunov (1892) proved this result for $X = \mathbb{R}^n$.

79.2. Proof of Theorem 79.A

Without loss of generality let $x_0 = 0$ and $t_0 = 0$. Let $X \neq \{0\}$.

PROOF OF THEOREM 79.A(a).

(I) Estimation of e^{tA}. Let $\mathbb{K} = \mathbb{C}$. Since the spectrum $\sigma(A)$ is compact, there exists a constant $c > 0$ with $\operatorname{Re} \lambda < -c$ for all $\lambda \in \sigma(A)$. We show

$$\|e^{tA}\| \leq Ce^{-ct} \quad \text{for all} \quad t \geq 0 \tag{8}$$

for fixed $C > 0$. In fact, from A_1(60b) it follows that

$$e^{tA} = (2\pi i)^{-1} \oint_{\partial U} e^{tz}(zI - A)^{-1} dz$$

if we choose an open disk U of radius R, which contains $\sigma(A)$ and where $\operatorname{Re}\lambda < -c$ for all $\lambda \in U$. We obtain

$$\|e^{tA}\| \le Re^{-ct}\sup_{z\in \partial U}\|(zI-A)^{-1}\| \quad \text{for all} \quad t\ge 0.$$

In the case of a real B-space X, i.e., for $\mathbb{K}=\mathbb{R}$ we use the complexification of A in $X_\mathbb{C}$. Then (8) holds in $X_\mathbb{C}$ and hence in X.

(II) *Representation formula.* For continuous g one immediately obtains by differentiation that

$$x(t) \stackrel{\text{def}}{=} e^{tA}a + \int_0^t e^{(t-s)A}g(s)\,ds \tag{9}$$

is a solution of the initial-value problem

$$x'(t) = Ax(t) + g(t), \quad x(0) = a. \tag{10}$$

(III) *Gronwall's lemma and an a priori estimate.* Let $x = x(t)$ be a solution of (5) on $[0, T]$. From Theorem 3.A it follows that this solution is unique and satisfies (10) with

$$g(t) = f(x(t), t). \tag{11}$$

Because of (6) there exists a closed ball B around the origin in X with

$$\|f(x,t)\| \le 2^{-1}c\|x\| \quad \text{for all} \quad x\in B,\ t\ge 0.$$

Let $x(t)\in B$ for all $t\in [0, T]$. It follows from (8) and (9) that

$$\|x(t)\| \le Ce^{-tc}\|a\| + 2^{-1}c\int_0^t e^{-c(t-s)}\|x(s)\|\,ds.$$

Gronwall's lemma of Section 3.5 with $f(t) = e^{tc}\|x(t)\|$ yields

$$\|x(t)\| \le Ce^{-ct/2}\|a\|. \tag{12}$$

Thus we have found the following *a priori* estimate:

There exists a neighborhood of zero $V\subset B$ which is independent of T and has the following property. A solution, defined for all $t\in [0, T]$, which starts at time $t=0$ in V and remains in B, does not leave $2^{-1}B$.

(IV) *Global existence of solutions.* Let $x(0)\in V$. To start out, Theorem 3.A implies that the initial-value problem (10) with (11) can be solved locally on $[0, T]$ for small $T > 0$. If one solves the initial-value problem again for $t = T$, then the *a priori* estimate shows that the solution can be continued for all $t \ge 0$. Thereby it remains in B. The asymptotic stability follows then immediately from (12). □

PROOF OF THEOREM 79.A(b). From uniqueness it follows that according to (9) every solution of

$$x' = Ax + f(x, t), \tag{13}$$

79.2. Proof of Theorem 79.A

with $x(0) = \delta b$ can be written as

$$x(t) = y(t) + z(t), \tag{14}$$

with

$$y(t) = \delta e^{tA}b, \quad z(t) = \int_0^t e^{(t-s)A} f(x(s), s)\, ds.$$

We now make essential use of the following two results in spectral theory:

(I) If $\lambda_0 > 0$ is the largest real part of points in the spectrum $\sigma(A)$, then from (8) there exists a constant C with

$$\|e^{tA}\| \leq C e^{(\lambda_0 + q/4)t} \quad \text{for all} \quad t \geq 0.$$

(II) For every $T > 0$ there exists a vector $b \in X$ with $0 < \|b\| \leq 1$ such that

$$4^{-1} e^{\lambda_0 T} \leq \|e^{TA}b\|, \tag{15}$$

$$\|e^{tA}b\| \leq 2 e^{\lambda_0 t} \quad \text{for all} \quad t \in [0, T]. \tag{16}$$

This will be proved in Problem 79.2. Since the numerical value of the constants γ, C, λ_0 is unimportant for the proof, we set $\gamma = C = \lambda_0 = 1$. Also let $q = 2$ in (7). For $q > 1$ one proceeds analogously.

(III) Suppose the equilibrium point $x_0 = 0$ is stable. For $\varepsilon = 10^{-3}$ there exists a $\delta > 0$ such that for every initial value $x(0)$ with $\|x(0)\| \leq \delta$ there exists a unique solution $x = x(t)$ of (13) for all $t \geq 0$ and

$$\|x(t)\| < 10^{-3} \quad \text{for all} \quad t \geq 0. \tag{17}$$

We will, however, construct a b with $\|b\| = 1$ such that the solution which satisfies $x(0) = \delta b$ violates condition (17) at some time $t = T$.

(IV) Construction of the contradiction. Let $R = 2.1$. We choose $T > 0$ such that

$$2 + 2\delta R^2 e^T = R. \tag{18}$$

Moreover, for $T > 0$ we choose b as in (II) and pose the initial-value problem for (13) with $x(0) = \delta b$. Because of the continuity of the solution there exists a τ with $0 < \tau < T$ and

$$\|x(t)\| \leq \delta R e^t \quad \text{for all} \quad t \in [0, \tau].$$

We want to show that τ can be increased up to T. With $q = 2$ it follows from (14) to (16) and (7) that for all $t \in [0, \tau]$:

$$\|y(t)\| \leq \|e^{tA} \delta b\| \leq 2\delta e^t,$$

and

$$\|z(t)\| \leq \int_0^t \|e^{(t-s)A}\| \, \|x(s)\|^2 \, ds \leq \int_0^t e^{3(t-s)/2} \delta^2 R^2 e^{2s} \, ds \leq 2\delta^2 R^2 e^{2t}$$

and hence

$$\|x(t)\| \leq \|y(t)\| + \|z(t)\| \leq \delta(2 + 2\delta R^2 e^\tau)e^t < \delta R e^t.$$

Therefore we can always increase τ for $\tau < T$. Using the continuity of the solution we may also choose $\tau = T$. From (15) and (18) it follows that

$$\|x(T)\| \geq \|y(T)\| - \|z(T)\| \geq 4^{-1}\delta e^T - 2\delta^2 R^2 e^{2T}$$
$$= 2(R-2)(9-4R)/R^2 > 10^{-3}.$$

This contradicts (17). □

79.3. Multipliers and the Fixed-Point Trick for Dynamical Systems

In order to give a unified description of the stability properties for equilibrium points, periodic solutions, and fixed points, we use the concept of multipliers. Those are suitably defined complex numbers.

Definition 79.4. A multiplier $\mu \in \mathbb{C}$ is called *asymptotically stable, critical, unstable* if and only if $|\mu| < 1, = 1, > 1$, respectively.

Critical multipliers usually create the most difficulties. If x_0 is a fixed point of the equation

$$x = Sx, \tag{19}$$

then, by definition, its *multipliers* are precisely the points in the spectrum $\sigma(S'(x_0))$ of the linearization. Theorem 4.C then takes the following form.

Proposition 79.5. *Let $S: U(x_0) \subseteq X \to X$ be C^1 in a B-space X over $\mathbb{K}$. If the fixed point x_0 of S has only asymptotically stable multipliers, then it is attracting.*

The behavior in the hyperbolic case, i.e., when x_0 is allowed to have unstable multipliers but no critical ones, will be studied in Chapter 80. There we will also examine critical multipliers (Center theorem).

If x_0 is an equilibrium point of the autonomous differential equation

$$x' = F(x), \tag{20}$$

then, by definition, its *multipliers* are precisely the points in the spectrum of $\exp F'(x_0)$. Therefore μ is a multiplier of x_0 if and only if there exists a $\lambda \in \sigma(F'(x_0))$ with $\mu = \exp \lambda$. Proposition 79.3 may then be formulated as follows.

79.3. Multipliers and the Fixed-Point Trick for Dynamical Systems

Proposition 79.6. *Let $F: U(x_0) \subseteq X \to X$ be C^2 in a B-space X over $\mathbb{K}$. If the equilibrium point x_0 of (20) has only asymptotically stable multipliers (or at least one unstable multiplier), then x_0 is asymptotically stable (or unstable).*

Critical multipliers in the context of dynamical systems will be discussed in Chapter 80 (Center theorem).

Now we explain an important *fixed-point trick*, through which equation (20) can be reduced to (19). Let $x = x(t)$ be the solution of (20) with $x(0) = a$. We let

$$\Phi_t(a) \stackrel{\text{def}}{=} x(t). \tag{21}$$

This defines the flow $\{\Phi_t\}$ at least locally for all a, t with $\|a - x_0\| < r$ and $-T \le t \le T$. For fixed t we let

$$S = \Phi_t.$$

Then the equilibrium points of (20) are fixed points of the so-called shift operator S. For the linear differential equation

$$x' = Ax \tag{22}$$

with $A \in L(X, X)$ the situation becomes particularly simple. Here we have $\Phi_t = \exp tA$ for all $t \in \mathbb{R}$. Letting $S = \exp A$ we obtain the following result.

Proposition 79.7. *Let $A \in L(X, X)$. The multipliers of the equilibrium point $x_0 = 0$ of (22) are precisely the multipliers of the fixed point $x_0 = 0$ for the shift operator $S = \exp A$.*

We now justify the important formula

$$\Phi_t'(x_0) = \exp tF'(x_0) \quad \text{for all} \quad t \in [-T, T], \tag{23}$$

i.e., the linearization of the flow at x_0 is equal in a natural way to the flow of the linearized differential equation at the point x_0.

Theorem 79.B (Structure of Flows for Autonomous Differential Equations). *Let $F: U(x_0) \subseteq X \to X$ be C^k, $k \ge 1$, in the open neighborhood $U(x_0)$ of a B-space X. Assume that the local flow $\{\Phi_t\}$ which corresponds to*

$$x' = F(x)$$

is defined for all a, t with $\|a - x_0\| < r$ and $-T \le t \le T$. Then:

(a) *The map $(a, t) \mapsto \Phi_t(a)$ is C^k.*
(b) *If x_0 is an equilibrium point, i.e., $F(x_0) = 0$, then (23) is satisfied.*
(c) *If one can choose $T = 1$, then the multipliers of the equilibrium point x_0 are equal to the multipliers of the fixed point x_0 of the shift operator $S = \Phi_1$.*

According to (23) we have

$$\zeta = \mu^T$$

for arbitrary $T > 0$. This shows the relation between the multipliers μ of the equilibrium point x_0 and the multipliers ζ of the fixed point x_0 of Φ_T. It follows that in the sense of Definition 79.4, μ and ζ always have the same stability properties, i.e., simultaneously $|\cdot| < 1$, $= 1$, > 0 is true for both.

PROOF. Ad(a). This follows immediately from Theorem 4.D.

Ad(b). Let $x = x(t, a)$ be the solution of (20) with $x(0, a) = a$, that is, $\Phi_t(a) = x(t, a)$. From Theorem 4.D we can differentiate with respect to a. For $a = x_0$ we obtain

$$x'_a(t, x_0)h = F'(x(t, x_0))x_a(t, x_0)h \quad \text{for all} \quad h \in X.$$

Because of $x(t, x_0) = x_0$ we have that $t \mapsto x_a(t, x_0)h$ is a solution of

$$x' = F'(x_0)x \quad \text{with} \quad x(0) = h,$$

and hence $x_a(t, x_0)h = (\exp t F'(x_0))h$. This is (23).

Ad(c). This follows from (23) with $t = 1$. □

79.4. Floquet Transformation Trick

In the following section we shall reduce the stability question for periodic solutions to Theorem 79.A. We will need the so-called Floquet transformation trick, and this section is a preparation for this. The important point is that the Floquet transformation yields differential equations where the principal part is time independent and hence Theorem 79.A can be applied.

We consider the linear differential equation

$$z' = B(t)z, \quad z(0) = z_0. \tag{24}$$

(H1) For every $t \in \mathbb{R}$ the map $B(t): X \to X$, defined on a B-space X over $\mathbb{K}$, is linear and continuous. Moreover, $t \mapsto B(t)$ is a continuous map from $\mathbb{R}$ into $L(X, X)$ with period $p > 0$.

For example, if $X = \mathbb{R}^n$ then $B(t)$ is an $(n \times n)$-matrix, where the elements are continuous, p-periodic functions of t.

From Corollary 3.8 it follows that (24) with (H1) has exactly one solution $z = z(t)$ which exists for all $t \in \mathbb{R}$. Let

$$S(t)z_0 = z(t).$$

From (24) we obtain the following differential equation for the shift operator S:

$$S'(t) = B(t)S(t), \quad S(0) = I. \tag{25}$$

Note that because of the integral representation

$$z(t) = z_0 + \int_0^t B(s)z(s)\,ds$$

79.4. Floquet Transformation Trick

(a)

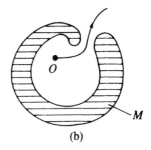
(b)

Figure 79.1

and (19*) of Chapter 3, the derivative $z'(t)$ exists as a uniform limit with respect to all z_0 in a ball. Thus $S'(t)$ exists in $L(X, X)$.

Especially, $S: \mathbb{R} \to L(X, X)$ is continuous. Since the initial-value problem for (24) has a unique global solution for any arbitrary initial time, it follows that $S(t): X \to X$ is bijective and continuous. From the open mapping theorem $A_1(36)$ it follows that $S(t)$ and $S(t)^{-1}$ belong to $L(X, X)$. Besides $z = z(t)$ also $z = z(t + p)$ is a solution of (24), that is, $z(t + p) = z(t)$. This implies the critical property

$$S(t + p) = S(t)S(p) \qquad \text{for all} \quad t \in \mathbb{R}. \tag{26}$$

We say that a compact set M in $\mathbb{C}$ does *not surround* the origin if there exists a half ray which originates at the origin and does not intersect M (Fig. 79.1(a)). This definition can be given in a more general form by replacing half rays with "reasonable" curves as shown in Figure 79.1(b). What we actually need is the fact that the set M contains an open neighborhood on which the function $z \mapsto \ln z$ is holomorphic.

(H2) The spectrum $\sigma(S(p))$ does not surround the origin. For example, this condition is always satisfied by $\dim X < \infty$, i.e., for differential equations in $\mathbb{R}^n$.

Proposition 79.8. *If* (H1), (H2) *hold, then the operator*

$$A = p^{-1} \ln S(p)$$

is well defined and $A \in L(X, X)$. *Furthermore, there exists a p-periodic, continuous operator function* $P: \mathbb{R} \to L(X, X)$ *with*

$$S(t) = P(t)e^{tA} \qquad \text{for all} \quad t \in \mathbb{R}. \tag{27}$$

For every $t \in \mathbb{R}$ *we find that* $P(t)^{-1} \in L(X, X)$.

PROOF. That A is well defined follows from (H2) and the operator calculus in A_1(60b). We have $S(p) = e^{pA}$. Let

$$P(t) = S(t)e^{-tA}.$$

From (26) it follows that

$$P(t + p) = S(t + p)e^{-(t+p)A} = S(t)S(p)e^{-pA}e^{-tA} = P(t).$$

Moreover, $P(t)^{-1} = e^{tA}S(t)^{-1}$. □

Definition 79.9. The points in the spectrum $\sigma(S(p))$ are called *Floquet multipliers* ζ. The transformation

$$z(t) = P(t)u(t) \tag{28}$$

is called Floquet transformation. Precisely the eigenvalues in $\sigma(S(p))$ are called Floquet eigenmultipliers.

An important approximation method to determine the Floquet multipliers goes as follows. One transforms (25) into the integral equation

$$S(p) = I + \int_0^p B(s)S(s)\,ds \tag{29}$$

which, according to Section 1.9, can be solved by successive approximation. This yields $S(p)$. A concrete example has already been discussed in Problem 4.8. In connection with the Hopf bifurcation of Section 79.10, it will be convenient to use the following criterion, which is based on the periodic differential equation

$$w'(t) - B(t)w(t) = -\lambda w(t), \qquad w(0) = w(p). \tag{30}$$

This is the original differential equation $w' = Bw$ in (24) in which the additional eigenvalue term $-\lambda w$ appears.

EXAMPLE 79.10. Assume (H1), (H2). Then $\lambda \in \mathbb{C}$ is an eigenvalue of (30) if and only if $\exp(p\lambda)$ is a Floquet eigenmultiplier.

PROOF. Let $S(p)a = e^{\lambda p}a$ with $a \neq 0$. Letting

$$z(t) = S(t)a \quad \text{and} \quad w(t) = e^{-\lambda t}z(t)$$

we obtain $w(0) = w(p) = a$ and

$$w' = -\lambda w + e^{-\lambda t}z' = -\lambda w + Bw$$

because of $z' = Bz$. This argument can be reversed. □

Now we present the key *trick*. When studying the stability of the equilibrium point $x_0 = 0$ of (24), we cannot apply Theorem 79.A right away, since B depends on t. However, if we use the Floquet transformation (28), then (24) passes to the new differential equation

$$u' = Au \tag{31}$$

where A does not depend on t. This follows from $P(t) = S(t)e^{-tA}$, hence

$$P'(t) = S'(t)e^{-tA} - S(t)Ae^{-tA} = B(t)P(t) - P(t)A$$

and $z' = P'u + Pu' = Bz$. Note that according to the operator calculus in A_1(60b) we may interchange functions of A.

Proposition 79.11 (Stability Criterium). *Assume* (H1), (H2). *If all Floquet multipliers of equation* (24) *are asymptotically stable (or at least one is unstable), then the equilibrium point* $x_0 = 0$ *of* (24) *is asymptotically stable (or unstable).*

PROOF. Because of $S(p) = e^{pA}$ we obtain the relation $\zeta = \mu^p$ between the Floquet multipliers ζ and the multipliers μ of (31). From $|\zeta| = |\mu|^p$ it follows that both have the same stability properties of Definition 79.4. Furthermore, these stability properties are not changed under Floquet transformation, since $P: [0, p] \to L(X, X)$ is continuous and thus Problem 1.7 implies that

$$\sup_{t \in \mathbb{R}} \|P(t)\| < \infty \quad \text{and} \quad \sup_{t \in \mathbb{R}} \|P(t)^{-1}\| < \infty.$$

An application of Theorem 79.A to the equilibrium point $u = 0$ of (31) yields the assertion. □

This transformation trick goes back to Floquet (1883).

79.5. Asymptotic Stability and Instability of Periodic Solutions

Consider the nonlinear differential equation

$$x' = F(x, t). \tag{32}$$

(H1) The map $F: X \times \mathbb{R} \to X$ is C^2, where X is a B-space over $\mathbb{K}$. For each $x \in X$ we assume that $t \mapsto F(x, t)$ has period $p > 0$.
(H2) Let $x = x_0(t)$ be a p-periodic solution of (32).
(H3) Let $B(t) = F_x(x_0(t), t)$ and let $S(p)$ denote the shift operator for $z' = Bz$. The spectrum of $S(p)$ does not surround the origin.

Definition 79.12. The *Floquet multipliers* of the periodic solution $x = x_0(t)$ are precisely the points of the spectrum $\sigma(S(p))$.

Theorem 79.C. *Assume* (H1)–(H3). *If all Floquet multipliers of* $x = x_0(t)$ *are asymptotically stable (or at least one Floquet multiplier is unstable), then* $x = x_0(t)$ *is asymptotically stable (or unstable).*

PROOF. We let $x = x_0(t) + z$ and obtain from (32) the equation

$$z' = B(t)z + g(z, t)$$

with

$$g(z, t) = F(x_0(t) + z, t) - F(x_0(t), t) - F_x(x_0(t), t)z.$$

As in (31) the Floquet transformation $z(t) = P(t)u(t)$ yields the new differential equation

$$u' = Au + P(t)^{-1}g(t, P(t)u) \tag{33}$$

with time-independent linear principal part A. As in the proof of Proposition 79.11 one then applies Theorem 79.A to (33). □

79.6. Orbital Stability

Unfortunately, Theorem 79.C cannot be used to study the asymptotic stability of autonomous differential equations

$$x'(t) = F(x(t)). \tag{34}$$

This follows because one is a Floquet eigenmultiplier if $x = x_0(t)$ is a nonconstant, p-periodic solution. Namely, besides $x = x_0(t)$ also $x = x_0(t + \tau)$ is a solution of (34), i.e.,

$$x_0'(t + \tau) = F(x_0(t + \tau)).$$

Differentiation with respect to τ gives

$$y'(t) = F_x(x_0(t))y(t) \tag{E}$$

with $y(t) = x_0'(t)$. Hence

$$y(p) = S(p)y(0).$$

The fact that $y(p) = y(0) \neq 0$ implies that $y(0)$ is an eigenvector of $S(p)$ for the eigenvalue one. Note that from $y(0) = 0$ and (E) it would immediately follow that $x_0(t) = $ const. This shows that the concept of Ljapunov stability is not practical for periodic solutions of autonomous differential equations. However, one can employ the concept of orbital stability. Recall that an orbit of $x = x_0(t)$ is the set $C = \{x_0(t): t \in \mathbb{R}\}$, i.e., the set of all solution points.

Definition 79.13. The periodic solution $x = x_0(t)$ of equation (34) is called *orbitally asymptotically stable* if and only if there exists an open neighborhood U of the orbit C of $x = x_0(t)$ such that every solution $x = x(t)$ of (34) with $x(t_0) \in U$ for fixed $t_0 \geq 0$ satisfies:

$$\lim_{t \to +\infty} \text{dist}(x(t), C) = 0.$$

The periodic solution $x = x_0(t)$ of equation (34) is called *orbitally stable* if and only if, for each neighborhood V of the orbit C of $x = x_0(t)$, there exists a neighborhood U of C such that every solution $x = x(t)$ of (34), with $x(t_0) \in U$ for fixed $t_0 \geq 0$, remains in V for all times $t \geq t_0$.

For a planet, this means that a sufficiently small perturbation at some fixed time leaves the new orbit in a neighborhood of the old orbit, but the course in time may change.

The periodic solution $x = x_0(t)$ is called *orbitally unstable* if and only if it is *not* orbitally stable.

Theorem 79.D. *Let $F: X \to X$ be C^1 on a B-space X over $\mathbb{K}$. Let*

$$x = x_0(t)$$

be a nonconstant, p-periodic solution of (34), *and assume that the set of all Floquet multipliers of $x = x_0(t)$ does not surround the origin. Then:*

(a) *If one is an algebraically simple Floquet eigenmultiplier of $x = x_0(t)$, and if all the remaining Floquet multipliers of $x = x_0(t)$ are asymptotically stable, then $x = x_0(t)$ is asymptotically orbitally stable.*
(b) *If $F: X \to X$ is C^2, and if there exists at least one unstable Floquet multiplier of $x = x_0(t)$, then $x = x_0(t)$ is orbitally unstable.*

The proof follows very simply from the Floquet transformation and the existence of a stable manifold. Since, for didactical reasons, we discuss such manifolds only in Chapter 80, we postpone the proof until then.

Assertion (b) of Theorem 79.D follows immediately from Theorem 79.C.

79.7. Perturbation of Simple Eigenvalues

The results of this section will mainly be used to examine the stability of bifurcation branches during the following sections. But actually, the results are of general interest. One of the main tools used by physicists for the solution of concrete problems is perturbation calculus. It has been applied with great success to celestial mechanics and quantum theory. In this direction we recommend the five volumes of Hagihara (1976) on celestial mechanics, Reed and Simon (1972, M), Vol. 4 on quantum mechanics, Bogoljubov and Širkov (1973, M), (1980, M), and Itzykson and Zuber (1980) on quantum field theory, and Kevorkian and Cole (1981, M) on boundary layers, for example, in the hydrodynamics of viscous fluids. The classical standard work on perturbation theory is Kato (1966, M). Modern methods which yield asymptotic expansions in connection with critical effects, e.g., in the presence of caustics in geometrical optics, may be found in Maslov (1972, M) and in the profound exposition of Leray (1978, M). As an introduction to these ideas we recommend Eckman and Sénéor (1976). Further literature may be found in the References to the Literature of this chapter.

The basic idea of perturbation calculus is the following. Knowing the exact solution of a particular problem one wants to compute the perturbed problem approximately by using expansions of small parameters. In quantum electrodynamics, for example, the interaction between electrons, positrons, and photons can be described by expansions with respect to Sommerfeld's fine structure constant $\varepsilon = 1/137$.

Experience shows that perturbed eigenvalue problems may have a very complicated structure. However, in the case of algebraically simple eigenvalues and certain generalizations the situation becomes transparent. Consider, for example, the eigenvalue problem

$$Ah = \lambda Bh, \qquad h \in X, \quad \lambda \in \mathbb{K}. \tag{35}$$

It is our goal to study the perturbation of the eigenvalue λ_0 of equation

$$A_0 x_1 = \lambda_0 B x_1, \qquad x_1 \neq 0. \tag{36}$$

(H) Let X and Y be B-spaces over $\mathbb{K}$ and $A_0, B \in L(X, Y)$.

Definition 79.14. Let $C = A_0 - \lambda_0 B$. Then λ_0 is called a B-simple eigenvalue of A_0 if and only if (36) holds for some x_1 and C is a Fredholm operator of index zero with $\dim N(C) = 1$ and the transversality condition

$$Bx_1 \notin R(C). \tag{37}$$

From Section 8.4 it follows that (37) is equivalent to the fact that there exists a $y_1^* \in Y^*$ with $C^* y_1^* = 0$ and

$$\langle y_1^*, Bx_1 \rangle = 1. \tag{38}$$

In the special case that $X = Y$, $B = I$, the identity, Proposition 8.18 shows that for compact A_0 and $\lambda_0 \neq 0$ the algebraically simple and I-simple eigenvalues coincide.

Proposition 79.15. *Assume* (H) *and let λ_0 be a B-simple eigenvalue of A_0.*
Then there exists a neighborhood $U(\lambda_0)$ in $\mathbb{K}$ and a number r such that equation (35) *has a unique eigenvalue*

$$\lambda(A) \quad \text{in } U(\lambda_0)$$

for every operator $A \in L(X, Y)$ with $\|A - A_0\| < r$.
This eigenvalue is simple and the map $A \mapsto \lambda(A)$ is analytic with $\lambda(A_0) = \lambda_0$. Moreover, we have

$$\lambda'(A_0)H = \langle y_1^*, Hx_1 \rangle \quad \text{for all} \quad H \in L(X, Y). \tag{39}$$

Corollary 79.16. *Let $U(\mu_0)$ be a neighborhood in $\mathbb{R}$ and let*

$$\mu \mapsto A(\mu)$$

be a C^k-map from $U(\mu_0)$ into $L(X, Y)$ with $A(\mu_0) = A_0$ and $k \geq 1$. If we write $\lambda(\mu)$ for $\lambda(A(\mu))$, then $\mu \mapsto \lambda(\mu)$ is a C^k-map with

$$\lambda'(\mu_0) = \langle y_1^*, A'(\mu_0) x_1 \rangle. \tag{40}$$

PROOF.

(I) Eigenvalue problem. We use the decomposition

$$X = N(C) \oplus N(C)^\perp$$

79.7. Perturbation of Simple Eigenvalues

and let $w = (\lambda, z)$, $W = \mathbb{K} \times N(C)^\perp$ as well as $H(A, \lambda, z) = A(x_1 + z) - \lambda B(x_1 + z)$, $(\lambda, z) \in W$, where $A \in L(X, Y)$.

In order to solve (35), we consider

$$H(A, \lambda, z) = 0, \qquad (\lambda, z) \in W. \tag{41}$$

The linearization is

$$H_w(A_0, \lambda_0, 0)w = A_0 z - \lambda_0 B z - \lambda B x_1 = Cz - \lambda B x_1.$$

The operator $H_w(A_0, \lambda_0, 0): W \to Y$ is bijective, because equation

$$Cz - \lambda B x_1 = y, \qquad (\lambda, z) \in W$$

has a unique solution for every $y \in Y$. In fact, from

$$\langle y_1^*, Cz \rangle = \langle C^* y_1^*, z \rangle = 0 \qquad \text{for all} \quad z \in N(C)^\perp$$

and (38) it follows that $\lambda = -\langle y_1^*, y \rangle$. Moreover, equation

$$Cz = y + \lambda B x_1, \qquad z \in N(C)^\perp$$

has a unique solution, since $\langle y_1^*, y + \lambda B x_1 \rangle = 0$.

The implicit function theorem (Theorem 4.B) implies that (41) can be uniquely solved for (λ, z) in a neighborhood of $(A_0, \lambda_0, 0)$. This gives the eigenvalues $\lambda(A)$.

(II) Simplicity of $\lambda(A)$. It remains to show that (35) has no eigensolutions $h \in N(C)^\perp$. We set $x_1 = 0$ in H and consider the equation

$$QH(A, \lambda, z) = 0, \qquad z \in N(C)^\perp, \tag{42}$$

where $Q: Y \to R(C)$ is a fixed projection operator onto $R(C)$. The linearization is

$$QH_z(A_0, \lambda_0, 0)z = A_0 z - \lambda_0 B z = Cz.$$

Therefore the operator

$$QH_z(A_0, \lambda_0, 0): N(C)^\perp \to R(C)$$

is bijective. The implicit function theorem implies that equation (42) can be uniquely solved for z in a neighborhood of $(A_0, \lambda_0, 0)$. This solution is $z = 0$.

Following this preparation we consider the equation

$$Az - \lambda B z = 0, \qquad z \in N(C)^\perp.$$

It follows that z is also a solution of (42), and hence it follows from the previous arguments that $z = 0$.

(III) Proof of (39). We have $z(A_0) = 0$. Differentiation of

$$A(x_1 + z(A)) - \lambda(A)B(x_1 + z(A)) = 0$$

at the point A_0 with respect to A gives

$$Hx_1 - (\lambda'(A_0)H)Bx_1 + C(z'(A_0)H) = 0$$

for all $H \in L(X, Y)$. Applying y_1^* we obtain (39). □

Important well-known results about the perturbation of nonsimple eigenvalues can be found in Problem 79.8.

79.8. Loss of Stability and the Main Theorem About Simple Curve Bifurcation

We now come to the main objective of this chapter. An important scientific phenomenon is that a state of a system looses its stability and thereby passes into a qualitatively different state. Mathematically, this is bifurcation through loss of stability. Consider, for example, the autonomous differential equation

$$z' = G(\mu, z) \tag{43}$$

in a real B-space X depending on a real parameter μ, which describes an outer perturbation of the system. Assume that the system has a family $\{z(\mu)\}$ of equilibrium states, i.e.,

$$G(\mu, z(\mu)) = 0 \quad \text{for all } \mu.$$

According to Theorem 79.A the stability of $z(\mu)$ is determined by the spectrum of $G_z(\mu, z(\mu))$. We consider two important possibilities which cause the loss of stability.

(i) *Simple curve bifurcation.* At $\mu = \mu_0$ a real eigenvalue $\lambda(\mu)$ of the linearization $G_z(\mu, z(\mu))$ passes over the imaginary axis with positive velocity. In Theorem 79.E below, we precisely state when, at the point $z(\mu_0)$, a new curve bifurcates from the curve $z = z(\mu)$ i.e., the system eventually passes into new equilibrium states (Fig. 79.2).
(ii) *Hopf bifurcation.* At $\mu = \mu_0$ a pair of conjugate-complex eigenvalues of $G_z(\mu, z(\mu))$ passes over the imaginary axis with positive velocity. In Theorem 79.F below, we precisely state when, in the neighborhood of $z(\mu_0)$, nonconstant, periodic solutions of (43) bifurcate from the curve $z = z(\mu)$ i.e., the system begins to oscillate.

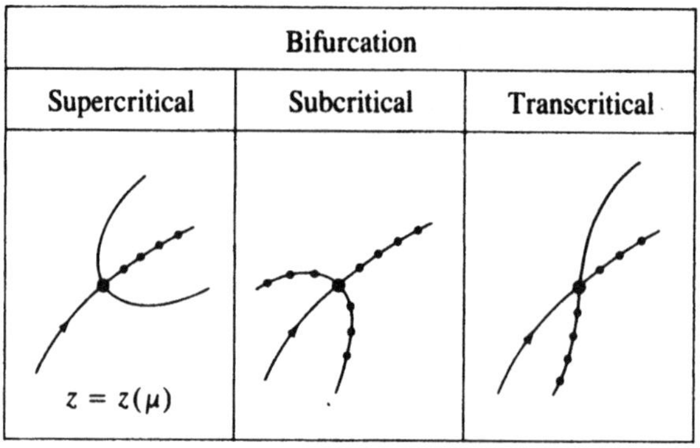

Bifurcation		
Supercritical	Subcritical	Transcritical
$z = z(\mu)$		

Figure 79.2

79.8. Loss of Stability and the Main Theorem About Simple Curve Bifurcation

In Section 79.11 we will show that the famous center theorem of Ljapunov is a special case of the main theorem about Hopf bifurcation (Theorem 79.F).

Whether or not the system passes into the bifurcation solution mainly depends on its stability. We assume that the known solution $z = z(\mu)$ is stable for $\mu < \mu_0$ and unstable for $\mu > \mu_0$. Roughly, we obtain the following picture for (i) and (ii). This corresponds to what one would naturally expect.

(S) *Stability principle.* The bifurcation solution is stable for $\mu > \mu_0$ and unstable for $\mu < \mu_0$.

In particular, if bifurcation solutions only appear for $\mu > \mu_0$ (supercritical bifurcation) or $\mu < \mu_0$ (subcritical bifurcation), then they are stable or unstable, respectively. This has schematically been pictured in Figure 79.2. The dotted lines represent the unstable solutions. In studying (i) and (ii) we will make essential use of the bunch theorem of Section 8.11. Our proofs will also give effective procedures for the construction of the solutions. Later on the stability concepts will be discussed in greater detail.

First we examine simple curve bifurcations for the stationary equation

$$G(\mu, z) = 0, \quad (\mu, z) \in \mathbb{R} \times X. \tag{44}$$

Our assumptions are as follows:

(H1) X and Y are real B-spaces with the continuous embedding $X \subseteq Y$ and corresponding embedding operator J.
(H2) (Trivial solution). The map $G: U(\mu_0, z(\mu_0)) \subseteq \mathbb{R} \times X \to Y$ is C^k, $k \geq 2$. There exists a C^k-map $z: U(\mu_0) \subseteq \mathbb{R} \to X$ with

$$G(\mu, z(\mu)) = 0 \quad \text{for all } \mu.$$

We call $z = z(\mu)$ the trivial solution branch.
(H3) (Loss of stability). There exists a C^1-map $\lambda: V(\mu_0) \subseteq \mathbb{R} \to \mathbb{R}$ such that $\lambda(\mu)$ is an eigenvalue of $G_z(\mu, z(\mu))$ for every μ with $\lambda(\mu_0) = 0$ and

$$\lambda'(\mu_0) > 0. \tag{45}$$

Moreover, $\lambda(\mu_0)$ is a J-simple eigenvalue of $G_z(\mu_0, z(\mu_0))$.

In (H3) note the following convention. We say that λ is an eigenvalue of $A \in L(X, Y)$ if the equation $Ax = \lambda Jx$ has a solution $x \neq 0$. Condition (45) elegantly indicates the change in stability. It states that at $\mu = \mu_0$ the real eigenvalue $\lambda(\mu)$ crosses the imaginary axis with nonvanishing velocity from the left to the right. As the proof will show, this will yield the important generic bifurcation condition.

From Section 79.7 it follows that (H3) is equivalent to the following condition (H3*), which is often easier to verify. Let $L = G_z(\mu_0, z(\mu_0))$.

(H3*) The operator $L: X \to Y$ is Fredholm of index zero with $\dim N(L) = 1$. There exists an element $x_1 \in N(L)$ and an element $y_1^* \in N(L^*)$ with $\langle y_1^*, Jx_1 \rangle = 1$ and the generic bifurcation condition

$$\langle y_1^*, G_{z\mu}(\mu_0, z(\mu_0))x_1 + G_{zz}(\mu_0, z(\mu_0))z'(\mu_0)x_1 \rangle > 0.$$

According to (40), the expression on the left-hand side is then equal to $\lambda'(\mu_0)$.

The weak stability concept used in the following will be defined precisely during the proof below. Moreover, the real numbers s vary in a sufficiently small neighborhood of zero.

Theorem 79.E (Main Theorem About Simple Curve Bifurcation of Crandall and Rabinowitz (1973)). *Assume* (H1)–(H3). *Then* $(\mu_0, z(\mu_0))$ *is a bifurcation point of the equation* (44).

In a neighborhood of the point $(\mu_0, z(\mu_0))$ *in* $\mathbb{R} \times X$, *all bifurcation solutions lie on a* C^{k-1}-*curve*

$$s \mapsto (\mu(s), z_0(s))$$

which passes at $s = 0$ *through the bifurcation point* $(\mu_0, z(\mu_0))$.

The trivial solution $z = z(\mu)$ *is weakly stable for* $\mu < \mu_0$ *and weakly unstable for* $\mu > \mu_0$.

If $\mu'(s) \neq 0$ *for all* $s \neq 0$, *then the bifurcation solution is weakly stable for* $\mu(s) > 0$ *and weakly unstable for* $\mu(s) < 0$.

If G *and* z *in* (H1), (H2) *are analytic, then also* $s \mapsto (\mu(s), z_0(s))$ *is analytic*.

PROOF. We use the bunch theorem (Theorem 8.B) of Section 8.11.

(I) Bifurcation branch. In order to simplify the notation let

$$\mu = \mu_0 + \varepsilon, \qquad F(\varepsilon, x) = G(\mu_0 + \varepsilon, z(\mu_0 + \varepsilon) + x). \qquad (46)$$

The original equation (44) then becomes

$$F(\varepsilon, x) = 0, \qquad (\varepsilon, x) \in \mathbb{R} \times X. \qquad (47)$$

Moreover, we have $F(\varepsilon, 0) \equiv 0$ and $L = F_x(0, 0)$. The important generic bifurcation condition

$$\langle y_1^*, F_{x\varepsilon}(0,0)x_1 \rangle > 0$$

of Theorem 8.B follows immediately from (H3*).

Theorem 8.B therefore yields the existence of a C^{k-1}-bifurcation branch $s \mapsto (\varepsilon(s), x(s))$ of (47) through the point $(0, 0)$ with

$$x(s) = s(x_1 + w(s)), \qquad w(s) \in N(L)^\perp$$

and $w(0) = 0$. The corresponding solution of (44) is

$$\mu(s) = \mu_0 + \varepsilon(s), \qquad z_0(s) = z(\mu(s)) + x(s).$$

(II) Weak stability of the trivial solution. We have

$$F_x(\varepsilon, x) = G_z(\mu_0 + \varepsilon, z(\mu) + x).$$

Since $\lambda = 0$ is a J-simple eigenvalue of $F_x(0, 0)$, it follows from Section 79.7 that there exists a neighborhood $U(0)$ in $\mathbb{C}$ such that the eigenvalue

79.8. Loss of Stability and the Main Theorem About Simple Curve Bifurcation

problem
$$F_x(\varepsilon, x)h = \lambda h, \qquad \lambda \in \mathbb{C}, \quad h \in X_\mathbb{C} \tag{48}$$

has exactly one eigenvalue λ in $U(0)$ for every (ε, x) in a neighborhood of zero. The solution (ε, x) of (47) is called *weakly stable* (or weakly unstable) if and only if $\operatorname{Re} \lambda < 0$ (or > 0). Because of (45) we have

$$\lambda(\mu) \lessgtr 0 \qquad \text{for} \quad \mu \lessgtr \mu_0.$$

This yields the stability result for the trivial solution.

(III) *Weak stability of the bifurcation solution.* We now consider

$$F_x(\varepsilon(s), x(s))h = \lambda(s)h.$$

Let $\varepsilon'(s) \neq 0$ for all $s \neq 0$ in a neighborhood of zero. In Problem 79.4 we prove

$$\lim_{s \to 0} \lambda(s)/\varepsilon'(s)s = -\langle y_1^*, F_{x\varepsilon}(0,0)x_1 \rangle < 0. \tag{49}$$

If, for example, $\varepsilon'(s) > 0$ for all small $s > 0$, then $\varepsilon(s) > 0$ and $\lambda(s) < 0$, i.e., the bifurcation solution is weakly stable for small $s > 0$. Analogously, one treats the other cases. □

Remark 79.17 (Construction of the Bifurcation Solution). We use (46). Furthermore, we let $N = N(F_x(0,0))$ and choose a decomposition $X = N \oplus N^\perp$. Equation

$$F_x(0,0)w = g, \qquad w \in N^\perp$$

has a unique solution for every $g \in Y$ with $\langle y_1^*, g \rangle = 0$, which will be denoted by $w = Sg$. Moreover, equation

$$F_x(0,0)w + \varepsilon F_{x\varepsilon}(0,0)x_1 = f, \qquad (w, \varepsilon) \in N^\perp \times \mathbb{R} \tag{50}$$

has also a unique solution for every $f \in X$, namely

$$\varepsilon = \langle y_1^*, f \rangle / \langle y_1^*, F_{x\varepsilon}(0,0)x_1 \rangle, \qquad w = S(f - \langle y_1^*, f \rangle). \tag{51}$$

The bifurcation solution has the form

$$x(s) = sx_1 + sw(s)$$

with $w(s) \in N^\perp$ and $w(0) = 0$. From $F(\varepsilon(s), x(s)) = 0$ follows (50) with

$$-f = s^{-1}F(\varepsilon(s), sx_1 + sw(s)) - F_x(0,0)w(s) - \varepsilon(s)F_{x\varepsilon}(0,0)x_1.$$

We insert this f into (51) and fix a number $s \neq 0$ in a sufficiently small neighborhood of zero. As starting value we choose $\varepsilon_0(s) = 0$, $w_0(s) = 0$. The bunch theorem of Section 8.11 implies that the corresponding iteration scheme for (51) converges to the bifurcation solution.

If F is analytic, then also $s \mapsto (\varepsilon(s), x(s))$ is analytic and one can use ansatz and comparison of coefficients in (51). Note that $f = O(s)$, $s \to 0$.

We now use the following condition (H4) in order to prove a stronger stability result.

(H4) Let $X = Y$. Again we set $L = G_z(\mu_0, z(\mu_0))$. Moreover, let $M = \sigma(L) - \{0\}$. We assume that M lies in the left half-plane with a positive distance to the imaginary axis. Moreover, we assume that there exists a neighborhood U of zero in $\mathbb{C}$ such that
$$\sigma(G(\mu, z)) \cap U$$
consists entirely of eigenvalues. More precisely, this should be true for all points (μ, z) in a sufficiently small neighborhood of the point $(\mu_0, z(\mu_0))$ in $\mathbb{R} \times X$.

Corollary 79.18 (Stability of Solutions in the Sense of Ljapunov). *If* (H1)–(H4) *holds, then we can replace weakly stable (or weakly unstable) with asymptotically stable (or unstable) in Theorem 79.E.*

PROOF. A well-known theorem in perturbation theory states that for small perturbations of L in the operator norm, the perturbation of M remains in the left open half-plane (see Problem 79.8). From (II) and (III) in the proof of Theorem 79.E we know the behavior of the perturbation of $\lambda = 0$. Theorem 79.A then yields the assertion. □

79.9. Loss of Stability and the Main Theorem About Hopf Bifurcation

We consider again the differential equation
$$z' = G(\mu, z). \tag{52}$$
Our assumptions are now as follows:

(H1) X and Y are real B-spaces with the continuous embedding $X \subseteq Y$ and corresponding embedding operator J.
(H2) (Trivial solution). The map $G: U(\mu_0, z(\mu_0)) \subseteq \mathbb{R} \times X \to Y$ is C^k, $k \geq 2$. There exists a C^k-map $z: U(\mu_0) \subseteq \mathbb{R} \to X$ with $G(\mu, z(\mu)) = 0$ for all μ.
(H3) (Loss of stability). There exists a C^1-map $\lambda: V(\mu_0) \subseteq \mathbb{R} \to \mathbb{R}$ such that $\lambda(\mu)$ is an eigenvalue of $G_z(\mu, z(\mu))$ for all μ with $\lambda(\mu_0) = i\omega_0$, $\omega_0 > 0$ and
$$\operatorname{Re} \lambda'(\mu_0) > 0. \tag{53}$$
Moreover, $\lambda(\mu_0)$ is a J-simple eigenvalue of $G_z(\mu_0, z(\mu_0))$ with respect to the complexification.

For $X = Y = \mathbb{R}^n$, J-simplicity means that $\lambda(\mu_0)$ is algebraically simple. The number $p_0 = 2\pi/\omega_0$ will arise as the limit period of the periodic solutions which are bifurcating from $z(\mu_0)$.

79.9. Loss of Stability and the Main Theorem About Hopf Bifurcation

Besides $\lambda(\mu)$ the complex conjugate number $\bar{\lambda}(\mu)$ is also an eigenvalue of $G_z(\mu, z(\mu))$. It follows from (53) that, for $\mu = \mu_0$, the pair $(\lambda(\mu), \bar{\lambda}(\mu))$ of complex conjugate eigenvalues crosses the imaginary axis with nonvanishing velocity from the left to the right. As the proof shows, this loss of stability yields the important generic bifurcation condition.

(H4) (Nonresonance condition). None of the numbers $ik\omega_0$ with $k = 0$ and $k = 2, 3, \ldots$ is an eigenvalue of $G_z(\mu_0, z(\mu_0))$.

Conditions (H3) and (H4) guarantee that the linearized equation $z' = G_z(\mu_0, z(\mu_0))z$ has periodic solutions of the form

$$z = (\cos k\omega_0 t)c + (\sin k\omega_0 t)d, \qquad c, d \in X$$

with $k = 1$, but not with $k = 0$ or $k = 2, 3, \ldots$.

Let $C_{2\pi}^m(\mathbb{R}, X)$ denote the set of all 2π-periodic, m-times continuously differentiable functions with values in X. Let

$$X_{2\pi} = C_{2\pi}^1(\mathbb{R}, X) \text{ and } Y_{2\pi} = C_{2\pi}(\mathbb{R}, Y).$$

(H5) (Technical condition). The operator $d/dt - \omega_0^{-1} G_z(\mu_0, z(\mu_0))$ from $X_{2\pi}$ into $Y_{2\pi}$ is Fredholm of index zero with two-dimensional null space.

For $X = Y = \mathbb{R}^n$ this condition follows automatically from (H1)–(H4) (see Problem 79.5). Let $A = G_z(\mu_0, z(\mu_0)) - i\omega_0 J$ with $\omega_0 > 0$. From Section 79.7 it follows that (H3) is equivalent to the following condition.

(H3*) The operator $A: X_C \to Y_C$ is Fredholm of index zero with dim $N(A) = 1$. There exists an element $a \in N(A)$ and an element $a^* \in N(A^*)$ with $\langle a^*, Ja \rangle = 1$ and

$$\text{Re}\langle a^*, G_{z\mu}(\mu_0, z(\mu_0))a + G_{zz}(\mu_0, z(\mu_0))z'(\mu_0)a \rangle > 0.$$

According to (40) the expression on the left-hand side is then equal to $\text{Re }\lambda'(\mu_0)$.

If $z = z_0(t)$ is a p-periodic solution of the original equation (52) with $p > 0$, then we use time scaling $\tau = 2\pi t/p$ to renorm it to period 2π. We let

$$x(\tau) = z_0(t)$$

and call the tuple

$$(\mu, p, x) \quad \text{in } \mathbb{R}^2 \times X_{2\pi}$$

a solution tuple of (52). By a phase shift we mean the transformation from

$$\tau \mapsto x(\tau) \quad \text{to} \quad \tau \mapsto x(\tau + \alpha).$$

A solution tuple (μ, p, x) is called nontrivial if and only if x does not coincide with the equilibrium point $z(\mu)$.

Our goal is to find a family

$$\mu = \mu(s), \qquad p = p(s), \qquad x = x_s \tag{54}$$

of solution tuples for all real s in a neighborhood $U(0)$ of zero with

$$(\mu(s), p(s), x_s) \to (\mu_0, p_0, z(\mu_0)) \quad \text{as} \quad s \to 0 \tag{55}$$

in $\mathbb{R}^2 \times X_{2\pi}$ as well as

$$\mu(s) = \mu(-s) \quad \text{and} \quad p(s) = p(-s) \quad \text{for all} \quad s \in U(0), \tag{56}$$

where $p_0 = 2\pi/\omega_0$. If we write the element $a \in X_C$ of (H3*) in the form $a = a_1 + ia_2$ with $a_1, a_2 \in X$, then we will obtain

$$x_s(\tau) = z(\mu_0) + s[(\cos \tau)a_1 + (\sin \tau)a_2] + o(s), \quad s \to 0. \tag{57}$$

Hence the tuple in (54) is nontrivial for $s \neq 0$, and we have nontrivial periodic solutions.

The following theorem is called the main theorem of Hopf bifurcation. The stability concepts mentioned will be given a precise form during the proof.

Theorem 79.F (Hopf (1942), Crandall and Rabinowitz (1975)). *Assume* (H1)–(H5).

(a) Existence of periodic solutions. *There exists a C^{k-1}-curve (54), whose points consist of solution tuples of equation (52). Furthermore, (55) through (57) hold.*
(b) Uniqueness. *There exists a neighborhood $U(\mu_0, p_0, z(\mu_0))$ in $\mathbb{R}^2 \times X_{2\pi}$ such that all nontrivial solution tuples of (52) in this neighborhood are given by (54) with $s \neq 0$ and by phase shifts of x_s.*
(c) Stability. *The trivial solution $z(\mu)$ is weakly stable for $\mu < \mu_0$ and weakly unstable for $\mu > \mu_0$. If $\mu'(s) \neq 0$ for all $s \neq 0$ in a neighborhood of zero, then (54) is weakly stable for $\mu(s) > \mu_0$ and weakly unstable for $\mu(s) < \mu_0$.*
(d) Analyticity. *If G and z in* (H1), (H2) *are analytic, then all functions in (54) depend analytically on s.*

The proof will yield an effective iteration scheme for the construction of (54). In the analytic case one can use ansatz and comparison of coefficients. If

$$\mu'(s) \neq 0$$

for all $s \neq 0$, it follows from (56) that nontrivial periodic solutions are possible only for $\mu > \mu_0$ or $\mu < \mu_0$, i.e., weakly stable supercritical bifurcation or weakly unstable subcritical bifurcation are the only bifurcations that occur. The natural scientist is interested in the former case.

Theorem 79.F can be applied to the case $X = \mathbb{R}^n$ (systems of ordinary differential equations), to parabolic partial differential equations, and to the time-dependent Navier–Stokes equations. In the last two cases one needs to modify the assumptions slightly. We again assume (H1) to (H5), but do not choose $X_{2\pi}$ and $Y_{2\pi}$ as above, but instead choose function spaces whose elements are Hölder continuous differentiable in an appropriate sense with respect to the space and time variable, and have period 2π with respect to time. Our proof will immediately apply to this situation. The appropriate spaces may be found in Joseph and Sattinger (1972) together with the *a priori* estimates that imply (H5).

In the following proof, the concept of weak stability and weak instability is based on the behavior of the *essential* Floquet multipliers (see Lemma 79.19 below). From the physical point of view, it is important to know the *orbital stability* or *orbital instability* of the bifurcating periodic solutions. However, similarly as in the proof of Corollary 79.18, it is not difficult to prove the orbital stability or orbital instability via Theorem 79.D by making *additional* natural assumptions about the spectrum of $G_z(\mu_0, z(\mu_0))$. An important result in this direction will be proved in Problem 79.9.

Further interesting results about Hopf bifurcation can be found in Problems 79.10 and 79.11, and in Chapter 80.

79.10. Proof of Theorem 79.F

We apply the bunch theorem (Theorem 8.B) of Section 8.11 to the operator equation (61) below. The important generic bifurcation condition

$$\det(\langle x_i^*, F_{x\varepsilon_j}(0,0)x_1 \rangle) \neq 0$$

of Theorem 8.B will follow from (58) below, and (58) is a consequence of the loss of stability $\operatorname{Re} \lambda'(\mu_0) > 0$.

Since the bunch theorem follows from the implicit function theorem, the same is true for the Hopf bifurcation (Theorem 79.F).

Step 1: Preparations.

By eventually passing to $f(\varepsilon, z) = G(\mu_0 + \varepsilon, z(\mu + \varepsilon) + z)$ we may assume right away that $\mu_0 = 0$ and $z(\mu) \equiv 0$. Also, for simplicity, we write x instead of Jx. Important is the generic bifurcation condition

$$\det \begin{pmatrix} (x_1^*|L_0 x_1) & (x_1^*|L_1 x_1) \\ (x_2^*|L_0 x_1) & (x_2^*|L_1 x_1) \end{pmatrix} \neq 0, \tag{58}$$

which will follow from $\operatorname{Re} \lambda'(\mu_0) \neq 0$, where we set

$$L_0 = \omega_0^{-1} G_z(0,0), \qquad L_1 = \omega_0^{-1} G_{z\mu}(0,0)$$

and $(f|g) = \int_{-\pi}^{\pi} \langle f(\tau), g(\tau) \rangle \, d\tau$. We have that $L_0, L_1 \in L(X, Y)$. Moreover, we define

$$x = e^{i\tau} a, \qquad x_1 = \operatorname{Re} x, \qquad x_2 = \operatorname{Im} x,$$
$$x^* = e^{-i\tau} a^*, \qquad x_1^* = \operatorname{Re} x^*, \qquad x_2^* = \operatorname{Im} x.$$

From (H3*) follows

$$L_0 a = ia, \qquad L_0^* a^* = ia^*, \qquad \langle a^*, a \rangle = \pi^{-1}, \tag{59}$$

where a has been renormed. Finally, we let

$$(T_\alpha f)(\tau) = f(\tau + \alpha).$$

Because of $a = a_1 + ia_2$ we have $x_1 = (\cos \tau)a_1 - (\sin \tau)a_2$, etc. The following statements are easily verified.

(I) Equations (59) imply that $(x_i^* | x_j) = \delta_{ij}$ for $i, j = 1, 2$ and

$$(x_1^* | L_0 x_1) = 0, \qquad (x_2^* | L_0 x_1) = 1.$$

(II) We prove (58). It is

$$\pi(x_1^* | L_1 x_1) = \langle a_1^*, L_1 a_1 \rangle - \langle a_2^*, L_1 a_2 \rangle$$
$$= \operatorname{Re}\langle a^*, L_1 a \rangle.$$

Thus (H3*) implies that

$$(x_1^* | L_1 x_1) = \omega_0^{-1} \operatorname{Re} \lambda'(\mu_0) > 0.$$

From (I) follows (58).

(III) $(T_\pi x_1)(\tau) = -x_1(\tau)$.

(IV) T_α is a linear map in the spaces span$\{x_1, x_2\}$ and span$\{x_1^*, x_2^*\}$.

(V) For each point $y \in \operatorname{span}\{x_1, x_2\}$ there exist real numbers β, r with

$$T_\beta y = r x_1,$$

because we have always $y = \operatorname{Re}(bx)$ for some $b \in \mathbb{C}$, hence $y = \operatorname{Re} e^{i\tau - i\beta} |b| a$.

(VI) Let $x \in X_{2\pi}$. With $\dot{x}$ we denote the derivative with respect to the time variable τ. It follows from (59) that x_1, x_2 and x_1^*, x_2^* are 2π-periodic solutions of the differential equations

$$\dot{x} - L_0 x = 0 \quad \text{and} \quad \dot{x}^* + L_0^* x^* = 0,$$

respectively. This way, we obtain all 2π-periodic solutions of the first differential equation in $X_{2\pi}$ as span$\{x_1, x_2\}$. For $X = Y = \mathbb{R}^n$ this follows from (H4) and Fourier expansions. In the general case it follows from (H5).

Step 2: Eigenvalue trick.

In order to eliminate the unknown period p, we let

$$x(\tau) = z(t) \quad \text{with} \quad \tau = 2\pi t/p.$$

From $z' = G(\mu, z)$ we obtain

$$\dot{x} = \frac{p}{2\pi} G(\mu, x).$$

By introducing a small parameter ρ through $p = 2\pi \omega_0^{-1}(1 + \rho)$ we obtain

$$\omega_0^{-1}(1 + \rho) G(\mu, x) - \dot{x} = 0. \tag{60}$$

79.10. Proof of Theorem 79.F

Setting $\varepsilon = (\rho, \mu)$ we can write (60) simply as

$$F(\varepsilon, x) = 0, \qquad \varepsilon \in \mathbb{R}^2, \quad x \in X_{2\pi} \qquad (61)$$

with $F: U(0, 0) \subseteq \mathbb{R}^2 \times X_{2\pi} \to Y_{2\pi}$.

Using this time scaling we can restrict ourselves to 2π-periodic solutions, where (61) now depends on *two* parameters ρ and μ.

Step 3: Existence proof.

For (61) we check the assumptions of the bunch theorem (Theorem 8.B) of Section 8.11.

From (60) it follows that

$$F_x(0, 0)x = L_0 x - \dot{x}.$$

According to (H5) this is a Fredholm operator of index zero. From (VI) the null space $N(F_x(0, 0))$ is spanned by x_1 and x_2. If we identify x_j^* with the linear, continuous functional $x \mapsto (x_j^* | x)$ on $X_{2\pi}$, then integration by parts gives

$$\langle x_j^*, F_x(0, 0)x \rangle = (x_j^* | L_0 x - \dot{x})$$
$$= (\dot{x}_j^* + L_0^* x_j^* | x) = 0 \qquad \text{for all} \quad x \in X_{2\pi},$$

hence $F_x(0, 0)^* x_j^* = 0$. Because of

$$\text{ind } F_x(0, 0) = 0 \quad \text{and} \quad \dim N(F_x(0, 0)) = 2$$

we obtain from Section 8.4 that $\dim N(F_x(0, 0)^*) = 2$. Therefore x_1^* and x_2^* span the null space $N(F_x(0, 0)^*)$.

Since the generic bifurcation condition

$$\det(\langle x_i^*, F_{x\varepsilon_j}(0, 0)x_1 \rangle) \neq 0$$

is identical with (58), Theorem 8.B implies the existence of a C^{k-1}-bifurcation branch $s \mapsto (\varepsilon(s), x_s)$ of (61) through the point $(0, 0)$ with

$$x_s = s x_1 + s w_s, \qquad w_s \in N(F_x(0, 0))^\perp$$

and $w_s = 0$ for $s = 0$. Time rescaling gives the existence result of Theorem 79.F.

Step 4: Uniqueness.

Let $N = N(F_x(0, 0))$. The operator

$$Px = \langle x_1^*, x \rangle x_1 + \langle x_2^*, x \rangle x_2$$

is a projection operator from $X_{2\pi}$ onto N. From Theorem 8.B it follows that the bifurcation branch is uniquely determined through

$$Px_s = s x_1.$$

Let $N^\perp = (I - P)X_{2\pi}$. From (IV) it follows that T_α leaves N as well as $N^\perp$ invariant. This implies $T_\alpha P = P T_\alpha$.

Let (ε, x) be a solution of (61) in a neighborhood of zero which has the form of a ball. Then also $(\varepsilon, T_\alpha x)$ is a solution which, because of $\|T_\alpha\| = 1$, remains in the neighborhood of zero. We have $T_\alpha Px \in N$. From (V) we can choose α such that $T_\alpha Px = rx_1$. This implies $PT_\alpha x = rx_1$. Hence we must have

$$T_\alpha x = x_s \quad \text{and} \quad \varepsilon = \varepsilon(s) \quad \text{with} \quad s = r,$$

i.e., x differs from x_r only by a phase shift.

Besides $(\varepsilon(s), x_s)$, also $(\varepsilon(s), T_\pi x_s)$ is a solution of (61). From (III) it follows that $PT_\pi x_s = sT_\pi x_1 = -sx_1$. Hence we must have $T_\pi x_s = x_{-s}$ and $\varepsilon(s) = \varepsilon(-s)$.

Step 5: Stability.

By definition the weak stability of the trivial solution has to be determined by the behavior of $\lambda(\mu)$. Because of (H3) we have

$$\operatorname{Re} \lambda(\mu) \lessgtr 0 \quad \text{for} \quad \mu \lessgtr \mu_0.$$

This implies the stability result of Theorem 79.F for the trivial solution.

In order to study the stability of the bifurcation solution, we consider the eigenvalue problem

$$F_x(\varepsilon(s), x_s)h = \kappa h, \quad h \in X_{2\pi}, \quad \kappa \in \mathbb{R} \tag{62}$$

and note that $\varepsilon = (\rho, \mu)$ and $p = 2\pi\omega_0^{-1}(1 + \rho)$.

Lemma 79.19. *Let $\mu'(s) \neq 0$ for all $s \neq 0$ in a neighborhood of zero. Then there exists a real C^1-function $s \mapsto \kappa(s)$ in a neighborhood of zero, where all $\kappa(s)$ are eigenvalues of* (62) *with $\kappa(0) = 0$ and*

$$\lim_{s \to 0} \kappa(s)/\mu'(s)s = -\omega_0^{-1} \operatorname{Re} \lambda'(\mu_0) < 0. \tag{63}$$

The proof will be given in Problem 79.6. Example 79.10 shows that $m(s) = e^{2\pi\kappa(s)}$ is a Floquet multiplier of the 2π-periodic solution x_s of

$$\dot{x} - \frac{p}{2\pi} G(\mu(s), x) = 0.$$

Moreover, for $s = 0$ equation (62) has the double eigenvalue $\kappa = 0$ with eigenfunctions x_1 and x_2. The index has nothing to do with x_s. Differentiation of $F(\varepsilon(s), x_s) = 0$ with respect to τ gives

$$F_x(\varepsilon(s), x_s)\dot{x}_s = 0.$$

Thus $\kappa = 0$ is an eigenvalue of (62) for $s \neq 0$. According to Lemma 79.19 we therefore may think of $(\kappa(s), 0)$ as a perturbation of the double eigenvalue $(0, 0)$. The corresponding Floquet multipliers are $(m(s), 1)$.

Motivated by Theorem 79.D about orbital stability, we define the weak stability of x_s according to the behavior of $m(s)$. The solution x_s is called *weakly stable* (or *weakly unstable*) if

$$|m(s)| < 1 \quad (\text{or} > 1).$$

This corresponds to $\kappa(s) < 0$ (or $\kappa(s) > 0$).

From (63) we obtain that $\kappa(s)$ has the opposite sign of $\mu'(s)s$. This implies the stability result of Theorem 79.F.

The considerations in the proof to Problem 79.9 justify the designation "weak" stability and "weak" instability in Theorem 79.F.

Step 6: Construction of the bifurcation solution.

The iteration scheme of Theorem 8.B to determine the solutions of (61) takes here the following specific form.

For every $y \in Y_{2\pi}$ the linear differential equation

$$\dot{x} - L_0 x = y + \rho L_0 x_1 + \mu L_1 x_1 \tag{64}$$

has a unique solution $(x, \rho, \mu) \in X_{2\pi} \times \mathbb{R}^2$ with

$$(x_1^* | x) = (x_2^* | x) = 0.$$

This solution is obtained by first solving the system of linear equations

$$\rho(x_1^* | L_0 x_1) + \mu(x_1^* | L_1 x_1) = -(x_1^* | y),$$
$$\rho(x_2^* | L_0 x_1) + \mu(x_2^* | L_1 x_1) = -(x_2^* | y),$$

and then finding x from (64). The solution of (64) will be denoted by $(x, \mu, \rho) = Sy$. We now set $x_s = sx_1 + sw_s$ and consider the equation

$$(w, \mu, \rho) = Sy \tag{65}$$

with

$$\omega_0 y = s^{-1}(1 + \rho)G(\mu, sx_1 + sw) - G_z(0,0)(x_1 + w) \\ - \rho G_z(0,0)x_1 - \mu G_{z\mu}(0,0)x_1. \tag{66}$$

The solution $(w_s, \mu(s), \rho(s))$ can then be determined from (65) and (66) by successive approximations with starting value $w = 0$, $\mu = \rho = 0$. In the analytic case one can also use ansatz and comparison of coefficients.

79.11. Applications to Ljapunov Bifurcation

We examine conditions under which there exist nonconstant, periodic solutions to the autonomous differential equation

$$z' = H(z) \tag{67}$$

in the neighborhood of the equilibrium point $z = 0$ (see Figure 79.3 for $\mathbb{R}^2$). In contrast to the Hopf bifurcation, no parameter occurs. We will, however, reduce this problem in a very simple fashion to the Hopf bifurcation for the equation

$$z' = H(z) + \mu E'(z) \tag{68}$$

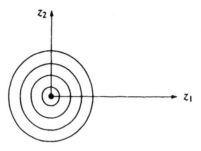

Figure 79.3

at the point $\mu = 0$. The *trick* is to choose the perturbation E' in such a way that every small periodic solution of (68) is also a solution of (67). We assume:

(H1) The map $H: U(0) \subseteq \mathbb{R}^n \to \mathbb{R}^n$ is C^2 with $H(0) = 0$.
(H2) (Conserved quantity). The C^3-function $E: \mathbb{R}^n \to \mathbb{R}$ is a conserved quantity for (67), i.e., we have $E(z(t)) = \text{const}$ along every solution of (67). Moreover, $E'(0) = 0$ and the matrix $E''(0)$ of the second-order partial derivatives of E at the point $z = 0$ is nonsingular.

Often, one can choose the energy as E.

(H3) (Nonresonance condition). $H'(0)$ has the algebraically simple eigenvalue $\omega_0 i$ with $\omega_0 > 0$ and no $k\omega_0 i$ with $k = 0$ or $k = 2, 3, \ldots$ is an eigenvalue of $H'(0)$. We set $p_0 = 2\pi/\omega_0$.

Theorem 79.G (Center Theorem of Ljapunov (1892)). *If* (H1)–(H3) *hold, then* (67) *has a family* $\{z_s\}$ *of nonconstant, periodic solutions of period* $p(s)$, *where*

$$\max_{t \in \mathbb{R}} |z_s(t)| \to 0, \qquad p(s) \to p_0$$

as $s \to 0$.

The name center theorem is used since the configuration in Figure 79.3 is called a center.

PROOF.

(I) Preparations. Recall the proof of the following two well-known results about conserved quantities:

$$E'(y)H(y) = 0 \quad \text{for all} \quad y \in \mathbb{R}^n, \tag{69}$$

$$A \stackrel{\text{def}}{=} E''(0)H'(0) + H'(0)^* E''(0) = 0. \tag{70}$$

Ad(69). For every $y \in \mathbb{R}^n$ there exists a solution z of (67) with $z(0) = y$. Because of

$$z'(t) = H(z(t)) = H'(0)z(t) + o(|z(t)|), \qquad t \to 0$$

we obtain from Taylor's theorem that

$$z(t) = y + tH'(0)y + o(t), \quad t \to 0. \tag{71}$$

Differentiation of $E(z(t)) = \text{const}$ gives $E'(z(t))z'(t) = 0$. For $t = 0$ we obtain (69).

Ad(70). From

$$E(z) = E(0) + 2^{-1}\langle z|E''(0)z\rangle + o(|z|^2), \quad z \to 0$$

and (71) it follows that

$$E(z(t)) = E(0) + 2^{-1}\langle y|E''(0)y\rangle + 2^{-1}t\langle y|Ay\rangle + o(t), \quad t \to 0.$$

This expression is constant in time, and hence $\langle y|Ay\rangle = 0$ for all $y \in \mathbb{R}^n$. Because of $A = A^*$ we thus have $A = 0$.

(II) Now we present the key trick: Every periodic solution $z(t) \not\equiv 0$ of (68), in a sufficiently small neighborhood of zero, is also a solution of (67). To see this, we compute

$$E(z(t))' = E'(z(t))z'(t) = E'(z(t))H(z(t)) + \mu|E'(z(t))|^2$$
$$= \mu|E'(z(t))|^2 = \mu\{|E''(0)z(t)|^2 + o(|z(t)|^2)\}.$$

Let $\mu > 0$ (or < 0). Because of $z(t) \not\equiv 0$ we have $z(t) \neq 0$ for all t. Hence we obtain $E''(0)z(t) \neq 0$ and therefore

$$E(z(t))' > 0 \quad \text{for all} \quad t \text{ (or } < 0).$$

But this is impossible for a periodic solution. Thus we must have $\mu = 0$.

(III) Hopf bifurcation for (68) at $\mu = 0$. We want to apply Theorem 79.F to (68). In doing so we only need to check condition (H3) of Section 79.9 about loss of stability.

Because of (H3) and (70) we can use a linear coordinate transformation to get at first $H'(0)$ and then $E''(0)$ into the following form:

$$H'(0) = \begin{pmatrix} 0 & \omega_0 & 0 \\ -\omega_0 & 0 & 0 \\ 0 & 0 & \omega \end{pmatrix}, \quad E''(0) = \begin{pmatrix} a & 0 & 0 \\ 0 & a & 0 \\ 0 & 0 & b \end{pmatrix}.$$

Because of $\det E''(0) \neq 0$ we have $a \neq 0$. Thus the matrix

$$H'(0) + \mu E''(0)$$

has the eigenvalue $\lambda(\mu) = \mu a + \omega_0 i$, and hence $\operatorname{Re} \lambda'(0) \neq 0$.

The bifurcation solutions of (68), which follow from Theorem 79.F, are the required solutions of (67). □

Problems

79.1. Quasi-eigenvectors. Let Y be a complex B-space and let $A \in L(Y, Y)$. Moreover, let λ be a boundary point of the spectrum $\sigma(A)$.

Prove:
(i) For every $\varepsilon > 0$ there exists a unit vector $y \in Y$ with $\|(A - \lambda I)y\| < \varepsilon$.
(ii) For every $T > 0$ and for every $\eta \in]0, 1[$ there exists a unit vector $y \in Y$ with

$$e^{t \operatorname{Re} \lambda}(1 - \eta) \leq \|e^{tA}y\| \leq e^{t \operatorname{Re} \lambda}(1 + \eta) \tag{72}$$

for all $t \in [0, T]$.

Solution: Ad(i). Let $\rho(A)$ be the resolvent set and let $R_\mu = (A - \mu I)^{-1}$ be the resolvent for $\mu \in \rho(A)$. We have

$$(A - \lambda I)R_\mu = (\mu - \lambda)R_\mu + I. \tag{73}$$

For $\lambda \in \sigma(A)$ and $\mu \in \rho(A)$ we have $\|R_\mu\| \geq |\mu - \lambda|^{-1}$. Otherwise we obtain $\|(A - \lambda I)R_\mu - I\| < 1$, i.e., because of the Neumann series (see A_1(57d)), $(A - \lambda I)R_\mu$ is invertible and hence also $A - \lambda I$, which contradicts $\lambda \in \sigma(A)$.

For the boundary point λ of $\sigma(A)$ we find a $\mu \in \rho(A)$ with $|\mu - \lambda| < \varepsilon/2$, and hence

$$\|R_\mu\| > 2/\varepsilon.$$

We choose a unit vector x with $\|R_\mu x\| > 2/\varepsilon$ and let $y = R_\mu(x/\|R_\mu x\|)$. From (73) it follows that

$$\|(A - \lambda I)y\| \leq |\mu - \lambda| + \|R_\mu x\|^{-1} < \varepsilon.$$

Ad(ii). Choose y as in (i). Power series expansion implies that

$$(e^{At} - e^{\lambda t}I)y = e^{\lambda t}(e^{(A - \lambda I)t} - I)y$$

$$= e^{\lambda t} \int_0^t e^{(A - \lambda I)s}(A - \lambda I)y \, ds$$

and thus

$$\|e^{At}y - e^{\lambda t}y\| \leq e^{t \operatorname{Re} \lambda} \int_0^T \|e^{(A - \lambda I)s}\| \varepsilon \, ds.$$

Because of $\|e^{\lambda t}y\| = e^{t \operatorname{Re} \lambda}$ we obtain (72) for sufficiently small $\varepsilon > 0$.

79.2. *Proof of (15) and (16).*

Solution: For a complex B-space X this immediately follows from (72). Now assume X is real. We choose Y equal to the complexification $X_{\mathbb{C}}$ (see A_1(23h)). Let $y = y_1 + iy_2$. From (72) it follows with $\eta = \frac{1}{2}$ that

$$\tfrac{1}{2}e^{T \operatorname{Re} \lambda} \leq \|e^{TA}y\| \leq \|e^{TA}y_1\| + \|e^{TA}y_2\|.$$

For $b = y_1$ or $b = y_2$ we therefore have

$$\tfrac{1}{4}e^{T \operatorname{Re} \lambda} \leq \|e^{TA}b\|.$$

This is (15). Let, for example, $b = y_1$. From (72) it follows that

$$\|e^{tA}b\| = \|\operatorname{Re}(e^{tA}y)\| \leq \|e^{tA}y\| \leq 2e^{t \operatorname{Re} \lambda}$$

for all $t \in [0, T]$. This is (16).

79.3. *A class of Fredholm operators.* Let X and Y be B-spaces over $\mathbb{K}$. Let M denote the set of all Fredholm operators $A \in L(X, Y)$ of index zero with $\dim N(A) = 1$. Prove that M is an analytic submanifold of $L(X, Y)$.

Hint: Use Proposition 79.15. See Crandall and Rabinowitz (1973).

79.4. Proof of (49).

Solution:

(I) We shall use the fact about real functions that
$$|r(t)| \leq o(1)a(t), \qquad t \to 0$$
and $a(t) \neq 0$ for $t \neq 0$ always implies that
$$r(t)/a(t) \to 0 \quad \text{as} \quad t \to 0.$$
Thus we have $r(t) = o(1)a(t)$.

(II) Furthermore, it always follows from
$$\|a(s)\| \leq o(1)\|a(s)\| + \|b\|, \qquad s \to 0$$
that $\|a(s)\|/2 \leq \|b\|$ for small $|s|$.

(III) Let $N^\perp = N(F_x(0,0))^\perp$. It is important that the linear equation
$$F_x(0,0)z - \varepsilon F_{x\varepsilon}(0,0)x_1 = y, \qquad (z, \varepsilon) \in N^\perp \times \mathbb{R} \qquad (74)$$
has a unique solution (z, ε) for every $y \in Y$ (see Remark 79.17). The open mapping theorem $A_1(36)$ therefore implies that there exists a constant c with
$$\|z\| + |\varepsilon| \leq c\|y\|. \qquad (75)$$

(IV) We write $F(s)$ for $F(\varepsilon(s), x(s))$. Let $\varepsilon'(s) \neq 0$ for $s \neq 0$. According to Section 79.7, equation
$$F_x(s)(x_1 + z(s)) = \lambda(s)J(x_1 + z(s))$$
has a C^1-solution $s \mapsto (\lambda(s), z(s))$ with $\lambda(s) \in \mathbb{R}$, $z(s) \in N^\perp$, and $\lambda(0) = 0$, $z(0) = 0$. From $F(s) \equiv 0$ follows
$$F_\varepsilon(s)\varepsilon'(s) + F_x(s)x'(s) = 0.$$
Subtraction gives
$$F_x(s)u(s) = \lambda(s)J(x_1 + z(s)) + F_\varepsilon(s)\varepsilon'(s) \qquad (76)$$
with $u(s) \stackrel{\text{def}}{=} x_1 + z(s) - x'(s)$. Because of $x'(s) - x_1 \in N^\perp$ we obtain $u(s) \in N^\perp$. Note that
$$F_x(s)u(s) = F_x(0,0)u(s) + o(1)u(s).$$
From $F(\varepsilon, 0) \equiv 0$ follows $F_\varepsilon(\varepsilon, 0) \equiv 0$. Therefore, Taylor's theorem of Section 4.6 implies
$$F_\varepsilon(s)\varepsilon'(s) = (F_\varepsilon(\varepsilon(s), x(s)) - F_\varepsilon(\varepsilon(s), 0))\varepsilon'(s)$$
$$= (F_{\varepsilon x}(0,0) + o(1))x(s)\varepsilon'(s)$$
$$= F_{\varepsilon x}(0,0)sx_1\varepsilon'(s) + o(1)s\varepsilon'(s).$$
Note that $x(s) = s(x_1 + o(1))$. Thus it follows from (76) that
$$F_x(0,0)u(s) - F_{x\varepsilon}(0,0)\varepsilon'(s)sx_1$$
$$= \lambda(s)(Jx_1 + o(1)) + o(1)u(s) + o(1)\varepsilon'(s)s. \qquad (77)$$
Equation (75) and (II) imply
$$\|u(s)\| + |\varepsilon'(s)s| \leq \text{const}\,|\lambda(s)|. \qquad (78)$$

Hence we have $\lambda(s) \neq 0$ for $s \neq 0$. An application of y_1^* to (77) gives

$$-\varepsilon'(s)s\langle y_1^*, F_{x\varepsilon}(0,0)x_1\rangle = \lambda(s)(1 + o(1)).$$

Here observe that $\langle y_1^*, F_x(0,0)x\rangle = 0$ for all $x \in X$ and note (78) and (I). For $s \to 0$ we obtain (49).

79.5. Special Fredholm operator. Prove that (H5) of Section 79.9 is always satisfied for $X = Y = \mathbb{R}^n$.

Solution: We let $Dx = \dot{x}$. Then $D: X_{2\pi} \to Y_{2\pi}$ is a Fredholm operator of index zero, because from $\dot{x} = 0$ follows $x = \text{const}$, and $\dot{x} = f$ means that

$$\int_0^{2\pi} f(t)\,dt = 0.$$

Thus we have $\dim N(D) = \operatorname{codim} R(D) = 1$. According to Section 8.4 the compact perturbation $D - \omega_0^{-1} G_z(\mu_0, z(\mu_0))$ of D is also a Fredholm operator of index zero.

79.6. Proof of Lemma 79.19.

Solution: We proceed analogously as in Problem 79.4. We set $N = N(F_x(0,0))$. From Theorem 8.B the bifurcation solution $x_s \in X_{2\pi}$ has the form

$$x_s = sx_1 + sw_s$$

with $w_s \in N^\perp$ and $w_s = 0$ for $s = 0$. We let $v_s = sw_s$. With x'_s or $\dot{x}_s$ we denote the derivatives with respect to the parameter s or time τ. Moreover let

$$F(s) = F(\varepsilon(s), x_s), \qquad F_\rho(s) = F_\rho(\varepsilon(s), x_s), \quad \text{etc.}$$

Note that $\varepsilon = (\rho, \mu)$.

(I) The equation

$$F_x(0,0)u - \kappa x_1 - \eta x_2 = y, \qquad (u, \kappa, \eta) \in N^\perp \times \mathbb{R}^2$$

has a unique solution for every $y \in Y_{2\pi}$. This follows as in Section 79.7 from

$$\langle x_i^*, x_j\rangle = \delta_{ij} \quad \text{and} \quad \langle x_i^*, F_x(0,0)x\rangle = 0 \quad \text{for all } x \in X_{2\pi}.$$

Hence, from the open mapping theorem $A_1(36)$ there exists a constant c with

$$\|u\| + |\kappa| + |\eta| \leq c\|y\|.$$

This estimate will now essentially be used.

(II) Because of (I), the implicit function theorem implies that equation

$$F_x(s)(x_1 + u_s) = \kappa(s)(x_1 + u_s) + \eta(s)(x_2 - \dot{w}_s) \tag{79}$$

has a C^1-solution curve $s \mapsto (u_s, \kappa(s), \eta(s))$ in $N^\perp \times \mathbb{R}^2$ which passes through $(0, 0, 0)$ at $s = 0$.

(III) Differentiation of $F(s) = s$ with respect to τ gives

$$F_x(s)\dot{x}_s = 0.$$

Moreover, it is $\dot{x}_s = -sx_2 + s\dot{w}_s$. Below we will show that $\kappa(s) \neq 0$ for

Problems 873

$s \neq 0$. From (II) we obtain

$$F_x(s)(x_1 + u_s + \kappa(s)^{-1}\eta(s)(x_2 - \dot{w}_s))$$
$$= \kappa(s)(x_1 + u_s + \kappa(s)^{-1}\eta(s)(x_2 - \dot{w}_s)),$$

i.e., $\kappa(s)$ is an eigenvalue of $F_x(s)$.

(IV) Differentiation of $F(s) = 0$ with respect to s gives

$$\rho'(s)F_\rho(s) + \mu'(s)F_\mu(s) + F_x(s)(x_1 + v'_s) = 0.$$

Subtraction of (79) yields

$$\rho'(s)F_\rho(s) + \mu'(s)F_\mu(s) + F_x(s)(v'_s - u_s)$$
$$+ \kappa(s)(x_1 + u_s) + \eta(s)(x_2 - \dot{w}_s) = 0.$$

From (60) and $G(\mu, 0) = 0$ it follows that

$$F_\mu(s) = s\omega_0^{-1}G_{\mu z}(0,0)x_1 + o(s) = sL_1 x_1 + o(s),$$
$$F_x(s) = F_x(0,0) + o(1),$$
$$F_\rho(s) = \omega_0^{-1}G(\mu(s), x_s) = (1 + \rho(s))^{-1}\dot{x}_s$$

as $s \to 0$.

Letting $\beta(s) = s\rho'(s)/(1 + \rho(s))$, we obtain

$$F_x(0,0)(v'_s - u_s) + \kappa(s)x_1 + (\eta(s) - \beta(s))x_2$$
$$= o(1)(v'_s - u_s) + o(1)\kappa(s) + o(1)(\eta(s) - \beta(s))$$
$$+ o(1)s\mu'(s) - s\mu'(s)L_1 x_1. \tag{80}$$

From (I) follows

$$\|v'_s - u_s\| + |\kappa(s)| + |\eta(s) - \beta(s)| \leq O(1)|s\mu'(s)|. \tag{81}$$

Note that $\langle x_1^*, L_1 x_1 \rangle = \omega_0^{-1}\operatorname{Re}\lambda'(\mu_0)$. An application of x_1^* to (80) together with (I) and (81) gives

$$|\kappa(s) + s\mu'(s)\omega_0^{-1}\operatorname{Re}\lambda'(\mu_0)| \leq o(1)s\mu'(s).$$

If $\mu'(s) \neq 0$ for $s \neq 0$, we may divide by $s\mu'(s)$ to obtain (63) for $s \to 0$.

79.7.** *Global Hopf bifurcation.* Similarly as in Section 79.9, we consider the autonomous differential equation

$$z' = G(\mu, z), \tag{82}$$

where $G: \mathbb{R} \times \mathbb{R}^n \to \mathbb{R}^n$ is C^1. We assume:
(i) $G(\mu, 0) = 0$ for all μ in a neighborhood of μ_0.
(ii) $G_z(\mu_0, 0)$ has $\omega_0 i$ as an algebraically simple eigenvalue with $\omega_0 > 0$, and we assume conditions (H3) and (H4) of Section 79.9 about loss of stability and nonresonance. We set $p_0 = 2\pi/\omega_0$.
(iii) For any parameter value $\mu \in \mathbb{R}$ there passes a unique solution curve of (82) through each point $z_0 \in \mathbb{R}^n$, which exists for all times t.

We define a set P in $\mathbb{R}^{n+2}$. The point (p, μ, z_0) then belongs to P precisely if there exists a nonconstant, p-periodic solution of (82) which passes through

z_0. Let S denote the component of $P \cup \{(p_0, \mu_0, 0)\}$ which contains the point $(p_0, \mu_0, 0)$. Prove that precisely one of the following two cases must occur:
(a) S is unbounded.
(b) S is bounded and contains a point (p, μ_1, z_1) different from $(p_0, \mu_0, 0)$, where z_1 is an equilibrium point of (82).

Interpretation. The points of S in a neighborhood of $(p_0, \mu_0, 0)$ correspond to p-periodic solutions with

$$\max_{t \in \mathbb{R}} |z(t)| + |p - p_0| + |\mu - \mu_0| \to 0.$$

This follows from Theorem 79.F (local Hopf bifurcation). In case (a), p-periodic solutions z of (82) lie on S with

$$\max_{t \in \mathbb{R}} |z(t)| + |p| + |\mu| \to \infty.$$

In case (b) there exists an equilibrium point z_1 for which a Hopf bifurcation occurs at $\mu = \mu_1$, which is different from the Hopf bifurcation at $z = 0$, $\mu = \mu_0$.

Pendulum as an example. In order to get an intuitive understanding of cases (a) and (b), we consider a pendulum under the influence of an outer force parameter μ which varies. Roughly, the following holds:

At $\mu = \mu_0$ the pendulum passes from a state of rest into periodic oscillations. In case (b) the pendulum returns to a state of rest.

If the pendulum does not return into a state of rest, then (a) implies that at least one of the following three situations occurs:
(α) For all μ there exist periodic oscillations.
(β) The period of the oscillations increases.
(γ) The amplitudes of the oscillations increase.

This behavior corresponds to what one intuitively expects.

Hint: This important theorem is due to Alexander and Yorke (1978). A relatively simple proof may be found in Ize (1976), p. 93. It uses the same idea as has been used in the proof of Theorem 15.C of Part I. Instead of the fixed-point index, a more sophisticated homotopy argument about essential maps is used. Moreover, we recommend Chow and Mallet-Paret (1978) (Fuller index) and Nussbaum (1978) (retarded functional differential equations).

79.8. *Two fundamental results about the perturbation of spectra.* Let $L(X, X)$ denote the set of all continuous linear operators $A: X \to X$ on the B-space X over $\mathbb{K} = \mathbb{R}, \mathbb{C}$ where $X \neq \{0\}$. Let $\sigma(A)$ denote the spectrum of A. Recall that, by definition, $\sigma(A)$ is equal to the spectrum of the complexification $A_\mathbb{C}$ if $\mathbb{K} = \mathbb{R}$. Moreover, in this case, λ is called an eigenvalue of A if and only if it is an eigenvalue of $A_\mathbb{C}$ (cf. A_1(23h), A_1(56)).

79.8a. *The upper continuity of the spectrum.* Show that the map

$$\sigma: L(X, X) \to 2^\mathbb{C} \qquad (83)$$

is upper semicontinuous, i.e., for each neighborhood U of $\sigma(A)$ in $\mathbb{C}$, there exists a neighborhood $V(A)$ of A in $L(X, X)$ such that

$$\sigma(B) \subset U \quad \text{for all} \quad B \in V(A).$$

In addition, for given $A \in L(X, X)$ and for each $\varepsilon > 0$, there exists a number

$\delta(A, \varepsilon) > 0$ such that

$$\text{dist}(\sigma(B), \sigma(A)) < \varepsilon$$

for all $B \in L(X, X)$ with $\|B - A\| < \delta$.

Hint: Use the resolvent $(A - \lambda I)^{-1}$ and the continuity of inverse formation from Problem 1.7 of Part I. Cf. Kato (1966, M), Chapter 4, 3.1.

Note that the map (83) is not necessarily lower semicontinuous, i.e., if an open set U is contained in $\sigma(A)$, then U is not necessarily contained in $\sigma(B)$, where B is a perturbation of A in $L(X, X)$. Indeed, there exist an infinite-dimensional B-space X and operators A and C such that

$$\sigma(A + \gamma C) = \begin{cases} \text{closed unit disk } D & \text{if } \gamma = 0, \\ \partial D & \text{if } \gamma \neq 0, \end{cases}$$

(cf. Kato (1966, M), Example 3.8 of Chapter 4). The following result shows that the situation is much better in finite-dimensional B-spaces X (e.g., $X = \mathbb{R}^n$ or $X = \mathbb{C}^n$).

79.8b. *Main theorem of perturbation theory in finite-dimensional B-spaces.* Let $A \in L(X, X)$ be given, where X is a finite-dimensional B-space over $\mathbb{K} = \mathbb{R}, \mathbb{C}$. Let $\lambda_1, \ldots, \lambda_r$ be the points in the spectrum $\sigma(A)$ of A with algebraic multiplicity $m_1, \ldots, m_r$, respectively, i.e., the $\lambda_1, \ldots, \lambda_r$ are the distinct eigenvalues of A, where λ_j is a solution of the characteristic equation

$$\det(A_\mathbb{C} - \lambda I) = 0$$

with multiplicity m_j for each j. Moreover, let $U_1, \ldots, U_r$ be open subsets of $\mathbb{C}$ with

$$\lambda_j \in U_j, \qquad j = 1, \ldots, r$$

and $\bar{U}_j \cap \bar{U}_k = \emptyset$ for all $j \neq k$. We set

$$U = \bigcup_{j=1}^{r} U_j.$$

Finally, let $V(A)$ be a connected neighborhood of A in $L(X, X)$ such that

$$\sigma(B) \subset U \qquad \text{for all} \quad B \in V(A).$$

According to Problem 79.8a, such a neighborhood always exists.

Show that

$$\sigma(B) \cap U_j \neq \emptyset \qquad \text{for all } j,$$

and show that the sum of the algebraic multiplicities of the eigenvalues of B contained in U_j is equal to m_j.

Solution: We use the properties of the mapping degree summarized in Section 13.6. We may assume that $\mathbb{K} = \mathbb{C}$. Otherwise, we replace X and A with the complexification $X_\mathbb{C}$ and $A_\mathbb{C}$, respectively. We set

$$f_B(\lambda) = \det(B - \lambda I)$$

and

$$\deg(f_B, U_j) = n_j.$$

Since $f_B(\lambda) \neq 0$ on ∂U_j, the mapping degree is well-defined. According to Proposition 14.2, the number n_j is equal to the sum of the multiplicities of the zeros of f_B in U_j. Since the function $B \mapsto f_B$ is continuous from $L(X, X)$ into $C(\bar{U}_j)$, it follows from the homotopy invariance of the mapping degree that the function

$$B \mapsto \deg(f_B, U_j)$$

is continuous on the *connected* set $V(A)$ and integer-valued. This implies

$$\deg(f_B, U_j) = \text{constant} \qquad \text{for all} \quad B \in V(A).$$

Letting $B = A$, we obtain $n_j = m_j$. Since $\deg(f_B, U_j) \neq 0$, we also obtain that f_B has at least one zero on U_j, i.e., $\sigma(B) \cap U_j \neq \emptyset$.

79.9. Hopf bifurcation in $\mathbb{R}^N$ and the orbital stability of the bifurcating periodic solutions. We consider the differential equation

$$z' = G(\mu, z) \tag{84}$$

depending on the real parameter μ. We are looking for nonconstant periodic solutions $z = z(t)$.

Our assumptions are as follows:

(H1) *Trivial solution (equilibrium state).* The map

$$G: U(\mu_0, 0) \subseteq \mathbb{R} \times \mathbb{R}^N \to \mathbb{R}^N$$

is C^2 with $G(\mu, 0) = 0$ for all μ in a neighborhood of μ_0.

(H2) *Loss of stability (transversality condition).* There exists an $\omega_0 > 0$ such that $i\omega_0$ is an algebraically simple eigenvalue of $G_z(\mu_0, 0)$.

According to Proposition 79.15, there exists a unique C^1-function $\mu \mapsto \lambda(\mu)$ on a neighborhood $V(\mu_0)$ of μ_0 such that $\lambda(\mu_0) = i\omega_0$, and $\lambda(\mu)$ is an algebraically simple eigenvalue of $G_z(\mu, 0)$ for all $\mu \in V(\mu_0)$. We assume that the so-called transversality condition

$$\operatorname{Re} \lambda'(\mu_0) > 0 \tag{85}$$

is satisfied.

(H3) *Nonresonance condition.* None of the numbers $\pm ik\omega_0$ with $k = 0$ and $k = 2, 3, \ldots$ is an eigenvalue of $G_z(\mu_0, 0)$.

(H3*) *Strong nonresonance condition.* All eigenvalues λ of $G_z(\mu_0, 0)$ different from $\pm i\omega_0$ satisfy the stability condition $\operatorname{Re} \lambda < 0$.

Conditions (H2) and (H3*) mean that the equilibrium point $z = 0$ of (84) is asymptotically stable for $\mu < \mu_0$ and unstable for $\mu > \mu_0$. This follows from Proposition 79.3 and Problem 79.8.

Show: If the assumptions (H1), (H2), and (H3) are satisfied, then the following are true.

(a) *Existence.* Let $W(0)$ be a sufficiently small neighborhood of $s = 0$ in $\mathbb{R}$. There exist two even C^1-functions

$$\mu, p: W(0) \to \mathbb{R} \tag{86}$$

and a family $\{z_s\}$ of nonconstant $p(s)$-periodic functions

$$z = z_s(t)$$

for each $s \neq 0$ in $W(0)$ such that $(\mu(s), z_s)$ is a solution of the original problem (84) for each $s \in W(0)$. Here, we have $z_s(t) \equiv 0$ for $s = 0$ and

$$\mu(s) = \mu_0 + O(s), \qquad\qquad s \to 0,$$

$$p(s) = \frac{2\pi}{\omega_0} + O(s), \qquad\qquad s \to 0,$$

$$z_s(t) = s\left(a\cos\frac{2\pi}{\omega_0}t + b\sin\frac{2\pi}{\omega_0}t\right) + o(s), \qquad s \to 0,$$

with

$$\lim_{s \to 0}\max_{t \in \mathbb{R}}|z_s(t)| = 0.$$

The real numbers a and b with $|a| + |b| \neq 0$ are independent of s.

If G is analytic in (H2), then $\mu(\cdot)$ and $p(\cdot)$ are also analytic.

(b) *Uniqueness.* There exists a neighborhood $U(0)$ in $\mathbb{R}^N$ of the equilibrium point $z = 0$, and there exist neighborhoods $U(\mu_0)$ and $U(p(0))$ in $\mathbb{R}$ such that the following holds. If C is the orbit of a p-periodic solution (μ, z) of (84) with

$$C \subset U(0), \qquad \mu \in U(\mu_0), \qquad \text{and} \qquad p \in U(p(0)),$$

then (μ, z) corresponds to one of the solutions in (a) up to time translation.

(c) *Stability.* If, in addition, condition (H3*) is satisfied, and if the bifurcation is *supercritical*, i.e.,

$$s\mu'(s) > 0 \qquad \text{for all} \quad s \neq 0 \quad \text{in } W(0), \tag{87a}$$

then the periodic solution $t \mapsto z_s(t)$ is asymptotically orbitally stable for all $s \neq 0$ in $W(0)$.

(d) *Instability.* If, in addition, condition (H3*) is satisfied, and if the bifurcation is *subcritical*, i.e.,

$$s\mu'(s) < 0 \qquad \text{for all} \quad s \neq 0 \quad \text{in } W(0), \tag{87b}$$

then $t \mapsto z_s(t)$ is orbitally unstable for all $s \neq 0$ in $W(0)$.

Solution: The existence and uniqueness of the solution $(\mu(s), z_s)$ follows from the main theorem on Hopf bifurcation (Theorem 79.F). We now investigate the stability of $t \mapsto z_s(t)$ by using Theorem 79.D.

(I) First, let $s = 0$. We set

$$L = e^{G_z(\mu_0, 0)p(0)}.$$

According to assumption (H3*), the operator $G_z(\mu_0, 0): \mathbb{R}^N \to \mathbb{R}^N$ has the eigenvalues

$$\lambda_1, \ldots, \lambda_r \qquad \text{with} \qquad \operatorname{Re}\lambda_j < 0 \qquad \text{for all } j$$

and $\lambda_\pm = \pm i\omega_0$, where λ_+ and λ_- are algebraically simple. Thus, the operator $L: \mathbb{R}^N \to \mathbb{R}^N$ has the eigenvalues

$$m_j = e^{\lambda_j p(0)}, \qquad j = 1, \ldots, r \qquad \text{with} \quad |m_j| < 1 \qquad \text{for all } j$$

and the eigenvalue

$$m = e^{\lambda_\pm p(0)} = e^{\pm 2\pi i} = 1,$$

where the algebraic multiplicity of m equals *two*.

(II) We now let $s \neq 0$. The Floquet multipliers of $t \mapsto z_s(t)$ are the *eigenvalues* of the shift operator $S(p(s))$ which corresponds to the differential equation

$$z'(t) = G_z(\mu(s), z_s(t))z(t).$$

According to (25), we obtain the integral equation

$$S(t) = I + \int_0^t G_z(\mu(s), z_s(\tau))S(\tau)\,d\tau.$$

Since $\max_{t \in \mathbb{R}} |z_s(t)| \to 0$ as $s \to 0$, according to Theorem 79.F, we obtain from Proposition 1.2 the crucial property

$$\lim_{s \to 0} S(p(s)) = L.$$

Therefore, we can use the methods of perturbation theory in order to study the structure of the eigenvalues of $S(p(s))$, which are the Floquet multipliers of z_s. Let $|s|$ be sufficiently small.

(II-1) From Problem 79.8a it follows that the perturbations of $m_1, \ldots, m_r$ remain within the open unit disk and outside a small neighborhood of the point $m = 1$.

(II-2) From Problem 79.8b it follows that the eigenvalue $m = 1$ of L splits into the eigenvalues $m_+(s)$ and $m_-(s)$ counted according to their algebraic multiplicity, i.e., the total algebraic multiplicity equals *two*. Moreover $m_+(s)$ and $m_-(s)$ lie in a small neighborhood of the point $m = 1$.

According to Example 79.10 and Lemma 79.19, there are two Floquet multipliers of z_s with

$$M_\pm(s) = e^{2\pi\kappa_\pm(s)}$$

and $\kappa_+(s) = \kappa(s)$, $\kappa_-(s) = 0$, where $\kappa(s) \to 0$ as $s \to 0$. From (II-1) and (II-2) we obtain

$$m_\pm(s) = M_\pm(s).$$

(III) It follows from Lemma 79.19 that

$$\operatorname{sgn} \kappa(s) = -\operatorname{sgn} s\mu'(s).$$

(III-1) For (87a), we obtain $\kappa(s) < 0$. Thus, all the eigenvalues of $S(p(s))$ (i.e., the Floquet multipliers of z_s) are contained in the open unit disk. According to Theorem 79.D, the periodic solution $t \mapsto z_s(t)$ is asymptotically orbitally stable.

(III-2) For (87b), we obtain $\kappa(s) > 0$. Therefore, the Floquet multiplier $m_+(s)$ satisfies $m_+(s) > 1$. According to Theorem 79.D, the solution z_s is orbitally unstable.

79.10.* *Degenerate Hopf bifurcation.*

79.10a. *Violation of the transversality condition* (H2) *in Problem* 79.9. Study Kielhöfer (1982).

79.10b. *Violation of the nonresonance condition* (H3) *in Problem* 79.9. Suppose that condition (H3) is not valid. A careful study of this so-called resonance case can be found in Recke (1987), (1988). In this connection it is very important that the four-dimensional system of branching equations can be essentially simplified by using the symmetry invariance of the problem under the action of the rotation group in $\mathbb{R}^2$.

Roughly speaking, one obtains the following result in the generic resonance case. If *two* pairs of complex conjugate eigenvalues of $G_z(\mu, 0)$ cross the imaginary axis at the parameter value $\mu = \mu_0$, then the equilibrium state $z = 0$ bifurcates into *four* families of periodic oscillations whose stability properties are *subtle*. In order to describe transparently these stability properties, it is necessary to consider multi-dimensional parameters μ in $\mathbb{R}^d$, where the dimension d of the parameter space is sufficiently large.

79.11. *Abstract parabolic equations and Hopf bifurcation.* In order to obtain a variant of Theorem 79.F, we consider the differential equation

$$z' + Lz + g(\mu, z) = 0, \tag{88}$$

where μ is a real parameter. We are looking for nonconstant periodic solutions $z = z(t)$ of (88).

In Section 19.26 we studied initial-value problems for abstract parabolic equations of the form (88) together with applications to parabolic differential equations. In this connection, sectorial operators L, fractional powers $(L + rI)^\alpha$, and abstract Sobolev spaces X_α played a fundamental role. The following results allow applications to Hopf bifurcation for broad classes of parabolic differential equations.

Recall that the assumption (H1a) below is satisfied, if, for example, the operator $L: D(L) \subseteq X \to X$ is linear, self-adjoint, and bounded below on the real H-space X, i.e., there is a real number c such that

$$(Lz|z) \geq c\|z\|^2 \quad \text{for all} \quad z \in D(L).$$

We make the following assumptions.

(H1a) *Sectorial operator L.* Let X be a real B-space. The operator

$$L: D(L) \subseteq X \to X$$

is linear, densely defined, graph closed, and the complexification $L_\mathbb{C}$ is sectorial on the complexification $X_\mathbb{C}$.

(H1b) *Compact resolvent.* The operator

$$(L_\mathbb{C} - \lambda I)^{-1}: X_\mathbb{C} \to X_\mathbb{C}$$

is compact for each $\lambda \in \mathbb{C}$ in the resolvent set of $L_\mathbb{C}$.

It follows from (H1a) that the operator $-L$ generates an analytic semigroup. Moreover, it follows that if $r > -\operatorname{Re} \lambda$ for all $\lambda \in \mathbb{C}$ in the spectrum of L, then the fractional powers $(L + rI)^\alpha$ are well defined for each $\alpha \geq 0$. In this connection, the B-spaces X_α (abstract Sobolev spaces) with norms $\|\cdot\|_\alpha$ are defined by

$$X_\alpha = D((L + rI)^\alpha)$$

and $\|z\|_\alpha = \|(L + rI)^\alpha z\|_X$ for all $z \in X_\alpha$. We have $X_\alpha \subseteq X$ for all $\alpha \geq 0$.

(H2) *Nonlinearity.* There is an α with $0 \leq \alpha < 1$ and a neighborhood U of $(\mu_0, 0)$ in $\mathbb{R} \times X_\alpha$ such that the function

$$g: U \to X$$

is C^m with $m \geq 2$. Henceforth the value of α is fixed. Furthermore, suppose that

$$g(\mu, 0) = 0 \quad \text{and} \quad g_z(0, 0) = 0$$

for all μ in a neighborhood of μ_0.

(H3) *Loss of stability (transversality condition).* There is an $\omega_0 > 0$ such that $\lambda = i\omega_0$ is an algebraically simple eigenvalue of L_C, i.e.,

$$\dim N(L_C - i\omega_0 I) = \operatorname{codim} R(L_C - i\omega_0 I) = 1$$

and $z \in N(L_C - i\omega_0 I)$ with $z \neq 0$ implies $z \notin R(L_C - i\omega_0 I)$.

This assumption tells us that $\lambda = i\omega_0$ is an I-simple eigenvalue of L_C regarded as a map from $(X_1)_C$ into X_C. Hence it follows from Section 79.7 that there exists a unique C^1-function $\mu \mapsto \lambda(\mu)$ on a neighborhood $U(\mu_0)$ of μ_0 such that $\lambda(\mu_0) = i\omega_0$ and $\lambda(\mu)$ is an eigenvalue of $L + g_z(\mu, 0)$ for each $\mu \in U(\mu_0)$. We assume that the transversality condition

$$\operatorname{Re} \lambda'(\mu_0) \neq 0 \tag{89}$$

is satisfied.

(H4) *Nonresonance condition.* None of the numbers $\pm ik\omega_0$ with $k = 0$ and $k = 2, 3, \ldots$ is in the spectrum of L.

(H4*) *Strong nonresonance condition.* All points in the spectrum of L different from $\pm i\omega_0$ satisfy $\operatorname{Re} \lambda > 0$, and $\operatorname{Re} \lambda'(\mu_0) < 0$.

By a *solution* of the original problem (88), we understand a continuous function

$$z: [0, \infty[\to X_\alpha$$

such that $z(t) \in D(L)$ for all $t > 0$ and the derivative

$$z':]0, \infty[\to X$$

is continuous and satisfies equation (88).

Show: If assumptions (H1) to (H4) are satisfied, then the following are true.

(a) *Existence.* Let $W(0)$ be a sufficiently small neighborhood of $s = 0$ in $\mathbb{R}$. There exist two even C^{m-1}-functions

$$\mu, p: W(0) \to \mathbb{R}$$

and a family $\{z_s\}$ of nonconstant $p(s)$-periodic functions

$$z = z_s(t)$$

for each $s \neq 0$ in $W(0)$ such that $(\mu(s), z_s)$ is a solution of the original problem (88) for each $s \in W(0)$. Here, we have $z_s(t) \equiv 0$ for $s = 0$ and

$$\mu(s) = \mu_0 + O(s), \qquad s \to 0,$$

$$p(s) = \frac{2\pi}{\omega_0} + O(s), \qquad s \to 0,$$

$$\lim_{s \to 0} \max_{t \geq 0} \|z_s(t)\|_\alpha = 0.$$

If g is analytic in (H2), then $\mu(\cdot)$ and $p(\cdot)$ are also analytic.

(b) *Uniqueness.* There exists a neighborhood $U(0)$ in X_α of the equilibrium point $z = 0$, and there exist neighborhoods $U(\mu_0)$ and $U(p(0))$ in $\mathbb{R}$ such that the following holds. If C is the orbit of a p-periodic solution (μ, z) of (88) with

$$C \subset U(0), \quad \mu \in U(\mu_0), \quad \text{and} \quad p \in U(p(0)),$$

then (μ, z) corresponds to one of the solutions in (a) up to time translation.

(c) *Stability.* If, in addition, condition (H4*) is satisfied, and if the bifurcation is *supercritical,* i.e.,

$$s\mu'(s) > 0 \quad \text{for all} \quad s \neq 0 \quad \text{in } W(0),$$

then the periodic solution $t \mapsto z_s(t)$ is asymptotically orbitally stable for all $s \neq 0$ in $W(0)$.

(d) *Instability.* If, in addition, condition (H4*) is satisfied, and if the bifurcation is *subcritical,* i.e.,

$$s\mu'(s) < 0 \quad \text{for all} \quad s \neq 0 \quad \text{in } W(0),$$

then $t \mapsto z_s(t)$ is orbitally unstable for all $s \neq 0$ in $W(0)$.

Hint: Use analogous arguments as in the proofs of Theorem 79.F, Problem 79.9, and Theorem 19.I on abstract parabolic equations. Replace the differential equation (88) with the equivalent integral equation

$$z(t) = S(t)z(0) - \int_0^t S(t-s)g(\mu, z(s))\,ds,$$

where $\{S(t)\}$ denotes the analytic semigroup generated by $-L$. Cf. Crandall and Rabinowitz (1977).

References to the Literature

Classical works: Floquet (1883), Ljapunov (1892), Poincaré (1892), Hopf (1942).
Introductory exposition in $\mathbb{R}^n$: Amann (1983, M).
Applications to mechanics: Abraham and Marsden (1978, M).
Stability theory in B-spaces: Daleckii and Krein (1970, M, B, H), Henry (1981, L).

Hopf bifurcation: Joseph and Sattinger (1972), Crandall and Rabinowitz (1975), (1977), Marsden and McCracken (1976, L), Sattinger (1979, L), Iooss (1979, M), Hassard (1981, M) (theory and applications), Chow and Hale (1982, M), Kielhöfer (1979), (1980), (1982), (1983), Arnold (1983, M), (1987, S), Vol. 5.

Hopf bifurcation at resonance: Recke (1987), (1988).

Global Hopf bifurcation: Alexander and Yorke (1978), Ize (1976), Chow and Mallet-Paret (1978), Nussbaum (1978), Fiedler (1986).

Stability and bifurcation: Iooss and Joseph (1980, M) (elementary introduction), Crandall and Rabinowitz (1973), Sattinger (1979, L), (1980, S) (see also the References to the Literature to Chapters 3, 80, and 81).

General References to the Literature on Perturbation Theory
Standard works: Kato (1966, M), Reed and Simon (1972, M), Vol. 4.
Introduction: Nayfeh (1973, M), Kevorkian and Cole (1981, M), Gitterman and Halpern (1981, M) (applications to physics), Kato (1982, M).

Matrices and operators: Baumgärtel (1985, M).

Singular perturbations for partial differential equations and control problems: Lions (1973, L).

Boundary layers for partial differential equations: Ljusternik and Višik (1957) (basic paper), Trenogin (1970, S).

Method of averaging: Bogoljubov and Mitropolskii (1965, M), Daleckii and Krein (1970, M), Hale (1980, M).

Celestial mechanics: Stumpf (1973, M), Vols. 1–3, Sternberg (1969, M), Hagihara (1976, M), Vols. 1–5.

Oscillating systems: Bogoljubov and Mitropolskii (1965, M), Kirchgraber and Stiefel (1978, M), Nayfeh and Mook (1979, M).

Quantum mechanics: Reed and Simon (1972, M), Vols. 1–4.

Quantum field theory: Bogoljubov and Širkov (1973, M), (1980, M), Itzykson and Zuber (1980, M), Lee (1981, M), Frampton (1987, M).

Reaction and diffusion: Fife (1978, S), (1979, L), Smoller (1983, M).

Maslov index and asymptotic expansions beyond the caustic: Eckman and Sénéor (1976) (introduction), Maslov (1972, M), Leray (1978, M).

Homogenization: Bensoussan, Lions, and Papanicolau (1978, M).

Regularization: Lions (1969, M).

Bifurcation: Chow and Hale (1982, M).

Geometric perturbation theory in physics: Omohundro (1986, M).

Appendix

Table 1 contains a survey of units from the international system of units. In addition, the following abbreviations are used:

> T   (tera, 10^{12}),
> G   (giga, 10^9),
> M   (mega, 10^6),
> k   (kilo, 10^3),
> f   (femto, 10^{-15}).

For example, 1 GeV (gigaelectron volt) is equal to 10^9 eV. The unit eV (electron volt) does not belong to the international system of units, but is frequently used in elementary particle physics. Because of Einstein's formula

$$E = mc^2,$$

which relates the energy E of a free particle, its mass m, and the velocity of light c, masses in elementary particle physics are measured in eV/c². The rest mass of the electron, for instance, is

$$m_e = 0.511 \cdot \text{MeV}/c^2.$$

The rest mass of the proton is

$$m_p = 1{,}836.1 m_e \sim 1 \text{ GeV}/c^2.$$

These two numbers form the basic scale for masses of elementary particles. Table 2 contains the values of important universal constants.

Table 1

_	Basic units		
length	m	meter	
time	s	second	
mass	kg	kilogram	
temperature	K	degrees Kelvin	
current strength	A	ampere	
amount of substance	mol	1 mol = $6.026 \cdot 10^{23}$ pieces	
luminous intensity	cd	candela	
	Derived units		
force	N	newton	N = kg m/s^2
energy, work	J	joule	J = N m = W s
	eV	electron volt	(1 eV = $1.6 \cdot 10^{-19}$ J)
velocity	m/s	—	—
acceleration	m/s^2	—	—
density	kg/m^3	—	—
pressure	Pa	pascal	Pa = N/m^2
power	W	watt	W = J/s = V A
action	Js	—	—
voltage	V	volt	V = W/A
charge	C	coulomb	C = A s
electric field strength	V/m	—	—
magnetic flow	Wb	weber	Wb = V s
magnetic field strength	T	tesla	T = Wb/m^2
electric resistance	Ω	ohm	Ω = V/A
inductance	H	henry	H = Wb/A
capacity	F	farad	F = C/V

Table 2

	Universal constants
velocity of light in the vacuum	$c = 2.99793 \cdot 10^8$ m/s
	$= 1/\sqrt{\varepsilon_0 \mu_0}$
Planck's action quantum	$h = 6.625 \cdot 10^{-34}$ J s
	($\hbar = h/2\pi$)
Boltzmann constant	$k = 1.380 \cdot 10^{-23}$ J/K
gravitational constant	$G = 6.674 \cdot 10^{-11}$ N m^2/kg^2
dielectric constant	$\varepsilon_0 = 8.854 \cdot 10^{-12}$ A s/V m
permeability constant	$\mu_0 = 4\pi \cdot 10^{-7}$ V s/A m
charge of the electron	$e = -1.602 \cdot 10^{-19}$ A s
rest mass of the electron	$m_e = 9.108 \cdot 10^{-31}$ kg
	$= 0.511$ MeV/c^2
rest mass of the proton	$m_p = 1.672 \cdot 10^{-27}$ kg
rest mass of the neutron	$m_n = 1.675 \cdot 10^{-27}$ kg

References

Abbott, L. and Pi, S. [eds.] (1985): *Inflationary Cosmology*. World Scientific, Singapore.

Ablowitz, M. and Segur, H. (1981): *Solitons and the Inverse Scattering Transform*. SIAM, Philadelphia.

Abraham, R. and Robbin, J. (1967): *Transversal Mappings and Flows*. Benjamin, Reading, MA.

Abraham, R. and Marsden, J. (1978): *Foundations of Mechanics*. Benjamin, Reading, MA. (Cf. also Marsden, Abraham, and Ratiu (1983).)

Abraham, R. (1983): *Dynamics—the Geometry of Behavior*. Part I: *Periodic Behavior*. Part II: *Chaotic Behavior*. Part III: *Global Behavior*. Part IV: *Bifurcation Behavior*. Birkhäuser, Basel.

Adams, J. (1962): *Vector fields on spheres*. Ann. of Math. 75, 603–632.

Agmon, S., Douglis, A., and Nirenberg, L. (1959): *Estimates near the boundary for solutions of elliptic partial differential equations*, I, II. Comm. Pure Appl. Math. 12 (1959), 623–727; 17 (1964), 35–92.

Ahrens, W. (1904): *Scherz und Ernst in der Mathematik*. Teubner, Leipzig.

Albers, D. and Alexanderson, A. [eds.] (1985): *Mathematical People*. Birkhäuser, Boston.

Alberti, P. and Uhlmann, A. (1981): *Dissipative Motion in State Space*. Teubner, Leipzig.

Albeverio, S., Fenstad, J., and Hoegh-Krohn, R. (1986): *Nonstandard Methods in Stochastic Analysis and Mathematical Physics*. Academic, New York.

Aleksandrov, P. [ed.] (1971): *Die Hilbertschen Probleme*. Geest & Portig, Leipzig.

Alexander, J. and Yorke, J. (1978): *Global bifurcation of periodic orbits*. Amer. J. Math. 100, 263–292.

Allgower, E. and Georg, K. (1980): *Simplicial and continuation methods for approximating fixed points*. SIAM Rev. 22, 28–85.

Allgower, E. and Georg, K. (1980a): *Homotopy methods for approximating several solutions to nonlinear systems of equations*. In: Forster, W. [ed.] (1980), 253–270.

Allgower, E. and Georg, K. (1990): *Numerical Continuation Methods*. Springer-Verlag, New York.

Alt, H. (1977): *A free boundary-value problem associated with the flow of ground water.* Arch. Rat. Mech. Anal. **64**, 111–126.

Alt, H. (1979): *Strömungen durch inhomogene poröse Medien mit freiem Rand.* J. Reine Angew. Math. **305**, 89–115.

Alt, H. and Caffarelli, L. (1981): *Existence and regularity for a minimum problem with free boundary.* J. Reine Angew. Math. **105**, 105–144.

Alt, H., Caffarelli, L., and Friedman, A. (1983): *Axially symmetric jet flows.* Arch. Rat. Mech. Anal. **81**, 97–149.

Amann, H. (1983): *Gewöhnliche Differentialgleichungen.* De Gruyter, Berlin.

Ambarzumjan, V. [ed.] (1980): *Probleme der modernen Kosmogonie.* Akademie-Verlag, Berlin.

American Institute of Physics Handbook (1972): McGraw-Hill, New York.

Amick, C. (1978): *Properties of steady Navier–Stokes solutions for certain unbounded channels and pipes.* Nonlinear Anal. **6**, 689–720.

Amick, C. and Toland, J. (1981): *On solitary water waves of finite amplitude.* Arch. Rat. Mech. Anal. **76**, 9–95.

Amick, C., Fraenkel, L., and Toland, J. (1982): *On the Stokes conjecture for the wave of extreme form.* Acta Math. **148**, 193–214.

Antman, S., (1972): *The theory of rods.* In: Flügge, S. [ed.] (1956), Vol. VIa/2, 641–703.

Antman, S. (1977): *Nonlinear elastic structures.* In: Rabinowitz, P. [ed.] (1977), 73–125.

Antman, S. and Rosenfeld, G. (1978): *Global behavior of buckled states of nonlinearly elastic rods.* SIAM Rev. **20**, 513–566.

Antman, S. (1980): *Buckled states of nonlinearly elastic plates.* Arch. Rat. Mech. Anal. **67**, 11–149.

Antman, S. and Nachman, A. (1980): *Large buckled states of rotating rods.* Nonlinear Anal. **4**, 303–327.

Antman, S. and Kenney, C. (1981): *Large buckled states of nonlinearly elastic rods under torsion, thrust, and gravity.* Arch. Rat. Mech. Anal. **76**, 289–300.

Antman, S. (1983): *The influence of elasticity on analysis: Modern developments.* Bull. Amer. Math. Soc. (N.S.) **9**, 267–291.

Antman, S. (1984): *Geometrical and analytical questions in nonlinear elasticity.* In: Chern, S. [ed.] (1984), 1–30.

Antman, S. and Ke-Gang Shih (1986): *Qualitative properties of large buckled states of spherical shells.* Arch. Rat. Mech. Anal. **93**, 357–384.

Antman, S. (1995): *Nonlinear Problems of Elasticity.* Springer-Verlag, New York.

Arms, J., Marsden, J., and Moncrief, V. (1982): *The structure of the space of solutions of Einstein's equations, II. Several Killing fields and the Einstein–Yang–Mills equations.* Ann. Physics **144**, 81–106.

Arnold, V. (1963): *Small divisors and problems of stability of motion in classical and celestial mechanics.* Uspekhi Mat. Nauk **18**, (6), 91–196 (Russian).

Arnold, V. (1966): *Sur la géometrie différentielle des groupes de Lie de dimension infini et ses applications a l'hydrodynamique des fluides parfait.* Ann. Inst. Fourier Grenoble **16** (1), 319–361.

Arnold, V. (1978): *Mathematical Methods of Classical Mechanics.* Springer-Verlag, Berlin.

Arnold, V. (1981): *Singularity Theory: Selected Papers.* University Press, Cambridge, England.

Arnold, V. (1983): *Geometrical Methods in the Theory of Ordinary Differential Equations.* Springer-Verlag, New York.

Arnold, V., Gusein-Zade, S., and Varčenko, A. (1985): *Singularities of Differentiable Maps*, Vols. 1, 2. Birkhäuser, Boston.

Arnold, V. (1986): *Catastrophe Theory*. Springer-Verlag, New York.

Arnold, V. et al. [eds.] (1987): *Dynamical Systems*, Vols. 1–8. Springer-Verlag, New York. (Classical mechanics, celestial mechanics, Kolmogorov–Arnold–Moser theory, bifurcation, catastrophe theory).

Arrow, K. and Debreu, G. (1954): *Existence of an equilibrium for a competitive economy*. Econometrica **22**, 265–290.

Arrow, K. and Intrilligator, A. (1982): *Handbook of Mathematical Economics*. North-Holland, New York.

Atiyah, M. (1979): *Geometry of Yang–Mills Fields* (Lezioni Fermiane). Acc. Naz. Lincei, Scuola Normale Superiore, Pisa.

Atiyah, M. (1984): *An interview*. Math. Intelligencer **6**, 9–19.

Aubin, J. (1977): *Applied Abstract Analysis*. Wiley, New York.

Aubin, J. (1979): *Applied Functional Analysis*. Wiley, New York.

Aubin, J. (1979a): *Mathematical Methods and Economic Theory*. North-Holland, New York.

Aubin, J. and Ekeland, I. (1983): *Applied Nonlinear Analysis*. Wiley, New York.

Aubin, T. (1982): *Nonlinear Analysis on Manifolds: Monge–Ampère Equations*. Springer-Verlag, New York.

Audouze, J. and Tran Thanh Van, J. (1985): *Fundamental Interactions and Cosmology*. Editions Frontières, Paris.

Aziz, A. and Na, T. (1984): *Perturbation Methods in Heat Transfer*. Springer-Verlag, New York.

Babin, A. and Višik, M. (1983): *Attractors of evolution equations and estimates of their dimensions*. Uspekhi Mat. Nauk **38** (4), 133–187 (Russian).

Babin, A. and Višik, M. (1983a): *On the dimension of attractors of the Navier–Stokes system and other evolution equations*. Dokl. Akad. Nauk SSSR **271** (6), 238–243.

Babin, A. and Višik, M. (1986): *Unstable invariant sets of semigroups of nonlinear operators and their perturbations*. Uspekhi Mat. Nauk **41** (4), 3–34.

Babuška, I., Rektorys, K., and Vyčichlo, F. (1960): *Mathematische Elastizitätstheorie der ebenen Probleme*. Akademie-Verlag, Berlin.

Baiocchi, C. (1972): *Su un problem a frontiera libera conesso a questioni di idraulica*. Ann. Mat. Pura Appl. **92**, 107–127.

Baiocchi, C. and Capelo, A. (1978): *Disequazioni variazionali e quasivariazionali*, Vols. 1, 2. Pitagora, Bologna.

Balian, R. (1982): *Du microscopique au macroscopique*. Cours de physique statistique de l'École Polytechnique, Paris.

Ball, J. (1977): *Convexity conditions and existence theorems in nonlinear elasticity*. Arch. Rat. Mech. Anal. **63**, 337–403.

Ball, J., Curie, J., and Olver, P. (1981): *Null Lagrangians, weak continuity, and variational problems of arbitrary order*. J. Funct. Anal. **41**, 135–174.

Ball, J. (1983): *Energy minimizing configurations in nonlinear elasticity*. In: Warsaw (1983), 1309–1314.

Ball, J. [ed.] (1983a): *Systems of Nonlinear Partial Differential Equations*. Reidel, Boston.

Bardos, C. (1982): *Introduction aux problèmes hyperbolique nonlinéaires*. Lecture Notes in Mathematics, Vol. 1047, 1–76. Springer-Verlag, Berlin.

Bartel, N. [ed.] (1987): *Supernovae as Distance Indicators*. Lecture Notes in Physics, Vol. 224. Springer-Verlag, New York.

Batchelor, G. (1967): *Introduction to Fluid Dynamics*. University Press, Cambridge, England.

Batchelor, G. (1982): *The Theory of Homogeneous Turbulence*. University Press, Cambridge, England.

Bauer, F., Garabedian, P., and Korn, D. (1972): *A Theory of Supercritical Wing Sections with Computer Programs and Examples*. Lecture Notes in Economics and Mathematical Systems, Vol. 66, Springer-Verlag, New York.

Baule, B. (1956): *Die Mathematik des Naturforschers und Ingenieurs*, Vols. 1–7. Hirzel, Leipzig.

Baumgärtel, H. (1985): *Analytic Perturbation Theory for Matrices and Operators*. Birkhäuser, Boston.

Baxter, R. (1982): *Exactly Solved Models in Statistical Mechanics*. Academic Press, New York.

Beale, J. (1979): *The existence of cnoidal water waves with surface tension*. J. Differential Equations **31**, 230–264.

Becher, P. and Böhme, M. (1981): *Eichtheorien der starken und elektroschwachen Wechselwirkung*. Teubner, Stuttgart.

Becker, E. (1965): *Gasdynamik*. Teubner, Stuttgart.

Beckert, H. (1972): *Zur Steuerung der Stabilität in elastischen Körpern*. Z. Angew. Math. Mech. **52**, 617–622.

Beckert, H. (1975): *Zur ersten Randwertaufgabe in der nichtlinearen Elastizitätstheorie*. Z. Angew. Math. Mech. **55**, 47–58.

Beckert, H. (1977): *Bemerkungen zur Theorie der Stabilität*. Ber. Sächs. Akad. Wiss. Leipzig, Math.-nat. Kl. **113**, 2. Akademie-Verlag, Berlin.

Beckert, H. (1982): *The initial-value problem for the general dynamical equations in nonlinear elasticity*. Z. Angew. Math. Mech. **62**, 357–369.

Beckert, H. (1984): *Nichtlineare Elastizitätstheorie*. Ber. Sächs. Akad. Wiss. Leipzig, Math.-nat. Kl. **117**, 2. Akademie-Verlag, Berlin.

Beckert, H. (1985): *Axiomatik-Mathematik und Erfahrung*. Ber. Sächs. Akad. Wiss. Leipzig, Math.-nat. Kl. **118**, 2. Akademie-Verlag, Berlin.

Beckert, H. (1986): *The bending of plates and their stability region*. Z. Angew. Math. Mech. **66**, 413–419.

Beem, J. and Ehrlich, P. (1981): *Global Lorentzian Geometry*. Marcel Dekker, New York.

Beju, I. (1972): *The place boundary-value problem in hyperelastostatics*, I, II. Bull. Math. Soc. Sci. Math. R. S. Roumanie (N.S.) **16**, 131–149; 283–314.

Beltrami, E. (1868): *Teoria fondamentale degli spazi di curvatura constante*. Ann. Mat. Pura Appl., Ser. 2 (2), 232–255.

Bénard, M. (1901): *Les tourbillons cellulaires dans une nappe liquide transportant de la chaleur convection en régime permanent*. Ann. Chem. Ser. 7 (23), 62–144.

Benkert, F. (1989): *On the number of stable local minima of some functionals*. Z. Anal. Anwendungen **8**, 89–96.

Bensoussan, A., Lions, J., and Papanicolao, G. (1978): *Asymptotic Methods in Periodic Structures*. North-Holland, Amsterdam.

Bergé, P., Pomeau, Y., and Vidal, C. (1984): *L'ordre dans le chaos*. Hermann, Paris. (English edition, 1986).

Berger, M. and Fife, P. (1967): *On von Kármán's equations*, I, II. Comm. Pure Appl. Math. **20** (1967), 687–719; **21** (1968), 227–241.

References

Berger, M. (1977): *Nonlinearity and Functional Analysis*. Academic Press, New York.
Bergmann, L. and Schaefer, C. (1979): *Lehrbuch der Experimentalphysik*, Vols. 1-4. De Gruyter, Berlin.
Berkeley (1983): *Proceedings of the AMS Summer Institute on Nonlinear Functional Analysis and its Applications*. Cf. Browder, F. [ed.] (1986).
Berkeley (1986): *Proceedings of the International Conference of Mathematicians* (to appear).
Bernadou, M. and Boisserie, J. (1982): *The Finite Element Method in Thin Shell Theory*. Birkhäuser, Boston.
Bernard, P. and Ratiu, T. [eds.] (1977): *Turbulence Seminar*. Lecture Notes in Mathematics, Vol. 615. Springer-Verlag, Berlin.
Bernoulli, J. and Euler, L. (1691/1744): *Abhandlungen über das Gleichgewicht und die Schwingungen der ebenen elastischen Kurven*. Ostwalds Klassiker, Vol. 175. Leipzig, 1910.
Bernoulli, J. (1705): *Véritable hypothèse de la résistance des solides*. (See *The Collected Works of J. Bernoulli*, Vol. 2. Geneva, 1744; and Bernoulli, J. and Euler, L. (1691/1744).)
Bernoulli, D. (1738): *Hydrodynamica*. Argentorati,
Bers, L. (1954): *Existence and uniqueness of a subsonic flow past a given profile*. Comm. Pure Appl. Math. **7** (1954), 441-504.
Bers, L. (1958): *Mathematical Aspects of Subsonic and Supersonic Gas Dynamics*. New York.
Besse, A. (1987): *Einstein Manifolds*. Springer-Verlag, New York.
Beyer, K. (1979): *Zur ersten Randwertaufgabe in der nichtlinearen Elastizitätstheorie*. Math. Nachr. **89**, 43-50.
Beyer, K. and Zeidler, E. (1979): *Existenz- und Eindeutigkeitsbeweis für Gezeitenwellen mit allgemeinen Wirbelverteilungen*. Math. Nachr. **88**, 227-254.
Birell, N. and Davies, P. (1982): *Quantum Fields in Curved Space*. University Press, Cambridge, England.
Birkhoff, G. (1923): *Relativity and Modern Physics*. University Press, Harvard, MA.
Birkhoff, G. (1950): *Hydrodynamics: A Study in Logic, Fact, and Similitude*. University Press, Princeton, NJ.
Birkhoff, G. and Zarantonello, E. (1957): *Jets, Wakes, and Cavities*. Academic Press, New York.
Bitsadze, A. (1964): *Equations of Mixed Type*. Macmillan, New York.
Blaschke, W. (1923): *Vorlesungen über Differentialgeometrie und geometrische Grundlagen von Einsteins Relativitätstheorie*, Vols. 1-3. Springer-Verlag, Berlin.
Blaschke, W. (1946): See Blaschke, W. (1957).
Blaschke, W. (1950): *Einführung in die Differentialgeometrie*. Springer-Verlag, Berlin.
Blaschke, W. (1957): *Reden und Reisen eines Geometers*. Verl. d. Wiss., Berlin.
Boffi, V. and Neunzert, H. [eds.] (1984): *Applications of Mathematics in Technology*. Teubner, Stuttgart.
Bogoljubov, N. and Mitropolskii, J. (1965): *Asymptotische Methoden in der Theorie nichtlinearer Schwingungen*. Akademie-Verlag, Berlin.
Bogoljubov, N. and Širkov, D. (1973): *Introduction to Quantum Field Theory*. Nauka, Moscow (Russian). (English edition: New York, 1979.)
Bogoljubov, N. and Širkov, D. (1980): *Quantum Fields*. Nauka, Moscow (Russian).
Bogoljubov, N. and Bogoljubov, N. Jr. (1980): *Introduction to Quantum Statistical Mechanics*. World Scientific, Singapore.

Bogoljubov, N. [ed.] (1983): *Mathematicians and Physicists*. Nauka, Moscow (Russian).

Bogoljubov, N. and Bogoljubov, N. Jr. (1984): *Quantum Statistical Mechanics*. Nauka, Moscow (Russian).

Böhm, J. and Reichhardt, H. [eds.] (1986): *Gauss–Riemann–Minkowski* (Selected Papers). Teubner, Leipzig.

Boltzmann, L. (1871): *Analytischer Beweis des zweiten Hauptsatzes der mechanischen Wärmetheorie aus dem Satz über das Gleichgewicht der lebendigen Kraft*. Wiener Berichte **63**, 712–733.

Boltzmann, L. (1872): *Weitere Studien über das Wärmegleichgewicht zwischen Gasmolekülen*. Wiener Berichte **66**, 275–361.

Boltzmann, L. (1873): Cf. Cohen, E. and Thirring, W. [eds.] (1973).

Boltzmann, L. (1909): *Wissenschaftliche Abhandlungen (Collected Works)*. Edited by F. Hasenöhrl, Vols. 1–3.

Bolzano, B. (1817): *Rein analytischer Beweis des Lehrsatzes, daß zwischen je zwei Werten, die ein entgegengesetztes Resultat gewähren, wenigstens eine reelle Wurzel der Gleichung liegt*. Prague. (Reprinted in: *Ostwalds Klassiker*, Vol. 153. Leipzig, 1905.)

Bona, J. and Smith, R. (1975): *The initial-value problem for the Korteweg–de Vries equation*. Philos. Trans. Roy. Soc. London **A278**, 555–601.

Bonnet, O. (1848): *Mémoire sur la théorie générale des surfaces*. J. École Polytechnique **19**, 1–146.

Borisovič, J. (1977): *Nonlinear Fredholm maps and the Leray–Schauder theory*. Uspekhi Mat. Nauk **32**, (4), 3–54.

Born, M. (1926): *Quantenmechanik der Stoßvorgänge*. Z. Phys. **37**, 803–827.

Born, M. (1957): *Physik im Wandel meiner Zeit*. Vieweg, Braunschweig. (English edition: *Physics in My Generation*. Springer-Verlag, Berlin, 1969.)

Born, M. and Infeld, L. (1967): *Erinnerungen an Einstein*. Union-Verlag, Berlin.

Börner, G. and Straumann, N. (1985): *Das Modell des inflationären Universums*. Physikalische Blätter **41**, 146–151.

Bothe, H. (1982): *The ambient structure of expanding attractors*. Math. Nachr. **107**, 327–348.

Bourbaki, N. (1960): *Eléments d'histoire des mathématiques*. Hermann, Paris.

Bratteli, C. and Robinson, D. (1979): *Operator Algebras and Quantum Statistical Mechanics*, Vols. 1, 2. Springer-Verlag, New York.

Brekhovskikh, L. (1985): *Mechanics of Continua and Wave Dynamics*. Springer-Verlag, New York.

Bresch, G. (1978): *Zwischenstufe Leben—Evolution ohne Ziel?* Piper, München.

Breuer, R. (1981): *Kontakt mit den Sternen*. Ullstein,

Brézis, H. (1972): *Multiplicateur de Lagrange en torsion elastoplastique*. Arch. Rat. Mech. Anal. **49**, 32–40.

Brézis, H. and Stampacchia, G. (1976): *The hodograph method in fluid dynamics in the light of variational inequalities*. Arch. Rat. Mech. Anal. **61**, 1–18.

Brezzi, F. (1978): *Finite element approximations of the von Kármán equations*. RAIRO Anal. Numér. **12**, 303–312.

Brockhaus ABC der Chemie (1965): Lexikon der Chemie, Vols. 1, 2. Brockhaus-Verlag, Leipzig.

Brockhaus ABC der Physik (1971): Lexikon der Physik, Vols. 1, 2. Brockhaus-Verlag, Leipzig.

Bronstein, I. and Semendjaev, K. (1979): *Taschenbuch der Mathematik*, Vols. 1, 2. Teubner, Leipzig. (English edition: *Handbook of Mathematics*, Van Nostrand, New York, 1985.)

Brouwer, L. (1912): *Über Abbildungen von Mannigfaltigkeiten*. Math. Ann. 71, 97–115.

Brouwer, L. (1975): *Collected Works*. North-Holland, Amsterdam.

Browder, F. (1954): *Strongly elliptic systems of differential equations*. In: Bers, L. [ed.] (1954): *Contributions to the Theory of Partial Differential Equations*, pp. 14–52. University Press, Princeton, NJ.

Browder, F. (1976): *Mathematical Developments Arising from Hilbert's Problems*. American Mathematical Society, New York.

Browder, F. (1976a): *Problems of present-day mathematics*. In: Browder, F. [ed.] (1976), 35–79.

Browder, F. [ed.] (1984): *The Mathematical Heritage of Henri Poincaré*. Proceedings of the Symposia in Pure Mathematics, Vol. 39, Parts I, II. American Mathematical Society, Providence, RI.

Browder, F. [ed.] (1986): *Nonlinear Functional Analysis and its Applications*, Vols. 1, 2. American Mathematical Society, Providence, RI.

Brown, A. (1935): *Functional dependence*. Trans. Amer. Math. Soc. 38, 379–394.

Buchner, M., Marsden, J., and Schecter, S. (1983): *Examples for the infinite-dimensional Morse lemma*. SIAM J. Math. Anal. 14, 1054–1055.

Buchner, M., Marsden, J., and Schecter, S. (1983a): *Applications of the blowing-up construction and algebraic geometry to bifurcation problems*. J. Differential Equations 48, 404–422.

Budó, A. (1956): *Theoretische Mechanik*. Verl. d. Wiss., Berlin.

Büdeler, W. (1981): *Faszinierendes Weltall*. Dt. Verlagsanstalt, Stuttgart.

Bühler, W. (1981): *Gauss: A Bibliographical Study*. Springer-Verlag, New York.

Bullough, R. and Caudrey, P. [eds.] (1980): *Solitons*. Springer-Verlag, New York.

Burger, E. (1959): *Einführung in die Theorie der Spiele*. De Gruyter, Berlin.

Busemann, A. and Föppl, O. (1928): *Physikalische Grundlagen der Elastomechanik*. In: Geiger, H. and Scheel, K. [eds.] (1926), Vol. 6, 1–46.

Caffarelli, L., Kohn, R., and Nirenberg, L. (1982): *Partial regularity of suitable weak solutions of the Navier–Stokes equations*. Comm. Pure Appl. Math. 35, 771–831.

Calogero, F. and Degasperis, A. (1982): *Spectral Transform and Solitons*. North-Holland, Amsterdam.

Carelli, A. et al. [ed.] (1979): *Astrofisica e cosmologia gravitazione quanti e relatività*. Giunti Barbéra, Firenze, Italia.

Carleman, T. (1921): *Über eine nichtlineare Randwertaufgabe bei der Gleichung $\Delta u = 0$*. Math. Z. 9, 35–43.

Carmeli, M. (1977): *Group Theory and General Relativity*. McGraw-Hill, New York.

Cartan, E. (1952): *Oeuvres complètes*, Vols. 1–4. Gauthier-Villars, Paris.

Cartan, E. (1955): *Calcul différentiel: Les systèmes différentiels extérieurs et leurs applications géométriques*. Hermann, Paris.

Cauchy, A. (1827): *De la pression ou tension dans un corpe solide*. Exercises de mathématique, Vol. 2, 42–56 = Cauchy, A. (1882), II. Série, Vol. 7, 60–78.

Cauchy, A. (1828): *Sur les équations qui experiment les conditions d'équilibre ou les lois de mouvement intérieur d'un corps solide*. Exercises de mathématique, Vol. 3, 160–187 = Cauchy, A. (1882), II Série: Vol. 8, 253–277.

Cauchy, A. (1882/1970): *Oeuvres complètes (Collected Works)*. I Série: Vols. 1–12; II Série: Vols. 1–15. Gauthier-Villars, Paris.

Cebeci, T. and Bradshaw, P. (1984): *Physical and Computational Aspects to Convective Heat Transfer*. Springer-Verlag, New York.

Cercignani, C. (1975): *Theory and Applications of the Boltzmann Equation*. Scottish Academic Press, Edinburgh.

Chandrasekhar, S. (1939): *An Introduction to the Study of Stellar Structure*. Dover, New York.

Chandrasekhar, S. (1961): *Hydrodynamic and Hydromagnetic Stability*. University Press, Oxford, England.

Chandrasekhar, S. (1969): *Ellipsoidal Figures of Equilibrium*. Yale University Press, New Haven, CT.

Chandrasekhar, S. (1983): *The Mathematical Theory of Black Holes*. Clarendon Press, Oxford, England.

Chang, K. and Howes, F. (1984): *Nonlinear Singular Perturbation Phenomena*. Springer-Verlag, New York.

Chavent, G. and Jaffré, J. (1986): *Mathematical Methods and Finite Elements for Reservoir Simulation*. North-Holland, Amsterdam.

Chern, S. (1945): *A simple intrinsic proof of the Gauss–Bonnet formula for closed Riemannian manifolds*. Ann. of Math. **45**, 747–752.

Chern, S. (1959): *Differentiable Manifolds*. Lecture Notes. University Press, Chicago, IL.

Chern, S. [ed.] (1984): *Seminar on Nonlinear Partial Differential Equations*. Springer-Verlag, New York.

Chernoff, P. and Marsden, J. (1974): *Properties of Infinite-Dimensional Hamiltonian Systems*. Lecture Notes in Mathematics, Vol. 425. Springer-Verlag, Berlin.

Chevalley, C. (1946): *Theory of Lie Groups*. University Press, Princeton, NJ.

Chillingworth, D., Marsden, J., and Wan, Y. (1983): *Symmetry and bifurcation in three-dimensional elasticity*, I, II. Arch. Rat. Mech. Anal. **80**, 295–331; **83**, 363–395.

Chipot, M. (1984): *Variational Inequalities and Flow in Porous Media*. Springer-Verlag, New York.

Choquet-Bruhat, Y. and Yorke, J. (1980): *The Cauchy problem*. In: Held, A. [ed.] (1980), Vol. 1, 99–136.

Choquet-Bruhat, Y. and Christodoulou, D. (1981): *Existence of global solutions of the Yang–Mills, Higgs, and spinor field equations in 3 + 1 dimensions*. Ann. Sci. École Norm. Sup. IV. Ser., **14**, 481–500.

Choquet-Bruhat, Y. DeWitt-Morette, C., and Dillard-Bleick, M. (1982): *Analysis, Manifolds, and Physics*. North-Holland, Amsterdam.

Choquet-Bruhat, Y. (1984): *Positive-energy theorems*. In: DeWitt, B., and Stora, R. [eds.] (1984), 739–784.

Chorin, A. (1973): *Numerical study of slightly viscous flow*. J. Fluid Mech. **57**, 785–796.

Chorin, A. (1975): *Lectures on Turbulence Theory*. Publish or Perish, Boston.

Chorin, A. et al. (1977): *Product formulas and numerical algorithms*. Comm. Pure Appl. Math. **31**, 205–256.

Chorin, A. (1978): *Vortex sheet approximation of boundary layers*. J. Comput. Phys. **27**, 428–442.

Chorin, A. and Marsden, J. (1979): *A Mathematical Introduction to Fluid Dynamics*. Springer-Verlag, New York.

Chorin, A. (1982): *The evolution of a turbulent vortex*. Comm. Math. Phys. **83**, 517–535.

Chow, S., Hale, J., and Mallet-Paret, J. (1975): *Applications of generic bifurcation*, I, II. Arch. Rat. Mech. Anal. **59**, 159–188; **62**, 209–235.
Chow, S. and Mallet-Paret, J. (1978): *The fuller index and global Hopf bifurcation*. J. Differential Equations **29**, 66–85.
Chow, S., Mallet-Paret, J., and Yorke, J. (1978): *Finding zeros of maps: homotopy methods that are constructive with probability one*. Math. Comput. **32**, 887–899.
Chow, S. and Hale, J. (1982): *Methods of Bifurcation Theory*. Springer-Verlag, New York.
Christodoulou, D. and Klainerman, S. (1988): Cf. additional references.
Ciarlet, P. (1977): *Numerical Analysis of the Finite Element Method*. North-Holland, Amsterdam.
Ciarlet, P. and Rabier, P. (1980): *Les équations de von Kármán*. Lecture Notes in Mathematics, Vol. 826. Springer-Verlag, Berlin.
Ciarlet, P. (1983): *Lectures on Three-Dimensional Elasticity*. Springer-Verlag, New York.
Ciarlet, P. and Nečas, J. (1984): *Unilateral problems in nonlinear three-dimensional elasticity (or how to carry a bottle of slivovitz)*. Arch. Rat. Mech. Anal. **87**, 319–338.
Claro, F. [ed.] (1985): *Nonlinear Phenomena in Physics*. Springer-Verlag, New York.
Clausius, R. (1865): *Über verschiedene für die Anwendung bequeme Formen der Hauptgleichungen der mechanischen Wärmetheorie*. Pogg. Ann. **125**, 353–400.
Cohen, E. and Thirring, W. [eds.] (1973): *The Boltzmann Equation*. Springer-Verlag, New York.
Cole, J. and Cook, L. (1986): *Transonic Aerodynamics*. North-Holland, Amsterdam.
Concus, P. and Finn, R. [eds.] (1987): *Variational Methods for Free Interfaces*. Springer-Verlag, New York.
Constantin, P. and Foias, C. (1985): *Global Ljapunov exponents, Kaplan–Yorke formulas, and the dimension of the attractors for two-dimensional Navier–Stokes equations*. Comm. Pure Appl. Math. **38**, 1–27.
Constantin, P., Foias, C., and Temam, R. (1985): *Attractors representing turbulent flows*. Memoirs Amer. Math. Soc., Vol. **53**. American Mathematical Society, Providence, RI.
Cornwell, J. (1984): *Group Theory in Physics*, Vols. 1, 2. Academic Press, New York.
Courant, R. and Robbins, H. (1941): *What is Mathematics?* University Press, Oxford, England.
Courant, R. and Friedrichs, K. (1948): *Supersonic Flow and Shock Waves*. Interscience, New York.
Courant, R. and Hilbert, D. (1959): *Methods of Mathematical Physics*, Vols. 1, 2. Interscience, New York.
Craigie, N. [ed.] (1986): *Theory and Detection of Magnetic Monopoles in Gauge Theories*. World Scientific, Singapore.
Crandall, M. and Rabinowitz, P. (1973): *Bifurcation, perturbation of simple eigenvalues, and linearized stability*. Arch. Rat. Mech. Anal. **52**, 161–180.
Crandall, M. and Rabinowitz, P. (1975): *The Hopf bifurcation theorem*. MRC Technical Report, No. 1604. University of Madison, Madison, WI.
Crandall, M. and Rabinowitz, P. (1977): *The Hopf bifurcation theorem*. Arch. Rat. Mech. Anal. **67**, 53–72.
Crandall, M. and Lions, P. (1983): *Viscosity solutions of Hamilton–Jacobi equations*. Trans. Amer. Math. Soc. **277**, 1–42.

Crank, J. (1984): *Free and Moving Boundary Problems.* Clarendon Press, Oxford, England.
Cronin, V. (1981): *The View from Planet Earth—Man Looks at the Cosmos.* Collins, Glasgow. (German edition: *Säulen des Himmels—die Weltbilder des Abendlandes.* Claasen, Düsseldorf.)
Curtis, W. and Miller, F. (1986): *Differentiable Manifolds and Theoretical Physics.* Academic Press, New York.
Cycon, R., Froese, R., Kirsch, W., and Simon, B. (1986): *Schrödinger Operators.* Springer-Verlag, New York.

Dacarogna, B. (1982): *Weak Continuity and Weak Lower Semicontinuity of Nonlinear Functionals.* Lecture Notes in Mathematics, Vol. 922. Springer-Verlag, Berlin.
Daleckii, J. and Krein, M. (1970): *Stability of Solutions of Differential Equations in Banach Spaces.* Nauka, Moscow (Russian). (English edition: American Mathematical Society, Providence, RI.)
Darcy, H. (1856): *Les fontaines publiques de la ville de Dijon.* Dalmont, Paris.
Davis, P. and Hersh, R. (1981): *The Mathematical Experience.* Birkhäuser, Boston.
Davis, P. and Chinn, W. (1985): *3.1416 and All That.* Birkhäuser, Boston.
Davydov, A. (1984): *Solitons in Molecular Systems.* Naukova Dumkova, Kiev (Russian).
Debnath, L. [ed.] (1985): *Advances in Nonlinear Waves.* Pitman, London.
Debreu, G. (1959): *Theory of Value.* Wiley, New York.
De Rham, G. (1955): *Variétés différentiables. Formes courants, formes harmoniques.* Hermann, Paris.
DeWitt, B. and Stora, R. [eds.] (1984): *Relativité, groupes et topologies,* II. North-Holland, Amsterdam.
DeWitt, B. (1984): *The space–time approach to quantum field theory.* In: DeWitt, B. and Stora, R. [eds.] (1984), 381–738.
DeWitt, B. (1984a): *Supermanifolds.* University Press, Cambridge, England.
Dias, J. (1975): *Variational inequalities for nonlinear maximal monotone operators in a Hilbert space.* Amer. J. Math. **97**, 905–914.
Dias, J. and Hernandez, J. (1975): *A Sturm–Liouville theorem for some odd multivalued maps.* Proc. Amer. Math. Soc. **53**, 72–74.
Dickerson, R., Gray, H., and Haight, G. (1974): *Chemical Priciples.* Benjamin, New York. (German edition: de Gruyter, Berlin, 1978.)
Dickey, R. (1976): *Bifurcation Problems in Nonlinear Elasticity.* Pitman, London.
Dictionary of Mathematics (1961): Cf. Mathematisches Wörterbuch (1961).
Dierker, E. (1974): *Topological Methods in Walrasian Economics.* Lecture Notes in Economics, Vol. 92, Springer-Verlag, New York.
Dieudonné, J. (1975): *Grundzüge der modernen Analysis,* Vols. 1–9. Verl. d. Wiss., Berlin. (English edition: *Foundations of Modern Analysis.* Academic Press, New York 1960ff. French edition: Gauthier-Villars, Paris, 1968ff.)
Dieudonné, J. (1981): *History of Functional Analysis.* North-Holland, Amsterdam.
Dieudonné, J. (1982): *The work of Bourbaki during the last thirty years.* Notices Amer. Math. Soc. **29**, 618–623.
Dionne, P. (1962): *Sur les problèmes de Cauchy bien posés.* J. Analyse Math. **10**, 1–90.
DiPerna, R. (1977): *Decay of solutions of hyperbolic systems of conservation laws with a convex extension.* Arch. Rat. Mech. Anal. **64**, 1–46.

DiPerna, R. (1979): *Uniqueness of solutions of hyperbolic conservation laws.* Indiana Univ. Math. J. **28**, 137–187.

DiPerna, R. (1983): *Convergence of the viscosity method for isentropic gas dynamics.* Comm. Math. Phys. **91**, 1–30.

DiPerna, R. and Majda, A. (1987): *Concentrations in regularizations for two-dimensional incompressible flow.* Comm. Pure Appl. Math. **40**, 301–346.

Dixon, W. (1978): *Special Relativity.* University Press, Cambridge, England.

Djubek, J., Kodnár, R., and Skaloud, M. (1983): *Limit State of the Plate Elements of Steel Structures.* Birkhäuser, Boston.

Do, C. (1975): *Problèmes de valeurs propres pur une inéquation variationelle sur un cône.* C. R. Acad. Sci. Paris, Ser. A–B, **280**, 45–48.

Do, C. (1976): *The buckling of a thin elastic plate subjected to unilateral conditions.* In: Germain, P. and Nayroles, B. [eds.] (1976), 307–316.

Do, C. (1977): *Bifurcation theory for elastic plates subjected to unilateral conditions.* J. Math. Anal. Appl. **60**, 435–448.

Donaldson, S. (1983): *An application of gauge theory to the topology of four-dimensional manifolds.* J. Diff. Geometry **18**, 279–315.

Drazin, P. (1983): *Solitons.* University Press, Cambridge, England.

Dubin, D. (1974): *Solvable Models in Algebraic Quantum Statistics.* Clarendon Press, Oxford, England.

Dubrovin, B., Novikov, S., and Fomenko, A. (1979): *Modern Geometry: Methods and Applications.* Nauka, Moscow (Russian). (English edition: Springer-Verlag, New York, 1984.)

Duffie, D. and Shafer, W. (1985): *Equilibrium in incomplete markets. I. A basic model of generic existence.* J. Math. Economics **14**, 285–300.

Dukas, H. and Hoffmann, B. (1972): *Albert Einstein—Creator and Rebel.* New York.

Dukas, H. and Hoffmann, B. (1979): *Albert Einstein—the Human Side.* University Press, Princeton, NJ.

Duvaut, G. and Lions, J. (1972): *Les inéquations en mécanique et en physique.* Dunod, Paris.

Dwoyer, D. et al. (1985): *Theoretical Approaches to Turbulence.* Springer-Verlag, New York.

Dyck, W. (1885): *Beiträge zur Analysis situs.* Berichte Sächs. Ges. Wiss. Leipzig **37**, 314–325.

Dyson, J. (1979): *Disturbing the Universe.* Harper & Row, New York. (German edition: *Innenansichten-Erinnerungen an die Zukunft.* Birkhäuser, Basel, 1981.)

Eavas, C. et al. [eds.] (1982): *Homotopy Methods and Global Convergence.* Plenum, New York.

Ebeling, W. (1976): *Strukturbildung bei irreversiblen Prozessen.* Teubner, Leipzig.

Ebeling, W. and Feistel, R. (1982): *Physik der Selbstorganisation und Evolution.* Akademie-Verlag, Berlin.

Ebeling, W. and Klimontowitsch, A. (1985): *Self-Organization and Turbulence.* Teubner, Leipzig.

Ebin, D. and Marsden, J. (1970): *Groups of diffeomorphisms and the motion of an incompressible fluid.* Ann. of Math. **92**, 102–163.

Ebin, D., Fischer, A., and Marsden, J. (1972): *Diffeomorphism groups, hydrodynamics and relativity.* In: Vanstone, J. [ed.], *Proceedings of the 13th Biennial Seminar of the Canadian Mathematical Congress*, pp. 135-279.

Ebin, D. (1979): *The initial-value problem for subsonic fluid motion*. Comm. Pure Appl. Math. **32**, 1-19.

Ebin, D. and Saxton, R. (1986): *The initial-value problem for elasto-dynamics of incompressible bodies*. Arch. Rat. Mech. Anal. **94**, 15-38.

Eckhaus, W. (1973): *Matched Asymptotic Expansions and Singular Perturbations*. North-Holland, Amsterdam.

Eckmann, J. and Sénéor, R. (1976): *The Maslov-WKB-method for the (an)harmonic oscillator*. Arch. Rat. Mech. Anal. **61**, 153-173.

Efimov, N. (1957): *Flächenverbiegung im Großen*. Akademie-Verlag, Berlin.

Einstein, A. (1905): *Zur Elektrodynamik bewegter Körper*. Ann. Physik **17**, 891-921.

Einstein, A. (1905a): *Ist die Trägheit eines Körpers von seinem Energieinhalt abhängig?* Ann. Physik **18**, 639-641.

Einstein, A. (1905b): *Über einen die Erzeugung und Verwandlung des Lichts betreffenden Gesichtspunkte*. Ann. Physik **17**, 132-148.

Einstein, A. (1905c): *Die von der molekular-kinetischen Theorie der Wärme geforderte Behandlung von in ruhenden Flüsssigkeiten suspendierten Teilchen*. Ann. Physik **17**, 549-560.

Einstein, A. (1915): *Zur allgemeinen Relativitätstheorie. Die Feldgleichungen der Gravitation*. Sitzungsber. Preuss. Akad. Wiss. Berlin vom 11.11.1915 und 2.12.1915.

Einstein, A. (1916): *Die Grundlagen der allgemeinen Relativitätstheorie*. Ann. Physik **49**, 769-822.

Einstein, A. (1920), (1921): Cf. Frank, P. (1949).

Einstein, A. (1953): *The Meaning of Relativity*. University Press, Princeton, NJ.

Einstein, A. (1955): Cf. Einstein, A. (1965) and Melcher, H. (1979).

Einstein, A. (1956): *Grundzüge der Relativitätstheorie*. Vieweg, Braunschweig.

Einstein, A. (1960): *Collected Writings of Albert Einstein*. Readex Microprint, New York. Cf. also the additional references.

Einstein, A. (1965): *Mein Weltbild*. Frankfurt/Main.

Eisenreich, G. and Sube, R. (1973): *Wörterbuch Physik* (Dictionary of Physics in English, French, German and Russian). Verlag-Technik, Berlin.

Eisenreich, G. and Sube, R. (1982): *Wörterbuch Mathematik* (Dictionary of Mathematics in English, French, German and Russian). Verlag-Technik, Berlin.

Ekeland, I. and Temam, R. (1974): *Analyse convexe et problèmes variationnels*. Dunod, Paris. (English edition: North-Holland, Amsterdam, 1976.)

Elliott, C. and Ockendon, J. (1982): *Weak and Variational Methods for Moving Boundaries*. Pitman, London.

Ellis, R. (1985): *Entropy, Large Deviations, and Statistical Physics*. Springer-Verlag, New York.

Emden, R. (1938): *Why we have heating?* Nature **141**, 908.

Encyclopedia of Astronomy and Space (1976): Edited by I. Ridpath. Macmillan, London.

Encyclopedia of Mathematics and Its Applications (1976): Edited by G. Rota, Vols. 1ff. Addison-Wesley, Reading, MA.

Encyclopedia of Mathematical Sciences (1987): Vols. 1ff. Springer-Verlag, New York.

Encyclopedic Dictionary of Mathematics (1977): Vols. 1, 2. MIT Press, Cambridge, MA.

Encyclopedic Dictionary of Physics (1977): Vols. 1-9. Supplementary Volumes 1-4. Edited by J. Thewlis. MIT Press, Cambridge, MA.

Enzyklopädie der mathematischen Wissenschaften (1904): Vols. 1ff. Teubner, Leipzig.
Euklid (325 B.C.): *Elemente.* Ostwalds Klassiker, Vols. 235, 236. Akad. Verlagsges., Leipzig.
Euler, L. (1744): Cf. Bernoulli, J. and Euler, L. (1691/1744).
Euler, L. (1755): *Principels généraux du mouvements des fluides.* Hist. de l'Acad. de Berlin.
Euler, L. (1911): *Opera omnia (Collected Works),* Vols. 1ff. Leipzig–Zürich since 1911, Lausanne since 1942.
Evans, L. (1986): *Quasiconvexity and partial regularity in the calculus of variations.* Arch. Rat. Mech. Anal. **95**, 227–252.

Faddeev, L. and Takhtadjan (1986): *The Hamiltonian Approach in the Theory of Solitons.* Nauka Moscow (Russian). (English edition: Springer-Verlag, New York, 1987.)
Fan, Ky (1952): *Fixed-point and minimax theorems in locally convex spaces.* Proc. Nat. Acad. Sci. U.S.A. **74**, 4749–4751.
Fan, Ky (1972): *A minimax inequality and applications.* In: Shisha, O. [ed.] (1972), *Inequalities,* pp. 103–114. Academic Press, New York.
Federer, H. (1969): *Geometric Measure Theory.* Springer-Verlag, Berlin.
Feigenbaum, M. (1980): *Universal behaviour in nonlinear systems.* Los Alamos Science **1**, 4–27.
Feistauer, M. (1984): *On irrotational flow through cascades of profiles in a layer of variable thickness.* Applikace Mathematiky **29**, 423–458.
Feistauer, M., Mandel, J., and Nečas, J. (1984): *Entropy regularization of the transonic potential flow problem.* Comm. Mat. Univ. Carolinae **25** (3), 431–443.
Feistauer, M. and Nečas, J. (1985): *On the solvability of transonic potential flow problems.* Z. Anal. Anwendungen **4**, 305–329.
Fermi, E., Pasta, J., and Ulam, S. (1955): *Studies of nonlinear problems.* Los Alamos Report, LA, 1940, 1955. Reproduced in: Newell, A. [ed.] (1974), *Nonlinear Wave Motion.* American Mathematical Society, Providence, RI.
Ferrara, S. (1983): *Aspects of supergravity theories.* In: Schmutzer, E. [ed.] (1983), 207–224.
Ferraris, M. and Kijowski, J. (1982): *Unified theory of electromagnetic and gravitational interactions.* Gen. Relativity Gravitation **14**, 37–47.
Ferraris, M. and Kijowski, J. (1982a): *On the equivalence of the relativistic theories of gravitation.* Gen. Relativity Gravitation **14**, 165–180.
Ferris, T. (1977): *The Red Limit.* New York. (German edition: Birkhäuser, Basel, 1982.)
Feynman, R., Leighton, R., and Sands, M. (1963): *The Feynman Lectures on Physics.* Addison-Wesley, Reading, MA. (German–English edition: Oldenbourg, München, 1974.)
Fichera, G. (1964): *Problemi elastostatici con vincoli unilateriali: il problema die Signorini con ambigue condizioni al contorno.* Mem. Accad. Naz. Lincei **8**, 91–140.
Fichera, G. (1972): *Existence theorems in elasticity.* In: Flügge, S. [ed.] (1956), Vol. VIa/2, 347–390.
Fichera, G. (1972a): *Boundary value problems of elasticity with unilateral constraints.* In: Flügge, S. [ed.] (1956), Vol. VIa/2, 391–424.
Fick, E. and Sauermann, G. (1982): *Quantenstatistik dynamischer Prozesse,* Vols. 1, 2. Geest & Portig, Leipzig.

Fiedler, B. (1986): *Global Hopf bifurcation of two-parameter flows*. Arch. Rat. Mech. Anal. **94**, 59–81.

Field, G. and Chaisson, E. (1986): *Das unsichtbare Universum*. Birkhäuser, Boston. (English edition: *The Invisible Universe*. Boston, 1985).

Fife, P. (1970): *The Bénard problem for general fluid dynamical equations and remarks on the Boussinesq approximation*. Indiana Univ. Math. J. **20**, 303–326.

Fife, P. (1978): *Asymptotic states for equations of reaction and diffusion*. Bull. Amer. Math. Soc. **84**, 693–726.

Fife, P. (1979): *Mathematical Aspects of Reacting and Diffusing Processes*. Lecture Notes in Biomathematics, Vol. 28. Springer-Verlag, Berlin.

Finn, R. and Gilbarg, D. (1957): *Three-dimensional subsonic flow, and asymptotic estimates for elliptic partial differential equations*. Acta Math. **98**, 265–296.

Finn, R. (1959): *On steady state solutions of the Navier–Stokes partial differential equations*. Arch. Rat. Mech. Anal. **3**, 381–396.

Finn, R. (1961): *On the steady-state solutions of the Navier–Stokes equations*. Acta Math. **105**, 197–244.

Finn, R. (1965): *Stationary solutions of Navier–Stokes equations*. Proc. Sympos. Appl. Math. **17**, 121–153.

Finn, R. (1985): *Equilibrium Capillary Surfaces*. Springer-Verlag, New York.

Fischer, A. and Marsden, J. (1972): *The Einstein evolution equations as a first-order symmetric hyperbolic system*. Comm. Math. Phys. **28**, 1–38.

Fischer, A. and Marsden, J. (1972a): *The Einstein equations of evolution—a geometric approach*. J. Math. Phys. **13**, 546–568.

Fischer, A., Marsden, J., and Moncrief, V. (1982): *The structure of the space of solutions of Einstein equations, II. Several Killing fields and the Einstein–Yang–Mills equations*. Ann. Physics **144**, 81–106.

Floquet, G. (1883): *Sur les équations différentielles linéaires à coefficients périodiques*. Ann. Sci. École Norm. Sup., Ser. 2, **12**, 47–89.

Flügge, S. [ed.] (1956): *Handbuch der Physik (Handbook of Physics)*, Vol. 1ff. Springer-Verlag, Berlin.

Fock, V. (1960): *Theorie von Raum, Zeit und Gravitation*. Akademie-Verlag, Berlin.

Foias, C. (1973): *Statistical study of Navier–Stokes equations, I. II*. Rend. Sem. Mat. Univ. Padova **48**, 219–348; **49**, 9–123.

Foias, C. and Temam, R. (1977): *Structure of the set of stationary solutions of the Navier–Stokes equations*. Comm. Pure Appl. Math. **30**, 149–164.

Foias, C. and Temam, R. (1982): *Homogeneous statistical solutions of Navier–Stokes equations*. Indiana Univ. Math. J. **29**, 913–957.

Forster, W. [ed.] (1980): *Numerical Solutions of Highly Nonlinear Problems*. North-Holland, New York.

Fortin, M. and Glowinski, R. [eds.] (1983): *Augmented Lagrangian's Methods: Applications to the Numerical Solutions of Boundary Value Problems*. North-Holland, New York.

Frampton, P. (1987): *Gauge Field Theories*. Benjamin, Reading, MA.

Frank, G. and Meyer-Spasche, R. (1981): *Computations of transitions in Taylor vortex flows*. Z. Angew. Math. Phys. **32**, 710–720.

Frank, P. (1949): *Einstein—sein Leben und seine Zeit*. München.

Frank, P. and Mises, R.v. (1962): *Die Differential- und Integralgleichungen der Mechanik und Physik*, Vols. 1, 2. Vieweg, Braunschweig.

Franke, H. (1969): *Lexikon der Physik*. Stuttgart.

Fredrickson, A. (1964): *Principles and Applications of Rheology.* Prentice-Hall, Englewood Cliffs, NJ.
Freed, D. and Uhlenbeck, K. (1984): *Instantons and Four-Manifolds.* Springer-Verlag, New York.
Fridman, A. and Polyachenko, V. (1984): *Physics of Gravitating Systems*, Vols. 1, 2. Springer-Verlag, New York.
Friedlander, S. (1980): *An Introduction to the Mathematical Theory of Geophysical Fluid Dynamics.* North-Holland, Amsterdam.
Friedman, A. (1982): *Variational Principles and Free Boundary Value Problems.* Wiley, New York.
Friedrich, H. (1981): *The analytic characteristic initial-value problem for Einstein's vacuum field equations as an initial-value problem for a first-order quasilinear symmetric hyperbolic system.* Proc. Roy. Soc. London **A378**, 401–421.
Friedrichs, K. (1927): *Rand- und Eigenwertprobleme aus der Theorie der elastischen Platten.* Math. Ann. **98**, 206–247.
Friedrichs, K. (1929): *Ein Verfahren das Minimum eines Integrals als das Maximum eines anderen Ausdrucks darzustellen.* Göttinger Nachrichten.
Friedrichs, K. and Stoker, J. (1942): *Buckling of the circular plate beyond the critical thrust.* J. Appl. Mech. **9**, A7–A14.
Friedrichs, K. and Hyers, D. (1954): *The existence of solitary waves.* Comm. Pure Appl. Math. **7**, 517–550.
Friedrichs, K. (1954): *Symmetric hyperbolic linear differential equations.* Comm. Pure Appl. Math. **7**, 345–392.
Fritzsch, H. (1982): *Quarks–Urstoff unserer Welt.* Piper, München.
Fritzsch, H. (1983): *Vom Urknall zum Verfall.* Piper, München.
Fröhlich, J. (1983): *Scaling and Self-Similarity in Physics. Renormalization in Statistical Mechanics and Dynamics.* Birkhäuser, Boston.
Frolov, V. (1976): *Black holes and quantum processes.* Uspekhi Fiz. Nauk **18**, 473–503 (Russian).
Frost, W. and Moulden, T. (1977): *Handbook of Turbulence.* Plenum, New York.
Fučik, S. and Kufner, A. [eds.] (1979): *Nonlinear Analysis. Function Spaces and Applications.* Teubner, Leipzig.
Fujita, H. and Kato, T. (1964): *On the Navier–Stokes initial value problem.* Arch. Rat. Mech. Anal. **16**, 269–315.

Gajewski, H. (1970): *Iterations- Projektions- und Projektions-Iterationsverfahren zur Berechnung visco-plastischer Strömungen.* Z. Angew. Math. Mech. **50**, 485–490.
Gajewski, H. (1970a): *Über einige Fehlerabschätzungen bei Gleichungen mit monotonen Potentialoperatoren in Banachräumen.* Monatsberichte Akad. d. Wiss. Berlin, 571–579.
Galdi, G. and Maremonti, P. (1986): *Monotonic decreasing and asymptotic behavior of the kinetic energy for weak solutions of the Navier–Stokes equations in external domains.* Arch. Rat. Mech. Anal. **94**, 253–266.
Gale, D. (1955): *The law of supply and demand.* Math. Scand. **3**, 155–169.
Galilei, G. (1638): *Discorsi e dimonstrazioni matematiche.* Firenze, Italia.
Galilei, G. (1890/1909): *Le opere di Galileo.* Edited by A. Favaro, Vols. 1–20. Firenze, Italia.
Gantmacher, F. and Krein, M. (1960): *Oszillationsmatrizen, Oszillationskerne und kleine Schwingungen mechanischer Systeme.* Akademie-Verlag, Berlin.

Garabedian, P. (1960): *Boundary Layer Theory*. McGraw-Hill, New York.
Garabedian, P. and Korn, D. (1971): *Analysis of transonic airfoils*. Comm. Pure Appl. Math. **24**, 841–851.
Garabedian, P. (1972): See Bauer, F., Garabedian, P., and Korn D. (1972).
Garcia, C. and Zangwill, W. (1983): *Pathways to Solutions, Fixed Points and Equilibria*. Prentice-Hall, Englewood Cliffs, NJ.
Gardiner, W. (1985): *Handbook of Stochastic Methods for Physics, Chemistry and the Natural Sciences*. Springer-Verlag, New York.
Gates, S. et al. (1983): *Superspace*. Benjamin, Reading, MA.
Gauss, C. (1827): *Disquisitiones generales circa superficies curvas*. In: Gauss, C. (1863), Vol. 5, 217–258, 341–347. (German translation: *Allgemeine Flächentheorie*. Oswalds Klassiker, Vol. 5, Leipzig, 1889.)
Gauss, C. (1863): *Werke (Collected Works)*, Vols. 1–12. Göttingen, 1863–1929.
Gautreau, R. and Savin, W. (1978): *Theory and Problems in Modern Physics*. McGraw-Hill, New York.
Geckeler, J. (1928): *Elastostatik*. In: Geiger, H. and Scheel, K. [eds.] (1926), Vol. 6, 141–308.
Geiger, H. and Scheel, K. [eds.] (1926): *Handbuch der Physik*, Vols. 1–24. Springer-Verlag, Berlin, 1926ff.
Geiringer, H. (1972): *Ideal plasticity*. In: Flügge, S. [eds.] (1956), Vol. VIa/3, 403–533.
Georg, K. (1981): *On tracing an implicitly defined curve by quasi-Newton steps and calculating bifurcation by local perturbations*. SIAM J. Sci. Statist. Comput. **2**, 35–50.
Georg, K. (1981a): *Numerical integration of the Davidenko equation*. In: Peitgen, H. and Walther, H. [eds.] (1981), 128–161.
Georgi, H. (1984): *Weak Interactions and Modern Particle Theory*. Benjamin, Reading, MA.
Gerhardt, C. (1976): *On the existence and uniqueness of a warpening function in the elasto-plastic torsion of a cylindrical bar with multiply connected cross section*. In: Germain, P. and Nayroles, B. [eds.] (1976), 328–342.
Germain, P. and Nayroles, B. [eds.] (1976): *Applications of Functional Analysis to Problems in Mechanics*. Lecture Notes in Mathematics, Vol. 503. Springer-Verlag, Berlin.
Gerthsen, C. (1982): *Physik* (14th edn). Springer-Verlag, New York.
Ghil, M. and Childress, S. (1987): *Topics in Geophysical Fluid Dynamics: Atmospheric Dynamics, Dynamo Theory, and Climate Dynamics*. University Press, Oxford, England.
Giacaglia, G. (1972): *Perturbation Methods in Nonlinear Systems*. Springer-Verlag, New York.
Giaquinta, M. and Hildebrandt, S. (1988): *Calculus of Variations*, Vols. 1–3 (to appear).
Gibbs, J. (1902): *Elementary Principles in Statistical Mechanics*. Yale University Press, New Haven, CT.
Gilbarg, D. (1960): *Jets and cavities*. In: Flügge, S. [ed.], (1958), Vol. IX, *Fluid Dynamics*, III, 311–445.
Gilkey, P. (1974): *The Index Theorem and the Heat Equation*. Publish or Perish, Boston.
Gilkey, P. (1984): *Invariance Theory, the Heat Equation and the Atiyah–Singer Index Theorem*. Publish or Perish, Boston.
Gilmore, R. (1981): *Catastrophe Theory for Scientists*. Wiley, New York.
Girault, V. and Raviart, P. (1981): *Finite Element Approximation of the Navier–Stokes*

Equations. Lecture Notes in Mathematics Vol. 749. Springer-Verlag, Berlin.
Girault, V. and Raviart, P. (1986): *Finite Element Methods for Navier–Stokes Equations.* Springer-Verlag, New York.
Gittel, H. (1987): *Studies on transonic problems by nonlinear variational inequalities.* Z. Anal. Anwendungen **6**, 449–458.
Gitterman, M. and Halpern, V. (1981): *Qualitative Analysis of Physical Problems.* Academic Press, New York.
Glansdorff, P. and Prigogine, I. (1971): *Thermodynamic Theory of Structure, Stability and Fluctuations.* Wiley, London.
Glicksberg, I. (1952): *A further generalization of the Kakutani fixed-point theorem with application to Nash equilibrium points.* Proc. Amer. Math. Soc. **3**, 170–174.
Glimm, J. (1965): *Solutions in the large for nonlinear systems of conservation laws.* Comm. Pure Appl. Math. **18**, 697–715.
Glowinski, R., Lions, J., and Trémolières, R. (1976): *Analyse numèrique des inéquations variationnelles,* Vols. 1, 2. Gauthier-Villars, Paris (English edition: North-Holland, Amsterdam, 1981.)
Glowinski, R. (1980): *Lectures on Numerical Methods for Nonlinear Variational Problems.* Springer-Verlag, New York.
Glowinski, R. (1983): *Numerical solution of nonlinear boundary value problems by variational methods. Applications.* In: Proceedings of the International Congress of Mathematicians in Warsaw (1983), Vol. 2, 1455–1508.
Glowinski, R. (1984): *Numerical Methods for Nonlinear Variational Problems.* Springer-Verlag, New York.
Goering, H. et al. (1983): *Singularly Perturbed Differential Equations.* Akademie-Verlag, Berlin.
Goldberg, S. (1984): *Understanding Relativity.* Birkhäuser, Boston.
Golubitsky, M. and Guillemin, V. (1973): *Stable Mappings and Their Singularities.* Springer-Verlag, New York.
Golubitsky, M. (1978): *An introduction to catastrophe theory and its applications.* SIAM Rev. **20**, 352–387.
Golubitsky, M. and Schaeffer, E. (1979): *A theory for imperfect bifurcation via singularity theory.* Comm. Pure Appl. Math. **32**, 21–98.
Golubitsky, M. and Marsden, J. (1983): *The Morse lemma in infinite dimensions via singularity theory.* SIAM J. Math. Anal. **14**, 1037–1044.
Golubitsky, M. and Schaeffer, D. (1984): *Singularities and Bifurcation Theory.* Springer-Verlag, New York.
Gompf, R. (1985): *An infinite set of exotic $\mathbb{R}^4$'s.* J. Diff. Geometry **21**, 317–328.
Green, G. (1839): *On the laws of reflection and refraction of light at the common surface of two non-crystallised media.* Trans. Cambridge Philos. Soc. **7**, 1–24.
Green, M. and Gross, D. [eds.] (1986): *Unified String Theories.* World Scientific, Singapore.
Green, M. (1986): *Superstrings and the unification of forces and particles.* In: Ruffini, R. [ed.] (1986), pp. 203–226.
Green, M., Schwarz, J., and Witten, E. (1987): *Superstrings,* Vols. 1, 2. University Press, Cambridge, England.
Greenberg, W. et al. (1986): *Boundary Value Problems in Abstract Kinetic Theory.* Birkhäuser, Boston.
Greiner, W. and Müller, B. (1984): *Theoretische Physik,* Vols. 1–10. Harri Deutsch, Frankfurt/Main, 1984ff. (Cf. also the additional references.)

Greiner, W. and Müller, B. (1986): *Die elektroschwache Wechselwirkung.* Cf. Greiner, W. and Müller, B. (1984), Vol. 8.

Greub, W., Halperin, S., and Vanstone, R. (1972): *Connections, Curvature, and Cohomology,* Vols. 1-3. Academic Press, New York.

Grib, A. et al. (1980): *Quantum Effects in Strong External Fields.* Atomizdat, Moscow (Russian).

Griffith, P. (1983): *Exterior Differential Systems and the Calculus of Variations.* Birkhäuser, Boston.

Grinčenko, V. and Ulitko, A. (1985): *Three-dimensional Problems in Elasticity and Plasticity,* Vols. 1-6. Naukova Dumka (Russian).

Gröger, K. (1979): *Initial-value problems for elastoplastic and elasto-viscoplastic systems.* In: Fučik, S. and Kufner, A. [eds.] (1979), 95-127.

Gromov, M. and Rohlin, V. (1970): *Embeddings and immersions in Riemannian geometry.* Uspekhi Mat. Nauk **25** (5), 3-62 (Russian).

Gromov, M. (1986): *Partial Differential Relations.* Springer-Verlag, New York.

Groot, S. de and Mazur, P. (1962): *Nonequilibrium Thermodynamics.* North-Holland, Amsterdam.

Guckenheimer, J. and Holmes, P. (1983): *Nonlinear Oscillations, Dynamical Systems, and Bifurcations.* Springer-Verlag, New York.

Guderley, G. (1957): *Theory of Transonic Flow.* Pergamon Press, New York.

Guillemin, V. and Pollack, A. (1974): *Differential Topology.* Prentice-Hall, Englewood Cliffs, NJ.

Günter, N. (1957): *Potentialtheorie.* Teubner, Leipzig.

Günther, M. (1987): *Ein einfacher Beweis für das nichtlineare Malodensky-Problem.* Math. Nachr. **130**, 251-265.

Günther, M. (1989): *Zum Einbettungssatz von J. Nash.* Math. Nachr. **144**, 165-187.

Gurtin, M. (1972): *The linear theory of elasticity.* In: Flügge, S. [ed.] (1956), Vol. VIa/2, 1-296.

Gurtin, M. (1981): *Topics in Finite Elasticity.* SIAM, Philadelphia.

Guth, A. (1981): *The inflationary universe.* Phys. Rev. **D23**, 347.

Gwinner, J. (1981): *On fixed points and variational inequalities: A circular tour.* Nonlinear Anal. **5**, 565-583.

Haar, A. and Kármán. T. v. (1909): *Theorie der Spannungszustände in plastischen und sandartigen Medien.* Göttinger Nachrichten **1909**, 204-218.

Haber, H. [ed.] (1987): *From the Planck Scale to the Weak Scale: Toward a Theory of the Universe.* World Scientific, Singapore.

Hagihara, Y. (1976): *Celestial Mechanics,* Vols. 1-5. MIT Press, Cambridge, MA.

Hajko, V. and Schilling, H. (1975): *Physik in Beispielen,* Vols. 1-6. Fachbuchverlag, Leipzig.

Haken, H. (1977): *Synergetics—an Introduction: Nonequilibrium, Phase Transitions and Self-Organisation in Physics, Chemistry and Biology.* Springer-Verlag, New York.

Haken, H. (1983): *Advanced Synergetics.* Springer-Verlag, New York.

Hale, J. (1980): *Ordinary Differential Equations.* Wiley, New York.

Hale, J. (1981): *Topics in Dynamic Bifurcation Theory.* American Mathematical Society, Providence, RI.

Halmos, P. (1983): *Selecta. Expository Writing.* Springer-Verlag, New York.

Halmos, P. (1985): *I Want To Be A Mathematician.* Springer-Verlag, New York.

Halphen, B. and Nguyen, Q. (1975): *Sur les matériaux standards généralisés.* Méc., Paris **14**, 39-63.
Haltiner, G. and Williams, R. (1980): *Numerical Weather Prediction and Dynamic Meteorology.* Wiley, New York.
Handbook of Applicable Mathematics (1980): Edited by W. Ledermann, Vols. 1-6. Wiley, Chichester, 1980ff.
Handbook of Chemistry and Physics (1980): The Chemical Rubber, Cleveland, OH.
Handbook of Mathematics (1961): Cf. Mathematisches Wörterbuch (1961).
Hänsel, H. and Neumann, W. (1972): *Physik: Eine Darstellung der Grundlagen,* Vols. 1-7. Verl. d. Wiss., Berlin.
Harrison, E. (1981): *Cosmology.* University Press, Cambridge, England.
Hartman, P. and Stampacchia, G. (1966): *On some nonlinear functional differential equations.* Acta Math. **115**, 271-310.
Hartmann, F. (1985): *The Mathematical Foundations of Structural Mechanics.* Springer-Verlag, New York.
Haslinger, J. (1983): *Approximations of the Signorini problem with friction obeying the Coulomb law.* Math. Methods Appl. Sci. **5**, 422-437.
Hassard, B., Kazarinoff, N., and Wang, Y. (1981): *Theory and Applications of Hopf Bifurcation.* University Press, Cambridge, England.
Hawking, S. (1972): *Black holes in general relativity.* Comm. Math. Phys. **25**, 152-166.
Hawking, S., Bardeen, J., and Carter, B. (1973): *The four laws of black hole mechanics.* Comm. Math. Phys. **31**, 161-170.
Hawking, S. and Ellis, G. (1973): *The Large Scale Structure of Space-Time.* University Press, Cambridge, England.
Hawking, S. (1975): *Particle creation by black holes.* Comm. Math. Phys. **43**, 199-220.
Hawking, S. and Israel, W. [eds.] (1979): *General Relativity: An Einstein Centenary Survey.* University Press, Cambridge, England.
Hawking, S. and Roček, M. [eds.] (1981): *Superspace and Supergravity.* University Press, Cambridge, England.
Hawking, S. (1984): *Quantum cosmology.* In: DeWitt, B. and Stora, R. [eds.] (1984), 330-380.
Heisenberg, W. (1925): *Quantenmechanische Umdeutung kinematischer und mechanischer Beziehungen.* Z. Phys. **33**, 879-884.
Heisenberg, W. (1927): *Anschaulicher Inhalt der quantentheoretischen Kinematik und Mechanik.* Z. Phys. **43**, 172-199.
Heisenberg, W. (1937): *Die physikalischen Prinzipien der Quantenmechanik.* Hirzel, Leipzig.
Heisenberg, W. (1968): *Nonlinear problems in physics.* In: Zabusky, N. [ed.] (1968), 1-18.
Heisenberg, W. (1977): *Schritte über Grenzen.* Piper, München.
Heisenberg, W. (1977a): *Tradition in der Wissenschaft.* Piper, München.
Heisenberg, W. (1978): *Physik und Philosophie.* Hirzel, Stuttgart.
Heisenberg, W. (1980): *Wandlungen in den Grundlagen der Wissenschaft.* Hirzel, Stuttgart.
Heisenberg, W. (1981): *Der Teil und das Ganze.* Piper, München.
Heisenberg, W. (1984): *Gesammelte Werke—Collected Works.* Springer-Verlag, New York.
Held, A. [ed.] (1980): *General Relativity and Gravitation,* Vols. 1, 2. Plenum, New York.

Helgason, S. (1962): *Differential Geometry and Symmetric Spaces.* Academic Press, New York.

Helmholtz, H. (1858): *Über Integrale der hydrodynamischen Gleichungen, welche den Wirbelbewegungen entsprechen.* Crelle Journal.

Henbest, N. and Marten, M. (1984): *Die neue Astronomie.* Birkhäuser, Basel.

Hencky, H. (1924): *Zur Theorie plastischer Deformationen.* Z. Angew. Math. Mech. **4**, 323–334.

Henry, D. (1981): *Geometric Theory of Semilinear Parabolic Equations.* Lecture Notes in Mathematics, Vol. 840. Springer-Verlag, Berlin.

Herrmann, D. (1978): *Entdecker des Himmels.* Urania, Leipzig.

Heywood, J. (1980): *The Navier–Stokes equations: on the existence, regularity and decay of solutions.* Indiana Univ. Math. J. **29**, 639–681.

Hilbert, D. (1897): *Die Theorie der algebraischen Zahlkörper.* In: Hilbert, D. (1932), Vol. 1, 63–527.

Hilbert, D. (1901): *Über die Flächen von konstanter Gaußscher Krümmung.* Trans. Amer. Math. Soc. **2**, 87–99.

Hilbert, D. (1903): *Über die Grundlagen der Geometrie.* Teubner, Leipzig. (12th edn, Teubner, Stuttgart, 1977.)

Hilbert, D. (1915): *Die Grundlagen der Physik.* Nachr. Akad. Wiss. Göttingen, Math.-phys. Kl. **1915**, 395–407; **1917**, 53–76.

Hilbert, D. (1930): *Naturerkenntnis und Logik.* Naturwissenschaften, 959–963.

Hilbert, D. (1932): *Gesammelte Abhandlungen (Collected Works),* Vols. 1–3. Springer-Verlag, Berlin.

Hilbig, H. (1964): *Existenzsätze für einige Totwasserprobleme der Hydrodynamik.* Ber. Sächs. Akademie Wiss. Leipzig, Math. Naturwiss. Klasse 105.

Hilbig, H. (1987): *Weitere Existenzsätze für Totwasserprobleme der Hydrodynamik.* Ber. Sächs. Akademie Wiss. Leipzig, Math.-Naturwiss. Klasse 119/5.

Hildebrandt, S. (1985): *Harmonic mappings of Riemannian manifolds.* In: Guisti, E. [ed.] (1985), *Harmonic Mappings and Minimal Immersions.* Lecture Notes Mathematics, Vol. 1161, pp. 1–117, Springer-Verlag, Berlin.

Hildebrandt, S. and Tromba, T. (1985): *Mathematics and Optimal Form.* Scientific American Books, New York. (German edition: *Panoptimum,* Spektrum der Wissenschaft, 1987.)

Hildebrandt, S. and Giaquinta, M. (1988): Cf. Giaquinta, M. (1988).

Hill, R. (1950): *The Mathematical Theory of Plasticity.* Clarendon Press, Oxford, England.

Hillel, D. (1980): *Fundamentals of Soil Physics.* Academic Press, New York.

Hilton, P. (1974): Cf. Otte, M. [ed.] (1974).

Hilton, P. and Young, G. [eds.] (1981): *New Directions in Applied Mathematics.* Springer-Verlag, New York.

Hincliffe, I. [ed.] (1987): *Cosmology and Particle Physics.* World Scientific, Singapore.

Hirsch, M. (1976): *Differential Topology.* Springer-Verlag, New York.

Hirsch, M., Pugh, C., and Shub, M. (1977): *Invariant Manifolds.* Lecture Notes in Mathematics, Vol. 583. Springer-Verlag, Berlin.

Hirzebruch, F. (1974): Cf. Otte, M. [ed.] (1974).

Hlaváček, I., Haslinger, J., Nečas, J., and Lovíšek, J. (1986): *Solution of Variational Inequalities in Mechanics.* Mir, Moscow (Russian).

Hodge, W. (1952): *The Theory and Applications of Harmonic Integrals.* University Press, Cambridge, England.

Hoffmann, K. (1981): *Fixpunktprinzipien und freie Randwertaufgaben.* In: Lecture Notes in Mathematics, Vol. 878, 169–181. Springer-Verlag, Berlin.

Hofmann, R. (1986): *Eine Klasse von Extremalproblemen für die Norm selbstadjungierter vollstetiger Operatoren im Hilbertraum mit Anwendung auf die Platte und die schwingende Saite.* Dissertation B, Karl-Marx-Universität, Leipzig.

Hofstadter, D. (1979): *Gödel, Escher, Bach: An Eternal Golden Braid.* Basic Books, New York.

Holmes, P. and Marsden, J. (1981): *Chaotic oscillations of a forced beam.* Arch. Rat. Mech. Anal. **76**, 135–166.

Holt, M. (1984): *Numerical Methods in Fluid Dynamics.* Springer-Verlag, New York.

Hooke, R. (1678): *Lectures de Potentia Restitutiva or of Spring Explaining the Power of Springing Bodies,* London = Gunther, R. (1931), Early Science in Oxford **8**, 331–356.

Hopf, E. (1942): *Abzweigung einer periodischen Lösung von einer stationären Lösung eines Differentialgleichungssystems.* Ber. Sächs. Akad. Wiss. Leipzig, Math.-phys. Kl. **94**, 1–22.

Hopf, E. (1948): *A mathematical example displaying the features of turbulence.* Comm. Pure Appl. Math. **1**, 303–322.

Hopf, E. (1950): *The partial differential equation $u_t + uu_x = u_{xx}$.* Comm. Pure Appl. Math. **3**, 201–230.

Hopf, E. (1951): *Über die Anfangswertaufgabe für die hydrodynamischen Grundgleichungen.* Math. Nachr. **4**, 213–231.

Hopf, E. (1952): *Statistical hydrodynamics and functional calculus.* J. Rat. Mech. Anal. **1**, 87–123.

Hörmander, L. (1967): *An Introduction to Complex Analysis in Several Variables.* Van Nostrand, Princeton, NJ.

Hörmander, L. (1976): *The boundary-value problems of physical geodesy.* Arch. Rat. Mech. Anal. **62**, 1–52.

Hornung, U. (1983): *A unilateral boundary value problem for unsteady water flow in porous media.* Meth. Verf. Math. Phys. **25**, 59–94.

Hoyle, F. and Wickramasinghe, N. (1979): *Life Cloud.* Sphere Books, New York.

Huang, K. (1963): *Statistical Mechanics.* Wiley, New York.

Huang, K. (1982): *Quarks, Leptons, and Gauge Fields.* World Scientific, Singapore.

Hughes, T. and Marsden (1976): *A Short Course in Hydrodynamics.* Publish or Perish, Boston.

Hughes, T., Kato, T., and Marsden, J. (1977): *Well-posed quasi-linear hyperbolic systems with applications to nonlinear elastodynamics and general relativity.* Arch. Rat. Mech. Anal. **63**, 272–294.

Hugoniot, H. (1889): *Sur la propagation du mouvement dans les corps et spécialement dans les gax parfaits,* J. École Polytechnique **58**, 1–125.

Hünlich, R. (1979): *On simultaneous torsion and tension of a circular cylindrical bar consisting of an elastoplastic material with linear hardening.* Z. Angew. Math. Mech. **59**, 509–516.

Hurt, N. (1983): *Geometric Quantization in Action.* Reidel, Boston.

Husseini, S., Lasry, J., and Magill, M. (1990): *Existence of equilibrium with incomplete markets.* J. Math. Economics **19**, 39–67.

Infeld, L. (1969): *Leben mit Einstein: Kontur einer Erinnerung.* Wien.

XXIIth International Conference on High Energy Physics in Leipzig (1984): *Proceedings*, Vols. 1, 2. Akad. d. Wiss., Berlin.

d'Inverno, R. (1983): *Computer methods in general relativity*. In: Schmutzer, E. [ed.] (1983), 93–113.

Iooss, G. (1979): *Bifurcation of Maps and Applications*. North-Holland, Amsterdam.

Iooss, G. and Joseph, D. (1980): *Elementary Stability and Bifurcation Theory*. Springer-Verlag, New York.

Itzykson, C. and Zuber, J. (1980): *Quantum Field Theory*. McGraw-Hill, New York.

Ivanov, V. (1983): *Gerade und Ungerade: die Asymmetrie des Gehirns und der Zeichenstruktur*. Hirzel, Stuttgart.

Ize, J. (1976): *Bifurcation Theory for Fredholm Operators*. American Mathematical Society, Providence, RI.

Jacob, C. (1959): *Introduction mathématique à la mécanique des fluides*. Gauthier-Villars, Paris.

Jaffe, A. (1984): *Ordering the universe: the role of mathematics*. Notices Amer. Math. Soc. **236**, 589–608.

Jäger, W., Moser, J., and Remmert, R. [eds.] (1984): *Perspectives in Mathematics*. Birkhäuser, Boston.

Jantzsch, E. (1982): *Die Selbstorganisation des Universums*. Deutscher Taschenbuchverlag, München.

Jentsch, L. (1977): *Zur Existenz von regulären Lösungen der Elastostatik stückweise homogener Körper*. Akademie-Verlag, Berlin.

Jentzsch, R. (1912): *Über Integralgleichungen mit positivem Kern*. J. Reine Angew. Math. **141**, 235–249.

John, F. (1965): *Estimates for the derivative of the stresses in a thin shell and interior shell equations*. Comm. Pure Appl. Math. **18**, 235–267.

John, F. (1971): *Refined interior equations for thin elastic shells*. Comm. Pure Appl. Math. **24**, 583–615.

John, F. (1972): *Uniqueness of nonlinear elastic equilibrium for prescribed boundary displacements and sufficiently small strains*. Comm. Pure Appl. Math. **25**, 627–634.

John, F. (1977): *Finite amplitude waves in homogeneous isotropic elastic solids*. Comm. Pure Appl. Math. **30**, 421–446.

John, F. (1982): *Partial Differential Equations*. Springer-Verlag, New York.

John, F. (1983): *Lower bounds for the life span of solutions of nonlinear wave equations in three-dimensional space*. Comm. Pure Appl. Math. **36**, 1–35.

John, F. (1985): *Collected Papers*. Edited by J. Moser, Vols. 1, 2. Birkhäuser, Boston.

John, J. (1985a): *Formation of singularities in elastic waves*. In: John, F. (1985), Vol. 1, 624–640.

John, F. (1987): *Existence for large times of strict solutions of nonlinear wave equations in three space dimensions for small initial data*. Comm. Pure Appl. Math. **40**, 79–110.

Jones, M. and Toland, J. (1986): *Symmetry and the bifurcation of capillary–gravity waves*. Arch. Rat. Mech. Anal. **96**, 79–110.

Joseph, D. (1965): *On the solvability of the Boussinesq equation*. Arch. Rat. Mech. Anal. **20**, 59–71.

Joseph, D. and Sattinger, D. (1972): *Bifurcating time periodic solutions and their stability*. Arch. Rat. Mech. Anal. **45**, 79–109.

Joseph, D. (1976): *Stability of Fluid Motions*, Vols. 1, 2. Springer-Verlag, New York.

Jost, R. (1984): *Mathematics and physics since 1800: discord and sympathy.* In: DeWitt, B. and Stora, R. [eds.] (1984), 4-50.

Judovič, V. (1966): *Secondary flows and fluid instability between rotating cylinders.* Prikl. Mat. Mekh. 30, 688-698 (Russian.)

Judovič, V. (1966a): *On the origin of convection.* Prikl. Mat. Mekh. 30, 1193-1199 (Russian).

Judovič, V. (1967): *Free convection and bifurcation.* Prikl. Mat. Mekh. 31, 101-111 (Russian).

Kačanov, L. (1969): *Foundations of Plasticity Theory.* Nauka, Moscow (Russian).

Kähler, E. (1934): *Einführung in die Theorie der Systeme von Differentialgleichungen.* Teubner, Leipzig.

Kähler, E. (1941): *Über die Beziehungen der Mathematik zu Astronomie und Physik.* Jahresber. Deutsche Math.-Verein. 51, 52-63.

Kähler, E. (1979): *Monadologie.* Privatdruck, Hamburg.

Kakutani, S. (1941): *A generalization of Brouwers fixed-point theorem.* Duke Math. J. 8, 457-459.

Karlin, S. (1959): *Mathematical Methods and Theory in Games, Programming and Economics.* Addison-Wesley, Reading, MA.

Karlin, S. (1967): *Total Positivity and Applications.* University Press, Stanford, CA.

Kármán, T. v. (1910): *Festigkeitsprobleme im Maschinenbau.* Enzyklopädie der mathematischen Wissenschaften, Vol. IV/4, 601-694.

Kato, T. (1966): *Perturbation Theory for Linear Operators.* Springer-Verlag, New York.

Kato, T. (1967): *On classical solutions of two-dimensional stationary Euler equations.* Arch. Rat. Mech. Anal. 25, 188-200.

Kato, T. (1972): *Nonstationary flow of viscous and ideal fluids in $\mathbb{R}^3$.* J. Funct. Anal. 9, 296-305.

Kato, T. (1975): *The Cauchy problem for quasilinear symmetric hyperbolic systems.* Arch. Rat. Mech. Anal. 58, 181-205.

Kato, T. (1975a): *Quasilinear equations of evolution with applications to partial Differential equations.* Lecture Notes in Mathematics, Vol. 448, 25-70. Springer-Verlag, New York.

Kato, T. (1982): *A Short Introduction to Perturbation Theory for Linear Operators.* Springer-Verlag, New York.

Kato, T. (1983): *On the Cauchy problem for the generalized Korteweg-de Vries equation.* In: Guillemin, V. [ed.] (1983), *Studies in Applied Mathematics,* pp. 93-128, Academic Press, New York.

Kato, T. and Lai, C. (1984): *Nonlinear evolution equations and the Euler flow.* J. Funct. Anal. 56, 15-28.

Kaufmann, W. (1979): *Black Holes and Warped Space-Time.* Freeman, San Francisco, CA.

Keller, H. and Meyer-Spasche, R. (1980): *Computation of the axialsymmetric flow between rotating cylinders.* J. Comput. Phys. 35, 100-109.

Kellog, R., Li, T., and Yorke, J. (1976): *A constructive proof of the Brouwer fixed-point theorem and computational results.* SIAM J. Numer. Anal. 13, 473-483.

Kellogg, O. (1929): *Foundations of Potential Theory.* Springer-Verlag, Berlin.

Kelly, A. (1967): *The stable, center-stable, center, center-unstable and unstable manifolds.* In: Abraham, R. and Robbin, J. (1967), 136-154.

Kelvin, Lord: See Thomson, W., Sir.
Kepler, J. (1609): *Astronomia nova*. (German translation by M. Caspar, München, 1939.)
Kepler, J. (1618): *Harmonice mundi*. (German translation by M. Caspar, München, 1939.)
Kepler, J. (1939): *Gesammelte Werke (Collected Works)*. Edited by W. v. Dyck and M. Caspar, München.
Kevorkian, J. and Cole, J. (1981): *Perturbation Methods in Applied Mathematics*. Springer-Verlag, New York.
Kielhöfer, H. (1979): *Hopf bifurcation at multiple eigenvalues*. Arch. Rat. Mech. Anal. **69**, 53–84.
Kielhöfer, H. (1980): *Degenerate bifurcation at simple eigenvalues and stability of bifurcating solutions*. J. Funct. Anal. **38**, 416–441.
Kielhöfer, H. (1982): *Floquet exponents of bifurcating periodic orbits*. Nonlinear Anal. **6**, 571–584.
Kielhöfer, H. and Lauterbach, R. (1983): *On the principle of reduced stability*. J. Funct. Anal. **53**, 99–111.
Kijowski, J. (1978): *On a new variational principle in general relativity and the energy of the gravitational field*. Gen. Relativity Gravitation **9**, 857–877.
Kijowski, J. (1985): *On positivity of energy of the graviational fied*. In: Ruffini, R. [ed.], Proceedings of the Forth Marcel Grossmann Meeting, Rome, 1985. North-Holland, Amsterdam.
Kinderlehrer, D. and Stampacchia, G. (1980): *An Introduction to Variational Inequalities and Their Applications*. Academic Press, New York.
Kinderlehrer, D. (1981): *Remarks about Signorinis problem in linear elasticity*. Ann. Scuola Norm. Sup. Pisa Cl. Sci. (4), **8**, 605–645.
Kippenhahn, R. (1980): *Geburt, Leben und Tod der Sterne*. Piper, München.
Kippenhahn, R. (1987): *Light from the Depth of Time*. Springer-Verlag, New York.
Kirchgässner, K. (1961): *Die Instabilität der Strömung zwischen zwei rotierenden Zylindern gegenüber Taylor–Wirbeln für beliebige Spaltbreiten*. Z. Angew. Math. Phys. **12**, 14–30.
Kirchgässner, K. and Sorger, P. (1969): *Branching analysis for the Taylor problem*. Quart. J. Mech. Appl. Math. **32**, 183–209.
Kirchgässner, K. and Kielhöfer, H. (1973): *Stability and bifurcation in fluid dynamics*. Rocky Mountain J. Math. **3**, 275–318.
Kirchgässner, K. (1975): *Bifurcation and nonlinear hydrodynamic stability*. SIAM Rev. **17**, 652–683.
Kirchgässner, K. (1975a): *Instability phenomena in fluid mechanics*. In: *SYNSPADE 1975*, Edited by J. Hubbard. Academic Press, New York.
Kirchgässner, K. (1981): *Periodic and nonperiodic solutions of reversible systems*. In: De Mottoni, P. and Salvadori, L. [eds.], *Nonlinear Differential Equations, Invariance, Stability, and Bifurcation*. Academic Press, New York, pp. 221–242.
Kirchgässner, K. (1988): *Nonlinearly resonant surface waves and homoclinic bifurcation* (to appear).
Kirchgraber, V. and Stiefel, E. (1978): *Methoden der analytischen Störungsrechnung und ihre Anwendungen*. Teubner, Stuttgart.
Kittel, C. *et al.* (1965): *Berkeley Physics: A Course in Physics*, Vols. 1–5. McGraw-Hill, New York. (German edition: Vieweg, Braunschweig.)
Kittel, C. (1969): *Thermal Physics*. Wiley, New York. (German edition: Geest & Portig, Leipzig, 1973.)

Klein, F. (1871): *Über die sogenannte nicht-euklidische Geometrie.* Math. Ann. **4**, 573–625. (Further papers on this topic may be found in Klein, F. (1921), Vol. 1, 241–410.)

Klein, F. (1872): *Erlangener Programm.* In: Klein, F. (1921), Vol. 1, 460–497.

Klein, F. (1921): *Gesammelte mathematische Abhandlungen (Collected Papers)*, Vols. 1–3. Springer-Verlag, Berlin.

Klein, F. (1926): *Vorlesungen über die Entwickung der Mathematik im 19. Jahrhundert*, Vols. 1, 2. Springer-Verlag, Berlin.

Klein, F. (1928): *Vorlesungen über nicht-euklidische Geometrie.* Springer-Verlag, Berlin.

Klein, M. (1973): *The development of Boltzmann's statistical ideas.* In: Cohen, E. and Thirring, W. [eds.] (1973), 53–106.

Kleinert, W. [ed.] (1987): *Gauge Theory of Stresses and Defects.* World Scientific, Singapore.

Kliesch, W. (1983): *Zur numerischen Bestimmung des Abbildungsgrades im $\mathbb{R}^n$ und zu seiner Anwendung bei der Lösung nichtlinearer Gleichungssysteme.* Dissertation. Universität, Leipzig.

Kliesch, W. (1984): *Zur numerischen Bestimmung des Abbildungsgrades im $\mathbb{R}^n$.* Z. Anal. Anwendungen **3**, 337–365; 489–502.

Kliesch, W. (1989): *A unified numerical approach to the topological degree in $\mathbb{R}^n$.* Math. Nachr. **142**, 181–213.

Kline, M. (1972): *Mathematical Thought from Ancient to Modern Times.* University Press, Oxford, England.

Klingenberg, W. (1978): *A Course in Differential Geometry.* Springer-Verlag, New York.

Klingenberg, W. (1983): *Riemannian Geometry.* De Gruyter, Berlin.

Knaster, B., Kuratowski, C., and Mazurkiewicz, S. (1929): *Ein Beweis des Fixpunktsatzes für n-dimensionale Simplexe.* Fund. Math. **14**, 132–137.

Knightly, G. and Sather, D. (1974): *Nonlinear buckled states of rectangular plates.* Arch. Rat. Mech. Anal. **54**, 356–372.

Knightly, G. and Sather, D. (1980): *Existence and stability of axial symmetric buckled states of spherical shells.* Arch. Rat. Mech. Anal. **63**, 305–319.

Knightly, G. and Sather, D. (1985): *A selection principle for Bénard-type convection.* Arch. Rat. Mech. Anal. **88**, 163–193.

Knops, R. [ed.] (1976): *Symposium on Nonlinear Analysis and Mechanics*, Vols. 1–4. Pitman, New York, 1976–79.

Knörrer, H. (1986): *Integrable Hamiltonsche Systeme und Algebraische Geometrie.* Jahresber. Deutsche Math.-Ver. **88**, 82–103.

Kobayashi, S. and Nomizu, K. (1963): *Foundation of Differential Geometry*, Vols. 1, 2. Interscience, New York.

Koiter, W. (1960): *General theorems for elastic-plastic solids.* In: *Progress in Solid Mechanics.* North-Holland, Amsterdam, pp. 165–221.

Koiter, W. and Simmonds, J. (1972): *Foundations of shell theory.* In: Proc. 13th Int. Congr. Appl. Mech. Moscow, 1972, pp. 150–176. Springer-Verlag, New York.

Koiter, W. (1980): *The intrinsic equations of shell theory with some applications.* In Nemat-Nasser, S. [ed.], *Mechanics Today*, Vol. 5. Pergamon Press, London.

Kompanejec, A. (1961): *Theoretical Physics.* Mir, Moscow (Russian).

Korn, A. (1907): *Sur les équations d'élasticité.* Ann. École Norm. **24**, 9–75.

Korneev, V. and Lange, U. (1984): *Approximate Solution of Plastic Flow Theory Problems.* Teubner, Leipzig.

Korteweg, D. and de Vries, G. (1895): *On the change of form of long waves and of a*

new type of long stationary waves. Philos. Mag. **39**, 422.
Kotschin, N. et al. (1954): *Theoretische Hydrodynamik.* Akademie-Verlag, Berlin.
Kounas, C. et al. [eds.] (1984): *Grand Unification With and Without Supersymmetry and Cosmological Implications.* World Scientific, Singapore.
Kowalewski, G. (1939): *Große Mathematiker.* Berlin.
Kramer, D. et al. (1980): *Exact Solutions of Einstein's Field Equations.* Verl. d. Wiss., Berlin.
Krasnoselskii, M. and Pokrovskii, A. (1983): *Systems with Hysteresis.* Nauka, Moscow. (English edition in preparation.)
Krasnoselskii, M. et al. (1985): *Positive Linear Systems: The Method of Positive Operators.* Nauka, Moscow (Russian).
Krein, M. (1964): *Lectures on Stability.* Kiev (Russian). (Cf. Daleckii, J. and Krein, M. (1970).)
Kreyszig, E. (1957): *Differentialgeometrie.* Geest & Portig, Leipzig.
Kruskal, M. (1960): *Maximal extension of the Schwarzschild metric.* Phys. Rev. **119**, 1743.
Kružkov, S. (1970): *Quasilinear equations of first order with several independent variables.* Mat. Sbornik **81**, 228–255.
Kubíček, M. and Marek, M. (1983): *Computational Methods in Bifurcation Theory and Dissipative Structures.* Springer-Verlag, New York.
Kubo, R., Toda, M., and Saitô, N. (1983): *Statistical Physics, I. Equilibrium Statistical Mechanics.* Springer-Verlag, New York.
Kubo, R., Toda, M., and Hashitsume, N. (1985): *Statistical Physics, II. Nonequilibrium Statistical Mechanics.* Springer-Verlag, New York.
Kučera, M., Nečas, J., and Souček, J. (1978): *The eigenvalue problem for variational inequalities and a new version of the Ljusternik–Schnirelman theory.* In: Cesari, L. [ed.] (1978), Nonlinear Analysis, 125–143, Academic Press, New York.
Kučera, M. (1982): *A new method for obtaining eigenvalues of variational inequalities.* Czechoslovak Math. J. **32**, 197–207.
Kučera, M. (1982a): *Bifurcation points of variational inequalities.* Czechoslovak Math. J. **32**, 208–226.
Kuhn, H. (1960): *Some combinatorical lemma in topology.* IBM J. Res. Develop. **4**, 518–524.
Kupradze, V. (1976): *Three-Dimensional Problems in Mathematical Elasticity and Thermoelasticity.* Nauka, Moscow (Russian). (English edition: North-Holland, Amsterdam, 1979.)
Ky Fan: Cf. Fan, Ky.

Ladyženskaja, O. (1959): *Solution "in the large" of the nonstationary boundary-value problem for the Navier–Stokes system with two space variables.* Comm. Pure Appl. Math. **12**, 427–433.
Ladyženskaja, O. (1970): *Mathematical Problems in the Dynamics of Viscous Incompressible Fluids.* Nauka, Moscow (Russian). (English edition: Gordon and Breach, New York, 1969.)
Ladyženskaja, O. and Solonnikov, V. (1977): *On the solvability of boundary-value problems and boundary initial-value problems for the Navier–Stokes equations with non-compact boundaries.* Vestnik Leningr. Univ., 1977, 39–47 (Russian).
Ladyženskaja, O. (1979): *On formulation and solvability of boundary-value problems for viscous incompressible fluids in domains with non-compact boundaries.* In: Equadiff

IV. Lecture Notes in Mathematics, Vol. 703, 233-240. Springer-Verlag, Berlin.
Ladyženskaja, O. and Solonnikov, V. (1980): *On solutions for the stationary Navier-Stokes equations with unbounded Dirichlet integral.* Zap. Naučn. Sem. Leningrad. Otdel. Mat. Inst. Steklov. **96**, 117-160 (Russian).
Ladyženskaja, O. (1982): *On the finite dimension of bounded invariant sets for the Navier-Stokes equations and other dissipative systems.* Zap. Naučn. Sem. Leningrad. Otdel. Mat. Inst. Steklov. **115**, 137-155 (Russian).
Ladyženskaja, O. (1986): *On some directions of the research in mathematical physics at the Steklov-Institute in Leningrad.* In: Vladimirov, V. (1986), 217-245 (Russian).
Lagrange, L. (1788): *La Mécanique Analytique*. Paris.
Lagrange, L. (1867/1892): *Oeuvres (Collected Works)*. Gauthier-Villars, Paris.
Lahaye, E. (1935): *Sur la représentation des racines systèmes d'équations transcendante.* Deuxième Congrès National des Sciences, Vol. 1, 141-146.
Lamb, G. (1980): *Elements of Soliton Theory*. Wiley, New York.
Lamb, H. (1924): *Hydrodynamics*. University Press, Cambridge, England. (German edition: Teubner, Leipzig, 1931.)
Lanchon, H: (1974): *Torsion élastoplastique d'une barre cylindrique de section simplement ou multiplement connexe.* J. Mech. **13**, 267-320.
Landau, L. and Lifšic, E. (1962): *Lehrbuch der Theoretischen Physik*, Vols. 1-10. Akademie-Verlag, Berlin. (English edition: Pergamon Press, Oxford, 1962ff.)
Landau, L., Achieser, A., and Lifšic, E. (1970): *Mechanik und Molekularphysik*. Akademie-Verlag, Berlin.
Lang, S. (1972): *Differential Manifolds*. Addison-Wesley, Reading, MA.
Langenbach, A. (1976): *Monotone Potentialoperatoren*. Verl. d. Wiss., Berlin.
Latal, H. and Mitter, H. [eds.] (1987): *Concepts and Trends in Particle Physics*. Springer-Verlag, New York.
Laue, M. v. (1950): *History of Physics*. Academic Press, New York. (German edition: Ullstein, Frankfurt/Main.)
Laugwitz, D. (1960): *Differentialgeometrie*. Teubner, Stuttgart. (English edition: Academic Press, New York, 1965.)
Lavrentjev, M. (1946): *On the theory of long waves.* Sbornik Inst. Mat. Akad. Nauk Ukrainsk. RSR **8**, 13-69 (Ukrainian). (English edition: Amer. Math. Soc. Transl., Vol. 102, 1954.)
Lax, P. (1957): *Hyperbolic systems of conservation laws.* Comm. Pure Appl. Math. **10**, 537-566.
Lax, P. and Wendroff, B. (1960): *Systems of conservation laws.* Comm. Pure Appl. Math. **13**, 217-237.
Lax, P. (1968): *Integrals of nonlinear equations of evolution and solitary waves.* Comm. Pure Appl. Math. **21**, 467-490.
Lax, P. (1973): *Hyperbolic Systems of Conservation Laws and the Mathematical Theory of Shock Waves*. SIAM, Philadelphia.
Lax, P. (1983): *Problems solved and unsolved concerning linear and nonlinear partial differential equations.* In: Warsaw (1983), 119-138.
LeBlond, P. and Mysak, L. (1978): *Waves in the Ocean*, Vols. 1, 2. Elsevier.
Lebovitz, N. (1977): *Bifurcation and stability problems in astrophysics.* In: Rabinowitz, P. [ed.] (1977), 259-284.
Lee, T. (1981): *Particle Physics and Introduction to Field Theory*. Harwood, New York.
Leites, D. (1980): *Introduction to the theory of supermanifolds.* Uspekhi Mat. Nauk **35** (1), 3-57 (Russian).

Leray, J. (1933): *Étude de diverses équations integrales non linéaires et de quelques problèmes que pose l'hydrodynamique.* J. Math. Pures Appl. **12**, 1–82.

Leray, J. (1934): *Essai sur les mouvements plans d'un liquide visceux que limitent des parois.* J. Math. Pures Appl. **13**, 331–418.

Leray, J. (1934a): *Sur le mouvement d'un liquide visqueux emplissant l'espace.* Acta Math. **63**, 193–248.

Leray, J. (1935): *Les problèmes de représentation conforme d'Helmholtz.* Comment. Math. Helv. **8**, 149–180.

Leray, J. (1978): *Analyse Lagrangienne et mécanique quantique.* Strasbourg, France. (English edition: MIT Press, 1981).

Les Houches (1951ff): *Summer Schools on Theoretical Physics.* North-Holland, Amsterdam.

Les Houches (1979): *Physical Cosmology.* North-Holland, Amsterdam.

Les Houches (1981): *Chaotic Behavior of Deterministic Systems.* North-Holland, Amsterdam.

Les Houches (1981a): *Gauge Theories in High Energy Physics.* North-Holland, Amsterdam.

Les Houches (1982): *Recent Developments in Field Theory and Statistical Mechanics.* North-Holland, Amsterdam.

Les Houches (1983): *Relativity, Groups and Topology, II.* North-Holland, Amsterdam.

Les Houches (1983a): *Birth and Infancy of Stars.* North-Holland, Amsterdam.

Levi-Civita, T. (1917): *Nozione di parallelismo in una varietà* Rend. Circ. Mat. Palermo **42**, 173–205.

Levi-Civita, T. (1925): *Détermination rigoreuse des ondes permanentes d'ampleur finie.* Math. Ann. **93**, 264–314.

Lévy, M. (1871): *Mémoire sur les équations des corps solides ductiles au-déla de la limite élastique.* J. Math. Pures Appl. **16**, 369–372.

Liang, E. and Sachs, R. (1980): *Cosmology.* In: Held, A. [ed.] (1980), Vol. 2, 329–357.

Lichtenstein, L. (1921): *Neuere Entwicklungen der Potentialtheorie.* In: *Enzyklopädie der mathematischen Wissenschaften*, Vol. II/3.1, 181–217.

Lichtenstein, L. (1923): *Astronomie und Mathematik in ihrer Wechselwirkung.* Hirzel, Leipzig.

Lichtenstein, L. (1924): *Über die erste Randwertaufgabe der Elastizitätstheorie.* Math. Z. **20**, 21–28.

Lichtenstein, L. (1929): *Grundlagen der Hydromechanik.* Springer-Verlag, Berlin.

Lichtenstein, L. (1930): *Über einige Hilfssätze der Potentialtheorie IV.* Ber. Sächs. Akad. Wiss. Leipzig **82**, 265–344.

Lichtenstein, L. (1931): *Vorlesungen über einige Klassen nichtlinearer Integralgleichungen und Integro-Differentialgleichungen nebst Anwendungen.* Springer-Verlag, Berlin.

Lichtenstein, L. (1933): *Gleichgewichtsfiguren rotierender Flüssigkeiten.* Springer-Verlag, Berlin.

Lie, S. (1934): *Gesammelte Abhandlungen (Collected Papers)*, Vols. 1–7. Teubner, Leipzig, 1934ff.

Lighthill, J. (1978): *Waves in Fluids.* University Press, Cambridge, England.

Lighthill, J. (1986): *An Informal Introduction to Theoretical Fluid Dynamics.* University Press, Oxford, England.

Lightman, A. et al. (1975): *Problem Book in Relativity and Gravitation.* University Press, Princeton, NJ.

Lin, C. and Reid, W. (1963): *Turbulent flow, theoretical aspects.* In: Flügge, S. [ed.] (1956), Vol. VIII/2, 438–523.

Linde, A. (1984): *Elementary particles and cosmology.* In: *International Conference on High Energy Physics in Leipzig (1984)*, Vol. 2, 125–148.

Lions, J. (1969): *Quelques méthodes de résolution des problémes aux limites non linéaires.* Dunod, Paris.

Lions, J. (1973): *Perturbation singulières dans les problèmes aux limites et en controle optimal.* Lecture Notes in Mathematics, Vol. 323. Springer-Verlag, Berlin.

Lions, P. (1982): *Generalized Solutions of Hamilton–Jacobi Equations.* Pitman, London.

Lions, P. (1983): *Hamilton–Jacobi–Bellman equations and the optimal control of stochastic systems.* In: Warsaw (1983), Vol. 2, 1403–1477.

Lions, P. (1984): *The concentration-compactness principle in the calculus of variations.* Ann. Inst. H. Poincaré. Anal. Non Linéaire **1**, 109–145.

List, S. (1978): *Generic bifurcation with applications to the von Kármán equations.* J. Differential Equations **30**, 89–118.

Ljapunov. A. (1882): *The general problem on the stability of motion.* In: Ljapunov, A. (1954), Vol. 2, 7–263 (Russian).

Ljapunov, A. (1906): *Sur les figures d'equilibre peu différentes d'une masse liquide homogène donée d'un mouvement rotation.* Mémories 1906. St Pétersburgh, 1–225. (See Ljapunov, A. (1954), Vol. 4.)

Ljapunov, A. (1954): *Collected Works*, Vols. 1–5. Nauka, Moscow (Russian).

Ljusternik, L. and Višik, M. (1957): *Regular degeneration and boundary layers for linear differential equations with small parameters.* Uspekhi Mat. Nauk **12** (5), 3–122 (Russian).

Lodge, A., McLeod, J., and Nohel, J. (1978): *A nonlinear singularly perturbed Volterra integrodifferential equation occurring in polymer rheology.* Proc. Roy. Soc. Edinburgh **A80**, 99–137.

Longair, M. [ed.] (1974): *Confrontation of Cosmological Theories with Observational Data.* Reidel, Boston.

Lopes, J. (1981): *Gauge Field Theory.* Pergamon Press, Oxford, England.

Lorenz, E. (1963): *Deterministic non-periodic flow.* J. Atomspheric Sci. **20**, 130–141.

Lotze, K. (1980): *Der Lebensweg der schwarzen Löcher.* Die Sterne **56**, 82–92, 149–159.

Love, A. (1906): *A Treatise on the Mathematical Theory of Elasticity.* University Press, Cambridge, England.

Ludwig, G. (1978): *Einführung in die Grundlagen der theoretischen Physik*, Vols. 1–4. Vieweg, Braunschweig.

Lurje, A. (1980): *Nonlinear Elasticity.* Nauka, Moscow (Russian).

Lüscher, E. (1980): *Pipers Buch der modernen Physik.* Piper, München.

Ma, S. (1982): *Modern Theory of Critical Phenomena.* Benjamin, London.

Macke, W. (1962): *Lehrbuch der theoretischen Physik*, Vols. 1–6. Geest & Portig, Leipzig.

Majda, A. (1982): *Smooth solutions for the equations of compressible and incompressible flow.* In: da Veiga, B. [ed.], *Fluid Dynamics.* Lecture Notes in Mathematics, Vol. 1047, 77–126. Springer-Verlag, Berlin.

Majda, A. (1983): *Systems of Conservation Laws in Several Space Variables.* In: Warsaw (1983), 1217–1224.

Majda, A. (1984): *Compressible Fluid Flow and Systems of Conservation Laws in Several*

Space Variables. Springer-Verlag, New York.
Mangoldt, H. v. and Knopp, K. (1957): *Einführung in die höhere Mathematik,* Vols. 1-3. Hirzel, Leipzig.
Manin, J. (1981): *Mathematics and Physics.* Birkhäuser, Boston.
Manin, Y. and Khenkin, G. (1982): *Yang-Mills-Dirac equations as Cauchy-Riemann equations in twistor space.* Soviet J. Nuclear Phys. 35, 941-950.
Manin, J. (1984): *Gauge Field Theory and Complex Manifolds.* Nauka, Moscow (Russian).
Manin, Y. (1985): *New exact solutions and cohomology analysis of ordinary and supersymmetric Yang-Mills equations.* Proc. Steklov Inst. Math. 165, 107-127.
Marlow, A. (1980): *Quantum Theory and Gravitation.* Academic Press, New York.
Marsden, J. (1968): *Generalized Hamiltonian mechanics.* Arch. Rat. Mech. Anal. 28, 323-361.
Marsden, J. (1968a): *Hamiltonian one-parameter groups.* Arch. Rat. Mech. Annal. 28, 362-396.
Marsden, J. (1974): *Applications of Global Analysis in Mathematical Physics.* Publish or Perish, Boston.
Marsden, J. and McCracken, M. (1976): *The Hopf Bifurcation and Its Applications.* Springer-Verlag, Berlin.
Marsden, J. and Tromba, A. (1976): *Vector Calculus.* Freeman, San Francisco, CA.
Marsden, J. (1980): *Lectures on Geometric Methods in Mathematical Physics.* SIAM, Philadelphia.
Marsden, J. (1983): *The initial-value problem and the dynamics of gravitational fields.* In: Schmutzer, E. [ed.] (1983), 115-126.
Marsden, J. and Hughes, T. (1983): *Mathematical Foundations of Elasticity.* Prentice-Hall, Englewood Cliffs, NJ.
Marsden, J., Abraham, R., and Ratiu, T. (1983): *Manifolds, Tensor Analysis, and Applications.* Addison-Wesley, Reading, MA.
Marsden, J. [ed.] (1984): *Fluids and Plasmas: Geometry and Dynamics.* Contemporary Mathematics, Vol. 28, American Mathematical Society, Providence, RI.
Martin, N. (1981): *Mathematical Theory of Entropy.* Addison-Wesley, London.
Maslov, V. (1972): *Théorie des perturbations et méthodes asymptotiques.* Dunod, Paris.
Mathematisches Wörterbuch (1961): Edited by J. Naas and H. Schmid. Akademie-Verlag, Berlin.
Massey, B. (1971): *Units, Dimensional Analysis, and Physical Similarity.* Van Nostrand, London.
Mathematics: The Unifying Thread in Science (1986): Notices Amer. Math. Soc. 33, 716-733.
Matsumara, A. and Nishida, T. (1979): *The initial-value problem for the equations of motion of compressible viscous and heat conductive fluids.* Proc. Japan Acad. A55, 337-342.
Maul, J. (1976): *Eine einheitliche Methode zur Lösung der ebenen Aufgaben der linearen Elastostatik.* Akademie-Verlag, Berlin.
Maurin, K. (1967): *Methods of Hilbert Spaces.* PWN, Warsaw.
Maurin, K. (1976): *Analysis,* Vols. 1, 2. Reidel, Boston.
Maurin, K. (1981): *Mathematik als Sprache und Kunst.* In: Maurin, K. *et al.* [eds.], *Offene Systeme,* Vol. 2, 118-241. Stuttgart,
Maurin, K. (1982): *Plato's cave parable and the development of modern physics.* Rend. Sem. Mat. Univ. Politec. Torino 40, 1-31.

Mehra, J. and Rechenberg, H. (1982): *The Historical Development of Quantum Theory*, Vols. 1-4. Springer-Verlag, New York.

Melcher, H. (1979): *Albert Einstein wider Vorurteile und Denkgewohnheiten*. Akademie-Verlag, Berlin.

Menzel, D. (1955): *Fundamental Formulas of Physics*. Prentice-Hall, New York.

Meyer, R. (1971): *Introduction to Mathematical Fluid Dynamics*. Wiley, New York.

Meyers, N. (1963): *An L_p-estimate for the gradient of solutions of second-order elliptic divergence equations*. Ann. Scuola Norm. Sup. Pisa 17 (3), 189-206.

Mezard, M. and Virusoro, M. [eds.] (1987): *Spin Glass Theory and Beyond*. World Scientific, Singapore.

Miersemann, E. (1975): *Verzweigungsprobleme für Variationsungleichungen*. Math. Nachr. 65, 187-209.

Miersemann, E. (1978): *Über höhere Verzweigungspunkte nichtlinearer Variationsungleichungen*. Math. Nachr. 85, 195-213.

Miersemann, E. (1979): *Über positive Lösungen von Eigenwertgleichungen mit Anwendungen auf ein Beulproblem für die Platte*. Z. Angew. Math. Mech. 59, 189-194.

Miersemann, E. (1980): *Zur Regularität der quasistatischen elastoviscoplastischen Verschiebungen und Spannungen*. Math. Nachr. 96, 293-299.

Miersemann, E. (1981): *Eigenvalue problems for variational inequalities*. Contemp. Math. 4, 25-43.

Miersemann, E. (1981a): *Eigenwertaufgaben für Variationsungleichungen*. Math. Nachr. 100, 221-228.

Miersemann, E. (1981b): *Zur Lösungsverzweigung bei Variationsungleichungen mit einer Anwendung auf den Knickstab mit begrenzter Durchbiegung*. Math. Nachr. 102, 7-15.

Miersemann, E. (1982): *Stabilitätsprobleme für Eigenwertaufgaben bei Beschränkungen für die Variationen mit einer Anwendung auf die Platte*. Math. Nachr. 106, 211-221.

Miller, J. and Sciama, D. (1980): *Gravitational collapse to the black hole states*. In: Held, A. [ed.] (1980), Vol. 2, 359-392.

Milne-Thomson, L. (1960): *Theoretical Hydrodynamics*. Macmillan, London.

Milnor, J. (1965): *Topology From the Differentiable Viewpoint*. University Press, Charlottesville, VA.

Milnor, J. (1983): *Hyperbolic geometry: the first 150 years*. In: Browder, F. [ed.] (1983), 25-40.

Minkowski, H. (1909): *Raum und Zeit*. Teubner, Leipzig. (English edition: *Space and time*. Calcutta Math. Soc. Bull. 1, 135-141.)

Miranda, C. (1941): *Un'osservazione su un teorema di Brouwer*. Boll. Un. Mat. Ital. Seconda Serie 3, 5-7.

Mises, R.v. (1913): *Methodik der festen Körper im plastisch-deformablen Zustand*. Nachr. Akad. Wiss. Göttingen, Math.-phys. Kl. 1913, 582-592.

Mises, R. v. (1962): Cf. Frank, P. and Mises, R. v. (1962).

Misner, C., Thorne, K., and Wheeler, J. (1973): *Gravitation*. Freeman, San Francisco, CA.

Miyoshi, T. (1985): *Foundations of the Numerical Analysis of Plasticity*. North-Holland, Amsterdam.

Mohapatra, R. (1986): *Unification and Supersymmetry*. Springer-Verlag, New York.

Monod, J. (1970): *Le hasard et la nécessité*. Paris. (German edition: *Zufall und Notwendigkeit*, München, 1971.)

Monastyrsky, M. (1987): *Riemann, Topology, and Physics*. Birkhäuser, Boston.

Montgomery, D. and Zippin, L. (1955): *Transformation Groups*. Interscience, New York.
Moore, F. (1977): *The Story of Astronomy*. MacDonald, London.
Morawetz, C. (1981): *Lectures on Nonlinear Waves and Shocks*. Springer-Verlag, New York.
Morawetz, C. (1982): *The mathematical approach to the sonic barrier*. Bull. Amer. Math. Soc. (N.S.) **6**, 127–145.
Moreau, J. (1968): *La notion de sur-potential et les liaisons unilatérales en élastostatique*. C. R. Acad. Sci. Paris Sér. **A267**, 954–957.
Moreau, J. (1976): *Applications of convex analysis to the treatment of elastoplastic systems*. In: Germain, P. and Nayroles, B. [eds.] (1976), 56–89.
Moritz, R. (1914): *Memorabilia Mathematics*. Macmillan, New York.
Morrey, C. (1952): *Quasi-convexity and the lower semicontinuity of multiple integrals*. Pacific J. Math. **2**, 25–53.
Morrey, C. (1966): *Multiple Integrals in the Calculus of Variations*. Springer-Verlag, New York.
Morrisson, P. (1984): *Zehn hoch. Dimensionen zwischen Quarks und Galaxien*. Spektrum der Wissenschaft, Heidelberg.
Morse, M. (1939): *The behaviour of a function on its critical set*. Ann. Math. **40**, 62–70.
Morse, P. and Feshbach, H. (1953): *Methods of Theoretical Physics*, Vols. 1, 2. McGraw-Hill, New York.
Mosco, U. (1976): *Implicit variational problems and quasivariational inequalities*. Lecture Notes in Mathematics, Vol. 543, 83–156. Springer-Verlag, Berlin.
Mouritsen, O. (1984): *Computer Studies of Phase Transitions and Critical Phenomena*. Springer-Verlag, New York.
Murat, F. (1978): *Compacticité par compensation*. Ann. Scuola Norm. Sup. Pisa Sci. Fis. Math. **5**, 489–507.
Murat, F. (1981): *L'injection du cône positif de H^{-1} dans $W^{-1,q}$ est compacte pour tout $1 < q < 2$*. J. Math. Pures Appl. **60**, 309–322.
Murat, F. (1987): *A survey on compensated compactness*. In: Cesari, L. [ed.] (1987), *Contributions to Modern Calculus of Variations*. Pitman, London, pp. 145–183.
Murman, E. [ed.] (1985): *Progress and Supercomputing in Computational Fluid Dynamics*. Birkhäuser, Boston.
Muskelišvili, N. (1954): *Fundamental Problems in Mathematical Elasticity*. Nauka, Moscow (Russian). (English edition: Noordhoff, Leyden, 1975.)

Naghdi, P. (1972): *Theory of plates and shells*. In: Flügge, S. [ed.] (1956), Vol. VIa/2, 425–640.
Narlikar, J. and Padmanabhan, T. (1982): *Quantum cosmology via path integrals*. Phys. Rep. **110**, 151–200.
Nash, J. (1951): *Non-cooperative games*. Ann. Math. **54**, 286–295.
Nash, J. (1956): *The embedding problem for Riemannian manifolds*. Ann. of Math. **63**, 20–63.
Naumann, J. (1982): *Zur Existenz und Regularität der Lösungen der Variationsungleichungen der Theorie visko-plastischer und starr-idealplastischer Flüssigkeiten*. Dissertation B, Universität Leipzig.
Naumann, J. (1984): *Parabolische Variationsungleichungen*. Teubner, Leipzig.
Navier, C. (1822): *Mémoire sur les lois du mouvement des fluides*. Mém. Acad. Sciences.

Nayfeh, A. (1973): *Perturbation Methods.* Wiley, New York.

Nayfeh, A. and Mook, D. (1979): *Nonlinear Oscillations.* Wiley, New York.

Nečas, J. et al. (1980): *On the solution of the variational inequality to the Signorini problem with small friction.* Boll. Un. Mat. Ital. (5) **178**, 796–811.

Nečas, J. and Hlávácek, I. (1981): *Mathematical Theory of Elastic and Elasto-plastic Bodies.* Elsevier, New York.

Nečas, J. (1983): *Introduction to the Theory of Nonlinear Elliptic Equations.* Teubner, Leipzig.

Nečas, J. and Hlaváček, I. (1986): Cf. Hlaváček, I. (1986).

Nekrasov, A. (1921): *On stationary waves,* I. II. Ivanova-Voznes. Bull. Inst. Polytechnic **3**, 52–65, **6**, 155–171 (Russian).

Neumann, J. v. (1928): *Zur Theorie der Gesellschaftsspiele.* Math. Ann. **100**, 295–320.

Neumann, J. v. (1932): *Mathematische Grundlagen der Quantenmechanik.* Springer-Verlag, Berlin. (English edition: *Mathematical Foundations of Quantum Mechanics.* University Press, Princeton, NJ, 1955).

Neumann, J. v. and Morgenstern, O. (1944): *Theory of Games and Economic Behaviour.* University Press, Princeton, NJ.

Neumann, J. v. (1947): *The Mathematician.* In: Neumann, J. v. (1961), Vol. 1, 1–9.

Neumann, J. v. and Richtmyer, R. (1950): *A method for the numerical calculation of hydrodynamics.* J. Appl. Phys. **21**, 232–237.

Neumann, J. v. (1961): *Collected Works.* Pergamon Press, New York.

Neumark, M. (1963): *Lineare Darstellungen der Lorentz-Gruppe.* Verl. d. Wiss., Berlin.

Newton, I. (1687): *Philosophiae Naturalis Principia Mathematica.* London.

Newton, I. (1797): *Opera.* Vols. 1–5. Edited by S. Horseley. London.

Newton, I. (1967): *The Mathematical Papers of Isaac Newton.* Edited by D. Whiteside, since 1967.

Nguyen, Q. (1973): *Matériaux élastoplastiques écrouissable.* Arch. Mech. Stos. **25**, 695–702.

Nguyen, Q. (1982): *Problèmes de plasticité et de rupture.* Publ. math. d'Orsay, No. 82.08. Université de Paris-Sud.

Ni, L. (1982): *A combinatorial approach to the mapping degree.* J. Math. Anal. Appl. **89**, 386–399.

Nickel, K. (1958): *Einige Eigenschaften von Lösungen der Prandtlschen Grenzschichtgleichung.* Arch. Rat. Mech. Anal. **2**, 1–31.

Nickel, K. (1963): *Die Prandtlschen Grenzschichtdifferentialgleichungen als asymptotischer Grenzfall der Navier–Stokesschen Differentialgleichungen und der Eulerschen Differentialgleichungen.* Arch. Rat. Mech. Anal. **13**, 1–14.

Nickel, K. (1984): *Minimal drag for wings with prescribed lift, roll moment and yaw moment, or how to fight adverse yaw.* In: Boffi, V. and Neunzert, H. [eds.] (1984), 7–50.

Nicolis, G. and Prigogine, I. (1977): *Self-organization in non-equilibrium systems. From Dissipative Structures to Order Through Fluctuations.* Wiley, New York.

Nieuwenhuizen, P. van (1984): *An introduction to simple supergravity and the Klein–Kaluza program.* In: DeWitt, B. and Stora, R. [eds.] (1984), 824–912.

Nikaido, H. (1956): *On the classical multilateral exchange problem.* Metroeconomica **8**, 135–145.

Nikaido, H. (1968): *Convex Structures and Economic Theory.* Academic Press, New York.

Nikaido, H. (1970): *Introduction to Sets and Mappings in Modern Economics.* North-Holland, New York.

Niordson, F. (1985): *Shell Theory.* North-Holland, Amsterdam.

Nirenberg, L. (1955): *Remarks on strongly elliptic systems.* Comm. Pure Appl. Math. **8**, 649–675.

Nobel Prizes (1954ff): *Nobel Lectures.* Edited by the Nobel Foundation, Stockholm.

Nonlinear Phenomena (1986): *Solitons and Coherent Structures.* Proceedings of a conference held at Santa Barbara, CA, 1985. Phys. D, **18**.

Novikov, S. et al. (1980): *Theory of Solitons. The Inverse Scattering Method.* Nauka, Moscow (Russian). (English edition: Plenum, New York, 1984.)

Novikov, I. and Frolov, V. (1986): *Physics of Black Holes.* Nauka, Moscow (Russian).

Nussbaum, R. (1978): *Differential Delay Equations With Two Time Lags.* American Mathematical Society, Providence, RI.

Oberdorfer, E. (1969): *Das internationale Maßsystem.* Springer-Verlag, Berlin.

Obuhov, A. (1983): *Kolmogorov flow and its realization in experiments.* Uspekhi Mat. Nauk **38** (4), 100–111 (Russian).

Ockendon, H. and Taylor, A. (1983): *Inviscid Fluid Flows.* Springer-Verlag, New York.

Oden, J. and Reddy, T. (1976): *Variational Methods in Theoretical Mechanics.* Springer-Verlag, New York.

Oden, J. (1979): *Existence theorems for a class of problems in nonlinear elasticity.* J. Math. Anal. Appl. **69**, 51–83.

Oden, J. (1980): *Computational Methods in Nonlinear Mechanics.* North-Holland, New York.

Odquist, F. (1930): *Über die Randwertaufgaben der Hydrodynamik zäher Flüssigkeiten.* Math. Z. **32**, 329–375.

Ogden, R. (1972): *Large deformation isotropic elasticity: on the correlation of theory and experiment for compressible rubberlike solids.* Proc. Roy. Soc. London, **A328**, 567–583.

Oleinik, O. (1957): *Discontinuous solutions of nonlinear equations.* Uspekhi Mat. Nauk **12** (3), 3–73 (Russian).

Oleinik, O., Kalašnikov, A., and Yui-Lin, C. (1958): *The Cauchy problem and boundary problems for equations of the type of nonstationary filtration.* Izv. Akad. Nauk SSSR Ser. Mat. **22**, 667–704 (Russian).

Oleinik, O. (1959): *On the construction of generalized solutions of the Cauchy problem for quasilinear equations via artificial viscosity,* Uspekhi Mat. Nauk **14** (2), 159–164 (Russian).

Oleinik, O. (1968): *Mathematical problems in the theory of boundary layers.* Uspekhi Mat. Nauk **23** (3), 4–65 (Russian).

Olfe, D. and Zakkay, V. (1964): *Supersonic Flow. Chemical Processes and Radiative Transfer.* Oxford.

Olszak, W. [ed.] (1980): *Thin Shell Theory: New Trends and Applications.* Springer-Verlag, Wien.

Omohundro, S. (1986): *Geometric Perturbation Theory in Physics.* World Scientific, Singapore.

O'Neill, B. (1983): *Semi-Riemannian Geometry.* Academic Press, New York.

Orear, J. (1966): *Fundamental Physics.* Wiley, New York. (German edition: Hanser-Verlag, München, 1971.)

Orlik, P. [ed.]. *Singularities.* Proc. Sympos. Pure Math., Vol. 40, Parts I, II. American Mathematical Society, Providence, RI.
Osborn, H. (1982): *Vector Bundles*, Vols. 1-3. Academic Press, New York.
Oswatitsch, K. (1976): *Grundlagen der Gasdynamik.* Springer-Verlag, Berlin.
Otte, M. [ed.] (1974): *Mathematiker über Mathematik.* Springer-Verlag, Berlin.
Owen, D. (1984): *A First Course in the Mathematical Foundations of Thermodynamics.* Springer-Verlag, New York.

Palais, R. (1969): *The Morse lemma on Banach spaces.* Bull. Amer. Math. Soc. **75**, 968–971.
Panagiotopoulos, P. (1985): *Inequality Problems in Mechanics and Applications.* Birkhäuser, Boston.
Pascali, D. (1986): *On critical points of nondifferentiable functions.* Libertas Math. **6**, 95–100.
Pauli, W. (1958): *Die allgemeinen Prinzipien der Wellenmechanik.* In: Flügge, S. [ed.] (1956), Vol. V/1, 1–168.
Pauli, W. (1973): *Lectures in Physics*, Vols. 1–6. MIT Press, Cambridge, MA.
Pazy, A. (1983): *Semigroups of Linear Operators and Applications to Partial Differential Equations.* Springer-Verlag, New York.
Peebles, P. (1980): *The Large Scale Structure of the Universe.* University Press, Princeton, NJ.
Peitgen, H. and Prüfer, M. (1979): *The Leray–Schauder continuation method is a constructive element in the numerical study of nonlinear eigenvalue and bifurcation problems.* In: Peitgen, H. and Walther, H. [eds.] (1979), 326–409.
Peitgen, H. and Walther, H. [eds.] (1979): *Functional Differential Equations and Approximation of Fixed Points.* Lecture Notes in Mathematics, Vol. 730. Springer-Verlag, Berlin.
Peitgen, H. and Walther, H. [eds.] (1981): *Numerical Solution of Nonlinear Equations.* Lecture Notes in Mathematics, Vol. 878. Springer-Verlag, New York.
Peitgen, H. (1984): *Harmonie in Chaos und Kosmos.* Bremen.
Peitgen, H. and Richter, P. (1985): *The Beauty of Fractals.* Springer-Verlag, New York.
Penrose, R. (1972): *Techniques of Differential Topology in Relativity.* SIAM, Philadelphia.
Penrose, R. (1977): *The twistor program.* Rep. Math. Phys. **12**, 65–76.
Penrose, R. (1979): *Singularities and time-asymmetry.* In: Hawking, S. and Israel, W. [eds.] (1979).
Penrose, R. and Ward, R. (1980): *Twistors for flat and curved space–time.* In: Held, A. [eds.] (1980), Vol. 2, 283–328.
Penrose, R. and Rindler, W. (1984): *Spinors and Space–Time*, Vols. 1, 2. University Press, Cambridge, England.
Perron, O. (1929): *Über Stabilität und asymptotisches Verhalten der Integrale von Differentialgleichungssystemen.* Math. Z. **29**, 129–160.
Perron, O. (1930): *Die Stabilitätsfrage bei Differentialgleichungen.* Math. Z. **32**, 703–728.
Petrina, D. and Gerasimenko, V. (1983): *A mathematical approach to the evolution of infinite systems in classical statistical mechanics.* Uspekhi Mat. Nauk **38** (5), 3–58 (Russian).
Peyret, R. and Taylor, T. (1985): *Computational Methods for Fluid Flow.* Springer-Verlag, New York.

Pflüger, A. (1965): *Stabilitätsprobleme der Elastostatik*. Springer-Verlag, Berlin.

Pflüger, A. (1967): *Elementare Schalenstatik*. Springer-Verlag, Berlin.

Phú, H. (1987): *Zur Lösung des linearisierten Knickstabproblems mit beschränkter Ausbiegung*. Z. Anal. Anwendungen (to appear).

Phú, H. (1987a): *On the optimal control of a hydroelectric power plant*. Systems Control Lett. 8, 281–288.

Pipkin, A. (1986): *Lectures on Viscoelastic Theory*. Springer-Verlag, New York.

Planck, M. (1900): *Zur Theorie des Gesetzes der Energieverteilung im Normalspektrum*. Verh. Dt. Physik. Ges. Berlin 2, 237–248.

Planck, M. (1909): *Gutachten zur Berufung Einsteins an die Universität Prag*. (Cf. Frank, P. (1949).)

Planck, M. (1913): *Vorlesungen über Thermodynamik*. De Gruyter, Berlin. (English edition: *Treatise on Thermodynamics*. Dover, New York, 1945.)

Planck, M. (1945): *Wissenschaftliche Autobiographie*. Barth, Leipzig. (English edition: *Scientific Autobiography*, Philosophical Library, 1949.)

Planck, M. (1967): *Der Kausalbegriff in der Physik*. Barth, Leipzig.

Pliss, V. (1977): *Solution Sets of Periodic Differential Equations*. Nauka, Moscow (Russian).

Poincaré, H. (1885): *Les figures equilibrium*. Acta Math. 7, 259–302.

Poincaré, H. (1892): *Les méthodes nouvelles de la mécanique céleste*, Vols. 1–3. Gauthier-Villars, Paris.

Poincaré, H. (1928): *Oeuvres (Collected Works)*, Vols. 1–10. Gauthier-Villars, Paris.

Pontrjagin, L. (1966): *Topological Groups*. Gordon and Breach, New York.

Poston, T. and Stewart, I. (1978): *Catastrophe Theory and Its Applications*. Pitman, London.

Prager, W. and Hodge, P. (1951): *Theory of Perfectly Plastic Solids*. Wiley, New York.

Prager, W. (1955): *Probleme der Plastizitätstheorie*. Birkhäuser, Basel. (English edition: *Problems in Plasticity*. Addison-Wesley, London, 1959.)

Prager, W. (1961): *Einführung in die Kontinuumsmechanik*. Birkhäuser, Basel.

Prandtl, L. (1904): *Über Flüssigkeitsbewegung bei sehr kleiner Reibung*. In: *Verh. d. III. Internat. Mathematikerkongresses, Heidelberg*.

Prandtl, L. (1924): *Spannungsverteilung in plastischen Körpern*. In: *Proceedings of the First Int. Congr. Appl. Mech., Delft*, pp. 43–54.

Prandtl, L. (1949): *Strömungslehre*. Vieweg, Braunschweig.

Press, H. et al. (1986): *Numerical Recipies: The Art of Scientific Computing*. University Press, Cambridge, England.

Pressley, A. and Segal, G. (1986): *Loop Groups*. Clarendon Press, Oxford, England.

Prigogine, I. (1979): *Vom Sein zum Werden*. Piper, München. (English edition: *From Being to Becoming*.)

Prigogine, I. and Stengers, I. (1981): *Dialog mit der Natur*. Piper, München.

Primas, H. (1983): *Chemistry, Quantum Mechanics, and Reductionism*. Perspectives in Theoretical Chemistry. Springer-Verlag, New York.

Prüfer, M. and Siegberg, H. (1979): *On computational aspects of topological degree in $\mathbb{R}^n$*. In: Peitgen, H. and Walther, H. [eds.] (1979), 410–433.

Pukhnačov, J. and Popov, J. (1985): *Mathematik ohne Formeln*. Urania-Verlag, Leipzig.

Pukhnačov, V. (1972): *The plane steady problem with free boundary for the Navier-Stokes equations*. Prikl. Mech. Techn. Fiz. 5, 126–134 (Russian).

Pukhnačov, V. (1975): *Nonclassical Problems in the Theory of a Boundary Layer*. University Press, Novosibirsk (Russian).

Quigg, C. (1983): *Gauge Theories of the Strong, Weak and Electromagnetic Interactions*. Benjamin, London.

Rabier, P. (1985): *Lectures on Topics in One-Parameter Bifurcation Problems*. Springer-Verlag, New York.

Rabier, P. (1986): *A general study of nonlinear problems with three solutions in Hilbert spaces*. Arch. Rat. Mech. Anal. **95**, 123–154.

Rabinowitz, P. (1968): *Existence and nonuniqueness of rectangular solutions of the Bénard problem*. Arch. Rat. Mech. Anal. **29**, 32–57.

Rabinowitz, P. [ed.] (1977): *Applications of Bifurcation Theory*. Academic Press, New York.

Ramanathan, R. (1983): *Introduction to the Theory of Economic Growth*. Lecture Notes in Economics, Vol. 205. Springer-Verlag, New York.

Rajaraman, R. (1982): *Solitons and Instantons*. North-Holland, Amsterdam.

Ramm, E. [ed.] (1982): *Buckling of Shells*. Springer-Verlag, New York.

Ranft, G. and Ranft, J. (1976): *Elementarteilchen*, Vols. 1, 2. Teubner, Leipzig.

Rankine, W. (1870): *On the thermodynamic theory of waves of finite longitudinal disturbance*. Trans. Roy. Soc. London **160**, 277–288.

Raschewskii, P. (1959): *Riemannsche Geometrie und Tensoranalysis*. Verl. d. Wiss., Berlin.

Rayleigh, J. (1910): *Aerial plane waves of finite amplitude*. Proc. Roy. Soc. London **84**, 247–284.

Reasenberg, R. and Shapiro, I. (1983): *Terrestrial and planetary relativity experiments*. In: Schmutzer, E. [ed.] (1983), 149–164.

Rebbi, C. and Soliani, G. (1984): *Solitons and Particles*. World Scientific, Singapore.

Recke, L. (1978): *Anwendung der Verzweigungstheorie auf geometrisch nichtlineare Schalengleichungen*. Dissertation, Humboldt-Universität, Berlin.

Recke, L. (1987): *Zur Überlagerung zweier Hopf-Bifurkationen*. Dissertation B, Humboldt-Universität Berlin (Seminarbericht Sektion Mathematik Nr. 79).

Reed, M. and Simon, B. (1972): *Methods of Modern Mathematical Physics*, Vols. 1–4. Academic Press, New York.

Reeken, M. (1977): *The equation of motion of a chain*. Math. Z. **155**, 219–237.

Reeken, M. (1979): *Classical solutions of the chain equations*. Math. Z. **165**, 143–169, **166**, 67–82.

Regge, T. (1984): *The group manifold approach to unified gravity*. In: DeWitt, B. and Stora, R. [eds.] (1984), 933–1006.

Reichardt, H. (1985): *Gauß und die Anfänge der nicht-euklidischen Geometrie*. With original papers by J. Bolyai, N. Lobačevskii, and F. Klein. Teubner-Verlag, Leipzig.

Reif, F. (1965): *Fundamentals of Statistical and Thermal Physics*. McGraw-Hill, New York. (German edition: De Gruyter, Berlin, 1976.)

Reiner, M. (1958): *Rheology*. In: Flügge, S. [ed.] (1956), Vol. VI, 434–550.

Reiss, E. (1977): *Imperfect bifurcation*. In: Rabinowitz, P. [ed.] (1977), 37–72.

Renardy, M. (1982): *Bifurcation from rotating waves*. Arch. Rat. Mech. Anal. **79**, 49–84.

Renardy, M. (1983): *A class of quasilinear parabolic equations with infinite delay and application to a problem of viscoelasticity*. J. Differential Equations **48**, 280–292.

Reuss, A. (1930): *Berücksichtigung der elastischen Formänderung in der Plastizitätstheorie.* Z. Angew, Math. Mech. **10**, 266–271.

Reynolds, O. (1885): *On the flow of gases.* Proc. Manch. Lit. Phil. Sci.

Rheinboldt, W. (1986): *Numerical Analysis of Parametrized Nonlinear Equations.* Wiley, New York.

Ricci, G. and Levi-Civita, T. (1901): *Méthodes de calcul différentiel absolu et leurs applications.* Math. Ann. **54**, 125–201.

Richtmyer, R. and Morton, K. (1967): *Difference Methods for Initial-Value Problems.* Interscience, New York.

Richtmyer, R. (1978): *Principles of Advanced Mathematical Physics*, Vols. 1, 2. Springer-Verlag, Berlin.

Riedl, R. (1976): *Die Strategie der Genesis.* Piper, München.

Riedrich, T. (1976): *Vorlesungen über nichtlineare Operatorgleichungen.* Teubner, Leipzig.

Riemann, B. (1854): *Über die Hypothesen, welche der Geometrie zugrunde liegen.* Habilitationsvortrag. Abh. Akad. Wiss. Göttingen **13**. (English translation: In: Spivak, M. (1979), Vol. 2.)

Riemann, B. (1860): *Über die Fortpflanzung ebener Luftwellen von endlicher Schwingungsweite.* Abh. Ges. Wiss. Göttingen. Math.-Naturwiss. Kl. **8**, p. 43.

Riemann, B. (1861): *Mathematical remarks answering a question asked by the famous Paris Academy.* In: Riemann, B. (1892), 391–404.

Riemann, B. (1892): *Gesammelte mathematische Werke (Collected Mathematical Works).* Teubner, Leipzig.

Riesz, F. and Nagy, B. (1978): *Functional Analysis.* Ungar, New York.

Rindler, W. (1977): *Essential Relativity. Special, General, Cosmological.* Springer-Verlag, New York.

Ritz, W. (1908): *Über eine neue Methode zur Lösung gewisser Randwertaufgaben.* Göttinger Nachr., Math.-Naturwiss. Kl. **1908**, 236–248.

Rivlin, R. and Ericksen, J. (1955): *Stress–deformation relations for isotropic materials.* J. Rat. Mech. Anal. **4**, 681–702.

Röpke, G. (1987): *Statistische Mechanik für das Nichtggleichgewicht.* Verl. d. Wiss., Berlin.

Ross, G. (1984): *Grand Unified Theories.* Benjamin, Reading, MA.

Roy, P. and Singh, V. [eds.] (1984): *Supersymmetry and Supergravity.* Lecture Notes in Physics, Vol. 208.

Rôzdestvenskii, B. and Janenko, N. (1978): *Systems of Quasilinear Equations.* Nauka, Moscow (Russian).

Ruelle, D. (1969): *Statistical Mechanics. Rigorous Results.* Benjamin, New York.

Ruelle, D. and Takens, F. (1971): *On the nature of turbulence.* Commun. Math. Phys. **20**, 167–192, **23**, 343–344.

Ruelle, D. (1978): *The Mathematical Structure of Classical Equilibrium Statistical Mechanics.* Addison-Wesley, Reading, MA.

Ruelle, D. (1980): *Strange attractors.* Math. Intelligencer **2**, 126–137.

Ruelle, D. (1981): *Differentiable dynamical systems and the problem of turbulence.* Bull. Amer. Math. Soc. (N.S.) **5**, 29–42.

Ruelle, D. (1983): *Turbulent dynamical systems.* In: Warsaw (1983), 271–283.

Ruffini, R. [ed.] (1987): *Proceedings of the Fourth Marcel Grossmann Meeting on General Relativity*, Vols. 1, 2. North-Holland, Amsterdam.

Russian Encyclopedia of Mathematics (1977): Edited by I. Vinogradov. Vol. 1ff. Sovetskaja Encyclopedia, Moscow (Russian).

Sabinina, E. (1961): *On the Cauchy problem for the equation of nonstationary gas filtration in several space variables.* Dokl. Akad. Nauk SSSR **136**, 1034–1037 (Russian).
Sachs, R. and Wu, H. (1977): *General Relativity for Mathematicians.* Springer-Verlag, New York.
Sagan, C. and Shklovsky, I. (1968): *Intelligent Life in the Universe.* Dell, New York.
Sagan, C. and Agel, J. (1973): *The Cosmic Connection: An Extraterrestrial Perspective.* New York. (German edition: München, 1978.)
Sagan, C. (1980): *Cosmos.* New York (German edition: München, 1982.)
Sagan, C. (1980a): *Signale der Erde.* Droemer, München, (English edition: *Murmurs of Earth*, New York, 1978.)
Sagan, C. (1982): *Aufbruch in den Kosmos.* Heyne, München. (English edition: *Broca's Brain*, New York, 1979.)
Saint-Venant, de M. (1871): *Sur les équations du mouvement intérieur des solides ductiles.* J. Math. Pures Appl. **16**, 373–382.
Salam, A. and Sezgin, E. [eds.] (1986): *Supergravity Theories, Anomalies, and Compactification,* Vols. 1, 2. World Scientific, Singapore.
Sard, A. (1942): *The measure of the critical points of differentiable maps.* Bull. Amer. Math. Soc. **48**, 883–890.
Sather, D. (1976): *Branching and stability for nonlinear shells.* In: Germain, P. and Nayroles, B. [eds.] (1976), 462–473.
Sattinger, D. (1977): *Selection mechanisms for pattern formation.* Arch. Rat. Mech. Anal. **66**, 31–42.
Sattinger, D. (1979): *Group-Theoretic Methods in Bifurcation Theory.* Lecture Notes in Mathematics, Vol. 762. Springer-Verlag, Berlin.
Sattinger, D. (1980): *Bifurcation and symmetry breaking in applied mathematics.* Bull. Amer. Math. Soc. (N.S.) **3**, 779–819.
Sattinger, D. (1980a): *Les symetries des équations et leurs applications dans la mécanique et la physique.* Publ. Math. d'Orsay No. 80.08. Université de Paris-Sud.
Sauer, R. (1960): *Gasdynamik.* Springer-Verlag, New York.
Sauer, R. (1966): *Nichtstationäre Probleme der Gasdynamik.* Springer-Verlag, Berlin.
Scarf, H. (1967): *The approximation of fixed points of continuous mappings.* SIAM J. Appl. Math. **15**, 1328–1343.
Schauder, J. (1978): *Oeuvres (Collected Works).* PWN, Warsaw.
Scheffer, V. (1980): *The Navier–Stokes equations on a bounded domain.* Commun. Math. Phys. **73**, 1–42.
Scheidegger, A. (1963): *Hydrodynamics in porous media.* In: Flügge, S. [ed.] (1956), Vol. VIII/2, 625–662.
Scheidt, J. v. and Purkert, W. (1983): *Random Eigenvalue Problems.* North-Holland. Amsterdam.
Schilling, K. (1986): *Simpliziale Algorithmen zur Berechnung von Fixpunkten mengenwertiger Operatoren.* Wissenschaftlicher Verlag, Trier.
Schlichting, S. (1960): *Boundary Layer Theory.* McGraw-Hill, New York.
Schmutzer, E. (1968): *Relativistische Physik.* Teubner, Leipzig.
Schmutzer, E. [ed.] (1983): *Proceedings of the 9th International Conference on General Relativity and Gravitation.* Verl. d. Wiss., Berlin.

Schochet, S. (1986): *The incompressible Euler equations in a bounded domain.* Comm. Math. Phys. **104**, 49–75.
Schoen, R. and Yau, S.: Cf. Yau.
Scholz, E. (1980): *Geschichte des Mannigfaltigkeitsbegriffes von Riemann bis Poincaré.* Birkhäuser, Basel.
Schouten, J. (1954): *Ricci Calculus.* Springer-Verlag, Berlin.
Schreier, S. (1982): *Compressible Flow.* Wiley, New York.
Schrödinger, E. (1926): *Quantisierung als Eigenwertproblem.* Ann. Physik **9**, 361–376.
Schrödinger, E. (1927): *Abhandlungen zur Wellenmechanik.* Barth, Leipzig.
Schumann, R. (1987): *Eine neue Methode zur Gewinnung starker Regularitätsaussagen für das Signorini-Problem in der linearen Elastizitätstheorie.* Dissertation B, Universität, Leipzig.
Schumann, R. (1988): *Regularity for Signorini's problem in linear elastostatics.* Manuscripta Math. **63**, 255–291.
Schuster, H. (1984): *Deterministic Chaos.* Physik-Verlag, Weinheim, Federal Republic of Germany.
Schwartz, J. (1968): *Differential Geometry and Topology.* Gordon and Breach, New York.
Schwartz, J. (1969): *Nonlinear Functional Analysis.* Gordon and Breach, New York.
Schwarz, J. [ed.] (1985): *Superstrings,* Vols. 1, 2. World Scientific, Singapore.
Schwarzschild, K. (1916): *Über das Gravitationsfeld eines Massenpunktes nach der Einsteinschen Theorie.* Sitzungsber. Preuss. Akad. Wiss. Berlin **1916**, 189–196.
Sedov, L. (1959): *Similarity and Dimensional Methods in Mechanics.* Cleaver-Hume Press, London.
Segrè, G. (1987): *Superstrings and four-dimensional physics.* In: Latal, H. and Mitter, H. [eds.] (1987), 101–150.
Seifert, H. (1983): *Black holes, singularities, and topology.* In: Schmutzer, E. [ed.] (1983), 133–150.
Serrin, J. (1952): *Existence theorems for some hydrodynamical free boundary-value problems.* J. Rat. Mech. Anal. **1**, 1–48.
Serrin, J. (1953): *On plane and axially symmetric free boundary problems.* J. Rat. Mech. Anal. **2**, 563–575.
Serrin, J. (1959): *Mathematical principles of classical fluid mechanics.* In: Flügge, S. [ed.] (1956), Vol. VIII/1, 125–246.
Serrin, J. (1959a): *On the stability of viscous fluid motions.* Arch. Rat. Mech. Anal. **3**, 1–13.
Serrin, J. (1963): *The initial-value problem for the Navier–Stokes equations.* In: Langer, R. [ed.] (1963), *Nonlinear Problems.* University of Wisconsin Press, pp. 69–98.
Serrin, J. (1979): *Conceptual analysis of the second law of thermodynamics.* Arch. Rat. Mech. Anal. **70**, 355–371.
Serrin, J. (1983): *The structure and laws of thermodynamics.* In: Warsaw (1983), 1717–1728.
Serrin, J. [ed.] (1986): *New Perspectives in Thermodynamics.* Springer-Verlag, New York.
Setti, L. and Van Hove, L. [eds.] (1984): *Large-Scale Structure of the Universe and Fundamental Physics.* First ESO–CERN Symposium.
Sexl, R. and Sexl, M. (1981): *Weisse Zwerge und schwarze Löcher.* Vieweg, Braunschweig.
Sexl, R. (1982): *Was die Welt zusammenhält: Physik auf der Suche nach dem Bauplan der Natur.* Dt. Verlagsanstalt, Stuttgart.

Sexl, R. and Urbantke, H. (1983): *Gravitation und Kosmologie*. Wissenschaftsverlag, Mannheim.
Shanahan, P. (1978): *The Atiyah–Singer Index Theorem*. Lecture Notes in Mathematics, Vol. 638. Springer-Verlag, Berlin.
Shih, T. (1984): *Numerical Heat Transfer*. Springer-Verlag, New York.
Shinbrot, M. (1973): *Lectures on Fluid Mechanics*. Gordon and Breach, New York.
Shinbrot, M. (1976): *The initial-value problem for surface waves under gravity*, I. II. Indiana Univ. Math. J. **25**, 281–300, 1049–1071.
Shinbrot, M. (1979): *The initial-value problem for surface waves under gravity*, III. J. Math. Anal. Appl. **67**, 340–391.
Showalter, W. (1978): *Mechanics of Non-Newtonian Fluids*. Pergamon Press, Oxford, England.
Siegel, C. and Moser, J. (1971): *Lectures on Celestial Mechanics*. Springer-Verlag, Berlin.
Signorini, A. (1959): *Questioni di elasticità nonlinearizzata e semilinearizzata*. Rend. Mat. **18**, 1–45.
Silk, J. (1980): *The Big Bang*. Freeman, San Francisco, CA.
Simon, B. (1984): *Fifteen problems in mathematical physics*. In: Jäger, W., Moser, J., and Remmert, R. [eds.] (1984), 423–450.
Simon, B. (1986): Cf. Cycon, R. (1986).
Simon, B. (1993): Cf. additional references.
Sinai, Ya. (1982): *Theory of Phase Transitions. Rigorous Results*. Pergamon Press, Oxford, England
Singh, V. (1983): *Grand unification and the Big Bang cosmology*. Progr. Phys. **31**, 569–590.
Smale, S. (1965): *An infinite-dimensional version of Sard's theorem*. Amer. J. Math. **87**, 861–866.
Smale, S. (1972): See Smale, S. (1980), pp. 95–105.
Smale, S. (1980): *The Mathematics of Time: Essays on Dynamical Systems, Economic Processes, and Related Topics*. Springer-Verlag, New York.
Smale, S. (1981): *The fundamental theorem of algebra and complexity theory*. Bull. Amer. Math. Soc. (N.S.) **4**, 1–36.
Smale, S. (1982): *Global analysis and economics*. In: Arrow, K. and Intrilligator, A. [eds.] (1982), Vol. 1, pp. 331–378.
Smirnow, W. (1956): *Lehrgang der höheren Mathematik*, Vols. 1–5. Verl. d. Wiss. Berlin. (English edition: *A Course in Higher Mathematics*. Addison-Wesley, Reading, MA, 1964.)
Smoller, J. (1983): *Shock Waves and Reaction–Diffusion Equations*. Springer-Verlag, New York.
Smoller, J. [ed.] (1983a): *Nonlinear partial differential equations*. Contemp. Math. **17**.
Socolescu, D. (1977): *Existenz- und Eindeutigkeitsbeweis für das Problem der Zusammenwirkung von Strahlen*. Indiana Univ. Math. J. **26**, 707–730.
Socolescu, D. (1980): *Existenz- und Eindeutigkeitsbeweis für ein freies Randwertproblem für die stationären Navier–Stokesschen Bewegungsgleichungen*. Arch. Rat. Mech. Anal. **73**, 191–242.
Sod, G. (1978): *A survey of several finite difference methods for systems of nonlinear hyperbolic conservation laws*. J. Comput. Phys. **29**, 1–31.
Sod, G. (1985): *Numerical Methods in Fluid Dynamics*, Vols. 1, 2. University Press, Cambridge, England.

Sokolovskii, V. (1955): *Plastizitätstheorie*. Verl. d. Wiss., Berlin,

Solonnikov, V. (1983): *Solvability of three-dimensional free boundary problems for the Navier–Stokes equations*. Banach Center Publ. 10, 361–403.

Solonnikov, V. (1984): *On the solvability of boundary and initial-boundary value problems of the Navier–Stokes system in a domain with noncompact boundaries*. Pacific J. Math. 93, 443–458.

Solomon, L. (1968): *Élasticité linéaires*. Masson, Paris.

Sommerfeld, A. (1944): See Sommerfeld, A. (1970).

Sommerfeld, A. (1954): *Vorlesungen über theoretische Physik*, Vols. 1–6. Geest & Portig, Leipzig.

Sommerfeld, A. (1970): *Mechanik der deformierbaren Medien*. Geest & Portig, Leipzig. (First edition, 1944.)

Sparrow, C. (1982): *The Lorenz Equation: Bifurcation, Chaos, and Strange Attractors*. Springer-Verlag, Berlin.

Sperner, E. (1928): *Neuer Beweis für die Invarianz der Dimensionszahl und des Gebietes*, Abh. Math. Sem. Univ. Hamburg 6, 265–272.

Spivak, M. (1979): *A Comprehensive Introduction to Differential Geometry*, Vols. 1–5. Publish or Perish, Boston.

Spohn, W. (1969): *Can mathematics be saved?* Notices Amer. Math. Soc. 16, 890–894.

Stephani, H. (1977): *Allgemeine Relativitätstheorie*. Verl. d. Wiss., Berlin. (English edition: *General Relativity*, Cambridge, 1981.)

Sternberg, S. (1964): *Lectures on Differential Geometry*. Prentice-Hall, Englewood Cliffs, NJ.

Sternberg, S. (1969): *Celestial Mechanics*, Vols. 1, 2. Benjamin, Reading, MA.

Stoker, J. (1957): *Water Waves*. Interscience, New York.

Stokes, G. (1845): *On the theories of the internal friction of fluids in motion*. Cam. Trans.

Stoppeli, F. (1954): *Un teorema di esistenza e di unicita relativo alle equazioni dell' elastostatica isoterma per deformazioni finite*. Recerche Mat. 3, 247–267.

Straumann, N. (1984): *General Relativity and Relativistic Astrophysics*. Springer-Verlag, New York.

Streater, R. and Wightman, A. (1964): *PCT, Spin, Statistics, and All That*. Benjamin, New York.

Strehlow, R. (1979): *Fundamentals of Combustion*. Krieger, New York.

Struik, D. (1926): *Détermination rigoreuse des ondes irrotationelles permanent dans un canal à profondeur finie*. Math. Ann. 95, 595–634.

Struik, D. (1948): *A Consise History of Mathematics*. Dover, New York.

Stuart, C. (1976): *Steadily rotating chains*. In: Germain, P. and Nayroles, B. [eds.] (1976), 490–499.

Stumpff, (1973): *Himmelsmechanik*, Vols. 1–3. Akademie-Verlag, Berlin.

Sulanke, R. and Wintgen, P. (1972): *Differentialgeometrie und Faserbündel*. Verl. d. Wiss., Berlin.

Sullivan, W. (1979): *Black Holes—the Edge of Space—the End of Time*. Anchor Press, New York. (German edition: Breidenstein, Frankfurt/Main, 1980.)

Swinney, H. and Gollub, J. (1978): *The transition to turbulence*. Physics Today 31 (8), 41–49.

Szabó, I. (1987): *Geschichte der mechanischen Prinzipien und ihrer wichtigsten Anwendungen*. Birkhäuser, Basel.

Szillard, R. (1974): *Theory and Analysis of Plates*. Prentice-Hall, Englewood Cliffs, NJ.

Szücs, E. (1980): *Similitude and Modelling*. Akadémiai Kiado, Budapest.

Ta-Pei Cheng and Ling-Fong Li (1984): *Gauge Theory of Elementary Particle Physics.* Clarendon Press, Oxford, England.

Tartar, L. (1979): *Compensated compactness and partial differential equations.* In: Knops, R. [ed.], Nonlinear Analysis and Mechanics, Vol. IV. Pitman, London, pp. 136–212.

Tartar, L. (1983): *The compensated compactness method applied to systems of conservation laws.* In: Ball, J. [ed.] (1983a).

Tassoul, J. (1978): *Theory of Rotating Stars.* University Press, Princeton, NJ.

Taube, M. (1985): *Evolution of Matter and Energy on a Cosmic and Planetary Scale.* Springer-Verlag, New York.

Taubes, C. (1986): *Physical and mathematical applications of gauge theories.* Notices Amer. Math. Soc. 33, 707–715.

Taylor, G. (1923): *Stability of a viscous liquid contained between two rotating cylinders.* Phil. Trans. Roy. Soc. London A223, 289–343.

Telionis D. (1981): *Unsteady Viscous Flow.* Springer-Verlag, New York.

Temam, R. (1975): *On the Euler equations of incompressible perfect fluids.* J. Funct. Anal. 20, 32–43.

Temam, R. (1977): *Navier–Stokes Equations: Theory and Numerical Analysis.* North-Holland, New York.

Temam, R. (1983): *Navier–Stokes Equation and Nonlinear Functional Analysis.* CBMS –NSF Regional Conference Series in Applied Mathematics. SIAM, Philadelphia.

Temam, R. (1983a): *Problèmes mathématiques en plasticité.* Gauthier-Villars, Paris. (English edition, Paris, 1985.)

Temam, R. (1986): *A generalized Norton–Hoff model and the Prandtl–Reuss law of plasticity.* Arch. Rat. Mech. Anal. 95, 137–183.

Thirring, W. (1983): *A Course in Mathematical Physics,* Vols. 1–4. Springer-Verlag, Wien.

Thom, R. (1972): *Stabilité structurelle et morphogénèse.* (English edition: Benjamin, New York, 1975.)

Thomasset, F. (1981): *Implementation of Finite Element Methods for Navier–Stokes Equations.* Springer-Verlag, New York.

Thomson, W., Sir, (1869): *On Vortex Motion.* Edin. Trans. XXV.

Ting, T. (1969): *Elastic–plastic torsion.* Arch. Rat. Mech. Anal. 34, 228–244.

Ting, T. (1972): *Topics in Mathematical Theory of Plasticity.* In: Flügge, S. [ed.] (1956), Vol. VIa/3, 535–623.

Tipler, F., Clarke, C., and Ellis, G. (1980): *Singularities and horizons.* In: Held, A. [ed.] (1980), Vol. 2, 97–206.

Todd, M. (1976): *The Computation of Fixed Points and Applications.* Lecture Notes in Economics, Vol. 124. Springer-Verlag, New York.

Treder, H. (1971): *Gravitationstheorie und Äquivalenzprinzip.* Akademie-Verlag, Berlin.

Treder, H. (1972): *Die Relativität der Trägheit.* Akademie-Verlag, Berlin.

Treder, H. (1974): *Über die Prinzipien der Dynamik von Einstein, Hertz, Mach und Poincaré.* Akademie-Verlag, Berlin.

Treder, H. (1975): *Elementare Kosmologie.* Akademie-Verlag, Berlin.

Treder, H. (1983): *Große Physiker und ihre Probleme.* Akademie-Verlag, Berlin.

Trefftz, E. (1927): *Ein Gegenstück zum Ritzschen Verfahren.* Zweiter Kongress für Technische Mechanik, Zürich.

Trefftz, E. (1928): *Mathematische Elastizitätstheorie.* In: Geiger, H. and Scheel, K. [eds.] (1926), Vol. 6, 47–140.

Trefil, J. (1983): *The Unexpected Vista: A Physicist's View of Nature.* Charles Scribner's Sons, New York.

Trefil, J. (1984): *The Moment of Creation: Big Bang Physics.* Charles Scribner's Sons, New York. (German edition: Birkhäuser, Basel, 1985.)

Trenogin, V. (1970): *The asymptotic method of Ljusternik–Višik.* Uspekhi Mat. Nauk **25** (4), 123–156 (Russian).

Tresca, H. (1864): C. R. Acad. Sci. Paris **59**, 754.

Triebel, H. (1972): *Höhere Analysis.* Verl. d. Wiss., Berlin.

Triebel, H. (1981): *Analysis und mathematische Physik.* Teubner, Leipzig. (English edition, Leipzig, 1985.)

Tromba, A. (1976): *Fredholm vector fields and transversality.* J. Funct. Anal. **23**, 362–368.

Tromba, A. (1976a): *Almost Riemannian structures on Banach manifolds, the Morse lemma, and the Darboux theorem.* Canad. J. Math. **28**, 640–652.

Tromba, A. (1978): *The Euler characteristic of vector fields on Banach manifolds and a globalization of Leray–Schauder degree.* Adv. Math. **28**, 148–173.

Tromba, A. (1983): *A sufficient condition for a critical point of a functional to be a minimum and its application to Plateau's problem.* Math. Ann. **263**, 303–312.

Truesdell, C. and Noll, W. (1965): *The non-linear field theories of mechanics.* In: Flügge, S. [ed.] (1956), Vol. III/3.

Truesdell, C. (1968): *Essay's in the History of Mechanics.* Springer-Verlag, New York.

Truesdell, C. (1977): *A First Course in Rational Mechanics,* Vols. 1, 2. Academic Press, New York.

Truesdell, C. and Muncaster, R. (1980): *Fundamentals of Maxwell's Kinetic Theory of a Simple Monatomic Gas.* Academic Press, New York.

Truesdell, C. (1983): *The influence of elasticity on analysis, the classical heritage.* Bull. Amer. Math. Soc. (N.S.) **9**, 293–310.

Turner, R. (1981): *Internal waves in fluids with rapidly varying density.* Ann. Scuola Norm. Sup. Pisa, Cl. Sci. Ser. 4, **8**, 513–573.

Tymoczko, T. [ed.] (1985): *New Directions in the Philosophy of Mathematics.* Birkhäuser, Boston.

Unsöld, A. and Baschek, B. (1981): *Der neue Kosmos.* Springer-Verlag, Berlin.

Uralceva, N. (1973): *On the solvability of the capillary problem.* Vestnik Leningrad. Univ. Ser. Math. (1973) **19**, 54–64; (1975) **1**, 143–149.

Van der Meer, C. (1985): *The Hamiltonian–Hopf Bifurcation.* Lecture Notes in Mathematics, Vol. 1160. Springer-Verlag, Berlin.

Van der Waerden, B. (1980): *Group Theory and Quantum Mechanics.* Springer-Verlag, New York.

Van Nostrand's Scientific Encyclopedia (1976): Vols. 1–5. Van Nostrand, New York. (German edition: *Enzyklopädie Naturwissenschaft und Technik,* Verlag moderne Industrie, München.)

Velte, W. (1966): *Stabilität und Verzweigung stationärer Lösungen der Navier–Stokesschen Gleichungen beim Taylorproblem.* Arch. Rat. Mech. Anal. **22**, 1–14.

Vinogradov, A. and Kuperschmidt, B. (1977): *Structure of Hamiltonian Mechanics.* Uspekhi Math. Nauk **32** (4), 175–236 (Russian).

Visconti, A. (1987): *Introductory Differential Geometry for Physicists*. World Scientific, Singapore.
Višik, M. and Fursikov, A. (1980): *Mathematical Problems in Statistical Hydrodynamics*. Nauka, Moscow (Russian). (German edition: Teubner, Leipzig, 1986.)
Vladimirov, V. (1976): *Einführung in die physikalische Theorie der Plastizität und Festigkeit*. Verlag für Grundstoffindustrie, Leipzig.
Vladimirov, V. [ed.] (1986): *Collection of Survey Articles on the Occasion of the Fifth Anniversary of the Steklov Institute*. Trudy Mat. Inst. Steklova **175**.
Vlasov, V. (1964): *General Theory of Shells and Its Applications in Engineering*. NASA, Washington, D.C.
Vogel, H. (1977): *Probleme aus der Physik*. Springer-Verlag, Berlin.
Volmir, A. (1972): *Nonlinear Dynamics of Plates and Shells*. Nauka, Moscow (Russian).
Vorovič, I. (1955): *On the existence of solutions in nonlinear shell theory*. Izv. Akad. Nauk SSSR Ser. mat. **19**, 173–186 (Russian).
Vul, E., Sinai, Ja, and Chanin, K. (1984): *Universality of Feigenbaum and thermodynamical formalism*. Uspekhi Mat. Nauk **39** (3), 3–37 (Russian).

Wahl, W. v. (1985): *The Equations of Navier–Stokes and Abstract Parabolic Equations*. Vieweg, Braunschweig.
Wald, R. (1984): *General Relativity*. University Press, Chicago, IL.
Walras, L. (1874): *Eléments d'economie politique pure*. Corbaz, Lausanne.
Walter, W. (1964): *Differential- und Integralungleichungen*. Springer-Verlag, New York.
Wang, C. (1979): *Mathematical Principles of Mechanics and Electromagnetism*, Vols. 1, 2. Plenum, New York.
Warner, F. (1971): *Foundations of Differentiable Manifolds and Lie Groups*. Scott-Foresman, Dallas, TX.
Warsaw (1983): *Proceedings of the International Conference of Mathematicians in Warsaw*, Vols. 1, 2. PWN, Warsaw and North-Holland, Amsterdam.
Washington, W. and Parkinson, C. (1987): *An Introduction to Three-Dimensional Climate Modelling*. University Press, Oxford, England.
Washizu, K. (1968): *Variational Methods in Elasticity and Plasticity*. Pergamon Press, Oxford, England.
Weber, J. (1980): *A search for gravitational radiation*. In: Held, A. [ed.] (1980), Vol. 1, 435–468.
Wehrl, A. (1978): *General properties of entropy*. Rev. Mod. Phys. **50**, 221–260.
Weierstrass, K. (1857): *Antrittsrede in der Berliner Akademie* (Inaugural speech). See Weierstrass, K. (1894/1927), Vol. 1, pp. 293–296.
Weierstrass, K. (1894/1927): *Mathematische Werke*. (*Mathematical Works*.), Vols. 1–7. Berlin.
Weinberg, S. (1972): *Gravitation and Cosmology*. Wiley, New York.
Weinberg, S. (1977): *The First Three Minutes: A Modern View of the Origin of the Universe*. Basic Books, New York. (German edition: Piper, München, 1977.)
Weinberg, S. [ed.] (1983): *Interaction Between Elementary Particle Physics and Cosmology*. Wiley, New York.
Weinberg, S. (1984): *Teile des Unteilbaren*. Spektrum der Wissenschaft, Heidelberg. (English edition: *The Discovery of Subatomic Particles*. Scientific American Books, New York, 1983.)

Weinberg, S. (1986): Cf. *Mathematics: The Unifying Thread* (1986).
Weinberg, S. [ed.] (1986a): *Physics in Higher Dimensions*. Wiley, New York.
Weinstein A. (1977): *Lectures on Symplectic Manifolds*. American Mathematical Society, Providence, RI.
Weizsäcker, C. v. (1973): *Die philosophische Interpretation der modernen Physik*. Deutsche Akademie d. Naturforscher Leopoldina, Halle.
Weizsäcker, C. v. (1976): *Die Tragweite der Wissenschaft*. Hirzel, Stuttgart.
Weizsäcker, C. v. (1976a): *Zum Weltbild der Physik*. Hirzel, Stuttgart.
Weizsäcker, C. v. (1979): *Die Einheit der Natur*. Hanser-Verlag, München.
Weizsäcker, C. v. (1979a): *Die Geschichte der Natur*. Vandenhoeck & Ruprecht, Göttingen.
Weller, W. and Winkler, H. (1974): *Grundkurs klassische Physik*, Vols. 1, 2. Teubner, Leipzig.
Wells, R. (1979): *Complex manifolds and mathematical physics*. Bull. Amer. Math. Soc. (N.S.) 1, 296–336.
Wells, R. (1980): *Differential Analysis on Complex Manifolds*. Springer-Verlag, New York.
Wess, J. and Bagger, J. (1983): *Supersymmetry and Supergravity*. University Press, Princeton, NJ.
West, P. (1986): *Introduction to Supersymmetry and Supergravity*. World Scientific, Singapore.
Westenholz, C. v. (1981): *Differential Forms in Mathematical Physics*. North-Holland, Amsterdam.
Weyl, H. (1923): *Raum, Zeit, Materie*. Springer-Verlag, Berlin. (First edition, 1918.)
Weyl, H. (1952): *Symmetry*. University Press, Princeton, NJ. (German edition: Birkhäuser, Stuttgart, 1955.)
Weyl, H. (1966): *Philosophie der Mathematik und der Naturwissenschaft*. Leibniz-Verlag, München. (English edition: *Philosophy of Mathematics and Natural Science*. University Press, Princeton, NJ., 1949.)
Weyl, H. (1968): *Gesammelte Werke (Collected Works)*, Vols. 1–4. Springer-Verlag, New York.
Whitham, G. (1974): *Linear and Nonlinear Waves*. Wiley, New York.
Whitney, H. (1936): *Differentiable manifolds*. Ann. of Math. 37, 645–680.
Whitney, H. (1944): *The self-intersection of a smooth n-manifold in 2n-space*. Ann. of Math. 45, 220–246.
Whitney, H. (1944a): *The singularities of a smooth n-manifold in $(2n-1)$-space*. Ann. of Math. 45, 247–293.
Wilkinson, W. (1960): *Non-Newtonian Fluids. Fluid Mechanics, Mixing and Heat Transfer*. Pergamon Press, New York.
Wille, F. (1982): *Humor in der Mathematik*. Vandenboeck & Ruprecht, Göttingen.
Williams, F. (1964): *Combustion Theory*. Addison-Wesley, Reading, MA.
Wilson, K. (1982): *The renormalization group and critical phenomena*. In: Nobel Prizes (1954ff), Vol. 1982, 57–87.
Wintner, A. (1947): *The Analytical Foundations of Celestial Mechanics*. University Press, Princeton, NJ.
Witten, E. (1986): *Topological tools in ten dimensions, and unification in ten dimensions*. In: Green, M. and Gross, D. [eds.] (1986), 400–458.
Worbs, E. (1955): *Carl Friedrich Gauss*, Koehler & Amelang, Leipzig.

Wussing, H. (1974). *Carl Friedrich Gauss.* Teubner, Leipzig.
Wussing, H. (1979): *Vorlesungen zur Geschichte der Mathematik.* Verl. d. Wiss., Berlin.
Wussing, H. [ed.] (1983): *Geschichte der Naturwissenschaften.* Edition Leipzig.

Yau, S. and Schoen, R. (1979): *On the proof of the positive mass conjecture in general relativity.* Comm. Math. Phys. **65** (1979), 45–76, **79** (1981), 231–260.
Yau, S. and Schoen, R. (1983): *The existence of a black hole due to condensation of matter.* Comm. Math. Phys. **90**, 575–579.
Yosida, K. (1965): *Functional Analysis.* Springer-Verlag, New York.
Young, L. (1981): *Mathematicians and Their Times.* North-Holland, Amsterdam.

Zabusky, N. [ed.] (1968): *Topics in Nonlinear Physics.* Springer-Verlag, New York.
Zee, A. [ed.] (1982): *Unity of Forces in the Universe,* Vols. 1, 2. World Scientific, Singapore.
Zeeman, F. (1976): *Euler Buckling.* Lecture Notes in Mathematics, Vol. 525, 373–395. Springer-Verlag, Berlin.
Zeidler, E. (1968): *Beiträge zur Theorie und Praxis freier Randwertaufgaben.* Habilitationsschrift, Universität Leipzig. Published as a monograph, Akademie-Verlag, Berlin, 1971.
Zeidler, E. (1971): *Existenzbeweis für cnoidal waves unter Berücksichtigung der Oberflächenspannung.* Arch. Rat. Mech. Anal. **41**, 81–107.
Zeidler, E. (1972): *Zur Bifurkationstheorie und zur Stabilitätstheorie der Navier-Stokesschen Gleichungen.* Math. Nachr. **52**, 167–205.
Zeidler, E. (1972a): *Existenz einer Gasblase in einer Parallel- und Zirkulationsströmung unter Berücksichtigung der Schwerkraft.* Beiträge Anal. **3**, 67–95.
Zeidler, E. (1972b): *Existenzbeweis für asymptotische Wirbelwellen.* Beiträge Anal. **3**, 109–134.
Zeidler, E. (1973): *Existenzbeweis für permanente Kapillar–Schwerewellen mit allgemeinen Wirbelverteilungen.* Arch. Rat. Mech. Anal. **50**, 34–72.
Zeidler, E. (1976): *Lokale und globale Bifurkationsresultate für Variationsungleichungen.* Math. Nachr. **71**, 37–63.
Zeidler, E. (1977): *Bifurcation theory and permanent waves.* In: Rabinowitz, P. [ed.] (1977), 203–224.
Zeidler, E. (1979): *Vektoranalysis, Differentialgeometrie, Tensoranalysis.* In: Bronstein, I. and Semendjaev, K. [eds.] (1979), Vol. 1, 605–658; Vol. 2, 70–87. (English edition: Bronshtein, I. and Semendjaev, K. [eds.]: *Handbook of Mathematics,* pp. 550–597; 808–825. Van Nostrand, New York, 1985.)
Zeldovič, J. and Novikov, I. (1971): *Theory of Gravitation and Evolution of Stars.* Nauka, Moscow (Russian).
Zeldovič, J. and Novikov, I. (1971a): *Relativistic Astrophysics.* University Press, Chicago, IL.
Zeytounian, R. (1987): *Les modèles asymptotiques de la mécanique des fluids,* Vols. 1, 2. Springer-Verlag, New York.
Ziegler, H. (1983): *Introduction to Thermomechanics.* North-Holland, Amsterdam.

Additional References

Cf. also the "Additional References" to the revised edition of Volume 1.

Adams, D. and Hedberg, L. (1996): *Function Spaces and Potential Theory.* Springer-Verlag, Berlin, Heidelberg (General Reference).

Aebischer, B. et al. (1994): *Symplectic Geometry: An Introduction.* Birkhäuser, Basel (Chapter 58).

Aldroubi, A. and Unser, M. (1996): *Wavelets in Medicine and Biology.* CRC Press, New York (General Reference).

Alexander, D. (1994): *A History of Complex Dynamics: From Schröder to Fatou und Julia.* Vieweg, Braunschweig (Chapter 79).

Alinhac, S. and Gérard, P. (1991): *Opérateurs pseudo-différentiels et théorème de Nash-Moser.* Intereditions, Paris (General Reference).

Allgower, E. and Georg, K. (1990): *Numerical Continuation Methods.* Springer-Verlag, New York (Chapter 78).

Allgower, E., Böhmer, K., and Golubitsky, M. (eds.) (1992): *Bifurcation and Symmetry: Cross Influence between Mathematics and Applications.* Birkhäuser, Basel (Chapter 79).

Amann, H. (1995): *Linear and Quasilinear Parabolic Problems. Vol. 1: Abstract Linear Theory. Vols. 2,3* (to appear). Birkhäuser, Basel (General Reference).

Ambrosetti, A. (1993): *A Primer of Nonlinear Analysis.* Cambridge University Press, Cambridge, UK (General Reference).

Ambrosetti, A. and Chang, K. (eds.) (1993): *Variational Methods in Nonlinear Analysis.* Gordon & Breach, Newark, NJ (General Reference).

Ambrosetti, A. and Coti-Zelati, V. (1993): *Periodic Solutions of Singular Lagrangian Systems.* Birkhäuser, Basel (Chapter 58).

Antes, H. and Panagiotopoulos, P. (1992): *The Boundary Integral Approach to Static and Dynamic Contact Problems: Equality and Inequality Methods.* Birkhäuser, Basel (Chapter 63).

Antman, S. (1995): *Nonlinear Elasticity.* Springer-Verlag, New York (Chapter 61).

Arbib, M. (ed.) (1995): *The Handbook of Brain Theory and Neural Networks.* MIT Press, Cambridge, MA (General Reference).

Arnold, (ed.) (1988/94): *Dynamical Systems*. Vols. 1–8. Encyclopedia of Mathematical Sciences. Springer-Verlag, New York (Chapter 79).
Aubin, J. (1991): *Viability Theory*. Birkhäuser, Basel (Chapter 77).
Aubin, J. (1993): *Optima and Equilibria*. Springer-Verlag, New York (Chapter 77).
Auerbach, A. (1994): *Interacting Electrons and Quantum Magnetism*. Springer-Verlag, New York (General Reference).

Baez, J. (1994): *Knots and Quantum Gravity*. Oxford University Press, Oxford (Chapter 76).
Baggett, L. (1992): *Functional Analysis: A Primer*. Marcel Dekker, New York (General Reference).
Bakelman, I. (1994): *Convex Analysis and Nonlinear Geometric Elliptic Equations*. Springer-Verlag, Berlin, Heidelberg (General Reference).
Banks, R. (1994): *Growth and Diffusion Phenomena*. Springer-Verlag, Berlin, Heidelberg (Chapter 69).
Bar'yakhtar, V., Chetkin, M., Ivanov, B., and Gadetskii, S. (1994): *Dynamics of Topological Magnetic Solitons: Experiment and Theory*. Springer-Verlag, Berlin, Heidelberg (Chapter 71).
Bartsch, T. (1993): *Topological Methods for Variational Problems with Symmetries*. Springer-Verlag, Berlin, Heidelberg (General Reference).
Beals, M. (1989): *Propagation and Interaction of Singularities in Nonlinear Hyperbolic Problems*. Birkhäuser, Basel (General Reference).
Beaulieu, L. (1997): *Nicolas Bourbaki: History and Legend*. Springer-Verlag, Berlin, Heidelberg (General Reference) (to appear).
Beem, J., Ehrlich, P., and Easley, K. (1996): *Global Lorentzian Geometry*. Marcel Dekker, New York (Chapter 76).
Bellissard, J. (1996): *Applications of C^*-Techniques to Modern Quantum Physics*. Springer-Verlag, Berlin, Heidelberg (Chapter 59).
Benatti, F. (1993): *Deterministic Chaos in Infinite Quantum Systems*. Springer-Verlag, Berlin, Heidelberg (Chapters 59 and 68).
Bensoussan, A., Da Prato, G., Delfour, M., and Mitter, S. (1993): *Representation and Control of Infinite-Dimensional Systems*. Vols. 1, 2. Birkhäuser, Basel (General Reference).
Berezin, F. and Shubin, M. (1991): *The Schrödinger Equation*. Kluwer, Dordrecht (Chapter 59).
Berezin, F. (1987): *Introduction to Superanalysis*. Reidel, Dordrecht (Chapter 76).
Bertil, G. and Kreiss, H. (1995): *Time Dependent Problems and Difference Methods*. Wiley, New York (General Reference).
Bertin, J., Glowinski, R., and Periaux, J. (eds.) (1989): *Hypersonics*. Vol. 1: *Defining the Hypersonic Environment*. Vol. 2: *Computation and Measurement of Hypersonic Flows*. Birkhäuser, Basel (Chapter 70).
Bertin, J., Periaux, J., and Ballmann, J. (1992): *Advances in Hypersonics*. Vols. 1–3. Birkhäuser, Basel (Chapter 70).
Bethuel, F., Brézis, H., and Hélein, F. (1994): *Ginzburg–Landau Vortices*. Birkhäuser, Basel (Chapter 67).
Binder, K. and Heermann, D. (1993): *Monte-Carlo Simulation in Statistical Physics*. Springer-Verlag, Berlin, Heidelberg (Chapter 68).
Binney, J. and Tremaine, S. (1988): *Galactic Dynamics*. Princeton University Press, Princeton, NJ (Chapter 76).

Additional References

Bishop, C. (1996): *Neural Networks of Pattern Recognition.* Oxford University Press, Oxford, UK (General Reference).

Bloom, F. (1993): *Mathematical Problems of Classical Nonlinear Electromagnetic Theory.* Longman, Harlow, UK (Chapter 76).

Bobylev, N., Burman, Yu., and Korovin, S. (1994): *Approximation Procedures in Nonlinear Oscillation Theory.* De Gruyter, Berlin (Chapter 79).

Boccaletti, D. and Pucacco, G. (1996): *Theory of Orbits.* Vols. 1, 2. Springer-Verlag, Berlin, Heidelberg (Chapter 79).

Bogoljubov, N. et al (1990): *General Principles of Quantum Field Theory.* Kluwer, Dordrecht (General Reference).

Border, K. (1985): *Fixed Point Theorems with Applications to Economics and Game Theory.* Cambridge University Press, Cambridge, UK (Chapter 78).

Bott, R. (1993): *Collected Works.* Vol. 1: *Topology and Lie Groups.* Vol. 2: *Differential Operators.* Vol. 3: *Foliations.* Vol. 4: *Mathematics Related to Physics.* Birkhäuser, Basel (General Reference).

Bott, R. and Tu, L. (1994): *Differential Forms in Algebraic Topology.* Springer-Verlag, New York (Chapter 74).

Bourbaki, N. (1994): *Elements of the History of Mathematics.* Springer-Verlag, Berlin, Heidelberg (General Reference).

Bourguignon, J. (1996): *Variational Calculus.* Springer-Verlag, Berlin, Heidelberg (General Reference).

Brand, H. (1995): *Spatial Structures in Systems Far From Equilibrium.* Springer-Verlag, Berlin, Heidelberg (Chapter 67).

Bredon, G. (1993): *Topology and Geometry.* Springer-Verlag, New York (General Reference).

Brody, T. (1993): *The Philosophy Behind Physics.* Springer-Verlag, Berlin, Heidelberg (General Reference).

Brokate, M. and Sprekels, J. (1996): *Hysteresis Phenomena in Phase Transitions.* Springer-Verlag, Berlin, Heidelberg (Chapter 67).

Browder, F. (ed.) (1992): *Nonlinear and Global Analysis.* Reprints from the Bulletin of the American Mathematical Society. Providence, RI (General Reference).

Brown, D. and Smith, K. (1991): *Frontiers of Mathematical Psychology.* Springer-Verlag, New York (General Reference).

Brown, L. (ed.) (1993): *Renormalization: From Lorentz to Landau and Beyond.* Springer-Verlag, New York (Chapter 59).

Brown, R. (1993): *A Topological Introduction to Nonlinear Analysis.* Birkhäuser, Basel (Chapter 77).

Brown, R. and Davis, S. (1994): *Free Boundaries in Viscous Flows.* Springer-Verlag, New York (Chapter 71).

Brumberg, V. (1995): *Analytical Techniques of Celestial Mechanics.* Springer-Verlag, Berlin, Heidelberg (Chapter 58).

Bruno, A. (1994): *The Restricted 3-Body Problem: Plane Periodic Orbits.* De Gruyter, Berlin (Chapter 58).

Buchheim, G. and Sonnemann, R. (eds.) (1989): *Lebensbilder von Ingenieurwissenschaftlern: Eine Sammlung von Biographien aus zwei Jahrhunderten.* Birkhäuser, Basel (General Reference).

Buechler, S. (1996): *Essential Stability Theory.* Springer-Verlag, Berlin, Heidelberg (Chapter 79).

Buttazzo, G. and Visintin, A. (eds.) (1994): *Motion by Mean Curvature and Related Topics.* De Gruyter, Berlin (cf. also Damlamian, Spruck, and Visintin (eds.) (1995)) (Chapter 74).

Caffarelli, L. and Cabré, X. (1995): *Fully Nonlinear Elliptic Equations*. American Mathematical Society, Providence, RI (General Reference).
Carmichael, H. (1997): *Quantum Statistical Methods in Quantum Optics*. Springer-Verlag, Berlin, Heidelberg (Chapter 59) (to appear).
Carmo, M. (1993): *Riemannian Geometry*. Birkhäuser, Boston, MA (Chapter 74).
Carmo, M. (1994): *Differential Forms*. Springer-Verlag, Berlin, Heidelberg (Chapter 74).
Cascuberta, C. and Castellet, M. (1992): *Mathematical Research Today and Tomorrow: Viewpoints of Seven Fields Medalists*. Springer-Verlag, Berlin, Heidelberg (General Reference).
Cercignani, C., Illner, R., and Pulvirenti, M. (1994): *The Theory of Dilute Gases*. Springer-Verlag, Berlin, Heidelberg (Chapter 68).
Chang, K. (1997): *Critical Point Theory and its Applications*. Springer-Verlag, Berlin, Heidelberg (General Reference) (to appear).
Choquet-Bruhat, Y., DeWitt-Morette, and Dillard-Bleick, M. (1988): *Analysis, Manifolds, and Physics*. Vol. 2. North-Holland, Amsterdam. (Chapter 73).
Chorin, A. (1994): *Vorticity and Turbulence*. Springer-Verlag, New York (Chapter 72).
Chossat, P. and Iooss, G. (1994): *The Couette-Taylor Flow*. Springer-Verlag, New York (Chapter 72).
Christodoulou, D. and Klainerman, S. (1993): *The Global Nonlinear Stability of the Minkowski Space*. Princeton University Press, Princeton, NJ (Chapter 76).
Chung, K. and Zhao, Z. (1994): *From Brownian Motion to Schrödinger's Equation*. Springer-Verlag, Berlin, Heidelberg (Chapter 59).
Ciarlet, P. (1990): *Plates and Junctions in Elastic Multi-Structures*. Springer-Verlag, New York (Chapter 65).
Clarke, C. (1994): *The Analysis of Space-Time Singularities*. Cambridge University Press, Cambridge, UK (Chapter 76).
Clément, P. and Lumer, G. (eds.) (1994): *Evolutione Equations, Control Theory, and Biomathematics*. Marcel Dekker, New York (General Reference).
Colombeau, J. (1992): *Multiplication of Distributions*. Lecture Notes in Mathematics Vol. 1532. Springer-Verlag, Berlin, Heidelberg (General Reference).
Colombini, F. et al. (1989): *Partial Differential Equations and the Calculus of Variations: Essays in Honor of Ennio de Giorgi*. Vols 1, 2. Birkhäuser, Basel (General Reference).
Colton, D. and Kress, R. (1992): *Inverse Acoustic and Electromagnetic Scattering*. Springer-Verlag, Berlin, Heidelberg (Chapter 59).
Companion Encyclopedia of the History and Philosophy of the Mathematical Sciences (1994): Edited by I. Grattan-Guiness. Rutledge, London (General Reference).
Conlon, L. (1992): *Differentiable Manifolds: A First Course*. Birkhäuser, Basel (Chapter 73).
Connes, A. (1994): *Noncommutative Geometry*. Academic Press, New York (General Reference).
Coughran, W., Cole, J., Lloyd, P., and White, J. (1994): *Semiconductors*. Springer-Verlag, Berlin, Heidelberg (Chapter 67).
Crandall, M., Benilan, P., and Pazy, A. (1997): *Nonlinear Evolution Governed by Accretive Operators*. Springer-Verlag, Berlin, Heidelberg (General Reference) (to appear).
Crandall, M., Hishii, M., and Lions, P. (1992): *A User's Guide to Viscosity Solutions of Second Order Partial Differential Equations*. Bull. AMS **27**, 1-67.
Cross, M. and Hohenberg, P. (1993): *Pattern Formation Outside of Equilibrium*. Rev. Mod. Physics **65**, 851-1112 (General Reference).

Czichos, H. (ed.) (1989): *Hütte—die Grundlagen der Ingenieurwissenschaften*. 29. völlig neubearbeitete Auflage. Springer-Verlag, Berlin, Heidelberg (General Reference).

Damlamian, A., Spruck, J., and Visintin, A. (1995): *Motion by Mean Curvature and Related Topics*. De Gruyter, Berlin (Chapter 74).

Dalen, D. van (ed.) (1996): *L.E.J. Brouwer Biography* (General Reference).

Dal Maso, G. (1993): *An Introduction to Γ-Convergence*. Birkhäuser, Basel (General Reference).

Dancer, E. (1994): *Weakly Nonlinear Dirichlet Problems on Long or Thin Domains*. American Mathematical Society, Providence, RI (General Reference).

Das, A. (1993): *The Special Theory of Relativity*. Springer-Verlag, New York (Chapter 75).

Dautray, R. and Lions, J. L. (1990/93): *Mathematical Analysis and Numerical Methods for Science and Technology*. Vol. 1: *Physical Origins and Classical Methods*. Vol. 2: *Functional and Variational Methods*. Vol. 3: *Spectral Theory and Applications*. Vol. 4: *Integral Equations and Numerical Methods*. Vol. 5: *Evolution Problems I*. Vol. 6: *Evolution Problems II—The Navier–Stokes Equations and Transport Equations and Numerical Methods*. Springer-Verlag, Berlin, Heidelberg (General Reference).

Davies, P. (ed.) (1989): *The New Physics*. Cambridge University Press, Cambridge, UK (General Reference).

Davis, P. (1993): *The Nature and Power of Mathematics*. Princeton University Press, Princeton, NJ (General Reference).

Day, W (1993): *Entropy and Partial Differential Equations*. Longman, Harlow, UK (General Reference).

De Gennes, P. and Prost, J. (1995): *The Physics of Liquid Crystals*. Clarendon Press, Oxford, UK (Chapter 70).

Deimling, K. (1992): *Multivalued Differential Equations*. De Gruyter, Berlin (Chapter 79).

Demazure, M. (1996): *Geometry: Catastrophes and Bifurcations*. Springer-Verlag, Berlin, Heidelberg (Chapter 79).

Deuflhard, P. and Bornemann, F. (1994): *Numerische Mathematik II: Integration gewöhnlicher Differentialgleichungen*. De Gruyter, Berlin (English edition in preparation) (General Reference).

Deuflhard, P. and Hohmann, A. (1993): *Numerische Mathematik I: Eine algorithmisch orientierte Einführung*. De Gruyter, Berlin (English edition: Numerical Analysis: A First Course in Scientific Computation, De Gruyter, Berlin, 1994) (General Reference).

Deuring, P. (1994): *The Stokes Problem in an Infinite Cone*. Akademie-Verlag, Berlin (Chapter 72).

DeWitt, B. (1992): *Supermanifolds*. Cambridge University Press, Cambridge, UK (Chapter 76).

DiBenedetto, E. (1993): *Degenerate Parabolic Equations*. Springer-Verlag, New York (General Reference).

Diekmann, O., Lunel, S., van Gils, A., and Walther, H. (1995): *Delay Equations: Functional Analysis, Complex Analysis, and Nonlinear Analysis*. Springer-Verlag, Berlin, Heidelberg.

Dieudonné, J. (1992): *Mathematics—the Music of Reason*. Springer-Verlag, Berlin, Heidelberg (General Reference).

Dittrich, W. and Reutter, M. (1994): *Classical and Quantum Dynamics from Classical Paths to Path Integrals*. Springer-Verlag, Berlin, Heidelberg (Chapter 59).
Dobrushin, R. and Kusnoka, S. (1993): *Statistical Mechanics and Fractals*. Springer-Verlag, Berlin, Heidelberg (Chapter 68).
Donoghue, J., Golowich, E., and Holstein, B. (1992): *The Dynamics of the Standard Model*. Cambridge University Press, Cambridge, UK (Chapters 59 and 76).
Duhem, P. (1991): *The Aim and Structure of Physical Theory*. Princeton University Press, Princeton, NJ (General Reference).
Duistermaat, H. and Kolk, J. (1996): *Lie Groups*. Springer-Verlag, Berlin, Heidelberg (General Reference).
Dunham, W. (1991): *Journey Through Genius: The Great Theorems of Mathematics*. Penguin Books, New York (General Reference).
Duren, P. and Zdravskoska, S. (1994): *Golden Years of Moscow Mathematics*. Oxford University Press, Oxford (General Reference).

Earman, J., Janssen, H., and Norton, J. (1993): *The Attraction of Gravitation: New Studies in the History of General Relativity*. Birkhäuser, Basel (Chapter 76).
Economou, E. (1990): *Green's Function in Quantum Physics*. Springer-Verlag, New York (Chapter 59).
Edwards, H. (1993): *Advanced Calculus: A Differential Forms Approach*. Birkhäuser, Basel (Chapter 74).
Efendiev, M. (1997): *Degree Theory for Nonlinear Pseudodifferential Operators and its Applications in Mathematical Physics* (Chapter 78) (to appear).
Egorov, Yu. and Shubin, M. (1991): *Partial Differential Equations*. Vols. 1–4. Encyclopedia of Mathematical Sciences. Springer-Verlag, New York (General Reference).
Ehlers, J. and Rindeler, W. (1996): *Grundzüge der Relativitätstheorie*. Springer-Verlag, Berlin, Heidelberg (Chapter 76).
Eigen, M. and Schuster, P. (1996): *The Hypercycle: A Principle of Natural Self-Organization*. Springer-Verlag, Berlin, Heidelberg (General Reference).
Eigen, M. and Winkler, R. (1993): *Laws of the Game: How the Principles of Nature Govern Chance*. Princeton University Press, Princeton, NJ (General Reference).
Einstein, A. (1992): *Mileva Marić: The Love Letters*. Edited by J. Renn and R. Schulmann. Princeton University Press, Princeton, NJ (General Reference).
Einstein, A. (1993): *Collected Papers*. Vol. 1: *The Early Years: 1879–1902*. Vol. 2: *The Swiss Years: Writings*, 1900–1909. Vol. 3: *The Swiss Years: Writings, 1909–1911*. Edited by M. Klein, A. Kox, J. Renn, and R. Schulmann. Princeton University Press, Princeton, NJ (Chapter 76).
Encrenaz, T. and Bibring, J. (1994): *The Solar System*. Springer-Verlag, Berlin, Heidelberg (Chapter 58).
Encyclopedia of Applied Physics. Edited by G. Trigg. Vol. 1. VCM Publishers, New York (General Reference).
Encyclopedia of Cosmology (1993): Edited by N. Hetherington. Garland Publishing, New York (Chapter 76).
Encyclopedia of Mathematics (1988): Vols. 1–10. Kluwer, Dordrecht (General Reference).
Encyclopedia of the Mathematical Sciences (1988): Vols. 1ff. Springer-Verlag, New York (General Reference).
Encyclopedia of Science and Technology (1992): Vols. 1–20. McGraw-Hill, New York (General Reference).

Encyclopedic Dictionary of Mathematics (1993): Edited by Kiyosi Ito. The MIT Press, Cambridge, MA (General Reference).

Enos, J. (ed.) (1993): *Dynamics and Control of Mechanical Systems: The Falling Cat and Related Problems*. American Mathematical Society, Providence, RI (Chapter 58).

Esposito, G. (1993): *Quantum Gravity, Quantum Cosmology, and Lorentzian Geometries*. Springer-Verlag, New York (Chapter 76).

Evans, L. (1990): *Weak Convergence Methods for Nonlinear Partial Differential Equations*. American Mathematical Society, Providence, RI (General Reference).

Evans, L. and Gariepy, R. (1996): *Measure Theory and Fine Properties of Functions*. CRC Press, Boca Raton, FL (General Reference).

Farkas, M: (1994): *Periodic Motions*. Springer-Verlag, Berlin, Heidelberg (Chapter 79).

Ferreyra, G., Goldstein, G., and Neubrander, F. (eds.) (1994): *Evolution Equations*. Marcel Dekker, New York (General Reference).

Finkelstein, D. (1995): *Quantum Relativity: A Synthesis of the Ideas of Einstein and Heisenberg*. Springer-Verlag, Berlin, Heidelberg (Chapter 76).

Fitzpatrick, P. and Furi, M. (eds.) (1993): *Topological Methods for Ordinary Differential Equations*. Lecture Notes Mathematics Vol. 1537. Springer-Verlag, Berlin, Heidelberg (Chapter 79).

Fletcher, C. (1991): *Computational Techniques for Fluid Dynamics*. Vols. 1, 2. Springer-Verlag, Berlin, Heidelberg (Chapter 70).

Fletcher, C. and Srinivas, K. (1992): *Computational Techniques for Fluid Dynamics: A Solutions Manual*. Springer-Verlag, Berlin, Heidelberg (Chapter 70).

Fomenko, A. (1994): *Visual Geometry and Topology*. Springer-Verlag, Berlin, Heidelberg (General Reference).

Fonseca, I., Gangbo, W. (1995): *Degree Theory in Analysis and Applications*. Clarendon Press, Oxford (Chapter 78).

Friedman, A. (1988/96): *Mathematics in Industrial Problems*. Vols. 1–7. Springer-Verlag, New York (General Reference).

Friedman, A. and Spruck, J. (eds.) (1992): *Variational and Free Boundary Problems*. Springer-Verlag, New York (General Reference)

Fröhlich, J. and Kerler, T. (1993): *Quantum Groups, Quantum Categories, and Quantum Field Theory*. Lecture Notes in Mathematics. Vol. 1542. Springer-Verlag, Berlin, Heidelberg (Chapters 59 and 76).

Fulde, P. (1995): *Electron Correlations in Molecules and Solids*. 3rd edition. Springer-Verlag, New York (General Reference).

Galdi, G (1994): *An Introduction to the Mathematical Theory of the Navier-Stokes Equations*. Vols. 1–4. Springer-Verlag, Berlin, Heidelberg (Vols. 3 and 4 to appear) (Chapter 72).

Gallot, S., Hulin, D., and Lafontaine, J. (1987): *Riemannian Geometry*. Springer-Verlag, Berlin, Heidelberg (Chapter 74).

Gamkrelidze, R. (ed.) (1991): *Geometry I–IV*. Encyclopedia of the Mathematical Sciences. Springer-Verlag, New York (General Reference).

Gardiner, C. (1993): *Handbook of Stochastic Methods for Physics, Chemistry and the Natural Sciences*. Springer-Verlag, New York (General Reference).

Gell-Mann, M. (1994): *The Quark and the Jaguar: Adventures in the Simple and the Complex*. Freeman, San Francisco, CA. (German edition: Das Quark und der Jaguar: eine neue Theorie erklärt die Welt, Piper, München, 1994) (General Reference).

Giaquinta, M. (1993): *Introduction to Regularity Theory for Nonlinear Elliptic Systems*. Birkhäuser, Basel (General Reference).

Giaquinta, M. and Hildebrandt, S. (1996): *Calculus of Variations*. Vols. 1, 2. Springer-Verlag, New York (General Reference).

Girvin, S. and Prange, R. (1990): *The Quantum Hall Effect*. Springer-Verlag, New York (Chapter 59).

Göckeler, M. and Schücker, T.: *Differential Geometry, Gauge Theories, and Gravity*. Cambridge University Press, Cambridge, UK (Chapter 74).

Godlewski, E. and Raviart, P. (1996): *Numerical Approximation of Hyperbolic Systems of Conservation Laws*. Springer-Verlag, New York (General Reference).

Goldberg, L. et al. (eds.) (1993): *Topological Methods in Modern Mathematics: A Symposium in Honor of John Milnor's 60th Birthday*. Publish or Perish, Huston, TX (General Reference).

Golub, G. and Ortega, J. (1993): *Scientific Computing: An Introduction with Parallel Computing*. Academic Press, New York (General Reference).

Gorshkov, V. (1994): *Physical and Biological Bases of Life Stability: Man, Biota, Environment*. Springer-Verlag, Berlin, Heidelberg (Chapter 79).

Greiner, W. (1994): *Classical Physics*. Vols. 1ff. Springer-Verlag, New York (General Reference).

Greiner, W. (1993/94): *Theoretical Physics*. Vols. 1–6. Cf. the following titles.

Greiner, W. (1994): *Quantum Mechanics: An Introduction*. Springer-Verlag, Berlin, Heidelberg (General Reference).

Greiner, W. (1993): *Relativistic Quantum Mechanics*. Springer-Verlag, Berlin, Heidelberg (General Reference).

Greiner, W. (1993): *Gauge Theory of Weak Interactions*. Springer-Verlag, Berlin, Heidelberg (General Reference).

Greiner, W. and Müller, B. (1994): *Quantum Mechanics: Symmetries*. Springer-Verlag, Berlin, Heidelberg (General Reference).

Greiner, W. and Reinhardt, J. (1994): *Quantum Electrodynamics*. Springer-Verlag, Berlin, Heidelberg (General Reference).

Greiner, W. and Reinhardt, J. (1996): *Field Quantization*. Springer-Verlag, New York (General Reference).

Greiner, W. and Schäfer, A. (1994): *Quantum Chromodynamics*. Springer-Verlag, Berlin, Heidelberg (General Reference).

Grisvard, P. (1992): *Singularities in Boundary Problems*. Masson, Paris (General Reference).

Grosche, C. and Steiner, F. (1996): *A Table of Feynman Path Integrals*. Springer-Verlag, Berlin, Heidelberg (Chapter 59).

Grosche, G., Ziegler, D., Ziegler, V., and Zeidler, E. (eds.) (1995): *Teubner-Taschenbuch der Mathematik II*. Teubner-Verlag, Stuttgart/Leipzig (General Reference).

Grosse, H. (1996): *Models in Statistical Physics and Quantum Field Theory*. Springer-Verlag, Berlin, Heidelberg (Chapter 68).

Gruber, P. and Wills, J. (1993): *Handbook of Convex Geometry*. Vols. 1, 2. North-Holland, Amsterdam (General Reference).

Gurtin, M. (1993): *Thermomechanics of Evolving Phase*. Clarendon Press, Oxford (Chapter 67).

Haag, R. (1993): *Local Quantum Physics: Fields, Particles, Algebras*, Springer-Verlag, Berlin, Heidelberg (General Reference).
Haken, H. (1996): *Principles of Brain Functioning*. Springer-Verlag, New York (Chapter 79).
Haken, H. and Wolf, H. (1992): *Molecular Physics and Elements of Quantum Chemistry*. Springer-Verlag, New York (General Reference).
Hale, J. and Koçak, H. (1991): *Dynamics of Bifurcations*. Springer-Verlag, Berlin, Heidelberg (cf. also Koçak (1989)) (Chapter 79).
Hatfield, B. (1992): *Quantum Field Theory of Point Particles and Strings*. Addison-Wesley, Redwood City, CA (Chapter 76).
Havin, V. and Jöricke, B. (1994): *The Uncertainty Principle in Harmonic Analysis*, Springer-Verlag, Berlin, Heidelberg (Chapter 59).
Heidmann, J. (1994): *Bioastronomie: Über irdisches Leben und außerirdische Intelligenz*. Springer-Verlag, Berlin, Heidelberg (Chapter 76).
Heidrich, D., Kliesch, W., and Quapp, W. (1991): *Properties of Chemically Interesting Potential Energy Surfaces*. Springer-Verlag, New York (Chapter 78).
Heisenberg, W. (1989): *Encounters with Einstein and Other Essays on People, Places, and Particles*. Princeton University Press, Princeton, NJ (General Reference).
Henneaux, M. and Teitelboim, C. (1993): *Quantization of Gauge Systems*. Princeton University Press, Princeton, NJ (Chapter 59).
Hermann, C. and Sapoval, B. (1994): *Physics of Semiconductors*. Springer-Verlag, New York (Chapter 67).
Hilbert, D. (1991): *Natur und mathematisches Erkennen*. Lectures given in 1919–1920. Edited by D. Rowe. Birkhäuser, Basel (General Reference).
Hiriart-Urruty, J. and Lemarchal, C. (1993): *Convex Analysis and Minimization Algorithms*. Vols. 1, 2. Springer-Verlag, Berlin, Heidelberg (General Reference).
Hirsch, M., Marsden, J., and Shub, M. (eds.) (1993): *From Topology to Computation. Proceedings of the Smalefest*. Springer-Verlag, New York (General Reference).
Hislop, P. and Sigal, I. (1996): *Introduction to Spectral Theory: With Applications to Schrödinger Operators*. Springer-Verlag, New York (Chapter 59).
Hofer, H. and Zehnder, E. (1994): *Symplectic Invariants and Hamiltonian Dynamics*. Birkhäuser, Basel (Chapter 79).
Holmes, M. (1995): *Introduction to Perturbation Methods*. Springer-Verlag, New York (General Reference).
Honerkamp, J. and Römer, H. (1993): *Theoretical Physics: A Classical Approach*. Springer-Verlag, New York (General Reference).
Hoppenstaedt, F. (1993): *Analysis and Simulation of Chaotic Systems*. Springer-Verlag, New York (Chapter 79).
Hoppenstaedt, F. and Peskin, C. (1994): *Mathematics in Medicine and in the Life Sciences*. Springer-Verlag, New York (General Reference).
Hubbard, J. and West, B. (1995): *Differential Equations: A Dynamical Systems Approach*. Vol. 2. Springer-Verlag, New York (Chapter 79).

Iagolnitzer, D. (1993): *Scattering in Quantum Field Theory*. Princeton University Press, Princeton, NJ (Chapter 59).
Ibach, H. and Lüth, H. (1993): *Solid-State Physics: An Introduction to Theory and Experiment*. Springer-Verlag, New York (General Reference).
Ibragimov, N. (1993): *CRC Handbook of Lie Group Analysis of Differential Equations*. CRC Press, Boca Raton, FL (General Reference).

Isham, C. (1989): *Modern Differential Geometry for Physicists.* World Scientific, Singapore (Chapter 74).

Ize, J. Massabo, I., and Vignoli, A. (1993): *Degree Theory for Equivariant Maps: the General S^1-Action.* American Mathematical Society, Providence, RI (General Reference).

Jacobs, K. (1992): *Invitation to Mathematics.* Princeton University Press, Princeton, NJ (General Reference).

Jaffe, A. and Quinn, F. (1993): *Theoretical Mathematics: Towards a Cultural Synthesis of Mathematics and Theoretical Physics.* Bull. AMS 29, 1–13. (Cf. also the controversial discussion of this article in Bull. AMS 30 (1994), 161–211.) (General Reference).

Jikov, V., Kozlov, S., and Oleinik, O. (1994): *Homogenization of Differential Operators.* Springer-Verlag, Berlin, Heidelberg (General Reference).

Joseph, D. (1990): *Fluid Dynamics of Viscoelastic Liquids.* Springer-Verlag, New York (Chapter 70).

Joseph, D. and Renardy, Y. (1993): *Fundamentals of Two-Fluid Dynamics.* Vols. 1, 2. Springer-Verlag, New York (Chapter 70).

Jost, J. (1994): *Differentialgeometrie und Minimalflächen.* Springer-Verlag, Berlin, Heidelberg (Chapter 74).

Jost, J. (1995): *Riemannian Geometry and Geometric Analysis.* Springer-Verlag, Berlin, Heidelberg (Chapter 74) (to appear).

Jost, J. (1996a): *Compact Riemann Surfaces.* Springer-Verlag, New York (Chapter 74).

Jost, J. (1997): *Postmodern Analysis.* Springer-Verlag, Berlin, Heidelberg (General Reference) (to appear).

Kac, M, Rota, G., and Schwartz, J. (1992): *Discrete Thoughts: Essays on Mathematics, Science, and Philosophy.* Birkhäuser, Basel (General Reference).

Kaiser, G. (1994): *A Friendly Guide to Wavelets.* Birkhäuser, Basel (General Reference).

Kaku, M. (1988): *Introduction to Superstring Theory.* Springer-Verlag, New York (Chapter 76).

Kaku, M. (1991): *Strings, Conformal Fields, and Topology.* Springer-Verlag, New York (Chapter 76).

Kaku, M. (1993): *Quantum Field Theory.* Oxford University Press, Oxford (Chapters 59 and 76).

Kaku, M. (1994): *Hyperspace: A Scientific Odyssey Through Parallel Universes, Time Warps, and the 10th Dimension.* Oxford University Press, Oxford (Chapter 76).

Kaku, M. and Trainer, J. (1987): *Beyond Einstein: The Cosmic Quest for the Theory of the Universe.* Bantam, New York (Chapter 76).

Karttunen et al. (1993): *Fundamental Astronomy.* Springer-Verlag, Berlin, Heidelberg (Chapter 76).

Kassel, C. (1994): *Quantum Groups.* Springer-Verlag, New York (General Reference).

Katok, A. and Hasselblatt, B. (1995): *Introduction to the Modern Theory of Dynamical Systems.* Cambridge University Press, Cambridge, UK (General Reference).

Kavian, O. (1993): *Introduction à la théorie des points critiques et applications aux problèmes elliptiques.* Springer-Verlag, New York (General Reference).

Kevorkian, J. and Cole, J. (1996): *Multiple Scale and Singular Perturbation Methods.* Springer-Verlag, New York (General Reference).

Kinderlehrer, D. et al. (eds.): *Microstructure and Phase Transition*. Springer-Verlag, New York (Chapter 61).
Kippenhahn, R. (1993): *100 Billion Suns*. Princeton University Press, Princeton, NJ (Chapter 76).
Kippenhahn, R. and Weigert, A. (1994): *Stellar Structure and Evolution*. Springer-Verlag, Berlin, Heidelberg (Chapter 76).
Kircher, R. and Bergner, W. (1991): *Three-Dimensional Simulation of Semiconductor Devices*. Birkhäuser, Basel (Chapter 67).
Kirsch, A. (1996): *An Introduction to the Mathematical Theory of Inverse Problems*. Springer-Verlag, New York (General Reference).
Knauf, A. and Sinai, Ya. (1997): *Classical Nonintegrability Quantum Chaos*. Birkhäuser, Basel (to appear).
Koçak, H. (1989): *Differential and Difference Equations Through Computer Experiments. With Diskettes*. Springer-Verlag, New York (cf. Hale and Koçak (1991)) (Chapter 79).
Kohonen, T. (1995): *Self-Organizing Maps*. Springer-Verlag, New York (Chapter 78).
Kolb, E. and Turner, M. (1990): *The Early Universe*. Addison-Wesley, Redwood City, CA (Chapter 76).
Kozlov, V. and Fedorov, U. (1996): *Memoirs on Integrable Systems*. Springer-Verlag, Berlin, Heidelberg (Chapter 58).
Kuchment, P. (1993): *Floquet Theory for Partial Differential Equations*. Birkhäuser, Basel (Chapter 79).
Kuksin, S. (1993): *Nearly Integrable Infinite-Dimensional Hamiltonian Systems*. Lecture Notes Mathematics Vol. 1556. Springer-Verlag, Berlin, Heidelberg (Chapter 58).
Kuperschmidt, B. (1992): *The Variational Principles of Dynamics*. World Scientific, Singapore (General Reference).
Kuzmin, A. (1992): *Non-Classical Equations of Mixed Type and Their Applications in Gas Dynamics*. Birkhäuser, Basel (Chapter 70).
Kuznetsov, Y. (1995): *Elements of Applied Bifurcation Theory*. Springer-Verlag, New York.

Lakshmikantham, V. (ed.) (1994): *First World Congress of Nonlinear Analysts*, Vols. 1-4. De Gruyter, Berlin (General Reference).
Lang, K. (1996): *Astrophysical Formulae*. Springer-Verlag, Berlin, Heidelberg (Chapter 76).
Lazutkin, V. (1993): *KAM-Theory and Semiclassical Approximations to Eigenfunctions*. Springer-Verlag, Berlin, Heidelberg (Chapter 58).
Leung, A. (1989): *Systems of Nonlinear Partial Differential Equations: Applications to Biology and Engineering*. Kluwer, Dordrecht (General Reference).
LeVeque, R. (1990): *Numerical Methods for Conservation Laws*. Birkhäuser, Basel (Chapter 70).
Li, M. and Vitányi, P. (1993): *An Introduction to Kolmogorov Complexity and its Applications*. Springer-Verlag, New York (General Reference).
Li, Ta-tsien (1994): *Global Classical Solutions for Quasilinear Hyperbolic Systems*. Wiley, New York (General Reference).
Lichtenberg, A. and Lieberman, M. (1992): *Regular and Chaotic Dynamics*. Springer-Verlag, New York (Chapter 76).
Lions, P. (1996): *Mathematical Topics in Fluid Dynamics. Vol. 1: Incompressible Models. Vol. 2: Compressible Models* (to appear). Oxford University Press, Oxford, UK (Chapter 72).

Louis, A. (1995): *Inverse and Ill-Posed Problems.* Springer-Verlag, New York (General Reference).

Lüst, D. and Theissen, S. (1989): *Lectures on String Theory.* Springer-Verlag, Berlin, Heidelberg (Chapter 76).

Lunardi, A. (1995): *Analytic Semigroups and Optimal Regularity in Parabolic Problems.* Birkhäuser, Basel (General Reference).

Lusztig, G. (1993): *Introduction to Quantum Groups.* Birkhäuser, Boston, MA (Chapter 76).

Mackey, G. (1992): *The Scope and History of Commutative and Noncommutative Harmonic Analysis.* American Mathematical Society, Providence, RI (General Reference).

Mackey, M. (1993): *Time's Arrow: The Origin of Thermodynamic Behavior.* Springer-Verlag, New York (Chapter 67).

Mainzer, K. (1994): *Thinking in Complexity: The Complex Dynamics of Matter, Mind, and Mankind.* Springer-Verlag, Berlin, Heidelberg (Chapter 79).

Málek, J., Nečas, M., Rokyta, and Růžička, M. (1996): *Weak and Measure-valued Solutions to Evolutionary Partial Differential Equations.* Chapman, London (General Reference).

Mandl, F. and Shaw, G. (1989): *Quantum Field Theory.* Wiley, New York (Chapter 59).

Mangel, M. and Segel, L. (eds.) (1996): *Classical Papers in Mathematical Biology.* Springer-Verlag, Berlin, Heidelberg.

Marathe, K. and Martucci, G. (1992): *The Mathematical Foundations of Gauge Theory.* North-Holland, Amsterdam (Chapter 59).

Marchioro, C. and Pulvirenti, M. (1994): *Mathematical Theory of Inviscid Fluids.* Springer-Verlag, New York (Chapter 70).

Markowich, P. (1986): *The Stationary Semiconductor Device Equations.* Springer-Verlag, Berlin, Heidelberg (Chapter 67).

Markowich, P. (1990): *Semiconductor Equations.* Springer-Verlag, Berlin, Heidelberg (Chapter 67).

Marsden, J. (1992): *Lectures in Mechanics.* Cambridge University Press, Cambridge, UK (Chapters 58 and 79).

Marsden, J. and Ratiu, T. (1994): *Introduction to Mechanics, and Symmetry: A Basic Exposition of Classical Mechanical Systems.* Springer-Verlag, New York (Chapter 58).

Marshak, R. (1993): *Conceptual Foundations of Modern Particle Physics.* World Scientific, Singapore (Chapter 59).

Maslova, N. (1993): *Nonlinear Evolution Equations: Kinetic Approach.* World Scientific, Singapore (General Reference).

Mason, L. and Hughstone, L. (1990): *Further Advances in Twistor Theory.* Vols. 1, 2. Longman, Essex, UK (Chapter 75).

Matveev, V. et al. (1994): *Algebro-Geometrical Approach to Nonlinear Evolution Equations.* Springer-Verlag, Berlin, Heidelberg (General Reference).

Mawhin, J. and Willem, M. (1989): *Critical Point Theory and Hamiltonian Systems.* Springer-Verlag, New York (General Reference).

McDuff, D. and Salamon, D. (1994): *Introduction to Symplectic Topology.* Oxford University Press, Oxford (General Reference).

Mehmeti, A. (1994): *Nonlinear Waves in Networks.* Akademie-Verlag, Berlin (Chapter 71).

Meirmanov, A. (1992): *The Stefan Problem.* De Gruyter, Berlin (Chapter 67).

Melo, W. de and van Strien, S. (1993): *One-Dimensional Dynamics*. Springer-Verlag, New York (Chapter 79).

Meyer, K. and Hall, G. (1992): *Introduction to Hamiltonian Dynamical Systems and the N-Body Problem*. Springer-Verlag, New York (Chapter 58).

Meyer, S. (1995): *Computersimulation von Gittermodellen*. Springer-Verlag, Berlin, Heidelberg (General Reference).

Meyer, Y. (1990): *Ondelettes et opérateurs*, Vols. 1-3. Hermann, Paris. (English edition: Wavelets and Operators, Cambridge University Press, Cambridge, UK.) (General Reference).

Mielke, A. (1991): *Hamiltonian and Lagrangian Flows on Center Manifolds with Applications to Elliptic Variational Problems*. Lecture Notes Mathematics, Vol. 1489. Springer-Verlag, Berlin, Heidelberg (Chapter 79).

Mikhailov, A. (1994): *Foundations of Synergetics*. Vols. 1, 2. Springer-Verlag, Berlin, Heidelberg (Chapter 67).

Milburn, G. and Walls, D. (1994): *Quantum Optics*. Springer-Verlag, Berlin, Heidelberg (Chapter 59).

Mitsuru Ikawa (ed.) (1993): *Spectral and Scattering Theory*. Marcel Dekker, New York (General Reference).

Monastirsky, M. (1993): *Topology of Gauge Fields and Condensed Matter*. Plenum Press, New York (General Reference).

Monteiro, M. (1993): *Differential Inclusions in Nonsmooth Mechanical Problems: Shocks and Dry Friction*. Birkhäuser, Basel (Chapter 63).

Moreau, J. (ed.) (1988): *Nonsmooth Mechanics and Applications*. Springer-Verlag, Wien (Chapter 60).

Müller, I. (1994): *Grundzüge der Thermodynamik*. Springer-Verlag, Berlin, Heidelberg (Chapter 67).

Müller, I. and Ruggeri, T. (1993): *Extended Thermodynamics*. Springer-Verlag, New York (Chapter 67).

Müller, W. et al. (1993): *Geometrie und Physik*. De Gruyter, Berlin (General Reference).

Murdock, J. (1991): *Perturbations*. Wiley, New York (General Reference).

Murray, J. (1989): *Mathematical Biology*. Springer-Verlag, Berlin, Heidelberg (Chapter 67).

Naber, G. (1992): *The Geometry of Minkowski Spacetime*. Springer-Verlag, New York (Chapter 75).

Nagasawa, M. (1993): *Schrödinger Equations and Diffusion Theory*. Birkhäuser, Basel (Chapter 59).

Nakahara, M. (1990): *Geometry, Topology, and Physics*. Hilger, Bristol (Chapter 74).

Narasimhan, M. (1993): *Principles of Continuum Mechanics*. Wiley, New York (Chapters 61 and 70).

Nettel, S. (1992): *Wave Physics*. Springer-Verlag, New York (General Reference).

Newton, R. (1988): *Scattering Theory of Waves and Particles*. Springer-Verlag, Berlin, Heidelberg (Chapter 59).

Nishikawa, K. and Wakatani, M. (1993): *Plasma Physics: Basic Theory with Fusion Applications*. Springer-Verlag, Berlin, Heidelberg (Chapter 68).

Nusse, H. and Yorke, J. (1994): *Dynamics: Numerical Explorations*. Springer-Verlag, New York (Chapter 79).

Oberguggenberger, M. (1992): *Multiplication of Distributions and Applications to Partial Differential Equations*. Harlow, Longman, UK (General Reference).

Oberguggenberger, M. and Rosinger, E. (1994): *Solution of Continuous Nonlinear Partial Differential Equations Through Order Completion.* North-Holland, New York (General Reference).

Ohya, M. and Petz, D. (1993): *Quantum Entropy and its Use.* Springer-Verlag, Berlin, Heidelberg (Chapter 68).

Ott, E. (1993): *Chaos in Dynamical Systems.* Cambridge University Press, Cambridge, UK (Chapter 79).

Pais, A. (1993): *Niels Bohr's Times in Physics, Philosophy, and Polity.* Oxford University Press, Oxford (Chapter 59).

Pauli, W. (1990): *Die allgemeinen Prinzipen der Wellenmechanik.* Neu herausgegeben und mit historischen Anmerkungen versehen von Norbert Straumann. (Revised edition by Norbert Straumann). Springer-Verlag, Berlin, Heidelberg (Chapter 59).

Pauli, W. (1992): *Scientific Correspondence with Bohr, Einstein, Heisenberg, and others.* Vols. 1–3. Edited by K. von Meyenn. Springer-Verlag, Berlin, Heidelberg (Chapter 59).

Pauli, W. (1994): *Writings on Physics and Philosophy.* Springer-Verlag, Berlin, Heidelberg (General Reference).

Peebles, P. (1991): *Quantum Mechanics.* Princeton University Press, Princeton, NJ (Chapter 59).

Peebles, P. (1993): *Principles of Physical Cosmology.* Princeton University Press, Princeton, NJ (Chapter 76).

Penrose, R. (1992): *The Emperor's New Mind Concerning Computers, Minds, and the Laws of Physics.* Oxford University Press, Oxford (General Reference).

Penrose, R. (1994): *Shadows of the Mind: The Search for the Missing Science of Conciousness.* Oxford University Press, Oxford (General Reference).

Peters, N. (1995): *Laminar and Turbulent Combustion.* Springer-Verlag, New York (Chapter 70) (to appear).

Peyard, M. (ed.) (1995): *Nonlinear Excitations in Biomolecules.* Springer-Verlag, New York (General Reference).

Pier, J. (ed.) (1995): *Development of Mathematics 1900–1950.* Birkhäuser, Basel (General Reference).

Plakida, N. (1994): *High-Temperature Superconductivity: Experiment and Theory.* Springer-Verlag, New York (Chapter 59).

Plessis, A. du and Wall, C. (1994): *The Geometry of Topological Stability.* Clarendon Press, Oxford (General Reference).

Polyakov, A. (1987): *Gauge Fields and Strings.* Academic Publishers, Harwood, NJ (Chapter 76).

Princeton Problems in Physics with Solutions. Edited by N. Newbury et al., Princeton University Press, Princeton, NJ (General Reference).

Proceedings of the International Congress of Mathematicians in Zurich. Birkhäuser, Basel (to appear).

Proto, A. and Plastino, A. (1995): *Maximum Entropy Principle.* Springer-Verlag, New York (Chapter 68) (to appear).

Prüss, J. (1993): *Evolutionary Integral Equations and Applications.* Birkhäuser, Basel (General Reference).

Quartapelle, L. (1993): *Numerical Solutions of the Incompressible Navier-Stokes Equations.* Birkhäuser, Basel (Chapter 72).

Quarteroni, A. and Valli, A. (1994): *Numerical Approximation of Partial Differential Equations.* Springer-Verlag, Berlin, Heidelberg (General Reference).

Ratcliffe, J. (1994): *Foundations of Hyperbolic Manifolds.* Springer-Verlag, New York (Chapter 74).
Ratner, V. et al. (1996): *Molecular Evolution.* Springer-Verlag, New York (General Reference).
Rauch, J. (1991). *Partial Differential Equations.* Springer-Verlag, New York (General Reference).
Raychaudhuri, A., Banerji, S., and Banerjee, A. (1992): *General Relativity, Astrophysics, and Cosmology.* Springer-Verlag, New York (Chapter 76).
Renardy, M. and Rogers, R. (1993): *Introduction to Partial Differential Equations.* Springer-Verlag, New York (General Reference).
Riemann, B. (1990): *Gesammelte Mathematische Werke, Wissenschaftlicher Nachlaß und Nachträge* (Collected Papers edited by R. Narasimhan). Teubner-Verlag, Leipzig and Springer-Verlag, Berlin, Heidelberg (Chapter 74).
Rivasseau, V. (1991): *From Perturbative to Constructive Renormalization.* Princeton University Press, Princeton, NJ (Chapter 59).
Roepstorff, G. (1994): *Path Integral Approach to Quantum Physics.* Springer-Verlag, Berlin, Heidelberg (Chapter 59).
Rolnick, W. (1994): *Fundamental Particles and Their Interactions.* Addison-Wesley, Reading, MA (Chapter 59).
Roubiček, T. (1997): *Relaxation in Optimization Theory and Variational Calculus.* De Gruyter, Berlin (to appear) (General Reference).
Ruder, H. et al. (1994): *Atoms in Strong Magnetic Fields: Quantum Mechanical Treatment and Applications in Astrophysics and Quantum Chaos.* Springer-Verlag, New York (Chapter 59).
Rudolph, E. (1994): *Philosophy, Mathematics, and Modern Physics: A Dialogue.* Springer-Verlag, Berlin, Heidelberg (General Reference).
Ruelle, D. (1989): *Chaotic Evolution and Strange Attractors.* Cambridge University Press, Cambridge, UK (Chapter 59).
Ruelle, D. (1993): *Chance and Chaos.* Princeton University Press, Princeton, NJ (Chapter 79).
Rychlik, M. and Yakobson, M. (1997): *One-dimensional Dynamical Systems.* Springer-Verlag, New York (General Reference).

Sanz, J., Martinez-Gonzalez, and Cayon, L. (eds.) (1994): *Present and Future of the Cosmic Microwave Background.* Springer-Verlag, New York (Chapter 76).
Sattinger, D. and Weaver, O. (1993): *Lie Groups, Lie Algebras, and Their Representations.* Springer-Verlag, New York (General Reference).
Schiepek, G. and Tschacher, W. (1997): *Synergetik in den Humanwissenschaften.* Springer-Verlag, Berlin, Heidelberg (General Reference) (to appear).
Schmutzer, E. (1989): *Grundlagen der theoretisehen Physik.* Vols. 1, 2. Deutscher Verlag der Wissenschaften, Berlin (General Reference).
Schneider, M. (1992): *Himmelsmechanik.* Vols. 1, 2. Bibliogr. Institut, Mannheim (Chapter 58).
Schneider, P., Ehlers, J., and Falco, E. (1992): *Gravitational Lenses.* Springer-Verlag, New York (Chapter 76).

Schulze, B. (1994): *Pseudo-Differential Boundary Value Problems, Conical Singularities, and Asymptotics.* Akademie-Verlag, Berlin (General Reference).

Schuster, R. (1996): *Grundkurs Biomathematik: Mathematische Modelle in Biologie, Biochemie, Medizin und Pharmazie mit Computerlösungen in Mathematica.* Teubner-Verlag, Stuttgart/Leipzig (General Reference).

Schwarz, A. (1993): *Quantum Field Theory and Topology.* Springer-Verlag, Berlin, Heidelberg (Chapter 59).

Schwarz, A. (1994): *Topology for Physicists.* Springer-Verlag, New York (General Reference).

Scott, G. and Davidson, K. (1993): *Wrinkles in Time.* Morrow, New York (Chapter 76).

Sell, G., Foias, C., and Temam, R. (1993): *Turbulence in Fluid Flows: A Dynamical Systems Approach.* Springer-Verlag, New York (Chapter 72).

Seydel, R. (1994): *Practical Bifurcation and Stability Analysis: From Equilibrium to Chaos.* Springer-Verlag, Berlin, Heidelberg (Chapter 79).

Shore, S. (1992): *An Introduction to Astrophysical Hydrodynamics.* Academic Press, San Diego, CA (Chapters 70 and 76).

Simon, B. (1993): *The Statistical Mechanics of Lattice Gases.* Princeton University Press, Princeton, NJ (Chapter 68).

Sirovich, L. (ed.) (1991): *New Perspectives in Turbulence.* Springer-Verlag, New York (Chapter 72).

Sirovich, L. (ed.) (1994): *Trends and Perspectives in Applied Mathematics.* Springer-Verlag, New York (General Reference).

Smoller, J. (1994): *Shock Waves and Reaction-Diffusion Equations.* 2nd enlarged edition. Springer-Verlag, New York (General Reference).

Smirnov, V. (1991): *Renormalization and Asymptotic Expansions.* Birkhäuser, Basel (General Reference).

Spohn, H. (1991): *Large Scale Dynamics of Interacting Particles.* Springer-Verlag, Berlin, Heidelberg (Chapter 68).

Stein, E. (1993): *Harmonic Analysis: Real-Variable Methods, Orthogonality, and Oscillatory Integrals.* Princeton University Press, Princeton, NJ (General Reference).

Stephani, H. (1989): *Differential Equations: Their Solution Using Symmetries.* Edited by M. MacCallum. Cambridge University Press, Cambridge, UK (Chapter 58).

Sterman, G. (1993): *An Introduction to Quantum Field Theory.* Cambridge University Press, Cambridge, UK (Chapter 59).

Stratonovich, R. (1992): *Nonlinear Equilibrium Thermodynamics.* Vols. 1, 2. Springer-Verlag, New York (Chapter 67).

Straub, D. (1989): *Thermofluiddynamics of Optimized Rocket Propulsions.* Birkhäuser. Basel (Chapter 70).

Straughan, B. (1992): *The Energy Method, Stability and Nonlinear Convection.* Springer-Verlag, New York (Chapter 69).

Strauss, W. (1989): *Nonlinear Wave Equations.* American Mathematical Society, Providence, RI (General Reference).

Struwe, M. (1988): *Semilinear wave equations.* Bull. AMS **26**, 53–85 (General Reference).

Struwe, M. (1990): *Variational Methods: Applications to Nonlinear Partial Differential Equations and Hamiltonian Systems.* Springer-Verlag, Berlin, Heidelberg (General Reference).

Sun, N. (1994): *Mathematical Modelling of Groundwater Pollution.* Springer-Verlag, Berlin, Heidelberg (Chapter 70).

Taylor, M. (1993): *Pseudodifferential Operators and Nonlinear Partial Differential Equations*. Birkhäuser, Boston (General Reference).

Taylor, M. (1996): *Partial Differential Equations*, Vols. 1–3. Springer-Verlag, New York (recommended as a standard text on the modern theory of linear and nonlinear partial differential equations).

Temam, R. (1988): *Infinite-Dimensional Dynamical Systems in Mechanics and Physics*. North-Holland, Amsterdam (General Reference).

Thaller, B. (1992): *The Dirac Equation*. Springer-Verlag, Berlin, Heidelberg (Chapter 59).

Thomas, J. (1995): *Numerical Partial Differential Equations*. Vols. 1,2. Springer-Verlag, New York (General Reference).

Thorne, K. (1994): *Black Holes and Time Warps: Einstein's Outrageous Legacy*. (German edition: Gekrümmter Raum und verbogene Zeit: Einstein's Vermächtnis, Droemer und Knaur, München, 1994.) (Chapter 76).

Tipler, P. (1991): *Physics for Scientists and Engineers*. World Publishers, New York (General Reference)

Toda, M. (1989): *Nonlinear Waves and Solitons*. Kluwer, Dordrecht (Chapter 71).

Tretter, C. (1993): *On λ-Nonlinear Eigenvalue Problems*. Akademie-Verlag, Berlin (General Reference).

Triebel, H. (1992): *Theory of Function Spaces*. Vol. 2. Birkhäuser, Basel (General Reference).

Troianello, G. (1987): *Elliptic Differential Equations and Obstacle Problems*. Plenum Press, New York (Chapter 63).

Tromba, T. (1993): *Teichmüller Theory in Riemannian Geometry*. Birkhäuser, Basel (Chapter 74).

Truhlar, G. (ed.) (1988): *Mathematical Frontiers in Computational Chemical Physics*. Springer-Verlag, New York (Chapter 67).

Urakawa, H. (1993): *Calculus of Variations and Harmonic Maps*. American Mathematical Society, Providence, RI (Chapter 74).

Van de Velde, E. (1994): *Concurrent Scientific Computing*. Springer-Verlag, New York (General Reference).

Vanhorn, W. (1994): *The Stokes Equation*. Akademie-Verlag, Berlin (Chapter 72).

Vilenkin, N. and Klimyk, A. (1992): *Representation of Lie Groups and Special Functions*. Vols. 1–4. Kluwer, Dordrecht (General Reference).

Vishik, M. and Fursikov, A. (1988): *Mathematical Problems of Statistical Hydromechanics*. Kluwer, Dordrecht (Chapter 72?).

Visintin, A. (1994): *Differentiable Models of Hysteresis*. Springer-Verlag, Berlin, Heidelberg (Chapter 60).

Volkenstein, M. (1994): *Physical Approaches to Biological Evolution*. Springer-Verlag, Berlin, Heidelberg (Chapter 79).

Wald, R. (1984): *General Relativity*. The University of Chicago Press, Chicago, IL (Chapter 76).

Waldschmidt, M. et al. (eds.) (1993): *From Number Theory to Physics*. Springer-Verlag, Berlin, Heidelberg (General Reference).

Weil, A. (1991): *Apprenticeship of a Mathematician*. Birkhäuser, Basel (General Reference).

Weinberg, S. (1989): *The Cosmological Constant Problem*. Rev. Mod. Phys. 61, 1–24 (Chapter 76).
Weinberg, S. (1992): *Dreams of a Final Theory*. Pantheon Books, New York (Chapter 76).
Weinberg, S. (1995): *The Quantum Theory of Fields*, Vols. 1,2. Cambridge University Press, Cambridge, UK (Chapter 59).
Wendland, W. (1997): *Integral Equation Methods for Boundary Value Problems*. Springer-Verlag, Berlin, Heidelberg (General Reference).
Wess, J. and Bagger, J. (1991): *Supersymmetry and Supergravity*. Second edition revised and expanded. Princeton University Press, Princeton, NJ (Chapter 76).
Wiedemann, H. (1993): *Particle Accelerator Physics*. Vol. 1: *Basic Principles and Linear Beam Dynamics*. Vol. 2: *Nonlinear and High Order Beam Dynamics*. Springer-Verlag, Berlin, Heidelberg (Chapter 59).
Wiggins, S. (1994): *Normally Hyperbolic Invariant Manifolds in Dynamical Systems*. Springer-Verlag, New York (Chapter 79).
Wiggins, S. (1994a): *Global Dynamics, Phase Space Transport, Orbits Homoclinic to Resonances, and Applications*. American Mathematical Society, Providence, RI (Chapter 79).
Wigner, E. (1993): *Collected Works*. Edited by A. Wightman. Vols. 1ff. Springer-Verlag, Berlin, Heidelberg (General Reference).
Willmore, T. (1993): *Riemannian Geometry*. Clarendon Press, Oxford (Chapter 74).
Winfree, A. (1990): *The Geometry of Biological Time*. Springer-Verlag, New York (Chapter 67).
Wu, J. (1996): *Theory and Applications of Partial Functional Equations*. Springer-Verlag, New York (General Reference).

Yndurain, F. (1993): *The Theory of Quark and Gluon Interactions*. Springer-Verlag, Berlin, Heidelberg (Chapter 59).

Zabczyk, J. (1992): *Optimal Control Theory*. Birkhäuser, Basel (General Reference).
Zeidler, E. (ed.) (1995): *Teubner-Taschenbuch der Mathematik I*. Teubner-Verlag, Stuttgart-Leipzig (General Reference).
Zeidler, E. (1995a): *Teubner-Taschenbuch der Mathematik II*, Chapters 10–19. Cf. Grosche, Ziegler, and Zeidler (eds.) (1995) (General Reference).
Zeidler, E. (1995): *Applied Functional Analysis: Applications to Mathematical Physics*. Springer-Verlag, New York (General Reference).
Zeidler, E. (1995): *Applied Functional Analysis: Main Principles and Their Applications*. Springer-Verlag, New York.
Zeldovich, Y. (1993): *Selected Works*. Vol. 1: *Chemical Physics and Hydrodynamics*. Vol. 2: *Particles, Nuclei, and the Universe*. Edited by G. Barenblatt and R. Sunyaev. Princeton University Press, Princeton, NJ (Chapter 67).
Ziemer, W. (1989): *Weakly Differentiable Functions*. Springer-Verlag, New York (General Reference).
Zinn-Justin, J. (1996): *Quantum Field Theory and Critical Phenomena*. Clarendon Press, Oxford, UK (General Reference).
Zwillinger, D. (1992): *Handbook of Differential Equations*. Academic Press, New York (General Reference).

List of Symbols

We use the following abbreviations:

B-space	Banach space
H-space	Hilbert space
M–S sequence	Moore–Smith sequence
F-derivative	Fréchet derivative
G-derivative	Gâteaux derivative

$A_i(10)$ means (10) in the Appendix to Part i.

General Notation

$\mathscr{A} \Rightarrow \mathscr{B}$	$\mathscr{A}$ implies $\mathscr{B}$
iff	if and only if
$\mathscr{A} \Leftrightarrow \mathscr{B}$	$\mathscr{A}$ iff $\mathscr{B}$
$f(x) \stackrel{\text{def}}{=} 2x$	$f(x) = 2x$ by definition
$x \in S$	x is an element of the set S
$x \notin S$	x is not an element of S
$\{x:\ldots\}$	set of all x with the property …
$S \subseteq T$	the set S is contained in the set T
$S \subset T$	S is properly contained in T
$S \subset\subset T$	S is a strongly proper subset of T, i.e., the closure of S is contained in T
$\cap, \cup, -$	intersection, union, difference
$\varnothing$	empty set
2^S	set of all all subsets of S, the power set of S
$X \times Y$	product set, $X \times Y = \{(x,y): x \in X, y \in Y\}$

I, id	identity mapping
$f: S \subseteq X \to Y$	mapping from the set S into the set Y with $S \subseteq X$
$D(f)$	domain of f, $D(f) = S$
$R(f)$	range of f, $R(f) = \{f(x): x \in S\}$
$G(f)$	graph of f, $G(f) = \{(x, f(x)): x \in S\}$
$N(f)$	null space of f, $N(f) = \{x: f(x) = 0\}$
Fix(f)	set of fixed points of f, Fix$(f) = \{x: f(x) = x\}$
dom(f), im(f)	identical with $D(f)$, $R(f)$, respectively
ker(f)	identical with $N(f)$
f surjective	mapping onto Y, i.e., $f(S) = Y$
f injective	one-to-one mapping
f bijective	one-to-one mapping onto Y, i.e., f is surjective and injective
$f(A)$	image of the set A, $f(A) = \{f(x): x \in A\}$
$f^{-1}(B)$	preimage of the set B, $f^{-1}(B) = \{x: f(x) \in B\}$
$f\|_A$	restriction of the map f to the set A
$f \circ g$	f applied to g, $(f \circ g)(x) = f(g(x))$
$f: S \to 2^Y$	multivalued mapping, $f(x)$ is a subset of Y
$R(f)$	range of the multivalued mapping f, $R(f) = \bigcup_{x \in S} f(x)$
$G(f)$	graph of the multivalued mapping f, $G(f) = \{(x, y): x \in S, y \in f(x)\}$
dom(f)	effective domain of the multivalued map f, dom$(f) = \{x: f(x) \neq \emptyset\}$
$\mathbb{N}$	set of the natural numbers $1, 2, \ldots$
$\mathbb{R}, \mathbb{C}, \mathbb{Q}, \mathbb{Z}$	set of the real, complex, rational, integer numbers
$\mathbb{K}$	$\mathbb{R}$ or $\mathbb{C}$
$\mathbb{R}_+$	nonnegative real numbers $\xi \geq 0$
$\mathbb{R}_>$	positive real numbers $\xi > 0$
$\mathbb{R}_-$	nonnegative real numbers $\xi \leq 0$
$\mathbb{R}^N$	set of all real N-tuples $x = (\xi_1, \ldots, \xi_N)$
$\mathbb{R}^N_+$	set of all $x \in \mathbb{R}^N$ with $\xi_i \geq 0$ for all i
$\mathbb{R}^N_>$, $\mathring{\mathbb{R}}^N_+$	set of all $x \in \mathbb{R}^N$ with $\xi_i > 0$ for all i
$H\mathbb{R}^N$	set of all $x \in \mathbb{R}^N$ with $\xi_1 \leq 0$ (special half-space)
Re z, Im z	real part of the complex number z, imaginary part of z
$D_i = \partial/\partial \xi_i$	partial derivative with respect to the ith coordinate
$\|x\|$	Euclidean norm of $x \in \mathbb{R}^N$, $\|x\| = (\sum_{i=1}^N \xi_i^2)^{1/2}$
$\langle x \| y \rangle$	Euclidean scalar product in $\mathbb{R}^N$, $\langle x \| y \rangle = \sum_{i=1}^N \xi_i \eta_i$, where $x = (\xi_1, \ldots, \xi_N)$ and $y = (\eta_1, \ldots, \eta_N)$
sgn a	signum of the real number a

List of Symbols

δ_{ij}	Kronecker symbol, $\delta_{ij} = 1$ if $i = j$, and $\delta_{ij} = 0$ if $i \neq j$
$g(x) = o(f(x))$, $x \to a$	$g(x)/f(x) \to 0$ as $x \to a$
$g(x) = O(f(x))$, $x \to a$	$|g(x)| \leq $ constant $|f(x)|$ for all x in a neighborhood of the point a
$[a,b]$, $]a,b[$, $[a,b[$	closed, open, half-open real interval
B^N	closed unit ball in $\mathbb{R}^N$, $B^N = \{x \in \mathbb{R}^N : |x| \leq 1\}$
S^N	N-dimensional unit sphere, $S^N = \{x \in \mathbb{R}^{N+1} : |x| = 1\}$
meas G	Lebesgue measure of the set G (see Part II)

Special Notation

The page number I123 refers to page 123 of Part I, etc., whereas the page number 123 refers to page 123 of the present volume.

		Page
X^*	dual space to X	I774
A^*	dual operator to the operator A	I775
F'	F-derivative of the operator F	I135
F_x	partial F-derivative of F with respect to x	I140
$d^n F(x; h_1, \ldots, h_n)$	nth F-differential of F at x in the directions $h_1, \ldots, h_n$	I143
$d^n F(x; h)$	identical with $d^n F(x; h, \ldots, h)$	
$\delta^n F(x; h_1, \ldots, h_n)$	nth variation of F at x in the directions $h_1, \ldots, h_n$	I134
$\delta^n F(x; h)$	identical with $\delta^n F(x; h, \ldots, h)$	
Nx^m	identical with $N(x, \ldots, x)$ where N is m-linear	I361
∂S	boundary of the set S	I751
$\overline{S}$	closure of S	I751
int S	interior of S	I751
$U(x)$	neighborhood of the point x, i.e., there exists an open set O such that $x \in O$ and $O \subseteq U(x)$	I751
supp f	support of the function f, supp f is the closure of the set $\{x : f(x) \neq 0\}$	I756
$\underline{\lim}$, $\overline{\lim}$	lower, upper limit	I761
diam S, $d(S)$	diameter of the set S	I762
$d(x, S)$	distance of the point x from the set S	I762
$d(S, T)$	distance between the sets S and T	I762
dist(x, S), dist(x, T)	identical with $d(x, S)$, $d(x, T)$, respectively	
$S + T$	sum of the sets S and T	I764
λS	product of the set S by the number λ	I764
span S	linear hull of the set S	I764
co S	convex hull of S	I764

$\overline{\operatorname{co}} S$	closed convex hull of S	I764
$\dim L$	dimension of the linear subspace L;	I765
	dimension of the manifold L	535
$\operatorname{codim} L$	codimension of the linear subspace L;	I765
	codimension of the submanifold L	557
X/Y	factor space (quotient space)	I765
$X \oplus Y$	direct sum or direct topological sum (the text always refers precisely to the momentary meaning)	I765, I766
$X^\perp$	complement of X with respect to a direct sum	I766
$X \times Y$	product space of two B-spaces (The norm on $X \times Y$ is given by $\|(x,y)\| = \|x\| + \|y\|$.)	I770
$\prod_\alpha X_\alpha$	general product space	I747, I755
$X_{\mathbb{C}}$	complexification of the real B-space X	I770
$A_{\mathbb{C}}$	complexification of the linear operator A	I770
$\operatorname{ind} A$	index of the linear operator	I767
$\operatorname{rank} A$	rank of the linear operator A, $\operatorname{rank} A = \dim R(A)$	
$\det A$	determinant of the linear operator A	I179
$\rho(A)$	resolvent set of the linear operator	I795
$\sigma(A)$	spectrum of the linear operator A	I795
$r(A)$	spectral radius of the linear operator A	I795

In real B-spaces, $\rho(A)$, $\sigma(A)$, and $r(A)$ always refer to the complexification of A, i.e., $\rho(A) = \rho(A_{\mathbb{C}})$, etc.

$(x\|y)$	inner (scalar) product in an H-space	I785
(x, y)	ordered pair, an element of the product set $X \times Y$, where $x \in X$ and $y \in Y$	
$\langle x\|y \rangle$	inner (scalar) product in $\mathbb{R}^N$ or $\mathbb{C}^N$	I771
$\langle f, x \rangle$	value of the linear functional f at the point x, $\langle f, x \rangle = f(x)$	
$\|x\|$	norm of x in a B-space	I768
$\|x\|$	Euclidean norm of x in $\mathbb{R}^N$ or $\mathbb{C}^N$, $\|x\| = \langle x\|x \rangle^{1/2}$	
$x_n \to x$	convergence (in norm)	I769
$x_n \rightharpoonup x$	weak convergence	I775
$x_n \overset{*}{\rightharpoonup} x$	weak* convergence	I775
$f_n \rightrightarrows f$	uniform convergence of functions	I801
$x \mapsto f(x)$	another notation for the mapping f	
$f(\cdot)$	another notation for the mapping f	
$(x_{n'})$	subsequence of the sequence (x_n)	
$\alpha = (\alpha_1, \ldots, \alpha_N)$	multi-index; the α_i are nonnegative integers	
$\|\alpha\|$	order of α, $\|\alpha\| = \alpha_1 + \cdots + \alpha_N$	
$D^\alpha = D_1^{\alpha_1} D_2^{\alpha_2} \ldots D_N^{\alpha_N}$	derivative in multi-index notation, $D_i = \partial/\partial \xi_i$	

List of Symbols

$\partial/\partial n$	derivative in the direction of the exterior normal	
$\Delta = D_1^2 + \cdots + D_N^2$	Laplace operator in $\mathbb{R}^N$	
V_3	real, three-dimensional, linear space	10
xy	scalar product of the vectors x and y in V_3	10
$x \times y$	vector product of the vectors x and y in V_3	10
b_i	vector of a basis $\{b_i\}$	11, 596, 617
b^i	vector of a dual basis $\{b^i\}$	13, 597, 620
e_i	vector of an orthonormal basis $\{e_i\}$ (basis vector of a Cartesian coordinate system)	11, 617

In the case of a finite-dimensional manifold, we write $b_i = \partial x/\partial u^i$ or briefly $b_i = \partial/\partial u^i$. Moreover, we have $b^i = du^i$, where $u^1, \ldots, u^n$ denote local coordinates. The precise definition of b_i and b^i on manifolds can be found in Section 73.23.

$\dot{x}(t)$	time derivative of the motion $x = x(t)$ at t	11
$U'(x)$	F-derivative of the potential U at the point x, $U'(x) = \operatorname{grad} U(x)$	13
E_{pot}	potential energy	33
E_{kin}	kinetic energy	32
$\partial F(x)$	subdifferential of the functional F at the point x	III385
$\boldsymbol{\omega}$	angular velocity vector	11
ω	angular velocity, $\omega = |\boldsymbol{\omega}|$; angular frequency, $\omega = \nu/2\pi$, where ν denotes the frequency	11 100
c	velocity of light	885
h	Planck's quantum of action	78
$\hbar$	$\hbar = h/2\pi$	
G	gravitational constant	36
k	Boltzmann constant	400
Re	Reynolds number	440
T^{ij}	energy-momentum tensor	725
$\mathscr{E}$	strain tensor	170
τ	stress tensor	177
σ	reduced stress tensor (first Piola–Kirchhoff tensor)	184
$L(V_3)$	space of all linear operators $\sigma: V_3 \to V_3$	166
$L_{\text{sym}}(V_3)$	space of all linear symmetric operators $\sigma: V_3 \to V_3$ on the H-space V_3	166
tr σ	trace of $\sigma \in L(V_3)$	166
det σ	determinant of $\sigma \in L(V_3)$	166
$[\sigma, \gamma]$	special product for $\sigma, \gamma \in L(V_3)$	166
$(\sigma|\gamma)$	scalar product on $L(V_3)$	166

div $\sigma(x)$	divergence of the tensor field $\sigma: V_3 \to L(V_3)$, i.e., $\sigma(x) \in L(V_3)$	167
$\sigma \circ \gamma$	dyadic product for $\sigma, \gamma \in L(V_3)$	167
adj σ	special operation for $\sigma \in L(V_3)$, adj $\sigma = (\det \sigma) \sigma^{-1}$	167
$J(x)$	Jacobian for $y = y(x)$, $J(x) = \det y'(x)$	
$w_{,i}$	identical with $D_i w$, where $D_i = \partial/\partial \xi_i$ (special notation in the theory of elasticity)	325
D^i	identical with $D_i = \partial/\partial \xi_i$ in a Cartesian coordinate system (This notation is used in connection with the Einstein summation convention. For example, $e_i D^i f$ means $\sum_{i=1}^{n} e_i D_i f$.)	

Special Notation for the Theory of Manifolds

x_φ	representative of the point x on the manifold M with respect to the chart φ, i.e., $x_\varphi = \varphi(x)$	534
v_φ	representative of the tangent vector v on the manifold M with respect to the chart φ, i.e., $v_\varphi = \dot{x}_\varphi(t)$	539
v_φ^*	representative of the cotangent vector v^* with respect to the chart φ	545
TM_x	tangent space of the manifold M at the point x	539
TM	tangent bundle of the manifold M	542
$T^2 M$	identical with $T(TM)$	542
TM_x^*	cotangent space of the manifold M at x (dual space to TM_x)	545
TM^*	cotangent bundle of M	545
$f'(x)$	tangent map $f'(x): TM_x \to TN_{f(x)}$ at the point x of the manifold M with respect to the map $f: M \to N$	540

In the literature, one also uses the following notations for $f'(x)$: $T_x f$, $Tf(x)$, $df(x)$, $Df(x)$, $d_x f$, df_x.

Tf	tangent map $Tf: TM \to TN$ of the map $f: M \to N$	542
$T^2 f$	identical with $T(Tf)$, i.e., $T^2 f$ maps $T^2 M$ into $T^2 N$	542
$df(x; h)$	differential of the map f at the point x in the direction of the tangent vector h, i.e., $df(x; h) = f'(x) h$	595

In the literature, one also uses the following notations for $df(x; h)$: $df_x(h)$ or $h(f)$. Note that $df(x; h)$ is identical with the directional derivative of f at the point x in the direction of h.

If $f: X \to Y$ is a C^1-map between the B-spaces X and Y over $\mathbb{K} = \mathbb{R}, \mathbb{C}$, then we have $TX_x = X$ and $TY_y = Y$ for all $x \in X$ and $y \in Y$, respectively. Moreover, the tangent map $f'(x)$ at x is identical with the F-derivative $f'(x): X \to X$. Finally, we have $df(x; h) = f'(x)h = \delta f(x; h)$ for all $x, h \in X$.

du^i	identical with $f'(x)$ for $f(x) = u^i(x)$, where $u^1, \ldots, u^n$ are local coordinates of x, i.e., $du^i(h) = du^i_x(h) = f'(x)h$	598
$j^k_x f(h)$	Taylor expansion of f at the point x up to kth order, i.e., $j^k_x f(h) = f(x) + \sum_{j=1}^k f^{(j)}(x) h^j / j!$	
$J^k f(x)$	k-jet of $f: M \to N$ at the point x	567

In the special case of a function $f: \mathbb{R} \to \mathbb{R}$, we have $J^k f(x) = (x, f(x), f'(x), \ldots, f^{(k)}(x))$, where $J^k f(x)$ is called the kth jet coordinate of f at the point x.

$J^k(M, N)$	k-jet manifold corresponding to smooth maps $f: M \to N$	567
$M \pitchfork N \bmod Y$	transversal intersection of the submanifolds M and N of Y	565
$f \pitchfork N \bmod Y$	the map $f: M \to Y$ is transversal to the submanifold N of Y	565
M_4, E_4	Minkowski, Einstein manifold	696, 731

Special Notation for the Tensor Calculus

$A^{i'}_i$	transformation coefficient, $A^{i'}_i = \partial u^{i'} / \partial u^i$	618
t^i	simply contravariant tensor	619
t_i	simply covariant tensor	619
$t^{j_1 \ldots j_r}_{i_1 \ldots i_s}$	r-fold contravariant and s-fold covariant tensor	619
$t_i t^i$	identical with $\sum_{i=1}^n t_i t^i$, according to Einstein's summation convention	616
δ^i_j	unit tensor	620
Alt t_{ij}	identical with $\frac{1}{2}(t_{ij} - t_{ji})$	622
Sym t_{ij}	identical with $\frac{1}{2}(t_{ij} + t_{ji})$	622
g_{ij}	metric tensor	620, 651
g	$\det(g_{ij})$	651
Γ^k_{ij}	Christoffel symbol	623
R^i_{jkm}	Riemann's curvature tensor	642
R_{jm}	Ricci tensor, $R_{jm} = R^k_{jkm}$	731
R	scalar curvature, $R = g^{jm} R_{jm}$	731
K	Gauss curvature of a surface, $K = R_{1212}/g$	637, 643
Ω^i_j	curvature form, $\Omega^i_j = \frac{1}{2} R^i_{jkm} du^k \wedge du^m$	689
$\nabla_i t_j$	covariant derivative of the tensor field t_j	623

$Dt_j/d\alpha$	absolute derivative of the tensor field t_j along the curve $u^i = u^i(\alpha)$, $$Dt_j/d\alpha = \dot{u}^i \nabla_i t_j$$	625
Dt_j	absolute differential of t_j, $$Dt_j = du^i \nabla_i t_j$$	625
grad f	gradient of a function f	626
div w	divergence of a vector field in $\mathbb{R}^n$	626
curl w	rotation of a vector field in $\mathbb{R}^3$	629
Curl w_i	rotation of a vector field in $\mathbb{R}^n$	630
$\varepsilon_{i_1\ldots i_n}$	Levi-Civita symbol, the sign of the permutation $\binom{1\ldots n}{i_1\ldots i_n}$	628
$E_{i_1\ldots i_n}$	standard n-fold covariant pseudotensor, $$E_{i_1\ldots i_n} = |g|^{1/2} \varepsilon_{i_1\ldots i_n}$$	628
$E^{i_1\ldots i_n}$	standard n-fold contravariant pseudotensor, $$E^{i_1\ldots i_n} = |g|^{-1/2} \varepsilon_{i_1\ldots i_n}$$	628
$d_i t_{i_1\ldots i_r}$	alternating differentiation of the antisymmetric tensor field $t_{\ldots}$, $$d_i t_{i_1\ldots i_r} = \text{Alt } D_i t_{i_1\ldots i_r}$$	664
ω	alternating differential form, $$\omega = t_{i_1\ldots i_r} du^{i_1} \wedge \cdots \wedge du^{i_r}$$	667
$d\omega$	derivative of the alternating differential form ω, $$d\omega = dt_{i_1\ldots i_r} \wedge du^{i_1} \wedge \cdots \wedge du^{i_r}$$	668
$*\omega$	dual differential form to ω	672
$i_v \omega$	inner product of the alternating differential form ω with the vector field v	671
$\delta\omega$	dual derivative of the alternating differential form ω	672
$L_a t^i$	Lie derivative of the tensor field t^i along a^i, $$L_a t^i = a^s D_s t^i - t^s D_s a^i$$	673
r_s	Schwarzschild radius	757

Function Spaces

The reader should also consult the List of Symbols to Parts I and II. In the following, G denotes a nonempty bounded open set in $\mathbb{R}^N$. Let $-\infty < a < b < \infty$. Moreover, let $k = 0, 1, \ldots, 1 \leq p < \infty$, and $0 < \alpha \leq 1$.

List of Symbols

$\partial G = \overline{\partial_1 G} \cup \overline{\partial_2 G}$	special decomposition of the boundary ∂G of G	242
$\partial G \in C^{k,\alpha}$	boundary property of the set G (If $k \geq 1$, then the boundary is smooth.)	I232
$\partial G \in C^{0,1}$	the set G has a piecewise smooth boundary	I232
$L(X, Y)$	space of linear continuous operators from X into Y	I135
$C(X, Y)$	space of continuous operators from X into Y	I148
$C^k(X, Y)$	space of k-times continuously F-differentiable operators from X into Y	I148
$C^0(X, Y)$	identical with $C(X, Y)$	
$H_\alpha(u)$	Hölder constant of u, i.e., $H_\alpha(u)$ is the smallest constant L with	

$$|u(x) - u(y)| \leq L|x - y|^\alpha$$

for all $x, y \in \overline{G}$

$\|u\|_{C[a,b]}$	identical with $\max_{a \leq x \leq b} |u(x)|$
$\|u\|_{C^k[a,b]}$	identical with $\sum_{j=0}^{k} \max_{a \leq x \leq b} |u^{(j)}(x)|$
$\|u\|_{C^{k,\alpha}[a,b]}$	identical with $\|u\|_{C^k[a,b]} + H_\alpha(u^{(k)})$, where the Hölder constant H_α refers to $\overline{G} = [a,b]$
$\|u\|_{C(\overline{G})}$	identical with $\max_{x \in \overline{G}} |u(x)|$
$\|u\|_{C^k(\overline{G})}$	identical with $\sum_{|\beta| \leq k} \max_{x \in \overline{G}} |D^\beta u(x)|$
$\|u\|_{C^{k,\alpha}(\overline{G})}$	identical with

$$\|u\|_{C^{k,\alpha}(\overline{G})} + \sum_{|\beta|=k} H_\alpha(D^\beta u)$$

$\|u\|_p$	identical with $(\int_G |u(x)|^p \, dx)^{1/p}$
$\|u\|_{p,\partial G}$	identical with $(\int_{\partial G} |u(x)|^p \, dO)^{1/p}$
$\|u\|_\infty$	identical with $\operatorname{ess\,sup}_{x \in G} |u(x)|$ (see Section 18.6 of Part II)
$\|u\|_{\infty,\partial G}$	identical with $\operatorname{ess\,sup}_{x \in \partial G} |u(x)|$
$\|u\|_{k,p}$	identical with $(\int_G \sum_{|\beta| \leq k} |D^\beta u(x)|^p \, dx)^{1/p}$
$\|u\|_{k,p,0}$	identical with $(\int_G \sum_{|\beta|=k} |D^\beta u(x)|^p \, dx)^{1/p}$
$C[a,b]$	real B-space of all continuous functions $u: [a,b] \to \mathbb{R}$ with the norm $\|u\|_{C[a,b]}$
$C^k[a,b]$	real B-space of all k-times continuously differentiable functions $u: [a,b] \to \mathbb{R}$ with the norm $\|u\|_{C^k[a,b]}$
$C(\overline{G})$	real B-space of all continuous functions $u: \overline{G} \to \mathbb{R}$ with the norm $\|u\|_{C(\overline{G})}$
$C^k(\overline{G})$	real B-space of all k-times continuously differentiable real functions on $\overline{G}$ with the norm $\|u\|_{C^k(\overline{G})}$. (More precisely, $C^k(\overline{G})$ consists of all continuous functions $u: \overline{G} \to \mathbb{R}$ which are k-times continuously differentiable on G and all of whose partial derivatives up to kth order can be continuously extended to the closure $\overline{G}$ of G.)

$C^{k,\alpha}(\bar{G})$	real B-space of all functions $u \in C^k(\bar{G})$ with $\|u\|_{C^{k,\alpha}(\bar{G})} < \infty$	
$C^{k,\alpha}(\partial G)$	space of $C^{k,\alpha}$-functions $u: \partial G \to \mathbb{R}$	1232
$C^\infty(G)$	space of infinitely differentiable functions $u: G \to \mathbb{R}$	
$C_0^\infty(G)$	space of all functions $u \in C^\infty(G)$ with compact support supp u, i.e., u vanishes outside a compact subset of G (see Section 18.1 of Part II)	
$L_p(G)$	Lebesgue space of all measurable functions $u: G \to \mathbb{R}$ with $\|u\|_p < \infty$ (see Section 18.6 of Part II)	
$L_p(\partial G)$	Lebesgue space of all measurable functions $u: \partial G \to \mathbb{R}$ with $\|u\|_{p,\partial G} < \infty$ (The measurability of u refers to the surface measure on ∂G.)	
$W_p^k(G)$	Sobolev space with the norm $\|u\|_{k,p}$ (see Section 21.2 of Part II)	
$\mathring{W}_p^k(G)$	closure of $C_0^\infty(G)$ in $W_p^k(G)$ (see Section 21.2 of Part II)	

The norms $\|\cdot\|_{k,p}$ and $\|\cdot\|_{k,p,0}$ are equivalent on the Sobolev space $\mathring{W}_p^k(G)$ (see Section 21.2 of Part II).

X^n	product space of X, i.e., X^n consists of all $u = (u_1, \ldots, u_n)$ with $u_i \in X$ for all i

If X is a B-space over $\mathbb{K} = \mathbb{R}, \mathbb{C}$, then X^n is also a B-space over $\mathbb{K}$ with the norm

$$\|u\| = \sum_{i=1}^n \|u_i\|_X.$$

In this way we obtain the real B-spaces $L_p(G)^n$, $W_p^k(G)^n$, etc. and the spaces $C^\infty(G)^n$, $C_0^\infty(G)^n$, etc.

$L_2(G; V_3)$	Lebesgue space of functions $u: G \to V_3$ with values in the real three-dimensional linear space V_3	243
$L_2(\partial G; V_3)$	Lebesgue space of functions $u: \partial G \to V_3$	243
$L_2(G; L_{\text{sym}}(V_3))$	Lebesgue space of functions $\sigma: G \to L_{\text{sym}}(V_3)$	243
$W_2^1(G; V_3)$	Sobolev space of functions $u: G \to V_3$	243
$\mathring{W}_2^1(G; V_3)$	Sobolev space of functions $u: G \to V_3$ with $u = 0$ on ∂G	242
$\mathring{W}_2^1(G, \partial_1 G; V_3)$	Sobolev space of functions $u: G \to V_3$ with $u = 0$ on $\partial_1 G$, where $\partial_1 G \subseteq \partial G$	242

List of Theorems

> The collection of all our experiences consists of what we know and what we have forgotten.
> Marie von Ebner-Eschenbach (1830–1916)

Theorem 58.A	(Balance and conservation laws)	34
Theorem 58.B	(The two-body problem and the three Kepler laws)	40
Theorem 58.C	(Stability principle of minimal potential energy)	51
Theorem 58.D	(Existence and uniqueness theorem for the motion of the rigid body)	54
Theorem 58.E	(Existence of Lagrange multipliers in Lagrangian mechanics)	68
Theorem 58.F	(Canonical transformations and the solution of the canonical equations via the Hamilton–Jacobi equation)	84
Theorem 58.G	(Lagrange brackets and the solution of the Hamilton–Jacobi equations via the canonical equations)	85
Theorem 59.A	(The uncertainty relation in quantum mechanics)	123
Theorem 60.A	(Existence and uniqueness theorem for the elastoplastic wire with linear hardening law)	152
Theorem 61.A	(The fundamental variational principle in nonlinear elasticity)	191
Theorem 61.B	(The Rivlin–Ericksen theorem on the most general constitutive law for homogeneous isotropic bodies)	207
Theorem 61.C	(The stored-energy function of a homogeneous elastic body)	208
Theorem 61.D	(Main theorem of linear elastostatics—generalized solutions)	211

Theorem 61.E	(Main theorem of linear elastodynamics—generalized solutions)	212
Theorem 61.F	(Local existence and uniqueness in nonlinear elasticity)	219
Theorem 61.G	(Classical solutions in linear elastostatics)	221
Theorem 61.H	(Convergence of a general approximation method in nonlinear elasticity)	228
Theorem 62.A	(The principle of minimal potential energy for convex material in nonlinear elasticity—existence and uniqueness of solutions)	244
Theorem 62.B	(The principle of maximal dual energy and duality in nonlinear elasticity)	247
Theorem 62.C	(Existence and uniqueness in linear quasi-statical plasticity)	262
Theorem 62.D	(Existence and uniqueness in linear statical plasticity)	263
Theorem 62.E	(Existence theorem for polyconvex material)	275
Theorem 62.F	(Korn's inequality)	279
Theorem 62.G	(Friedrichs' duality)	287
Theorem 63.A	(Existence and uniqueness for the Signorini problem in elasticity)	298
Theorem 64.A	(Main theorem of bifurcation theory for variational inequalities)	308
Theorem 64.B	(Supported beams, variational inequalities, and bifurcation)	313
Theorem 65.A	(Existence theorem for the von Kármán plate equation)	332
Theorem 65.B	(Bifurcation theorem for the von Kármán plate equation)	333
Theorem 65.C	(Supported plates, variational inequalities, and bifurcation)	341
Theorem 66.A	(Existence and uniqueness theorem for a general model in plasticity including internal state variables and hardening effects)	355
Theorem 67.A	(Solution of the Gibbs equation for gases and liquids)	377
Theorem 69.A	(Existence and uniqueness theorem for Carleman's radiation problem)	426
Theorem 71.A	(Complex function theory and hydrodynamics)	453
Theorem 71.B	(Permanent gravity waves and bifurcation)	461
Theorem 72.A	(Existence and uniqueness theorem for the stationary Navier–Stokes equations)	490
Theorem 72.B	(Existence and uniqueness theorem for the instationary Navier–Stokes equations)	494
Theorem 72.C	(The Taylor problem for viscous flow and bifurcation)	499
Theorem 72.D	(The Bénard problem for viscous flow and bifurcation)	510
Theorem 73.A	(Existence and uniqueness theorem for ordinary differential equations on Banach manifolds)	547
Theorem 73.B	(Linearization principle for diffeomorphisms)	552

List of Theorems

Theorem 73.C	(Construction of manifolds via submersions—the preimage theorem)	556
Theorem 73.D	(Construction of manifolds via subimmersions)	558
Theorem 73.E	(Construction of manifolds via embeddings)	559
Theorem 73.F	(Construction of diffeomorphisms via ordinary differential equations and the generalized Morse lemma)	561
Theorem 73.G	(Construction of manifolds via transversality)	565
Theorem 73.H	(Sard's theorem)	587
Theorem 73.I	(Whitney's embedding theorem)	588
Theorem 74.A	(Parallel transport of vector fields and covariant differentiation)	627
Theorem 74.B	(Main theorem of surface theory of Bonnet)	640
Theorem 74.C	(Theorema egregium of Gauss)	643
Theorem 74.D	(Riemann's curvature tensor and Riemann's theorem on the local flatness of Riemannian manifolds)	654
Theorem 76.A	(The fundamental variational principle for the motion of light and matter in general relativity)	733
Theorem 76.B	(Friedman's model of the universe)	739
Theorem 76.C	(The Schwarzschild solution in general relativity)	757
Theorem 76.D	(The motion of the Perihelion of planets in general relativity)	759
Theorem 76.E	(The Kruskal solution in general relativity and black holes)	769
Theorem 77.A	(The fixed-point theorem of Brouwer)	799
Theorem 77.B	(The inequality of Fan)	801
Theorem 77.C	(The main theorem about n-person games of Nash)	802
Theorem 77.D	(The fixed-point theorem of Fan and Glicksberg)	805
Theorem 77.E	(The main theorem of mathematical economics of Gale, Nikaido, and Debreu)	807
Theorem 78.A	(The Sard–Smale theorem for Banach manifolds)	829
Theorem 78.B	(Parametrized version of the Sard–Smale theorem)	833
Theorem 78.C	(The main theorem about the generic finiteness of the solution set of operator equations)	834
Theorem 79.A	(Ljapunov's main theorem of stability theory in B-spaces)	843
Theorem 79.B	(Structure of flows for autonomous differential equations in B-spaces)	847
Theorem 79.C	(Asymptotic stability and instability of periodic solutions in B-spaces)	851
Theorem 79.D	(Orbital stability)	853
Theorem 79.E	(Loss of stability and the main theorem about the bifurcation of equilibrium points)	858
Theorem 79.F	(Loss of stability and the main theorem about Hopf bifurcation)	862
Theorem 79.G	(Center theorem of Ljapunov)	868

List of the Most Important Definitions

International system of units 883
Dimension of important physical quantities...................... 884
Universal constants.. 884
   Planck's quantum of action h ($\hbar = h/2\pi$)................... 102, 884
   velocity of light c 700, 884
   Boltzmann constant k 400, 884
   gravitational constant G 36, 884
Hubble constant.. 57
Critical density of the universe.................................. 61
Critical temperature of the universe for the existence of specific elementary particles 62
Reynolds number Re and turbulence............................ 440
Elementary units 750
   Planck length, Planck time, Planck temperature, Planck energy, Planck mass, elementary charge (charge of the proton)........ 751

Energy
   total.. 34
   kinetic.. 32
   potential 33
   inner 378, 387
   free... 387
   elastic.. 190, 198
   dual elastic 236
Energy
   of a free particle 721
   of a photon 102

List of the Most Important Definitions

Energy-momentum tensor.	725
current density vector	422
conservation law	34, 422, 724
Entropy	387, 397
Thermodynamical potential	385, 387
inner energy	387
free energy	387
enthalpy	387
free enthalpy	387
statistical potential of Gibbs	387
Temperature of a radiation	58
The fundamental partition function in statistical physics	400
temperature	398
chemical potential	398
Velocity	11
virtual velocity	49
Acceleration	11
Momentum	28, 32
momentum of a photon	102
generalized momentum in Hamiltonian mechanics	73
Angular momentum	32
spin	125
Mass	
in classical mechanics	26, 28
in special relativity	721
in general relativity	731, 736
rest mass	721
Center of mass	33
Force	
conservative force	33
centrifugal force	30
Coriolis force	30
constraining force	46
the four fundamental forces in the universe	135
Torque	32
Potential of a force	33
gauge invariance in classical mechanics	33
gauge invariance in modern physics (see Part V)	
Potential of a velocity	454
Work (force times displacement)	32
Power (work divided by time)	32
principle of virtual power	16, 19, 49
Action (energy times time)	82
action along a motion	82

principle of stationary action 70
Lagrange function (Lagrangian)... 21, 71
Hamilton function (Hamiltonian)... 21, 73
Lagrange multipliers and constraining forces............................. 46, 67
Legendre transformation
    in mechanics 65
    in thermodynamics 385

Wave
    wave length λ 100
    wave vector... 100
    frequency ν 100
    angular frequency ω. 100
    phase displacement 101
    phase velocity... 100
    group velocity... 101
    dispersion relation 100
    polarization... 101
    plane waves.. 100
    spherical waves.. 103
Damped oscillations
    mean life-time... 104
    frequency-time uncertainty relation.................................... 104
Probability of presence for particles in
    quantum mechanics... 114
Heisenberg's uncertainty relation... 123
    position-momentum uncertainty 124
    energy-time uncertainty for quasi-stable quantum states 130
Measurements in quantum mechanics
    expectation value.. 114
    dispersion and mean error 114
Decay probability for particles... 106
Reaction probability for particles.. 107
Cross section for particle reactions 106

Equilibrium point (state)
    of a dynamical system 841
    of a mechanical system 16, 50
    of a thermodynamical system (see also KMS states in Part V)......... 373
    in game theory 802
    in mathematical economics (Walras equilibrium)....................... 806
Stability of an equilibrium point in the sense of Ljapunov
    asymptotically stable.. 842
    stable... 841
    unstable... 842

List of the Most Important Definitions 967

Orbital stability of periodic processes 852
Statical stability of mechanical systems....................... 16, 50
Strongly stable states in elastostatics........................... 218
Stability of thermodynamical equilibria via thermodynamical potentials 389
Multipliers for equilibrium states
    asymptotically stable.. 846
    critical.. 846
    unstable... 846
Floquet multipliers for periodic processes........................ 851
Bifurcation point (see page 428 of Part I)

Deformation... 176
    displacement... 176
Strain tensor $\mathscr{E}$ 170
Stress tensor τ.. 177
Reduced stress tensor σ (first Piola–Kirchhoff tensor)............... 184
Principal strain... 169
Piola transformation.. 175
Piola identity 175
Constitutive law.. 176
    dual constitutive law 246
Yield condition in plasticity 155, 200
Stored energy function (density of the elastic potential energy of a body) 190
Dual energy of an elastic body.................................. 236
Friedrichs' duality 287
Trefftz' duality 288

Flow of fluids
    inviscid (ideal)... 439
    viscous.. 438
    incompressible 438
    stationary 438
    irrotational 438
Circulation 438
Viscosity 437
Tensor of inner friction 436
Pressure (force divided by surface).............................. 435

Inertial system
    in classical mechanics 31
    in the theory of relativity................................... 700
Proper time.. 718
Lorentz transformation.................................... 706, 711
Poincaré group... 712
Minkowski space–time manifold M_4........................... 713

Einstein space–time manifold E_4	730
Friedman metric	
of the closed universe	739
of the open universe	741
Schwarzschild metric of the sun	756
Kruskal metric and black holes	768
Banach manifold	535
tangent vector	538
tangent space	539
tangent bundle	542
cotangent vector (covector or also differential form)	545
cotangent space	545
cotangent bundle	545
Chart	533
chart map	534
chart space	534
local coordinates	534
admissible chart	535
Atlas	534
equivalent atlas	536
maximal atlas	535
Differentiable structure	536
C^k-manifold	535
topological manifold	535
analytical manifold	535
Submanifold	556
Manifold with boundary	584
Dimension of a manifold	535
codimension of a submanifold	557
Orientation of a manifold	582
Fredholm mapping (see also Section 8.4 of Part I)	552
double splitting map	531
the linear subspace L splits the space X (see page 766 of Part I)	
projection operator (see page 766 of Part I)	
direct topological sum (see page 766 of Part I)	
Mappings between manifolds	
tangent map at a point	541
tangent map on the tangent bundle	543
C^k-map	537
diffeomorphism	537
regular and singular (critical) value	552
regular and singular (critical) point	552
etale map	551

List of the Most Important Definitions

submersion 551
immersion 551
subimmersion 551
embedding 559
proper map 551
closed map 551
k-jet $J^k f(x)$ at the point x 567
k-jet manifold $J^k(M, N)$ 567
Vector field on a manifold 543
Flow on a manifold 547
Bundles
    abstract bundle 544
    vector bundle 589
    fiber bundle and principal fiber bundle (see Part V)

Singularities and catastrophe theory
    equivalence of maps 571
    unfolding 575
        versal 575
        universal 575
    k-determination 574
    transversality 565
    structural stability 579
    genericity 579
    Whitney topology 569

Derivative $f'(x): TM_x \to TN_{f(x)}$ of a map $f: M \to N$ at the point x 541
Differential $df(x; h) = f'(x)h$ 595
Directional derivative 595
Derivation 600
Subgradient ∂F of a functional (see page 385 of Part III)
Covariant derivative of a tensor field 623
Absolute derivative 625
Absolute differential 625
Parallel transport of a tensor field 626, 650
Alternating differentiation of antisymmetric tensor fields ... 664
Differentiation of alternating differential forms 668
Lie derivative 673
Lie group 677
Lie algebra 676

Tensor 13
    covariant 618
    contravariant 619
    symmetric 622

    antisymmetric (skew-symmetric or alternating) 622
    tensor in V_3 166
Tensor density 630
Pseudotensor 628
Pseudotensor density... 630
Unit tensor 620
Metric tensor 651
    signature of the metric tensor 654
Curvature tensor 642
Gaussian curvature 637
Affine connected manifold...................................... 649
Riemannian manifold
    general... 651
    proper 651
Local flatness of Riemannian manifolds 654
Geodesic 653
    affine geodesic.. 650

Sperner simplex 798
Regular solution curve... 819
Homotopy method 818
Mapping degree (see also Chapter 12 of Part I) 670, 826
Fixed-point index (see also Chapter 12 of Part I).................. 825
Residual set (massive set)....................................... 570

List of Basic Equations in Mathematical Physics

Classical mechanics
    Newton's fundamental equation in inertial systems............... 26
    Newton's fundamental equation in arbitrary systems of reference... 30
    Gauss' principle of least constraint and the general basic equations
        of point mechanics with side conditions 45
    Lagrange's equation and the fundamental variational principle of
        stationary action 70, 71
    Hamilton's canonical equation................................. 72
    Hamilton–Jacobi equation...................................... 82
    Poisson's equation.. 77
    the principle of virtual power (virtual work)............. 17, 18, 19, *49*
    the statical stability principle for mechanical equilibrium states
        and the principle of minimal potential energy 16, *50, 51*
    the general dynamical stability principle of Ljapunov.......... 20, *843*
    symplectic manifolds and classical mechanics................. Part V
    algebraic approach to classical mechanics via operator algebras .. Part V
Nonlinear elasticity
    Cauchy's fundamental equation in nonlinear elasticity............ 176
    hyperelasticity and the principle of minimal potential energy....... 190
    the principle of dual elastic energy............................ 245
    linear material and linear Hooke's law 201
    convex material and nonlinear Hooke's law..................... 235
    polyconvex material.. 209
Plasticity
    statical plasticity.. 262
    quasi-statical plasticity 259
    quasi-dynamical plasticity.................................... 350

Dynamical plasticity and rheology
   ideal plastic von Mises liquid .... Part V
   viscoplastic Bingham liquid.................................... Part V
   viscous Williamson liquid .... Part V
   non-Newtonian liquids .... Part V

Hydrodynamics
   the fundamental equations for smooth flow 434
   the Newton equation in hydrostatics............................ 445
   the Euler equation for inviscid flow............................ 439
   the Navier–Stokes equation for viscous flow 438
   the Prandtl boundary layer equation and singular perturbation theory 520
   the Bernoulli equation and conservation of energy 438
   complex function theory and inviscid flow in the plane........... 453
   basic equation in filtration theory .... Part V
   basic equation for non-Newtonian fluids....................... Part V

Gas dynamics and shock waves
   the fundamental equations for nonsmooth flow of gases and liquids
      and the Rankine–Hugoniot jump conditions .... Part V
   the fundamental equations for isentropic flow and the Helmholtz
      vorticity equation..................................... Part V

Phenomenological thermodynamics
   the three laws of thermodynamics 378
   Gibbs' fundamental equation 374
   the stability of thermodynamical equilibria via thermodynamical potentials... 389
   heat conduction 424

Statistical physics
   the fundamental model 397
      Bose statistics... 402
      Fermi statistics.. 403
      classical Boltzmann statistics 420
      quasi-classical statistics in phase space 417
   symplectic manifolds and the classical Gibbs statistics in phase
      space .... Part V
   the fundamental Hilbert space approach to quantum statistics and
      von Neumann's density matrix......................... Part V
   the modern algebraic approach to quantum statistics via operator
      algebras (equilibrium states correspond to KMS functionals) Part V
   operator algebras, diagonalization of the Hamilton operator, and
      quasi-particles in superfluidity .... Part V

Irreversible thermodynamics
   the fundamental equation for the entropy production (Onsager's
      relations) .... Part V
   the Boltzmann equation...................................... Part V
   plasma physics .... Part V

    physical kinetics and superfluidity .... Part V
    phase transitions... Part V
Chemistry
    the fundamental equation of chemical kinetics .... Part V
    the fundamental equation of quantum chemistry.............. Part V
Theory of special relativity
    Einstein's principle of relativity 699
    principle of constant velocity of light........................ 700
    motion of free particles 719
    motion of photons.. 722
    motion of relativistic inviscid fluids and cosmology........... 726, 739
    relativistic electromagnetism................................ Part V
    energy-momentum tensor and relativistic field theories........ Part V
    Dirac's equation for the relativistic electron .... Part V
Theory of general relativity
    Einstein's fundamental equation 730
    Hilbert's variational principle for the Einstein equation 734
    the fundamental equation for the motion of matter and light in the
        universe... 730
Electromagnetism
    Maxwell's fundamental equations .... Part V
    quantum electrodynamics .... Part V
    electromagnetism and gauge field theory..................... Part V
Quantum mechanics
    Schrödinger's fundamental equation of nonrelativistic quantum mechanics... 112
    Dirac's equation for the relativistic electron .... Part V
    relativistic field theories.................................... Part V
    the fundamental Hilbert space model and the dualism between
        waves and particles..................................... Part V
    algebraic approach to quantum mechanics via operator algebras   Part V
Quantum field theory
    approach via canonical quantization......................... Part V
    approach via Feynman integral .... Part V
    axiomatic approach via operator algebras .... Part V
Free quantum fields... Part V
Quantum electrodynamics, scattering processes, and the S-matrix .. Part V
Gauge field theory and the curvature of principal fiber bundles.... Part V
Elementary particle physics and gauge field theory
    the Weinberg–Salam model for the electroweak interaction and
        the group $SU(2) \times U(1)$............................. Part V
    the quark model for the strong interaction and the groups $SU(3)$
        (color group) and $SU(n)$ (flavor group). .... Part V
    the grand unification of electromagnetic, weak, and strong interaction and the group $SU(5)$................................ Part V

supersymmetry and graded Lie algebras .... Part V
superstring theory and the unification of all fundamental interactions in nature including gravitation .... Part V

Important Principles

Conservation laws
   conserved quantities in classical mechanics. 32
   current density vectors and conservation laws 422
   energy-momentum tensor and conservation laws in relativistic field theories 723
   Noether theorem and general conservation laws via global symmetries .... Part V
   conservation laws and the Rankine–Hugoniot jump conditions in gas dynamics (shock waves). .... Part V
   conservation laws and the fundamental equations of irreversible thermodynamics. .... Part V
Propagation of discontinuities along characteristics (wave propagation) .... Part V
   propagation of light in electromagnetism .... Part V
   transversal and longitudinal elastic waves. .... Part V
   propagation of sound. .... Part V
   shock waves in gas dynamics .... Part V
Symmetries in nature correspond to groups .... Part V
Stability
   dynamical stability. 20, 841
   statical stability and the principle of minimal potential energy   16, 50, 51
   statical stability in elasticity 193, 304
   orbital stability of periodic processes. 852
   stability of thermodynamical equilibria. 389
Loss of stability leads to bifurcation
   equilibrium states bifurcate into new equilibrium states 856
   equilibrium states bifurcate into periodic processes (Hopf bifurcation)   860
Similarity and the structure of physical laws
   basic idea. 89
   turbulence and the Reynolds number 440
   turbulence and the Kolmogorov law. 513
Singular limits and singular perturbation theory
   boundary layers in hydrodynamics 518
   incompressible flow as a singular limit of compressible flow for $c_s \to \infty$ (c_s = speed of sound). .... Part V
   inviscid flow as a singular limit of viscous flow for $\eta \to 0$ (η = viscosity)   520
   geometrical optics as a singular limit of electromagnetic waves for $\lambda \to 0$ (λ = wavelength of light). .... Part V

classical mechanics as a singular limit of the theory of relativity
for $c \to \infty$ (c = velocity of light) 706, 735
classical mechanics as a singular limit of quantum mechanics for
$h \to 0$ (h = Planck's quantum of action). 129
simultaneous biological oscillations with completely different periods (a basic problem in mathematical biology). .... Part V
The fundamental algebraic approach to modern physics via operator algebras
  observables correspond to operators .... Part V
  states correspond to positive functionals .... Part V
  the dynamics of physical systems corresponds to one-parameter automorphism groups of the algebra .... Part V
  thermodynamical states correspond to special functionals called KMS states .... Part V
Applications of the algebraic approach to physics
  classical mechanics .... Part V
  classical statistical physics .... Part V
  quantum statistics .... Part V
  quantum field theory .... Part V
Heisenberg's uncertainty principle
  position-momentum uncertainty 124
  energy-time uncertainty for quasi-stable quantum states 130
  general formulation (see also Part V) 123
Spin and statistics
  Fermions (i.e., elementary particles with half-numberly spin) satisfy the Fermi statistics 126
  Bosons (i.e., elementary particles with integer spin) satisfy the Bose statistics 126
Principle of indistinguishability of identical elementary particles .. 401, 417
Pauli principle: in a system of Fermions, two particles can never be in the same quantum state 126
Fundamental principles in modern physics
  the field equations are the Euler equations to the variational principle of stationary action .... Part V
  global symmetries yield conservation laws via the Noether theorem .... Part V
  local symmetries yield the fundamental interactions in nature via gauge field theory .... Part V
  interactions correspond to the curvature of appropriate manifolds Part V
  elementary particles correspond to irreducible representations of appropriate Lie groups .... Part V
Probability plays a fundamental role in nature .... 113, 366, Part V

Index

The reader should also consult the detailed Contents of this volume and the index material (List of Theorems, etc.). Moreover, the reader may also consult the Indices of Parts I through III. If several page numbers belong to the same catch word, then the primary reference is *italized*. The page number 1345 refers to page 345 of Part I, etc.

absolute
  derivative 625
  differential 625
  space 29, 31, 702, *704ff*
  temperature (*see* temperature)
  time *31*, 702
acceleration 11
accessory quadratic variational problem 218
action *81*, 884
  propagation of 81
addition theorem for velocities in relativity 708ff
adiabatic process *380*, 394
admissible chart 535
admissible coordinate system 632
affine connected manifolds 649
Airy's stress function 328, 337, *345*
algebraically simple eigenvalue *1374*, 853ff
alternating differential forms 664ff (*see also* Part V)
  basic ideas of the calculus for 665

differentiation of 665, *668*
integration of 665, *669*
amplitude 101
analytic map 1362
angular
  frequency 28, *99*
  momentum 32
  momentum in quantum mechanics 115
  velocity 12
angular momentum tensor 725ff
antisymmetric (skew-symmetric) 622
approximation methods in
  elasticity *224ff*, 252
  general relativity 735
  hydrodynamics 483
  quantum mechanics 129
approximation models in elasticity 163, *240ff*
area preserving maps 644
asymptotically stable 842
atlas 534
  abstract 602

977

atlas (*continued*)
  equivalent 536
  maximal 535
  oriented 583
autonomous differential equations 548, 841

Baire category I802
Baire space I801
balance of
  angular momentum 34
  energy 34
  momentum 34
Banach manifold (*see* manifold)
Banach space (*see* B-space)
baryon number 416
basic equations in mathematical physics, list of 953
basis space of a bundle 544
beam equation 315ff
Bénard problem 505*ff*, 518
bending of beams and rods 311ff
Bernoulli constant 453
Bernoulli equation 433, *438*, 453
Betti number 688
bifurcation
  and loss of stability 221*ff*, 227, 841, 856*ff*, 860*ff*
  for beams and rods 311*ff*, 320
  for dynamical systems 856ff (*see also* Part V)
  for plates 323, *333*
  for variational inequalities 303ff
  in elasticity · 221, 227, 303*ff*, 311ff, 332, 346
  in hydrodynamics 448, *495, 505*
bifurcation point
  of an operator equation I358
  of a variational inequality 304
bifurcation principle 821, 837, *841*
big bang 57*ff*, 742*ff*
bilinear form I141
  compact (*see* Part II)
  nondegenerate 561
  strictly positive (*see* Part II)
  strongly positive (*see* Part II)
  weakly nondegenerate 561
black-body radiation 410ff

black hole 768, 771*ff*, 778ff
black–white dipole hole 768, 775*ff*
blowing-up lemma 606
Bohr's atomic model 137
Boltzmann constant 59, 366, *400, 884*
Boltzmann statistics 420
Bose statistics *402*, 408
  basic ideas of 401
boson *126*, 751
boundary layers 518ff
boundary of stability 222
Boussinesq approximation 506, *512*
Brouwer's fixed-point theorem I51
  elementary proof of 799
  equivalent statements to the 795, *810*
Brouwer's principle 797
B-space (Banach space) I769ff
buckling force *304*, 314
bundle 544
  abstract 544
  cotangent 545
  fiber (*see* Part V)
  isomorphism 544
  morphism 544
  normal 591
  space 544
  tangent 542
  vector 590

canonical equation 21, 72, 83ff
canonical transformation 83
Carnot's cycle 394
catastrophe theory 572ff
  basic ideas of 572
causal structure of manifolds 787
causality 714ff
cavities 471
center
  of mass 33
  theorem, general (*see* Part V)
  theorem of Ljapunov 868
center, stable, and unstable manifolds (*see* Part V)
centrifugal force 30
chain rule for tangent maps 543
Chandrasekhar mass, critical *412*, 778
characteristic classes 690

Index

chart   533
  abstract   536
  admissible   535
  admissible oriented   584
  compatible   534
  image   534
  map   535
  space   534
Chebyshev inequality   115, *398*
chemical potential   *374, 398*
  as a Lagrange multiplier   398
  of an ideal gas   *377, 406*
  of a photon gas   *409*
  of elementary particles   *416*
chemical reactions   373, *389ff*
Christoffel symbols   623
  effective calculation of the   734
  intuitive meaning of the   682
circulation   438
closed map   551
cnoidal waves   466ff
cobasis   576
coboundary   671
cocycle   671
codimension   I765
  of a map   576
  of a submanifold   557
coherently oriented   585
cohomology   671
  class   671
  group   671
compact operator   I53
compact set   *I756*, I769
compact spacelike support   724
compensated compactness   264ff
complex flow potential   454
complex velocity   453
complexification   I770
component of a set   I757
cone   I276
configuration space   22, *48*
conformal maps   645
conjugate functional   65ff
connected set   I757
connection in a principal fiber bundle   690 (*see also* Part V)
conservation laws   34, *422ff*, 724
  basic ideas of   422
  differential forms and   782
  in relativity   724 (*see also* Part V)
conservation of
  angular momentum   35
  energy   34
  mass   423
  momentum   35
conservative force   33
conserved quantities in mechanics   32
constitutive laws
  basic ideas of   147ff
  convex material   237, *244*ff
  dual   237, *246*
  elastic energy of the cuboid and   *198*, 230
  elasto-viscoplastic material   151ff, *348ff*, 354
  for homogeneous bodies   205ff
  for homogeneous isotropic bodies   206
  for the friction tensor   436ff
  Fourier's law   424
  hardening and   148, 151ff, *348ff*, 353
  Hencky material   *202*, 256
  Hooke's law   *148, 201*, 233ff, 255
  ideal plastic material   149
  ideal plastic von Mises liquid   157 (*see also* Part V)
  in heat conduction   424
  in hydrodynamics   434ff
  in hyperelasticity   190
  in linear elasticity   *201*, 208, 255
  in nonlinear elasticity   *176, 185*
  in plasticity   148ff, *154ff*, 200ff, 257ff, 348ff
  inverse (dual)   237, *246*
  linear material   *201*, 208, 255, 259
  Mooney–Rivlin material   209
  Navier–Stokes liquid   438
  non-Newtonian liquid   157 (*see also* Part V)
  Ogden's material   209
  polyconvex material   *209*, 241, 273
  rubberlike material   *208*, 277
  Saint Venant–Kirchhoff material   208
  theory of invariants and   202ff
  viscoplastic Bingham liquid   157 (*see also* Part V)
  viscoplastic material   *149ff*, 348ff

constitutive laws (*continued*)
  viscous material   150
  viscous Williamson liquid   157 (*see also* Part V)
constraining force   46ff
  of a pendulum   93
continuation method
  in heat conduction   422ff
  in nonlinear elasticity   224ff
  in numerical mathematics   818ff
continuity equation   434
continuous operator   I755, I770
contraction of tensors   622
contravariant tensor   619
convex
  functional   III245, III380
  set   III245
Coriolis force   30
cosmos
  early   63, 747*ff*
  future of the   745ff
cotangent bundle   545
cotangent vector   545, 595
Couette flow   497
coulomb (unit of charge)   884
Coulomb's law   38
countable basis   I754, *536*
covariance   720
covariant differentiation   623
  motivation of the   626
  on surfaces   646
covariant tensor   618
covector field on a manifold   546
critical mass density of the universe   60, *744*
critical temperature for elementary particles   62
cross section for elementary particle processes   *106*, 137, 139
current density vector   422*ff*, 725
curvature scalar   731
curvature tensor   642*ff*, 650, 653, 655, 683, 731, 734, *784*
  analytic meaning of the   650
  definition of the   642
  geometric meaning of the   654, 683
  important properties of the   643, *784*
  in general relativity   730*ff*, 734
  theorema egregium of Gauss and the   642
  Riemann's theorem on local flatness and the   654
curve following algorithm   812ff, *822*, 837

de Broglie matter wave   112
de Rham cohomology   671
dead water problem   471
decay of particles   105
decay probability   106
deflection of light in general relativity   765
deformation   177
  of a cuboid   198
degree of a tensor   619
derivation   600
dielectricity constant   38, 884
diffeomorphism   537
  construction of a   560
  local   538
differentiable structure   536
differential   595
  absolute   625
  calculus on manifolds   648, *663ff*
differential forms   599
  and conservation laws   782
  in mechanics (*see* Part V)
  in thermodynamics   383ff
differential topology   688ff
  and numerical mathematics   817ff
  and the fixed-point index   825
  and the mapping degree   825
differentiation
  absolute   *625*, 650
  alternating   664
  covariant   *623*, 650
  Lie   673
  of differential forms   665ff
  on manifolds   540*ff*, 663ff
dimension analysis
  and boundary layers   521
  and turbulence   440, *515*
  basic ideas of   89
dimension of a manifold   535
direct sum   I766
directional derivative   595, 600, 673

dispersion
  in quantum mechanics   114
  in statistical physics   398, 407
  in turbulence   515
  relation   100, 102, 449
divergence of a tensor   166
door-in/door-out principle   813, 820
Doppler effect   91
  and the red shift of galaxies   57
double splitting maps   531
dual elastic energy   245
dualism between waves and particles   107ff
duality
  in elasticity   235, 237, 245ff, 289
  in plasticity   259, 262
  in the calculus of variations   284
  of Friedrichs   284ff
  of Trefftz   288 (see also Part III)
dyadic product   167
dynamical stability   841ff
  basic ideas of   20
dynamical systems (see ordinary differential equations)

e-homeomorphism   559
eddies in turbulence   513
Einstein's
  principle of relativity   32, 695
  space-time manifold   730ff
  summation convention   11, 616
Einstein's equations   730
  and bifurcation   788
  initial-value problem for   787
  light quantum hypothesis   102
elastic energy of the cuboid   198
elastic potential energy   190
elasticity
  basic ideas in   159ff, 234ff
  linear   239, 255
  module   148
  nonlinear   158, 176
  plane   345
  typical difficulties in   179
elastodynamics, linear   212
elastostatics   180
elasto-viscoplastic wire   151

electron volt   883ff
elementary particles   130ff
  critical temperature for   62
  cross section for   106, 137, 139
  lifetime of   130
elliptic geometry   656ff
embedding   559
embedding theorem of
  Nash   661
  Whitney   588
energy   19, 34
  and the first law of thermodynamics   379
  dissipation   513ff
  dual elastic   245
  elastic potential   190
  fluctuations   399, 406
  free   387
  in mechanics   34
  inner   374, 379, 386ff
  kinetic   32
  of a free particle   721
  of a photon   102, 722
  of the present universe   61
  potential   33
  total   34
energy–momentum tensor   725, 727, 732, 734
energy–time uncertainty   104, 130
enthalpy   387, 391
entropy   363ff, 380, 397
  and equilibrium states   388
  and Gibbs' fundamental equation   374
  and information   364, 419
  and statistical weights   366, 420
  and the second law of thermodynamics   380
  as a thermodynamical potential   387
  balance   434
  basic ideas of   363ff
  in classical statistics   417
  in hydrodynamics   434
  in quasi-classical statistics   417
  in statistical physics   397
  of radiation   59, 409
equation of state   372, 377, 414
equilibrium forms of rotating fluids   476ff

equilibrium point
  asymptotically stable 842
  in economics 806
  of a differential equation 841
  of dynamical systems 841
  of Nash 802
  of Walras 806
  stable 841
  unstable 842
equilibrium states
  computation of 387
  in mechanics 50
  in thermodynamics 373, 387ff
  necessary condition for 35
equivalence of maps 571
equivalent system of reference 29
etale mapping 551
Euler
  characteristic 686, 688ff, 692
  class 689
  equation in hydrodynamics 439
events
  lightlike 715
  spacelike 715
  timelike 715
expansion of the universe 58ff, 742ff
expectation value (see mean value)
extremal principles in thermodynamics 387

Fan's inequality 801
Feigenbaum bifurcation 522
Fermi statistics 402, 408
  basic ideas of 401
fermion 126, 751
fiber 544
field quantum 131ff
fixed-point index 824 (see also Part I)
  and differential topology 825
fixed-point theorem of
  Brouwer 799
  Fan–Glicksberg 805
  Kakutani 804
fixed-point theorems 795ff
Floquet
  eigenmultiplier 850
  multiplier 850ff
  transformation 850

  trick 846
flow
  around a body 474
  in tubes 439
  irrotational 438
  laminar 440
  on a manifold 547ff, 847
  parallel 437
  planar 453
  stationary 438
  turbulent 440
fluid
  compressible 437
  ideal (see inviscid)
  incompressible 438
  inviscid 439
  viscous 438
flux 450
force 26
  and Newton's basic equation 26
  centrifugal 30
  conservative 33
  constraining 46
  Coriolis 30
  Coulomb 38
  four fundamental forces in nature 135
  gravitational 35
  inertial 30
  unit of 884
form invariance 720
four momentum 723
four velocity 723
Fourier's law 424
Fredholm operator I365, 552
free
  boundary-value problem 450
  energy 387, 390
  enthalpy 387, 391
  fall 26
  particle 719
frequency 28, 99
frequency–time uncertainty relation 104
friction 436ff
Friedman model
  closed 736, 745
  open 741, 747
function spaces, notations for 940

Index

fundamental
  equation of Gibbs *374*, 382
  equations of surface theory 639
  forms of Gauss 633
  interactions in nature 134ff
future of our cosmos 745ff

Galileian principle of relativity 31
game theory 802
gas constant 377
gas dynamics 444 (*see also* Part V)
gauge invariance 33
gauge theory 612, 690 (*see also* Part V)
Gauss
  basic equations in mechanics of 48
  map 689
  principle of least constraint of 48
Gaussian curvature 611, *637*, *643*, 656, 659, 682ff, 686, 689
generic finiteness of the solution set 521, *834*
genericity *570*, *573*, *579*
genus 686
geodesic coordinates, local 784
geodesic curvature 685
geodesics *646ff*, 653
  affine 650
  in classical mechanics 684
  in general relativity 731
geometrization of mechanics 21
geometry
  elliptic *656ff*, 659
  Euclidean 656, 659
  hyperbolic *656ff*, 659
  non-Euclidean *656ff*, 659
  Riemannian 634, *651*, 730ff
Gibbs' fundamental equation *374*, 382
Gibbs' phase rule 391
gradient method 255
grand unification theory *135*, 748
  (*see also* Part V)
gravitational
  acceleration 26, 37
  collapse of stars 790
  constant *36ff*, 884
  force 35
  force, first-order approximation 27, 36

force in a mine 91
law 35
potential 36
potential of a body 90
potential of a spherical shell 91
group velocity *101*, 110

half-life period 104
Hamilton–Jacobi equation *82ff*
  for a free particle 723
Hamiltonian equation (*see* canonical equation)
hardening of material *151ff*, 348ff
harmonic oscillator 27, 127
  in quantum mechanics 79, *122*, 126
Hausdorff dimension (*see* Part V)
Hausdorff measure 521
heat 370, *379*
  capacity 378
  capacity, specific *376ff*, 424
  conduction 423ff
  conductivity number 424
Heisenberg's uncertainty relation 123ff
Hencky material, nonlinear 202, *256*
Hénon attractor 522
Hertzsprung–Russel diagram 777
Hilbert's variational problem in general relativity 734, 784
Hölder inequality (*see* Part II)
Hölder spaces I230
holonomic constraints 48
homotopy methods 817ff
  basic ideas of 818
Hooke's law, linear 148, 199, 201
Hooke's law, nonlinear 233
Hopf bifurcation 840, *862ff*, 873, 876
  and abstract parabolic equations 879
  degenerate 878
  global 873
H-space (Hilbert space) I784ff (*see also* Part II)
Hubble constant 57, *743*
Hubble's law 57, *743*
hydrodynamics 433ff
hydrogen atom *118ff*, 121, 137
hydrostatics 445
hyperbolic geometry 656ff

hyperelasticity 190ff
  general strategy of 196
hysteresis *148ff*, 348ff

ideal
  fluid (*see* inviscid fluid)
  gas 377, 394, *403ff*
  plastic material 149
immersion 551
incompressible fluid 437
index
  of a Fredholm operator 552
  picture of tensors 623
  principle for tensors 616, *623*, *625*, 629, 679ff
  principle of mathematical physics *625*, 681
  principle, inverse 681
inequality
  of Chebyshev 115, *398*
  of Fan 801
  of Gårding 214 (*see also* Part II)
  of Hölder (*see* Part II)
  of Korn 248, *279ff*
  of Poincaré-Friedrichs (*see* Part II)
  quasi-variational 807
  variational *296ff*, *303ff*
inertia tensor 53
inertial
  charts 713
  force 30
  system 28, 30, *699ff*, 702ff, 782
infinitesimal motion 12
infinitesimal rotation 12
infinitesimally small rigid motion *292*, 344
inflationary universe 749
information III294, III307
inner friction 436ff
integrability conditions 343, 640, 643, 654, 667, 669
interactions in nature, four fundamental 134ff
international system of units 883
invariants 619
inviscid fluid 439
inviscid (ideal) relativistic fluid 726ff
irreversible 381

irrotational flow 438
isotropic tensor functions 203ff

jet 567
  coordinates 566
joule 883

k-determined map 574, 576, 602
k-equivalent maps 567
Kepler's laws 22, 39, 41
Kerr–Newman solution 779
kinetic energy 19, *32*
Kolmogorov's laws in turbulence 513ff
Korn's inequality 248, *279ff*
Korteweg–de Vries equation 468ff
  (*see also* Part II)
Kruskal solution 767
K-theory 594 (*see also* Part V)
Kutta–Jukovski formula 474

Lagrange
  brackets 85
  equation 21, *70*
  function 21, *71*
  manifold 87
Lagrange multiplier
  and chemical potential 398
  and temperature 398
  in mechanics 67ff
  rule in variational inequalities 306
Lamé constants 199
  dual 256
Laplace operator on manifolds 673
law of
  equipartition 62, 378, *417ff*
  Hagen–Poisseuille 439
  Kepler 22, *39*, 41
  Kolmogorov 513ff
  mass action 392
  thermodynamics, first 366, 369ff, *379ff*
  thermodynamics, second 363, 365ff, 369ff, *380ff*
  thermodynamics, third 385
  thermodynamics, zero-th 380
  Walras 806

Lax pair 470
Lebesgue spaces (see Part II)
left invariant vector field 678
Legendre transformation
  and conjugate functionals 65ff
  basic ideas of 66
  in elasticity 238, *286ff*
  in mechanics 74
  in the calculus of variations 284ff
  in thermodynamics 385
Legendre–Hadamard condition 218
lemma of
  Knaster–Kuratowski–Mazurkiewicz 798
  Morse 560, 603
  Morse–Tromba 605
  Ricci 642, 654, *679*
  Sperner 797
  Sperner, cubic 815
length contraction in relativity 708
length preserving maps 644
lepton number 416
Leray–Schauder principle, constructive 823
lever principle 14
Lie algebra 676ff
Lie derivative 673ff
  motivation for the 674
Lie group 677
lifetime of
  black holes 780
  elementary particles 130
lifting of indices 680
light 107
  cone 715
  dualism between wave and particle 107ff
  quantum hypothesis 102
  trap 773
lightlike events 715
linearization principle for
  differential equations 842
  flows 842
  maps 550ff
list of
  important principles I866, *956*
  symbols 933
  the most important definitions 946
  theorems 943

Ljapunov
  bifurcation 867
  center theorem 868
  stability 20, *841*
local coordinates (see representatives)
locally convex space 797
locally flat 654, 684
longitudinal waves 101
Lorentz group 712
Lorentz transformation
  general 711
  proper 711
  special 706
Lorenz attractor 523
loss of stability and bifurcation 221ff, 227, *856ff*, *860ff*
lower semicontinuous I456
  multivalued mapping I450
lowering of indices 680

magnetic monopoles 750
manifolds 535
  affine connected 649
  basic strategy of the theory of 535
  center, stable, and unstable (see Part V)
  in general relativity 730
  in special relativity 713
  metrizable 537
  one-dimensional 538, 585, 817ff
  orientation of 582
  principles for constructing *555ff*, 563ff
  Riemannian 651
  with boundary 584
  with countable basis 536
mapping degree 824ff (see also Part I)
  and differential topology 824ff
  and Sperner simplices 814
  for maps on manifolds 670
maps (see operator properties)
  area preserving 644
  conformal 645
  length preserving 644
  orientation preserving 582
mass 26, *28*, 33, 37
  balance 434
  Chandrasekhar 412

mass (*continued*)
  in general relativity 731, *736*
  in special relativity 721
  of black holes *773*, 779
  of neutron stars 778
  of white dwarf stars, maximal *412*, 778
mass–energy equivalence 719ff
mathematical economics 806
mathematics and physics 1ff
matrix mechanics in quantum theory 78ff
maximal monotone operator (*see* Part II)
maximum (*see* minimum)
Maxwell's velocity distribution 406
meager set I802
mean curvature 637
  and minimal surfaces 682
mean lifetime *104*, 135, 137
mean value
  in quantum mechanics 114
  in statistical physics 398
  of velocity in turbulent flows 515
mechanics 9ff
  basic ideas of 14
  Gaussian 45
  Hamiltonian 72
  Lagrangian 70
  Newtonian 25
  Poissonian 77
  quantum 112ff
mesons 131
metric *620*, 651
  Eddington 774
  Friedman 738ff
  Kerr–Newman 779
  Kruskal 768ff
  Newton 735
  Schwarzschild 756ff
  space I761
  tensor 620, *634*, 651, 731
Michelson experiment 703
minimal surfaces 682
minimax theorem 803
minimum
  bound III276
  free III193
  local III193
  regular 551
  strict III193
  strictly stable 193
  strongly stable 218
Minkowski space–time manifold 696
modern mathematical physics 752, 952, 955
momentum 28, *32*
  angular 32
  balance *34*, 434
  generalized 73
  of a photon 102, *722*
monotone operator (*see* Part II)
Mooney–Rivlin material 209
Morse index 653
Morse lemma *1110*, I343
  generalized *560*, 603
Morse–Tromba lemma 605
multilinearization of maps 572
multiplier *846*, 851
  asymptotically stable 846
  critically 846
  Floquet 850
  of a fixed point 846
  of an equilibrium point 846
  of a periodic solution 850
  unstable 846

Nash equilibrium point 802
natural
  basis 597, 617, *632*, 649
  boundary condition *192*, 299
  coordinates 597
  projection 542, *544*, 545
  system of units 751
Navier–Stokes equations 438, *479ff*
  basic ideas of the 480ff
n-body problem *38*, 42
negative retract principle *808*, 810
neutron stars 415, *778*
newton (unit of force) 884
Newton's basic equation *25ff*, 32
  in general systems of reference 28ff
nondegenerate singular point 561
non-Euclidean geometry 655ff
nonresonance condition *861*, 876, 880
normal bundle 591

normal coordinates 96
normal forms
  and catastrophe theory 573ff, 578
  for immersions 554, 559
  for subimmersions 552, 558
  for submersions 553, 556
  of oscillating systems 96
nowhere dense set 1751
nuclear forces 132

Ogden's material 209
operator properties, important 1889
  (see also the Indices to Parts II and III)
orbit 852
orbitally
  asymptotically stable 852
  stable 852, 877, 881
  unstable 853
ordinary differential equations
  bifurcation theory for 856
  center, stable, and unstable manifolds for (see Part V)
  on manifolds and flows 548
  stability theory for 841
orientation 582ff
  coherent 585
  of a manifold 585
  of a manifold with boundary 585
  preserving map 582
oscillating systems, normal form of 96

parallel axiom 655ff
parallel transport of tensors 626ff, 645ff
  basic ideas of the 627
  geometric interpretation of the 645
  in the sense of Levi-Civita 626ff, 645ff, 650, 682
  in the sense of Lie 675
partial regularity of differential equations 521
partition function in statistical physics 400
pascal (unit of pressure) 884
Pauli principle 125ff, 413

PCT-invariance in nature 712
pendulum 92, 318, 874
Penrose transformation via twistors 789
Perihelion of Mercury, motion of the 758ff
period of oscillation 99
periodic system of the elements 127
permanent waves 448ff
perpetuum mobile of the second kind 394
perturbation
  of orbits 759ff
  of simple eigenvalues 853ff
  of the spectrum 874ff
  theory 853, 874
  theory, singular 519
phase 100
  space 73
  transition (see Part V)
  transition in the cosmos 748
photon 108ff, 722
  equation of motion of a 731
  gas 59, 408
Piola identity 167, 175
Piola–Kirchhoff tensor
  first 165, 186, 189
  second 186
Piola transformation 175
Planck's quantum of action 78, 102, 108, 126, 366, 884
Planck's radiation law 58, 366, 408ff
  and the big bang 58
plane wave 100
plastic torsion (see Part V)
plasticity
  basic ideas of 147ff, 154ff
  condition of von Mises (yield condition) 156, 201
  dynamical (plastic liquids) 156 (see also Part V)
  historical remarks on 155ff
  quasi-dynamical 154, 348ff
  quasi-statical 154, 257ff
  statical 154, 262ff
plates 322ff
  basic ideas for 322
  with obstacles 339

Poincaré (*see also* theorem of Poincaré)
  group 712, 781
  model 657
  transformation 711
Poisson
  brackets 77
  mechanics 77
  number 148
polarization 101
polyconvex material of Ball 209, 241, 273*ff*
positive-energy theorem in general relativity 791
potential 33
  energy 17, *33*
  of a force 33
  of a velocity 454
  operator III229
power 32, 884
Prandtl number 508
Prandtl's boundary layer equation 520
pressure 394, 435, 884
  in the weak sense 487
pricing system 806
principal axes of
  inertia 53
  strain 169
  stress 178
principal moments of inertia 53
principle
  bifurcation *821*, 837, *841*
  door-in/door-out 813, 820
  for constructing diffeomorphisms 560
  for constructing manifolds 555ff, 563ff
  negative retract *808*, 810
  of Brouwer 797
  of causality 716
  of constant enthalpy 391
  of constant velocity of light 700
  of least constraint 45ff
  of linearization for dynamical system 842
  of linearization for maps 550
  of maximal dual energy 245, 259ff, 262
  of maximal entropy 388
  maximal plastic work 345
  of maximal signal velocity 716
  of minimal free energy 390
  of minimal free enthalpy 391
  of minimal potential energy 19, *51*, 244
  of multilinearization for maps 560, *572ff*, 604ff
  of relativity 695
  of relativity, classical *31*, 703
  of stationary action 69ff
  stationary potential energy 190
  turning point 821
principle of virtual power 16ff, *50*
  basic ideas of the 17
  in elastostatics 188
  in hyperelasticity 192
principle of virtual work *18*, 345
  and the principle of virtual power 18
principles, list of important 956
probability of presence for particles 114
production of
  entropy 435
  mass 423
  momentum 435
projection–iteration method 254
projection operator I766
propagation of action 81
propagation velocity 100
proper map *1173ff*, 551
proper time
  in general relativity 732
  in special relativity 718
pseudomonotone operators
  definition of (*see* Part II)
  in elasticity 322ff
  in hydrodynamics 481ff
pseudosphere 660
pseudotensor 628
  density 630
  dual 629
pull-back 670
pulsars 778

quantization 98ff
  general approach to (*see* Part V)
  of classical mechanics in the sense of Heisenberg 78

of classical mechanics in the sense of
    Schrödinger 112
  of the phase space 126
quantum
  cosmology 750
  of action 78, 102, 108, 126, 366,
    884
quantum statistics (see also Part V)
  basic ideas of 401
  Bose statistics 402, 408
  Fermi statistics 402, 408
quasars 778
quasi-classical approximations
  in general relativity 735
  in quantum mechanics 129
quasi-concave I456
quasi-convex I456
quasi-eigenvector 869
quasi-equilibrium states 372ff, 374,
    379ff
quasi-statical plastic material 257
quasi-variational inequality 807

radiation problem of Carleman 422ff
Rankine–Hugoniot jump condition
  in gas dynamics 444 (see also
    Part V)
Rayleigh number 508
reaction probability 107
reaction velocity 389
red limit of the cosmos 778
red shift
  and the expansion of the universe
    57, 742
  in gravitational fields 766
  in the spectrum of galaxies 57, 742
region I758
regular
  minimum 51
  point 552
  solution curve 819
  space I596, 537
  state space 380
  value 552
relatively compact set I756
relativity principle
  of Galilei 31, 703
  of Einstein 32, 695

representatives (local coordinates)
  of a cotangent vector 545
  of a map 537
  of a point of a manifold 534
  of a point of the tangent bundle 543
  of a tangent vector 539
residual set 570, 829
retract I50
reversible process 381
Reynolds number 440, 479, 485, 495,
    520
Ricci tensor 731
Riemannian manifold 651
  proper 651
  pseudo- 651
rigid body 52ff, 54
Ritz method 252
rod equation 315ff
rubberlike material of Ogden 208
Rutherford's scattering formula 139

Saint Venant–Kirchhoff material 208
scalar curvature 731
scales
  equilibrium of a pair of 14ff
  motion of a pair of 19ff
  stability of a pair of 14ff
scattering of particles 106, 138ff
Schauder estimates 460
Schrödinger equation 112ff
Schwarzschild solution 756
section
  of a bundle 544
  of a vector bundle 590
sectorial operator (see Part II)
semiflow 547
semigroup (see Part II)
sets, fundamental properties of I893
shell theory 325
shift operator 847ff
similarity in mathematical physics
    439, 484
simple
  contravariant tensor 619
  covariant tensor 619
simplicial algorithms 812
simplicial methods 794ff
  basic ideas of 798

singular
  homology groups 688 (*see also* Part V)
  perturbation problems 519
  point 552
  point, nondegenerate 561
  value 552
sinking of a space ship 773
sinusoidal waves 466ff
skew-symmetric (*see* antisymmetric)
slow deformation processes 349
small oscillations 96
Sobolev spaces (*see* Part II)
solitary waves 466ff
solitons 468ff
spacelike events 715
specific
  densities 442
  entropy 376
  heat capacity 376
  inner energy 376
spectrum I795
  does not surround the origin 849
Sperner simplex 798, 814
spherical waves 103
spin 125
  and statistics 125
splitting of spaces I766, 531, 550
  of maps 531
stability
  basic ideas of 16, 20
  dynamical (in the sense of Ljapunov) 20, 842
  in elasticity 193, 217
  of dynamical systems 841
  of equilibrium points 841
  of mechanical systems 16, 50
  of orbits 852
  of periodic solutions 851
  of solutions of ordinary differential equations 841
  of the planetary system 42
  of thermodynamical states 389
  orbital 852
  statical 16, 50
stability principle in mechanics 16, 50ff
  basic idea of the 16

stable equilibrium
  dynamically 20, 842
  statically 16, 50
star, death of a 777ff
star models
  classical 412
  relativistic 790
state 371
  equation of 372, 377, 414
  equilibrium 373, 387ff
  quasi-equilibrium 372ff, 387ff
  strictly stable 193
  strongly stable 218, 244
stationary flow 438
statistical physics (*see also* Part V)
  basic ideas of 396ff
  Boltzmann statistics in 420
  Bose statistics in 402, 408
  classical 417, 420
  classification of states and 401
  Fermi statistics in 403, 408
  ideal gas in 403ff
  photon gas in 408ff
  pure energy statistics in 402
  quasi-classical 417
statistical potential 387, 400
statistical solution of the Navier–Stokes equations 522
Stefan–Boltzmann law 409ff, 781
stored energy function 165, 190
  general structure of the 207
  dual 246, 256, 294
strain tensor 165, 170
  geometrical meaning of the 171ff
  linearized 170
strange attractor 522 (*see also* Part V)
stream line 454
stress force 165, 177, 181
stress tensor 165, 177, 181ff
  basic formulas for the 189
  for the inner friction 434, 436ff
  in hyperelasticity 190
strict local minimum III193
strictly stable 193
strings 753ff
strongly
  continuous operator I498
  elliptic systems 213

Index   991

k-determined map   574
stable solution   193, *218*, 244
structural stability   579
subgradient   III385
subimmersion   551
submanifold   556
  with boundary   556
submersion   551
sun, history of the   777
supergravity   751
super-Lie groups   752
supermanifolds   752
superstring theory   752
supersymmetry   751
surface maps   644
surface theory
  curvature properties in   636ff
  metric properties in   631ff
symplectic geometry and mechanics   22
  (*see also* Part V)

tangent bundle   542
tangent map   543
  at a point   541
tangent vector   538
  abstract   539
  concrete   538
  local coordinates of a   539
  transformation rule for a   539
Taylor problem   495ff
Taylor vortices   495
temperature
  as a Lagrange multiplier   398
  basic ideas for   363ff
  compensation   388
  Gibbs' fundamental equation and   374
  in statistical physics   398
  of radiation   *58*, 408
  regular states and   380
  zero-th law of thermodynamics and   380
tensor   618ff
  antisymmetric (skew-symmetric)   622
  construction of a   681
  contraction of a   622

  contravariant   619
  covariant   618ff
  curvature (*see* curvature tensor)
  degree of a   619
  density   630
  inertia (*see* inertia tensor)
  in three-dimensional space   *13*, 166
  metric (*see* metric tensor)
  parallel transport of a (*see* parallel transport)
  strain (*see* strain tensor)
  stress (*see* stress tensor)
  symmetric   622
  torsion (*see* torsion tensor)
  trace of a   166
  type of a   679
  unit (*see* unit tensor)
tensor calculus   615ff
  basic ideas of the   615
  on manifolds   648ff
tensor field
  constant in the sense of Levi-Civita   627
  constant in the sense of Lie   675
  curl of a   630
  divergence of a   *166*, 626, 629
theorem of (*see also* lemma of)
  Adams   692
  Birkhoff   786
  Bolzano–Poincaré–Miranda   808
  Bonnet (main theorem of surface theory)   640
  Brouwer   799
  Chern   689
  Crandall and Rabinowitz   858, 862, 879
  de Rham   688
  Euler (polyeder theorem)   687
  Fan and Glicksberg   805
  Gale, Nikaido, and Debreu   807
  Gauss (angular sum theorem)   686
  Gauss (integral theorem)   665
  Gauss (theorema egregium)   643
  Gauss–Bonnet–Chern   685ff
  Hartman–Stampacchia   803
  Hodge   672
  Hopf   862, 876
  Kakutani   804

theorem of (*continued*)
  Ljapunov (center theorem) 868
  Ljapunov (stability theorem) 843
  Morse 691
  Murat 267
  Nash 661
  Poincaré (differential forms) 669
  Poincaré (duality theorem) 688
  Poincaré–Hopf 691
  Riemann 654
  Rivlin–Ericksen 207
  Sard 587
  Sard (parametrized) 823
  Sard–Smale 829ff
  Sard–Smale (parametrized) 832
  Stokes (integral theorem) 660, 670
  Thom (transversality theorem) 570
  Whitney 588
theorema egregium of Gauss 643
theory of relativity
  general 730ff
  special 694ff
thermodynamical
  equilibrium states 373, 387ff
  potential 385, 387
  quasi-equilibrium states 372*ff*, 374, 379ff
  states 371*ff*, 379ff
thermodynamical process 371ff, *379*
  adiabatic 380
  closed 380
  irreversible 365, *381*
  isothermal 387
  reversible 365, *381*
thermodynamics 363ff
  basic ideas of *363, 369*
  deviation from reversibility in 381
three-body problem 42ff
time dilation in relativity 709
timelike events 715
topological direct sum I766
topological vector space 797
topology and analysis 685ff
torque 32
torsion tensor 650
  geometric meaning of the 683
total energy 34
total momentum 32

trace of a tensor 166
transition functions of a vector bundle 590
transport theorem 441
transversal
  linear subspaces 564
  maps 565
  submanifolds 565
  waves 101
transversality 563ff
  basic ideas of 563
  in catastrophe theory 577
  theorem, general 565
  theorem of Thom 570
Trefftz method 253
turbulence 439*ff*, 479ff, 495
  and Kolmogorov's laws 513ff
  and stochastic velocities 515
  and strange attractors 522ff
  basic ideas of 439
  classical Landau–Hopf theory of 524
turning point principle 821
twin paradox 718
twistors 789
two-body problem 39
type of a tensor 679

uncertainty
  of energy and time 130
  of frequency and time 104
  of momentum and position 124
  of the angular–momentum components 124
  principle of Heisenberg 123
unfolding
  universal 575
  versal 575
unit tensor 620
units
  elementary 751
  international system of 883ff
universal constants 750, *884*
universe at a temperature of 10^{11} K 63
unstable 842
upper semicontinuous I456
  multivalued map I450

vaporization of black holes 780
variational inequalities
  bifurcation problems for 303ff
  in elasticity *296ff, 303ff*
  in mathematical economics 806
vector
  axial 629
  polar 629
vector bundle 589
  isomorphism 594
  morphism 593
vector fields on manifolds 543
velocity
  complex 453
  of light 884
  potential 454
  vector 11
violation of parity in weak interaction 701
virtual
  displacement 18
  motion 49
  power *16*, 50
  singularities in general relativity 767
  velocity 49
  work *20*, 50
viscoplastic material 150
viscous fluid 438
viscous material 150
volt 884
von Kármán's plate equations *326ff*, 334

Walras equilibrium 806
Walras law 806
water waves *448ff*, 466ff
watt (unit of power) 884
wave
  cnoidal 466
  length 100
  longitudinal 101
  matter 112
  number 100
  permanent 448ff
  plane 100
  sinusoidal 466
  solitary 466ff
  spherical 103
  transversal 101
  vector 100
weakly stable *860*, 862, 866
white dwarf stars *412ff*, 778
white hole 768, *775*
Whitney topology 569
Wien's displacement law 418, 718
WKB method 129
work
  in mechanics 16, *32*
  in thermodynamics 370, *379*, 393
world line 717
world sheet 753

yield condition in plasticity 200
Yukawa potential 132

PGMO 06/01/2018